LITERATURE

For additional information on Intel products in the U.S. or Canada, call Intel's Literature Center at (800) 548-4725 or write to:

Intel Literature
P.O. Box 7641
Mt. Prospect, Il 60056-7641

To order literature outside of the U.S. and Canada contact your local international sales office.

CURRENT DATABOOKS

Product line databooks contain datasheets, application notes, article reprints, and other design information. Databooks can be ordered in the U.S. and Canada by calling TAB/McGraw-Hill at 1-800-822-8158; outside of the U.S. and Canada contact your local international sales office.

Title	Intel Order Number	ISBN
Automotive Products	231792	N/A
Embedded Applications (2 vol. set)	270648	1-55512-242-6
Embedded Microcontrollers	270646	1-55512-230-2
Embedded Microprocessors	272396	1-55512-231-0
Flash Memory (2 vol. set)	210830	1-55512-232-9
Intel486™ Microprocessors and Related Products	241731	1-55512-235-3
i960® Processors and Related Products	272084	1-55512-234-5
Military and Special Products	210461	N/A
Networking	297360	1-55512-236-1
OEM Boards, Systems and Software	280407	1-55512-237-X
Packaging	240800	1-55512-238-8
Pentium™ Processors and Related Products	241732	1-55512-239-6
Peripheral Components	296467	1-55512-240-X

A complete set of this information is available on CD-ROM through Intel's Data on Demand program, order number 240897. For information about Intel's Data on Demand ask for item number 240952.

intel®

24-HOUR AUTOMATED TECHNICAL SUPPORT*

Intel's Application Bulletin Board System (BBS) and FaxBack System are at your service, 24-hours a day, at no charge, and the information is updated frequently.

FaxBack SYSTEM

Technical and product information are available 24-hours a day! Order documents containing:

- Product Announcements
- Product Literature
- Intel Device Characteristics
- Design/Application Recommendations
- Stepping/Change Notifications
- Quality and Reliability Information

Information on the following subjects is also available:

- Microcontroller and Flash
- OEM Branded Systems
- Multibus/BBS Listing
- Multimedia
- Development Tools
- Quality and Reliability/Change Notification
- Microprocessor/PCI/Peripheral
- Intel Architecture Lab

To use FaxBack for Intel components and systems, dial **(800) 628-2283** or (916) 356-3105 (U.S. and Canada) *or +44{0} 1793-496646 (Europe)* and follow the automated voice-prompt menu. Document orders will be faxed to the fax number you specify. Catalogs are updated twice a month, so call for the latest information!

BULLETIN BOARD SYSTEM

Intel's Application Bulletin Board System (BBS) enables file retrieval 24-hours a day. The following can be located on the BBS:

- Software Drivers
- Tool Information
- Software/Application Utilities
- Product/Technical Documentation
- Firmware Upgrades
- Quality and Reliability Data

To use the Intel Application BBS (components and systems), dial **(916) 356-3600** for download access (U.S. and Canada) *or +44{0} 1793-496340 (Europe).* The BBS will support 1200–19200 baud rate modem. *Typical modem configuration: 9600 baud rate, No Parity, 8 Data Bits, 1 Stop Bit.* A directory listing of BBS files is also available through FaxBack or our 800 BBS (800-897-2536).

Retail Products

Information on Intel's retail products (Coprocessors and wireless, video, personal conferencing and network products) is available through the following services:

Internet : ftp.intel.com (143.185.65.2)
CompuServe : GO INTELFORUM (modem settings: E-7-1, up to 14.4 Kbps)

Country	BBS (N-8-1, up to 14.4 Kbps)	FaxBack
North America	(503) 264-7999	(800) 525-3019 or (503) 264-6835
Europe	+44 1 793-432955	+44 1 793-432509
Australia	+61 2 975-3066	+61 2 975-3922
Taiwan	+886 2 718-6422	+886 2 514-0815
Singapore	+65 256-4776	+65 256-5350
Hong Kong	+852 530-4116	+852 844-4448
Korea	+822 784-3430	+822 767-2594

*Support services provided courtesy of Intel Application Support

intel®

Embedded Microcontrollers

MCS® 48 Microcontrollers, MCS 51 Microcontrollers, MCS 251 Microcontrollers, MCS 96 Microcontrollers

1995

Additional copies of this document or other Intel literature may be obtained from:

Intel Corporation
Literature Sales
P.O. Box 7641
Mt. Prospect, IL 60056-7641

or call 1-800-879-4683

DATASHEET DESIGNATIONS

Intel uses various datasheet markings to designate each phase of the document as it relates to the product. The markings appear in the lower inside corner of each datasheet page. Following are the definitions of each marking:

Datasheet Marking	Description
Product Preview	Contains information on products in the design phase of development. Do not finalize a design with this information. Revised information will be published when the product becomes available.
Advanced Information	Contains information on products being sampled or in the initial production phase of development.*
Preliminary	Contains preliminary information on new products in production.*
No Marking	Contains information on products in full production.*

* Specifications within these datasheets are subject to change without notice. Verify with your local Intel sales office that you have the latest datasheet before finalizing a design.

intel®

Table of Contents

Table of Contents (Continued)

Table of Contents (Continued)

Alphanumeric Index

intel®

1

1

MCS® 48 Microcontroller Family

SINGLE COMPONENT MCS®-48 SYSTEM

1.0 INTRODUCTION

Sections 2 through 5 describe in detail the functional characteristics of the 8748H and 8749H EPROM. 8048AH/8049AH/8050AH ROM, and 8035AHL/8039AHL/8040-AHL CPU only single component microcomputers. Unless otherwise noted, details within these sections apply to all versions. This chapter is limited to those functions useful in single-chip implementations of the MCS®-48. The Chapter on the Expanded MCS®-48 System discusses functions which allow expansion of program memory, data memory, and input output capability.

2.0 ARCHITECTURE

The following sections break the MCS-48 Family into functional blocks and describe each in detail. The following description will use the 8048AH as the representative product for the family. See Figure 1.

2.1 Arithmetic Section

The arithmetic section of the processor contains the basic data manipulation functions of the 8048AH and can be divided into the following blocks:

- Arithmetic Logic Unit (ALU)
- Accumulator
- Carry Flag
- Instruction Decoder

In a typical operation data stored in the accumulator is combined in the ALU with data from another source on the internal bus (such as a register or I/o port) and the result is stored in the accumulator or another register.

The following is more detailed description of the function of each block.

INSTRUCTION DECODER

The operation code (op code) portion of each program instruction is stored in the Instruction Decoder and converted to outputs which control the function of each of the blocks of the Arithmetic Section. These lines control the source of data and the destination register as well as the function performed in the ALU.

ARITHMETIC LOGIC UNIT

The ALU accepts 8-bit data words from one or two sources and generates an 8-bit result under control of the Instruction Decoder. The ALU can perform the following functions:

- Add With or Without Carry
- AND, OR, Exclusive OR
- Increment/Decrement
- Bit Complement
- Rotate Left, Right
- Swap Nibbles
- BCD Decimal Adjust

If the operation performed by the ALU results in a value represented by more than 8 bits (overflow of most significant bit), a Carry Flag is set in the Program Status Word.

ACCUMULATOR

The accumulator is the single most important data register in the processor, being one of the sources of input to the ALU and often the destination of the result of operations performed in the ALU. Data to and from I/O ports and memory also normally passes through the accumulator.

2.2 Program Memory

Resident program memory consists of 1024, 2048, or 4096 words eight bits wide which are addressed by the program counter. In the 8748H and the 8749H this memory is user programmable and erasable EPROM; in the 8048AH/8049AH/8050AH the memory is ROM which is mask programmable at the factory. The 8035AHL/8039AHL/8040AHL has no internal program memory and is used with external memory devices. Program code is completely interchangeable among the various versions. To access the upper 2K of program memory in the 8050AH, and other MCS-48 devices, a select memory bank and a JUMP or CALL instruction must be executed to cross the 2K boundary.

There are three locations in Program Memory of special importance as shown in Figure 2.

LOCATION 0
Activating the Reset line of the processor causes the first instruction to be fetched from location 0.

LOCATION 3
Activating the Interrupt input line of the processor (if interrupt is enabled) causes a jump to subroutine at location 3.

LOCATION 7
A timer/counter interrupt resulting from timer counter overflow (if enabled) causes a jump to subroutine at location 7.

Therefore, the first instruction to be executed after initialization is stored in location 0, the first word of an external interrupt service subroutine is stored in location 3, and the first word of a timer/counter service routines

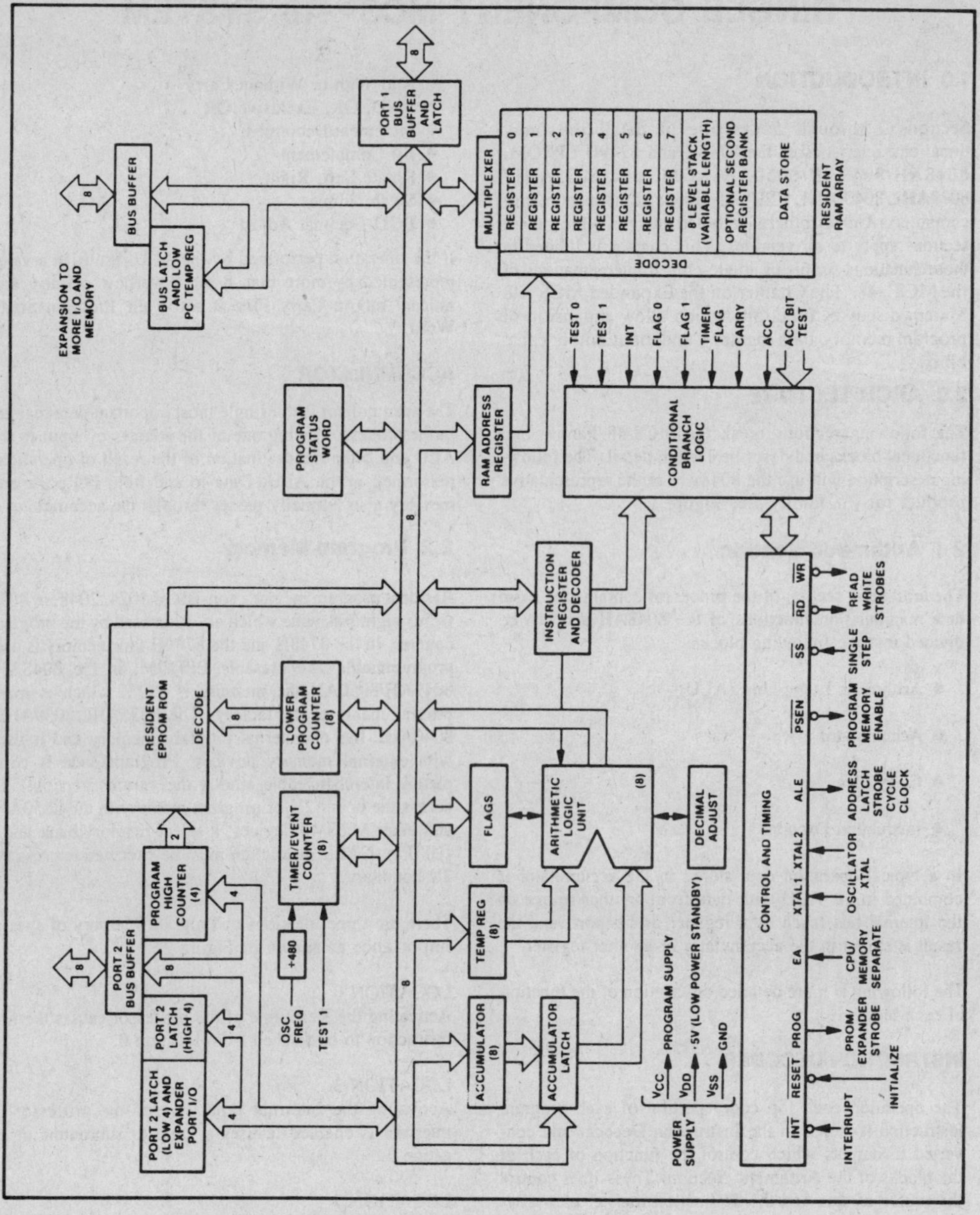

Figure 1. 8748H/8048H/8749AH/8050AH Block Diagram

is stored in location 7. Program memory can be used to store constants as well as program instructions. Instructions such as MOVP and MOVP3 allow easy access to data "lookup" tables.

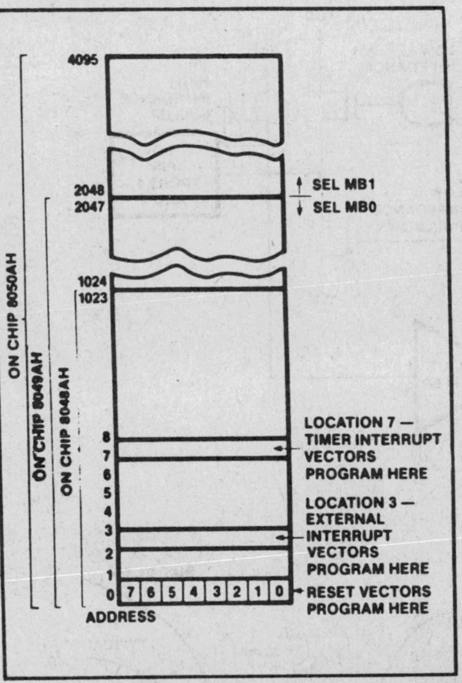

Figure 2. Program Memory Map

2.3 Data Memory

Resident data memory is organized as 64, 128, or 256 by 8-bits wide in the 8048AH, 8049AH and 8050AH. All locations are indirectly addressable through either of two RAM Pointer Registers which reside at address 0 and 1 of the register array. In addition, as shown in Figure 3, the first 8 locations (0–7) of the array are designated as working registers and are directly addressable by several instructions. Since these registers are more easily addressed, they are usually used to store frequently accessed intermediate results. The DJNZ instruction makes very efficient use of the working registers as program loop counters by allowing the programmer to decrement and test the register in a single instruction.

By executing a Register Bank Switch instruction (SEL RB) RAM locations 24–31 are designated as the working registers in place of locations 0–7 and are then directly addressable. This second bank of working registers may be used as an extension of the first bank or reserved for use during interrupt service subroutines allowing the registers of Bank 0 used in the main program to be instantly "saved" by a Bank Switch. Note that if this second bank is not used, locations 24–31 are still addressable as general purpose RAM. Since the two RAM pointer Registers R0 and R1 are a part of the working register array, bank switching effectively creates two more pointer registers (R0/and R1/) which can be used with R0 and R1 to easily access up to four separate working areas in RAM at one time. RAM locations (8–23) also serve a dual role in that they contain the program counter stack as explained in Section 2.6. These locations are addressed by the Stack Pointer during subroutine calls as well as by RAM Pointer Registers R0 and R1. If the level of subroutine nesting is less than 8, all stack registers are not required and can be used as general purpose RAM locations. Each level of subroutine nesting not used provides the user with two additional RAM locations.

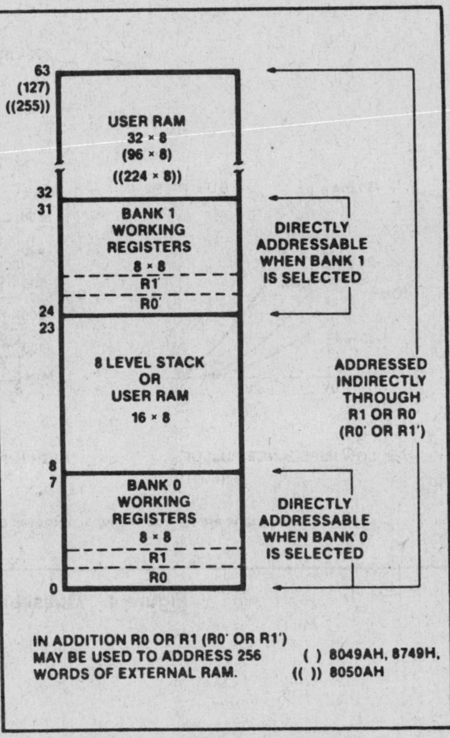

Figure 3. Data Memory Map

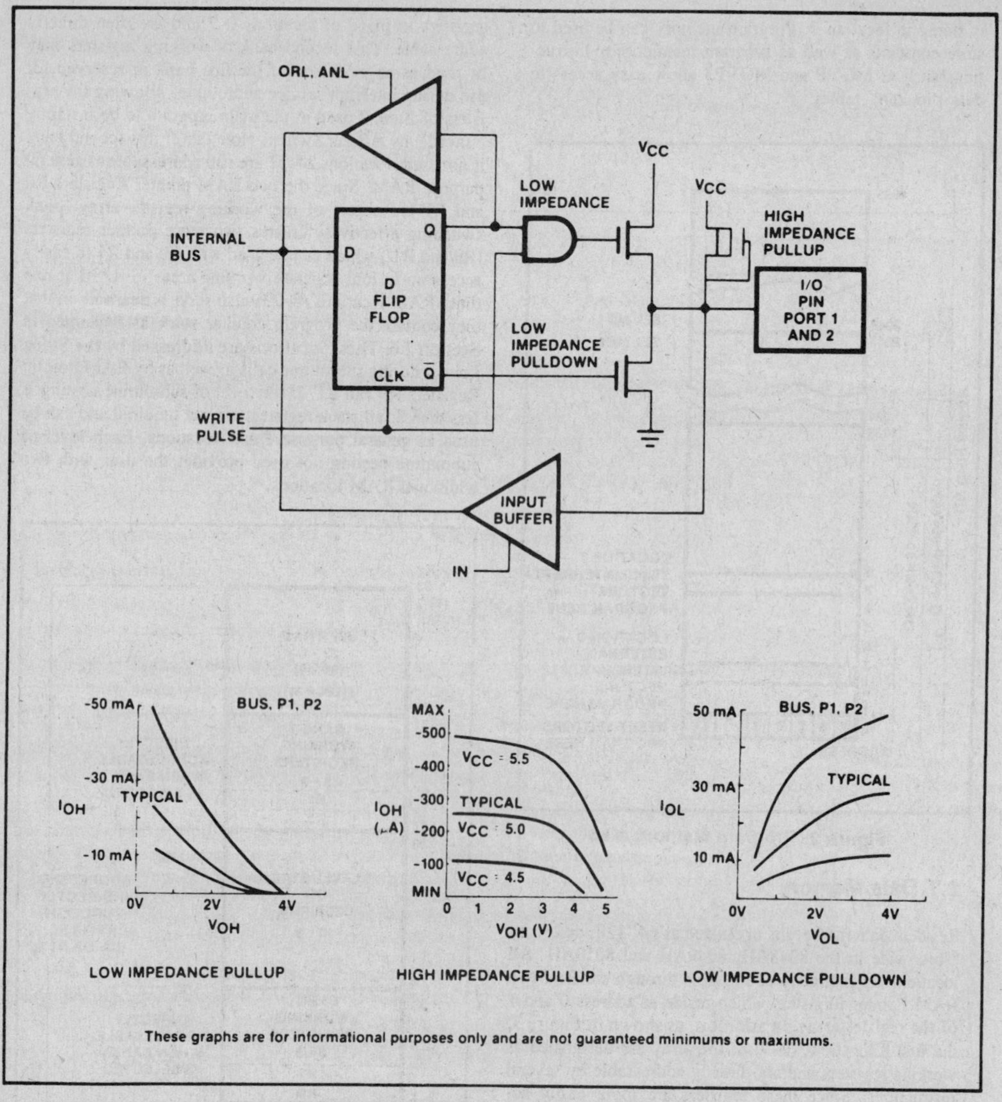

Figure 4. "Quasi-bidirectional" Port Structure

2.4 Input/Output

The 8048AH has 27 lines which can be used for input or output functions. These lines are grouped as 3 ports of 8 lines each which serve as either inputs, outputs or bidirectional ports and 3 "test" inputs which can alter program sequences when tested by conditional jump instructions.

PORTS 1 AND 2

Ports 1 and 2 are each 8 bits wide and have identical characteristics. Data written to these ports is statically latched and remains unchanged until rewritten. As input ports these lines are non-latching, i.e., inputs must be present until read by an input instruction. Inputs are fully TTL compatible and outputs will drive one standard TTL load.

The lines of ports 1 and 2 are called quasi-bidirectional because of a special output circuit structure which allows each line to serve as an input, and output, or both even though outputs are statically latched. Figure 4 shows the circuit configuration in detail. Each line is continuously pulled up to V_{CC} through a resistive device of relatively high impedance.

This pullup is sufficient to provide the source current for a TTL high level yet can be pulled low by a standard TTL gate thus allowing the same pin to be used for both input and output. To provide fast switching times in a "0" to "1" transition a relatively low impedance device is switched in momentarily ($\approx$ 1/5 of a machine cycle) whenever a "1" is written to the line. When a "0" is written to the line a low impedance device overcomes the light pullup and provides TTL current sinking capability. Since the pulldown transistor is a low impedance device a "1" must first be written to any line which is to be used as an input. Reset initializes all lines to the high impedance "1" state.

It is important to note that the ORL and the ANL are read/write operations. When executed, the μC "reads" the port, modifies the data according to the instruction, then "writes" the data back to the port. The "writing" (essentially an OUTL instruction) enables the low impedance pull-up momentarily again even if the data was unchanged from a "1." This specifically applies to configurations that have inputs and outputs mixed together on the same port. See also section 8 in the Expanded MCS-48 System chapter.

BUS

Bus is also an 8-bit port which is a true bidirectional port with associated input and output strobes. If the bidirectional feature is not needed, Bus can serve as either a statically latched output port or non-latching input port. Input and output lines on this port cannot be mixed however.

As a static port, data is written and latched using the OUTL instruction and inputted using the INS instruction. The INS and OUTL instructions generate pulses on the corresponding $\overline{RD}$ and $\overline{WR}$ output strobe lines; however, in the static port mode they are generally not used. As a bidirectional port the MOVX instructions are used to read and write the port. A write to the port generates a pulse on the $\overline{WR}$ output line and output data is valid at the trailing edge of $\overline{WR}$. A read of the port generates a pulse on the $\overline{RD}$ output line and input data must be valid at the trailing edge of $\overline{RD}$. When not being written or read, the BUS lines are in a high impedance state. See also sections 7 and 8 in the Expanded MCS-48 System chapter.

2.5 Test and INT Inputs

Three pins serve as inputs and are testable with the conditional jump instruction. These are T0, T1, and $\overline{INT}$. These pins allow inputs to cause program branches without the necessity to load an input port into the accumulator. The T0, T1, and $\overline{INT}$ pins have other possible functions as well. See the pin description in Section 3.

2.6 Program Counter and Stack

The Program Counter is an independent counter while the Program Counter Stack is implemented suing pairs of registers in the Data Memory Array. Only 10, 11, or 12 bits of the Program Counter are used to address the 1024, 2048, or 4096 words of on-board program memory of the 8048AH, 8049AH, or 8050AH, while the most significant bits can be used for external Program Memory fetches. See Figure 5. The Program Counter is initialized to zero by activating the Reset line.

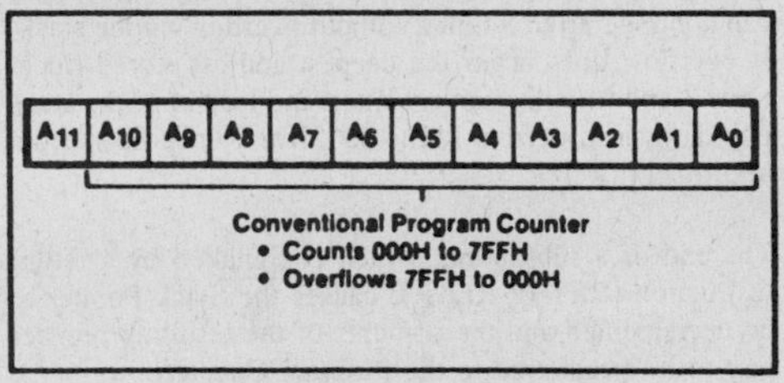

Figure 5. Program Counter

An interrupt or CALL to a subroutine causes the contents of the program counter to be stored in one of the 8 register pairs of the Program Counter Stack as shown in Figure 6. The pair to be used is determined by a 3-bit Stack Pointer which is part of the Program Status Word (PSW).

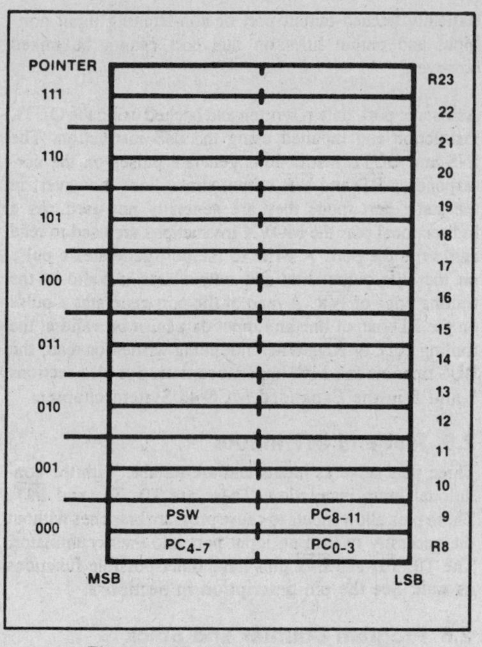

Figure 6. Program Counter Stack

Data RAM locations 8-23 are available as stack registers and are used to store the Program Counter and 4 bits of PSW as shown in Figure 6. The Stack Pointer when initialized to 000 points to RAM locations 8 and 9. The first subroutine jump or interrupt results in the program counter contents being transferred to locations 8 and 9 of the RAM array. The stack pointer is then incremented by one to point to locations 10 and 11 in anticipation of another CALL. Nesting of subroutines wihtin subroutines can continue up to 8 times without overflowing the stack. If overflow does occur the deepest address stored (locations 8 and 9) will be overwritten and lost since the stack pointer overflows from 111 to 000. It also underflows from 000 to 111.

The end of a subroutine, which is signalled by a return instruction (RET or RETR), causes the Stack Pointer to be decremented and the contents of the resulting register pair to be transferred to the Program Counter.

2.7 Program Status Word

An 8-bit status word which can be loaded to and from the accumulator exists called the Program Status Word (PSW). Figure 7 shows the information available in the word. The Program Status Word is actually a collection of flip-flops throughout the machine which can be read or written as a whole. The ability to write to PSW allows for easy restoration of machine status after a power down sequence.

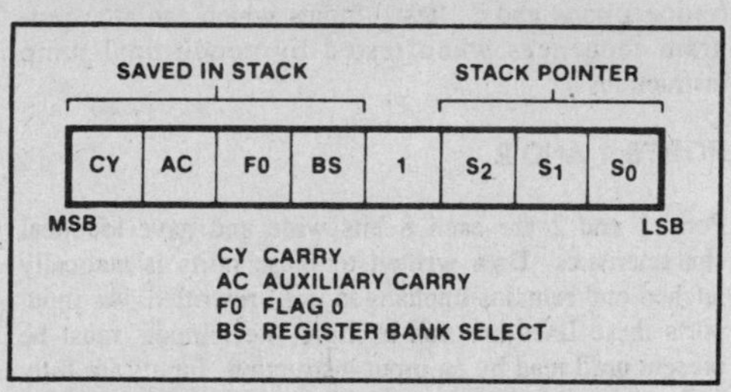

Figure 7. Program Status Word (PSW)

The upper four bits of PSW are stored in the Program Counter Stack with every call to subroutine or interrupt vector and are optionally restored upon return with the RETR instruction. The RET return instruction does not update PSW.

The PSW bit definitions are as follows:

Bits 0–2: Stack Pointer bits (S_0, S_1, S_2)

Bit 3: Not used ("1" level when read)

Bit 4: Working Register Bank Switch Bit (BS)
0 = Bank 0
1 = Bank 1

Bit 5: Flag 0 bit (F0) user controlled flag which can be complemented or cleared, and tested with the conditional jump instruction JF0.

Bit 6: Auxiliary Carry (AC) carry bit generated by an ADD instruction and used by the decimal adjust instruction DA A.

Bit 7: Carry (CY) carry flag which indicates that the previous operation has resulted in overflow of the accumulator.

2.8 Conditional Branch Logic

The conditional branch logic within the processsor enables several conditions internal and external to the processor to be tested by the users program. By using the conditional jump instruction the conditions that are listed in Table 1 can effect a change in the sequence of the program execution.

Table 1

Device Testable	Jump Conditions (Jump On)	
		not all
Accumulator	All zeros	zeros
Accumulator Bit	—	1
Carry Flag	0	1
User Flags (F0, F1)	—	1
Timer Overflow Flag	—	1
Test Inputs (T0, T1)	0	1
Interrupt Input ($\overline{INT}$)	0	—

2.9 Interrupt

An interrupt sequence is initiated by applying a low "0" level input to the INT pin. Interrupt is level triggered and active low to allow "WIRE ORing" of several interrupt sources at the input pin. Figure 8 shows the interrupt logic of the 8048AH. The Interrupt line is sampled every instruction cycle and when detected causes a "call to subroutine" at location 3 in program memory as soon as all cycles of the current instruction are complete. On 2-cycle instructions the interrupt line is sampled on the 2nd cycle only. $\overline{INT}$ must be held low for at least 3 machine cycles to ensure proper interrupt operations. As in any CALL to subroutine, the Program Counter and Program Status word are saved in the stack. For a description of this operation see the previous section, Program Counter and Stack. Program Memory location 3 usually contains an unconditional jump to an interrupt service subroutine elsewhere in program memory. The end of an interrupt service subroutine is signalled by the execution of a Return and Restore Status instruction RETR. The interrupt system is single level in that once an interrupt is detected all further interrupt requests are ignored until execution of an RETR reenables the interrupt input logic. This occurs at the beginning of the second cycle of the RETR instruction. This sequence holds true also for an internal interrupt generated by timer overflow. If an internal timer/counter generated interrupt and an external interrupt are detected at the same time, the external source will be recognized. See the following Timer/Counter section for a description of timer interrupt. If needed, a second external interrupt can be created by enabling the timer/counter interrupt, loading FFH in the Counter (ones less than terminal count), and enabling the event counter mode. A "1" to "0" transition on the T1 input will then cause an interrupt vector to location 7.

INTERRUPT TIMING

The interrupt input may be enabled or disabled under Program Control using the EN I and DIS I instructions. Interrupts are disabled by Reset and remain so until enabled by the users program. An interrupt request must be removed before the RETR instruction is executed upon return from the service routine otherwise the processor will re-enter the service routine immediately. Many peripheral devices prevent this situation by resetting their interrupt request line whenever the processor accesses (Reads or Writes) the peripherals data buffer register. If the interrupting device does not require access by the processor, one output line of the 8048AH may be designated as an "interrupt acknowledge" which is activated by the service subroutine to reset the interrupt request. The $\overline{INT}$ pin may also be tested using the conditional jump instruction JNI. This instruction may be used to detect the presence of a pending interrupt before interrupts are enabled. If interrupt is left disabled, $\overline{INT}$ may be used as another test input like T0 and T1.

1

2.10 Timer/Counter

The 8048AH contains a counter to aid the user in counting external events and generating accurate time delays without placing a burden on the processor for these functions. In both modes the counter operation is the same, the only difference being the source of the input to the counter. The timer/event counter is shown in Figure 9.

COUNTER

The 8-bit binary counter is presettable and readable with two MOV instructions which transfer the contents of the accumulator to the counter and vice versa. The counter content may be affected by Reset and should be initialized by software. The counter is stopped by a Reset or STOP TCNT instruction and remains stopped until started as a timer by a START T instruction or as an event counter by a START CNT instruction. Once started the counter will increment to this maximum count (FF) and overflow to zero continuing its count until stopped by a STOP TCNT instruction or Reset.

The increment from maximum count to zero (overflow) results in the setting of an overflow flag flip-flop and in the generation of an interrupt request. The state of the overflow flag is testable with the conditional jump instruction JTF. The flag is reset by executing a JTF or by Reset. The interrupt request is stored in a latch and then ORed with the external interrupt input INT. The timer interrupt may be enabled or disabled independently of external interrupt by the EN TCNTI and DIS TCNTI instructions. If enabled, the counter overflow will cause a subroutine call to location 7 where the timer or counter service routine may be stored.

If timer and external interrupts occur simultaneously, the external source will be recognized and the Call will be to

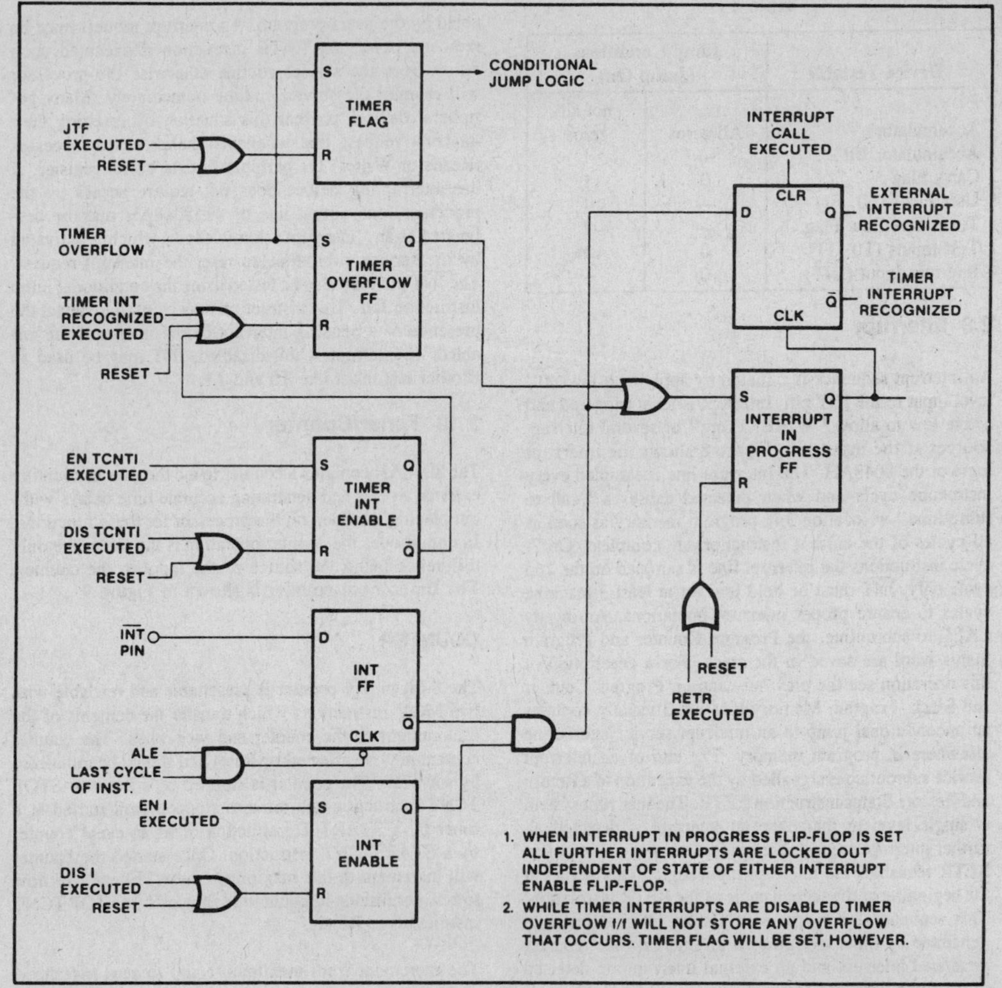

Figure 8. Interrupt Logic

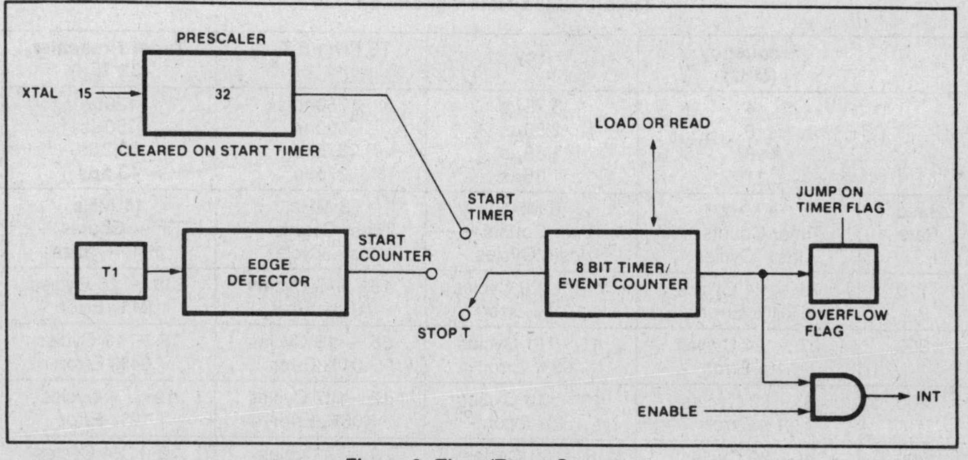

Figure 9. Timer/Event Counter

location 3. Since the timer interrupt is latched it will remain pending until the external device is serviced and immediately be recognized upon return from the service routine. The pending timer interrupt is reset by the Call to location 7 or may be removed by executing a DIS TCNT1 instruction.

AS AN EVENT COUNTER

Execution of a START CNT instruction connects the T1 input pin to the counter input and enables the counter. The T1 input is sampled at the beginning of state 3 or in later MCS-48 devices in state time 4. Subsequent high to low transitions on T1 will cause the counter to increment. T1 must be held low for at least 1 machine cycle to insure it won't be missed. The maximum rate at which the counter may be incremented is once per three instruction cycles (every 5.7 μsec when using an 8 MHz crystal) — there is no minimum frequency. T1 input must remain high for at least 1/5 machine cycle after each transition.

AS A TIMER

Eexcution of a START T instruction connects an internal clock to the counter input and enables the counter. The internal clock is derived bypassing the basic machine cycle clock through a ÷32 prescaler. The prescaler is reset during the START T instruction. The resulting clock increments the counter every 32 machine cycles. Various delays from 1 to 256 counts can be obtained by presetting the counter and detecting overflow. Times longer than 256 counts may be achieved by accumulating multiple overflows in a register under software control. For time resolution less than 1 count an external clock can be applied to the T1 input and the counter operated in the event counter mode. ALE divided by 3 or more can serve as this external clock. Very small delays or "fine tuning" of larger delays can be easily accomplished by software delay loops.

Often a serial link is desirable in an MCS-48 family member. Table 2 lists the timer counts and cycles needed for a specific baud rate given a crystal frequency.

2.11 Clock and Timing Circuits

Timing generation for the 8048AH is completely selfcontained with the exeception of a frequency reference which can be XTAL, ceramic resonator, or external clock source. The Clock and Timing circuitry can be divided into the following functional blocks.

OSCILLATOR

The on-board oscillator is a high gain parallel resonant circuit with a frequency range of 1 to 11 MHz. The X1 external pin is the input to the amplifier stage while X2 is the output. A crystal or ceramic resonator connected between X1 and X2 provides the feedback and phase shift required for oscillation. If an accurate frequency reference is not required, ceramic resonator may be used in place of the crystal.

For accurate clocking, a crystal should be used. An externally generated clock may also be applied to X1–X2 as the frequency source. See the data sheet for more information.

Table 2. Baud Rate Generation

	Frequency (MHz)	T_{cy}	T0 Prr(1/5 T_{cy})	Timer Prescaler (32 T_{cy})
	4 6 8 11	3.75μs 2.50μs 1.88μs 1.36μs	750ns 500ns 375ns 275ns	120μs 80μs 60.2μs 43.5μs
Baud Rate	4 MHz Timer Counts + Instr. Cycles	6 MHz Timer Counts + Instr. Cycles	8 MHz Timer Counts + Instr. Cycles	11 MHz Timer Counts + Instr. Cycles
110	75 + 24 Cycles .01% Error	113 + 20 Cycles .01% Error	151 + 3 Cycles .01% Error	208 + 28 Cycles .01% Error
300	27 + 24 Cycles .1% Error	41 + 21 Cycles .03% Error	55 + 13 Cycles .01% Error	76 + 18 Cycles. .04% Error
1200	6 + 30 Cycles .1% Error	10 + 13 Cycles .1% Error	12 + 27 Cycles .06% Error	19 + 4 Cycles .12% Error
1800	4 + 20 Cycles .1% Error	6 + 30 Cycles .1% Error	9 + 7 Cycles .17% Error	12 + 24 Cycles .12% Error
2400	3 + 15 Cycles .1% Error	5 + 6 Cycles .4% Error	6 + 24 Cycles .29% Error	9 + 18 Cycles .12% Error
4800	1 + 23 Cycles 1.0% Error	2 + 19 Cycles .4% Error	3 + 14 Cycles .74% Error	4 + 25 Cycles .12% Error

STATE COUNTER

The output of the oscillator is divided by 3 in the State Counter to create a clock which defines the state times of the machine (CLK). CLK can be made available on the external pin T0 by executing an ENT0 CLK instruction. The output of CLK on T0 is disabled by Reset of the processor.

CYCLE COUNTER

CLK is then divided by 5 in the Cycle Counter to provide a clock which defines a machine cycle consisting of 5 machine states as shown in Figure 10. Figure 11 shows the different internal operations as divided into the machine states. This clock is called Address Latch Enable (ALE) because of its function in MCS-48 systems with external memory. It is provided continuously on the ALE output pin.

2.12 Reset

The reset input provides a means for initialization for the processor. This Schmitt-trigger input has an internal pull-up device which in combination with an external 1 μ fd capacitor provides an internal reset pulse of sufficient length to guarantee all circuitry is reset, as shown in Figure 12. If the reset pulse is generated externally the RESET pin must be held low for at least 10 milliseconds after the power supply is within tolerance. Only 5 machine cycles (6.8 μs @ 11 MHz) are required if power is already on and the oscillator has stabilized. ALE and PSEN (if EA = 1) are active while in Reset.

Reset performs the following functions:

1) Sets program counter to zero.

2) Sets stack pointer to zero.

3) Selects register bank 0.

4) Selects memory bank 0.

5) Sets BUS to high impedance state (except when EA = 5V).

6) Sets Ports 1 and 2 to input mode.

7) Disables interrupts (timer and external).

8) Stops timer.

9) Clears timer flag.

10) Clears F0 and F1.

11) Disables clock output from T0.

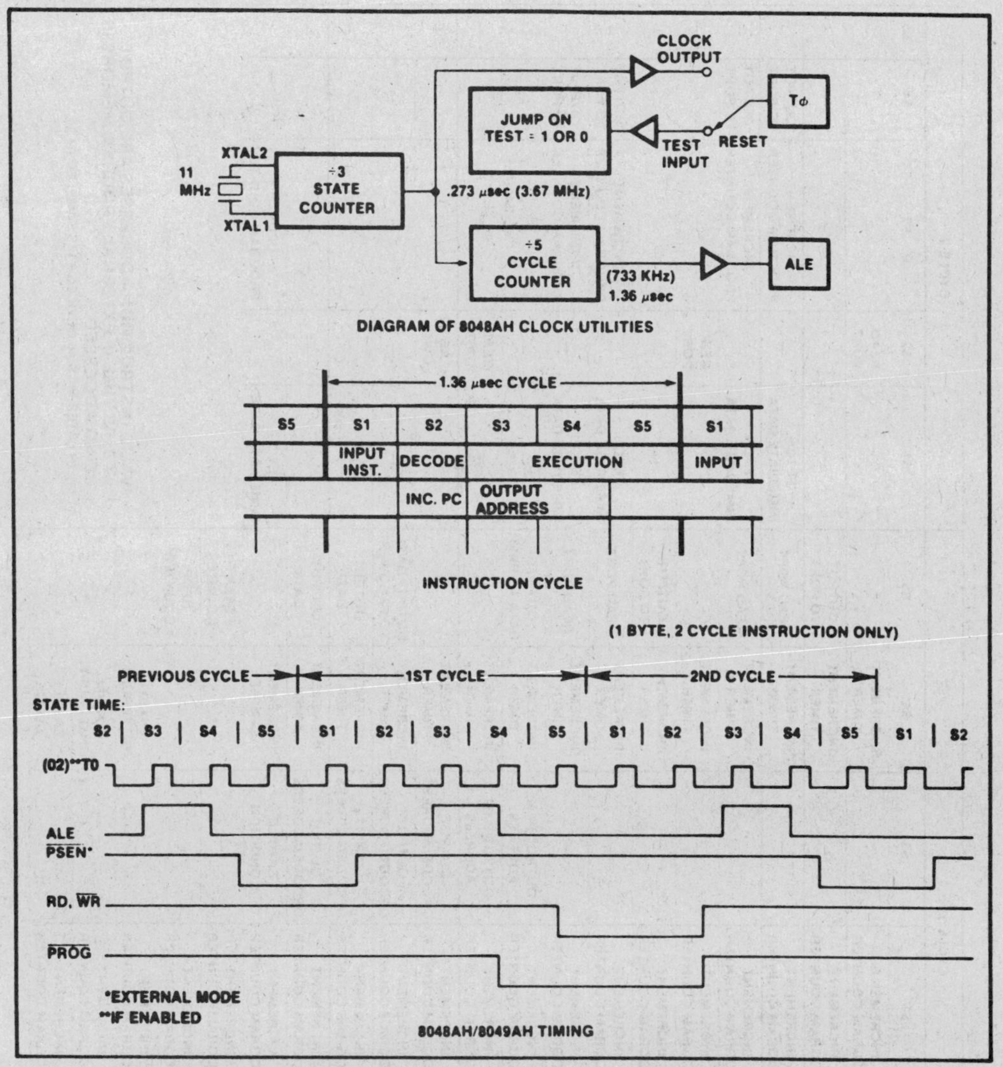

Figure 10. MCS®-48 Timing Generation and Cycle Timing

2.13 Single-Step

This feature, as pictured in Figure 13, provides the user with a debug capability in that the processor can be stepped through the program one instruction at a time. While stopped, the address of the next instruction to be fetched is available concurrently on BUS and the lower half of Port 2. The user can therefore follow the program through each of the instruction steps. A timing diagram, showing the interaction between output ALE and input $\overline{SS}$, is shown. The BUS buffer contents are lost during single step; however, a latch may be added to reestablish the lost I/O capability if needed. Data is valid at the leading edge of ALE.

	CYCLE 1					CYCLE 2				
INSTRUCTION	S1	S2	S3	S4	S5	S1	S2	S3	S4	S5
IN A,P	FETCH INSTRUCTION	INCREMENT PROGRAM COUNTER	—	*INCREMENT TIMER	—	—	READ PORT	—	• —	—
OUTL P,A	FETCH INSTRUCTION	INCREMENT PROGRAM COUNTER	—	*INCREMENT TIMER	OUTPUT TO PORT	—	—	—	• —	—
ANL P, # DATA	FETCH INSTRUCTION	INCREMENT PROGRAM COUNTER	—	*INCREMENT TIMER	READ PORT	FETCH IMMEDIATE DATA	—	INCREMENT PROGRAM COUNTER	*OUTPUT TO PORT	—
ORL P, # DATA	FETCH INSTRUCTION	INCREMENT PROGRAM COUNTER	—	*INCREMENT TIMER	READ PORT	FETCH IMMEDIATE DATA	—	INCREMENT PROGRAM COUNTER	*OUTPUT TO PORT	—
INS A, BUS	FETCH INSTRUCTION	INCREMENT PROGRAM COUNTER	—	INCREMENT TIMER	—	—	READ PORT	—	• —	—
OUTL BUS, A	FETCH INSTRUCTION	INCREMENT PROGRAM COUNTER	—	INCREMENT TIMER	OUTPUT TO PORT	—	—	—	• —	—
ANL BUS, # DATA	FETCH INSTRUCTION	INCREMENT PROGRAM COUNTER	—	*INCREMENT TIMER	READ PORT	FETCH IMMEDIATE DATA	—	INCREMENT PROGRAM COUNTER	*OUTPUT TO PORT	—
ORL BUS, # DATA	FETCH INSTRUCTION	INCREMENT PROGRAM COUNTER	—	*INCREMENT TIMER	READ PORT	FETCH IMMEDIATE DATA	—	INCREMENT PROGRAM COUNTER	*OUTPUT TO PORT	—
MOVX @ R,A	FETCH INSTRUCTION	INCREMENT PROGRAM COUNTER	OUTPUT RAM ADDRESS	INCREMENT TIMER	OUTPUT DATA TO RAM	—	—	—	• —	—
MOVX A,@R	FETCH INSTRUCTION	INCREMENT PROGRAM COUNTER	OUTPUT RAM ADDRESS	INCREMENT TIMER	—	—	READ DATA	—	• —	—
MOVD A,P_i	FETCH INSTRUCTION	INCREMENT PROGRAM COUNTER	OUTPUT OPCODE/ADDRESS	INCREMENT TIMER	—	—	READ P2 LOWER	—	• —	—
MOVD P_i,A	FETCH INSTRUCTION	INCREMENT PROGRAM COUNTER	OUTPUT OPCODE/ADDRESS	INCREMENT TIMER	OUTPUT DATA TO P2 LOWER	—	—	—	• —	—
ANLD P,A	FETCH INSTRUCTION	INCREMENT PROGRAM COUNTER	OUTPUT OPCODE/ADDRESS	INCREMENT TIMER	OUTPUT DATA	—	—	—	• —	—
ORLD P,A	FETCH INSTRUCTION	INCREMENT PROGRAM COUNTER	OUTPUT OPCODE/ADDRESS	INCREMENT TIMER	OUTPUT DATA	—	—	—	• —	—
J(CONDITIONAL)	FETCH INSTRUCTION	INCREMENT PROGRAM COUNTER	SAMPLE CONDITION	*INCREMENT SAMPLE	—	FETCH IMMEDIATE DATA	—	UPDATE PROGRAM COUNTER	• —	—
STRT T STRT CNT	FETCH INSTRUCTION	INCREMENT PROGRAM COUNTER	—	• —	START COUNTER					
STOP TCNT	FETCH INSTRUCTION	INCREMENT PROGRAM COUNTER	—	• —	STOP COUNTER					
ENI	FETCH INSTRUCTION	INCREMENT PROGRAM COUNTER	—	• ENABLE INTERRUPT	—					
DIS I	FETCH INSTRUCTION	INCREMENT PROGRAM COUNTER	—	• DISABLE INTERRUPT	—					
ENTO CLK	FETCH INSTRUCTION	INCREMENT PROGRAM COUNTER		• ENABLE CLOCK	—					

*VALID INSTRUCTION ADDRESSES ARE OUTPUT AT THIS TIME IF EXTERNAL PROGRAM MEMORY IS BEING ACCESSED.

(1) IN LATER MCS-48 DEVICES T1 IS SAMPLED IN S4.

Figure 11. 8048AH/8049AH Instruction Timing Diagram

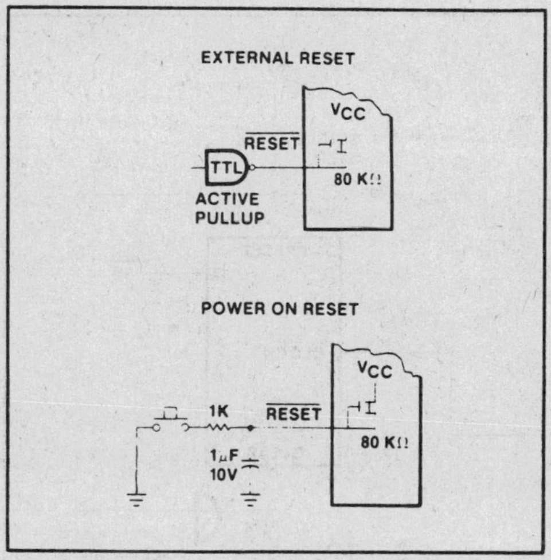

Figure 12.

TIMING

The 8048AH operates in a single-step mode as follows:

1) The processor is requested to stop by applying a low level on $\overline{SS}$.
2) The processor responds by stopping during the address fetch portion of the next instruction. If a double cycle instruction is in progress when the single step command is received, both cycles will be completed before stopping.
3) The processor acknowledges it has entered the stopped state by raising ALE high. In this state (which can be maintained indefinitely) the address of the next instruction to be fetched is present on BUS and the lower half of port 2.
4) $\overline{SS}$ is then raised high to bring the processor out of the stopped mode allowing it to fetch the next instruction. The exit from stop is indicated by the processor bringing ALE low.
5) To stop the processor at the next instruction $\overline{SS}$ must be brought low again soon after ALE goes low. If $\overline{SS}$ is left high the processor remains in a "Run" mode.

A diagram for implementing the single-step function of the 8748H is shown in Figure 13. D-type flip-flop with preset and clear is used to generate $\overline{SS}$. In the run mode SS is held high by keeping the flip-flop preset (preset has precedence over the clear input). To enter single step, preset is removed allowing ALE to bring $\overline{SS}$ low via the clear input. ALE should be buffered since the clear input of an SN7474 is the equivalent of 3 TTL loads. The processor is now in the stopped state. The next instruction is initiated by clocking a "1" into the flip-flop. This "1" will not appear on $\overline{SS}$ unless ALE is high removing clear from the flip-flop. In response to $\overline{SS}$ going high the processor begins an instruction fetch which brings ALE low resetting $\overline{SS}$ through the clear input and causing the processor to again enter the stopped state.

2.14 Power Down Mode (8048AH, 8049AH, 8050AH, 8039AHL, 8035AHL, 8040AHL)

Extra circuitry has been added to the 8048AH/8049AH/8050AH ROM version to allow power to be removed from all but the data RAM array for low power standby operation. In the power down mode the contents of data RAM can be maintained while drawing typically 10% to 15% of normal operating power requirements.

V_{CC} serves as the 5V supply pin for the bulk of circuitry while the V_{DD} pin supplies only the RAM array. In normal operation both pins are a 5V while in standby, V_{CC} is at ground and V_{DD} is maintained at its standby value. Applying Reset to the processor through the $\overline{RESET}$ pin inhibits any access to the RAM by the processor and guarantees that RAM cannot be inadvertently altered as power is removed from V_{CC}.

A typical power down sequence (Figure 14) occurs as follows:

1) Imminent power supply failure is detected by user defined circuitry. Signal must be early enough to allow 8048AH to save all necessary data before V_{CC} falls below normal operating limits.
2) Power fail signal is used to interrupt processor and vector it to a power fail service routine.
3) Power fail routine saves all important data and machine status in the internal data RAM array. Routine may also initiate transfer of backup supply to the V_{DD} pin and indicate to external circuitry that power fail routine is complete.
4) Reset is applied to guarantee data will not be altered as the power supply falls out of limits. Reset must be held low until V_{CC} is at ground level.

Recovery from the Power Down mode can occur as any other power-on sequence with an external capacitor on the Reset input providing the necessary delay. See the previous section on Reset.

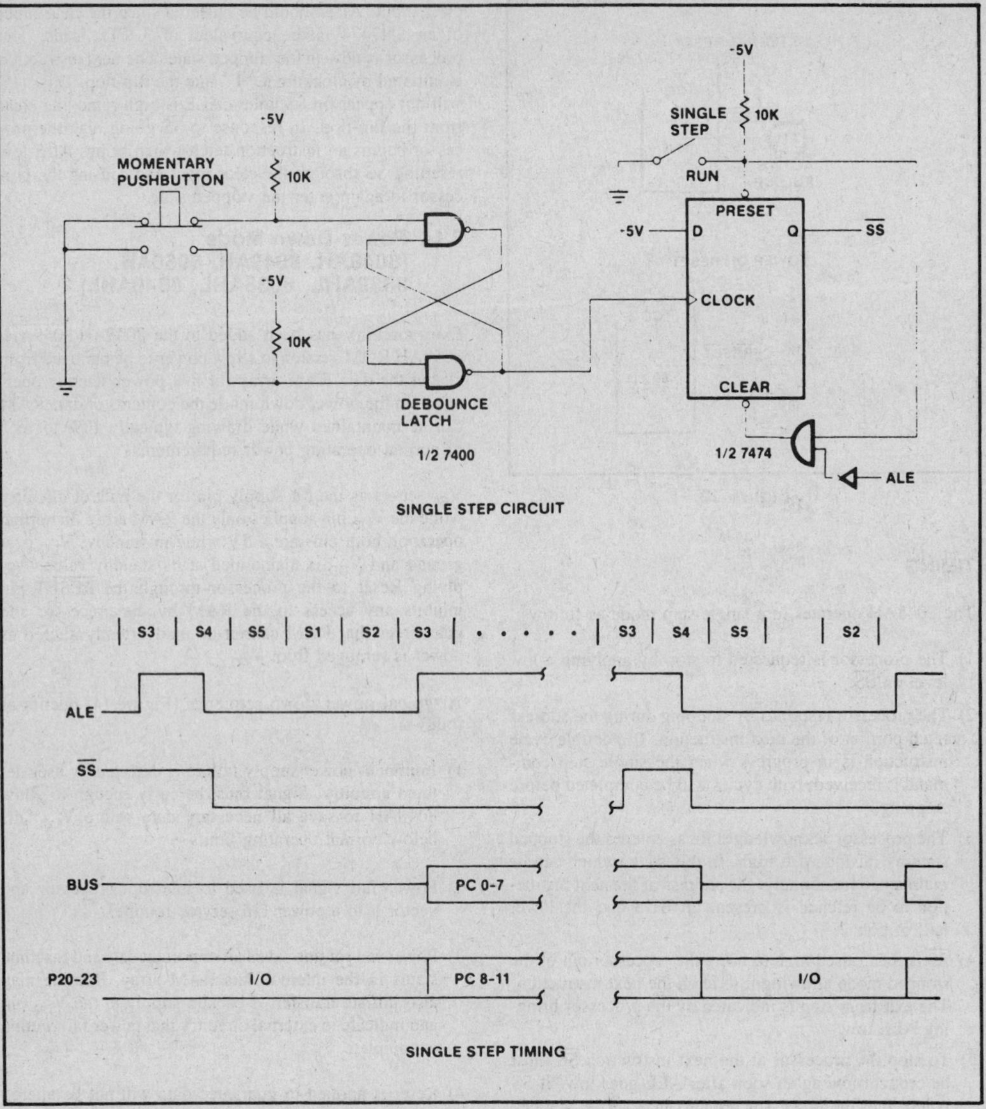

Figure 13. Single Step Operation

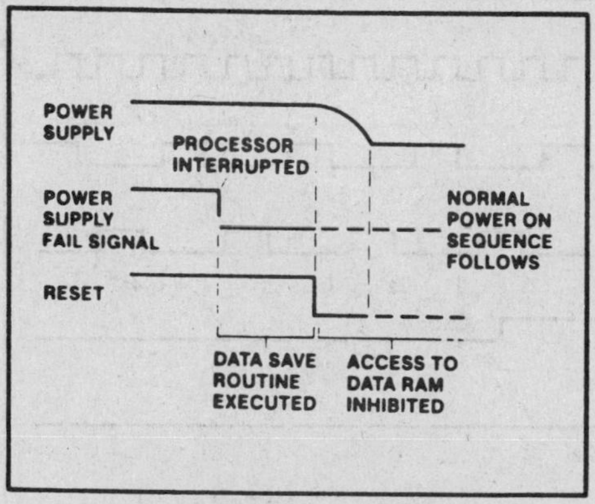

Figure 14. Power Down Sequence

2.15 External Access Mode

Normally the first 1K (8048AH), 2K (8049AH), or 4K (8050AH) words of program memory are automatically fetched from internal ROM or EPROM. The EA input pin however allows the user to effectively disable internal program memory by forcing all program memory fetches to reference external memory. The following chapter explains how access to external program memory is accomplished.

The External Access mode is very useful in system test and debug because it allows the user to disable his internal applications program and substitute an external program of his choice — a diagnostic routine for instance. In addition, the date sheet shows how internal program memory can be read externally, independent of the processor. A "1" level on EA initiates the external accesss mode. For proper operation, Reset should be applied while the EA input is changed.

2.16 Sync Mode

The 8048AH, 8049AH, 8050AH has incorporated a new SYNC mode. The Sync mode is provided to ease the design of multiple controller circuits by allowing the designer to force the device into known phase and state time. The SYNC mode may also be utilized by automatic test equipment (ATE) for quick, easy, and efficient synchronizing between the tester and the DUT (device under test).

SYNC mode is enabled when SS' pin is raised to high voltage level of +12 volts. To begin synchronization, T0 is raised to 5 volts at least four clocks cycles after SS'. T0 must be high for at least four X1 clock cycles to fully reset the prescaler and time state generators. T0 may then be brought down with the rising edge of X1. Two clock cycles later, with the rising edge of X1, the device enters into Time State 1, Phase 1, SS' is then brought down to 5 volts 4 clocks later after T0. RESET' is allowed to go high 5 tCY (75 clocks) later for normal execution of code. See Figure 15.

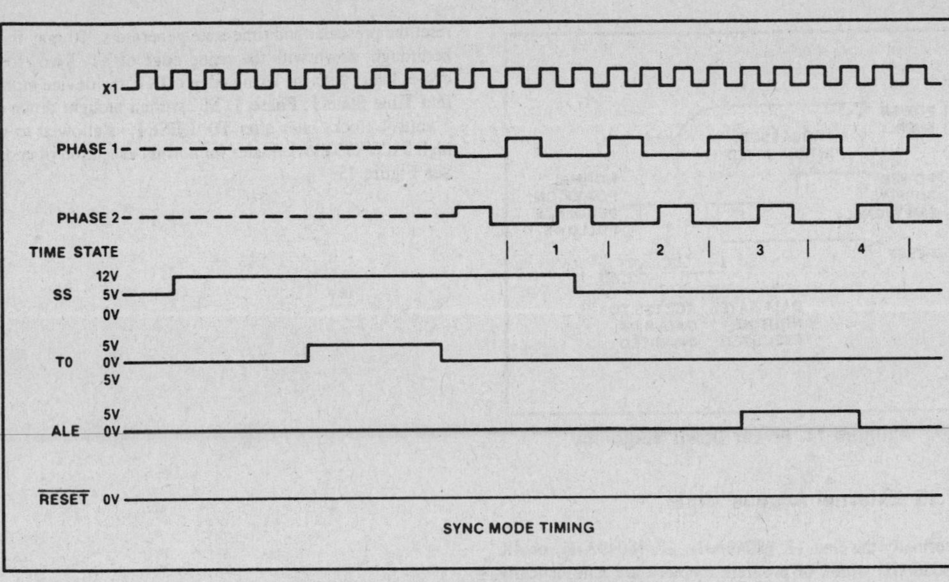

Figure 15. Sync Mode Timing

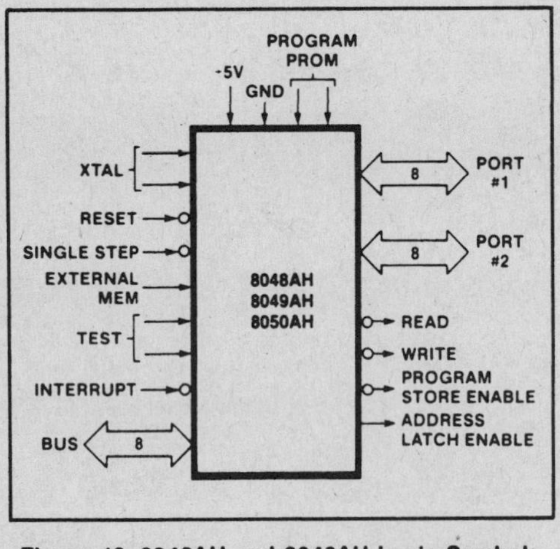

Figure 16. 8048AH and 8049AH Logic Symbol

3.0 PIN DESCRIPTION

The MCS-48 processors are packaged in 40 pin Dual In-Line Packages (DIP's). Table 3 is a summary of the functions of each pin. Figure 16 is the logic symbol for the 8048AH product family. Where it exists, the second paragraph describes each pin's function in an expanded MCS-48 system. Unless otherwise specified, each input is TTL compatible and each output will drive one standard TTL load.

Table 3. Pin Description

Designation	Pin Number*	Function
V_{SS}	20	Circuit GND potential
V_{DD}	26	Programming power supply; 21V during program for the 8748H/8749H: +5V during operation for both ROM and EPROM. Low power standby pin in 8048AH and 8049AH/8050AH ROM versions.
V_{CC}	40	Main power supply; +5V during operation and during 8748H and 8749H programming.
PROG	25	Program pulse; +18V input pin during 8748H/8749H programming. Output strobe for 8243 I/O expander.
P10-P17 (Port 1)	27-34	8-bit quasi-bidirectional port. (Internal Pullup ≈ 50KΩ)
P20-P27 (Port 2)	21-24 35-38	8-bit quasi-bidirectional port. (Internal Pullup ≈ 50KΩ)
		P20-P23 contain the four high order program counter bits during an external program memory fetch and serve as a 4-bit I/O expander bus for 8243.
D0-D7 (BUS)	12-19	True bidirectional port which can be written or read synchronously using the $\overline{RD}$, $\overline{WR}$ strobes. The port can also be statically latched.
		Contains the 8 low order program counter bits during an external program memory fetch, and receives the addressed instruction under the control of $\overline{PSEN}$. Also contains the address and data during an external RAM data store instruction, under control of ALE, $\overline{RD}$, and $\overline{WR}$.
T0	1	Input pin testable using the conditional transfer instructions JT0 and JNT0. T0 can be designated as a clock output using ENT0 CLK instruction. T0 is also used during programming and sync mode.
T1	39	Input pin testable using the JT1, and JNT1 instructions. Can be designated the event counter input using the STRT CNT instruction. (See Section 2.10).
$\overline{INT}$	6	Interrupt input. Initiates an interrupt if interrupt is enabled. Interrupt is disabled after a reset. (Active low)
		Interrupt must remain low for at least 3 machine cycles to ensure proper operation.
$\overline{RD}$	8	Output strobe activated during a BUS read. Can be used to enable data onto the BUS from an external device. (Active low)
		Used as a Read Strobe to External Data Memory.
$\overline{RESET}$	4	Input which is used to initialize the processor. Also used during EPROM programming and verification. (Active low) (Internal pullup ≈ 80K Ω)
$\overline{WR}$	10	Output strobe during a BUS write. (Active low) Used as write strobe to external data memory.
ALE	11	Address Latch Enable. This signal occurs once during each cycle and is useful as a clock output.
		The negative edge of ALE strobes address into external data and program memory.

Designation	Pin Number*	Function
$\overline{PSEN}$	9	Program Store Enable. This output occurs only during a fetch to external program memory. (Active low)
$\overline{SS}$	5	Single step input can be used in conjunction with ALE to "single step" the processor through each instruction. (Active low) (Internal pullup ≈ 300KΩ) +12V for sync modes (See 2.16).
EA	7	External Access input which forces all program memory fetches to reference external memory. Useful for emulation and debug, and essential for testing and program verification. (Active high) +12V for 8048AH 8049AH 8050AH program verification and +18V for 8748H 8749H program verification (Internal pullup ≈ 10MΩ on 8048AH 8049AH 8035AHL 8039AHL 8050AH 8040AHL)
XTAL1	2	One side of crystal input for internal oscillator. Also input for external source.
XTAL2	3	Other side of crystal external source input.

*Unless otherwise stated, inputs do not have internal pullup resistors. 8048AH, 8748H, 8049AH, 8050AH, 8040AHL

4.0 PROGRAMMING, VERIFYING AND ERASING EPROM

The internal Program Memory of the 8748H and the 8749H may be erased and reprogrammed by the user as explained in the following sections. See also the 8748H and 8749H data sheets.

4.1 Programming/Verification

In brief, the programming process consists of: activating the program mode, applying an address, latching the address, applying data, and applying a programming pulse. This programming algorithm applies to both the 8748H and 8749H. Each word is programmed completely before moving on to the next and is followed by a verification step. The following is a list of the pins used for programming and a descsription of their functions:

Pin	Function
XTAL 1	Clock Input (3 to 4 MHz)
Reset	Initialization and Address Latching
Test 0	Selection of Program (0V) or Verify (5V) Mode
EA	Activation of Program/Verify Modes
BUS	Address and Data Input Data Output During Verify
P20-1	Address Input for 8748H
P20-2	Address Input for 8749H
V_{DD}	Programming Power Supply
PROG	Program Pulse Input
P10–P11	Tied to ground (8749H only)

8748H AND 8749H ERASURE CHARACTERISTICS

The erasure characteristics of the 8748H and 8749H are such that erasure begins to occur when exposed to light with wavelengths shorter than approximately 4000 Angstroms (A). It should be noted that sunlight and certain types of fluorescent lamps have wavelengths in the 3000–4000A range. Data show that constant exposure to room level fluorescent lighting could erase the typical 8748H and 8749H in approximately 3 years while it would take approximately 1 week to cause erasure when exposed to direct sunlight. If the 8748H or 8749H is to be exposed to these types of lighting conditions for extended periods of time, opaque labels should be placed over the 8748H window to prevent unintentional erasure.

When erased, bits of the 8748H and 8749H Program Memory are in the logic "0" state.

The recommended erasure procedure for the 8748H and 8749H is exposure to shortwave ultraviolet light which has a wavelength of 2537 Angstroms (A). The integrated dose (i.e., UV intensity X exposure time) for erasure should be a minimum of 15W-sec/cm^2. The erasure time with this dosage is approximately 15 to 20 minutes using an ultraviolet lamp with a 12000μW/cm^2 power rating. The 8748H and 8749H should be placed within one inch from the lamp tubes during erasure. Some lamps have a filter in their tubes and this filter should be removed before erasure.

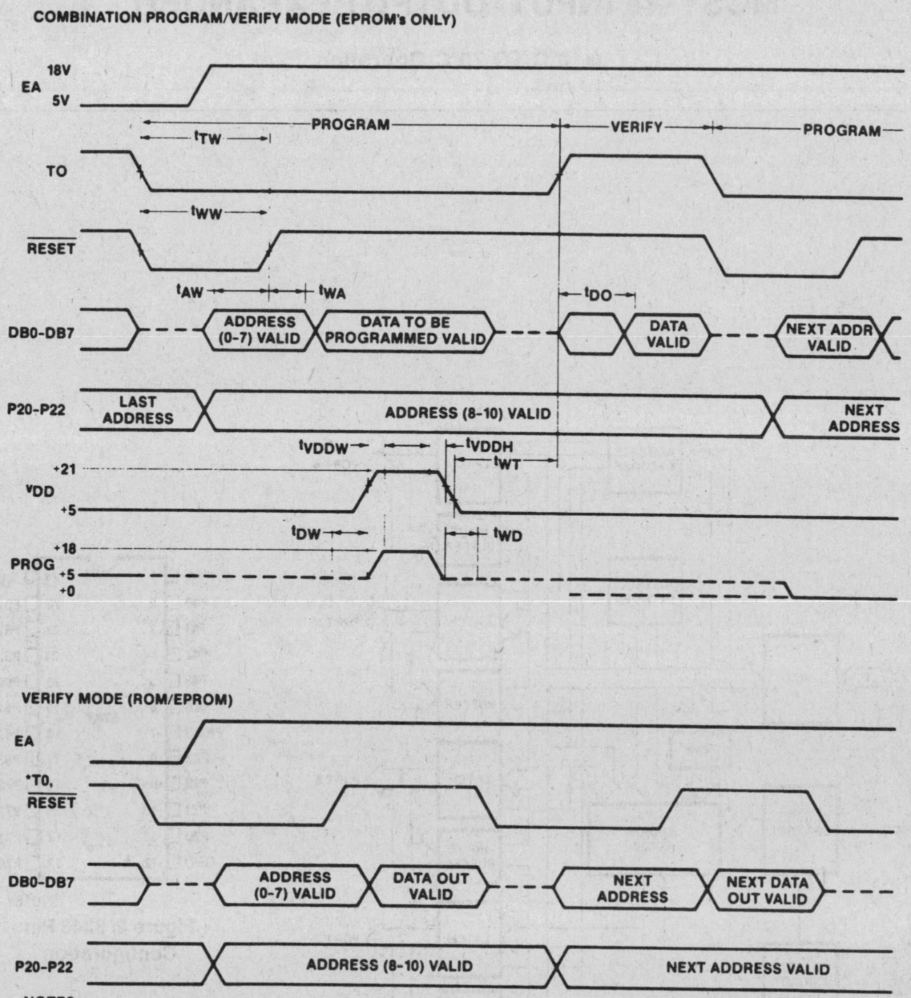

NOTES:
1. PROG MUST FLOAT IF EA IS LOW (I.E., ≠ 18V).

*T0 ON EPROM ONLY.

Figure 17. Program/Verify Sequence for 8749H/8748H

intel®

PRELIMINARY

8243
MCS®-48 INPUT/OUTPUT EXPANDER

■ 0°C TO 70°C Operation

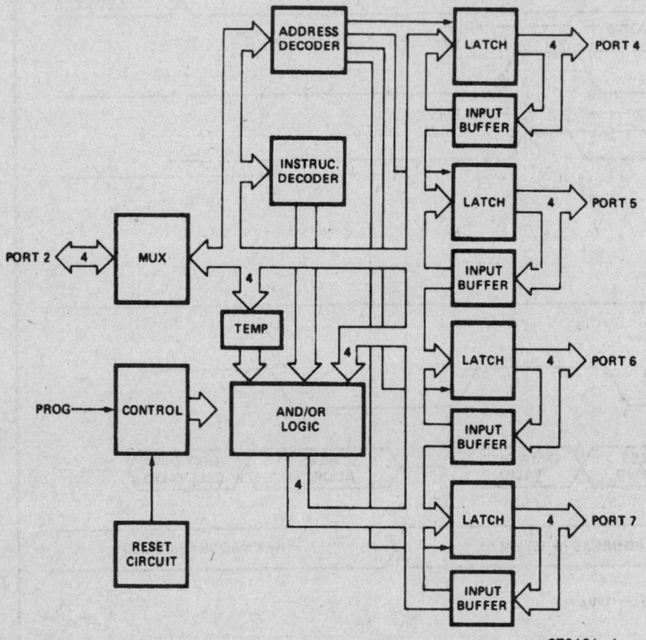

Figure 1. 8243 Block Diagram

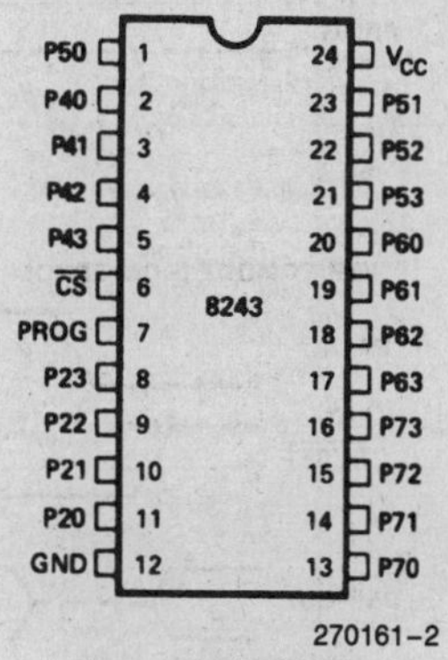

Figure 2. 8243 Pin Configuration

November 1990
Order Number: 270161-001

Table 1. Pin Description

Symbol	Pin No.	Function
PROG	7	Clock Input. A high to low transition on PROG signifies that address and control are available on P20–P23, and a low to high transition signifies that data is available on P20–P23.
$\overline{CS}$	6	Chip Select Input. A high on CS inhibits any change of output or internal status.
P20–P23	11–8	Four (4) bit bi-directional port contains the address and control bits on a high to low transition of PROG. During a low to high transition, P2 contains the data for a selected output port if a write operation, or the data from a selected port before the low to high transition if a read operation.
GND	12	0V supply.
P40–P43	2–5	Four (4) bit bi-directional I/O ports.
P50–P53 P60–P63 P70–P73	1, 23–21 20–17 13–16	May be programmed to be input (during read), low impedance latched output (after write), or a tri-state (after read). Data on pins P20–P23 may be directly written, ANDed or ORed with previous data.
V_{CC}	24	+5V supply.

FUNCTIONAL DESCRIPTION

General Operation

The 8243 contains four 4-bit I/O ports which serve as an extension of the on-chip I/O and are addressed as Ports 4–7. The following operations may be performed on these ports:

- Transfer Accumulator to Port.
- Transfer Port to Accumulator.
- AND Accumulator to Port.
- OR Accumulator to Port.

All communication between the 8048 and the 8243 occurs over Port 2 (P20–P23) with timing provided by an output pulse on the PROG pin of the processor. Each transfer consists of two 4-bit nibbles:

The first containing the "op code" and port address and the second containing the actual 4-bits of data. A high to low transition of the PROG line indicates that address is present while a low to high transition indicates the presence of data. Additional 8243's may be added to the 4-bit bus and chip selected using additional output lines from the 8048/8748/8035.

Power On Initialization

Initial application of power to the device forces input/output Ports 4, 5, 6, and 7 to the tri-state and Port 2 to the input mode. The PROG pin may be either high or low when power is applied. The first high to low transition of PROG causes the device to exit power on mode. The power on sequence is initiated if V_{CC} drops below 1V.

P21	P20	Address Code	P23	P22	Instruction Code
0	0	Port 4	0	0	Read
0	1	Port 5	0	1	Write
1	0	Port 6	1	0	ORLD
1	1	Port 7	1	1	ANLD

Write Modes

The device has three write modes. MOVD Pi, A directly writes new data into the selected port and old data is lost. ORLD Pi, A takes new data, OR's it with the old data and then writes it to the port. ANLD Pi, A takes new data, AND's it with the old data and then writes it to the port. Operation code and port address are latched from the input Port 2 on the high to low transition of the PROG pin. On the low to high transition of PROG data on Port 2 is transferred to the logic block of the specified output port.

After the logic manipulation is performed, the data is latched and outputed. The old data remains latched until new valid outputs are entered.

Read Mode

The device has one read mode. The operation code and port address are latched from the input Port 2 on the high to low transition of the PROG pin. As soon as the read operation and port address are decoded, the appropriate outputs are tri-stated, and the input buffers switched on. The read operation is terminated by a low to high transition of the PROG pin. The port (4, 5, 6 or 7) that was selected is switched to the tri-stated mode while Port 2 is returned to the input mode.

Normally, a port will be in an output (write mode) or input (read mode). If modes are changed during operation, the first read following a write should be ignored; all following reads are valid. This is to allow the external driver on the port to settle after the first read instruction removes the low impedance drive from the 8243 output. A read of any port will leave that port in a high impedance state.

ABSOLUTE MAXIMUM RATINGS*

Ambient Temperature Under Bias ...... 0°C to 70°C
Storage Temperature .... −65°C to +150°C
Voltage on Any Pin
with Respect to Ground.......... −0.5V to +7V
Power Dissipation .... 1 Watt

NOTICE: This data sheet contains preliminary information on new products in production. It is valid for the devices indicated in the revision history. The specifications are subject to change without notice.

**WARNING: Stressing the device beyond the "Absolute Maximum Ratings" may cause permanent damage. These are stress ratings only. Operation beyond the "Operating Conditions" is not recommended and extended exposure beyond the "Operating Conditions" may affect device reliability.*

D.C. CHARACTERISTICS T_A = 0°C to 70°C, V_{CC} = 5V ± 10%

Symbol	Parameter	Min	Typ	Max	Units	Test Conditions
V_{IL}	Input Low Voltage	−0.5		0.8	V	
V_{IH}	Input High Voltage	2.0		V_{CC} + 0.5	V	
V_{OL1}	Output Low Voltage Ports 4–7			0.45	V	I_{OL} = 4.5 mA*
V_{OL2}	Output Low Voltage Port 7			1	V	I_{OL} = 20 mA
V_{OH1}	Output High Voltage Ports 4–7	2.4			V	I_{OH} = 240 μA
I_{IL1}	Input Leakage Ports 4–7	−10		20	μA	V_{in} = V_{CC} to 0V
I_{IL2}	Input Leakage Port 2, CS, PROG	−10		10	μA	V_{in} = V_{CC} to 0V
V_{OL3}	Output Low Voltage Port 2			0.45	V	I_{OL} = 0.6 mA
I_{CC}	V_{CC} Supply Current		10	20	mA	(Note 1)
V_{OH2}	Output Voltage Port 2	2.4				I_{OH} = 100 μA
I_{OL}	Sum of All I_{OL} From 16 Outputs			72	mA	4.5 mA Each Pin

*Refer to Figure 3 for additional sink current capability.

A.C. CHARACTERISTICS $T_A = 0°C$ to $70°C$, $V_{CC} = 5V \pm 10\%$

Symbol	Parameter	Min	Max	Units	Test Conditions
t_A	Code Valid before PROG	50		ns	80 pF Load
t_B	Code Valid after PROG	60		ns	20 pF Load
t_C	Data Valid before PROG	200		ns	80 pF Load
t_D	Data Valid after PROG	20		ns	20 pF Load
t_H	Floating after PROG	0	150	ns	20 pF Load
t_K	PROG Negative Pulse Width	700		ns	
t_{CS}	CS Valid before/after PROG	50		ns	
t_{PO}	Ports 4–7 Valid after PROG		700	ns	100 pF Load
t_{LP1}	Ports 4–7 Valid before/after PROG	100		ns	
t_{ACC}	Port 2 Valid after PROG		650	ns	80 pF Load

NOTE:
1. I_{CC} (−40°C to 85°C EXPRESS options) 15 mA typical/25 mA maximum.

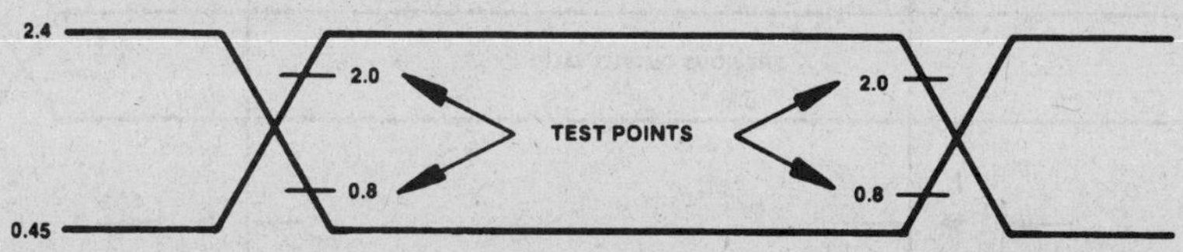

270161–3

A.C. Testing: Inputs are driven at 2.4V for a Logic "1" and 0.45V for a Logic "0". Output timing measurements are made at 2.0V for Logic "1" and 0.8V for a Logic "0".

WAVEFORMS

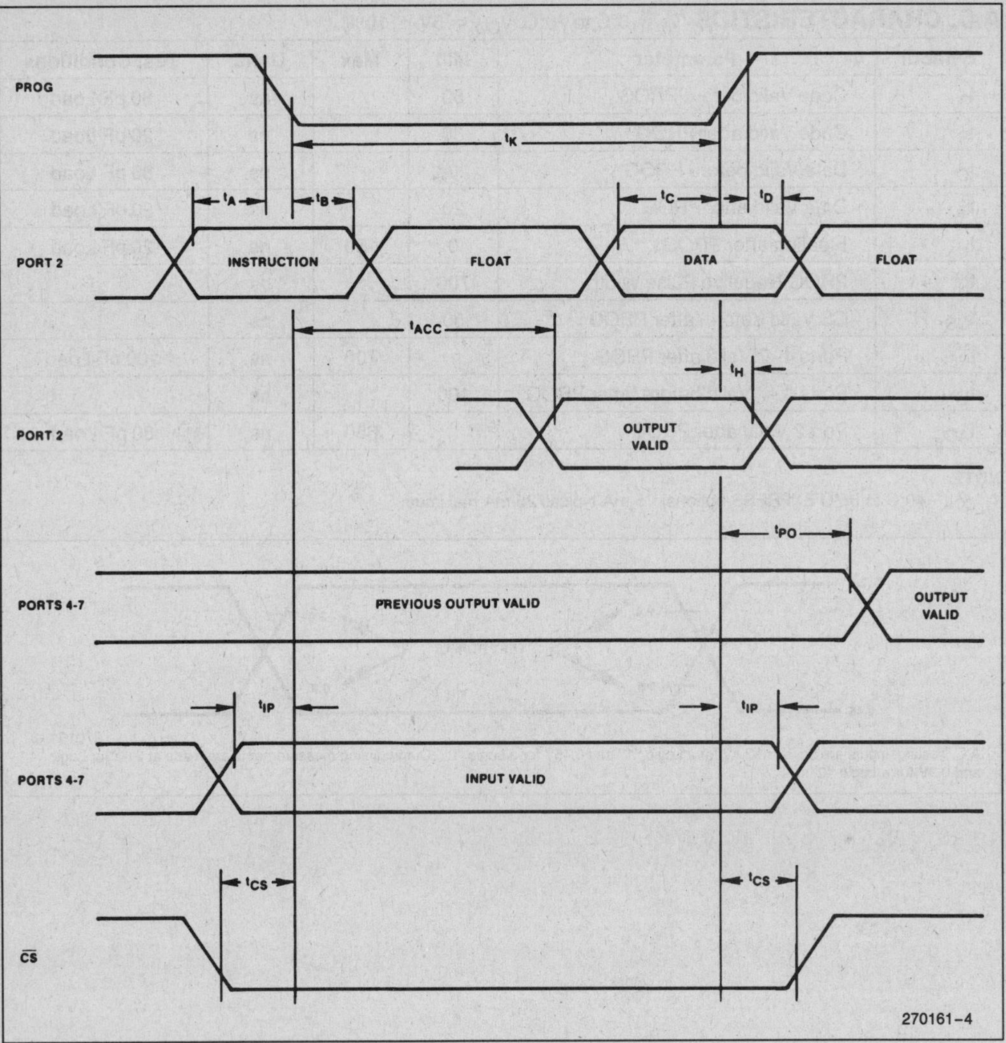

PRELIMINARY

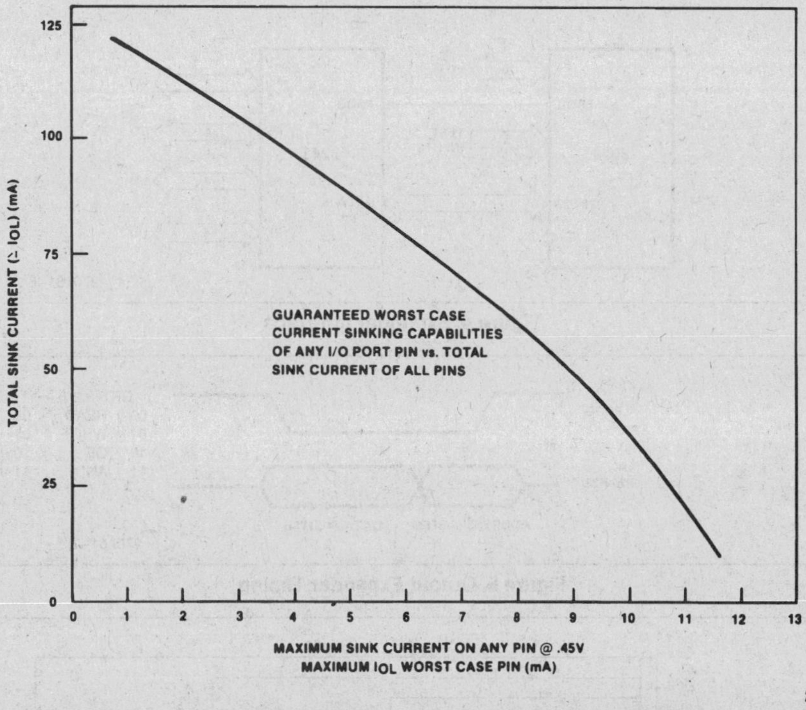

Figure 3. 8243 Current Sink Capability

Sink Capability

The 8243 can sink 5 mA @ 0.45V on each of its 16 I/O lines simultaneously. If, however, all lines are not sinking simultaneously or all lines are not fully loaded, the drive capability of any individual line increases as is shown by the accompanying curve.

For example, if only 5 of the 16 lines are to sink current at one time, the curve shows that each of those 5 lines is capable of sinking 9 mA @ 0.45V (if any lines are to sink 9 mA the total I_{OL} must not exceed 45 mA or five 9 mA loads).

Example: How many pins can drive 5 TTL loads (1.6 mA) assuming remaining pins are unloaded?

$I_{OL} = 5 \times 1.6 \text{ mA} = 8 \text{ mA}$

$\epsilon I_{OL} = 60 \text{ mA from curve}$

$\# \text{ pins} = 60 \text{ mA} \div 8 \text{ mA/pin} = 7.5 = 7$

In this case, 7 lines can sink 8 mA for a total of 56 mA. This leaves 4 mA sink current capability which can be divided in any way among the remaining 8 I/O lines of the 8243.

NOTE:

A10 to 50 KΩ pullup resistor to +5V should be added to 8243 outputs when driving to 5V CMOS directly.

Example: This example shows how the use of the 20 mA sink capability of Port 7 affects the sinking capability of the other I/O lines.

An 8243 will drive the following loads simultaneously.

2 loads—20 mA @ 1V (Port 7 only)

8 loads—4 mA @ 0.45V

6 loads—3.2 mA @ 0.45V

Is this within the specified limits?

$\epsilon I_{OL} = (2 \times 20) + (8 \times 4) + (6 \times 3.2) = 91.2 \text{ mA}.$

From the curve: for $I_{OL} = 4$ mA, $\epsilon I_{OL} \cong 93$ mA. Since 91.2 mA < 93 mA the loads are within specified limits.

Although the 20 mA @ 1V loads are used in calculating ϵI_{OL}, it is the largest current required @ 0.45V which determines the maximum allowable ϵI_{OL}.

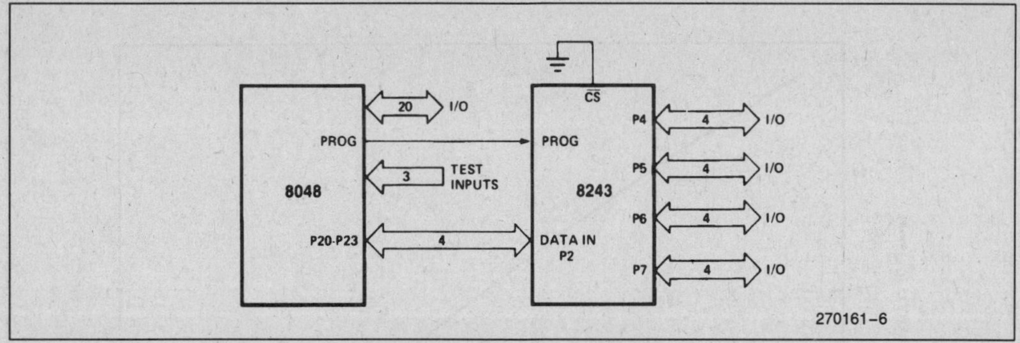

Figure 4. Expander Interface

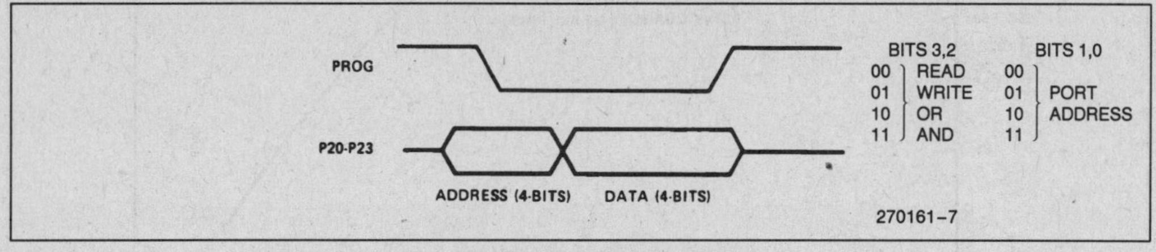

Figure 5. Output Expander Timing

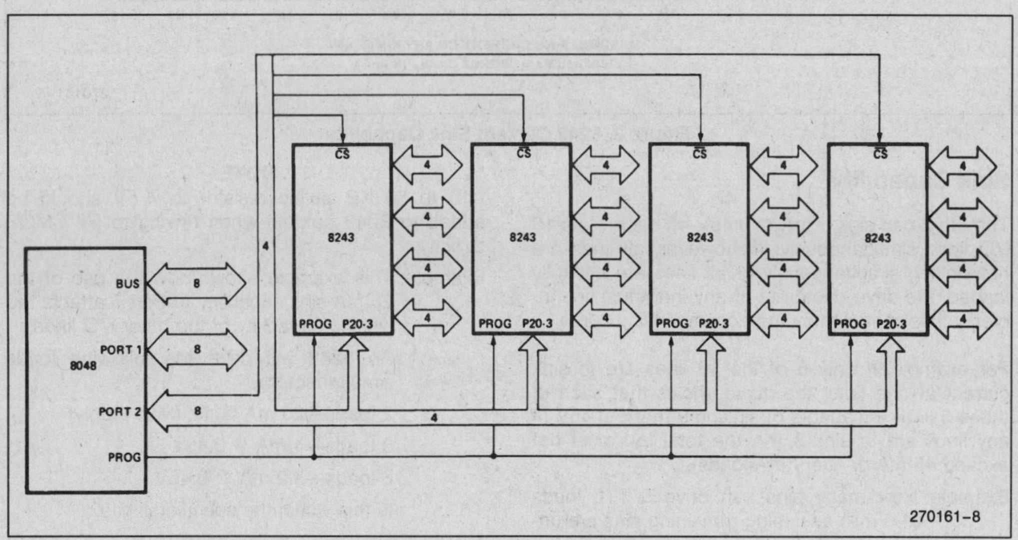

Figure 6. Using Multiple 8243's

intel®

P8748H/P8749H
8048AH/8035AHL/8049AH/8039AHL/8050AH/8040AHL
HMOS SINGLE-COMPONENT 8-BIT MICROCONTROLLER

- **High Performance HMOS II**
- **Interval Time/Event Counter**
- **Two Single Level Interrupts**
- **Single 5-Volt Supply**
- **Over 96 Instructions; 90% Single Byte**
- **Programmable ROMs Using 21V**
- **Easily Expandable Memory and I/O**
- **Up to 1.36 μs Instruction Cycle All Instructions 1 or 2 Cycles**

1

The Intel MCS®-48 family are totally self-sufficient, 8-bit parallel computers fabricated on single silicon chips using Intel's advanced N-channel silicon gate HMOS process.

The family contains 27 I/O lines, an 8-bit timer/counter, and on-board oscillator/clock circuits. For systems that require extra capability, the family can be expanded using MCS®-80/MCS®-85 peripherals.

These microcontrollers are available in both masked ROM and ROMless versions as well as a new version, The Programmable ROM. The Programmable ROM provides the user with the capability of a masked ROM while providing the flexibility of a device that can be programmed at the time of requirement and to the desired data. Programmable ROM's allow the user to lower inventory levels while at the same time decreasing delay times and code risks.

These microcomputers are designed to be efficient controllers as well as arithmetic processors. They have extensive bit handling capability as well as facilities for both binary and BCD arithmetic. Efficient use of program memory results from an instruction set consisting of mostly single byte instructions and no instructions over 2 bytes in length.

Device	Internal	Memory	RAM STANDBY
8050AH	4K x 8 ROM	256 x 8 RAM	yes
8049AH	2K x 8 ROM	128 x 8 RAM	yes
8048AH	1K x 8 ROM	64 x 8 RAM	yes
8040AHL	None	256 x 8 RAM	yes
8039AHL	None	128 x 8 RAM	yes
8035AHL	None	64 x 8 RAM	yes
P8749H	2K x 8 Programmable ROM	128 x 8 RAM	no
P8748H	1K x 8 Programmable ROM	64 x 8 RAM	no

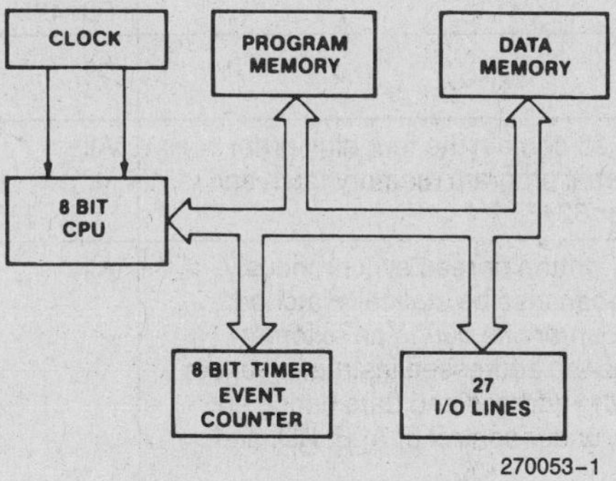

Figure 1. Block Diagram

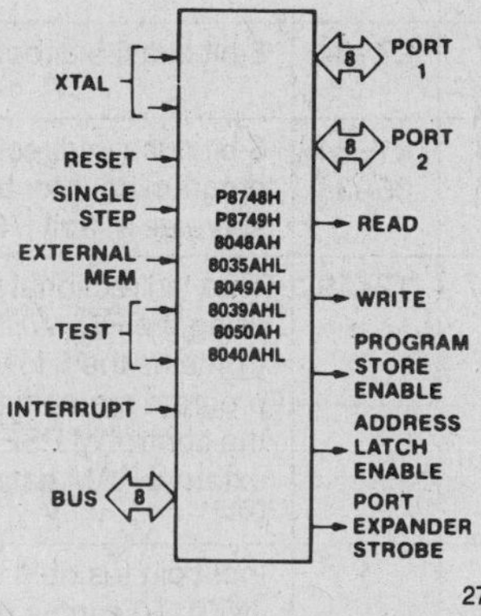

Figure 2. Logic Symbol

August 1989
Order Number: 270053-003

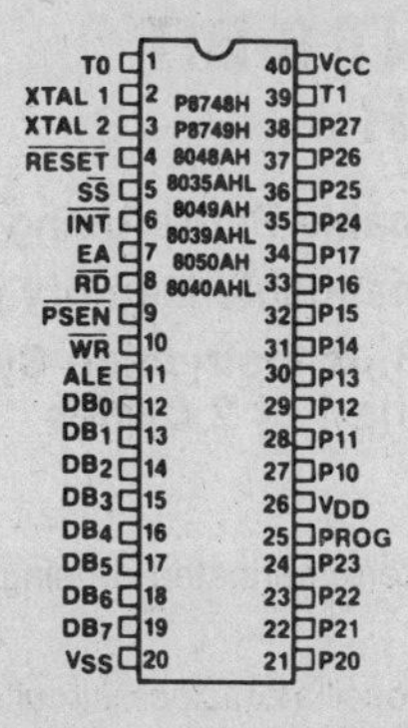

Figure 3. Pin Configuration

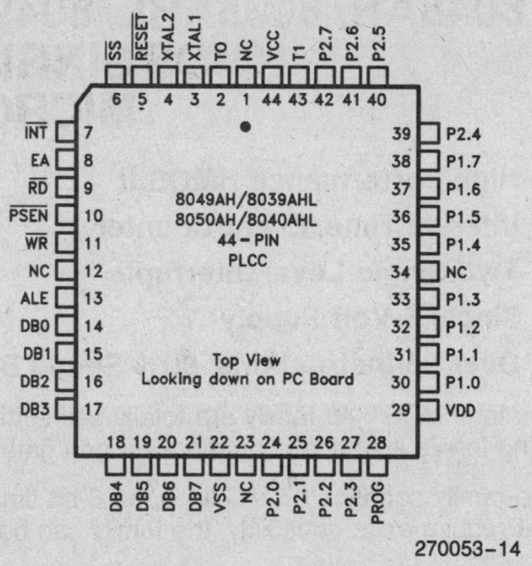

Figure 4. Pad Configuration

Table 1. Pin Description

Symbol	Pin No.	Function	Device
V_{SS}	20	Circuit GND potential.	All
V_{DD}	26	+5V during normal operation.	All
		Low power standby pin.	8048AH 8035AHL 8049AH 8039AHL 8050AH 8040AHL
		Programming power supply (+21V).	P8748H P8749H
V_{CC}	40	Main power supply; +5V during operation and programming.	All
PROG	25	Output strobe for 8243 I/O expander.	All
		Program pulse (+18V) input pin During Programming.	P8748H P8749H
P10–P17 Port 1	27–34	8-bit quasi-bidirectional port.	All
P20–P23 P24–P27 Port 2	21–24 35–38	8-bit quasi-bidirectional port. P20–P23 contain the four high order program counter bits during an external program memory fetch and serve as a 4-bit I/O expander bus for 8243.	All
DB0–DB7 BUS	12–19	True bidirectional port which can be written or read synchronously using the $\overline{RD}$, $\overline{WR}$ strobes. The port can also be statically latched. Contains the 8 low order program counter bits during an external program memory fetch, and receives the addressed instruction under the control of $\overline{PSEN}$. Also contains the address and data during an external RAM data store instruction, under control of ALE, $\overline{RD}$, and $\overline{WR}$.	All
T0	1	Input pin testable using the conditional transfer instruction JT0 and JNT0. T0 can be designated as a clock output using ENT0 CLK instruction.	All
		Used during programming.	P8748H P8749H

Table 1. Pin Description (Continued)

Symbol	Pin No.	Function	Device
T1	39	Input pin testable using the JT1, and JNT1 instructions. Can be designated the timer/counter input using the STRT CNT instruction.	All
$\overline{INT}$	6	Interrupt input. Initiates an interrupt if interrupt is enabled. Interrupt is disabled after a reset. Also testable with conditional jump instruction. (Active low) interrupt must remain low for at least 3 machine cycles for proper operation.	All
$\overline{RD}$	8	Output strobe activated during a BUS read. Can be used to enable data onto the bus from an external device. Used as a read strobe to external data memory. (Active low)	All
$\overline{RESET}$	4	Input which is used to initialize the processor. (Active low) (Non TTL V_{IH})	All
		Used during power down.	8048AH 8035AHL 8049AH 8039AHL 8050AH 8040AHL
		Used during programming.	P8748H P8749H
		Used during ROM verification.	8048AH P8748H 8049AH P8749H 8050AH
$\overline{WR}$	10	Output strobe during a bus write. (Active low) Used as write strobe to external data memory.	All
ALE	11	Address latch enable. This signal occurs once during each cycle and is useful as a clock output. The negative edge of ALE strobes address into external data and program memory.	All
$\overline{PSEN}$	9	Program store enable. This output occurs only during a fetch to external program memory. (Active low)	All
$\overline{SS}$	5	Single step input can be used in conjunction with ALE to "single step" the processor through each instruction.	All
		(Active low) Used in sync mode.	8048AH 8035AHL 8049AH 8039AHL 8050AH 8040AHL
EA	7	External access input which forces all program memory fetches to reference external memory. Useful for emulation and debug. (Active high)	All
		Used during (18V) programming.	P8748H P8749H
		Used during ROM verification (12V).	8048AH 8049AH 8050AH
XTAL1	2	One side of crystal input for internal oscillator. Also input for external source. (Non TTL V_{IH})	All
XTAL2	3	Other side of crystal input.	All

Table 2. Instruction Set

Accumulator

Mnemonic	Description	Bytes	Cycles
ADD A, R	Add register to A	1	1
ADD A, @R	Add data memory to A	1	1
ADD A, #data	Add immediate to A	2	2
ADDC A, R	Add register with carry	1	1
ADDC A, @R	Add data memory with carry	1	1
ADDC A, #data	Add immediate with carry	2	2
ANL A, R	And register to A	1	1
ANL A, @R	And data memory to A	1	1
ANL A, #data	And immediate to A	2	2
ORL A, R	Or register to A	1	1
ORL A, @R	Or data memory to A	1	1
ORL A, #data	Or immediate to A	2	2
XRL A, R	Exclusive or register to A	1	1
XRL A, @R	Exclusive or data memory to A	1	1
XRL A, #data	Exclusive or immediate to A	2	2
INC A	Increment A	1	1
DEC A	Decrement A	1	1
CLR A	Clear A	1	1
CPL A	Complement A	1	1
DA A	Decimal adjust A	1	1
SWAP A	Swap nibbles of A	1	1
RL A	Rotate A left	1	1
RLC A	Rotate A left through carry	1	1
RR A	Rotate A right	1	1
RRC A	Rotate A right through carry	1	1

Input/Output

Mnemonic	Description	Bytes	Cycles
IN A, P	Input port to A	1	2
OUTL P, A	Output A to port	1	2
ANL P, #data	And immediate to port	2	2
ORL P, #data	Or immediate to port	2	2
INS A, BUS	Input BUS to A	1	2
OUTL BUS, A	Output A to BUS	1	2
ANL BUS, #data	And immediate to BUS	2	2
ORL BUS, #data	Or immediate to BUS	2	2
MOVD A, P	Input expander port to A	1	2
MOVD P, A	Output A to expander port	1	2
ANLD P, A	And A to expander port	1	2
ORLD P, A	Or A to expander port	1	2

Registers

Mnemonic	Description	Bytes	Cycles
INC R	Increment register	1	1
INC @R	Increment data memory	1	1
DEC R	Decrement register	1	1

Branch

Mnemonic	Description	Bytes	Cycles
JMP addr	Jump unconditional	2	2
JMPP @A	Jump indirect	1	2
DJNZ R, addr	Decrement register and skip	2	2
JC addr	Jump on carry = 1	2	2
JNC addr	Jump on carry = 0	2	2
JZ addr	Jump on A zero	2	2
JNZ addr	Jump on A not zero	2	2
JT0 addr	Jump on T0 = 1	2	2
JNT0 addr	Jump on T0 = 0	2	2
JT1 addr	Jump on T1 = 1	2	2
JNT1 addr	Jump on T1 = 0	2	2
JF0 addr	Jump on F0 = 1	2	2
JF1 addr	Jump on F1 = 1	2	2
JTF addr	Jump on timer flag	2	2
JNI addr	Jump on $\overline{INT}$ = 0	2	2
JBb addr	Jump on accumulator bit	2	2

Table 2. Instruction Set (Continued)

Subroutine

Mnemonic	Description	Bytes	Cycles
CALL addr	Jump to subroutine	2	2
RET	Return	1	2
RETR	Return and restore status	1	2

Flags

Mnemonic	Description	Bytes	Cycles
CLR C	Clear carry	1	1
CPL C	Complement carry	1	1
CLR F0	Clear flag 0	1	1
CPL F0	Complement flag 0	1	1
CLR F1	Clear flag 1	1	1
CPL F1	Complement flag 1	1	1

Data Moves

Mnemonic	Description	Bytes	Cycles
MOV A, R	Move register to A	1	1
MOV A, @R	Move data memory to A	1	1
MOV A, #data	Move immediate to A	2	2
MOV R, A	Move A to register	1	1
MOV @R, A	Move A to data memory	1	1
MOV R, #data	Move immediate to register	2	2
MOV @R, #data	Move immediate to data memory	2	2
MOV A, PSW	Move PSW to A	1	1
MOV PSW, A	Move A to PSW	1	1
XCH A, R	Exchange A and register	1	1
XCH A, @R	Exchange A and data memory	1	1
XCHD A, @R	Exchange nibble of A and data memory	1	1
MOVX A, @R	Move external data memory to A	1	2
MOVX @R, A	Move A to external data memory	1	2
MOVP A, @A	Move to A from current page	1	2
MOVP3 A, @A	Move to A from page 3	1	2

Timer/Counter

Mnemonic	Description	Bytes	Cycles
MOV A, T	Read timer/counter	1	1
MOV T, A	Load timer/counter	1	1
STRT T	Start timer	1	1
STRT CNT	Start counter	1	1
STOP TCNT	Stop timer/counter	1	1
EN TCNTI	Enable timer/counter interrupt	1	1
DIS TCNTI	Disable timer/counter interrupt	1	1

Control

Mnemonic	Description	Bytes	Cycles
EN I	Enable external interrupt	1	1
DIS I	Disable external interrupt	1	1
SEL RB0	Select register bank 0	1	1
SEL RB1	Select register bank 1	1	1
SEL MB0	Select memory bank 0	1	1
SEL MB1	Select memory bank 1	1	1
ENT0 CLK	Enable clock output on T0	1	1

Mnemonic	Description	Bytes	Cycles
NOP	No operation	1	1

1

ABSOLUTE MAXIMUM RATINGS*

Case Temperature Under Bias .... 0°C to +70°C
Storage Temperature .... −65°C to +150°C
Voltage on any Pin with Respect to Ground .... −0.5V to +7V
Power Dissipation 1.5W

NOTICE: This is a production data sheet. The specifications are subject to change without notice.

******WARNING: Stressing the device beyond the "Absolute Maximum Ratings" may cause permanent damage. These are stress ratings only. Operation beyond the "Operating Conditions" is not recommended and extended exposure beyond the "Operating Conditions" may affect device reliability.*

D.C. CHARACTERISTICS T_A = 0°C to +70°C; $V_{CC} = V_{DD}$ = 5V ±10%; V_{SS} = 0V

Symbol	Parameter	Limits			Unit	Test Conditions	Device
		Min	Typ	Max			
V_{IL}	Input Low Voltage (All Except $\overline{RESET}$, X1, X2)	−0.5		0.8	V		All
V_{IL1}	Input Low Voltage ($\overline{RESET}$, X1, X2)	−0.5		0.6	V		All
V_{IH}	Input High Voltage (All Except XTAL1, XTAL2, $\overline{RESET}$)	2.0		V_{CC}	V		All
V_{IH1}	Input High Voltage (X1, X2, $\overline{RESET}$)	3.8		V_{CC}	V		All
V_{OL}	Output Low Voltage (BUS)			0.45	V	I_{OL} = 2.0 mA	All
V_{OL1}	Output Low Voltage ($\overline{RD}$, $\overline{WR}$, $\overline{PSEN}$, ALE)			0.45	V	I_{OL} = 1.8 mA	All
V_{OL2}	Output Low Voltage (PROG)			0.45	V	I_{OL} = 1.0 mA	All
V_{OL3}	Output Low Voltage (All Other Outputs)			0.45	V	I_{OL} = 1.6 mA	All
V_{OH}	Output High Voltage (BUS)	2.4			V	I_{OH} = −400 μA	All
V_{OH1}	Output High Voltage ($\overline{RD}$, $\overline{WR}$, $\overline{PSEN}$, ALE)	2.4			V	I_{OH} = −100 μA	All
V_{OH2}	Output High Voltage (All Other Outputs)	2.4			V	I_{OH} = −40 μA	All

D.C. CHARACTERISTICS T_A = 0°C to +70°C; V_{CC} = V_{DD} = 5V ±10%; V_{SS} = 0V (Continued)

Symbol	Parameter	Limits			Unit	Test Conditions	Device
		Min	Typ	Max			
I_{L1}	Leakage Current (T1, $\overline{INT}$)			±10	μA	$V_{SS} \leq V_{IN} \leq V_{CC}$	All
I_{LI1}	Input Leakage Current (P10–P17, P20–P27, EA, $\overline{SS}$)			−500	μA	$V_{SS} + 0.45 \leq V_{IN} \leq V_{CC}$	All
I_{LI2}	Input Leakage Current $\overline{RESET}$	−10		−300	μA	$V_{SS} \leq V_{IN} \leq 3.8$	All
I_{LO}	Leakage Current (BUS, T0) (High Impedance State)			±10	μA	$V_{SS} \leq V_{IN} \leq V_{CC}$	All
I_{DD}	V_{DD} Supply Current (RAM Standby)		3	5	mA		8048AH 8035AHL
			4	7	mA		8049AH 8039AHL
			5	10	mA		8050AH 8040AHL
I_{DD} + I_{CC}	Total Supply Current*		30	65	mA		8048AH 8035AHL
			35	70	mA		8049AH 8039AHL
			40	80	mA		8050AH 8040AHL
			30	100	mA		P8748H
			50	110	mA		P8749H
V_{DD}	RAM Standby Voltage	2.2		5.5	V	Standby Mode Reset $\leq V_{IL1}$	8048AH 8035AH
		2.2		5.5	V		8049AH 8039AH
		2.2		5.5	V		8050AH 8040AHL

*I_{CC} + I_{DD} are measured with all outputs in their high impedance state; $\overline{RESET}$ low; 11 MHz crystal applied; $\overline{INT}$, $\overline{SS}$, and EA floating.

A.C. CHARACTERISTICS $T_A = 0°C$ to $+70°C$; $V_{CC} = V_{DD} = 5V \pm 10\%$; $V_{SS} = 0V$

Symbol	Parameter	f (t) (Note 3)	11 MHz		Unit	Conditions (Note 1)
			Min	Max		
t	Clock Period	1/xtal freq	90.9	1000	ns	(Note 3)
t_{LL}	ALE Pulse Width	3.5t−170	150		ns	
t_{AL}	Addr Setup to ALE	2t−110	70		ns	(Note 2)
t_{LA}	Addr Hold from ALE	t−40	50		ns	
t_{CC1}	Control Pulse Width ($\overline{\text{RD}}$, $\overline{\text{WR}}$)	7.5t−200	480		ns	
t_{CC2}	Control Pulse Width ($\overline{\text{PSEN}}$)	6t−200	350		ns	
t_{DW}	Data Setup before $\overline{\text{WR}}$	6.5t−200	390		ns	
t_{WD}	Data Hold after $\overline{\text{WR}}$	t−50	40		ns	
t_{DR}	Data Hold ($\overline{\text{RD}}$, $\overline{\text{PSEN}}$)	1.5t−30	0	110	ns	
t_{RD1}	$\overline{\text{RD}}$ to Data in	6t−170		375	ns	
t_{RD2}	$\overline{\text{PSEN}}$ to Data in	4.5t−170		240	ns	
t_{AW}	Addr Setup to $\overline{\text{WR}}$	5t−150	300		ns	
t_{AD1}	Addr Setup to Data ($\overline{\text{RD}}$)	10.5t−220		730	ns	
t_{AD2}	Addr Setup to Data ($\overline{\text{PSEN}}$)	7.5t−200		460	ns	
t_{AFC1}	Addr Float to $\overline{\text{RD}}$, $\overline{\text{WR}}$	2t−40	140		ns	(Note 2)
t_{AFC2}	Addr Float to $\overline{\text{PSEN}}$	0.5t−40	10		ns	(Note 2)
t_{LAFC1}	ALE to Control ($\overline{\text{RD}}$, $\overline{\text{WR}}$)	3t−75	200		ns	
t_{LAFC2}	ALE to Control ($\overline{\text{PSEN}}$)	1.5t−75	60		ns	
t_{CA1}	Control to ALE ($\overline{\text{RD}}$, $\overline{\text{WR}}$, PROG)	t−65	25		ns	
t_{CA2}	Control to ALE ($\overline{\text{PSEN}}$)	4t−70	290		ns	
t_{CP}	Port Control Setup to PROG	1.5t−80	50		ns	
t_{PC}	Port Control Hold to PROG	4t−260	100		ns	
t_{PR}	PROG to P2 Input Valid	8.5t−120		650	ns	
t_{PF}	Input Data Hold from PROG	1.5t	0	140	ns	
t_{DP}	Output Data Setup	6t−290	250		ns	
t_{PD}	Output Data Hold	1.5t−90	40		ns	
t_{PP}	PROG Pulse Width	10.5t−250	700		ns	
t_{PL}	Port 2 I/O Setup to ALE	4t−200	160		ns	
t_{LP}	Port 2 I/O Hold to ALE	0.5t−30	15		ns	
t_{PV}	Port Output from ALE	4.5t+100		5.0	ns	
t_{0PRR}	T0 Rep Rate	3t	270		ns	
t_{CY}	Cycle Time	15t	1.36	15.0	μs	

NOTES:
1. Control outputs: $C_L = 80$ pF. BUS Outputs: $C_L = 150$ pF.
2. BUS High Impedance Load 20 pF
3. f(t) assumes 50% duty cycle on X1, X2. Max clock period is for a 1 MHz crystal input.

WAVEFORMS

INSTRUCTION FETCH FROM PROGRAM MEMORY

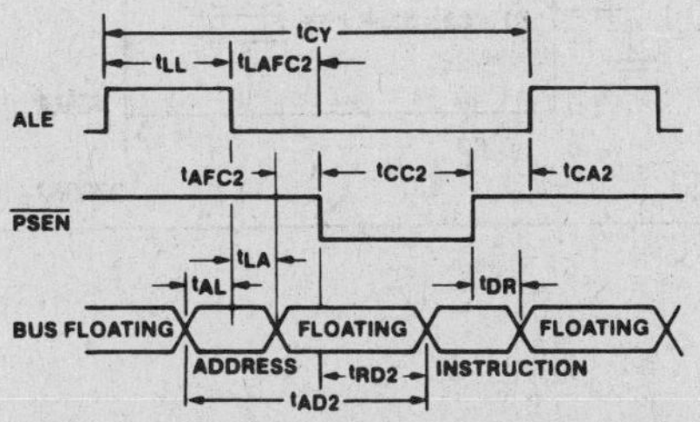

270053-4

READ FROM EXTERNAL DATA MEMORY

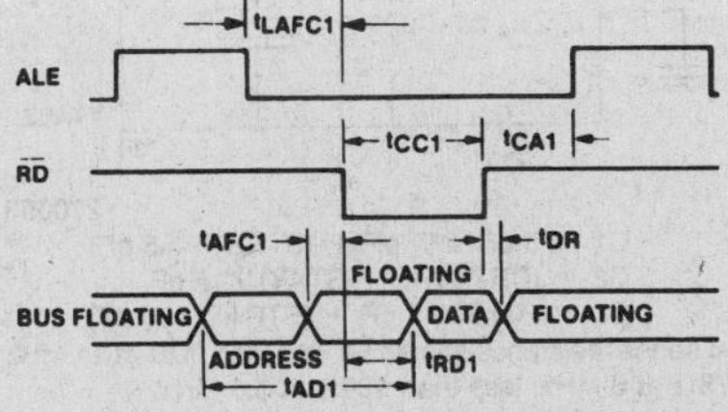

270053-5

WRITE TO EXTERNAL DATA MEMORY

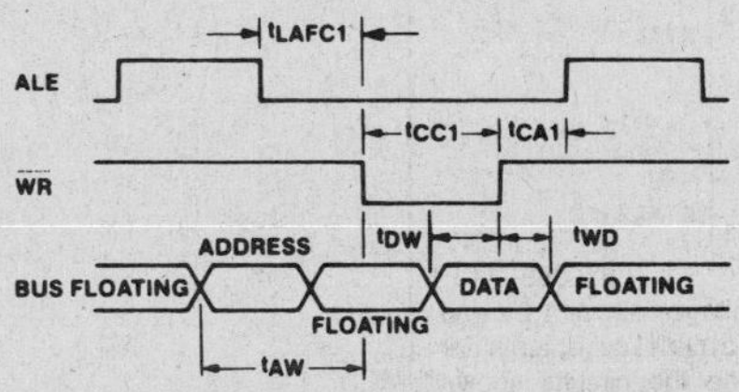

270053-6

INPUT AND OUTPUT FOR A.C. TESTS

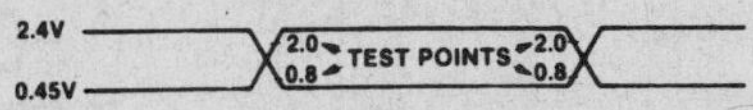

270053-7

A.C. testing inputs are driven at 2.4V for a logic "1" and 0.45V for a logic "0". Output timing measurements are made at 2.0V for a logic "1" and 0.8V for a logic "0".

PORT 1/PORT 2 TIMING

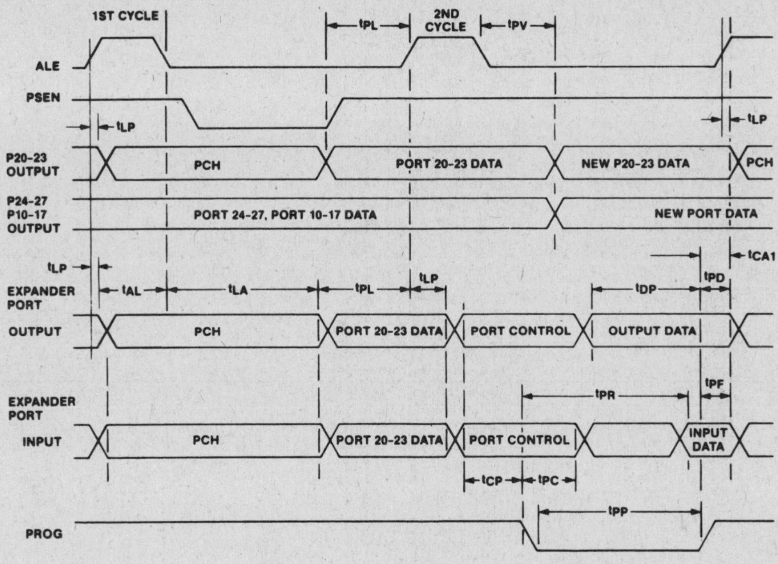

270053-8

CRYSTAL OSCILLATOR MODE

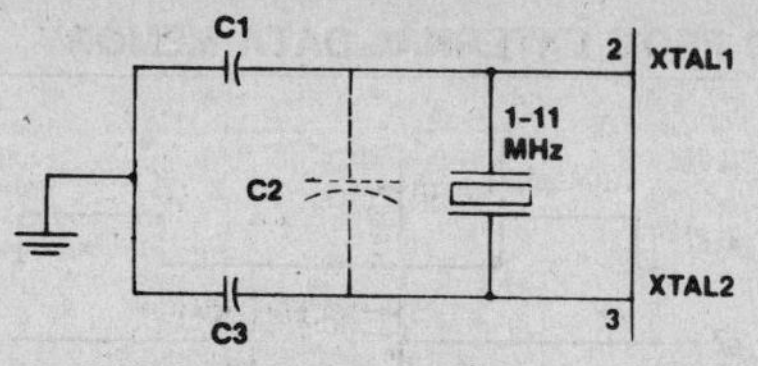

270053-9

C1 = 5 pF ±½ pF + (STRAY < 5 pF)
C2 = (CRYSTAL + STAY) < 8 pF
C3 = 20 pF ±1 pF + (STRAY < 5 pF)

Crystal series resistance should be less than 30Ω at 11 MHz; less than 75Ω at 6 MHz; less than 180Ω at 3.6 MHz.

CERAMIC RESONATOR MODE

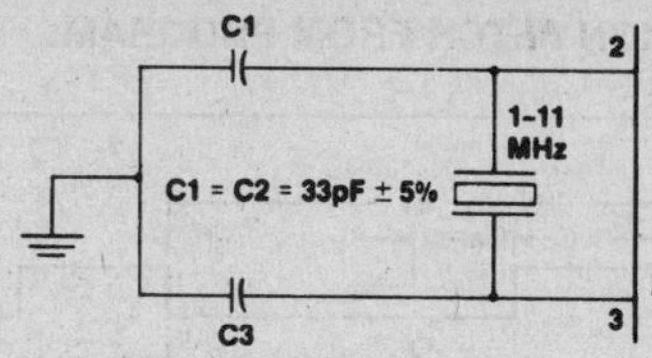

270053-10

DRIVING FROM EXTERNAL SOURCE

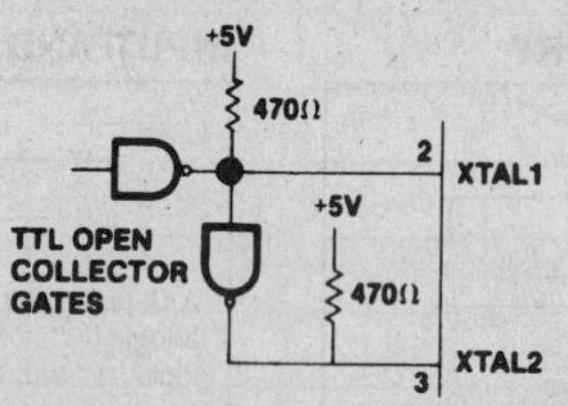

270053-11

For XTAL1 and XTAL2 define "high" as voltages above 1.6V and "low" as voltages below 1.6V. The duty cycle requirements for externally driving XTAL1 and XTAL2 using the circuits shown above are as follows: XTAL1 must be high 35–65% of the period and XTAL2 must be high 35–65% of the period. Rise and fall times must be faster than 20 ns.

PROGRAMMING AND VERIFYING THE P8749H/48H PROGRAMMABLE ROM

Programming Verification

In brief, the programming process consists of: activating the program mode, applying an address, latching the address, applying data, and applying a programming pulse. Each word is programmed completely before moving on to the next and is followed by a verification step. The following is a list of the pins used for programming and a description of their functions:

Pin	Function
XTAL1 XTAL2	Clock Input (3 to 4.0 MHz)
$\overline{\text{RESET}}$	Initialization and Address Latching
T0	Selection of Program or Verifying Mode
EA	Activation of Program/Verify Modes
BUS	Address and Data Input Data Output During Verify
P20–P22	Address Input
V_{DD}	Programming Power Supply
PROG	Program Pulse Input

WARNING:

An attempt to program a missocketed P8749H/48H will result in severe damage to the part. An indication of a properly socketed part is the appearance of the ALE clock output. The lack of this clock may be used to disable the programmer.

The Program/Verify sequence is:

1. V_{DD} = 5V, Clock applied or internal oscillator operating, $\overline{\text{RESET}}$ = 0V, T0 = 5V, EA = 5V, BUS and PROG floating. P10 and P11 must be tied to ground.
2. Insert P8749H/48H in programming socket
3. T0 = 0V (select program mode)
4. EA = 18V (activate program mode)
5. Address applied to BUS and P20–22
6. $\overline{\text{RESET}}$ = 5V (latch address)
7. Data applied to BUS
8. V_{DD} = 21V (programming power)
9. PROG = V_{CC} or float followed by one 50 ms pulse to 18V
10. V_{DD} = 5V
11. T0 = 5V (verify mode)
12. Read and verify data on BUS
13. T0 = 0V
14. $\overline{\text{RESET}}$ = 0V and repeat from step 5
15. Programmer should be at conditions of step 1 when P8749H/48H is removed from socket.

NOTE:
Once programmed the P8749H/48H cannot be erased.

1

A.C. TIMING SPECIFICATION FOR PROGRAMMING P8748H/P8749H ONLY

T_A = 25°C ±5°C; V_{CC} = 5V ±5%; V_{DD} = 21 ±0.5V

Symbol	Parameter	Min	Max	Unit	Test Conditions
t_{AW}	Address Setup Time to $\overline{RESET}$	$4t_{CY}$			
t_{WA}	Address Hold Time After $\overline{RESET}$	$4t_{CY}$			
t_{DW}	Data in Setup Time to PROG	$4t_{CY}$			
t_{WD}	Data in Hold Time After PROG	$4t_{CY}$			
t_{PH}	$\overline{RESET}$ Hold Time to Verify	$4t_{CY}$			
t_{VDDW}	V_{DD} Hold Time Before PROG	0	1.0	ms	
t_{VDDH}	V_{DD} Hold Time After PROG	0	1.0	ms	
t_{PW}	Program Pulse Width	50	60	ms	
t_{TW}	T0 Setup Time for Program Mode	$4t_{CY}$			
t_{WT}	T0 Hold Time After Program Mode	$4t_{CY}$			
t_{DO}	T0 to Data Out Delay		$4t_{CY}$		
t_{WW}	$\overline{RESET}$ Pulse Width to Latch Address	$4t_{CY}$			
t_r, t_f	V_{DD} and PROG Rise and Fall Times	0.5	100	μs	
t_{CY}	CPU Operation Cycle Time	3.75	5	μs	
t_{RE}	$\overline{RESET}$ Setup Time before EA	$4t_{CY}$			

NOTE:
If Test 0 is high, t_{DO} can be triggered by $\overline{RESET}$.

D.C. CHARACTERISTICS FOR PROGRAMMING P8748H/P8749H ONLY

T_A = 25°C ±5°C; V_{CC} = 5V ±5%; V_{DD} = 21 ±0.5V

Symbol	Parameter	Min	Max	Unit	Test Conditions
V_{DDH}	V_{DD} Program Voltage High Level	20.5	21.5	V	
V_{DDL}	V_{DD} Voltage Low Level	4.75	5.25	V	
V_{PH}	PROG Program Voltage High Level	17.5	18.5	V	
V_{PL}	PROG Voltage Low Level	4.0	V_{CC}	V	
V_{EAH}	EA Program or Verify Voltage High Level	17.5	18.5	V	
I_{DD}	V_{DD} High Voltage Supply Current		20.0	mA	
I_{PROG}	PROG High Voltage Supply Current		1.0	mA	
I_{EA}	EA High Voltage Supply Current		1.0	mA	

SUGGESTED ROM VERIFICATION ALGORITHM FOR ROM DEVICE ONLY

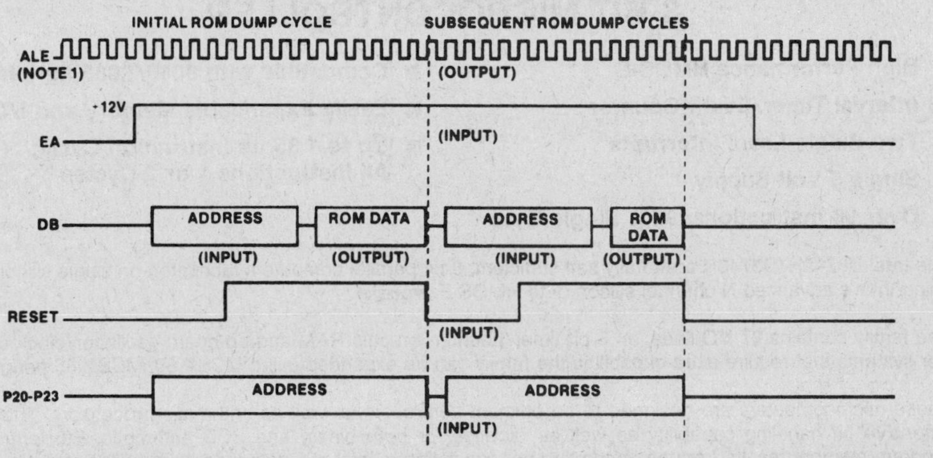

270053-12

	48H	49H	50H
A10	0	ADDR	ADDR
A11	0	0	ADDR

$V_{CC} = V_{DD} = +5V$
$V_{SS} = 0V$

NOTE:
ALE is function of X1, X2 inputs.

COMBINATION PROGRAM/VERIFY MODE (PROGRAMMABLE ROMS ONLY)

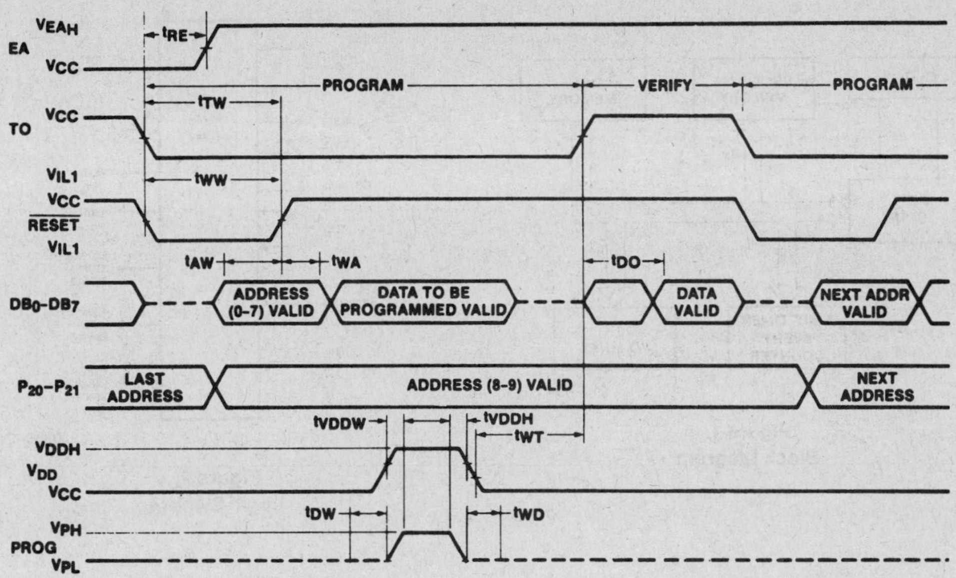

270053-13

intel®

D8748H/D8749H HMOS-E SINGLE-COMPONENT 8-BIT MICROCONTROLLER

- **High Performance HMOS-E**
- **Interval Timer/Event Counter**
- **Two Single Level Interrupts**
- **Single 5-Volt Supply**
- **Over 96 Instructions; 90% Single Byte**
- **Compatible with 8080/8085 Peripherals**
- **Easily Expandable Memory and I/O**
- **Up to 1.35 μs Instruction Cycle; All Instructions 1 or 2 Cycles**

The Intel D8749H/D8748H are totally self-sufficient, 8-bit parallel computers fabricated on single silicon chips using Intel's advanced N-channel silicon gate HMOS-E process.

The family contains 27 I/O lines, an 8-bit timer/counter, on-chip RAM and on-board oscillator/clock circuits. For systems that require extra capability, the family can be expanded using MCS®-80/MCS®-85 peripherals.

These microcomputers are designed to be efficient controllers as well as arithmetic processors. They have extensive bit handling capability as well as facilities for both binary and BCD arithmetic. Efficient use of program memory results from an instruction set consisting mostly of single byte instructions and no instructions over 2 bytes in length.

Device	Internal Memory	
D8749H	2K x 8 EPROM	128 x 8 RAM
D8748H	1K x 8 EPROM	64 x 8 RAM

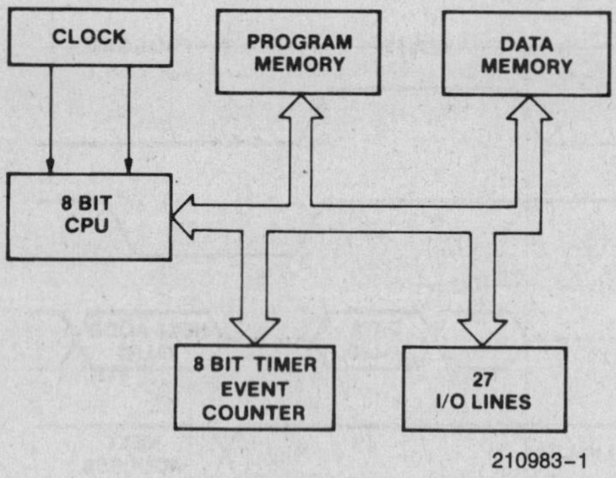

Figure 1. Block Diagram

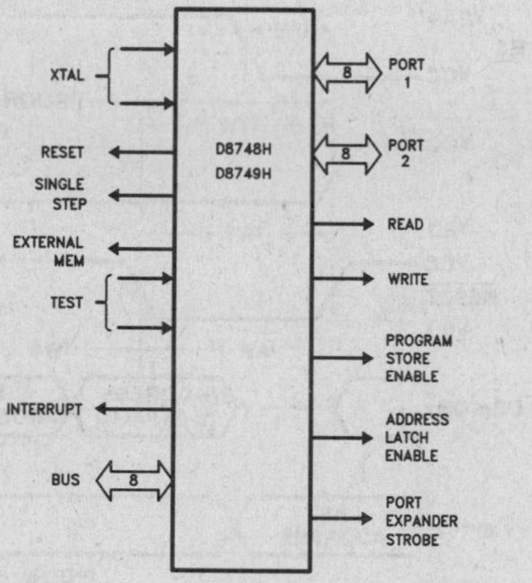

Figure 2. Logic Symbol

September 1992
Order Number: 210983-004

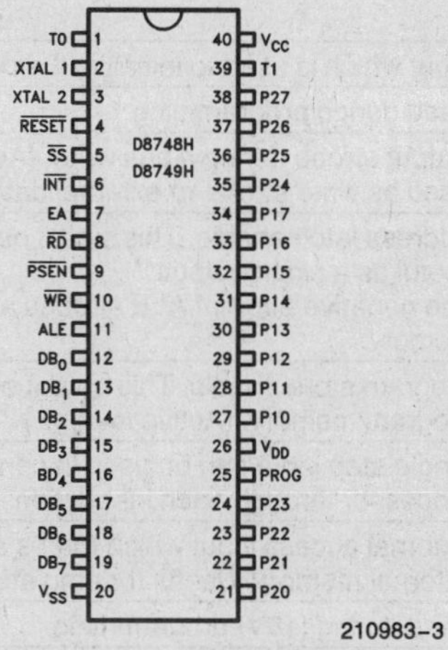

Figure 3. Pin Configuration

Table 1. Pin Description (40-Pin DIP)

Symbol	Pin No.	Function
V_{SS}	20	Circuit GND potential.
V_{DD}	26	+5V during normal operation.
		Programming power supply (+21V).
V_{CC}	40	Main power supply; +5V during operation and programming.
PROG	25	Output strobe for 8243 I/O expander.
		Program pulse (+18V) input pin during programming.
P10–P17 Port 1	27–34	8-bit quasi-bidirectional port.
P20–P23	21–24	8-bit quasi-bidirectional port. P20–P23 contain the four high order program counter bits during an external program memory fetch and serve as a 4-bit I/O expander bus for 8243.
P24–P27 Port 2	35–38	
DB0–DB7 BUS	12–19	True bidirectional port which can be written or read synchronously using the $\overline{RD}$, $\overline{WR}$ strobes. The port can also be statically latched. Contains the 8 low order program counter bits during an external program memory fetch, and receives the addressed instruction under the control of $\overline{PSEN}$. Also contains the address and data during an external RAM data store instruction, under control of ALE, $\overline{RD}$, and $\overline{WR}$.
T0	1	Input pin testable using the conditional transfer instructions JT0 and JNT0. T0 can be designated as a clock output using ENT0 CKL instruction.
		Used during programming.
T1	39	Input pin testable using the JT1, and JNT1 instructions. Can be designated the timer/counter input using the STRT CNT instruction.
$\overline{INT}$	6	Interrupt input. Initiates an interrupt if interrupt is enabled. Interrupt is disabled after a reset. Also testable with conditional jump instruction. (Active low) interrupt must remain low for at least 3 machine cycles for proper operation.
$\overline{RD}$	8	Output strobe activated during a BUS read. Can be used to enable data onto the bus from an external device. Used as a read strobe to external data memory. (Active low)

Table 1. Pin Description (40-Pin DIP) (Continued)

Symbol	Pin No.	Function
$\overline{\text{RESET}}$	4	Input which is used to initialize the processor. (Active low) (Non TTL V_{IH})
		Used during programming.
$\overline{\text{WR}}$	10	Output strobe during a bus write. (Active low) Used as write strobe to external data memory.
ALE	11	Address latch enable. This signal occurs once during each cycle and is useful as a clock output. The negative edge of ALE strobes address into external data and program memory.
$\overline{\text{PSEN}}$	9	Program store enable. This output occurs only during a fetch to external program memory. (Active low.)
$\overline{\text{SS}}$	5	Single step input can be used in conjunction with ALE to "single step" the processor through each instruction.
EA	7	External access input which forces all program memory fetches to reference external memory. Useful for emulation and debug. (Active high.)
		Used during (18V) programming.
XTAL1	2	One side of crystal input for internal oscillator. Also input for external source. (Non TTL V_{IH}.)
XTAL2	3	Other side of crystal input.

Table 2. Instruction Set

Mnemonic	Description	Bytes	Cycles
ACCUMULATOR			
ADD A, R	Add register to A	1	1
ADD A,@R	Add data memory to A	1	1
ADD A, #data	Add immediate to A	2	2
ADDC A, R	Add register with carry	1	1
ADDC A, @R	Add data memory with carry	1	1
ADDC A, #data	Add immediate with carry	2	2
ANL A, R	And register to A	1	1
ANL A, @R	And data memory to A	1	1
ANL A, #data	And immediate to A	2	2
ORL A, R	Or register to A	1	1
ORL A, @R	Or data memory to A	1	1
ORL A, #data	Or immediate to A	2	2
XRL A, R	Exclusive or register to A	1	1
XRL A, @R	Exclusive or data memory to A	1	1
XRL A, #data	Exclusive or immediate to A	2	2

Mnemonic	Description	Bytes	Cycles
ACCUMULATOR (Continued)			
INC A	Increment A	1	1
DEC A	Decrement A	1	1
CLR A	Clear A	1	1
CPL A	Complement A	1	1
DA A	Decimal adjust A	1	1
SWAP A	Swap nibbles of A	1	1
RL A	Rotate A left	1	1
RLC A	Rotate A left through carry	1	1
RR A	Rotate A right	1	1
RRC A	Rotate A right through carry	1	1
INPUT/OUTPUT			
IN A, P	Input port to A	1	2
OUTL P, A	Output A to port	1	2
ANL P, #data	And immediate to port	2	2
ORL P, #data	Or immediate to port	2	2
INS A, BUS	Input BUS to A	1	2
OUTL BUS, A	Output A to BUS	1	2
ANL BUS, #data	And immediate to BUS	2	2
ORL BUS, #data	Or immediate to BUS	2	2
MOVD A, P	Input expander port to A	1	2

Table 2. Instruction Set (Continued)

Mnemonic	Description	Bytes	Cycles
INPUT/OUTPUT (Continued)			
MOVD P, A	Output A to expander port	1	2
ANLD P, A	And A to expander port	1	2
ORLD P, A	Or A to expander port	1	2
REGISTERS			
INC R	Increment register	1	1
INC @R	Increment data memory	1	1
DEC R	Decrement register	1	1
BRANCH			
JMP addr	Jump unconditional	2	2
JMPP @A	Jump indirect	1	2
DJNZ R, addr	Decrement register and skip	2	2
JC addr	Jump on carry = 1	2	2
JNC addr	Jump on carry = 0	2	2
JZ addr	Jump on A zero	2	2
JNZ addr	Jump on A not zero	2	2
JT0 addr	Jump on T0 = 1	2	2
JNT0 addr	Jump on T0 = 0	2	2
JT1 addr	Jump on T1 = 1	2	2
JNT1 addr	Jump on T1 = 0	2	2
JF0 addr	Jump on F0 = 1	2	2
JF1 addr	Jump on F1 = 1	2	2
JTF addr	Jump on timer flag	2	2
JNI addr	Jump on INT = 0	2	2
JBb addr	Jump on accumulator bit	2	2
SUBROUTINE			
CALL addr	Jump to subroutine	2	2
RET	Return	1	2
RETR	Return and restore status	1	2
FLAGS			
CLR C	Clear carry	1	1
CPL C	Complement carry	1	1
CLR F0	Clear flag 0	1	1
CPL F0	Complement flag 0	1	1
CLR F1	Clear flag 1	1	1
CPL F1	Complement flag 1	1	1
DATA MOVES			
MOV A, R	Move register to A	1	1
MOV A, @R	Move data memory to A	1	1
MOV A, #data	Move immediate to A	2	2

Mnemonic	Description	Bytes	Cycles
DATA MOVES (Continued)			
MOV R, A	Move A to register	1	1
MOV @R, A	Move A to data memory	1	1
MOV R, #data	Move immediate to register	2	2
MOV @R, #data	Move immediate to data memory	2	2
MOV A, PSW	Move PSW to A	1	1
MOV PSW, A	Move A to PSW	1	1
XCH A, R	Exchange A and register	1	1
XCH A, @R	Exchange A and data memory	1	1
XCHD A, @R	Exchange nibble of A and register	1	1
MOVX A, @R	Move external data memory to A	1	2
MOVX @R, A	Move A to external data memory	1	2
MOVP A, @A	Move to A from current page	1	2
MOVP3 A, @A	Move to A from page 3	1	2
TIMER/COUNTER			
MOV A, T	Read timer/counter	1	1
MOV T, A	Load timer/counter	1	1
STRT T	Start timer	1	1
STRT CNT	Start counter	1	1
STOP TCNT	Stop timer/counter	1	1
EN TCNTI	Enable timer/counter interrupt	1	1
DIS TCNTI	Disable timer/counter interrupt	1	1
CONTROL			
EN I	Enable external interrupt	1	1
DIS I	Disable external interrupt	1	1
SEL RB0	Select register bank 0	1	1
SEL RB1	Select register bank 1	1	1
SEL MB0	Select memory bank 0	1	1
SEL MB1	Select memory bank 1	1	1
ENT0 CLK	Enable clock output on T0	1	1
NOP	No operation	1	1

ABSOLUTE MAXIMUM RATINGS*

Ambient Temperature Under Bias . . . . 0°C to +70°C
Storage Temperature . . . . −65°C to +150°C
Voltage On Any Pin With Respect to Ground. . . . . −0.5V to +7V
Power Dissipation . . . . 1.0 Watt

NOTICE: This is a production data sheet. The specifications are subject to change without notice.

**WARNING: Stressing the device beyond the "Absolute Maximum Ratings" may cause permanent damage. These are stress ratings only. Operation beyond the "Operating Conditions" is not recommended and extended exposure beyond the "Operating Conditions" may affect device reliability.*

DC CHARACTERISTICS T_A = 0°C to +70°C; $V_{CC} = V_{DD}$ = 5V ±10%; V_{SS} = 0V

Symbol	Parameter	Limits			Unit	Test Conditions	Device
		Min	Typ	Max			
V_{IL}	Input Low Voltage (All Except RESET, X1, X2)	−0.5		0.8	V		All
V_{IL1}	Input Low Voltage ($\overline{\text{RESET}}$, X1, X2)	−0.5		0.6	V		All
V_{IH}	Input High Voltage (All Except XTAL1, XTAL2, $\overline{\text{RESET}}$	2.0		V_{CC}	V		All
V_{IH1}	Input High Voltage (X1, X2, $\overline{\text{RESET}}$)	3.8		V_{CC}	V		All
V_{OL}	Output Low Voltage (BUS)			0.45	V	I_{OL} = 2.0 mA	All
V_{OL1}	Output Low Voltage ($\overline{\text{RD}}$, $\overline{\text{WR}}$, $\overline{\text{PSEN}}$, ALE)			0.45	V	I_{OL} = 1.8 mA	All
V_{OL2}	Output Low Voltage (PROG)			0.45	V	I_{OL} = 1.0 mA	All
V_{OL3}	Output Low Voltage (All Other Outputs)			0.45	V	I_{OL} = 1.6 mA	All
V_{OH}	Output High Voltage (BUS)	2.4			V	I_{OH} = −400 μA	All
V_{OH1}	Output High Voltage ($\overline{\text{RD}}$, $\overline{\text{WR}}$, $\overline{\text{PSEN}}$, ALE)	2.4			V	I_{OH} = −100 μA	All
V_{OH2}	Output High Voltage (All Other Outputs)	2.4			V	I_{OH} = −40 μA	All
I_{L1}	Leakage Current (T1, $\overline{\text{INT}}$)			±10	μA	$V_{SS} \leq V_{IN} \leq V_{CC}$	All
I_{LI1}	Input Leakage Current (P10–P17, P20–P27, EA, $\overline{\text{SS}}$)			−500	μA	$V_{SS} + 0.45 \leq V_{IN} \leq V_{CC}$	All
I_{LI2}	Input Leakage Current $\overline{\text{RESET}}$	−10		−300	μA	$V_{SS} \leq V_{IN} \leq 3.8V$	All
I_{LO}	Leakage Current (BUS, T0) (High Impedance State)			±10	μA	$V_{SS} \leq V_{IN} \leq V_{CC}$	All
$I_{DD} + I_{CC}$	Total Supply Current*		80	100	mA		8748H
			95	110	mA		8749H

NOTE:
*$I_{CC} + I_{DD}$ is measured with all outputs disconnected; $\overline{\text{SS}}$, $\overline{\text{RESET}}$, and $\overline{\text{INT}}$ equal to V_{CC}; EA equal to V_{SS}.

AC CHARACTERISTICS $T_A = 0°C$ to $+70°C$; $V_{CC} = V_{DD} = 5V \pm 10\%$; $V_{SS} = 0V$

Symbol	Parameter	f(t) (Note 3)	11 MHz		Unit	Conditions (Note 1)
			Min	Max		
t	Clock Period	1/xtal freq	90.9	1000	ns	(Note 3)
t_{LL}	ALE Pulse Width	3.5t − 170	150		ns	
t_{AL}	Addr Setup to ALE	2t − 110	70		ns	(Note 2)
t_{LA}	Addr Hold from ALE	t − 40	50		ns	
t_{CC1}	Control Pulse Width ($\overline{RD}$, $\overline{WR}$)	7.5t − 200	480		ns	
t_{CC2}	Control Pulse Width ($\overline{PSEN}$)	6t − 200	350		ns	
t_{DW}	Data Setup before $\overline{WR}$	6.5t − 200	390		ns	
t_{WD}	Data Hold after $\overline{WR}$	t − 50	40		ns	
t_{DR}	Data Hold ($\overline{RD}$, $\overline{PSEN}$)	1.5t − 30	0	110	ns	
t_{RD1}	$\overline{RD}$ to Data In	6t − 170		375	ns	
t_{RD2}	$\overline{PSEN}$ to Data In	4.5t − 170		240	ns	
t_{AW}	Addr Setup to $\overline{WR}$	5t − 150	300		ns	
t_{AD1}	Addr Setup to Data ($\overline{RD}$)	10.5t − 220		730	ns	
t_{AD2}	Addr Setup to Data ($\overline{PSEN}$)	7.5t − 200		460	ns	
t_{AFC1}	Addr Float to $\overline{RD}$, $\overline{WR}$	2t − 40	140		ns	(Note 2)
t_{AFC2}	Addr Float to $\overline{PSEN}$	0.5t − 40	10		ns	(Note 2)
t_{LAFC1}	ALE to Control ($\overline{RD}$, $\overline{WR}$)	3t − 75	200		ns	
t_{LAFC2}	ALE to Control ($\overline{PSEN}$)	1.5t − 75	60		ns	
t_{CA1}	Control to ALE ($\overline{RD}$, $\overline{WR}$, PROG)	t − 65	25		ns	
t_{CA2}	Control to ALE ($\overline{PSEN}$)	4t − 70	290		ns	
t_{CP}	Port Control Setup to PROG	1.5t − 80	50		ns	
t_{PC}	Port Control Hold to PROG	4t − 260	100		ns	
t_{PR}	PROG to P2 Input Valid	8.5t − 120		650	ns	
t_{PF}	Input Data Hold from PROG	1.5t	0	140	ns	
t_{DP}	Output Data Setup	6t − 290	250		ns	
t_{PD}	Output Data Hold	1.5t − 90	40		ns	
t_{PP}	PROG Pulse Width	10.5t − 250	700		ns	
t_{PL}	Port 2 I/O Setup to ALE	4t − 200	160		ns	
t_{LP}	Port 2 I/O Hold to ALE	0.5t − 30	15		ns	
t_{PV}	Port Output from ALE	4.5t + 100		510	ns	
t_{OPRR}	T0 Rep Rate	3t	270		ns	
t_{CY}	Cycle Time	15t	1.36	15.0	μs	

NOTES:
1. Control outputs CL = 80 pF; BUS outputs CL = 150 pF.
2. BUS High Impedance Load 20 pF.
3. f(t) assumes 50% duty cycle on X1, X2. Max clock period is for a 1 MHz crystal input.

WAVEFORMS

INSTRUCTION FETCH FROM PROGRAM MEMORY

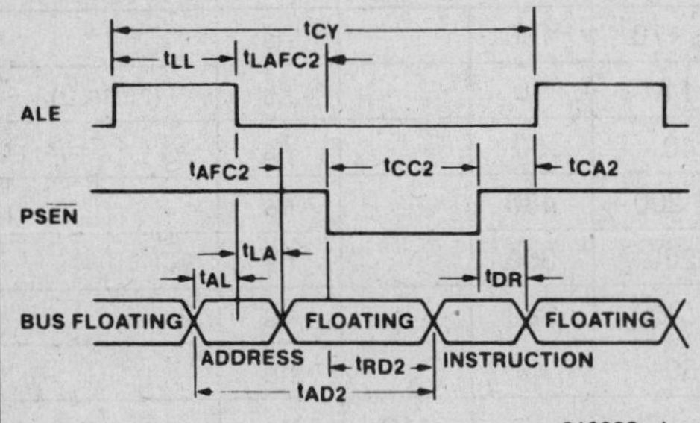

210983-4

WRITE TO EXTERNAL DATA MEMORY

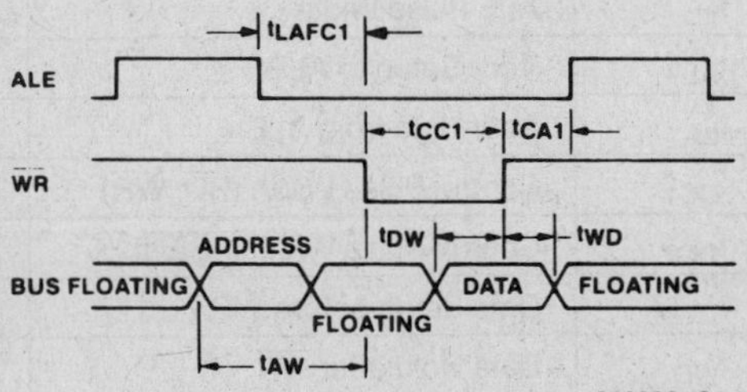

210983-6

READ FROM EXTERNAL DATA MEMORY

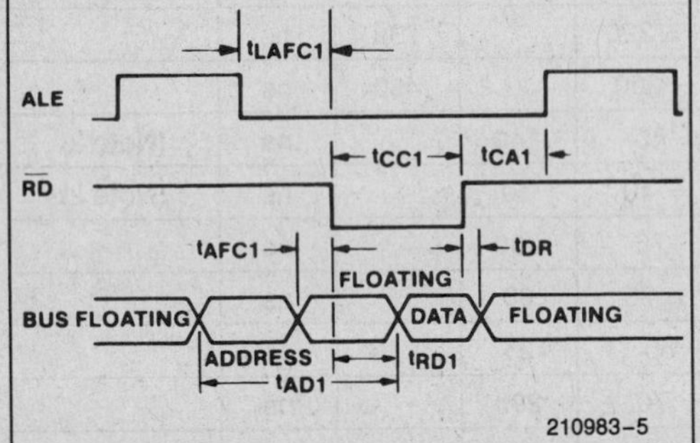

210983-5

INPUT AND OUTPUT FOR AC TESTS

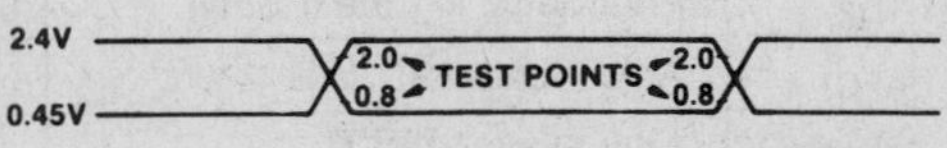

210983-7

AC testing inputs are driven at 2.4V for a Logic "1" and 0.45V for a Logic "0". Output timing measurements are made at 2.0V for a Logic "1" and 0.8V for a Logic "0."

PORT 1/PORT 2 TIMING

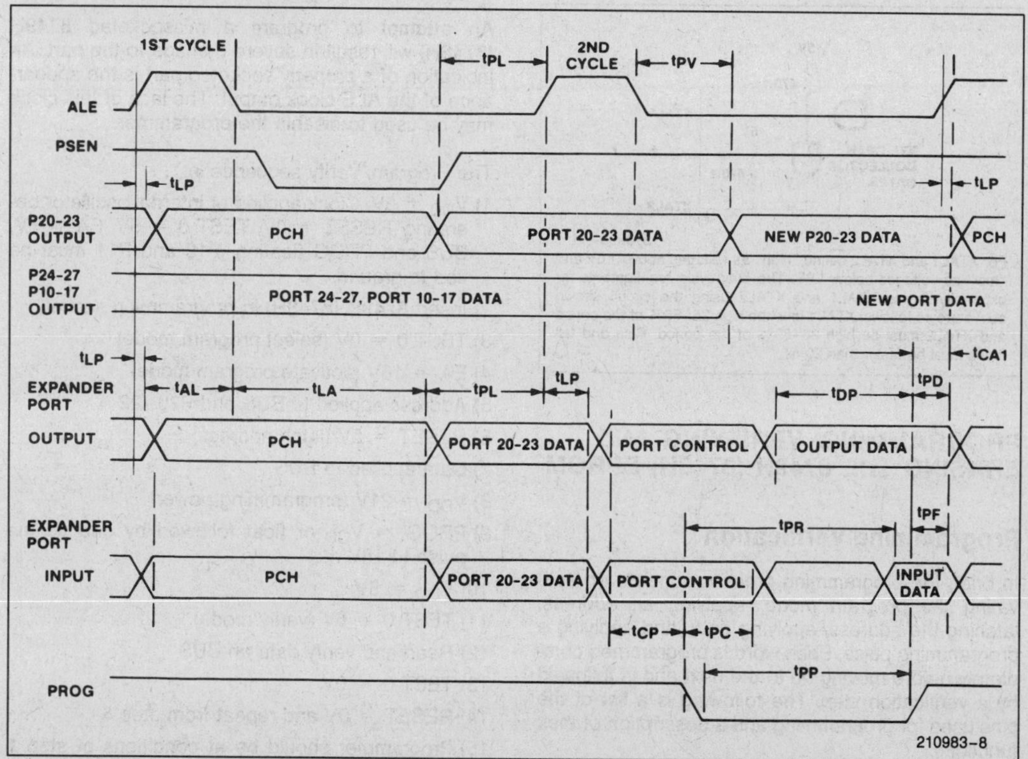

210983-8

CRYSTAL OSCILLATOR MODE

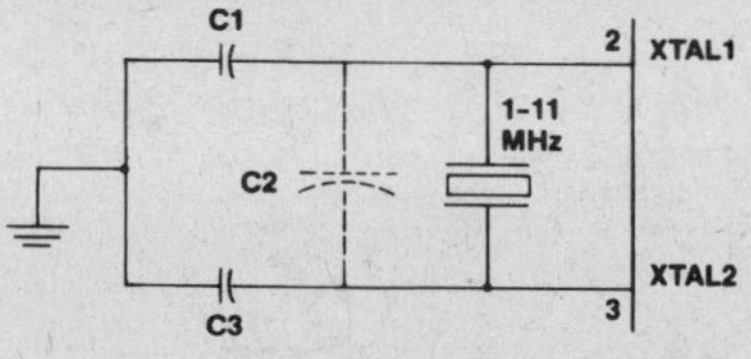

210983-9

C1 = 5 pF ±½ pF + (STRAY < 5 pF)
C2 = (CRYSTAL + STRAY) < 8 pF
C3 = 20 pF ±1 pF (STRAY < 5 pF)

Crystal series resistance should be less than 30Ω at 11 MHz; less than 75Ω at 6 MHz; less than 180Ω at 3.6 MHz.

CERAMIC RESONATOR MODE

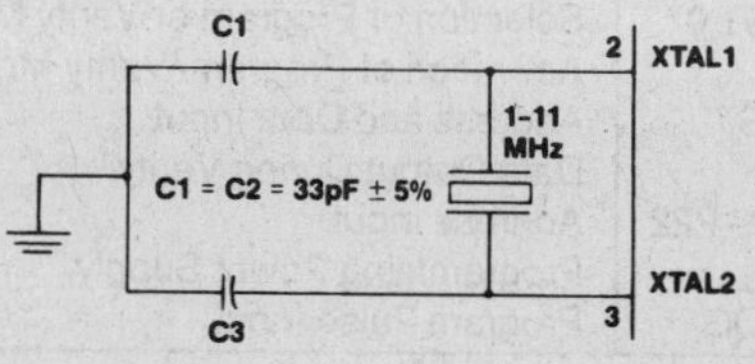

210983-10

DRIVING FROM EXTERNAL SOURCE

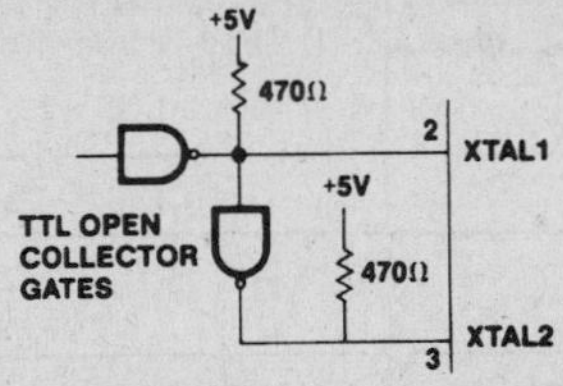

210983-11

For XTAL1 and XTAL2 define "high" as voltages above 1.6V and "low" as voltages below 1.6V. The duty cycle requirements for externally driving XTAL1 and XTAL2 using the circuit shown above are as follows: XTAL1 must be high 35–65% of the period and XTAL2 must be high 36–65% of the period. Rise and fall times must be faster than 20 ns.

PROGRAMMING, VERIFYING AND ERASING THE 8749H (8748H) EPROM

Programming Verification

In brief, the programming process consists of: activating the program mode, applying an address, latching the address, applying data, and applying a programming pulse. Each word is programmed completely before moving on to the next and is followed by a verification step. The following is a list of the pins used for programming and a description of their functions:

Pin	Function
XTAL 1	Clock Input (3 to 4.0 MHz)
XTAL 2	
RESET	Initialization and Address Latching
TEST 0	Selection of Program or Verify Mode
EA	Activation of Program/Verify Modes
BUS	Address and Data Input Data Output During Verify
P20–P22	Address Input
V_{DD}	Programming Power Supply
PROG	Program Pulse Input

WARNING

An attempt to program a missocketed 8749H (8748H) will result in severe damage to the part. An indication of a properly socketed part is the appearance of the ALE clock output. The lack of this clock may be used to disable the programmer.

The Program/Verify sequence is:

1) V_{DD} = 5V, Clock applied or internal oscillator operating. RESET = 0V, TEST 0 = 5V, EA = 5V, BUS and PROG floating. P10 and P11 must be tied to ground.
2) Insert 8749H (8748H) in programming socket.
3) TEST 0 = 0V (select program mode)
4) EA = 18V (activate program mode)
5) Address applied to BUS and P20–22
6) RESET = 5V (latch address)
7) Data applied to BUS
8) V_{DD} = 21V (programming power)
9) PROG = V_{CC} or float followed by one 50 ms pulse to 18V
10) V_{DD} = 5V
11) TEST 0 = 5V (verify mode)
12) Read and verify data on BUS
13) TEST 0 = 0V
14) RESET = 0V and repeat from step 5
15) Programmer should be at conditions of step 1 when 8749H (8748H) is removed from socket.

AC TIMING SPECIFICATION FOR PROGRAMMING 8748H/8749H

T_A = 25°C ±5°C; V_{CC} = 5V ±5%; V_{DD} = 21V ±0.5V

Symbol	Parameter	Min	Max	Unit	Test Conditions
t_{AW}	Address Setup Time to $\overline{RESET}$ ↑	$4t_{CY}$			
t_{WA}	Address Hold Time after $\overline{RESET}$ ↑	$4t_{CY}$			
t_{DW}	Data in Setup Time to PROG ↑	$4t_{CY}$			
t_{WD}	Data in Hold Time after PROG ↓	$4t_{CY}$			
t_{PH}	$\overline{RESET}$ Hold Time to Verify	$4t_{CY}$			
t_{VDDW}	V_{DD} Hold Time before PROG ↑	0	1.0	ms	
t_{VDDH}	V_{DD} Hold Time after PROG ↓	0	1.0	ms	
t_{PW}	Program Pulse Width	50	60	ms	
t_{TW}	TEST 0 Setup Time for Program Mode	$4t_{CY}$			
t_{WT}	TEST 0 Hold Time after Program Mode	$4t_{CY}$			
t_{DO}	TEST 0 to Data Out Delay		$4t_{CY}$		
t_{WW}	$\overline{RESET}$ Pulse Width to Latch Address	$4t_{CY}$			
t_r, t_f	V_{DD} and PROG Rise and Fall Times	0.5	100	μs	
t_{CY}	CPU Operation Cycle Time	3.75	5	μs	
t_{RE}	$\overline{RESET}$ Setup Time before EA ↑	$4t_{CY}$			

NOTE:
If TEST 0 is high, t_{DO} can be triggered by $\overline{RESET}$ ↑.

DC SPECIFICATION FOR PROGRAMMING 8748H/8749H

T_A = 25°C ±5°C; V_{CC} = 5V ±5%; V_{DD} = 21V ±0.5V

Symbol	Parameter	Min	Max	Unit	Test Conditions
V_{DDH}	V_{DD} Program Voltage High Level	20.5	21.5	V	
V_{DDL}	V_{DD} Voltage Low Level	4.75	5.25	V	
V_{PH}	PROG Program Voltage High Level	17.5	18.5	V	
V_{PL}	PROG Voltage Low Level	4.0	V_{CC}	V	
V_{EAH}	EA Program or Verify Voltage High Level	17.5	18.5	V	
I_{DD}	V_{DD} High Voltage Supply Current		20.0	mA	
I_{PROG}	PROG High Voltage Supply Current		1.0	mA	
I_{EA}	EA High Voltage Supply Current		1.0	mA	

WAVEFORMS

COMBINATION PROGRAM/VERIFY MODE (EPROMs ONLY)

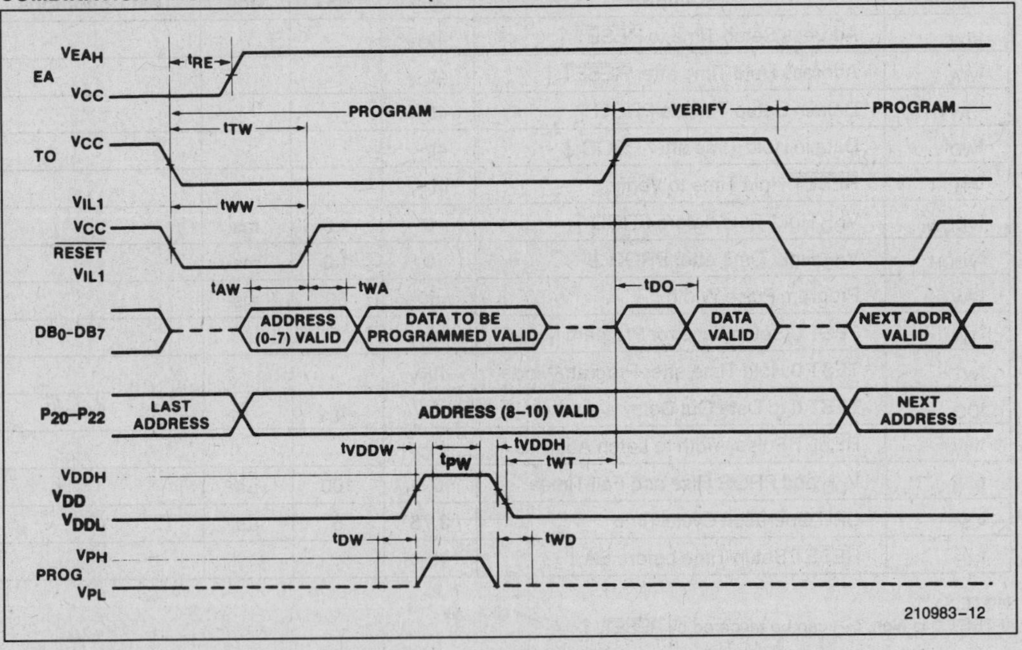

VERIFY MODE

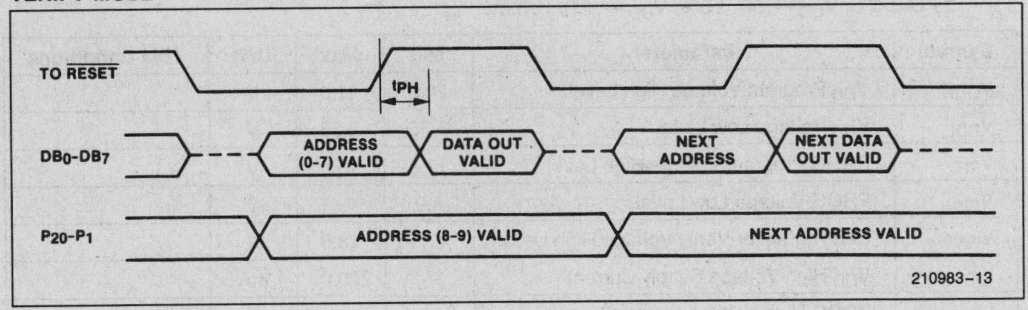

SUGGESTED EPROM VERIFICATION ALGORITHM FOR HMOS-E DEVICE ONLY

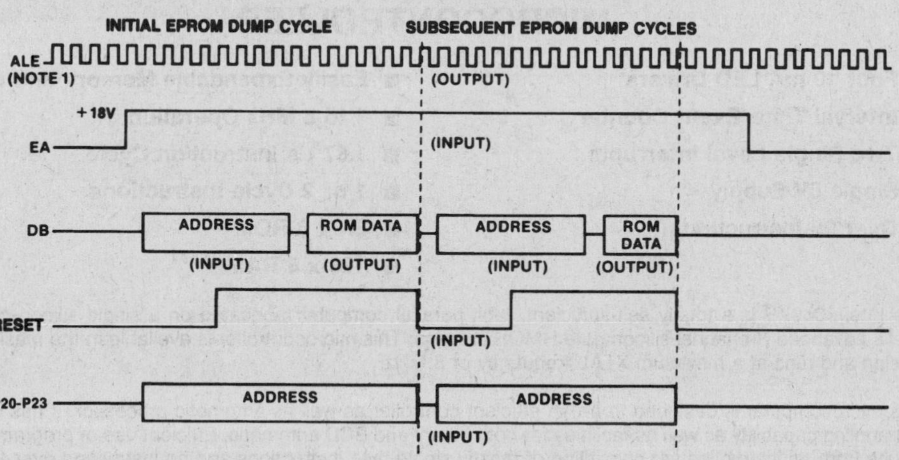

210983-14

$V_{CC} = V_{DD} = +5V$
$V_{SS} = 0V$

	48H	49H
A10	0	ADDR
A11	0	0

NOTE:
ALE is function of X1, X2 inputs.

P8049KB
HMOS SINGLE-COMPONENT 8-BIT MICROCONTROLLER

- Four 10 mA LED Drivers
- Interval Time/Event Counter
- Two Single Level Interrupts
- Single 5V Supply
- Over 96 Instructions
- Easily Expandable Memory and I/O
- 1 to 8 MHz Operation
- 1.87 μs Instruction Cycle
- 1 or 2 Cycle Instructions
- 2K x 8 ROM
- 128 x 8 RAM

The Intel 8049KB is a totally self-sufficient, 8-bit parallel computer fabricated on a single silicon chip using Intel's advanced N-channel silicon gate HMOS process. This microcontroller is available in the masked ROM version and runs at a maximum XTAL frequency of 8 MHz.

This microcomputer is designed to be an efficient controller as well as arithmetic processor. It has extensive bit handling capability as well as facilities for both binary and BCD arithmetic. Efficient use of program memory results from an instruction set consisting of mostly single byte instructions and no instruction over 2 bytes in length.

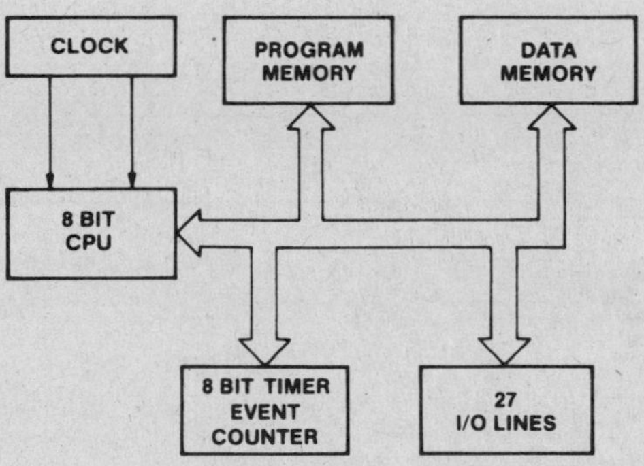

270790–1

Figure 1. Block Diagram

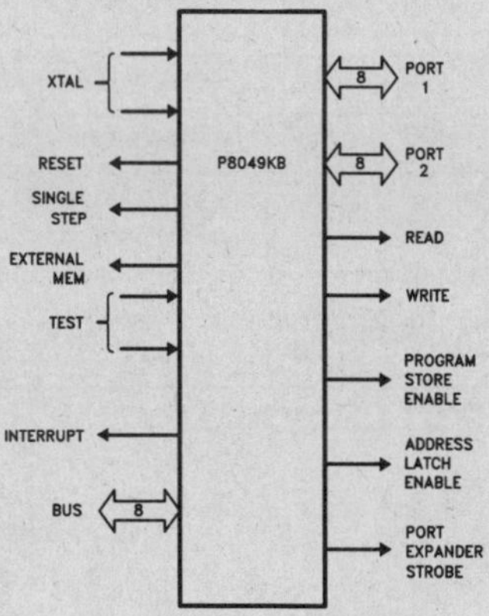

270790–2

Figure 2. Logic Symbol

February 1990
Order Number: 270790-002

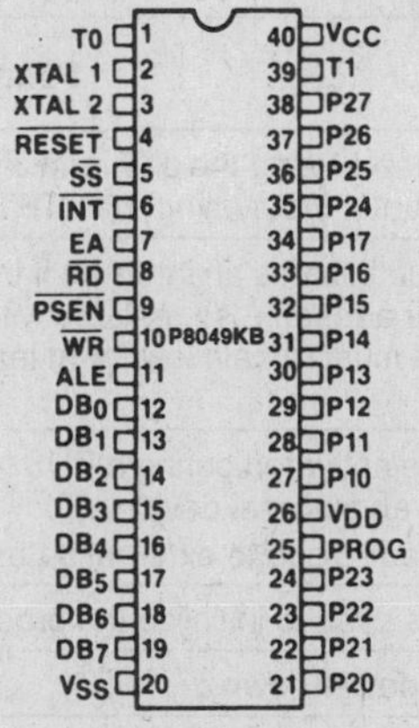

Pin Configuration

Table 1. Pin Description

Symbol	Pin No.	Function
V_{SS}	20	Circuit GND potential.
V_{DD}	26	+5V during normal operation.
		Low power standby pin.
		Programming power supply (+21V).
V_{CC}	40	Main power supply; +5V during operation and programming.
P10–P17 Port 1	27–34	8-bit quasi-bidirectional port.
P20–P23 P24–P27 Port 2	21–24 35–38	8-bit quasi-bidirectional port. P20–P23 contain the four high order program counter bits during an external program memory fetch and serve as a 4-bit I/O expander bus for 8243.
DB0–DB7 BUS	12–19	True bidirectional port which can be written or read synchronously using the $\overline{RD}$, $\overline{WR}$ strobes. The port can also be statically latched. Contains the 8 low order program counter bits during an external program memory fetch, and receives the addressed instruction under the control of $\overline{PSEN}$. Also contains the address and data during an external RAM data store instruction, under control of ALE, $\overline{RD}$, and $\overline{WR}$.
T0	1	Input pin testable using the conditional transfer instruction JT0 and JNT0. T0 can be designated as a clock output using ENT0 CLK instruction.
		Used during programming.

Table 1. Pin Description (Continued)

Symbol	Pin No.	Function
T1	39	Input pin testable using the JT1, and JNT1 instructions. Can be designated the timer/counter input using the STRT CNT instruction.
$\overline{INT}$	6	Interrupt input. Initiates an interrupt if interrupt is enabled. Interrupt is disabled after a reset. Also testable with conditional jump instruction. (Active low) interrupt must remain low for at least 3 machine cycles for proper operation.
$\overline{RD}$	8	Output strobe activated during a BUS read. Can be used to enable data onto the bus from an external device. Used as a read strobe to external data memory. (Active low)
$\overline{RESET}$	4	Input which is used to initialize the processor. (Active low) (Non TTL V_{IH})
		Used during power down.
		Used during programming.
		Used during ROM verification.
$\overline{WR}$	10	Output strobe during a bus write. (Active low) Used as write strobe to external data memory.
ALE	11	Address latch enable. This signal occurs once during each cycle and is useful as a clock output. The negative edge of ALE strobes address into external data and program memory.
$\overline{PSEN}$	9	Program store enable. This output occurs only during a fetch to external program memory. (Active low)
$\overline{SS}$	5	Single step input can be used in conjunction with ALE to "single step" the processor through each instruction.
		(Active low) Used in sync mode.
EA	7	External access input which forces all program memory fetches to reference external memory. Useful for emulation and debug. (Active high)
		Used during (18V) programming.
		Used during ROM verification (12V).
XTAL1	2	One side of crystal input for internal oscillator. Also input for external source. (Non TTL V_{IH})
XTAL2	3	Other side of crystal input.

Table 2. Instruction Set

Accumulator

Mnemonic	Description	Bytes	Cycles
ADD A, R	Add register to A	1	1
ADD A, @R	Add data memory to A	1	1
ADD A, #data	Add immediate to A	2	2
ADDC A, R	Add register with carry	1	1
ADDC A, @R	Add data memory with carry	1	1
ADDC A, #data	Add immediate with carry	2	2
ANL A, R	And register to A	1	1
ANL A, @R	And data memory to A	1	1
ANL A, #data	And immediate to A	2	2
ORL A, R	Or register to A	1	1
ORL A, @R	Or data memory to A	1	1
ORL A, #data	Or immediate to A	2	2
XRL A, R	Exclusive or register to A	1	1
XRL A, @R	Exclusive or data memory to A	1	1
XRL A, #data	Exclusive or immediate to A	2	2
INC A	Increment A	1	1
DEC A	Decrement A	1	1
CLR A	Clear A	1	1
CPL A	Complement A	1	1
DA A	Decimal adjust A	1	1
SWAP A	Swap nibbles of A	1	1
RL A	Rotate A left	1	1
RLC A	Rotate A left through carry	1	1
RR A	Rotate A right	1	1
RRC A	Rotate A right through carry	1	1

Input/Output

Mnemonic	Description	Bytes	Cycles
IN A, P	Input port to A	1	2
OUTL P, A	Output A to port	1	2
ANL P, #data	And immediate to port	2	2
ORL P, #data	Or immediate to port	2	2
INS A, BUS	Input BUS to A	1	2
OUTL BUS, A	Output A to BUS	1	2
ANL BUS, #data	And immediate to BUS	2	2
ORL BUS, #data	Or immediate to BUS	2	2
MOVD A, P	Input expander port to A	1	2
MOVD P, A	Output A to expander port	1	2
ANLD P, A	And A to expander port	1	2
ORLD P, A	Or A to expander port	1	2

Registers

Mnemonic	Description	Bytes	Cycles
INC R	Increment register	1	1
INC @R	Increment data memory	1	1
DEC R	Decrement register	1	1

Branch

Mnemonic	Description	Bytes	Cycles
JMP addr	Jump unconditional	2	2
JMPP @A	Jump indirect	1	2
DJNZ R, addr	Decrement register and skip	2	2
JC addr	Jump on carry = 1	2	2
JNC addr	Jump on carry = 0	2	2
JZ addr	Jump on A zero	2	2
JNZ addr	Jump on A not zero	2	2
JT0 addr	Jump on T0 = 1	2	2
JNT0 addr	Jump on T0 = 0	2	2
JT1 addr	Jump on T1 = 1	2	2
JNT1 addr	Jump on T1 = 0	2	2
JF0 addr	Jump on F0 = 1	2	2
JF1 addr	Jump on F1 = 1	2	2
JTF addr	Jump on timer flag	2	2
JNI addr	Jump on $\overline{INT}$ = 0	2	2
JBb addr	Jump on accumulator bit	2	2

1

Table 2. Instruction Set (Continued)

Subroutine

Mnemonic	Description	Bytes	Cycles
CALL addr	Jump to subroutine	2	2
RET	Return	1	2
RETR	Return and restore status	1	2

Flags

Mnemonic	Description	Bytes	Cycles
CLR C	Clear carry	1	1
CPL C	Complement carry	1	1
CLR F0	Clear flag 0	1	1
CPL F0	Complement flag 0	1	1
CLR F1	Clear flag 1	1	1
CPL F1	Complement flag 1	1	1

Data Moves

Mnemonic	Description	Bytes	Cycles
MOV A, R	Move register to A	1	1
MOV A, @R	Move data memory to A	1	1
MOV A, #data	Move immediate to A	2	2
MOV R, A	Move A to register	1	1
MOV @R, A	Move A to data memory	1	1
MOV R, #data	Move immediate to register	2	2
MOV @R, #data	Move immediate to data memory	2	2
MOV A, PSW	Move PSW to A	1	1
MOV PSW, A	Move A to PSW	1	1
XCH A, R	Exchange A and register	1	1
XCH A, @R	Exchange A and data memory	1	1
XCHD A, @R	Exchange nibble of A and data memory	1	1
MOVX A, @R	Move external data memory to A	1	2
MOVX @R, A	Move A to external data memory	1	2
MOVP A, @A	Move to A from current page	1	2
MOVP3 A, @A	Move to A from page 3	1	2

Timer/Counter

Mnemonic	Description	Bytes	Cycles
MOV A, T	Read timer/counter	1	1
MOV T, A	Load timer/counter	1	1
STRT T	Start timer	1	1
STRT CNT	Start counter	1	1
STOP TCNT	Stop timer/counter	1	1
EN TCNTI	Enable timer/counter interrupt	1	1
DIS TCNTI	Disable timer/counter interrupt	1	1

Control

Mnemonic	Description	Bytes	Cycles
EN I	Enable external interrupt	1	1
DIS I	Disable external interrupt	1	1
SEL RB0	Select register bank 0	1	1
SEL RB1	Select register bank 1	1	1
SEL MB0	Select memory bank 0	1	1
SEL MB1	Select memory bank 1	1	1
ENT0 CLK	Enable clock output on T0	1	1

Mnemonic	Description	Bytes	Cycles
NOP	No operation	1	1

ABSOLUTE MAXIMUM RATINGS*

Case Temperature Under Bias ....0°C to +70°C
Storage Temperature ....−65°C to +150°C
Voltage on any Pin with Respect to Ground......................−0.5V to +7V
Power Dissipation..........................1.5W

NOTICE: This data sheet contains preliminary information on new products in production. The specifications are subject to change without notice. Verify with your local Intel Sales office that you have the latest data sheet before finalizing a design.

**WARNING: Stressing the device beyond the "Absolute Maximum Ratings" may cause permanent damage. These are stress ratings only. Operation beyond the "Operating Conditions" is not recommended and extended exposure beyond the "Operating Conditions" may affect device reliability.*

1

D.C. CHARACTERISTICS $T_A = 0°C$ to $+70°C$; $V_{CC} = V_{DD} = 5V \pm 10\%$; $V_{SS} = 0V$

Symbol	Parameter	Limits			Unit	Test Conditions
		Min	Typ	Max		
V_{IL}	Input Low Voltage (All Except $\overline{RESET}$, X1, X2)	−0.5		0.8	V	
V_{IL1}	Input Low Voltage ($\overline{RESET}$, X1, X2)	−0.5		0.6	V	
V_{IH}	Input High Voltage (All Except XTAL1, XTAL2, $\overline{RESET}$)	2.0		V_{CC}	V	
V_{IH1}	Input High Voltage (X1, X2, $\overline{RESET}$)	3.8		V_{CC}	V	
V_{OL}	Output Low Voltage (BUS)			0.45	V	$I_{OL} = 2.0$ mA
V_{OL1}	Output Low Voltage ($\overline{RD}$, $\overline{WR}$, $\overline{PSEN}$, ALE)			0.45	V	$I_{OL} = 1.8$ mA
V_{OL2}	Output Low Voltage (PROG)			0.45	V	$I_{OL} = 1.0$ mA
V_{OL3}	Output Low Voltage (All Other Outputs)			0.45	V	$I_{OL} = 1.6$ mA
V_{OL4}	Output Low Voltage (Any Four Port Outputs)			0.45	V	$I_{OL} = 10$ mA
V_{OH}	Output High Voltage (BUS)	2.4			V	$I_{OH} = -400\ \mu A$
V_{OH1}	Output High Voltage ($\overline{RD}$, $\overline{WR}$, $\overline{PSEN}$, ALE)	2.4			V	$I_{OH} = -100\ \mu A$
V_{OH2}	Output High Voltage (All Other Outputs)	2.4			V	$I_{OH} = -40\ \mu A$

D.C. CHARACTERISTICS $T_A = 0°C$ to $+70°C$; $V_{CC} = V_{DD} = 5V \pm 10\%$; $V_{SS} = 0V$ (Continued)

Symbol	Parameter	Limits			Unit	Test Conditions
		Min	Typ	Max		
I_{L1}	Leakage Current (T1, $\overline{INT}$)			±10	μA	$V_{SS} \leq V_{IN} \leq V_{CC}$
I_{LI1}	Input Leakage Current (P10–P17, P20–P27, EA, $\overline{SS}$)			−500	μA	$V_{SS} + 0.45 \leq V_{IN} \leq V_{CC}$
I_{LI2}	Input Leakage Current $\overline{RESET}$	−10		−300	μA	$V_{SS} \leq V_{IN} \leq 3.8$
I_{LO}	Leakage Current (BUS, T0) (High Impedance State)			±10	μA	$V_{SS} \leq V_{IN} \leq V_{CC}$
I_{DD}	V_{DD} Supply Current (RAM Standby)		3	5	mA	
			4	7	mA	
			5	10	mA	
$I_{DD} + I_{CC}$	Total Supply Current*		30	65	mA	
			35	70	mA	
			40	80	mA	
			30	100	mA	
			50	110	mA	
V_{DD}	RAM Standby Voltage	2.2		5.5	V	Standby Mode Reset $\leq V_{IL1}$
		2.2		5.5	V	
		2.2		5.5	V	

*$I_{CC} + I_{DD}$ are measured with all outputs in their high impedance state; $\overline{RESET}$ low; 8 MHz crystal applied; $\overline{INT}$, $\overline{SS}$, and EA floating.

†Any four Port Outputs can be loaded to a 10 mA maximum. Excessive heating and dissipation will result if more than four outputs are loaded to 10 mA.

1

A.C. CHARACTERISTICS $T_A = 0°C$ to $+70°C$; $V_{CC} = V_{DD} = 5V \pm 10\%$; $V_{SS} = 0V$

Symbol	Parameter	f (t) (Note 3)	8 MHz		Unit	Conditions (Note 1)
			Min	Max		
t	Clock Period	1/xtal freq	125	1000	ns	(Note 3)
t_{LL}	ALE Pulse Width	3.5t−170	268		ns	
t_{AL}	Addr Setup to ALE	2t−110	140		ns	(Note 2)
t_{LA}	Addr Hold from ALE	t−40	85		ns	
t_{CC1}	Control Pulse Width ($\overline{RD}$, $\overline{WR}$)	7.5t−200	675		ns	
t_{CC2}	Control Pulse Width ($\overline{PSEN}$)	6t−200	550		ns	
t_{DW}	Data Setup before $\overline{WR}$	6.5t−200	613		ns	
t_{WD}	Data Hold after $\overline{WR}$	t−50	75		ns	
t_{DR}	Data Hold ($\overline{RD}$, $\overline{PSEN}$)	1.5t−30	0	158	ns	
t_{RD1}	$\overline{RD}$ to Data in	6t−170		580	ns	
t_{RD2}	$\overline{PSEN}$ to Data in	4.5t−170		393	ns	
t_{AW}	Addr Setup to $\overline{WR}$	5t−150	475		ns	
t_{AD1}	Addr Setup to Data ($\overline{RD}$)	10.5t−220		1093	ns	
t_{AD2}	Addr Setup to Data ($\overline{PSEN}$)	7.5t−200		738	ns	
t_{AFC1}	Addr Float to $\overline{RD}$, $\overline{WR}$	2t−40	210		ns	(Note 2)
t_{AFC2}	Addr Float to $\overline{PSEN}$	0.5t−40	23		ns	(Note 2)
t_{LAFC1}	ALE to Control ($\overline{RD}$, $\overline{WR}$)	3t−75	300		ns	
t_{LAFC2}	ALE to Control ($\overline{PSEN}$)	1.5t−75	113		ns	
t_{CA1}	Control to ALE ($\overline{RD}$, $\overline{WR}$, PROG)	t−65	60		ns	
t_{CA2}	Control to ALE ($\overline{PSEN}$)	4t−70	430		ns	
t_{CP}	Port Control Setup to PROG	1.5t−80	108		ns	
t_{PC}	Port Control Hold to PROG	4t−260	240		ns	
t_{PR}	PROG to P2 Input Valid	8.5t−120		943	ns	
t_{PF}	Input Data Hold from PROG	1.5t	0	188	ns	
t_{DP}	Output Data Setup	6t−290	460		ns	
t_{PD}	Output Data Hold	1.5t−90	98		ns	
t_{PP}	PROG Pulse Width	10.5t−250	1063		ns	
t_{PL}	Port 2 I/O Setup to ALE	4t−200	300		ns	
t_{LP}	Port 2 I/O Hold to ALE	0.5t−30	33		ns	
t_{PV}	Port Output from ALE	4.5t+100		663	ns	
t_{OPRR}	T0 Rep Rate	3t	375		ns	
t_{CY}	Cycle Time	15t	1.87	28	μs	

NOTES:
1. Control outputs: C_L = 80 pF. BUS Outputs: C_L = 150 pF.
2. BUS High Impedance Load 20 pF
3. f(t) assumes 50% duty cycle on X1, X2. Max clock period is for a 1 MHz crystal input.

WAVEFORMS

INSTRUCTION FETCH FROM PROGRAM MEMORY

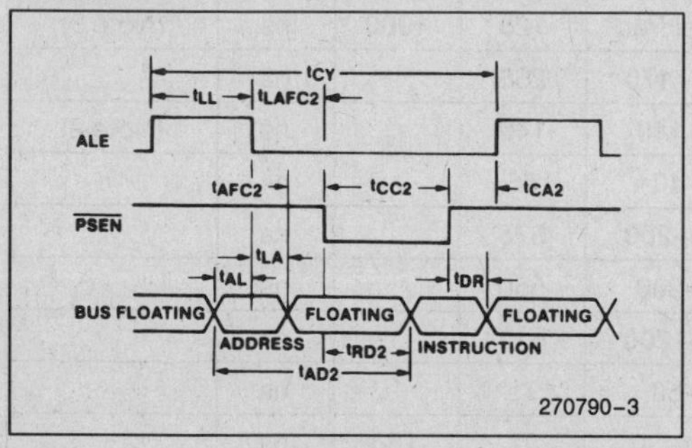

270790-3

READ FROM EXTERNAL DATA MEMORY

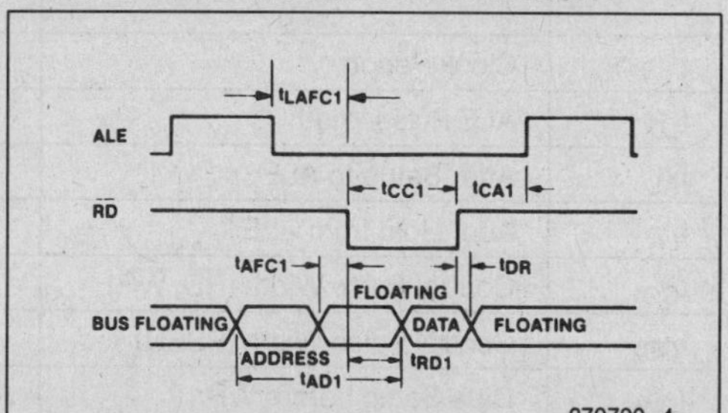

270790-4

WRITE TO EXTERNAL DATA MEMORY

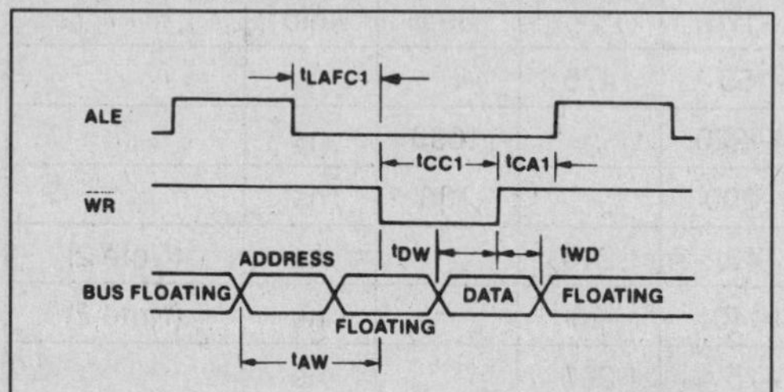

270790-5

INPUT AND OUTPUT FOR A.C. TESTS

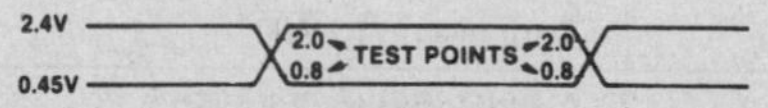

270790-6

A.C. testing inputs are driven at 2.4V for a logic "1" and 0.45V for a logic "0". Output timing measurements are made at 2.0V for a logic "1" and 0.8V for a logic "0".

PORT 1/PORT 2 TIMING

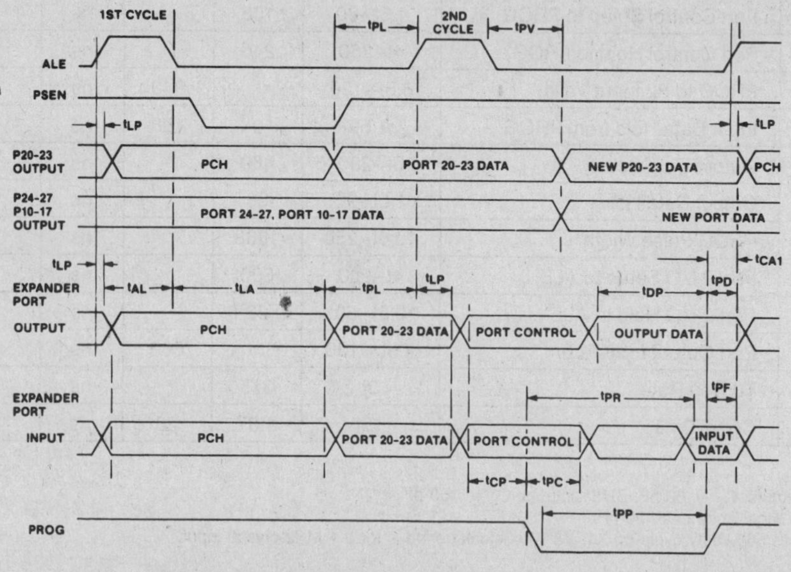

270790-7

CRYSTAL OSCILLATOR MODE

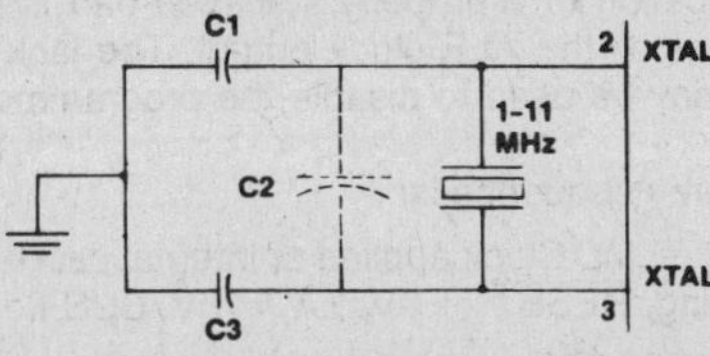

270790-8

C1 = 5 pF ±½ pF + (STRAY < 5 pF)
C2 = (CRYSTAL + STAY) < 8 pF
C3 = 20 pF ±1 pF + (STRAY < 5 pF)

Crystal series resistance should be less than 50Ω at 8 MHz; less than 75Ω at 6 MHz; less than 180Ω at 3.6 MHz.

CERAMIC RESONATOR MODE

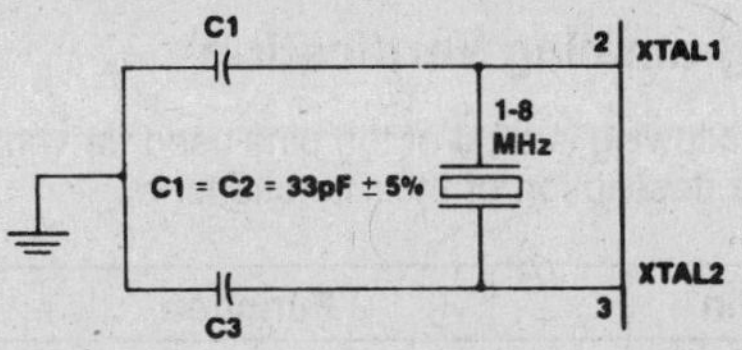

270790-9

DRIVING FROM EXTERNAL SOURCE

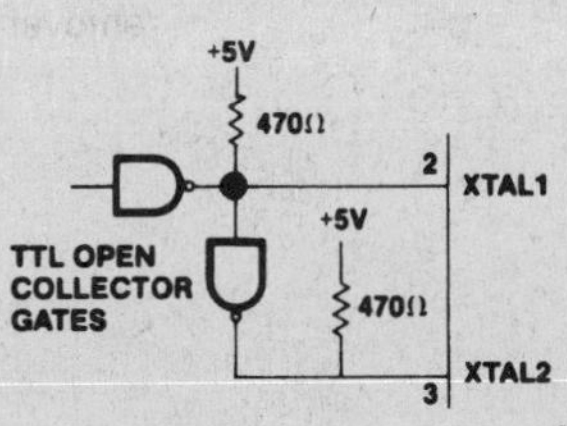

270790-10

For XTAL1 and XTAL2 define "high" as voltages above 1.6V and "low" as voltages below 1.6V. The duty cycle requirements for externally driving XTAL1 and XTAL2 using the circuits shown above are as follows: XTAL1 must be high 35–65% of the period and XTAL2 must be high 35–65% of the period. Rise and fall times must be faster than 20 ns.

VERIFYING THE 8049KB ROM

Programming Verification

The following is a list of the pins used for verification and a description of their functions:

Pin	Function
XTAL1	Clock Input (3 to 4.0 MHz)
XTAL2	
$\overline{RESET}$	Initialization and Address Latching
T0	Selection of Program or Verifying Mode
EA	Activation Verify Modes
BUS	Address and Data Output During Verify
P20–P22	Address Input

WARNING:

An indication of a properly socketed part is the appearance of the ALE clock output. The lack of this clock may be used to disable the programmer.

The Verify sequence is:

1. V_{DD} = 5V, Clock applied or internal oscillator operating, $\overline{RESET}$ = 0V, EA = 5V, BUS floating.
2. Insert 8049KB in verify socket
3. EA = 12V (activate verify mode)
4. Address applied to BUS and P20–23
5. $\overline{RESET}$ = 5V (latch address)
6. Read and verify Data on BUS
7. $\overline{RESET}$ = 0V and repeat from step 4
8. Verify socket should be at conditions of step 1 for removal from socket.

SUGGESTED ROM VERIFICATION ALGORITHM FOR ROM DEVICE ONLY

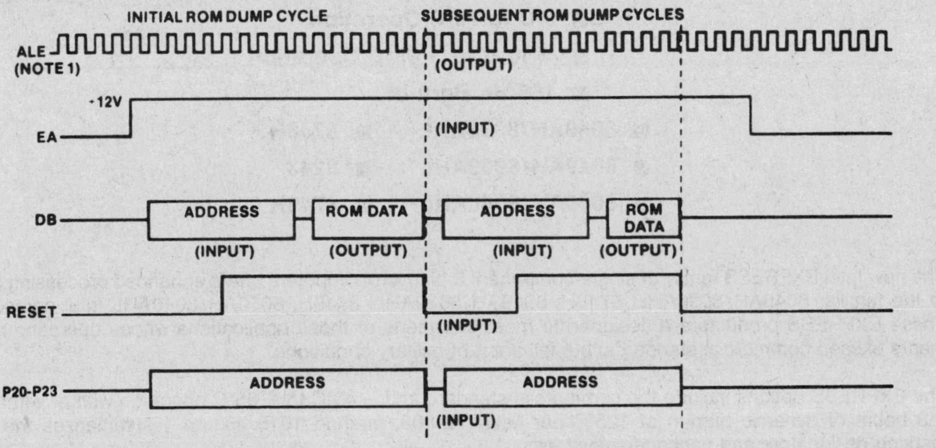

270790-11

	48H	49H	50H
A10	0	ADDR	ADDR
A11	0	0	ADDR

$V_{CC} = V_{DD} = +5V$
$V_{SS} = 0V$

NOTE:
ALE is function of X1, X2 inputs.

intel®

PRELIMINARY

MCS®-48
EXPRESS

- 0°C to 70°C Operation
- −40°C to +85°C Operation
- 168 Hr. Burn-In

- 8048AH/8035AHL
- 8049AH/8039AHL
- 8050AH/8040AHL
- 8748H
- 8243
- 8749H

The new Intel EXPRESS family of single-component 8-bit microcomputers offers enhanced processing options to the familiar 8048AH/8035AHL, 8748H, 8049AH/8039AHL, 8749H, 8050AH/8040AHL Intel components. These EXPRESS products are designed to meet the needs of those applications whose operating requirements exceed commercial standards, but fall short of military conditions.

The EXPRESS options include the commercial standard and −40°C to +85°C operation with or without 168 ±8 hours of dynamic burn-in at 125°C per MIL-STD-883, method 1015. Figure 1 summarizes the option marking designators and package selections.

For a complete description of 8048AH/8035AHL, 8748H, 8049AH/8309AHL, 8749H, 8040AHL and 8050AH features and operating characteristics, refer to the respective standard commercial grade data sheet. This document highlights only the electrical specifications which differ from the respective commercial part.

Temp Range °C	0–70	−40–+85	0–70	−40–+85
Burn In	**0 Hrs**	**0 Hrs**	**168 Hrs**	**168 Hrs**
	P8048AH	TP8048AH	QP8048AH	LP8048AH
	D8048AH	TD8048AH	QD8048AH	LD8048AH
	D8748H	TD8748H	QD8748H	LD8748H
	P8035AHL	TP8035AHL	QP8035AHL	LP8035AHL
	D8035AHL	TD8035AHL	QD8035AHL	LD8035AHL
	P8049AH	TP8049AH	QP8049AH	LP8049AH
	D8049AH	TD8049AH	QD8049AH	LD8049AH
	D8749H	TD8749AH	QD8749H	LD8749AH
	P8039AHL	TP8039AHL	QP8039AHL	LP8039AHL
	D8039AHL	TD8039AHL	QD8039AHL	LD8039AHL
	P8050AH	TP8050AH	QP8050AH	LP8050AH
	D8050AH	TD8050AH	QD8050AH	LD8050AH
	P8040AHL	TP8040AHL	QP8040AHL	LP8040AHL
	D8040AHL	TD8040AHL	QD8040AHL	LD8040AHL
	P8243	TP8243	QP8243	—
	D8243	TD8243	QD8243	LD8243

* Commercial Grade
P Plastic Package
D Cerdip Package

September 1987
Order Number: 270225-002

Extended Temperature Electrical Specification Deviations*

1

TP8048AH/TP8035AHL/LP8048AH/LP8035AHL
TD8048AH/TD8035AHL/LD8048AH/LD8035AHL

D.C. CHARACTERISTICS $T_A = -40°C$ to $+85°C$; $V_{CC} = V_{DD} = 5V \pm 10\%$; $V_{SS} = 0V$

Symbol	Parameter	Limits			Unit	Test Conditions
		Min	Typ	Max		
V_{IH}	Input High Voltage (All Except XTAL1, XTAL2, $\overline{RESET}$)	2.2		V_{CC}	V	
I_{DD}	V_{DD} Supply Current		4	8	mA	
$I_{DD} + I_{CC}$	Total Supply Current		40	80	mA	

TP8049AH/TP8039AHL/LP8049AH/LP8039AHL
TD8049AH/TD8039AHL/LD8049AH/LD8039AHL

D.C. CHARACTERISTICS $T_A = -40°C$ to $+85°C$; $V_{CC} = V_{DD} = 5V \pm 10\%$; $V_{SS} = 0V$

Symbol	Parameter	Limits			Unit	Test Conditions
		Min	Typ	Max		
V_{IH}	Input High Voltage (All Except XTAL1, XTAL2, $\overline{RESET}$)	2.2		V_{CC}	V	
I_{DD}	V_{DD} Supply Current		5	10	mA	
$I_{DD} + I_{CC}$	Total Supply Current		50	100	mA	

TP8050AH/TP8040AHL/LP8050AHL/LP8040AHL
TD8050AH/TD8040AHL/LD8050AH/LD8040AHL

D.C. CHARACTERISTICS $T_A = -40°C$ to $+85°C$; $V_{CC} = V_{DD} = 5V \pm 10\%$; $V_{SS} = 0V$

Symbol	Parameter	Limits			Unit	Test Conditions
		Min	Typ	Max		
V_{IH}	Input High Voltage (All Except XTAL1, XTAL2, $\overline{RESET}$)	2.2		V_{CC}	V	
I_{DD}	V_{DD} Supply Current		10	20	mA	
$I_{DD} + I_{CC}$	Total Supply Current		75	120	mA	

*Extended Temperature Electrical Specification Deviations**

TD8748H/LD8748H

D.C. CHARACTERISTICS $T_A = -40°C$ to $+85°C$; $V_{CC} = V_{DD} = 5V \pm 10\%$; $V_{SS} = 0V$

Symbol	Parameter	Limits			Unit	Test Conditions
		Min	Typ	Max		
V_{IH}	Input High Voltage (All Except XTAL1, XTAL2, $\overline{RESET}$)	2.2		V_{CC}	V	
$I_{DD} + I_{CC}$	Total Supply Current		50	130	mA	

TD8749H/LD8749H

D.C. CHARACTERISTICS $T_A = -40°C$ to $+85°C$; $V_{CC} = V_{DD} = 5V \pm 10\%$; $V_{SS} = 0V$

Symbol	Parameter	Limits			Unit	Test Conditions
		Min	Typ	Max		
V_{IH}	Input High Voltage (All Except XTAL1, XTAL2, $\overline{RESET}$)	2.2		V_{CC}	V	
$I_{DD} + I_{CC}$	Total Supply Current		75	150	mA	

TP8743/TD8243/LD8243

D.C. CHARACTERISTICS $T_A = -40°C$ to $+85°C$; $V_{CC} = 5V \pm 10\%$; $V_{SS} = 0V$

Symbol	Parameter	Limits			Unit	Test Conditions
		Min	Typ	Max		
I_{CC}	V_{CC} Supply Current		15	25	mA	

*Refer to individual commercial grade data sheet for complete operating characteristics.

2

MCS® 51 Microcontroller Family

2

MCS® 51 Family of Microcontrollers Architectural Overview

September 1993

Order Number: 270251-004

MCS®-51 Family of Microcontrollers Architectural Overview

CONTENTS

INTRODUCTION

The 8051 is the original member of the MCS®-51 family, and is the core for all MCS-51 devices. The features of the 8051 core are:

- 8-bit CPU optimized for control applications
- Extensive Boolean processing (single-bit logic) capabilities
- 64K Program Memory address space
- 64K Data Memory address space
- On-chip Program Memory
- 128 bytes of on-chip Data RAM
- 32 bidirectional and individually addressable I/O lines
- Two 16-bit timer/counters
- Full duplex UART
- 6-source/5-vector interrupt structure with two priority levels
- On-chip clock oscillator

The basic architectural structure of this 8051 core is shown in Figure 1.

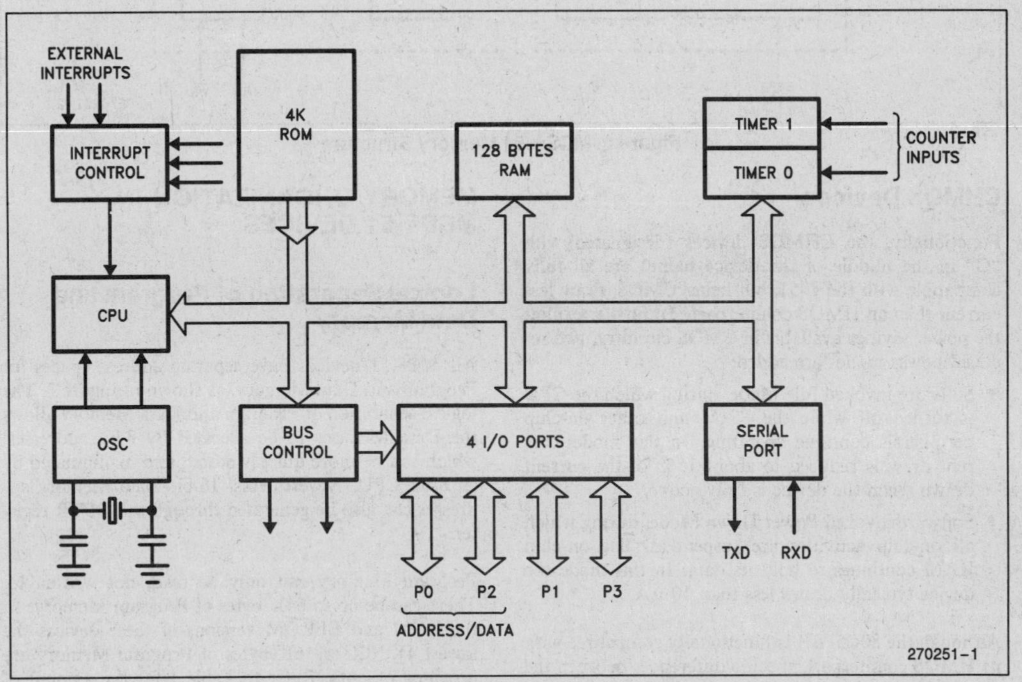

Figure 1. Block Diagram of the 8051 Core

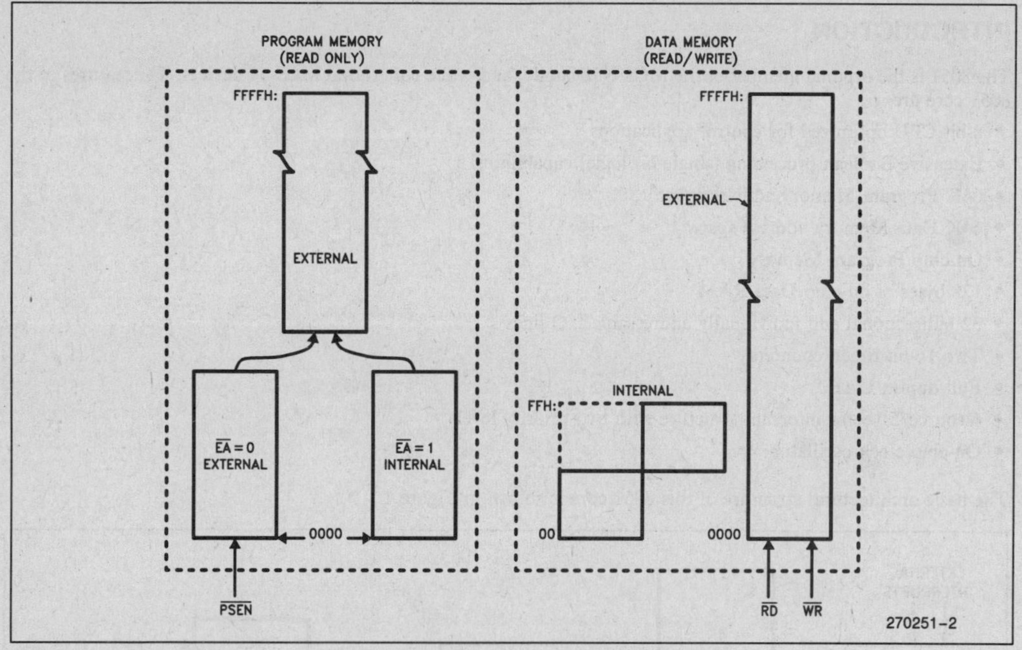

Figure 2. MCS®-51 Memory Structure

CHMOS Devices

Functionally, the CHMOS devices (designated with "C" in the middle of the device name) are all fully compatible with the 8051, but being CMOS, draw less current than an HMOS counterpart. To further exploit the power savings available in CMOS circuitry, two reduced power modes are added:

- Software-invoked Idle Mode, during which the CPU is turned off while the RAM and other on-chip peripherals continue operating. In this mode, current draw is reduced to about 15% of the current drawn when the device is fully active.
- Software-invoked Power Down Mode, during which all on-chip activities are suspended. The on-chip RAM continues to hold its data. In this mode the device typically draws less than 10 μA.

Although the 80C51BH is functionally compatible with its HMOS counterpart, specific differences between the two types of devices must be considered in the design of an application circuit if one wishes to ensure complete interchangeability between the HMOS and CHMOS devices. These considerations are discussed in the Application Note AP-252, "Designing with the 80C51BH".

For more information on the individual devices and features, refer to the Hardware Descriptions and Data Sheets of the specific device.

MEMORY ORGANIZATION IN MCS®-51 DEVICES

Logical Separation of Program and Data Memory

All MCS-51 devices have separate address spaces for Program and Data Memory, as shown in Figure 2. The logical separation of Program and Data Memory allows the Data Memory to be accessed by 8-bit addresses, which can be more quickly stored and manipulated by an 8-bit CPU. Nevertheless, 16-bit Data Memory addresses can also be generated through the DPTR register.

Program Memory can only be read, not written to. There can be up to 64K bytes of Program Memory. In the ROM and EPROM versions of these devices the lowest 4K, 8K or 16K bytes of Program Memory are provided on-chip. Refer to Table 1 for the amount of on-chip ROM (or EPROM) on each device. In the ROMless versions all Program Memory is external. The read strobe for external Program Memory is the signal $\overline{\text{PSEN}}$ (Program Store Enable).

Data Memory occupies a separate address space from Program Memory. Up to 64K bytes of external RAM can be addressed in the external Data Memory space. The CPU generates read and write signals, $\overline{RD}$ and $\overline{WR}$, as needed during external Data Memory accesses.

External Program Memory and external Data Memory may be combined if desired by applying the $\overline{RD}$ and $\overline{PSEN}$ signals to the inputs of an AND gate and using the output of the gate as the read strobe to the external Program/Data memory.

Program Memory

Figure 3 shows a map of the lower part of the Program Memory. After reset, the CPU begins execution from location 0000H.

As shown in Figure 3, each interrupt is assigned a fixed location in Program Memory. The interrupt causes the CPU to jump to that location, where it commences execution of the service routine. External Interrupt 0, for example, is assigned to location 0003H. If External Interrupt 0 is going to be used, its service routine must begin at location 0003H. If the interrupt is not going to be used, its service location is available as general purpose Program Memory.

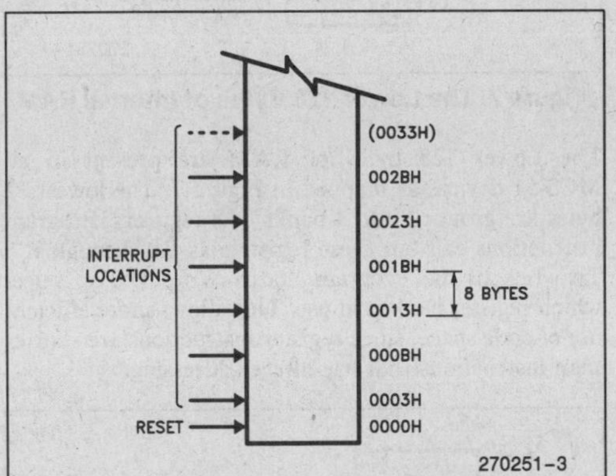

Figure 3. MCS®-51 Program Memory

The interrupt service locations are spaced at 8-byte intervals: 0003H for External Interrupt 0, 000BH for Timer 0, 0013H for External Interrupt 1, 001BH for Timer 1, etc. If an interrupt service routine is short enough (as is often the case in control applications), it can reside entirely within that 8-byte interval. Longer service routines can use a jump instruction to skip over subsequent interrupt locations, if other interrupts are in use.

The lowest 4K (or 8K or 16K or 32K) bytes of Program Memory can be either in the on-chip ROM or in an external ROM. This selection is made by strapping the $\overline{EA}$ (External Access) pin to either V_{CC} or V_{SS}.

In the 4K byte ROM devices, if the $\overline{EA}$ pin is strapped to V_{CC}, then program fetches to addresses 0000H through 0FFFH are directed to the internal ROM. Program fetches to addresses 1000H through FFFFH are directed to external ROM.

In the 8K byte ROM devices, $\overline{EA} = V_{CC}$ selects addresses 0000H through 1FFFH to be internal, and addresses 2000H through FFFFH to be external.

In the 16K byte ROM devices, $\overline{EA} = V_{CC}$ selects addresses 0000H through 3FFFH to be internal, and addresses 4000H through FFFFH to be external.

If the $\overline{EA}$ pin is strapped to V_{SS}, then all program fetches are directed to external ROM. The ROMless parts must have this pin externally strapped to V_{SS} to enable them to execute properly.

The read strobe to external ROM, $\overline{PSEN}$, is used for all external program fetches. $\overline{PSEN}$ is not activated for internal program fetches.

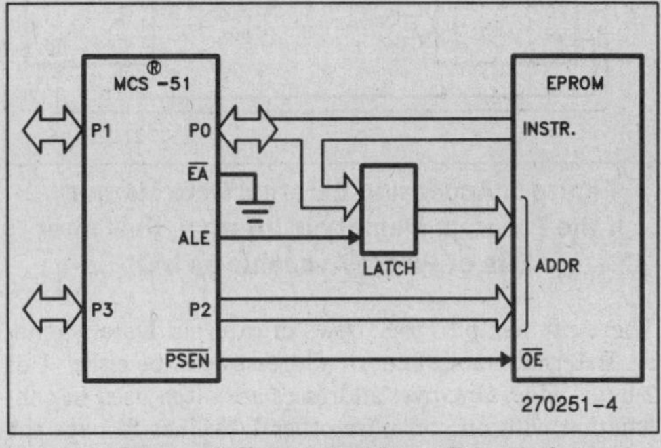

Figure 4. Executing from External Program Memory

The hardware configuration for external program execution is shown in Figure 4. Note that 16 I/O lines (Ports 0 and 2) are dedicated to bus functions during external Program Memory fetches. Port 0 (P0 in Figure 4) serves as a multiplexed address/data bus. It emits the low byte of the Program Counter (PCL) as an address, and then goes into a float state awaiting the arrival of the code byte from the Program Memory. During the time that the low byte of the Program Counter is valid on P0, the signal ALE (Address Latch Enable) clocks this byte into an address latch. Meanwhile, Port 2 (P2 in Figure 4) emits the high byte of the Program Counter (PCH). Then $\overline{PSEN}$ strobes the EPROM and the code byte is read into the microcontroller.

Program Memory addresses are always 16 bits wide, even though the actual amount of Program Memory used may be less than 64K bytes. External program execution sacrifices two of the 8-bit ports, P0 and P2, to the function of addressing the Program Memory.

Data Memory

The right half of Figure 2 shows the internal and external Data Memory spaces available to the MCS-51 user.

Figure 5 shows a hardware configuration for accessing up to 2K bytes of external RAM. The CPU in this case is executing from internal ROM. Port 0 serves as a multiplexed address/data bus to the RAM, and 3 lines of Port 2 are being used to page the RAM. The CPU generates $\overline{RD}$ and $\overline{WR}$ signals as needed during external RAM accesses.

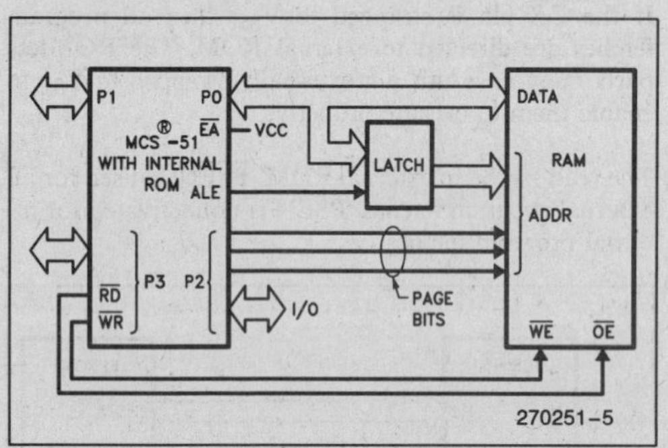

Figure 5. Accessing External Data Memory. If the Program Memory is Internal, the Other Bits of P2 are Available as I/O.

There can be up to 64K bytes of external Data Memory. External Data Memory addresses can be either 1 or 2 bytes wide. One-byte addresses are often used in conjunction with one or more other I/O lines to page the RAM, as shown in Figure 5. Two-byte addresses can also be used, in which case the high address byte is emitted at Port 2.

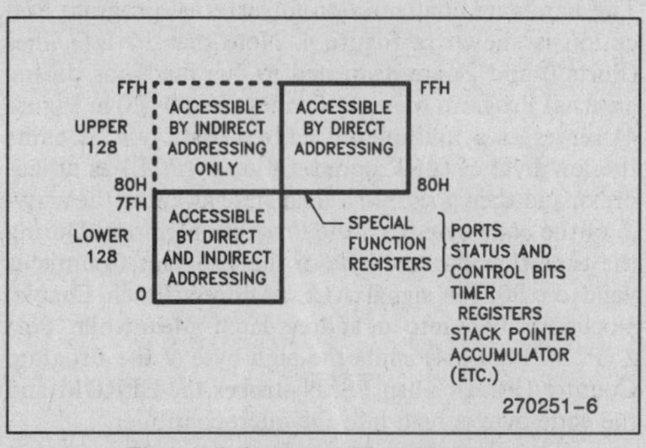

Figure 6. Internal Data Memory

Internal Data Memory is mapped in Figure 6. The memory space is shown divided into three blocks, which are generally referred to as the Lower 128, the Upper 128, and SFR space.

Internal Data Memory addresses are always one byte wide, which implies an address space of only 256 bytes. However, the addressing modes for internal RAM can in fact accommodate 384 bytes, using a simple trick. Direct addresses higher than 7FH access one memory space, and indirect addresses higher than 7FH access a different memory space. Thus Figure 6 shows the Upper 128 and SFR space occupying the same block of addresses, 80H through FFH, although they are physically separate entities.

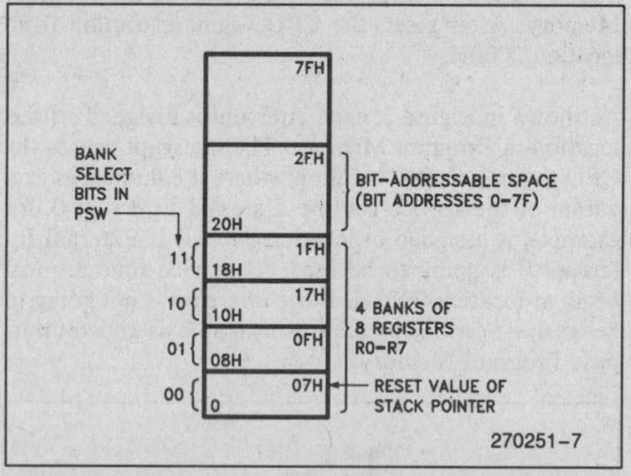

Figure 7. The Lower 128 Bytes of Internal RAM

The Lower 128 bytes of RAM are present in all MCS-51 devices as mapped in Figure 7. The lowest 32 bytes are grouped into 4 banks of 8 registers. Program instructions call out these registers as R0 through R7. Two bits in the Program Status Word (PSW) select which register bank is in use. This allows more efficient use of code space, since register instructions are shorter than instructions that use direct addressing.

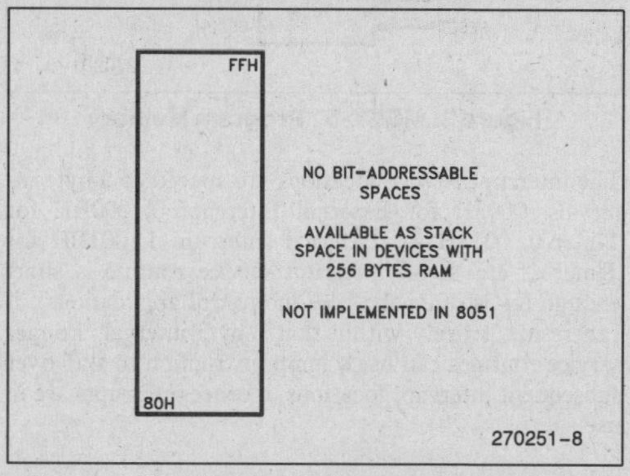

Figure 8. The Upper 128 Bytes of Internal RAM

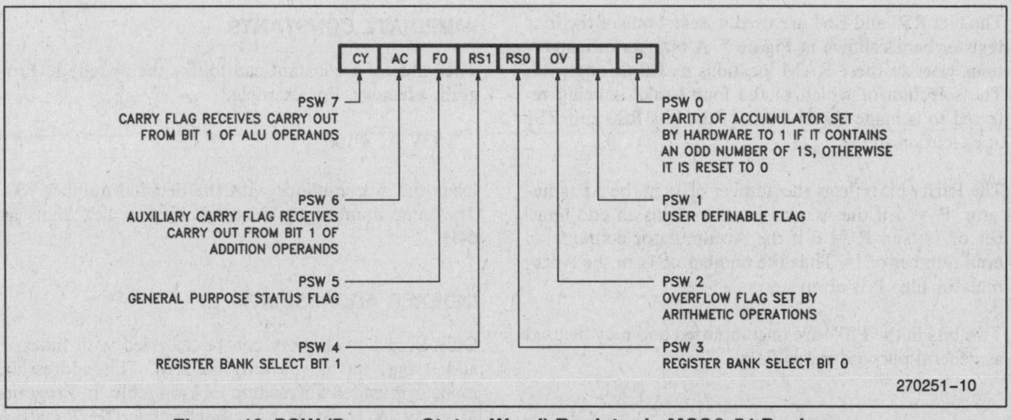

Figure 10. PSW (Program Status Word) Register in MCS®-51 Devices

The next 16 bytes above the register banks form a block of bit-addressable memory space. The MCS-51 instruction set includes a wide selection of single-bit instructions, and the 128 bits in this area can be directly addressed by these instructions. The bit addresses in this area are 00H through 7FH.

All of the bytes in the Lower 128 can be accessed by either direct or indirect addressing. The Upper 128 (Figure 8) can only be accessed by indirect addressing. The Upper 128 bytes of RAM are not implemented in the 8051, but are in the devices with 256 bytes of RAM. (See Table 1).

Figure 9 gives a brief look at the Special Function Register (SFR) space. SFRs include the Port latches, timers, peripheral controls, etc. These registers can only be accessed by direct addressing. In general, all MCS-51 microcontrollers have the same SFRs as the 8051, and at the same addresses in SFR space. However, enhancements to the 8051 have additional SFRs that are not present in the 8051, nor perhaps in other proliferations of the family.

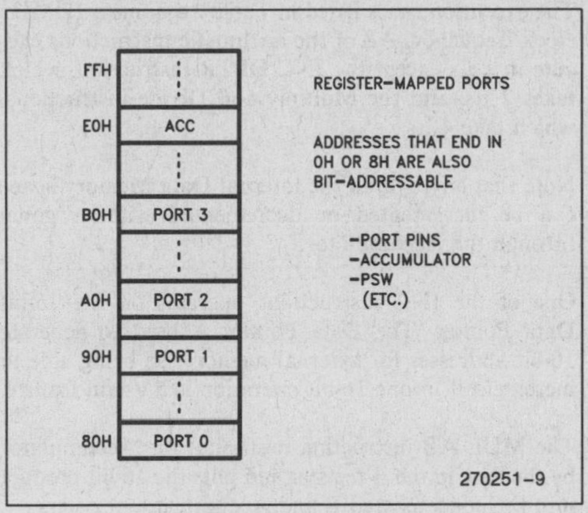

Figure 9. SFR Space

Sixteen addresses in SFR space are both byte- and bit-addressable. The bit-addressable SFRs are those whose address ends in 000B. The bit addresses in this area are 80H through FFH.

THE MCS®-51 INSTRUCTION SET

All members of the MCS-51 family execute the same instruction set. The MCS-51 instruction set is optimized for 8-bit control applications. It provides a variety of fast addressing modes for accessing the internal RAM to facilitate byte operations on small data structures. The instruction set provides extensive support for one-bit variables as a separate data type, allowing direct bit manipulation in control and logic systems that require Boolean processing.

An overview of the MCS-51 instruction set is presented below, with a brief description of how certain instructions might be used. References to "the assembler" in this discussion are to Intel's MCS-51 Macro Assembler, ASM51. More detailed information on the instruction set can be found in the MCS-51 Macro Assembler User's Guide (Order No. 9800937 for ISIS Systems, Order No. 122752 for DOS Systems).

Program Status Word

The Program Status Word (PSW) contains several status bits that reflect the current state of the CPU. The PSW, shown in Figure 10, resides in SFR space. It contains the Carry bit, the Auxiliary Carry (for BCD operations), the two register bank select bits, the Overflow flag, a Parity bit, and two user-definable status flags.

The Carry bit, other than serving the functions of a Carry bit in arithmetic operations, also serves as the "Accumulator" for a number of Boolean operations.

The bits RS0 and RS1 are used to select one of the four register banks shown in Figure 7. A number of instructions refer to these RAM locations as R0 through R7. The selection of which of the four banks is being referred to is made on the basis of the bits RS0 and RS1 at execution time.

The Parity bit reflects the number of 1s in the Accumulator: P = 1 if the Accumulator contains an odd number of 1s, and P = 0 if the Accumulator contains an even number of 1s. Thus the number of 1s in the Accumulator plus P is always even.

Two bits in the PSW are uncommitted and may be used as general purpose status flags.

Addressing Modes

The addressing modes in the MCS-51 instruction set are as follows:

DIRECT ADDRESSING

In direct addressing the operand is specified by an 8-bit address field in the instruction. Only internal Data RAM and SFRs can be directly addressed.

INDIRECT ADDRESSING

In indirect addressing the instruction specifies a register which contains the address of the operand. Both internal and external RAM can be indirectly addressed.

The address register for 8-bit addresses can be R0 or R1 of the selected register bank, or the Stack Pointer. The address register for 16-bit addresses can only be the 16-bit "data pointer" register, DPTR.

REGISTER INSTRUCTIONS

The register banks, containing registers R0 through R7, can be accessed by certain instructions which carry a 3-bit register specification within the opcode of the instruction. Instructions that access the registers this way are code efficient, since this mode eliminates an address byte. When the instruction is executed, one of the eight registers in the selected bank is accessed. One of four banks is selected at execution time by the two bank select bits in the PSW.

REGISTER-SPECIFIC INSTRUCTIONS

Some instructions are specific to a certain register. For example, some instructions always operate on the Accumulator, or Data Pointer, etc., so no address byte is needed to point to it. The opcode itself does that. Instructions that refer to the Accumlator as A assemble as accumulator-specific opcodes.

IMMEDIATE CONSTANTS

The value of a constant can follow the opcode in Program Memory. For example,

```
MOV A, #100
```

loads the Accumulator with the decimal number 100. The same number could be specified in hex digits as 64H.

INDEXED ADDRESSING

Only Program Memory can be accessed with indexed addressing, and it can only be read. This addressing mode is intended for reading look-up tables in Program Memory. A 16-bit base register (either DPTR or the Program Counter) points to the base of the table, and the Accumulator is set up with the table entry number. The address of the table entry in Program Memory is formed by adding the Accumulator data to the base pointer.

Another type of indexed addressing is used in the "case jump" instruction. In this case the destination address of a jump instruction is computed as the sum of the base pointer and the Accumulator data.

Arithmetic Instructions

The menu of arithmetic instructions is listed in Table 2. The table indicates the addressing modes that can be used with each instruction to access the <byte> operand. For example, the ADD A,<byte> instruction can be written as:

```
ADD    A,7FH     (direct addressing)
ADD    A,@R0     (indirect addressing)
ADD    A,R7      (register addressing)
ADD    A,#127    (immediate constant)
```

The execution times listed in Table 2 assume a 12 MHz clock frequency. All of the arithmetic instructions execute in 1 μs except the INC DPTR instruction, which takes 2 μs, and the Multiply and Divide instructions, which take 4 μs.

Note that any byte in the internal Data Memory space can be incremented or decremented without going through the Accumulator.

One of the INC instructions operates on the 16-bit Data Pointer. The Data Pointer is used to generate 16-bit addresses for external memory, so being able to increment it in one 16-bit operation is a useful feature.

The MUL AB instruction multiplies the Accumulator by the data in the B register and puts the 16-bit product into the concatenated B and Accumulator registers.

Table 2. A List of the MCS®-51 Arithmetic Instructions

Mnemonic	Operation	Addressing Modes				Execution Time (μs)
		Dir	Ind	Reg	Imm	
ADD A,<byte>	A = A + <byte>	X	X	X	X	1
ADDC A,<byte>	A = A + <byte> + C	X	X	X	X	1
SUBB A,<byte>	A = A − <byte> − C	X	X	X	X	1
INC A	A = A + 1	Accumulator only				1
INC <byte>	<byte> = <byte> + 1	X	X	X		1
INC DPTR	DPTR = DPTR + 1	Data Pointer only				2
DEC A	A = A − 1	Accumulator only				1
DEC <byte>	<byte> = <byte> − 1	X	X	X		1
MUL AB	B:A = B x A	ACC and B only				4
DIV AB	A = Int [A/B] B = Mod [A/B]	ACC and B only				4
DA A	Decimal Adjust	Accumulator only				1

The DIV AB instruction divides the Accumulator by the data in the B register and leaves the 8-bit quotient in the Accumulator, and the 8-bit remainder in the B register.

Oddly enough, DIV AB finds less use in arithmetic "divide" routines than in radix conversions and programmable shift operations. An example of the use of DIV AB in a radix conversion will be given later. In shift operations, dividing a number by 2^n shifts its n bits to the right. Using DIV AB to perform the division completes the shift in 4 μs and leaves the B register holding the bits that were shifted out.

The DA A instruction is for BCD arithmetic operations. In BCD arithmetic, ADD and ADDC instructions should always be followed by a DA A operation, to ensure that the result is also in BCD. Note that DA A will not convert a binary number to BCD. The DA A operation produces a meaningful result only as the second step in the addition of two BCD bytes.

Table 3. A List of the MCS®-51 Logical Instructions

Mnemonic	Operation	Addressing Modes				Execution Time (μs)
		Dir	Ind	Reg	Imm	
ANL A,<byte>	A = A .AND. <byte>	X	X	X	X	1
ANL <byte>,A	<byte> = <byte> .AND. A	X				1
ANL <byte>,#data	<byte> = <byte> .AND. #data	X				2
ORL A,<byte>	A = A .OR. <byte>	X	X	X	X	1
ORL <byte>,A	<byte> = <byte> .OR. A	X				1
ORL <byte>,#data	<byte> = <byte> .OR. #data	X				2
XRL A,<byte>	A = A .XOR. <byte>	X	X	X	X	1
XRL <byte>,A	<byte> = <byte> .XOR. A	X				1
XRL <byte>,#data	<byte> = <byte> .XOR. #data	X				2
CRL A	A = 00H	Accumulator only				1
CPL A	A = .NOT. A	Accumulator only				1
RL A	Rotate ACC Left 1 bit	Accumulator only				1
RLC A	Rotate Left through Carry	Accumulator only				1
RR A	Rotate ACC Right 1 bit	Accumulator only				1
RRC A	Rotate Right through Carry	Accumulator only				1
SWAP A	Swap Nibbles in A	Accumulator only				1

Logical Instructions

Table 3 shows the list of MCS-51 logical instructions. The instructions that perform Boolean operations (AND, OR, Exclusive OR, NOT) on bytes perform the operation on a bit-by-bit basis. That is, if the Accumulator contains 00110101B and <byte> contains 01010011B, then

```
ANL     A,<byte>
```

will leave the Accumulator holding 00010001B.

The addressing modes that can be used to access the <byte> operand are listed in Table 3. Thus, the ANL A,<byte> instruction may take any of the forms

```
ANL     A,7FH     (direct addressing)
ANL     A,@R1     (indirect addressing)
ANL     A,R6      (register addressing)
ANL     A,#53H    (immediate constant)
```

All of the logical instructions that are Accumulator-specific execute in 1μs (using a 12 MHz clock). The others take 2 μs.

Note that Boolean operations can be performed on any byte in the lower 128 internal Data Memory space or the SFR space using direct addressing, without having to use the Accumulator. The XRL <byte>, #data instruction, for example, offers a quick and easy way to invert port bits, as in

```
XRL     P1,#0FFH
```

If the operation is in response to an interrupt, not using the Accumulator saves the time and effort to stack it in the service routine.

The Rotate instructions (RL A, RLC A, etc.) shift the Accumulator 1 bit to the left or right. For a left rotation, the MSB rolls into the LSB position. For a right rotation, the LSB rolls into the MSB position.

The SWAP A instruction interchanges the high and low nibbles within the Accumulator. This is a useful operation in BCD manipulations. For example, if the Accumulator contains a binary number which is known to be less than 100, it can be quickly converted to BCD by the following code:

```
MOV     B,#10
DIV     AB
SWAP    A
ADD     A,B
```

Dividing the number by 10 leaves the tens digit in the low nibble of the Accumulator, and the ones digit in the B register. The SWAP and ADD instructions move the tens digit to the high nibble of the Accumulator, and the ones digit to the low nibble.

Data Transfers

INTERNAL RAM

Table 4 shows the menu of instructions that are available for moving data around within the internal memory spaces, and the addressing modes that can be used with each one. With a 12 MHz clock, all of these instructions execute in either 1 or 2 μs.

The MOV <dest>, <src> instruction allows data to be transferred between any two internal RAM or SFR locations without going through the Accumulator. Remember the Upper 128 byes of data RAM can be accessed only by indirect addressing, and SFR space only by direct addressing.

Note that in all MCS-51 devices, the stack resides in on-chip RAM, and grows upwards. The PUSH instruction first increments the Stack Pointer (SP), then copies the byte into the stack. PUSH and POP use only direct addressing to identify the byte being saved or restored,

Table 4. A List of the MCS®-51 Data Transfer Instructions that Access Internal Data Memory Space

Mnemonic	Operation	Addressing Modes				Execution Time (μs)
		Dir	Ind	Reg	Imm	
MOV A,<src>	A = <src>	X	X	X	X	1
MOV <dest>,A	<dest> = A	X	X	X		1
MOV <dest>, <src>	<dest> = <src>	X	X	X	X	2
MOV DPTR,#data16	DPTR = 16-bit immediate constant.				X	2
PUSH <src>	INC SP : MOV "@SP",<src>	X				2
POP <dest>	MOV <dest>, "@SP" : DEC SP	X				2
XCH A,<byte>	ACC and <byte> exchange data	X	X	X		1
XCHD A,@Ri	ACC and @Ri exchange low nibbles		X			1

but the stack itself is accessed by indirect addressing using the SP register. This means the stack can go into the Upper 128, if they are implemented, but not into SFR space.

In devices that do not implement the Upper 128, if the SP points to the Upper 128, PUSHed bytes are lost, and POPped bytes are indeterminate.

The Data Transfer instructions include a 16-bit MOV that can be used to initialize the Data Pointer (DPTR) for look-up tables in Program Memory, or for 16-bit external Data Memory accesses.

The XCH A, <byte> instruction causes the Accumulator and addressed byte to exchange data. The XCHD A,@Ri instruction is similar, but only the low nibbles are involved in the exchange.

To see how XCH and XCHD can be used to facilitate data manipulations, consider first the problem of shifting an 8-digit BCD number two digits to the right. Figure 11 shows how this can be done using direct MOVs, and for comparison how it can be done using XCH instructions. To aid in understanding how the code works, the contents of the registers that are holding the BCD number and the content of the Accumulator are shown alongside each instruction to indicate their status after the instruction has been executed.

	2A	2B	2C	2D	2E	ACC
MOV A,2EH	00	12	34	56	78	78
MOV 2EH,2DH	00	12	34	56	56	78
MOV 2DH,2CH	00	12	34	34	56	78
MOV 2CH,2BH	00	12	12	34	56	78
MOV 2BH,#0	00	00	12	34	56	78

(a) Using direct MOVs: 14 bytes, 9 μs

	2A	2B	2C	2D	2E	ACC
CLR A	00	12	34	56	78	00
XCH A,2BH	00	00	34	56	78	12
XCH A,2CH	00	00	12	56	78	34
XCH A,2DH	00	00	12	34	78	56
XCH A,2EH	00	00	12	34	56	78

(b) Using XCHs: 9 bytes, 5 μs

Figure 11. Shifting a BCD Number Two Digits to the Right

After the routine has been executed, the Accumulator contains the two digits that were shifted out on the right. Doing the routine with direct MOVs uses 14 code bytes and 9 μs of execution time (assuming a 12 MHz clock). The same operation with XCHs uses less code and executes almost twice as fast.

To right-shift by an odd number of digits, a one-digit shift must be executed. Figure 12 shows a sample of code that will right-shift a BCD number one digit, using the XCHD instruction. Again, the contents of the registers holding the number and of the Accumulator are shown alongside each instruction.

	2A	2B	2C	2D	2E	ACC
MOV R1,#2EH	00	12	34	56	78	XX
MOV R0,#2DH	00	12	34	56	78	XX
loop for R1 = 2EH:						
LOOP: MOV A,@R1	00	12	34	56	78	78
XCHD A,@R0	00	12	34	58	78	76
SWAP A	00	12	34	58	78	67
MOV @R1,A	00	12	34	58	67	67
DEC R1	00	12	34	58	67	67
DEC R0	00	12	34	58	67	67
CJNE R1,#2AH,LOOP						
loop for R1 = 2DH:	00	12	38	45	67	45
loop for R1 = 2CH:	00	18	23	45	67	23
loop for R1 = 2BH:	08	01	23	45	67	01
CLR A	08	01	23	45	67	00
XCH A,2AH	00	01	23	45	67	08

Figure 12. Shifting a BCD Number One Digit to the Right

First, pointers R1 and R0 are set up to point to the two bytes containing the last four BCD digits. Then a loop is executed which leaves the last byte, location 2EH, holding the last two digits of the shifted number. The pointers are decremented, and the loop is repeated for location 2DH. The CJNE instruction (Compare and Jump if Not Equal) is a loop control that will be described later.

The loop is executed from LOOP to CJNE for R1 = 2EH, 2DH, 2CH and 2BH. At that point the digit that was originally shifted out on the right has propagated to location 2AH. Since that location should be left with 0s, the lost digit is moved to the Accumulator.

EXTERNAL RAM

Table 5 shows a list of the Data Transfer instructions that access external Data Memory. Only indirect addressing can be used. The choice is whether to use a one-byte address, @Ri, where Ri can be either R0 or R1 of the selected register bank, or a two-byte address, @DPTR. The disadvantage to using 16-bit addresses if only a few K bytes of external RAM are involved is that 16-bit addresses use all 8 bits of Port 2 as address bus. On the other hand, 8-bit addresses allow one to address a few K bytes of RAM, as shown in Figure 5, without having to sacrifice all of Port 2.

All of these instructions execute in 2 μs, with a 12 MHz clock.

Table 5. A List of the MCS®-51 Data Transfer Instructions that Access External Data Memory Space

Address Width	Mnemonic	Operation	Execution Time (μs)
8 bits	MOVX A,@Ri	Read external RAM @Ri	2
8 bits	MOVX @Ri,A	Write external RAM @Ri	2
16 bits	MOVX A,@DPTR	Read external RAM @DPTR	2
16 bits	MOVX @DPTR,A	Write external RAM @DPTR	2

Note that in all external Data RAM accesses, the Accumulator is always either the destination or source of the data.

The read and write strobes to external RAM are activated only during the execution of a MOVX instruction. Normally these signals are inactive, and in fact if they're not going to be used at all, their pins are available as extra I/O lines. More about that later.

LOOKUP TABLES

Table 6 shows the two instructions that are available for reading lookup tables in Program Memory. Since these instructions access only Program Memory, the lookup tables can only be read, not updated. The mnemonic is MOVC for "move constant".

If the table access is to external Program Memory, then the read strobe is $\overline{\text{PSEN}}$.

Table 6. The MCS®-51 Lookup Table Read Instructions

Mnemonic	Operation	Execution Time (μs)
MOVC A,@A+DPTR	Read Pgm Memory at (A+DPTR)	2
MOVC A,@A+PC	Read Pgm Memory at (A+PC)	2

The first MOVC instruction in Table 6 can accommodate a table of up to 256 entries, numbered 0 through 255. The number of the desired entry is loaded into the Accumulator, and the Data Pointer is set up to point to beginning of the table. Then

```
MOVC    A,@A+DPTR
```

copies the desired table entry into the Accumulator.

The other MOVC instruction works the same way, except the Program Counter (PC) is used as the table base, and the table is accessed through a subroutine. First the number of the desired entry is loaded into the Accumulator, and the subroutine is called:

```
        MOV     A,ENTRY_NUMBER
        CALL    TABLE
```

The subroutine "TABLE" would look like this:

```
TABLE:  MOVC    A,@A+PC
        RET
```

The table itself immediately follows the RET (return) instruction in Program Memory. This type of table can have up to 255 entries, numbered 1 through 255. Number 0 can not be used, because at the time the MOVC instruction is executed, the PC contains the address of the RET instruction. An entry numbered 0 would be the RET opcode itself.

Boolean Instructions

MCS-51 devices contain a complete Boolean (single-bit) processor. The internal RAM contains 128 addressable bits, and the SFR space can support up to 128 other addressable bits. All of the port lines are bit-addressable, and each one can be treated as a separate single-bit port. The instructions that access these bits are not just conditional branches, but a complete menu of move, set, clear, complement, OR, and AND instructions. These kinds of bit operations are not easily obtained in other architectures with any amount of byte-oriented software.

Table 7. A List of the MCS®-51 Boolean Instructions

Mnemonic		Operation	Execution Time (μs)
ANL	C,bit	C = C .AND. bit	2
ANL	C,/bit	C = C .AND. .NOT. bit	2
ORL	C,bit	C = C .OR. bit	2
ORL	C,/bit	C = C .OR. .NOT. bit	2
MOV	C,bit	C = bit	1
MOV	bit,C	bit = C	2
CLR	C	C = 0	1
CLR	bit	bit = 0	1
SETB	C	C = 1	1
SETB	bit	bit = 1	1
CPL	C	C = .NOT. C	1
CPL	bit	bit = .NOT. bit	1
JC	rel	Jump if C = 1	2
JNC	rel	Jump if C = 0	2
JB	bit,rel	Jump if bit = 1	2
JNB	bit,rel	Jump if bit = 0	2
JBC	bit,rel	Jump if bit = 1; CLR bit	2

The instruction set for the Boolean processor is shown in Table 7. All bit accesses are by direct addressing. Bit addresses 00H through 7FH are in the Lower 128, and bit addresses 80H through FFH are in SFR space.

Note how easily an internal flag can be moved to a port pin:

```
        MOV     C,FLAG
        MOV     P1.0,C
```

In this example, FLAG is the name of any addressable bit in the Lower 128 or SFR space. An I/O line (the LSB of Port 1, in this case) is set or cleared depending on whether the flag bit is 1 or 0.

The Carry bit in the PSW is used as the single-bit Accumulator of the Boolean processor. Bit instructions that refer to the Carry bit as C assemble as Carry-specific instructions (CLR C, etc). The Carry bit also has a direct address, since it resides in the PSW register, which is bit-addressable.

Note that the Boolean instruction set includes ANL and ORL operations, but not the XRL (Exclusive OR) operation. An XRL operation is simple to implement in software. Suppose, for example, it is required to form the Exclusive OR of two bits:

```
    C = bit1 .XRL. bit2
```

The software to do that could be as follows:

```
          MOV     C,bit1
          JNB     bit2,OVER
          CPL     C
OVER:     (continue)
```

First, bit1 is moved to the Carry. If bit2 = 0, then C now contains the correct result. That is, bit1 .XRL. bit2 = bit1 if bit2 = 0. On the other hand, if bit2 = 1 C now contains the complement of the correct result. It need only be inverted (CPL C) to complete the operation.

This code uses the JNB instruction, one of a series of bit-test instructions which execute a jump if the addressed bit is set (JC, JB, JBC) or if the addressed bit is not set (JNC, JNB). In the above case, bit2 is being tested, and if bit2 = 0 the CPL C instruction is jumped over.

JBC executes the jump if the addressed bit is set, and also clears the bit. Thus a flag can be tested and cleared in one operation.

All the PSW bits are directly addressable, so the Parity bit, or the general purpose flags, for example, are also available to the bit-test instructions.

RELATIVE OFFSET

The destination address for these jumps is specified to the assembler by a label or by an actual address in Program Memory. However, the destination address assembles to a **relative** offset byte. This is a signed (two's complement) offset byte which is added to the PC in two's complement arithmetic if the jump is executed.

The range of the jump is therefore −128 to +127 Program Memory bytes relative to the first byte following the instruction.

Jump Instructions

Table 8 shows the list of unconditional jumps.

Table 8. Unconditional Jumps in MCS®-51 Devices

Mnemonic	Operation	Execution Time (μs)
JMP addr	Jump to addr	2
JMP @A+DPTR	Jump to A+DPTR	2
CALL addr	Call subroutine at addr	2
RET	Return from subroutine	2
RETI	Return from interrupt	2
NOP	No operation	1

The Table lists a single "JMP addr" instruction, but in fact there are three—SJMP, LJMP and AJMP—which differ in the format of the destination address. JMP is a generic mnemonic which can be used if the programmer does not care which way the jump is encoded.

The SJMP instruction encodes the destination address as a relative offset, as described above. The instruction is 2 bytes long, consisting of the opcode and the relative offset byte. The jump distance is limited to a range of −128 to +127 bytes relative to the instruction following the SJMP.

The LJMP instruction encodes the destination address as a 16-bit constant. The instruction is 3 bytes long, consisting of the opcode and two address bytes. The destination address can be anywhere in the 64K Program Memory space.

The AJMP instruction encodes the destination address as an 11-bit constant. The instruction is 2 bytes long, consisting of the opcode, which itself contains 3 of the 11 address bits, followed by another byte containing the low 8 bits of the destination address. When the instruction is executed, these 11 bits are simply substituted for the low 11 bits in the PC. The high 5 bits stay the same. Hence the destination has to be within the same 2K block as the instruction following the AJMP.

In all cases the programmer specifies the destination address to the assembler in the same way: as a label or as a 16-bit constant. The assembler will put the destination address into the correct format for the given instruction. If the format required by the instruction will not support the distance to the specified destination address, a "Destination out of range" message is written into the List file.

The JMP @A+DPTR instruction supports case jumps. The destination address is computed at execution time as the sum of the 16-bit DPTR register and the Accumulator. Typically, DPTR is set up with the address of a jump table, and the Accumulator is given an index to the table. In a 5-way branch, for example, an integer 0 through 4 is loaded into the Accumulator. The code to be executed might be as follows:

```
MOV     DPTR,#JUMP_TABLE
MOV     A,INDEX_NUMBER
RL      A
JMP     @A+DPTR
```

The RL A instruction converts the index number (0 through 4) to an even number on the range 0 through 8, because each entry in the jump table is 2 bytes long:

```
JUMP_TABLE:
        AJMP    CASE_0
        AJMP    CASE_1
        AJMP    CASE_2
        AJMP    CASE_3
        AJMP    CASE_4
```

Table 8 shows a single "CALL addr" instruction, but there are two of them—LCALL and ACALL—which differ in the format in which the subroutine address is given to the CPU. CALL is a generic mnemonic which can be used if the programmer does not care which way the address is encoded.

The LCALL instruction uses the 16-bit address format, and the subroutine can be anywhere in the 64K Program Memory space. The ACALL instruction uses the 11-bit format, and the subroutine must be in the same 2K block as the instruction following the ACALL.

In any case the programmer specifies the subroutine address to the assembler in the same way: as a label or as a 16-bit constant. The assembler will put the address into the correct format for the given instructions.

Subroutines should end with a RET instruction, which returns execution to the instruction following the CALL.

RETI is used to return from an interrupt service routine. The only difference between RET and RETI is that RETI tells the interrupt control system that the interrupt in progress is done. If there is no interrupt in progress at the time RETI is executed, then the RETI is functionally identical to RET.

Table 9 shows the list of conditional jumps available to the MCS-51 user. All of these jumps specify the destination address by the relative offset method, and so are limited to a jump distance of −128 to +127 bytes from the instruction following the conditional jump instruction. Important to note, however, the user specifies to the assembler the actual destination address the same way as the other jumps: as a label or a 16-bit constant.

Table 9. Conditional Jumps in MCS®-51 Devices

Mnemonic	Operation	Addressing Modes				Execution Time (μs)
		Dir	Ind	Reg	Imm	
JZ rel	Jump if A = 0	Accumulator only				2
JNZ rel	Jump if A ≠ 0	Accumulator only				2
DJNZ <byte>,rel	Decrement and jump if not zero	X		X		2
CJNE A,<byte>,rel	Jump if A ≠ <byte>	X			X	2
CJNE <byte>,#data,rel	Jump if <byte> ≠ #data		X	X		2

There is no Zero bit in the PSW. The JZ and JNZ instructions test the Accumulator data for that condition.

The DJNZ instruction (Decrement and Jump if Not Zero) is for loop control. To execute a loop N times, load a counter byte with N and terminate the loop with a DJNZ to the beginning of the loop, as shown below for N = 10:

```
       MOV     COUNTER,#10
LOOP:  (begin loop)
          *
          *
          *
        (end loop)
       DJNZ    COUNTER,LOOP
        (continue)
```

The CJNE instruction (Compare and Jump if Not Equal) can also be used for loop control as in Figure 12. Two bytes are specified in the operand field of the instruction. The jump is executed only if the two bytes are not equal. In the example of Figure 12, the two bytes were the data in R1 and the constant 2AH. The initial data in R1 was 2EH. Every time the loop was executed, R1 was decremented, and the looping was to continue until the R1 data reached 2AH.

Another application of this instruction is in "greater than, less than" comparisons. The two bytes in the operand field are taken as unsigned integers. If the first is less than the second, then the Carry bit is set (1). If the first is greater than or equal to the second, then the Carry bit is cleared.

CPU TIMING

All MCS-51 microcontrollers have an on-chip oscillator which can be used if desired as the clock source for the CPU. To use the on-chip oscillator, connect a crystal or ceramic resonator between the XTAL1 and XTAL2 pins of the microcontroller, and capacitors to ground as shown in Figure 13.

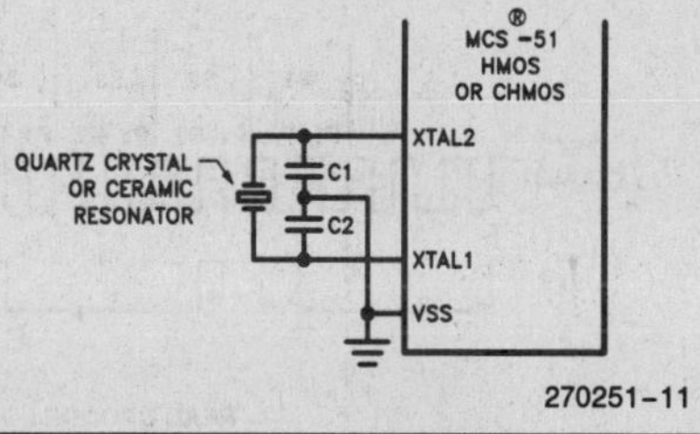

Figure 13. Using the On-Chip Oscillator

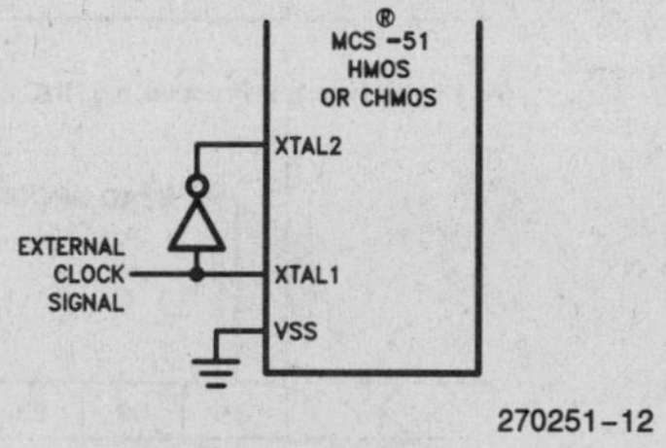

A. HMOS or CHMOS

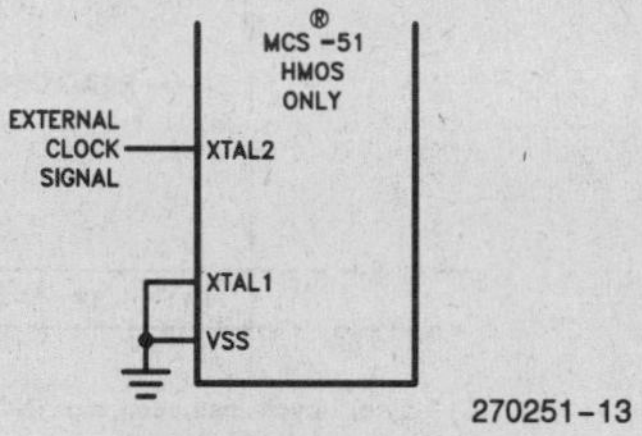

B. HMOS Only

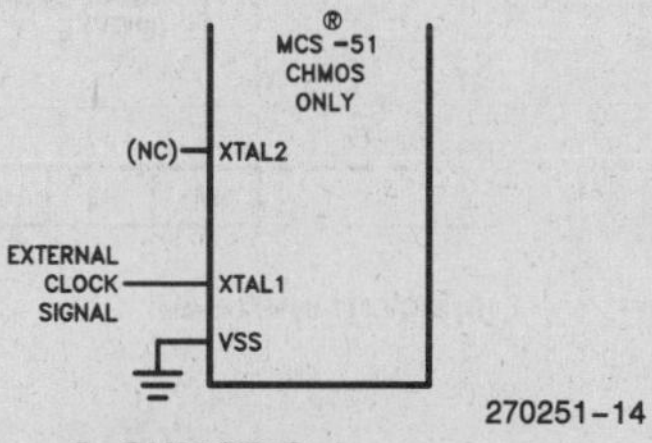

C. CHMOS Only

Figure 14. Using an External Clock

Examples of how to drive the clock with an external oscillator are shown in Figure 14. Note that in the HMOS devices (8051, etc.) the signal at the XTAL2 pin actually drives the internal clock generator. In the CHMOS devices (80C51BH, etc.) the signal at the XTAL1 pin drives the internal clock generator. If only one pin is going to be driven with the external oscillator signal, make sure it is the right pin.

The internal clock generator defines the sequence of states that make up the MCS-51 machine cycle.

Machine Cycles

A machine cycle consists of a sequence of 6 states, numbered S1 through S6. Each state time lasts for two oscillator periods. Thus a machine cycle takes 12 oscillator periods or 1 μs if the oscillator frequency is 12 MHz.

Each state is divided into a Phase 1 half and a Phase 2 half. Figure 15 shows the fetch/execute sequences in

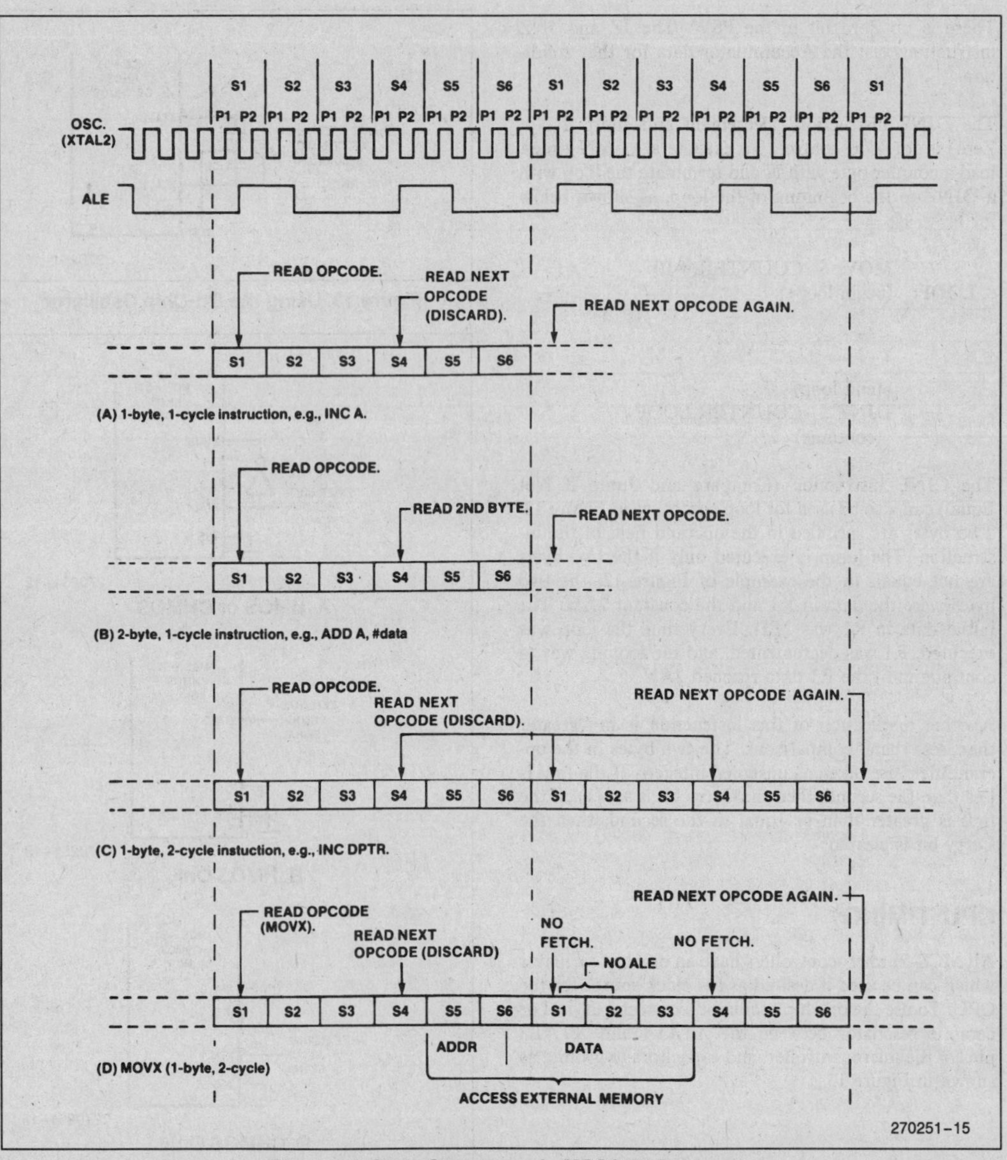

Figure 15. State Sequences in MCS®-51 Devices

states and phases for various kinds of instructions. Normally two program fetches are generated during each machine cycle, even if the instruction being executed doesn't require it. If the instruction being executed doesn't need more code bytes, the CPU simply ignores the extra fetch, and the Program Counter is not incremented.

Execution of a one-cycle instruction (Figure 15A and B) begins during State 1 of the machine cycle, when the opcode is latched into the Instruction Register. A second fetch occurs during S4 of the same machine cycle. Execution is complete at the end of State 6 of this machine cycle.

The MOVX instructions take two machine cycles to execute. No program fetch is generated during the second cycle of a MOVX instruction. This is the only time program fetches are skipped. The fetch/execute sequence for MOVX instructions is shown in Figure 15(D).

The fetch/execute sequences are the same whether the Program Memory is internal or external to the chip. Execution times do not depend on whether the Program Memory is internal or external.

Figure 16 shows the signals and timing involved in program fetches when the Program Memory is external. If Program Memory is external, then the Program Memory read strobe $\overline{PSEN}$ is normally activated twice per machine cycle, as shown in Figure 16(A).

If an access to external Data Memory occurs, as shown in Figure 16(B), two $\overline{PSEN}$s are skipped, because the address and data bus are being used for the Data Memory access.

Note that a Data Memory bus cycle takes twice as much time as a Program Memory bus cycle. Figure 16 shows the relative timing of the addresses being emitted at Ports 0 and 2, and of ALE and $\overline{PSEN}$. ALE is used to latch the low address byte from P0 into the address latch.

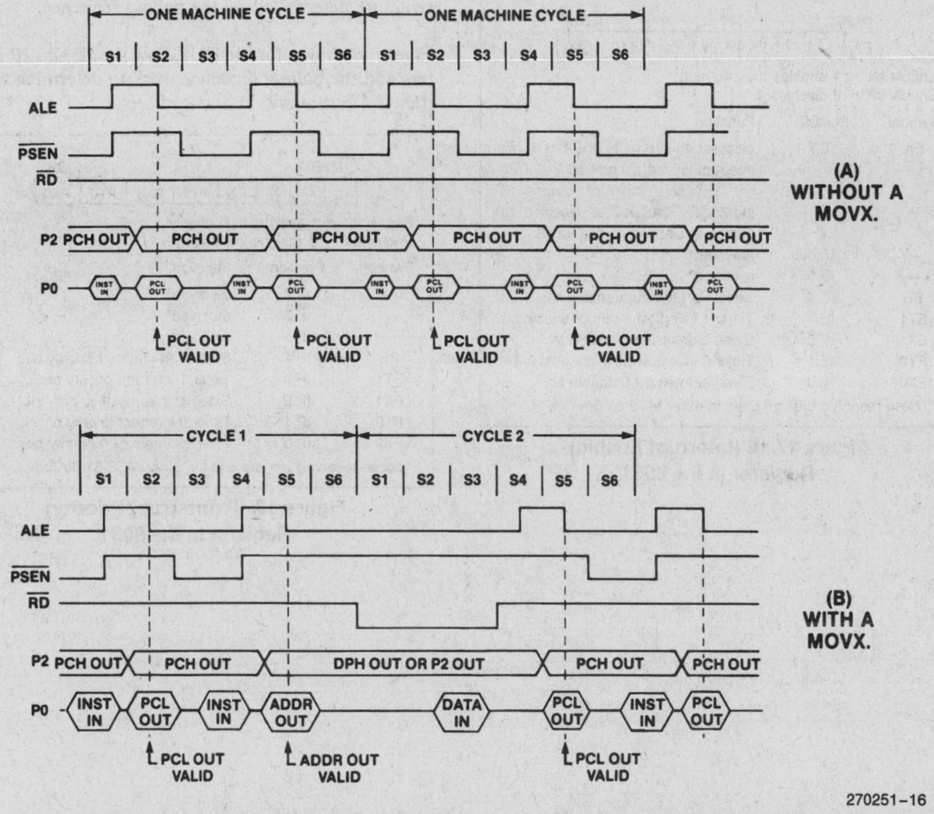

Figure 16. Bus Cycles in MCS®-51 Devices Executing from External Program Memory

When the CPU is executing from internal Program Memory, $\overline{PSEN}$ is not activated, and program addresses are not emitted. However, ALE continues to be activated twice per machine cycle and so is available as a clock output signal. Note, however, that one ALE is skipped during the execution of the MOVX instruction.

Interrupt Structure

The 8051 core provides 5 interrupt sources: 2 external interrupts, 2 timer interrupts, and the serial port interrupt. What follows is an overview of the interrupt structure for the 8051. Other MCS-51 devices have additional interrupt sources and vectors as shown in Table 1. Refer to the appropriate chapters on other devices for further information on their interrupts.

INTERRUPT ENABLES

Each of the interrupt sources can be individually enabled or disabled by setting or clearing a bit in the SFR named IE (Interrupt Enable). This register also contains a global disable bit, which can be cleared to disable all interrupts at once. Figure 17 shows the IE register for the 8051.

(MSB) (LSB)

EA	—	—	ES	ET1	EX1	ET0	EX0

Enable bit = 1 enables the interrupt.
Enable bit = 0 disables it.

Symbol	Position	Function
EA	IE.7	disables all interrupts. If EA = 0, no interrupt will be acknowledged. If EA = 1, each interrupt source is individually enabled or disabled by setting or clearing its enable bit.
—	IE.6	reserved*
—	IE.5	reserved*
ES	IE.4	Serial Port Interrupt enable bit.
ET1	IE.3	Timer 1 Overflow Interrupt enable bit.
EX1	IE.2	External Interrupt 1 enable bit.
ET0	IE.1	Timer 0 Overflow Interrupt enable bit.
EX0	IE.0	External Interrupt 0 enable bit.

*These reserved bits are used in other MCS-51 devices.

Figure 17. IE (Interrupt Enable) Register in the 8051

INTERRUPT PRIORITIES

Each interrupt source can also be individually programmed to one of two priority levels by setting or clearing a bit in the SFR named IP (Interrupt Priority). Figure 18 shows the IP register in the 8051.

A low-priority interrrupt can be interrupted by a high-priority interrupt, but not by another low-priority interrupt. A high-priority interrupt can't be interrupted by any other interrupt source.

If two interrupt requests of different priority levels are received simultaneously, the request of higher priority level is serviced. If interrupt requests of the same priority level are received simultaneously, an internal polling sequence determines which request is serviced. Thus within each priority level there is a second priority structure determined by the polling sequence.

Figure 19 shows, for the 8051, how the IE and IP registers and the polling sequence work to determine which if any interrupt will be serviced.

(MSB) (LSB)

—	—	—	PS	PT1	PX1	PT0	PX0

Priority bit = 1 assigns high priority.
Priority bit = 0 assigns low priority.

Symbol	Position	Function
—	IP.7	reserved*
—	IP.6	reserved*
—	IP.5	reserved*
PS	IP.4	Serial Port interrupt priority bit.
PT1	IP.3	Timer 1 interrupt priority bit.
PX1	IP.2	External Interrupt 1 priority bit.
PT0	IP.1	Timer 0 interrupt priority bit.
PX0	IP.0	External Interrupt 0 priority bit.

*These reserved bits are used in other MCS-51 devices.

Figure 18. IP (Interrupt Priority) Register in the 8051

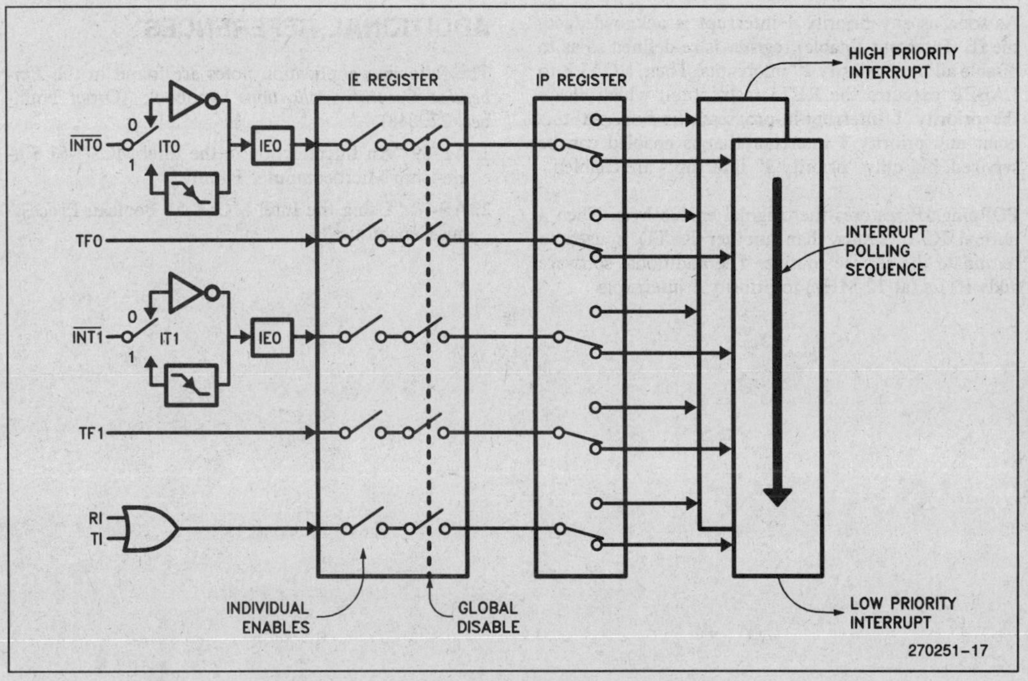

Figure 19. 8051 Interrupt Control System

In operation, all the interrupt flags are latched into the interrupt control system during State 5 of every machine cycle. The samples are polled during the following machine cycle. If the flag for an enabled interrupt is found to be set (1), the interrupt system generates an LCALL to the appropriate location in Program Memory, unless some other condition blocks the interrupt. Several conditions can block an interrupt, among them that an interrupt of equal or higher priority level is already in progress.

The hardware-generated LCALL causes the contents of the Program Counter to be pushed onto the stack, and reloads the PC with the beginning address of the service routine. As previously noted (Figure 3), the service routine for each interrupt begins at a fixed location.

Only the Program Counter is automatically pushed onto the stack, not the PSW or any other register. Having only the PC be automatically saved allows the programmer to decide how much time to spend saving which other registers. This enhances the interrupt response time, albeit at the expense of increasing the programmer's burden of responsibility. As a result, many interrupt functions that are typical in control applications—toggling a port pin, for example, or reloading a timer, or unloading a serial buffer—can often be completed in less time than it takes other architectures to commence them.

SIMULATING A THIRD PRIORITY LEVEL IN SOFTWARE

Some applications require more than the two priority levels that are provided by on-chip hardware in MCS-51 devices. In these cases, relatively simple software can be written to produce the same effect as a third priority level.

First, interrupts that are to have higher priority than 1 are assigned to priority 1 in the IP (Interrupt Priority) register. The service routines for priority 1 interrupts that are supposed to be interruptible by "priority 2" interrupts are written to include the following code:

```
        PUSH    IE
        MOV     IE,#MASK
        CALL    LABEL
          *******
   (execute service routine)
          *******
        POP     IE
        RET
LABEL:  RETI
```

As soon as any priority 1 interrupt is acknowledged, the IE (Interrupt Enable) register is re-defined so as to disable all but "priority 2" interrupts. Then, a CALL to LABEL executes the RETI instruction, which clears the priority 1 interrupt-in-progress flip-flop. At this point any priority 1 interrupt that is enabled can be serviced, but only "priority 2" interrupts are enabled.

POPping IE restores the original enable byte. Then a normal RET (rather than another RETI) is used to terminate the service routine. The additional software adds 10 μs (at 12 MHz) to priority 1 interrupts.

ADDITIONAL REFERENCES

The following application notes are found in the *Embedded Control Applications* handbook. (Order Number: 270648)

1. AP-69 "An Introduction to the Intel MCS®-51 Single-Chip Microcomputer Family"
2. AP-70 "Using the Intel MCS®-51 Boolean Processing Capabilities"

intel®

MCS® 51 8-BIT CONTROL-ORIENTED MICROCONTROLLERS

Commercial/Express

8031AH/8051AH/8051AHP
8032AH/8052AH
8751H/8751H-8
8751BH/8752BH

- **High Performance HMOS Process**
- **Internal Timers/Event Counters**
- **2-Level Interrupt Priority Structure**
- **32 I/O Lines (Four 8-Bit Ports)**
- **64K External Program Memory Space**
- **Security Feature Protects EPROM Parts Against Software Piracy**
- **Boolean Processor**
- **Bit-Addressable RAM**
- **Programmable Full Duplex Serial Channel**
- **111 Instructions (64 Single-Cycle)**
- **64K External Data Memory Space**
- **Extended Temperature Range (−40°C to +85°C)**

2

The MCS® 51 controllers are optimized for control applications. Byte-processing and numerical operations on small data structures are facilitated by a variety of fast addressing modes for accessing the internal RAM. The instruction set provides a convenient menu of 8-bit arithmetic instructions, including multiply and divide instructions. Extensive on-chip support is provided for one-bit variables as a separate data type, allowing direct bit manipulation and testing in control and logic systems that require Boolean processing.

The 8751H is an EPROM version of the 8051AH. It has 4 Kbytes of electrically programmable ROM which can be erased with ultraviolet light. It is fully compatible with the 8051AH but incorporates one additional feature: a Program Memory Security bit that can be used to protect the EPROM against unauthorized readout. The 8751H-8 is identical to the 8751H but only operates up to 8 MHz.

The 8051AHP is identical to the 8051AH with the exception of the Protection Feature. To incorporate this Protection Feature, program verification has been disabled and external memory accesses have been limited to 4K.

The 8052AH is an enhanced version of the 8051AH. It is backwards compatible with the 8051AH and is fabricated with HMOS II technology. The 8052AH enhancements are listed in the table below. Also refer to this table for the ROM, ROMless and EPROM versions of each product.

Device	Internal Memory		Timers/ Event Counters	Interrupts
	Program	Data		
8031AH	none	128 x 8 RAM	2 x 16-Bit	5
8051AH	4K x 8 ROM	128 x 8 RAM	2 x 16-Bit	5
8051AHP	4K x 8 ROM	128 x 8 RAM	2 x 16-Bit	5
8751H	4K x 8 EPROM	128 x 8 RAM	2 x 16-Bit	5
8751H-8	4K x 8 EPROM	128 x 8 RAM	2 x 16-Bit	5
8751BH	4K x 8 EPROM	128 x 8 RAM	2 x 16-Bit	5
8032AH	none	256 x 8 RAM	3 x 16-Bit	6
8052AH	8K x 8 ROM	256 x 8 RAM	3 x 16-Bit	6
8752BH	8K x 8 EPROM	256 x 8 RAM	3 x 16-Bit	6

October 1994
Order Number: 272318-002

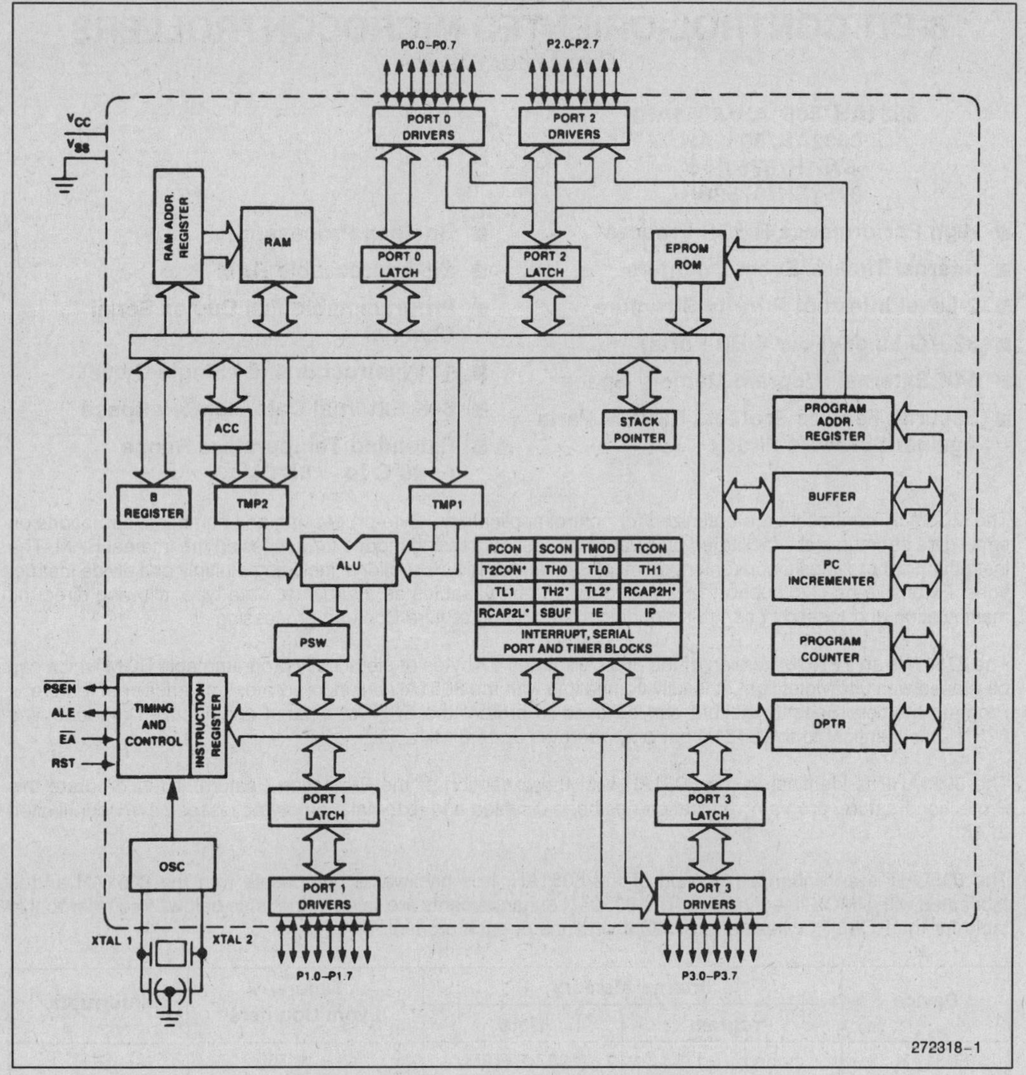

Figure 1. MCS® 51 Controller Block Diagram

PROCESS INFORMATION

The 8031AH/8051AH and 8032AH/8052AH devices are manufactured on P414.1, an HMOS II process. The 8751H/8751H-8 devices are manufactured on P421.X, an HMOS-E process. The 8751BH and 8752BH devices are manufactured on P422. Additional process and reliability information is available in Intel's *Components Quality and Reliability Handbook,* Order No. 210997.

PACKAGES

Part	Prefix	Package Type	θ_{ja}	θ_{jc}
8051AH 8031AH 8052AH 8032AH 8752BH*	P D N	40-Pin Plastic DIP 40-Pin CERDIP 44-Pin PLCC	45°C/W 45°C/W 46°C/W	16°C/W 15°C/W 16°C/W
8751H 8751H-8	D	40-Pin CERDIP	45°C/W	15°C/W
8051AHP	P D	40-Pin Plastic DIP 40-Pin CERDIP	45°C/W 45°C/W	16°C/W 15°C/W
8751BH	P N	40-Pin Plastic DIP 44-Pin PLCC	36°C/W 47°C/W	12°C/W 16°C/W

NOTE:
*8752BH is 36°/10° for D, and 38°/22° for N.

All thermal impedance data is approximate for static air conditions at 1W of power dissipation. Values will change depending on operating conditions and application. See the Intel *Packaging Handbook* (Order Number 240800) for a description of Intel's thermal impedance test methodology.

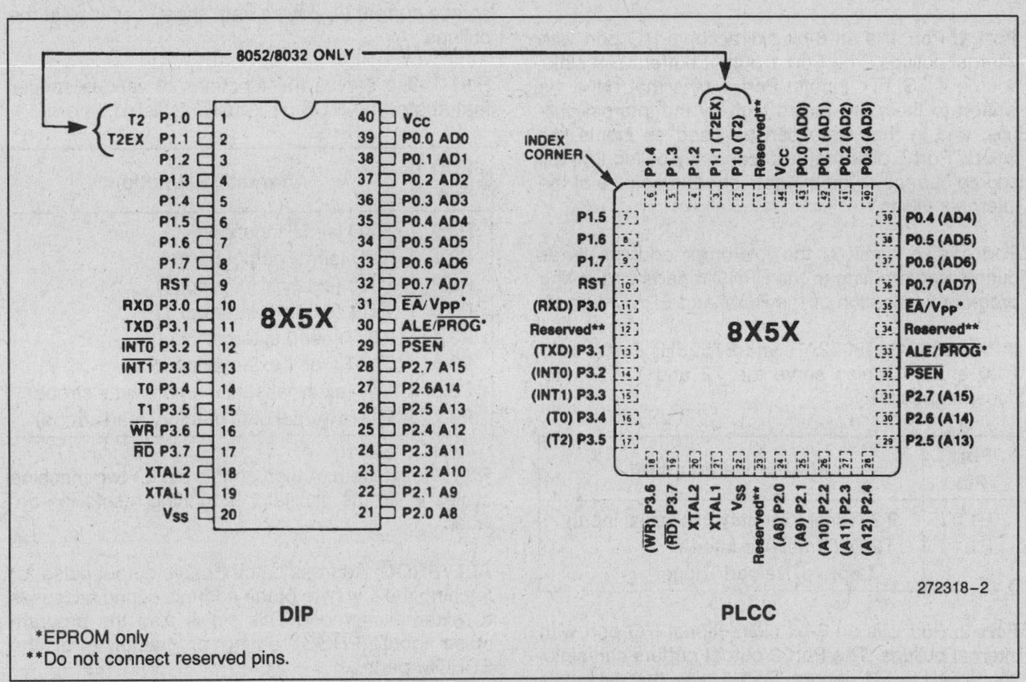

Figure 2. MCS® 51 Controller Connections

PIN DESCRIPTIONS

V_{CC}: Supply voltage.

V_{SS}: Circuit ground.

Port 0: Port 0 is an 8-bit open drain bidirectional I/O port. As an output port each pin can sink 8 LS TTL inputs.

Port 0 pins that have 1's written to them float, and in that state can be used as high-impedance inputs.

Port 0 is also the multiplexed low-order address and data bus during accesses to external Program and Data Memory. In this application it uses strong internal pullups when emitting 1's and can source and sink 8 LS TTL inputs.

Port 0 also receives the code bytes during programming of the EPROM parts, and outputs the code bytes during program verification of the ROM and EPROM parts. External pullups are required during program verification.

Port 1: Port 1 is an 8-bit bidirectional I/O port with internal pullups. The Port 1 output buffers can sink/source 4 LS TTL inputs. Port 1 pins that have 1's written to them are pulled high by the internal pullups, and in that state can be used as inputs. As inputs, Port 1 pins that are externally pulled low will source current (I_{IL} on the data sheet) because of the internal pullups.

Port 1 also receives the low-order address bytes during programming of the EPROM parts and during program verification of the ROM and EPROM parts.

In the 8032AH, 8052AH and 8752BH, Port 1 pins P1.0 and P1.1 also serve the T2 and T2EX functions, respectively.

Port Pin	Alternative Function
P1.0	T2 (Timer/Counter 2 External Input)
P1.1	T2EX (Timer/Counter 2 Capture/Reload Trigger)

Port 2: Port 2 is an 8-bit bidirectional I/O port with internal pullups. The Port 2 output buffers can sink/source 4 LS TTL inputs. Port 2 pins that have 1's written to them are pulled high by the internal pullups, and in that state can be used as inputs. As inputs, Port 2 pins that are externally pulled low will source current (I_{IL} on the data sheet) because of the internal pullups.

Port 2 emits the high-order address byte during fetches from external Program Memory and during accesses to external Data Memory that use 16-bit addresses (MOVX @DPTR). In this application it uses strong internal pullups when emitting 1's. During accesses to external Data Memory that use 8-bit addresses (MOVX @Ri), Port 2 emits the contents of the P2 Special Function Register.

Port 2 also receives the high-order address bits during programming of the EPROM parts and during program verification of the ROM and EPROM parts.

The protection feature of the 8051AHP causes bits P2.4 through P2.7 to be forced to 0, effectively limiting external Data and Code space to 4K each during external accesses.

Port 3: Port 3 is an 8-bit bidirectional I/O port with internal pullups. The Port 3 output buffers can sink/source 4 LS TTL inputs. Port 3 pins that have 1's written to them are pulled high by the internal pullups, and in that state can be used as inputs. As inputs, Port 3 pins that are externally pulled low will source current (I_{IL} on the data sheet) because of the pullups.

Port 3 also serves the functions of various special features of the MCS 51 Family, as listed below:

Port Pin	Alternative Function
P3.0	RXD (serial input port)
P3.1	TXD (serial output port)
P3.2	$\overline{\text{INT0}}$ (external interrupt 0)
P3.3	$\overline{\text{INT1}}$ (external interrupt 1)
P3.4	T0 (Timer 0 external input)
P3.5	T1 (Timer 1 external input)
P3.6	$\overline{\text{WR}}$ (external data memory write strobe)
P3.7	$\overline{\text{RD}}$ (external data memory read strobe)

RST: Reset input. A high on this pin for two machine cycles while the oscillator is running resets the device.

ALE/$\overline{\text{PROG}}$: Address Latch Enable output pulse for latching the low byte of the address during accesses to external memory. This pin is also the program pulse input ($\overline{\text{PROG}}$) during programming of the EPROM parts.

In normal operation ALE is emitted at a constant rate of 1/6 the oscillator frequency, and may be used for external timing or clocking purposes. Note, however, that one ALE pulse is skipped during each access to external Data Memory.

$\overline{PSEN}$: Program Store Enable is the read strobe to external Program Memory.

When the device is executing code from external Program Memory, $\overline{PSEN}$ is activated twice each machine cycle, except that two $\overline{PSEN}$ activations are skipped during each access to external Data Memory.

$\overline{EA}/V_{PP}$: External Access enable $\overline{EA}$ must be strapped to V_{SS} in order to enable any MCS 51 device to fetch code from external Program memory locations starting at 0000H up to FFFFH. $\overline{EA}$ must be strapped to V_{CC} for internal program execution.

Note, however, that if the Security Bit in the EPROM devices is programmed, the device will not fetch code from any location in external Program Memory.

This pin also receives the programming supply voltage (VPP) during programming of the EPROM parts.

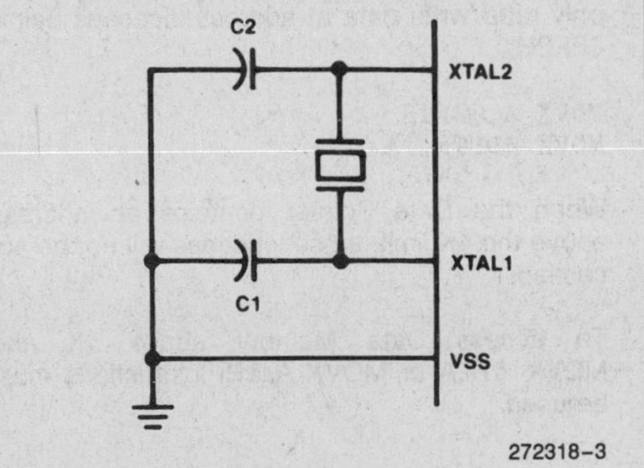

C1, C2 = 30 pF ±10 pF for Crystals
For Ceramic Resonators contact resonator manufacturer.

Figure 3. Oscillator Connections

XTAL1: Input to the inverting oscillator amplifier.

XTAL2: Output from the inverting oscillator amplifier.

OSCILLATOR CHARACTERISTICS

XTAL1 and XTAL2 are the input and output, respectively, of an inverting amplifier which can be configured for use as an on-chip oscillator, as shown in Figure 3. Either a quartz crystal or ceramic resonator may be used. More detailed information concerning the use of the on-chip oscillator is available in Application Note AP-155, "Oscillators for Microcontrollers," Order No. 230659.

To drive the device from an external clock source, XTAL1 should be grounded, while XTAL2 is driven, as shown in Figure 4. There are no requirements on the duty cycle of the external clock signal, since the input to the internal clocking circuitry is through a divide-by-two flip-flop, but minimum and maximum high and low times specified on the data sheet must be observed.

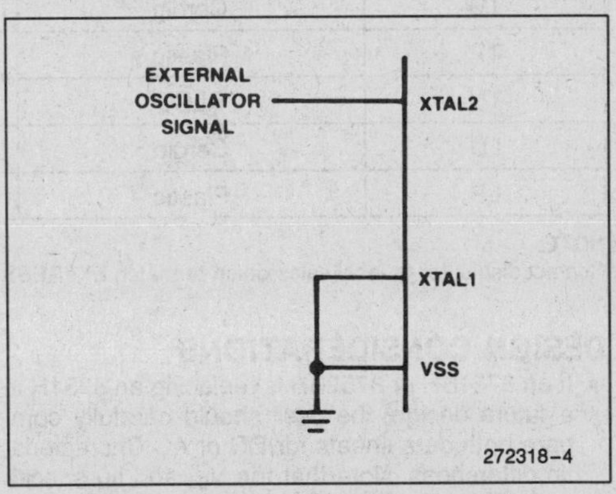

Figure 4. External Drive Configuration

EXPRESS Version

The Intel EXPRESS system offers enhancements to the operational specifications of the MCS 51 family of microcontrollers. These EXPRESS products are designed to meet the needs of those applications whose operating requirements exceed commercial standards.

The EXPRESS program includes the commercial standard temperature range with burn-in, and an extended temperature range with or without burn-in.

With the commercial standard temperature range, operational characteristics are guaranteed over the temperature range of 0°C to +70°C. With the extended temperature range option, operational characteristics are guaranteed over a range of −40°C to +85°C.

The optional burn-in is dynamic, for a minimum time of 160 hours at 125°C with V_{CC} = 5.5V ±0.25V, following guidelines in MIL-STD-883, Method 1015.

Package types and EXPRESS versions are identified by a one- or two-letter prefix to the part number. The prefixes are listed in Table 1.

For the extended temperature range option, this data sheet specifies the parameters which deviate from their commercial temperature range limits.

Table 1. EXPRESS Prefix Identification

Prefix	Package Type	Temperature Range	Burn-In
P	Plastic	Commercial	No
D	Cerdip	Commercial	No
N	PLCC	Commercial	No
TD	Cerdip	Extended	No
TP	Plastic	Extended	No
TN	PLCC	Extended	No
LD	Cerdip	Extended	Yes
LP	Plastic	Extended	Yes

NOTE:
Contact distributor or local sales office to match EXPRESS prefix with proper device.

DESIGN CONSIDERATIONS

- If an 8751BH or 8752BH is replacing an 8751H in a future design, the user should carefully compare both data sheets for DC or AC Characteristic differences. Note that the V_{IH} and I_{IH} specifications for the $\overline{EA}$ pin differ significantly between the devices.
- Exposure to light when the EPROM device is in operation may cause logic errors. For this reason, it is suggested that an opaque label be placed over the window when the die is exposed to ambient light.
- The 8051AHP cannot access external Program or Data memory above 4K. This means that the following instructions that use the Data Pointer only read/write data at address locations below 0FFFH:

```
MOVX A,@DPTR
MOVX @DPTR, A
```

When the Data Pointer contains an address above the 4K limit, those locations will not be accessed.

To access Data Memory above 4K, the MOVX @Ri,A or MOVX A,@Ri instructions must be used.

ABSOLUTE MAXIMUM RATINGS*

Ambient Temperature Under Bias . −40°C to +85°C

Storage Temperature .... −65°C to +150°C

Voltage on $\overline{EA}/V_{PP}$ Pin to V_{SS}

8751H .... −0.5V to +21.5V

8751BH/8752BH .... −0.5V to +13.0V

Voltage on Any Other Pin to V_{SS} .... −0.5V to +7V

Power Dissipation.......................... 1.5W

NOTICE: This is a production data sheet. It is valid for the devices indicated in the revision history. The specifications are subject to change without notice.

**WARNING: Stressing the device beyond the "Absolute Maximum Ratings" may cause permanent damage. These are stress ratings only. Operation beyond the "Operating Conditions" is not recommended and extended exposure beyond the "Operating Conditions" may affect device reliability.*

OPERATING CONDITIONS

Symbol	Description	Min	Max	Units
T_A	Ambient Temperature Under Bias Commercial Express	 0 −40	 +70 +85	 °C °C
V_{CC}	Supply Voltage	4.5	5.5	V
F_{OSC}	Oscillator Frequency	3.5	12	MHz

DC CHARACTERISTICS (Over Operating Conditions)

All parameter values apply to all devices unless otherwise indicated

Symbol	Parameter	Min	Max	Units	Test Conditions
V_{IL}	Input Low Voltage (Except $\overline{EA}$ Pin of 8751H and 8751H-8)	−0.5	0.8	V	
V_{IL1}	Input Low Voltage to $\overline{EA}$ Pin of 8751H and 8751H-8	0	0.7	V	
V_{IH}	Input High Voltage (Except XTAL2, RST)	2.0	V_{CC} + 0.5	V	
V_{IH1}	Input High Voltage to XTAL2, RST	2.5	V_{CC} + 0.5	V	XTAL1 = V_{SS}
V_{IH2}	Input High Voltage to $\overline{EA}$ pin of 8751BH and 8752BH	4.5	5.5V		
V_{OL}	Output Low Voltage (Ports 1, 2, 3)*		0.45	V	I_{OL} = 1.6 mA
V_{OL1}	Output Low Voltage (Port 0, ALE, $\overline{PSEN}$)* 8751H, 8751H-8 All Others		 0.60 0.45 0.45	 V V V	 I_{OL} = 3.2 mA I_{OL} = 2.4 mA I_{OL} = 3.2 mA
V_{OH}	Output High Voltage (Ports 1, 2, 3, ALE, $\overline{PSEN}$)	2.4		V	I_{OH} = −80 μA
V_{OH1}	Output High Voltage (Port 0 in External Bus Mode)	2.4		V	I_{OH} = −400 μA
I_{IL}	Logical 0 Input Current (Ports 1, 2, 3, and RST)		−500	μA	V_{IN} = 0.45V
I_{IL1}	Logical 0 Input Current ($\overline{EA}$) 8751H and 8751H-8 8751BH 8752BH	 −10	 −15 −10 0.5	 mA mA mA mA	 V_{IN} = 0.45V V_{IN} = V_{SS} V_{IN} = V_{SS}

DC CHARACTERISTICS (Over Operating Conditions)

All parameter values apply to all devices unless otherwise indicated (Continued)

Symbol	Parameter	Min	Max	Units	Test Conditions
I_{IL2}	Logical 0 Input Current (XTAL2)		−3.2	mA	$V_{IN} = 0.45V$
I_{LI}	Input Leakage Current (Port 0) 8751H and 8751H-8 All Others		 ±100 ±10	 μA μA	 $0.45 \leq V_{IN} \leq V_{CC}$ $0.45 \leq V_{IN} \leq V_{CC}$
I_{IH}	Logical 1 Input Current ($\overline{EA}$) 8751H and 8751H-8 8751BH/8752BH		 500 1	 μA mA	 $V_{IN} = 2.4V$ $4.5V < V_{IN} < 5.5V$
I_{IH1}	Input Current to RST to Activate Reset		500	μA	$V_{IN} < (V_{CC} - 1.5V)$
I_{CC}	Power Supply Current: 8031AH/8051AH/8051AHP 8032AH/8052AH/8751BH/8752BH 8751H/8751H-8		 125 175 250	 mA mA mA	 All Outputs Disconnected; $\overline{EA} = V_{CC}$
C_{IO}	Pin Capacitance		10	pF	Test freq = 1 MHz

NOTES:

1. Capacitive loading on Ports 0 and 2 may cause spurious noise pulses to be superimposed on the V_{OL}s of ALE/$\overline{PROG}$ and Ports 1 and 3. The noise is due to external bus capacitance discharging into the Port 0 and Port 2 pins when these pins make 1-to-0 transitions during bus operations. In the worst cases (capacitive loading > 100 pF), the noise pulse on the ALE/$\overline{PROG}$ pin may exceed 0.8V. In such cases it may be desirable to qualify ALE with a Schmitt Trigger, or use an address latch with a Schmitt Trigger STROBE input.
2. ALE/$\overline{PROG}$ refers to a pin on the 8751BH. ALE refers to a timing signal that is output on the ALE/$\overline{PROG}$ pin.
3. Under steady state (non-transient) conditions, I_{OL} must be externally limited as follows:

Maximum I_{OL} per port pin:	10 mA
Maximum I_{OL} per 8-bit port -	
Port 0:	26 mA
Ports 1, 2, and 3:	15 mA
Maximum total I_{OL} for all output pins:	71 mA

If I_{OL} exceeds the test condition, V_{OL} may exceed the related specification. Pins are not guaranteed to sink current greater than the listed test conditions.

EXPLANATION OF THE AC SYMBOLS

Each timing symbol has 5 characters. The first character is always a 'T' (stands for time). The other characters, depending on their positions, stand for the name of a signal or the logical status of that signal. The following is a list of all the characters and what they stand for.

A: Address
C: Clock
D: Input Data
H: Logic level HIGH
I: Instruction (program memory contents)
L: Logic level LOW, or ALE
P: $\overline{\text{PSEN}}$
Q: Output data
R: $\overline{\text{RD}}$ signal
T: Time
V: Valid
W: $\overline{\text{WR}}$ signal
X: No longer a valid logic level
Z: Float

For example,

TAVLL = Time from Address Valid to ALE Low.
TLLPL = Time from ALE Low to PSEN Low.

AC CHARACTERISTICS (Under Operating Conditions; Load Capacitance for Port 0, ALE/$\overline{\text{PROG}}$, and $\overline{\text{PSEN}}$ = 100 pF; Load Capacitance for All Other Outputs = 80 pF)

2

EXTERNAL PROGRAM MEMORY CHARACTERISTICS

Symbol	Parameter	12 MHz Oscillator		Variable Oscillator		Units
		Min	Max	Min	Max	
1/TCLCL	Oscillator Frequency			3.5	12.0	MHz
TLHLL	ALE Pulse Width	127		2TCLCL − 40		ns
TAVLL	Address Valid to ALE Low	43		TCLCL − 40		ns
TLLAX	Address Hold after ALE Low	48		TCLCL − 35		ns
TLLIV	ALE Low to Valid Instr In					
	8751H		183		4TCLCL − 150	ns
	All Others		233		4TCLCL − 100	ns
TLLPL	ALE Low to $\overline{\text{PSEN}}$ Low	58		TCLCL − 25		ns
TPLPH	$\overline{\text{PSEN}}$ Pulse Width					
	8751H	190		3TCLCL − 60		ns
	All Others	215		3TCLCL − 35		ns
TPLIV	$\overline{\text{PSEN}}$ Low to Valid Instr In					
	8751H		100		3TCLCL − 150	ns
	All Others		125		3TCLCL − 125	ns
TPXIX	Input Instr Hold after $\overline{\text{PSEN}}$	0		0		ns
TPXIZ	Input Instr Float after $\overline{\text{PSEN}}$		63		TCLCL − 20	ns
TPXAV	$\overline{\text{PSEN}}$ to Address Valid	75		TCLCL − 8		ns
TAVIV	Address to Valid Instr In					
	8751H		267		5TCLCL − 150	ns
	All Others		302		5TCLCL − 115	ns
TPLAZ	$\overline{\text{PSEN}}$ Low to Address Float		20		20	ns
TRLRH	$\overline{\text{RD}}$ Pulse Width	400		6TCLCL − 100		ns
TWLWH	$\overline{\text{WR}}$ Pulse Width	400		6TCLCL − 100		ns
TRLDV	$\overline{\text{RD}}$ Low to Valid Data In		252		5TCLCL − 165	ns
TRHDX	Data Hold after $\overline{\text{RD}}$	0		0		ns
TRHDZ	Data Float after $\overline{\text{RD}}$		97		2TCLCL − 70	ns
TLLDV	ALE Low to Valid Data In		517		8TCLCL − 150	ns
TAVDV	Address to Valid Data In		585		9TCLCL − 165	ns

EXTERNAL PROGRAM MEMORY CHARACTERISTICS (Continued)

Symbol	Parameter	12 MHz Oscillator		Variable Oscillator		Units
		Min	Max	Min	Max	
TLLWL	ALE Low to $\overline{RD}$ or $\overline{WR}$ Low	200	300	3TCLCL−50	3TCLCL+50	ns
TAVWL	Address to $\overline{RD}$ or $\overline{WR}$ Low	203		4TCLCL−130		ns
TQVWX	Data Valid to $\overline{WR}$ Transition 8751H All Others	 13 23		 TCLCL−70 TCLCL−60		 ns ns
TQVWH	Data Valid to $\overline{WR}$ High	433		7TCLCL−150		ns
TWHQX	Data Hold after $\overline{WR}$	33		TCLCL−50		ns
TRLAZ	$\overline{RD}$ Low to Address Float		20		20	ns
TWHLH	$\overline{RD}$ or $\overline{WR}$ High to ALE High 8751H All Others	 33 43	 133 123	 TCLCL−50 TCLCL−40	 TCLCL+50 TCLCL+40	 ns ns

NOTE:
*The 8751H-8 is identical to the 8751H but only operates up to 8 MHz. When calculating the AC Characteristics for the 8751H-8, use the 8751H formula for variable oscillators.

EXTERNAL PROGRAM MEMORY READ CYCLE

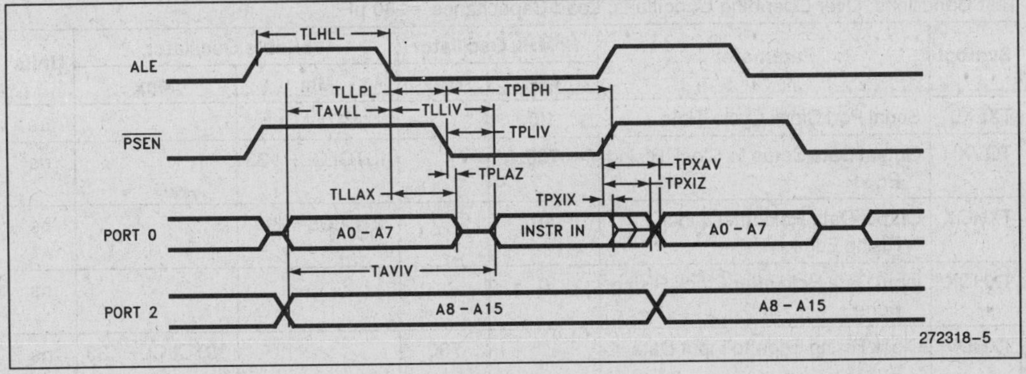

272318–5

EXTERNAL DATA MEMORY READ CYCLE

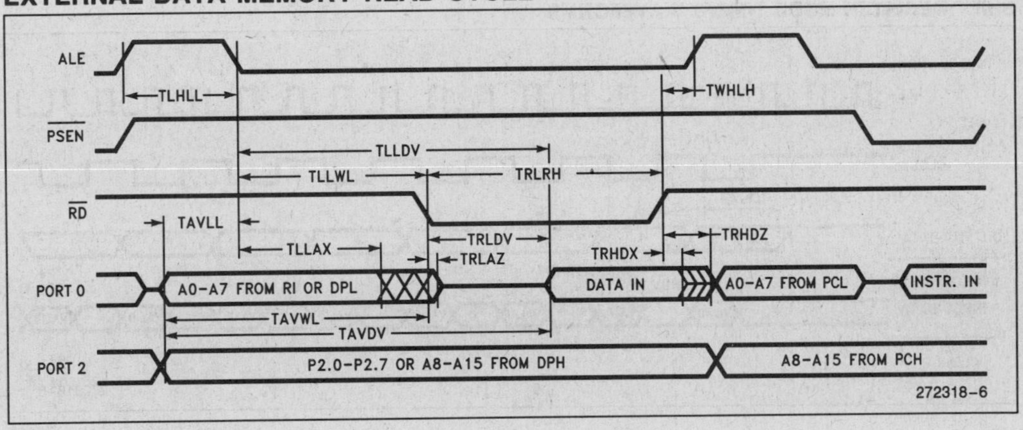

272318–6

EXTERNAL DATA MEMORY WRITE CYCLE

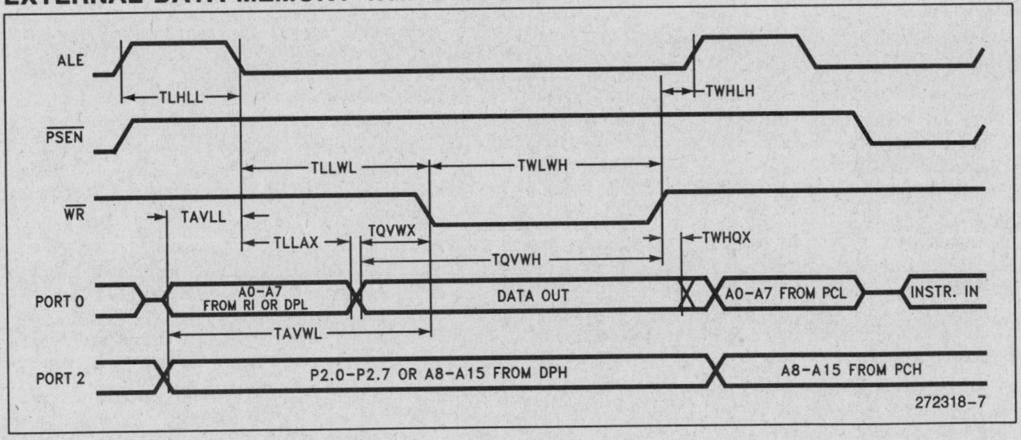

272318–7

SERIAL PORT TIMING—SHIFT REGISTER MODE

Test Conditions: Over Operating Conditions; Load Capacitance = 80 pF

Symbol	Parameter	12 MHz Oscillator		Variable Oscillator		Units
		Min	Max	Min	Max	
TXLXL	Serial Port Clock Cycle Time	1.0		12TCLCL		μs
TQVXH	Output Data Setup to Clock Rising Edge	700		10TCLCL − 133		ns
TXHQX	Output Data Hold after Clock Rising Edge	50		2TCLCL − 117		ns
TXHDX	Input Data Hold after Clock Rising Edge	0		0		ns
TXHDV	Clock Rising Edge to Input Data Valid		700		10TCLCL − 133	ns

SHIFT REGISTER MODE TIMING WAVEFORMS

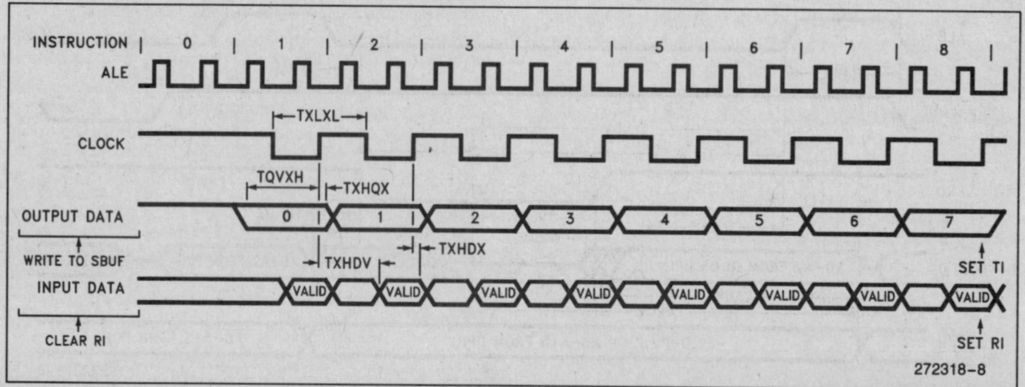

EXTERNAL CLOCK DRIVE

Symbol	Parameter	Min	Max	Units
1/TCLCL	Oscillator Frequency (except 8751H-8) 8751H-8	3.5 3.5	12 8	MHz MHz
TCHCX	High Time	20		ns
TCLCX	Low Time	20		ns
TCLCH	Rise Time		20	ns
TCHCL	Fall Time		20	ns

EXTERNAL CLOCK DRIVE WAVEFORM

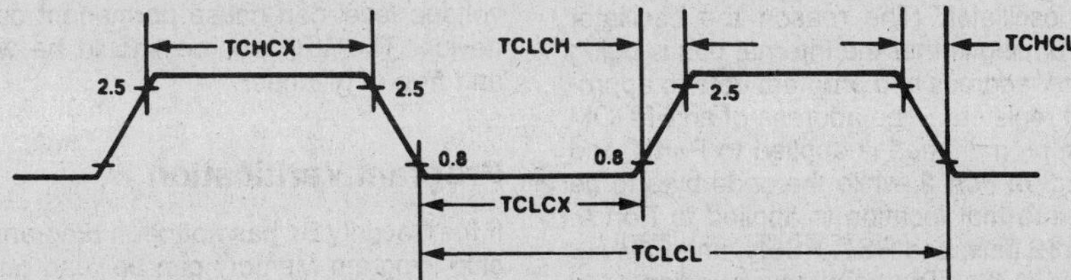

272318-9

AC TESTING INPUT, OUTPUT WAVEFORM

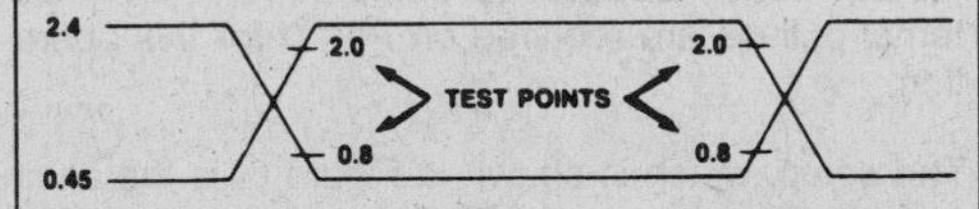

272318-10

AC Testing: Inputs are driven at 2.4V for a Logic "1" and 0.45V for a Logic "0". Timing measurements are made at 2.0V for a Logic "1" and 0.8V for a Logic "0".

EPROM CHARACTERISTICS

Table 3. EPROM Programming Modes

Mode	RST	PSEN	ALE	EA	P2.7	P2.6	P2.5	P2.4
Program	1	0	0*	VPP	1	0	X	X
Verify	1	0	1	1	0	0	X	X
Security Set	1	0	0*	VPP	1	1	X	X

NOTE:
"1" = logic high for that pin
"0" = logic low for that pin
"X" = "don't care"
"VPP" = +21V ±0.5V
*ALE is pulsed low for 50 ms.

PROGRAMMING THE 8751H

To be programmed, the part must be running with a 4 to 6 MHz oscillator. (The reason the oscillator needs to be running is that the internal bus is being used to transfer address and program data to appropriate internal registers.) The address of an EPROM location to be programmed is applied to Port 1 and pins P2.0–P2.3 of Port 2, while the code byte to be programmed into that location is applied to Port 0. The other Port 2 pins, and RST, PSEN, and EA/V_{PP} should be held at the "Program" levels indicated in Table 3. ALE/PROG is **pulsed** low for 50 ms to program the code byte into the addressed EPROM location. The setup is shown in Figure 5.

Normally EA/V_{PP} is held at a logic high until just before ALE/PROG is to be pulsed. Then EA/V_{PP} is raised to +21V, ALE/PROG is pulsed, and then EA/V_{PP} is returned to a logic high. Waveforms and detailed timing specifications are shown in later sections of this data sheet.

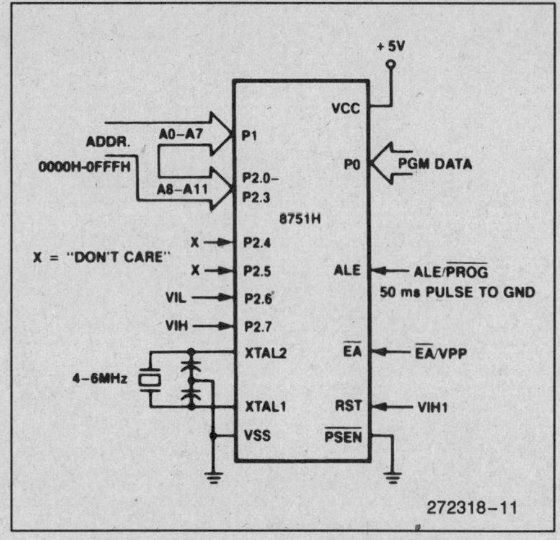

Figure 5. Programming Configuration

Note that the EA/VPP pin must not be allowed to go above the maximum specified VPP level of 21.5V for any amount of time. Even a narrow glitch above that voltage level can cause permanent damage to the device. The VPP source should be well regulated and free of glitches.

Program Verification

If the Security Bit has not been programmed, the on-chip Program Memory can be read out for verification purposes, if desired, either during or after the programming operation. The address of the Program Memory location to be read is applied to Port 1 and pins P2.0–P2.3. The other pins should be held at the "Verify" levels indicated in Table 3. The contents of the addressed location will come out on Port 0. External pullups are required on Port 0 for this operation.

The setup, which is shown in Figure 6, is the same as for programming the EPROM except that pin P2.7 is held at a logic low, or may be used as an active-low read strobe.

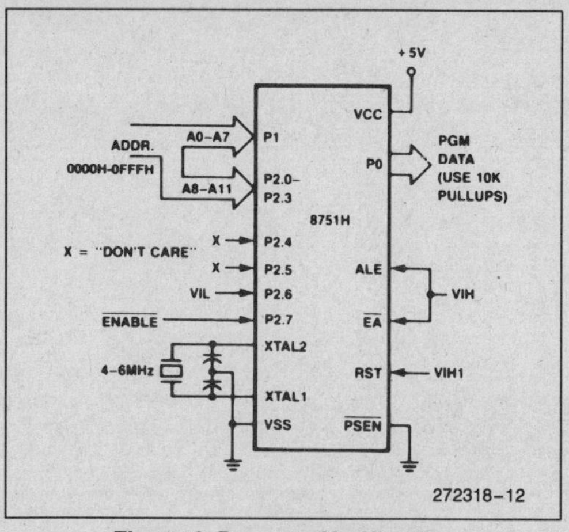

Figure 6. Program Verification

EPROM Security

The security feature consists of a "locking" bit which when programmed denies electrical access by any external means to the on-chip Program Memory. The bit is programmed as shown in Figure 7. The setup and procedure are the same as for normal EPROM programming, except that P2.6 is held at a logic high. Port 0, Port 1 and pins P2.0–P2.3 may be in any state. The other pins should be held at the "Security" levels indicated in Table 3.

Once the Security Bit has been programmed, it can be cleared only by full erasure of the Program Memory. While it is programmed, the internal Program Memory can not be read out, the device can not be further programmed, and it **can not execute out of external program memory**. Erasing the EPROM, thus clearing the Security Bit, restores the device's full functionality. It can then be reprogrammed.

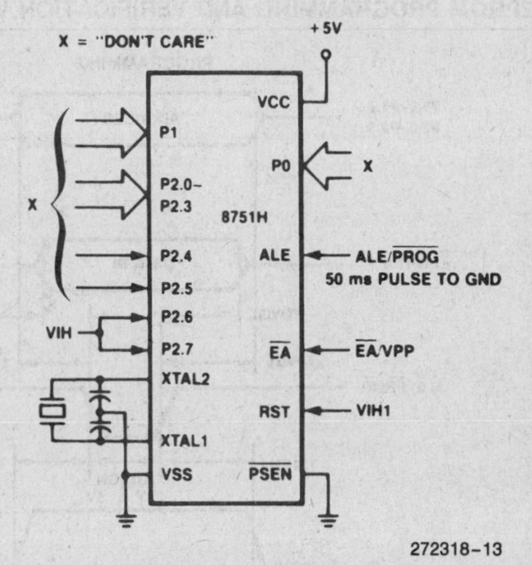

Figure 7. Programming the Security Bit

Erasure Characteristics

Erasure of the EPROM begins to occur when the device is exposed to light with wavelengths shorter than approximately 4,000 Angstroms. Since sunlight and fluorescent lighting have wavelengths in this range, exposure to these light sources over an extended time (about 1 week in sunlight, or 3 years in room-level fluorescent lighting) could cause inadvertent erasure. If an application subjects the device to this type of exposure, it is suggested that an opaque label be placed over the window.

The recommended erasure procedure is exposure to ultraviolet light (at 2537 Angstroms) to an integrated dose of at least 15 W-sec/cm^2. Exposing the EPROM to an ultraviolet lamp of 12,000 μW/cm^2 rating for 20 to 30 minutes, at a distance of about 1 inch, should be sufficient.

Erasure leaves the array in an all 1's state.

EPROM PROGRAMMING AND VERIFICATION CHARACTERISTICS

T_A = 21°C to 27°C; VCC = 5V ±10%; VSS = 0V

Symbol	Parameter	Min	Max	Units
VPP	Programming Supply Voltage	20.5	21.5	V
IPP	Programming Supply Current		30	mA
1/TCLCL	Oscillator Frequency	4	6	MHz
TAVGL	Address Setup to $\overline{\text{PROG}}$ Low	48TCLCL		
TGHAX	Address Hold after $\overline{\text{PROG}}$	48TCLCL		
TDVGL	Data Setup to $\overline{\text{PROG}}$ Low	48TCLCL		
TGHDX	Data Hold after $\overline{\text{PROG}}$	48TCLCL		
TEHSH	P2.7 ($\overline{\text{ENABLE}}$) High to VPP	48TCLCL		
TSHGL	VPP Setup to $\overline{\text{PROG}}$ Low	10		μs
TGHSL	VPP Hold after $\overline{\text{PROG}}$	10		μs
TGLGH	$\overline{\text{PROG}}$ Width	45	55	ms
TAVQV	Address to Data Valid		48TCLCL	
TELQV	$\overline{\text{ENABLE}}$ Low to Data Valid		48TCLCL	
TEHQZ	Data Float after $\overline{\text{ENABLE}}$	0	48TCLCL	

EPROM PROGRAMMING AND VERIFICATION WAVEFORMS

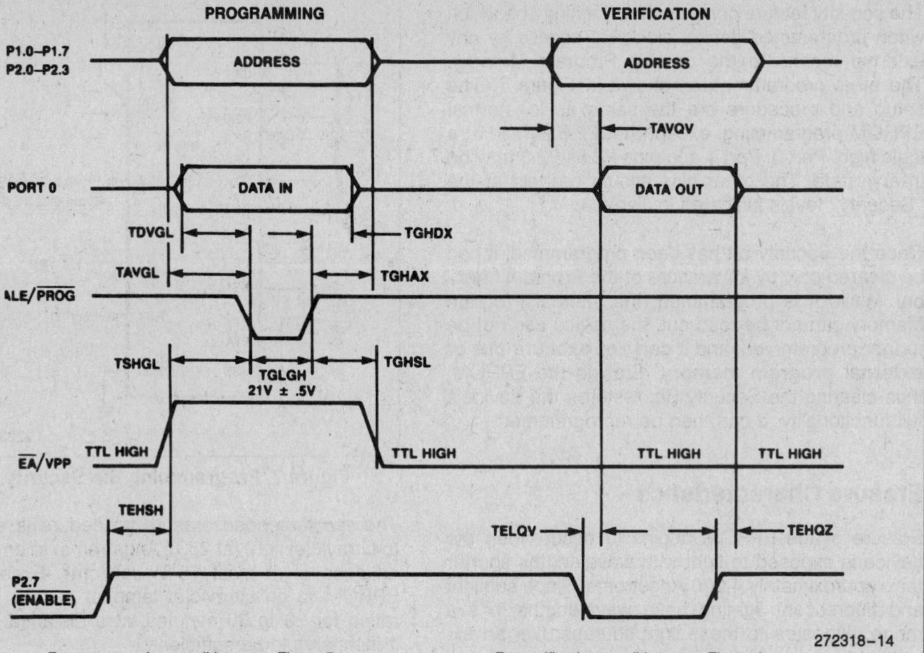

For programming conditions see Figure 5. For verification conditions see Figure 6.

Programming the 8751BH/8752BH

To be programmed, the 875XBH must be running with a 4 to 6 MHz oscillator. (The reason the oscillator needs to be running is that the internal bus is being used to transfer address and program data to appropriate internal registers.) The address of an EPROM location to be programmed is applied to Port 1 and pins P2.0 - P2.4 of Port 2, while the code byte to be programmed into that location is applied to Port 0. The other Port 2 and 3 pins, and RST, $\overline{PSEN}$, and $\overline{EA}/V_{PP}$ should be held at the "Program" levels indicated in Table 1. ALE/$\overline{PROG}$ is pulsed low to program the code byte into the addressed EPROM location. The setup is shown in Figure 8.

Normally $\overline{EA}/V_{PP}$ is held at a logic high until just before ALE/$\overline{PROG}$ is to be pulsed. Then $\overline{EA}/V_{PP}$ is raised to V_{PP}, ALE/$\overline{PROG}$ is pulsed low, and then $\overline{EA}/V_{PP}$ is returned to a valid high voltage. The voltage on the $\overline{EA}/V_{PP}$ pin must be at the valid $\overline{EA}/V_{PP}$ high level before a verify is attempted. Waveforms and detailed timing specifications are shown in later sections of this data sheet.

Note that the $\overline{EA}/V_{PP}$ pin must not be allowed to go above the maximum specified V_{PP} level for any amount of time. Even a narrow glitch above that voltage level can cause permanent damage to the device. The V_{PP} source should be well regulated and free of glitches.

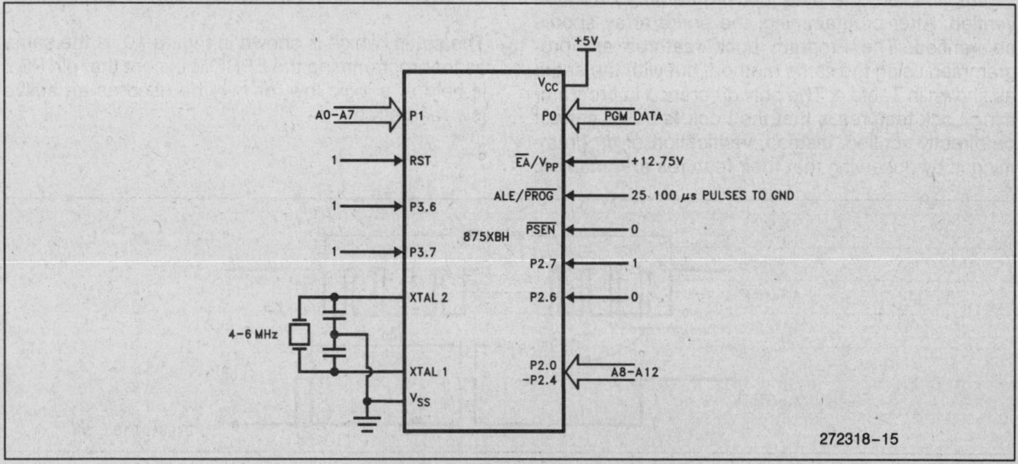

Figure 8. Programming the EPROM

Table 4. EPROM Programming Modes for 875XBH

MODE	RST	$\overline{PSEN}$	ALE/ $\overline{PROG}$	$\overline{EA}$/ V_{PP}	P2.7	P2.6	P3.6	P3.7
Program Code Data	1	0	0*	V_{PP}	1	0	1	1
Verify Code Data	1	0	1	1	0	0	1	1
Program Encryption Table Use Addresses 0-1FH	1	0	0*	V_{PP}	1	0	0	1
Program Lock Bits (LBx) x=1	1	0	0*	V_{PP}	1	1	1	1
x=2	1	0	0*	V_{PP}	1	1	0	0
Read Signature	1	0	1	1	0	0	0	0

NOTES:
"1" = Valid high for that pin
"0" = Valid low for that pin
"V_{PP}" = +12.75V ±0.25V
*ALE/$\overline{PROG}$ is pulsed low for 100 uS for programming. (Quick-Pulse Programming)

QUICK-PULSE PROGRAMMING ALGORITHM

The 875XBH can be programmed using the Quick-Pulse Programming Algorithm for microcontrollers. The features of the new programming method are a lower V_{PP} (12.75 volts as compared to 21 volts) and a shorter programming pulse. For example, it is possible to program the entire 8 Kbytes of 875XBH EPROM memory in less than 25 seconds with this algorithm!

To program the part using the new algorithm, V_{PP} must be 12.75 ±0.25 Volts. ALE/$\overline{PROG}$ is pulsed low for 100 μseconds, 25 times as shown in Figure 9. Then, the byte just programmed may be verified. After programming, the entire array should be verified. The Program Lock features are programmed using the same method, but with the setup as shown in Table 4. The only difference in programming Lock features is that the Lock features cannot be directly verified. Instead, verification of programming is by observing that their features are enabled.

PROGRAM VERIFICATION

If the Lock Bits have not been programmed, the on-chip Program Memory can be read out for verification purposes, if desired, either during or after the programming operation. The address of the Program Memory location to be read is applied to Port 1 and pins P2.0 - P2.4. The other pins should be held at the "Verify" levels indicated in Table 1. The contents of the addressed location will come out on Port 0. External pullups are required on Port 0 for this operation. (If the Encryption Array in the EPROM has been programmed, the data present at Port 0 will be Code Data XNOR Encryption Data. The user must know the Encryption Array contents to manually "unencrypt" the data during verify.)

The setup, which is shown in Figure 10, is the same as for programming the EPROM except that pin P2.7 is held at a logic low, or may be used as an active low read strobe.

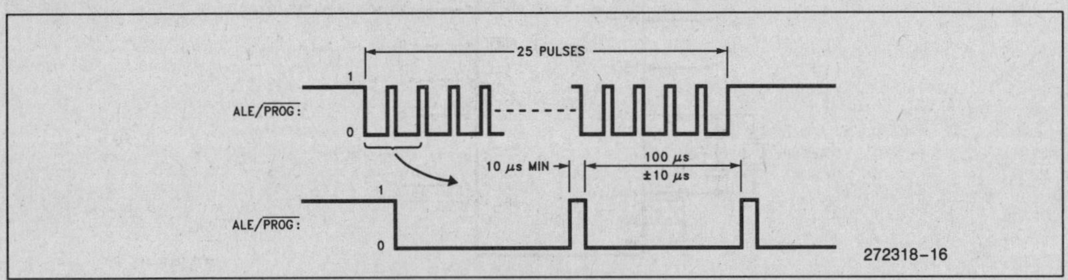

Figure 9. $\overline{PROG}$ Waveforms

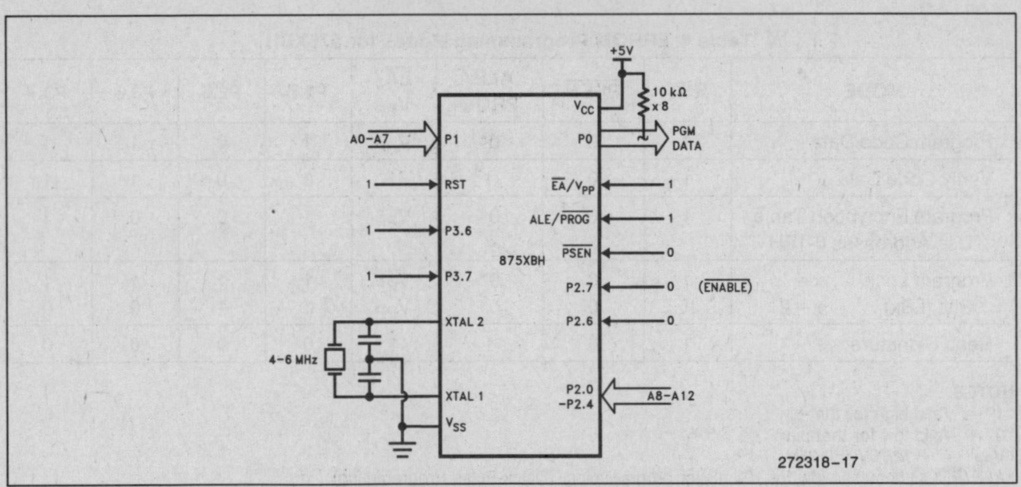

Figure 10. Verifying the EPROM

PROGRAM MEMORY LOCK

The two-level Program Lock system consists of 2 Lock bits and a 32-byte Encryption Array which are used to protect the program memory against software piracy.

ENCRYPTION ARRAY

Within the EPROM array are 32 bytes of Encryption Array that are initially unprogrammed (all 1s). Every time that a byte is addressed during a verify, 5 address lines are used to select a byte of the Encryption Array. This byte is then exclusive-NORed (XNOR) with the code byte, creating an Encrypted Verify byte. The algorithm, with the array in the unprogrammed state (all 1s), will return the code in its original, unmodified form.

It is recommended that whenever the Encryption Array is used, at least one of the Lock Bits be programmed as well.

LOCK BITS

Also included in the EPROM Program Lock scheme are two Lock Bits which function as shown in Table 5.

Erasing the EPROM also erases the Encryption Array and the Lock Bits, returning the part to full unlocked functionality.

To ensure proper functionality of the chip, the internally latched value of the $\overline{EA}$ pin must agree with its external state.

Table 5. Lock Bits and their Features

Lock Bits		Logic Enabled
LB1	**LB2**	
U	U	Minimum Program Lock features enabled. (Code Verify will still be encrypted by the Encryption Array)
P	U	MOVC instructions executed from external program memory are disabled from fetching code bytes from internal memory, $\overline{EA}$ is sampled and latched on reset, and further programming of the EPROM is disabled
P	P	Same as above, but Verify is also disabled
U	P	Reserved for Future Definition

P = Programmed
U = Unprogrammed

READING THE SIGNATURE BYTES

The signature bytes are read by the same procedure as a normal verification of locations 030H and 031H, except that P3.6 and P3.7 need to be pulled to a logic low. The values returned are:

(030H) = 89H indicates manufactured by Intel
(031H) = 51H indicates 8751BH
52H indicates 8752BH

ERASURE CHARACTERISTICS

Erasure of the EPROM begins to occur when the 8752BH is exposed to light with wavelengths shorter than approximately 4,000 Angstroms. Since sunlight and fluorescent lighting have wavelengths in this range, exposure to these light sources over an extended time (about 1 week in sunlight, or 3 years in room-level fluorescent lighting) could cause inadvertent erasure. If an application subjects the device to this type of exposure, it is suggested that an opaque label be placed over the window.

The recommended erasure procedure is exposure to ultraviolet light (at 2537 Angstroms) to an integrated dose of at lease 15 W-sec/cm. Exposing the EPROM to an ultraviolet lamp of 12,000 μW/cm rating for 30 minutes, at a distance of about 1 inch, should be sufficient.

Erasure leaves the array in an all 1s state.

EPROM PROGRAMMING AND VERIFICATION CHARACTERISTICS

(T_A = 21°C to 27°C, V_{CC} = 5.0V ±10%, V_{SS} = 0V)

Symbol	Parameter	Min	Max	Units
V_{PP}	Programming Supply Voltage	12.5	13.0	V
I_{PP}	Programming Supply Current		50	mA
1/TCLCL	Oscillator Frequency	4	6	MHz
TAVGL	Address Setup to $\overline{\text{PROG}}$ Low	48TCLCL		
TGHAX	Address Hold After $\overline{\text{PROG}}$	48TCLCL		
TDVGL	Data Setup to $\overline{\text{PROG}}$ Low	48TCLCL		
TGHDX	Data Hold After $\overline{\text{PROG}}$	48TCLCL		
TEHSH	P2.7 ($\overline{\text{ENABLE}}$) High to V_{PP}	48TCLCL		
TSHGL	V_{PP} Setup to $\overline{\text{PROG}}$ Low	10		μs
TGHSL	V_{PP} Hold After $\overline{\text{PROG}}$	10		μs
TGLGH	$\overline{\text{PROG}}$ Width	90	110	μs
TAVQV	Address to Data Valid		48TCLCL	
TELQV	$\overline{\text{ENABLE}}$ Low to Data Valid		48TCLCL	
TEHQZ	Data Float After $\overline{\text{ENABLE}}$	0	48TCLCL	
TGHGL	$\overline{\text{PROG}}$ High to $\overline{\text{PROG}}$ Low	10		μs

EPROM PROGRAMMING AND VERIFICATION WAVEFORMS

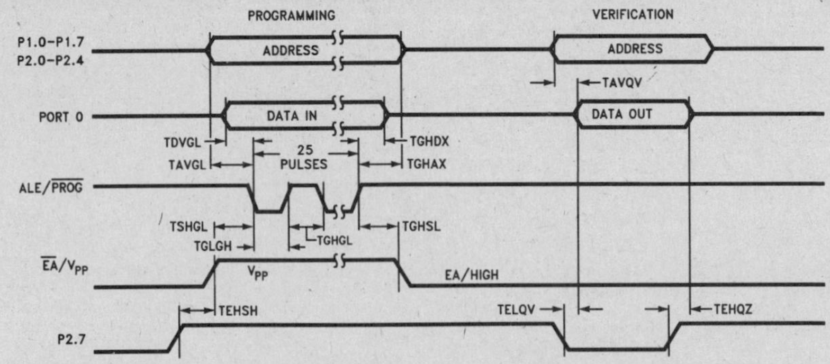

272318–18

DATA SHEET REVISION HISTORY

Datasheets are changed as new device information becomes available. Verify with your local Intel sales office that you have the latest version before finalizing a design or ordering devices.

The following differences exist between this datasheet (272318-002) and the previous version (272318-001):

1. Removed QP and QD (commercial with extended burn-in) from Table 1. EXPRESS Prefix Identification.

This datasheet (272318-001) replaces the following datasheets:

MCS® 51 Controllers (270048-007)
8051AHP (270279-004)
8751BH (270248-005)
8751BH EXPRESS (270708-001)
8752BH (270429-004)
8752BH EXPRESS (270650-002)

87C51/80C51BH/80C31BH
CHMOS SINGLE-CHIP 8-BIT MICROCONTROLLER

Commercial/Express

87C51/80C51BH/80C51BHP/80C31BH
***See Table 1 for Proliferation Options**

- **High Performance CHMOS EPROM**
- **24 MHz Internal Operation (-3 Only)**
- **Improved Quick-Pulse Programming Algorithm**
- **3-Level Program Memory Lock**
- **Boolean Processor**
- **128-Byte Data RAM**
- **32 Programmable I/O Lines**
- **Two 16-Bit Timer/Counters**
- **Extended Temperature Range (−40°C to +85°C)**
- **5 Interrupt Sources**
- **Programmable Serial Port**
- **TTL- and CMOS-Compatible Logic Levels**
- **64K External Program Memory Space**
- **64K External Data Memory Space**
- **ONCE Mode Facilitates System Testing**
- **Power Control Modes**
  - **Idle**
  - **Power Down**

MEMORY ORGANIZATION

PROGRAM MEMORY: Up to 4 Kbytes of the program memory can reside on-chip (except 80C31BH). In addition the device can address up to 64K of program memory external to the chip.

DATA MEMORY: This microcontroller has a 128 x 8 on-chip RAM. In addition it can address up to 64 Kbytes of external data memory.

The Intel 87C51 is a single-chip control-oriented microcontroller which is fabricated on Intel's reliable CHMOS III-E technology. The Intel 80C51BH/80C31BH is fabricated on CHMOS III technology. Being a member of the MCS® 51 controller family, the 87C51/80C51BH/80C31BH uses the same powerful instruction set, has the same architecture, and is pin-for-pin compatible with the existing MCS 51 controller family of products.

The 80C51BHP is identical to the 80C51BH with the exception of the Protection Feature. To incorporate this Protection Feature, program verification has been disabled and external memory accesses have been limited to 4K.

The 87C51-3 has the same 3.5 MHz to 20 MHz frequency range as the 87C51-20 when operating out of external program/data memory. When running out of internal program/data memory, the 87C51-3 can operate up to 24 MHz.

The extremely low operating power, along with the two reduced power modes, Idle and Power Down, make this part very suitable for low power applications. The Idle mode freezes the CPU while allowing the RAM, timer/counters, serial port and interrupt system to continue functioning. The Power Down mode saves the RAM contents but freezes the oscillator, causing all other chip functions to be inoperative.

For the remainder of this document, the 87C51, 80C51BH, and 80C31BH will be referred to as the 87C51/BH, unless information applies to a specific device.

October 1994
Order Number: 272335-002

Table 1. Proliferation Options

	*Standard	-1	-2	-3	-20
80C31BH	X	X	X		
80C51BH	X	X	X		
80C51BHP	X	X	X		
87C51	X	X	X	X	X

NOTES:
* —3.5 MHz to 12 MHz; V_{CC} = 5V ±20%
-1 —3.5 MHz to 16 MHz; V_{CC} = 5V ±20%
-2 —0.5 MHz to 12 MHz; V_{CC} = 5V ±20%
-3 —3.5 MHz to 20 MHz external program execution; V_{CC} = 5V ±20%
24 MHz internal program execution
-20 —3.5 MHz to 20 MHz; V_{CC} = 5V ±20%

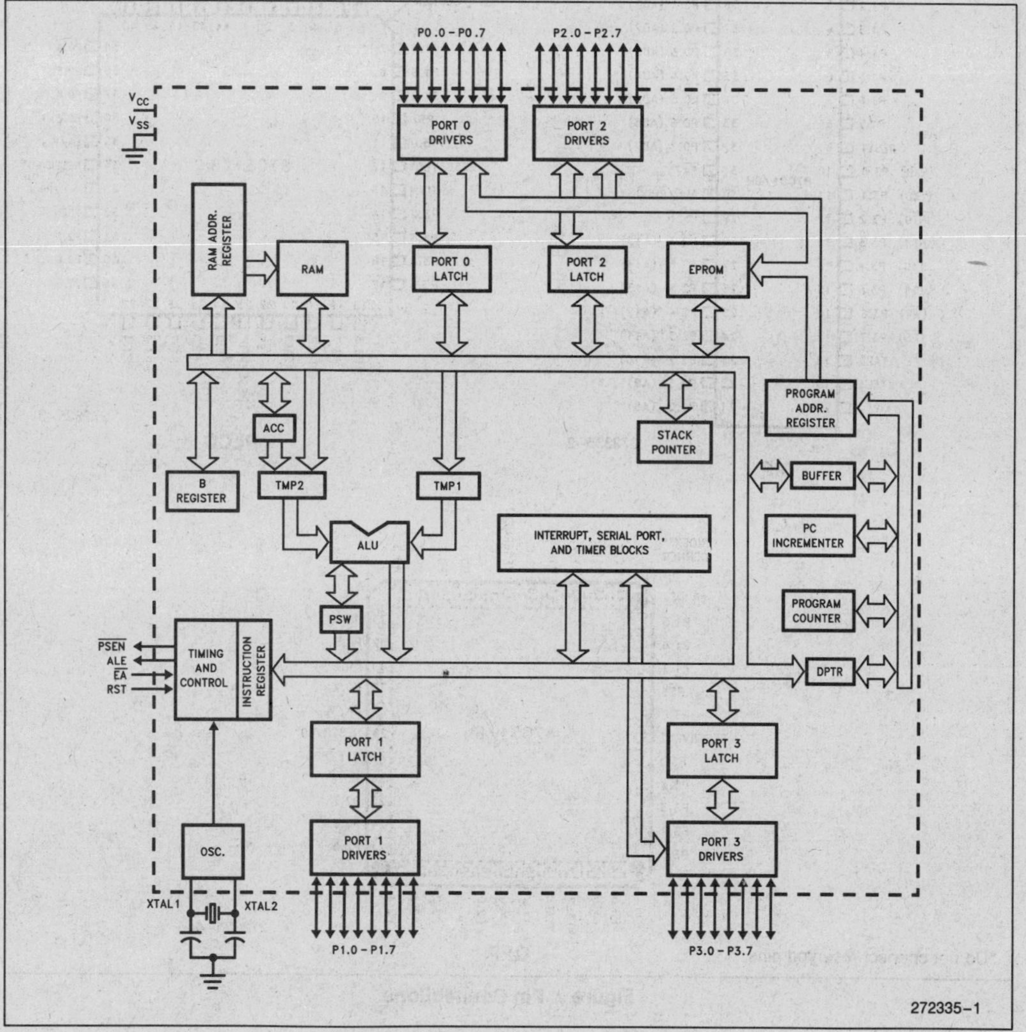

Figure 1. 87C51/BH Block Diagram

PROCESS INFORMATION

The 87C51 is manufactured on P629.0, a CHMOS III-E process. The 80C51BH/80C31BH are manufactured on P645, a CHMOS III process. Additional process and reliability information is available in Intel's *Components Quality and Reliability Handbook,* Order No. 210997.

PACKAGES

Part	Prefix	Package Type
87C51/BH	P	40-Pin Plastic DIP (OTP)
	D	40-Pin CERDIP (EPROM)
	N	44-Pin PLCC (OTP)
	S	44-Pin QFP (OTP)
80C51BHP	P	40-Pin Plastic DIP
	N	44-Pin PLCC

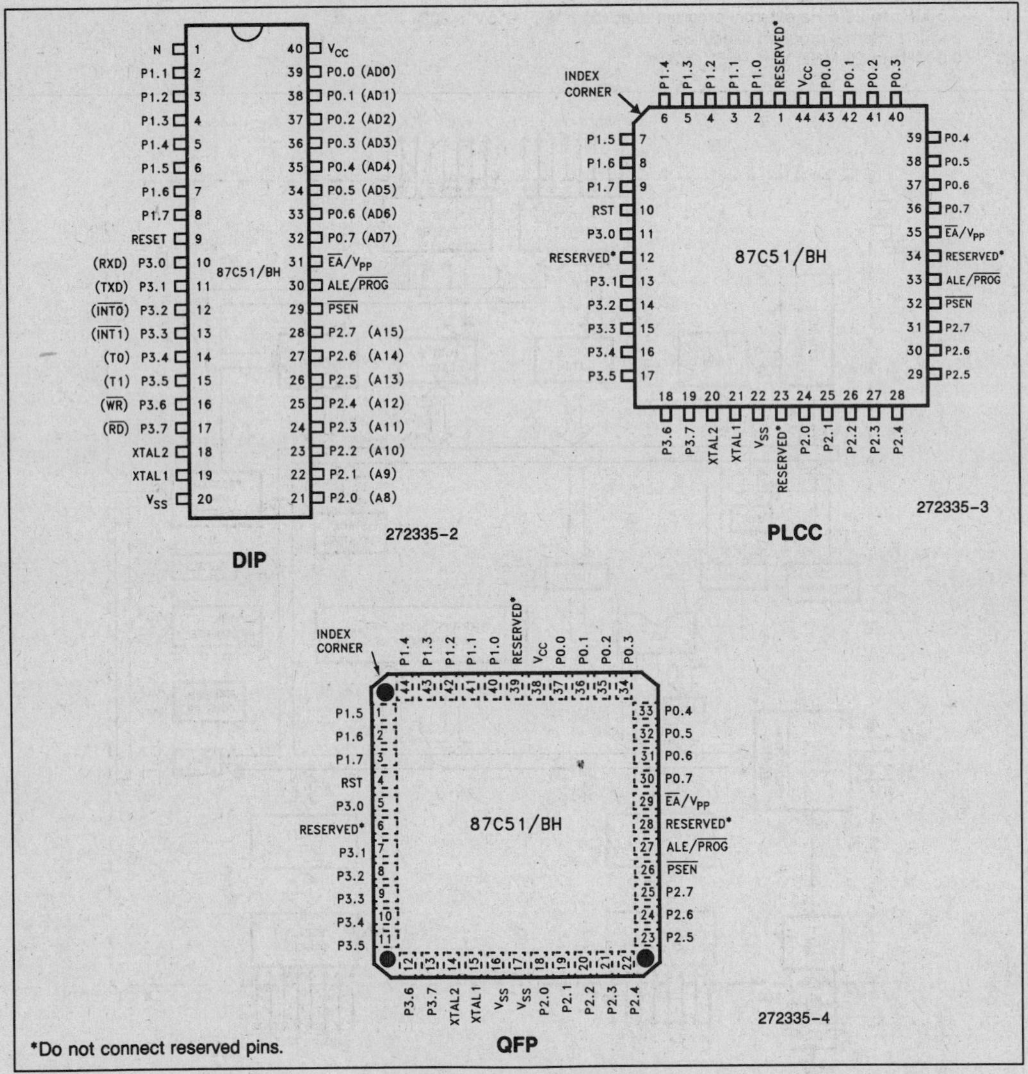

Figure 2. Pin Connections

PIN DESCRIPTION

V_{CC}: Supply voltage during normal, Idle and Power Down operations.

V_{SS}: Circuit ground.

Port 0: Port 0 is an 8-bit open drain bidirectional I/O port. As an output port each pin can sink 8 LS TTL inputs. Port 0 pins that have 1's written to them float, and in that state can be used as high-impedance inputs.

Port 0 is also the multiplexed low-order address and data bus during accesses to external memory. In this application it uses strong internal pullups when emitting 1's.

Port 0 also receives the code bytes during EPROM programming, and outputs the code bytes during program verification. External pullups are required during program verification.

Port 1: Port 1 is an 8-bit bidirectional I/O port with internal pullups. Port 1 pins that have 1's written to them are pulled high by the internal pullups, and in that state can be used as inputs. As inputs, Port 1 pins that are externally pulled low will source current (I_{IL}, on the data sheet) because of the internal pullups.

Port 1 also receives the low-order address bytes during EPROM programming and program verification.

Port 2: Port 2 is an 8-bit bidirectional I/O port with internal pullups. Port 2 pins that have 1's written to them are pulled high by the internal pullups, and in that state can be used as inputs. As inputs, Port 2 pins that are externally pulled low will source current (I_{IL}, on the data sheet) because of the internal pullups.

Port 2 emits the high-order address byte during fetches from external Program memory and during accesses to external Data Memory that use 16-bit address (MOVX @DPTR). In this application it uses strong internal pullups when emitting 1's.

During accesses to external Data Memory that use 8-bit addresses (MOVX @Ri), Port 2 emits the contents of the P2 Special Function Register.

Port 2 also receives some control signals and the high-order address bits during EPROM programming and program verification.

In the 80C51BHP, bits 2.4–2.7 are forced to 0, effectively limiting external data and code space to 4K each during external accesses (see Design Considerations).

Port 3: Port 3 is an 8-bit bidirectional I/O port with internal pullups. Port 3 pins that have 1's written to them are pulled high by the internal pullups, and in that state can be used as inputs. As inputs, Port 3 pins that are externally pulled low will source current (I_{IL}, on the data sheet) because of the pullups.

Port 3 also serves the functions of various special features of the MCS-51 Family, as listed below:

Pin	Name	Alternate Function
P3.0	RXD	Serial input line
P3.1	TXD	Serial output line
P3.2	$\overline{INT0}$	External Interrupt 0
P3.3	$\overline{INT1}$	External Interrupt 1
P3.4	T0	Timer 0 external input
P3.5	T1	Timer 1 external input
P3.6	$\overline{WR}$	External Data Memory Write strobe
P3.7	$\overline{RD}$	External Data Memory Read strobe

Port 3 also receives some control signals for EPROM programming and program verification.

RST: Reset input. A high on this pin for two machine cycles while the oscillator is running resets the device. The port pins will be driven to their reset condition when a minimum V_{IH1} voltage is applied whether the oscillator is running or not (87C51 only). An internal pulldown resistor permits a power-on reset with only a capacitor connected to V_{CC}.

In order to reset the 80C51BH/80C31BH port pins, the oscillator must be running. At least 19 oscillator periods must occur after a logic 1 is applied to the RST input before the port pins are driven to their reset state.

ALE/$\overline{PROG}$: Address Latch Enable output signal for latching the low byte of the address during accesses to external memory. This pin is also the program pulse input ($\overline{PROG}$) during EPROM programming.

If desired, ALE operation can be disabled by setting bit 0 of SFR location 8EH. With this bit set, the pin is weakly pulled high. However, the ALE disable feature will be suspended during a MOVX or MOVC instruction, idle mode, power down mode and ICE mode. The ALE disable feature will be terminated by reset. When the ALE disable feature is suspended or terminated, the ALE pin will no longer be pulled up weakly. Setting the ALE-disable bit has no effect if the microcontroller is in external execution mode.

In normal operation ALE is emitted at a constant rate of 1/6 the oscillator frequency, and may be used for external timing or clocking purposes. Note, however, that one ALE pulse is skipped during each access to external Data Memory.

$\overline{\text{PSEN}}$: Program Store Enable is the Read strobe to External Program Memory. When the 87C51/BH is executing from Internal Program Memory, $\overline{\text{PSEN}}$ is inactive (high). When the device is executing code from External Program Memory, $\overline{\text{PSEN}}$ is activated twice each machine cycle, except that two $\overline{\text{PSEN}}$ activations are skipped during each access to External Data Memory.

$\overline{\text{EA}}/V_{PP}$: External Access enable. $\overline{\text{EA}}$ must be strapped to V_{SS} in order to enable the 87C51/BH to fetch code from External Program Memory locations starting at 0000H up to FFFFH. Note, however, that if either of the Lock Bits is programmed, the logic level at $\overline{\text{EA}}$ is internally latched during reset.

$\overline{\text{EA}}$ must be strapped to V_{CC} for internal program execution.

This pin also receives the 12.75V programming supply voltage (V_{PP}) during EPROM programming.

XTAL1: Input to the inverting oscillator amplifier.

XTAL2: Output from the inverting oscillator amplifier.

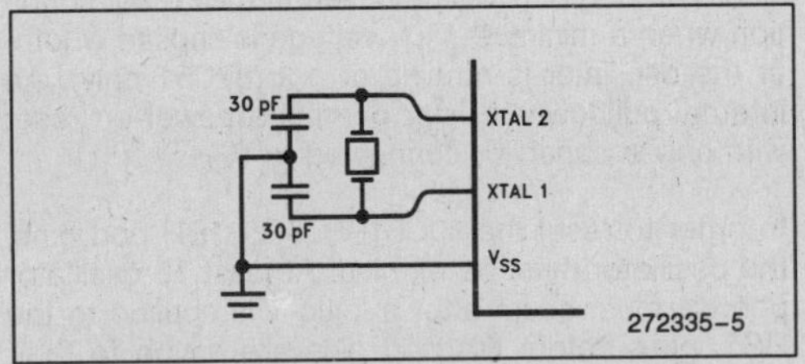

Figure 3. Using the On-Chip Oscillator

OSCILLATOR CHARACTERISTICS

XTAL1 and XTAL2 are the input and output, respectively, of an inverting amplifier which can be configured for use as an on-chip oscillator, as shown in Figure 3.

To drive the device from an external clock source, XTAL1 should be driven, while XTAL2 is left unconnected, as shown in Figure 4. There are no requirements on the duty cycle of the external clock signal, since the input to the internal clocking circuitry is through a divide-by-two flip-flop, but minimum and maximum high and low times specified on the data sheet must be observed.

An external oscillator may encounter as much as a 100 pF load at XTAL1 when it starts up. This is due to interaction between the amplifier and its feedback capacitance. Once the external signal meets the V_{IL} and V_{IH} specifications the capacitance will not exceed 20 pF.

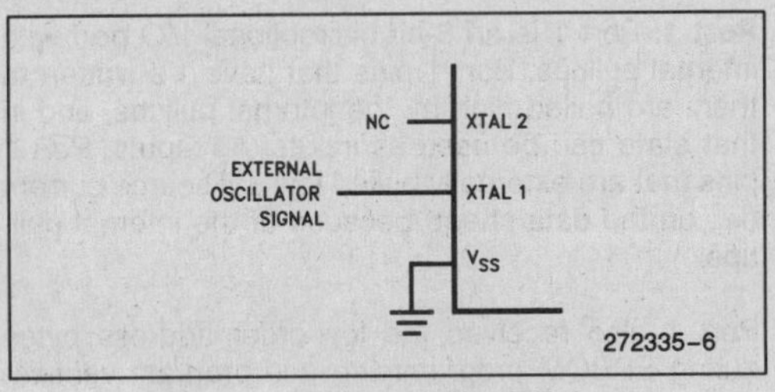

Figure 4. External Clock Drive

IDLE MODE

In Idle Mode, the CPU puts itself to sleep while all the on-chip peripherals remain active. The mode is invoked by software. The content of the on-chip RAM and all the Special Functions Registers remain unchanged during this mode. The Idle Mode can be terminated by any enabled interrupt or by a hardware reset.

It should be noted that when Idle is terminated by a hardware reset, the device normally resumes program execution, from where it left off, up to two machine cycles before the internal reset algorithm takes control. On-chip hardware inhibits access to internal RAM in this event, but access to the port pins is not inhibited. To eliminate the possibility of an unexpected write to a port pin when Idle is terminated by reset, the instruction following the one that invokes Idle should not be one that writes to a port pin or to external memory.

POWER DOWN MODE

In the Power Down mode the oscillator is stopped, and the instruction that invokes Power Down is the last instruction executed. The on-chip RAM and Special Function Registers retain their values until the Power Down mode is terminated.

The only exit from Power Down is a hardware reset. Reset redefines the SFRs but does not change the on-chip RAM. The reset should not be activated before V_{CC} is restored to its normal operating level and must be held active long enough to allow the oscillator to restart and stabilize.

DESIGN CONSIDERATIONS

- Exposure to light when the device is in operation may cause logic errors. For this reason, it is suggested that an opaque label be placed over the window when the die is exposed to ambient light.
- When running out of internal program/data memory, the 87C51-3 can be operated using a 24 MHz clock. If the 87C51-3 is running out of external program/data memory, the operating frequency must be between 3.5 MHz to 20 MHz. The 87C51-3 will not function properly at 24 MHz when running out of external program/data memory.
- The 80C51BHP cannot access external Program or Data memory above 4K. This means that the following instructions that use the Data Pointer only read/write data at address locations below 0FFFH:

```
MOVX A,@DPTR
MOVX @DPTR,A
```

  When the Data Pointer contains an address above the 4K limit, those locations will not be accessed. To access Data Memory above 4K, the `MOVX @Ri,A` or `MOVX A,@Ri` instructions must be used.
- The 87C51 has some additional features that are not available on the 80C51BH/80C31BH. The features are: asynchronous port reset, 4 interrupt priority levels, power off flag, ALE disable, serial port automatic address recognition, serial port framing error detection, 64-byte encryption array, and 3 program lock bits. These features cannot be used with the 80C51BH/80C31BH.

Table 2. Status of the External Pins during Idle and Power Down

Mode	Program Memory	ALE	$\overline{PSEN}$	PORT0	PORT1	PORT2	PORT3
Idle	Internal	1	1	Data	Data	Data	Data
Idle	External	1	1	Float	Data	Address	Data
Power Down	Internal	0	0	Data	Data	Data	Data
Power Down	External	0	0	Float	Data	Data	Data

NOTE:
For more detailed information on the reduced power modes refer to current Embedded Microcontrollers and Processors Handbook Volume I (Order No. 270645), and Application Note AP-252 (Embedded Applications Handbook, Order No. 270648), "Designing with the 80C51BH."

ONCE MODE

The ONCE ("On-Circuit Emulation") mode facilitates testing and debugging of systems using the 87C51/BH without the 87C51/BH having to be removed from the circuit. The ONCE mode is invoked by:

1. Pull ALE low while the device is in reset and $\overline{PSEN}$ is high;
2. Hold ALE low as RST is deactivated.

While the device is in ONCE mode, the Port 0 pins float, and the other port pins and ALE and $\overline{PSEN}$ are weakly pulled high. The oscillator circuit remains active. While the 87C51BH is in this mode, an emulator or test CPU can be used to drive the circuit. Normal operation is restored when a normal reset is applied.

87C51/BH EXPRESS

The Intel EXPRESS system offers enhancements to the operational specifications of the MCS-51 family of microcontrollers. These EXPRESS products are designed to meet the needs of those applications whose operating requirements exceed commercial temperature.

The EXPRESS program includes the commercial standard temperature range with burn-in and an extended temperature range with or without burn-in.

With the commercial standard temperature range, operational characteristics are guaranteed over the temperature range of 0°C to 70°C. With the extended temperature range option, operational characteristics are guaranteed over the range of −40°C to +85°C.

The optional burn-in is dynamic for a minimum time of 160 hours at 125°C with V_{CC} = 6.9V ±0.25V, following guidelines in MIL-STD-883, Method 1015.

Package types and EXPRESS versions are identified by a one- or two-letter prefix to the part number. The prefixes are listed in Table 3.

For the extended temperature range option, this data sheet specifies the parameters which deviate from their commercial temperature range limits.

Table 3. Prefix Identification

Prefix	Package Type	Temperature Range	Burn-in
P	Plastic	Commercial	No
D	Cerdip	Commercial	No
N	PLCC	Commercial	No
S	QFP	Commercial	No
TP	Plastic	Extended	No
TD	Cerdip	Extended	No
TN	PLCC	Extended	No
TS	QFP	Extended	No
LP	Plastic	Extended	Yes
LD	Cerdip	Extended	Yes
LN	PLCC	Extended	Yes

NOTE:
Contact distributor or local sales office to match EXPRESS prefix to proper device.

Examples:

P87C51 indicates 87C51 in a plastic package and specified for commercial temperature range, without burn-in.

LD87C51 indicates 87C51 in a cerdip package and specified for extended temperature range with burn-in.

ABSOLUTE MAXIMUM RATINGS*

Ambient Temperature Under Bias . −40°C to +85°C
Storage Temperature .... −65°C to +150°C
Voltage on $\overline{EA}/V_{PP}$ Pin to V_{SS} .... 0V to +13.0V
Voltage on Any Other Pin to V_{SS} .. −0.5V to +6.5V
Maximum I_{OL} per I/O Pin .... 15 mA
Power Dissipation............................ 1.5W
(Based on package heat transfer limitations, not device power consumption.)

NOTICE: This data sheet contains preliminary information on new products in production. It is valid for the devices indicated in the revision history. The specifications are subject to change without notice.

**WARNING: Stressing the device beyond the "Absolute Maximum Ratings" may cause permanent damage. These are stress ratings only. Operation beyond the "Operating Conditions" is not recommended and extended exposure beyond the "Operating Conditions" may affect device reliability.*

OPERATING CONDITIONS

Symbol	Description	Min	Max	Unit
T_A	Ambient Temperature Under Bias Commercial Express	 0 −40	 +70 +85	 °C °C
V_{CC}	Supply Voltage	4.5	5.5	V
f_{OSC}	Oscillator Frequency 87C51/BH 87C51-1/BH-1 87C51-2/BH-2 87C51-20	 3.5 3.5 0.5 3.5	 12 16 12 20	MHz

DC CHARACTERISTICS (Over Operating Conditions)

All parameter values apply to all devices unless otherwise indicated.

Symbol	Parameter	Min	Typ(1)	Max	Unit	Test Conditions
V_{IL}	Input Low Voltage Commercial Express	 −0.5 −0.5		 $0.2\ V_{CC} - 0.1$ $0.2\ V_{CC} - 0.15$	 V V	
V_{IL1}	Input Low Voltage $\overline{EA}$ Commercial Express	 0 −0.5		 $0.2\ V_{CC} - 0.3$ $0.2\ V_{CC} - 0.35$	 V V	
V_{IH}	Input High Voltage (Except XTAL1, RST) Commercial Express	 $0.2\ V_{CC} + 0.9$ $0.2\ V_{CC} + 1$		 $V_{CC} + 0.5$ $V_{CC} + 0.5$	 V V	
V_{IH1}	Input High Voltage (XTAL1, RST) Commercial Express	 $0.7\ V_{CC}$ $0.7\ V_{CC} + 0.1$		 $V_{CC} + 0.5$ $V_{CC} + 0.5$	 V V	
V_{OL}	Output Low Voltage(6) (Ports 1, 2, 3)			0.3	V	$I_{OL} = 100\ \mu A$(2)
				0.45	V	$I_{OL} = 1.6$ mA(2)
				1.0	V	$I_{OL} = 3.5$ mA(2)
V_{OL1}	Output Low Voltage(6) (Port 0, ALE, $\overline{PSEN}$)			0.3	V	$I_{OL} = 200\ \mu A$(2)
				0.45	V	$I_{OL} = 3.2$ mA(2)
				1.0	V	$I_{OL} = 7.0$ mA(2)

DC CHARACTERISTICS (Over Operating Conditions) (Continued)

Symbol	Parameter	Min	Typ(1)	Max	Unit	Test Conditions
V_{OH}	Output High Voltage (Ports 1, 2, 3, ALE, $\overline{PSEN}$) 87C51	$V_{CC} - 0.3$			V	$I_{OH} = -10\ \mu A^{(3)}$
		$V_{CC} - 0.7$			V	$I_{OH} = -30\ \mu A^{(3)}$
		$V_{CC} - 1.5$			V	$I_{OH} = -60\ \mu A^{(3)}$
V_{OH}	Output High Voltage (Ports 1, 2, 3, ALE, $\overline{PSEN}$) 80C51BH/31BH	$0.9\ V_{CC}$			V	$I_{OH} = -10\ \mu A^{(3)}$ $V_{CC} = 5V \pm 10\%$
		$0.75\ V_{CC}$			V	$I_{OH} = -25\ \mu A$
		2.4			V	$I_{OH} = -60\ \mu A^{(3)}$
V_{OH1}	Output High Voltage (Port 0 in External Bus Mode) 87C51	$V_{CC} - 0.3$			V	$I_{OH} = -200\ \mu A^{(3)}$
		$V_{CC} - 0.7$			V	$I_{OH} = -3.2\ mA^{(3)}$
		$V_{CC} - 1.5$			V	$I_{OH} = -7.0\ mA^{(3)}$
V_{OH1}	Output High Voltage (Port 0 in External Bus Mode) 80C51BH/31BH	$0.9\ V_{CC}$			V	$I_{OH} = -80\ \mu A$ $V_{CC} = 5V \pm 10\%$
		$0.75\ V_{CC}$			V	$I_{OH} = -300\ \mu A$
		2.4			V	$I_{OH} = -800\ \mu A$
I_{IL}	Logical 0 Input Current (Ports 1, 2, 3)					$V_{IN} = 2V$ (87C51) $V_{IN} = 0.45V$
	Commercial			−50	μA	
	Express			−75	μA	
I_{LI}	Input Leakage Current (Port 0)			±10	μA	$0.45 < V_{IN} < V_{CC}$
I_{TL}	Logical 1-to-0 Transition Current (Ports 1, 2, 3)					$V_{IN} = 2V$
	Commercial			−650	μA	
	Express			−750	μA	
RRST	RST Pulldown Resistor					
	80C51BH/31BH	50		300	KΩ	
	87C51	40		225	KΩ	
C_{IO}	Pin Capacitance		10		pF	@ 1 MHz, 25°C
I_{CC}	Power Supply Current					(Note 4)
	Active Mode					
	@ 12 MHz (Figure 5)		11.5	20	mA	
	@ 16 MHz			26	mA	
	@ 20 MHz			32	mA	
	@ 24 MHz			38	mA	
	Idle Mode					
	@ 12 MHz (Figure 5)		1.7	5	mA	
	@ 16 MHz			6	mA	
	@ 20 MHz			7.6	mA	
	@ 24 MHz			9	mA	
	87C51Express					
	@ 12 MHz			7	mA	
	@ 16 MHz			8	mA	
	@ 20 MHz			9	mA	
	@ 24 MHz			10	mA	
	Power Down Mode		5	50	μA	

NOTES:

1. "Typicals" are based on a limited number of samples taken from early manufacturing lots and are not guaranteed. The values listed are at room temp, 5V.
2. Capacitive loading on Ports 0 and 2 may cause noise pulses above 0.4V to be superimposed on the V_{OL}s of ALE and Ports 1, 2 and 3. The noise is due to external bus capacitance discharging into the Port 0 and Port 2 pins when these pins change from 1 to 0. In applications where capacitive loading exceeds 100 pF, the noise pulses on these signals may exceed 0.8V. It may be desirable to qualify ALE or other signals with a Schmitt Trigger, or CMOS-level input logic.
3. Capacitive loading on Ports 0 and 2 may cause the V_{OH} on ALE and $\overline{PSEN}$ to momentarily fall below the $0.9V_{CC}$ specification when the address bits are stabilizing.
4. See Figures 6 through 8 for I_{CC} test conditions. Minimum V_{CC} for Power Down is 2V.
5. Under steady state (non-transient) conditions, I_{OL} must be externally limited as follows:

Maximum I_{OL} per port pin:	10 mA
Maximum I_{OL} per 8-bit port—	
Port 0:	26 mA
Ports 1, 2, and 3:	15 mA
Maximum total I_{OL} for all output pins:	71 mA

If I_{OL} exceeds the test condition, V_{OL} may exceed the related specification. Pins are not guaranteed to sink greater than the listed test conditions.

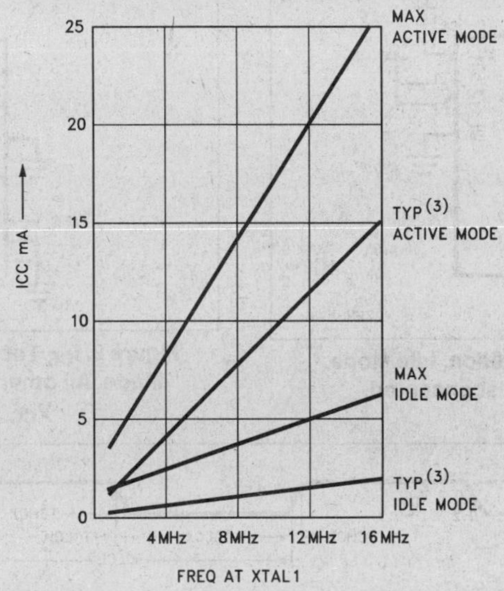

272335-26

I_{CC}MAX at other frequencies is given by:

Active Mode: $I_{CC}MAX = 1.47 \times FREQ + 2.35$

Idle Mode: $I_{CC}MAX = 0.33 \times FREQ + 1.05$

where FREQ is the external oscillator frequency in MHz. I_{CC}MAX is given in mA.

Figure 5. I_{CC} vs FREQ. Valid only within Frequency Specifications of the Device under Test

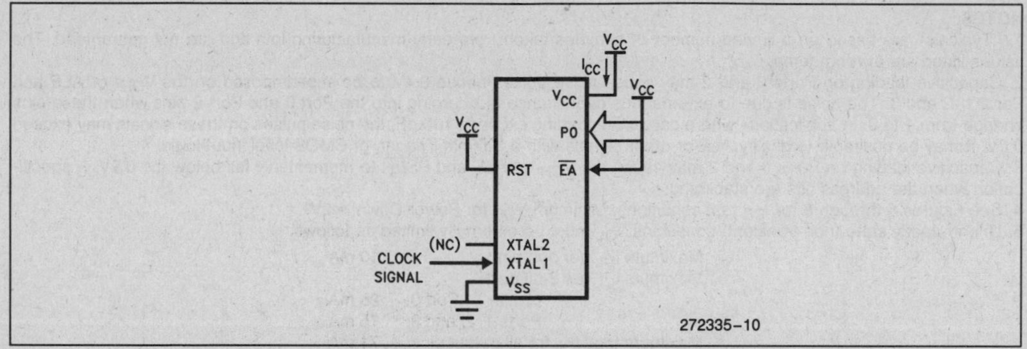

Figure 6. I_{CC} Test Condition, Active Mode. All other pins are disconnected.

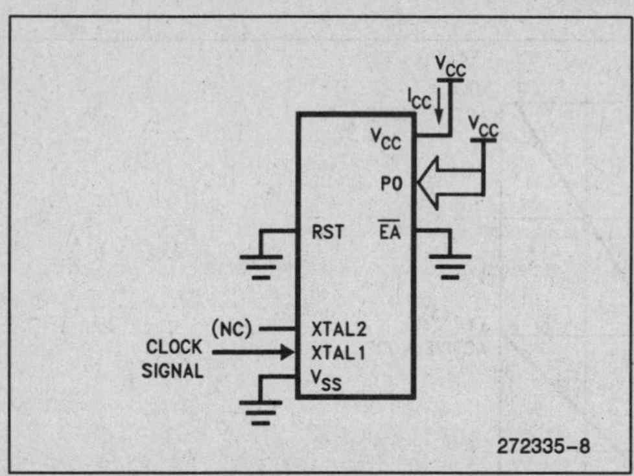

Figure 7. I_{CC} Test Condition, Idle Mode. All other pins are disconnected.

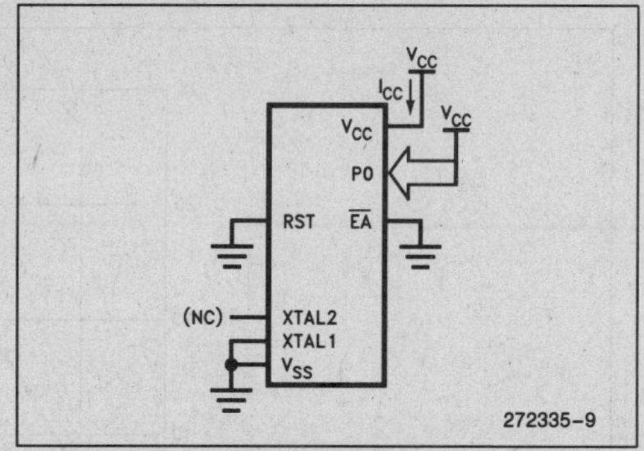

Figure 9. I_{CC} Test Condition, Power Down Mode. All other pins are disconnected. V_{CC} = 2V to 5.5V.

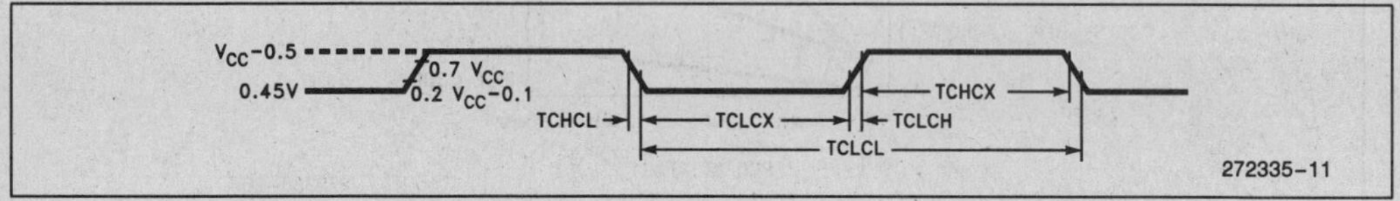

Figure 8. Clock Signal Waveform for I_{CC} Tests in Active and Idle Modes TCLCH = TCHCL = 5 ns

EXPLANATION OF THE AC SYMBOLS

Each timing symbol has 5 characters. The first character is always a 'T' (stands for time). The other characters, depending on their positions, stand for the name of a signal or the logical status of that signal. The following is a list of all the characters and what they stand for.

A:Address.
C:Clock.
D:Input data.
H:Logic level HIGH.
I:Instruction (program memory contents).
L:Logic level LOW, or ALE.
P:$\overline{PSEN}$.
Q:Output data.
R:$\overline{RD}$ signal.
T:Time.
V:Valid.
W:$\overline{WR}$ signal.
X:No longer a valid logic level.
Z:Float.

For example,

TAVLL = Time from Address Valid to ALE Low.
TLLPL = Time from ALE Low to PSEN Low.

2

AC CHARACTERISTICS: (Over Operating Conditions; Load Capacitance for Port 0, ALE, and $\overline{PSEN}$ = 100 pF; Load Capacitance for All Other Outputs = 80 pF)

EXTERNAL MEMORY CHARACTERISTICS

Symbol	Parameter	12 MHz Oscillator		20 MHz Oscillator		Variable Oscillator		Units
		Min	Max	Min	Max	Min	Max	
1/TCLCL	Oscillator Frequency 87C51/BH 87C51-1/BH-1 87C51-2/BH-2 87C51-20					 3.5 3.5 0.5 3.5	 12 16 12 20	MHz
TLHLL	ALE Pulse Width 87C51/BH/-20	127		60		2TCLCL−40		ns
TAVLL	Address Valid to ALE Low 87C51/-20 80C51BH/C31BH	 43 28		 10		 TCLCL−40 TCLCL−55		 ns ns
TLLAX	Address Hold After ALE Low 87C51/-20 80C51BH/C31BH	 53 48		 20		 TCLCL−30 TCLCL−35		 ns ns
TLLIV	ALE Low to Valid Instr In 87C51/BH 87C51-20		 234		 125		 4TCLCL−100 4TCLCL−75	 ns ns
TLLPL	ALE Low to $\overline{PSEN}$ Low 87C51/-20 80C51BH/C31BH	 53 43		 20		 TCLCL−30 TCLCL−40		 ns ns
TPLPH	$\overline{PSEN}$ Pulse Width 87C51/BH/-20	205		105		3TCLCL−45		ns
TPLIV	$\overline{PSEN}$ Low to Valid Instr In 87C51/BH 87C51-20		 145		 60		 3TCLCL−105 3TCLCL−90	 ns ns

EXTERNAL MEMORY CHARACTERISTICS (Continued)

Symbol	Parameter	12 MHz Oscillator		20 MHz Oscillator		Variable Oscillator		Units
		Min	Max	Min	Max	Min	Max	
TPXIX	Input Instr Hold After $\overline{PSEN}$							
	87C51/BH/-20	0		0		0		ns
TPXIZ	Input Instr Float After $\overline{PSEN}$							
	87C51/BH		59				TCLCL − 25	ns
	87C51-20				30		TCLCL − 20	ns
TAVIV	Address to Valid Instr In							
	87C51/BH/-20		312		145		5TCLCL − 105	ns
TPLAZ	$\overline{PSEN}$ Low to Address Float							
	87C51/BH/-20		10		10		10	ns
TRLRH	$\overline{RD}$ Pulse Width							
	87C51/BH/-20	400		200		6TCLCL − 100		ns
TWLWH	$\overline{WR}$ Pulse Width							
	87C51/BH/-20	400		200		6TCLCL − 100		ns
TRLDV	$\overline{RD}$ Low to Valid Data In							
	87C51/BH		252				5TCLCL − 165	ns
	87C51-20				155		5TCLCL − 95	ns
TRHDX	Data Hold After $\overline{RD}$							
	87C51/BH/-20	0		0		0		ns
TRHDZ	Data Float After $\overline{RD}$							
	87C51/-20		107		40		2TCLCL − 60	ns
	80C51BH/C31BH		97				2TCLCL − 70	ns
TLLDV	ALE Low to Valid Data In							
	87C51/BH		517				8TCLCL − 150	ns
	87C51-20				310		8TCLCL − 90	ns
TAVDV	Address to Valid Data In							
	87C51/BH		585				9TCLCL − 165	ns
	87C51-20				360		9TCLCL − 90	ns
TLLWL	ALE Low to $\overline{RD}$ or $\overline{WR}$ Low							
	87C51/BH/-20	200	300	100	200	3TCLCL − 50	3TCLCL + 50	ns
TAVWL	Address to $\overline{RD}$ or $\overline{WR}$ Low							
	87C51/BH	203				4TCLCL − 130		ns
	87C51-20			110		4TCLCL − 90		ns
TQVWX	Data Valid to $\overline{WR}$ Transition							
	87C51	33				TCLCL − 50		ns
	80C51BH/C31BH	23				TCLCL − 60		ns
	87C51-20			15		TCLCL − 35		ns

EXTERNAL MEMORY CHARACTERISTICS (Continued)

Symbol	Parameter	12 MHz Oscillator		20 MHz Oscillator		Variable Oscillator		Units
		Min	Max	Min	Max	Min	Max	
TWHQX	Data Hold After $\overline{WR}$							
	87C51/BH	33				TCLCL − 50		ns
	87C51-20			10		TCLCL − 40		ns
TQVWH	Data Valid to $\overline{WR}$ High							
	87C51/BH	433				7TCLCL − 150		ns
	87C51-20			280		7TCLCL − 70		ns
TRLAZ	$\overline{RD}$ Low to Address Float							
	87C51/BH/-20		0		0		0	ns
TWHLH	$\overline{RD}$ or $\overline{WR}$ High to ALE High							
	87C51/BH/-20	43	123	10	90	TCLCL − 40	TCLCL + 40	ns

2

EXTERNAL PROGRAM MEMORY READ CYCLE

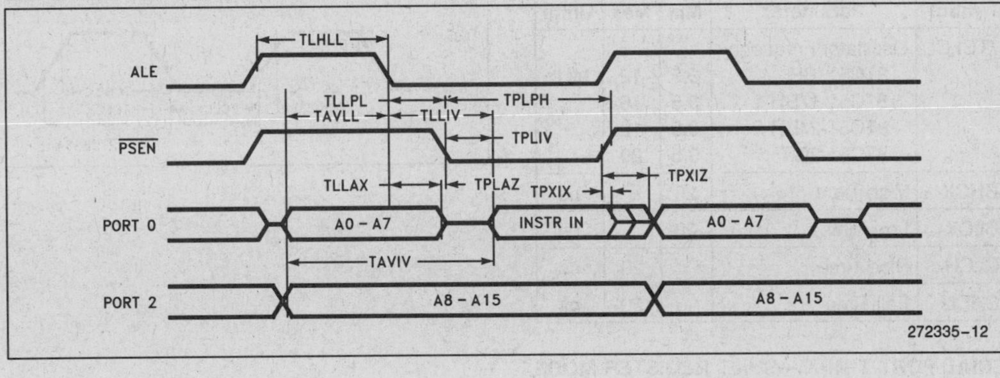

EXTERNAL DATA MEMORY READ CYCLE

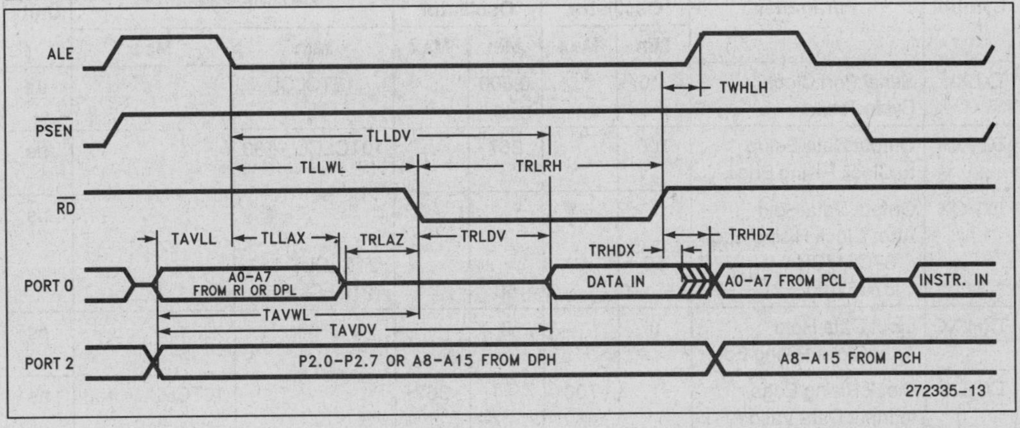

EXTERNAL DATA MEMORY WRITE CYCLE

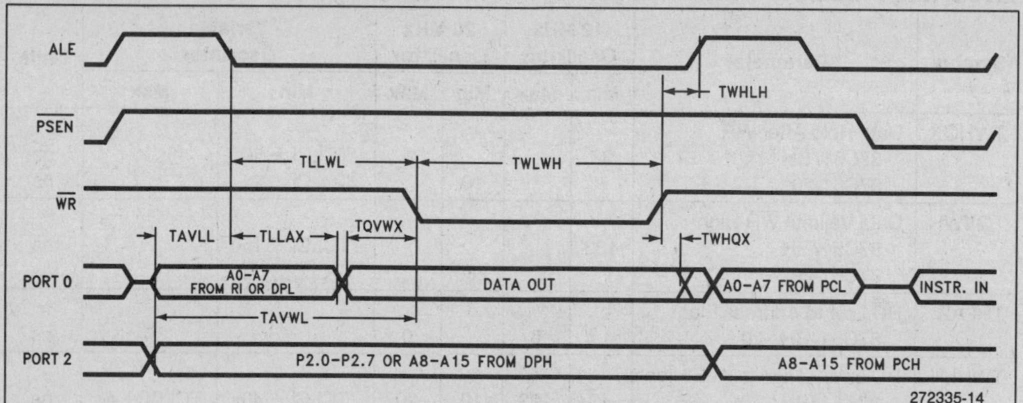

272335-14

EXTERNAL CLOCK DRIVE

Symbol	Parameter	Min	Max	Units
1/TCLCL	Oscillator Frequency			
	87C51/BH	3.5	12	MHz
	87C51-1/BH-1	3.5	16	
	87C51-2/BH-2	0.5	12	
	87C51-20	3.5	20	
TCHCX	High Time	20		ns
TCLCX	Low Time	20		ns
TCLCH	Rise Time		20	ns
TCHCL	Fall Time		20	ns

EXTERNAL CLOCK DRIVE WAVEFORM

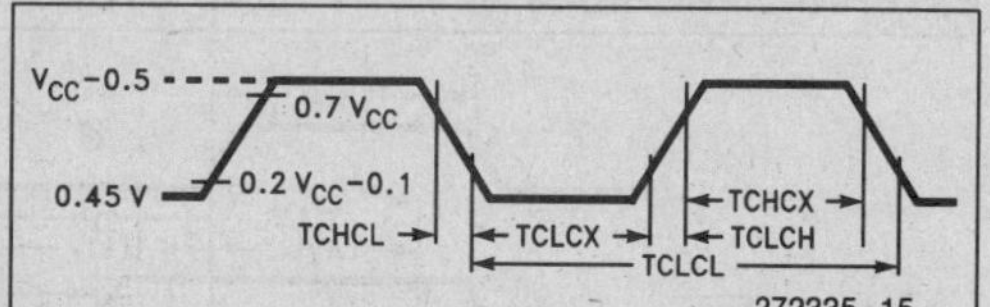

272335-15

SERIAL PORT TIMING—SHIFT REGISTER MODE

Symbol	Parameter	12 MHz Oscillator		20 MHz Oscillator		Variable Oscillator		Units
		Min	Max	Min	Max	Min	Max	
TXLXL	Serial Port Clock Cycle Time	1.0		0.600		12TCLCL		μs
TQVXH	Output Data Setup to Clock Rising Edge	700		367		10TCLCL − 133		ns
TXHQX	Output Data Hold After Clock Rising Edge							ns
	87C51/BH	50				2TCLCL − 117		
	87C51-20			50		2TCLCL − 50		
TXHDX	Input Data Hold After Clock Rising Edge	0		0		0		ns
TXHDV	Clock Rising Edge to Input Data Valid		700		367		10TCLCL − 133	ns

SHIFT REGISTER MODE TIMING WAVEFORMS

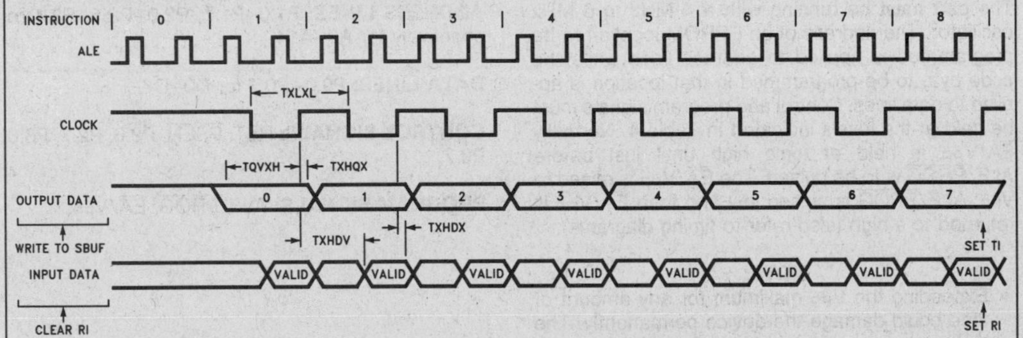

272335–18

AC TESTING INPUT, OUTPUT WAVEFORMS

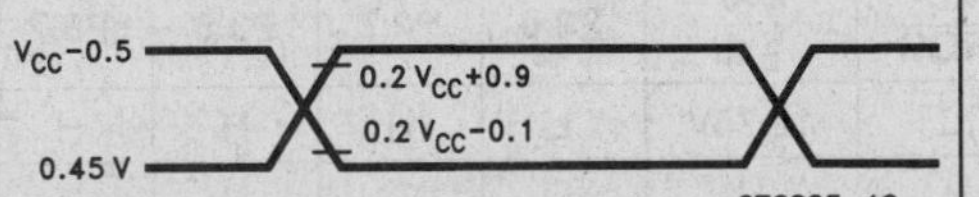

272335–19

AC inputs during testing are driven at $V_{CC} - 0.5$ for a Logic "1" and 0.45V for a Logic "0." Timing measurements are made at V_{IH} min for a Logic "1" and V_{IL} max for a Logic "0".

FLOAT WAVEFORMS

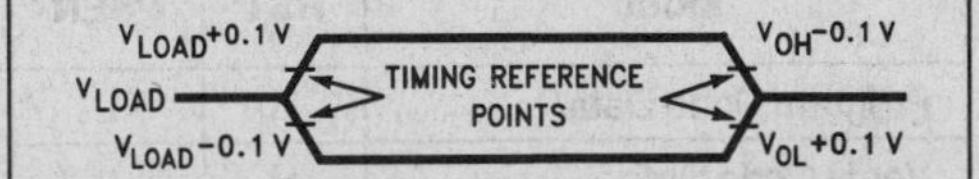

272335–20

For timing purposes a port pin is no longer floating when a 100 mV change from load voltage occurs, and begins to float when a 100 mV change from the loaded V_{OH}/V_{OL} level occurs. $I_{OL}/I_{OH} = \pm 20$ mA.

PROGRAMMING THE 87C51

The part must be running with a 4 MHz to 6 MHz oscillator. The address of an EPROM location to be programmed is applied to address lines while the code byte to be programmed in that location is applied to data lines. Control and program signals must be held at the levels indicated in Table 4. Normally $\overline{EA}/V_{PP}$ is held at logic high until just before ALE/$\overline{PROG}$ is to be pulsed. The $\overline{EA}/V_{PP}$ is raised to V_{PP}, ALE/$\overline{PROG}$ is pulsed low and then $\overline{EA}/V_{PP}$ is returned to a high (also refer to timing diagrams).

NOTE:

- Exceeding the V_{PP} maximum for any amount of time could damage the device permanently. The V_{PP} source must be well regulated and free of glitches.

DEFINITION OF TERMS

ADDRESS LINES: P1.0–P1.7, P2.0–P2.5, P3.4 respectively for A0–A14.

DATA LINES: P0.0–P0.7 for D0–D7.

CONTROL SIGNALS: RST, $\overline{PSEN}$, P2.6, P2.7, P3.6, P3.7.

PROGRAM SIGNALS: ALE/$\overline{PROG}$, $\overline{EA}/V_{PP}$.

Table 4. EPROM Programming Modes

Mode		RST	$\overline{PSEN}$	ALE/ $\overline{PROG}$	$\overline{EA}$/ V_{PP}	P2.6	P2.7	P3.6	P3.7
Program Code Data		H	L	⊔	12.75V	L	H	H	H
Verify Code Data		H	L	H	H	L	L	H	H
Program Encryption Array Address 0–3F		H	L	⊔	12.75V	L	H	L	H
Program Lock Bits	Bit 1	H	L	⊔	12.75V	H	H	H	H
	Bit 2	H	L	⊔	12.75V	H	H	L	L
	Bit 3	H	L	⊔	12.75V	H	L	H	L
Read Signature Byte		H	L	H	H	L	L	L	L

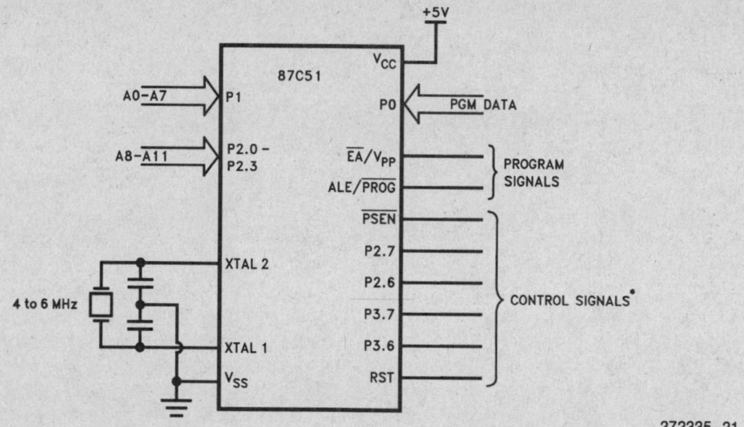

*See Table 4 for proper input on these pins

Figure 10. Programming the EPROM

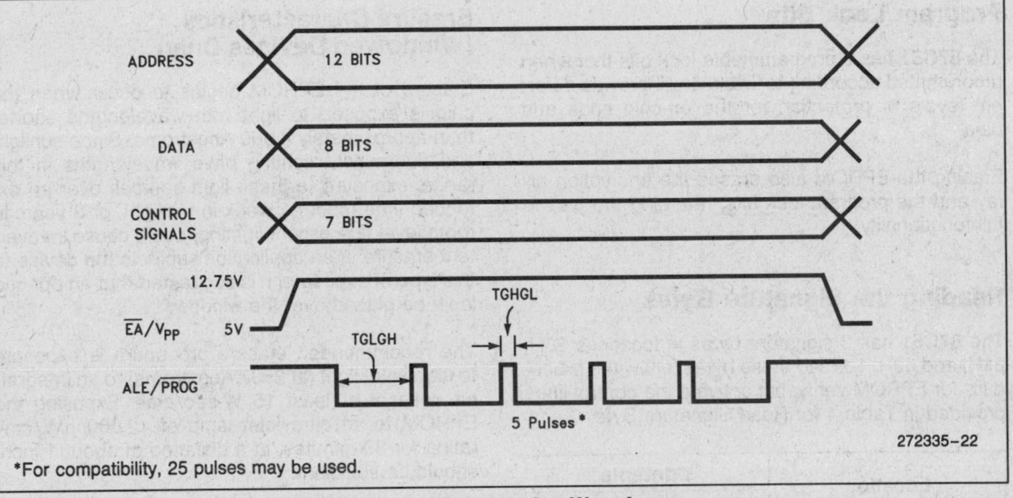

Figure 11. Programming Waveforms

PROGRAMMING ALGORITHM

Refer to Table 4 and Figures 10 and 11 for address, data, and control signals set up. To program the 87C51 the following sequence must be exercised.

1. Input the valid address on the address lines.
2. Input the appropriate data byte on the data lines.
3. Activate the correct combination of control signals.
4. Raise $\overline{EA}/V_{PP}$ from V_{CC} to 12.75V ±0.25V.
5. Pulse ALE/$\overline{PROG}$ 5 times* for the EPROM array, and 25 times for the encryption table and the lock bits.

Repeat 1 through 5 changing the address and data for the entire array or until the end of the object file is reached.

Program Verify

Verification may be done after programming either one byte or a block of bytes. In either case a complete verify of the array will ensure reliable programming of the 87C51.

The lock bits cannot be directly verified. Verification of the lock bits is done by observing that their features are enabled.

EPROM Lock System

The 87C51 program lock system, when programmed, protects the onboard program against software piracy.

The 87C51 has a 3-level program lock system and a 64-byte encryption array. Since this is an EPROM device, all locations are user-programmable. See Table 5.

Encryption Array

Within the EPROM array are 64 bytes of Encryption Array that are initially unprogrammed (all 1's). Every time that a byte is addressed during a verify, 6 address lines are used to select a byte of the Encryption Array. This byte is then exclusive-NOR'ed (XNOR) with the code byte, creating an Encryption Verify byte. The algorithm, with the array in the unprogrammed state (all 1's), will return the code in its original, unmodified form. For programming the Encryption Array, refer to Table 4 (Programming the EPROM).

When using the encryption array, one important factor needs to be considered. If a code byte has the value 0FFH, verifying the byte will produce the encryption byte value. If a large block (>64 bytes) of code is left unprogrammed, a verification routine will display the contents of the encryption array. For this reason all unused code bytes should be programmed with some value other than 0FFH, and not all of them the same value. This will ensure maximum program protection.

Program Lock Bits

The 87C51 has 3 programmable lock bits that when programmed according to Table 5 will provide different levels of protection for the on-chip code and data.

Erasing the EPROM also erases the encryption array and the program lock bits, returning the part to full functionality.

Reading the Signature Bytes

The 87C51 has 3 signature bytes in locations 30H, 31H, and 60H. To read these bytes follow the procedure for EPROM verify, but activate the control lines provided in Table 4 for Read Signature Byte.

Location	Contents
	87C51
30H	89H
31H	58H
60H	51H

Erasure Characteristics (Windowed Devices Only)

Erasure of the EPROM begins to occur when the chip is exposed to light with wavelengths shorter than approximately 4,000 Angstroms. Since sunlight and fluorescent lighting have wavelengths in this range, exposure to these light sources over an extended time (about 1 week in sunlight, or 3 years in room level fluorescent lighting) could cause inadvertent erasure. If an application subjects the device to this type of exposure, it is suggested that an opaque label be placed over the window.

The recommended erasure procedure is exposure to ultraviolet light (at 2537 Angstroms) to an integrated dose of at least 15 W-sec/cm^2. Exposing the EPROM to an ultraviolet lamp of 12,000 μW/cm^2 rating for 30 minutes, at a distance of about 1 inch, should be sufficient.

Erasure leaves the array in an all 1's state.

Table 5. Program Lock Bits and the Features

Program Lock Bits				Protection Type
	LB1	LB2	LB3	
1	U	U	U	No program lock features enabled. (Code verify will still be encrypted by the encryption array if programmed.)
2	P	U	U	MOVC instructions executed from external program memory are disabled from fetching code bytes from internal memory, EA is sampled and latched on reset, and further programming of the EPROM is disabled.
3	P	P	U	Same as 2, also verify is disabled.
4	P	P	P	Same as 3, also external execution is disabled.

EPROM PROGRAMMING AND VERIFICATION CHARACTERISTICS:

(T_A = 21°C to 27°C, V_{CC} = 5V ±10%, V_{SS} = 0V)

Symbol	Parameter	Min	Max	Units
V_{PP}	Programming Supply Voltage	12.5	13.0	V
I_{PP}	Programming Supply Current		75	mA
1/TCLCL	Oscillator Frequency	4	6	MHz
TAVGL	Address Setup to $\overline{\text{PROG}}$ Low	48TCLCL		
TGHAX	Address Hold After $\overline{\text{PROG}}$	48TCLCL		
TDVGL	Data Setup to $\overline{\text{PROG}}$ Low	48TCLCL		
TGHDX	Data Hold After $\overline{\text{PROG}}$	48TCLCL		
TEHSH	P2.7 ($\overline{\text{ENABLE}}$) High to V_{PP}	48TCLCL		
TSHGL	V_{PP} Setup to $\overline{\text{PROG}}$ Low	10		μs
TGHSL	V_{PP} Hold After $\overline{\text{PROG}}$	10		μs
TGLGH	$\overline{\text{PROG}}$ Width	90	110	μs
TAVQV	Address to Data Valid		48TCLCL	
TELQV	$\overline{\text{ENABLE}}$ Low to Data Valid		48TCLCL	
TEHQZ	Data Float After $\overline{\text{ENABLE}}$	0	48TCLCL	
TGHGL	$\overline{\text{PROG}}$ High to $\overline{\text{PROG}}$ Low	10		μs

2

EPROM PROGRAMMING AND VERIFICATION WAVEFORMS

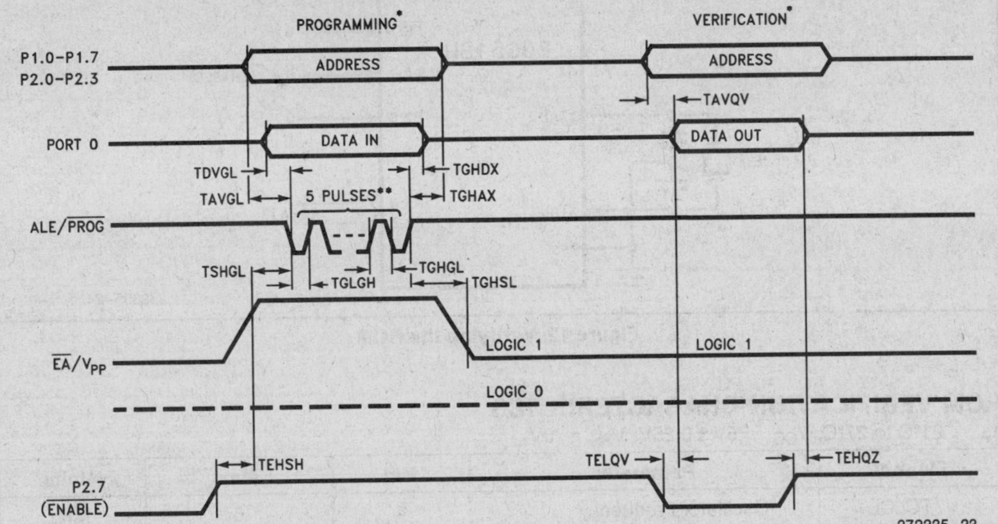

272335–23

*For programming conditions see Figure 10.
**5 pulses for the EPROM array, 25 pulses for the encryption table and lock bits.

ROM CHARACTERISTICS

Table 6 shows the logic levels for verifying the code data and reading the signature bytes on the 80C51BH/80C31BH.

Table 6. ROM Modes

Mode	RST	$\overline{PSEN}$	ALE	$\overline{EA}$	P2.7	P2.6
Verify Code Data	1	0	1	1	0	0

NOTES:
"1" = Valid high for that pin
"0" = Valid low for that pin

Program Verification

The address of the Program Memory location to be read is applied to Port 1 and pins P2.0–P2.3. The other pins should be held at the "Verify" levels indicated in Table 2. The contents of the addressed locations will come out on Port 0. External pullups are required on Port 0 for this operation.

Figure 12 shows the setup for verifying the program memory.

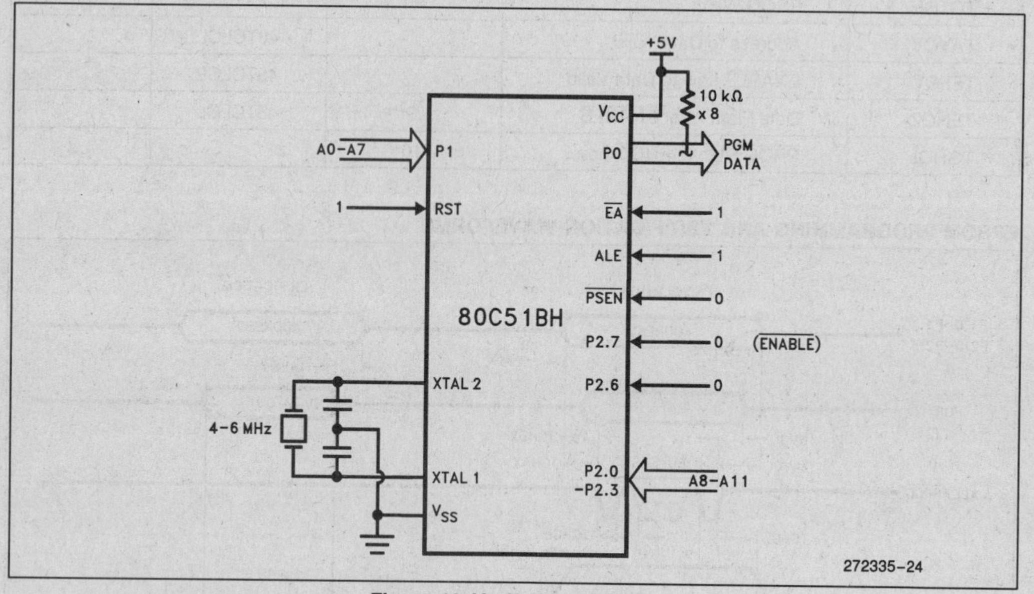

Figure 12. Verifying the ROM

ROM VERIFICATION CHARACTERISTICS

T_A = 21°C to 27°C; V_{CC} = 5V ± 0.25V; V_{SS} = 0V

Symbol	Parameter	Min	Max	Units
1/TCLCL	Oscillator Frequency	4	6	MHz
TAVQV	Address to Data Valid		48TCLCL	
TELQV	ENABLE Low to Data Valid		48TCLCL	
TEHQZ	Data Float after ENABLE	0	48TCLCL	

ROM VERIFICATION WAVEFORMS

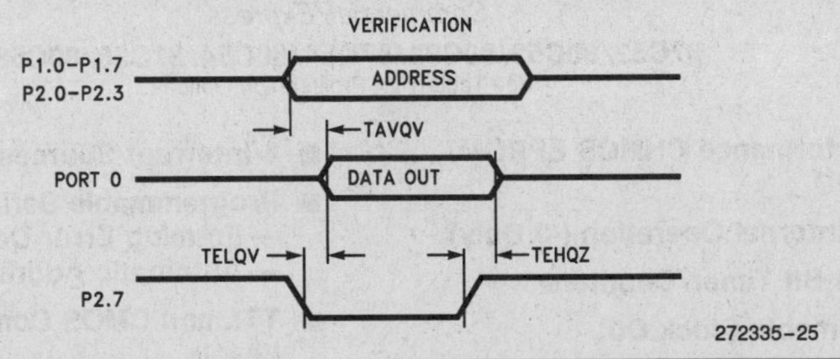

Thermal Impedance

All thermal impedance data is approximate for static air conditions at 1W of power dissipation. Values will change depending on operating conditions and applications. See the Intel Packaging Handbook (Order No. 240800) for a description of Intel's thermal impedance test methodology.

Package	θ_{JA}	θ_{JC}	Device
P	45°C/W	16°C/W	87C51
	75°C/W	23°C/W	BH
D	45°C/W	15°C/W	87C51
	36°C/W	13°C/W	BH
N	46°C/W	16°C/W	All
S	98°C/W	24°C/W	All

DATA SHEET REVISION HISTORY

Data sheets are changed as new device information becomes available. Verify with your local Intel sales office that you have the latest version before finalizing a design or ordering devices.

The following differences exist between this datasheet (272335-002) and the previous version (272335-001):

1. Removed −L, I_{OL} = ±10 mA from Float Waveforms figure.

2. Removed QP, QD and QN (commercial with extended burn-in) from Table 3. Prefix Identification.

This data sheet (272335-001) replaces the following:

80C51BH/80C31BH Express	270218-003
80C51BHP	270603-004
87C51/80C51BH/80C31BH	270147-008
87C51 Express	270430-002
87C51-20/-3	272082-002

8XC52/54/58
CHMOS SINGLE-CHIP 8-BIT MICROCONTROLLER

Commercial/Express

87C52/80C52/80C32/87C54/80C54/87C58/80C58

*See Table 1 for Proliferation Options

- **High Performance CHMOS EPROM/ROM/CPU**
- **24 MHz Internal Operation (-3 Only)**
- **Three 16-Bit Timer/Counters**
- **Programmable Clock Out**
- **Up/Down Timer/Counter**
- **Three Level Program Lock System**
- **8K/16K/32K On-Chip Program Memory**
- **256 Bytes of On-Chip Data RAM**
- **Improved Quick Pulse Programming Algorithm**
- **Boolean Processor**
- **32 Programmable I/O Lines**
- **6 Interrupt Sources**
- **Programmable Serial Channel with:**
  - **—Framing Error Detection**
  - **—Automatic Address Recognition**
- **TTL and CMOS Compatible Logic Levels**
- **64K External Program Memory Space**
- **64K External Data Memory Space**
- **MCS® 51 Compatible Instruction Set**
- **Power Saving Idle and Power Down Modes**
- **ONCE (On-Circuit Emulation) Mode**
- **Four-Level Interrupt Priority**
- **Extended Temperature Range (−40°C to +85°C)**

MEMORY ORGANIZATION

ROM Device	EPROM Version	ROMless Version	ROM/EPROM Bytes	RAM Bytes
80C52	87C52	80C32	8K	256
80C54	87C54	80C32	16K	256
80C58	87C58	80C32	32K	256

These devices can address up to 64 Kbytes of external program/data memory.

The Intel 8XC52/8XC54/8XC58 is a single-chip control-oriented microcontroller which is fabricated on Intel's reliable CHMOS III-E technology. Being a member of the MCS 51 family of controllers, the 8XC52/8XC54/8XC58 uses the same powerful instruction set, has the same architecture, and is pin-for-pin compatible with the existing MCS 51 family of products. The 8XC52/8XC54/8XC58 is an enhanced version of the 87C51/80C51BH/80C31BH. The added features make it an even more powerful microcontroller for applications that require clock output, and up/down counting capabilities such as motor control. It also has a more versatile serial channel that facilitates multi-processor communications.

The 8XC5X-3 has the same 3.5 MHz to 20 MHz frequency range as the 8XC5X-20 when operating out of external program/data memory. When running out of internal program/data memory, the 8XC5X-3 can operate up to 24 MHz.

Throughout this document 8XC5X will refer to the 8XC52, 80C32, 8XC54 and 8XC58 unless information applies to a specific device.

October 1994
Order Number: 272336-002

2

Table 1. Proliferations Options

	Standard*1	-1	-2	-3	-20
80C32	X*2	X*2	X		X
80C52	X*2	X*2	X	X	X
87C52	X	X	X	X	X
80C54	X	X	X	X	X
87C54	X	X	X	X	X
80C58	X	X	X	X	X
87C58	X	X	X	X	X

NOTES:
*1 3.5 MHz to 12 MHz; 5V ±20%
*2 5V ±20% Commercial
5V ±10% Express
-1 3.5 MHz to 16 MHz; 5V ±20%
-2 0.5 MHz to 12 MHz; 5V ±20%
-3 3.5 MHz to 20 MHz; 5V ±20% External Execution
3.5 MHz to 24 MHz; 5V ±20% Internal Execution
-20 3.5 MHz to 20 MHz; 5V ±20%

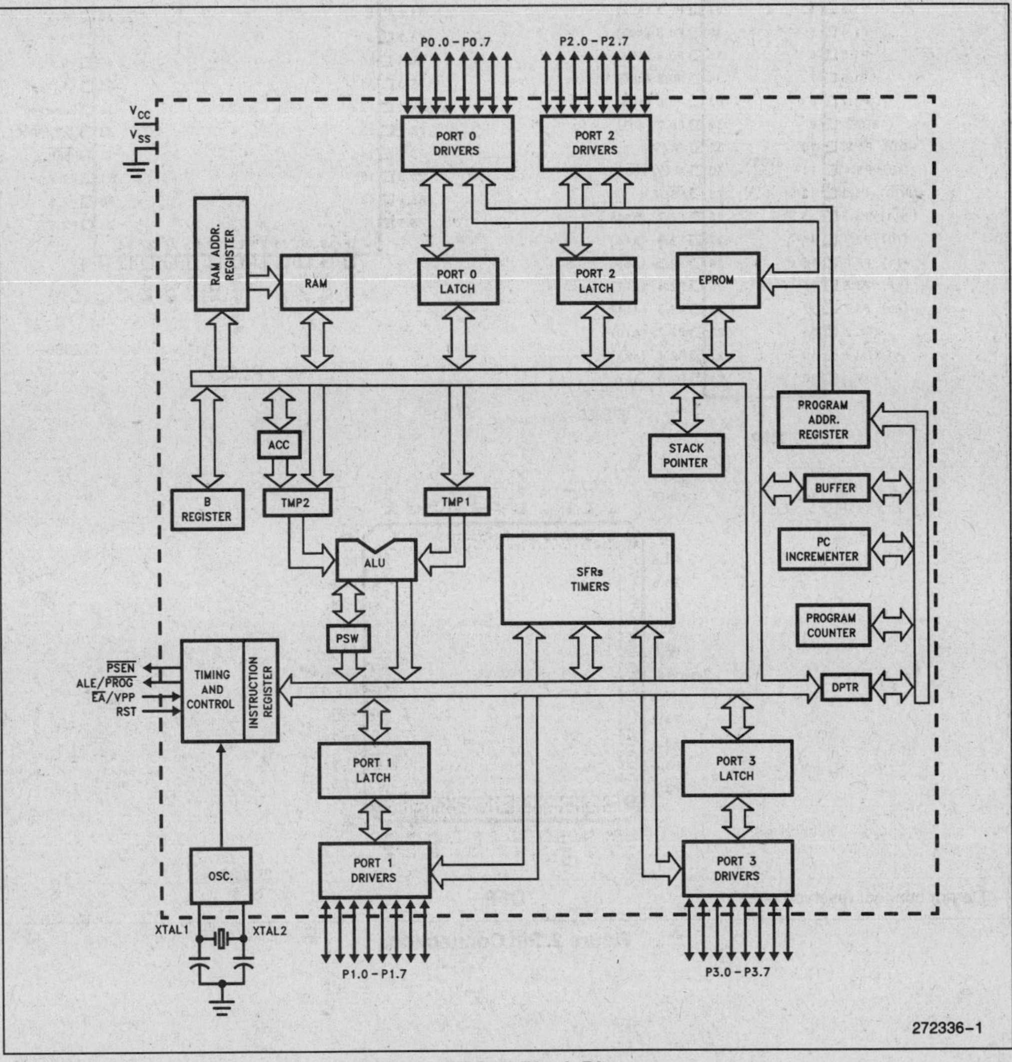

Figure 1. 8XC5X Block Diagram

PROCESS INFORMATION

This device is manufactured on P629.0, a CHMOS III-E process. The 80C32-Standard, -1 and -2, and 80C52-Standard, -1 and -2, are manufactured on P645, a CHMOS III process. Additional process and reliability information is available in Intel's *Components Quality and Reliability Handbook,* Order No. 210997.

PACKAGES

Part	Prefix	Package Type
8XC5X	P	40-Pin Plastic DIP (OTP)
87C5X	D	40-Pin CERDIP (EPROM)
8XC5X	N	44-Pin PLCC (OTP)
8XC5X	S	44-Pin QFP (OTP)

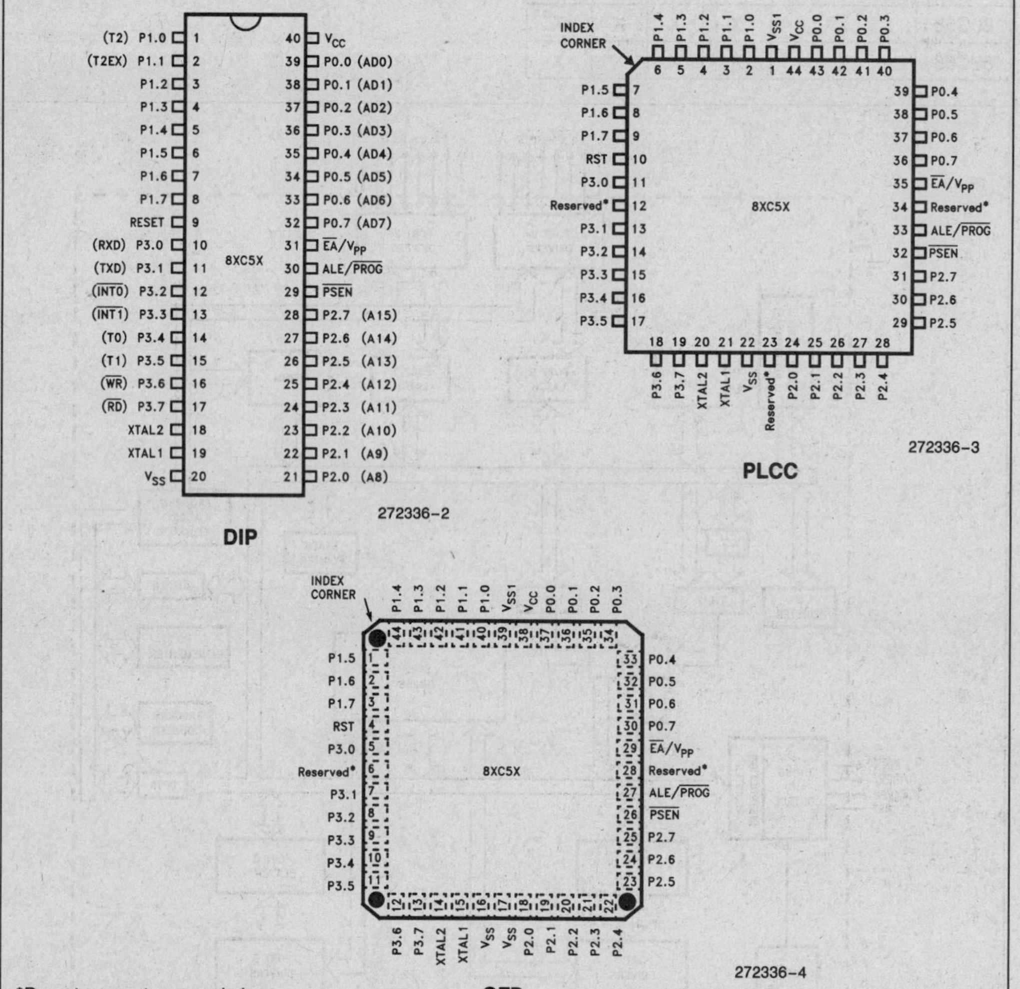

*Do not connect reserved pins.

Figure 2. Pin Connections

PIN DESCRIPTIONS

V_{CC}: Supply voltage.

V_{SS}: Circuit ground.

V_{SS1}: Secondary ground (not on DIP). Provided to reduce ground bounce and improve power supply by-passing.

NOTE:
This pin is not a substitute for the V_{SS} pin (pin 22). (Connection not necessary for proper operation.)

Port 0: Port 0 is an 8-bit, open drain, bidirectional I/O port. As an output port each pin can sink several LS TTL inputs. Port 0 pins that have 1's written to them float, and in that state can be used as high-impedance inputs.

Port 0 is also the multiplexed low-order address and data bus during accesses to external Program and Data Memory. In this application it uses strong internal pullups when emitting 1's, and can source and sink several LS TTL inputs.

Port 0 also receives the code bytes during EPROM programming, and outputs the code bytes during program verification. External pullup resistors are required during program verification.

Port 1: Port 1 is an 8-bit bidirectional I/O port with internal pullups. The Port 1 output buffers can drive LS TTL inputs. Port 1 pins that have 1's written to them are pulled high by the internal pullups, and in that state can be used as inputs. As inputs, Port 1 pins that are externally pulled low will source current (I_{IL}, on the data sheet) because of the internal pullups.

In addition, Port 1 serves the functions of the following special features of the 8XC5X:

Port Pin	Alternate Function
P1.0	T2 (External Count Input to Timer/Counter 2), Clock-Out
P1.1	T2EX (Timer/Counter 2 Capture/Reload Trigger and Direction Control)

Port 1 receives the low-order address bytes during EPROM programming and verifying.

Port 2: Port 2 is an 8-bit bidirectional I/O port with internal pullups. The Port 2 output buffers can drive LS TTL inputs. Port 2 pins that have 1's written to them are pulled high by the internal pullups, and in that state can be used as inputs. As inputs, Port 2 pins that are externally pulled low will source current (I_{IL}, on the data sheet) because of the internal pullups.

Port 2 emits the high-order address byte during fetches from external Program Memory and during accesses to external Data Memory that use 16-bit addresses (MOVX @DPTR). In this application it uses strong internal pullups when emitting 1's. During accesses to external Data Memory that use 8-bit addresses (MOVX @Ri), Port 2 emits the contents of the P2 Special Function Register.

Some Port 2 pins receive the high-order address bits during EPROM programming and program verification.

Port 3: Port 3 is an 8-bit bidirectional I/O port with internal pullups. The Port 3 output buffers can drive LS TTL inputs. Port 3 pins that have 1's written to them are pulled high by the internal pullups, and in that state can be used as inputs. As inputs, Port 3 pins that are externally pulled low will source current (I_{IL}, on the data sheet) because of the pullups.

Port 3 also serves the functions of various special features of the 8051 Family, as listed below:

Port Pin	Alternate Function
P3.0	RXD (serial input port)
P3.1	TXD (serial output port)
P3.2	$\overline{INT0}$ (external interrupt 0)
P3.3	$\overline{INT1}$ (external interrupt 1)
P3.4	T0 (Timer 0 external input)
P3.5	T1 (Timer 1 external input)
P3.6	$\overline{WR}$ (external data memory write strobe)
P3.7	$\overline{RD}$ (external data memory read strobe)

RST: Reset input. A high on this pin for two machine cycles while the oscillator is running resets the device. The port pins will be driven to their reset condition when a minimum V_{IHI} voltage is applied whether the oscillator is running or not. An internal pulldown resistor permits a power-on reset with only a capacitor connected to V_{CC}.

ALE: Address Latch Enable output pulse for latching the low byte of the address during accesses to external memory. This pin (ALE/$\overline{PROG}$) is also the program pulse input during EPROM programming for the 87C5X.

In normal operation ALE is emitted at a constant rate of 1/6 the oscillator frequency, and may be used for external timing or clocking purposes. Note, however, that one ALE pulse is skipped during each access to external Data Memory.

2

If desired, ALE operation can be disabled by setting bit 0 of SFR location 8EH. With this bit set, the pin is weakly pulled high. However, the ALE disable feature will be suspended during a MOVX or MOVC instruction, idle mode, power down mode and ICE mode. The ALE disable feature will be terminated by reset. When the ALE disable feature is suspended or terminated, the ALE pin will no longer be pulled up weakly. Setting the ALE-disable bit has no affect if the microcontroller is in external execution mode.

Throughout the remainder of this data sheet, ALE will refer to the signal coming out of the ALE/$\overline{PROG}$ pin, and the pin will be referred to as the ALE/$\overline{PROG}$ pin.

$\overline{PSEN}$: Program Store Enable is the read strobe to external Program Memory.

When the 8XC5X is executing code from external Program Memory, $\overline{PSEN}$ is activated twice each machine cycle, except that two $\overline{PSEN}$ activations are skipped during each access to external Data Memory.

$\overline{EA}$/V_{PP}: External Access enable. $\overline{EA}$ must be strapped to VSS in order to enable the device to fetch code from external Program Memory locations 0000H to 0FFFFH. Note, however, that if any of the Lock bits are programmed, $\overline{EA}$ will be internally latched on reset.

$\overline{EA}$ should be strapped to V_{CC} for internal program executions.

This pin also receives the programming supply voltage (V_{PP}) during EPROM programming.

XTAL1: Input to the inverting oscillator amplifier.

XTAL2: Output from the inverting oscillator amplifier.

OSCILLATOR CHARACTERISTICS

XTAL1 and XTAL2 are the input and output, respectively, of a inverting amplifier which can be configured for use as an on-chip oscillator, as shown in Figure 3. Either a quartz crystal or ceramic resonator may be used. More detailed information concerning the use of the on-chip oscillator is available in Application Note AP-155, "Oscillators for Microcontrollers", Order No. 230659.

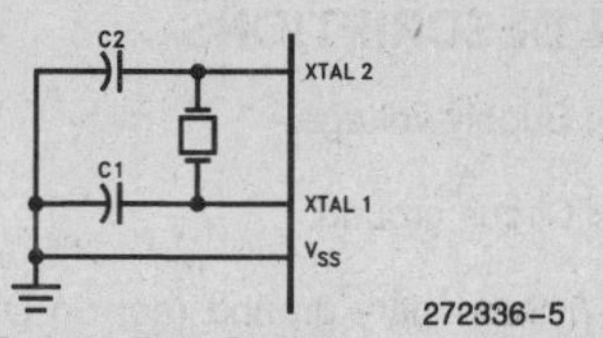

272336-5

C1, C2 = 30 pF ±10 pF for Crystals
For Ceramic Resonators, contact resonator manufacturer.

Figure 3. Oscillator Connections

To drive the device from an external clock source, XTAL1 should be driven, while XTAL2 floats, as shown in Figure 4. There are no requirements on the duty cycle of the external clock signal, since the input to the internal clocking circuitry is through a divide-by-two flip-flop, but minimum and maximum high and low times specified on the data sheet must be observed.

An external oscillator may encounter as much as a 100 pF load at XTAL1 when it starts up. This is due to interaction between the amplifier and its feedback capacitance. Once the external signal meets the V_{IL} and V_{IH} specifications the capacitance will not exceed 20 pF.

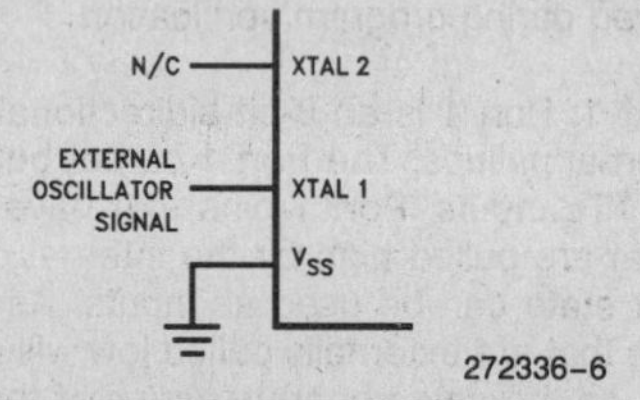

272336-6

Figure 4. External Clock Drive Configuration

IDLE MODE

The user's software can invoke the Idle Mode. When the microcontroller is in this mode, power consumption is reduced. The Special Function Registers and the onboard RAM retain their values during Idle, but the processor stops executing instructions. Idle Mode will be exited if the chip is reset or if an enabled interrupt occurs.

Table 2. Status of the External Pins during Idle and Power Down

Mode	Program Memory	ALE	$\overline{PSEN}$	PORT0	PORT1	PORT2	PORT3
Idle	Internal	1	1	Data	Data	Data	Data
Idle	External	1	1	Float	Data	Address	Data
Power Down	Internal	0	0	Data	Data	Data	Data
Power Down	External	0	0	Float	Data	Data	Data

POWER DOWN MODE

To save even more power, a Power Down mode can be invoked by software. In this mode, the oscillator is stopped and the instruction that invoked Power Down is the last instruction executed. The on-chip RAM and Special Function Registers retain their values until the Power Down mode is terminated.

On the 8XC5X either a hardware reset or an external interrupt can cause an exit from Power Down. Reset redefines all the SFRs but does not change the on-chip RAM. An external interrupt allows both the SFRs and on-chip RAM to retain their values.

To properly terminate Power Down, the reset or external interrupt should not be executed before V_{CC} is restored to its normal operating level, and must be held active long enough for the oscillator to restart and stabilize (normally less than 10 ms).

With an external interrupt, INT0 and INT1 must be enabled and configured as level-sensitive. Holding the pin low restarts the oscillator but bringing the pin back high completes the exit. Once the interrupt is serviced, the next instruction to be executed after RETI will be the one following the instruction that put the device into Power Down.

DESIGN CONSIDERATION

- The 80C32-Standard, -1 and -2, and 80C52-Standard, -1 and -2 do not have the following features: Timer 2 clockout, 4 interrupt priority levels, asynchronous port reset, 64-byte encryption array, and 3 program lock bits.
- When running out of internal program/data memory, the 8XC5X-3 can be operated using a 24 MHz clock. If the 8XC5X-3 is running out of external program/data memory, the operating frequency must be between 3.5 MHz to 20 MHz. The 8XC5X-3 will not function properly at 24 MHz when running out of external program/data memory.
- The window on the D87C5X must be covered by an opaque label. Otherwise, the DC and AC characteristics may not be met, and the device may be functionally impaired.
- When the idle mode is terminated by a hardware reset, the device normally resumes program execution, from where it left off, up to two machine cycles before the internal reset algorithm takes control. On-chip hardware inhibits access to internal RAM in this event, but access to the port pins is not inhibited. To eliminate the possibility of an unexpected write when Idle is terminated by reset, the instruction following the one that invokes Idle should not be one that writes to a port pin or to external memory.

ONCE MODE

The ONCE ("On-Circuit Emulation") Mode facilitates testing and debugging of systems using the 8XC5X without the 8XC5X having to be removed from the circuit. The ONCE Mode is invoked by:

1) Pull ALE low while the device is in reset and $\overline{PSEN}$ is high;

2) Hold ALE low as RST is deactivated.

While the device is in ONCE Mode, the Port 0 pins float and the other port pins and ALE and $\overline{PSEN}$ are weakly pulled high. The oscillator circuit remains active. While the 8XC5X is in this mode, an emulator or test CPU can be used to drive the circuit. Normal operation is restored when a normal reset is applied.

NOTE:
For more detailed information on the reduced power modes refer to current Embedded Microcontrollers and Processors Handbook Volume I, (Order No. 270645) and Application Note AP-252 (Embedded Applications Handbook, Order No. 270648), "Designing with the 80C51BH."

8XC5X EXPRESS

The Intel EXPRESS system offers enhancements to the operational specifications of the MCS 51 family of microcontrollers. These EXPRESS products are designed to meet the needs of those applications whose operating requirements exceed commercial standards.

The EXPRESS program includes the commercial standard temperature range with burn-in and an extended temperature range with or without burn-in.

With the commercial standard temperature range, operational characteristics are guaranteed over the temperature range of 0°C to +70°C. With the extended temperature range option, operational characteristics are guaranteed over the range of −40°C to +85°C.

The optional burn-in is dynamic for a minimum time of 168 hours at 125°C with V_{CC} = 6.9V ±0.25V, following guidelines in MIL-STD-883, Method 1015.

Package types and EXPRESS versions are identified by a one- or two-letter prefix to the part number. The prefixes are listed in Table 3.

For the extended temperature range option, this data sheet specifies the parameters which deviate from their commercial temperature range limits.

Table 3. Prefix Identification

Prefix	Package Type	Temperature Range	Burn-In
P	Plastic	Commercial	No
D	Cerdip	Commercial	No
N	PLCC	Commercial	No
S	QFP	Commercial	No
TP	Plastic	Extended	No
TD	Cerdip	Extended	No
TN	PLCC	Extended	No
TS	QFP	Extended	No
LP	Plastic	Extended	Yes
LD	Cerdip	Extended	Yes
LN	PLCC	Extended	Yes
LS	QFP	Extended	Yes

NOTE:
Contact distributor or local sales office to match EXPRESS prefix with proper device.

EXAMPLES:
P80C52 indicates 80C52 in a plastic package and specified for commercial temperature range, without burn-in. TD80C52 indicates 80C52 in a Cerdip package and specified for extended temperature range, without burn-in.

ABSOLUTE MAXIMUM RATINGS*

Ambient Temperature Under Bias . −40°C to +85°C

Storage Temperature .... −65°C to +150°C

Voltage on EA/V_{PP} Pin to V_{SS} .... 0V to +13.0V

Voltage on Any Other Pin to V_{SS} .. −0.5V to +6.5V

I_{OL} Per I/O Pin .... 15 mA

Power Dissipation.......................... 1.5W
(based on PACKAGE heat transfer limitations, not device power consumption)

NOTICE: This data sheet contains preliminary information on new products in production. The specifications are subject to change without notice. Verify with your local Intel Sales office that you have the latest data sheet before finalizing a design.

**WARNING: Stressing the device beyond the "Absolute Maximum Ratings" may cause permanent damage. These are stress ratings only. Operation beyond the "Operating Conditions" is not recommended and extended exposure beyond the "Operating Conditions" may affect device reliability.*

2

OPERATING CONDITIONS

Symbol	Description	Min	Max	Units
T_A	Ambient Temperature Under Bias Commercial Express	 0 −40	 +70 +85	 °C °C
V_{CC}	Supply Voltage	4.0	6.0	V
f_{OSC}	Oscillator Frequency 8XC5X 8XC5X-1 8XC5X-20	 3.5 3.5 3.5	 12 16 20	 MHz MHz MHz

DC CHARACTERISTICS (Over Operating Conditions)

All parameter values apply to all devices unless otherwise indicated.

Symbol	Parameter	Min	Typ (Note 4)	Max	Unit	Test Conditions
V_{IL}	Input Low Voltage	−0.5		0.2 V_{CC}−0.1	V	
V_{IL1}	Input Low Voltage $\overline{EA}$	0		0.2 V_{CC}−0.3	V	
V_{IH}	Input High Voltage (Except XTAL1, RST)	0.2 V_{CC}+0.9		V_{CC}+0.5	V	
V_{IH1}	Input High Voltage (XTAL1, RST)	0.7 V_{CC}		V_{CC}+0.5	V	
V_{OL}	Output Low Voltage **(Note 5)** (Ports 1, 2 and 3)			0.3	V	I_{OL} = 100 μA (Note 1)
				0.45	V	I_{OL} = 1.6 mA (Note 1)
				1.0	V	I_{OL} = 3.5 mA (Note 1)
V_{OL1}	Output Low Voltage **(Note 5)** (Port 0, ALE, $\overline{PSEN}$)			0.3	V	I_{OL} = 200 μA (Note 1)
				0.45	V	I_{OL} = 3.2 mA (Note 1)
				1.0	V	I_{OL} = 7.0 mA (Note 1)
V_{OH}	Output High Voltage (Ports 1, 2 and 3, ALE, $\overline{PSEN}$)	V_{CC}−0.3			V	I_{OH} = −10 μA
		V_{CC}−0.7			V	I_{OH} = −30 μA
		V_{CC}−1.5			V	I_{OH} = −60 μA

DC CHARACTERISTICS (Over Operating Conditions) (Continued)

All parameter values apply to all devices unless otherwise indicated.

Symbol	Parameter	Min	Typ (Note 4)	Max	Unit	Test Conditions
V_{OH1}	Output High Voltage (Port 0 in External Bus Mode)	$V_{CC}-0.3$			V	$I_{OH} = -200\ \mu A$
		$V_{CC}-0.7$			V	$I_{OH} = -3.2$ mA
		$V_{CC}-1.5$			V	$I_{OH} = -7.0$ mA
I_{IL}	Logical 0 Input Current (Ports 1, 2 and 3)			−50	μA	$V_{IN} = 0.45V$
I_{LI}	Input leakage Current (Port 0)			±10	μA	$V_{IN} = V_{IL}$ or V_{IH}
I_{TL}	Logical 1 to 0 Transition Current (Ports 1, 2 and 3)					
	Commercial			−650	μA	$V_{IN} = 2V$
	Express			−750	μA	
RRST	RST Pulldown Resistor	40		225	KΩ	
CIO	Pin Capacitance		10		pF	@1 MHz, 25°C
I_{CC}	Power Supply Current:					(Note 3)
	Active Mode					
	at 12 MHz (Figure 5)		15	30	mA	
	at 16 MHz			38	mA	
	at 20 MHz			47	mA	
	at 24 MHz			56	mA	
	Idle Mode					
	at 12 MHz (Figure 5)		5	7.5	mA	
	at 16 MHz			9.5	mA	
	at 20 MHz			11.5	mA	
	at 24 MHz			13.5	mA	
	Power Down Mode		5	75	μA	

NOTES:

1. Capacitive loading on Ports 0 and 2 may cause noise pulses above 0.4V to be superimposed on the V_{OL}s of ALE and Ports 1, 2 and 3. The noise is due to external bus capacitance discharging into the Port 0 and Port 2 pins when these pins change from 1 to 0. In applications where capacitive loading exceeds 100 pF, the noise pulses on these signals may exceed 0.8V. It may be desirable to qualify ALE or other signals with a Schmitt Triggers, or CMOS-level input logic.
2. Capacitive loading on Ports 0 and 2 cause the V_{OH} on ALE and $\overline{PSEN}$ to drop below the 0.9 V_{CC} specification when the address lines are stabilizing.
3. See Figures 6–9 for test conditions. Minimum V_{CC} for Power Down is 2V.
4. Typicals are based on a limited number of samples and are not guaranteed. The values listed are at room temperature and 5V.
5. Under steady state (non-transient) conditions, I_{OL} must be externally limited as follows:
   Maximum I_{OL} per port pin: 10mA
   Maximum I_{OL} per 8-bit port—
   Port 0: 26 mA
   Ports 1, 2 and 3: 15 mA
   Maximum total I_{OL} for all output pins: 71 mA

If I_{OL} exceeds the test condition, V_{OL} may exceed the related specification. Pins are not guaranteed to sink current greater than the listed test conditions.

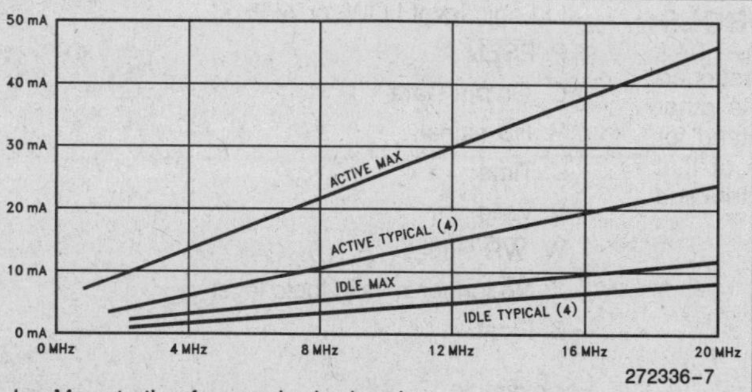

I_{CC} Max at other frequencies is given by:
Active Mode
I_{CC} Max = 2.2 × Freq + 3.1
Idle Mode
I_{CC} Max = 0.5 × Freq + 1.5
Where Freq is in MHz, I_{CC} Max is given in mA.

Figure 5. I_{CC} vs Frequency

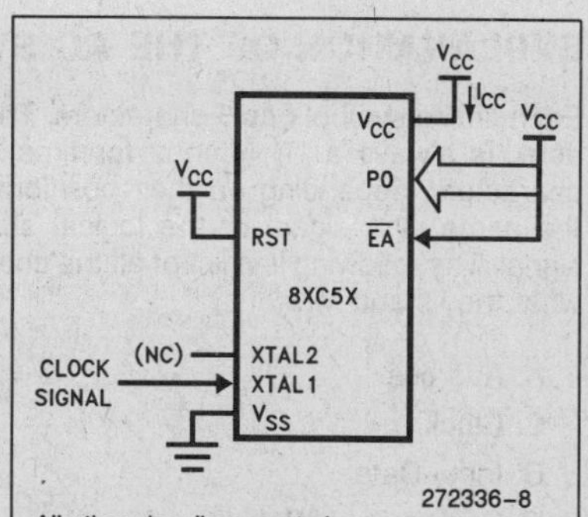

All other pins disconnected
TCLCH = TCHCL = 5 ns

Figure 6. I_{CC} Test Condition, Active Mode

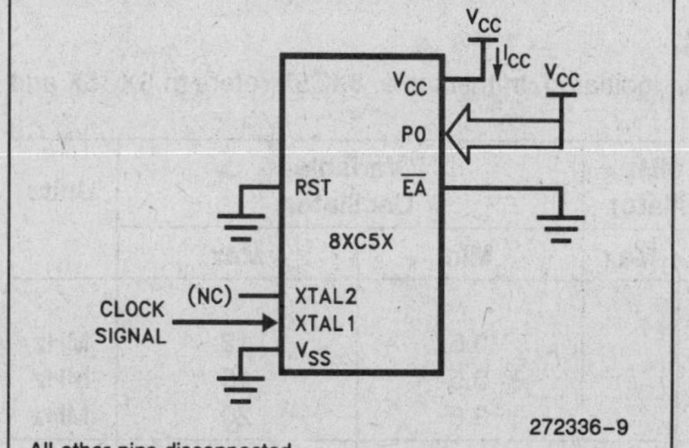

All other pins disconnected
TCLCH = TCHCL = 5 ns

Figure 7. I_{CC} Test Condition Idle Mode

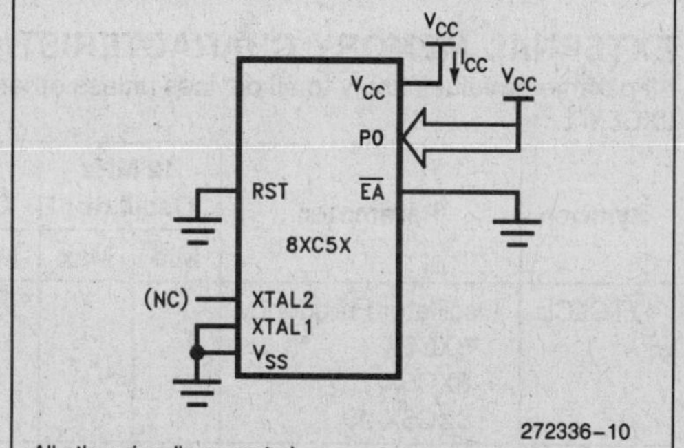

All other pins disconnected

Figure 8. I_{CC} Test Condition, Power Down Mode V_{CC} = 2.0V to 6.0V

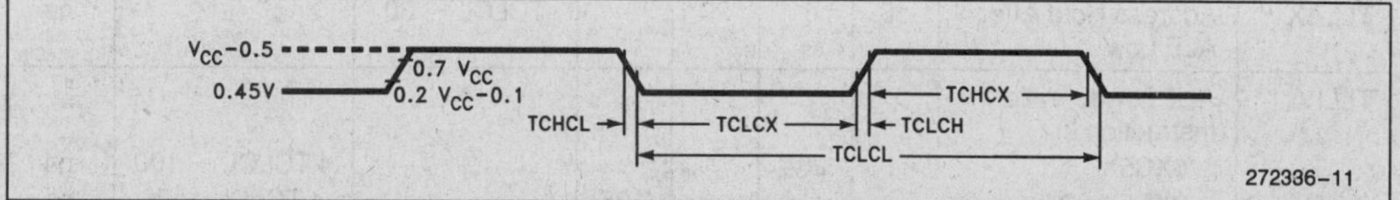

Figure 9. Clock Signal Waveform for I_{CC} Tests in Active and Idle Modes. TCLCH = TCHCL = 5 ns

2

EXPLANATION OF THE AC SYMBOLS

Each timing symbol has 5 characters. The first character is always a 'T' (stands for time). The other characters, depending on their positions, stand for the name of a signal or the logical status of that signal. The following is a list of all the characters and what they stand for.

A: Address
C: Clock
D: Input Data
H: Logic level HIGH
I: Instruction (program memory contents)
L: Logic level LOW, or ALE
P: $\overline{\text{PSEN}}$
Q: Output Data
R: $\overline{\text{RD}}$ signal
T: Time
V: Valid
W: $\overline{\text{WR}}$ signal
X: No longer a valid logic level
Z: Float

For example,

TAVLL = Time from Address Valid to ALE Low
TLLPL = Time from ALE Low to $\overline{\text{PSEN}}$ Low

AC CHARACTERISTICS (Over Operating Conditions, Load Capacitance for Port 0, ALE/$\overline{\text{PROG}}$ and $\overline{\text{PSEN}}$ = 100 pF, Load Capacitance for All Other Outputs = 80 pF)

EXTERNAL MEMORY CHARACTERISTICS

All parameter values apply to all devices unless otherwise indicated. In this table, 8XC5X refers to 8XC5X and 8XC5X-1.

Symbol	Parameter	12 MHz Oscillator		20 MHz Oscillator		Variable Oscillator		Units
		Min	Max	Min	Max	Min	Max	
1/TCLCL	Oscillator Frequency 8XC5X 8XC5X-1 8XC5X-20					 3.5 3.5 3.5	 12 16 20	 MHz MHz MHz
TLHLL	ALE Pulse Width	127		60		2 TCLCL − 40		ns
TAVLL	Address Valid to ALE Low	43		10		TCLCL − 40		ns
TLLAX	Address Hold After ALE Low	53		20		TCLCL − 30		ns
TLLIV	ALE Low to Valid Instruction In 8XC5X 8XC5X-20		 234 		 125		 4 TCLCL − 100 4 TCLCL − 75	 ns ns
TLLPL	ALE Low to $\overline{\text{PSEN}}$ Low	53		20		TCLCL − 30		ns
TPLPH	$\overline{\text{PSEN}}$ Pulse Width	205		105		3 TCLCL − 45		ns
TPLIV	$\overline{\text{PSEN}}$ Low to Valid Instruction In 8XC5X 8XC5X-20		 145 		 60		 3 TCLCL − 105 3 TCLCL − 90	 ns ns
TPXIX	Input Instruction Hold After $\overline{\text{PSEN}}$	0		0		0		ns

EXTERNAL MEMORY CHARACTERISTICS (Continued)

All parameter values apply to all devices unless otherwise indicated.

Symbol	Parameter	12 MHz Oscillator		20 MHz Oscillator		Variable Oscillator		Units
		Min	Max	Min	Max	Min	Max	
TPXIZ	Input Instruction Float After $\overline{PSEN}$ 8XC5X 8XC5X-20		59		30		TCLCL − 25 TCLCL − 20	ns ns
TAVIV	Address to Valid Instruction In		312		145		5 TCLCL − 105	ns
TPLAZ	$\overline{PSEN}$ Low to Address Float		10		10		10	ns
TRLRH	$\overline{RD}$ Pulse Width	400		200		6 TCLCL − 100		ns
TWLWH	$\overline{WR}$ Pulse Width	400		200		6 TCLCL − 100		ns
TRLDV	$\overline{RD}$ Low to Valid Data In 8XC5X 8XC5X-20		252		155		5 TCLCL − 165 5 TCLCL − 95	ns ns
TRHDX	Data Hold After $\overline{RD}$	0		0		0		ns
TRHDZ	Data Float After $\overline{RD}$		107		40		2 TCLCL − 60	ns
TLLDV	ALE Low to Valid Data In 8XC5X 8XC5X-20		517		310		8 TCLCL − 150 8 TCLCL − 90	ns ns
TAVDV	Address to Valid Data In 8XC5X 8XC5X-20		585		360		9 TCLCL − 165 9 TCLCL − 90	ns ns
TLLWL	ALE Low to $\overline{RD}$ or $\overline{WR}$ Low	200	300	100	200	3 TCLCL − 50	3 TCLCL + 50	ns
TAVWL	Address Valid to $\overline{WR}$ Low 8XC5X 8XC5X-20	203		110		4 TCLCL − 130 4 TCLCL − 90		ns ns
TQVWX	Data Valid before $\overline{WR}$ 8XC5X 8XC5X-20	33		15		TCLCL − 50 TCLCL − 35		ns ns
TWHQX	Data Hold after $\overline{WR}$ 8XC5X 8XC5X-20	33		10		TCLCL − 50 TCLCL − 40		ns ns
TQVWH	Data Valid to $\overline{WR}$ High 8XC5X 8XC5X-20	433		280		7 TCLCL − 150 7 TCLCL − 70		ns ns
TRLAZ	$\overline{RD}$ Low to Address Float		0		0		0	ns
TWHLH	$\overline{RD}$ or $\overline{WR}$ High to ALE High	43	123	10	90	TCLCL − 40	TCLCL + 40	ns

EXTERNAL PROGRAM MEMORY READ CYCLE

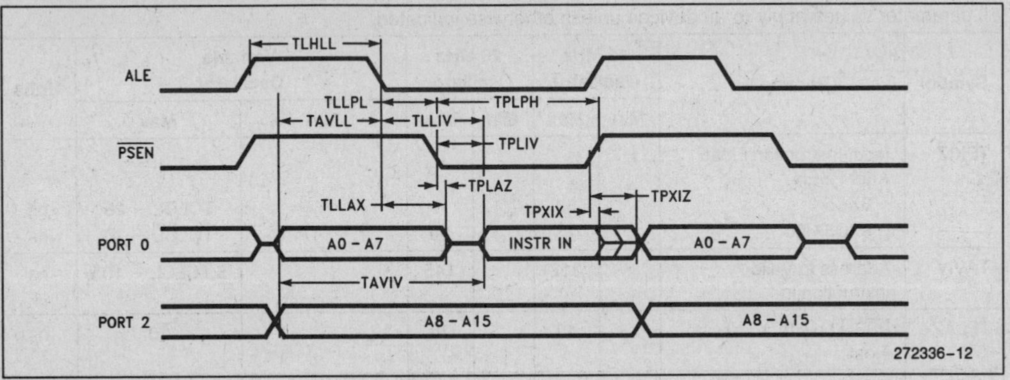

EXTERNAL DATA MEMORY READ CYCLE

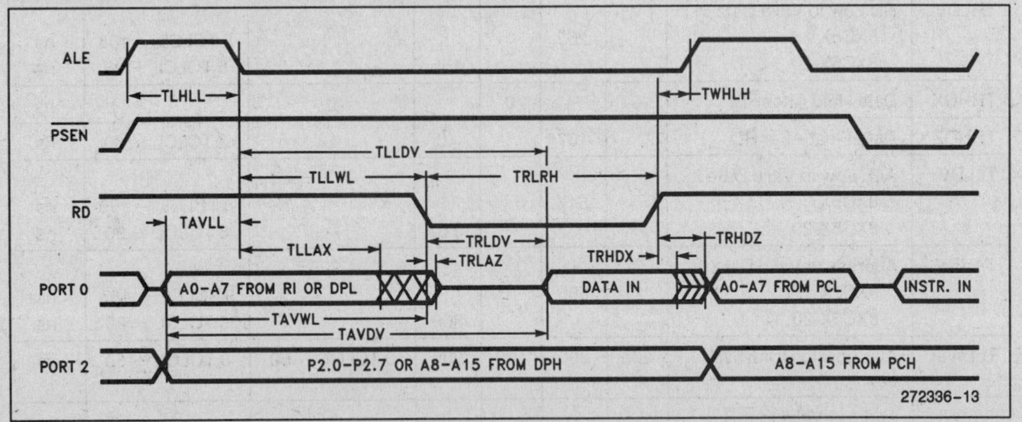

EXTERNAL DATA MEMORY WRITE CYCLE

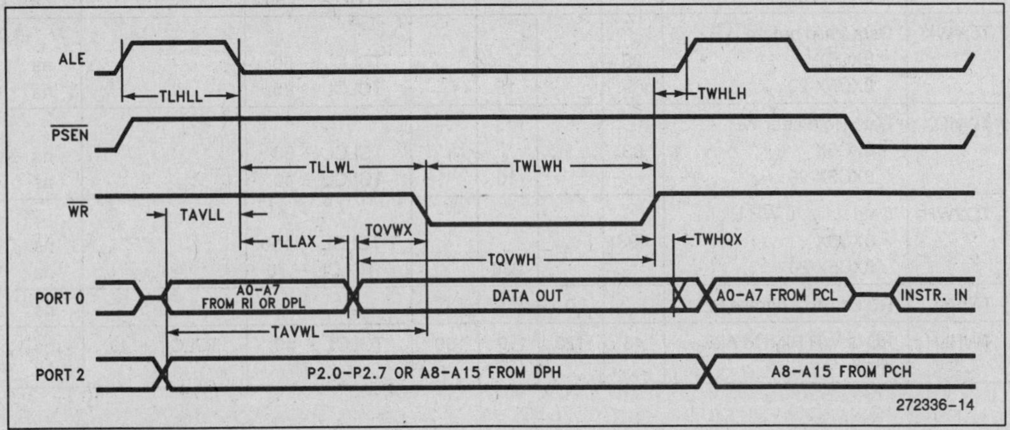

SERIAL PORT TIMING - SHIFT REGISTER MODE

Test Conditions: Over Operating Conditions; Load Capacitance = 80 pF

Symbol	Parameter	12 MHz Oscillator		20 MHz Oscillator		Variable Oscillator		Units
		Min	Max	Min	Max	Min	Max	
TXLXL	Serial Port Clock Cycle Time	1		0.600		12 TCLCL		μs
TQVXH	Output Data Setup to Clock Rising Edge	700		367		10 TCLCL − 133		ns
TXHQX	Output Data Hold after Clock Rising Edge							
	8XC5X	50				2 TCLCL − 117		ns
	8XC5X-20			50		2 TCLCL − 50		ns
TXHDX	Input Data Hold After Clock Rising Edge	0		0		0		ns
TXHDV	Clock Rising Edge to Input Data Valid		700		367		10 TCLCL − 133	ns

SHIFT REGISTER MODE TIMING WAVEFORMS

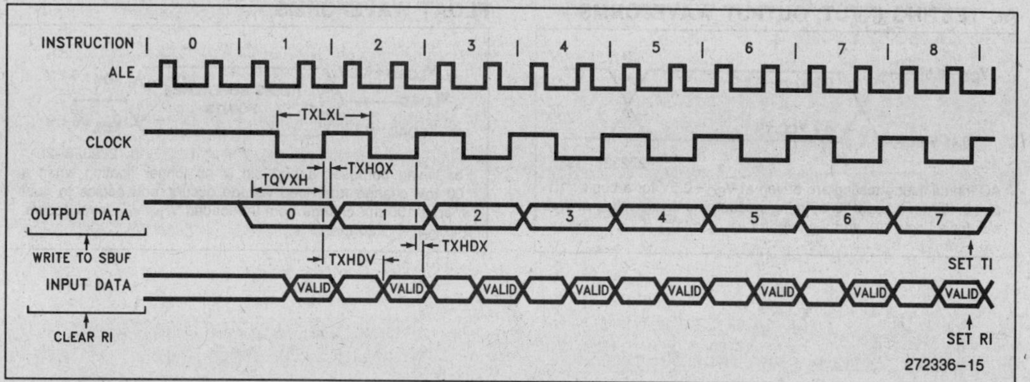

272336-15

EXTERNAL CLOCK DRIVE

Symbol	Parameter	Min	Max	Units
1/TCLCL	Oscillator Frequency			
	8XC5X	3.5	12	MHz
	8XC5X-1	3.5	16	
	8XC5X-20	3.5	20	
TCHCX	High Time	20		ns
TCLCX	Low Time	20		ns
TCLCH	Rise Time		20	ns
TCHCL	Fall Time		20	ns

EXTERNAL CLOCK DRIVE WAVEFORM

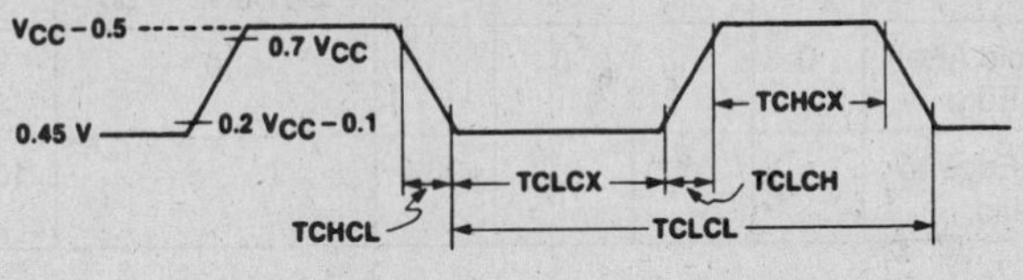

272336–16

AC TESTING INPUT, OUTPUT WAVEFORMS

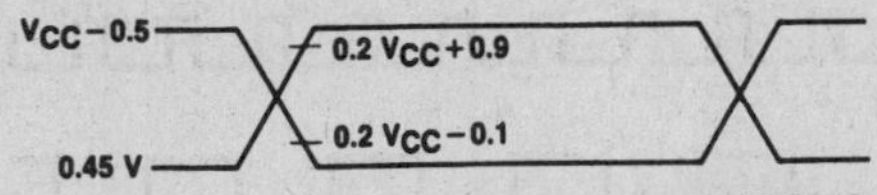

272336–19

AC Inputs during testing are driven at $V_{CC}-0.5V$ for a Logic "1" and 0.45V for a Logic "0". Timing measurements are made at V_{IH} min for a Logic "1" and V_{IL} max for a Logic "0".

FLOAT WAVEFORMS

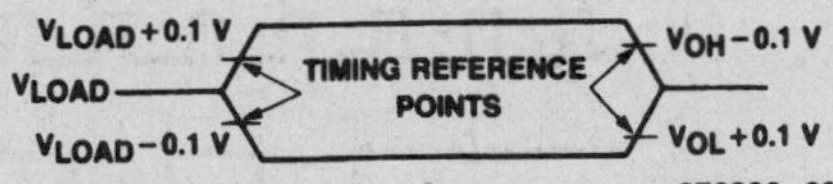

272336–20

For timing purposes a port pin is no longer floating when a 100 mV change from load voltage occurs, and begins to float when a 100 mV change from the loaded V_{OH}/V_{OL} level occurs. $I_{OL}/I_{OH} = \pm 20$ mA.

PROGRAMMING THE EPROM

The part must be running with a 4 MHz to 6 MHz oscillator. The address of an EPROM location to be programmed is applied to address lines while the code byte to be programmed in that location is applied to data lines. Control and program signals must be held at the levels indicated in Table 4. Normally $\overline{EA}/V_{PP}$ is held at logic high until just before ALE/$\overline{PROG}$ is to be pulsed. The $\overline{EA}/V_{PP}$ is raised to V_{PP}, ALE/$\overline{PROG}$ is pulsed low and then $\overline{EA}/V_{PP}$ is returned to a high (also refer to timing diagrams).

NOTES:

- Exceeding the V_{PP} maximum for any amount of time could damage the device permanently. The V_{PP} source must be well regulated and free of glitches.

DEFINITION OF TERMS

ADDRESS LINES: P1.0–P1.7, P2.0–P2.5 respectively for A0–A13.

DATA LINES: P0.0–P0.7 for D0–D7.

CONTROL SIGNALS: RST, $\overline{PSEN}$, P2.6, P2.7, P3.3, P3.6, P3.7

PROGRAM SIGNALS: ALE/$\overline{PROG}$, $\overline{EA}/V_{PP}$

Table 4. EPROM Programming Modes

Mode		RST	$\overline{PSEN}$	ALE/$\overline{PROG}$	$\overline{EA}/V_{PP}$	P2.6	P2.7	P3.3	P3.6	P3.7
Program Code Data		H	L	⊔	12.75V	L	H	H	H	H
Verify Code Data		H	L	H	H	L	L	L	H	H
Program Encryption Array Address 0–3FH		H	L	⊔	12.75V	L	H	H	L	H
Program Lock Bits	Bit 1	H	L	⊔	12.75V	H	H	H	H	H
	Bit 2	H	L	⊔	12.75V	H	H	H	L	L
	Bit 3	H	L	⊔	12.75V	H	L	H	H	L
Read Signature Byte		H	L	H	H	L	L	L	L	L

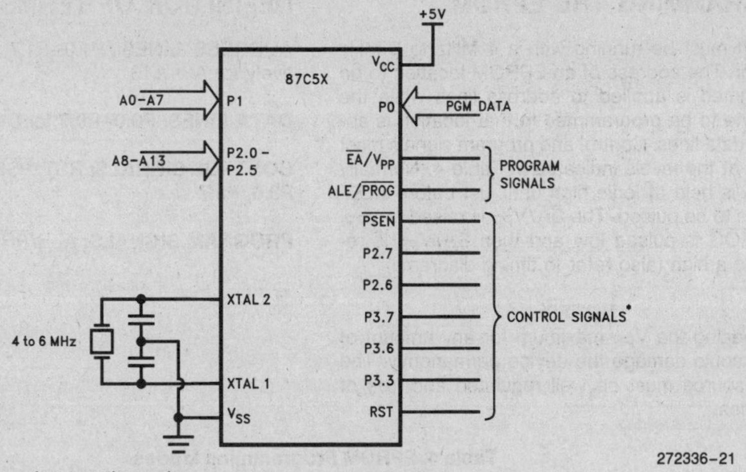

*See Table 4 for proper input on these pins

Figure 10. Programming the EPROM

PROGRAMMING ALGORITHM

Refer to Table 4 and Figures 10 and 11 for address, data, and control signals set up. To program the 87C5X the following sequence must be exercised.

1. Input the valid address on the address lines.
2. Input the appropriate data byte on the data lines.
3. Activate the correct combination of control signals.
4. Raise $\overline{EA}/V_{PP}$ from V_{CC} to 12.75V ±0.25V.
5. Pulse ALE/$\overline{PROG}$ 5 times for the EPROM array, and 25 times for the encryption table and the lock bits.

Repeat 1 through 5 changing the address and data for the entire array or until the end of the object file is reached.

PROGRAM VERIFY

Program verify may be done after each byte or block of bytes is programmed. In either case a complete verify of the programmed array will ensure reliable programming of the 87C5X.

The lock bits cannot be directly verified. Verification of the lock bits is done by observing that their features are enabled.

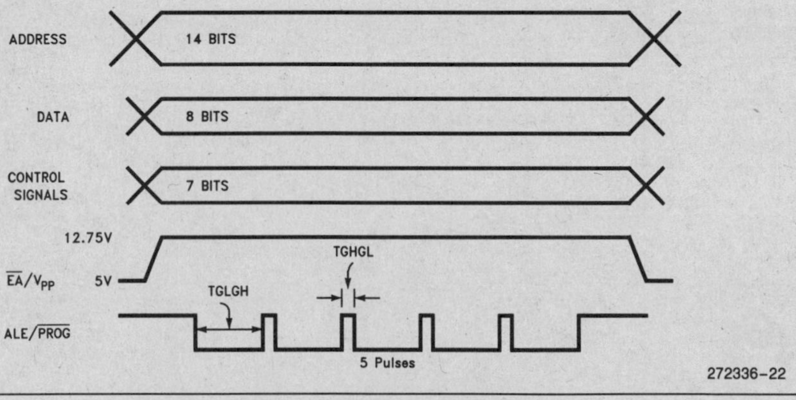

Figure 11. Programming Signal's Waveforms

ROM and EPROM Lock System

The program lock system, when programmed, protects the onboard program against software piracy.

The 80C5X has a one-level program lock system and a 64-byte encryption table. See line 2 of Table 5. If program protection is desired. the user submits the encryption table with their code. and both the lock-bit and encryption array are programmed by the factory. The encryption array is not available without the lock bit. For the lock bit to be programmed, the user must submit an encryption table.

The 87C5X has a 3-level program lock system and a 64-byte encryption array. Since this is an EPROM device, all locations are user-programmable. See Table 5.

Encryption Array

Within the EPROM array are 64 bytes of Encryption Array that are initially unprogrammed (all 1's). Every time that a byte is addressed during a verify, 6 address lines are used to select a byte of the Encryption Array. This byte is then exclusive-NOR'ed (XNOR) with the code byte, creating an Encryption Verify byte. The algorithm, with the array in the unprogrammed state (all 1's), will return the code in its original, unmodified form. For programming the Encryption Array, refer to Table 4 (Programming the EPROM).

When using the encryption array, one important factor needs to be considered. If a code byte has the value 0FFH, verifying the byte will produce the encryption byte value. If a large block (>64 bytes) of code is left unprogrammed, a verification routine will display the contents of the encryption array. For this reason all unused code bytes should be programmed with some value other than 0FFH, and not all of them the same value. This will ensure maximum program protection.

Program Lock Bits

The 87C5X has 3 programmable lock bits that when programmed according to Table 5 will provide different levels of protection for the on-chip code and data.

Erasing the EPROM also erases the encryption array and the program lock bits, returning the part to full functionality.

Reading the Signature Bytes

The 8XC5X has 3 signature bytes in locations 30H, 31H, and 60H. To read these bytes follow the procedure for EPROM verify, but activate the control lines provided in Table 4 for Read Signature Byte.

Location	Device	Contents
30H	All	89H
31H	80C52 All Others	58H/53H 58H
60H	80C52	52H/12H
	87C52	52H
	80C54	54H/14H
	87C54	54H
	80C58	58H/18H
	87C58	58H

Erasure Characteristics (Windowed Packages Only)

Erasure of the EPROM begins to occur when the chip is exposed to light with wavelength shorter than approximately 4,000 Angstroms. Since sunlight and fluorescent lighting have wavelengths in this range, exposure to these light sources over an extended time (about 1 week in sunlight, or 3 years in room-level fluorescent lighting) could cause inadvertent erasure. If an application subjects the device to this type of exposure, it is suggested that an opaque label be placed over the window.

The recommended erasure procedure is exposure to ultraviolet light (at 2537 Angstroms) to an integrated dose of at least 15 W-sec/cm^2. Exposing the EPROM to an ultraviolet lamp of 12,000 μW/cm^2 rating for 30 minutes, at a distance of about 1 inch, should be sufficient.

Erasure leaves all the EPROM Cells in a 1's state.

2

Table 5. Program Lock Bits and the Features

Program Lock Bits				Protection Type
	LB1	LB2	LB3	
1	U	U	U	No Program Lock features enabled. (Code verify will still be encrypted by the Encryption Array if programmed.)
2	P	U	U	MOVC instructions executed from external program memory are disabled from fetching code bytes from internal memory, EA is sampled and latched on Reset, and further programming of the EPROM is disabled.
3	P	P	U	Same as 2, also verify is disabled.
4	P	P	P	Same as 3, also external execution is disabled.

NOTE:
Any other combination of the lock bits is not defined.

EPROM PROGRAMMING AND VERIFICATION CHARACTERISTICS

(T_A = 21°C to 27°C; V_{CC} = 5V ±20%; V_{SS} = 0V)

Symbol	Parameter	Min	Max	Units
V_{PP}	Programming Supply Voltage	12.5	13.0	V
I_{PP}	Programming Supply Current		75	mA
1/TCLCL	Oscillator Frequency	4	6	MHz
TAVGL	Address Setup to $\overline{\text{PROG}}$ Low	48TCLCL		
TGHAX	Address Hold after $\overline{\text{PROG}}$	48TCLCL		
TDVGL	Data Setup to $\overline{\text{PROG}}$ Low	48TCLCL		
TGHDX	Data Hold after $\overline{\text{PROG}}$	48TCLCL		
TEHSH	(Enable) High to V_{PP}	48TCLCL		
TSHGL	V_{PP} Setup to $\overline{\text{PROG}}$ Low	10		µs
TGHSL	V_{PP} Hold after $\overline{\text{PROG}}$	10		µs
TGLGH	$\overline{\text{PROG}}$ Width	90	110	µs
TAVQV	Address to Data Valid		48TCLCL	
TELQV	ENABLE Low to Data Valid		48TCLCL	
TEHQZ	Data Float after ENABLE	0	48TCLCL	
TGHGL	$\overline{\text{PROG}}$ High to $\overline{\text{PROG}}$ Low	10		µs

EPROM PROGRAMMING AND VERIFICATION WAVEFORMS

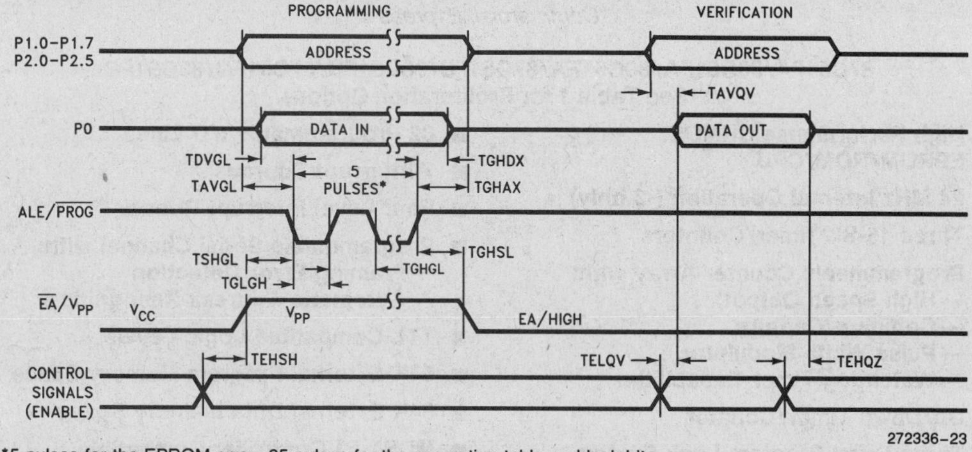

*5 pulses for the EPROM array, 25 pulses for the encryption table and lock bits.

Thermal Impedance

All thermal impedance data is approximate for static air conditions at 1W of power dissipation. Values will change depending on operating conditions and applications. See the Intel Packaging Handbook (Order Number 240800) for a description of Intel's thermal impedance test methodology.

Package	θ_{JA}	θ_{JC}	Device
P	45°C/W	16°C/W	All
D	45°C/W	15°C/W	All
N	46°C/W	16°C/W	All
S	87°C/W	18°C/W	52
	96°C/W	24°C/W	54
	90°C/W	22°C/W	58

DATA SHEET REVISION HISTORY

Data sheets are changed as new device information becomes available. Verify with your local Intel sales office that you have the latest version before finalizing a design or ordering devices.

The following differences exist between this datasheet (272336-002) and the previous version (272336-001):

1. Removed 8XC5X-L from the data sheet.
2. Included features not available in 80C32-Standard, -1 and -2, and 80C52-Standard, -1 and -2 devices.

This 8XC5X datasheet (272336-001) replaces the following datasheets:

87C52/80C52/80C32	270757-003
87C52/80C52/80C32 EXPRESS	270868-002
87C52-20/80C52-20/80C32-20	272272-001
87C54/80C54	270816-004
87C54/80C54 EXPRESS	270901-001
87C54-20/-3 80C54-20/-3	270941-003
87C54/80C58	270900-003
87C58/80C58 EXPRESS	270902-001
87C58-20/-3 80C58-20/-3	272029-002

8XC51FX
CHMOS SINGLE-CHIP 8-BIT MICROCONTROLLERS

Commercial/Express

87C51FA/83C51FA/80C51FA/87C51FB/83C51FB/87C51FC/83C51FC
***See Table 1 for Proliferation Options**

- High Performance CHMOS EPROM/ROM/CPU
- 24 MHz Internal Operation (-3 only)
- Three 16-Bit Timer/Counters
- Programmable Counter Array with:
  - High Speed Output,
  - Compare/Capture,
  - Pulse Width Modulator,
  - Watchdog Timer Capabilities
- Up/Down Timer/Counter
- Three Level Program Lock System
- 8K/16K/32K On-Chip Program Memory
- 256 Bytes of On-Chip Data RAM
- Improved Quick Pulse Programming Algorithm
- Boolean Processor
- 32 Programmable I/O Lines
- 7 Interrupt Sources
- Four Level Interrupt Priority
- Programmable Serial Channel with:
  - Framing Error Detection
  - Automatic Address Recognition
- TTL Compatible Logic Levels
- 64K External Program Memory Space
- 64K External Data Memory Space
- MCS® 51 Controller Compatible Instruction Set
- Power Saving Idle and Power Down Modes
- ONCE (On-Circuit Emulation) Mode
- Extended Temperature Range (−40°C to +85°C)

MEMORY ORGANIZATION

ROM Device	EPROM Version	ROMLESS Version	ROM/ EPROM Bytes	RAM Bytes
83C51FA	87C51FA	80C51FA	8K	256
83C51FB	87C51FB	80C51FA	16K	256
83C51FC	87C51FC	80C51FA	32K	256

These devices can address up to 64 Kbytes of external program/data memory.

The Intel 87C51FA/8XC51FB/8XC51FC is a single-chip control oriented microcontroller which is fabricated on Intel's reliable CHMOS III-E technology. The Intel 83C51FA/80C51FA is fabricated on CHMOS III technology. Being a member of the MCS® 51 controller family, the 8XC51FA/8XC51FB/8XC51FC uses the same powerful instruction set, has the same architecture, and is pin-for-pin compatible with the existing MCS 51 controller products. The 8XC51FA/8XC51FB/8XC51FC is an enhanced version of the 8XC52/8XC54/8XC58. Its added features make it an even more powerful microcontroller for applications that require Pulse Width Modulation, High Speed I/O and up/down counting capabilities such as motor control.

The 8XC51FX-3 has the same 3.5 MHz to 20 MHz frequency range as the 8XC51FX-20 when operating out of external program/data memory. When running out of internal program/data memory, the 8XC51FX-3 can operate up to 24 MHz.

For the remainder of this document, the 8XC51FA, 8XC51FB, 8XC51FC will be referred to as the 8XC51FX, unless information applies to a specific device.

October 1994
Order Number: 272322-002

Table 1. Proliferation Options

	Standard*1	-1	-2	-3	-20
80C51FA	X	X	X		
83C51FA	X	X	X		
87C51FA	X	X	X	X	X
83C51FB	X	X	X	X	X
87C51FB	X	X	X	X	X
83C51FC	X	X	X	X	X
87C51FC	X	X	X	X	X

NOTES:
*1 3.5 MHz to 12 MHz; 5V ±20%
-1 3.5 MHz to 16 MHz; 5V ±20%
-2 0.5 MHz to 12 MHz; 5V ±20%
-3 3.5 MHz to 20 MHz; 5V ±20% External Execution
3.5 MHz to 24 MHz; 5V ±20% Internal Execution
-20 3.5 MHz to 20 MHz; 5V ±20%

2

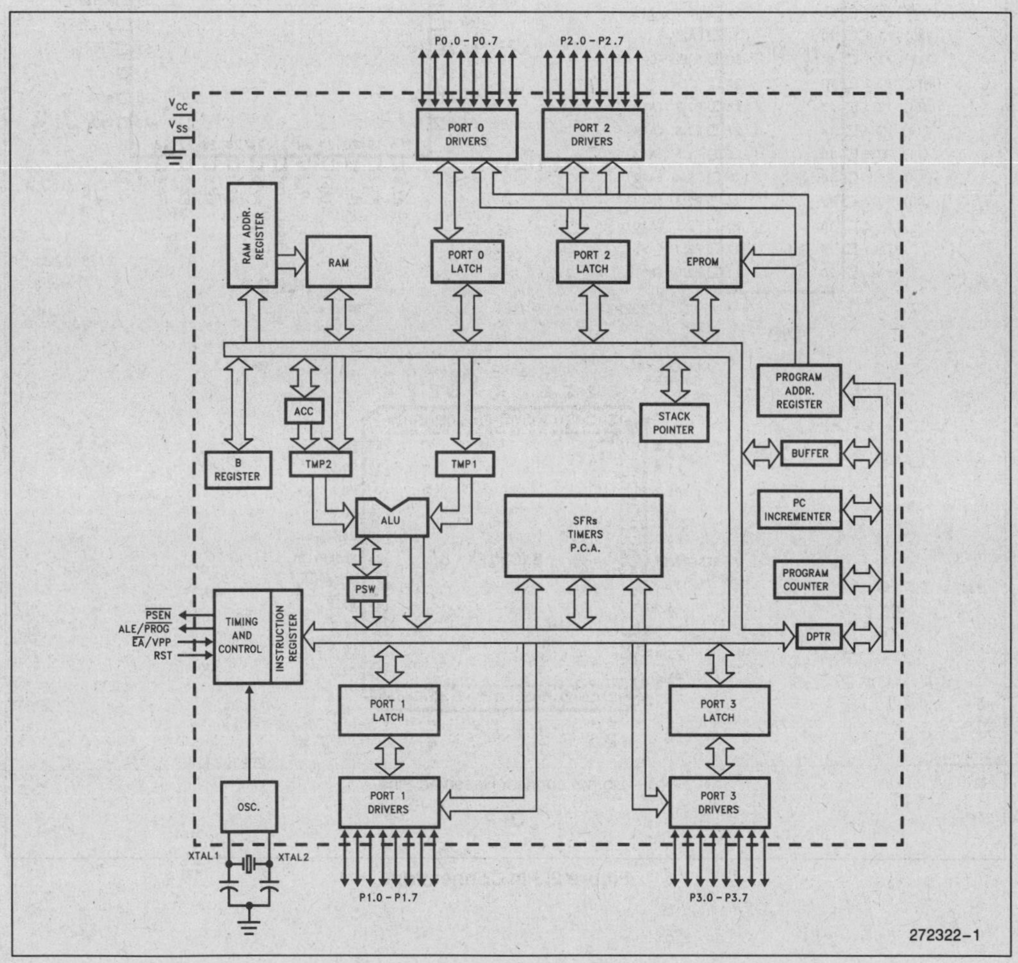

Figure 1. 8XC51FX Block Diagram

PROCESS INFORMATION

The 87C51FA/8XC51FB/8XC51FC is manufactured on P629.0, a CHMOS III-E process. The 83C51FA/80C51FA are manufactured on P645, a CHMOS III process. Additional process and reliability information is available in Intel's *Components Quality and Reliability Handbook,* Order No. 210997.

PACKAGES

Part	Prefix	Package Type
8XC51FX	P	40-Pin Plastic DIP
	D	40-Pin CERDIP
	N	44-Pin PLCC
	S	44-Pin QFP

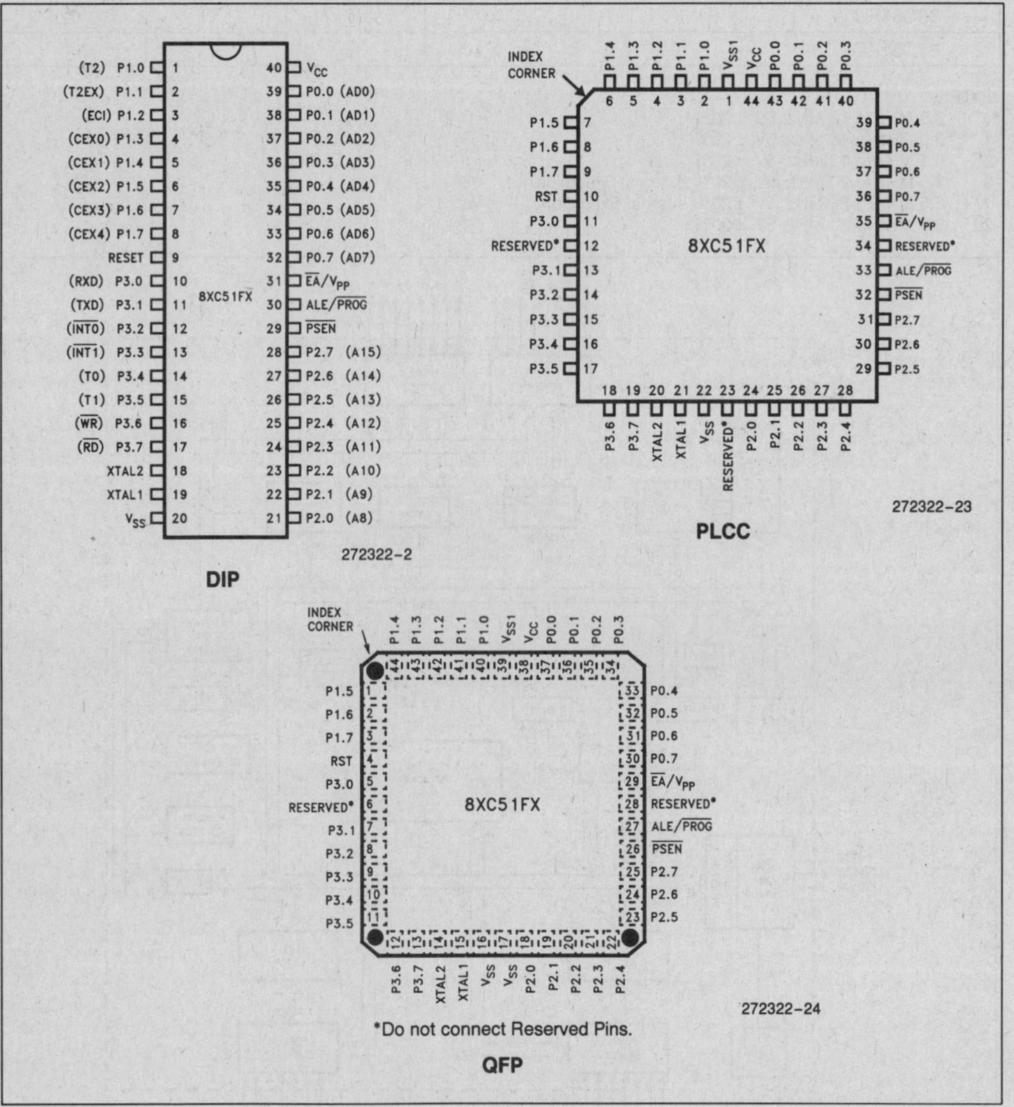

Figure 2. Pin Connections

PIN DESCRIPTIONS

V_{CC}: Supply voltage.

V_{SS}: Circuit ground.

V_{SS1}: Secondary ground (not on DIP devices or any 83C51FA/80C51FA device). Provided to reduce ground bounce and improve power supply by-passing.

NOTE:
This pin is not a substitution for the V_{SS} pin. (Connection not necessary for proper operation.)

Port 0: Port 0 is an 8-bit, open drain, bidirectional I/O port. As an output port each pin can sink several LS TTL inputs. Port 0 pins that have 1's written to them float, and in that state can be used as high-impedance inputs.

Port 0 is also the multiplexed low-order address and data bus during accesses to external Program and Data Memory. In this application it uses strong internal pullups when emitting 1's, and can source and sink several LS TTL inputs.

Port 0 also receives the code bytes during EPROM programming, and outputs the code bytes during program verification. External pullup resistors are required during program verification.

Port 1: Port 1 is an 8-bit bidirectional I/O port with internal pullups. The Port 1 output buffers can drive LS TTL inputs. Port 1 pins that have 1's written to them are pulled high by the internal pullups, and in that state can be used as inputs. As inputs, Port 1 pins that are externally pulled low will source current (I_{IL}, on the data sheet) because of the internal pullups.

In addition, Port 1 serves the functions of the following special features of the 8XC51FX:

Port Pin	Alternate Function
P1.0	T2 (External Count Input to Timer/Counter 2), Clock Out
P1.1	T2EX (Timer/Counter 2 Capture/Reload Trigger and Direction Control)
P1.2	ECI (External Count Input to the PCA)
P1.3	CEX0 (External I/O for Compare/Capture Module 0)
P1.4	CEX1 (External I/O for Compare/Capture Module 1)
P1.5	CEX2 (External I/O for Compare/Capture Module 2)
P1.6	CEX3 (External I/O for Compare/Capture Module 3)
P1.7	CEX4 (External I/O for Compare/Capture Module 4)

Port 1 receives the low-order address bytes during EPROM programming and verifying.

Port 2: Port 2 is an 8-bit bidirectional I/O port with internal pullups. The Port 2 output buffers can drive LS TTL inputs. Port 2 pins that have 1's written to them are pulled high by the internal pullups, and in that state can be used as inputs. As inputs, Port 2 pins that are externally pulled low will source current (I_{IL}, on the data sheet) because of the internal pullups.

Port 2 emits the high-order address byte during fetches from external Program Memory and during accesses to external Data Memory that use 16-bit addresses (MOVX @DPTR). In this application it uses strong internal pullups when emitting 1's. During accesses to external Data Memory that use 8-bit addresses (MOVX @Ri), Port 2 emits the contents of the P2 Special Function Register.

Some Port 2 pins receive the high-order address bits during EPROM programming and program verification.

Port 3: Port 3 is an 8-bit bidirectional I/O port with internal pullups. The Port 3 output buffers can drive LS TTL inputs. Port 3 pins that have 1's written to them are pulled high by the internal pullups, and in that state can be used as inputs. As inputs, Port 3 pins that are externally pulled low will source current (I_{IL}, on the data sheet) because of the pullups.

Port 3 also serves the functions of various special features of the MCS-51 Family, as listed below:

Port Pin	Alternate Function
P3.0	RXD (serial input port)
P3.1	TXD (serial output port)
P3.2	$\overline{INT0}$ (external interrupt 0)
P3.3	$\overline{INT1}$ (external interrupt 1)
P3.4	T0 (Timer 0 external input)
P3.5	T1 (Timer 1 external input)
P3.6	$\overline{WR}$ (external data memory write strobe)
P3.7	$\overline{RD}$ (external data memory read strobe)

RST: Reset input. A high on this pin for two machine cycles while the oscillator is running resets the device. The port pins will be driven to their reset condition when a minimum V_{IH1} voltage is applied whether the oscillator is running or not (except 83C51FA/80C51FA). An internal pulldown resistor permits a power-on reset with only a capacitor connected to V_{CC}.

In order to reset the 83C51FA/80C51FA port pins, the oscillator must be running. At least 19 oscillator periods must occur after a logic 1 is applied to the RST input before the port pins are driven to their reset state.

ALE: Address Latch Enable output pulse for latching the low byte of the address during accesses to external memory. This pin (ALE/$\overline{PROG}$) is also the program pulse input during EPROM programming for the 87C51FX.

In normal operation ALE is emitted at a constant rate of 1/6 the oscillator frequency, and may be used for external timing or clocking purposes. Note, however, that one ALE pulse is skipped during each access to external Data Memory.

If desired, ALE operation can be disabled by setting bit 0 of SFR location 8EH. With this bit set, the pin is weakly pulled high. However, the ALE disable feature will be suspended during a MOVX or MOVC instruction, idle mode, power down mode and ICE mode. The ALE disable feature will be terminated by reset. When the ALE disable feature is suspended or terminated, the ALE pin will no longer be pulled up weakly. Setting the ALE-disable bit has no affect if the microcontroller is in external execution mode.

Throughout the remainder of this data sheet, ALE will refer to the signal coming out of the ALE/$\overline{PROG}$ pin, and the pin will be referred to as the ALE/$\overline{PROG}$ pin.

$\overline{PSEN}$: Program Store Enable is the read strobe to external Program Memory.

When the 8XC51FX is executing code from external Program Memory, $\overline{PSEN}$ is activated twice each machine cycle, except that two $\overline{PSEN}$ activations are skipped during each access to external Data Memory.

$\overline{EA}$/V_{PP}: External Access enable. $\overline{EA}$ must be strapped to VSS in order to enable the device to fetch code from external Program Memory locations 0000H to 0FFFH. Note, however, that if either of the Program Lock bits are programmed, $\overline{EA}$ will be internally latched on reset.

$\overline{EA}$ should be strapped to V_{CC} for internal program executions.

This pin also receives the programming supply voltage (V_{PP}) during EPROM programming.

XTAL1: Input to the inverting oscillator amplifier.

XTAL2: Output from the inverting oscillator amplifier.

OSCILLATOR CHARACTERISTICS

XTAL1 and XTAL2 are the input and output, respectively, of a inverting amplifier which can be configured for use as an on-chip oscillator, as shown in Figure 3. Either a quartz crystal or ceramic resonator may be used. More detailed information concerning the use of the on-chip oscillator is available in Application Note AP-155, "Oscillators for Microcontrollers."

To drive the device from an external clock source, XTAL1 should be driven, while XTAL2 floats, as shown in Figure 4. There are no requirements on the duty cycle of the external clock signal, since the input to the internal clocking circuitry is through a divide-by-two flip-flop, but minimum and maximum high and low times specified on the data sheet must be observed.

An external oscillator may encounter as much as a 100 pF load at XTAL1 when it starts up. This is due to interaction between the amplifier and its feedback capacitance. Once the external signal meets the V_{IL} and V_{IH} specifications the capacitance will not exceed 20 pF.

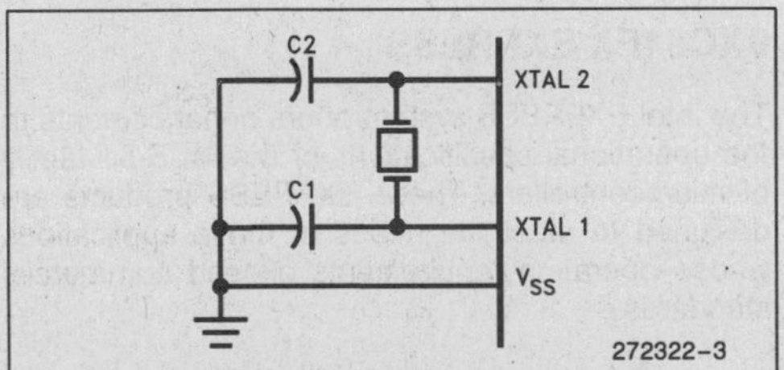

Figure 3. Oscillator Connections

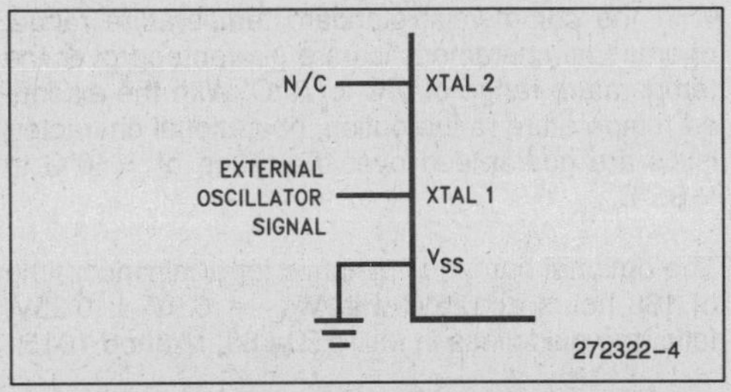

Figure 4. External Clock Drive Configuration

IDLE MODE

The user's software can invoke the Idle Mode. When the microcontroller is in this mode, power consumption is reduced. The Special Function Registers and the onboard RAM retain their values during Idle, but the processor stops executing instructions. Idle Mode will be exited if the chip is reset or if an enabled interrupt occurs. The PCA timer/counter can optionally be left running or paused during Idle Mode.

POWER DOWN MODE

To save even more power, a Power Down mode can be invoked by software. In this mode, the oscillator is stopped and the instruction that invoked Power Down is the last instruction executed. The on-chip RAM and Special Function Registers retain their values until the Power Down mode is terminated.

On the 8XC51FX either hardware reset or external interrupt can cause an exit from Power Down. Reset redefines all the SFRs but does not change the on-chip RAM. An external interrupt allows both the SFRs and the on-chip RAM to retain their values.

To properly terminate Power Down the reset or external interrupt should not be executed before V_{CC} is restored to its normal operating level and must be held active long enough for the oscillator to restart and stabilize (normally less than 10 ms).

With an external interrupt, INT0 or INT1 must be enabled and configured as level-sensitive. Holding the pin low restarts the oscillator but bringing the pin back high completes the exit. Once the interrupt is serviced, the next instruction to be executed after RETI will be the one following the instruction that put the device into Power Down.

2

DESIGN CONSIDERATION

- When running out of internal program/data memory, the 8XC51FX-3 can be operated using a 24 MHz clock. If the 8XC51FX-3 is running out of external program/data memory, the operating frequency must be between 3.5 MHz to 20 MHz. The 8XC51FX-3 will not function properly at 24 MHz when running out of external program/data memory.
- Ambient light is known to affect the internal RAM contents during operation. If the 87C51FX application requires the part to be run under ambient lighting, an opaque label should be placed over the window to exclude light.
- When the idle mode is terminated by a hardware reset, the device normally resumes program execution, from where it left off, up to two machine cycles before the internal reset algorithm takes control. On-chip hardware inhibits access to internal RAM in this event, but access to the port pins is not inhibited. To eliminate the possibility of an unexpected write when Idle is terminated by reset, the instruction following the one that invokes Idle should not be one that writes to a port pin or to external memory.

Table 2. Status of the External Pins during Idle and Power Down

Mode	Program Memory	ALE	$\overline{\text{PSEN}}$	PORT0	PORT1	PORT2	PORT3
Idle	Internal	1	1	Data	Data	Data	Data
Idle	External	1	1	Float	Data	Address	Data
Power Down	Internal	0	0	Data	Data	Data	Data
Power Down	External	0	0	Float	Data	Data	Data

NOTE:
For more detailed information on the reduced power modes refer to current Embedded Microcontrollers and Processors Handbook Volume I, and Application Note AP-252 (Embedded Applications Handbook), "Designing with the 80C51BH."

- The 83C51FA/80C51FA do not have the following features: Timer 2 clockout, 4 interrupt priority levels, asynchronous port reset, 64-byte encryption array, and 3 program lock bits. These features cannot be used with the 83C51FA/80C51FA.

ONCE MODE

The ONCE ("On-Circuit Emulation") Mode facilitates testing and debugging of systems using the 8XC51FX without the 8XC51FX having to be removed from the circuit. The ONCE Mode is invoked by:

1) Pull ALE low while the device is in reset and $\overline{PSEN}$ is high;
2) Hold ALE low as RST is deactivated.

While the device is in ONCE Mode, the Port 0 pins float, and the other port pins and ALE and $\overline{PSEN}$ are weakly pulled high. The oscillator circuit remains active. While the 8XC51FX is in this mode, an emulator or test CPU can be used to drive the circuit. Normal operation is restored when a normal reset is applied.

8XC51FX EXPRESS

The Intel EXPRESS system offers enhancements to the operational specifications of the MCS-51 family of microcontrollers. These EXPRESS products are designed to meet the needs of those applications whose operating requirements exceed commercial standards.

The EXPRESS program includes the commercial standard temperature range with burn-in and an extended temperature range with or without burn-in.

With the commercial standard temperature range, operational characteristics are guaranteed over the temperature range of 0°C to 70°C. With the extended temperature range option, operational characteristics are guaranteed over the range of −40°C to +85°C.

The optional burn-in is dynamic for a minimum time of 168 hours at 125°C with V_{CC} = 6.9V ± 0.25V, following guidelines in MIL-STD-883, Method 1015.

Package types and EXPRESS versions are identified by a one- or two-letter prefix to the part number. The prefixes are listed in Table 3.

For the extended temperature range option, this data sheet specifies the parameters which deviate from their commercial temperature range limits.

Table 3. Prefix Identification

Prefix	Package Type	Temperature Range	Burn-In
D	Cerdip	Commercial	No
N	PLCC	Commercial	No
P	Plastic	Commercial	No
S	QFP	Commercial	No
LD	Cerdip	Extended	Yes
LN	PLCC	Extended	Yes
LP	Plastic	Extended	Yes
LS	QFP	Extended	Yes
TD	Cerdip	Extended	No
TN	PLCC	Extended	No
TP	Plastic	Extended	No
TS	QFP	Extended	No

NOTE:
Contact distributor or local sales office to match EXPRESS prefix with proper device.

EXAMPLES:
P87C51FC indicates 87C51FC in a plastic package and specified for commercial temperature range, without burn-in.
LD87C51FC indicates 87C51FC in a cerdip package and specified for extended temperature range with burn-in.

ABSOLUTE MAXIMUM RATINGS*

Ambient Temperature Under Bias . −40°C to +85°C
Storage Temperature .... −65°C to +150°C
Voltage on EA/V_{PP} Pin to V_{SS} .... 0V to +13.0V
Voltage on Any Other Pin to V_{SS} .. −0.5V to +6.5V
I_{OL} per I/O Pin .... 15 mA
Power Dissipation.......................... 1.5W
(based on PACKAGE heat transfer limitations, not device power consumption)

NOTICE: This data sheet contains preliminary information on new products in production. It is valid for the devices indicated in the revision history. The specifications are subject to change without notice.

**WARNING: Stressing the device beyond the "Absolute Maximum Ratings" may cause permanent damage. These are stress ratings only. Operation beyond the "Operating Conditions" is not recommended and extended exposure beyond the "Operating Conditions" may affect device reliability.*

OPERATING CONDITIONS

Symbol	Description	Min	Max	Units
T_A	Ambient Temperature Under Bias Commercial Express	 0 −40	 +70 +85	°C
V_{CC}	Supply Voltage All Others	 4.0	 6.0	V
f_{OSC}	Oscillator Frequency 8XC51FX 8XC51FX-1 8XC51FX-2 8XC51FX-20	 3.5 3.5 0.5 3.5	 12 16 12 20	MHz

DC CHARACTERISTICS (Over Operating Conditions)

All parameter values apply to all devices unless otherwise indicated.

Symbol	Parameter	Min	Typical (Note 4)	Max	Units	Test Conditions
V_{IL}	Input Low Voltage	−0.5		0.2 V_{CC} − 0.1	V	
V_{IL1}	Input Low Voltage $\overline{EA}$	0		0.2 V_{CC} − 0.3	V	
V_{IH}	Input High Voltage (Except XTAL1, RST)	0.2 V_{CC} + 0.9		V_{CC} + 0.5	V	
V_{IH1}	Input High Voltage (XTAL1, RST)	0.7 V_{CC}		V_{CC} + 0.5	V	
V_{OL}	Output Low Voltage (Note 5) (Ports 1, 2 and 3)			0.3 0.45 1.0	V	I_{OL} = 100 μA I_{OL} = 1.6 mA (Note 1) I_{OL} = 3.5 mA
V_{OL1}	Output Low Voltage (Note 5) (Port 0, ALE/$\overline{PROG}$, $\overline{PSEN}$)			0.3 0.45 1.0	V	I_{OL} = 200 μA I_{OL} = 3.2 mA (Note 1) I_{OL} = 7.0 mA
V_{OH}	Output High Voltage (Ports 1, 2 and 3 ALE/$\overline{PROG}$ and $\overline{PSEN}$)	V_{CC} − 0.3 V_{CC} − 0.7 V_{CC} − 1.5			V	I_{OH} = −10 μA I_{OH} = −30 μA (Note 2) I_{OH} = −60 μA
V_{OH1}	Output High Voltage (Port 0 in External Bus Mode) 83C51FA/80C51FA (Express)	V_{CC} − 0.3 V_{CC} − 0.7 V_{CC} − 1.5			V	I_{OH} = −200 μA I_{OH} = −3.2 mA (Note 2) I_{OH} = −7.0 mA I_{OH} = −6.0 mA
I_{IL}	Logical 0 Input Current (Ports 1, 2 and 3) Express (83C51FA/80C51FA) All Others			 −75 −50	μA	V_{IN} = 0.45V

DC CHARACTERISTICS (Over Operating Conditions)

All parameter values apply to all devices unless otherwise indicated. (Continued)

Symbol	Parameter	Min	Typical (Note 4)	Max	Units	Test Conditions
I_{LI}	Input leakage Current (Port 0) Express (83C51FA/80C51FA) All Others			±15 ±10	μA	$V_{IN} = V_{IL}$ or V_{IH}
I_{TL}	Logical 1 to 0 Transition Current (Ports 1, 2 and 3) Express Commercial			−750 −650	μA	$V_{IN} = 2V$
RRST	RST Pulldown Resistor 83C51FA/80C51FA All Others	50 40		300 225	KΩ	
CIO	Pin Capacitance		10		pF	@1MHz, 25°C
I_{CC}	Power Supply Current: Active Mode At 12 MHz (Figure 5) At 16 MHz At 20 MHz At 24 MHz Idle Mode At 12 MHz (Figure 5) At 16 MHz At 20 MHz At 24 MHz Power Down Mode		 15 5 5	 30 38 47 56 7.5 9.5 11.5 13.5 75	 mA mA mA mA mA mA mA mA μA	(Note 3)

NOTES:

1. Capacitive loading on Ports 0 and 2 may cause noise pulses above 0.4V to be superimposed on the V_{OL}s of ALE and Ports 1, 2 and 3. The noise is due to external bus capacitance discharging into the Port 0 and Port 2 pins when these pins change from 1 to 0. In applications where capacitance loading exceeds 100 pF, the noise pulses on these signals may exceed 0.8V. It may be desirable to qualify ALE or other signals with a Schmitt Trigger, or CMOS-level input logic.
2. Capacitive loading on Ports 0 and 2 cause the V_{OH} on ALE and $\overline{PSEN}$ to drop below the 0.9 V_{CC} specification when the address lines are stabilizing.
3. See Figures 6–9 for test conditions. Minimum V_{CC} for power down is 2V.
4. Typicals are based on limited number of samples, and are not guaranteed. The values listed are at room temperature and 5V.
5. Under steady state (non-transient) conditions, I_{OL} must be externally limited as follows:
   Maximum I_{OL} per port pin: 10 mA
   Maximum I_{OL} per 8-bit port -
   Port 0: 26 mA
   Ports 1, 2, and 3: 15 mA
   Maximum total I_{OL} for all output pins: 71 mA

If I_{OL} exceeds the test condition, V_{OL} may exceed the related specification. Pins are not guaranteed to sink current greater than the listed test conditions.

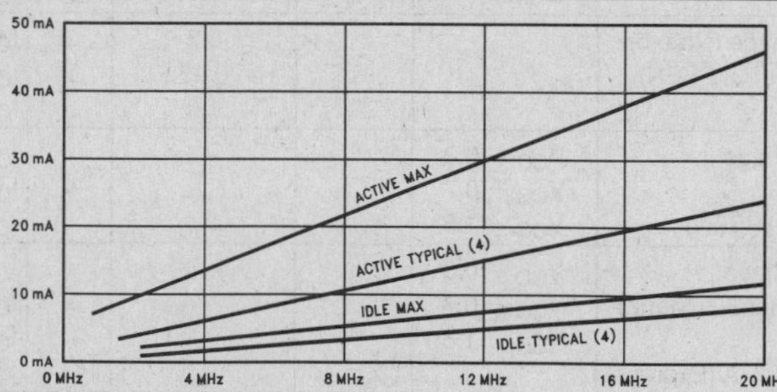

272322–5

I_{CC} Max at other frequencies is given by:

Active Mode I_{CC} MAX = 2.2 × FREQ + 3.1

Idle Mode I_{CC} MAX = 0.5 × FREQ + 1.5

Where FREQ is in MHz, I_{CC}MAX is given in mA.

Figure 5. I_{CC} vs Frequency

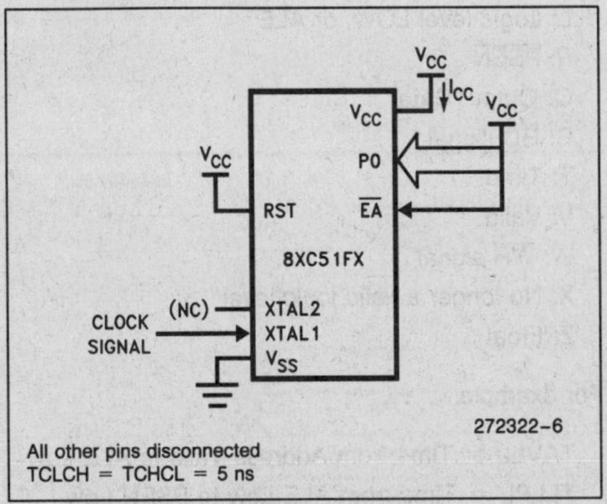

Figure 6. I_{CC} Test Condition, Active Mode

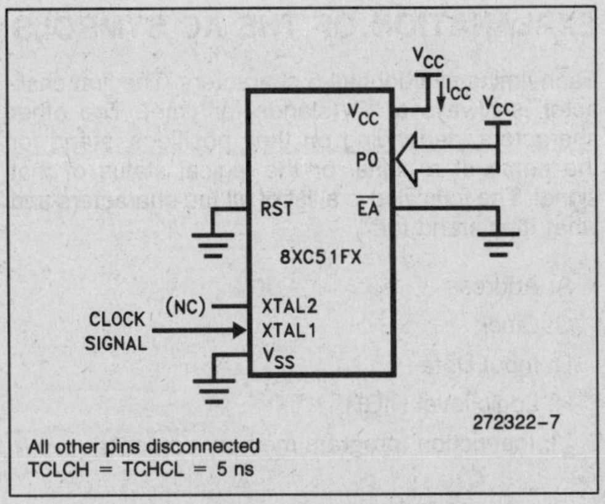

Figure 7. I_{CC} Test Condition Idle Mode

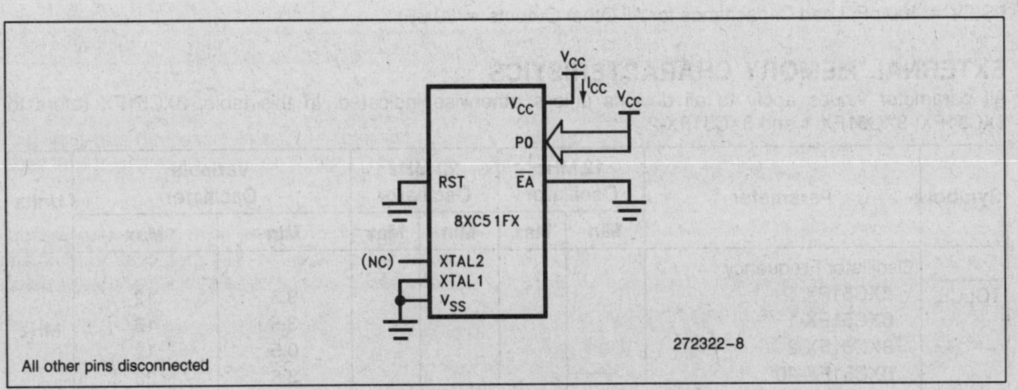

Figure 8. I_{CC} Test Condition, Power Down Mode. V_{CC} = 2.0V to 6.0V.

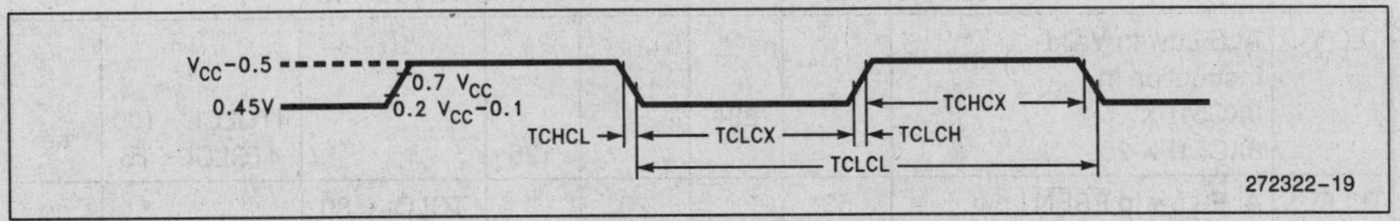

Figure 9. Clock Signal Waveform for I_{CC} Tests in Active and Idle Modes. TCLCH = TCHCL = 5 ns.

EXPLANATION OF THE AC SYMBOLS

Each timing symbol has 5 characters. The first character is always a 'T' (stands for time). The other characters, depending on their positions, stand for the name of a signal or the logical status of that signal. The following is a list of all the characters and what they stand for.

A: Address
C: Clock
D: Input Data
H: Logic level HIGH
I: Instruction (program memory contents)
L: Logic level LOW, or ALE
P: $\overline{PSEN}$
Q: Output Data
R: $\overline{RD}$ signal
T: Time
V: Valid
W: $\overline{WR}$ signal
X: No longer a valid logic level
Z: Float

For example,

TAVLL = Time from Address Valid to ALE Low
TLLPL = Time from ALE Low to PSEN Low

AC CHARACTERISTICS (Over Operating Conditions, Load Capacitance for Port 0, ALE/$\overline{PROG}$ and PSEN = 100 pF, Load Capacitance for All Other Outputs = 80 pF)

EXTERNAL MEMORY CHARACTERISTICS

All parameter values apply to all devices unless otherwise indicated. In this table, 8XC51FX refers to 8XC51FX, 8XC51FX-1 and 8XC51FX-2.

Symbol	Parameter	12 MHz Oscillator		20 MHz Oscillator		Variable Oscillator		Units
		Min	Max	Min	Max	Min	Max	
$\frac{1}{TCLCL}$	Oscillator Frequency 8XC51FX 8XC51FX-1 8XC51FX-2 8XC51FX-20					 3.5 3.5 0.5 3.5	 12 16 12 20	MHz
TLHLL	ALE Pulse Width	127		60		2TCLCL−40		ns
TAVLL	Address Valid to ALE Low	43		10		TCLCL−40		ns
TLLAX	Address Hold After ALE Low	53		20		TCLCL−30		ns
TLLIV	ALE Low to Valid Instruction In 8XC51FX 8XC51FX-20		234		125		4TCLCL−100 4TCLCL−75	ns
TLLPL	ALE Low to $\overline{PSEN}$ Low	53		20		TCLCL−30		ns
TPLPH	$\overline{PSEN}$ Pulse Width	205		105		3TCLCL−45		ns
TPLIV	$\overline{PSEN}$ Low to Valid Instruction In 8XC51FX 8XC51FX-20		145		60		3TCLCL−105 3TCLCL−90	ns
TPXIX	Input Instruction Hold After $\overline{PSEN}$	0		0		0		ns
TPXIZ	Input Instruction Float After $\overline{PSEN}$ 8XC51FX 8XC51FX-20		59		30		TCLCL-25 TCLCL-20	ns

EXTERNAL MEMORY CHARACTERISTICS (Continued)

All parameter values apply to all devices unless otherwise indicated

Symbol	Parameter	12 MHz Oscillator		20 MHz Oscillator		Variable Oscillator		Units
		Min	Max	Min	Max	Min	Max	
TAVIV	Address to Valid Instruction In		312		145		5TCLCL − 105	ns
TPLAZ	$\overline{\text{PSEN}}$ Low to Address Float		10		10		10	ns
TRLRH	$\overline{\text{RD}}$ Pulse Width	400		200		6TCLCL − 100		ns
TWLWH	$\overline{\text{WR}}$ Pulse Width	400		200		6TCLCL − 100		ns
TRLDV	$\overline{\text{RD}}$ Low to Valid Data In 8XC51FX 8XC51FX-20		 252 		 155		 5TCLCL − 165 5TCLCL − 95	ns
TRHDX	Data Hold After $\overline{\text{RD}}$	0		0		0		ns
TRHDZ	Data Float After $\overline{\text{RD}}$		107		40		2TCLCL − 60	ns
TLLDV	ALE Low to Valid Data In 8XC51FX 8XC51FX-20		 517 		 310		 8TCLCL − 150 8TCLCL − 90	ns
TAVDV	Address to Valid Data In 8XC51FX 8XC51FX-20		 585 		 360		 9TCLCL − 165 9TCLCL − 90	ns
TLLWL	ALE Low to $\overline{\text{RD}}$ or $\overline{\text{WR}}$ Low	200	300	100	200	3TCLCL − 50	3TCLCL + 50	ns
TAVWL	Address Valid to $\overline{\text{WR}}$ Low 8XC51FX 8XC51FX-20	 203 		 110		 4TCLCL − 130 4TCLCL − 90		ns
TQVWX	Data Valid to $\overline{\text{WR}}$ Transition 8XC51FX 8XC51FX-20	 33 		 15		 TCLCL − 50 TCLCL − 35		ns
TWHQX	Data Hold after $\overline{\text{WR}}$ 8XC51FX 8XC51FX-20	 33 		 10		 TCLCL − 50 TCLCL − 40		ns
TQVWH	Data Valid to $\overline{\text{WR}}$ High 8XC51FX 8XC51FX-20	 433 		 280		 7TCLCL − 150 7TCLCL − 70		ns
TRLAZ	$\overline{\text{RD}}$ Low to Address Float		0		0		0	ns
TWHLH	$\overline{\text{RD}}$ or $\overline{\text{WR}}$ High to ALE High	43	123	10	90	TCLCL − 40	TCLCL + 40	ns

EXTERNAL PROGRAM MEMORY READ CYCLE

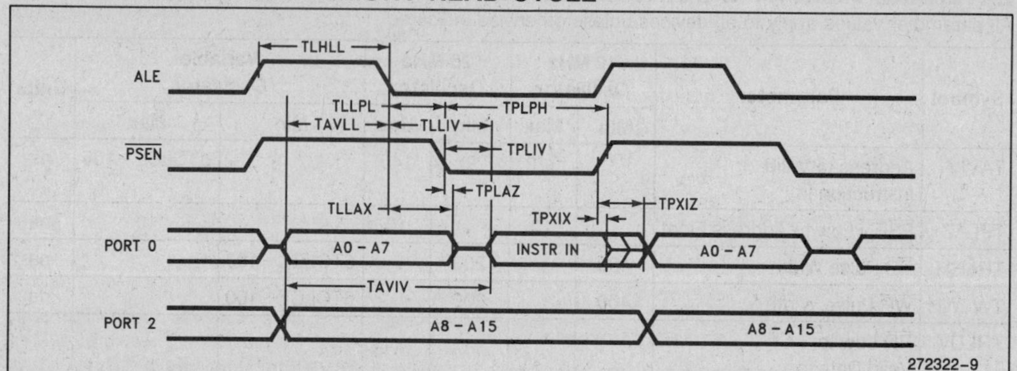

272322-9

EXTERNAL DATA MEMORY READ CYCLE

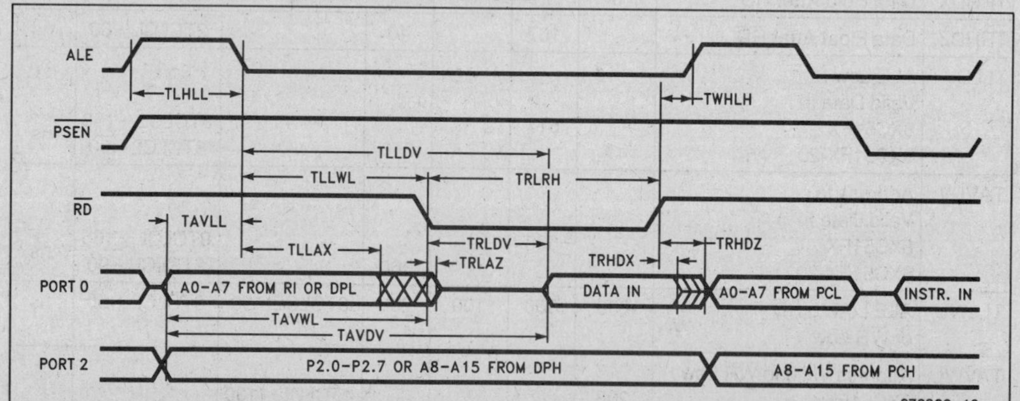

272322-10

EXTERNAL DATA MEMORY WRITE CYCLE

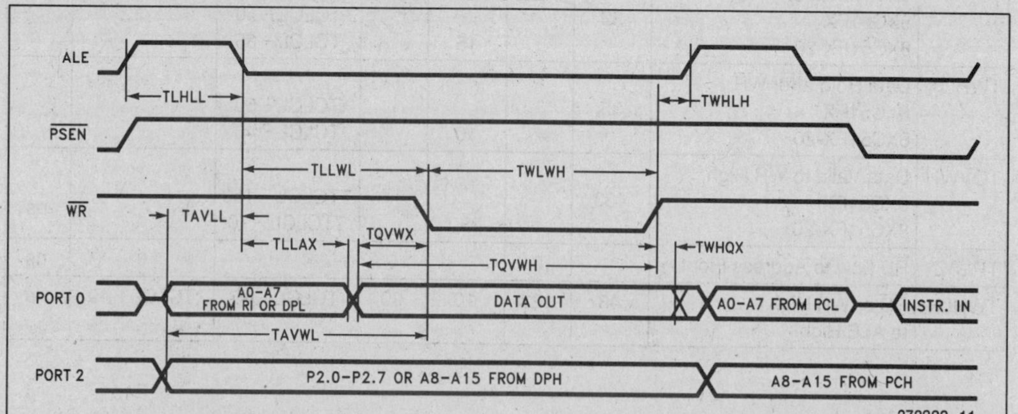

272322-11

SERIAL PORT TIMING—SHIFT REGISTER MODE

Test Conditions: Over Operating Conditions; Load Capacitance = 80 pF

Symbol	Parameter	12 MHz Oscillator		20 MHz Oscillator		Variable Oscillator		Units
		Min	Max	Min	Max	Min	Max	
TXLXL	Serial Port Clock Cycle Time	1		0.600		12TCLCL		μs
TQVXH	Output Data Setup to Clock Rising Edge	700		367		10TCLCL − 133		ns
TXHQX	Output Data Hold after Clock Rising Edge 8XC51FX 8XC51FX-20	 50		 50		 2TCLCL − 117 2TCLCL − 50		ns
TXHDX	Input Data Hold After Clock Rising Edge	0		0		0		ns
TXHDV	Clock Rising Edge to Input Data Valid		700		367		10TCLCL − 133	ns

SHIFT REGISTER MODE TIMING WAVEFORMS

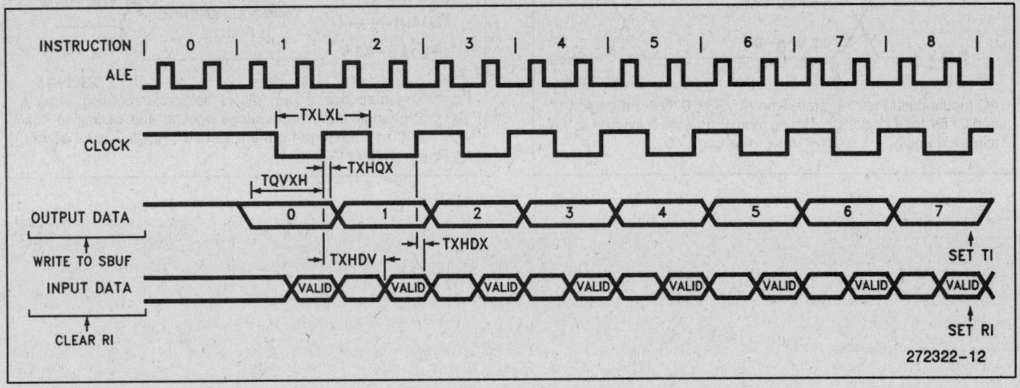

EXTERNAL CLOCK DRIVE

Symbol	Parameter	Min	Max	Units
1/TCLCL	Oscillator Frequency 8XC51FX 8XC51FX-1 8XC51FX-2 8XC51FX-20	 3.5 3.5 0.5 3.5	 12 16 12 20	MHz
TCHCX	High Time	20		ns
TCLCX	Low Time	20		ns
TCLCH	Rise Time		20	ns
TCHCL	Fall Time		20	ns

EXTERNAL CLOCK DRIVE WAVEFORM

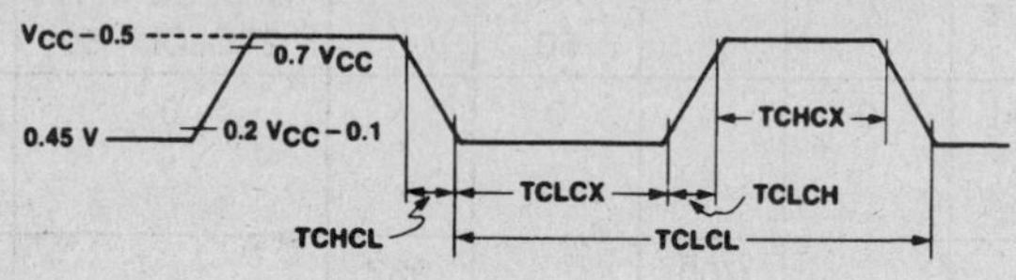

272322–13

AC TESTING INPUT, OUTPUT WAVEFORMS

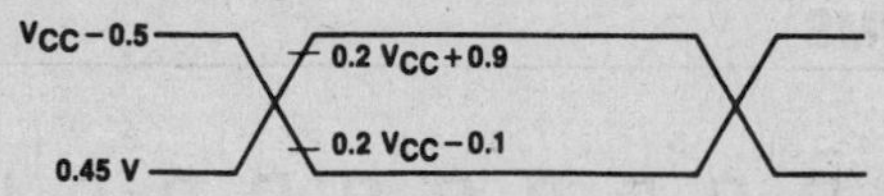

272322–14

AC Inputs during testing are driven at $V_{CC}-0.5V$ for a Logic "1" and 0.45V for a Logic "0". Timing measurements are made at V_{IH} min for a Logic "1" and V_{OL} max for a Logic "0".

FLOAT WAVEFORMS

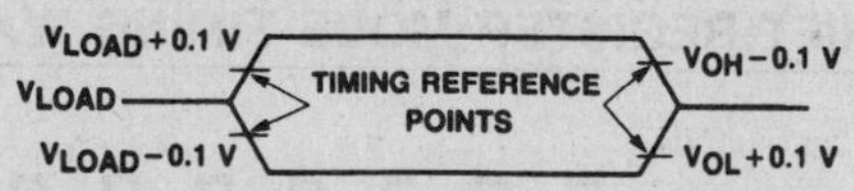

272322–15

For timing purposes a port pin is no longer floating when a 100 mV change from load voltage occurs, and begins to float when a 100 mV change from the loaded V_{OH}/V_{OL} level occurs. $I_{OL}/I_{OH} = \pm 20$ mA.

PROGRAMMING THE EPROM

To be programmed, the part must be running with a 4 to 6 MHz oscillator. (The reason the oscillator needs to be running is that the internal bus is being used to transfer address and program data to appropriate internal EPROM locations.) The address of an EPROM location to be programmed is applied to Port 1 and pins P2.0 - P2.4 of Port 2, while the code byte to be programmed into that location is applied to Port 0. The other Port 2 and 3 pins, RST $\overline{\text{PSEN}}$, and $\overline{\text{EA}}/V_{PP}$ should be held at the "Program" levels indicated in Table 4. ALE/$\overline{\text{PROG}}$ is pulsed low to program the code byte into the addressed EPROM location. The setup is shown in Figure 10.

Normally $\overline{\text{EA}}/V_{PP}$ is held at logic high until just before ALE/$\overline{\text{PROG}}$ is to be pulsed. Then $\overline{\text{EA}}/V_{PP}$ is raised to V_{PP}, ALE/$\overline{\text{PROG}}$ is pulsed low, and then $\overline{\text{EA}}/V_{PP}$ is returned to a valid high voltage. The voltage on the $\overline{\text{EA}}/V_{PP}$ pin must be at the valid $\overline{\text{EA}}/V_{PP}$ high level before a verify is attempted. Waveforms and detailed timing specifications are shown in later sections of this data sheet.

NOTE:

- $\overline{\text{EA}}/V_{PP}$ pin must not be allowed to go above the maximum specified V_{PP} level for any amount of time. Even a narrow glitch above that voltage level can cause permanent damage to the device. The V_{PP} source should be well regulated and free of glitches.

Table 4. EPROM Programming Modes

Mode		RST	$\overline{\text{PSEN}}$	ALE/$\overline{\text{PROG}}$	$\overline{\text{EA}}/V_{PP}$	P2.6	P2.7	P3.3	P3.6	P3.7
Program Code Data		H	L	⊔	12.75V	L	H	H	H	H
Verify Code Data		H	L	H	H	L	L	L	H	H
Program Encryption Array Address 0–3FH		H	L	⊔	12.75V	L	H	H	L	H
Program Lock Bits	Bit 1	H	L	⊔	12.75V	H	H	H	H	H
	Bit 2	H	L	⊔	12.75V	H	H	H	L	L
	Bit 3	H	L	⊔	12.75V	H	L	H	H	L
Read Signature Byte		H	L	H	H	L	L	L	L	L

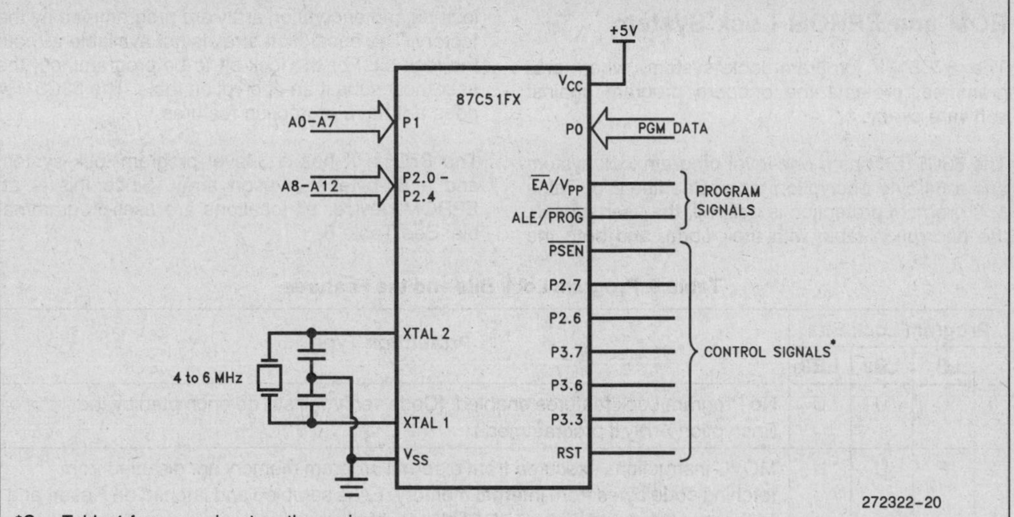

*See Table 4 for proper input on these pins

Figure 10. Programming the EPROM

PROGRAMMING ALGORITHM

Refer to Table 4 and Figures 10 and 11 for address, data, and control signals set up. To program the 87C51FX the following sequence must be exercised.

1. Input the valid address on the address lines.
2. Input the appropriate data byte on the data lines.
3. Activate the correct combination of control signals.
4. Raise $\overline{EA}/V_{PP}$ from V_{CC} to 12.75V $\pm$0.25V.
5. Pulse, ALE/$\overline{PROG}$ 5 times for the EPROM array, and 25 times for the encryption table and the lock bits.

Repeat 1 through 5 changing the address and data for the entire array or until the end of the object file is reached.

PROGRAM VERIFY

Program verify may be done after each byte or block of bytes is programmed. In either case a complete verify of the programmed array will ensure reliable programming of the 87C51FX.

The lock bits cannot be directly verified. Verification of the lock bits is done by observing that their features are enabled.

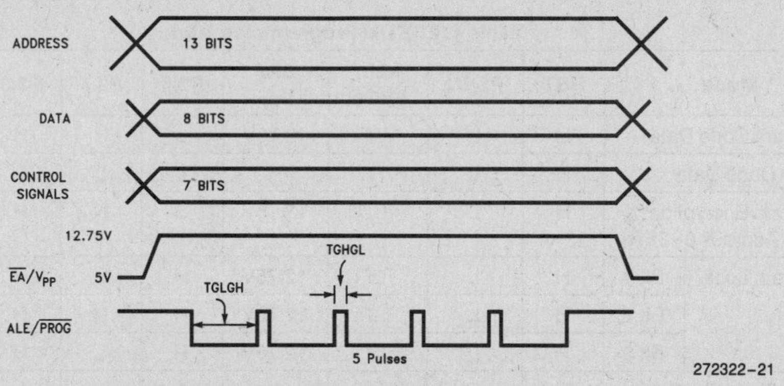

Figure 11. Programming Signals Waveforms

ROM and EPROM Lock System

The 87C51FX program lock system, when programmed, protects the onboard program against software piracy.

The 83C51FX has a one-level program lock system and a 64-byte encryption table. See line 2 of Table 5. If program protection is desired, the user submits the encryption table with their code, and both the lock-bit and encryption array are programmed by the factory. The encryption array is not available without the lock bit. For the lock bit to be programmed, the user must submit an encryption table. The 83C51FA does not have protection features.

The 87C51FX has a 3-level program lock system and a 64-byte encryption array. Since this is an EPROM device, all locations are user-programmable. See Table 5.

Table 5. Program Lock Bits and the Features

Program Lock Bits				Protection Type
	LB1	LB2	LB3	
1	U	U	U	No Program Lock features enabled. (Code verify will still be encrypted by the Encryption Array if programmed.)
2	P	U	U	MOVC instructions executed from external program memory are disabled from fetching code bytes from internal memory, EA is sampled and latched on Reset, and further programming of the EPROM is disabled.
3	P	P	U	Same as 2, also verify is disabled.
4	P	P	P	Same as 3, also external execution is disabled.

Any other combination of the lock bits is not defined.

Encryption Array

Within the EPROM array are 64 bytes of Encryption Array that are initially unprogrammed (all 1's). Every time that a byte is addressed during a verify, 6 address lines are used to select a byte of the Encryption Array. This byte is then exclusive-NOR'ed (XNOR) with the code byte, creating an Encryption Verify byte. The algorithm, with the array in the unprogrammed state (all 1's), will return the code in its original, unmodified form. For programming the Encryption Array, refer to Table 4 (Programming the EPROM).

When using the encryption array, one important factor needs to be considered. If a code byte has the value 0FFH, verifying the byte will produce the encryption byte value. If a large block (>64 bytes) of code is left unprogrammed, a verification routine will display the contents of the encryption array. For this reason all unused code bytes should be programmed with some value other than 0FFH, and not all of them the same value. This will ensure maximum program protection.

Program Lock Bits

The 87C51FX has 3 programmable lock bits that when programmed according to Table 5 will provide different levels of protection for the on-chip code and data.

Erasing the EPROM also erases the encryption array and the program lock bits, returning the part to full functionality.

Reading the Signature Bytes

The 87C51FX has 3 signature bytes in locations 30H, 31H, and 60H. The 83C51FA has 2 signature bytes in locations 30H and 31H. To read these bytes follow the procedure for EPROM verify, but activate the control lines provided in Table 4 for Read Signature Byte.

Location	Device	Contents
30H	All	89H
31H	83C51FA All Others	53H 58H
60H	87C51FA	FAH
	83C51FB	7BH/FBH
	87C51FB	FBH
	83C51FC	7CH/FCH
	87C51FC	FCH

Erasure Characteristics (Windowed Packages Only)

Erasure of the EPROM begins to occur when the chip is exposed to light with wavelength shorter than approximately 4,000 Angstroms. Since sunlight and fluorescent lighting have wavelengths in this range, exposure to these light sources over an extended time (about 1 week in sunlight, or 3 years in room-level fluorescent lighting) could cause inadvertent erasure. If an application subjects the device to this type of exposure, it is suggested that an opaque label be placed over the window.

The recommended erasure procedure is exposure to ultraviolet light (at 2537 Angstroms) to an integrated dose of at least 15 W-sec/cm. Exposing the EPROM to an ultraviolet lamp of 12,000 μW/cm rating for 30 minutes, at a distance of about 1 inch, should be sufficient.

Erasure leaves all the EPROM Cells in a 1's state.

EPROM PROGRAMMING AND VERIFICATION CHARACTERISTICS

(T_A = 21°C to 27°C; V_{CC} = 5V ±20%; V_{SS} = 0V)

Symbol	Parameter	Min	Max	Units
V_{PP}	Programming Supply Voltage	12.5	13.0	V
I_{PP}	Programming Supply Current		75	mA
1/TCLCL	Oscillator Frequency	4	6	MHz
TAVGL	Address Setup to $\overline{PROG}$ Low	48TCLCL		
TGHAX	Address Hold after $\overline{PROG}$	48TCLCL		
TDVGL	Data Setup to $\overline{PROG}$ Low	48TCLCL		
TGHDX	Data Hold after $\overline{PROG}$	48TCLCL		
TEHSH	P2.7 (ENABLE) High to V_{PP}	48TCLCL		
TSHGL	V_{PP} Setup to $\overline{PROG}$ Low	10		μs
TGHSL	V_{PP} Hold after $\overline{PROG}$	10		μs
TGLGH	$\overline{PROG}$ Width	90	110	μs
TAVQV	Address to Data Valid		48TCLCL	
TELQV	ENABLE Low to Data Valid		48TCLCL	
TEHQZ	Data Float after ENABLE	0	48TCLCL	
TGHGL	$\overline{PROG}$ High to $\overline{PROG}$ Low	10		μs

EPROM PROGRAMMING AND VERIFICATION WAVEFORMS

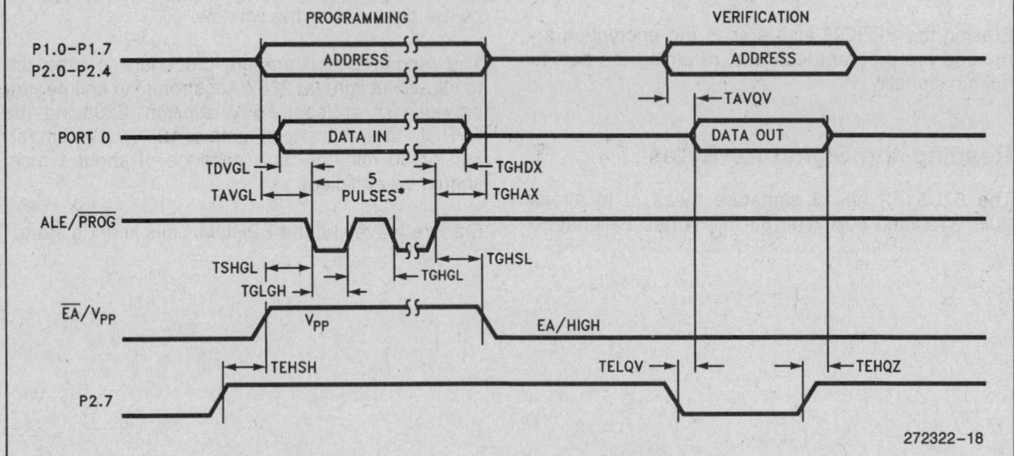

NOTE:
*5 pulses for the EPROM array, 25 pulses for the encryption table and lock bits.

83C51FA ROM CHARACTERISTICS

Table 6 shows the logic levels for verifying the code data and reading the signature bytes on the 83C51FA.

Table 6. ROM Modes

Mode	RST	$\overline{PSEN}$	ALE	$\overline{EA}$	P2.7	P2.6	P3.6	P3.7
Verify Code Data	1	0	1	1	0	0	1	1
Read Signature	1	0	1	1	0	0	0	0

NOTES:
"1" = Valid high for that pin
"0" = Valid low for that pin

Program Verification

The on-chip Program Memory can be read out for verification purposes, if desired. The address of the Program Memory location to be read is applied to Port 1 and pins P2.0–P2.4. The other pins should be held at the "Verify" levels indicated in Table 6. The contents of the addressed locations will come out on Port 0. External pullups are required on Port 0 for this operation.

Figure 12 shows the setup for verifying the program memory.

2

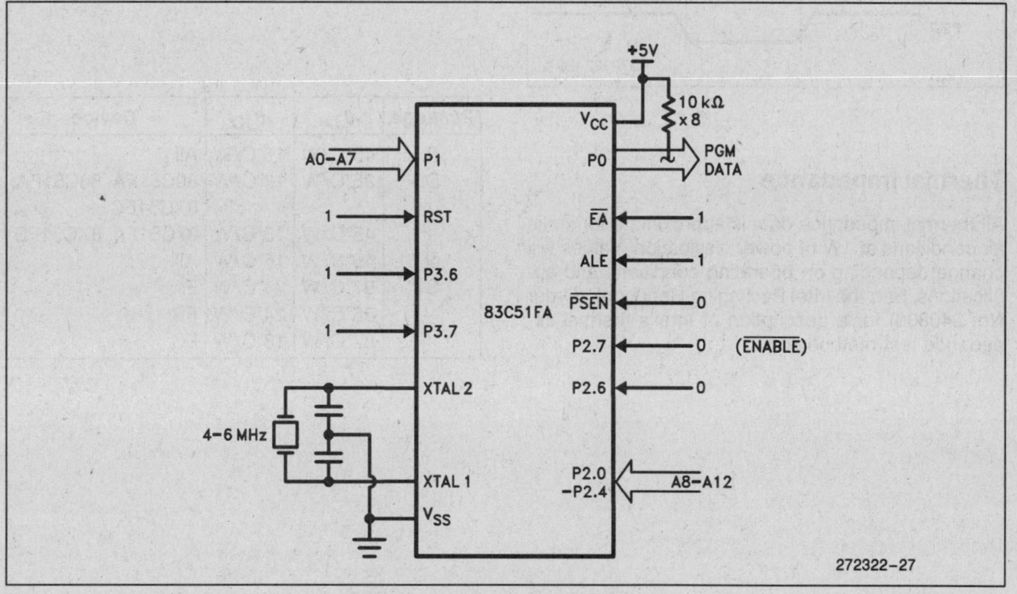

Figure 12. Verifying the ROM

ROM VERIFICATION CHARACTERISTICS

T_A = 21°C to 27°c; V_{CC} = 5V ±0.25V; V_{SS} = 0V

Symbol	Parameter	Min	Max	Units
1/TCLCL	Oscillator Frequency	4	6	MHz
TAVQV	Address to Data Valid		48TCLCL	
TELQV	ENABLE Low to Data Valid		48TCLCL	
TEHQZ	Data Float after ENABLE	0	48TCLCL	

ROM VERIFICATION WAVEFORMS

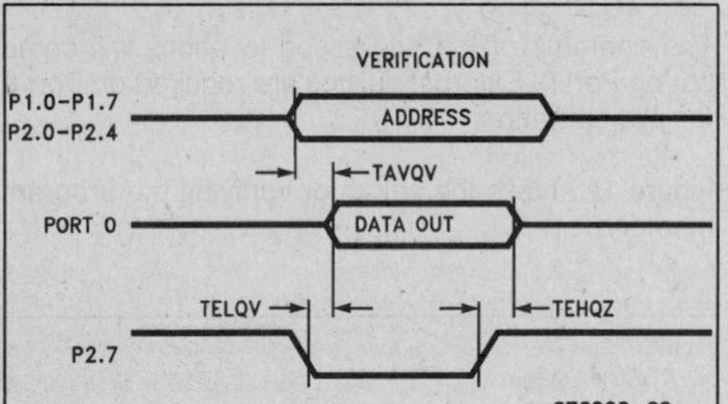

272322–28

Reading the Signature Bytes

The signature bytes are read by the same procedure as a normal verification of locations 030H and 031H, except that P3.6 and P3.7 need to be pulled to a logic low. The values returned are:

(030H) = 89H indicates manufacture by Intel
(031H) = 53H indicates 83C51FA

Thermal Impedance

All thermal impedance data is approximate for static air conditions at 1W of power dissipation. Values will change depending on operating conditions and applications. See the Intel Packaging Handbook (Order No. 240800) for a description of Intel's thermal impedance test methodology.

Package	θ_{JA}	θ_{JC}	Device
P	45°C/W	16°C/W	All
D	36°C/W	13°C/W	80C51FA, 83C51FA, 8XC51FC
	45°C/W	15°C/W	87C51FA, 8XC51FB
N	46°C/W	16°C/W	All
S	97°C/W	24°C/W	FA
	96°C/W	24°C/W	FB
	87°C/W	18°C/W	FC

DATA SHEET REVISION HISTORY

Data sheets are changed as new device information becomes available. Verify with your local Intel sales office that you have the latest version before finalizing a design or ordering devices.

The following differences exist between this datasheet (272322-002) and the previous version (272322-001):

1. Removed 8XC51FX-L from datasheet.
2. Include V_{OH1} for 83C51FA (Express)/80C51FA (Express).

This 8XC51FX datasheet (272322-001) replaces the following datasheets:

87C51FA/83C51FA/80C51FA	270258-007
83C51FA/80C51FA EXPRESS	270620-001
87C51FA EXPRESS	270619-001
87C51FA-20/-3	272081-002
87C51FB/83C51FB	270563-005
87C51FB-20/-3 83C51FB-20/-3	272080-002
87C51FB/83C51FB EXPRESS	270767-002
87C51FC/83C51FC	270789-004
87C51FC/83C51FC EXPRESS	270903-001
87C51FC-20/-3 83C51FC-20/-3	272028-002

2

8XL52/54/58
LOW VOLTAGE
CHMOS SINGLE-CHIP 8-BIT MICROCONTROLLERS

Commercial/Express

87L52/80L52/87L54/80L54/87L58/80L58

- **High Performance CHMOS OTP ROM**
- **Low Voltage Operation**
- **20 MHz Commercial/16 MHz Express Operation**
- **Three 16-Bit Timer/Counters**
- **Up/Down Timer/Counter**
- **Three Level Program Lock System**
- **8K/16K/32K On-Chip Program Memory**
- **256 Bytes of On-Chip Data RAM**
- **Improved Quick Pulse Programming Algorithm**
- **Boolean Processor**
- **32 Programmable I/O Lines**
- **6 Interrupt Sources**
- **Four Level Interrupt Priority**
- **Programmable Serial Channel with:**
  - **—Framing Error Detection**
  - **—Automatic Address Recognition**
- **64K External Program Memory Space**
- **64K External Data Memory Space**
- **MCS® 51 Microcontroller Compatible Instruction Set**
- **Power Saving Idle and Power Down Modes**
- **ONCE (On-Circuit Emulation) Mode**
- **Extended Temperature Range (−40°C to +85°C)**

MEMORY ORGANIZATION

OTP ROM Version	ROM Version	ROM/ OTP ROM Bytes	RAM Bytes
87L52	80L52	8K	256
87L54	80L54	16K	256
87L58	80L58	32K	256

These devices can address up to 64 Kbytes of external program/data memory.

The Intel 8XL52/8XL54/8XL58 is a single-chip control oriented microcontroller which is fabricated on Intel's reliable CHMOS III-E technology. Being a member of the MCS 51 microcontroller family, the 8XL52/8XL54/8XL58 uses the same powerful instruction set, has the same architecture, and is pin-for-pin compatible with the existing MCS 51 microcontroller products.

The 8XL5X is a 3V version of current 8XC5X and will operate from 2.7V to 3.6V at a frequency range of 3.5 MHz to 16 MHz (Express)/20 MHz (Commercial).

For the remainder of this document, the 8XL52, 8XL54, 8XL58 will be referred to as the 8XL5X, unless information applies to a specific device.

November 1994
Order Number: 272468-002

	Standard	-1	-20*
80L52	X	X	X
87L52	X	X	X
80L54	X	X	X
87L54	X	X	X
80L58	X	X	X
87L58	X	X	X

NOTE:

Standard	3.5 MHz to 12 MHz; 2.7V to 3.6V
-1	3.5 MHz to 16 MHz; 2.7V to 3.6V
-20*	3.5 MHz to 20 MHz; 2.7V to 3.6V

*Only available for commercial standard temperature range, not available at express temperature range.

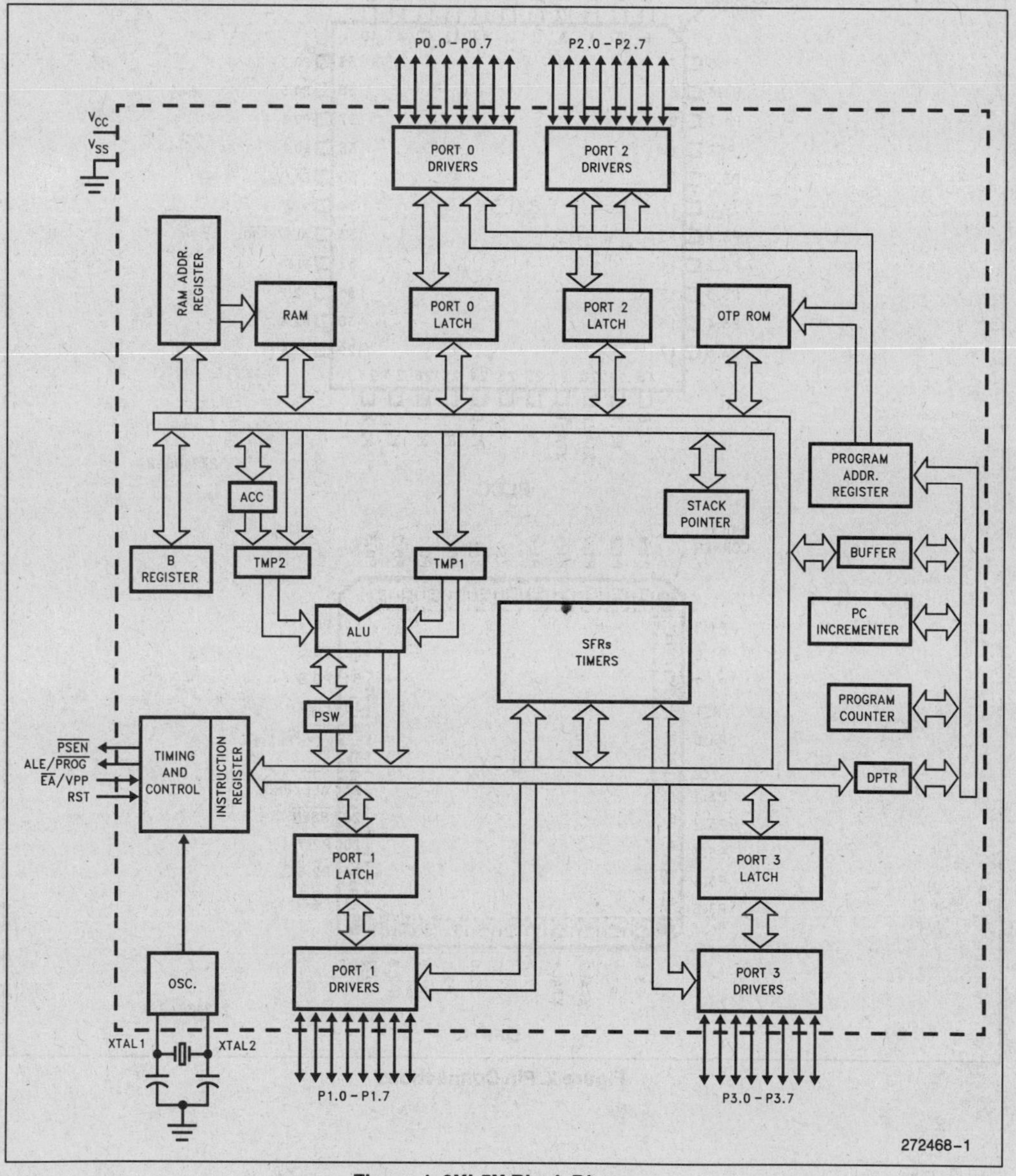

Figure 1. 8XL5X Block Diagram

PROCESS INFORMATION

The 8XL52/8XL54/8XL58 is manufactured on P629.5, a CHMOS III-E process. Additional process and reliability information is available in Intel's *Components Quality and Reliability Handbook,* Order Number 210997.

PACKAGES

Part	Prefix	Package Type
8XL5X	N	44-Pin PLCC (OTP)
	S	44-Pin QFP (OTP)

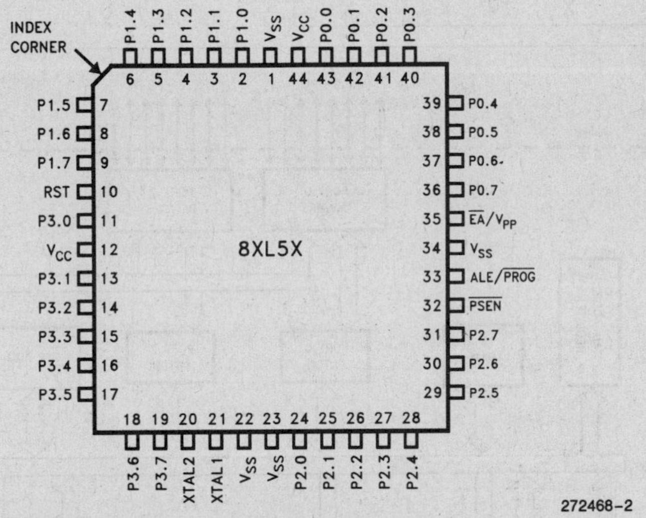

PLCC

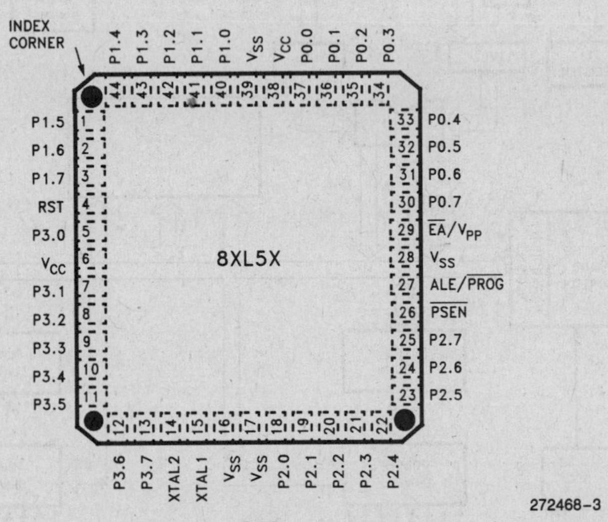

QFP

Figure 2. Pin Connections

ADVANCE INFORMATION

PIN DESCRIPTIONS

V_{CC}: Supply voltage.

V_{SS}: Circuit ground.

Port 0: Port 0 is an 8-bit, open drain, bidirectional I/O port. As an output port each pin can sink several inputs. Port 0 pins that have 1's written to them float, and in that state can be used as high-impedance inputs.

Port 0 is also the multiplexed low-order address and data bus during accesses to external Program and Data Memory. In this application it uses strong internal pullups when emitting 1's, and can source and sink several inputs.

Port 0 also receives the code bytes during OTP ROM programming, and outputs the code bytes during program verification. External pullup resistors are required during program verification.

Port 1: Port 1 is an 8-bit bidirectional I/O port with internal pullups. The Port 1 output buffers can drive several inputs. Port 1 pins that have 1's written to them are pulled high by the internal pullups, and in that state can be used as inputs. As inputs, Port 1 pins that are externally pulled low will source current (I_{IL}, on the data sheet) because of the internal pullups.

In addition, Port 1 serves the functions of the following special features of the 8XL5X:

Port Pin	Alternate Function
P1.0	T2 (External Count Input to Timer/ Counter 2), Clock Out
P1.1	T2EX (Timer/ Counter 2 Capture/ Reload Trigger and Direction Control)

Port 1 receives the low-order address bytes during OTP ROM programming and verifying.

Port 2: Port 2 is an 8-bit bidirectional I/O port with internal pullups. The Port 2 output buffers can drive several inputs. Port 2 pins that have 1's written to them are pulled high by the internal pullups, and in that state can be used as inputs. As inputs, Port 2 pins that are externally pulled low will source current (I_{IL}, on the data sheet) because of the internal pullups.

Port 2 emits the high-order address byte during fetches from external Program Memory and during accesses to external Data Memory that use 16-bit addresses (MOVX @DPTR). In this application it uses strong internal pullups when emitting 1's. During accesses to external Data Memory that use 8-bit addresses (MOVX @Ri), Port 2 emits the contents of the P2 Special Function Register.

Some Port 2 pins receive the high-order address bits during OTP ROM programming and program verification.

Port 3: Port 3 is an 8-bit bidirectional I/O port with internal pullups. The Port 3 output buffers can drive several inputs. Port 3 pins that have 1's written to them are pulled high by the internal pullups, and in that state can be used as inputs. As inputs, Port 3 pins that are externally pulled low will source current (I_{IL}, on the data sheet) because of the pullups.

Port 3 also serves the functions of various special features of the MCS-51 Family, as listed below:

Port Pin	Alternate Function
P3.0	RXD (serial input port)
P3.1	TXD (serial output port)
P3.2	$\overline{\text{INT0}}$ (external interrupt 0)
P3.3	$\overline{\text{INT1}}$ (external interrupt 1)
P3.4	T0 (Timer 0 external input)
P3.5	T1 (Timer 1 external input)
P3.6	$\overline{\text{WR}}$ (external data memory write strobe)
P3.7	$\overline{\text{RD}}$ (external data memory read strobe)

RST: Reset input. A high on this pin for two machine cycles while the oscillator is running resets the device. The port pins will be driven to their reset condition when a minimum V_{IH2} voltage is applied whether the oscillator is running or not. An internal pulldown resistor permits a power-on reset with only a capacitor connected to V_{CC}.

ALE: Address Latch Enable output pulse for latching the low byte of the address during accesses to external memory. This pin (ALE/$\overline{PROG}$) is also the program pulse input during OTP ROM programming for the 87L5X.

In normal operation ALE is emitted at a constant rate of 1/6 the oscillator frequency, and may be used for external timing or clocking purposes. Note, however, that one ALE pulse is skipped during each access to external Data Memory.

If desired, ALE operation can be disabled by setting bit 0 of SFR location 8EH. With this bit set, the pin is weakly pulled high. However, the ALE disable feature will be suspended during a MOVX or MOVC instruction, idle mode, power down mode and ICE mode. The ALE disable feature will be terminated by reset. When the ALE disable feature is suspended or terminated, the ALE pin will no longer be pulled up weakly. Setting the ALE-disable bit has no affect if the microcontroller is in external execution mode.

Throughout the remainder of this data sheet, ALE will refer to the signal coming out of the ALE/$\overline{PROG}$ pin, and the pin will be referred to as the ALE/$\overline{PROG}$ pin.

$\overline{PSEN}$: Program Store Enable is the read strobe to external Program Memory.

When the 8XL5X is executing code from external Program Memory, $\overline{PSEN}$ is activated twice each machine cycle, except that two $\overline{PSEN}$ activations are skipped during each access to external Data Memory.

$\overline{EA}$/V_{PP}: External Access enable. $\overline{EA}$ must be strapped to VSS in order to enable the device to fetch code from external Program Memory locations 0000H to 0FFFH. Note, however, that if either of the Program Lock bits are programmed, $\overline{EA}$ will be internally latched on reset.

$\overline{EA}$ must be strapped to V_{CC} for internal program executions.

This pin also receives the programming supply voltage (V_{PP}) during OTP ROM programming.

XTAL1: Input to the inverting oscillator amplifier.

XTAL2: Output from the inverting oscillator amplifier.

OSCILLATOR CHARACTERISTICS

XTAL1 and XTAL2 are the input and output, respectively, of a inverting amplifier which can be configured for use as an on-chip oscillator, as shown in Figure 3. Either a quartz crystal or ceramic resonator may be used. More detailed information concerning the use of the on-chip oscillator is available in Application Note AP-155, "Oscillators for Microcontrollers."

To drive the device from an external clock source, XTAL1 should be driven, while XTAL2 floats, as shown in Figure 4. There are no requirements on the duty cycle of the external clock signal, since the input to the internal clocking circuitry is through a divide-by-two flip-flop, but minimum and maximum high and low times specified on the data sheet must be observed.

An external oscillator may encounter as much as a 100 pF load at XTAL1 when it starts up. This is due to interaction between the amplifier and its feedback capacitance. Once the external signal meets the V_{IL} and V_{IH} specifications the capacitance will not exceed 20 pF.

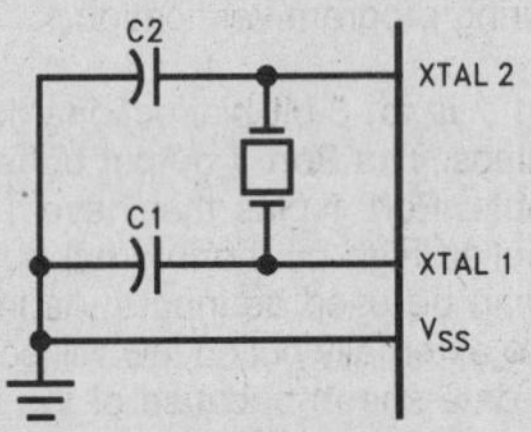

C1, C2 = 30 pF ±10 pF for Crystals
For Ceramic Resonators, contact resonator manufacturer.

Figure 3. Oscillator Connections

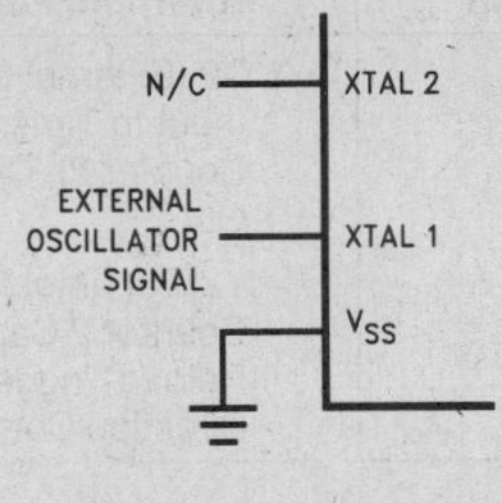

Figure 4. External Clock Drive Configuration

IDLE MODE

The user's software can invoke the Idle Mode. When the microcontroller is in this mode, power consumption is reduced. The Special Function Registers and the onboard RAM retain their values during Idle, but the processor stops executing instructions. Idle Mode will be exited if the chip is reset or if an enabled interrupt occurs. The PCA timer/counter can optionally be left running or paused during Idle Mode.

POWER DOWN MODE

To save even more power, a Power Down mode can be invoked by software. In this mode, the oscillator is stopped and the instruction that invoked Power Down is the last instruction executed. The on-chip RAM and Special Function Registers retain their values until the Power Down mode is terminated.

On the 8XL5X either hardware reset or external interrupt can cause an exit from Power Down. Reset redefines all the SFRs but does not change the on-chip RAM. An external interrupt allows both the SFRs and the on-chip RAM to retain their values.

To properly terminate Power Down the reset or external interrupt should not be executed before V_{CC} is restored to its normal operating level and must be held active long enough for the oscillator to restart and stabilize (normally less than 10 ms).

With an external interrupt, INT0 or INT1 must be enabled and configured as level-sensitive. Holding the pin low restarts the oscillator but bringing the pin back high completes the exit. Once the interrupt is serviced, the next instruction to be executed after RETI will be the one following the instruction that put the device into Power Down.

DESIGN CONSIDERATION

- The 8XL5X will operate from 2.7V to 3.6V with a frequency range of 3.5 MHz to 16 MHz (Express)/20 MHz (Commercial). Operating beyond these specifications could cause improper device functionality.
- All V_{CC} and V_{SS} pins must be connected. Please refer to Figure 2, Pin Connections, for the specific pins.
- When the idle mode is terminated by a hardware reset, the device normally resumes program execution, from where it left off, up to two machine cycles before the internal reset algorithm takes control. On-chip hardware inhibits access to internal RAM in this event, but access to the port pins is not inhibited. To eliminate the possibility of an unexpected write when Idle is terminated by reset, the instruction following the one that invokes Idle should not be one that writes to a port pin or to external memory.

ONCE MODE

The ONCE ("On-Circuit Emulation") Mode facilitates testing and debugging of systems using the 8XL5X without the 8XL5X having to be removed from the circuit. The ONCE Mode is invoked by:

1) Pull ALE low while the device is in reset and $\overline{PSEN}$ is high;
2) Hold ALE low as RST is deactivated.

While the device is in ONCE Mode, the Port 0 pins float, and the other port pins and ALE and $\overline{PSEN}$ are weakly pulled high. The oscillator circuit remains active. While the 8XL5X is in this mode, an emulator or test CPU can be used to drive the circuit. Normal operation is restored when a normal reset is applied.

2

Table 1. Status of the External Pins during Idle and Power Down

Mode	Program Memory	ALE	$\overline{PSEN}$	PORT0	PORT1	PORT2	PORT3
Idle	Internal	1	1	Data	Data	Data	Data
Idle	External	1	1	Float	Data	Address	Data
Power Down	Internal	0	0	Data	Data	Data	Data
Power Down	External	0	0	Float	Data	Data	Data

NOTE:
For more detailed information on the reduced power modes refer to current Embedded Microcontrollers and Processors Handbook Volume I, #270646, and Application Note AP-252 (Embedded Applications Handbook), #270648, "Designing with the 80C51BH."

8XL5X EXPRESS

The Intel EXPRESS system offers enhancements to the operational specifications of the MCS-51 family of microcontrollers. These EXPRESS products are designed to meet the needs of those applications whose operating requirements exceed commercial standards.

The EXPRESS program includes the commercial standard temperature range with burn-in and an extended temperature range with or without burn-in.

With the commercial standard temperature range, operational characteristics are guaranteed over the temperature range of 0°C to 70°C. With the extended temperature range option, operational characteristics are guaranteed over the range of −40°C to +85°C.

Package types and EXPRESS versions are identified by a one- or two-letter prefix to the part number. The prefixes are listed in Table 2.

For the extended temperature range option, this data sheet specifies the parameters which deviate from their commercial temperature range limits.

Table 2. Prefix Identification

Prefix	Package Type	Temperature Range
N	PLCC	Commercial
S	QFP	Commercial
TN	PLCC	Extended
TS	QFP	Extended

NOTE:
Contact your distributor or local sales office to match the EXPRESS prefix with the proper device.

EXAMPLES:
N87L51FC indicates 87L51FC in a PLCC package and specified for commercial temperature range, without burn-in.

TN87L51FC indicates 87L51FC in a PLCC package and specified for extended temperature range with burn-in.

ABSOLUTE MAXIMUM RATINGS*

Ambient Temperature Under Bias . −40°C to +85°C

Storage Temperature .... −65°C to +150°C

Voltage on EA/V_{PP} Pin to V_{SS} .... 0V to +13.0V

Voltage on Any Other Pin to V_{SS} .. −0.5V to +6.5V

I_{OL} per I/O Pin .... 15 mA

Power Dissipation.......................... 1.5W
(based on PACKAGE heat transfer limitations, not device power consumption)

NOTICE: This data sheet contains information on products in the sampling and initial production phases of development. It is valid for the devices indicated in the revision history. The specifications are subject to change without notice.

******WARNING: Stressing the device beyond the "Absolute Maximum Ratings" may cause permanent damage. These are stress ratings only. Operation beyond the "Operating Conditions" is not recommended and extended exposure beyond the "Operating Conditions" may affect device reliability.*

2

OPERATING CONDITIONS

Symbol	Description	Min	Max	Units
T_A	Ambient Temperature Under Bias			
	Commercial	0	+70	°C
	Express	−40	+85	°C
V_{CC}	Supply Voltage	2.7	3.6	V

DC CHARACTERISTICS (Over Operating Conditions)

All parameter values apply to all devices unless otherwise indicated.

Symbol	Parameter	Min	Max	Units	Test Conditions
V_{IL}	Input Low Voltage (except XTAL1, RST)	−0.5	0.8	V	
V_{IL1}	Input Low Voltage (XTAL1, RST)	−0.5	0.2 V_{CC} − 0.1	V	
V_{IH}	Input High Voltage (Except XTAL1, RST, $\overline{EA}$)	2.0	V_{CC} + 0.5	V	
V_{IH1}	Input High Voltage ($\overline{EA}$)	V_{CC} − 1.0	V_{CC} + 0.5	V	
V_{IH2}	Input High Voltage (XTAL1, RST)	0.7 V_{CC}	V_{CC} + 0.5	V	
V_{OL}	Output Low Voltage (Note 4) (Ports 1, 2 and 3)		0.4	V	I_{OL} = 1.6 mA (Note 1)
V_{OL1}	Output Low Voltage (Note 4) (Port 0, ALE/$\overline{PSEN}$)		0.4	V	I_{OL} = 3.2 mA (Note 1)
V_{OH}	Output High Voltage (Ports 1, 2 and 3, ALE, $\overline{PSEN}$	V_{CC} − 0.7		V	I_{OH} = −30 μA (Note 2)
V_{OH1}	Output High Voltage (Port 0 in External Bus Mode)	2.4		V	I_{OH} = −1.0 mA (Note 2)
I_{IL}	Logical 0 Input Current (Ports 1, 2 and 3)		−50	μA	V_{IN} = 0.4V
I_{LI}	Input Leakage Current (Port 0)		±10	μA	0 < V_{IN} < V_{CC}

DC CHARACTERISTICS (Over Operating Conditions)

All parameter values apply to all devices unless otherwise indicated. (Continued)

Symbol	Parameter	Min	Max	Units	Test Conditions
I_{TL}	Logical 1 to 0 Transition Current (Ports 1, 2 and 3)		−350	μA	$V_{IN} = 1.4V$
RRST	RST Pulldown Resistor	40	225	KΩ	
I_{CC}	Power Supply Current				(Note 3)
	Active Mode at 16 MHz		25	mA	
	Idle Mode at 16 MHz		8	mA	
	Power-Down Mode		30	μA	

NOTES:

1. Capacitive loading on Ports 0 and 2 may cause noise pulses above 0.4V to be superimposed on the V_{OL}s of ALE and Ports 1, 2 and 3. The noise is due to external bus capacitance discharging into the Port 0 and Port 2 pins when these pins change from 1 to 0. In applications where capacitance loading exceeds 100 pF, the noise pulses on these signals may exceed 0.8V. It may be desirable to qualify ALE or other signals with a Schmitt Trigger, or CMOS-level input logic.
2. Capacitive loading on Ports 0 and 2 cause the V_{OH} on ALE and $\overline{PSEN}$ to drop below the 0.9 V_{CC} specification when the address lines are stabilizing.
3. See Figures 6-9 for test conditions. Minimum V_{CC} for power down is 2V.
4. Under steady state (non-transient) conditions, I_{OL} must be externally limited as follows:

Maximum I_{OL} per port pin: 10 mA
Maximum I_{OL} per 8-bit port -
Port 0: 26 mA
Ports 1, 2, and 3: 15 mA
Maximum total I_{OL} for all output pins: 71 mA

If I_{OL} exceeds the test condition, V_{OL} may exceed the related specification. Pins are not guaranteed to sink current greater than the listed test conditions.

Running the device with $\overline{EA}$ at a higher voltage than V_{CC} sinks additional currrent.

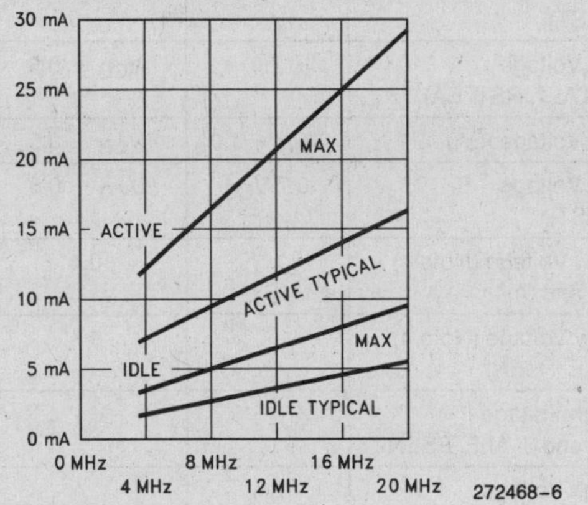

I_{CC} Max at other frequencies (3.5 MHz to 20 MHz) is given by:
Active Mode
I_{CC} MAX = 1.1 × FREQ + 7.6
Idle Mode
I_{CC} MAX = 0.4 × FREQ + 1.8
Where FREQ is in MHz, I_{CC}MAX is given in mA.

Figure 5. I_{CC} vs Frequency

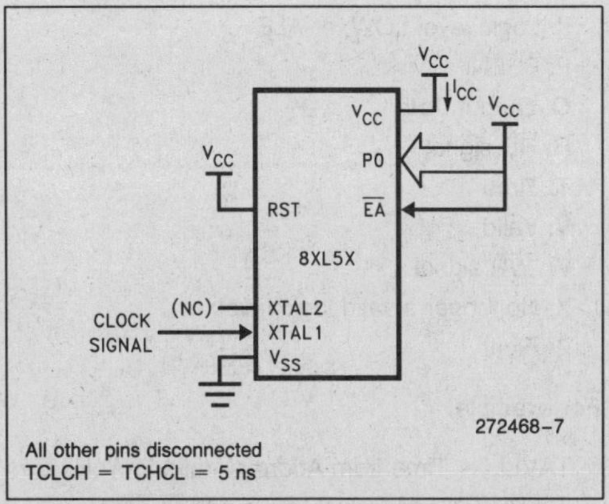

Figure 6. I_{CC} Test Condition, Active Mode

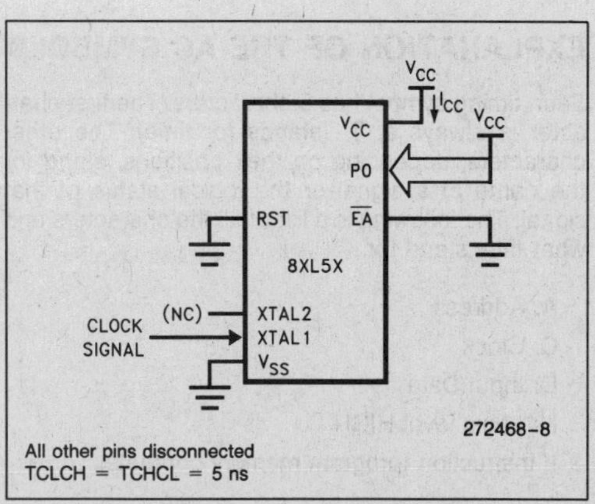

Figure 7. I_{CC} Test Condition Idle Mode

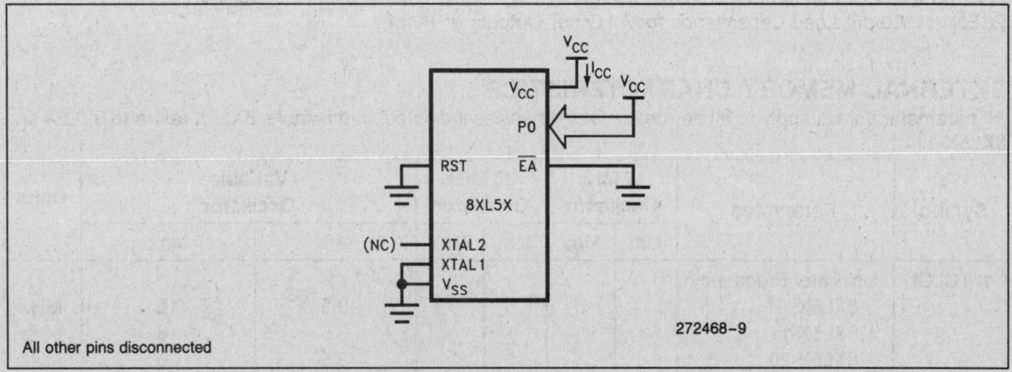

Figure 8. I_{CC} Test Condition, Power Down Mode.
V_{CC} = 2.7V to 3.6V.

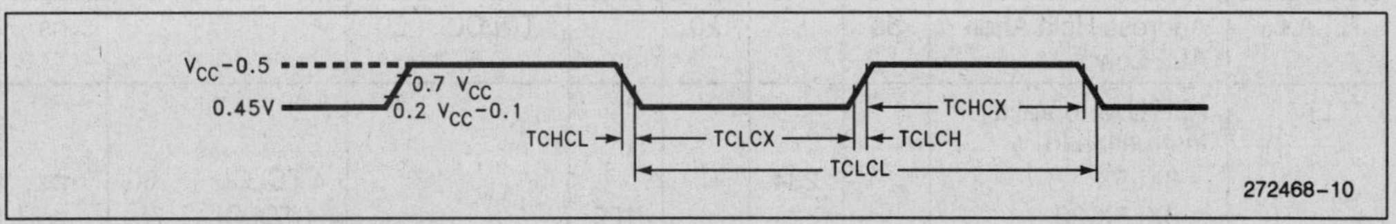

Figure 9. Clock Signal Waveform for I_{CC} Tests in Active and Idle Modes. TCLCH = TCHCL = 5 ns.

EXPLANATION OF THE AC SYMBOLS

Each timing symbol has 5 characters. The first character is always a 'T' (stands for time). The other characters, depending on their positions, stand for the name of a signal or the logical status of that signal. The following is a list of all the characters and what they stand for.

A: Address
C: Clock
D: Input Data
H: Logic level HIGH
I: Instruction (program memory contents)
L: Logic level LOW, or ALE
P: $\overline{\text{PSEN}}$
Q: Output Data
R: $\overline{\text{RD}}$ signal
T: Time
V: Valid
W: $\overline{\text{WR}}$ signal
X: No longer a valid logic level
Z: Float

For example,

TAVLL = Time from Address Valid to ALE Low
TLLPL = Time from ALE Low to $\overline{\text{PSEN}}$ Low

AC CHARACTERISTICS (Over Operating Conditions, Load Capacitance for Port 0, ALE/$\overline{\text{PROG}}$ and $\overline{\text{PSEN}}$ = 100 pF, Load Capacitance for All Other Outputs = 80 pF)

EXTERNAL MEMORY CHARACTERISTICS

All parameter values apply to all devices unless otherwise indicated. In this table, 8XL5X refers to 8XL5X and 8XL5X-1.

Symbol	Parameter	12 MHz Oscillator		20 MHz Oscillator		Variable Oscillator		Units
		Min	Max	Min	Max	Min	Max	
1/TCLCL	Oscillator Frequency 8XL5X 8XL5X-1 8XL5X-20					 3.5 3.5 3.5	 12 16 20	 MHz MHz MHz
TLHLL	ALE Pulse Width	127		60		2 TCLCL − 40		ns
TAVLL	Address Valid to ALE Low	43		10		TCLCL − 40		ns
TLLAX	Address Hold After ALE Low	53		20		TCLCL − 30		ns
TLLIV	ALE Low to Valid Instruction In 8XL5X 8XL5X-20		 234		 125		 4 TCLCL − 100 4 TCLCL − 75	 ns ns
TLLPL	ALE Low to $\overline{\text{PSEN}}$ Low	53		20		TCLCL − 30		ns
TPLPH	$\overline{\text{PSEN}}$ Pulse Width	205		105		3 TCLCL − 45		ns
TPLIV	$\overline{\text{PSEN}}$ Low to Valid Instruction In 8XL5X 8XL5X-20		 145		 60		 3 TCLCL − 105 3 TCLCL − 90	 ns ns
TPXIX	Input Instruction Hold After $\overline{\text{PSEN}}$	0		0		0		ns

EXTERNAL MEMORY CHARACTERISTICS (Continued)

All parameter values apply to all devices unless otherwise indicated.

Symbol	Parameter	12 MHz Oscillator		20 MHz Oscillator		Variable Oscillator		Units
		Min	Max	Min	Max	Min	Max	
TPXIZ	Input Instruction Float After $\overline{PSEN}$							
	8XL5X		59				TCLCL − 25	ns
	8XL5X-20				30		TCLCL − 20	ns
TAVIV	Address to Valid Instruction In		312		145		5 TCLCL − 105	ns
TPLAZ	$\overline{PSEN}$ Low to Address Float		10		10		10	ns
TRLRH	$\overline{RD}$ Pulse Width	400		200		6 TCLCL − 100		ns
TWLWH	$\overline{WR}$ Pulse Width	400		200		6 TCLCL − 100		ns
TRLDV	$\overline{RD}$ Low to Valid Data In							
	8XL5X		252				5 TCLCL − 165	ns
	8XL5X-20				155		5 TCLCL − 95	ns
TRHDX	Data Hold After $\overline{RD}$	0		0		0		ns
TRHDZ	Data Float After $\overline{RD}$		107		40		2 TCLCL − 60	ns
TLLDV	ALE Low to Valid Data In							
	8XL5X		517				8 TCLCL − 150	ns
	8XL5X-20				310		8 TCLCL − 90	ns
TAVDV	Address to Valid Data In							
	8XL5X		585				9 TCLCL − 165	ns
	8XL5X-20				360		9 TCLCL − 90	ns
TLLWL	ALE Low to $\overline{RD}$ or $\overline{WR}$ Low	200	300	100	200	3 TCLCL − 50	3 TCLCL + 50	ns
TAVWL	Address Valid to $\overline{WR}$ Low							
	8XL5X	203				4 TCLCL − 130		ns
	8XL5X-20			110		4 TCLCL − 90		ns
TQVWX	Data Valid before $\overline{WR}$							
	8XL5X	33				TCLCL − 50		ns
	8XL5X-20			15		TCLCL − 35		ns
TWHQX	Data Hold after $\overline{WR}$							
	8XL5X	33				TCLCL − 50		ns
	8XL5X-20			10		TCLCL − 40		ns
TQVWH	Data Valid to $\overline{WR}$ High							
	8XL5X	433				7 TCLCL − 150		ns
	8XL5X-20			280		7 TCLCL − 70		ns
TRLAZ	$\overline{RD}$ Low to Address Float		0		0		0	ns
TWHLH	$\overline{RD}$ or $\overline{WR}$ High to ALE High	43	123	10	90	TCLCL − 40	TCLCL + 40	ns

EXTERNAL PROGRAM MEMORY READ CYCLE

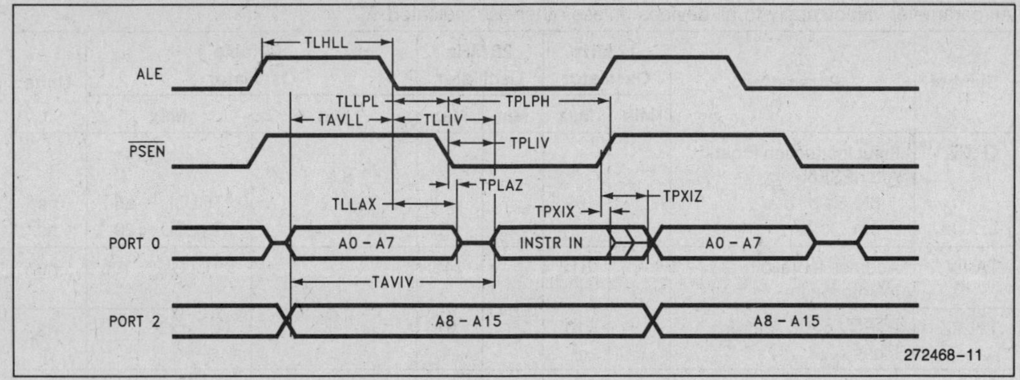

EXTERNAL DATA MEMORY READ CYCLE

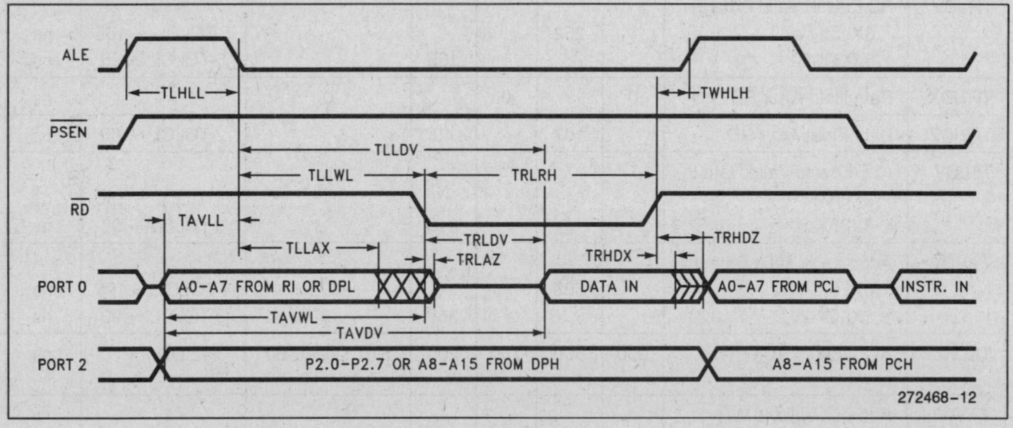

EXTERNAL DATA MEMORY WRITE CYCLE

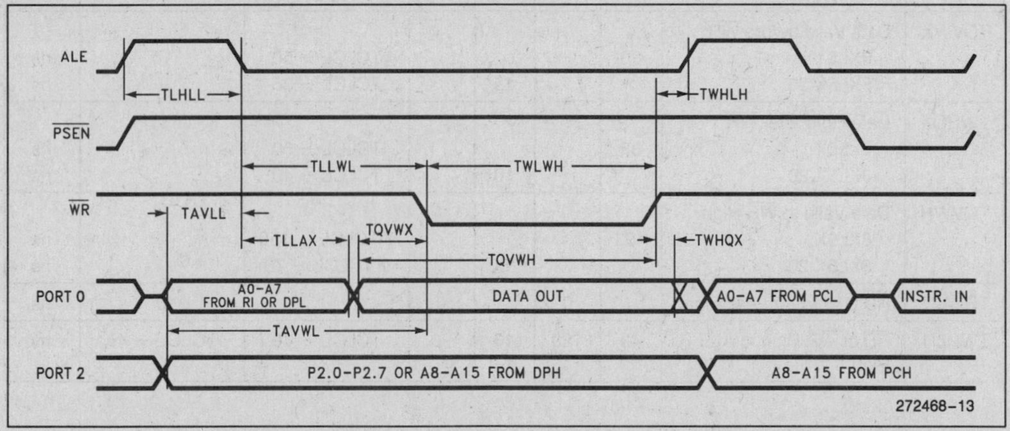

SERIAL PORT TIMING - SHIFT REGISTER MODE

Test Conditions: Over Operating Conditions; Load Capacitance = 80 pF

Symbol	Parameter	12 MHz Oscillator		20 MHz Oscillator		Variable Oscillator		Units
		Min	Max	Min	Max	Min	Max	
TXLXL	Serial Port Clock Cycle Time	1		0.600		12 TCLCL		μs
TQVXH	Output Data Setup to Clock Rising Edge	700		367		10 TCLCL − 133		ns
TXHQX	Output Data Hold after Clock Rising Edge							
	8XL5X	50				2 TCLCL − 117		ns
	8XL5X-20			50		2 TCLCL − 50		ns
TXHDX	Input Data Hold After Clock Rising Edge	0		0		0		ns
TXHDV	Clock Rising Edge to Input Data Valid		700		367		10 TCLCL − 133	ns

SHIFT REGISTER MODE TIMING WAVEFORMS

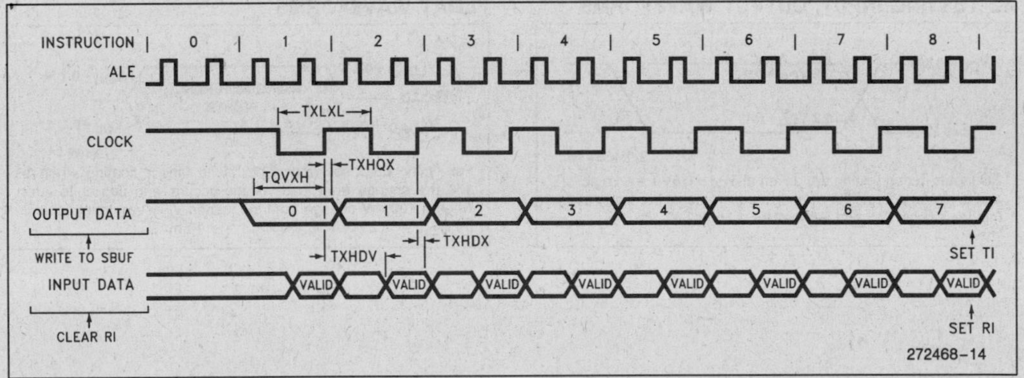

272468-14

EXTERNAL CLOCK DRIVE

Symbol	Parameter	Min	Max	Units
1/TCLCL	Oscillator Frequency 8XL5X 8XL5X-1 8XL5X-20	 3.5 3.5 3.5	 12 16 20	MHz
TCHCX	High Time	20		ns
TCLCX	Low Time	20		ns
TCLCH	Rise Time		20	ns
TCHCL	Fall Time		20	ns

EXTERNAL CLOCK DRIVE WAVEFORM

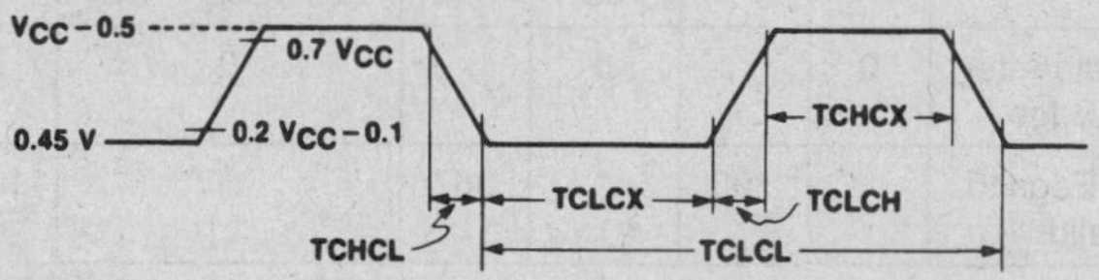

272468-15

AC TESTING INPUT, OUTPUT WAVEFORMS

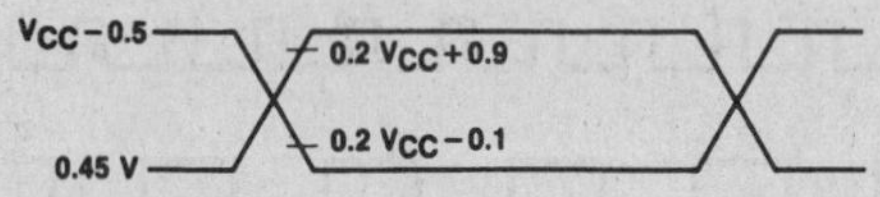

272468-16

AC Inputs during testing are driven at $V_{CC}-0.5V$ for a Logic "1" and 0.45V for a Logic "0". Timing measurements are made at V_{IH} min for a Logic "1" and V_{IL} max for a Logic "0".

FLOAT WAVEFORMS

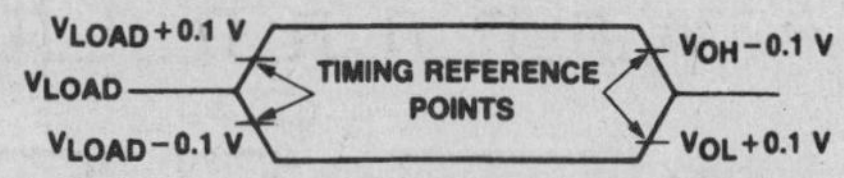

272468-17

For timing purposes a port pin is no longer floating when a 100 mV change from load voltage occurs, and begins to float when a 100 mV change from the loaded V_{OH}/V_{OL} level occurs. $I_{OL}/I_{OH} = \pm 20$ mA (-L, $I_{OL}/I_{OH} = \pm 10$ mA).

PROGRAMMING THE OTP ROM

To be programmed, the part must be running with a 4 to 6 MHz oscillator. (The reason the oscillator needs to be running is that the internal bus is being used to transfer address and program data to appropriate internal OTP ROM locations.) The address of an OTP ROM location to be programmed is applied to Port 1 and pins P2.0 - P2.4 of Port 2, while the code byte to be programmed into that location is applied to Port 0. The other Port 2 and 3 pins, RST $\overline{PSEN}$, and $\overline{EA}/V_{PP}$ should be held at the "Program" levels indicated in Table 3. ALE/$\overline{PROG}$ is pulsed low to program the code byte into the addressed OTP ROM location. The setup is shown in Figure 10.

Normally $\overline{EA}/V_{PP}$ is held at logic high until just before ALE/$\overline{PROG}$ is to be pulsed. Then $\overline{EA}/V_{PP}$ is raised to V_{PP}, ALE/$\overline{PROG}$ is pulsed low, and then $\overline{EA}/V_{PP}$ is returned to a valid high voltage. The voltage on the $\overline{EA}/V_{PP}$ pin must be at the valid $\overline{EA}/V_{PP}$ high level before a verify is attempted. Waveforms and detailed timing specifications are shown in later sections of this data sheet.

NOTE:

- $\overline{EA}/V_{PP}$ pin must not be allowed to go above the maximum specified V_{PP} level for any amount of time. Even a narrow glitch above that voltage level can cause permanent damage to the device. The V_{PP} source should be well regulated and free of glitches.

Table 3. OTP ROM Programming Modes
(H = 2.7V to 3.6V; H1 = 5V ±10%)

Mode		RST	$\overline{PSEN}$	ALE/$\overline{PROG}$	$\overline{EA}/V_{PP}$	P2.6	P2.7	P3.3	P3.6	P3.7	V_{CC}
Program Code Data		H1	L	⊔	12.75V	L	H1	H1	H1	H1	H1
Verify Code Data		H	L	H	H	L	L	L	H	H	H
Program Encryption Array Address 0–3FH		H1	L	⊔	12.75V	L	H1	H1	L	H1	H1
Program Lock Bits	Bit 1	H1	L	⊔	12.75V	H1	H1	H1	H1	H1	H1
	Bit 2	H1	L	⊔	12.75V	H1	H1	H1	L	L	H1
	Bit 3	H1	L	⊔	12.75V	H1	L	H1	H1	L	H1
Read Signature Byte		H	L	H	H	L	L	L	L	L	H

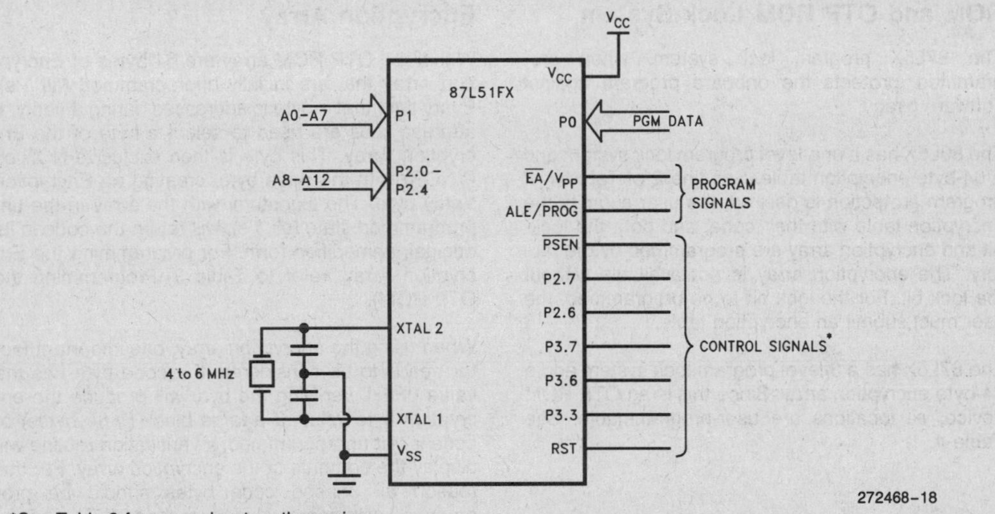

*See Table 2 for proper input on these pins

Figure 10. Programming the OTP ROM

PROGRAMMING ALGORITHM

Refer to Table 3 and Figures 10 and 11 for address, data, and control signals set up. To program the 87L5X the following sequence must be exercised.

1. Input the valid address on the address lines.
2. Input the appropriate data byte on the data lines.
3. Activate the correct combination of control signals.
4. Raise $\overline{EA}/V_{PP}$ from V_{CC} to 12.75V ±0.25V.
5. Pulse, ALE/$\overline{PROG}$ 5 times for the OTP ROM array, and 25 times for the encryption table and the lock bits.

Repeat 1 through 5 changing the address and data for the entire array or until the end of the object file is reached.

PROGRAM VERIFY

Program verify may be done after each byte or block of bytes is programmed. In either case a complete verify of the programmed array will ensure reliable programming of the 8XL5X.

The lock bits cannot be directly verified. Verification of the lock bits is done by observing that their features are enabled.

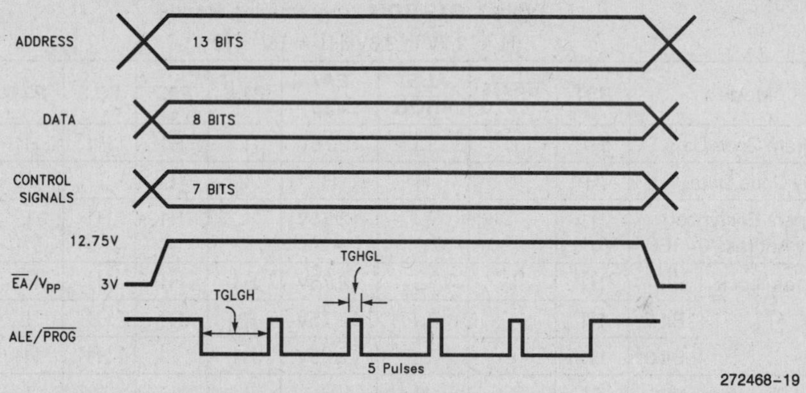

Figure 11. Programming Signals Waveforms

ROM and OTP ROM Lock System

The 87L5X program lock system, when programmed, protects the onboard program against software piracy.

The 80L5X has a one-level program lock system and a 64-byte encryption table. See line 2 of Table 4. If program protection is desired, the user submits the encryption table with their code, and both the lock-bit and encryption array are programmed by the factory. The encryption array is not available without the lock bit. For the lock bit to be programmed, the user must submit an encryption table.

The 87L5X has a 3-level program lock system and a 64-byte encryption array. Since this is an OTP ROM device, all locations are user-programmable. See Table 4.

Encryption Array

Within the OTP ROM array are 64 bytes of Encryption Array that are initially unprogrammed (all 1's). Every time that a byte is addressed during a verify, 6 address lines are used to select a byte of the Encryption Array. This byte is then exclusive-NOR'ed (XNOR) with the code byte, creating an Encryption Verify byte. The algorithm, with the array in the unprogrammed state (all 1's), will return the code in its original, unmodified form. For programming the Encryption Array, refer to Table 3 (Programming the OTP ROM).

When using the encryption array, one important factor needs to be considered. If a code byte has the value 0FFH, verifying the byte will produce the encryption byte value. If a large block (>64 bytes) of code is left unprogrammed, a verification routine will display the contents of the encryption array. For this reason all unused code bytes should be programmed with some value other than 0FFH, and not all of them the same value. This will ensure maximum program protection.

Table 4. Program Lock Bits and the Features

Program Lock Bits				Protection Type
	LB1	LB2	LB3	
1	U	U	U	No Program Lock features enabled. (Code verify will still be encrypted by the Encryption Array if programmed.)
2	P	U	U	MOVC instructions executed from external program memory are disabled from fetching code bytes from internal memory, EA is sampled and latched on Reset, and further programming of the OTP ROM is disabled.
3	P	P	U	Same as 2, also verify is disabled.
4	P	P	P	Same as 3, also external execution is disabled.

Any other combination of the lock bits is not defined.

Program Lock Bits

The 8XL5X has 3 programmable lock bits that when programmed according to Table 4 will provide different levels of protection for the on-chip code and data.

Reading the Signature Bytes

The 87L5X/80L5X has 3 signature bytes in locations 30H, 31H, and 60H. To read these bytes follow the procedure for OTP ROM verify, but activate the control lines provided in Table 3 for Read Signature Byte.

Location	Device	Contents
30H	All	89H
31H	All	58H
60H	80L52	30
	87L52	B0
	80L54	31
	87L54	B1
	80L58	32
	87L58	B2

OTP ROM PROGRAMMING AND VERIFICATION CHARACTERISTICS

(T_A = 21°C to 27°C; V_{CC} = 2.7V to 3.6V; V_{SS} = 0V)

Symbol	Parameter	Min	Max	Units
V_{PP}	Programming Supply Voltage	12.5	13.0	V
I_{PP}	Programming Supply Current		75	mA
1/TCLCL	Oscillator Frequency	4	6	MHz
TAVGL	Address Setup to $\overline{PROG}$ Low	48TCLCL		
TGHAX	Address Hold after $\overline{PROG}$	48TCLCL		
TDVGL	Data Setup to $\overline{PROG}$ Low	48TCLCL		
TGHDX	Data Hold after $\overline{PROG}$	48TCLCL		
TEHSH	P2.7 (ENABLE) High to V_{PP}	48TCLCL		
TSHGL	V_{PP} Setup to $\overline{PROG}$ Low	10		μs
TGHSL	V_{PP} Hold after $\overline{PROG}$	10		μs
TGLGH	$\overline{PROG}$ Width	90	100	μs
TAVQV	Address to Data Valid		48TCLCL	
TELQV	ENABLE Low to Data Valid		48TCLCL	
TEHQZ	Data Float after ENABLE	0	48TCLCL	
TGHGL	$\overline{PROG}$ High to $\overline{PROG}$ Low	10		μs

OTP ROM PROGRAMMING AND VERIFICATION WAVEFORMS

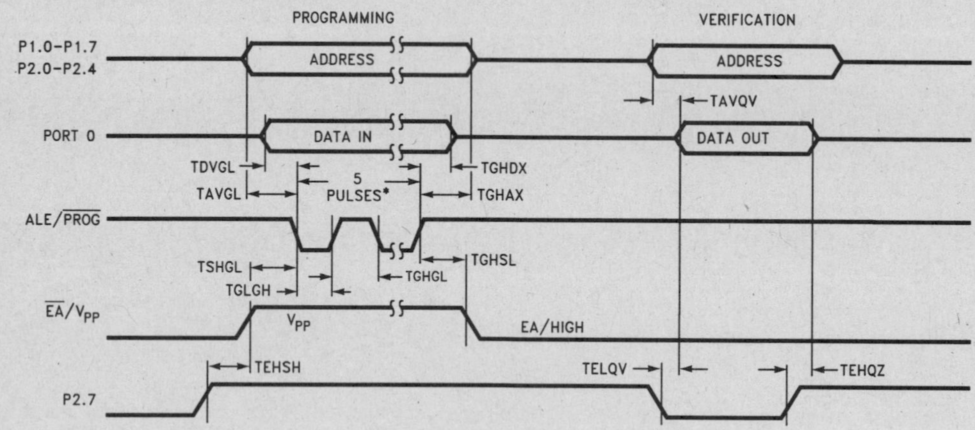

272468–20

NOTE:
*5 pulses for the OTP ROM array, 25 pulses for the encryption table and lock bits.

Thermal Impedance

All thermal impedance data is approximate for static air conditions at 1W of power dissipation. Values will change depending on operating conditions and applications. See the Intel Packaging Handbook (Order Number 240800) for a description of Intel's thermal impedance test methodology.

Package	θ_{JA}	θ_{JC}	Device
N	46°C/W	16°C/W	All
S	87°C/W	18°C/W	52
	96°C/W	24°C/W	54
	90°C/W	22°C/W	58

DATA SHEET REVISION HISTORY

This is the first issue of this data sheet.

ADVANCE INFORMATION

8XL51FA/FB/FC LOW VOLTAGE CHMOS SINGLE-CHIP 8-BIT MICROCONTROLLERS

Commercial/Express

87L51FA/83L51FA/80L51FA/87L51FB/83L51FB/87L51FC/83L51FC

- High Performance CHMOS OTP ROM/ROM/CPU
- Low Voltage Operation
- 20 MHz Commercial/16 MHz Express Operation
- Three 16-Bit Timer/Counters
- Programmable Counter Array with:
  - High Speed Output,
  - Compare/Capture,
  - Pulse Width Modulator,
  - Watchdog Timer Capabilities
- Up/Down Timer/Counter
- Three Level Program Lock System
- 8K/16K/32K On-Chip Program Memory
- 256 Bytes of On-Chip Data RAM
- Improved Quick Pulse Programming Algorithm
- Boolean Processor
- 32 Programmable I/O Lines
- 7 Interrupt Sources
- Four Level Interrupt Priority
- Programmable Serial Channel with:
  - Framing Error Detection
  - Automatic Address Recognition
- 64K External Program Memory Space
- 64K External Data Memory Space
- MCS® 51 Microcontroller Compatible Instruction Set
- Power Saving Idle and Power Down Modes
- ONCE (On-Circuit Emulation) Mode
- Extended Temperature Range (−40°C to +85°C)

MEMORY ORGANIZATION

ROM Device	OTP ROM Version	ROMLESS Version	ROM/ OTP ROM Bytes	RAM Bytes
83L51FA	87L51FA	80L51FA	8K	256
83L51FB	87L51FB	80L51FA	16K	256
83L51FC	87L51FC	80L51FA	32K	256

These devices can address up to 64 Kbytes of external program/data memory.

The Intel 8XL51FA/8XL51FB/8XL51FC is a single-chip control oriented microcontroller which is fabricated on Intel's reliable CHMOS III-E technology. Being a member of the MCS® 51 microcontroller family, the 8XL51FA/8XL51FB/8XL51FC uses the same powerful instruction set, has the same architecture, and is pin-for-pin compatible with the existing MCS 51 microcontroller products.

The 8XL51FX is a 3V version of current 8XC51FX and will operate from 2.7V to 3.6V at a frequency range of 3.5 MHz to 16 MHz (Express)/20 MHz (Commercial).

For the remainder of this document, the 8XL51FA, 8XL51FB, 8XL51FC will be referred to as the 8XL51FX, unless information applies to a specific device.

October 1994
Order Number: 272356-003

	Standard	−1	−20*
80L51FA	X	X	X
83L51FA	X	X	X
87L51FA	X	X	X
83L51FB	X	X	X
87L51FB	X	X	X
83L51FC	X	X	X
87L51FC	X	X	X

NOTE:
Standard 3.5 MHz to 12 MHz; 2.7V to 3.6V
−1 3.5 MHz to 16 MHz; 2.7V to 3.6V
−20* 3.5 MHz to 20 MHz; 2.7V to 3.6V
*Only available for commercial standard temperature range, not available at express temperature range.

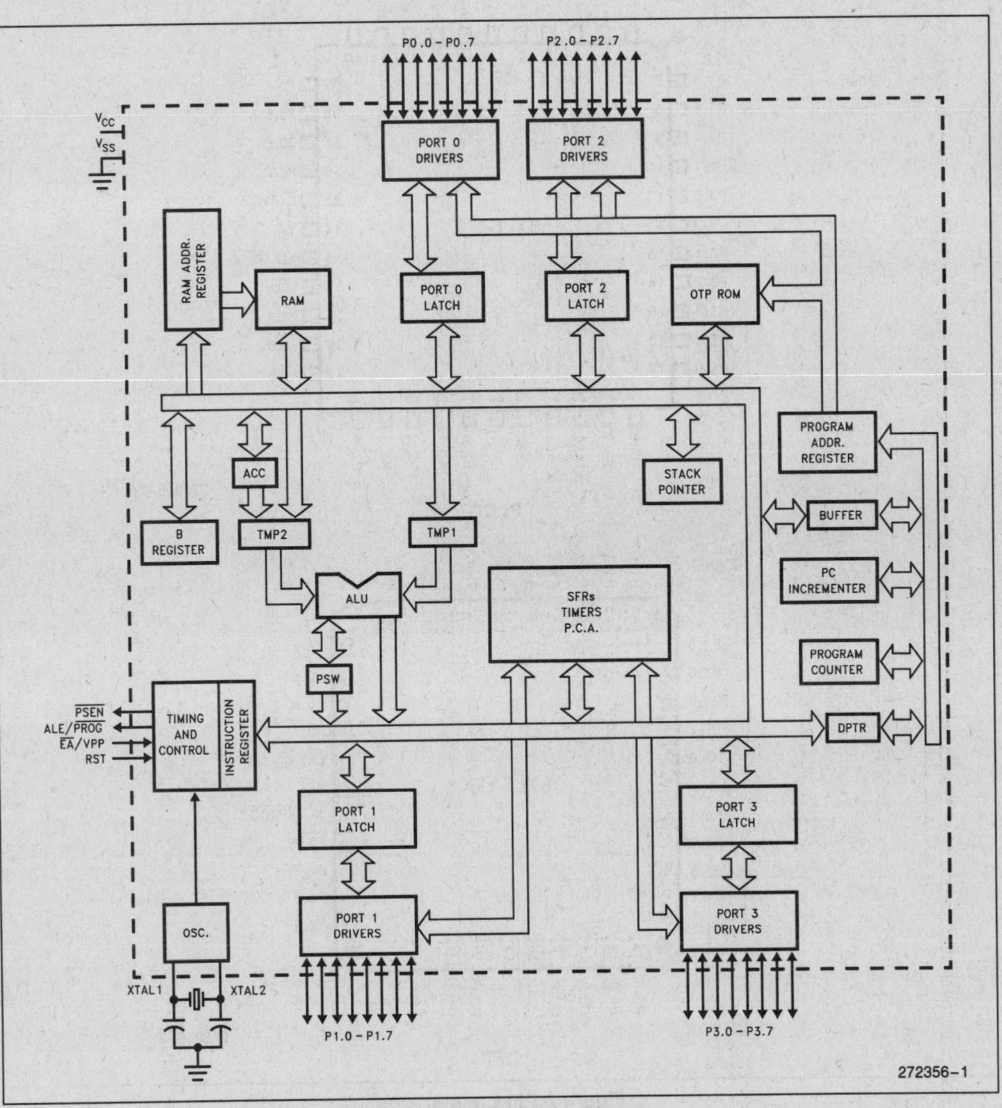

Figure 1. 8XL51FX Block Diagram

2

PROCESS INFORMATION

The 8XL51FA/8XL51FB/8XL51FC is manufactured on P629.5, a CHMOS III-E process. Additional process and reliability information is available in Intel's *Components Quality and Reliability Handbook*, Order Number 210997.

PACKAGES

Part	Prefix	Package Type
8XL51FX	N	44-Pin PLCC (OTP)
	S	44-Pin QFP (OTP)

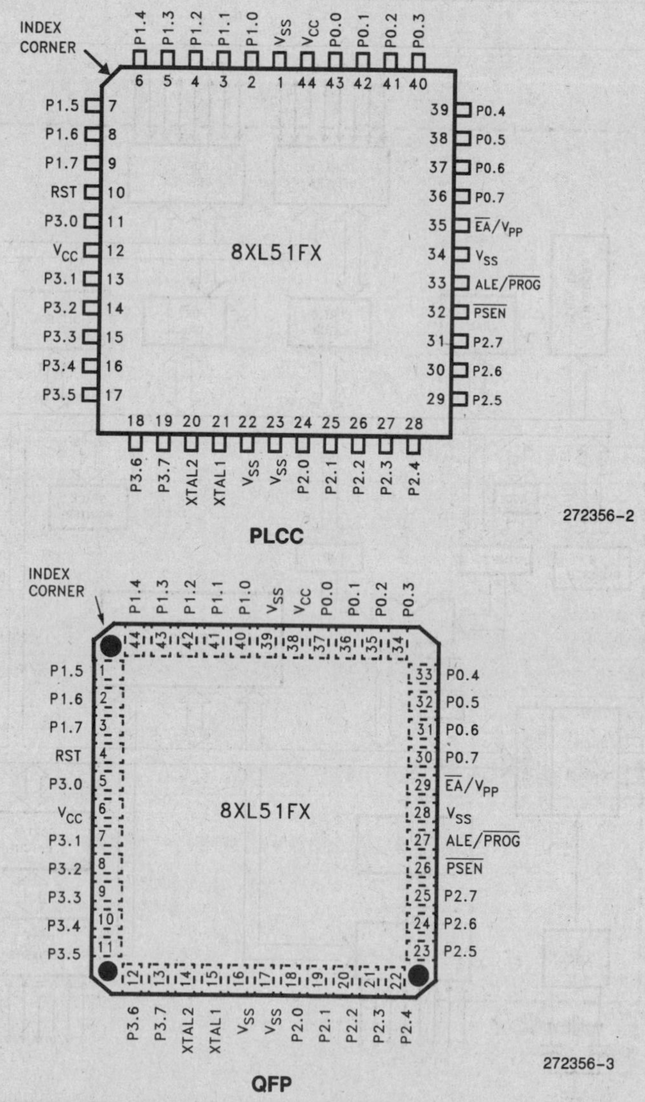

Figure 2. Pin Connections

ADVANCE INFORMATION

PIN DESCRIPTIONS

V_{CC}: Supply voltage.

V_{SS}: Circuit ground.

Port 0: Port 0 is an 8-bit, open drain, bidirectional I/O port. As an output port each pin can sink several inputs. Port 0 pins that have 1's written to them float, and in that state can be used as high-impedance inputs.

Port 0 is also the multiplexed low-order address and data bus during accesses to external Program and Data Memory. In this application it uses strong internal pullups when emitting 1's, and can source and sink several inputs.

Port 0 also receives the code bytes during OTP ROM programming, and outputs the code bytes during program verification. External pullup resistors are required during program verification.

Port 1: Port 1 is an 8-bit bidirectional I/O port with internal pullups. The Port 1 output buffers can drive several inputs. Port 1 pins that have 1's written to them are pulled high by the internal pullups, and in that state can be used as inputs. As inputs, Port 1 pins that are externally pulled low will source current (I_{IL}, on the data sheet) because of the internal pullups.

In addition, Port 1 serves the functions of the following special features of the 8XL51FX:

Port Pin	Alternate Function
P1.0	T2 (External Count Input to Timer/Counter 2), Clock Out
P1.1	T2EX (Timer/Counter 2 Capture/Reload Trigger and Direction Control)
P1.2	ECI (External Count Input to the PCA)
P1.3	CEX0 (External I/O for Compare/Capture Module 0)
P1.4	CEX1 (External I/O for Compare/Capture Module 1)
P1.5	CEX2 (External I/O for Compare/Capture Module 2)
P1.6	CEX3 (External I/O for Compare/Capture Module 3)
P1.7	CEX4 (External I/O for Compare/Capture Module 4)

Port 1 receives the low-order address bytes during OTP ROM programming and verifying:

Port 2: Port 2 is an 8-bit bidirectional I/O port with internal pullups. The Port 2 output buffers can drive several inputs. Port 2 pins that have 1's written to them are pulled high by the internal pullups, and in that state can be used as inputs. As inputs, Port 2 pins that are externally pulled low will source current (I_{IL}, on the data sheet) because of the internal pullups.

Port 2 emits the high-order address byte during fetches from external Program Memory and during accesses to external Data Memory that use 16-bit addresses (MOVX @DPTR). In this application it uses strong internal pullups when emitting 1's. During accesses to external Data Memory that use 8-bit addresses (MOVX @Ri), Port 2 emits the contents of the P2 Special Function Register.

Some Port 2 pins receive the high-order address bits during OTP ROM programming and program verification.

Port 3: Port 3 is an 8-bit bidirectional I/O port with internal pullups. The Port 3 output buffers can drive several inputs. Port 3 pins that have 1's written to them are pulled high by the internal pullups, and in that state can be used as inputs. As inputs, Port 3 pins that are externally pulled low will source current (I_{IL}, on the data sheet) because of the pullups.

Port 3 also serves the functions of various special features of the MCS 51 microcontroller family, as listed below:

Port Pin	Alternate Function
P3.0	RXD (serial input port)
P3.1	TXD (serial output port)
P3.2	$\overline{INT0}$ (external interrupt 0)
P3.3	$\overline{INT1}$ (external interrupt 1)
P3.4	T0 (Timer 0 external input)
P3.5	T1 (Timer 1 external input)
P3.6	$\overline{WR}$ (external data memory write strobe)
P3.7	$\overline{RD}$ (external data memory read strobe)

RST: Reset input. A high on this pin for two machine cycles while the oscillator is running resets the device. The port pins will be driven to their reset condition when a minimum V_{IH2} voltage is applied whether the oscillator is running or not. An internal pulldown resistor permits a power-on reset with only a capacitor connected to V_{CC}.

ALE: Address Latch Enable output pulse for latching the low byte of the address during accesses to external memory. This pin (ALE/$\overline{PROG}$) is also the program pulse input during OTP ROM programming for the 87L51FX.

In normal operation ALE is emitted at a constant rate of 1/6 the oscillator frequency, and may be used for external timing or clocking purposes. Note, however, that one ALE pulse is skipped during each access to external Data Memory.

If desired, ALE operation can be disabled by setting bit 0 of SFR location 8EH. With this bit set, the pin is weakly pulled high. However, the ALE disable feature will be suspended during a MOVX or MOVC instruction, idle mode, power down mode and ICE mode. The ALE disable feature will be terminated by reset. When the ALE disable feature is suspended or terminated, the ALE pin will no longer be pulled up weakly. Setting the ALE-disable bit has no affect if the microcontroller is in external execution mode.

Throughout the remainder of this data sheet, ALE will refer to the signal coming out of the ALE/$\overline{PROG}$ pin, and the pin will be referred to as the ALE/$\overline{PROG}$ pin.

$\overline{PSEN}$: Program Store Enable is the read strobe to external Program Memory.

When the 8XL51FX is executing code from external Program Memory, $\overline{PSEN}$ is activated twice each machine cycle, except that two $\overline{PSEN}$ activations are skipped during each access to external Data Memory.

$\overline{EA}/V_{PP}$: External Access enable. $\overline{EA}$ must be strapped to VSS in order to enable the device to fetch code from external Program Memory locations 0000H to 0FFFH. Note, however, that if either of the Program Lock bits are programmed, $\overline{EA}$ will be internally latched on reset.

$\overline{EA}$ must be strapped to V_{CC} for internal program executions.

This pin also receives the programming supply voltage (V_{PP}) during OTP ROM programming.

XTAL1: Input to the inverting oscillator amplifier.

XTAL2: Output from the inverting oscillator amplifier.

OSCILLATOR CHARACTERISTICS

XTAL1 and XTAL2 are the input and output, respectively, of a inverting amplifier which can be configured for use as an on-chip oscillator, as shown in Figure 3. Either a quartz crystal or ceramic resonator may be used. More detailed information concerning the use of the on-chip oscillator is available in Application Note AP-155, "Oscillators for Microcontrollers."

To drive the device from an external clock source, XTAL1 should be driven, while XTAL2 floats, as shown in Figure 4. There are no requirements on the duty cycle of the external clock signal, since the input to the internal clocking circuitry is through a divide-by-two flip-flop, but minimum and maximum high and low times specified on the data sheet must be observed.

An external oscillator may encounter as much as a 100 pF load at XTAL1 when it starts up. This is due to interaction between the amplifier and its feedback capacitance. Once the external signal meets the V_{IL} and V_{IH} specifications the capacitance will not exceed 20 pF.

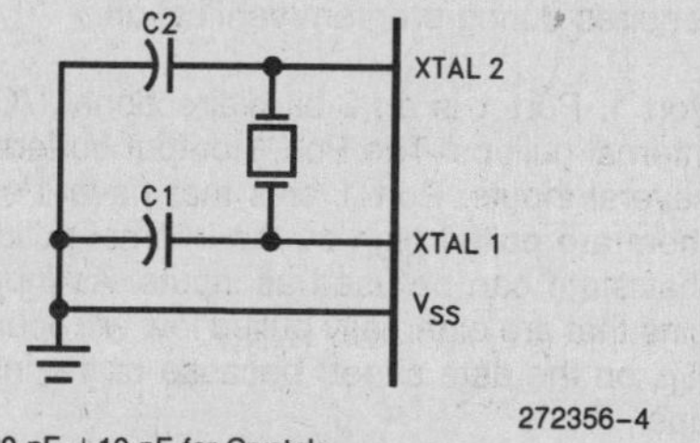

C1, C2 = 30 pF ±10 pF for Crystals
For Ceramic Resonators, contact resonator manufacturer.

Figure 3. Oscillator Connections

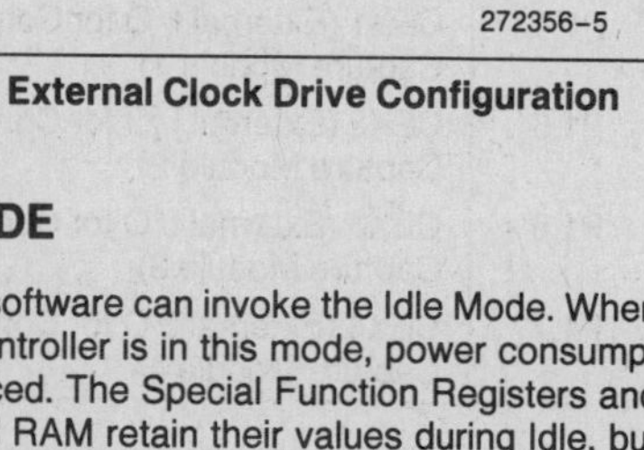

Figure 4. External Clock Drive Configuration

IDLE MODE

The user's software can invoke the Idle Mode. When the microcontroller is in this mode, power consumption is reduced. The Special Function Registers and the onboard RAM retain their values during Idle, but the processor stops executing instructions. Idle Mode will be exited if the chip is reset or if an enabled interrupt occurs. The PCA timer/counter can optionally be left running or paused during Idle Mode.

POWER DOWN MODE

To save even more power, a Power Down mode can be invoked by software. In this mode, the oscillator is stopped and the instruction that invoked Power Down is the last instruction executed. The on-chip RAM and Special Function Registers retain their values until the Power Down mode is terminated.

On the 8XL51FX either hardware reset or external interrupt can cause an exit from Power Down. Reset redefines all the SFRs but does not change the on-chip RAM. An external interrupt allows both the SFRs and the on-chip RAM to retain their values.

To properly terminate Power Down the reset or external interrupt should not be executed before V_{CC} is restored to its normal operating level and must be held active long enough for the oscillator to restart and stabilize (normally less than 10 ms).

With an external interrupt, INT0 or INT1 must be enabled and configured as level-sensitive. Holding the pin low restarts the oscillator but bringing the pin back high completes the exit. Once the interrupt is serviced, the next instruction to be executed after RETI will be the one following the instruction that put the device into Power Down.

DESIGN CONSIDERATION

- The 8XL51FX will operate from 2.7V to 3.6V with a frequency range of 3.5 MHz to 16 MHz (Express)/20 MHz (Commercial). Operating beyond these specifications could cause improper device functionality.
- All V_{CC} and V_{SS} pins must be connected. Please refer to Figure 2, Pin Connections, for the specific pins.
- When the idle mode is terminated by a hardware reset, the device normally resumes program execution, from where it left off, up to two machine cycles before the internal reset algorithm takes control. On-chip hardware inhibits access to internal RAM in this event, but access to the port pins is not inhibited. To eliminate the possibility of an unexpected write when Idle is terminated by reset, the instruction following the one that invokes Idle should not be one that writes to a port pin or to external memory.

ONCE MODE

The ONCE ("On-Circuit Emulation") Mode facilitates testing and debugging of systems using the 8XL51FX without the 8XL51FX having to be removed from the circuit. The ONCE Mode is invoked by:

1) Pull ALE low while the device is in reset and $\overline{\text{PSEN}}$ is high;
2) Hold ALE low as RST is deactivated.

While the device is in ONCE Mode, the Port 0 pins float, and the other port pins and ALE and $\overline{\text{PSEN}}$ are weakly pulled high. The oscillator circuit remains active. While the 8XL51FX is in this mode, an emulator or test CPU can be used to drive the circuit. Normal operation is restored when a normal reset is applied.

2

Table 1. Status of the External Pins during Idle and Power Down

Mode	Program Memory	ALE	$\overline{\text{PSEN}}$	PORT0	PORT1	PORT2	PORT3
Idle	Internal	1	1	Data	Data	Data	Data
Idle	External	1	1	Float	Data	Address	Data
Power Down	Internal	0	0	Data	Data	Data	Data
Power Down	External	0	0	Float	Data	Data	Data

NOTE:
For more detailed information on the reduced power modes refer to current Embedded Microcontrollers and Processors Handbook Volume I, #270646, and Application Note AP-252 (Embedded Applications Handbook), #270648, "Designing with the 80C51BH."

8XL51FX EXPRESS

The Intel EXPRESS system offers enhancements to the operational specifications of the MCS-51 family of microcontrollers. These EXPRESS products are designed to meet the needs of those applications whose operating requirements exceed commercial standards.

The EXPRESS program includes the commercial standard temperature range with burn-in and an extended temperature range with or without burn-in.

With the commercial standard temperature range, operational characteristics are guaranteed over the temperature range of 0°C to 70°C. With the extended temperature range option, operational characteristics are guaranteed over the range of −40°C to +85°C.

Package types and EXPRESS versions are identified by a one- or two-letter prefix to the part number. The prefixes are listed in Table 2.

For the extended temperature range option, this data sheet specifies the parameters which deviate from their commercial temperature range limits.

Table 2. Prefix Identification

Prefix	Package Type	Temperature Range
N	PLCC	Commercial
S	QFP	Commercial
TN	PLCC	Extended
TS	QFP	Extended

NOTE:
Contact your distributor or local sales office to match the EXPRESS prefix with the proper device.

EXAMPLES:
N87L51FC indicates 87L51FC in a PLCC package and specified for commercial temperature range, without burn-in.

TN87L51FC indicates 87L51FC in a PLCC package and specified for extended temperature range with burn-in.

ADVANCE INFORMATION

ABSOLUTE MAXIMUM RATINGS*

Ambient Temperature Under Bias . −40°C to +85°C
Storage Temperature .... −65°C to +150°C
Voltage on EA/V_{PP} Pin to V_{SS} .... 0V to +13.0V
Voltage on Any Other Pin to V_{SS} .. −0.5V to +6.5V
I_{OL} per I/O Pin .... 15 mA
Power Dissipation......................... 1.5W
(based on PACKAGE heat transfer limitations, not device power consumption)

NOTICE: This data sheet contains information on products in the sampling and initial production phases of development. It is valid for the devices indicated in the revision history. The specifications are subject to change without notice.

**WARNING: Stressing the device beyond the "Absolute Maximum Ratings" may cause permanent damage. These are stress ratings only. Operation beyond the "Operating Conditions" is not recommended and extended exposure beyond the "Operating Conditions" may affect device reliability.*

2

OPERATING CONDITIONS

Symbol	Description	Min	Max	Units
T_A	Ambient Temperature Under Bias			
	Commercial	0	+70	°C
	Express	−40	+85	°C
V_{CC}	Supply Voltage	2.7	3.6	V

DC CHARACTERISTICS (Over Operating Conditions)

All parameter values apply to all devices unless otherwise indicated.

Symbol	Parameter	Min	Max	Units	Test Conditions
V_{IL}	Input Low Voltage (except XTAL1, RST)	−0.5	0.8	V	
V_{IL1}	Input Low Voltage (XTAL1, RST)	−0.5	0.2 V_{CC} − 0.1	V	
V_{IH}	Input High Voltage (Except XTAL1, RST, $\overline{EA}$)	2.0	V_{CC} + 0.5	V	
V_{IH1}	Input High Voltage ($\overline{EA}$)	V_{CC} − 1.0	V_{CC} + 0.5	V	
V_{IH2}	Input High Voltage (XTAL1, RST)	0.7 V_{CC}	V_{CC} + 0.5	V	
V_{OL}	Output Low Voltage (Note 4) (Ports 1, 2 and 3)		0.4	V	I_{OL} = 1.6 mA (Note 1)
V_{OL1}	Output Low Voltage (Note 4) (Port 0, ALE/$\overline{PSEN}$)		0.4	V	I_{OL} = 3.2 mA (Note 1)
V_{OH}	Output High Voltage (Ports 1, 2 and 3, ALE, $\overline{PSEN}$	V_{CC} − 0.7		V	I_{OH} = −30 μA (Note 2)
V_{OH1}	Output High Voltage (Port 0 in External Bus Mode)	2.4		V	I_{OH} = −1.0 mA (Note 2)
I_{IL}	Logical 0 Input Current (Ports 1, 2 and 3)		−50	μA	V_{IN} = 0.4V
I_{LI}	Input Leakage Current (Port 0)		±10	μA	$0 < V_{IN} < V_{CC}$

DC CHARACTERISTICS (Over Operating Conditions)

All parameter values apply to all devices unless otherwise indicated. (Continued)

Symbol	Parameter	Min	Max	Units	Test Conditions
I_{TL}	Logical 1 to 0 Transition Current (Ports 1, 2 and 3)		−350	μA	$V_{IN} = 1.4V$
RRST	RST Pulldown Resistor	40	225	KΩ	
I_{CC}	Power Supply Current Active Mode at 16 MHz Idle Mode at 16 MHz Power-Down Mode		 25 8 30	 mA mA μA	(Note 3)

NOTES:

1. Capacitive loading on Ports 0 and 2 may cause noise pulses above 0.4V to be superimposed on the V_{OL}s of ALE and Ports 1, 2 and 3. The noise is due to external bus capacitance discharging into the Port 0 and Port 2 pins when these pins change from 1 to 0. In applications where capacitance loading exceeds 100 pF, the noise pulses on these signals may exceed 0.8V. It may be desirable to qualify ALE or other signals with a Schmitt Trigger, or CMOS-level input logic.
2. Capacitive loading on Ports 0 and 2 cause the V_{OH} on ALE and $\overline{PSEN}$ to drop below the 0.9 V_{CC} specification when the address lines are stabilizing.
3. See Figures 6–9 for test conditions. Minimum V_{CC} for power down is 2V.
4. Under steady state (non-transient) conditions, I_{OL} must be externally limited as follows:

Maximum I_{OL} per port pin: 10 mA
Maximum I_{OL} per 8-bit port -
Port 0: 26 mA
Ports 1, 2, and 3: 15 mA
Maximum total I_{OL} for all output pins: 71 mA

If I_{OL} exceeds the test condition, V_{OL} may exceed the related specification. Pins are not guaranteed to sink current greater than the listed test conditions.

Running the device with $\overline{EA}$ at a higher voltage than V_{CC} sinks additional currrent.

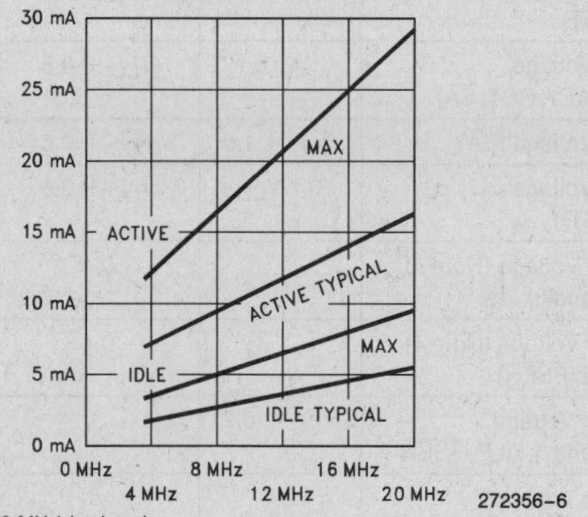

I_{CC} Max at other frequencies (3.5 MHz to 20 MHz) is given by:
Active Mode
I_{CC} MAX = 1.1 × FREQ + 7.6
Idle Mode
I_{CC} MAX = 0.4 × FREQ + 1.8
Where FREQ is in MHz, I_{CC} MAX is given in mA.

Figure 5. I_{CC} vs Frequency

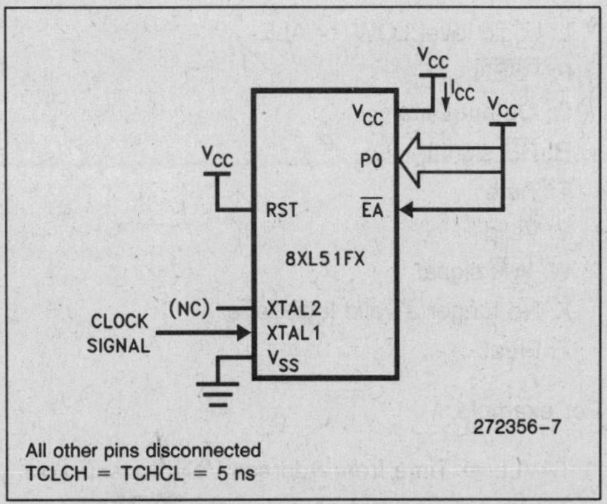

Figure 6. I_{CC} Test Condition, Active Mode

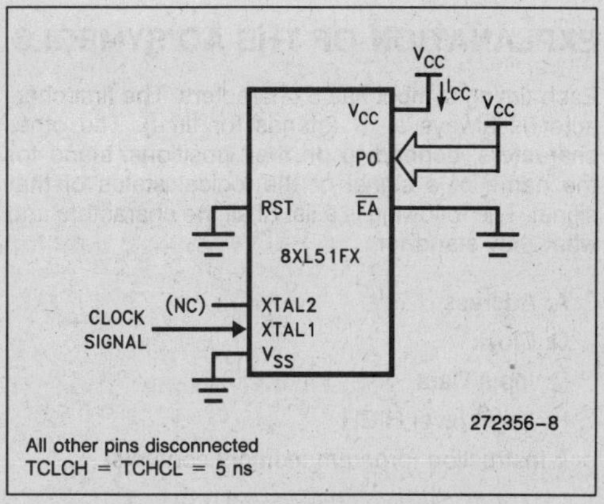

Figure 7. I_{CC} Test Condition Idle Mode

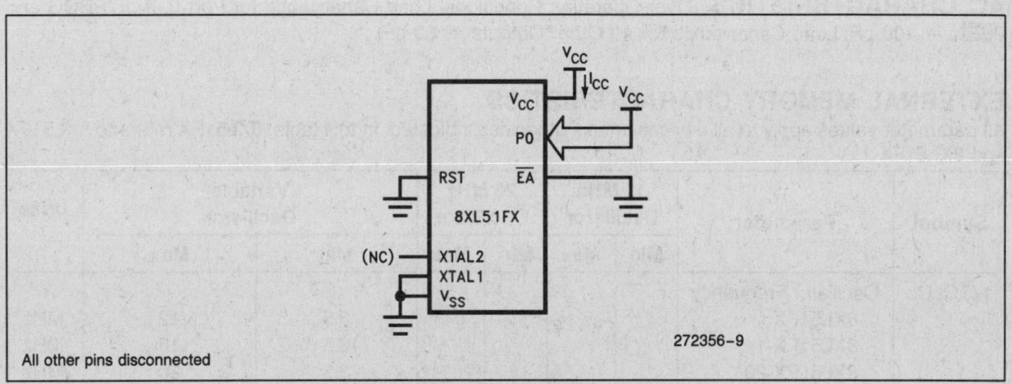

Figure 8. I_{CC} Test Condition, Power Down Mode.
V_{CC} = 2.7V to 3.6V.

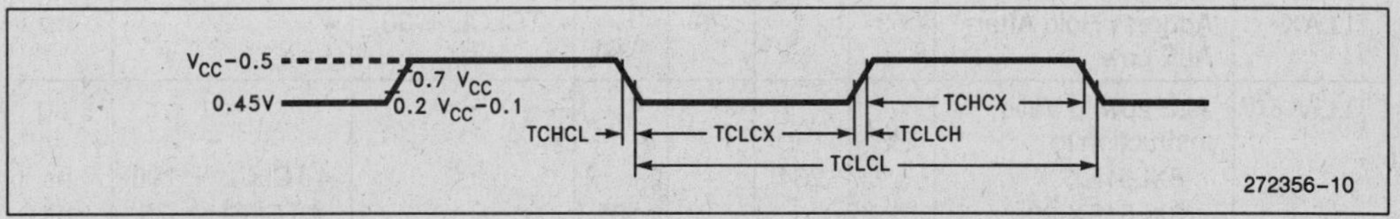

Figure 9. Clock Signal Waveform for I_{CC} Tests in Active and Idle Modes. TCLCH = TCHCL = 5 ns.

EXPLANATION OF THE AC SYMBOLS

Each timing symbol has 5 characters. The first character is always a 'T' (stands for time). The other characters, depending on their positions, stand for the name of a signal or the logical status of that signal. The following is a list of all the characters and what they stand for.

A: Address
C: Clock
D: Input Data
H: Logic level HIGH
I: Instruction (program memory contents)
L: Logic level LOW, or ALE
P: $\overline{\text{PSEN}}$
Q: Output Data
R: $\overline{\text{RD}}$ signal
T: Time
V: Valid
W: $\overline{\text{WR}}$ signal
X: No longer a valid logic level
Z: Float

For example,

TAVLL = Time from Address Valid to ALE Low
TLLPL = Time from ALE Low to $\overline{\text{PSEN}}$ Low

AC CHARACTERISTICS (Over Operating Conditions, Load Capacitance for Port 0, ALE/$\overline{\text{PROG}}$ and $\overline{\text{PSEN}}$ = 100 pF, Load Capacitance for All Other Outputs = 80 pF)

EXTERNAL MEMORY CHARACTERISTICS

All parameter values apply to all devices unless otherwise indicated. In this table, 8XL51FX refers to 8XL51FX and 8XL51FX-1.

Symbol	Parameter	12 MHz Oscillator		20 MHz Oscillator		Variable Oscillator		Units
		Min	Max	Min	Max	Min	Max	
1/TCLCL	Oscillator Frequency							
	8XL51FX					3.5	12	MHz
	8XL51FX-1					3.5	16	MHz
	8XL51FX-20					3.5	20	MHz
TLHLL	ALE Pulse Width	127		60		2 TCLCL − 40		ns
TAVLL	Address Valid to ALE Low	43		10		TCLCL − 40		ns
TLLAX	Address Hold After ALE Low	53		20		TCLCL − 30		ns
TLLIV	ALE Low to Valid Instruction In							
	8XL51FX		234				4 TCLCL − 100	ns
	8XL51FX-20				125		4 TCLCL − 75	ns
TLLPL	ALE Low to $\overline{\text{PSEN}}$ Low	53		20		TCLCL − 30		ns
TPLPH	$\overline{\text{PSEN}}$ Pulse Width	205		105		3 TCLCL − 45		ns
TPLIV	$\overline{\text{PSEN}}$ Low to Valid Instruction In							
	8XL51FX		145				3 TCLCL − 105	ns
	8XL51FX-20				60		3 TCLCL − 90	ns
TPXIX	Input Instruction Hold After $\overline{\text{PSEN}}$	0		0		0		ns

EXTERNAL MEMORY CHARACTERISTICS (Continued)

All parameter values apply to all devices unless otherwise indicated.

Symbol	Parameter	12 MHz Oscillator		20 MHz Oscillator		Variable Oscillator		Units
		Min	Max	Min	Max	Min	Max	
TPXIZ	Input Instruction Float After $\overline{PSEN}$							
	8XL51FX		59				TCLCL − 25	ns
	8XL51FX-20				30		TCLCL − 20	ns
TAVIV	Address to Valid Instruction In		312		145		5 TCLCL − 105	ns
TPLAZ	$\overline{PSEN}$ Low to Address Float		10		10		10	ns
TRLRH	$\overline{RD}$ Pulse Width	400		200		6 TCLCL − 100		ns
TWLWH	$\overline{WR}$ Pulse Width	400		200		6 TCLCL − 100		ns
TRLDV	$\overline{RD}$ Low to Valid Data In							
	8XL51FX		252				5 TCLCL − 165	ns
	8XL51FX-20				155		5 TCLCL − 95	ns
TRHDX	Data Hold After $\overline{RD}$	0		0		0		ns
TRHDZ	Data Float After $\overline{RD}$		107		40		2 TCLCL − 60	ns
TLLDV	ALE Low to Valid Data In							
	8XL51FX		517				8 TCLCL − 150	ns
	8XL51FX-20				310		8 TCLCL − 90	ns
TAVDV	Address to Valid Data In							
	8XL51FX		585				9 TCLCL − 165	ns
	8XL51FX-20				360		9 TCLCL − 90	ns
TLLWL	ALE Low to $\overline{RD}$ or $\overline{WR}$ Low	200	300	100	200	3 TCLCL − 50	3 TCLCL + 50	ns
TAVWL	Address Valid to $\overline{WR}$ Low							
	8XL51FX	203				4 TCLCL − 130		ns
	8XL51FX-20			110		4 TCLCL − 90		ns
TQVWX	Data Valid before $\overline{WR}$							
	8XL51FX	33				TCLCL − 50		ns
	8XL51FX-20			15		TCLCL − 35		ns
TWHQX	Data Hold after $\overline{WR}$							
	8XL51FX	33				TCLCL − 50		ns
	8XL51FX-20			10		TCLCL − 40		ns
TQVWH	Data Valid to $\overline{WR}$ High							
	8XL51FX	433				7 TCLCL − 150		ns
	8XL51FX-20			280		7 TCLCL − 70		ns
TRLAZ	$\overline{RD}$ Low to Address Float		0		0		0	ns
TWHLH	$\overline{RD}$ or $\overline{WR}$ High to ALE High	43	123	10	90	TCLCL − 40	TCLCL + 40	ns

EXTERNAL PROGRAM MEMORY READ CYCLE

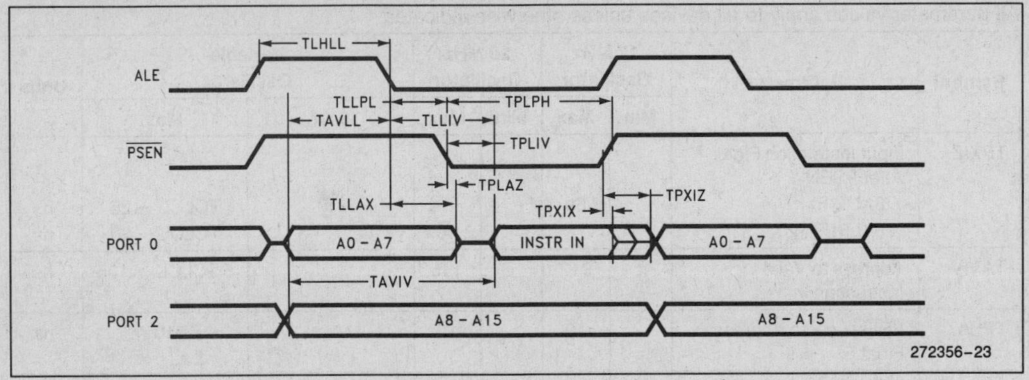

272356-23

EXTERNAL DATA MEMORY READ CYCLE

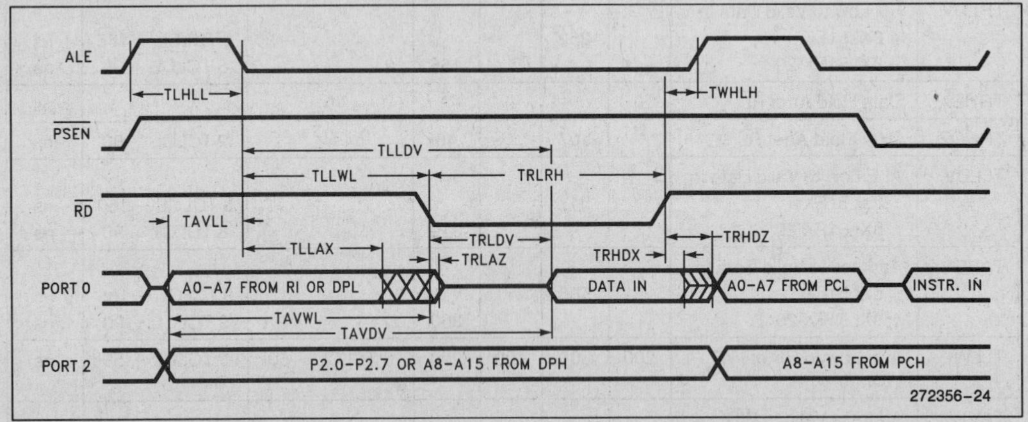

272356-24

EXTERNAL DATA MEMORY WRITE CYCLE

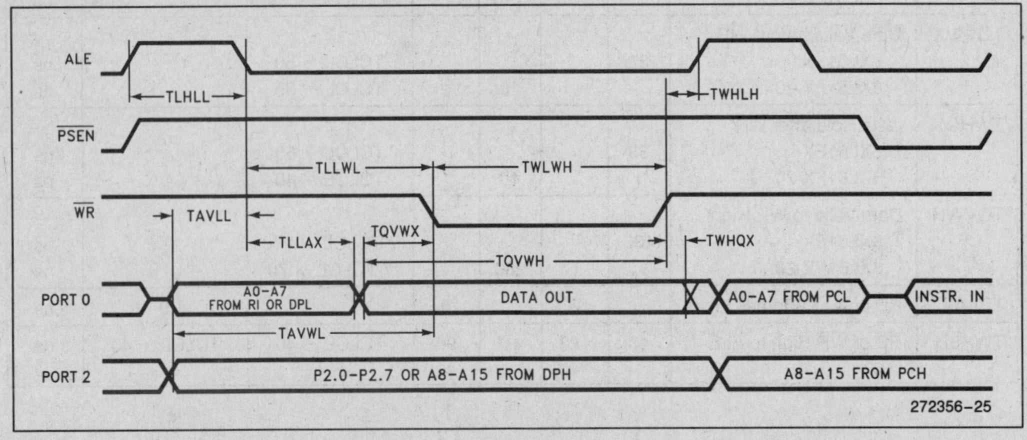

272356-25

SERIAL PORT TIMING - SHIFT REGISTER MODE

Test Conditions: Over Operating Conditions; Load Capacitance = 80 pF

Symbol	Parameter	12 MHz Oscillator		20 MHz Oscillator		Variable Oscillator		Units
		Min	Max	Min	Max	Min	Max	
TXLXL	Serial Port Clock Cycle Time	1		0.600		12 TCLCL		μs
TQVXH	Output Data Setup to Clock Rising Edge	700		367		10 TCLCL − 133		ns
TXHQX	Output Data Hold after Clock Rising Edge							
	8XC5X	50				2 TCLCL − 117		ns
	8XC5X-20			50		2 TCLCL − 50		ns
TXHDX	Input Data Hold After Clock Rising Edge	0		0		0		ns
TXHDV	Clock Rising Edge to Input Data Valid		700		367		10 TCLCL − 133	ns

SHIFT REGISTER MODE TIMING WAVEFORMS

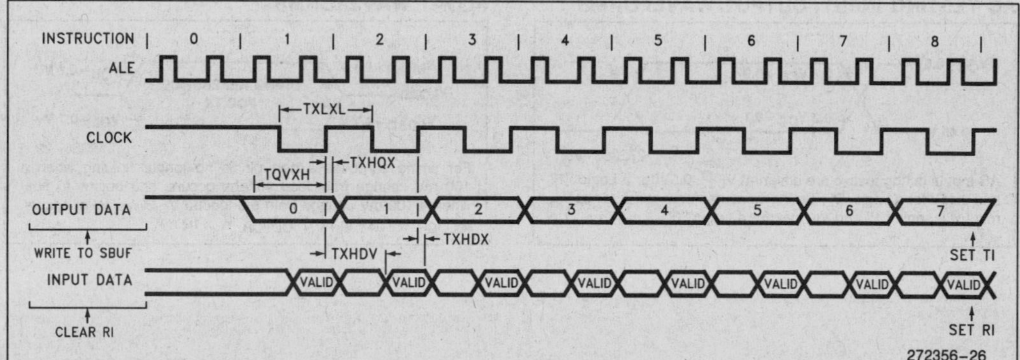

272356-26

EXTERNAL CLOCK DRIVE

Symbol	Parameter	Min	Max	Units
1/TCLCL	Oscillator Frequency 8XL51FX 8XL51FX-1 8XL51FX-20	 3.5 3.5 3.5	 12 16 20	MHz
TCHCX	High Time	20		ns
TCLCX	Low Time	20		ns
TCLCH	Rise Time		20	ns
TCHCL	Fall Time		20	ns

EXTERNAL CLOCK DRIVE WAVEFORM

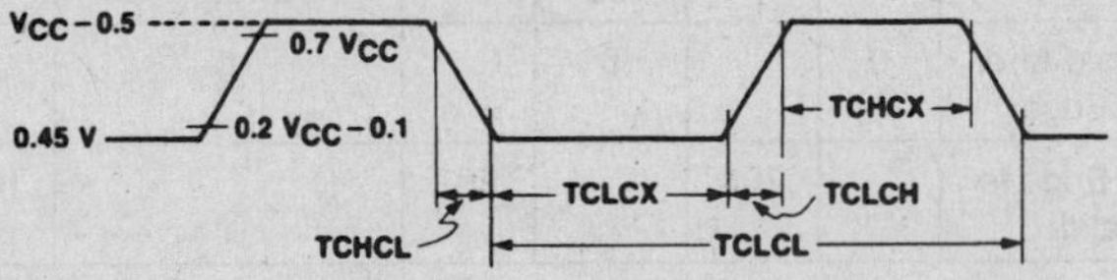

272356-27

AC TESTING INPUT, OUTPUT WAVEFORMS

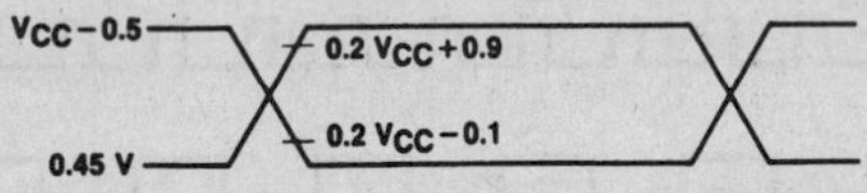

272356-28

AC Inputs during testing are driven at $V_{CC}-0.5V$ for a Logic "1" and 0.45V for a Logic "0". Timing measurements are made at V_{IH} min for a Logic "1" and V_{IL} max for a Logic "0".

FLOAT WAVEFORMS

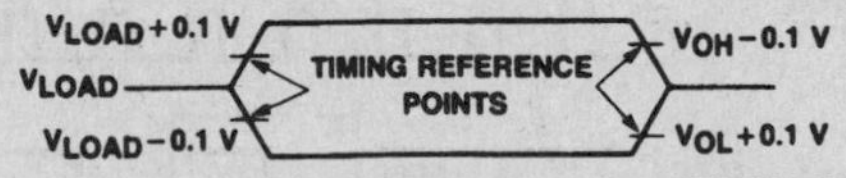

272356-29

For timing purposes a port pin is no longer floating when a 100 mV change from load voltage occurs, and begins to float when a 100 mV change from the loaded V_{OH}/V_{OL} level occurs. $I_{OL}/I_{OH} = \pm 20$ mA (-L, $I_{OL}/I_{OH} = \pm 10$ mA).

PROGRAMMING THE OTP ROM

To be programmed, the part must be running with a 4 to 6 MHz oscillator. (The reason the oscillator needs to be running is that the internal bus is being used to transfer address and program data to appropriate internal OTP ROM locations.) The address of an OTP ROM location to be programmed is applied to Port 1 and pins P2.0 - P2.4 of Port 2, while the code byte to be programmed into that location is applied to Port 0. The other Port 2 and 3 pins, RST $\overline{PSEN}$, and $\overline{EA}/V_{PP}$ should be held at the "Program" levels indicated in Table 3. ALE/$\overline{PROG}$ is pulsed low to program the code byte into the addressed OTP ROM location. The setup is shown in Figure 10.

Normally $\overline{EA}/V_{PP}$ is held at logic high until just before ALE/$\overline{PROG}$ is to be pulsed. Then $\overline{EA}/V_{PP}$ is raised to V_{PP}, ALE/$\overline{PROG}$ is pulsed low, and then $\overline{EA}/V_{PP}$ is returned to a valid high voltage. The voltage on the $\overline{EA}/V_{PP}$ pin must be at the valid $\overline{EA}/V_{PP}$ high level before a verify is attempted. Waveforms and detailed timing specifications are shown in later sections of this data sheet.

NOTE:

- $\overline{EA}/V_{PP}$ pin must not be allowed to go above the maximum specified V_{PP} level for any amount of time. Even a narrow glitch above that voltage level can cause permanent damage to the device. The V_{PP} source should be well regulated and free of glitches.

Table 3. OTP ROM Programming Modes

(H = 2.7V to 3.6V; H1 = 5V ±10%)

Mode		RST	$\overline{PSEN}$	ALE/ $\overline{PROG}$	$\overline{EA}$/ V_{PP}	P2.6	P2.7	P3.3	P3.6	P3.7	V_{CC}
Program Code Data		H1	L	⊔	12.75V	L	H1	H1	H1	H1	H1
Verify Code Data		H	L	H	H	L	L	L	H	H	H
Program Encryption Array Address 0–3FH		H1	L	⊔	12.75V	L	H1	H1	L	H1	H1
Program Lock Bits	Bit 1	H1	L	⊔	12.75V	H1	H1	H1	H1	H1	H1
	Bit 2	H1	L	⊔	12.75V	H1	H1	H1	L	L	H1
	Bit 3	H1	L	⊔	12.75V	H1	L	H1	H1	L	H1
Read Signature Byte		H	L	H	H	L	L	L	L	L	H

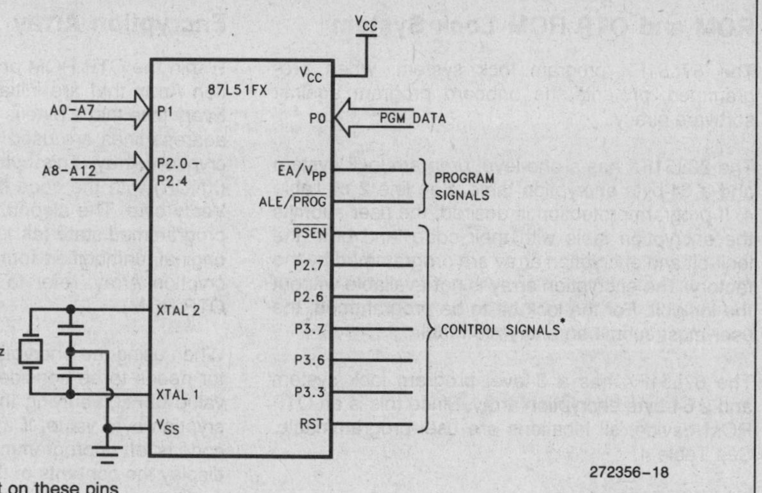

*See Table 2 for proper input on these pins

Figure 10. Programming the OTP ROM

PROGRAMMING ALGORITHM

Refer to Table 3 and Figures 10 and 11 for address, data, and control signals set up. To program the 87L51FX the following sequence must be exercised.

1. Input the valid address on the address lines.
2. Input the appropriate data byte on the data lines.
3. Activate the correct combination of control signals.
4. Raise $\overline{EA}/V_{PP}$ from V_{CC} to 12.75V ±0.25V.
5. Pulse, ALE/$\overline{PROG}$ 5 times for the OTP ROM array, and 25 times for the encryption table and the lock bits.

Repeat 1 through 5 changing the address and data for the entire array or until the end of the object file is reached.

PROGRAM VERIFY

Program verify may be done after each byte or block of bytes is programmed. In either case a complete verify of the programmed array will ensure reliable programming of the 87L51FX.

The lock bits cannot be directly verified. Verification of the lock bits is done by observing that their features are enabled.

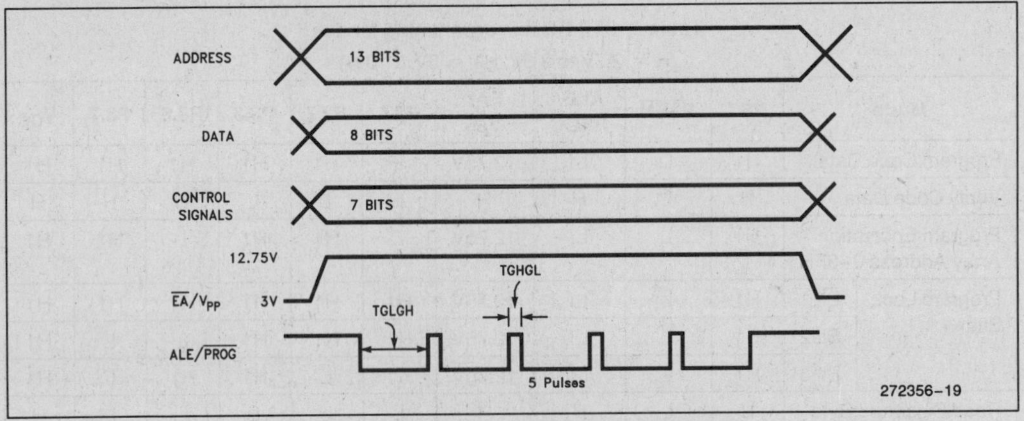

Figure 11. Programming Signals Waveforms

ROM and OTP ROM Lock System

The 87L51FX program lock system, when programmed, protects the onboard program against software piracy.

The 83L51FX has a one-level program lock system and a 64-byte encryption table. See line 2 of Table 4. If program protection is desired, the user submits the encryption table with their code, and both the lock-bit and encryption array are programmed by the factory. The encryption array is not available without the lock bit. For the lock bit to be programmed, the user must submit an encryption table.

The 87L51FX has a 3-level program lock system and a 64-byte encryption array. Since this is an OTP ROM device, all locations are user-programmable. See Table 4.

Encryption Array

Within the OTP ROM array are 64 bytes of Encryption Array that are initially unprogrammed (all 1's). Every time that a byte is addressed during a verify, 6 address lines are used to select a byte of the Encryption Array. This byte is then exclusive-NOR'ed (XNOR) with the code byte, creating an Encryption Verify byte. The algorithm, with the array in the unprogrammed state (all 1's), will return the code in its original, unmodified form. For programming the Encryption Array, refer to Table 3 (Programming the OTP ROM).

When using the encryption array, one important factor needs to be considered. If a code byte has the value 0FFH, verifying the byte will produce the encryption byte value. If a large block (>64 bytes) of code is left unprogrammed, a verification routine will display the contents of the encryption array. For this reason all unused code bytes should be programmed with some value other than 0FFH, and not all of them the same value. This will ensure maximum program protection.

Table 4. Program Lock Bits and the Features

Program Lock Bits				Protection Type
	LB1	LB2	LB3	
1	U	U	U	No Program Lock features enabled. (Code verify will still be encrypted by the Encryption Array if programmed.)
2	P	U	U	MOVC instructions executed from external program memory are disabled from fetching code bytes from internal memory, EA is sampled and latched on Reset, and further programming of the OTP ROM is disabled.
3	P	P	U	Same as 2, also verify is disabled.
4	P	P	P	Same as 3, also external execution is disabled.

Any other combination of the lock bits is not defined.

Program Lock Bits

The 87L51FX has 3 programmable lock bits that when programmed according to Table 4 will provide different levels of protection for the on-chip code and data.

Reading the Signature Bytes

The 87L51FX/83L51FX has 3 signature bytes in locations 30H, 31H, and 60H. To read these bytes follow the procedure for OTP ROM verify, but activate the control lines provided in Table 3 for Read Signature Byte.

Location	Device	Contents
30H	All	89H
31H	All	58H
60H	83L51FA	70H
	87L51FA	F0H
	83L51FB	71H
	87L51FB	F1H
	83L51FC	72H
	87L51FC	F2H

2

OTP ROM PROGRAMMING AND VERIFICATION CHARACTERISTICS

(T_A = 21°C to 27°C; V_{CC} = 2.7V to 3.6V; V_{SS} = 0V)

Symbol	Parameter	Min	Max	Units
V_{PP}	Programming Supply Voltage	12.5	13.0	V
I_{PP}	Programming Supply Current		75	mA
1/TCLCL	Oscillator Frequency	4	6	MHz
TAVGL	Address Setup to $\overline{PROG}$ Low	48TCLCL		
TGHAX	Address Hold after $\overline{PROG}$	48TCLCL		
TDVGL	Data Setup to $\overline{PROG}$ Low	48TCLCL		
TGHDX	Data Hold after $\overline{PROG}$	48TCLCL		
TEHSH	P2.7 (ENABLE) High to V_{PP}	48TCLCL		
TSHGL	V_{PP} Setup to $\overline{PROG}$ Low	10		μs
TGHSL	V_{PP} Hold after $\overline{PROG}$	10		μs
TGLGH	$\overline{PROG}$ Width	90	100	μs
TAVQV	Address to Data Valid		48TCLCL	
TELQV	ENABLE Low to Data Valid		48TCLCL	
TEHQZ	Data Float after ENABLE	0	48TCLCL	
TGHGL	$\overline{PROG}$ High to $\overline{PROG}$ Low	10		μs

OTP ROM PROGRAMMING AND VERIFICATION WAVEFORMS

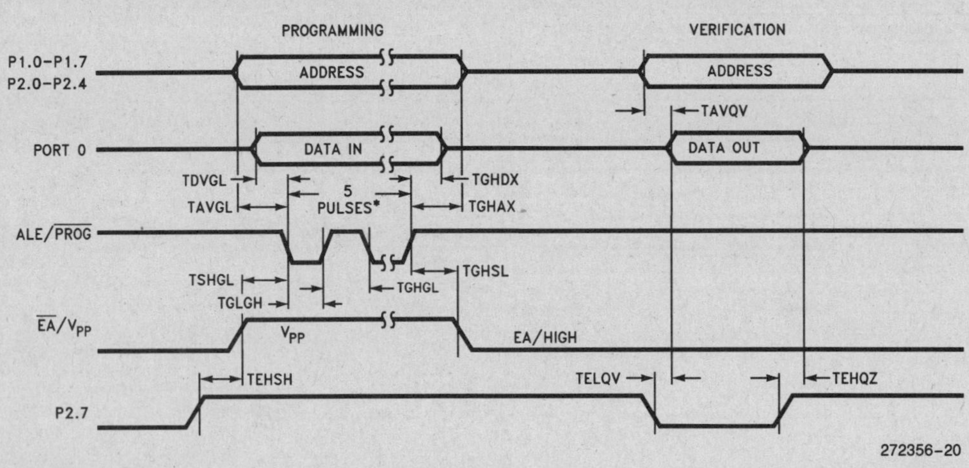

NOTE:
*5 pulses for the OTP ROM array, 25 pulses for the encryption table and lock bits.

Thermal Impedance

The thermal impedance data is approximate for static air conditions at 1W of power dissipation. Values will change depending on operating conditionis and applications. See the Intel Packaging Handbook (Order Number 240800) for a description of Intel's thermal impedance test methodology.

Package	θ_{JA}	θ_{JC}	Device
N	46°C/W	16°C/W	All
S	87°CW	18°C/W	52
	96°C/W	24°C/W	54
	90°C/W	22°C/W	58

Data Sheet Revision History

This 8XL51FA/FB/FC data sheet (Advanced Information) replaces 272356-001 (Product Preview).

2

intel® PRELIMINARY

8XC51GB
CHMOS SINGLE-CHIP 8-BIT MICROCONTROLLER

Commercial/Express

87C51GB—8 Kbytes OTP/8 Kbytes Internal Program Memory
83C51GB—8 Kbytes Factory Programmable ROM
80C51GB—CPU with RAM and I/O
8XC51GB—3.5 MHz to 12 MHz ±20% V_{CC}
8XC51GB-1—3.5 MHz to 16 MHz ±20% V_{CC}

- **8 Kbytes On-Chip ROM/OTP ROM**
- **256 Bytes of On-Chip Data RAM**
- **Two Programmable Counter Arrays with:**
  - **2 x 5 High Speed Input/Output Channels Compare/Capture**
  - **Pulse Width Modulators**
  - **Watchdog Timer Capabilities**
- **Three 16-Bit Timer/Counters with**
  - **Four Programmable Modes:**
  - **Capture, Baud Rate Generation (Timer 2)**
- **Dedicated Watchdog Timer**
- **8-Bit, 8-Channel A/D with:**
  - **Eight 8-Bit Result Registers**
  - **Four Programmable Modes**
- **Programmable Serial Channel with:**
  - **Framing Error Detection**
  - **Automatic Address Recognition**
- **Serial Expansion Port**
- **Programmable Clock Out**
- **Extended Temperature Range: (−40°C to +85°C)**
- **48 Programmable I/O Lines with 40 Schmitt Trigger Inputs**
- **15 Interrupt Sources with:**
  - **7 External, 8 Internal Sources**
  - **4 Programmable Priority Levels**
- **Pre-Determined Port States on Reset**
- **High Performance CHMOS Process**
- **TTL and CHMOS Compatible Logic Levels**
- **Power Saving Modes**
- **64K External Data Memory Space**
- **64K External Program Memory Space**
- **Three Level Program Lock System**
- **ONCE (ON-Circuit Emulation) Mode**
- **Quick Pulse Programming Algorithm**
- **MCS® 51 Microcontroller Fully Compatible Instruction Set**
- **Boolean Processor**
- **Oscillator Fail Detect**
- **Available in 68-Pin PLCC**

MEMORY ORGANIZATION

PROGRAM MEMORY: Up to 8 Kbytes of the program memory can reside in the on-chip ROM. Also, the device can address up to 64K of program memory external to the chip.

DATA MEMORY: This microcontroller has a 256 x 8 on-chip RAM. In addition it can address up to 64 Kbytes of external data memory.

The Intel 8XC51GB is a single-chip control oriented microcontroller which is fabricated on Intel's CHMOS III-E technology. The 8XC51GB is an enhanced version of the 8XC51FA and uses the same powerful instruction set and architecture as existing MCS 51 microcontroller products. Added features make it an even more powerful microcontroller for applications that require On-Chip A/D, Pulse Width Modulation, High Speed I/O, up/down counting capabilities and memory protection features. It also has a more versatile serial channel that facilitates multi-processor communications.

November 1994
Order Number: 272337-002

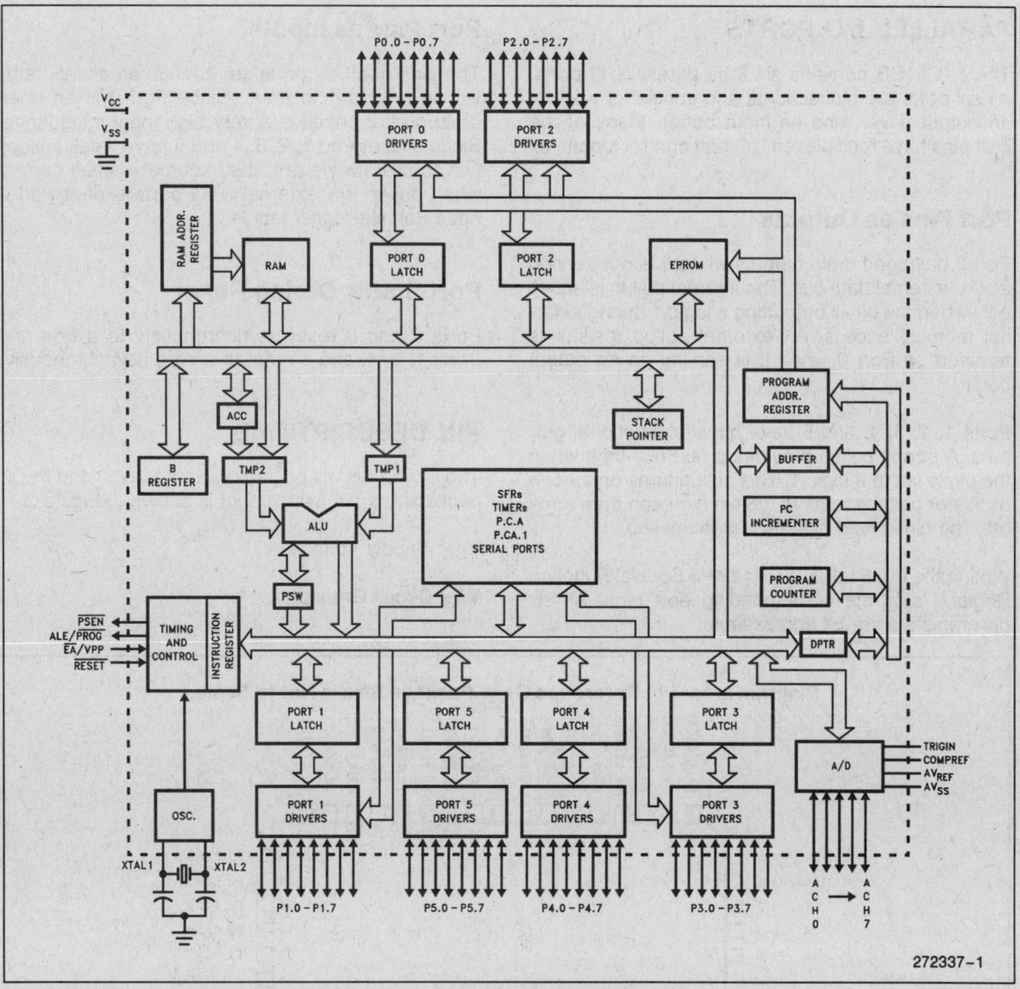

Figure 1. 8XC51GB Block Diagram

PROCESS INFORMATION

This device is manufactured on P629.0, a CHMOS III-E process. Additional process and reliability information is available in Intel's Components Quality and Reliability Handbook, Order No. 210997.

PACKAGES

Part	Prefix	Package Type
8XC51GB	N	68-Pin PLCC

PARALLEL I/O PORTS

The 8XC51GB contains six 8-bit parallel I/O ports. All six ports are bidirectional and consist of a latch, an output driver, and an input buffer. Many of the port pins have multiplexed I/O and control functions.

Port Pins as Outputs

Port 0 has open drain outputs when it is not serving as the external data bus. The internal pullup is active only when the pin is outputting a logic 1 during external memory access. An external pullup resistor is required on Port 0 when it is serving as an output port.

Ports 1, 2, 3, 4, and 5 have quasi-bidirectional outputs. A strong pullup provides a fast rise time when the pin is set to a logic 1. This pullup turns on for two oscillator periods to drive the pin high and then turns off. The pin is held high by a weak pullup.

Writing the P0, P1, P2, P3, P4 or P5 Special Function Register sets the corresponding port pins. All six port registers are bit addressable.

Port Pins as Inputs

The pins of all six ports are configured as inputs by writing a logic 1 to them. Since Port 0 is an open drain port, it provides a very high input impedance. Since pins of Port 1, 2, 3, 4 and 5 have weak pullups (which are always on), they source a small current when driven low externally. All ports except Port 0 have Schmitt trigger inputs.

Port States During Reset

Ports 0 and 3 reset asynchronously to a one and Ports 1, 2, 4, and 5 reset to a zero asynchronously.

PIN DESCRIPTIONS

The 8XC51GB will be packaged in the 68-lead PLCC package. Its pin assignment is shown in Figure 2.

V_{CC}: Supply Voltage.

V_{SS}: Circuit Ground.

Diagram is for Pin Reference Only. Package Size is Not to Scale.

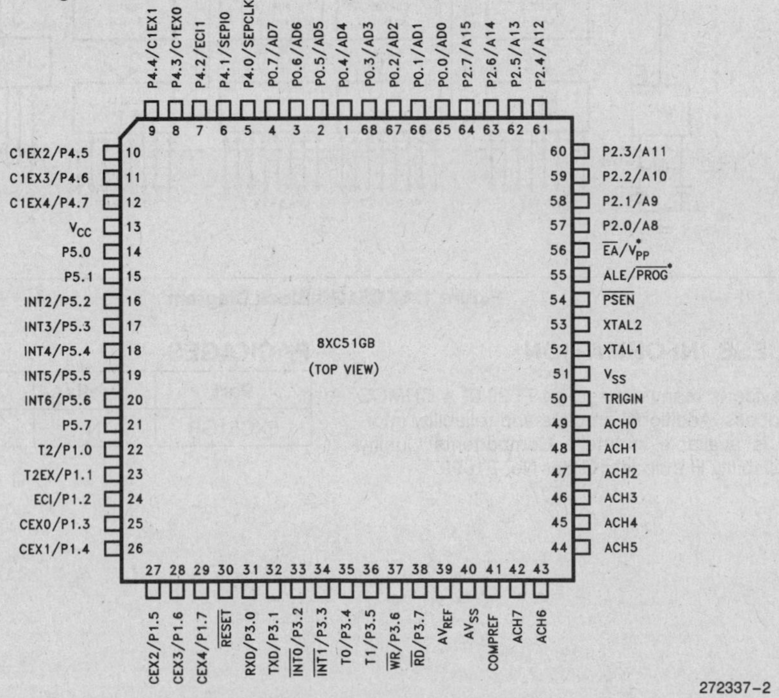

*OTP only

Figure 2. Pin Connections

ALTERNATE PORT FUNCTIONS

Ports 0, 1, 2, 3, 4 and 5 have alternate functions as well as their I/O function as described below.

Port Pin	Alternate Function
P0.0/ADO–P0.7/AD7	Multiplexed Address/Data for External Memory
P1.0/T2	Timer 2 External Clock Input/Clock-Out
P1.1/T2EX	Timer 2 Reload/Capture/Direction Control
P1.2/ECI	PCA External Clock Input
P1.3/CEXO–P1.7/CEX4	PCA Capture Input, Compare/PWM Output
P2.0/A8–P2.7/A15	High Byte of Address for External Memory
P3.0/RXD	Serial Port Input
P3.1/TXD	Serial Port Output
P3.2/$\overline{\text{INT0}}$	External Interrupt 0
P3.3/$\overline{\text{INT1}}$	External Interrupt 1
P3.4/T0	Timer 0 External Clock Input
P3.5/T1	Timer 1 External Clock Input
P3.6/$\overline{\text{WR}}$	Write Strobe for External Memory
P3.7/$\overline{\text{RD}}$	Read Strobe for External Memory
P4.0/SEPCLK	Clock Source for Serial Expansion Port
P4.1/SEPDAT	Data I/O for the Serial Expansion Port
P4.2/ECI1	PCA1 External Clock Input
P4.3/C1EX0–P4.7/C1EX4	PCA1 Capture Input, Compare/PWM Output
P5.2/INT2–P5.6/INT6	External Interrupt INT2–INT6

RST: Reset input. A low on this pin for two machine cycles while the oscillator is running resets the device. The port pins will be driven to their reset condition when a voltage below V_{IL} max voltage is applied, whether the oscillator is running or not. An internal pullup resistor permits a power-on reset with only a capacitor connected to V_{SS}.

ALE/$\overline{\text{PROG}}$: Address Latch Enable output pulse for latching the low byte of the address during accesses to external memory. This pin (ALE/$\overline{\text{PROG}}$) is also the program pulse input during programming of the 87C51GB.

In normal operation ALE is emitted at a constant rate of 1/6 the oscillator frequency, and may be used for external timing or clocking purposes. Note, however, that one ALE pulse is skipped during each access to external Data Memory.

If desired, ALE operation can be disabled by setting bit 0 of SFR location 8EH. With this bit set, the pin is weakly pulled high. However, the ALE disable feature will be suspended during a MOVX or MOVC instruction, idle mode, power down mode and ICE mode. The ALE disable feature will be terminated by reset. When the ALE disable feature is suspended or terminated, the ALE pin will no longer be pulled up weakly. Setting the ALE-disable bit has no affect if the microcontroller is in external execution mode.

Throughout the remainder of this data sheet, ALE will refer to the signal coming out of the ALE/$\overline{\text{PROG}}$ pin, and the pin will be referred to as the ALE/$\overline{\text{PROG}}$ pin.

$\overline{\text{PSEN}}$: Program Store Enable is the read strobe to external Program Memory.

When the 8XC51GB is executing code from external Program Memory, $\overline{\text{PSEN}}$ is activated twice each machine cycle, except that two $\overline{\text{PSEN}}$ activations are skipped during each access to external Data Memory.

$\overline{\text{EA}}$/V_{PP}: External Access enable. $\overline{\text{EA}}$ must be strapped to V_{SS} in order to enable the device to fetch code from external Program Memory locations 0000H to 1FFFH. Note, however, that if either of the Program Lock bits are programmed, $\overline{\text{EA}}$ will be internally latched on reset.

$\overline{\text{EA}}$ should be strapped to V_{CC} for internal program executions.

This pin also receives the 12.75V programming supply voltage (V_{PP}) during programming (OTP only).

XTAL1: Input to the inverting oscillator amplifier.

XTAL2: Output from the inverting oscillator amplifier.

A/D CONVERTER

The 8XC51GB A/D converter has a resolution of 8 bits and an accuracy of ±1 LSB (±2 LSB for channels 0 and 1). The conversion time for a single channel is 20 µs at a clock frequency of 16 MHz with the sample and hold function included. Independent supply voltages are provided for the A/D. Also, the A/D operates both in Normal Mode or in Idle Mode.

The A/D has 8 analog input pins; ACH0 (A/D CHannel 0) ... ACH7, 1 reference input pin; COMPREF (COMParison REFerence), 1 control input pin; TRIGIN (TRIGger IN), and 2 power pins; AVREF (Voltage REFerence) and analog ground (ANalog GrouND). In addition, the A/D has 8 conversion result registers; ADRES0 (A/D result for channel 0) ... ADRES7, 1 comparison result register; ACMP (Analog Comparison), and 1 control register; ACON (A/D Control).

The control bit ACE (A/D Conversion Enable) in ACON controls whether the A/D is in operation or not. ACE = 0 idles the A/D. ACE = 1 enables A/D conversion. The control bit AIM (A/D Input mode) in ACON controls the mode of channel selection. AIM = 0 is the Scan Mode, and AIM = 1 is the Select Mode. The result registers ADRES4 ... ADRES7 always contain the result of a conversion from the corresponding channels ACH4 ... CH7. However, the result registers ADRES0 ... ADRES3 depend on the mode selected. In the scan mode, ADRES0 ... ADRES3 contain the values from ACH0 ... ACH3. In the Select Mode, one of the four channels ACH0 ... ACH3 is converted four times, and the four values are stored sequentially in locations ADRES0 ... ADRES3. Its channel is selected by bits ACS1 and ACS0 (A/D Channel Select 1 and 0) in ACON.

PROGRAMMABLE COUNTER ARRAYS

The Programmable Counter Arrays (PCA–PCA1) are each made up of a Counter Module and five Register/Comparator Modules as shown below. The 16-bit output of the counter module is available to all five Register/Comparator Modules, providing one common timing reference. Each Register/Comparator Module is associated with a pin of Port 1 or Port 4 and is capable of performing input capture, output compare and pulse width modulation functions. The PCAs are exactly the same in function except for the addition of clock input sources on PCA1.

The PCA Counter and five Register/Comparator Modules each have a status bit in the CCON/C1CON Special Function Registers. These six status bits are set according to the selected modes of operation described below. The CCON/C1CON Register provides a convenient means to determine which of the six PCA/PCA1 interrupts has occurred. The EC Bit in the IE (Interrupt Enable) Special Function Register is a global interrupt enable for the PCA.

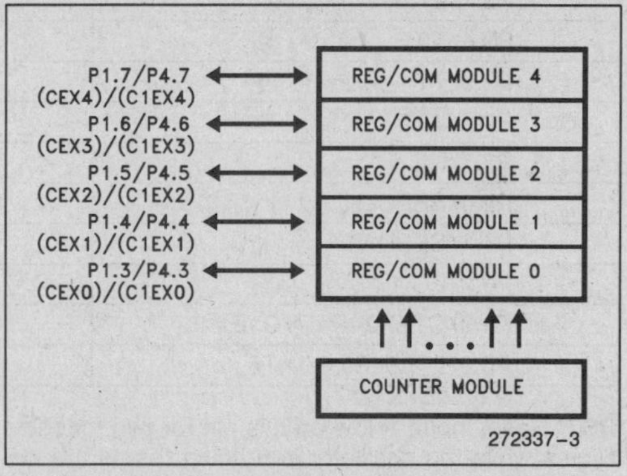

Figure 3. Programmable Counter Arrays

OSCILLATOR CHARACTERISTICS

XTAL1 and XTAL2 are the input and output, respectively, of an inverting amplifier which can be configured for use as an on-chip oscillator, as shown in Figure 4. Either a quartz crystal or ceramic resonator may be used. More detailed information concerning the use of the on-chip oscillator is available in Application Note AP-155, "Oscillators for Microcontrollers," Order No. 230659.

To drive the device from an external clock source, XTAL should be driven, while XTAL2 floats, as shown in Figure 5. There are no requirements on the duty cycle of the external clock signal, since the input to the internal clocking circuitry is through a divide-by-two flip-flop, but minimum and maximum high and low times specified on the data sheet must be observed.

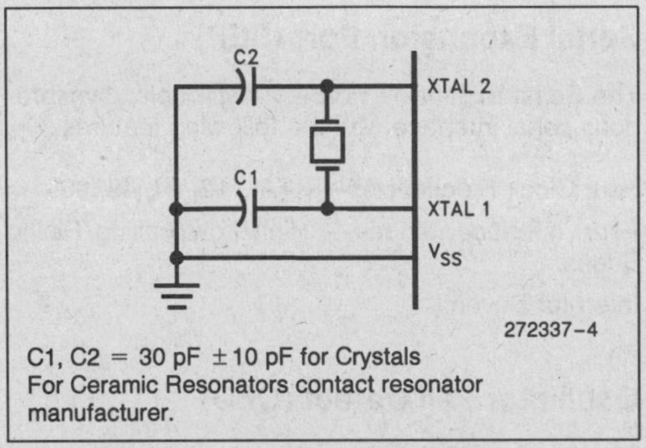

Figure 4. Oscillator Connections

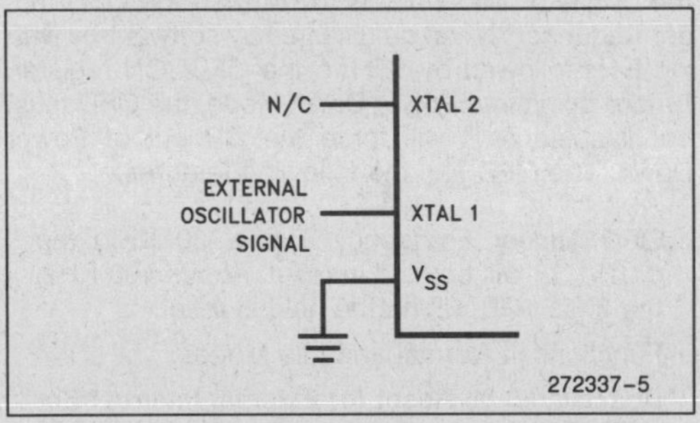

Figure 5. External Clock Drive Configuration

IDLE MODE

The user's software can invoke the Idle Mode. When the microcontroller is in this mode, power consumption is reduced. The Special Function Registers and the onboard RAM retain their values during idle, peripherals continue to operate, but the processor stops executing instructions. Idle Mode will be exited if the chip is reset or if an enabled interrupt occurs. The PCA timer/counter can optionally be left running or paused during Idle Mode. The Watchdog Timer continues to count in Idle Mode and must be serviced to prevent a device RESET while in Idle.

POWER DOWN MODE

To save even more power, a Power Down mode can be invoked by software. In this mode, the oscillator is stopped and the instruction that invoked Power Down is the last instruction executed. The on-chip RAM and Special Function Registers retain their values until the Power Down mode is terminated.

On the 8XC51GB either a hardware reset or an external interrupt can cause an exit from Power Down. Reset redefines all the SFRs but does not change the on-chip RAM. An external interrupt does not redefine the SFR's or change the on-chip RAM. An external interrupt will modify the interrupt associated SFR's in the same way an interrupt will in all other modes. The interrupt must be enabled and configured as level sensitive. To properly terminate Power Down the reset or external interrupt should not be executed before V_{CC} is restored to its normal operating level. The reset or external interrupt must be held active long enough for the oscillator to restart and stabilize. The Oscillator Fail Detect must be disabled prior to entering Power Down.

DESIGN CONSIDERATIONS

- When the idle mode is terminated by a hardware reset, the device normally resumes program execution, from where it left off, up to two machine cycles before the internal reset algorithm takes control. On-chip hardware inhibits access to internal RAM in this event, but access to the port pins is not inhibited. To eliminate the possibility of an unexpected write when Idle is terminated by reset, the instruction following the one that invokes Idle should not be one that writes to a port pin or to external memory.
- As $\overline{RESET}$ rises, the 8XC51GB will remain in reset for up to 5 machine cycles (60 oscillator periods) after $\overline{RESET}$ reaches V_{IH1}.

Table 1. Status of the External Pins during Idle and Power Down

Mode	Program Memory	ALE	$\overline{PSEN}$	PORT0	PORT1	PORT2	PORT3
Idle	Internal	1	1	Data	Data	Data	Data
Idle	External	1	1	Float	Data	Address	Data
Power Down	Internal	0	0	Data	Data	Data	Data
Power Down	External	0	0	Float	Data	Data	Data

NOTE:
For more detailed information on the reduced power modes refer to current Embedded Microcontrollers and Processors Handbook Volume I (Order No. 270645), and Application Note AP-252 (Embedded Applications Handbook, Order No. 270648), "Designing with the 80C51BH."

2

ONCE MODE

The ONCE ("On-Circuit Emulation") Mode facilitates testing and debugging of systems using the 8XC51GB without removing it from the circuit. The ONCE Mode is invoked by:

1) Pulling ALE low while the device is in reset and $\overline{PSEN}$ is high;
2) Holding ALE low as $\overline{RESET}$ is deactivated.

While the device is in ONCE Mode, the Port 0 pins float, and the other port pins and ALE and $\overline{PSEN}$ are weakly pulled high. The oscillator circuit remains active. While the 8XC51GB is in this mode, an emulator or test CPU can be used to drive the circuit. Normal operation is restored when a normal reset is applied.

Watchdog Timer (WDT)

The 8XC51GB contains a dedicated Watchdog Timer (WDT) to allow recovery from a software or hardware upset. The WDT consists of a 14-bit counter which is cleared on Reset, and subsequently incremented every machine cycle. While the oscillator is running, the WDT will be incrementing and cannot be disabled. The counter may be reset by writing 1EH and E1H in sequence to the WDTRST Special Function Register. If the counter is not reset before it reaches 3FFFH (16383D), the chip will be forced into a reset sequence by the WDT. This works out to 12.28 ms @ 16 MHz. WDTRST is a write only register. The WDT does not force the external reset pin low.

While in Idle mode the WDT continues to count. If the user does not wish to exit Idle with a reset, then the processor must be periodically "woken up" to service the WDT. In Power Down mode, the WDT stops counting and holds its current value.

Serial Expansion Port (SEP)

The Serial Expansion Port is a half-duplex synchronous serial interface with the following features:

Four Clock Frequencies— XTAL/12, 24, 48, 96.

Four Interface Modes— High/Low/Falling/Rising Edges.

Interrupt Driven.

Oscillator Fail Detect (OFD)

The Oscillator Fail Detect circuitry triggers a reset if the oscillator frequency is lower than the OFD trigger frequency. It can be disabled by software by writing E1H followed by 1EH to the OFDCON register. Before going into Power Down Mode, the OFD must be disabled or it will force the GB out of Power Down. The OFD has the following features.

OFD Trigger Frequency: Below 20 KHz, the 8XC51GB will be held in reset. Above 400 KHz, the 8XC51GB will not be held is reset.

Functions in Normal and Idle Modes.

Reactivated by Reset (or External Interrupt Zero/One Pins) after Software Disable.

8XC51GB EXPRESS

The Intel EXPRESS products are designed to meet the needs of those applications whose operating requirements exceed commercial standards.

With the commercial standard temperature range, operational characteristics are guaranteed over the temperature range of 0°C to +70°C. With the extended temperature range option, operational characteristics are guaranteed over the range of −40°C to +85°C. The 87C51GB EXPRESS is packaged in the 68-lead PLCC package. In order to designate a part as an EXPRESS part, a "T" is added as a prefix to the part number. TN87C51GB denotes an EXPRESS part in a PLCC package.

All AC and DC parameters in this data sheet apply to the EXPRESS devices.

ABSOLUTE MAXIMUM RATINGS*

Ambient Temperature under Bias ....0°C to +70°C

Storage Temperature ....−65°C to +150°C

Voltage on EA/V_{PP} Pin to V_{SS} ....0V to +13.0V*

I_{OL} per I/O Pin ....15 mA

Voltage on Any Other Pin to V_{SS} ....−0.5V to +6.5V

Power Dissipation..........................1.5W
(Based on Package heat transfer limitations, not device power consumption)

*OTP only.

NOTICE: This data sheet contains preliminary information on new products in production. The specifications are subject to change without notice. Verify with your local Intel Sales office that you have the latest data sheet before finalizing a design.

**WARNING: Stressing the device beyond the "Absolute Maximum Ratings" may cause permanent damage. These are stress ratings only. Operation beyond the "Operating Conditions" is not recommended and extended exposure beyond the "Operating Conditions" may affect device reliability.*

2

OPERATING CONDITIONS

Symbol	Description	Min	Max	Units
T_A	Ambient Temperature Under Bias			
	Commercial	0	+70	°C
	Express	−40	+85	°C
V_{CC}	Supply Voltage	4.0	6.0	V
f_{OSC}	Oscillator Frequency			
	8XC51GB	3.5	12	MHz
	8XC51GB-1	3.5	16	MHz

DC CHARACTERISTICS (Over Operating Conditions)

Symbol	Parameter	Min	Typ(1)	Max	Unit	Test Conditions
V_{IL}	Input Low Voltage (except Port 2 and $\overline{EA}$)	−0.5		$0.2\ V_{CC} - 0.1$	V	
V_{IL1}	Input Low Voltage (Port 2)	−0.5		$0.2\ V_{CC} - 0.3$	V	
V_{IL2}	Input Low Voltage ($\overline{EA}$)	0		$0.2\ V_{CC} - 0.3$	V	
V_{IH}	Input High Voltage (except XTAL1 and $\overline{RST}$)	$0.2\ V_{CC} + 0.9$		$V_{CC} + 0.5$	V	
V_{IH1}	Input High Voltage (XTAL1, $\overline{RST}$)	$0.7\ V_{CC}$		$V_{CC} + 0.5$	V	
V_{OL}	Output Low Voltage (Ports 1, 2, 3, 4 and 5)			0.3	V	$I_{OL} = 100\ \mu A$ (2,3)
				0.45	V	$I_{OL} = 1.6$ mA (2,3)
				1.0	V	$I_{OL} = 3.5$ mA (2,3)
V_{OL1}	Output Low Voltage (Port 0, $\overline{PSEN}$, ALE)			0.3	V	$I_{OL} = 200\ \mu A$ (2,3)
				0.45	V	$I_{OL} = 3.2$ mA (2,3)
				1.0	V	$I_{OL} = 7.0$ mA (2,3)

DC CHARACTERISTICS (Over Operating Conditions) (Continued)

Symbol	Parameter	Min	Typ(1)	Max	Unit	Test Conditions
V_{OH}	Output High Voltage (Ports 1, 2, 3, 4 and 5, ALE, $\overline{PSEN}$)	$V_{CC} - 0.3$			V	$I_{OH} = -10\ \mu A$ (4)
		$V_{CC} - 0.7$			V	$I_{OH} = -30\ \mu A$ (4)
		$V_{CC} - 1.5$			V	$I_{OH} = -60\ \mu A$ (4)
V_{OH1}	Output High Voltage (Port 0 in External Bus Mode)	$V_{CC} - 0.3$			V	$I_{OH} = -200\ \mu A$
		$V_{CC} - 0.7$			V	$I_{OH} = -3.2$ mA
		$V_{CC} - 1.5$			V	$I_{OH} = -7.0$ mA
I_{IL}	Logical 0 Input Current (Ports 1, 2, 3, 4, 5)			−50	μA	$V_{IN} = 0.45V$
I_{TL}	Logical 1-to-0 Transition Current (Ports 1, 2, 3, 4, 5)			−650	μA	$V_{IN} = 2.0V$
I_{LI}	Input Leakage Current (Port 0)			±10	μA	$0.45 < V_{IN} < V_{CC}$
RRST	RST Pullup Resistor	50		300	kΩ	
C_{IO}	Pin Capacitance		10		pF	Freq = 1 MHz $T_A = 25°C$
I_{PD}	Power Down Current			50	μA	(5)
I_{DL}	Idle Mode Current			18	mA	(5)
I_{CC}	Operating Current @16 MHz			50	mA	(5)
I_{REF}	A/D Converter Reference Current			5	mA	

NOTES:

1. Typical values are obtained using $V_{CC} = 5.0V$, $T_A = 25°C$, and are not guaranteed.
2. Under steady state (non-transient) conditions, I_{OL} must be externally limited as follows:
   Maximum I_{OL} per Port Pin: 10 mA
   Maximum I_{OL} per 8-Bit Port—
   Port 0: 26 mA
   Ports 1–5: 15 mA
   Maximum Total I_{OL} for All Outputs Pins: 101 mA
   If I_{OL} exceeds the test conditions, V_{OL} may exceed the related specification. Pins are not guaranteed to sink current greater than the listed test conditions.
3. Capacitive loading on Ports 0 and 2 may cause spurious noise pulses above 0.4V on the low level outputs of ALE and Ports 1, 2 and 3. The noise is due to external bus capacitance discharging into the Port 0 and Port 2 pins when these pins change from 1 to 0. In applications where capacitive loading exceeds 100 pF, the noise pulses on these signals may exceed 0.8V. It may be desirable to qualify ALE or other signals with a Schmitt Trigger, or CMOS-level input logic.
4. Capacitive loading on Ports 0 and 2 cause the V_{OH} on ALE and $\overline{PSEN}$ to drop below the 0.9 V_{CC} specification when the address lines are stabilizing.
5. See Figures 6–10 for test conditions. Minimum V_{CC} for Power Down is 2V.

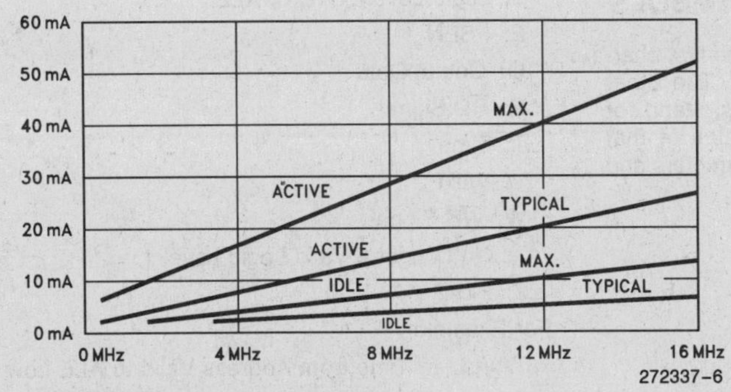

I_{CC} Max at other frequencies is given by:
Active Mode
I_{CC} Max = (Osc Freq × 3) + 4
Idle Mode
I_{CC} Max = (Osc Freq × 0.5) + 4
Where Osc Freq is in MHz, I_{CC} is in mA. $T_{CLCH} = T_{CHCL}$ = 5 ns

Figure 6. I_{CC} vs Frequency

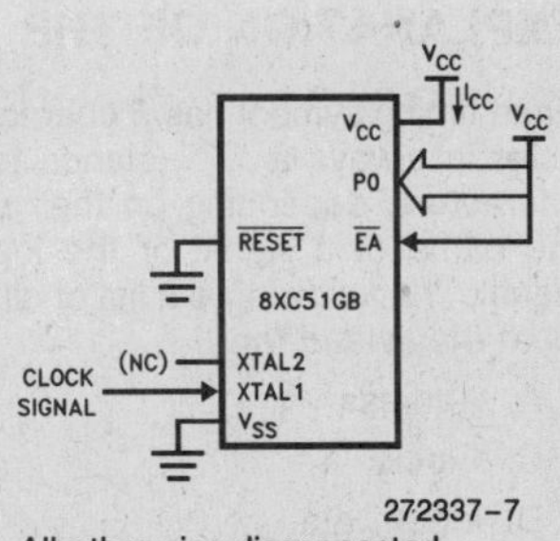

All other pins disconnected.
TCLCH = TCHCL = 5 ns

Figure 7. I_{CC} Test Condition, Active Mode

2

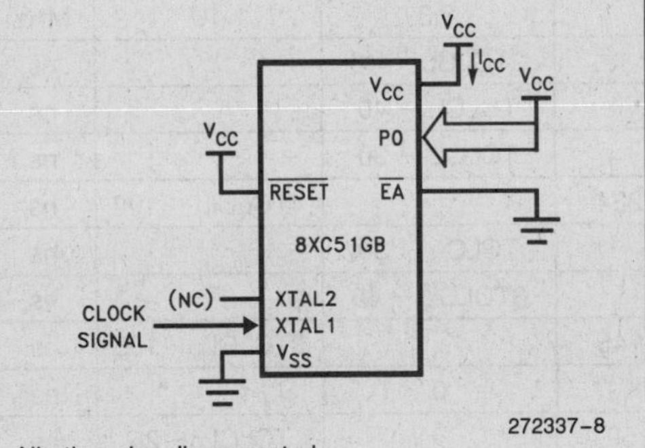

All other pins disconnected.
TCLCH = TCHCL = 5 ns

Figure 8. I_{CC} Test Condition Idle Mode

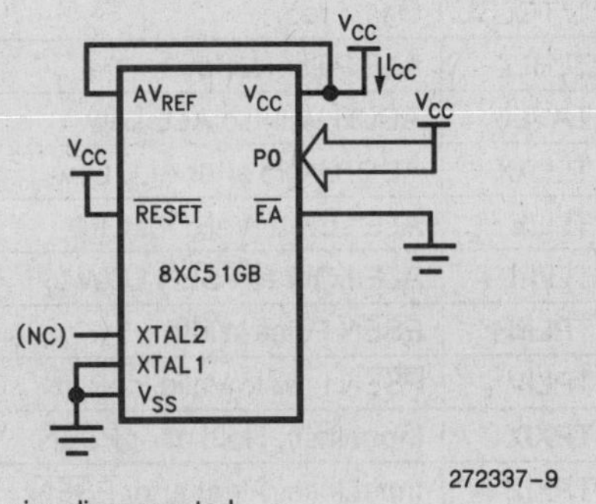

All other pins disconnected.

Figure 9. I_{CC} Test Condition, Power Down Mode V_{CC} = 2.0V to 5.5V

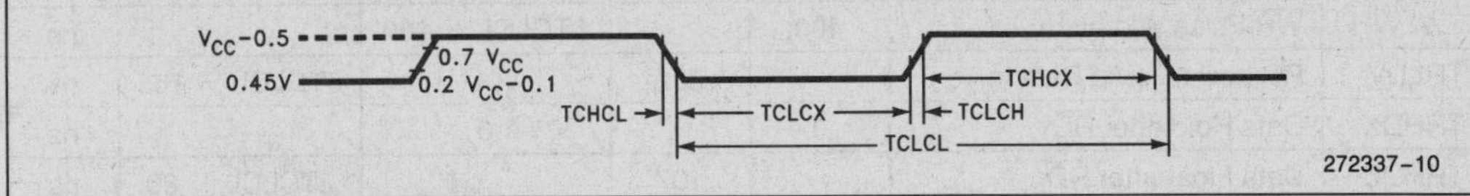

Figure 10. Clock Signal Waveform for I_{CC} Tests in Active and Idle Modes. TCLCH = TCHCL = 5 ns.

EXPLANATION OF THE AC SYMBOLS

Each timing symbol has 5 characters. The first character is always a "T" (stands for time). The other characters, depending on their positions, stand for the name of a signal or the logical status of that signal. The following is a list of all the characters and what they stand for:

A: Address

C: Clock

D: Input Data

H: Logic Level HIGH

I: Instruction (Program Memory Contents)

L: Logic Level LOW, or ALE

P: $\overline{PSEN}$

Q: Output Data

R: $\overline{RD}$ Signal

T: Time

V: Valid

W: $\overline{WR}$ Signal

X: No Longer a Valid Logic Level

Z: Float

For Example:

TAVLL = Time from Address Valid to ALE Low

TLLPL = Time from ALE Low to $\overline{PSEN}$ Low

AC SPECIFICATIONS

Over Operating Conditions, Load Capacitance on Port 0, ALE, and $\overline{PSEN}$ = 100 pF, Load Capacitance on all other outputs = 80 pF

EXTERNAL PROGRAM AND DATA MEMORY CHARACTERISTICS

Symbol	Parameter	12 MHz Osc.		Variable Osc.		Units
		Min	Max	Min	Max	
1/TCLCL	Osc. Freq.			3.5	16	MHz
TLHLL	ALE Pulse Width	127		2TCLCL − 40		ns
TAVLL	ADDR Valid to ALE Low	43		TCLCL − 40		ns
TLLAX	ADDR Hold after ALE Low	53		TCLCL − 30		ns
TLLIV	ALE Low to Valid Inst. IN		234		4TCLCL − 100	ns
TLLPL	ALE LOW to $\overline{PSEN}$ LOW	53		TCLCL − 30		ns
TPLPH	$\overline{PSEN}$ Pulse Width	205		3TCLCL − 45		ns
TPLIV	$\overline{PSEN}$ Low to Valid Instr In		145		3TCLCL − 105	ns
TPXIX	Input Instr. Hold after $\overline{PSEN}$	0		0		ns
TPXIZ	Input Instr. Float after $\overline{PSEN}$		59		TCLCL − 25	ns
TAVIV	ADDR to Valid Instr. In		312		5TCLCL − 105	ns
TPLAZ	$\overline{PSEN}$ Low to ADDR Float		10		10	ns
TRLRH	$\overline{RD}$ Pulse Width	400		6TCLCL − 100		ns
TWLWH	$\overline{WR}$ Pulse Width	400		6TCLCL − 100		ns
TRLDV	$\overline{RD}$ Low to Valid Data In		252		5TCLCL − 165	ns
TRHDX	Data Hold after $\overline{RD}$	0		0		ns
TRHDZ	Data Float after $\overline{RD}$		107		2TCLCL − 60	ns
TLLDV	ALE Low to Valid Data In		517		8TCLCL − 150	ns
TAVDV	ADDR to Valid Data In		585		9TCLCL − 165	ns
TLLWL	ALE Low to $\overline{RD}$ or $\overline{WR}$ Low	200	300	3TCLCL − 50	3TCLCL + 50	ns
TAVWL	ADDR Valid to $\overline{RD}$ or $\overline{WR}$ Low	203		4TCLCL − 130		ns
TQVWX	Data Valid to $\overline{WR}$ Transition	33		TCLCL − 50		ns
TWHQX	Data Hold after $\overline{WR}$	33		TCLCL − 50		ns
TQVWH	Data Valid to $\overline{WR}$ High	433		7 TCLCL − 150		ns
TRLAZ	$\overline{RD}$ Low to Addr Float		0		0	ns
TWHLH	$\overline{RD}$ or $\overline{WR}$ High to ALE High	43	123	TCLCL − 40	TCLCL + 40	ns

EXTERNAL PROGRAM MEMORY READ CYCLE

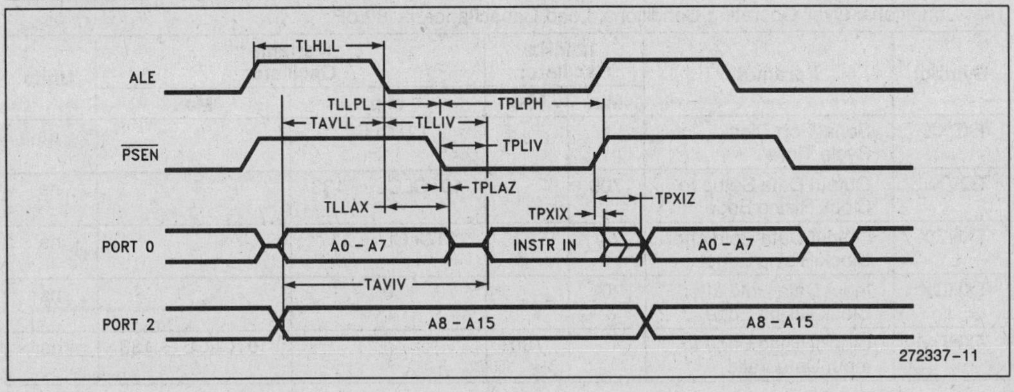

272337-11

EXTERNAL DATA MEMORY READ CYCLE

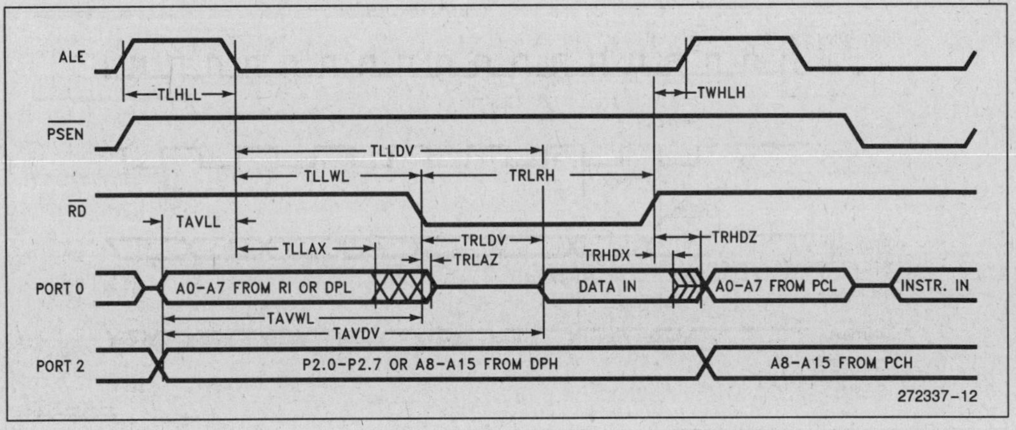

272337-12

EXTERNAL DATA MEMORY WRITE CYCLE

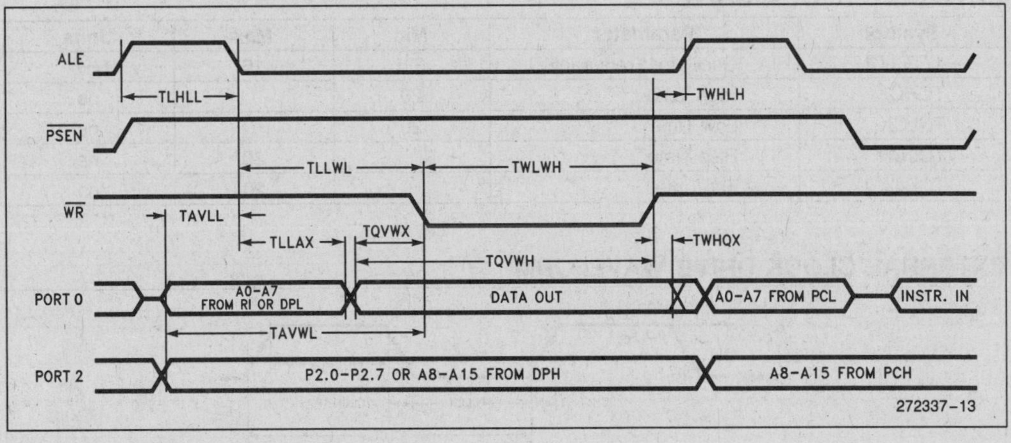

272337-13

SERIAL PORT TIMING—SHIFT REGISTER MODE

Test Conditions: Over Operating Conditions, Load Capacitance = 80 pF

Symbol	Parameter	12 MHz Oscillator		Variable Oscillator		Units
		Min	Max	Min	Max	
TXLXL	Serial Port Clock Cycle Time	1		12TCLCL		μs
TQVXH	Output Data Setup to Clock Rising Edge	700		10TCLCL − 133		ns
TXHQX	Output Data Hold after Clock Rising Edge	50		2TCLCL − 117		ns
TXHDX	Input Data Hold after Clock Rising Edge	0		0		ns
TXHDV	Clock Rising Edge to Input Data Valid		700		10TCLCL − 133	ns

SHIFT REGISTER MODE TIMING WAVEFORMS

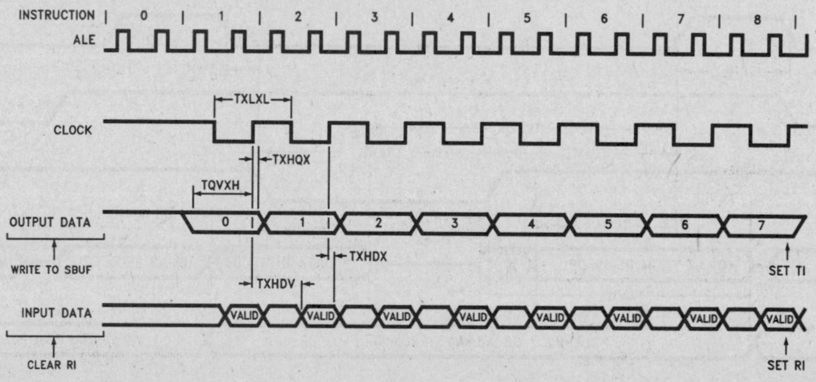

272337–14

EXTERNAL CLOCK DRIVE

Symbol	Parameter	Min	Max	Units
1/TCLCL	Oscillator Frequency	3.5	16	MHz
TCHCX	High Time	20		ns
TCLCX	Low Time	20		ns
TCLCH	Rise Time		20	ns
TCHCL	Fall Time		20	ns

EXTERNAL CLOCK DRIVE WAVEFORM

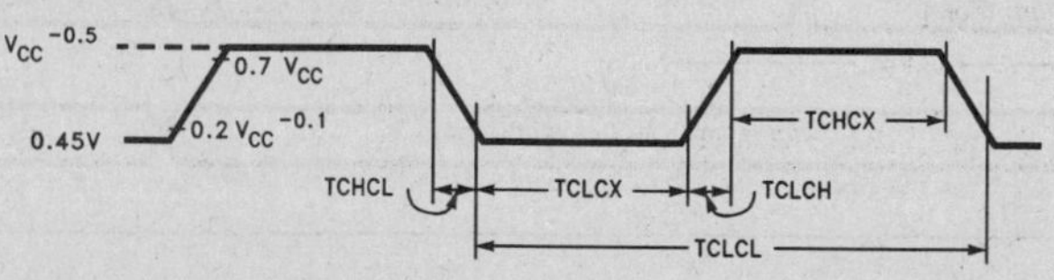

272337–15

SEP AC TIMING SPECIFICATIONS

Test Conditions: Over Operating Conditions, Load Capacitance = 80 pF

Symbol	Parameter	12 MHz Oscillator		Variable Oscillator		Units
		Min	Max	Min	Max	
TXSXL	SEPCLK Cycle Time	1		12 TCLCL		μs
TXSST	Output Data Setup to SEPCLK	435		6 TCLCL − 65		ns
TXSOH	Output Data Hold after SEPCLK	445		6 TCLCL − 55		ns
TXSIH	Input Data Hold after SEPCLK Sampling Edge	210		2 TCLCL + 43		ns
TXSDV	Input Data Valid to SEPCLK Sampling Edge		947		12 TCLCL − 53	ns

SEP Waveform (SEPS1 = 0; SEPS0 = 0; CLKPOL = 0; CLKPH = 0)

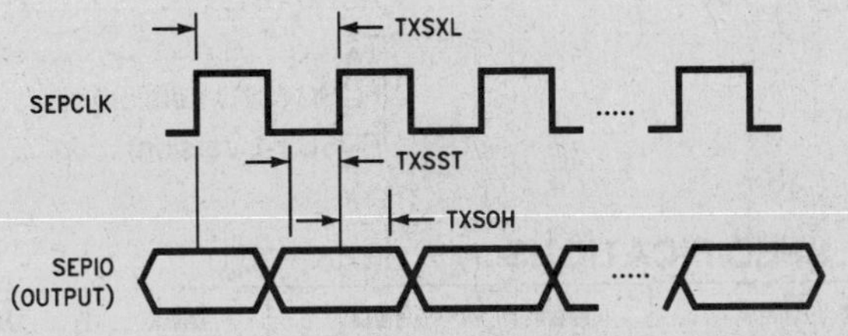

272337-16

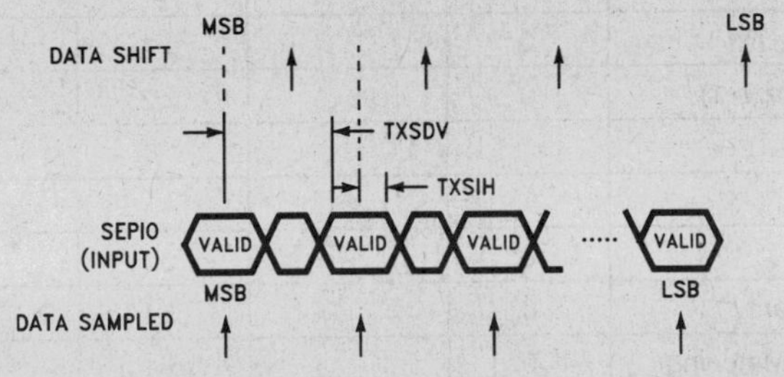

272337-17

AC TESTING INPUT, OUTPUT WAVEFORMS

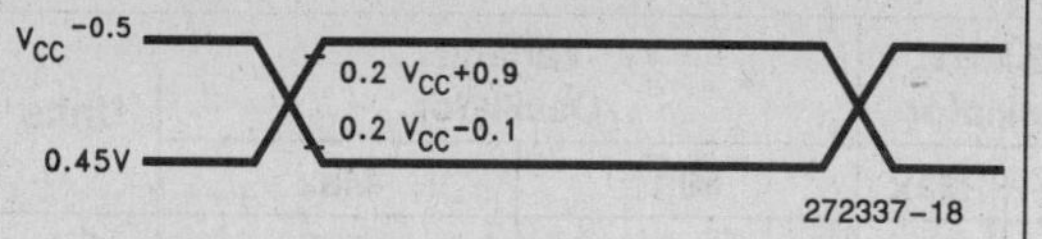

AC inputs during testing are driven at $V_{CC}-0.5V$ for a Logic "1" and 0.5V for a Logic "0". Timing measurements are made at V_{IH} for a Logic "1" and V_{OL} max for a Logic "0".

FLOAT WAVEFORMS

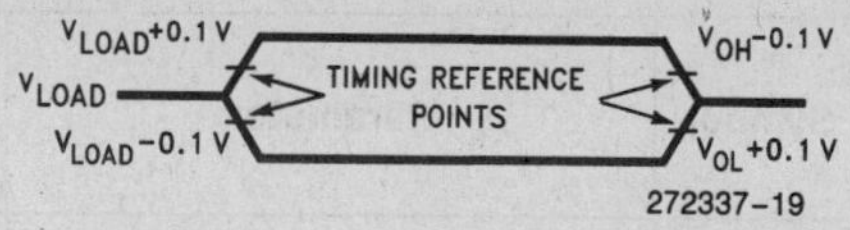

For timing purposes a port pin is no longer floating when a 100 mV change from load voltage occurs, and begins to float when a 100 mV change from the loaded V_{OH}/V_{OL} level occurs. $I_{OL}/I_{OH} \geq \pm 20$ mA.

A TO D CHARACTERISTICS

The absolute conversion accuracy is dependent on the accuracy of AV_{REF}. The specifications given below assume adherence to the Operating Conditions section of this data sheet. Testing is done at $AV_{REF} = 5.12V$, and $V_{CC} = 5.0V$.

OPERATING CONDITIONS

V_{CC} .... 4.0V to 6.0V
AV_{REF} .... 4.5V to 5.5V
V_{SS}, AV_{SS} 0V
ACH0–7 .... AV_{SS} to V_{REF}
T_A .... 0°C to +70°C Ambient
FOSC (STD Version) .... 3.5 MHz to 12 MHz
FOSC (-1 Version) .... 3.5 MHz to 16 MHz

A/D CONVERTER SPECIFICATIONS T_A = 0°C to +70°C

Parameter	Min	Typ*	Max	Units**	Notes
Resolution	256 8		256 8	Levels Bits	
Absolute Error (Ch 2–7)	0		±1	LSB	
Absolute Error (Ch 0 and 1)	0		±2	LSB	
Full Scale Error		±1		LSB	
Zero Offset Error		±1		LSB	
Non-Linearity	0		±1	LSB	
Differential Non-Linearity	0		±1	LSB	
Channel-to-Channel Matching	0		±1	LSB	
Repeatability		±0.25		LSB	

A/D CONVERTER SPECIFICATIONS $T_A = 0°C$ to $+70°C$ (Continued)

Parameter	Min	Typ*	Max	Units**	Notes
Temperature Coefficients: Offset Full Scale Differential Non-Linearity		 0.003 0.003 0.003		 LSB/°C LSB/°C LSB/°C	
Input Capacitance		3		pF	
Off Isolation	−60			dB	(8, 9)
Feedthrough		−60		dB	(8)
V_{CC} Power Supply Rejection		−60		dB	(8)
Input Resistance to Sample-and-Hold Capacitor	750		1.2K	Ω	
DC Input Leakage	0		3.0	μA	

NOTES:
*These values are expected for most parts at 25°C
**AN "LSB" as used here, has a value of approximately 20 mV.
8. DC to 100 KHz
9. Multiplexer Break-Before-Make Guaranteed.
10. There is no indication when a single A/D conversion is complete. Please refer to the 8XC51GB Hardware Description on how to read a single A/D conversion.
11. $T_{CY} = 12$ TCLCL

A/D Conversion Time		Notes
Per Channel	26 T_{CY}	(10, 11)
8 Conversions	208 T_{CY}	(11)

PROGRAMMING THE OTP

The part must be running with a 4 MHz to 6 MHz oscillator. The address of a location to be programmed is applied to address lines while the code byte to be programmed in that location is applied to data lines. Control and program signals must be held at the levels indicated in Table 2. Normally $\overline{EA}/V_{PP}$ is held at logic high until just before ALE/$\overline{PROG}$ is to be pulsed. The $\overline{EA}/V_{PP}$ is raised to V_{PP}, ALE/$\overline{PROG}$ is pulsed low and then $\overline{EA}/V_{PP}$ is returned to a high (also refer to timing diagrams).

NOTE:
Exceeding the V_{PP} maximum for any amount of time could damage the device permanently. The V_{PP} source must be well regulated and free of glitches.

DEFINITION OF TERMS

ADDRESS LINES: P1.0–P1.7, P2.0–P2.4, respectively for A0–A12.

DATA LINES: P0.0–P0.7 for D0–D7.

CONTROL SIGNALS: $\overline{RST}$, $\overline{PSEN}$, P2.6, P2.7, P3.3, P3.6, P3.7

PROGRAM SIGNALS: ALE/$\overline{PROG}$, $\overline{EA}/V_{PP}$

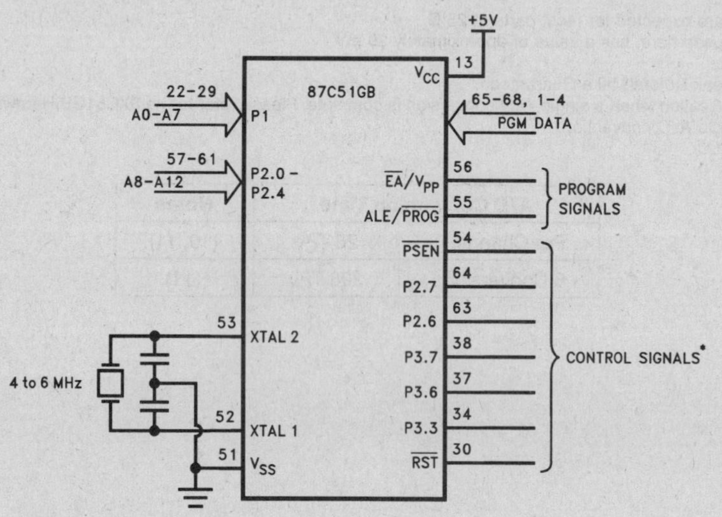

272337-20

*See Table 2 for proper input on these pins.

Figure 11. Programming the OTP

Table 2. OTP Programming Modes

Mode		$\overline{RST}$	$\overline{PSEN}$	ALE/$\overline{PROG}$	$\overline{EA}/V_{PP}$	P2.6	P2.7	P3.3	P3.6	P3.7
Program Code Data		L	L	⊔	12.75V	L	H	H	H	H
Verify Code Data		L	L	H	H	L	L	L	H	H
Program Encryption Array Address 0–3FH		L	L	⊔	12.75V	L	H	H	L	H
Program Lock Bits	Bit 1	L	L	⊔	12.75V	H	H	H	H	H
	Bit 2	L	L	⊔	12.75V	H	H	H	L	L
	Bit 3	L	L	⊔	12.75V	H	L	H	H	L
Read Signature Byte		L	H	H	H	L	L	L	L	L

PROGRAMMING ALGORITHM

Refer to Table 2 and Figures 11 and 12 for address, data, and control signals set up. To program the 87C51GB the following sequence must be exercised.

1. Input the valid address on the address lines.
2. Input the appropriate data byte on the data lines.
3. Activate the correct combination of control signals.
4. Raise $\overline{EA}/V_{PP}$ from V_{CC} to 12.75V ±0.25V.
5. Pulse ALE/$\overline{PROG}$ 5 times for the OTP array, and 25 times for the encryption table and the lock bits.

Repeat 1 through 5 changing the address and data for the entire array or until the end of the object file is reached.

PROGRAM VERIFY

Program verify may be done after each byte that is programmed, or after a block of bytes that is programmed. In either case a complete verify of the array will ensure that it has been programmed correctly.

The lock bits cannot be directly verified. Verification of the lock bits is done by observing that their features are enabled. Refer to the Program Lock section in this data sheet.

2

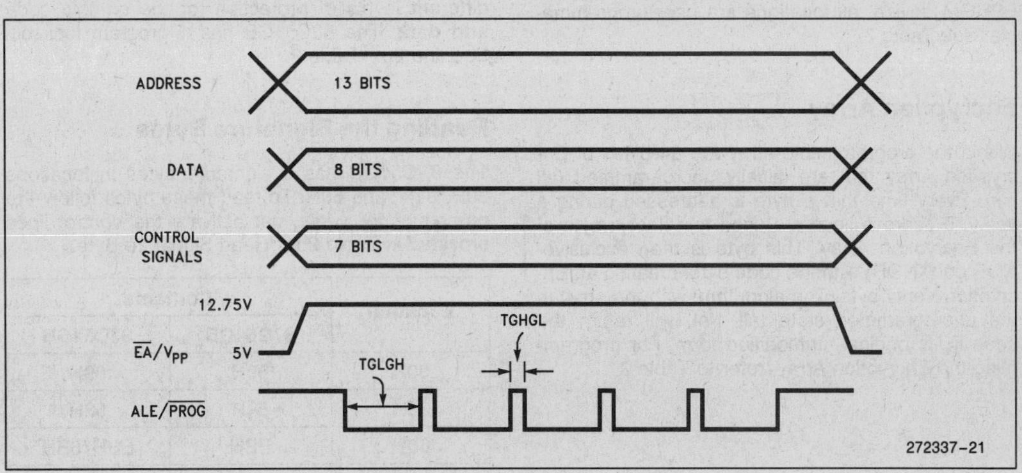

Figure 12. Programming Signal's Waveforms

ROM and EPROM Lock System

The 87C51GB and the 83C51GB program lock systems, when programmed, protect the on-board program against software piracy.

The 83C51GB has a one-level program lock system and a 64-byte encryption table. See line 2 of Table 3. If program protection is desired, the user submits the encryption table with their code, and both the lock bit and encryption array are programmed by the factory. The encryption array is not available without the lock bit. For the lock bit to be programmed, the user must submit an encryption table.

The 87C51GB has a 3-level program lock system and a 64-byte encryption array. Since this is an EPROM device, all locations are user programmable. See Table 3.

Encryption Array

Within the programmable array are 64 bytes of Encryption Array that are initially unprogrammed (all 1's). Every time that a byte is addressed during a verify, 5 address lines are used to select a byte of the Encryption Array. This byte is then exclusive-NOR'ed (XNOR) with the code byte, creating an Encryption Verify byte. The algorithm, with the array in the unprogrammed state (all 1's), will return the code in its original, unmodified form. For programming the Encryption Array, refer to Table 2.

When using the encryption array feature, one important factor needs to be considered. If a code byte has the value 0FFH, verification of the byte will produce the encryption byte value. If a large block (>64 bytes) of code is left unprogrammed, a verification routine will display the contents of the encryption array. For this reason it is strongly recommended that all unused code bytes be programmed with some value other than 0FFH, and not all of them the same value. This practice will ensure the maximum possible program protection.

Program Lock Bits

The 87C51GB has 3 programmable lock bits that when programmed according to Table 3 will provide different levels of protection for the on-chip code **and data**. The 83C51GB has 1 program lock bit. See line 2 of Table 3.

Reading the Signature Bytes

The 8XC51GB has 3 signature bytes in locations 30H, 31H, and 60H. To read these bytes follow the procedure for verify, but activate the control lines provided in Table 2 for Read Signature Byte.

Location	Contents	
	87C51GB	83C51GB
30H	89H	89H
31H	58H	58H
60H	EBH	EBH/6BH

Table 3. Program Lock Bits and the Features

*Program Lock Bits				Protection Type
	LB1	LB2	LB3	
1	U	U	U	No Program Lock features enabled. (Code verify will still be encrypted by the Encryption Array if programmed).
2	P	U	U	MOVC instructions executed from external program memory are disabled from fetching code bytes from internal memory, EA is sampled and latched on Reset, and further programming of the EPROM is disabled.
3	P	P	U	Same as 2, also verify is disabled.
4	P	P	P	Same as 3, also external execution is disabled.

*Any other combination of lock bits is not defined.

OTP PROGRAMMING AND VERIFICATION CHARACTERISTICS

(T_A = 21°C to 27°C; V_{CC} = 5V ± 20%; V_{SS} = 0V)

Symbol	Parameter	Min	Max	Units
V_{PP}	Programming Supply Voltage	12.5	13.0	V
I_{PP}	Programming Supply Current		75	mA
1/TCLCL	Oscillator Frequency	4	6	MHz
TAVGL	Address Setup to $\overline{\text{PROG}}$ Low	48TCLCL		
TGHAX	Address Hold after $\overline{\text{PROG}}$	48TCLCL		
TDVGL	Data Setup to $\overline{\text{PROG}}$ Low	48TCLCL		
TGHDX	Data Hold after $\overline{\text{PROG}}$	48TCLCL		
TEHSH	(Enable) High to V_{PP}	48TCLCL		
TSHGL	V_{PP} Setup to $\overline{\text{PROG}}$ Low	10		µs
TGHSL	V_{PP} Hold after $\overline{\text{PROG}}$	10		µs
TGLGH	$\overline{\text{PROG}}$ Width	90	110	µs
TAVQV	Address to Data Valid		48TCLCL	
TELQV	ENABLE Low to Data Valid		48TCLCL	
TEHQZ	Data Float after ENABLE	0	48TCLCL	
TGHGL	$\overline{\text{PROG}}$ High to $\overline{\text{PROG}}$ Low	10		µs

PROGRAMMING AND VERIFICATION WAVEFORMS

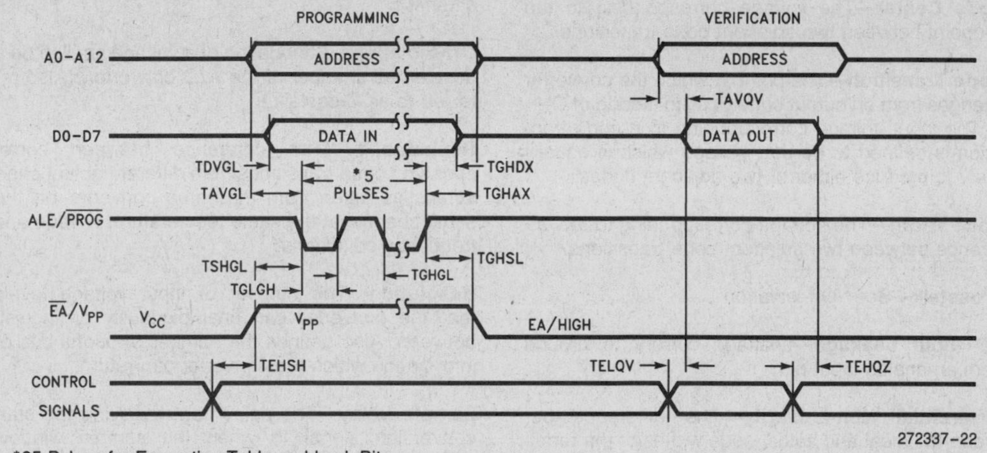

*25 Pulses for Encryption Table and Lock Bits.

A/D Glossary of Terms

Absolute Error—The maximum difference between corresponding actual and ideal code transitions. Absolute Error accounts for all deviations of an actual converter from an ideal converter.

Actual Characteristic—The characteristic of an actual converter. The characteristic of a given converter may vary over temperature, supply voltage, and frequency conditions. An actual characteristic rarely has ideal first and last transition locations or ideal code widths. It may even vary over multiple conversions under the same conditions.

Break-Before-Make—The property of a multiplexer which guarantees that a previously selected channel will be deselected before a new channel is selected (e.g., the converter will not short inputs together).

Channel-to-Channel Matching—The difference between corresponding code transitions of actual characteristics taken from different channels under the same temperature, voltage and frequency conditions.

Characteristic—A graph of input voltage versus the resultant output code for an A/D converter. It describes the transfer function of the A/D converter.

Code—The digital value output by the converter.

Code Center—The voltage corresponding to the midpoint between two adjacent code transitions.

Code Transition—The point at which the converter changes from an output code of Q, to a code of Q + 1. The input voltage corresponding to a code transition is defined to be that voltage which is equally likely to produce either of two adjacent codes.

Code Width—The voltage corresponding to the difference between two adjacent code transitions.

Crosstalk—See "Off-Isolation".

DC Input Leakage—Leakage current to ground from an analog input pin.

Differential Non-Linearity—The difference between the ideal and actual code widths of the terminal based characteristic.

Feedthrough—Attenuation of a voltage applied on the selected channel of the A/D Converter after the sample window closes.

Full Scale Error—The difference between the expected and actual input voltage corresponding to the full scale code transition.

Ideal Characteristic—A characteristic with its first code transition at $V_{IN} = 0.5$ LSB, its last code transition at $V_{IN} = (V_{REF} - 1.5$ LSB) and all code widths equal to one LSB.

Input Resistance—The effective series resistance from the analog input pin to the sample capacitor.

LSB—Least Significant Bit—The voltage corresponding to the full scale voltage divided by 2^n, where n is the number of bits of resolution of the converter. For an 8-bit converter with a reference voltage of 5.12V, one LSB is 20 mV. Note that this is different than digital LSBs since an uncertainty of two LSBs, when referring to an A/D converter, equals 40 mV. (This has been confused with an uncertainty of two digital bits, which would mean four counts, or 80 mV).

Monotonic—The property of successive approximation converters which guarantees that increasing input voltages produce adjacent codes of increasing value, and that decreasing input voltages produce adjacent codes of decreasing value.

No Missed Codes—For each and every output code, there exists a unique input voltage range which produces that code only.

Non-Linearity—The maximum deviation of code transitions of the terminal based characteristic from the corresponding code transitions of the ideal characteristic.

Off-Isolation—Attenuation of a voltage applied on a deselected channel of the A/D converter. (Also referred to as Crosstalk.)

Repeatability—The difference between corresponding code transitions from different actual characteristics taken from the same converter on the same channel at the same temperature, voltage and frequency conditions.

Resolution—The number of input voltage levels that the converter can unambiguously distinguish between. Also defines the number of useful bits of information which the converter can return.

Sample Delay—The delay from receiving the start conversion signal to when the sample window opens.

Sample Delay Uncertainty—The variation in the sample delay.

Sample Time—The time that the sample window is open.

Sample Time Uncertainty—The variation in the sample time.

Sample Window—Begins when the sample capacitor is attached to a selected channel and ends when the sample capacitor is disconnected from the selected channel.

Successive Approximation—An A/D conversion method which uses a binary search to arrive at the best digital representation of an analog input.

Temperature Coefficients—Change in the stated variable per degree centigrade temperature change. Temperature coefficients are added to the typical values of a specification to see the effect of temperature drift.

Terminal Based Characteristic—An actual characteristic which has been rotated and translated to remove zero offset and full scale error.

V_{CC} Rejection—Attenuation of noise on the V_{CC} line to the A/D converter.

Zero Offset—The difference between the expected and actual input voltage corresponding to the first code transition.

DATA SHEET REVISION SUMMARY

The following differences exist between this datasheet and the previous version (270869-003):

1. Merged 87C51GB Express (270889-001).
2. New order number 272337-001.

The following differences exist between the 270869-003 data sheet and the previous version (270869-002):

1. Changed data sheet status from "Advance Information" to "Preliminary" and updated associated notices.
2. Added 83C51GB throughout.
3. Added Package and Process Information.
4. Clarified ±2 LSB accuracy for channels 0 and 1 in A/D Converter Section.
5. Added "ROM and EPROM Lock System" section and added 83C51GB to "Program Lock Bits" section.
6. Modified Signature Bytes Table.

The following differences exist between the 270869-002 data sheet and the previous version (270869-001):

1. Changed data sheet status from "Product Preview" to "Advance Information" and updated associated notices.
2. Asynchronous port reset was added to $\overline{\text{RESET}}$ pin description.
3. ALE disable paragraph was added to ALE pin description.
4. C_1, C_2 guidelines clarified in Figure 4.
5. Operating Conditions heading was added.
6. Maximum I_{OL} per I/O pin was added to Absolute Maximum Ratings.
7. VT_+, VT_-, V_{HYS}, V_{OL2}, and V_{TL} removed.
8. V_{OL} value for ALE included with V_{OL1}.
9. V_{IL1} and V_{IL2} added.
10. RRST minimum changed from 40K to 50K. RRST maximum changed from 225K to 300K.
11. I_{PD} maximum changed from 200 μA to 50 μA.
12. I_{DL} maximum changed from 15 mA to 18 mA.
13. Typical values for I_{PD}, I_{DL}, I_{CC}, and I_{REF} removed.
14. Note 3 (page 9) was reworded.
15. SEP AC Timings added.
16. A/D Absolute Error for Channels 0 and 1 changed to ±2 LSB.
17. T_{CY} clarified.
18. Encryption array paragraph was added.
19. Corrected pin numbers on Figure 11 to reflect PLCC package.

8XC51SL/LOW VOLTAGE 8XC51SL KEYBOARD CONTROLLER

80C51SL — CPU with RAM and I/O; V_{CC} = 5V ±10%

81C51SL — 16K ROM Preprogrammed with SystemSoft Keyboard Controller and Scanner Firmware. V_{CC} = 5V ±10%.

83C51SL — 16K Factory Programmed ROM. V_{CC} = 5V ±10%.

87C51SL — 16K OTP ROM. V_{CC} = 5V ±10%.

Low Voltage 80C51SL — CPU with RAM and I/O; V_{CC} = 3.3V ±0.3V

Low Voltage 81C51SL — 16K ROM Preprogrammed with SystemSoft Keyboard Controller and Scanner Firmware. V_{CC} = 3.3V ±0.3V.

Low Voltage 83C51SL — 16K Factory Programmed ROM. V_{CC} = 3.3V ±0.3V.

Low Voltage 87C51SL — 16K OTP ROM. V_{CC} = 3.3V ±0.3V.

- **Proliferation of 8051 Architecture**
- **Complete 8042 Keyboard Control Functionality**
- **8042 Style Host Interface**
- **Optional Hardware Speedup of GATEA20 and RCL**
- **Local 16 x 8 Keyboard Switch Matrix Support**
- **Two Industry Standard Serial Keyboard Interfaces; Supported via Four High Drive Outputs**
- **5 LED Drivers**
- **Low Power CHMOS Technology**
- **4-Channel, 8-Bit A/D**
- **Interface for up to 32 Kbytes of External Memory**
- **Slew Rate Controlled I/O Buffers Used to Minimize Noise**
- **256 Bytes Data RAM**
- **Three Multifunction I/O Ports**
- **10 Interrupt Sources with 6 User-Definable External Interrupts**
- **2 MHz–16 MHz Clock Frequency**
- **100-Pin PQFP (8XC51SL)
100-Pin SQFP (Low Voltage 8XC51SL)**

The 8XC51SL, based on Intel's industry-standard MCS® 51 microcontroller family, is designed for keyboard control in laptop and notebook PCs. The highly integrated keyboard controller incorporates an 8042-style UPI host interface with expanded memory, keyboard scan, and power management. The 8XC51SL supports both serial and scanned keyboard interfaces and is available in pre-programmed versions to reduce time to market. The Low Voltage 8XC51SL is the 3.3V version optimized for even further power savings. Throughout the remainder of this document, both devices will generally be referred to as 51SL.

The 8XC51SL is a pin-for-pin compatible replacement for the 8XC51SL-BG. It does, however have some additional functionality. Those additional functions are as follows:

1. 16K OTP ROM: The 8XC51SL-BG had only 8K of ROM.
2. New Register Set: The 8XC51SL adds a second set of host interface registers available for use in supporting power management. This required an additional address line (A1) for decoding. To accommodate this, one V_{CC} pin was removed. However, in order to maintain compatibility with the -BG version, an enable bit for this new register set was added in configuration register 1. This allows the 8XC51SL to be drop in compatible to existing 8XC51SL-BG designs; no software modifications required.

NOTE:
The changes made to the V_{CC} pins require that all three V_{CC} pins be properly connected. Failing to do so could result in high leakage current and possible damage to the device.

October 1993
Order Number: 272271-002

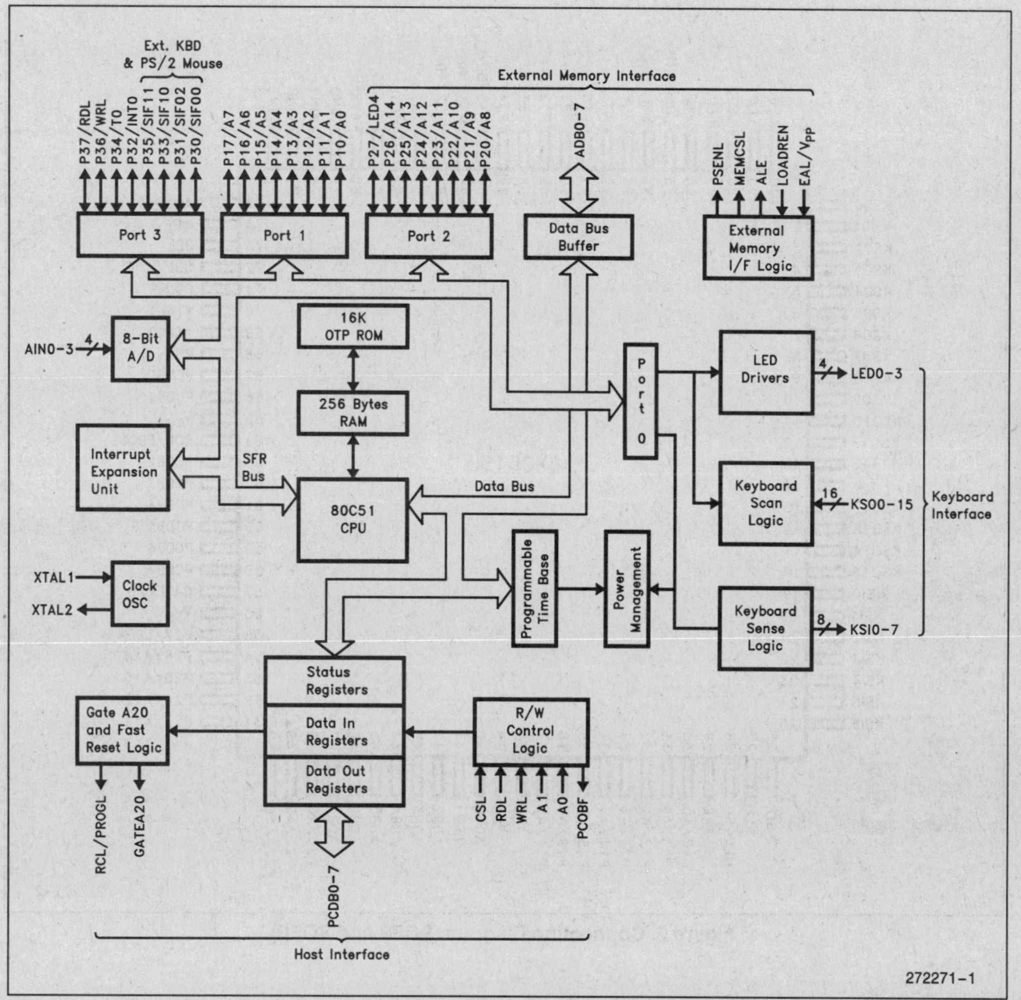

Figure 1. Block Diagram

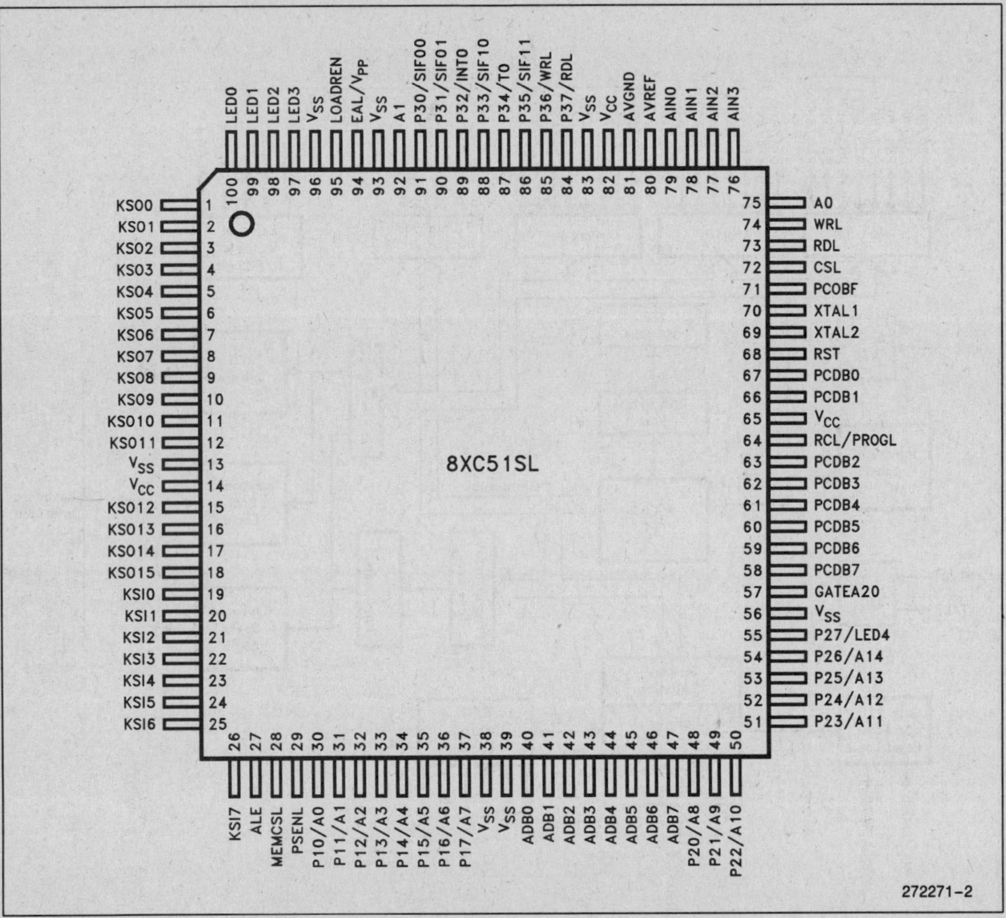

Figure 2. Connection Diagram (PQFP and SQFP)

PACKAGES

Part	Prefix	Suffix	Package Type
8XC51SL	KU	AH	100-Pin PQFP
Low Voltage 8XC51SL	SB	AL	100-Pin SQFP

PIN DESCRIPTIONS

Table 1. Pin Descriptions

Symbol	Type	Description
V_{SS}		Circuit ground potential.
V_{CC}		Supply voltage during normal, Idle, and Power-Down operation; nominally +5V ±10% for 8XC51SL, +3.3V ± 0.3V for Low Voltage 8XC51SL.
PCDB0–7	I/O	Host interface data bus. An 8-bit bidirectional port for data transfers between the host processor and the keyboard controller.
WRL	I	The active-low, host-interface write signal.
RDL	I	The active-low, host-interface read signal.
CSL	I	The active-low, host-interface chip select.
A0–A1	I	Host-Interface Address select inputs.
PCOBF	O	The active-high, host-interface Output Buffer Full interrupt.
GATEA20	O	Gate A20 control signal output.
RCL/PROGL	O	Host reset—active low. This pin is also the program pulse input during EPROM programming.
LED0–3	O	LED output drivers.
KSI0–7	I	Keyboard input scan lines (input Port 0). Schmitt inputs with 5K–20K pull-up resistors.
KSO0–15	O	Keyboard output scan lines.
PORT 1 P10/A0– P17/A7	I/O	Port 1 is a general-purpose, 8-bit bidirectional port with internal pull-ups. It also supports the following user-selectable functions: P10–P16 are available for connection to dedicated keyboard inputs. A0–A7 output the low-order address byte (refer to LOADREN signal).
LOADREN	I	Low address enable. When set high, address bits A0–A7 are output on P10–P17.
PORT2 P20–6/A8–14 P27/LED4	I/O	Port 2 is a general-purpose, 8-bit bidirectional port with internal pull-ups on P20–6/A8–14. It also supports the following user-selectable functions: P20–6/A8–14 output the high-order address byte. P27/LED4 is available as a fifth LED output driver (by writing to the port bit 7).
PORT 3 P30/SIF00 P31/SIF01 P32/INT0 P33/SIF10 P34/T0 P35/SIF11 P36/WRL P37/RDL	I/O	Port 3 is a general-purpose, 8-bit bidirectional port. P32/INT0, P34/T0, P36/WRL, and P37/RDL have internal pull-ups. P30/SIF00, P31/SIF01, P33/SIF10, and P35/SIF11 are high-drive open-drain outputs. It also supports the following user-selectable functions: A high-drive, open-drain output to support an external serial keyboard interface (typically CLK); RXD (8051 UART serial input port); SIF0INTL (serial interface interrupt 0). A high-drive, open-drain output to support an external serial keyboard interface (typically DATA); TXD (8051 UART serial output port). INT0L (external interrupt 0). A high-drive, open-drain output to support an external serial keyboard interface (typically mouse CLK); SIF1INTL (external interrupt 1). AUXOBF1 (output buffer full—mouse support); T0 (Timer/Counter 0 external input). A high-drive, open-drain output to support an external serial keyboard interface (typically mouse DATA); T1 (Timer/Counter 1 external input). WRL (external data memory write strobe); inactive at addresses 7FF0–7FFFH. AUXOBF2 (output buffer full interrupt); INT2L (external interrupt); RDL (external data memory read strobe); inactive at addresses 7FF0–FFFFH.

PIN DESCRIPTIONS (Continued)

Table 1. Pin Descriptions (Continued)

Symbol	Type	Description
XTAL1	I	Input to the on-chip oscillator.
XTAL2	O	Output from the on-chip oscillator.
AVGND		Analog ground potential.
AVREF		Analog supply voltage; nominally +5V ±10% for 8XC51SL, +3.3V ±0.3V for Low Voltage 8XC51SL.
AIN0–3	I	A/D Analog input channels.
ADB0–7	I/O	External address/data bus. Multiplexes the low-address byte and data during external memory accesses.
EAL/V_{PP}	I	External address input. When held high, the 51SL CPU executes out of internal Program Memory unless the program counter exceeds 3FFFH. When held low, the 51SL CPU always executes out of external memory. EAL is latched on the falling edge of RST. This pin also receives the programming supply voltage (V_{PP}) during EPROM programming.
ALE	O	Address Latch Enable output pulse latches the low address byte during external memory access. ALE is output at a constant rate of 1/6 the oscillator frequency, whether or not there are accesses to external memory. One ALE pulse is skipped during the execution of a MOVX instruction. ALE is disabled during Idle mode and can also be disabled via Configuration register 1 control.
PSENL	O	Program Store Enable is the read strobe to external program memory. PSENL is qualified with RDL and A15 for use with an external Flash memory. PSENL is not active when the device executes out of internal program memory.
MEMCSL	I/O	External Memory Chip Select for code space address 4000H and above, when EAL is inactive (i.e., high). For EAL low, MEMCSL is active. Goes inactive during Idle mode and Power-Down mode. If external memory interfacing is not required, MEMCSL can be configured as a general purpose I/O (controlled via Configuration register 1).
RST	I	Resets the keyboard controller. Hold RST high for two machine cycles.

8XC51SL/LOW VOLTAGE 8XC51SL PIN CHARACTERISTICS

Table 2. Pin Characteristics

Pin No.	Pin Name	Type	Term	Reset	PD Mode
1	KSO0	O	OD	TRI	HOLD
2	KSO1	O	OD	TRI	HOLD
3	KSO2	O	OD	TRI	HOLD
4	KSO3	O	OD	TRI	HOLD
5	KSO4	O	OD	TRI	HOLD
6	KSO5	O	OD	TRI	HOLD
7	KSO6	O	OD	TRI	HOLD
8	KSO7	O	OD	TRI	HOLD
9	KSO8	O	OD	TRI	HOLD
10	KSO9	O	OD	TRI	HOLD
11	KSO10	O	OD	TRI	HOLD
12	KSO11	O	OD	TRI	HOLD
13	V_{SS}				
14	V_{CC}				
15	KSO12	O	OD	TRI	HOLD
16	KSO13	O	OD	TRI	HOLD
17	KSO14	O	OD	TRI	HOLD
18	KSO15	O	OD	L	HOLD
19	KSI0	I	5K–20K PU		NC
20	KSI1	I	5K–20K PU		NC
21	KSI2	I	5K–20K PU		NC
22	KSI3	I	5K–20K PU		NC
23	KSI4	I	5K–20K PU		NC
24	KSI5	I	5K–20K PU		NC
25	KSI6	I	5K–20K PU		NC
26	KSI7	I	5K–20K PU		NC
27	ALE	O		L	L
28	MEMCSL	O		L (EAL = 0)	H
29	PSENL	O		L	L
30	P10/A0	I/O	PU	WH	HOLD
31	P11/A1	I/O	PU	WH	HOLD
32	P12/A2	I/O	PU	WH	HOLD
33	P13/A3	I/O	PU	WH	HOLD
34	P14/A4	I/O	PU	WH	HOLD
35	P15/A5	I/O	PU	WH	HOLD
36	P16/A6	I/O	PU	WH	HOLD
37	P17/A7	I/O	PU	WH	HOLD

8XC51SL/LOW VOLTAGE 8XC51SL PIN CHARACTERISTICS (Continued)

Table 2. Pin Characteristics (Continued)

Pin No.	Pin Name	Type	Term	Reset	PD Mode
38	V_{SS}				
39	V_{SS}				
40	ADB0	I/O		TRI	TRI
41	ADB1	I/O		TRI	TRI
42	ADB2	I/O		TRI	TRI
43	ADB3	I/O		TRI	TRI
44	ADB4	I/O		TRI	TRI
45	ADB5	I/O		TRI	TRI
46	ADB6	I/O		TRI	TRI
47	ADB7	I/O		TRI	TRI
48	P20/A8	I/O	PU	WH	HOLD
49	P21/A9	I/O	PU	WH	HOLD
50	P22/A10	I/O	PU	WH	HOLD
51	P23/A11	I/O	PU	WH	HOLD
52	P24/A12	I/O	PU	WH	HOLD
53	P25/A13	I/O	PU	WH	HOLD
54	P26/A14	I/O	PU	WH	HOLD
55	P27/LED4	I/O	OD	TRI	HOLD
56	V_{SS}				
57	GATEA20	O		WH	HOLD
58	PCDB7	I/O		TRI	TRI
59	PCDB6	I/O		TRI	TRI
60	PCDB5	I/O		TRI	TRI
61	PCDB4	I/O		TRI	TRI
62	PCDB3	I/O		TRI	TRI
63	PCDB2	I/O		TRI	TRI
64	RCL/PROGL	O		WH	HOLD
65	V_{CC}				
66	PCDB1	I/O		TRI	TRI
67	PCDB0	I/O		TRI	TRI
68	RST	I			
69	XTAL2	O			H
70	XTAL1	I			
71	PCOBF	O		L	HOLD
72	CSL	I			
73	RDL	I			
74	WRL	I			

8XC51SL/LOW VOLTAGE 8XC51SL PIN CHARACTERISTICS (Continued)

Table 2. Pin Characteristics (Continued)

Pin No.	Pin Name	Type	Term	Reset	PD Mode
75	A0	I			
76	AIN3	I			
77	AIN2	I			
78	AIN1	I			
79	AIN0	I			
80	AVREF				
81	AVGND				
82	V_{CC}				
83	V_{SS}				
84	P37/RDL	I/O	PU	WH	HOLD
85	P36/WRL	I/O	PU	WH	HOLD
86	P35/SIF11	I/O	OD	TRI	HOLD
87	P34/T0	I/O	PU	WH	HOLD
88	P33/SIF10	I/O	OD	L	HOLD
89	P32/INT0	I/O	PU	WH	HOLD
90	P31/SIF01	I/O	OD	TRI	HOLD
91	P30/SIF00	I/O	OD	L	HOLD
92	A1	I			
93	V_{SS}				
94	EAL	I			
95	LOADREN	I			
96	V_{SS}				
97	LED3	O	OD	TRI	HOLD
98	LED2	O	OD	TRI	HOLD
99	LED1	O	OD	TRI	HOLD
100	LED0	O	OD	TRI	HOLD

NOTES:
1. During Power Down mode all floating I/O pins or inputs without internal pullups should be driven.
2. PU = Pulled Up, OD = Open Drain, WH = Weak High, TRI = Tri-State.

PORT STRUCTURES AND OPERATION

All three 51SL ports are bidirectional. Each consists of a latch (Special Function Registers P1 through P3), an output driver, and an input buffer. Port 0 of the 51SL CPU does not connect to the package pins. It is used internally to drive the keyboard scan logic.

The output drivers of ports 1 and 2 can be used in accesses to external memory. The 51SL provides the LOADREN signal to facilitate external memory interfaces. When the LOADREN signal is high, Port 1 outputs the low byte of the external memory address. If LOADREN is tied low, then the Port 1 signals continue to emit the P1 SFR content. Port 2 outputs the upper seven bits of the high byte of the external address when the address is 15 bits wide and either EAL is tied low or EAL is tied high and Bit 0 (ADDREN) of configuration register 1 is set. Otherwise, the Port 2 pins continue to emit the P2 SFR content.

I/O Configurations

All port pins with the exception of P27/LED4, P30/SIF00, P31/SIF01, P33/SIF10, and P35/SIF11 have fixed internal pullups and therefore are called "quasi-bidirectional ports". When configured as inputs, the pins are pulled high by the pullups and will source current when externally pulled low.

During a 15-bit external program memory access, Port 2 outputs the high address byte. In the 80C51 the Port 2 drivers use the strong pullup during the entire time that they are emitting a "1" on a Port 2 bit. In this instance, the 80C51 weak quasi-bidirectional pullup condition that normally occurs after two oscillator periods does not occur. Port 1 and Port 2 of the 51SL emulate the quasi-bidirectional pullup condition during program memory access, not this extended strong pullup condition.

POWER MANAGEMENT

The 51SL uses low power CHMOS and provides for two further power savings modes, available when inactive: Idle mode, typically between keystrokes; and Power Down mode, upon command from the host. A four channel, eight-bit A/D converter is also included for power management (i.e., battery voltage/temperature monitoring, etc.).

Idle Mode

Idle mode is initiated by an instruction that sets the PCON.0 bit (SFR address 87H) in the 51SL. In Idle mode, the internal clock signal to the 51SL CPU is gated off, but not to the interrupt timer and Serial Port functions. The 51SL status is preserved in its entirety: the Stack Pointer, Program Counter, Program Status Word, Accumulator, and all other registers maintain their data. The port pins hold the logic levels they had when Idle mode was activated. ALE and PSENL are held high. If an A/D conversion is in process when Idle mode is entered, any conversion results may contain erroneous data. Idle mode is exited via a hardware reset, or an enable interrupt.

Power Down Mode

Power Down mode is initiated by an instruction that sets bit PCON.1 in the 51SL CPU. When the 51SL enters Power Down mode, all internal clocks, including the 51SL core clock, are turned off. If an external crystal is used, the internal oscillator is turned off. MEMCSL, the external memory select signal, goes inactive unless it is configured as a general purpose I/O (i.e., unless bit 3 of configuration register 1 is a "1"). ALE and PSENL are both forced low. RAM contents are preserved.

Power Down mode can only be exited via a reset. This reset may occur either from the RST pin, or an internally generated reset. See the 51SL Hardware Description (Order No. #272268) for a detailed description of this reset.

HOST INTERFACE

The 51SL host interface is functionally compatible with the 8042 style UPI interface. It consists of the PCDB0–7 data bus; the RDL, WRL, A0 and CSL control signals; and the Keyboard Status register, Input Data register, and Output Data register. In addition, a second address line, A1, has been added to decode a second set of registers for power management functions. These registers are identical to the keyboard registers. The host interface also includes a PCOBF interrupt, GATEA20, and host reset (RCL) outputs. Two additional OBF signals, AUXOBF1 and AUXOBF2 are available through firmware configuration of P34/T0 and P37/RDL respectively.

KEYBOARD SCAN

The interface to the keyboard scan logic includes 16 slew-rate-controlled, open drain scan out lines (KSO0–15) and eight Schmitt trigger sense lines (KSI0–7) with internal pullup resistors. KSI0–7 connect directly to Port 0 of the 51SL CPU. The 16 scan out lines are controlled by the four low order bits of Port 0. Together KSO0–15 and KSI0–7 form a keyboard matrix.

EXTERNAL KEYBOARD AND MOUSE INTERFACE

Industry standard PC-AT compatible keyboards employ a two wire, bidirectional TTL interface for data transmission. Several sources also supply PS/2 mouse products that employ the same type of interface. To facilitate system expansion, the 51SL provides four signal pins that may be used to implement this interface directly for an external keyboard and mouse.

The 51SL has four high-drive, open-drain, bidirectional port pins that can be used for external serial interfaces, such as ISA external keyboard and PS/2-type mouse interfaces. They are P30/SIF00, P31/SIF01, P33/SIF10, and P35/SIF11. P33/SIF10 is connected to the firmware configurable level/edge sensitive INTL interrupt pin of the 51SL CPU. P30/SIF00 is connected to the edge sensitive SIF0INTL interrupt pin of the 51SL CPU. Note that on the Low Voltage 8XC51SL these inputs are protected to 5.5V in order to provide compatibility with as many external keyboard and PS/2 mouse devices as possible.

DESIGN CONSIDERATIONS

The low voltage characteristics of the Low Voltage 8XC51SL have indicated that additional care should be taken in selection of the crystal used in the oscillator circuit. In particular, series resistance of a crystal seems to have the largest effect on start-up time and steady state amplitude. Consequently, the lower the series resistance the better, although medium to better quality crystals are generally more than adequate.

ELECTRICAL SPECIFICATIONS

ABSOLUTE MAXIMUM RATINGS*

Ambient Temperature
Under Bias .... −40°C to +85°C

Storage Temperature .... −65°C to +150°C

Voltage on Any Pin to V_{SS} ... −0.5V to V_{CC} + 0.5V

Power Dissipation........................1.0W**

**This value is based on the maximum allowable die temperature and the thermal resistance of the package.

NOTICE: This data sheet contains information on products in the sampling and initial production phases of development. It is valid for the devices indicated in the revision history. The specifications are subject to change without notice.

**WARNING: Stressing the device beyond the "Absolute Maximum Ratings" may cause permanent damage. These are stress ratings only. Operation beyond the "Operating Conditions" is not recommended and extended exposure beyond the "Operating Conditions" may affect device reliability.*

OPERATING CONDITIONS

8XC51SL: T_A (Under Bias) = 0°C to +70°C, V_{CC} = +5V ± 10%, V_{SS} = 0V
Low Voltage 8XC51SL: T_A (Under Bias) = 0°C to +70°C, V_{CC} = +3.3V ± 0.3V, V_{SS} = 0V

8XC51SL DC Characteristics (Over Operating Conditions)

Symbol	Parameter	Min	Max	Units	Test Conditions
V_{IL}	Input Low Voltage (Except XTAL1, RST)	−0.5	0.8	V	
V_{IL1}	Input Low Voltage (XTAL1, RST)	−0.5	0.2 V_{CC} − 0.1		
V_{IH}	Input High Voltage (Except EAL, PCDB0−7, ADB0−7, XTAL1, RST, CSL, RDL, WRL, LOADREN, A0, A1)	2.4	V_{CC} + 0.5	V	
V_{IH1}	Input High Voltage (EAL)	V_{CC} − 1.5	V_{CC} + 0.5	V	
V_{IH2}	Input High Voltage (PCDB0−7, ADB00-7, XTAL1, RST, CSL, RDL, WRL, LOADREN, A0, A1)	0.7 V_{CC}	V_{CC} + 0.5	V	
R_P	Internal Port Resistors KSI0−7	5	20	KΩ	
V_{OL}	Output Low Voltage BP Pins(1) (Except P27/LED4)	−0.5	0.4	V	I_{OL} = 16 mA
V_{OL1}	Output Low Voltage P27/LED4, LED0−3	−0.5	0.8	V	I_{OL} = 12 mA
V_{OL2}	QB Pins(2), PCDB0−7, RCL, ADB0−7, GATEA20, KSO0−15, MEMCSL, ALE, PSENL, PCOBF	−0.5	0.4	V	I_{OL} = 4 mA
V_{OH}	Output High Voltage QB Pins, ALE, PSENL, PCOBF	2.4	V_{CC} + 0.5	V	I_{OH} = −60 μA
V_{OH1}	Outut High Voltage MEMCSL, PCDB0−7, ADB0−7	4.0	V_{CC} + 0.5	V	I_{OH} = −2.0 mA
V_{OH2}	Output High Voltage RCL, GATEA20	4.0	V_{CC} +0.5	V	I_{OH} = 60 μA

8XC51SL DC Characteristics (Over Operating Conditions) (Continued)

Symbol	Parameter	Min	Max	Units	Test Conditions
I_{IL}	Logical 0 Input Current QB(2) Pins		−50	μA	$V_{IN} = 0.4V$
I_{LI}	Input Leakage Current (BP and Pure Input Pins except for KSI0–7, XTAL1, and EAL)		±10	μA	$0 < V_{IN} < V_{CC}$
I_{TL}	Logical 1 to 0 Transition Current QB(2) Pins		−1	mA	$V_{IN} = 2.0V$
I_{CC}	Power Supply Current Active Mode at 16 MHz Idle Mode at 16 MHz Power-Down Mode		 38 15 TBD	 mA mA μA	

Low Voltage 8XC51SL DC Characteristics (Over Operating Conditions)

Symbol	Parameter	Min	Max	Units	Test Conditions
V_{IL}	Input Low Voltage (Except XTAL1, RST, KSI0–7)	−0.5	0.8	V	
V_{IL1}	Input Low Voltage (XTAL1, RST)	−0.5	0.2 V_{CC} − 0.1		
V_{IL2}	Input Low Voltage (KSI0–7)	−0.5	0.6		
V_{IH}	Input High Voltage (Except EAL, PCDB0–7, ADB0–7, XTAL1, RST) P30, P31, P33, P35)	2.0	V_{CC} + 0.5	V	
V_{IH1}	Input High Voltage (EAL)	V_{CC} − 1	V_{CC} + 0.5	V	
V_{IH2}	Input High Voltage (PCDB0–7, ADB0–7, XTAL1, RST)	0.7 V_{CC}	V_{CC} + 0.5	V	
V_{IH3}	Input High Voltage (P30, P31, P33, P35)	2.0	5.5	V	
R_P	Internal Port Resistors KSI0–7	5	20	KΩ	
V_{OL}	Output Low Voltage BP Pins(1) (Except P27/LED4)	−0.5	0.4	V	I_{OL} = 16 mA
V_{OL1}	Output Low Voltage P27/LED4, LED0–3	−0.5	0.8	V	I_{OL} = 12 mA
V_{OL2}	Output Low Voltage QB Pins(2), PCDB0–7, RCL, ADB0–7, GATEA20, KSO0–15, MEMCSL, ALE, PSENL, PCOBF	−0.5	0.4	V	I_{OL} = 4 mA
V_{OH}	Output High Voltage QB Pins, ALE, PSENL, PCOBF	V_{CC} − 0.7	V_{CC} + 0.5	V	I_{OH} = −60 μA
V_{OH1}	Output High Voltage MEMCSL, PCDB0–7, ADB0–7	2.4	V_{CC} + 0.5	V	I_{OH} = −2.0 mA
V_{OH2}	Output High Voltage RCL, GATEA20	2.4	V_{CC} +0.5	V	I_{OH} = 60 μA
I_{IL}	Logical 0 Input Current QB(2) Pins		−50	μA	$V_{IN} = 0.4V$
I_{LI}	Input Leakage Current (BP and Pure Input Pins except for KSI0–7, XTAL1, and EAL)		±10	μA	$0 < V_{IN} < V_{CC}$

Low Voltage 8XC51SL DC Characteristics (Over Operating Conditions)

Symbol	Parameter	Min	Max	Units	Test Conditions
I_{TL}	Logical 1 to 0 Transition Current QB(2) Pins		−650	μA	V_{IN} = 1.5V
I_{CC}	Power Supply Current Active Mode at 16 MHz Idle Mode at 16 MHz Power-Down Mode		 25 10 175	 mA mA μA	

NOTES:
1. Bidirectional (BP) pins include P27/LED4, P30/SIF00, P31/SIF01, P33/SIF10, P36/SIF11, MEMCSL, PCDB0–7, and ADB0–7.
2. Quasi-bidirectional (QB) pins include P20–6/A8–A14, P32/INT0, P34/T0, P36/WRL, P37/RDL and P10–7/A0–7.
3. Pure input pins include LOADREN, EAL, A0, A1, CSL, RDL, WRL, RST, AIN0–3, and XTAL1.

AC Characteristics

EXPLANATION OF THE AC SYMBOLS

Each timing symbol has three or five characters. The first character is always "T" (for time). The other characters, depending on their positions, stand for the name of a signal or the logical status of that signal. Table 3 lists the characters and their meanings.

Example

TAVLL = Time for Address Valid to ALE Low.

TLLPL = Time for ALE Low to PSEN Low.

Table 3. AC Symbol Characters

Char.	Meaning
A	Address
C	Clock
D	Input Data
H	Logic Level HIGH
I	Instruction (Program Memory Contents)
L	Logic Level LOW, or ALE
P	PSENL
Q	Output Data
R	RDL Signal
T	Time
V	Valid
W	WRL Signal
X	No Longer a Valid Logic Level
Z	Float

HOST-INTERFACE TIMING

All Outputs Loaded with 50 pF

Symbol	Parameter	Min	Max	Units
TAR	CSL, A0/A1 Setup to RD Low	0		ns
TRA	CSL, A0/A1 Hold after RDL High	0		ns
TAD	CSL, A0/A1 to Data Out Delay		50	ns
TAW	CSL, A0/A1 Setup to WRL Low	0		ns
TWA	CSL, A0/A1 Hold after WRL High	10		ns
TDW	Data Setup to WRL High	60		ns
TWD	Data Hold after WRL High	5		ns
TWW	Minimum Pulse Width of WRL	50		ns
TRR	RDL Pulse Width	50		ns
TRD	RDL Low to Data Out Delay		50	ns
TDF	RDL High to Data Float Delay		50	ns

EXTERNAL MEMORY TIMING

TCLCL = 1 Clock Period, All Outputs Loaded with 50 pF

Symbol	Parameter	Min	Max	Units
1/TCLCL	Oscillator Frequency	2	16	MHz
TLHLL	ALE Pulse Width	2TCLCL − 40		ns
TAVLL	Address Valid to ALE Low	TCLCL − 40		ns
TLLAX	Address Hold after ALE Low	TCLCL − 30		ns
TLLIV	ALE Low to Valid Instruction In		4TCLCL − 100	ns
TLLPL	ALE Low to PSENL Low	TCLCL − 30		ns
TPLPH	PSENL Pulse Width	3TCLCL − 45		ns
TPLIV	PSENL Low to Valid Instruction In		3TCLCL − 105	ns
TPXIX	Input Instruction Hold after PSENL High	0		ns
TPXIZ	Input Instruction Float after PSENL High		TCLCL − 25	ns
TAVIV	Address to Valid Instruction In		5TCLCL − 105	ns
TPLAZ	PSENL Low to Address Float		10	ns
TRLRH	P37/RDL Pulse Width	6TCLCL − 50		ns
TWLWH	P36/WRL Pulse Width	6TCLCL − 50		ns
TRLDV	P37RDL Low to Valid Data In		5TCLCL − 100	ns
TRHDX	Data Hold after P37/RDL	0		ns
TRHDZ	Data Float after P37/RDL		2TCLCL − 50	ns
TLLDV	ALE Low to Valid Data In		8TCLCL − 100	ns
TAVDV	Address to Valid Data In		9TCLCL − 100	ns
TLLWL	ALE Low to P37/RDL or P36/WRL Low	3TCLCL − 25	3TCLCL + 25	ns
TAVWL	Address Valid to P36/WRL Low	4TCLCL − 50		ns
TQVWX	Data Valid before P36/WRL	TCLCL − 25		ns
TWHQX	Data Hold after P36/WRL	TCLCL − 25		ns
TQVWH	Data Valid to P36/WRL High	7TCLCL − 50		ns
TRLAZ	P37/RDL Low to Address Float		0	ns
TWHLH	P37/RDL or P36/WRL High to ALE High	TCLCL − 25	TCLCL + 25	ns

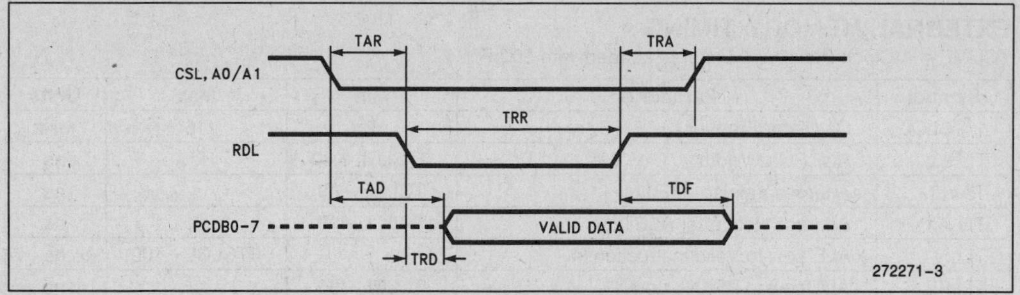

Figure 3. Host-Interface Read

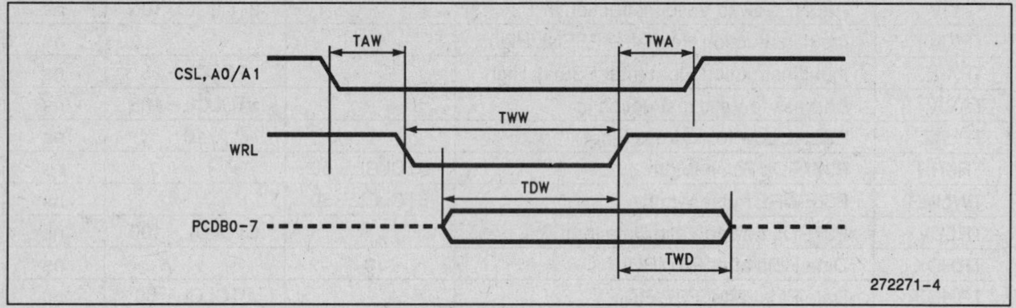

Figure 4. Host-Interface Write

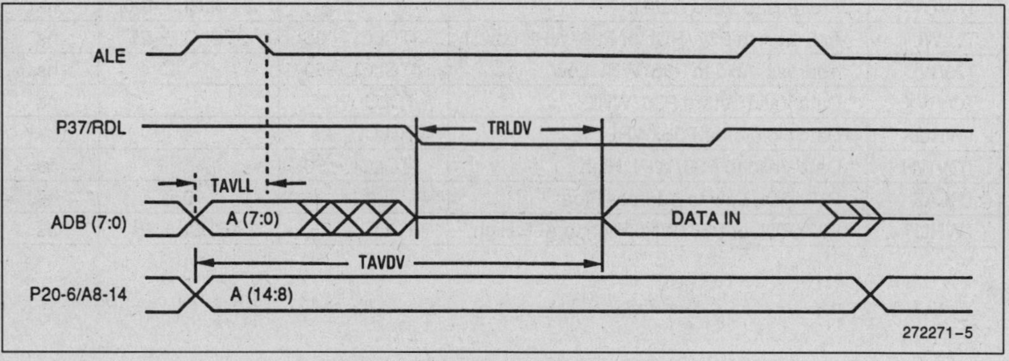

Figure 5. External Data Memory Read

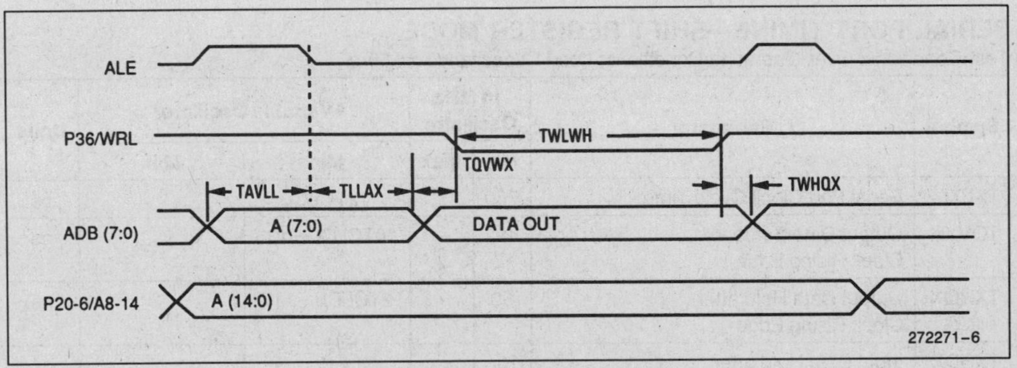

Figure 6. External Data Memory Write

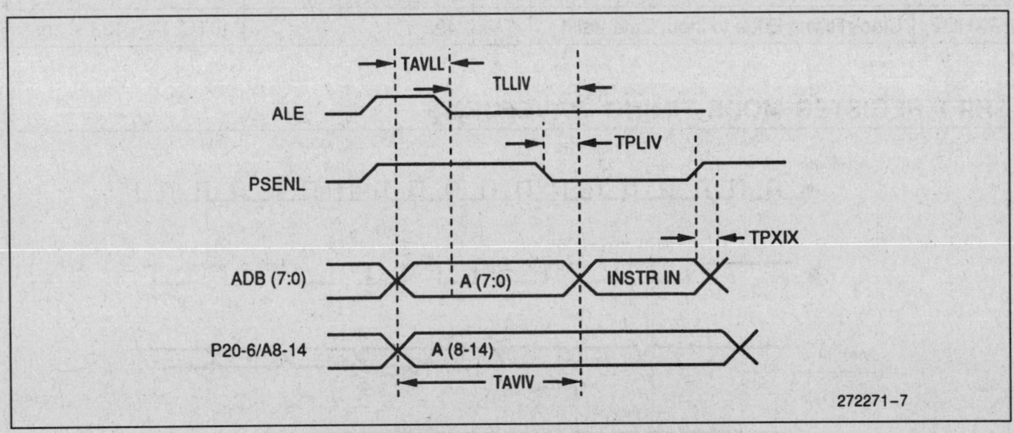

Figure 7. External Program Memory Read

SERIAL PORT TIMING—SHIFT REGISTER MODE

Test Conditions: Over Operating Conditions, Load Capacitance = 50 pF

Symbol	Parameter	16 MHz Oscillator		Variable Oscillator		Units
		Min	Max	Min	Max	
TXLXL	Serial Port Clock Cycle Time	750		12TCLCL		ns
TQVXH	Output Data Setup to Clock Rising Edge	492		10TCLCL − 133		ns
TXHQX	Output Data Hold after Clock Rising Edge	50		2TCLCL − 117		ns
TXHDX	Input Data Hold after Clock Rising Edge	0		0		ns
TXHDV	Clock Rising Edge to Input Data Valid		492		10TCLCL − 133	ns

SHIFT REGISTER MODE TIMING WAVEFORMS

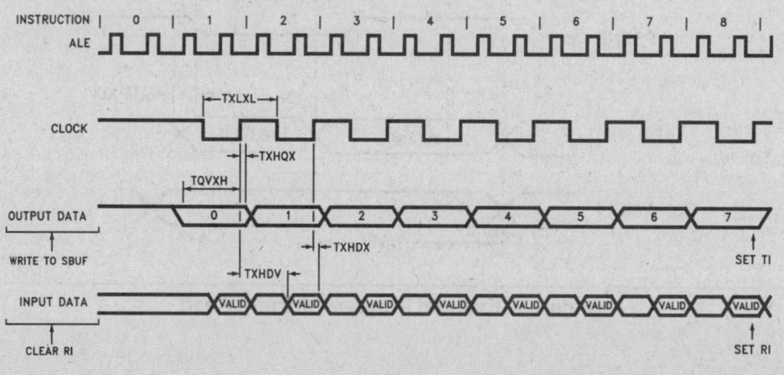

272271-8

EXTERNAL CLOCK DRIVE

Symbol	Parameter	Min	Max	Units
1/TCLCL	Oscillator Frequency	2.0	16	MHz
TCHCX	High Time	20		ns
TCLCX	Low Time	20		ns
TCLCH	Rise Time		20	ns
TCHCL	Fall Time		20	ns

EXTERNAL CLOCK DRIVE WAVEFORM

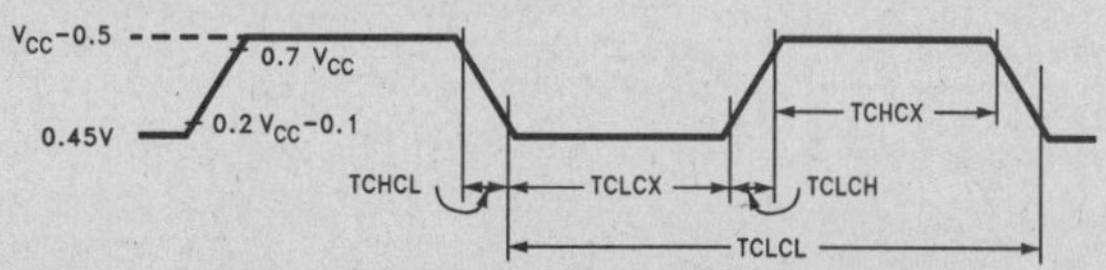

272271-9

PROGRAMMING THE OTP

The part must be running with a 4 MHz to 6 MHz oscillator. The address of a location to be programmed is applied to address lines, while the code byte to be programmed in that location is applied to data lines. Control and program signals must be held at the levels indicated in Table 4. Normally EAL/V_{PP} is held at a logic high until just before RCL/PROGL is to be pulsed. The EAL/V_{PP} is raised to V_{PP}, RCL/PROGL is pulsed low and then EAL/V_{PP} is returned to V_{CC} (also refer to timing diagrams). Also, the LOADREN signal must be grounded when programming or verifying.

NOTE:
Exceeding the V_{PP} maximum for any amount of time could damage the device permanently. The V_{PP} source must be well regulated and free of glitches.

DEFINITION OF TERMS

ADDRESS LINES: P10–P17, P20–P25, respectively for A0–A13.

DATA LINES: ADB0–7.

CONTROL SIGNALS: RST, GATEA20, P26, P27, P32, P36, P37.

PROGRAM SIGNALS: RCL/PROGL, EAL/V_{PP}.

PROGRAMMING ALGORITHM

Refer to Table 4 and Figures 8 and 9 for address, data and control signals setup. To program the 87C51SL the following sequence must be exercised.

1. Input the valid address on the address lines.
2. Input the appropriate data byte on the data lines.
3. Activate the correct combination of control signals.
4. Raise EAL/V_{PP} from V_{CC} to 12.75V ±0.25V.
5. Pulse RCL/PROGL 5 times.

Repeat 1 through 5 changing the address and data for the entire array or until the end of the object file is reached.

2

Table 4. OTP Programming Modes

Mode	RST	GATEA20	RCL/PROGL	EAL/V_{PP}	P26	P27	P32	P36	P37
Program Code Data	H	L	⊔ (pulse low)	12.75V	L	H	H	H	H
Verify Code Data	H	L	H	H	L	L	L	H	H
Read Signature Byte	H	L	H	H	L	L	L	L	L

Note that in the above table, to program code data on the Low Voltage 87C51SL V_{CC} must be raised to 5V ±10%. In addition, all address lines, data lines, and control signals being driven to a "High" level must be raised to 5V ±10%. The RCL/PROGL signal must pulse between 0V and 5V ±10%.

To verify code data or read the signature bytes of the Low Voltage 87C51SL V_{CC} must be set to 3.3V ±0.3V. In addition, all address lines and control signals being driven to a "High" level must be raised to 3.3V ±0.3V.

For the standard (5V version) of the 87C51SL V_{CC} must always be at 5V ±10%, and all "High" voltages must meet the DC specs indicated in the DC Characteristics section of this document.

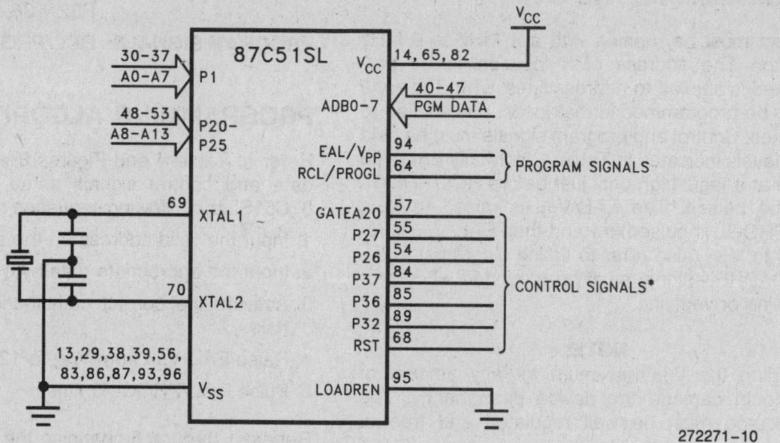

*See Table 4 for proper input on these pins.

Figure 8. Programming/Verifying the OTP

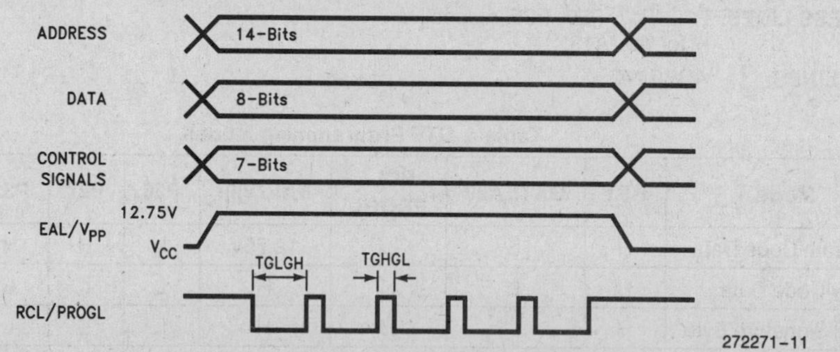

Figure 9. Programming Signal's Waveforms

PROGRAM VERIFY

Program verify may be done after each byte that is programmed, or after a block of bytes that is programmed. In either case a complete verify of the array will ensure that it has been programmed correctly.

READING THE SIGNATURE BYTES

The 8XC51SL and Low Voltage 8XC51SL each have three signature bytes in locations 30H, 31H, and 60H. To read these bytes, follow the procedure for EPROM verify, but activate the control lines provided in Table 4 for Read Signature Byte.

Location	Contents			
	87C51SL	83C51SL	Low Voltage 87C51SL	Low Voltage 83C51SL
30H	89H	89H	89H	89H
31H	58H	58H	58H	58H
60H	BBH	3BH	ABH	2BH

OTP PROGRAMMING AND VERIFICATION CHARACTERISTICS

T_A = 21°C to 27°C; V_{CC} = 5V ± 10% for 87C51SL, 3.3V ± 0.3V for Low Voltage 87C51SL (verification only). V_{CC} for programming the Low Voltage 87C51SL must be 5.0V ±10%. V_{SS} = 0V

Symbol	Parameter	Min	Max	Units
V_{PP}	Programming Supply Voltage	12.5	13.0	V
I_{PP}	Programming Supply Current		75	mA
1/TCLCL	Oscillator Frequency	4	6	MHz
TAVGL	Address Setp to PROGL Low	48TCLCL		
TGHAX	Address Hold after PROGL	48TCLCL		
TDVGL	Data Setup to PROGL Low	48TCLCL		
TGHDX	Data Hold after PROGL	48TCLCL		
TEHSH	(Enable) High to V_{PP}	48TCLCL		
TSHGL	V_{PP} Setup to PROGL Low	10		µs
TGHSL	V_{PP} Hold after PROGL	10		µs
TGLGH	PROGL Width	90	110	µs
TAVQV	Address to Data Valid		48TCLCL	
TELQV	ENABLE Low to Data Valid		48TCLCL	
TEHQZ	Data Float after Enable	0	48TCLCL	
TGHGL	PROGL High to PROGL Low	10		µs

PROGRAMMING AND VERIFICATION WAVEFORMS

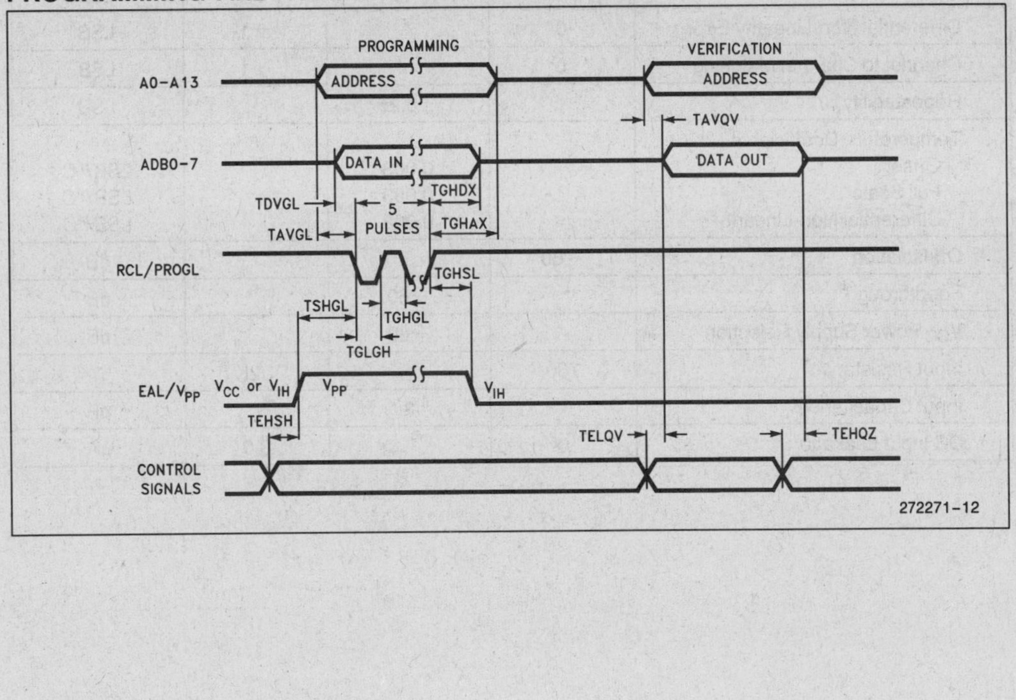

A/D CHARACTERISTICS

The 51SL includes a four-channel, 8-bit A/D converter. This A/D, with eight bits of accuracy, uses successive approximation with a switch capacitor comparator. It is designed to be used for sampling static analog signals (i.e., ideally suited for power management tasks such as battery voltage monitoring, etc.). The nominal conversion rate is 20 μs at 16 MHz. The analog high and low voltage references are connected to AVREF and AVGND, respectively. The four input channels, AIN0–3 are connected from the package pins, unbuffered, to an analog multiplexer (on-chip). The absolute conversion accuracy is dependent upon the accuracy of AVREF. The specifications given assume adherence to the operating conditions section of this data sheet.

Testing is done at AVREF = 5.12V and V_{CC} = 5.0V for the 8XC51SL, and at AVREF = 3.2V and V_{CC} = 3.3V for the Low Voltage 8XC51SL.

OPERATING CONDITIONS

V_{CC}
- 8XC51SL .... 4.5V to 5.5V
- Low Voltage 8XC51SL .... 3.0V to 3.6V

AVREF
- 8XC51SL .... 4.5V to 5.5V
- Low Voltage 8XC51SL .... 3.0V to 3.6V

V_{SS}, AVSS 0V

AIN0–3 .... AVSS to AVREF

T_A .... 0°C to +70°C Ambient

F_{OSC} .... 2 MHz to 16 MHz

A/D CONVERTER SPECIFICATIONS (Over Operating Conditions)

Parameter	Min	Typ	Max	Units
Resolution	255 8		256 8	Levels Bits
Absolute Error	0		±1	LSB
Full Scale Error		±1		LSB
Zero Offset Error		±1		LSB
Non-Linearity Error	0		±1	LSB
Differential Non-Linearity Error	0		±1	LSB
Channel to Channel Matching	0		±1	LSB
Repeatability		±0.25		LSB
Temperature Coefficients Offset Full Scale Differential Non-Linearity		 0.003 0.003 0.003		 LSB/°C LSB/°C LSB/°C
Off Isolation	−60			dB
Feedthrough		−60		dB
V_{CC} Power Supply Rejection		−60		dB
Input Resistance	750		1.2K	Ω
Input Capacitance		3		pF
DC Input Leakage	0		3.0	μA

A/D Glossary of Terms

Absolute Error—The maximum difference between corresponding actual and ideal code transitions. Absolute Error accounts for all deviations of an actual converter from an ideal converter.

Actual Characteristic—The characteristic of an actual converter. The characteristic of a given converter may vary over temperature, supply voltage, and frequency conditions. An actual characteristic rarely has ideal first and last transition locations or ideal code widths. It may even vary over multiple conversions under the same conditions.

Break-Before-Make—The property of a multiplexer which guarantees that a previously selected channel will be deselected before a new channel is selected (e.g., the converter will not short inputs together).

Channel-to-Channel Matching—The difference between corresponding code transitions of actual characteristics taken from different channels under the same temperature, voltage and frequency conditions.

Characteristic—A graph of input voltage versus the resultant output code for an A/D converter. It describes the transfer function of the A/D converter.

Code—The digital value output by the converter.

Code Center—The voltage corresponding to the midpoint between two adjacent code transitions.

Code Transition—The point at which the converter changes from an output code of Q, to a code of Q + 1. The input voltage corresponding to a code transition is defined to be that voltage which is equally likely to produce either of two adjacent codes.

Code Width—The voltage corresponding to the difference between two adjacent code transitions.

Crosstalk—See "Off-Isolation".

DC Input Leakage—Leakage current to ground from an analog input pin.

Differential Non-Linearity—The difference between the ideal and actual code widths of the terminal based characteristic.

Feedthrough—Attenuation of a voltage applied on the selected channel of the A/D Converter after the sample window closes.

Full Scale Error—The difference between the expected and actual input voltage corresponding to the full scale code transition.

Ideal Characteristic—A characteristic with its first code transition at V_{IN} = 0.5 LSB, its last code transition at V_{IN} = (V_{REF} − 1.5 LSB) and all code widths equal to one LSB.

Input Resistance—The effective series resistance from the analog input pin to the sample capacitor.

LSB—Least Significant Bit—The voltage corresponding to the full scale voltage divided by 2^n, where n is the number of bits of resolution of the converter. For an 8-bit converter with a reference voltage of 5.12V, one LSB is 20 mV. Note that this is different than digital LSBs since an uncertainty of two LSBs, when referring to an A/D converter, equals 40 mV. (This has been confused with an uncertainty of two digital bits, which would mean four counts, or 80 mV).

Monotonic—The property of successive approximation converters which guarantees that increasing input voltages produce adjacent codes of increasing value, and that decreasing input voltages produce adjacent codes of decreasing value.

No Missed Codes—For each and every output code, there exists a unique input voltage range which produces that code only.

Non-Linearity—The maximum deviation of code transitions of the terminal based characteristic from the corresponding code transitions of the ideal characteristic.

Off-Isolation—Attenuation of a voltage applied on a deselected channel of the A/D converter. (Also referred to as Crosstalk.)

Repeatability—The difference between corresponding code transitions from different actual characteristics taken from the same converter on the same channel at the same temperature, voltage and frequency conditions.

Resolution—The number of input voltage levels that the converter can unambiguously distinguish between. Also defines the number of useful bits of information which the converter can return.

Sample Delay—The delay from receiving the start conversion signal to when the sample window opens.

Sample Delay Uncertainty—The variation in the sample delay.

Sample Time—The time that the sample window is open.

Sample Time Uncertainty—The variation in the sample time.

Sample Window—Begins when the sample capacitor is attached to a selected channel and ends when the sample capacitor is disconnected from the selected channel.

Successive Approximation—An A/D conversion method which uses a binary search to arrive at the best digital representation of an analog input.

Temperature Coefficients—Change in the stated variable per degree centrigrade temperature change. Temperature coefficients are added to the typical values of a specification to see the effect of temperature drift.

Terminal Based Characteristic—An actual characteristic which has been rotated and translated to remove zero offset and full scale error.

V_{CC} Rejection—Attenuation of noise on the V_{CC} line to the A/D converter.

Zero Offset—The difference between the expected and actual input voltage corresponding to the first code transition.

DATA SHEET REVISION SUMMARY

The following differences exist between this data sheet (272271-002) and the previous version (272271-001).

1. Data sheet status changed from "Product Preview" to "Advance Information".
2. Title page item number three describing the global interrupt enable change was removed.
3. Title page item number two was corrected to read "... was added in configuration register 1."
4. In the 8XC51SL DC Characteristics section:

   The V_{OH} test condition (I_{OH}) changed from −0.8 mA to −60 μA.

   The V_{OH1} test condition (I_{OH}) changed from −4.0 mA to −2.0 mA.

   V_{OH2} was added.

   The XTAL1 and EAL pins were added to the I_{LI} spec.

   The I_{TL} spec changed from −650 μA to −1 mA.

   The I_{CC} idle spec changed from 10 mA to 15 mA.

   The I_{CC} Power Down spec changed from 100 μA to TBD.
5. In the Low Voltage 8XC51SL DC Characteristics section:

   The V_{OH} spec changed from 2.4V to V_{CC} −0.7

   The V_{OH} test condition (I_{OH}) changed from −0.8 mA to −60 μA.

   V_{OH2} was added.

   Pins were clarified in the I_{LI} spec.

   The I_{TL} test condition (V_{IN}) was changed from TBD to 1.5V.

   The I_{CC} Power Down spec changed from 100 μA to 175 μA.
6. The load capacitance for all timing tables was changed to 50 pF.
7. In the Host Interface Timing Section TWD changed from 0 ns to 5 ns.
8. The External Memory Timing table changed as follows:

Spec.	Old	New
TLLIV	4TCLCL-50	4TCLCL-100
TPLIV	3TCLCL-50	3TCLCL-105
TPXIZ	TCLCL-15	TCLCL-25
TAVIV	5TCLCL-50	5TCLCL-105
TRLDV	5TCLCL-50	5TCLCL-100
TLLDV	8TCLCL-50	8TCLCL-100
TAVDV	9TCLCL-50	9TCLCL-100
TMVDV	9TCLCL-50	Removed
TMVIV	5TCLCL-50	Removed

9. In Figures 5 and 7 the MEMCSL waveforms were removed.
10. Clarification was added in the Programming Algorithm section.
11. In the A/D Converter Specifications section the minimum resolution was changed from 256 levels to 255 levels.
12. The Data Sheet Revision Summary was added.

8XC152JA/JB/JC/JD UNIVERSAL COMMUNICATION CONTROLLER 8-BIT MICROCONTROLLER

- **8K Factory Mask Programmable ROM Available**

- **Superset of 80C51 Architecture**
- **Multi-Protocol Serial Communication I/O Port (2.048 Mbps/2.4 Mbps Max)**
  - **—SDLC/HDLC Only**
  - **—CSMA/CD and SDLC/HDLC**
  - **—User Definable Protocols**
- **Full Duplex/Half Duplex**
- **MCS®-51 Compatible UART**
- **16.5 MHz Maximum Clock Frequency**
- **Multiple Power Conservation Modes**
- **64KB Program Memory Addressing**
- **64KB Data Memory Addressing**
- **256 Bytes On-Chip RAM**
- **Dual On-Chip DMA Channels**
- **Hold/Hold Acknowledge**
- **Two General Purpose Timer/Counters**
- **5 or 7 I/O Ports**
- **56 Special Function Registers**
- **11 Interrupt Sources**
- **Available in 48 Pin Dual-in-Line Package and 68 Pin Surface Mount PLCC Package**
  (See Packaging Spec. Order #231369)

The 80C152, which is based on the MCS®-51 CPU, is a highly integrated single-chip 8-bit microcontroller designed for cost-sensitive, high-speed, serial communications. It is well suited for implementing Integrated Services Digital Networks (ISDN), emerging Local Area Networks, and user defined serial backplane applications. In addition to the multi-protocol communication capability, the 80C152 offers traditional microcontroller features for peripheral I/O interface and control.

Silicon implementations are much more cost effective than multi-wire cables found in board level parallel-to-serial and serial-to-parallel converters. The 83C152 contains, in silicon, all the features needed for the serial-to-parallel conversion. Other 83C152 benefits include: 1) better noise immunity through differential signaling or fiber optic connections, 2) data integrity utilizing the standard, designed in CRC checks, and 3) better modularity of hardware and software designs. All of these—cost, network parameter and real estate improvements—apply to 83C152 serial links between boards or systems and 83C152 serial links on a single board.

October 1989
Order Number: 270431-003

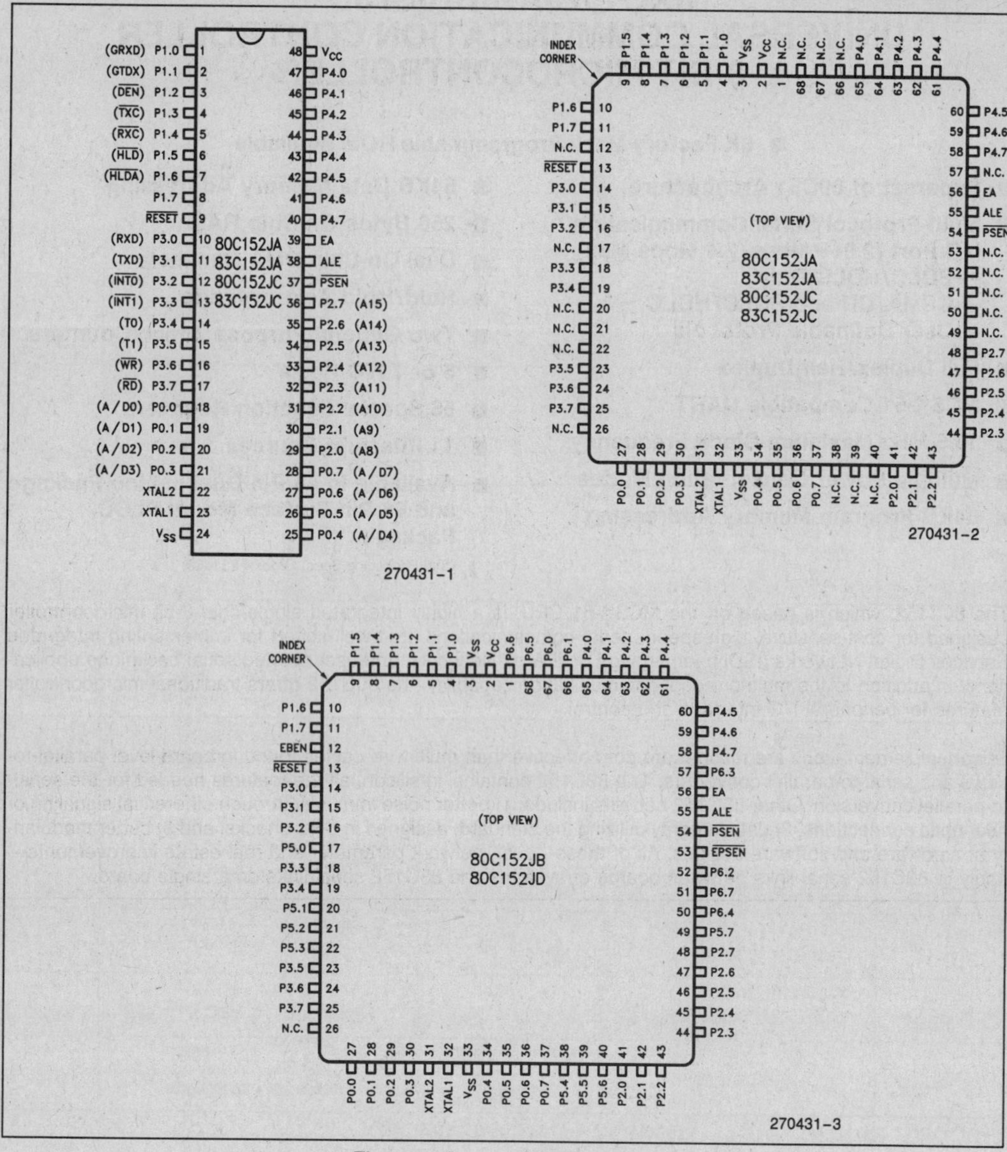

Figure 1. Connection Diagrams

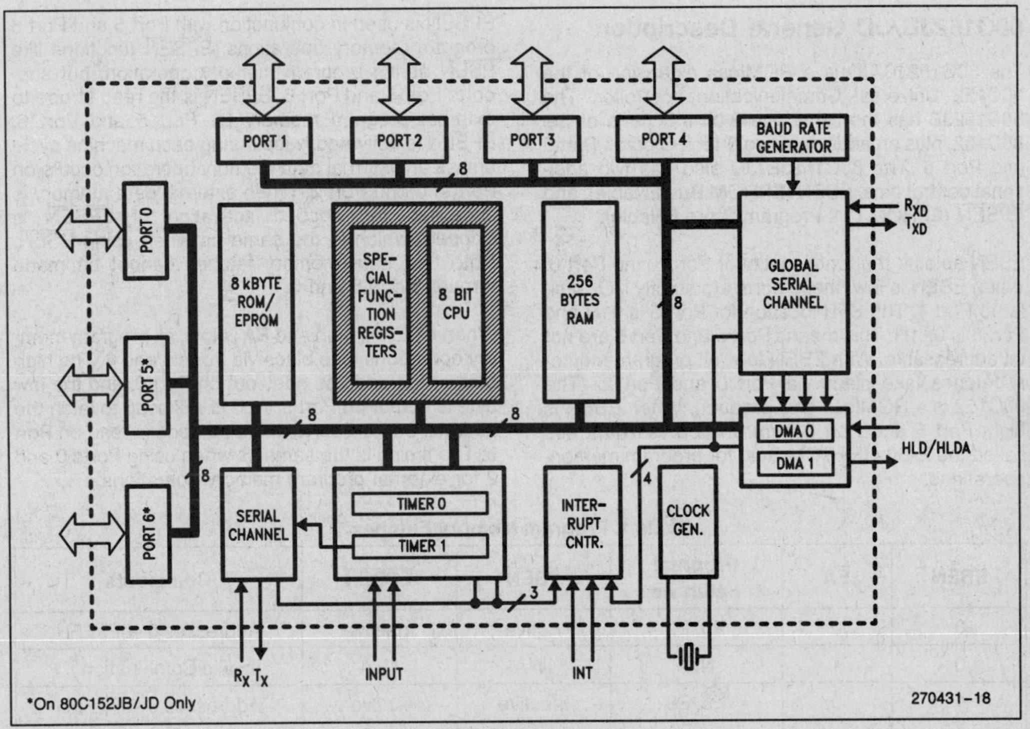

Figure 2. Block Diagram

80C152JB/JD General Description

The 80C152JB/JD is a ROMless extension of the 80C152 Universal Communication controller. The 80C152JB has the same five 8-bit I/O ports of the 80C152, plus an additional two 8-bit I/O ports, Port 5 and Port 6. The 80C152JB/JD also has two additional control pins, EBEN (**E**PROM **B**us **EN**able), and $\overline{\text{EPSEN}}$ (**E**PROM bus **P**rogram **S**tore **EN**able).

EBEN selects the functionality of Port 5 and Port 6. When EBEN is low, these ports are strictly I/O, similar to Port 4. The SFR location for Port 5 is 91H and Port 6 is 0A1H. This means Port 5 and Port 6 are not bit addressable. With EBEN low, all program memory fetches take place via Port 0 and Port 2. (The 80C152 is a ROMless only product). When EBEN is high, Port 5 and Port 6 form an address/data bus called the E-Bus (EPROM-Bus) for program memory operations.

$\overline{\text{EPSEN}}$ is used in conjunction with Port 5 and Port 6 program memory operations. $\overline{\text{EPSEN}}$ functions like $\overline{\text{PSEN}}$ during program memory operation, but supports Port 5 and Port 6. $\overline{\text{EPSEN}}$ is the read strobe to external program memory for Port 5 and Port 6. $\overline{\text{EPSEN}}$ is activated twice during each machine cycle unless an external data memory operation occurs on Port(s) 0 and Port 2. When external data memory is accessed the second activation of $\overline{\text{EPSEN}}$ is skipped, which is the same as when using $\overline{\text{PSEN}}$. Note that data memory fetches cannot be made through Ports 5 and 6.

When EBEN is high and EA is low, all program memory operations take place via Ports 5 and 6. The high byte of the address goes out on Port 6, and the low byte is output on Port 5. ALE is still used to latch the address on Port 5. Next, the op code is read on Port 5. The timing is the same as when using Ports 0 and 2 for external program memory operations.

Table 1. Program Memory Fetches

EBEN	$\overline{\text{EA}}$	Program Fetch via	$\overline{\text{PSEN}}$	$\overline{\text{EPSEN}}$	Comments
0	0	P0, P2	Active	Inactive	Addresses 0–0FFFFH
0	1	N/A	N/A	N/A	Invalid Combination
1	0	P5, P6	Inactive	Active	Addresses 0–0FFFFH
1	1	P5, P6 P0, P2	Inactive Active	Active Inactive	Addresses 0–1FFFH Addresses ≥ 2000H

Table 2. 8XC152 Product Differences

ROMless Version	CSMA/CD and HDLC/SDLC	HDLC/SDLC Only	ROM Version Available	PLCC and DIP	PLCC Only	5 I/O Ports	7 I/0 Ports
80C152JA	*		*(83C152JA)	*		*	
80C152JB	*				*		*
80C152JC		*	*(83C152JC)	*		*	
80C152JD		*			*		*

NOTES:
* = options available
0 standard frequency range 3.5 MHz to 12 MHz
0 "–1" frequency range 3.5 MHz to 16.5 MHz

<table>
<tr><th colspan="2">Pin #</th><th rowspan="2">Pin Description</th></tr>
<tr><th>DIP</th><th>PLCC(1)</th></tr>
<tr><td>48</td><td>2</td><td>V_{CC}—Supply voltage.</td></tr>
<tr><td>24</td><td>3,33(2)</td><td>V_{SS}—Circuit ground.</td></tr>
<tr><td>18-21,
25-28</td><td>27-30,
34-37</td><td>Port 0—Port 0 is an 8-bit open drain bidirectional I/O port. As an output port each pin can sink 8 LS TTL inputs. Port 0 pins that have 1s written to them float, and in that state can be used as high-impedance inputs.
Port 0 is also the multiplexed low-order address and data bus during accesses to external program memory if EBEN is pulled low. During accesses to external Data Memory, Port 0 always emits the low-order address byte and serves as the multiplexed data bus. In these applications it uses strong internal pullups when emitting 1s.
Port 0 also outputs the code bytes during program verification. External pullups are required during program verification.</td></tr>
<tr><td>1-8</td><td>4-11</td><td>Port 1—Port 1 is an 8-bit bidirectional I/O port with internal pullups. Port 1 pins that have 1s written to them are pulled high by the internal pullups, and in that state can be used as inputs. As inputs, Port 1 pins that are externally being pulled low will source current (I_{IL}, on the data sheet) because of the internal pullups.
Port 1 also serves the functions of various special features of the 8XC152, as listed below:
<table>
<tr><th>Pin</th><th>Name</th><th>Alternate Function</th></tr>
<tr><td>P1.0</td><td>GRXD</td><td>GSC data input pin</td></tr>
<tr><td>P1.1</td><td>GTXD</td><td>GSC data output pin</td></tr>
<tr><td>P1.2</td><td>$\overline{DEN}$</td><td>GSC enable signal for an external driver</td></tr>
<tr><td>P1.3</td><td>$\overline{TXC}$</td><td>GSC input pin for external transmit clock</td></tr>
<tr><td>P1.4</td><td>$\overline{RXC}$</td><td>GSC input pin for external receive clock</td></tr>
<tr><td>P1.5</td><td>$\overline{HLD}$</td><td>DMA hold input/output</td></tr>
<tr><td>P1.6</td><td>$\overline{HLDA}$</td><td>DMA hold acknowledge input/output</td></tr>
</table></td></tr>
<tr><td>29-36</td><td>41-48</td><td>Port 2—Port 2 is an 8-bit bidirectional I/O port with internal pullups. Port 2 pins that have 1s written to them are pulled high by the internal pullups, and in that state can be used as inputs. As inputs, Port 2 pins that are externally being pulled low will source current (I_{IL}, on the data sheet) because of the internal pullups.
Port 2 emits the high-order address byte during fetches from external Program Memory if EBEN is pulled low. During accesses to external Data Memory that use 16-bit addresses (MOVX @ DPTR and DMA operations), Port 2 emits the high-order address byte. In these applications it uses strong internal pullups when emitting 1s.
During accesses to external Data Memory that use 8-bit addresses (MOVX @ Ri), Port 2 emits the contents of the P2 Special Function Register.
Port 2 also receives the high-order address bits during program verification.</td></tr>
<tr><td>10- 17</td><td>14-16,
18, 19,
23-25</td><td>Port 3—Port 3 is an 8-bit bidirectional I/O port with internal pullups. Port 3 pins that have 1s written to them are pulled high by the internal pullups, and in that state can be used as inputs. As inputs, Port 3 pins that are externally being pulled low will source current (I_{IL}, on the data sheet) because of the pullups.
Port 3 also serves the functions of various special features of the MCS-51 Family, as listed below:
<table>
<tr><th>Pin</th><th>Name</th><th>Alternate Function</th></tr>
<tr><td>P3.0</td><td>RXD</td><td>Serial input line</td></tr>
<tr><td>P3.1</td><td>TXD</td><td>Serial output line</td></tr>
<tr><td>P3.2</td><td>$\overline{INT0}$</td><td>External Interrupt 0</td></tr>
<tr><td>P3.3</td><td>$\overline{INT1}$</td><td>External Interrupt 1</td></tr>
<tr><td>P3.4</td><td>T0</td><td>Timer 0 external input</td></tr>
<tr><td>P3.5</td><td>T1</td><td>Timer 1 external input</td></tr>
<tr><td>P3.6</td><td>$\overline{WR}$</td><td>External Data Memory Write strobe</td></tr>
<tr><td>P3.7</td><td>$\overline{RD}$</td><td>External Data Memory Read strobe</td></tr>
</table></td></tr>
</table>

Pin Description (Continued)

Pin #		Pin Description
47-40	65-58	**Port 4**—Port 4 is an 8-bit bidirectional I/O port with internal pullups. Port 4 pins that have 1s written to them are pulled high by the internal pullups, and in that state can be used as inputs. As inputs, Port 4 pins that are externally being pulled low will source current (I_{IL}, on the data sheet) because of the internal pullups. In addition, Port 4 also receives the low-order address bytes during program verification.
9	13	**$\overline{RST}$**—Reset input. A logic low on this pin for three machine cycles while the oscillator is running resets the device. An internal pullup resistor permits a power-on reset to be generated using only an external capacitor to V_{SS}. Although the GSC recognizes the reset after three machine cycles, data may continue to be transmitted for up to 4 machine cycles after Reset is first applied.
38	55	**ALE**—Address Latch Enable output signal for latching the low byte of the address during accesses to external memory. In normal operation ALE is emitted at a constant rate of ⅙ the oscillator frequency, and may be used for external timing or clocking purposes. Note, however, that one ALE pulse is skipped during each access to external Data Memory. While in Reset, ALE remains at a constant high level.
37	54	**$\overline{PSEN}$**—Program Store Enable is the Read strobe to External Program Memory. When the 8XC152 is executing from external program memory, $\overline{PSEN}$ is active (low). When the device is executing code from External Program Memory, $\overline{PSEN}$ is activated twice each machine cycle, except that two $\overline{PSEN}$ activations are skipped during each access to External Data Memory. While in Reset, $\overline{PSEN}$ remains at a constant high level.
39	56	**$\overline{EA}$**—External Access enable. $\overline{EA}$ must be externally pulled low in order to enable the 8XC152 to fetch code from External Program Memory locations 0000H to 0FFFH. $\overline{EA}$ must be connected to V_{CC} for internal program execution.
23	32	**XTAL1**—Input to the inverting oscillator amplifier and input to the internal clock generating circuits.
22	31	**XTAL2**—Output from the inverting oscillator amplifier.
N/A	17, 20 21, 22 38, 39 40, 49	**Port 5**—Port 5 is an 8-bit bidirectional I/O port with internal pullups. Port 5 pins that have 1s written to them are pulled high by the internal pullups, and in that state can be used as inputs. As inputs, Port 5 pins that are externally being pulled low will source current (I_{IL}, on the data sheet) because of the internal pullups. Port 5 is also the multiplexed low-order address and data bus during accesses to external program memory if EBEN is pulled high. In this application it uses strong pullups when emitting 1s.
N/A	67, 66 52, 57 50, 68 1, 51	**Port 6**—Port 6 is an 8-bit bidirectional I/O port with internal pullups. Port 6 pins that have 1s written to them are pulled high by the internal pullups, and in that state can be used as inputs. As inputs, Port 6 pins that are externally pulled low will source current (I_{IL}, on the data sheet) because of the internal pullups. Port 6 emits the high-order address byte during fetches from external Program Memory if EBEN is pulled high. In this application it uses strong pullups when emitting 1s.
N/A	12	**EBEN**—E-Bus Enable input that designates whether program memory fetches take place via Ports 0 and 2 or Ports 5 and 6. Table 1 shows how the ports are used in conjunction with EBEN.
N/A	53	**$\overline{EPSEN}$**—E-bus Program Store Enable is the Read strobe to external program memory when EBEN is high. Table 2 shows when $\overline{EPSEN}$ is used relative to $\overline{PSEN}$ depending on the status of EBEN and $\overline{EA}$.

OSCILLATOR CHARACTERISTICS

XTAL1 and XTAL2 are the input and output, respectively, of an inverting amplifier which can be configured for use as an on-chip oscillator, as shown in Figure 3.

To drive the device from an external clock source, XTAL1 should be driven, while XTAL2 is left unconnected, as shown in Figure 4. There are no requirements on the duty cycle of the external clock signal, since the input to the internal clocking circuitry is through a divide-by-two flip-flop, but minimum and maximum high and low times specified on the Data Sheet must be observed.

An external oscillator may encounter as much as a 100 pF load at XTAL1 when it starts-up. This is due to interaction between the amplifier and its feedback capacitance. Once the external signal meets the V_{IL} and V_{IH} specifications the capacitance will not exceed 20 pF.

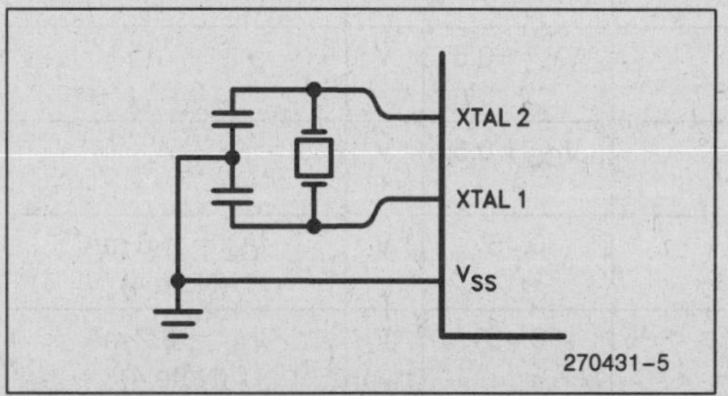

Figure 3. Using the On-Chip Oscillator

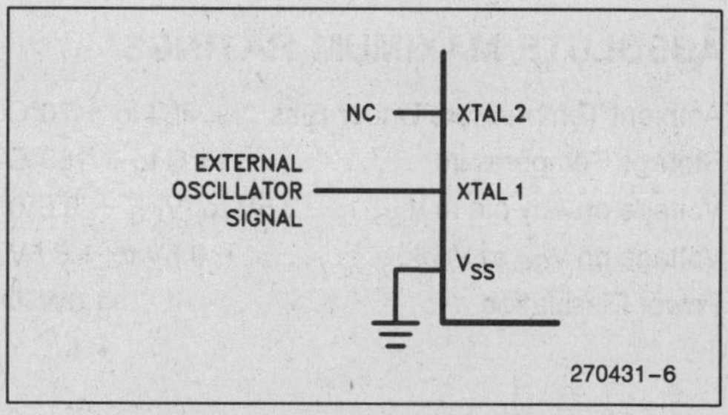

Figure 4. External Clock Drive

IDLE MODE

In Idle Mode, the CPU puts itself to sleep while most of the on-chip peripherals remain active. The major peripherals that do not remain active during Idle, are the DMA channels. The Idle Mode is invoked by software. The content of the on-chip RAM and all the Special Function Registers remain unchanged during this mode. The Idle Mode can be terminated by any enabled interrupt or by a hardware reset.

POWER DOWN MODE

In Power Down Mode, the oscillator is stopped and all on-chip functions cease except that the on-chip RAM contents are maintained. The mode Power Down is invoked by software. The Power Down Mode can be terminated only by a hardware reset.

2

Table 3. Status of the External Pins During Idle and Power Down Modes

80C152JA/83C152JA/80C152JC/83C152JC

Mode	Program Memory	ALE	$\overline{PSEN}$	Port 0	Port 1	Port 2	Port 3	Port 4
Idle	Internal	1	1	Data	Data	Data	Data	Data
Idle	External	1	1	Float	Data	Address	Data	Data
Power Down	Internal	0	0	Data	Data	Data	Data	Data
Power Down	External	0	0†	Float	Data	Data	Data	Data

80C152JB/80C152JD

Mode	Instruction Bus	ALE	$\overline{PSEN}$	$\overline{EPSEN}$	Port 0	Port 1	Port 2	Port 3	Port 4	Port 5	Port 6
Idle	P0, P2	1	1	1	Float	Data	Address	Data	Data	0FFH	0FFH
Idle	P5, P6	1	1	1	Data	Data	Data	Data	Data	0FFH	Address
Power Down	P0, P2	0	0	1	Float	Data	Data	Data	Data	0FFH	0FFH
Power Down	P5, P6	0	1†	0	Data	Data	Data	Data	Data	0FFH	0FFH

NOTE:
For more detailed information on the reduced power modes refer to the Embedded Controller Handbook, and Application Note AP-252, "Designing with the 80C51BH."
†Note difference of logic level of $\overline{PSEN}$ during Power Down for ROM JA/JC and ROM emulation mode for JC/JD.

ABSOLUTE MAXIMUM RATINGS*

Ambient Temperature Under Bias .... 0°C to +70°C

Storage Temperature .... −65°C to +150°C

Voltage on Any pin to V_{SS} .. −0.5V to (V_{CC} + 0.5V)

Voltage on V_{CC} to VSS .... −0.5V to +6.5V

Power Dissipation .... 1.0W(9)

NOTICE: This data sheet contains preliminary information on new products in production. The specifications are subject to change without notice. Verify with your local Intel Sales office that you have the latest data sheet before finalizing a design.

**WARNING: Stressing the device beyond the "Absolute Maximum Ratings" may cause permanent damage. These are stress ratings only. Operation beyond the "Operating Conditions" is not recommended and extended exposure beyond the "Operating Conditions" may affect device reliability.*

D.C. CHARACTERISTICS (T_A = 0°C to +70°C; V_{CC} = 5V ±10%; V_{SS} = 0V)

Symbol	Parameter	Min	Typ (Note 3)	Max	Unit	Test Conditions
V_{IL}	Input Low Voltage (All Except $\overline{EA}$, EBEN)	−0.5		$0.2V_{CC}-0.1$	V	
V_{IL1}	Input Low Voltage ($\overline{EA}$, EBEN)	−0.5		$0.2V_{CC}-0.3$	V	
V_{IH}	Input High Voltage (Except XTAL1, $\overline{RST}$)	$0.2V_{CC}+0.9$		$V_{CC}+0.5$	V	
V_{IH1}	Input High Voltage (XTAL1, $\overline{RST}$)	$0.7V_{CC}$		$V_{CC}+0.5$	V	
V_{OL}	Output Low Voltage (Ports 1, 2, 3, 4, 5, 6)			0.45	V	I_{OL} = 1.6 mA (Note 4)
V_{OL1}	Output Low Voltage (Port 0, ALE, $\overline{PSEN}$, $\overline{EPSEN}$)			0.45	V	I_{OL} = 3.2 mA (Note 4)
V_{OH}	Output High Voltage (Ports 1, 2, 3, 4, 5, 6 COMM9 ALE, $\overline{PSEN}$, $\overline{EPSEN}$)	2.4			V	I_{OH} = −60 μA V_{CC} = 5V ±10%
		$0.9V_{CC}$			V	I_{OH} = −10 μA
V_{OH1}	Output High Voltage (Port 0 in External Bus Mode)	2.4			V	I_{OH} = −400 μA V_{CC} = 5V ±10%
		$0.9V_{CC}$			V	I_{OH} = −40 μA (Note 5)
I_{IL}	Logical 0 Input Current (Ports 1, 2, 3, 4, 5, 6)			−50	μA	V_{IN} = 0.45V
I_{TL}	Logical 1 to 0 Transition Current (Ports 1, 2, 3, 4, 5, 6)			−650	μA	V_{IN} = 2V
I_{LI}	Input Leakage (Port 0, $\overline{EA}$)			±10	μA	$0.45<V_{IN}<V_{CC}$
RRST	Reset Pullup Resistor	40			kΩ	
I_{IH}	Logical 1 Input Current (EBEN)			+60	μA	
I_{CC}	Power Supply Current:					
	Active (16.5 MHz)		31	41.1	mA	(Note 6)
	Idle (16.5 MHz)		8	15.4	mA	(Note 6)
	Power Down Mode		10		μA	V_{CC} = 2.0V to 5.5V

MAX I_{CC} (ACTIVE) = (2.24 × FREQ) + 4.16 (Note 6)
MAX I_{CC} (IDLE) = (0.8 × FREQ) + 2.2 (Note 6)

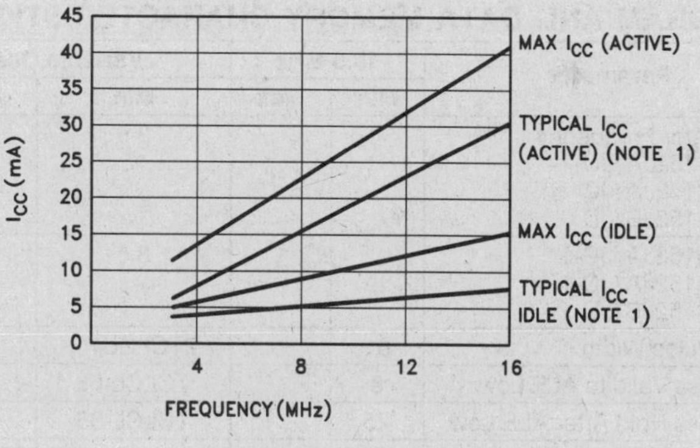

Figure 5. I_{CC} vs Frequency

EXPLANATION OF THE AC SYMBOLS

Each timing symbol has 5 characters. The first character is always a 'T' (stands for time). The other characters, depending on their positions, stand for the name of a signal or the logical status of that signal. The following is a list of all the characters and what they stand for.

A: Address.
C: Clock
D: Input data.
H: Logic level HIGH.
I: Instruction (program memory contents).
L: Logic level LOW, or ALE.
P: $\overline{PSEN}$.
Q: Output data.
R: $\overline{READ}$ signal.
T: Time.
V: Valid.
W: $\overline{WRITE}$ signal.
X: No longer a valid logic level.
Z: Float.

For example,

TAVLL = Time for Address Valid to ALE Low.
TLLPL = Time for ALE Low to $\overline{PSEN}$ Low.

A.C. CHARACTERISTICS (T_A = 0°C to +70°C; V_{CC} = 5V ±10%; V_{SS} = 0V; Load Capacitance for Port 0, ALE, and $\overline{PSEN}$ = 100 pF; Load Capacitance for All Other Outputs = 80 pF)

EXTERNAL PROGRAM AND DATA MEMORY CHARACTERISTICS (Note 7, 10)

Symbol	Parameter	16.5 MHz		Variable Oscillator		Unit
		Min	Max	Min	Max	
1/TCLCL	Oscillator Frequency 80C152JA/JC 83C152JA/JC 80C152JB/JD			3.5	12	MHz
	80C152JA/JC-1 83C152JA/JC-1 80C152JB/JD-1			3.5	16.5	MHz
TLHLL	ALE Pulse Width	81		2TCLCL-40		ns
TAVLL	Address Valid to ALE Low	5		TCLCL-55		ns
TLLAX	Address Hold After ALE Low	25		TCLCL-35		ns
TLLIV	ALE Low to Valid Instruction In		142		4TCLCL-100	ns
TLLPL	ALE Low to $\overline{PSEN}$ Low	20		TCLCL-40		ns
TPLPH	$\overline{PSEN}$ Pulse Width	137		3TCLCL-45		ns
TPLIV	$\overline{PSEN}$ Low to Valid Instruction In		77		3TCLCL-105	ns
TPXIX	Input Instruction Hold After $\overline{PSEN}$	0		0		ns
TPXIZ	Input Instruction Float After $\overline{PSEN}$		35		TCLCL-25	ns
TAVIV	Address to Valid Instruction In		198		5TCLCL-105	ns
TPLAZ	$\overline{PSEN}$ Low to Address Float		10		10	ns
TRLRH	$\overline{RD}$ Pulse Width	263		6TCLCL-100		ns
TWLWH	$\overline{WR}$ Pulse Width	263		6TCLCL-100		ns
TRLDV	$\overline{RD}$ Low to Valid Data In		138		5TCLCL-165	ns
TRHDX	Data Hold After $\overline{RD}$	0		0		ns
TRHDZ	Data Float After $\overline{RD}$		51		2TCLCL-70	ns
TLLDV	ALE Low to Valid Data In		335		8TCLCL-150	ns
TAVDV	Address to Valid Data In		380		9TCLCL-165	ns
TLLWL	ALE Low to $\overline{RD}$ or $\overline{WR}$ Low	132	232	3TCLCL-50	3TCLCL+50	ns
TAVWL	Address to $\overline{RD}$ or $\overline{WR}$ Low	112		4TCLCL-130		ns
TQVWX(8)	Data Valid to $\overline{WR}$ Transition	196		6TCLCL-167		ns
TWHQX	Data Hold After $\overline{WR}$	10		TCLCL-50		ns
TRLAZ	$\overline{RD}$ Low to Address Float		0		0	ns
TWHLH	$\overline{RD}$ or $\overline{WR}$ High to ALE High	20	100	TCLCL-40	TCLCL+40	ns

EXTERNAL PROGRAM MEMORY READ CYCLE

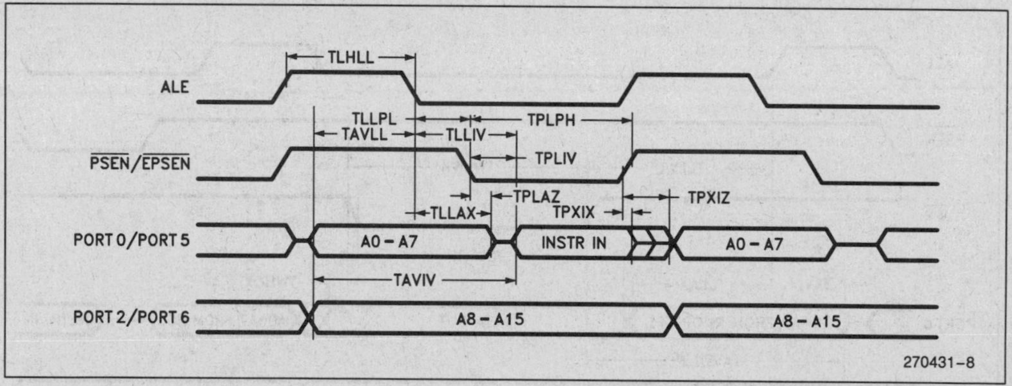

EXTERNAL DATA MEMORY READ CYCLE

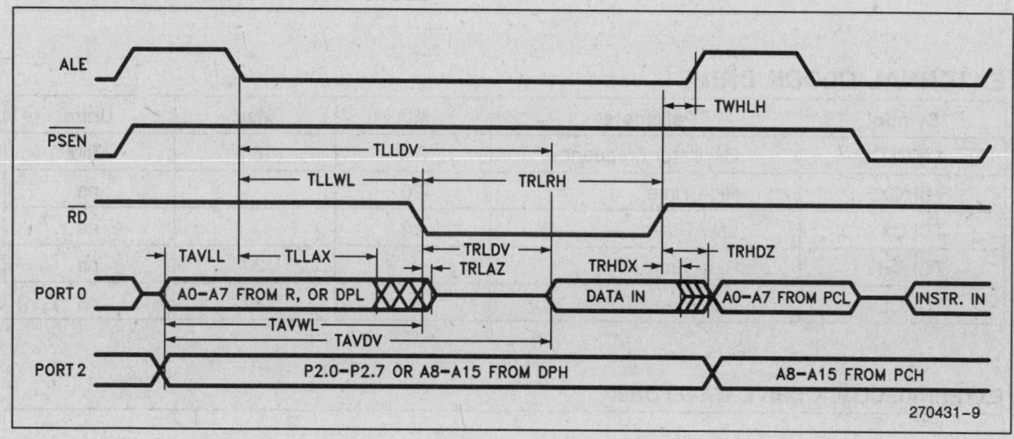

EXTERNAL DATA MEMORY WRITE CYCLE

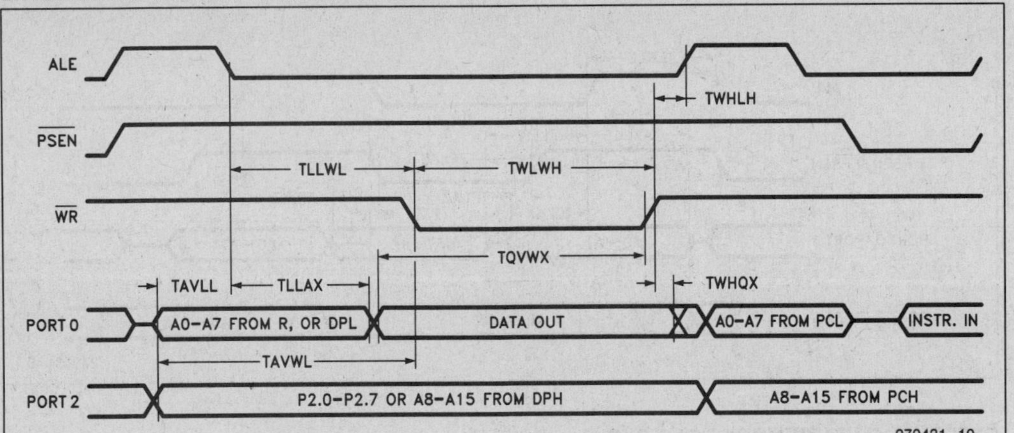

270431–10

EXTERNAL CLOCK DRIVE

Symbol	Parameter	Min	Max	Units
1/TCLCL	Oscillator Frequency	3.5	16.5	MHz
TCHCX	High Time	20		ns
TCLCX	Low Time	20		ns
TCLCH	Rise Time		20	ns
TCHCL	Fall Time		20	ns

EXTERNAL CLOCK DRIVE WAVEFORM

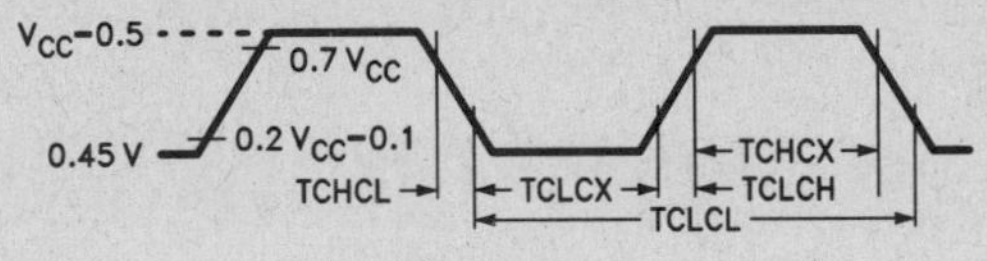

270431–11

LOCAL SERIAL CHANNEL TIMING—SHIFT REGISTER MODE

Symbol	Parameter	16.5 MHz		Variable Oscillator		Units
		Min	Max	Min	Max	
TXLXL	Serial Port Clock Cycle Time	727		12TCLCL		ns
TQVXH	Output Data Setup to Clock Rising Edge	473		10TCLCL-133		ns
TXHQX	Output Data Hold After Clock Rising Edge	4		2TCLCL-117		ns
TXHDX	Input Data Hold After Clock Rising Edge	0		0		ns
TXHDV	Clock Rising Edge to Input Data Valid		473		10TCLCL-133	ns

2

SHIFT REGISTER MODE TIMING WAVEFORMS

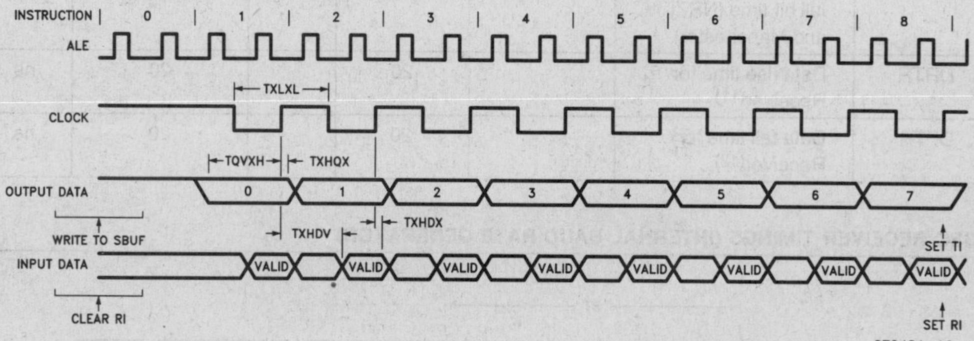

270431-12

A.C. TESTING:

INPUT, OUTPUT WAVEFORMS

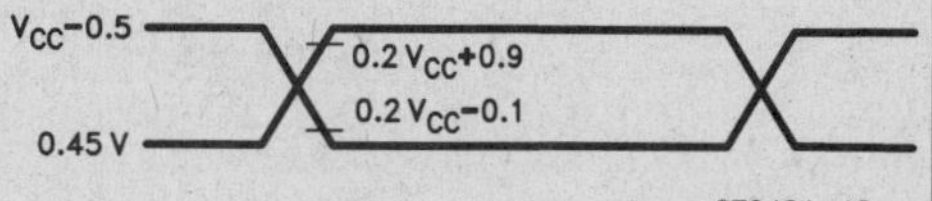

270431-13

AC Inputs During Testing are Driven at $V_{CC}-0.5$ for a Logic "1" and 0.45V for a Logic "0". Timing Measurements are made at V_{IH} Min for a Logic "1" and V_{IL} Max for a Logic "0".

FLOAT WAVEFORM

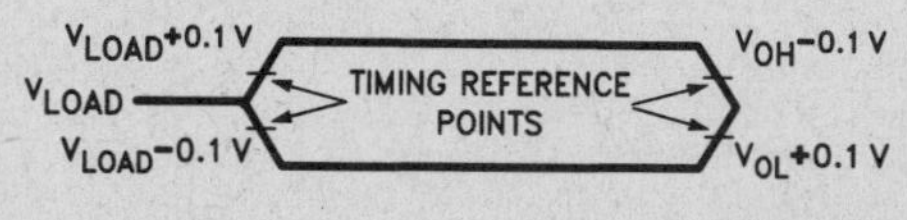

270431-14

For Timing Purposes a Port Pin is no Longer Floating when a 100 mV change from Load Voltage Occurs, and Begins to Float when a 100 mV change from the Loaded V_{OH}/V_{OL} Level occurs $I_{OL}/I_{OH} \geq \pm 20$ mA.

GLOBAL SERIAL PORT TIMINGS—Internal Baud Rate Generator

Symbol	Parameter	16.5 MHz (BAUD = 0)		Variable Oscillator		Unit
		Min	Max	Min	Max	
HBTJR	Allowable jitter on the Receiver for ½ bit time (Manchester encoding only)		0.0375		(0.125 × (BAUD + 1) × 8TCLCL) − 25 ns	μs
FBTJR	Allowable jitter on the Receiver for one full bit time (NRZI and Manchester)		0.10		(0.25 × (BAUD + 1) × 8TCLCL) − 25 ns	μs
HBTJT	Jitter of data from Transmitter for ½ bit time (Manchester encoding only)		±10		±10	ns
FBTJT	Jitter of data from Transmitter for one full bit time (NRZI and Manchester)		±10		±10	ns
DRTR	Data rise time for Receiver(11)		20		20	ns
DFTR	Data fall time for Receiver(12)		20		20	ns

GSC RECEIVER TIMINGS (INTERNAL BAUD RATE GENERATOR)

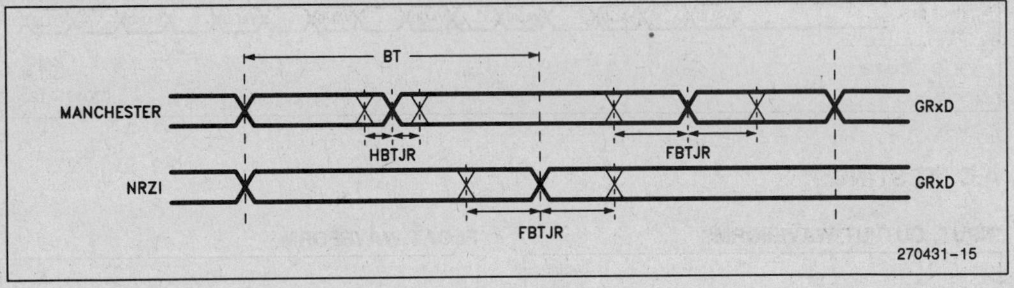

GSC TRANSMIT TIMINGS (INTERNAL BAUD RATE GENERATOR)

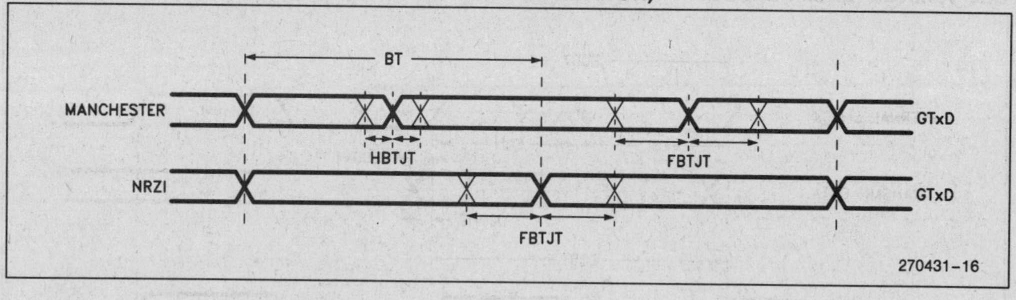

GLOBAL SERIAL PORT TIMINGS—External Clock

Symbol	Parameter	16.5 MHz		Variable Oscillator		Unit
		Min	Max	Min	Max	
1/ECBT	GSC Frequency with an External Clock		2.4	0.009	$F_{OSC} \times 0.145$	MHz
ECH	External Clock High	170		2TCLCL + 45 ns		ns
ECL[13]	External Clock Low	170		2TCLCL + 45 ns		ns
ECRT	External Clock Rise Time[11]		20		20	ns
ECFT	External Clock Fall Time[12]		20		20	ns
ECDVT	External Clock to Data Valid Out - Transmit (to External Clock Negative Edge)		150		150	ns
ECDHT	External Clock Data Hold - Transmit (to External Clock Negative Edge)	0		0		ns
ECDSR	External Clock Data Set-up - Receiver (to External Clock Positive Edge)	45		45		ns
ECDHR	External Clock to Data Hold - Receiver (to External Clock Positive Edge)	50		50		ns

GSC TIMINGS (EXTERNAL CLOCK)

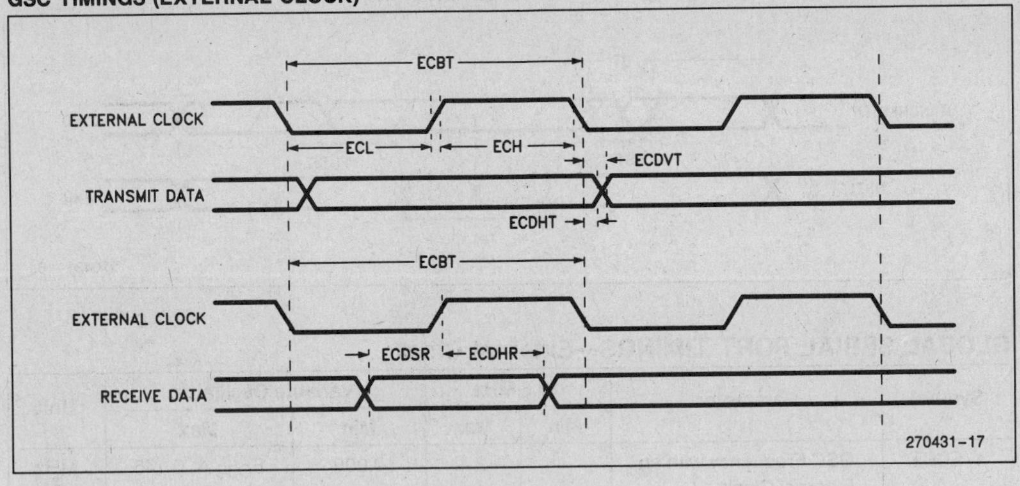

270431-17

NOTES:

1. N.C. pins on PLCC package may be connected to internal die and should not be used in customer applications.
2. It is recommended that both Pin 3 and Pin 33 be grounded for PLCC devices.
3. "Typicals" are based on samples taken from early manufacturing lots and are not guaranteed. The measurements were made with V_{CC} = 5V at room temperature.
4. Capacitive loading on Ports 0 and 2 may cause spurious noise pulses to be superimposed on the V_{OL}s of ALE and Ports 1 and 3. The noise is due to external bus capacitance discharging into the Port 0 and Port 2 pins when these pins make 1-to-0 transitions during bus operations. In the worst cases (capacitive loading > 100 pF), the noise pulse on the ALE pin may exceed 0.8V. In such cases it may be desirable to qualify ALE with a Schmitt Trigger, or use an address latch with a Schmitt Trigger STROBE input.
5. Capacitive loading on Ports 0 and 2 may cause the V_{OH} on ALE and $\overline{PSEN}$ to momentarily fall below the $0.9V_{CC}$ specification when the address bits are stabilizing.
6. I_{CC} is measured with all output pins disconnected; XTAL1 driven with TCLCH, TCHCL = 5 ns, $V_{IL} = V_{SS} + 0.5V$, $V_{IH} = V_{CC} - 0.5V$; XTAL2 N.C.; Port 0 pins connected to V_{CC}. "Operating" current is measured with $\overline{EA}$ connected to V_{CC} and $\overline{RST}$ connected to V_{SS}. "Idle" current is measured with $\overline{EA}$ connected to V_{SS}, $\overline{RST}$ connected to V_{CC} and GSC inactive.
7. The specifications relating to external data memory characteristics are also applicable to DMA operations.
8. TQVWX should not be confused with TQVWX as specified for 80C51BH. On 80C152, TQVWX is measured from data valid to rising edge of $\overline{WR}$. On 80C51BH, TQVWX is measured from data valid to falling edge of $\overline{WR}$. See timing diagrams.
9. This value is based on the maximum allowable die temperature and the thermal resistance of the package.
10. All specifications relating to external program memory characteristics are applicable to:
    $\overline{EPSEN}$ for $\overline{PSEN}$
    Port 5 for Port 0
    Port 6 for Port 2
    when EBEN is at a Logical 1 on the 80C152JB/JD.
11. Same as TCLCH, use External Clock Drive Waveform.
12. Same as TCHCL, use External Clock Drive Waveform.
13. When using the same external clock to drive both the receiver and transmitter, the minimum ECL spec effectively becomes 195 ns at all frequencies (assuming 0 ns propagation delay) because ECDVT (150 ns) plus ECDSR (45 ns) requirements must also be met (150 + 45 = 195 ns). The 195 ns requirement would also increase to include the maximum propagation delay between receivers and transmitters.

DESIGN NOTES

Within the 8XC152 there exists a race condition that may set both the RDN and AE bits at the end of a valid reception. This will not cause a problem in the application as long as the following steps are followed:

—Never give the receive error interrupt a higher priority than the valid reception interrupt

—Do not leave the valid reception interrupt service routine when AE is set by using a RETI instruction until AE is cleared. To clear AE set the GREN bit, this enables the receiver. If the user desires that the receiver remain disabled, clear GREN after setting it before leaving the interrupt service routine.

—If the AE bit is checked by user software in response to a valid reception interrupt, the status of AE should be considered invalid.

The race condition is dependent upon both the temperature that the device is currently operating at and the processing the device received during the wafer fabrication.

When the idle mode is terminated by a hardware reset, the device normally resumes program execution, from where it left off, up to two machine cycles before the internal reset algorithm takes control. On-chip hardware inhibits access to internal RAM in this event, but access to the port pins is not inhibited. To eliminate the possibility of an unexpected write when Idle is terminated by reset, the instruction following the one that invokes Idle should not be one that writes to a port pin or to external memory.

DATA SHEET REVISION SUMMARY

The following represent the key differences between the "-003" and the "-002" version of the 80C152/83C152 data sheet. Please review this summary carefully.

1. Removed minimum GSC frequency spec when used with an external clock.
2. Change figure "External Program Memory Read Cycle" to show Port 0/Port 5 address floating after PSEN goes low.
3. Added design note on terminating idle with reset.
4. Added status of PSEN during Power Down mode to Table 3.
5. Moved all notes to back of data sheet.
6. Changed microcomputer to microcontroller.
7. Added External Oscillator start-up capacitance note.

The following represent the key differences between the "-002" and the "-001" version of the 80C152/83C152 data sheet. Please review this summary carefully.

1. Status of data sheet changed from "ADVANCED" to "PRELIMINARY".
2. 80C152JC, 83C152JC, and 80C152JD were added.
3. Added AE/RDN design note.
4. This revision summary was added.
5. Note #13 was added (Effective ECL spec at higher clock rates).
6. Table #2 changed to Table #3 (Status of pins during Idle/Power Down).
7. Current Table #2 was added (JA vs. JB vs. JC vs. JD matrix).
8. Transmit jitter spec changed from ±35 ns and ±70 ns to ±10 ns.

3

MCS® 251 Microcontroller

3

intel®

MCS® 251 Architecture Overview

- Binary Code Compatible with MCS 51® Architecture
- 5 to 15 Times Increase in Performance Compared to MCS 251 Microcontroller at the Same Clock Speed
- 3-Stage Pipeline CPU Architecture
- 2 Clocks (1 State) per Instruction
- 16-Bit Internal Code Bus
  - 2 Bytes/State Code Fetch
- 8-Bit ALU
  - High Speed 8-Bit Source and 16-Bit Destination Bus
- 40 Bytes General Purpose Register File
  - Accessible as 16 8-Bit, 16 16-Bit or 10 32-Bit Registers
  - With Accumulator Functionality and Data Indexing Capability
- 24-Bit Linear Code and Data Addressing
- 64KB Stack Space
- New Instructions and Addressing Modes
  - 8, 16 and Limited 32-Bit Data Transfer, Arithmetic and Logical Instructions
  - Supports Register, Immediate, Direct, Indirect, Displacement, Relative and Bit Addressing
- Support 64 Interrupt Sources
  - 1 TRAP Instruction Interrupt (Highest Priority)
  - 1 Non-Maskable Interrupt (2nd Highest Priority)
  - 62 Maskable Interrupt Sources
  - 4 Interrupt Priority Levels
- Complete Tools Support
  - Assemblers and C Compilers
  - In-Circuit Emulator
  - OTP/EPROM Programmer
  - Debuggers

3

MCS 251 architecture is the next generation of Intel MCS 51 architecture that increases system performance by a factor of five using existing MCS 51 microcontroller code. By rewriting code using the MCS 251 architecture instructions, designers can increase performance up to 15 times.

The new MCS 251 architecture brings high performance, an increased memory mix and addressing, low power, low noise, efficient high level language support, enhanced instruction set, integrated features and functionality to the 8-bit embedded control market segment. All products in the family will be built around the new MCS 251 architecture core. They will have various peripherals, memories, input/ouput (I/O) ports, and a BIU (Bus Interface Unit). Most importantly, new proliferations will be binary code-compatible with existing 8051 components. Software investment is protected while providing an easy performance upgrade path. (See table 1 for features and benefits listing.)

The first product based on the new architecture will be 80C51 pin-compatible and a direct "plug-in" replacement. Anticipated availability of general samples is Q1 of 1995 with additional proliferations to follow later in the year.

Development tools for the new architecture will be available from some of the leading tools vendors. The vendors, including BSO Tasking; Data I/O Corporation; Keil Elektronik GmbH/Franklin Software Inc.; Metalink Corporation; Needham; Nohau Corporation; Production Languages Corporation (PLC); SMS Mikrocomputer-Systeme GmbH; and System General Corporation are planning to supply a full range of hardware and software development tools such as emulators, compilers, programmers and debuggers.

November 1994
Order Number: 272620-001

Table 1. Features and Benefits

Features	Benefits
• 3-stage pipeline CPU architecture • 1 state (2 clocks) per machine cycle (vs 6 states (12 clocks) per machine cycle for MCS® 51 microcontroller) • 16-bit internal code bus	• High performance. 5 to 15 times increase in performance compared to MCS 51 microcontroller at the same clock speed • High instruction throughput at low clock speed reduces power consumption and RFI

• Enhanced MCS 51 instruction set with:
  • 16-bit and limited 32-bit data transfer, arithmetic and logic instructions
  • Register-to-register operations
  • Extended addressing modes
  • Improved control instructions
  • Bigger bit addressable space
	• Increased performance and programming flexibility.
	• Reduced code size.

• Binary code-compatible with MCS 51 microcontroller	• Protects software investment • Easy performance upgrade from MCS 51 microcontroller applications
• Register-based machine with 40 register bytes accessible as 16 8-bit registers, 16 16-bit registers, 10 32-bit registers or a combination of all. All registers are general-purpose with accumulator functionality and data indexing capability.	• Increased performance and programming flexibility • Increased efficiency for C code
• 24-bit linear addressing for up to 16 MB memory space	• Support bigger code and data memory requirement.
• 64 KB extended stack space and additional stack instructions	• Increased efficiency for C code

CPU Overview

The central processing unit (CPU) represents the heart of the MCS 251 architecture and consists of the ALU, instruction sequencer, program counter and the register file, connected by high-speed source and destination buses. The CPU performs all instructions as a series of 8-bit microcontroller operations. The CPU is implemented using typical pipelining techniques, and is built around a three-stage pipeline. The pipe stages are instruction fetch or decode, address generation or data fetch, and execution or write back. The three-stage pipeline implementation offers the best trade-off between performance and design complexity.

Memory Interface

The CPU interfaces with the peripherals, memories and other chip units via internal instruction and data buses. The MCS 251 architecture has an internal 16-bit wide instruction bus capable of sustained 2 bytes/state code fetches. It also has an 8-bit data bus capable of one byte/state data transfer rate.

The MCS 251 architecture has one contiguous 16MB address space that is used for both code and data. The 16MB address space will be partitioned for internal and external access, depending on the amount of on-chip memory.

Code memory can reside anywhere in the address space except for reserved areas. Further restrictions may prevent code execution from certain locations that may vary from product to product. The code memory resides outside the CPU and will be partitioned as internal and external memory, depending on the amount of on-chip code memory.

Data memory can reside anywhere in the address space except for reserved areas. The lower 32 bytes of the data memory space can be accessed as both data memory and as four banks of general-purpose registers. All MCS 251 controllers will have this memory as part of the CPU; the amount of additional on-chip data memory may vary from product to product.

To provide flexibility to designers during development and production, the MCS 251 microcontroller will have one-time programmable (OTP), ROM and CPU-only versions available. Flash memory versions will be available in the future.

The MCS 251 architecture also supports an extra 32 bytes of general-purpose registers which reside in the CPU as register file. This is in addition to the 4 banks of 8 registers found on the MCS 51 microcontroller architecture. The register file can be addressed in the following ways, depending upon the registers to be accessed: registers 0–15 can be addressed as either byte, word or double word (Dword) registers; registers 16–31 can be addressed as either word or Dword registers; and registers 56–63 can be addressed only as Dword registers. There are 16 possible byte registers, 16 possible word registers and 10 possible Dword registers depending on the combinations. (See Figure 1 for register file diagram.)

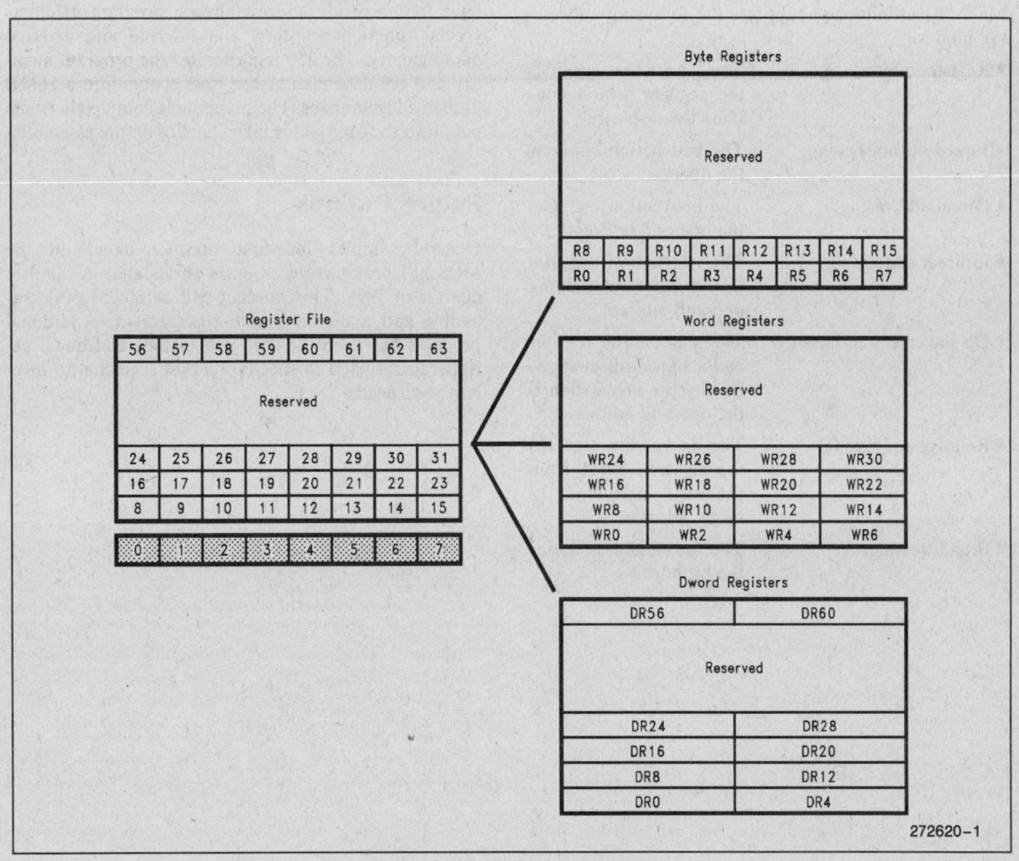

Figure 1. Register File Diagram

New Instruction Set

The MCS 251 architecture instruction set provides the user with newly-defined instructions that take advantage of the new architecture while maintaining all existing MCS 51 microcontroller instructions. Many of the MCS 251 microcontroller instructions can operate on either 8-, 16-, or limited 32-bit operands. This capability allows the MCS 251 architecture proliferations to be used more easily and efficiently with high-level programming languages such as "C."

The instruction set consists of data instructions, bit instructions and control instructions. Data instructions process 8-, 16-, and limited 32-bit data; bit instructions manipulate bits; and control instructions manage program flow. Each instruction type has a unique set of addressing modes; for example, not all data addressing modes apply to control instructions and vice-versa. The MCS 251 architecture supports the following addressing modes:

- Register addressing: The instruction specifies the register which contains the operand.
- Immediate addressing: The instruction contains the operand.
- Direct addressing: The instruction contains the operand address.
- Indirect addressing: The instruction specifies the register containing operand address.
- Displacement addressing: The sum of the register and a signed offset specified by the instruction is the operand address.
- Relative addressing: The instruction contains the relative offset from the next instruction to target address.
- Bit addressing: The instruction contains the bit address.

Interrupt Overview

The MCS 251 architecture supports one non-maskable interrupt, one TRAP (technical return analysis program) instruction interrupt and up to 62 maskable interrupt sources. The user can select each interrupt individually as well as control its priority level. The non-maskable interrupt is fixed in hardware to always have the second highest priority (after TRAP) and is always enabled. The exact number of interrupt sources, both internal and external, depends on the specific MCS 251 architecture proliferations.

Compatibility

The MCS 251 architecture is code-compatible with the MCS 51 architecture. And, all 51 microcontroller instructions are available on the 251. The 51 controllers have four separate address spaces: program memory, special function registers, and internal and external data memory. The 251 incorporates the program memory and the data memory address spaces into a 16MB unified address space. The mapping is completely transparent to the user and is taken care of by the assembler.

Future Outlook

General samples for first product based on the MCS 251 architecture core are anticipated in the first quarter of 1995. This product will be 80C51 pin-compatible and a direct plug-in replacement. Additional products based on the core will include additional peripherals targeted to specific market segments to meet customer needs.

NCS® 251 Architecture Block Diagram

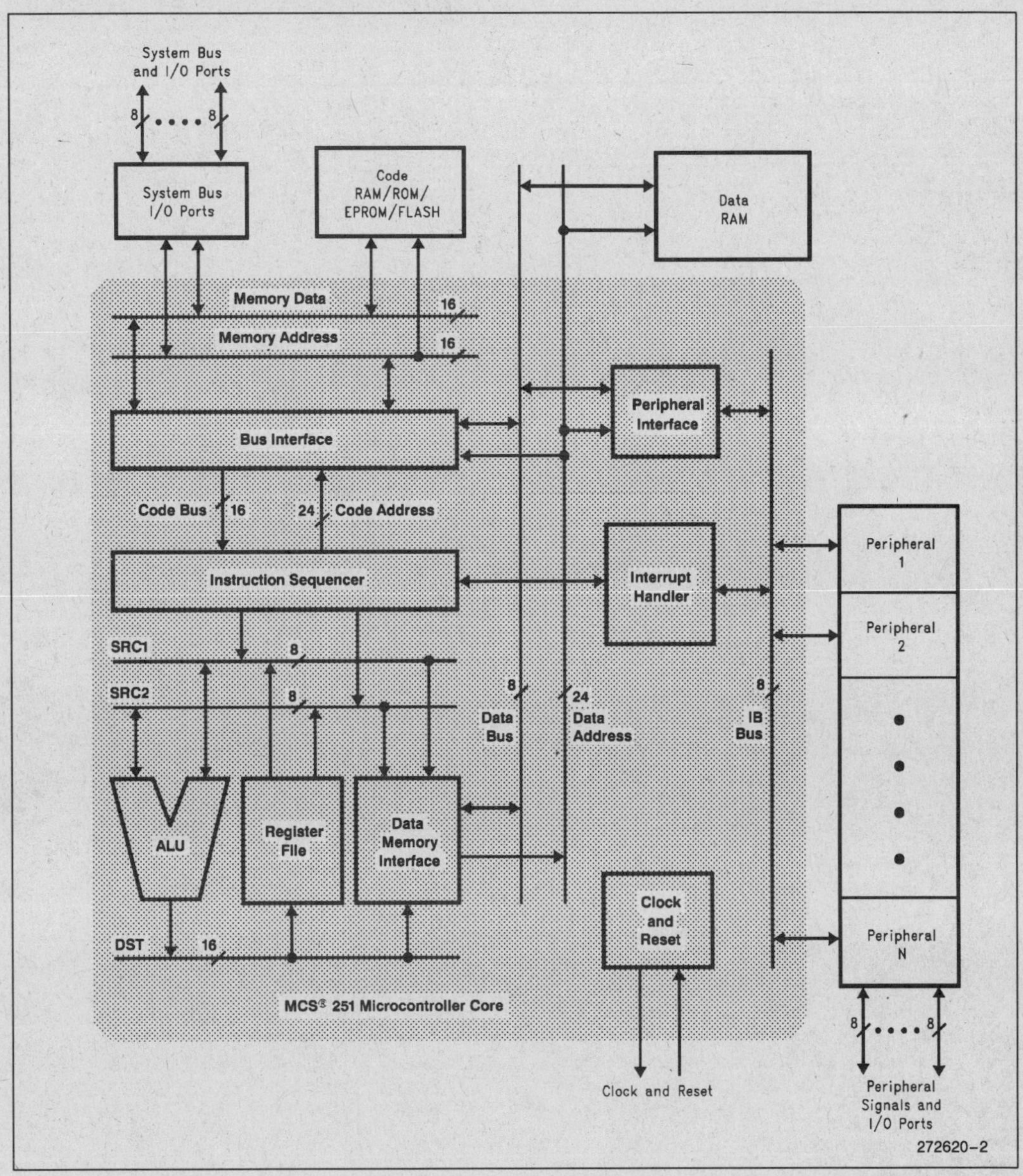

intel®

4

MCS® 96 Microcontroller Family

4

MCS®-96 Architectural Overview

September 1992

Order Number: 272109-002

MCS-96 Architectural Overview

CONTENTS PAGE

CONTENTS PAGE

1.0 INTRODUCTION

The MCS-96 family members are all high performance microcontrollers with a 16-bit CPU and at least 230 bytes of on-chip RAM. The Intel MCS-96 family easily handles high speed calculations and fast input/output (I/O) operations. Typical applications include closed-loop control and mid-range digital signal processing. Modems, motor control system, printers, engine control system, photocopiers, anti-lock brakes, air conditioner control systems, disk drives and medical instrumentation all use MCS-96 products.

All of the MCS-96 components share a common instruction set and architecture. However, the CHMOS components have enhancements to provide higher performance with lower power consumption. To further decrease power usage, idle and power-down modes are available on these devices. These microcontrollers contain dedicated I/O subsystems and perform 16-bit arithmetic instructions including multiply and divide operations.

This overview briefly describes the MCS-96 instruction set and architecture and provides descriptions for the 8X9X, 80C196KB, 80C196KC and 80C196KR key features. Comprehensive user's guides that contain more information about these devices are available. Figure 1.1 shows a block diagram of the MCS-96 architecture.

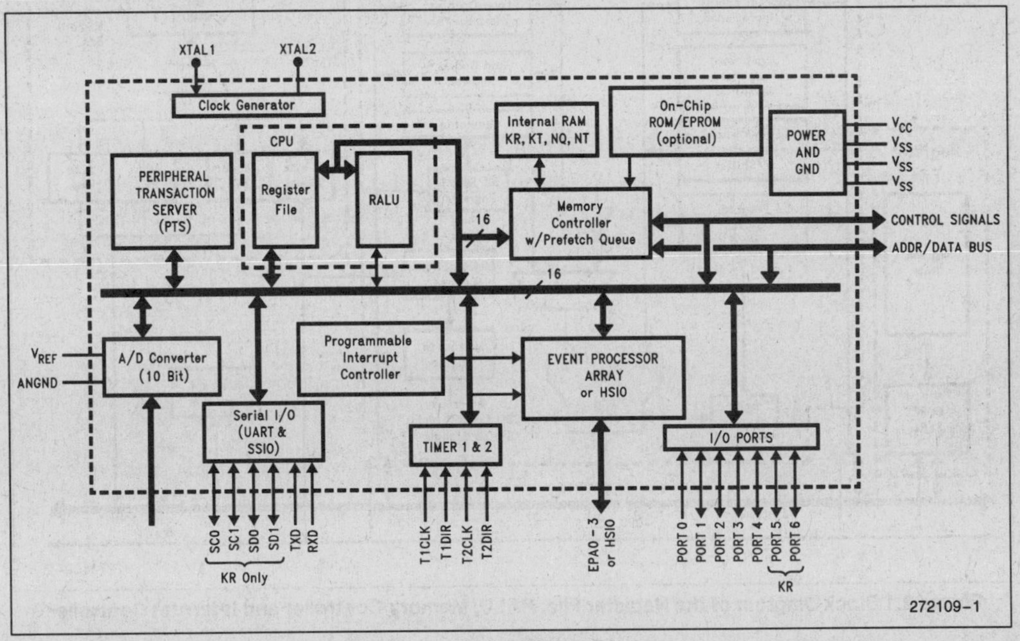

Figure 1.1. MCS-96 Block Diagram

2.0 THE CPU

The major components of the MCS-96 CPU are the Register File and the Register/Arithmetic Logic Unit (RALU). Locations 00H through 17H are the I/O control registers or Special Function Registers (SFRs). Locations 18H and 19H contain the stack pointer, which can serve as general purpose RAM when not performing stack operations. The remaining bytes of the register file serve as general purpose RAM, accessible as bytes, words or double-words.

Calculations performed by the CPU take place in the RALU. The RALU shown in Figure 2.1 contains a 17-bit ALU, the Program Status Word (PSW), the Program Counter (PC), a loop counter and three temporary registers. The RALU operates directly on the Register File, thus eliminating accumulator bottleneck and providing for direct control of I/O operations through the SFRs.

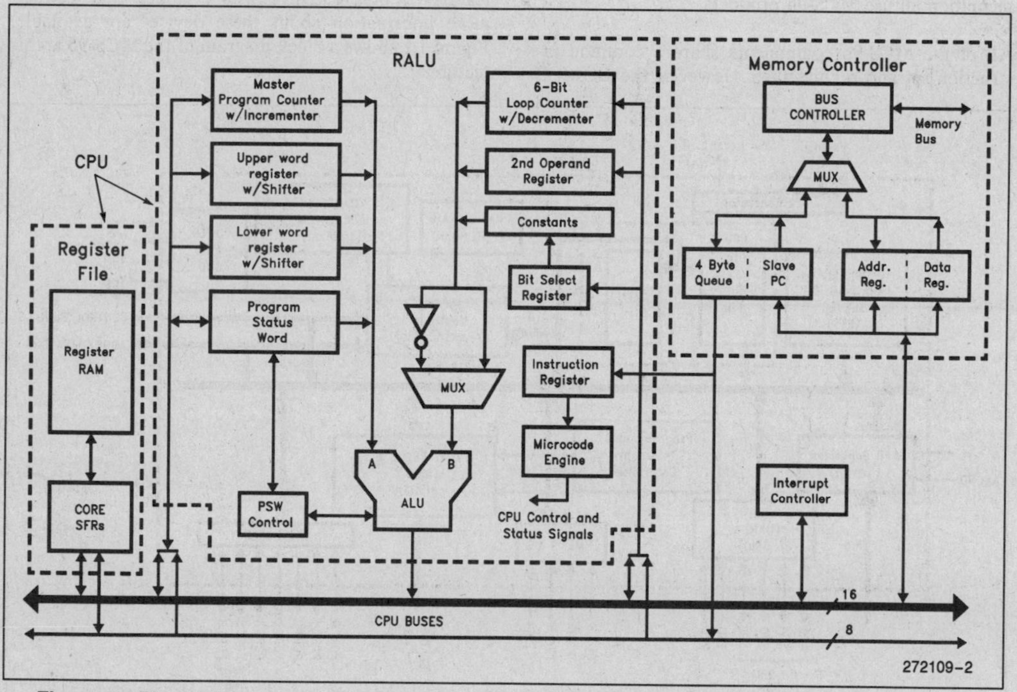

Figure 2.1 Block Diagram of the Register File, RALU, Memory Controller and Interrupt Controller

3.0 THE ARCHITECTURE

The MCS-96 supports a complete instruction set which includes bit operations, byte operations, word operations, double-word operations (unsigned 32-bit), long operations (signed 32-bit), flag manipulations as well as jump and call instructions. All the standard logical and arithmetic instructions function as both byte and word operations. The Jump Bit Set and Jump Bit Clear instructions can operate on any of the SFRs or bytes in the lower register file. These fast bit manipulations allow for rapid I/O functions.

Byte and word operations make-up most of the instruction set. The assembly language ASM-96 uses a "B" suffix on a mnemonic for a byte operation, otherwise the mnemonic refers to a word operation. One, two or three operand forms exist for many of the instructions.

Long and double-word operations include shifts, normlize, multiply and divide. The divide instruction functions as a 32-bit by 16-bit divide that generates a 16-bit quotient and 16-bit remainder. The word multiply operates as a 16-bit by 16-bit multiply with a 32-bit result. Both operations can function in either the signed or unsigned mode. The normalize instruction and sticky bit flag provide hardware support for the software floating point package (FPAL-96).

3.1 Addressing Modes

The MCS-96 instruction set supports the following addressing modes: register-direct, indirect, indirect with auto-increment, immediate, short-indexed and long-indexed. These modes increase the flexibility and overall execution speed of the MCS-96 devices. Each instruction uses at least one of the addressing modes. These modes and formats are shown in Figure 3.1.

```
Mnem    Dest or Src1              ;One operand direct
Mnem    Dest, Src1                ;Two operand direct
Mnem    Dest, Src1, Src2          ;Three operand direct

Mnem    #Src1                     ;One operand immediate
Mnem    Dest, #Src1               ;Two operand immediate
Mnem    Dest, Src1, #Src2         ;Three operand immediate

Mnem    [addr]                    ;One operand indirect
Mnem    [addr]+                   ;One operand indirect auto-increment
Mnem    Dest, [addr]              ;Two operand indirect
Mnem    Dest, [addr]+             ;Two operand indirect auto-increment
Mnem    Dest, Src1, [addr]        ;Three operand indirect
Mnem    Dest, Src1, [addr]+       ;Three operand indirect auto-increment

Mnem    Dest, offs[addr]          ;Two operand indexed (short or long)
Mnem    Dest, Src1, offs[addr]    ;Three operand indexed (short or long)

Where:
Mnem = instruction mnemonic
Dest = destination register
Src1, Src2 = source registers
addr = word register used in computing the address of an operand
offs = offset used in computing the address of an operand
```

Figure 3.1 Instruction Format

4

The register-direct and immediate addressing modes execute faster than the other addressing modes. The register-direct addressing mode provides access to the addresses in the register file and the SFRs. The indexed modes provide for direct access to the remainder of the 64K address space. Immediate addressing uses the data following the opcode as the operand.

Both of the indirect addressing modes use the value in a word register as the address of the operand. The indirect auto-increment mode increments a word address by one after a byte operation and two after a word operation. This addressing mode provides easy access into look-up tables.

The long-indexed addressing mode provides direct access to any of the locations in the 64K address space. This mode forms the address of the operand by adding a 16-bit 2's complement value to the contents of a word register. Indexing with the zero register allows "direct" addressing to any location. The short-indexed addressing mode forms the address of the operand by adding an 8-bit 2's complement value to the contents of a word register.

The 8XC196NT has 9 new instructions which have been implemented to support addressing the extended 1 Mbyte address space of the 8XC196NT family. Four extended load and store instructions using indirect, indirect auto increment, or extended indexed addressing, can be used to address the 1 Mbyte address space. Three instructions are for extended calls, branches, and jumps. An extended version of the interruptible and non-interruptible block moves have also been implemented.

The multiple addressing modes of the MCS-96 make it easy to program in assembly language and provide an excellent interface to high-level languages. The instructions accepted by the assembler consist of mnemonics followed by either addresses or data. Refer to the Quick Reference section for Instruction Summary tables for each device. The MCS-96 Macro Assembler Users Guide contains additional ASM96 information.

4.0 8X9X PERIPHERALS

Standard I/O Ports—The 8X9X has five 8-bit I/O ports. Port 0 is an input port that is also the anlog input for the A/D converter. Port 1 is a quasi-bidirectional port. Port 2 contains three types of port lines: quasi-bidirectional, input and output. Other functions on the 8X9X share the input and output lines with Port 2. Ports 3 and 4 are open-drain bidirectional ports that share their pins with the address/data bus.

Timers—The 8X9X has two 16-bit timers, Timer1 and Timer2. An internal clock increments the Timer1 value every 8 state times. (A state time is 3 oscillator periods.) An external clock increments Timer2 on every positive and negative transition. Either an internal or external source can reset Timer2. Timer1 and Timer2 can generate an interrupt when crossing the 0FFFFH/0000H boundary. The 8X9X also includes separate, dedicated timers for the Serial Port baud rate generator and Watchdog Timer. The Watchdog Timer is an internal timer that resets the system if the software fails to operate properly.

Table 1. MCS-96 Family Devices

Product	Timers	HSIO /EPA	A/D CHS	Serial Port	Synch. Serial Port	PWMS	PTS	Slave Port	3-Phase Waveform Generator
8098	2	HSIO	4	YES		1			
8097BH	2	HSIO	8	YES		1			
8097JF	2	HSIO	8	YES		1			
80C198	2	HSIO	4	YES		1			
80C196KB	2	HSIO	8	YES		1			
80C196KC	2	HSIO	8	YES		3	YES		
80C196KD	2	HSIO	8	YES		3	YES		
80C196KR	2	EPA	8	YES	YES		YES	YES	
80C196KT	2	EPA	8	YES	YES		YES	YES	
80C196NT	2	EPA	4	YES	YES		YES	YES	
80C196MC	2	EPA	13	YES			YES		YES

High Speed Input Unit (HSI)—The 8X9X HSI unit can record times of external events with a 9 state time resolution. It can monitor four independently configurable HSI lines and capture the value of Timer1 when an event(s) takes place. The four types of events that can trigger captures include: rising edges only, falling edges only, rising or falling edges, or every eighth rising edge. The HSI unit can store up to 8 entries (Timer1 values). Reading the HSI holding register unloads the earliest entry placed in the FIFO. The HSI unit can generate an interrupt when loading an entry into the HSI holding register or loading the sixth entry into the FIFO.

High Speed Output Unit (HSO)—The 8X9X HSO unit can trigger events at specified times based on Timer1 or Timer2. These programmable events include: starting an A/D conversion, resetting Timer2, generating up to four software time delays, and setting or clearing one or more of the 6 HSO output lines. The HSO unit stores pending events and the specified times in a Content Addressable Memory (CAM) file. This file stores up to eight commands. Each command specifies the action time, the nature of the action, whether an interrupt is to occur, and whether Timer1 or Timer2 is the reference timer. Every 8 state times the HSO compares the CAM locations for time matches. The HSO unit triggers the specified event when it finds a time match. A command is cleared from the CAM as soon as it executes.

The Serial Port—The Serial Port on the 8X9X has one synchronous (Mode 0) and three asynchronous modes (Modes 1, 2 and 3). The asynchronous modes are full duplex, meaning they can transmit and receive data simultaneously. The receiver on the 8X9X is buffered so the reception of a second byte may begin before the first byte is read. The most common use of Mode 0, the synchronous mode, is to expand the I/O capability of the 8X9X using shift registers. Mode 1 is the standard asynchronous mode used for normal serial communication. The data frame for Mode 1 consists of 10 bits: a start bit, 8 data bits (LSB first) and a stop bit. If parity is enabled (PEN = 1), an even parity bit is sent instead of the 8th data bit. Modes 2 and 3 are 9-bit modes commonly used for multiprocessor communications. The data frame used in these modes consists of 11 bits: a start bit, nine data bits (LSB first) and a stop bit. Devices in Mode 2 will interrupt upon reception only if the 9th data bit is set. Devices in Mode 3 will always interrupt upon reception. Mode 3 also allows transmission of 8 data bits plus an even parity bit.

Pulse Width Modulator (PWM)—The PWM output waveform is a variable duty cycle pulse that repeats every 256 state times. The Pulse Width Modulator of the 8X9X can provide useful signals for a variety of applications. The PWM output can perform digital to analog conversions and drive several types of motors that require a PWM waveform for more efficient operation.

A/D Converter—The 8X9X A/D Converter converts an analog input to a 10-bit digital equivalent. The main components of the A/D Converter are: 8 analog inputs, an 8 to 1 multiplexer, a sample and hold capacitor and the resistor ladder. The A/D Quick Reference section defines the A/D terms. The A/D Converter can start a conversion immediately or the High Speed Output unit can trigger a conversion at a preprogrammed time. The A/D Converter performs a conversion in 88 state times. Upon completion of each conversion the converter can generate a conversion complete interrupt. The 8X9X provides separate V_{REF} and ANGND supply pins to isolate noise on the V_{CC} or V_{SS} lines.

Interrupts—There are 21 interrupt sources and 8 interrupt vectors on the 8X9X. When the interrupt controller detects one of the 8 interrupts it sets the corresponding bit in the interrupt pending register. Individual interrupts are enabled or disabled by setting or clearing bits in the interrupt mask register. When the interrupt controller decides to process an interrupt, it executes a "call" to an Interrupt Service Routine (ISR). The corresponding interrupt vector contains the address of the ISR. The interrupt controller then clears the associated pending bit.

5.0 8XC196KB PERIPHERALS

Standard I/O Ports—The 8XC196KB has five 8-bit I/O ports. Port 0 is an input port that is also the analog input for the A/D converter. Port 1 is a quasi-bidirectional port. Port 2 contains three types of port lines: quasi-bidirectional, input and output. Other functions on the 8XC196KB share the input and output lines with Port 2. Ports 3 and 4 are open-drain bidirectional ports that share their pins with the address/data bus.

4

Timers—The 8XC196KB has two 16-bit timers, Timer1 and Timer2. An internal clock increments the Timer1 value every 8 state times. (A state time is 2 oscillator periods.) An external clock increments or decrements Timer2 on every positive and negative transition. Either an internal or external source can reset Timer2. Timer1 can generate an interrupt when crossing the 0FFFFH/0000H boundary. Timer2 can generate an interrupt when crossing the 0FFFFH/0000H boundary or the 7FFFH/8000H boundary. The 8XC196KB also includes separate, dedicated timers for the baud rate generator and Watchdog Timer. The Watchdog Timer is an internal timer that resets the system if the software fails to operate properly.

High Speed Input Unit (HSI)—The 8XC196KB HSI unit can record times of external events with a 9 state time resolution. It can monitor four independently configurable HSI lines and capture the value of Timer1 when an event(s) takes place. The four types of events that can trigger captures include: rising edges only, falling edges only, rising or falling edges, or every eighth rising edge. The HSI unit can store up to 8 entries (Timer1 values), 7 in the 7-level FIFO and 1 in the HSI holding register. Reading the HSI holding register unloads the earliest entry placed in the FIFO. The HSI unit can generate an interrupt when: loading an entry into the HSI holding register, loading the fourth entry into the FIFO or loading the sixth entry into the FIFO.

High Speed Output Unit (HSO)—The 8XC196KB HSO unit can trigger events at specified times based on Timer1 or Timer2. These programmable events include: starting an A/D conversion, resetting Timer2, generating up to four software time delays, and setting or clearing one or more of the 6 HSO output lines. The HSO unit stores pending events and the specified times in a Content Addressable Memory (CAM) file. This file stores up to eight commands. Each command specifies the action time, the nature of the action, whether an interrupt is to occur, and whether Timer1 or Timer2 is the reference timer. Every 8 state times the HSO compares the CAM locations for time matches. The HSO unit triggers the specified event when it finds a time match. A command can either clear from the CAM as soon as it executes or remain in the CAM as a locked CAM entry and continue to execute whenever its time tag matches the reference timer. Locked entries are useful in applications requiring periodic or repetitive events to occur such as multiple PWMs.

The Serial Port—The Serial Port on the 8XC196KB has one synchronous (Mode 0) and three asynchronous modes (Modes 1, 2 and 3). The asynchronous modes are full duplex, meaning they can transmit and receive data simultaneously. The receiver on the 8XC196KB is buffered so the reception of a second byte may begin before the first byte is read. The transmitter is also double buffered and can generate continual transmissions. The most common use of Mode 0, the synchronous mode, is to expand the I/O capability of the 8XC196KB using shift registers. Mode 1 is the standard asynchronous mode used for normal serial communication. The data frame for Mode 1 consists of 10 bits: a start bit, 8 data bits (LSB first) and a stop bit. If parity is enabled (PEN = 1), an even parity bit is sent instead of the 8th data bit. Modes 2 and 3 are 9-bit modes commonly used for multiprocessor communications. The data frame used in these modes consists of 11 bits: a start bit, nine data bits (LSB first) and a stop bit. Devices in Mode 2 will interrupt upon reception only if the 9th data bit is set. Devices in Mode 3 will always interrupt upon reception. Mode 3 also allows transmission of 8 data bits plus an even parity bit.

Pulse Width Modulator (PWM)—The Pulse Width Modulator of the 8XC196KB can provide useful signals for a variety of applications. The PWM output can perform digital to analog conversions and drive several types of motors that require a PWM waveform for more efficient operaton. The PWM output waveform is a variable duty cycle pulse that repeats every 256 state times or 512 state times.

A/D Converter—The 8XC196KB A/D Converter converts an analog input to a 10-bit digital equivalent. The main components of the A/D Converter are: 8 analog inputs, an 8 to 1 multiplexer, a sample and hold capacitor and the resistor ladder. Refer to the data sheet for all specifications on A/D performance. The A/D Quick Reference section defines the A/D terms. The A/D Converter can start a conversion immediately or the High Speed Output unit can trigger a conversion at a preprogrammed time. The A/D Converter can perform a conversion in either 91 state times for low crystal frequencies and 158 state times for higher crystal frequencies. Upon completion of each conversion the converter generates a conversion complete interrupt. The 8XC196KB provides separate V_{REF} and ANGND supply pins to isolate noise on the V_{CC} or V_{SS} lines.

Interrupts—There are 28 interrupt sources and 16 interrupt vectors on the 8XC196KB. Additionally, there are 2 special interrupt vectors for Software Trap and Unimplemented Opcodes. When the interrupt controller detects one of the 16 interrupts it sets the corresponding bit in one of two interrupt pending registers. Individual interrupts are enabled or disabled by setting or clearing bits in the interrupt mask registers. When the interrupt controller decides to process an interrupt, it executes a "call" to an Interrupt Service Routine (ISR). The corresponding interrupt vector contains the address of the ISR. The interrupt controller then clears the associated pending bit.

6.0 8XC196KC and 8XC196KD PERIPHERALS

Standard I/O Ports—The 8XC196KC/KD has five 8-bit I/O ports. Port 0 is an input port that is also the analog input for the A/D converter. Port 1 is a quasi-bidirectional port that shares pins with two PWM outputs. Port 2 contains three types of port lines: quasi-bidirectional, input and output. Other functions on the 8XC196KC/KD share the input and output lines with Port 2. Ports 3 and 4 are open-drain bidirectional ports that share their pins with the address/data bus.

Timers—The 8XC196KC/KD has two 16-bit timers, Timer1 and Timer2. An internal clock increments the Timer1 value every 8 state times. (A state time is 2 oscillator periods.) An internal clock or an external clock can drive Timer2. When clocked internally Timer2 can increment every 1 or 8 state times. When clocked externally Timer2 increments or decrements on every positive and negative transition. Either an internal or external source can reset Timer2. Timer1 can generate an interrupt when crossing the 0FFFFH/0000H boundary. Timer2 can generate an interrupt when crossing the 0FFFFH/0000H boundary or the 7FFFH/8000H boundary. The 8XC196KC/KD also includes separate, dedicated timers for the baud rate generator and Watchdog Timer. The Watchdog Timer is an internal timer that resets the system if the software fails to operate properly.

High Speed Input Unit (HSI)—The 8XC196KC/KD HSI unit can record times of external events with a 9 state time resolution. It can monitor four independently configurable HSI lines and capture the value of Timer1 when an event(s) takes place. The four types of events that can trigger captures include: rising edges only, falling edges only, rising or falling edges, or every eighth rising edge. The HSI unit can store up to 8 entries (Timer1 values), 7 in the 7-level FIFO and 1 in the HSI holding register. Reading the HSI holding register unloads the earliest entry placed in the FIFO. The HSI unit can generate an interrupt when: loading an entry into the HSI holding register, loading the fourth entry into the FIFO or loading the sixth entry into the FIFO.

High Speed Output Unit (HSO)—The 8XC196KC/KD HSO unit can trigger events at specified times based on Timer1 or Timer2. These programmable events include: starting an A/D conversion, resetting Timer2, generating up to four software timers, and setting or clearing one or more of the 6 HSO output lines. The HSO unit stores pending events and the specified times in a Content Addressable Memory (CAM) file. This file stores up to eight commands. Each command specifies the action time, the nature of the action, whether an interrupt is to occur, and whether Timer1 or Timer2 is the reference timer. Every 8 state times the HSO compares the CAM locations for time matches. The HSO unit triggers the specified event when it finds a time match. A command can either clear from the CAM as soon as it executes or remain in the CAM as a locked CAM entry and continue to execute whenever its time tag matches the reference timer. Locked entries are useful in applications requiring periodic or repetitive events to occur such as multiple PWMs.

The Serial Port—The Serial Port on the 8XC196KC/KD has one synchronous (Mode 0) and three asynchronous modes (Modes 1, 2 and 3). The asynchronous modes are full duplex, meaning they can transmit and receive data simultaneously. The receiver on the 8XC196KC/KD is buffered so the reception of a second byte may begin before the first byte is read. The transmitter is also double bufered and can generate continual transmissions. The most common use of Mode 0, the synchronous mode, is to expand the I/O capability of the 8XC196KC/KD using shift registers. Mode 1 is the standard asynchronous mode used for normal serial communication. The data frame for Mode 1 consists of 10 bits: a start bit, 8 data bits (LSB first) and a stop bit. If parity is enabled (PEN = 1), an even parity bit is sent instead of the 8th data bit. Modes 2 and 3 are 9-bit modes commonly used for multiprocessor communications. The data frame used in these modes consists of 11 bits: a start bit, nine data bits (LSB first) and a stop bit. Devices in Mode 2 will interrupt upon reception only if the 9th data bit is set. Devices in Mode 3 will always interrupt upon reception. Mode 3 also allows transmission of 8 data bits plus an even parity bit.

Pulse Width Modulator (PWM)—The 8CX196KC/KD has 3 PWM outputs. The output waveform is a variable duty cycle pulse which is selectable to repeat every 256 state times or 512 state times. Several types of motors require a PWM waveform for most efficient operation. Additionally, filtering this waveform will produce a DC level that can change in 256 steps by varying the duty cycle.

A/D Converter—The 8XC196KC/KD A/D Converter converts an analog input to a digital equivalent. Resolution is either 8 or 10 bits with programmable sample and convert times. The main components of the A/D Converter are: a sample and hold, an 8-channel multiplexer, and an 8-bit or 10-bit successive approximation analog-to-digital converter. Refer to the data sheet for all specifications on A/D performance. The A/D Quick Reference section defines the A/D terms. The converter can start a conversion immediately or the High Speed Output unit can trigger a conversion at a preprogrammed time. Upon completion of each conversion the converter generates a conversion complete interrupt. The 8XC196KC/KD provides separate V_{REF} and ANGND supply pins to isolate noise on the V_{CC} or V_{SS} lines.

Interrupts—There are 28 interrupt sources and 16 interrupt vectors on the 8XC196KC/KD. In addition there are 2 special interrupt vectors (Software Trap and Unimplemented Opcode) used in Intel development tools or evaluation boards. When the interrupt controller detects one of the 16 interrupts it sets the corresponding bit in one of two interrupt pending registers. Individual interrupts are enabled or disabled by setting or clearing bits in the interrupt mask registers. When the interrupt controller decides to process an interrupt, it executes a "call" to an Interrupt Service Routine (ISR). The corresponding interrupt vector contains the address of the ISR. The interrupt controller then clears the associated pending bit.

Peripheral Transaction Server (PTS)—The PTS is a microcoded hardware interrupt processor. It responds to interrupts with a fixed set of actions. These actions consist of: transferring data, starting an A/D conversion, reading the HSI FIFO and loading HSO events. The PTS completes these tasks much faster than using interrupt driven service routines. The PTS can service all interrupts except NMI, Trap and Unimplemented Opcode. Each interrupt managed by the PTS requires a block of data called the PTS Control Block (PTSCB). Each PTSCB requires 8 data bytes in register RAM. The PTSCB determines: the type of PTS, the number of PTS responses (if applicable), the source for data and the destination (if applicable). PTS cycles have a higher priority than interrupts and may temporarily suspend interrupt service routines.

7.0 8XC196KR and 8XC196KT PERIPHERALS

Standard I/O Ports—The 8XC196KR/KT has six 8-bit I/O ports. Each pin operates as a dedicated input or output. Most pins also have an alternate function. The KR/KT does not use the quasi-bidirectional port pins found on previous MCS-96 devices. As an input, the pin is a true high impedance with no pull-ups or pull-downs. Most ports (Ports 1, 2, 5 and 6) have direction registers (Px_DIR), mode registers (Px_MODE), data input registers (Px_PIN) and data output registers (Px_REG). This allows the user to configure each port pin as input, output, open-drain output or alternate function. Ports 3 and 4 have Px_PIN and Px_REG registers and lack internal pull-ups. As standard outputs, these pins can only function in open-drain mode and need external pull-ups. Ports 3 and 4 also are the multiplexed address/data bus. When emitting the address, an internal pull-up device is active and does not need external pull-ups. Port 0 is the analog input port, and only has a Px_PIN register because there are no output drivers. As a digital port, Port 0 pins can only function as inputs.

Event Processor Array (EPA)—The EPA performs input event capture and output event generation functions using Timer1 and Timer2. It consists of 10 capture/compare modules, 2 compare only modules and the 2 timers. In capture mode, when an external event occurs the EPA stores the value of the timer, generates an interrupt or both. A rising, falling, or any edge can trigger a capture. All captures are double buffered. In compare mode, when the timer matches the value in the compare register the EPA changes the state of an output pin, generates an interrupt, or both. The EPA sets, resets or toggles the pin when the compare occurs. The timers can count up or count down. The clock source to the timers can be internal or external. The clock also goes through a programmable prescaler. The prescaler divides the oscillator frequency within a range of 1 to 64. The EPA also allows two channels to control a single output pin that is useful for high-speed PWM generation.

Serial I/O Port (SIO)—The SIO (also known as the UART) supports 8- or 9-bit data frames with one synchronous mode and 3 asynchronous modes. The synchronous mode transmits or receives 8 bits of data without start or stop bits and generates a shift clock. All other devices must synchronize to the 8XC196KR/KT's shift clock. The asynchronous frames contain a start and stop bit, making them either 10 or 11 bits long. The 11-bit frames allow implementation of specialized multiprocessor communication interfaces. Two of the asynchronous modes support parity error detection. All three asynchronous modes support full or half duplex operation. Also included is a dedicated baud rate generator. The SIO on the 8XC196KR/KT is compatible with all MCS-96 and MCS-51 devices.

Synchronous Serial I/O Port (SSIO)—The SSIO includes two Serial I/O communication ports with separate data and clock pins. The data format is eight data bits only. The clock and data pins can be inputs or outputs. This peripheral supports several standard synchronous serial protocols. A handshake mode allows two serial channels to transfer data without requiring extra lines to convey their status. The handshake mode also permits servicing of the SSIO by the PTS. The serial channel includes a dedicated baud rate generator. Each channel has a single byte buffer. If clocked externally, both channels can simultaneously operate at different frequencies. Maximum baud rate is $\frac{1}{8}$ the oscillator frequency. The transmission or reception of a byte sets an interrupt pending flag.

A/D Converter—Converts analog inputs to a digital equivalent. Resolution is either 8 or 10 bits with programmable sample and convert times. The main components of the A/D converter are: 8 analog inputs, 8 to 1 multiplexer, sample and hold capacitor and the resistor ladder. Refer to the data sheet for all specifications on A/D performance. The A/D Quick Reference section defines the A/D terms. Another function implemented with the A/D converter is threshold detection. The converter generates an interrupt when the analog input is greater than or less than a programmed digital value. The KR/KT provides separate V_{REF} and ANGND supply pins to isolate noise on the V_{CC} or V_{SS} lines.

Interrupts—There are 37 interrupt sources and 18 interrupt vectors on the 8XC196KR/KT. With so many more sources than vectors, the KR/KT implements indirect interrupts. 17 of the interrupts are direct, which means each interrupt has one source and a dedicated vector location. The remaining 20 interrupt sources are the indirect interrupts. The term indirect is used because they share the same interrupt vector and another register identifies the interrupt source. The register, EPAIPV, contains the highest ending interrupt. EPAIPV is read to determine the interrupt needing sevice. The TIJMP instruction with EPAIPV simplifies the servicing of indirect interrupts. The direct interrupts include: NMI, External Interrupt, Trap, Unimplemented Opcode, SIO interrupts, SSIO interrupts, slave port interrupts, A/D converter and the lower 4 EPA channels. The indirect interrupts include: the upper 6 EPA channels, the 2 compare channels, all 10 EPA overruns and both timer overflows.

Peripheral Transaction Server (PTS)—The PTS is a microcoded hardware interrupt processor. It responds to interrupts with a fixed set of actions. These actions consist of: transferring data, starting an A/D conversion or generating PWM outputs. The PTS completes these tasks much faster than using interrupt driven service routines. The PTS can service all interrupts except NMI, Trap and Unimplemented Opcode. The register PTSSEL selects the interrupts handled by the PTS. Each interrupt managed by the PTS requires a block of data called the PTS Control Block (PTSCB). Each PTSCB requires 8 data bytes in register RAM. The PTSCB determines: the type of PTS, the number of PTS responses (if applicable), the source for data and the destination (if applicable). PTS cycles have a higher priority than interrupts and may temporarily suspend interrupt service routines.

Slave Port—The slave port is an interface between the KR/KT and a microprocessor. The KR/KT sits on the address/data bus of the processor and is accessed as a memory mapped peripheral. The slave port includes: a chip select input, 8-bit bidirectional data bus, an address input line, ALE input (to latch the address), $\overline{WR}$ and $\overline{RD}$ inputs to input/output data and an interrupt output. The address line and the $\overline{RD}/\overline{WR}$ select which registers are accessed (Output data, Status output, Input data or Command input). The various control signals and port structure allow the KR/KT and the processor to communicate with each other without having to be synchronized.

8.0 8XC196NT PERIPHERALS

Extended Address Port (EPORT)—The 80C196NT is the first member of the MCS-96 family to offer addressing that exceeds 64 Kbytes. The 80C196NT has a 1 Mbyte liner address space which is implemented through 4 address lines added by the EPORT. EPORT lines are individually assigned to function as either address or I/O. When assigned as I/O, they have the same functionality as a standard I/O port. As an input, the pin is a true high-impedance with no pull-ups or pull-downs. As an output, the pin is either complementary or open-drain. When assigned as address, the EPORT outputs address A16–A19. The address is strongly driven through the entire bus cycle, eliminating the need for an address latch.

Standard I/O Ports—The 8XC196NT has six 8-bit I/O ports. Each pin operates as a dedicated input or output. Most pins also have an alternate function. The NT does not use the quasi-bidirectional port pins found on previous MCS-96 devices. As an input, the pin is a true high impedance with no pull-ups or pull-downs. Most ports (Ports 1, 2, 5 and 6) have direction registers (Px_DIR), mode registers (Px_MODE), data input registers (Px_PIN) and data output registers (Px_REG). This allows the user to configure each port pin as input, output, open-drain output or alternate function. Ports 3 and 4 have Px_PIN and Px_REG registers and lack internal pull-ups. As standard outputs, these pins can only function in open-drain mode and need external pull-ups. Ports 3 and 4 also are the multiplexed address/data bus. When emitting the address, an internal pull-up device is active and does not need external pull-ups. Port 0 is the analog input port, and only has a Px_PIN register because there are no output drivers. As a digital port, Port 0 pins can only function as inputs.

Event Processor Array (EPA)—The EPA performs input event capture and output event generation functions using Timer1 and Timer2. It consists of 10 capture/compare modules, 2 compare only modules and the 2 timers. In capture mode, when an external event occurs the EPA stores the value of the timer, generates an interrupt or both. A rising, falling, or any edge can trigger a capture. All captures are double buffered. In compare mode, when the timer matches the value in the compare register the EPA changes the state of an output pin, generates an interrupt, or both. The EPA sets, resets or toggles the pin when the compare occurs. The timers can count up or count down. The clock source to the timers can be internal or external. The clock also goes through a programmable prescaler. The prescaler divides the oscillator frequency within a range of 1 to 64. The EPA also allows two channels to control a single output pin that is useful for high-speed PWM generation.

Serial I/O Port (SIO)—The SIO (also known as the UART) supports 8- or 9-bit data frames with one synchronous mode and 3 asynchronous modes. The synchronous mode transmits or receives 8 bits of data without start or stop bits and generates a shift clock. All other devices must synchronize to the 8XC196NT's shift clock. The asynchronous frames contain a start and stop bit, making them either 10 or 11 bits long. The 11-bit frames allow implementation of specialized multiprocessor communication interfaces. Two of the asynchronous modes support parity error detection. All three asynchronous modes support full or half duplex operation. Also included is a dedicated baud rate generator. The SIO on the 8XC196NT is compatible with all MCS-96 and MCS-51 devices.

Synchronous Serial I/O Port (SSIO)—The SSIO includes two Serial I/O communication ports with separate data and clock pins. The data format is eight data bits only. The clock and data pins can be inputs or outputs. This peripheral supports several standard synchronous serial protocols. A handshake mode allows two serial channels to transfer data without requiring extra lines to convey their status. The handshake mode also permits servicing of the SSIO by the PTS. The serial channel includes a dedicated baud rate generator. Each channel has a single byte buffer. If clocked externally, both channels can simultaneously operate at different frequencies. Maximum baud rate is $\frac{1}{8}$ the oscillator frequency. The transmission or reception of a byte sets an interrupt pending flag.

A/D Converter—Converts analog inputs to a digital equivalent. Resolution is either 8 or 10 bits with programmable sample and convert times. The main components of the A/D converter are: 4 analog inputs, 4 to 1 multiplexer, sample and hold capacitor and the resistor ladder. Refer to the data sheet for all specifications on A/D performance. The A/D Quick Reference section defines the A/D terms. Another function implemented with the A/D converter is threshold detection. The converter generates an interrupt when the analog input is greater than or less than a programmed digital value. The NT provides separate V_{REF} and ANGND supply pins to isolate noise on the V_{CC} or V_{SS} lines.

Interrupts—There are 37 interrupt sources and 18 interrupt vectors on the 8XC196NT. With so many more sources than vectors, the NT implements indirect interrupts. 17 of the interrupts are direct, which means each interrupt has one source and a dedicated vector location. The remaining 20 interrupt sources are the indirect interrupts. The term indirect is used because they share the same interrupt vector and another register identifies the interrupt source. The register, EPAIPV, contains the highest ending interrupt. EPAIPV is read to determine the interrupt needing service. The TIJMP instruction with EPAIPV simplifies the servicing of indirect interrupts. The direct interrupts include: NMI, External Interrupt, Trap, Unimplemented Opcode, SIO interrupts, SSIO interrupts, slave port interrupts, A/D converter and the lower 4 EPA channels. The indirect interrupts include: the upper 6 EPA channels, the 2 compare channels, all 10 EPA overruns and both timer overflows.

Peripheral Transaction Server (PTS)—The PTS is a microcoded hardware interrupt processor. It responds to interrupts with a fixed set of actions. These actions consist of: transferring data, starting an A/D conversion or generating PWM outputs. The PTS completes these tasks much faster than using interrupt driven service routines. The PTS can service all interrupts except NMI, Trap and Unimplemented Opcode. The register PTSSEL selects the interrupts handled by the PTS. Each interrupt managed by the PTS requires a block of data called the PTS Control Block (PTSCB). Each PTSCB requires 8 data bytes in register RAM. The PTSCB determines: the type of PTS, the number of PTS responses (if applicable), the source for data and the destination (if applicable). PTS cycles have a higher priority than interrupts and may temporarily suspend interrupt service routines.

Slave Port—The slave port is an interface between the NT and a microprocessor. The NT sits on the address/data bus of the processor and is accessed as a memory mapped peripheral. The slave port includes: a chip select input, 8-bit bidirectional data bus, an address input line, ALE input (to latch the address), $\overline{WR}$ and $\overline{RD}$ inputs to input/output data and an interrupt output. The address line and the $\overline{RD}/\overline{WR}$ select which registers are accessed (Output data, Status output, Input data or Command input). The various control signals and port structure allow the NT and the processor to communicate with each other without having to be synchronized.

9.0 8XC196MC PERIPHERALS

On-Chip Peripherals—The 8XC196MC's on-chip peripherals provide special functions useful in a variety of applications. The peripherals are monitored and controlled via special function registers (SFRs) that can be accessed indirectly or windowed and thereby treated as CPU "accumulators."

I/O Ports—The 8XC196MC has 7 I/O ports, labeled 0–6. Individual port pins are multiplexed to serve for standard I/O or to carry special signals. All ports are 8-bit except port 1 which is a 5-bit port.

Ports 0, 1, 2 and 6 are controlled by SFRs that can be directly addressed by the RALU through a window in the register file. Ports 0 and 1 serve as input to the 13-channel A/D, and can also be read as digital inputs. Port 2 can be configured either as standard I/O ports or to serve special functions. Port 6 is the output port for the PWM and WG units.

Ports 3, 4 and 5 are memory mapped and cannot be windowed. These ports are accessed only via 16-bit addresses. Ports 3 and 4 also serve as the 16-bit external address/data bus. The Port 5 lines can be selected for standard I/O or to serve as system bus control pins.

Timers and the Event Processor Array (EPA)—The Event Processor Array (EPA) performs input and output functions associated with timers 1 and 2. In the input mode the EPA monitors an input pin for signal transitions and records the timer value when the event occurs. The "captured" event is thus tagged with its time. In the output mode the EPA waits until the timer matches a stored time value and then sets, clears or toggles an output pin. This is a "compare" event. Both capture and compare events initiate interrupts which can be handled by a normal service routine or the PTS. The 8XC196MC has 4 capture/compare modules and 4 compare-only modules.

The two 16-bit timers can be clocked by the internal clock generator or by external sources. An external "quadrature clocking" mode is available for monitoring speed and direction from a position encoder.

Pulse Width Modulation Unit—The 8XC196MC has a PWM module that provides two PWM outputs. This module is in addition to the waveform generator. The duty cycle and the period of each output is programmable through a respective 8-bit register. The module has an 8-bit counter, two 8-bit PWM compare registers and an 8-bit period register. The PWM output pins are controlled with bits in the output control register of the waveform generator.

A/D Converter—The 13-channel A/D converter can perform 10-bit conversions or faster 8-bit conversions. Automated A/D conversions and result storage are facilitated by the A/D scan mode of the PTS. The sample-and-hold times and the conversion times are programmable. The A/D can also act as a programmable comparator and issue an interrupt when the input crosses a threshold. Conversions can be performed on the analog ground and reference voltage, and the results can be used to calculate gain and zero offset errors. The zero offset compensation circuit is also programmable, enabling automatic offset adjustment.

Interrupt Controller and Peripheral Transaction Server (PTS)—The 8XC196MC's flexible interrupt handling system has two main components: the programmable interrupt controller and the Peripheral Transaction Server (PTS). The interrupt controller has a hardware priority scheme that can be modified by user software. These interrupts are serviced by user-written interrupt service routines. The user can select most interrupts to be serviced by the PTS instead of the programmable interrupt controller. The PTS has several microcoded hardware interrupt service routines whose execution is interleaved with normal instruction execution. The result is high-speed, low-overhead interrupt handling. The PTS can perform single and burst transfers of bytes or 16-bit words between any memory locations, manage multiple analog-to-digital (A/D) conversions and control a software serial channel which allows either synchronous or asynchronous operations.

Waveform Generator—The Waveform Generator (WG) produces 3 pairs of complimentary PWM signals. This peripheral is optimized for controlling 3-phase induction AC motors. It can also control brushless DC motors and DC to AC inverters. A dead-time generator and phase inverter circuit provide non-overlapping on-timers for each PWM output pair. Each signal is independently programmable.

MCS®-96 A/D Converter Quick Reference

October 1991

Order Number: 272115-001

1.0 The MCS-96 A/D Converter

Analog inputs to the MCS-96 family are handled by the A/D converter system. As shown in Figure 1, the converter system has an 8 channel multiplexer, a sample and hold, and a 10-bit successive approximation A/D converter. Conversions can be performed on one of 8 channels, the inputs of which share pins with port 0.

There are various versions of the A/D converter, depending on the specific device type. The 8X9X family offers a 10-bit fixed conversion time. The 8XC196KB family offers a 10-bit conversion with either a fast or slow conversion time. The 8XC196KC family offers an 8- or 10-bit conversion with programmable sample and convert times. The 8XC196KR has all of the KC features, with the addition of offset correction and internal conversion of V_{ref} and ANGND. The 8XC196MC includes all of the KR features, and the multiplexer has been expanded to 13 analog input channels.

This chapter describes the basic operation and terminology of the A/D converter. The different devices control the A/D in different ways, but the principals of operation remain the same throughout.

1.1 A/D Conversion Process

The conversion process is initiated by an HSO or EPA command, or by writing a one to the GO Bit in the A/D Control Register. Either activity causes a start conversion signal to be sent to the A/D converter control logic.

Once the A/D unit receives a start conversion signal, there is a one state time delay before sampling (Sample Delay) while the successive approximation register is reset and the proper multiplexer channel is selected. After the sample delay, the multiplexer output is connected to the sample capacitor and remains connected for the sample time. After the "sample window" closes, the input to the sample capacitor is disconnected from the multiplexer so that changes on the input pin will not alter the stored charge while the conversion is in progress. The comparator is then auto-zeroed and the conversion begins. The sample delay and sample time uncertainties are each approximately ±50 ns, independent of clock speed.

To perform the actual analog-to-digital conversion the MCS-96 implements a successive approximation algorithm. The converter hardware consists of a 256-resistor ladder, a comparator, coupling capacitors and a 10-bit successive approximation register (SAR) with logic that guides the process. The resistor ladder provides 20 mV steps (V_{REF} = 5.12V), while capacitive coupling creates 5 mV steps within the 20 mV ladder voltages. Therefore, 1024 internal reference voltages are available for comparison against the analog input to generate a 10-bit conversion result.

A successive approximation conversion is performed by comparing a sequence of reference voltages, to the analog input, in a binary search for the reference voltage that most closely matches the input. The ½ full scale reference voltage is the first tested. This corresponds to a 10-bit result where the most significant bit is zero, and all other bits are ones (0111.1111.11b). If the analog input was less than the test voltage, bit 10 of the SAR is left a zero, and a new test voltage of ¼ full scale (0011.1111.11b) is tried. If this test voltage was lower than the analog input, bit 9 of the SAR is set and bit 8 is cleared for the next test (0101.1111.11b). This binary search continues until 10 tests have occurred, at which time the valid 10-bit conversion result resides in the SAR where it can be read by software.

1.2 A/D Interface Suggestions

The external interface circuitry to an analog input is highly dependent upon the application, and can impact converter characteristics. In the external circuit's design, important factors such as input pin leakage, sample capacitor size and multiplexer series resistance from the input pin to the sample capacitor must be considered.

These factors are idealized in Figure 1. The external input circuit must be able to charge a sample capacitor (C_S) through a series resistance (R_I) to an accurate voltage given a DC leakage (I_L). Typically C_S is around 2 pF, R_I is around 5 KΩ and I_L is specified as 3 μA. In determining the necessary source impedance R_S, the value of V_{BIAS} is not important.

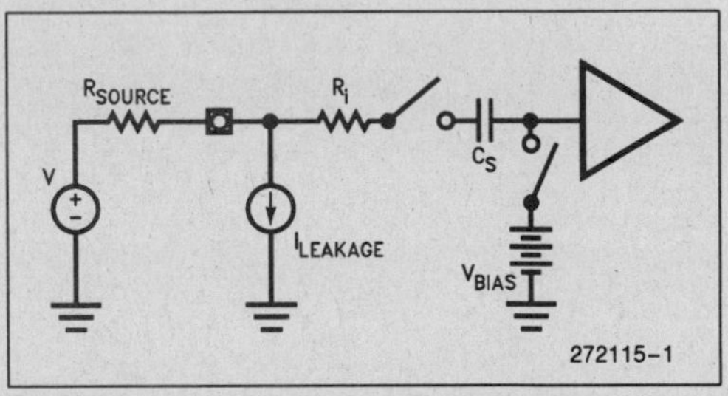

Figure 1. Idealized A/D Sampling Circuitry

External circuits with source impedances of 1 KΩ or less will be able to maintain an input voltage within a tolerance of about ±0.61 LSB (1.0 KΩ × 3.0 μA = 3.0 mV) given the DC leakage. Source impedances above 2 KΩ can result in an external error of at least one LSB due to the voltage drop caused by the 3 μA leakage. In addition, source impedances above 25 KΩ may degrade converter accuracy as a result of the internal sample capacitor not being fully charged during the sample window.

If large source impedances degrade converter accuracy because the sample capacitor is not charged during the sample time, an external capacitor connected to the pin compensates for this. Since the sample capacitor is 2 pF, a 0.005 μF capacitor (2048 * 2 pF) will charge the sample capacitor to an accurate input voltage of ±0.5 LSB. An external capacitor does not compensate for the voltage drop across the source resistance, but charges the sample capacitor fully during the sample time.

Placing an external capacitor on each analog input will also reduce the sensitivity to noise, as the capacitor combines with series resistance in the external circuit to form a low-pass filter. In practice, one should include a small series resistance prior to the external capacitor on the analog input pin and choose the largest capacitor value practical, given the frequency of the signal being converted. This provides a low-pass filter on the input, while the resistor will also limit input current during over-voltage conditions.

Figure 2 shows a simple analog interface circuit based upon the discussion above. The circuit in the figure also provides limited protection against over-voltage conditions on the analog input. Should the input voltage inappropriately drop significantly below ground, diode D2 will forward bias at about 0.8 DCV. Since the specification of the pin on most devices has an absolute maximum low voltage of −0.3V, this will leave about 0.5V across the 270Ω resistor, or about 2 mA of current. This should limit the current to a safe amount. Note that if any input pins are driven much beyond V_{REF} or below ANGND, the accuracy of all analog input channels may be adversely affected. This is because the input protection circuit will start to conduct, thus injecting current into the internal reference circuitry and upsetting the reference voltage. Refer to the data sheet for exact device specifications.

However, before any circuit is used in an actual application, it should be thoroughly analyzed for applicability to the specific problem at hand.

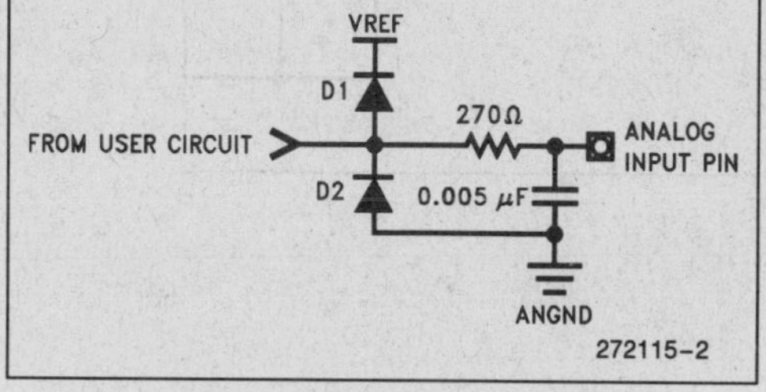

Figure 2. Suggested A/D Input Circuit

ANALOG REFERENCES

Reference supply levels and noise strongly influence the absolute accuracy of the conversion. For this reason, it is recommended that the ANGND pin be tied to the V_{SS} pins close to the device. Bypass capacitors should also be used between V_{REF} and ANGND. ANGND should be within about a tenth of a volt of V_{SS}. V_{REF} should be well regulated and used only for the A/D converter. The V_{REF} supply needs to be able to source around 5 mA.

Note that if only ratiometric information is desired, V_{REF} can be connected to V_{CC}. In addition, V_{REF} and ANGND must be connected even if the A/D converter is not being used. Remember that Port 0 receives its power from the V_{REF} and ANGND pins even when it is used as digital I/O.

1.3 The A/D Transfer Function

The conversion result is a 8- or 10-bit ratiometric representation of the input voltage, so the numerical value obtained from the conversion will be:

$$\mathrm{INT}\ [255 \times (V_{IN} - \mathrm{ANGND})/(V_{REF} - \mathrm{ANGND})]\ \text{or}$$

$$\mathrm{INT}\ [1023 \times (V_{IN} - \mathrm{ANGND})/(V_{REF} - \mathrm{ANGND})]$$

This produces a stair-stepped transfer function when the output code is plotted versus input voltage (see Figure 3). The resulting digital codes can be taken as simple ratiometric information, or they provide information about absolute voltages or relative voltage changes on the inputs. The more demanding the application is on the A/D converter, the more important it is to fully understand the converter's operation. For simple applications, knowing the absolute error of the converter is sufficient. However, closing a servo-loop with analog inputs necessitates a detailed understanding of an A/D converter's operation and errors.

The errors inherent in an analog-to-digital conversion process are many: quantizing error, zero offset, full-scale error, differential non-linearity and non-linearity. These are "transfer function" errors related to the A/D converter. In addition, converter temperature drift, V_{CC} rejection, sample-hold feedthrough, multiplexer off-isolation, channel-to-channel matching and random noise should be considered. Fortunately, one "Absolute Error" specification is available which describes the sum total of all deviations between the actual conversion process and an ideal converter. However, the various sub-components of error are important in many applications. These error components are described in the text below where ideal and actual converters are compared.

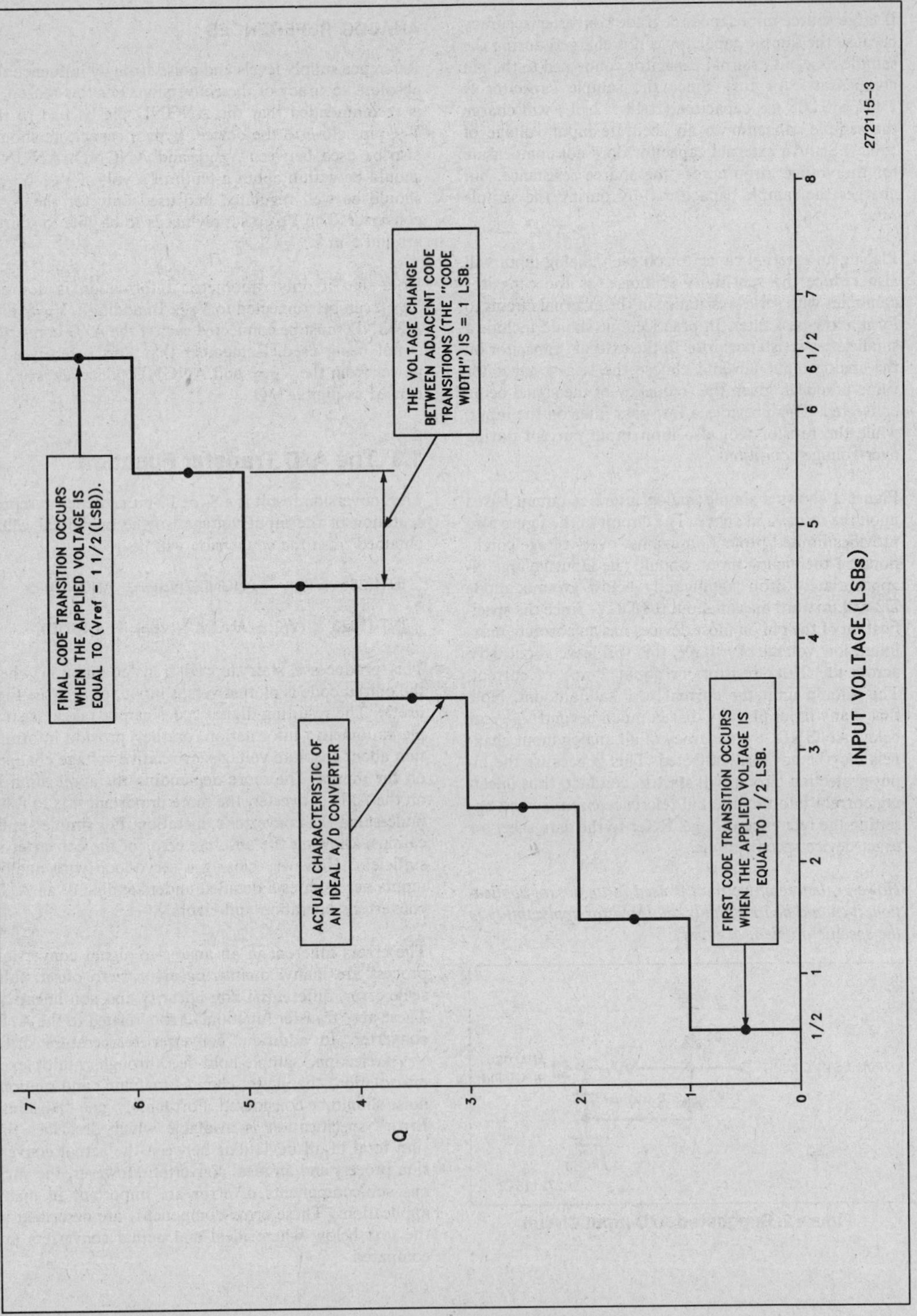

Figure 3. Ideal A/D Characteristic

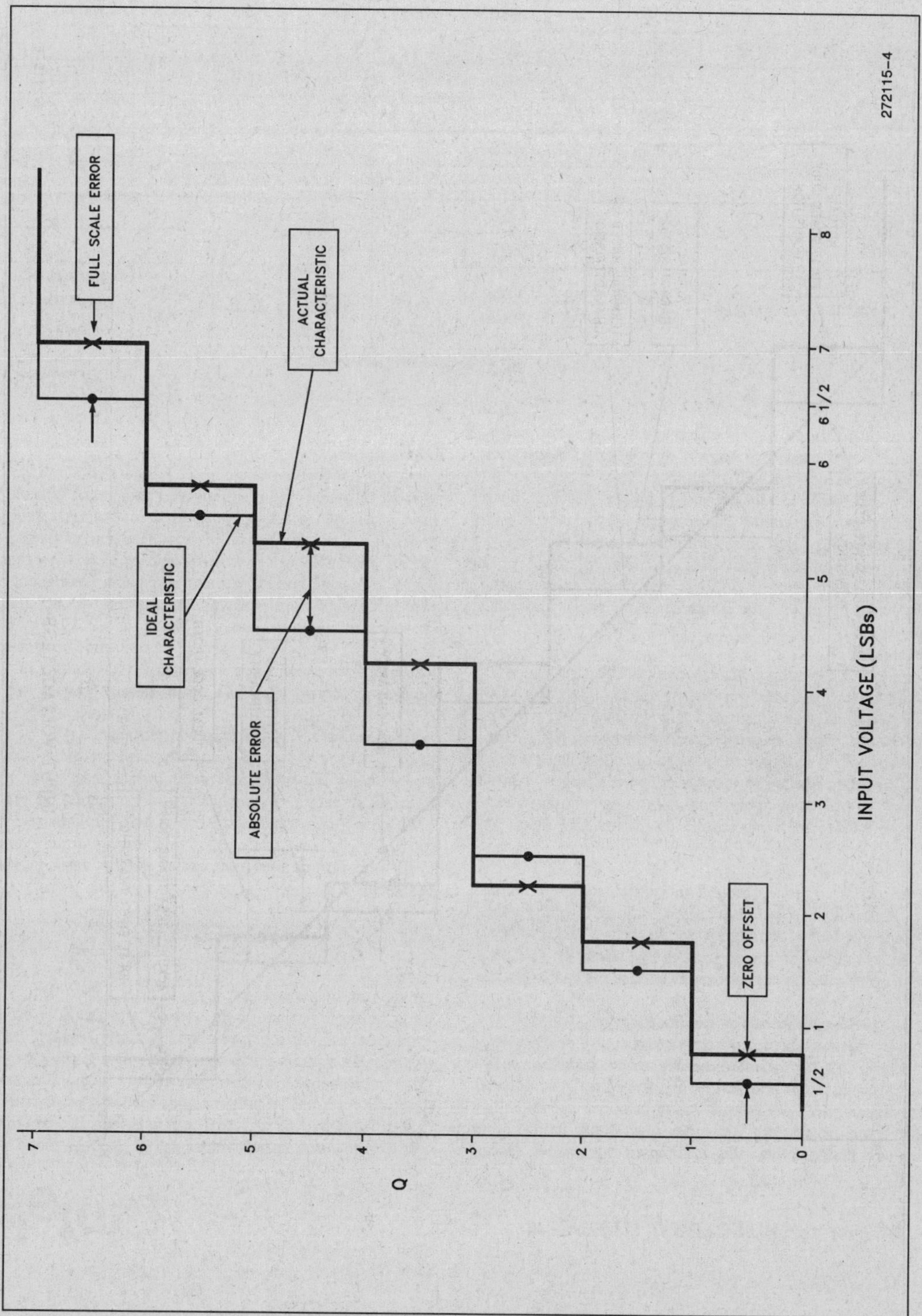

Figure 4. Actual and Ideal Characteristics

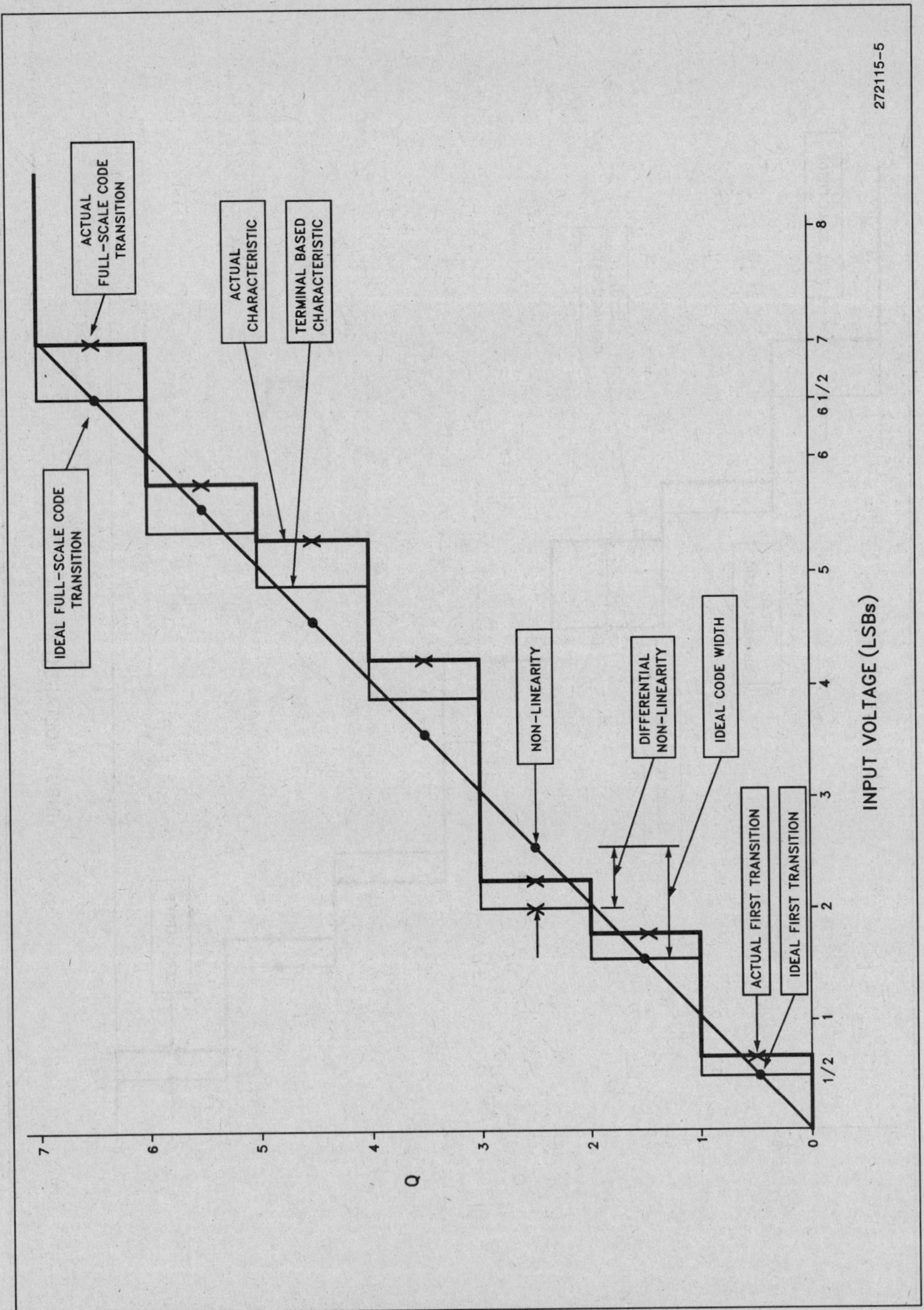

Figure 5. Terminal Based Characteristic

An unavoidable error simply results from the conversion of a continuous voltage to an integer digital representation. This error is called quantizing error, and is always ±0.5 LSB. Quantizing error is the only error seen in a perfect A/D converter, and is obviously present in actual converters. Figure 3 shows the transfer function for an ideal 3-bit A/D converter (i.e., the Ideal Characteristic).

Note that in Figure 3 the Ideal Characteristic possesses unique qualities: its first code transition occurs when the input voltage is 0.5 LSB; its full-scale code transition occurs when the input voltage equals the full-scale reference minus 1.5 LSB; and its code widths are all exactly one LSB. These qualities result in a digitization without offset, full-scale or linearity errors. In other words, a perfect conversion.

Figure 4 shows an Actual Characteristic of a hypothetical 3-bit converter, which is not perfect. When the Ideal Characteristic is overlaid with the imperfect characteristic, the actual converter is seen to exhibit errors in the location of the first and final code transitions and code widths. The deviation of the first code transition from ideal is called "zero offset", and the deviation of the final code transition from ideal is "full-scale error". The deviation of the code widths from ideal causes two types of errors. Differential Non-Linearity and Non-Linearity. Differential Non-Linearity is a local linearity error measurement, whereas Non-Linearity is an overall linearity error measure.

Differential Non-Linearity is the degree to which actual code widths differ from the ideal one LSB width. It gives the user a measure of how much the input voltage may have changed in order to produce a one count change in the conversion result. Non-Linearity is the worst case deviation of code transitions from the corresponding code transitions of the Ideal Characteristic. Non-Linearity describes how much Differential Non-Linearities could add up to produce an overall maximum departure from a linear characteristic. If the Differential Non-Linearity errors are too large, it is possible for an A/D converter to miss codes or exhibit non-monotonicity. Neither behavior is desirable in a closed-loop system. A converter has no missed codes if there exists for each output code a unique input voltage range that produces that code only. A converter is monotonic if every subsequent code change represents an input voltage change in the same direction.

Differential Non-Linearity and Non-Linearity are quantified by measuring the Terminal Based Linearity Errors. A Terminal Based Characteristic results when an Actual Characteristic is shifted and rotated to eliminate zero offset and full-scale error (see Figure 5). The Terminal Based Characteristic is similar to the Actual Characteristic that would be seen if zero offset and full-scale error were externally trimmed away. In practice, this is done by using input circuits which include gain and offset trimming. In addition, V_{REF} could also be closely regulated and trimmed within the specified range to affect full-scale error.

Other factors that affect a real A/D Converter system include sensitivity to temperature, failure to completely reject all unwanted signals, multiplexer channel dissimilarities and random noise. Fortunately these effects are small.

Temperature sensitivities are described by the rate at which typical specifications change with a change in temperature.

Undesired signals come from three main sources. First, noise on $V_{CC}-V_{CC}$ Rejection. Second, input signal changes on the channel being converted after the sample window has closed–Feedthrough. Third, signals applied to channels not selected by the multiplexer–Off-Isolation.

Finally, multiplexer on-channel resistances differ slightly from one channel to the next causing Channel-to-Channel Matching errors, and random noise in general results in Repeatability errors.

1.4 A/D Glossary of Terms

Figures 3, 4 and 5 display many of these terms. Refer to AP-406 'MCS-96 Analog Acquisition Primer' for additional information on the A/D terms.

ABSOLUTE ERROR–The maximum difference between corresponding actual and ideal code transitions. Absolute Error accounts for all deviations of an actual converter from an ideal converter.

ACTUAL CHARACTERISTIC–The characteristic of an actual converter. The characteristic of a given converter may vary over temperature, supply voltage, and frequency conditions. An Actual Characteristic rarely has ideal first and last transition locations or ideal code widths. It may even vary over multiple conversion under the same conditions.

BREAK-BEFORE-MAKE–The property of a multiplexer which guarantees that a previously selected channel will be deselected before a new channel is selected. (e.g., the converter will not short inputs together.)

CHANNEL-TO-CHANNEL MATCHING–The difference between corresponding code transitions of actual characteristics taken from different channels under the same temperature, voltage and frequency conditions.

CHARACTERISTIC–A graph of input voltage versus the resultant output code for an A/D converter. It describes the transfer function of the A/D converter.

CODE–The digital value output by the converter.

CODE CENTER–The voltage corresponding to the midpoint between two adjacent code transitions.

CODE TRANSITION–The point at which the converter changes from an output code of Q, to a code of Q+1. The input voltage corresponding to a code transition is defined to be that voltage which is equally likely to produce either of two adjacent codes.

CODE WIDTH–The voltage corresponding to the difference between two adjacent code transitions.

CROSSTALK–See "Off-Isolation".

DC INPUT LEAKAGE–Leakage current to ground from an analog input pin.

DIFFERENTIAL NON-LINEARITY–The difference between the ideal and actual code widths of the terminal based characteristic of a converter.

FEEDTHROUGH–Attenuation of a voltage applied on the selected channel of the A/D converter after the sample window closes.

FULL SCALE ERROR–The difference between the expected and actual input voltage corresponding to the full scale code transition.

IDEAL CHARACTERISTIC–A characteristic with its first code transition at $V_{IN} = 0.5$ LSB, its last code transition at $V_{IN} = (V_{REF} - 1.5$ LSB) and all code widths equal to one LSB.

INPUT RESISTANCE–The effective series resistance from the analog input pin to the sample capacitor.

LSB (LEAST SIGNIFICANT BIT)–The voltage value corresponding to the full scale voltage divided by 2^n, where n is the number of bits of resolution of the converter. For a 10-bit converter with a reference voltage of 5.12 volts, one LSB is 5.0 mV. Note that this is different than digital LSBs, since an uncertainty of two LSBs, when referring to an A/D converter, equals 10 mV. (This has been confused with an uncertainty of two digital bits, which would mean four counts, or 20 mV.)

MONOTONIC–The property of successive approximation converters which guarantees that increasing input voltages produce adjacent codes of increasing value, and that decreasing input voltages produce adjacent codes of decreasing value.

NO MISSED CODES–For each and every output code, there exists a unique input voltage range which produces that code only.

NON-LINEARITY–The maximum deviation of code transitions of the terminal based characteristic from the corresponding code transitions of the ideal characteristics.

OFF-ISOLATION–Attenuation of a voltage applied on a deselected channel of the A/D converter. (Also referred to as Crosstalk.)

REPEATABILITY–The difference between corresponding code transitions from different actual characteristics taken from the same converter on the same channel at the same temperature, voltage and frequency conditions.

RESOLUTION–The number of input voltage levels that the converter can unambiguously distinguish between. Also defines the number of useful bits of information which the converter can return.

SAMPLE DELAY–The delay from receiving the start conversion signal to when the sample window opens.

SAMPLE DELAY UNCERTAINTY–The variation in the Sample Delay.

SAMPLE TIME–The time that the sample window is open.

SAMPLE TIME UNCERTAINTY–The variation in the sample time.

SAMPLE WINDOW–Begins when the sample capacitor is attached to a selected channel and ends when the sample capacitor is disconnected from the selected channel.

SUCCESSIVE APPROXIMATION–An A/D conversion method which uses a binary search to arrive at the best digital representation of an analog input.

TEMPERATURE COEFFICIENTS–Change in the stated variable per degree centigrade temperature change. Temperature coefficients are added to the typical values of a specification to see the effects of temperature drift.

TERMINAL BASED CHARACTERISTIC–An Actual Characteristic which has been rotated and translated to remove zero offset and full-scale error.

V_{CC} REJECTION–Ratio of the change in the A/D characteristic to the change in V_{CC}.

ZERO OFFSET–The difference between the expected and actual input voltage corresponding to the first code transition.

intel® PRELIMINARY

809XBH/839XBH/879XBH COMMERCIAL/EXPRESS HMOS MICROCONTROLLER

- **879XBH: an 809XBH with 8 Kbytes of On-Chip EPROM**
- **839XBH: an 809XBH with 8 Kbytes of On-Chip ROM**

- **232 Byte Register File**
- **Register-to-Register Architecture**
- **10-Bit A/D Converter with S/H**
- **Five 8-Bit I/O Ports**
- **20 Interrupt Sources**
- **Pulse-Width Modulated Output**
- **ROM/EPROM Lock**
- **Run-Time Programmable EPROM**
- **Extended Temperature Available**
- **High Speed I/O Subsystem**
- **Full Duplex Serial Port**
- **Dedicated Baud Rate Generator**
- **6.25 μs 16 x 16 Multiply**
- **6.25 μs 32/16 Divide**
- **16-Bit Watchdog Timer**
- **Four 16-Bit Software Timers**
- **Two 16-Bit Counter/Timers**
- **Extended Burn-In Available**

The MCS®-96 family of 16-bit microcontrollers consists of many members, all of which are designed for high-speed control functions. The MCS-96 family members produced using Intel's HMOS-III process are described in this data sheet.

The CPU supports bit, byte, and word operations. Thirty-two bit double-words are supported for a subset of the instruction set. With a 12 MHz input frequency the 8096BH can do a 16-bit addition in 1.0 μs and a 16 x 16-bit multiply or 32/16 divide in 6.25 μs. Instruction execution times average 1 to 2 μs in typical applications.

Four high-speed trigger inputs are provided to record the times at which external events occur. Six high-speed pulse generator outputs are provided to trigger external events at preset times. The high-speed output unit can simultaneously perform software timer functions. Up to four 16-bit software timers can be in operation at once.

The on-chip A/D converter includes a Sample and Hold, and converts up to 8 multiplexed analog input channels to 10-bit digital values. With a 12 MHz crystal, each conversion takes 22 μs. This feature is only available on the 8X95BHs and 8X97BHs, with the 8X95BHs having 4 multiplexed analog inputs.

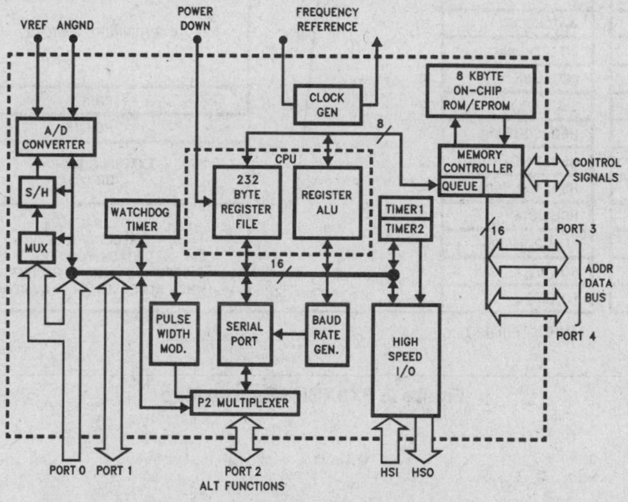

Figure 1. 8X9XBH Block Diagram

July 1992
Order Number: 270090-010

Also provided on-chip are a serial port, a Watchdog Timer and a pulse-width modulated output signal.

With the commercial (standard) temperature option, operational characteristics are guaranteed over the temperature range of 0°C to +70°C. With the extended temperature range option, operational characteristics are guaranteed over the temperature range of −40°C to +85°C. Unless otherwise noted, the specifications are the same for both options.

With the extended burn-in option, the burn-in is dynamic for a minimum time of 160 hours at 125°C with V_{CC} = 5.5V ±0.5V, following the guidelines in MIL-STD-883, Method 1015.

See the Packaging Information for extended temperature and extended burn-in designators.

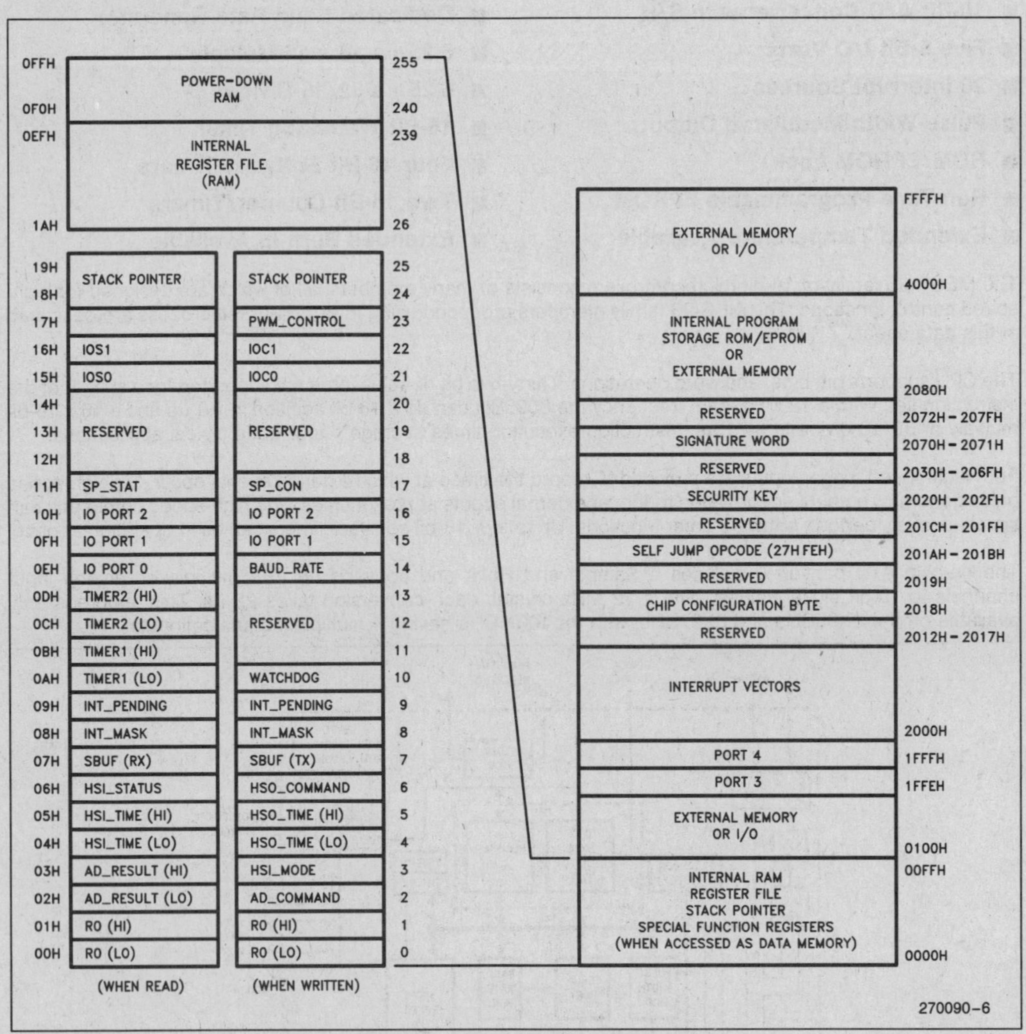

Figure 2. 8X9XBH Memory Map

PACKAGING

The 8096BH is available in 48-pin, 64-pin and 68-pin packages, with and without A/D, and with and without on-chip ROM or EPROM. The 8096BH numbering system is shown in Figure 3. Figures 5–10 show the pinouts for the 48-, 64- and 68-pin packages. The 48-pin version is offered in a Dual-In-Line package while the 68-pin versions come in a Plastic Leaded Chip Carrier (PLCC), a Pin Grid Array (PGA) or a Type "B" Leadless Chip Carrier.

	Factory Masked ROM			CPU			User Programmable					
							EPROM			OTP		
	68-Pin	64-Pin	48-Pin	68-Pin	64-Pin	48-Pin	68-Pin	64-Pin	48-Pin	68-Pin	64-Pin	48-Pin
ANALOG	8397BH	8397BH	8395BH	8097BH	8097BH	8095BH	8797BH		8795BH	8797BH	8797BH	
NO ANALOG	8396BH			8096BH								

Figure 3. 8X9X Packaging

Package Designators:
- N = PLCC
- C = Ceramic DIP
- A = Ceramic Pin Grid Array
- P = Plastic DIP
- R = Ceramic LCC
- U = Shrink DIP

Prefix Designators:
- T = Extended Temperature
- L = Extended Temperature with 160 Hours Burn-in

Package Type	θ_{ja}	θ_{jc}
68L PGA	35°C/W	10°C/W
68L PLCC	37°C/W	13°C/W
68L LCC	28°C/W	14°C/W
64L Shrink DIP	56°C/W	—
48L Plastic DIP	38°C/W	19°C/W
48L Ceramic DIP	26°C/W	6.5°C/W

Figure 4. 8X9XBH Thermal Characteristics

All thermal impedance data is approximate for static air conditions at 1W of power dissipation. Values will change depending on operating conditions and application. See the Intel *Packaging Handbook* (Order Number 240800) for a description of Intel's thermal impedance test methodology.

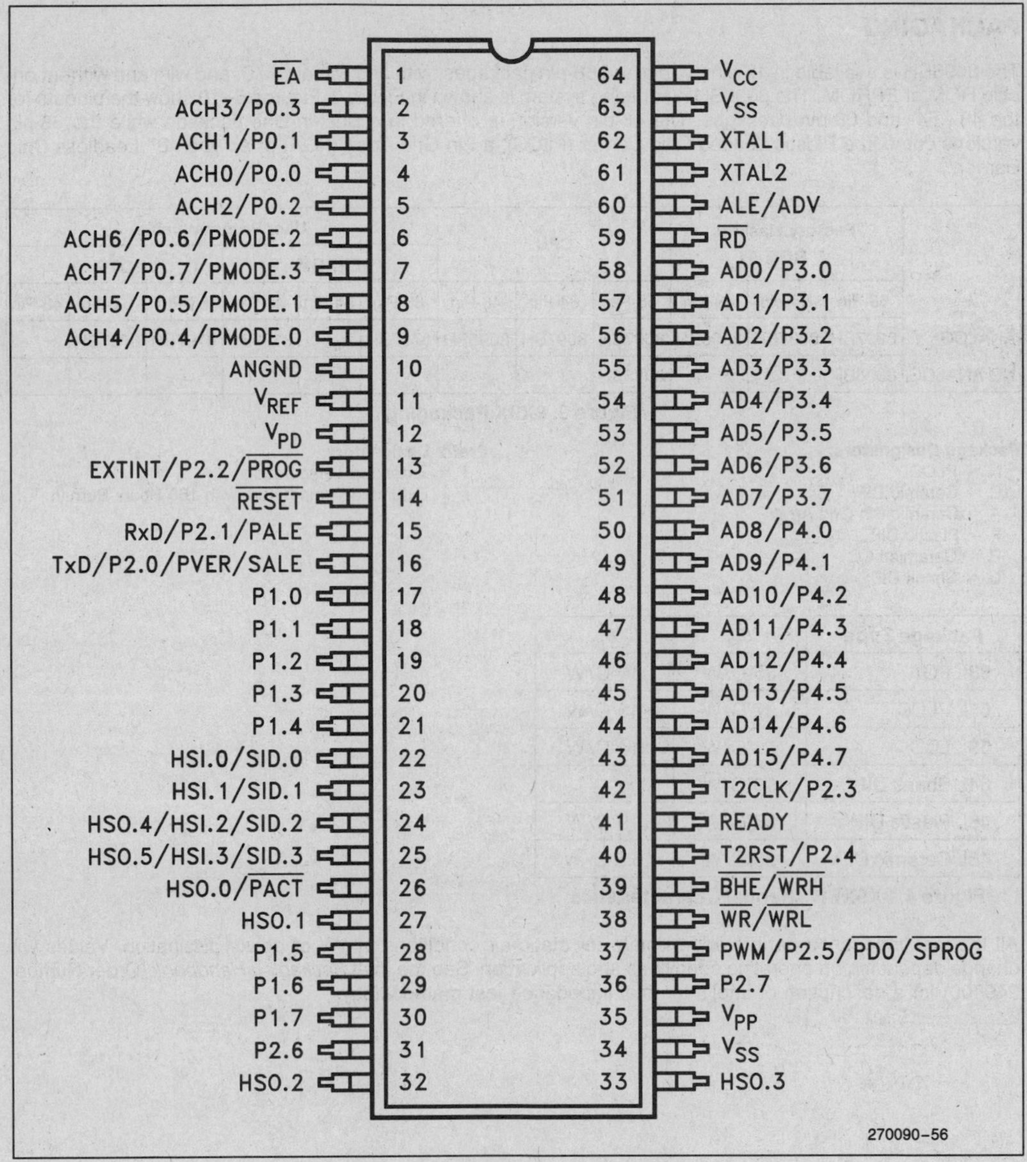

Figure 5. 64-Pin Shrink-DIP Package

Pins Facing Down

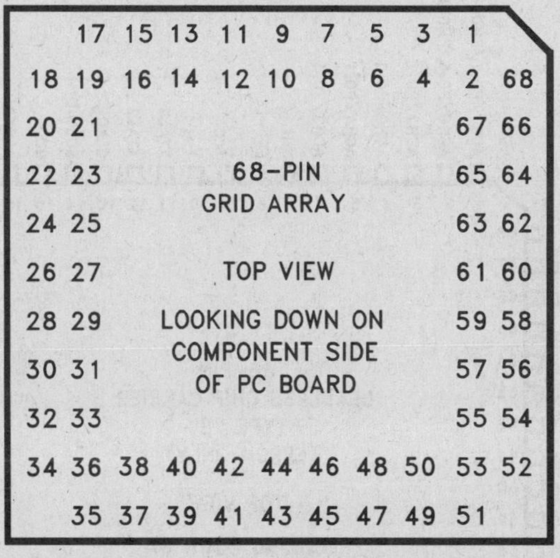

270090–4

Figure 6. 68-Pin PGA Package

PGA	Description	PGA	Description	PGA	Description
1	ACH7/P0.7/PMODE.3	24	AD6/P3.6	47	P1.6
2	ACH6/P0.6/PMODE.2	25	AD7/P3.7	48	P1.5
3	ACH2/P0.2	26	AD8/P4.0	49	HSO.1
4	ACH0/P0.0	27	AD9/P4.1	50	HSO.0/$\overline{PACT}$
5	ACH1/P0.1	28	AD10/P4.2	51	HSO.5/HSI.3/SID.3
6	ACH3/P0.3	29	AD11/P4.3	52	HSO.4/HSI.2/SID.2
7	NMI	30	AD12/P4.4	53	HSI.1/SID.1
8	$\overline{EA}$	31	AD13/P4.5	54	HSI.0/SID.0
9	V_{CC}	32	AD14/P4.6	55	P1.4
10	V_{SS}	33	AD15/P4.7	56	P1.3
11	XTAL1	34	T2CLK/P2.3	57	P1.2
12	XTAL2	35	READY	58	P1.1
13	CLKOUT	36	T2RST/P2.4	59	P1.0
14	BUSWIDTH	37	$\overline{BHE}$/$\overline{WRH}$	60	TXD/P2.0/PVER/SALE
15	INST	38	$\overline{WR}$/$\overline{WRL}$	61	RXD/P2.1/$\overline{PALE}$
16	ALE/$\overline{ADV}$	39	PWM/P2.5/$\overline{PDO}$/$\overline{SPROG}$	62	$\overline{RESET}$
17	$\overline{RD}$	40	P2.7	63	EXTINT/P2.2/$\overline{PROG}$
18	AD0/P3.0	41	V_{PP}	64	V_{PD}
19	AD1/P3.1	42	V_{SS}	65	V_{REF}
20	AD2/P3.2	43	HSO.3	66	ANGND
21	AD3/P3.3	44	HSO.2	67	ACH4/P0.4/PMODE.0
22	AD4/P3.4	45	P2.6	68	ACH5/P0.5/PMODE.1
23	AD5/P3.5	46	P1.7		

Figure 7. PGA Function Pinouts

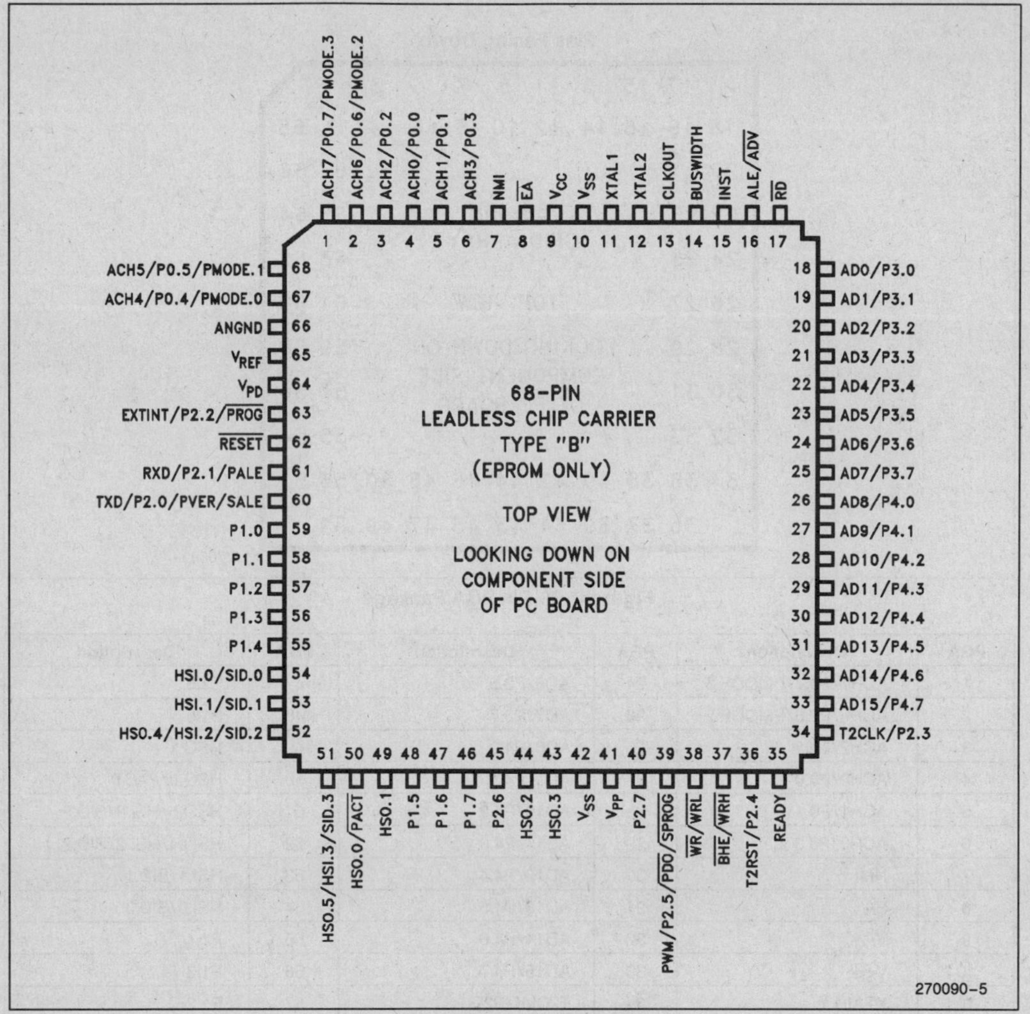

Figure 8. 68-Pin LCC Package

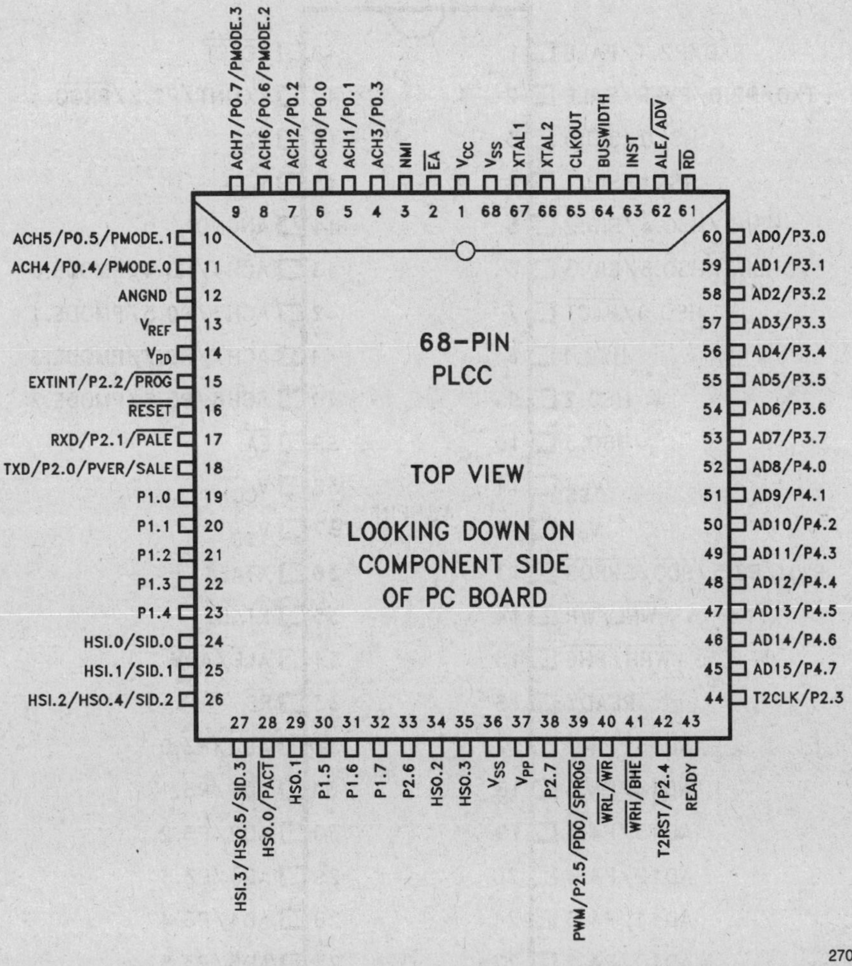

270090-3

Figure 9. 68-Pin PLCC Package

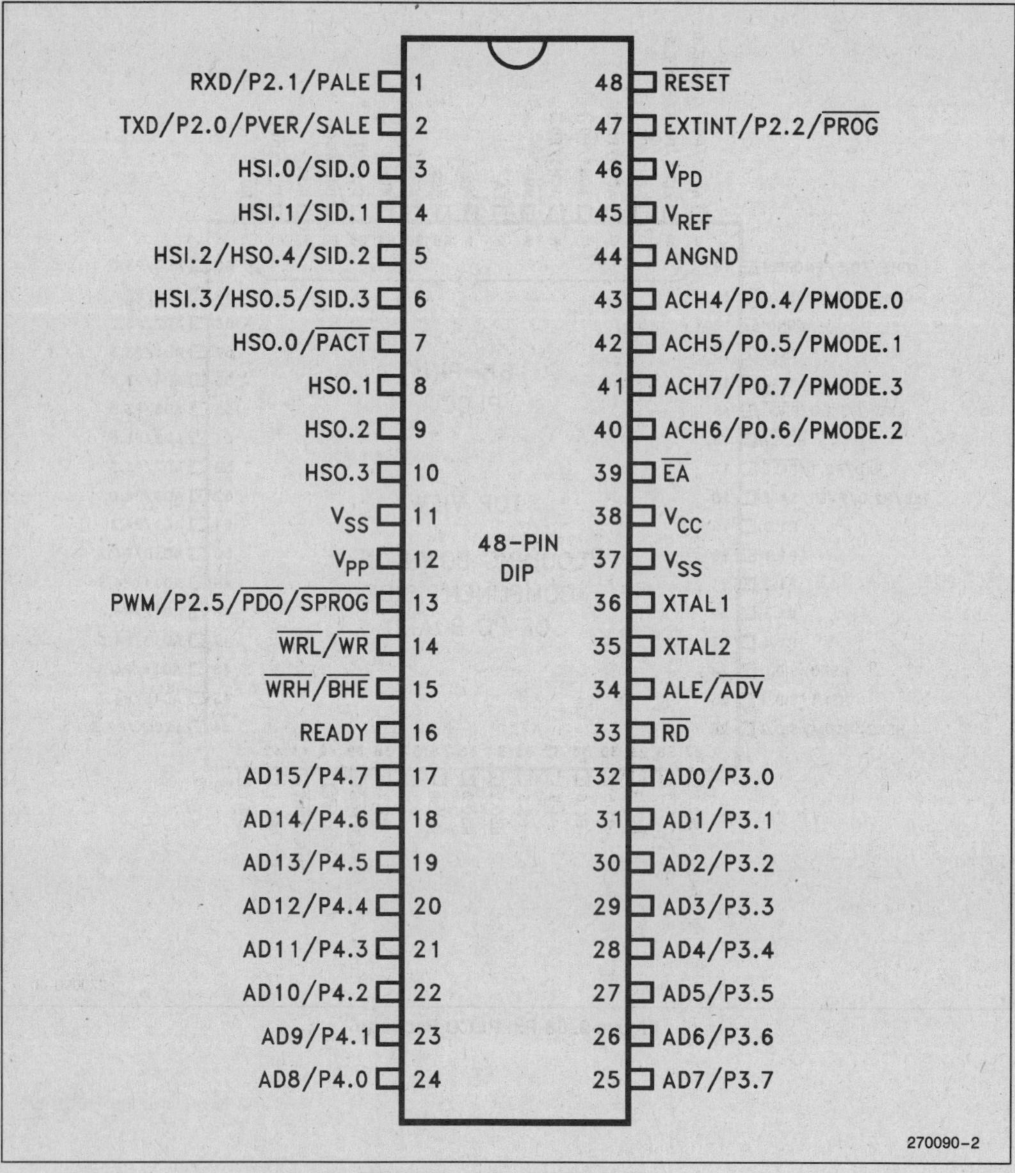

Figure 10. 48-Pin DIP Package

PIN DESCRIPTIONS

Symbol	Name and Function
V_{CC}	Main supply voltage (5V).
V_{SS}	Digital circuit ground (0V). There are two V_{SS} pins, both of which must be connected.
V_{PD}	RAM standby supply voltage (5V). This voltage must be present during normal operation. In a Power Down condition (i.e. V_{CC} drops to zero), if $\overline{RESET}$ is activated before V_{CC} drops below spec and V_{PD} continues to be held within spec., the top 16 bytes in the Register File will retain their contents.
V_{REF}	Reference voltage for the A/D converter (5V). V_{REF} is also the supply voltage to the analog portion of the A/D converter and the logic used to read Port 0. Must be connected to use A/D or Port 0.
ANGND	Reference ground for the A/D converter. Must be held at nominally the same potential as V_{SS}.
V_{PP}	Programming voltage for the EPROM devices. It should be +12.75V for programming and will float to 5V otherwise. The pin should not be above V_{CC} for ROM and CPU devices. This pin must be left floating in the application circuit for EPROM devices.
XTAL1	Input of the oscillator inverter and of the internal clock generator.
XTAL2	Output of the oscillator inverter.
CLKOUT*†	Output of the internal clock generator. The frequency of CLKOUT is ⅓ the oscillator frequency. It has a 33% duty cycle.
$\overline{RESET}$	Reset input to the chip. Input low for a minimum 10 XTAL1 cycles to reset the chip. The subsequent low-to-high transition re-synchronizes CLKOUT and commences a 10-state-time RESET sequence.
BUSWIDTH*†	Input for bus width selection. If CCR bit 1 is a one, this pin selects the bus width for the bus cycle in progress. If BUSWIDTH is a 1, a 16-bit bus cycle occurs. If BUSWIDTH is a 0 an 8-bit cycle occurs. If CCR bit 1 is a 0, the bus is always an 8-bit bus. If this pin is left unconnected, it will rise to V_{CC}.
NMI*†	A positive transition causes a vector to external memory location 0000H.
INST*†	Output high during an external memory read indicates the read is an instruction fetch. INST is valid throughout the bus cycle.
$\overline{EA}$	Input for memory select (External Access). $\overline{EA}$ equal to a TTL-high causes memory accesses to locations 2000H through 3FFFH to be directed to on-chip ROM/EPROM. $\overline{EA}$ equal to a TTL-low causes accesses to these locations to be directed to off-chip memory. $\overline{EA}$ equal to 12.75V causes the device to enter the Programming Mode.
ALE/$\overline{ADV}$	Address Latch Enable or Address Valid output, as selected by CCR. Both pin options provide a latch to demultiplex the address from the address/data bus. When the pin is $\overline{ADV}$, it goes high at the end of the bus cycle. ALE/$\overline{ADV}$ is activated only during external memory accesses.
$\overline{RD}$	Read signal output to external memory. $\overline{RD}$ is activated only during external memory reads.
$\overline{WR}$/$\overline{WRL}$	Write and Write Low output to external memory, as selected by the CCR. $\overline{WR}$ will go low for every external write, while $\overline{WRL}$ will go low only for external writes where an even byte is being written. $\overline{WR}$/$\overline{WRL}$ is activated only during external memory writes.
$\overline{BHE}$/$\overline{WRH}$	Bus High Enable or Write High output to external memory, as selected by the CCR. $\overline{BHE}$ will go low for external writes to the high byte of the data bus. $\overline{WRH}$ will go low for external writes where an odd byte is being written. $\overline{BHE}$/$\overline{WRH}$ is activated only during external memory writes.

PIN DESCRIPTIONS (Continued)

Symbol	Name and Function
READY	Ready input to lengthen external memory cycles. If the pin is low prior to the falling edge of CLKOUT, the memory controller goes into a wait mode until the next positive transition of CLKOUT occurs with READY high. When the external memory is not being used, READY has no effect. Internal control of the number of wait states inserted into a bus cycle held not ready is available in the CCR.
HSI	Inputs to High Speed Input Unit. Four HSI pins are available: HSI.0, HSI.1, HSI.2 and HSI.3. Two of them (HSI.2 and HSI.3) are shared with the HSO Unit.
HSO	Outputs from High Speed Output Unit. Six HSO pins are available: HSO.0, HSO.1, HSO.2, HSO.3, HSO.4 and HSO.5. Two of them (HSO.4 and HSO.5) are shared with the HSI Unit.
Port 0‡	8-bit high impedance input-only port. These pins can be used as digital inputs and/or as analog inputs to the on-chip A/D converter.
Port 1†	8-bit quasi-bidirectional I/O port.
Port 2†	8-bit multi-functional port. Six of its pins are shared with other functions in the 8096BH, the remaining 2 are quasi-bidirectional.
Ports 3 and 4	8-bit bidirectional I/O ports with open drain outputs. These pins are shared with the multiplexed address/data bus which has strong internal pullups. Ports 3 and 4 are also used as a command, address and data path by EPROM devices operating in the Programming Mode.*
PMODE	Determines the EPROM programming mode.
$\overline{\text{PACT}}$	A low signal in Auto Programming Mode indicates that programming is in progress. A high signal indicates programming is complete.
PVAL	A low signal in Auto Programming Mode indicates that the device programmed correctly.
SALE	A falling edge in Auto Programming Mode indicates that Ports 3 and 4 contain valid programming address/command information (output from master).
$\overline{\text{SPROG}}$	A falling edge in Auto Programming Mode indicates that Ports 3 and 4 contain valid programming data (output from master).
SID	Assigns a pin of Ports 3 and 4 to each slave to pass programming verification.
PALE	A falling edge in Slave Programming Mode and Auto Configuration Byte Programming Mode indicates that Ports 3 and 4 contain valid programming address/command information (input to slave).
$\overline{\text{PROG}}$	A falling edge in Slave Programming Mode indicates that Ports 3 and 4 contain valid programming data (input to slave).
PVER	A high signal in Slave Programming Mode and Auto Configuration Byte Programming Mode indicates the byte programmed correctly.
PVAL	A high signal in Slave Programming Mode indicates the device programmed correctly.
$\overline{\text{PDO}}$	A low signal in Slave Programming Mode indicates that the PROG pulse was applied for longer than allowed.

*Not available on Shrink-DIP package
†Not available on 48-pin device
‡Port 0.0.1.2.3 not available on 48-pin device

ELECTRICAL CHARACTERISTICS
ABSOLUTE MAXIMUM RATINGS*

Ambient Temperature Under Bias −55°C to +125°C

Storage Temperature .... −60°C to +150°C

Voltage from $\overline{EA}$ or V_{PP} to V_{SS} or ANGND .... −0.3V to +13.0V

Voltage from Any Other Pin to V_{SS} or ANGND .... −0.3V to +7.0V(1)

Average Output Current from Any Pin .... 10 mA

Power Dissipation(2) 1.5W

NOTES:
1. This includes V_{PP} and $\overline{EA}$ on ROM and CPU only devices.
2. Power dissipation is based on package heat transfer limitations, not device power consumption.

NOTICE: This data sheet contains preliminary information on new products in production. It is valid for the devices indicated in the revision history. The specifications are subject to change without notice.

**WARNING: Stressing the device beyond the "Absolute Maximum Ratings" may cause permanent damage. These are stress ratings only. Operation beyond the "Operating Conditions" is not recommended and extended exposure beyond the "Operating Conditions" may affect device reliability.*

OPERATING CONDITIONS

(All characteristics in this data sheet apply to these operating conditions unless otherwise noted.)

Symbol	Parameter	Min	Max	Units
T_A	Ambient Temperature Under Bias **Commercial** Temp.	0	+70	°C
T_A	Ambient Temperature Under Bias **Extended** Temp.	−40	+85	°C
V_{CC}	Digital Supply Voltage	4.50	5.50	V
V_{REF}	Analog Supply Voltage	4.50	5.50	V
F_{OSC}	Oscillator Frequency	6.0	12	MHz
V_{PD}	Power-Down Supply Voltage	4.50	5.50	V

NOTE:
ANGND and V_{SS} should be nominally at the same potential.

DC CHARACTERISTICS

Symbol	Parameter	Min	Max	Units	Test Conditions
I_{CC}	V_{CC} Supply Current **Commercial** Temp.		240	mA	All Outputs Disconnected.
I_{CC}	V_{CC} Supply Current **Extended** Temp.		270	mA	
I_{CC1}	V_{CC} Supply Current ($T_A \geq 70°C$)		185	mA	
I_{PD}	V_{PD} Supply Current		1	mA	Normal operation and Power-Down.
I_{REF}	V_{REF} Supply Current **Commercial** Temp.		8	mA	
I_{REF}	V_{REF} Supply Current **Extended** Temp.		10	mA	
V_{IL}	Input Low Voltage	−0.3	+0.8	V	

DC CHARACTERISTICS (Continued)

Symbol	Parameter	Min	Max	Units	Test Conditions
V_{IH}	Input High Voltage (Except $\overline{RESET}$, NMI, XTAL1)	2.0	V_{CC} + 0.5	V	
V_{IH1}	Input High Voltage, $\overline{RESET}$ Rising	2.4	V_{CC} + 0.5	V	
V_{IH2}	Input High Voltage, $\overline{RESET}$ Falling (Hysteresis)	2.1	V_{CC} + 0.5	V	
V_{IH3}	Input High Voltage, NMI, XTAL1	2.2	V_{CC} + 0.5	V	
I_{LI}	Input Leakage Current to each pin of HSI, P3, P4, and to P2.1.		±10	μA	V_{in} = 0 to V_{CC}
I_{LI1}	DC Input Leakage Current to each pin of P0		+3	μA	V_{in} = 0 to V_{CC}
I_{IH}	Input High Current to $\overline{EA}$		100	μA	V_{IH} = 2.4V
I_{IL}	Input Low Current to each pin of P1, and to P2.6, P2.7 **Commercial** Temp.		−125	μA	V_{IL} = 0.45V
I_{IL}	Input Low Current to each pin of P1, and to P2.6, P2.7 **Extended** Temp.		−150	μA	V_{IL} = 0.45V
I_{IL1}	Input Low Current to $\overline{RESET}$	−0.25	−2	mA	V_{IL} = 0.45V
I_{IL2}	Input Low Current P2.2, P2.3, P2.4, READY, BUSWIDTH		−50	μA	V_{IL} = 0.45V
V_{OL}	Output Low Voltage on Quasi-Bidirectional port pins and P3, P4 when used as ports		0.45	V	I_{OL} = 0.8 mA (Note 1)
V_{OL1}	Output Low Voltage on Quasi-Bidirectional port pins and P3, P4 when used as ports		0.75	V	I_{OL} = 2.0 mA (Notes 1, 2, 3)
V_{OL2}	Output Low Voltage on Standard Output pins, $\overline{RESET}$ and Bus/Control Pins		0.45	V	I_{OL} = 2.0 mA (Notes 1, 2, 3)
V_{OH}	Output High Voltage on Quasi-Bidirectional pins	2.4		V	I_{OH} = −20 μA (Note 1)
V_{OH1}	Output High Voltage on Standard Output pins and Bus/Control pins	2.4		V	I_{OH} = −200 μA (Note 1)
I_{OH3}	Output High Current on $\overline{RESET}$	−50		μA	V_{OH} = 2.4V
C_S	Pin Capacitance (Any Pin to V_{SS})		10	pF	F_{TEST} = 1.0 MHz

NOTES:
1. Quasi-bidirectional pins include those on P1, for P2.6 and P2.7. Standard Output Pins include TXD, RXD (Mode 0 only), PWM, and HSO pins. Bus/Control pins include CLKOUT, ALE, $\overline{BHE}$, $\overline{RD}$, $\overline{WR}$, INST and AD0–15.
2. Maximum current per pin must be externally limited to the following values if V_{OL} is held above 0.45V.
   - I_{OL} on quasi-bidirectional pins and Ports 3 and 4 when used as ports: 4.0 mA
   - I_{OL} on standard output pins and $\overline{RESET}$: 8.0 mA
   - I_{OL} on Bus/Control pins: 2.0 mA
3. During normal (non-transient) operation the following limits apply:
   - Total I_{OL} on Port 1 must not exceed 8.0 mA.
   - Total I_{OL} on P2.0, P2.6, $\overline{RESET}$ and all HSO pins must not exceed 15 mA.
   - Total I_{OL} on Port 3 must not exceed 10 mA.
   - Total I_{OL} on P2.5, P2.7, and Port 4 must not exceed 20 mA.

AC CHARACTERISTICS

Test Conditions: Load Capacitance on Output Pins = 80 pF

TIMING REQUIREMENTS (The system must meet these specifications to work with the 8X9XBH.)

Symbol	Parameter	Min	Max	Units
T_{CLYX} (2, 3)	READY Hold after CLKOUT Edge	0 (1)		ns
T_{LLYV}	End of ALE/$\overline{ADV}$ to READY Valid		$2\,T_{OSC} - 70$	ns
T_{LLYH}	End of ALE/$\overline{ADV}$ to READY High	$2\,T_{OSC} + 40$	$4\,T_{OSC} - 80$	ns
T_{YLYH}	Non-Ready Time		1000	ns
T_{AVDV} (4)	Address Valid to Input Data Valid		$5\,T_{OSC} - 120$ (5)	ns
T_{RLDV}	$\overline{RD}$ Active to Input Data Valid		$3\,T_{OSC} - 100$ (5)	ns
T_{RHDX}	Data Hold after $\overline{RD}$ Inactive	0		ns
T_{RHDZ}	$\overline{RD}$ Inactive to Input Data Float	0	$T_{OSC} - 25$	ns
T_{AVGV} (2, 4)	Address Valid to BUSWIDTH Valid		$2\,T_{OSC} - 125$	ns
T_{LLGX} (2, 3)	BUSWIDTH Hold after ALE/$\overline{ADV}$ Low	$T_{OSC} + 40$		ns
T_{LLGV} (2, 3)	ALE/$\overline{ADV}$ Low to BUSWIDTH Valid		$T_{OSC} - 75$	ns
T_{RLPV}	Reset Low to Ports Valid		$10\,T_{OSC}$	ns

NOTES:

1. If the 48-pin or 64-pin device is being used then this timing can be generated by assuming that the CLKOUT falling edge has occurred at $2\,T_{OSC} + 55$ (TLLCH(max) + TCHCL(max)) after the falling edge of ALE.
2. Pins not bonded out on 64-pin devices.
3. Pins not bonded out on 48-pin devices.
4. The term "Address Valid" applies to AD0–15, $\overline{BHE}$ and INST.
5. If wait states are used, add $3\,T_{OSC} * N$ where N = number of wait states.

TIMING RESPONSES (MCS-96 devices meet these specs.)

Symbol	Parameter	Min	Max	Units
F_{XTAL}	Oscillator Frequency	6.0	12.0	MHz
T_{OSC}	Oscillator Period	83	166	ns
T_{OHCH}	XTAL1 Rising Edge to Clockout Rising Edge	0[4]	120[4]	ns
T_{CHCH}[1, 4]	CLKOUT Period[3]	3 T_{OSC}[3]	3 T_{OSC}[3]	ns
T_{CHCL}[1, 4]	CLKOUT High Time	$T_{OSC} - 35$	$T_{OSC} + 10$	ns
T_{CLLH}[1, 4]	CLKOUT Low to ALE High	−30	+15	ns
T_{LLCH}[4]	ALE/$\overline{\text{ADV}}$ Low to CLKOUT High[1]	$T_{OSC} - 25$	$T_{OSC} + 45$	ns
T_{LHLL}	ALE/$\overline{\text{ADV}}$ High Time	$T_{OSC} - 30$	$T_{OSC} + 35$[5]	ns
T_{AVLL}[6]	Address Setup to End of ALE/$\overline{\text{ADV}}$	$T_{OSC} - 50$		ns
T_{RLAZ}[7]	$\overline{\text{RD}}$ or $\overline{\text{WR}}$ Low to Address Float **Commercial** Temp.	Typ. = 0	10	ns
T_{RLAZ}[7]	$\overline{\text{RD}}$ or $\overline{\text{WR}}$ Low to Address Float **Extended** Temp.		25	ns
T_{LLRL}	End of ALE/$\overline{\text{ADV}}$ to $\overline{\text{RD}}$ or $\overline{\text{WR}}$ Active	$T_{OSC} - 40$		ns
T_{LLAX}[7]	Address Hold after End of ALE/$\overline{\text{ADV}}$	$T_{OSC} - 40$		ns
T_{WLWH}	$\overline{\text{WR}}$ Pulse Width	3 $T_{OSC} - 35$[2]		ns
T_{QVWH}	Output Data Valid to End of $\overline{\text{WR}}$/$\overline{\text{WRL}}$/$\overline{\text{WRH}}$	3 $T_{OSC} - 60$[2]		ns
T_{WHQX}	Output Data Hold after $\overline{\text{WR}}$/$\overline{\text{WRL}}$/$\overline{\text{WRH}}$	$T_{OSC} - 50$		ns
T_{WHLH}	End of $\overline{\text{WR}}$/$\overline{\text{WRL}}$/$\overline{\text{WRH}}$ to ALE/$\overline{\text{ADV}}$ High	$T_{OSC} - 75$		ns
T_{RLRH}	$\overline{\text{RD}}$ Pulse Width	3 $T_{OSC} - 30$[2]		ns
T_{RHLH}	End of $\overline{\text{RD}}$ to ALE/$\overline{\text{ADV}}$ High	$T_{OSC} - 45$		ns
T_{CLLL}[4]	CLOCKOUT Low[1] to ALE/$\overline{\text{ADV}}$ Low	$T_{OSC} - 40$	$T_{OSC} + 35$	ns
T_{RHBX}[4]	$\overline{\text{RD}}$ High to INST[1], $\overline{\text{BHE}}$, AD8-15 Inactive	$T_{OSC} - 25$	$T_{OSC} + 30$	ns
T_{WHBX}[4]	$\overline{\text{WR}}$ High to INST[1], $\overline{\text{BHE}}$, AD8-15 Inactive	$T_{OSC} - 50$	$T_{OSC} + 100$	ns
T_{HLHH}	$\overline{\text{WRL}}$, $\overline{\text{WRH}}$ Low to $\overline{\text{WRL}}$, $\overline{\text{WRH}}$ High	2 $T_{OSC} - 35$	2 $T_{OSC} + 40$	ns
T_{LLHL}	ALE/$\overline{\text{ADV}}$ Low to $\overline{\text{WRL}}$, $\overline{\text{WRH}}$ Low	2 $T_{OSC} - 30$	2 $T_{OSC} + 55$	ns
T_{QVHL}	Output Data Valid to $\overline{\text{WRL}}$, $\overline{\text{WRH}}$ Low	$T_{OSC} - 60$		ns

NOTES:
1. Pins not bonded out on 64-pin devices.
2. If more than one wait state is desired, add 3 T_{OSC} for each additional wait state.
3. CLKOUT is directly generated as a divide by 3 of the oscillator. The period will be 3 T_{OSC} ±10 ns if T_{OSC} is constant and the rise and fall times on XTAL1 are less than 10 ns.
4. CLKOUT, INST, and $\overline{\text{BHE}}$ pins not bonded out on 48-pin and 64-pin devices.
5. Max spec applies only to ALE. Min spec applies to both ALE and $\overline{\text{ADV}}$.
6. The term "Address Valid" applies to AD0–15, $\overline{\text{BHE}}$ and INST.
7. The term " Address" in this definition applies to AD0–7 for 8-bit cycles, and AD0–15 for 16-bit cycles.

WAVEFORM—SYSTEM BUS TIMINGS

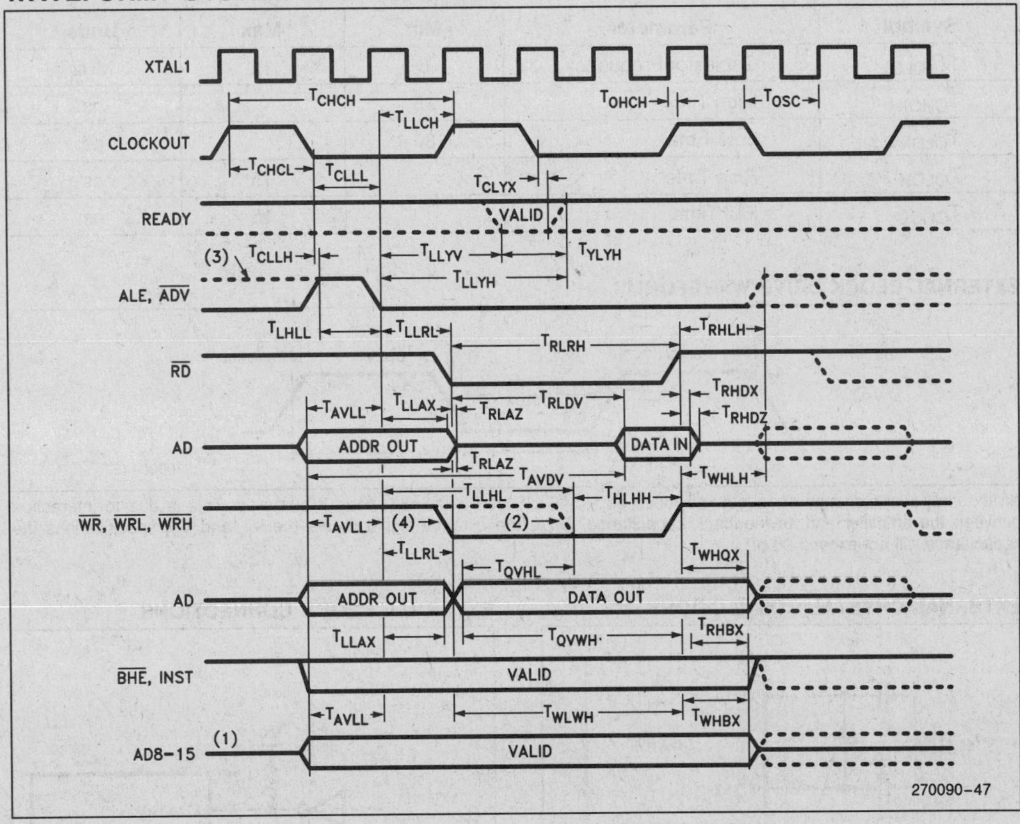

NOTES:
1. 8-bit bus only.
2. 8-bit or 16-bit bus and write strobe mode selected.
3. When $\overline{ADV}$ selected.
4. 8- or 16-bit bus and no write strobe mode selected.

WAVEFORM—BUSWIDTH* TIMINGS

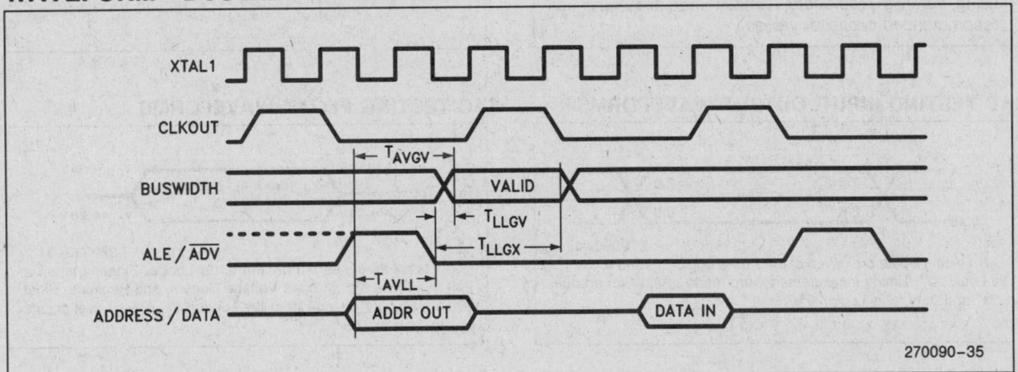

*Buswidth is not bonded out on 48- and 64-pin devices.

EXTERNAL CLOCK DRIVE

Symbol	Parameter	Min	Max	Units
$1/T_{OLOL}$	Oscillator Frequency	6	12	MHz
T_{OHOX}	High Time	25		ns
T_{OLOX}	Low Time	30		ns
T_{OLOH}	Rise Time		15	ns
T_{OHOL}	Fall Time		15	ns

EXTERNAL CLOCK DRIVE WAVEFORMS

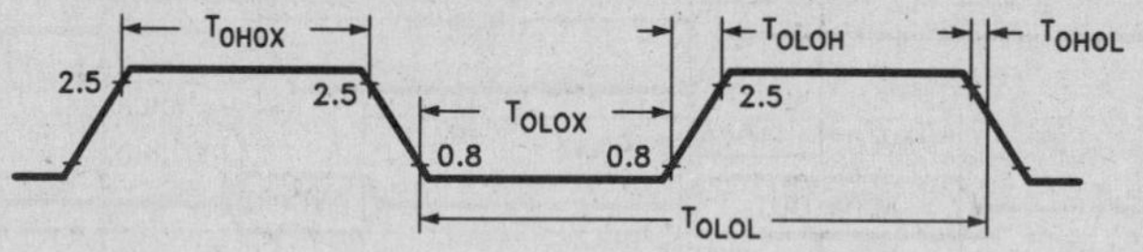

270090–48

An external oscillator may encounter as much as a 100 pF load at XTAL1 when it starts up. This is due to interaction between the amplifier and its feedback capacitance. Once the external signal meets the V_{IL} and V_{IH} specifications the capacitance will not exceed 20 pF.

EXTERNAL CRYSTAL CONNECTIONS

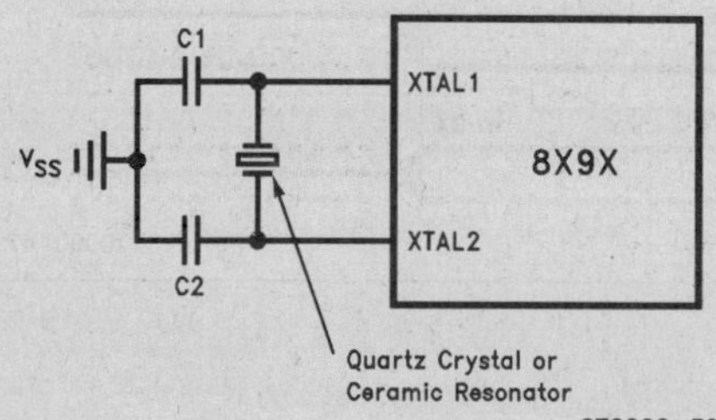

270090–58

NOTE:
Keep oscillator components close to chip and use short, direct traces to XTAL1, XTAL2 and V_{SS}. When using crystals, C1 = 30 pF and C2 = 30 pF. When using ceramic resonators, consult manufacturerer for recommended capacitor values.

EXTERNAL CLOCK CONNECTIONS

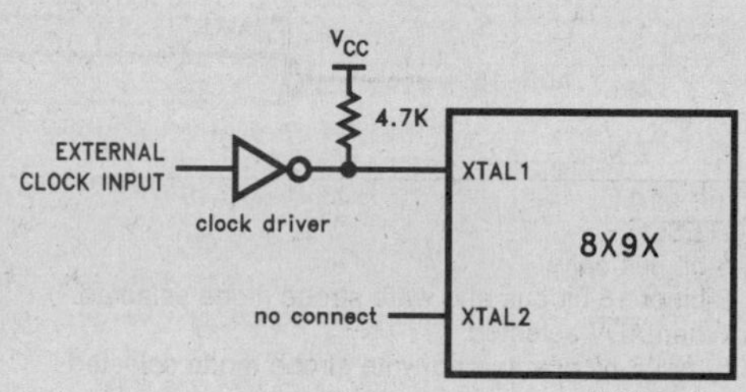

270090–59

AC TESTING INPUT, OUTPUT WAVEFORMS

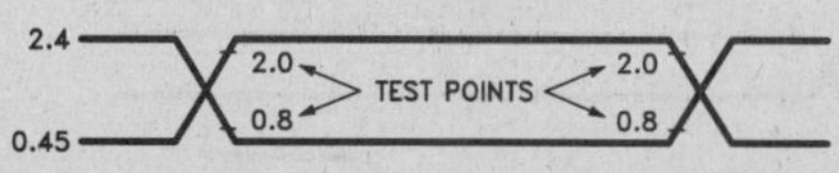

270090–49

AC Testing inputs are driven at 2.4V for a Logic "1" and 0.45V for a Logic "0". Timing measurements are made at 2.0V for a Logic "1" and 0.8V for a Logic "0".

AC TESTING FLOAT WAVEFORMS

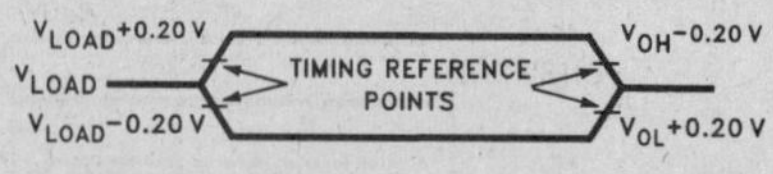

270090–51

For Timing Purposes a Port Pin is no Longer Floating when a 200 mV change from Load Voltage Occurs, and Begins to Float when a 200 mV change from the Loaded V_{OH}/V_{OL} Level occurs $I_{OL}/I_{OH} \geq \pm 8$ mA.

MINIMUM HARDWARE CONFIGURATION CIRCUITS

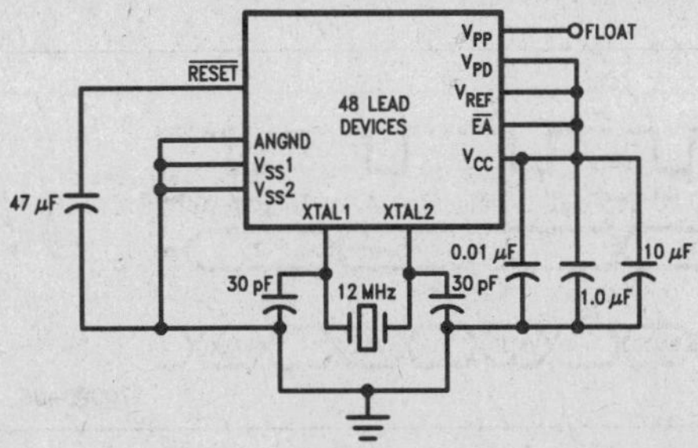

270090-52

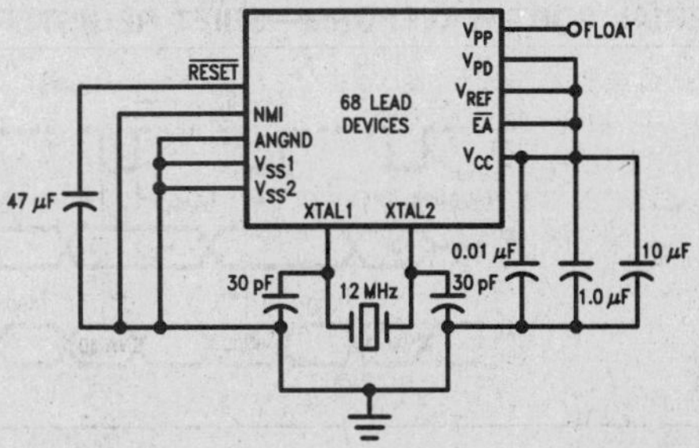

270090-53

AC CHARACTERISTICS—SERIAL PORT—SHIFT REGISTER MODE

SERIAL PORT TIMING—SHIFT REGISTER MODE

Test Conditions: Load Capacitance = 80 pF

Symbol	Parameter	Min	Max	Units
T_{XLXL}	Serial Port Clock Period	8 T_{OSC}		ns
T_{XLXH}	Serial Port Clock Falling Edge to Rising Edge	4 T_{OSC} − 50	4 T_{OSC} + 50	ns
T_{QVXH}	Output Data Setup to Clock Rising Edge	3 T_{OSC}		ns
T_{XHQX}	Output Data Hold After Clock Rising Edge	2 T_{OSC} − 70		ns
T_{XHQV}	Next Output Data Valid After Clock Rising Edge		2 T_{OSC} + 50	ns
T_{DVXH}	Input Data Setup to Clock Rising Edge	2 T_{OSC} + 200		ns
T_{XHDX}	Input Data Hold After Clock Rising Edge	0		ns
T_{XHQZ}	Last Clock Rising to Output Float		5 T_{OSC}	ns

WAVEFORM—T_{RLPV}

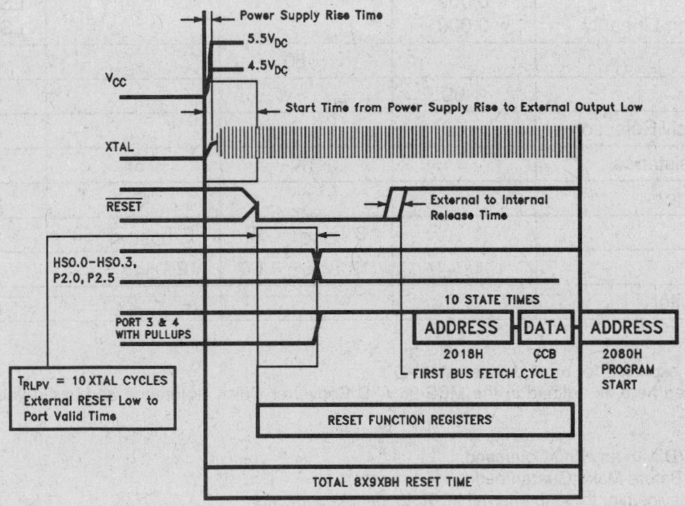

270090-60

WAVEFORM—SERIAL PORT—SHIFT REGISTER MODE

SERIAL PORT WAVEFORM—SHIFT REGISTER MODE

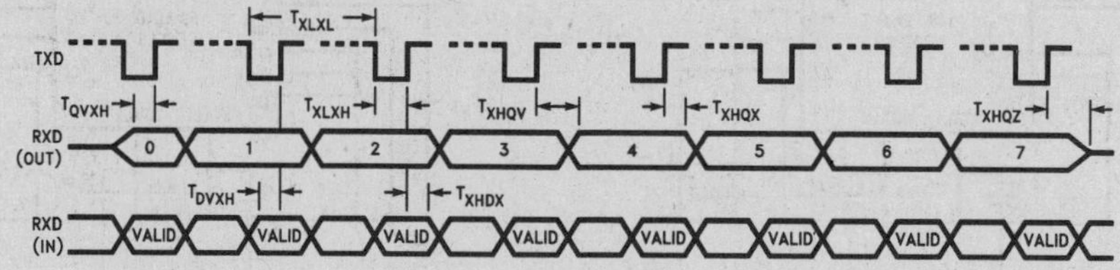

270090–36

A/D CONVERTER SPECIFICATIONS

A/D Converter operation is verified only on the 8097BH, 8397BH, 8095BH, 8395BH, 8797BH, 8795BH.

The absolute conversion accuracy is dependent on the accuracy and stability of V_{REF}.

See the MCS-96 A/D Converter Quick Reference for definitions of A/D Converter terms.

Parameter	Typical*	Minimum	Maximum	Units**	Notes
Resolution		1024 10	1024 10	Levels Bits	
Absolute Error		0	±4	LSBs	
Full Scale Error	−0.5 ±0.5			LSBs	
Zero Offset Error	±0.5			LSBs	
Non-Linearity		0	±4	LSBs	
Differential Non-Linearity		> −1	+2	LSBs	
Channel-to-Channel Matching		0	±1	LSBs	
Repeatability	±0.25			LSBs	
Temperature Coefficients: Offset Full Scale Differential Non-Linearity	 0.009 0.009 0.009			 LSB/°C LSB/°C LSB/°C	
Off Isolation		−60		dB	1, 3
Feedthrough	−60			dB	1
V_{CC} Power Supply Rejection	−60			dB	1
Input Series Resistance		1K	5K	Ω	4
DC Input Leakage		0	3.0	μA	
Sample Delay		$3 T_{OSC} - 50$	$3 T_{OSC} + 50$	ns	2
Sample Time		$12 T_{OSC} - 50$	$12 T_{OSC} + 50$	ns	
Sampling Capacitor			2	pF	

NOTES:
* These values are expected for most devices at 25°C.
** An "LSB", as used here, is defined in the MCS-96 A/D Converter Quick Reference and has a value of approximately 5 mV.
1. DC to 100 KHz.
2. For starting the A/D with an HSO Command.
3. Multiplexer Break-Before-Make Guaranteed.
4. Resistance from device pin, through internal MUX, to sample capacitor.

EPROM SPECIFICATIONS

EPROM PROGRAMMING OPERATING CONDITIONS

Symbol	Parameter	Min	Max	Units
T_A	Ambient Temperature during Programming	20	30	°C
V_{CC}, V_{PD}, V_{REF}(1)	Supply Voltages during Programming	4.5	5.5	V
V_{EA}	Programming Mode Supply Voltage	9.0	13.0	V(2)
V_{PP}	EPROM Programming Supply Voltage	12.50	13.0	V(2)
V_{SS}, ANGND(3)	Digital and Analog Ground	0	0	V
$F_{OSC}1$	Oscillator Frequency during Auto and Slave Programming	6.0	6.0	MHz
$F_{OSC}2$	Oscillator Frequency during Run-Time Programming	6.0	12.0	MHz

NOTES:
1. V_{CC}, V_{PD} and V_{REF} should nominally be at the same voltage during programming.
2. V_{EA} and V_{PP} must never exceed the maximum voltage for any amount of time or the device may be damaged.
3. V_{SS} and ANGND should nominally be at the same voltage (0V) during programming.

AC EPROM PROGRAMMING CHARACTERISTICS

Symbol	Parameter	Min	Max	Units
T_{AVLL}	ADDRESS/COMMAND Valid to PALE Low	0		T_{OSC}
T_{LLAX}	ADDRESS/COMMAND Hold After PALE Low	80		T_{OSC}
T_{DVPL}	Output Data Setup Before $\overline{PROG}$ Low	0		T_{OSC}
T_{PLDX}	Data Hold After $\overline{PROG}$ Falling	80		T_{OSC}
T_{LLLH}	PALE Pulse Width	180		T_{OSC}
T_{PLPH}	$\overline{PROG}$ Pulse Width	250 T_{OSC}	100 μs + 144 T_{OSC}	
T_{LHPL}	PALE High to $\overline{PROG}$ Low	250		T_{OSC}
T_{PHLL}	$\overline{PROG}$ High to Next PALE Low	600		T_{OSC}
T_{PHDX}	Data Hold After $\overline{PROG}$ High	30		T_{OSC}
T_{PHVV}	$\overline{PROG}$ High to PVER/$\overline{PDO}$ Valid	500		T_{OSC}
T_{LLVH}	PALE Low to PVER/$\overline{PDO}$ High	100		T_{OSC}
T_{PLDV}	$\overline{PROG}$ Low to VERIFICATION/DUMP Data Valid	100		T_{OSC}
T_{SHLL}	RESET High to First PALE Low (not shown)	2000		T_{OSC}

DC EPROM PROGRAMMING CHARACTERISTICS

Symbol	Parameter	Min	Max	Units
I_{PP}	V_{PP} Supply Current (Whenever Programming)		100	mA

WAVEFORM—EPROM PROGRAMMING

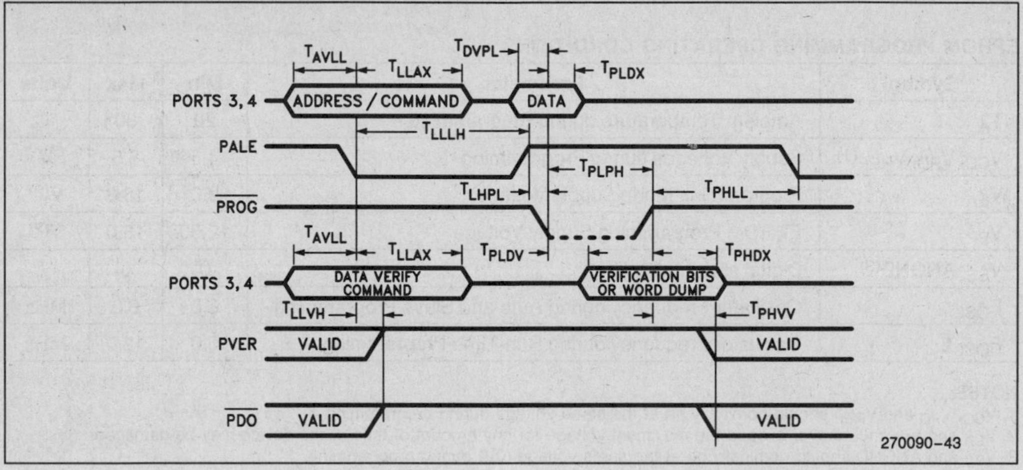

8X9XBH ERRATA

Devices covered by this data sheet (see Revision History) have the following errata.

1. INDEXED, 3 OPERAND MULTIPLY

The displacement portion of an indexed, three operand (byte or word) multiply may not be in the range of 200H thru 17FFH inclusive. If you must use these displacements, execute an indexed, two operand multiply and a move if necessary.

2. HSI FIFO OPERATION

The High Speed Input (HSI) has three deviations from the specifications. Note that "events" are defined as one or more pin transitions. "Entries" are defined as the recording of one or more events.

A. The resolution is nine states instead of eight states. Events occurring on the same pin more frequently than once every nine states may be lost.

B. A mismatch between the nine state HSI resolution and the eight state hardware timer causes one time-lag value to be skipped every nine timer counts. Events may receive a time-tag one count later than expected.

C. If the FIFO and Holding Register are empty, the first event will transfer into the Holding Register, leaving the FIFO empty again. The next event that occurs will be the first event loaded into the empty FIFO. If the first two events into an **empty** FIFO (not counting the Holding Register) occur coincident with each other, both are recorded as one entry with one time-tag. If the second event occurs within 9 states after the first, the events will be entered separately with time-tags at least one count apart. If the second event enters the FIFO coincident with the "skipped" time-tag situation (see B above) the time tags will be at least two counts apart.

3. RESERVED LOCATION 2019H

The 1990 Architectural Overview recommended that address 2019H be loaded with 0FFH. The recommendation is now 20H.

4. RESERVED LOCATION 201CH

Reading reserved location 201CH, either internally or externally, will return "201C" as data.

5. SERIAL PORT SECTION

Serial Port Flags—Reading SP_STAT may not clear the TI or RI flag if that flag was set within two state times prior to the read. In addition, the parity error bit (RPE/RB8) may not be correct if it is read within two state times after RI is set. Use the following code to replace ORB sp_image, SP_STAT.

```
SP_READ: LDB TEMP, SP_STAT
         ORB SP_IMAGE, TEMP
         JBS TEMP,5,SP_READ; if TI is
         set then read again
         JBS TEMP,6,SP_READ; if RI is
         set then read again
         ANDB SP_IMAGE,#7FH; clear
         false RB8/RPE
         ORB SP_IMAGE, TEMP; load
         correct RB8/RPE
```

DATA SHEET REVISION HISTORY

This data sheet (270090-010) is valid for devices marked with an "E" at the end of the topside tracking number. Data sheets are changed as new device information becomes available. Verify with your local Intel sales office that you have the latest version before finalizing a design or ordering devices.

The difference between this data sheet (-010) and the previous one (-009) is the I_{OL}/I_{OH} for float waveform testing changed from ±15 mA to ±8 mA (this data sheet).

The following differences exist between (-009) data sheet and (-008).

1. The Express (extended temperature and burn-in options) were added to this data sheet. The 8X9XBH Express data sheet (270433-004) is now obsolete.
2. Changes were made to the format of the data sheet and the SFR descriptions were removed. No specification changes made.
3. Added Reserved Location 201CH errata.

The following differences exist between the -008 data sheet and the -007 data sheet.

1. The -007 data sheet was valid for devices marked with a "D" at the end of the top side tracking number.
2. The following errata were removed: RESET and the Quasi-Bidirectional Ports, Software RESET Timing, and Using T2CLK as the source for Timer2.
3. The HSI FIFO Operation errata definition was changed to match change in the HSI FIFO Operation.

The following differences exist between the -007 data sheet and the -006 data sheet.

1. T_{CCLH} changed from Min = −20 ns, Max = +25 ns to Min = −30 ns, Max = +15 ns.
2. T_{XHQX} changed from Min = 2 T_{OSC} − 50 ns to Min = 2 T_{OSC} − 70 ns.
3. T_{OLOX} changed from Min = 25 ns to Min = 30 ns.
4. An errata was added changing the recommendation for address 2019H from 0FFH to 20H.
5. The power supply sequencing section has been deleted. The information is in the Hardware Design Information.
6. The method of identifying the current change indicator was added to the differences between the -005 and -004 data sheets.
7. A bug was not documented in the -004 data sheet and was fixed before the -005 data sheet. Information on the bug was added to the difference between the -005 and -004 data sheets.

Differences between -006 and -005 data sheets.

1. All EPROM programming mode information has been deleted and moved to the Hardware Design Information chapter.
2. Shrink-DIP package information has been added.
3. A new RESET timing specification has been added for clarity.
4. Software Reset pin timing information has been added.
5. HSO I_{OL} specifications have been improved so that all HSO pins have the same drive capability.
6. Port 3 and Port 4 pin descriptions were clarified, indicating the necessity of pullup if the pins are used as ports.
7. HSI FIFO overflow description added.

Differences between the -005 and the -004 data sheets.

1. The -005 data sheet corresponds to devices marked with a "D" at the end of the topside tracking number. The -004 data sheet corresponded to devices which are not marked with a "D".
2. Much of the description of device functionality has been deleted. All of this information is already in the MCS-96 Architectural Overview.
3. The A/D converter specification for Differential Non-linearity has been changed to be a minimum of > −1 lsbs to a maximum of +2 LSBs.
4. 8X9XBH errata section. The JBS and JBC on Port 0 errata has been fixed on the latest device stepping.
5. 8X9XBH errata section. The errata for the 48-pin devices has been fixed on the latest device stepping. This errata caused the upper 8 bits on the Address/Data bus to be latched when resetting into an 8-bit external memory system.
6. 8X9XBH errata section. An errata existed which caused the device to be held in RESET for extended periods of time with the internal RESET pin pulled down internally. The condition occurred when the XTAL inputs were driven before V_{CC} was stable and within the data sheet specification. The condition was worse at cold. This errata was not documented in the -004 data sheet. It has been fixed on the latest device stepping.
7. 8X9XBH errata section. Errata 3 and 4 have been added to the errata list. These errata exist for all steppings of the device.

Differences between the -004 and the -003 data sheets.

1. The bus control figures and bus timing diagrams were modified to more accurately describe their operation. In particular the 8-bit bus modes now reflect the use of Write Strobe Mode.
2. Additional text was added to the Analog/Digital description of the conversion process to clarify its operation and usefulness.
3. Text was added to the interrupt description section to indicate the maximum transition speed of the input signal relative to the CPU's state timing. A figure was included to graphically demonstrate the interrupt response timing.
4. The pin descriptions were modified to indicate that V_{PP} must normally float in the application.
5. The input low voltage specification (V_{IL1}) was deleted and is covered by the V_{IL} specification.
6. A suggested minimum configuration circuit was added to the material.
7. The A/D Converter Specifications for Differential Non-Linearity has been corrected to be a maximum of +2 LSB's.
8. The EPROM programming section figures were corrected to indicate the correct interface to a 2764A-2. A reset circuit was added to these figures and the signal $\overline{PVAL}$ (Port 3.X and Port 4.X) is now identified as the valid signal for program verification in the Auto Programming Mode. Text was added to this section to reference the requirement of using the Auto Configuration Byte Programming Mode for 48-lead devices. Figure 22A was edited for corrections to the text, and now indicates PVER (Port 2.0). The EPROM circuits were corrected to show 6 MHz operation for programming devices from internal microcode.
9. The protected memory section was edited to indicate that the CPU will enter a "JUMP ON SELF" condition when ROM/EPROM dump mode is complete.
10. An 8X9XBH ERRATA section was added.
11. This REVISION HISTORY was added.

8097JF/8397JF/8797JF COMMERCIAL/EXPRESS HMOS MICROCONTROLLER

8797JF: an 8097JF with 16 Kbytes of On-Chip EPROM
8397JF: an 8097JF with 16 Kbytes of On-Chip ROM

- 232 Byte Register File
- 256 Bytes XRAM for Code
- 10-Bit A/D Converter with S/H
- Five 8-Bit I/O Ports
- 20 Interrupt Sources
- Pulse-Width Modulated Output
- ROM/EPROM Lock
- Run-Time Programmable EPROM (OTP)
- Extended Temperature Available
- High Speed I/O Subsystem
- Full Duplex Serial Port
- Dedicated Baud Rate Generator
- 6.25 μs 16 x 16 Multiply
- 6.25 μs 32/16 Divide
- 16-Bit Watchdog Timer
- Four 16-Bit Software Timers
- Two 16-Bit Counter/Timers
- Extended Burn-In Available

The MCS 96 microcontroller family of 16-bit microcontrollers consists of many members, all of which are designed for high-speed control functions. The MCS-96 family members produced using Intel's HMOS-III process are described in this data sheet.

The CPU supports bit, byte, and word operations. Thirty-two bit double-words are supported for a subset of the instruction set. With a 12 MHz input frequency the 8097JF can do a 16-bit addition in 1.0 μs and a 16 x 16-bit multiply or 32/16 divide in 6.25 μs. Instruction execution times average 1 to 2 μs in typical applications.

4

Four high-speed trigger inputs are provided to record the times at which external events occur. Six high-speed pulse generator outputs are provided to trigger external events at preset times. The high-speed output unit can simultaneously perform software timer functions. Up to four 16-bit software timers can be in operation at once.

The on-chip A/D converter includes a Sample and Hold, and converts up to 8 multiplexed analog input channels to 10-bit digital values. With a 12 MHz crystal, each conversion takes 22 μs.

Also provided on-chip are a serial port, a Watchdog Timer and a pulse-width modulated output signal.

With the commercial (standard) temperature option, operational characteristics are guaranteed over the temperature range of 0°C to +70°C. With the extended temperature range option, operational characteristics are guaranteed over the temperature range of −40°C to +85°C. Unless otherwise noted, the specifications are the same for both options.

With the extended burn-in option, the burn-in is dynamic for a minimum time of 160 hours at 125°C with V_{CC} = 5.5V ± 0.5V, following the guidelines in MIL-STD-883, Method 1015.

See the Packaging information for extended temperature and extended burn-in designators.

November 1994
Order Number: 270795-006

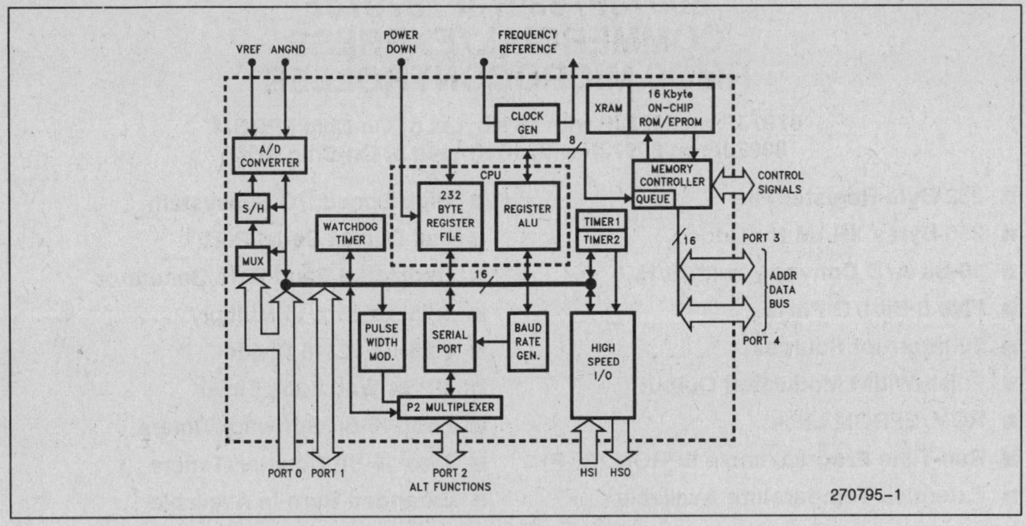

Figure 1. 8X97JF Block Diagram

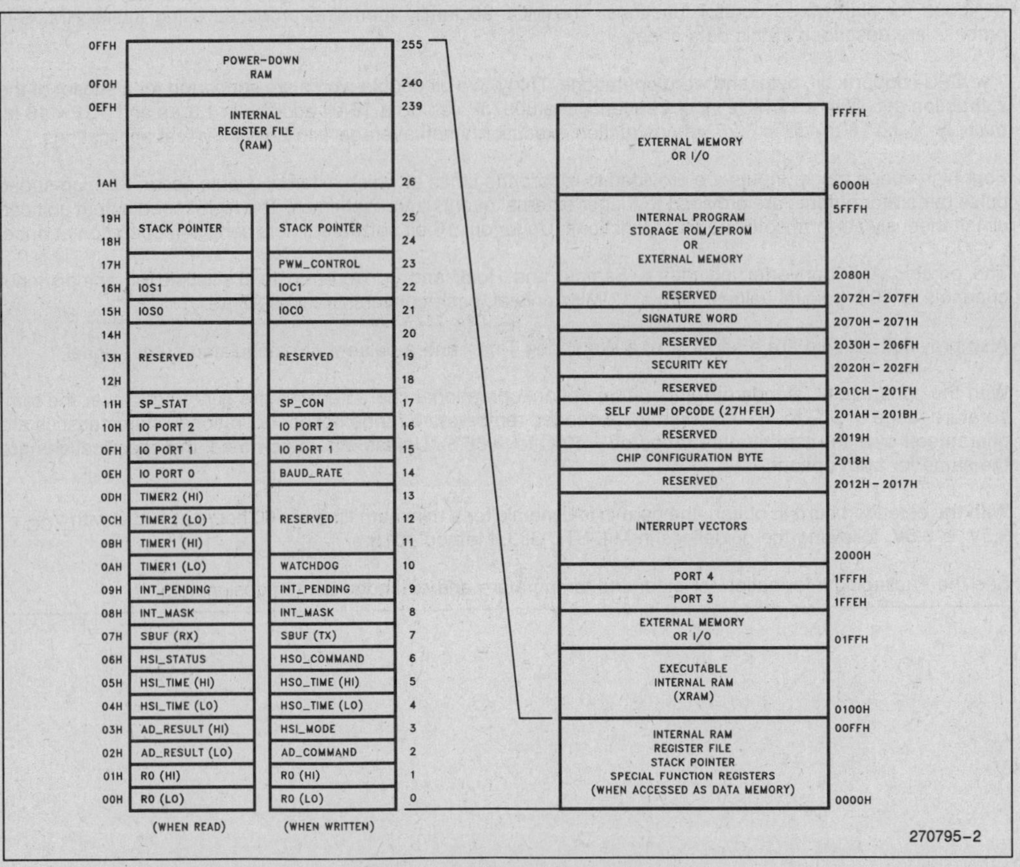

Figure 2. 8X97JF Memory Map

PACKAGING

The 8097JF is available in 64-pin and 68-pin packages, with and without on-chip ROM or EPROM. The 8097JF numbering system is shown in Figure 3. Figures 5–6 show the pinouts for the 64- and 68-pin packages. The 64-pin version is offered in a Shrink-DIP package while the 68-pin versions come in a Plastic Leaded Chip Carrier (PLCC).

8X97JF PACKAGING

Factory Masked ROM		CPU		User Programmable OTP	
68-Pin	64-Pin	68-Pin	64-Pin	68-Pin	64-Pin
8397JF	8397JF	8097JF	8097JF	8797JF	8797JF

Figure 3. The 8097JF Family Nomenclature

Package Designators:
N = PLCC
U = Shrink DIP

Prefix Designators:
T = Extended Temperature
L = Extended Temperature with 160 hrs Burn-in

Package Type	θ_{ja}	θ_{jc}
68L PLCC	37°C/W	13°C/W
64L Shrink DIP	56°C/W	—

Figure 4. 8X97JF Thermal Characteristics

All thermal impedance data is approximate for static air conditions a 1W of power dissipation. Values will change depending on operating conditions and application. See the Intel *Packaging Handbook* (Order Number 240800) for a description of Intel's thermal impedance test methodology.

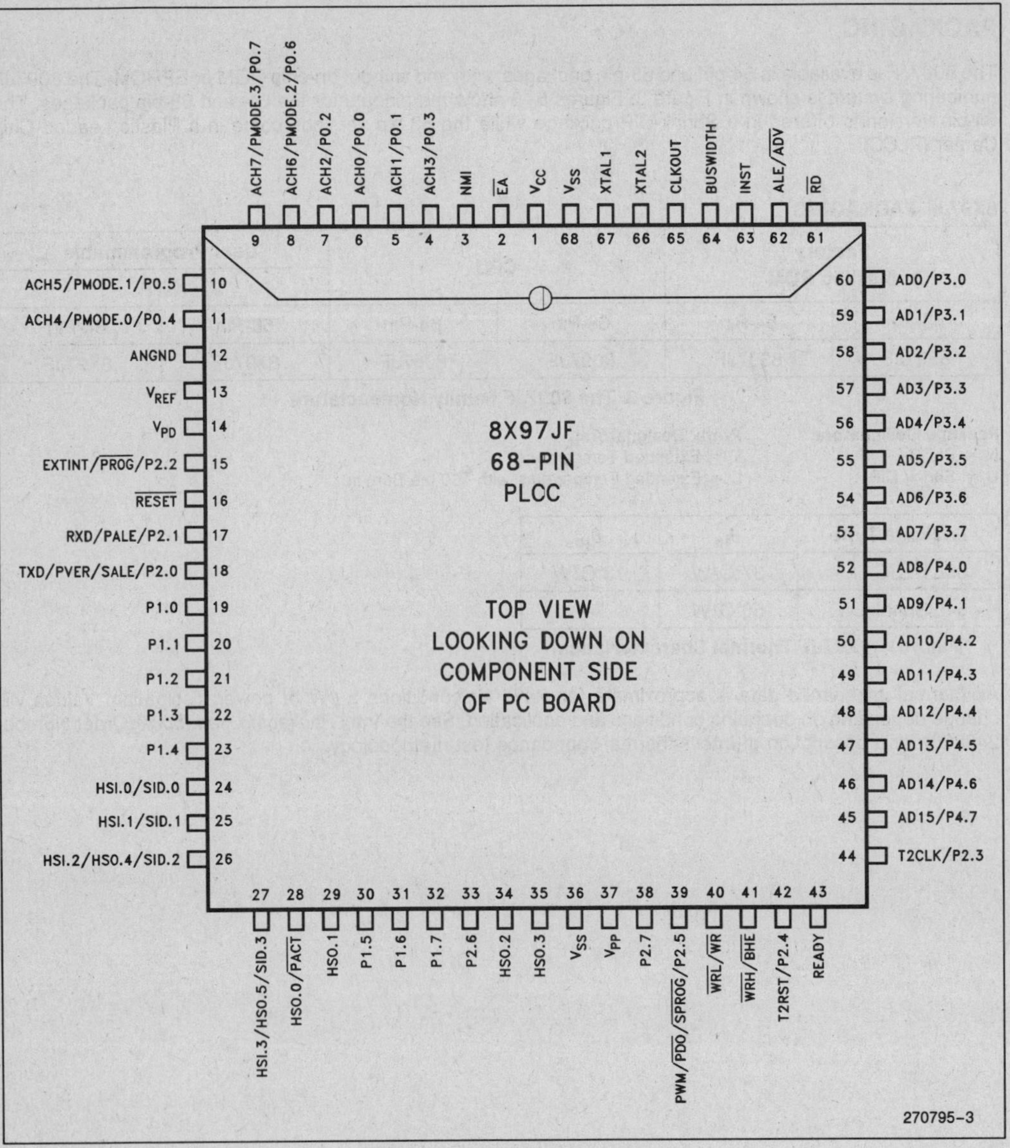

Figure 5. 68-Pin PLCC Package

PRELIMINARY

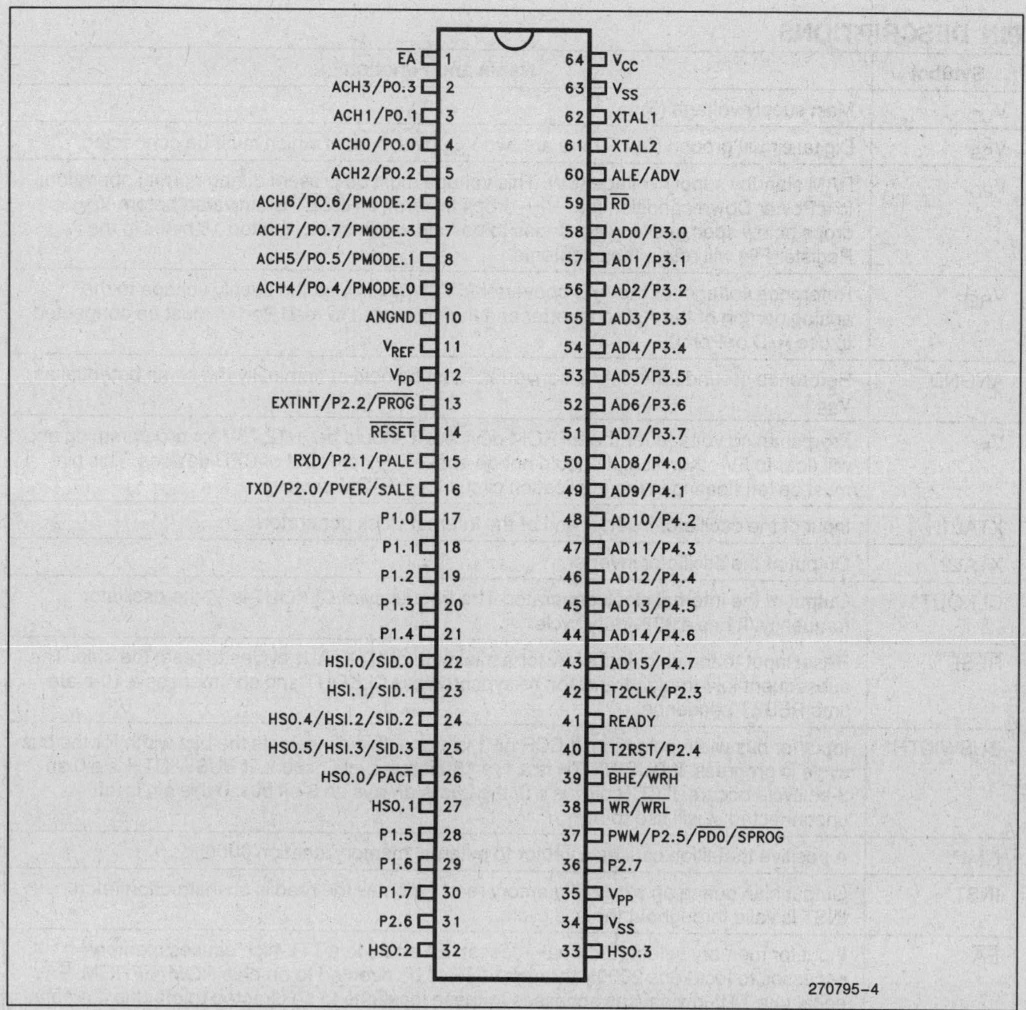

Figure 6. Shrink-DIP Package

PIN DESCRIPTIONS

Symbol	Name and Function
V_{CC}	Main supply voltage (5V).
V_{SS}	Digital circuit ground (0V). There are two V_{SS} pins, both of which must be connected.
V_{PD}	RAM standby supply voltage (5V). This voltage must be present during normal operation. In a Power Down condition (i.e., V_{CC} drops to zero), if $\overline{RESET}$ is activated before V_{CC} drops below spec and V_{PD} continues to be held within spec., the top 16 bytes in the Register File will retain their contents.
V_{REF}	Reference voltage for the A/D converter (5V). V_{REF} is also the supply voltage to the analog portion of the A/D converter and the logic used to read Port 0. Must be connected to use A/D or Port 0.
ANGND	Reference ground for the A/D converter. Must be held at nominally the same potential as V_{SS}.
V_{PP}	Programming voltage for the EPROM devices. It should be +12.75V for programming and will float to 5V otherwise. It should not be above V_{CC} for ROM or CPU devices. This pin must be left floating in the application circuit for EPROM devices.
XTAL1	Input of the oscillator inverter and of the internal clock generator.
XTAL2	Output of the oscillator inverter.
CLKOUT*	Output of the internal clock generator. The frequency of CLKOUT is ⅓ the oscillator frequency. It has a 33% duty cycle.
$\overline{RESET}$	Reset input to the chip. Input low for a minimum of 10 XTAL1 cycles to reset the chip. The subsequent low-to-high transition re-synchronizes CLKOUT and commences a 10-state-time RESET sequence.
BUSWIDTH*	Input for bus width selection. If CCR bit 1 is a one, this pin selects the bus width for the bus cycle in progress. If BUSWIDTH is a 1, a 16-bit bus cycle occurs. If BUSWIDTH is a 0 an 8-bit cycle occurs. If CCR bit 1 is a 0, the bus is always an 8-bit bus. If this pin is left unconnected, it will rise to V_{CC}.
NMI*	A positive transition causes a vector to external memory location 0000H.
INST*	Output high during an external memory read indicates the read is an instruction fetch. INST is valid throughout the bus cycle.
$\overline{EA}$	Input for memory select (External Access). $\overline{EA}$ equal to a TTL-high causes memory accesses to locations 2000H through 5FFF to be directed to on-chip ROM/EPROM. $\overline{EA}$ equal to a TTL-low causes accesses to these locations to be directed to off-chip memory. $\overline{EA}$ = +12.75V causes the device to enter the Programming Mode.
ALE/$\overline{ADV}$	Address Latch Enable or Address Valid output, as selected by CCR. Both pin options provide a latch to demultiplex the address from the address/data bus. When the pin is $\overline{ADV}$, it goes inactive high at the end of the bus cycle. ALE/$\overline{ADV}$ is activated only during external memory accesses.
$\overline{RD}$	Read signal output to external memory. $\overline{RD}$ is activated only during external memory reads.
$\overline{WR}$/$\overline{WRL}$	Write and Write Low output to external memory, as selected by the CCR. $\overline{WR}$ will go low for every external write, while $\overline{WRL}$ will go low only for external writes where an even byte is being written. $\overline{WR}$/$\overline{WRL}$ is activated only during external memory writes.
$\overline{BHE}$/$\overline{WRH}$	Bus High Enable or Write High output to external memory, as selected by the CCR. $\overline{BHE}$ will go low for external writes to the high byte of the data bus. $\overline{WRH}$ will go low for external writes where an odd byte is being written. $\overline{BHE}$/$\overline{WRH}$ is activated only during external memory writes.

*Not available on Shrink-DIP Package

PIN DESCRIPTIONS (Continued)

Symbol	Name and Function
READY	Ready input to lengthen external memory cycles. If the pin is low prior to the falling edge of CLKOUT, the memory controller goes into a wait mode until the next positive transition of CLKOUT occurs with READY high. When the external memory is not being used, READY has no effect. Internal control of the number of wait states inserted into a bus cycle held not ready is available in the CCR.
HSI	Inputs to High Speed Input Unit. Four HSI pins are available: HSI.0, HSI.1, HSI.2 and HSI.3. Two of them (HSI.2 and HSI.3) are shared with the HSO Unit.
HSO	Outputs from High Speed Output Unit. Six HSO pins are available: HSO.0, HSO.1, HSO.2, HSO.3, HSO.4 and HSO.5. Two of them (HSO.4 and HSO.5) are shared with the HSI Unit.
Port 0	8-bit high impedance input-only port. These pins can be used as digital inputs and/or as analog inputs to the on-chip A/D converter.
Port 1	8-bit quasi-bidirectional I/O port.
Port 2	8-bit multi-functional port. Six of its pins are shared with other functions in the 8096JF, the remaining 2 are quasi-bidirectional. These pins are also used to input and output control signals on EPROM devices in Programming Mode.
Ports 3 and 4	8-bit bidirectional I/O ports with open drain outputs. These pins are shared with the multiplexed address/data bus, which has strong internal pull-ups. Ports 3 and 4 are also used as a command, address and data path by EPROM devices operating in the Programming Mode.
PMODE	Determines the EPROM programming mode.
$\overline{\text{PACT}}$	A low signal in Auto Programming Mode indicates that programming is in progress. A high signal indicates programming is complete.
PVAL	A low signal in Auto Programming Mode indicates that the device programmed correctly.
SALE	A falling edge in Auto Programming Mode indicates that Ports 3 and 4 contain valid programming address/command information (output from master).
$\overline{\text{SPROG}}$	A falling edge in Auto Programming Mode indicates that Ports 3 and 4 contain valid programming data (output from master).
SID	Assigns a pin of Ports 3 and 4 to each slave to pass programming verification.
PALE	A falling edge in Slave Programming Mode and Auto Configuration Byte Programming Mode indicates that Ports 3 and 4 contain valid programming address/command information (input to slave).
$\overline{\text{PROG}}$	A falling edge in Slave Programming Mode indicates that Ports 3 and 4 contain valid programming data (input to slave).
PVER	A high signal in Slave Programming Mode and Auto Configuration Byte Programming Mode indicates the byte programmed correctly.
PVAL	A high signal in Slave Programming Mode indicates the device programmed correctly.
$\overline{\text{PDO}}$	A low signal in Slave Programming Mode indicates that the PROG pulse was applied for longer than allowed.

4

ELECTRICAL CHARACTERISTICS
ABSOLUTE MAXIMUM RATINGS*

Ambient Temperature
Under Bias .... −55°C to +125°C

Storage Temperature .... −60°C to +150°C

Voltage from $\overline{EA}$ or V_{PP}
to V_{SS} or ANGND .... −0.3V to +13.0V

Voltage from Any Other Pin to
V_{SS} or ANGND .... −0.3V to +7.0V(1)

Average Output Current from Any Pin .... 10 mA

Power Dissipation(2) 1.5W

NOTES:

1. This includes V_{PP} and $\overline{EA}$ on ROM and CPU only devices.
2. Power dissipation is based on package heat transfer characteristics, not device power consumption.

NOTICE: This data sheet contains preliminary information on new products in production. It is valid for the devices indicated in the revision history. The specifications are subject to change without notice.

**WARNING: Stressing the device beyond the "Absolute Maximum Ratings" may cause permanent damage. These are stress ratings only. Operation beyond the "Operating Conditions" is not recommended and extended exposure beyond the "Operating Conditions" may affect device reliability.*

OPERATING CONDITIONS

(All characteristics specified in this data sheet apply to these operating conditions unless otherwise noted.)

Symbol	Parameter	Min	Max	Units
T_A	Ambient Temperature Under Bias **Commercial** Temp.	0	+70	°C
T_A	Ambient Temperature Under Bias **Extended** Temp.	−40	+85	°C
V_{CC}	Digital Supply Voltage	4.50	5.50	V
V_{REF}	Analog Supply Voltage	4.50	5.50	V
F_{OSC}	Oscillator Frequency	6.0	12	MHz
V_{PD}	Power-Down Supply Voltage	4.50	5.50	V

NOTE:
ANGND and V_{SS} should be nominally at the same potential.

DC CHARACTERISTICS

Symbol	Parameter	Min	Max	Units	Test Conditions
I_{CC}	V_{CC} Supply Current **Commercial** Temp.		300	mA	All Outputs Disconnected
I_{CC}	V_{CC} Supply Current **Extended** Temp.		330	mA	
I_{CC}	V_{CC} Supply Current ($T_A \geq 70°C$)		245	mA	
I_{PD}	V_{PD} Supply Current		1	mA	Normal operation and Power-Down
I_{REF}	V_{REF} Supply Current **Commercial** Temp.		8	mA	
I_{REF}	V_{REF} Supply Current **Extended** Temp.		10	mA	
V_{IL}	Input Low Voltage	−0.3	+0.8	V	
V_{IL1}	Input Low Voltage, $\overline{RESET}$ **Commercial** Temp.	−0.3	+0.8	V	
V_{IL1}	Input Low Voltage, $\overline{RESET}$ **Extended** Temp.	−0.3	+0.7	V	

DC CHARACTERISTICS (Continued)

Symbol	Parameter	Min	Max	Units	Test Conditions
V_{IH}	Input High Voltage (Except $\overline{RESET}$, NMI, XTAL1)	2.0	V_{CC} + 0.5	V	
V_{IH1}	Input High Voltage, $\overline{RESET}$ Rising	2.4	V_{CC} + 0.5	V	
V_{IH2}	Input High Voltage, $\overline{RESET}$ Falling (Hysteresis)	2.1	V_{CC} + 0.5	V	
V_{IH3}	Input High Voltage, NMI, XTAL1 **Commercial** Temp.	2.2	V_{CC} + 0.5	V	
V_{IH3}	Input High Voltage, NMI, XTAL1 **Extended** Temp.	2.3	V_{CC} + 0.5	V	
I_{LI}	Input Leakage Current to each pin of HSI, P3, P4, and to P2.1		±10	μA	V_{IN} = 0 to V_{CC}
I_{LI1}	DC Input Leakage Current to each pin of P0		+3	μA	V_{IN} = 0 to V_{CC}
I_{IH}	Input High Current to $\overline{EA}$		100	μA	V_{IH} = 2.4V
I_{IL}	Input Low Current to each pin of P1, and to P2.6, P2.7 **Commercial** Temp.		−125	μA	V_{IL} = 0.45V
I_{IL}	Input Low Current to each pin of P1, and to P2.6, P2.7 **Extended** Temp.		−150	μA	V_{IL} = 0.45V
I_{IL1}	Input Low Current to $\overline{RESET}$	−0.25	−2	mA	V_{IL} = 0.45V
I_{IL2}	Input Low Current P2.2, P2.3, P2.4, READY, BUSWIDTH		−50	μA	V_{IL} = 0.45V
V_{OL}	Output Low Voltage on Quasi-Bidirectional port pins and P3, P4 when used as ports		0.45	V	I_{OL} = 0.8 mA (Note 1)
V_{OL1}	Output Low Voltage on Quasi-Bidirectional port pins and P3, P4 when used as ports		0.75	V	I_{OL} = 2.0 mA (Notes 1, 2, 3)
V_{OL2}	Output Low Voltage on Standard Output pins, $\overline{RESET}$ and Bus/Control Pins		0.45	V	I_{OL} = 2.0 mA (Notes 1, 2, 3)
V_{OH}	Output High Voltage on Quasi-Bidirectional pins	2.4		V	I_{OH} = −20 μA (Note 1)
V_{OH1}	Output High Voltage on Standard Output pins and Bus/Control pins	2.4		V	I_{OH} = −200 μA (Note 1)
I_{OH3}	Output High Current on $\overline{RESET}$	−50		μA	V_{OH} = 2.4V
C_S	Pin Capacitance (Any Pin to V_{SS})		10	pF	F_{TEST} = 1.0 MHz

NOTES:
1. Quasi-bidirectional pins include those on P1, for P2.6 and P2.7. Standard Output Pins include TXD, RXD (Mode 0 only), PWM and HSO pins. Bus/Control pins include CLKOUT, ALE, $\overline{BHE}$, $\overline{RD}$, $\overline{WR}$, INST and AD0–15.
2. Maximum current per pin must be externally limited to the following values if V_{OL} is held above 0.45V.
   I_{OL} on quasi-bidirectional pins and Ports 3 and 4 when used as ports: 4.0 mA
   I_{OL} on standard output pins and $\overline{RESET}$: 8.0 mA
   I_{OL} on Bus/Control pins: 2.0 mA
3. During normal (non-transient) operation the following limits apply:
   Total I_{OL} on Port 1 must not exceed 8.0 mA.
   Total I_{OL} on P2.0, P2.6, $\overline{RESET}$ and all HSO pins must not exceed 15 mA.
   Total I_{OL} on Port 3 must not exceed 10 mA.
   Total I_{OL} on P2.5, P2.7, and Port 4 must not exceed 20 mA.

AC CHARACTERISTICS

Test Conditions: Load Capacitance on Output Pins = 80 pF

TIMING REQUIREMENTS (The system must meet these specifications to work with the 8X97JF)

Symbol	Parameter	Min	Max	Units
T_{CLYX}(3)	READY Hold after CLKOUT Edge	0(1)		ns
T_{LLYV}	End of ALE/$\overline{ADV}$ to READY Valid		2 T_{OSC} − 70	ns
T_{LLYH}	End of ALE/$\overline{ADV}$ to READY High	2 T_{OSC} + 40	4 T_{OSC} − 80	ns
T_{YLYH}	Non-Ready Time		1000	ns
T_{AVDV}(2)	Address Valid to Input Data Valid		5 T_{OSC} − 120(4)	ns
T_{RLDV}	$\overline{RD}$ Active to Input Data Valid		3 T_{OSC} − 100(4)	ns
T_{RHDX}	Data Hold after $\overline{RD}$ Inactive	0		ns
T_{RHDZ}	$\overline{RD}$ Inactive to Input Data Float	0	T_{OSC} − 25	ns
T_{AVGV}(2,3)	Address Valid to BUSWIDTH Valid		2 T_{OSC} − 125	ns
T_{LLGX}(3)	BUSWIDTH Hold after ALE/$\overline{ADV}$ Low	T_{OSC} + 40		ns
T_{LLGV}(3)	ALE/$\overline{ADV}$ Low to BUSWIDTH Valid		T_{OSC} − 100	ns
T_{RLPV}	Reset Low to Ports Valid		10 T_{OSC}	ns

NOTES:
1. If the 64-pin device is being used then this timing can be generated by assuming that the CLKOUT falling edge has occurred at 2 T_{OSC} + 55 (TLLCH(max) + TCHCL(max)) after the falling edge of ALE.
2. The term "Address Valid" applies to AD0–15, $\overline{BHE}$ and INST.
3. Pins not bonded out on 64-pin devices.
4. If wait states are used, add 3 T_{OSC} * N where N = number of wait states.

TIMING RESPONSES (8X97JF devices meet these specs.)

Symbol	Parameter	Min	Max	Units
F_{XTAL}	Oscillator Frequency	6.0	12.0	MHz
T_{OSC}	Oscillator Period	83	166	ns
T_{OHCH}(3)	XTAL1 Rising Edge to Clockout Rising Edge	0	120	ns
T_{CHCH}(3)	CLKOUT Period	3 T_{OSC}(2)	3 T_{OSC}(2)	ns
T_{CHCL}(3)	CLKOUT High Time	$T_{OSC} - 35$	$T_{OSC} + 10$	ns
T_{CLLH}(3)	CLKOUT Low to ALE High	− 30	+ 15	ns
T_{LLCH}(3)	ALE/$\overline{ADV}$ Low to CLKOUT High	$T_{OSC} - 25$	$T_{OSC} + 45$	ns
T_{LHLL}	ALE/$\overline{ADV}$ High Time	$T_{OSC} - 30$	$T_{OSC} + 35$(4)	ns
T_{AVLL}(5)	Address Setup to End of ALE/$\overline{ADV}$	$T_{OSC} - 50$		ns
T_{RLAZ}(6)	$\overline{RD}$ or $\overline{WR}$ Low to Address Float **Commercial** Temp.	Typ. = 0	10	ns
T_{RLAZ}(6)	$\overline{RD}$ or $\overline{WR}$ Low to Address Float **Extended** Temp.		25	ns
T_{LLRL}	End of ALE/$\overline{ADV}$ to $\overline{RD}$ or $\overline{WR}$ Active	$T_{OSC} - 40$		ns
T_{LLAX}(6)	Address Hold after End of ALE/$\overline{ADV}$	$T_{OSC} - 40$		ns
T_{WLWH}	$\overline{WR}$ Pulse Width	3 $T_{OSC} - 35$(1)		ns
T_{QVWH}	Output Data Valid to End of $\overline{WR}$/$\overline{WRL}$/$\overline{WRH}$	3 $T_{OSC} - 60$(1)		ns
T_{WHQX}	Output Data Hold after $\overline{WR}$/$\overline{WRL}$/$\overline{WRH}$	$T_{OSC} - 50$		ns
T_{WHLH}	End of $\overline{WR}$/$\overline{WRL}$/$\overline{WRH}$ to ALE/$\overline{ADV}$ High	$T_{OSC} - 75$		ns
T_{RLRH}	$\overline{RD}$ Pulse Width	3 $T_{OSC} - 30$(1)		ns
T_{RHLH}	End of $\overline{RD}$ to ALE/$\overline{ADV}$ High	$T_{OSC} - 45$		ns
T_{CLLL}(3)	CLOCKOUT Low to ALE/$\overline{ADV}$ Low	$T_{OSC} - 40$	$T_{OSC} + 35$	ns
T_{RHBX}(3)	$\overline{RD}$ High to INST, $\overline{BHE}$, AD8-15 Inactive	$T_{OSC} - 25$	$T_{OSC} + 30$	ns
T_{WHBX}(3)	$\overline{WR}$ High to INST, $\overline{BHE}$, AD8-15 Inactive	$T_{OSC} - 50$	$T_{OSC} + 100$	ns
T_{HLHH}	$\overline{WRL}$, $\overline{WRH}$ Low to $\overline{WRL}$, $\overline{WRH}$ High	2 $T_{OSC} - 35$	2 $T_{OSC} + 40$	ns
T_{LLHL}	ALE/$\overline{ADV}$ Low to $\overline{WRL}$, $\overline{WRH}$ Low	2 $T_{OSC} - 30$	2 $T_{OSC} + 55$	ns
T_{QVHL}	Output Data Valid to $\overline{WRL}$, $\overline{WRH}$ Low	$T_{OSC} - 60$		ns

NOTES:
1. If more than one wait state is desired, add 3 T_{OSC} for each additional wait state.
2. CLKOUT is directly generated as a divide by 3 of the oscillator. The period will be 3 T_{OSC} ±10 ns if T_{OSC} is constant and the rise and fall times on XTAL1 are less than 10 ns.
3. CLKOUT, INST and $\overline{BHE}$ pins not bonded out on 64-lead package.
4. Max spec applies only to ALE. Min spec applies to both ALE and $\overline{ADV}$.
5. The term "Address Valid" applies to AD0–15, $\overline{BHE}$ and INST.
6. The term " Address" in this specification applies to AD0–7 for 8-bit cycles, and AD0–15 for 16-bit cycles.

WAVEFORM—SYSTEM BUS TIMINGS

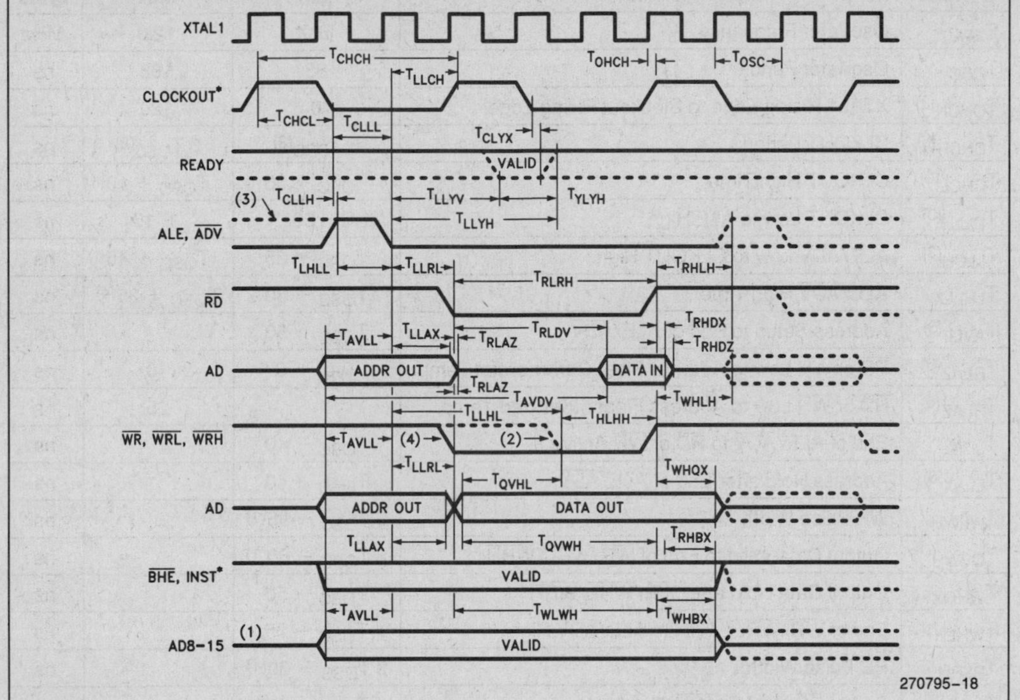

NOTES:
1. 8-bit bus only.
2. 8-bit or 16-bit bus and write strobe mode selected.
3. When $\overline{ADV}$ selected.
4. 8- or 16-bit bus and no write strobe mode selected.

WAVEFORM—BUSWIDTH* TIMINGS

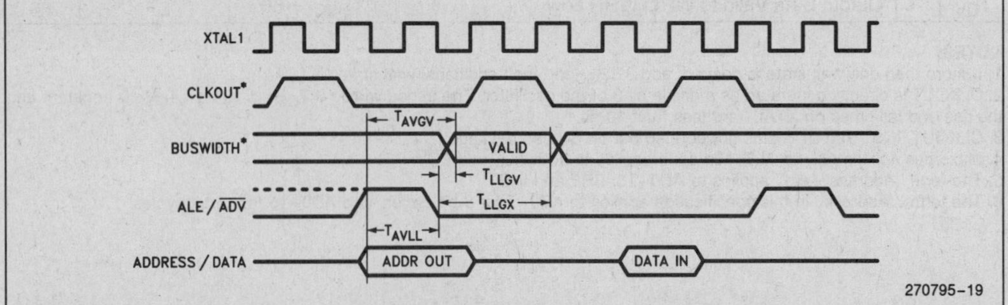

*Not available on 64-lead package.

WAVEFORM—T_{RLPV}

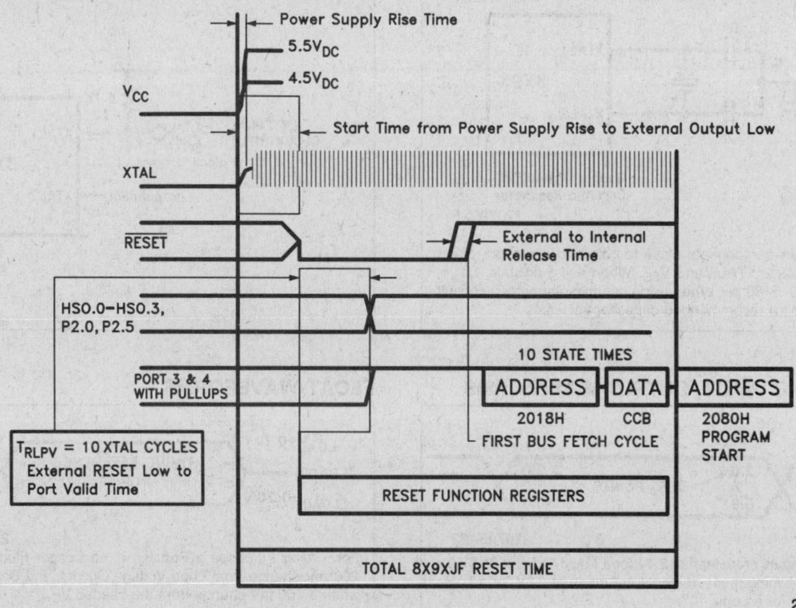

270795–20

EXTERNAL CLOCK DRIVE

Symbol	Parameter	Min	Max	Units
$1/T_{OLOL}$	Oscillator Frequency	6	12	MHz
T_{OHOX}	High Time	25		ns
T_{OLOX}	Low Time	30		ns
T_{OLOH}	Rise Time		15	ns
T_{OHOL}	Fall Time		15	ns

EXTERNAL CLOCK DRIVE WAVEFORMS

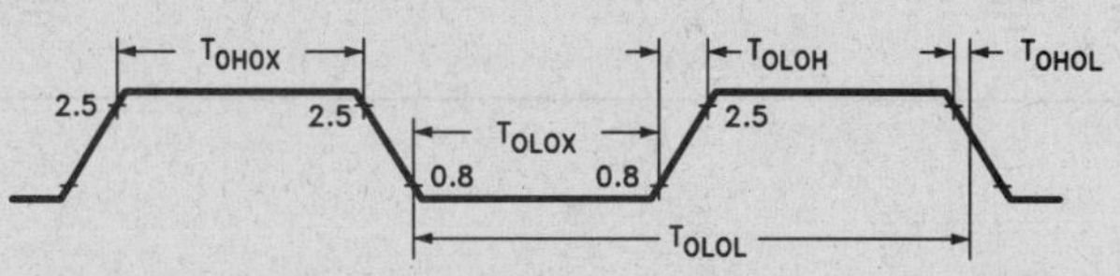

270795–22

An external oscillator may encounter as much as a 100 pF load at XTAL1 when it starts-up. This is due to interaction between the amplifier and its feedback capacitance. Once the external signal meets the V_{IL} and V_{IH} specifications the capacitance will not exceed 20 pF.

EXTERNAL CRYSTAL CONNECTIONS

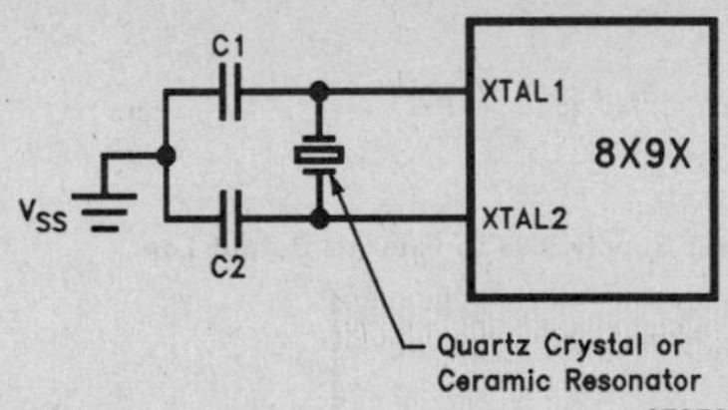

270795-5

NOTE:
Keep oscillator components close to chip and use short, direct traces to XTAL1, XTAL2 and V_{SS}. When using crystals, C1 = 30 pF and C2 = 30 pF. When using ceramic resonators, consult manufacturer for recommended capacitor values.

EXTERNAL CLOCK CONNECTIONS

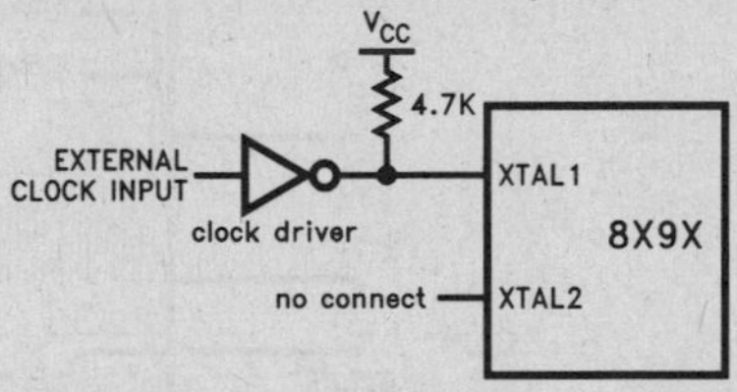

270795-6

AC TESTING INPUT, OUTPUT WAVEFORMS

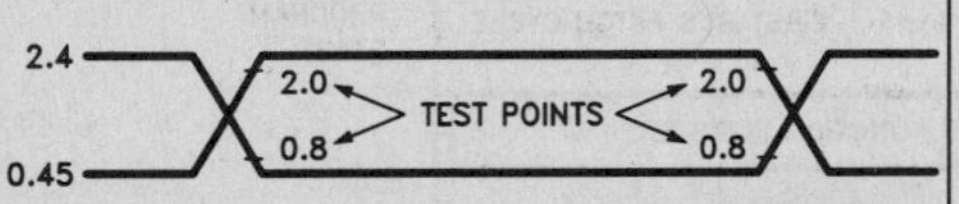

270795-23

AC Testing inputs are driven at 2.4V for a Logic "1" and 0.45V for a Logic "0". Timing measurements are made at 2.0V for a Logic "1" and 0.8V for a Logic "0".

FLOAT WAVEFORMS

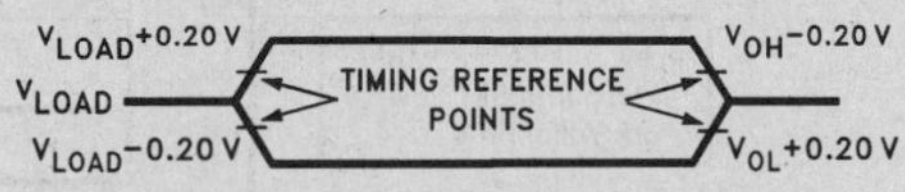

270795-24

For Timing Purposes a Port Pin is no Longer Floating when a 200 mV change from Load Voltage Occurs, and Begins to Float when a 200 mV change from the Loaded V_{OH}/V_{OL} Level occurs $I_{OL}/I_{OH} \geq \pm 8$ mA.

MINIMUM HARDWARE CONFIGURATION CIRCUITS

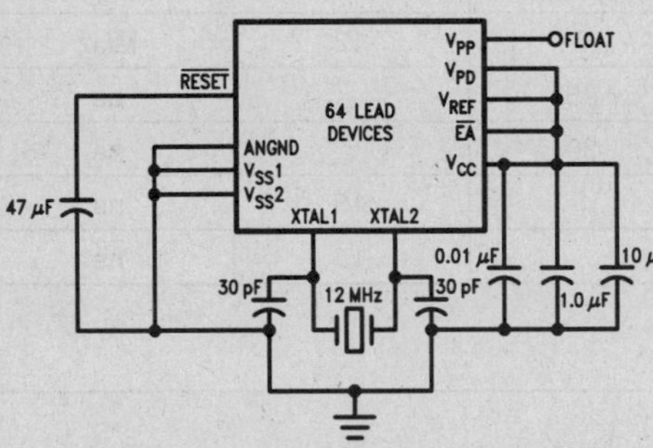

270795-25

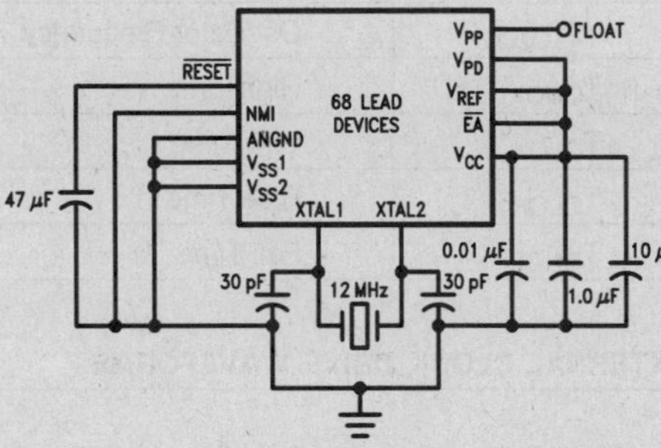

270795-26

AC CHARACTERISTICS—SERIAL PORT—SHIFT REGISTER MODE

SERIAL PORT TIMING—SHIFT REGISTER MODE

Test Conditions: Load Capacitance = 80 pF

Symbol	Parameter	Min	Max	Units
T_{XLXL}	Serial Port Clock Period	$8\ T_{OSC}$		ns
T_{XLXH}	Serial Port Clock Falling Edge to Rising Edge	$4\ T_{OSC} - 50$	$4\ T_{OSC} + 50$	ns
T_{QVXH}	Output Data Setup to Clock Rising Edge	$3\ T_{OSC}$		ns
T_{XHQX}	Output Data Hold After Clock Rising Edge	$2\ T_{OSC} - 70$		ns
T_{XHQV}	Next Output Data Valid After Clock Rising Edge		$2\ T_{OSC} + 50$	ns
T_{DVXH}	Input Data Setup to Clock Rising Edge	$2\ T_{OSC} + 200$		ns
T_{XHDX}	Input Data Hold After Clock Rising Edge	0		ns
T_{XHQZ}	Last Clock Rising to Output Float		$5\ T_{OSC}$	ns

WAVEFORM—SERIAL PORT—SHIFT REGISTER MODE

SERIAL PORT WAVEFORM—SHIFT REGISTER MODE

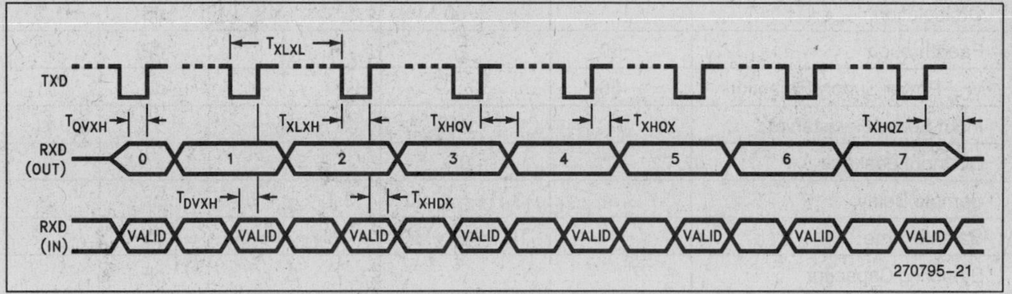

A/D CONVERTER SPECIFICATIONS

The absolute conversion accuracy is dependent on the accuracy and stability of V_{REF}.

See the MCS-96 A/D Converter Quick Reference for definitions of A/D Converter terms.

Parameter	Typical*	Minimum	Maximum	Units**	Notes
Resolution		1024 10	1024 10	Levels Bits	
Absolute Error		0	±4	LSBs	
Full Scale Error	−0.5 ±0.5			LSBs	
Zero Offset Error	±0.5			LSBs	
Non-Linearity		0	±4	LSBs	
Differential Non-Linearity		> −1	+2	LSBs	
Channel-to-Channel Matching		0	±1	LSBs	
Repeatability	±0.25			LSBs	
Temperature Coefficients: Offset Full Scale Differential Non-Linearity	 0.009 0.009 0.009			 LSB/°C LSB/°C LSB/°C	
Off Isolation		−60		dB	1, 3
Feedthrough	−60			dB	1
V_{CC} Power Supply Rejection	−60			dB	1
Input Series Resistance		1K	5K	Ω	4
DC Input Leakage		0	3.0	μA	
Sample Delay		$3\ T_{OSC} - 50$	$3\ T_{OSC} + 50$	ns	2
Sample Time		$12\ T_{OSC} - 50$	$12\ T_{OSC} + 50$	ns	
Sampling Capacitor			2	pF	

NOTES:

* These values are expected for most devices at 25°C.

** An "LSB", as used here, is defined in the MCS-96 A/D Converter Quick Reference and has a value of approximately 5 mV.

1. DC to 100 KHz.
2. For starting the A/D with an HSO Command.
3. Multiplexer Break-Before-Make Guaranteed.
4. Resistance from device pin, through internal MUX, to sample capacitor.

OTP EPROM SPECIFICATIONS

EPROM PROGRAMMING OPERATING CONDITIONS

Symbol	Parameter	Min	Max	Units
T_A	Ambient Temperature during Programming	20	30	C
V_{CC}, V_{PD}, V_{REF}(1)	Supply Voltages during Programming	4.5	5.5	V
V_{EA}	Programming Mode Supply Voltage	9.0	13.0	V(2)
V_{PP}	EPROM Programming Supply Voltage	12.50	13.0	V(2)
V_{SS}, ANGND(3)	Digital and Analog Ground	0	0	V
F_{OSC}(1)	Oscillator Frequency during Auto and Slave Programming	6.0	6.0	MHz
F_{OSC}(2)	Oscillator Frequency during Run-Time Programming	6.0	12.0	MHz

NOTES:
1. V_{CC}, V_{PD} and V_{REF} should nominally be at the same voltage during programming.
2. V_{EA} and V_{PP} must never exceed the maximum voltage for any amount of time or the device may be damaged.
3. V_{SS} and ANGND should nominally be at the same voltage (0V) during programming.

AC EPROM PROGRAMMING CHARACTERISTICS

Symbol	Parameter	Min	Max	Units
T_{AVLL}	ADDRESS/COMMAND Valid to PALE Low	0		T_{OSC}
T_{LLAX}	ADDRESS/COMMAND Hold After PALE Low	80		T_{OSC}
T_{DVPL}	Output Data Setup Before $\overline{PROG}$ Low	0		T_{OSC}
T_{PLDX}	Data Hold After $\overline{PROG}$ Falling	80		T_{OSC}
T_{LLLH}	PALE Pulse Width	180		T_{OSC}
T_{PLPH}	$\overline{PROG}$ Pulse Width	250 T_{OSC}	100 μs + 144 T_{OSC}	
T_{LHPL}	PALE High to $\overline{PROG}$ Low	250		T_{OSC}
T_{PHLL}	$\overline{PROG}$ High to Next PALE Low	600		T_{OSC}
T_{PHDX}	Data Hold After $\overline{PROG}$ High	30		T_{OSC}
T_{PHVV}	$\overline{PROG}$ High to PVER/$\overline{PDO}$ Valid	500		T_{OSC}
T_{LLVH}	PALE Low to PVER/$\overline{PDO}$ High	100		T_{OSC}
T_{PLDV}	$\overline{PROG}$ Low to VERIFICATION/DUMP Data Valid	100		T_{OSC}
T_{SHLL}	RESET High to First PALE Low (not shown)	2000		T_{OSC}

DC EPROM PROGRAMMING CHARACTERISTICS

Symbol	Parameter	Min	Max	Units
I_{PP}	V_{PP} Supply Current (Whenever Programming)		100	mA

WAVEFORM—EPROM PROGRAMMING (OTP)

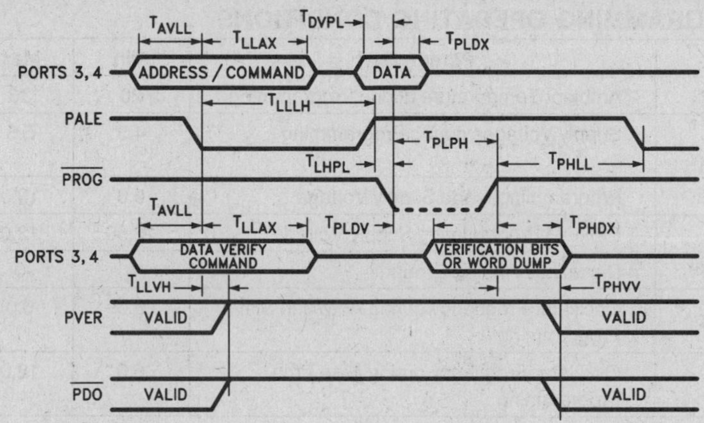

270795–27

REVISION HISTORY

This data sheet (270795-006) is valid for devices with a "B" at the end of the topside tracking number. Data sheets are changed as new device information becomes available. Verify with your local Intel sales office that you have the latest version before finalizing a design or ordering devices.

The following difference exists between this -006 data sheet and the previous one -005.

1. The I_{OL}/I_{OH} for float waveform testing changed from ±15 mA to ±8 mA (this data sheet).

The following differences exist between -005 and -004.

1. The Express (extended temperature and burn-in options) were added to this data sheet. The 8X9XJF EXPRESS data sheet (270796-001) is now obsolete.
2. Changes were made to the format of the data sheet and the SFR descriptions were removed. No spec changes were made.
3. Added Reserved Location 201CH errata.

The following differences exist between the -004 data sheet and the -003 data sheet.

1. The -003 data sheet was valid only for devices marked with an "A" at the end of the top side tracking number.
2. Added V_{IL1} (Input Low Voltage, $\overline{\text{RESET}}$)

The following differences exist between the -003 data sheet and the -002 data sheet.

1. The reserved location section and the power supply sequencing section has been deleted. This information is in the Hardware Design Information.
2. The Software Reset Timing bug was removed from the Functional Deviations. The RESET pin will pull down for at least 2 states if a software reset or watchdog timer overflow occurs.

Differences between the -002 and -001 data sheets.

1. The TLLGV spec has been changed from Max = T_{OSC} − 75 ns to Max = T_{OSC} = 100 ns.
2. The TCLLH spec has been changed from Min = −20 ns and Max = +25 ns to Min = −30 ns and Max = +15 ns.
3. The TXHQX spec has been changed from Min = 2 T_{OSC} − 50 ns to 2T_{OSC} − 70 ns.
4. The TOLOX spec has been changed from Min = 25 ns to Min = 30 ns.
5. Added "20" recommendation for reserved address 2019H to EPROM specification.
6. Added errata.

8X97JF ERRATA

Devices covered by this data sheet (see Revision History) have the following errata.

1. INDEXED, 3 OPERAND MULTIPLY

   The displacement portion of an indexed, three operand (byte or word) multiply may not be in the range of 200H thru 17FFH inclusive. If you must use these displacements, execute an indexed, two operand multiply and a move if necessary.

2. 8X97JF HIGH SPEED INPUTS

   The High Speed Input (HSI) has three deviations from the specifications. Note that "events" are defined as one or more pin transitions. "Entries" are defined as the recording of one or more events.

   A. The resolution is nine states instead of eight states. Events occurring on the same pin more frequently than once every nine state times may be lost.

   B. A mismatch between the nine state HSI resolution and the eight state hardware timer causes one time-tag value to be skipped every nine timer counts. Events may receive a time tag one count later than expected.

   C. If the FIFO and Holding Register are empty, the first event will transfer into the Holding Register, leaving the FIFO empty again. The next event that occurs will be the first event loaded into the empty FIFO. If the first two events into any **empty** FIFO (not counting the Holding Register) occur coincident with each other, both are recorded as one entry with one time-tag. If the second event occurs within 9 states after the first, the events will be entered separately with time-tags at least one count apart. If the second event enters the FIFO coincident with the "skipped" time-tag situation (see B above) the time-tags will be at least two counts apart.

3. RESERVED LOCATION 2019H

   The 1990 Architectural Overview recommends that reserved location 2019H be filled with hex value FFH. The recommendation is now to fill 2019H with hex value 20H.

4. RESERVED LOCATION 201CH

   Reading reserved location 201CH, either internally or externally, will return "201C" as data.

5. SERIAL PORT SECTION

   Serial Port Flags—Reading SP_STAT may not clear the TI or RI flag if that flag was set within two state times prior to the read. In addition, the parity error bit (RPE/RB8) may not be correct if it is read within two state times after RI is set. Use the following code to replace ORB sp_image, SP_STAT.

```
SP_READ: LDB TEMP, SP_STAT
         ORB SP_IMAGE, TEMP
         JBS TEMP,5,SP_READ; if TI
         is set then read again
         JBS TEMP,6,SP_READ; if RI
         is set then read again
         ANDB SP_IMAGE,#7FH; clear
         false RB8/RPE
         ORB SP_IMAGE, TEMP; load
         correct RB8/RPE
```

4

intel®

PRELIMINARY

8098/8398/8798 COMMERCIAL/EXPRESS HMOS MICROCONTROLLER

- 8798: an 8098 with 8 Kbytes of On-Chip EPROM
- 8398: an 8098 with 8 Kbytes of On-Chip ROM

- 232 Byte Register File
- Register-to-Register Architecture
- 10-Bit A/D Converter with S/H
- Two 8-Bit and Two 4-Bit I/O Ports
- 20 Interrupt Sources
- Pulse-Width Modulated Output
- ROM/EPROM Lock
- High Speed I/O Subsystem
- Extended Temperature Available
- Full Duplex Serial Port
- Dedicated Baud Rate Generator
- 6.25 μs 16 x 16 Multiply
- 6.25 μs 32/16 Divide
- 16-Bit Watchdog Timer
- Four 16-Bit Software Timers
- Two 16-Bit Counter/Timers
- Run-Time Programmable EPROM
- Extended Burn-In Available

The MCS®-96 family of 16-bit microcontrollers consists of many members, all of which are designed for high-speed control functions. The 8X98 members were designed specifically for those applications that require the speed of a 16-bit microcontroller but are limited by board space and cost requirements to an 8-bit external bus. The 8X98 members are produced using Intel's HMOS-III process.

The CPU supports bit, byte, and word operations. Thirty-two bit double-words are supported for a subset of the instruction set. With a 12 MHz input frequency the 8098 can do a 16-bit addition in 1.0 μs and a 16 x 16-bit multiply or 32/16 divide in 6.25 μs. Instruction execution times average 1 to 2 μs in typical applications.

Four high-speed trigger inputs are provided to record the times at which external events occur. Six high-speed pulse generator outputs are provided to trigger external events at preset times. The high-speed output unit can simultaneously perform software timer functions. Up to four 16-bit software timers can be in operation at once.

The on-chip A/D converter includes a Sample and Hold, and converts up to 4 multiplexed analog input channels to 10-bit digital values. With a 12 MHz crystal, each conversion takes 22 μs.

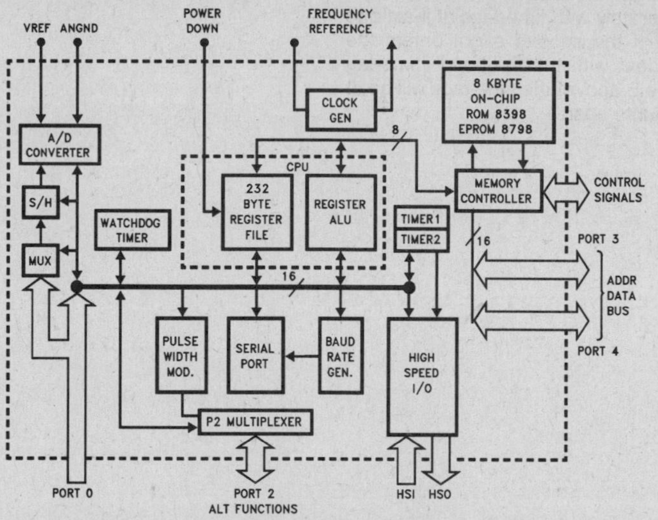

Figure 1. 8X98 Block Diagram

August 1992
Order Number: 270532-008

Also provided on-chip are a serial port, a Watchdog Timer and a pulse-width modulated output signal.

With the commercial (standard) temperature option, operational characteristics are guaranteed over the temperature range of 0°C to +70°C. With the extended temperature range option, operational characteristics are guaranteed over the temperature range of −40°C to +85°C. Unless otherwise noted, the specifications are the same for both options.

With the extended burn-in option, the burn-in is dynamic for a minimum time of 160 hours at 125°C with V_{CC} = 5.5V ±0.5V, following the guidelines in MIL-STD-883, Method 1015.

See the Packaging information for extended temperature and extended burn-in designators.

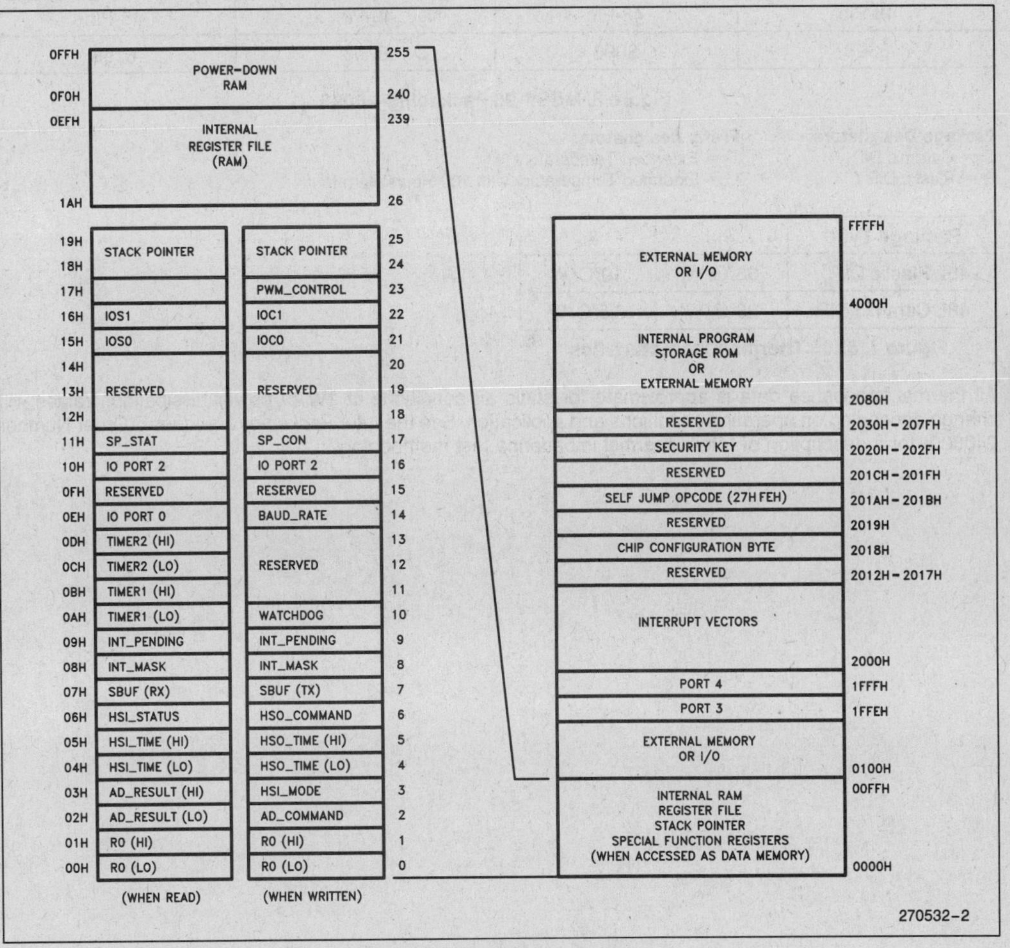

Figure 2. 8X98 Memory Map

PACKAGING

The 8098 is available in a 48-pin package with and without on-chip ROM or EPROM. The MCS-96 numbering system for the 8X98 devices is shown in Figure 4. Figure 6 shows the pinout for the 48-pin package. The 48-pin version is offered in a Dual-In-Line package.

Factory Masked ROM	CPU	User Programmable	
		EPROM	OTP
48-Pin	48-Pin	48-Pin	48-Pin
8398	8098	8798	8798

Figure 3. MCS®-96 Packaging—8098

Package Designators:
C = Ceramic DIP
P = Plastic DIP

Prefix Designators:
T = Extended Temperature
L = Extended Temperature with 160 Hours Burn-In

Package Type	θ_{ja}	θ_{jc}
48L Plastic DIP	38°C/W	19°C/W
48L Ceramic DIP	26°C/W	6.5°C/W

Figure 4. 8X98 Thermal Characteristics

All thermal impedance data is approximate for static air conditions at 1W of power dissipation. Values will change depending on operating conditions and application. See the Intel *Packaging Handbook* (Order Number 240800) for a description of Intel's thermal impedance test methodology.

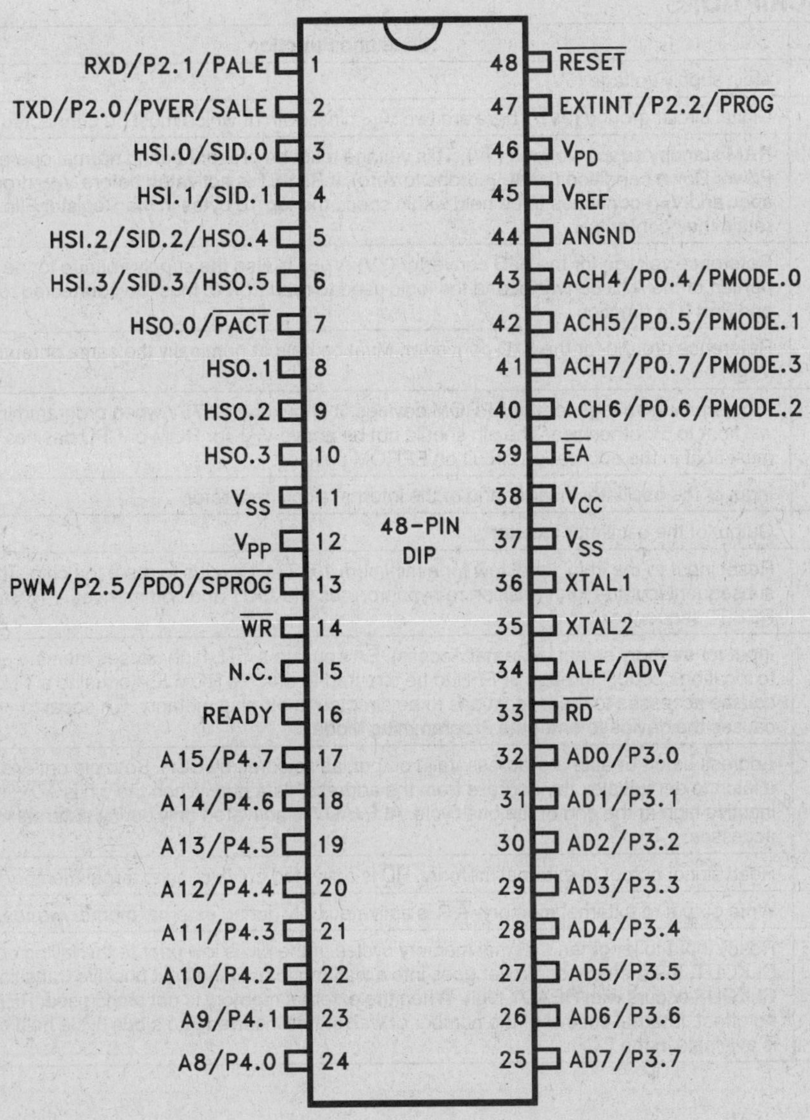

270532-9

Figure 5. 48-Pin Package

PIN DESCRIPTIONS

Symbol	Name and Function
V_{CC}	Main supply voltage (5V).
V_{SS}	Digital circuit ground (0V). There are two V_{SS} pins, both of which must be connected.
V_{PD}	RAM standby supply voltage (5V). This voltage must be present during normal operation. In a Power Down condition (i.e. V_{CC} drops to zero), if $\overline{RESET}$ is activated before V_{CC} drops below spec and V_{PD} continues to be held within spec., the top 16 bytes in the Register File will retain their contents.
V_{REF}	Reference voltage for the A/D converter (5V). V_{REF} is also the supply voltage to the analog portion of the A/D converter and the logic used to read Port 0. Must be connected for A/D and Port 0 to function.
ANGND	Reference ground for the A/D converter. Must be held at nominally the same potential as V_{SS}.
V_{PP}	Programming voltage for the EPROM devices. It should be 12.75V when programming and will float to 5V otherwise. The pin should not be above V_{CC} for ROM or CPU devices. This pin must float in the application circuit on EPROM devices.
XTAL1	Input of the oscillator inverter and of the internal clock generator.
XTAL2	Output of the oscillator inverter.
$\overline{RESET}$	Reset input to the chip. Input low for a minimum 10XTAL1 cycles to reset the chip. The subsequent low-to-high transition re-synchronizes CLKOUT and commences a 10-state-time RESET sequence.
$\overline{EA}$	Input for memory select (External Access). $\overline{EA}$ equal to a TTL-high causes memory accesses to locations 2000H through 3FFFH to be directed to on-chip ROM. $\overline{EA}$ equal to a TTL-low causes accesses to these locations to be directed to off-chip memory. $\overline{EA}$ equal to +12.75V causes the device to enter the Programming Mode.
ALE/$\overline{ADV}$	Address Latch Enable or Address Valid output, as selected by CCR. Both pin options provide a latch to demultiplex the address from the address/data bus. When the pin is $\overline{ADV}$, it goes inactive high at the end of the bus cycle. ALE/$\overline{ADV}$ is activated only during external memory accesses.
$\overline{RD}$	Read signal output to external memory. $\overline{RD}$ is activated only during external memory reads.
$\overline{WR}$	Write output to external memory. $\overline{WR}$ is activated only during external memory writes.
READY	Ready input to lengthen external memory cycles. If the pin is low prior to the falling edge of CLKOUT, the memory controller goes into a wait mode until the next positive transition of CLKOUT occurs with READY high. When the external memory is not being used, READY has no effect. Internal control of the number of wait states inserted into a bus cycle held not ready is available in the CCR.

PIN DESCRIPTIONS (Continued)

Symbol	Name and Function
HSI	Inputs to High Speed Input Unit. Four HSI pins are available: HSI.0, HSI.1, HSI.2 and HSI.3. Two of them (HSI.2 and HSI.3) are shared with the HSO Unit.
HSO	Outputs from High Speed Output Unit. Six HSO pins are available: HSO.0, HSO.1, HSO.2, HSO.3, HSO.4 and HSO.5. Two of them (HSO.4 and HSO.5) are shared with the HSI Unit.
Port 0	4-bit high impedance input-only port. These pins can be used as digital inputs and/or as analog inputs to the on-chip A/D converter.
Port 2	4-bit multi-functional port. Its pins are shared with other functions in the 8098.
Ports 3 and 4	8-bit bidirectional I/O ports with open drain outputs. These pins are shared with the multiplexed address/data bus which has strong internal pullups. Ports 3 and 4 are also used as a command, address and data path by EPROM devices in the Programming Mode.
PMODE	Determines the EPROM programming mode.
$\overline{PACT}$	A low signal in Auto Programming Mode indicates that programming is in progress. A high signal indicates programming is complete.
PVAL	A low signal in Auto Programming Mode indicates that the device programmed correctly.
SALE	A falling edge in Auto Programming Mode indicates that Ports 3 and 4 contain valid programming address/command information (output from master).
$\overline{SPROG}$	A falling edge in Auto Programming Mode indicates that Ports 3 and 4 contain valid programming data (output from master).
SID	Assigns a pin of Ports 3 and 4 to each slave to pass programming verification.
PALE	A falling edge in Slave Programming Mode and Auto Configuration Byte Programming Mode indicates that Ports 3 and 4 contain valid programming address/command information (input to slave).
$\overline{PROG}$	A falling edge in Slave Programming Mode indicates that Ports 3 and 4 contain valid programming data (input to slave).
PVER	A high signal in Slave Programming Mode and Auto Configuration Byte Programming Mode indicates the byte programmed correctly.
PVAL	A high signal in Slave Programming Mode indicates the device programmed correctly.
$\overline{PDO}$	A low signal in Slave Programming Mode indicates that the PROG pulse was applied for longer than allowed.

4

ELECTRICAL CHARACTERISTICS
ABSOLUTE MAXIMUM RATINGS*

Ambient Temperature Under Bias −55°C to +125°C

Storage Temperature .... −60°C to +150°C

Voltage from $\overline{EA}$ or V_{PP} to V_{SS} or ANGND .... −0.3V to +13.0V

Voltage from Any Other Pin to V_{SS} or ANGND .... −0.3V to +7.0V(1)

Average Output Current from Any Pin .... 10 mA

Power Dissipation(2) 1.5W

NOTES:

1. This includes V_{PP} on ROM and CPU only devices.
2. Power dissipation is based on package heat transfer limitations, not device power consumption.

NOTICE: This data sheet contains preliminary information on new products in production. It is valid for the devices indicated in the revision history. The specifications are subject to change without notice.

**WARNING:* *Stressing the device beyond the "Absolute Maximum Ratings" may cause permanent damage. These are stress ratings only. Operation beyond the "Operating Conditions" is not recommended and extended exposure beyond the "Operating Conditions" may affect device reliability.*

OPERATING CONDITIONS

(All characteristics specified in this data sheet apply to these operating conditions unless otherwise noted.)

Symbol	Parameter	Min	Max	Units
T_A	Ambient Temperature Under Bias **Commercial** Temp.	0	+70	C
T_A	Ambient Temperature Under Bias **Extended** Temp.	−40	+85	C
V_{CC}	Digital Supply Voltage	4.50	5.50	V
V_{REF}	Analog Supply Voltage	4.50	5.50	V
F_{OSC}	Oscillator Frequency	6.0	12	MHz
V_{PD}	Power-Down Supply Voltage	4.50	5.50	V

NOTE:
ANGND and V_{SS} should be nominally at the same potential.

DC CHARACTERISTICS

Symbol	Parameter	Min	Max	Units	Test Conditions
I_{CC}	V_{CC} Supply Current **Commercial** Temp.		240	mA	All Outputs Disconnected.
I_{CC}	V_{CC} Supply Current **Extended** Temp.		270	mA	
I_{CC}	V_{CC} Supply Current ($T_A \geq 70°C$)		185	mA	
I_{PD}	V_{PD} Supply Current		1	mA	Normal operation and Power-Down.
I_{REF}	V_{REF} Supply Current **Commercial** Temp.		8	mA	
I_{REF}	V_{REF} Supply Current **Extended** Temp.		10	mA	
V_{IL}	Input Low Voltage	−0.3	+0.8	V	

DC CHARACTERISTICS (Continued)

Symbol	Parameter	Min	Max	Units	Test Conditions
V_{IH}	Input High Voltage (Except $\overline{RESET}$, NMI, XTAL1)	2.0	V_{CC} +0.5	V	
V_{IH1}	Input High Voltage, $\overline{RESET}$ Rising	2.4	V_{CC} +0.5	V	
V_{IH2}	Input High Voltage, $\overline{RESET}$ Falling Hysteresis	2.1	V_{CC} +0.5	V	
V_{IH3}	Input High Voltage, NMI, XTAL1	2.2	V_{CC} +0.5	V	
I_{LI}	Input Leakage Current to each pin of HSI, P3, P4, and to P2.1.		±10	μA	V_{in} = 0 to V_{CC}
I_{LI1}	DC Input Leakage Current to each pin of P0		+3	μA	V_{in} = 0 to V_{CC}
I_{IH}	Input High Current to $\overline{EA}$		100	μA	V_{IH} = 2.4V
I_{IL}	Input Low Current to each pin of P1 and to P2.6, P2.7 **Commercial** Temp.		−125	μA	V_{IL} = 0.45V
I_{IL}	Input Low Current to each pin of P1 and to P2.6, P2.7 **Extended** Temp.		−150	μA	
I_{IL1}	Input Low Current to $\overline{RESET}$	−0.25	−2	mA	V_{IL} = 0.45V
I_{IL2}	Input Low Current P2.2		−50	μA	V_{IL} = 0.45V
V_{OL}	Output Low Voltage on P3, P4 when used as ports		0.45	V	I_{OL} = 0.8 mA (Note 1)
V_{OL1}	Output Low Voltage on P3, P4 when used as ports		0.75	V	I_{OL} = 2.0 mA (Notes 1, 2, 3)
V_{OL2}	Output Low Voltage on Standard Output pins, $\overline{RESET}$ and Bus/Control Pins		0.45	V	I_{OL} = 2.0 mA (Notes 1, 2, 3)
V_{OH1}	Output High Voltage on Standard Output pins and Bus/Control pins	2.4		V	I_{OH} = −200 μA (Note 1)
I_{OH3}	Output High Current on $\overline{RESET}$	−50		μA	V_{OH} = 2.4V
C_S	Pin Capacitance (Any Pin to V_{SS})		10	pF	F_{TEST} = 1.0 MHz

NOTES:
1. Standard Output Pins include TXD, RXD (Mode 0 only), PWM, and HSO pins. Bus/Control pins include ALE, $\overline{RD}$, $\overline{WR}$, AD0–AD7 and A8–A15.
2. Maximum current per pin must be externally limited to the following values if V_{OL} is held above 0.45V.
   - I_{OL} on Ports 3 and 4 when used as ports: 4.0 mA
   - I_{OL} on standard output pins and $\overline{RESET}$: 8.0 mA
   - I_{OL} on Bus/Control pins: 2.0 mA
3. During normal (non-transient) operation the following limits apply:
   - Total I_{OL} on P2.0, $\overline{RESET}$ and all HSO pins must not exceed 15 mA.
   - Total I_{OL} on Port 3 must not exceed 10 mA.
   - Total I_{OL} on P2.5 and Port 4 must not exceed 20 mA.

AC CHARACTERISTICS Test Conditions: Load Capacitance on Output Pins = 80 pF

TIMING REQUIREMENTS (The system must meet these specifications to work with the 8X98.)

Symbol	Parameter	Min	Max	Units
T_{LLYV}	End of ALE/$\overline{ADV}$ to READY Valid		$2\ T_{OSC} - 70$	ns
T_{LLYH}	End of ALE/$\overline{ADV}$ to READY High	$2\ T_{OSC} + 40$	$4\ T_{OSC} - 80$	ns
T_{YLYH}	Non-Ready Time		1000	ns
T_{AVDV}(1)	Address Valid to Input Data Valid		$5\ T_{OSC} - 120$(2)	ns
T_{RLDV}	$\overline{RD}$ Active to Input Data Valid		$3\ T_{OSC} - 100$(2)	ns
T_{RHDX}	Data Hold after $\overline{RD}$ Inactive	0		ns
T_{RHDZ}	$\overline{RD}$ Inactive to Input Data Float	0	$T_{OSC} - 25$	ns

NOTE:
1. The term "Address Valid" applies to A0–A15.
2. If wait states are used, add 3 Tosc * N where N = number of states.

TIMING RESPONSES (8X98 devices meet these specs.)

Symbol	Parameter	Min	Max	Units
F_{XTAL}	Oscillator Frequency	6.0	12.0	MHz
T_{OSC}	Oscillator Period	83	166	ns
T_{LHLL}	ALE/$\overline{ADV}$ High Time	$T_{OSC} - 30$	$T_{OSC} + 35$(3)	ns
T_{AVLL}(4)	Address Setup to End of ALE/$\overline{ADV}$	$T_{OSC} - 50$		ns
T_{RLAZ}(5)	$\overline{RD}$ or $\overline{WR}$ Low to Address Float **Commercial** Temp.	Typ. = 0	10	ns
T_{RLAZ}(5)	$\overline{RD}$ or $\overline{WR}$ Low to Address Float **Extended** Temp.		25	ns
T_{LLRL}	End of ALE/$\overline{ADV}$ to $\overline{RD}$ Active	$T_{OSC} - 40$		ns
T_{LLAX}(5)	Address Hold after End of ALE/$\overline{ADV}$	$T_{OSC} - 40$		ns
T_{WLWH}(6)	$\overline{WR}$ Pulse Width	$3\ T_{OSC} - 35$(2)		ns
T_{WLWH}(7)	$\overline{WR}$ Pulse Width	$2\ T_{OSC} - 35$(2)	$2\ T_{OSC} + 40$	ns
T_{QVWH}	Output Data Valid to End of $\overline{WR}$	$3\ T_{OSC} - 60$(2)		ns
T_{WHQX}	Output Data Hold after $\overline{WR}$	$T_{OSC} - 50$		ns
T_{WHLH}	End of $\overline{WR}$ to ALE/$\overline{ADV}$ High	$T_{OSC} - 75$		ns
T_{RLRH}	$\overline{RD}$ Pulse Width	$3\ T_{OSC} - 30$(2)		ns

TIMING RESPONSES (8X98 devices meet these specs.) (Continued)

Symbol	Parameter	Min	Max	Units
T_{RHLH}	End of $\overline{RD}$ to ALE/$\overline{ADV}$ High	$T_{OSC} - 45$		ns
T_{RHBX}	$\overline{RD}$ High to A8–A15 Inactive	$T_{OSC} - 25$	$T_{OSC} + 30$	ns
T_{WHBX}	$\overline{WR}$ High to A8–A15 Inactive	$T_{OSC} - 50$	$T_{OSC} + 100$	ns
T_{LLWL}(6)	ALE/$\overline{ADV}$ Low to $\overline{WR}$ Low	$T_{OSC} - 40$		ns
T_{LLWL}(7)	ALE/$\overline{ADV}$ Low to $\overline{WR}$ Low	$2\,T_{OSC} - 30$	$2\,T_{OSC} + 55$	ns
T_{QVWL}(6)	Output Data Valid to $\overline{WR}$ Low	$T_{OSC} - 60$		ns
T_{QVWL}(7)	Output Data Valid to $\overline{WR}$ Low	$T_{OSC} - 30$		ns

NOTES:
2. If more than one wait state is desired, add 3 Tosc for each additional wait state.
3. Max spec applies only to ALE. Min spec applies to both ALE and $\overline{ADV}$.
4. The term "Address Valid" applies to AD0–AD7, A8–A15.
5. The term "Address" in this definition applies to AD0–AD7.
6. Write Strobe Mode is not selected.
7. Write Strobe Mode is selected.

WAVEFORM—SYSTEM BUS TIMINGS

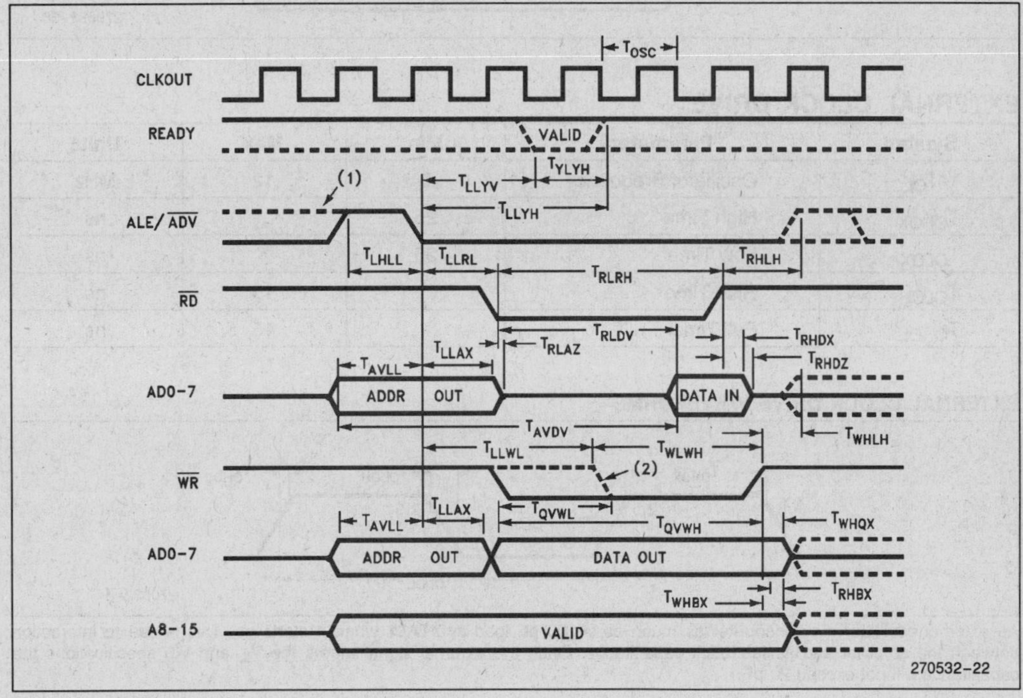

NOTES:
1. When $\overline{ADV}$ selected.
2. When Write Strobe Mode selected

WAVEFORM—T_{RLPV}

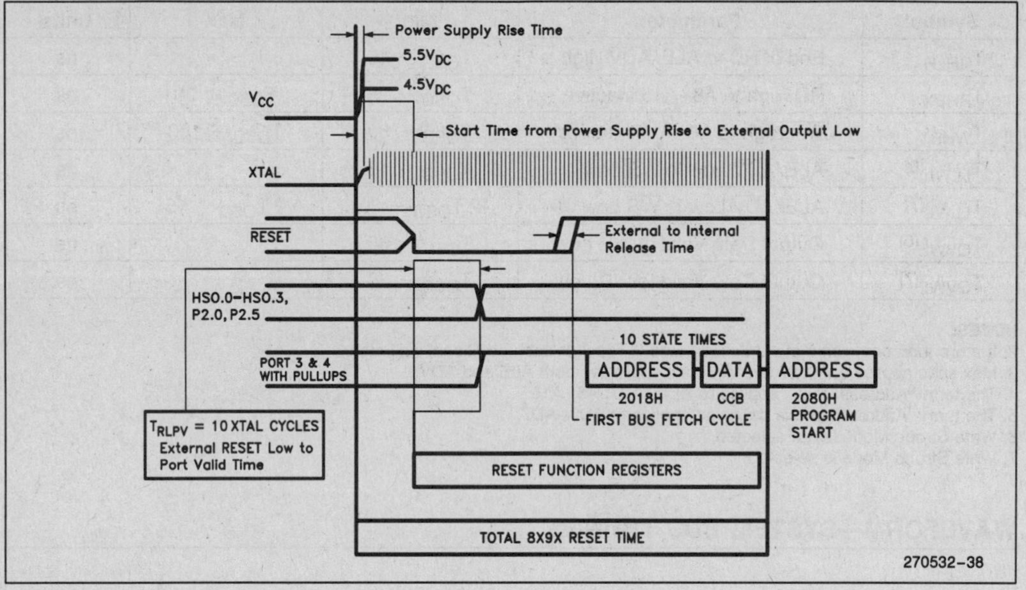

EXTERNAL CLOCK DRIVE

Symbol	Parameter	Min	Max	Units
$1/T_{OLOL}$	Oscillator Frequency	6	12	MHz
T_{OHOX}	High Time	25		ns
T_{OLOX}	Low Time	30		ns
T_{OLOH}	Rise Time		15	ns
T_{OHOL}	Fall Time		15	ns

EXTERNAL CLOCK DRIVE WAVEFORMS

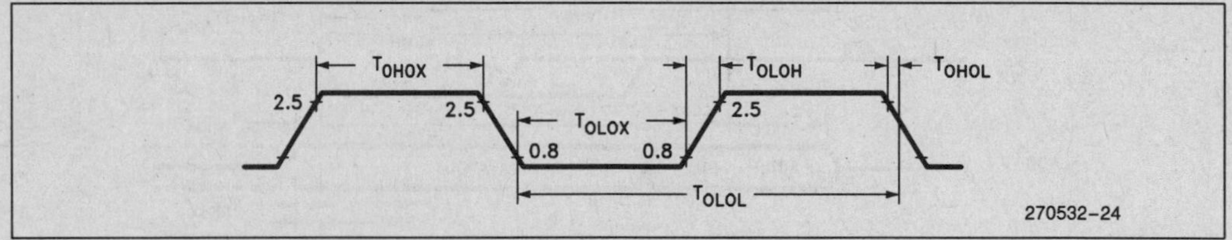

An external oscillator may encounter as much as a 100 pF load at XTAL1 when it starts up. This is due to interaction between the amplifier and its feedback capacitance. Once the external signal meets the V_{IL} and V_{IH} specifications the capacitance will not exceed 20 pF.

PRELIMINARY

EXTERNAL CRYSTAL CONNECTIONS

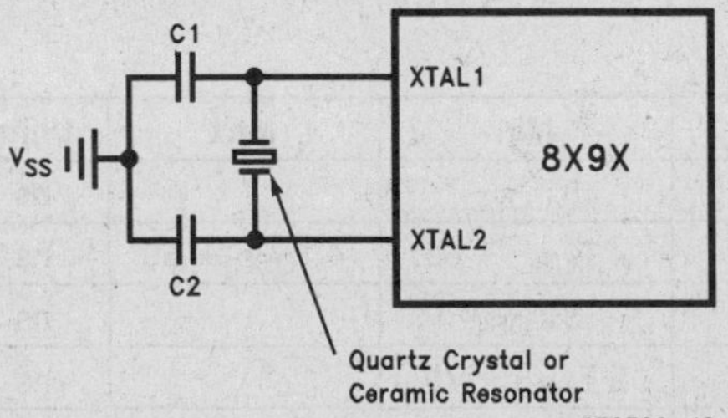

270532–40

NOTE:
Keep oscillator components close to chip and use short, direct traces to XTAL1, XTAL2 and V_{SS}. When using crystals, C1 = 30 pF, C2 = 30 pF. When using ceramic resonators, consult manufacturer for recommended capacitor values.

EXTERNAL CLOCK CONNECTIONS

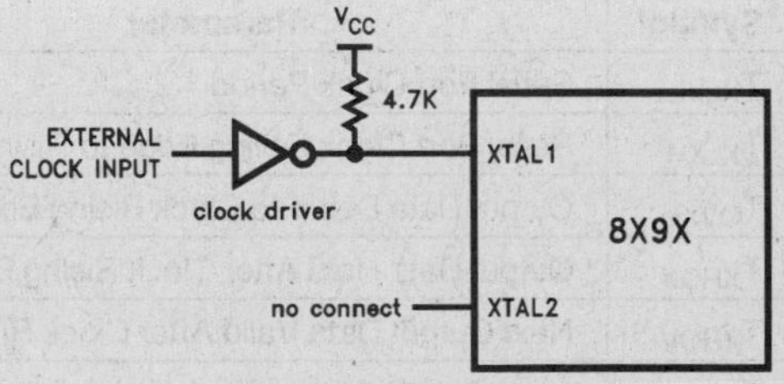

270532–41

AC TESTING INPUT, OUTPUT WAVEFORMS

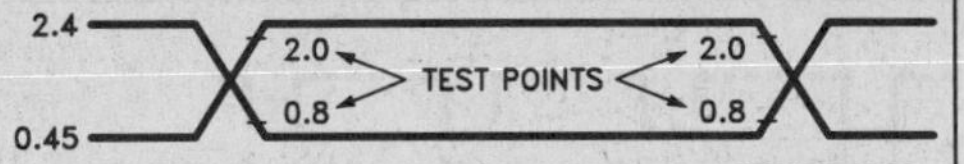

270532–25

AC Testing inputs are driven at 2.4V for a Logic "1" and 0.45V for a Logic "0". Timing measurements are made at 2.0V for a Logic "1" and 0.8V for a Logic "0".

FLOAT WAVEFORMS

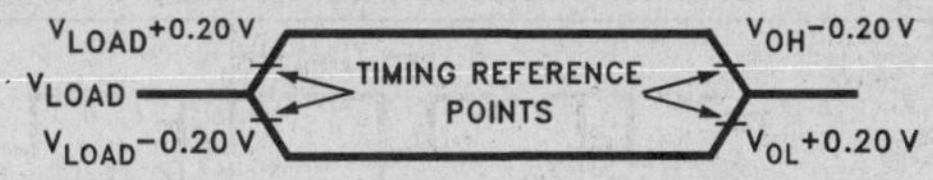

270532–26

For Timing Purposes a Port Pin is no Longer Floating when a 200 mV change from Load Voltage Occurs, and Begins to Float when a 200 mV change from the Loaded V_{OH}/V_{OL} Level occurs $I_{OL}/I_{OH} \geq \pm 8$ mA.

MINIMUM HARDWARE CONFIGURATION CIRCUIT

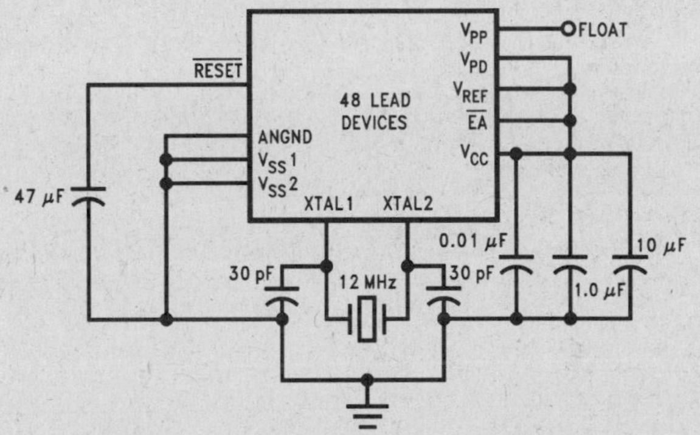

270532–42

AC CHARACTERISTICS—SERIAL PORT—SHIFT REGISTER MODE

SERIAL PORT TIMING—SHIFT REGISTER MODE

Test Conditions: Load Capacitance = 80 pF

Symbol	Parameter	Min	Max	Units
T_{XLXL}	Serial Port Clock Period	8 T_{OSC}		ns
T_{XLXH}	Serial Port Clock Falling Edge to Rising Edge	4 T_{OSC} − 50	4 T_{OSC} + 50	ns
T_{QVXH}	Output Data Setup to Clock Rising Edge	3 T_{OSC}		ns
T_{XHQX}	Output Data Hold After Clock Rising Edge	2 T_{OSC} − 70		ns
T_{XHQV}	Next Output Data Valid After Clock Rising Edge		2 T_{OSC} + 50	ns
T_{DVXH}	Input Data Setup to Clock Rising Edge	2 T_{OSC} + 200		ns
T_{XHDX}	Input Data Hold After Clock Rising Edge	0		ns
T_{XHQZ}	Last Clock Rising to Output Float		5 T_{OSC}	ns

WAVEFORM—SERIAL PORT—SHIFT REGISTER MODE

SERIAL PORT WAVEFORM—SHIFT REGISTER MODE

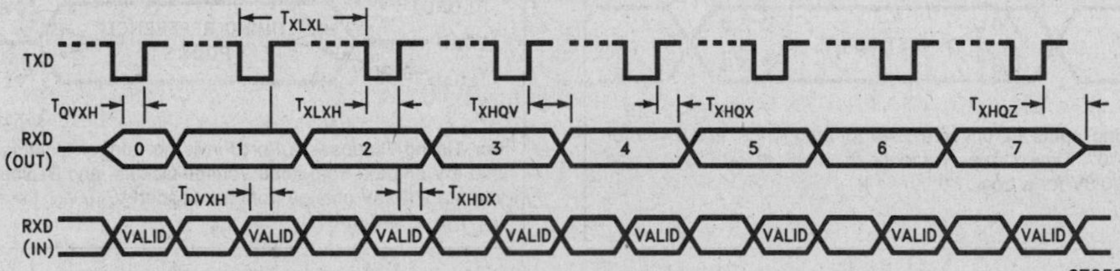

270532–23

A/D CONVERTER SPECIFICATIONS

The absolute conversion accuracy is dependent on the accuracy and stability of V_{REF}.

See the MCS-96 A/D Converter Quick Reference for definition of A/D Converter Terms.

Parameter	Typical*(1)	Minimum	Maximum	Units**	Notes
Resolution		1024 10	1024 10	Levels Bits	
Absolute Error		0	±4	LSBs	
Full Scale Error	−0.5 ±0.5			LSBs	
Zero Offset Error	±0.5			LSBs	
Non-Linearity		0	±4	LSBs	
Differential Non-Linearity		> −1	+2	LSBs	
Channel-to-Channel Matching		0	±1	LSBs	
Repeatability	±0.25			LSBs	
Temperature Coefficients: Offset Full Scale Differential Non-Linearity	 0.009 0.009 0.009			 LSB/°C LSB/°C LSB/°C	
Off Isolation		−60		dB	1, 3
Feedthrough	−60			dB	1
V_{CC} Power Supply Rejection	−60			dB	1
Input Series Resistance		1K	5K	Ω	4
DC Input Leakage		0	3.0	μA	
Sample Delay		$3\,T_{OSC} - 50$	$3\,T_{OSC} + 50$	ns	2
Sample Time		$12\,T_{OSC} - 50$	$12\,T_{OSC} + 50$	ns	
Sampling Capacitor	2			pF	

NOTES:

* These values are expected for most parts at 25°C.

** An "LSB", as used here, is defined in the MCS-96 A/D Converter Quick Reference and has a value of approximately 5 mV.

1. DC to 100 KHz.
2. For starting the A/D with an HSO Command.
3. Multiplexer Break-Before-Make Guaranteed.
4. Resistance from device pin, through internal MUX, to sample capacitor.

EPROM SPECIFICATIONS

EPROM PROGRAMMING OPERATING CONDITIONS

Symbol	Parameter	Min	Max	Units
T_A	Ambient Temperature during Programming	20	30	°C
V_{CC}, V_{PD}, V_{REF}(1)	Supply Voltages during Programming	4.5	5.5	V
V_{EA}	Programming Mode Supply Voltage	9.0	13.0	V(2)
V_{PP}	EPROM Programming Supply Voltage	12.50	13.0	V(2)
V_{SS}, ANGND(3)	Digital and Analog Ground	0	0	V
$F_{OSC}1$	Oscillator Frequency during Auto and Slave Programming	6.0	6.0	MHz
$F_{OSC}2$	Oscillator Frequency during Run-Time Programming	6.0	12.0	MHz

NOTES:
1. V_{CC}, V_{PD} and V_{REF} should nominally be at the same voltage during programming.
2. V_{EA} and V_{PP} must never exceed the maximum voltage for any amount of time or the device may be damaged.
3. V_{SS} and ANGND should nominally be at the same voltage (0V) during programming.

AC EPROM PROGRAMMING CHARACTERISTICS

Symbol	Parameter	Min	Max	Units
T_{AVLL}	ADDRESS/COMMAND Valid to PALE Low	0		T_{OSC}
T_{LLAX}	ADDRESS/COMMAND Hold After PALE Low	80		T_{OSC}
T_{DVPL}	Output Data Setup Before $\overline{PROG}$ Low	0		T_{OSC}
T_{PLDX}	Data Hold After $\overline{PROG}$ Falling	80		T_{OSC}
T_{LLLH}	PALE Pulse Width	180		T_{OSC}
T_{PLPH}	$\overline{PROG}$ Pulse Width	250 T_{OSC}	100 μs + 144 T_{OSC}	
T_{LHPL}	PALE High to $\overline{PROG}$ Low	250		T_{OSC}
T_{PHLL}	$\overline{PROG}$ High to Next PALE Low	600		T_{OSC}
T_{PHDX}	Data Hold After $\overline{PROG}$ High	30		T_{OSC}
T_{PHVV}	$\overline{PROG}$ High to PVER/$\overline{PDO}$ Valid	500		T_{OSC}
T_{LLVH}	PALE Low to PVER/$\overline{PDO}$ High	100		T_{OSC}
T_{PLDV}	$\overline{PROG}$ Low to VERIFICATION/DUMP Data Valid	100		T_{OSC}
T_{SHLL}	RESET High to First PALE Low (not shown)	2000		T_{OSC}

DC EPROM PROGRAMMING CHARACTERISTICS

Symbol	Parameter	Min	Max	Units
I_{PP}	V_{PP} Supply Current (Whenever Programming)		100	mA

WAVEFORM—EPROM PROGRAMMING

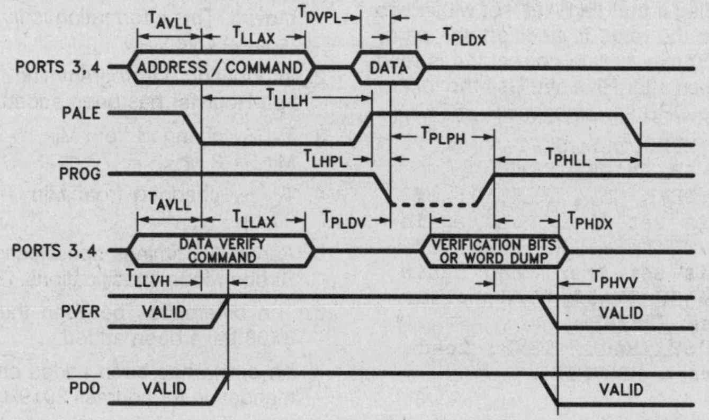

270532–37

DIFFERENCES BETWEEN THE 8X9XBH AND 8X98

1. CCB.1 must be set to a logical 0 on the 8X98.
2. The following 8X9XBH pins and corresponding functions are not available on the 8X98:

   BUSWIDTH

   CLKOUT

   INST

   NMI

   Port 0.0–0.3 (ACH0–3)

   Port 1.0–1.7

   Port 2.6

   Port 2.7

   P2.3 (T2CLK)

   P2.4 (T2RST).

8X98 ERRATA

Devices covered by this data sheet (see Revision History) have the following errata.

1. INDEXED, 3 OPERAND MULTIPLY

The displacement portion of an indexed, three operand (byte or word) multiply may not be in the range of 200H thru 17FFH inclusive. If you must use these displacements, execute an indexed, two operand multiply and a move if necessary.

2. HSI FIFO OPERATION

The High Speed Input (HSI) has three deviations from the specifications. Note that "events" are defined as one or more pin transitions. "Entries" are defined as the recording of one or more events.

A. The resolution is nine states instead of eight states. Events occurring on the same pin more frequently than once every nine states may be lost.

B. A mismatch between the nine state HSI resolution and the eight state hardware timer causes one time-tag value to be skipped every nine timer counts. Events may receive a time-tag one count later than expected.

C. If the FIFO and Holding Register are empty, the first event will transfer into the Holding Register, leaving the FIFO empty again. The next event that occurs will be the first event loaded into the empty FIFO. If the first two events into an **empty** FIFO (not counting the Holding Register) occur coincident with each other, both are recorded as one entry with one time-tag. If the second event occurs within nine states after the first, the events will be entered separately with time-tags at least one count apart. If the second event enters the FIFO coincident with the "skipped" time-tag situation (see B above) the time-tags will be at least two counts apart.

3. RESERVED LOCATION 2019H

The 1990 Architectural Overview recommended that address 2019H be filled with hex value 0FFH. The recommendation is now 20H.

4. RESERVED LOCATION 201CH

Reading reserved location 201CH, either internally or externally, will return "201C" as data.

5. SERIAL PORT SECTION

Serial Port Flags—Reading SP_STAT may not clear the TI or RI flag if that flag was set within two state times prior to the read. In addition, the parity error bit (RPE/RB8) may not be correct if it is read within two state times after RI is set. Use the following code to replace ORB sp_image, SP_STAT.

```
SP_READ:  LDB TEMP, SP_STAT
          ORB SP_IMAGE, TEMP
          JBS TEMP, 5, SP_READ; if
          TI is set then read again
          JBS TEMP, 6, SP_READ; if
          RI is set then read again
          ANDB SP_IMAGE,#7FH; clear
          false RB8/RPE
          ORB SP_IMAGE, TEMP; load
          correct RB8/RPE
```

DATA SHEET REVISION HISTORY

This data sheet (270532-008) is valid for devices with an "E" at the end of the topside tracking number. Data sheets are changed as new device information becomes available. Verify with your local Intel sales office that you have the latest version before finalizing a design or ordering devices.

The difference between -007 and -008 is the I_{OL}/I_{OH} for the float waveform testing changed from ±15 mA to ±8 mA.

The following differences exist between (-007) data sheet and the (-006).

1. The Express (extended temperature and burn-in options) were added to this data sheet. The 8X98 Express data sheet (270914-002) is now obsolete.
2. Changes were made to the format of the data sheet and the SFR descriptions were removed. No specification changes were made.
3. Added Reserved Location 201CH errata.

The following differences exist between the -006 data sheet and the -005 data sheet.

1. The -005 data sheet was valid for devices marked with a "D" at the end of the top side tracking number.
2. The following errata were removed: RESET and the Quasi-Bidirectional Ports, Software RESET Timing, and Using T2CLK as the source for Timer2.
3. The HSI FIFO Operation errata definition was changed to match a change in the HSI FIFO operation.

The following differences exist between the -005 data sheet and the -004 data sheet.

1. Most of the functional description has been removed. This information is in the MCS-96 Architectural Overview.
2. Information on programming the Chip Configuration Register has been added.
3. T_{XHQX} changed from Min = 2 T_{OSC} − 50 ns to Min = 2 T_{OSC} − 70 ns.
4. T_{OLOX} changed from Min = 25 ns to Min = 30 ns.
5. Added AC timings specifications to clarify Write Strobe Mode specifications.
6. The differences between the 8X9XBH and the 8X98 have been added.
7. An errata has been added changing the recommendation for address 2019H from 0FFH to 20H.

Differences between the -004 and -003 data sheets.

1. All EPROM programming mode information has been moved to the Hardware Design Information Chapter.
2. CCB RESET FETCH and JBS/JBC on Port 0 anamolies have been corrected on the current steppings of the 8X98.
3. New information regarding T2CLK and new information about RESET of the Quasi Ports have been added to the Errata section.
4. The Extended Reset errata has been eliminated on the silicon and in the data sheet.
5. HSI Mode register is undefined until the user code initializes this register.
6. Minimum DNL us now > −1 LSB.
7. HSI FIFO overflow description added.

Differences between the -002 and -003 data sheets.

1. All 8798 EPROM information has been added as a complete section after the Analog Section.
2. The chip configuration byte values now indicate the use of WRITE STROBE with 8-bit systems. Write Strobe design text was added to the explanation.
3. The interrupt information now includes a worst case timing diagram.
4. The EPROM 8798 was added as necessary throughout the text.
5. NMI pin information was deleted.
6. Reset Register Status was added and the state of the HSO pins after RESET.
7. A diagram of the Interrupt Pending Register is now included.

8. A diagram of the PSW Register was added.
9. V_{IL1} was deleted. This was a RESET pin characteristic that has been improved to match the other characteristics.
10. The Differential Non-Linearity specification in the A/D converter specifications was corrected to read +2 LSBs.
11. Power On Reset — New information on Extended Reset Time was added to the Errata Section.

80C196KB10/83C196KB10/80C196KB12/83C196KB12 COMMERCIAL/EXPRESS CHMOS MICROCONTROLLER

83C196KB — 8 Kbytes of Factory Mask-Programmed ROM
80C196KB — ROMless

- 8 Kbytes of On-Chip ROM Available
- 232 Byte Register File
- Register-to-Register Architecture
- 28 Interrupt Sources/16 Vectors
- 2.3 μs 16 x 16 Multiply (12 MHz)
- 4.0 μs 32/16 Divide (12 MHz)
- Powerdown and Idle Modes
- Five 8-Bit I/O Ports
- 16-Bit Watchdog Timer
- Dynamically Configurable 8-Bit or 16-Bit Buswidth
- Extended Temperature Available
- Full Duplex Serial Port
- High Speed I/O Subsystem
- 16-Bit Timer
- 16-Bit Up/Down Counter with Capture
- Pulse-Width-Modulated Output
- Four 16-Bit Software Timers
- 10-Bit A/D Converter with Sample/Hold
- $\overline{\text{HOLD}}$/$\overline{\text{HLDA}}$ Bus Protocol
- 10 MHz and 12 MHz Available
- Extended Burn-In Available

The 80C196KB 16-bit microcontroller is a high performance member of the MCS® 96 microcontroller family. The 80C196KB is compatible with the 8096BH and uses a true superset of the 8096BH instructions. Intel's CHMOS process provides a high performance processor along with low power consumption. To further reduce power requirements, the processor can be placed into Idle or Powerdown Mode.

The 80C196KB has a 232-byte register file and an optional 8 Kbyte of on-chip ROM. Bit, byte, word and some 32-bit operations are available on the 80C196KB. With a 12 MHz oscillator a 16-bit addition takes 0.66 μs, and the instruction times average 0.5 μs to 1.5 μs in typical applications.

Four high-speed capture inputs are provided to record times when events occur. Six high-speed outputs are available for pulse or waveform generation. The high-speed output can also generate four software timers or start an A/D conversion. Events can be based on the timer or up/down counter. Also provided on-chip are an A/D converter, serial port, watchdog timer and a pulse-width-modulated output signal.

The 80C196KB10 and 83C196KB10 have a maximum guaranteed frequency of 10 MHz. The 80C196KB12 and 83C196KB12 have a maximum guaranteed frequency of 12 MHz. All references to the 80C196KB also refer to the 80C196KB10, 83C196KB10, 80C196KB12 and 83C196KB12 unless otherwise noted.

With the commercial (standard) temperature option, operational characteristics are guaranteed over the temperature range of 0°C to +70°C. With the extended temperature range option, operational characteristics are guaranteed over the temperature range of −40°C to +85°C. With the extended burn-in option, the burn-in is dynamic for a minimum time of 160 hours at 125°C with V_{CC} = 5.5V ±0.5V, following the guidelines in MIL-STD-883, Method 1015. The specifications which are different for the extended temperature and extended burn-in devices are listed in this data sheet. Otherwise, the commercial specifications apply for both.

November 1994
Order Number: 270918-002

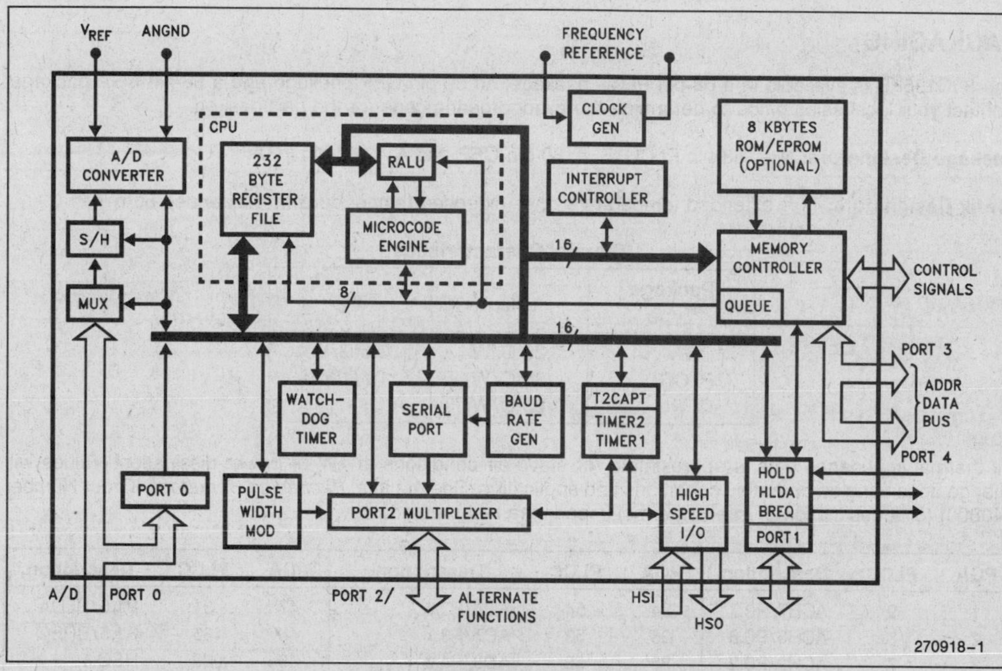

Figure 1. 80C196KB Block Diagram

EXTERNAL MEMORY OR I/O	0FFFFH – 4000H
INTERNAL ROM/EPROM OR EXTERNAL MEMORY	4000H – 2080H
RESERVED	2080H – 2040H
UPPER 8 INTERRUPT VECTORS	2040H – 2030H
ROM/EPROM SECURITY KEY	2030H – 2020H
RESERVED	2020H – 2019H
CHIP CONFIGURATION BYTE	2019H – 2018H
RESERVED	2018H – 2014H
LOWER 8 INTERRUPT VECTORS PLUS 2 SPECIAL INTERRUPTS	2014H – 2000H
PORT 3 AND PORT 4	2000H – 1FFEH
EXTERNAL MEMORY OR I/O	1FFEH – 0100H
INTERNAL DATA MEMORY - REGISTER FILE (STACK POINTER, RAM AND SFRS) EXTERNAL PROGRAM CODE MEMORY	0100H – 0000H

Figure 2. Memory Map

PACKAGING

The 80C196KB is available in a 68-pin PLCC package, an 80-pin QFP package and a 68-pin PGA package. Contact your local sales office to determine the exact ordering code for the part desired.

Package Designators: N = 68-pin PLCC, S = 80-pin QFP and A = 68-pin PGA.

Prefix Designators: T = extended temperature, L = extended temperature with extended burn-in.

Thermal Characteristics

Package Type	θ_{ja}	θ_{jc}
PGA	28°C/W	3.5°C/W
PLCC	35°C/W	12°C/W
QFP	85°C/W	—

All thermal impedance data is approximate for static air conditions at 1W of power dissipation. Values will change depending on operating conditions and application. See the Intel *Packaging Handbook* (Order Number 240800) for a description of Intel's thermal impedance test methodology.

PGA	PLCC	Description	PGA	PLCC	Description	PGA	PLCC	Description
1	9	ACH7/P0.7	24	54	AD6/P3.6	47	31	P1.6/$\overline{\text{HLDA}}$
2	8	ACH6/P0.6	25	53	AD7/P3.7	48	30	P1.5/$\overline{\text{BREQ}}$
3	7	ACH2/P0.2	26	52	AD8/P4.0	49	29	HSO.1
4	6	ACH0/P0.0	27	51	AD9/P4.1	50	28	HSO.0
5	5	ACH1/P0.1	28	50	AD10/P4.2	51	27	HSO.5/HSI.3
6	4	ACH3/P0.3	29	49	AD11/P4.3	52	26	HSO.4/HSI.2
7	3	NMI	30	48	AD12/P4.4	53	25	HSI.1
8	2	$\overline{\text{EA}}$	31	47	AD13/P4.5	54	24	HSI.0
9	1	V_{CC}	32	46	AD14/P4.6	55	23	P1.4
10	68	V_{SS}	33	45	AD15/P4.7	56	22	P1.3
11	67	XTAL1	34	44	T2CLK/P2.3	57	21	P1.2
12	66	XTAL2	35	43	READY	58	20	P1.1
13	65	CLKOUT	36	42	T2RST/P2.4	59	19	P1.0
14	64	BUSWIDTH	37	41	$\overline{\text{BHE}}$/$\overline{\text{WRH}}$	60	18	TXD/P2.0
15	63	INST	38	40	$\overline{\text{WR}}$/$\overline{\text{WRL}}$	61	17	RXD/P2.1
16	62	ALE/$\overline{\text{ADV}}$	39	39	PWM/P2.5	62	16	$\overline{\text{RESET}}$
17	61	$\overline{\text{RD}}$	40	38	P2.7/T2CAPTURE	63	15	EXTINT/P2.2
18	60	AD0/P3.0	41	37	V_{PP}	64	14	V_{SS}[1]
19	59	AD1/P3.1	42	36	V_{SS}	65	13	V_{REF}
20	58	AD2/P3.2	43	35	HSO.3/SID3	66	12	ANGND
21	57	AD3/P3.3	44	34	HSO.2/SID2	67	11	ACH4/P0.4
22	56	AD4/P3.4	45	33	P2.6/T2UP-DN	68	10	ACH5/P0.5
23	55	AD5/P3.5	46	32	P1.7/$\overline{\text{HOLD}}$			

NOTE:
1. This pin was formerly the Clock Detect Enable pin. This function is not guaranteed to work. This pin must be directly connected to V_{SS}.

PRELIMINARY

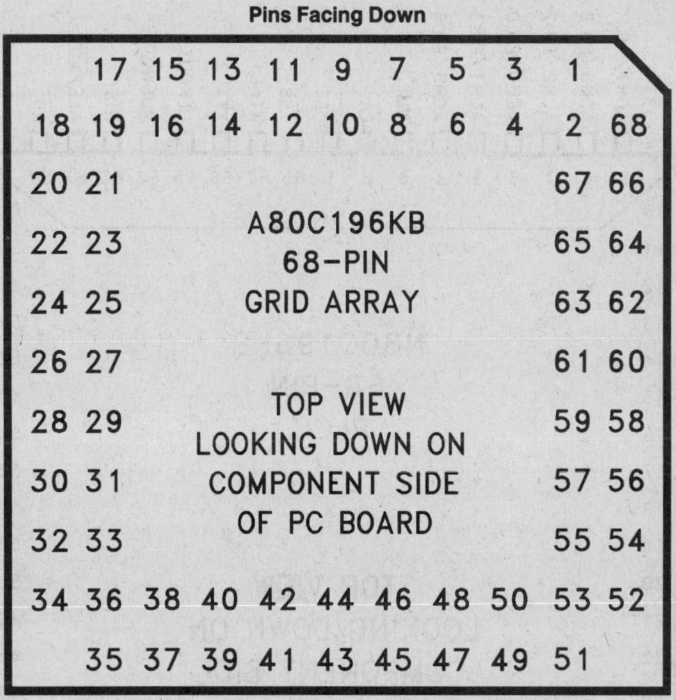

270918-2

Figure 3. 68-Pin Package (Pin Grid Array—Top View)

4

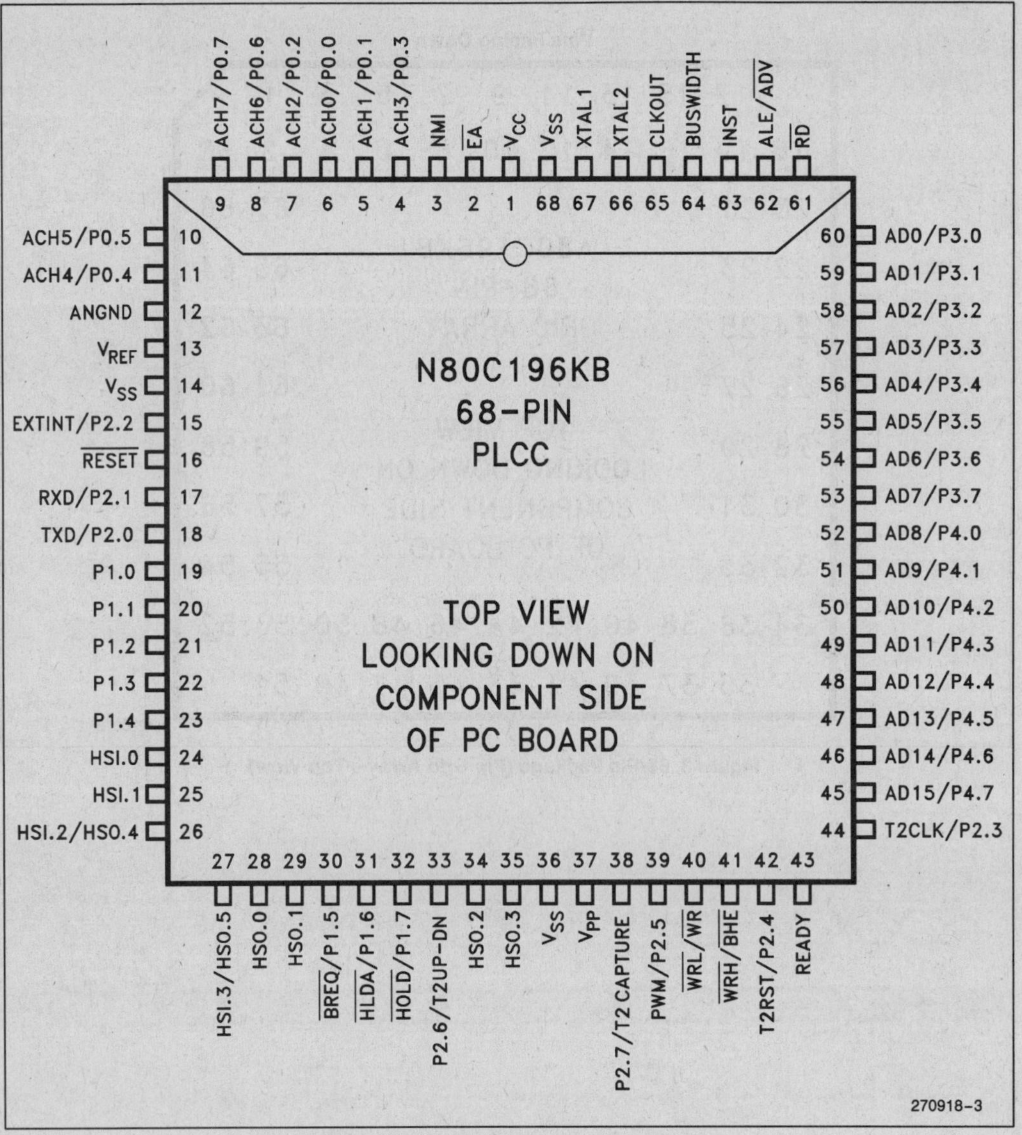

Figure 4. 68-Pin Package (PLCC—Top View)

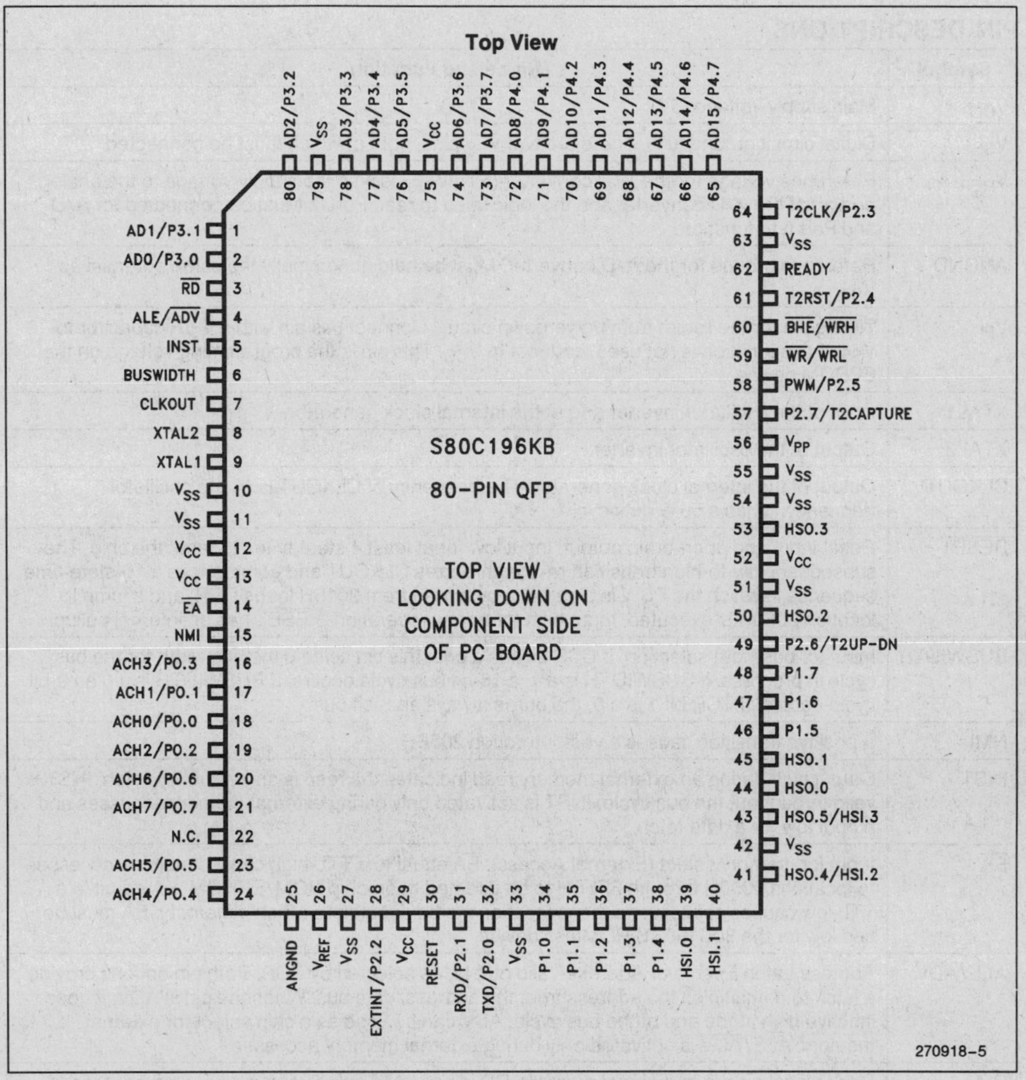

Figure 5. 80-Pin Quad Flat Pack (QFP)

PIN DESCRIPTIONS

Symbol	Name and Function
V_{CC}	Main supply voltage (5V).
V_{SS}	Digital circuit ground (0V). There are two V_{SS} pins, both of which must be connected.
V_{REF}	Reference voltage for the A/D converter (5V). V_{REF} is also the supply voltage to the analog portion of the A/D converter and the logic used to read Port 0. Must be connected for A/D and Port 0 to function.
ANGND	Reference ground for the A/D converter. Must be held at nominally the same potential as V_{SS}.
V_{PP}	Timing pin for the return from powerdown circuit. Connect this pin with a 1 μF capacitor to V_{SS}. If this function is not used, connect to V_{CC}. This pin is the programming voltage on the EPROM device.
XTAL1	Input of the oscillator inverter and of the internal clock generator.
XTAL2	Output of the oscillator inverter.
CLKOUT	Output of the internal clock generator. The frequency of CLKOUT is ½ the oscillator frequency. It has a 50% duty cycle.
$\overline{RESET}$	Reset input and open-drain output. Input low for at least 4 state times to reset the chip. The subsequent low-to-high transition re- synchronizes CLKOUT and commences a 10-state-time sequence in which the PSW is cleared, a byte read from 2018H loads CCR, and a jump to location 2080H is executed. Input high for normal operation. $\overline{RESET}$ has an internal pullup.
BUSWIDTH	Input for buswidth selection. If CCR bit 1 is a one, this pin selects the bus width for the bus cycle in progress. If BUSWIDTH is a 1, a 16-bit bus cycle occurs. If BUSWIDTH is a 0 an 8-bit cycle occurs. If CCR bit 1 is a 0, the bus is always an 8-bit bus.
NMI	A positive transition causes a vector through 203EH.
INST	Output high during an external memory read indicates the read is an instruction fetch. INST is valid throughout the bus cycle. INST is activated only during external memory accesses and output low for a data fetch.
$\overline{EA}$	Input for memory select (External Access). $\overline{EA}$ equal to a TTL-high causes memory accesses to locations 2000H through 3FFFH to be directed to on-chip ROM/EPROM. $\overline{EA}$ equal to a TTL-low causes accesses to these locations to be directed to off-chip memory. $\overline{EA}$ must be tied low for the 80C196KB ROMless device.
ALE/$\overline{ADV}$	Address Latch Enable or Address Valid output, as selected by CCR. Both pin options provide a latch to demultiplex the address from the address/data bus. When the pin is $\overline{ADV}$, it goes inactive high at the end of the bus cycle. $\overline{ADV}$ can be used as a chip select for external memory. ALE/$\overline{ADV}$ is activated only during external memory accesses.
$\overline{RD}$	Read signal output to external memory. $\overline{RD}$ is activated only during external memory reads.
$\overline{WR}$/$\overline{WRL}$	Write and Write Low output to external memory, as selected by the CCR. $\overline{WR}$ will go low for every external write, while $\overline{WRL}$ will go low only for external writes where an even byte is being written. $\overline{WR}$/$\overline{WRL}$ is activated only during external memory writes.
$\overline{BHE}$/$\overline{WRH}$	Bus High Enable or Write High output to external memory, as selected by the CCR. $\overline{BHE}$ = 0 selects the bank of memory that is connected to the high byte of the data bus. A0 = 0 selects the bank of memory that is connected to the low byte of the data bus. Thus accesses to a 16-bit wide memory can be to the low byte only (A0 = 0, $\overline{BHE}$ = 1), to the high byte only (A0 = 1, $\overline{BHE}$ = 0), or both bytes (A0 = 0, $\overline{BHE}$ = 0). If the $\overline{WRH}$ function is selected, the pin will go low if the bus cycle is writing to an odd memory location. $\overline{BHE}$/$\overline{WRH}$ is valid only during 16-bit external memory write cycles.

PIN DESCRIPTIONS (Continued)

Symbol	Name and Function
READY	Ready input to lengthen external memory cycles, for interfacing to slow or dynamic memory, or for bus sharing. If the pin is high, CPU operation continues in a normal manner. If the pin is low prior to the falling edge of CLKOUT, the memory controller goes into a wait mode until the next positive transition in CLKOUT occurs with READY high. When the external memory is not being used, READY has no effect. Internal control of the number of wait states inserted into a bus cycle held not ready is available through configuration of CCR.
HSI	Inputs to High Speed Input Unit. Four HSI pins are available: HSI.0, HSI.1, HSI.2 and HSI.3. Two of them (HSI.2 and HSI.3) are shared with the HSO Unit. The HSI pins are also used as the SID in Slave Programming Mode on the EPROM device.
HSO	Outputs from High Speed Output Unit. Six HSO pins are available: HSO.0, HSO.1, HSO.2, HSO.3, HSO.4 and HSO.5. Two of them (HSO.4 and HSO.5) are shared with the HSI Unit.
Port 0	8-bit high impedance input-only port. These pins can be used as digital inputs and/or as analog inputs to the on-chip A/D converter. These pins set the Programming Mode on the EPROM device.
Port 1	8-bit quasi-bidirectional I/O port.
Port 2	8-bit multi-functional port. All of its pins are shared with other functions in the 80C196KB.
Ports 3 and 4	8-bit bidirectional I/O ports with open drain outputs. These pins are shared with the multiplexed address/data bus which has strong internal pullups. Available as I/O only on the ROM and EPROM devices.
$\overline{\text{HOLD}}$	Bus Hold input requesting control of the bus. Enabled by setting WSR.7.
$\overline{\text{HLDA}}$	Bus Hold acknowledge output indicating release of the bus. Enabled by setting WSR.7.
$\overline{\text{BREQ}}$	Bus Request output activated when the bus controller has a pending external memory cycle. Enabled by setting WSR.7.
TxD	The TxD pin is used for serial port transmission in Modes 1, 2 and 3. The TxD function is enabled by setting IOC1 5. In mode 0 the pin is used as the serial clock output.
RxD	Serial Port Receive pin used for serial port reception. The RxD function is enabled by setting SPCON.3. In mode 0 the pin functions as input or output data.
EXTINT	A rising edge on the EXTINT pin will generate an external interrupt. EXTINT is selected as the external interrupt source by setting IOC1.1 high.
T2CLK	The T2CLK pin is the Timer2 clock input or the serial port baud rate generator input.
T2RST	A rising edge on the T2RST pin will reset Timer2. The external reset function is enabled by setting IOCO.03 T2RST is enabled as the reset source by clearing IOCO.5.
PWM	Port 2.5 can be enabled as a PWM output by setting IOC1.O The duty cycle of the PWM is determined by the value loaded into the PWM-CONTROL register (17H).
T2UP-DN	The T2UP-DN pin controls the direction of Timer2 as an up or down counter. The Timer2 up/down function is enabled by setting IOC2.1.
T2CAP	A rising edge on P2.7 will capture the value of Timer2 in the T2CAPTURE register (location 0CH in Window 15).

4

ELECTRICAL CHARACTERISTICS
ABSOLUTE MAXIMUM RATINGS*

Ambient Temperature
Under Bias .... −55°C to +125°C

Storage Temperature .... −65°C to +150°C

Voltage On Any Pin to V_{SS} .... −0.5V to +7.0V

Power Dissipation(1) 1.5W

NOTE:
1. Power Dissipation is based on package heat transfer, not device power consumption.

NOTICE: This data sheet contains preliminary information on new products in production. It is valid for the devices indicated in the revision history. The specifications are subject to change without notice.

**WARNING:* Stressing the device beyond the "Absolute Maximum Ratings" may cause permanent damage. These are stress ratings only. Operation beyond the "Operating Conditions" is not recommended and extended exposure beyond the "Operating Conditions" may affect device reliability.*

OPERATING CONDITIONS

(All characteristics in this data sheet apply to these operating conditions unless otherwise noted.)

Symbol	Description	Min	Max	Units
T_A	Ambient Temperature Under Bias	0	+70	°C
V_{CC}	Digital Supply Voltage	4.50	5.50	V
V_{REF}	Analog Supply Voltage	4.50	5.50	V
F_{OSC}	Oscillator Frequency **12 MHz**	3.5	12	MHz
F_{OSC}	Oscillator Frequency **10 MHz**	3.5	10	MHz

NOTE:
ANGND and V_{SS} should be nominally at the same potential.

DC CHARACTERISTICS

Symbol	Description	Min	Typ(7)	Max	Units	Test Conditions
V_{IL}	Input Low Voltage	−0.5		0.8	V	
V_{IH}	Input High Voltage (Note 1)	$0.2\ V_{CC} + 0.9$		$V_{CC} + 0.5$	V	
V_{IH1}	Input High Voltage on XTAL 1	$0.7\ V_{CC}$		$V_{CC} + 0.5$	V	
V_{IH2}	Input High Voltage on RESET	2.6		$V_{CC} + 0.5$	V	
V_{OL}	Output Low Voltage			0.3 0.45 1.5	V V V	$I_{OL} = 200\ \mu A$ $I_{OL} = 3.2\ mA$ $I_{OL} = 7\ mA$
V_{OH}	Output High Voltage (Standard Outputs)	$V_{CC} - 0.3$ $V_{CC} - 0.7$ $V_{CC} - 1.5$			V V V	$I_{OH} = -200\ \mu A$ $I_{OH} = -3.2\ mA$ $I_{OH} = -7\ mA$
V_{OH1}	Output High Voltage (Quasi-bidirectional Outputs)	$V_{CC} - 0.3$ $V_{CC} - 0.7$ $V_{CC} - 1.5$			V V V	$I_{OH} = -10\ \mu A$ $I_{OH} = -30\ \mu A$ $I_{OH} = -60\ \mu A$

NOTES:
1. All pins except RESET and XTAL1.
2. Holding these pins below V_{IH} in Reset may cause the part to enter test modes.

DC CHARACTERISTICS (Continued)

Symbol	Description	Min	Typ(7)	Max	Units	Test Conditions
I_{LI}	Input Leakage Current (Std. Inputs)			±10	μA	$0 < V_{IN} < V_{CC} - 0.3V$
I_{LI1}	Input Leakage Current (Port 0)			+3	μA	$0 < V_{IN} < V_{REF}$
I_{TL}	1 to 0 Transition Current (QBD Pins)			−650	μA	$V_{IN} = 2.0V$
I_{IL}	Logical 0 Input Current (QBD Pins)			−50	μA	$V_{IN} = 0.45V$
I_{IL1}	Logical 0 Input Current in Reset (Note 2) (ALE, $\overline{RD}$, $\overline{WR}$, $\overline{BHE}$, INST, P2.0)			−1.2	mA	$V_{IN} = 0.45V$
Hyst	Hysteresis on RESET Pin	300			mV	
I_{CC}	Active Mode Current in Reset		40	55	mA	XTAL1 = 12 MHz $V_{CC} = V_{PP} = V_{REF} = 5.5V$
I_{REF}	A/D Converter Reference Current		2	5	mA	
I_{IDLE}	Idle Mode Current		10	22	mA	
I_{CC1}	Active Mode Current		15	22	mA	XTAL1 = 3.5 MHz
I_{PD}	Powerdown Mode Current		5	50	μA	$V_{CC} = V_{PP} = V_{REF} = 5.5V$
R_{RST}	Reset Pullup Resistor	6K		50K	Ω	
C_S	Pin Capacitance (Any Pin to V_{SS})			10	pF	$F_{TEST} = 1.0$ MHz

NOTES:
(Notes apply to all specifications)
1. QBD (Quasi-bidirectional) pins include Port 1, P2.6 and P2.7.
2. Standard Outputs include AD0–15, $\overline{RD}$, $\overline{WR}$, ALE, $\overline{BHE}$, INST, HSO pins, PWM/P2.5, CLKOUT, RESET, Ports 3 and 4, TXD/P2.0 and RXD (in serial mode 0). The V_{OH} specification is not valid for RESET. Ports 3 and 4 are open-drain outputs.
3. Standard Inputs include HSI pins, CDE, $\overline{EA}$, READY, BUSWIDTH, NMI, RXD/P2.1, EXTINT/P2.2, T2CLK/P2.3 and T2RST/P2.4.
4. Maximum current per pin must be externally limited to the following values if V_{OL} is held above 0.45V or V_{OH} is held below $V_{CC} - 0.7V$:
   I_{OL} on Output pins: 10 mA
   I_{OH} on quasi-bidirectional pins: self limiting
   I_{OH} on Standard Output pins: 10 mA
5. Maximum current per bus pin (data and control) during normal operation is ±3.2 mA.
6. During normal (non-transient) conditions the following total current limits apply:

Port 1, P2.6	I_{OL}: 29 mA	I_{OH} is self limiting
HSO, P2.0, RXD, $\overline{RESET}$	I_{OL}: 29 mA	I_{OH}: 26 mA
P2.5, P2.7, $\overline{WR}$, $\overline{BHE}$	I_{OL}: 13 mA	I_{OH}: 11 mA
AD0–AD15	I_{OL}: 52 mA	I_{OH}: 52 mA
$\overline{RD}$, ALE, INST–CLKOUT	I_{OL}: 13 mA	I_{OH}: 13 mA

7. Typicals are based on a limited number of samples and are not guaranteed. The values listed are at room temperature and $V_{REF} = V_{CC} = 5V$.

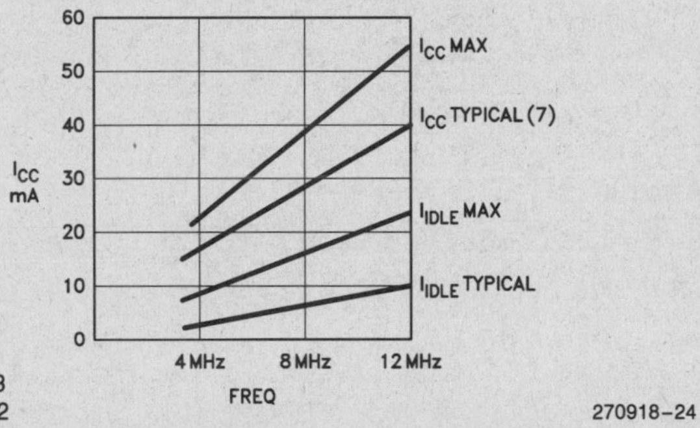

I_{CC} Max = 3.88 × FREQ + 8.43
I_{IDLE} Max = 1.65 × FREQ + 2.2

Figure 6. I_{CC} and I_{IDLE} vs Frequency

AC CHARACTERISTICS

Test Conditions: Capacitive load on all pins = 100 pF, Rise and fall times = 10 ns, F_{OSC} = 10/12 MHz

The system must meet these specifications to work with the 80C196KB:

Symbol	Description	Min	Max	Units	Notes
T_{AVYV}	Address Valid to READY Setup				
	80C196KB10/83C196KB10		$2\,T_{OSC} - 90$	ns	
	80C196KB12/83C196KB12		$2\,T_{OSC} - 85$	ns	
T_{LLYV}	ALE Low to READY Setup				
	80C196KB10/83C196KB10		$T_{OSC} - 80$	ns	
	80C196KB12/83C196KB12		$T_{OSC} - 72$	ns	
T_{YLYH}	Non READY Time	No upper limit		ns	
T_{CLYX}	READY Hold after CLKOUT Low	0	$T_{OSC} - 30$	ns	(Note 1)
T_{LLYX}	READY Hold after ALE Low	$T_{OSC} - 15$	$2\,T_{OSC} - 40$	ns	(Note 1)
T_{AVGV}	Address Valid to Buswidth Setup		$2\,T_{OSC} - 85$	ns	
T_{LLGV}	ALE Low to Buswidth Setup		$T_{OSC} - 70$	ns	
T_{CLGX}	Buswidth Hold after CLKOUT Low	0		ns	
T_{AVDV}	Address Valid to Input Data Valid				
	80C196KB10/83C196KB10		$3\,T_{OSC} - 70$	ns	(Note 2)
	80C196KB12/83C196KB12		$3\,T_{OSC} - 67$	ns	
T_{RLDV}	$\overline{RD}$ Active to Input Data Valid				
	80C196KB10/83C196KB10		$T_{OSC} - 30$	ns	(Note 2)
	80C196KB12/83C196KB12		$T_{OSC} - 23$	ns	
T_{CLDV}	CLKOUT Low to Input Data Valid		$T_{OSC} - 50$	ns	
T_{RHDZ}	End of $\overline{RD}$ to Input Data Float		$T_{OSC} - 20$	ns	
T_{RXDX}	Data Hold after $\overline{RD}$ Inactive	0		ns	

NOTES:
1. If max is exceeded, additional wait states will occur.
2. When using wait states, add $2\,T_{OSC} \times n$, where n = number of wait states.

AC CHARACTERISTICS

Test Conditions: Capacitive load on all pins = 100 pF, Rise and fall times = 10 ns, F_{OSC} = 10/12 MHz

The 80C196KB will meet these specifications:

Symbol	Description	Min	Max	Units	Notes
F_{XTAL}	Frequency on $XTAL_1$ 80C196KB10/83C196KB10 80C196KB12/83C196KB12	 3.5 3.5	 10 12	 MHz MHz	 (Note 2) (Note 2)
T_{OSC}	I/F_{XTAL} 80C196KB10/83C196KB10 80C196KB12/83C196KB12	 100 83	 286 286	 ns ns	
T_{XHCH}	XTAL1 High to CLKOUT High or Low	40	110	ns	(Note 3)
T_{CLCL}	CLKOUT Cycle Time	$2\ T_{OSC}$		ns	
T_{CHCL}	CLKOUT High Period	$T_{OSC} - 10$	$T_{OSC} + 10$	ns	
T_{CLLH}	CLKOUT Falling Edge to ALE Rising	−5	15	ns	
T_{LLCH}	ALE Falling Edge to CLKOUT Rising	−15	15	ns	
T_{LHLH}	ALE Cycle Time	$4\ T_{OSC}$		ns	(Note 5)
T_{LHLL}	ALE High Period	$T_{OSC} - 10$	$T_{OSC} + 10$	ns	
T_{AVLL}	Address Setup to ALE Falling Edge	$T_{OSC} - 20$			
T_{LLAX}	Address Hold after ALE Falling Edge	$T_{OSC} - 40$		ns	
T_{LLRL}	ALE Falling Edge to $\overline{RD}$ Falling Edge	$T_{OSC} - 40$		ns	
T_{RLCL}	$\overline{RD}$ Low to CLKOUT Falling Edge	5	30	ns	
T_{RLRH}	$\overline{RD}$ Low Period	$T_{OSC} - 5$	$T_{OSC} + 25$	ns	(Note 5)
T_{RHLH}	$\overline{RD}$ Rising Edge to ALE Rising Edge	T_{OSC}	$T_{OSC} + 25$	ns	(Note 4)
T_{RLAZ}	$\overline{RD}$ Low to Address Float		10	ns	
T_{LLWL}	ALE Falling Edge to $\overline{WR}$ Falling Edge	$T_{OSC} - 10$		ns	
T_{CLWL}	CLKOUT Low to $\overline{WR}$ Falling Edge	0	25	ns	
T_{QVWH}	Data Stable to $\overline{WR}$ Rising Edge 80C196KB10/83C196KB10 80C196KB12/83C196KB12	 $T_{OSC} - 30$ $T_{OSC} - 23$		 ns ns	 (Note 5)
T_{CHWH}	CLKOUT High to $\overline{WR}$ Rising Edge	−10	10	ns	
T_{WLWH}	$\overline{WR}$ Low Period	$T_{OSC} - 30$	$T_{OSC} + 5$	ns	(Note 5)
T_{WHQX}	Data Hold after $\overline{WR}$ Rising Edge	$T_{OSC} - 10$		ns	
T_{WHLH}	$\overline{WR}$ Rising Edge to ALE Rising Edge	$T_{OSC} - 10$	$T_{OSC} + 15$	ns	(Note 4)
T_{WHBX}	$\overline{BHE}$, INST Hold after $\overline{WR}$ Rising Edge	$T_{OSC} - 10$		ns	
T_{RHBX}	$\overline{BHE}$, INST Hold after $\overline{RD}$ Rising Edge	$T_{OSC} - 10$		ns	
T_{WHAX}	AD8–15 Hold after $\overline{WR}$ Rising Edge	$T_{OSC} - 50$		ns	
T_{RHAX}	AD8–15 Hold after $\overline{RD}$ Rising Edge	$T_{OSC} - 25$		ns	

NOTES:

T_{OSC} = 83.3 ns at 12 MHz; T_{OSC} = 100 ns at 10 MHz.

1. Customers whose applications require an 83C196KB to meet the 80C196KB specifications listed above should contact an Intel Field Sales Representative.
2. Testing performed at 3.5 MHz. However, the part is static by design and will typically operate below 1 Hz.
3. Typical specification, not guaranteed.
4. Assuming back-to-back bus cycles.
5. When using wait states, add $2\ T_{OSC} \times n$, where n = number of wait states.

System Bus Timings

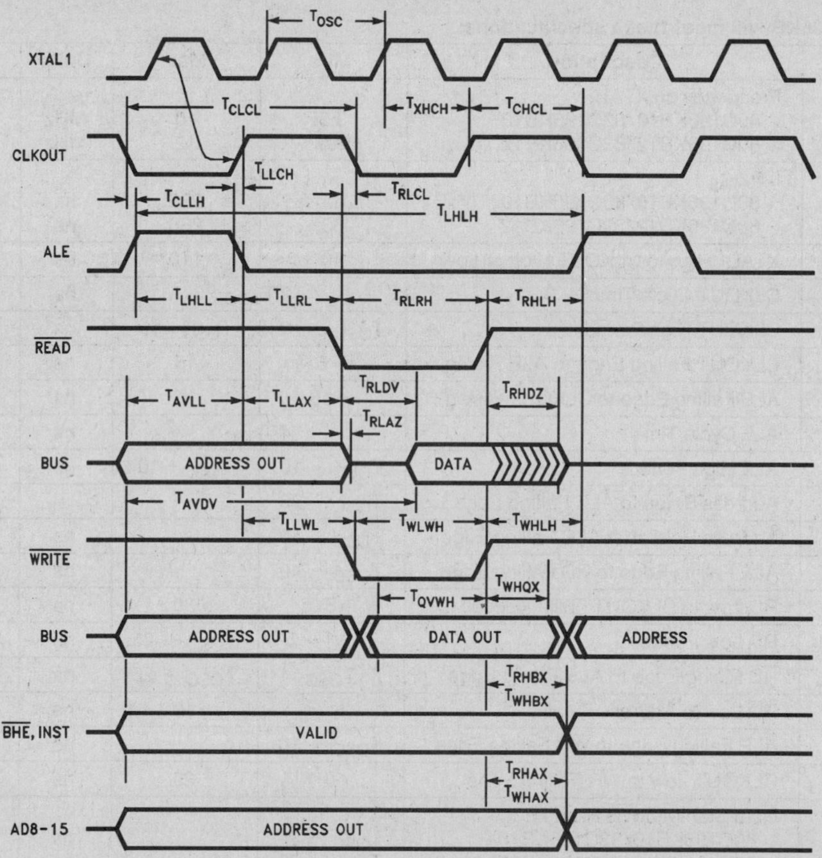

270918-25

READY Timings (One Wait State)

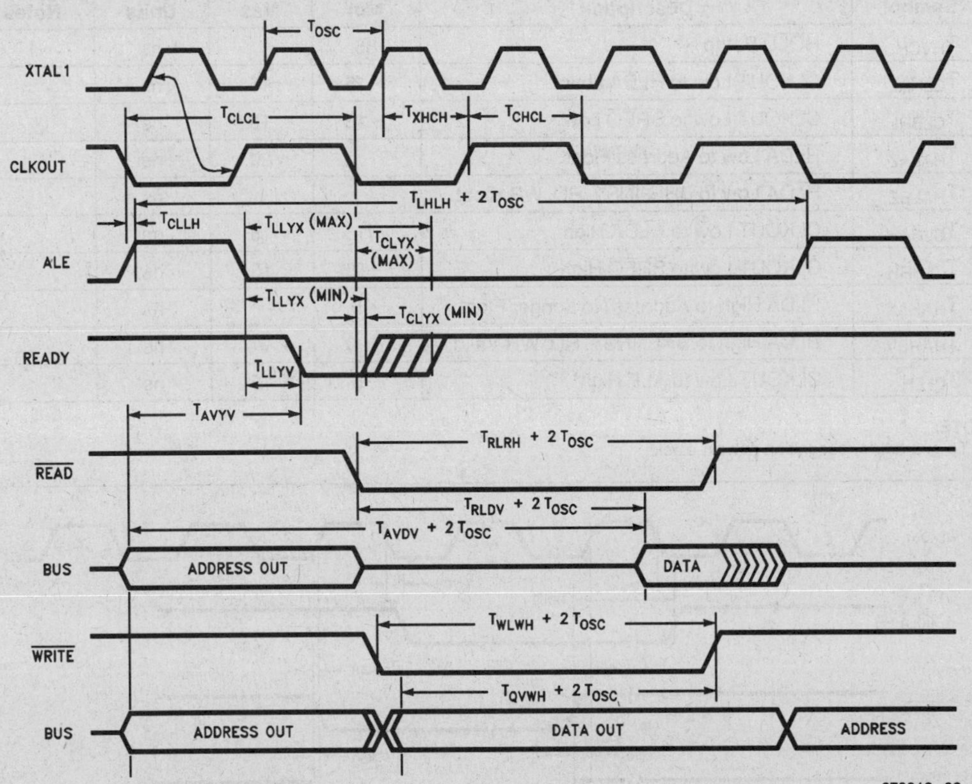

270918–26

Buswidth Timings

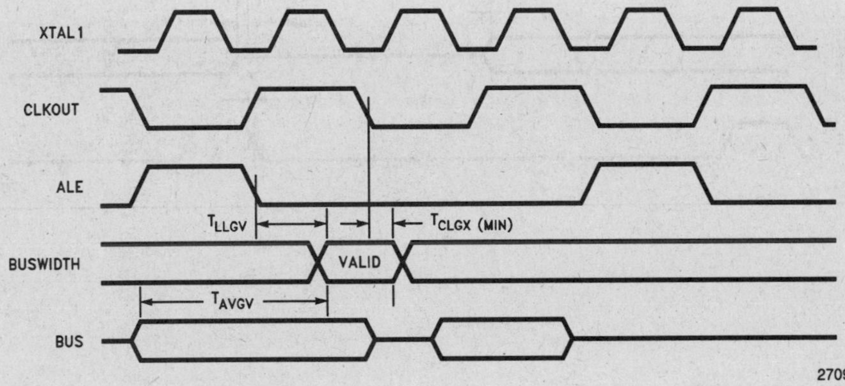

270918–27

$\overline{HOLD}$/$\overline{HLDA}$ TIMINGS

Symbol	Description	Min	Max	Units	Notes
T_{HVCH}	$\overline{HOLD}$ Setup	85		ns	1
T_{CLHAL}	CLKOUT Low to $\overline{HLDA}$ Low	−15	15	ns	
T_{CLBRL}	CLKOUT Low to $\overline{BREQ}$ Low	−15	15	ns	
T_{HALAZ}	$\overline{HLDA}$ Low to Address Float		20	ns	
T_{HALBZ}	$\overline{HLDA}$ Low to BHE, INST, RD, WR Float			ns	
T_{CLHAH}	CLKOUT Low to $\overline{HLDA}$ High	−15	15	ns	
T_{CLBRH}	CLKOUT Low to $\overline{BREQ}$ High	−15	15	ns	
T_{HAHAX}	$\overline{HLDA}$ High to Address No Longer Float	−5		ns	
T_{HAHBV}	$\overline{HLDA}$ High to $\overline{BHE}$, INST, $\overline{RD}$, $\overline{WR}$ Valid	−20		ns	
T_{CLLH}	CLKOUT Low to ALE High	−5	15	ns	

NOTE:
1. To guarantee recognition at next clock.

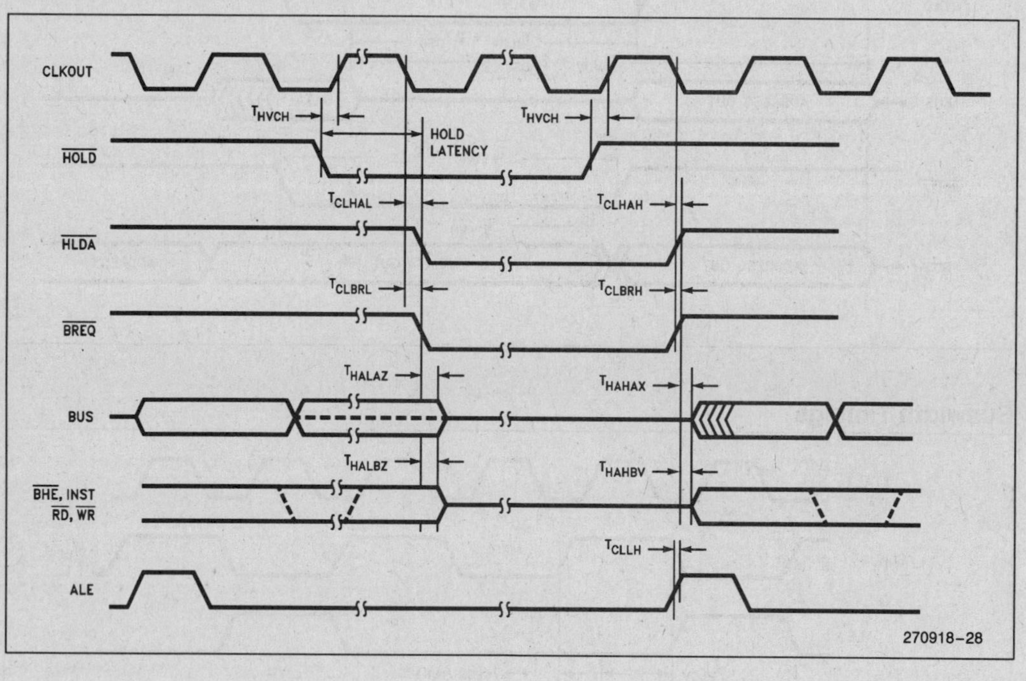

EXTERNAL CLOCK DRIVE

Symbol	Parameter	Min	Max	Units
$1/T_{XLXL}$	Oscillator Frequency 80C196KB10 80C196KB12	 3.5 3.5	 10.0 12.0	 MHz MHz
T_{XLXL}	Oscillator Frequency 80C196KB10 80C196KB12	 100 83	 286 286	 ns ns
T_{XHXX}	High Time	32		ns
T_{XLXX}	Low Time	32		ns
T_{XLXH}	Rise Time		10	ns
T_{XHXL}	Fall Time		10	ns

EXTERNAL CLOCK DRIVE WAVEFORMS

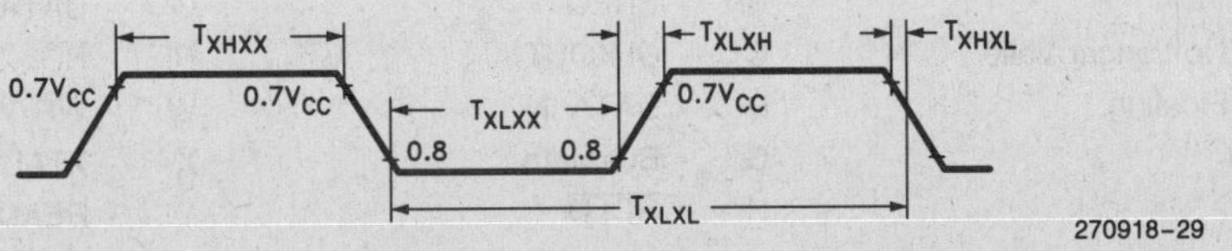

270918-29

An external oscillator may encounter as much as a 100 pF load at XTAL1 when it starts-up. This is due to interaction between the amplifier and its feedback capacitance. Once the external signal meets the V_{IL} and V_{IH} specifications the capacitance will not exceed 20 pF.

EXTERNAL CRYSTAL CONNECTIONS

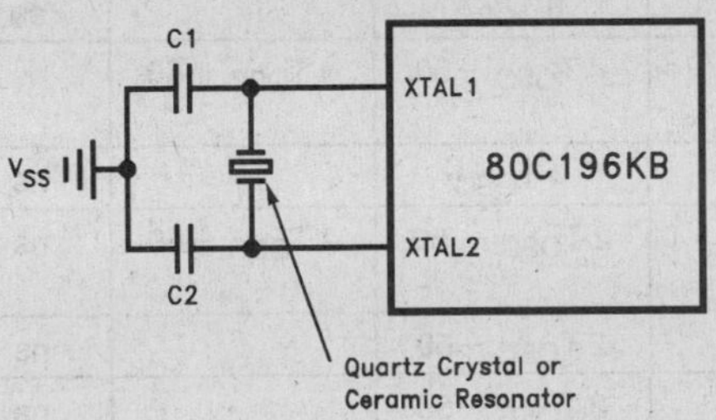

270918-36

NOTE:
Keep oscillator components close to chip and use short, direct traces to XTAL1, XTAL2 and V_{SS}. When using crystals, C1 = C2 ≈ 20 pF. When using ceramic resonators, consult manufacturer for recommended capacitor values.

EXTERNAL CLOCK CONNECTIONS

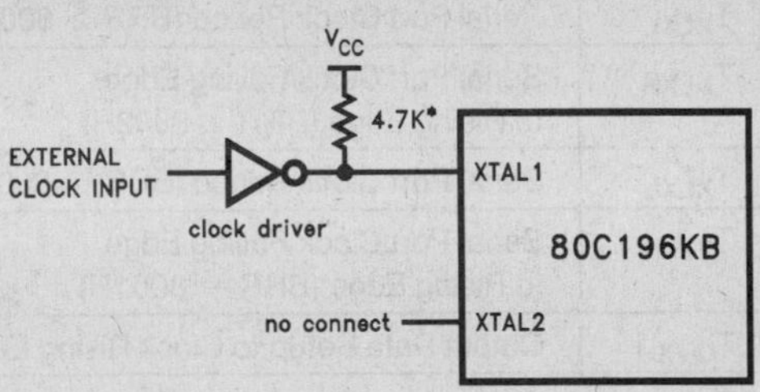

270918-37

* Required if TTL driver used.
Not needed if CMOS driver is used.

AC TESTING INPUT, OUTPUT WAVEFORMS

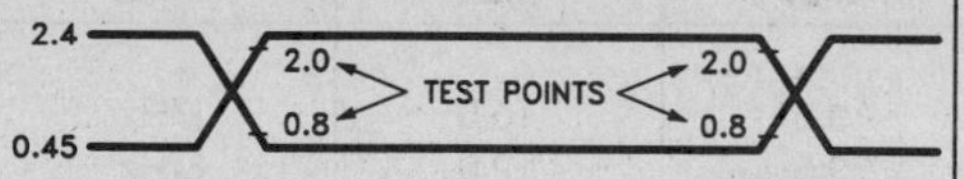

270918-30

AC Testing inputs are driven at 2.4V for a Logic "1" and 0.45V for a Logic "0" Timing measurements are made at 2.0V for a Logic "1" and 0.8V for a Logic "0".

FLOAT WAVEFORMS

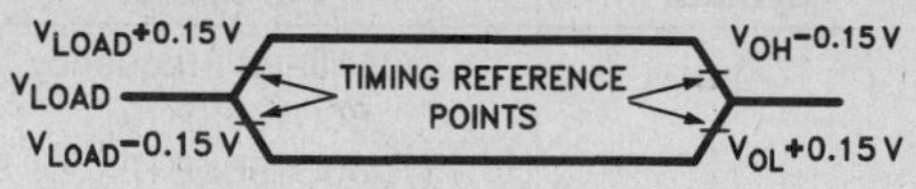

270918-31

For Timing Purposes a Port Pin is no Longer Floating when a 150 mV change from Load Voltage Occurs and Begins to Float when a 150 mV change from the Loaded V_{OH}/V_{OL} Level occurs $I_{OL}/I_{OH} = \pm 15$ mA.

EXPLANATION OF AC SYMBOLS

Each symbol is two pairs of letters prefixed by "T" for time. The characters in a pair indicate a signal and its condition, respectively. Symbols represent the time between the two signal/condition points.

Conditions:

- H - High
- L - Low
- V - Valid
- X - No Longer Valid
- Z - Floating

Signals:

- A - Address
- B - $\overline{BHE}$
- BR - $\overline{BREQ}$
- C - CLKOUT
- D - DATA IN
- G - Buswidth
- H - $\overline{HOLD}$
- HA - $\overline{HLDA}$
- L - ALE/$\overline{ADV}$
- Q - DATA OUT
- R - $\overline{RD}$
- W - $\overline{WR}/\overline{WRH}/\overline{WRL}$
- X - XTAL1
- Y - READY

AC CHARACTERISTICS—SERIAL PORT—SHIFT REGISTER MODE

SERIAL PORT TIMING—SHIFT REGISTER MODE

Symbol	Parameter	Min	Max	Units
T_{XLXL}	Serial Port Clock Period (BRR ≥ 8002H)	6 T_{OSC}		ns
T_{XLXH}	Serial Port Clock Falling Edge to Rising Edge (BRR ≥ 8002H)	4 T_{OSC} −50	4 T_{OSC} +50	ns
T_{XLXL}	Serial Port Clock Period (BRR = 8001H)	4 T_{OSC}		ns
T_{XLXH}	Serial Port Clock Falling Edge to Rising Edge (BRR = 8001H)	2 T_{OSC} − 50	2 T_{OSC} +50	ns
T_{QVXH}	Output Data Setup to Clock Rising Edge	2 T_{OSC} −50		ns
T_{XHQX}	Output Data Hold after Clock Rising Edge	2 T_{OSC} −50		ns
T_{XHQV}	Next Output Data Valid after Clock Rising Edge		2 T_{OSC} +50	ns
T_{DVXH}	Input Data Setup to Clock Rising Edge	T_{OSC} +50		ns
T_{XHDX}	Input Data Hold after Clock Rising Edge	0		ns
T_{XHQZ}	Last Clock Rising to Output Float		1 T_{OSC}	ns

WAVEFORM—SERIAL PORT—SHIFT REGISTER MODE

SERIAL PORT WAVEFORM—SHIFT REGISTER MODE

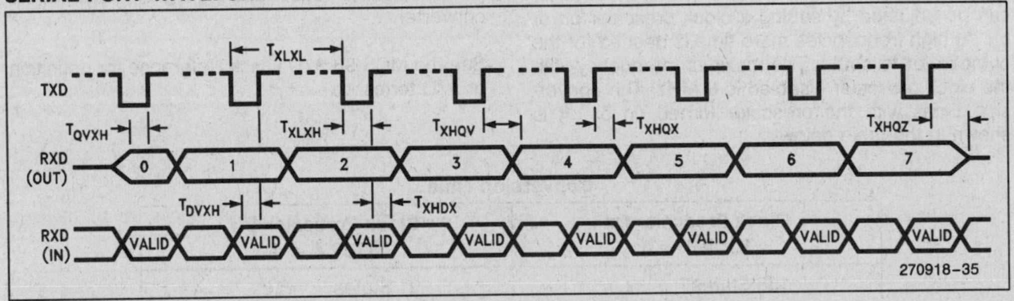

4

A TO D CHARACTERISTICS

There are two modes of A/D operation: with or without clock prescaler. The speed of the A/D converter can be adjusted by setting a clock prescaler on or off. At high frequencies more time is needed for the comparator to settle. The maximum frequency with the clock prescaler disabled is 8 MHz. The conversion times with the prescaler turned on or off is shown in the table below.

The converter is ratiometric, so the absolute accuracy is directly dependent on the accuracy and stability of V_{REF}. V_{REF} must be close to V_{CC} since it supplies both the resistor ladder and the digital section of the converter.

See the MCS-96 A/D Quick Reference for definition of A/D terms.

Conversion Time

Clock Prescaler On IOC2.4 = 0	Clock Prescaler Off IOC2.4 = 1
158 States 26.33 μs @ 12 MHz	91 States 22.75 μs @ 8 MHz

A/D CONVERTER SPECIFICATIONS

Parameter	Typical(1)	Minimum	Maximum	Units*	Notes
Resolution		512 9	1024 10	Levels Bits	
Absolute Error		0	±4	LSBs	
Full Scale Error	0.25 ±0.50			LSBs	
Zero Offset Error	−0.25 ±0.50			LSBs	
Non-Linearity Error	1.5 ±2.5	0	±4	LSBs	
Differential Non-Linearity Error		> −1	+2	LSBs	
Channel-to-Channel Matching	±0.1	0	±1	LSBs	
Repeatability	±0.25			LSBs	
Temperature Coefficients: Offset Full Scale Differential Non-Linearity	 0.009 0.009 0.009			 LSB/°C LSB/°C LSB/°C	
Off Isolation		−60		dB	2, 3
Feedthrough	−60			dB	2
V_{CC} Power Supply Rejection	−60			dB	2
Input Series Resistance		1K	5K	Ω	4
DC Input Leakage		0	3.0	μA	
Sample Time: Prescaler On Prescaler Off	15 8			States States	
Sampling Capacitor	3			pF	

NOTES:

*An "LSB", as used here, has a value of approximately 5 mV.

1. Typical values are expected for most devices at 25°C but are not tested or guaranteed.
2. DC to 100 KHz.
3. Multiplexer Break-Before-Make Guaranteed.
4. Resistance from device pin, through internal MUX, to sample capacitor.

EXTENDED TEMPERATURE/EXTENDED BURN-IN ONLY SPECIFICATIONS

Symbols	Description	Min	Max	Units
RESET Hysteresis	Hysteresis on RESET Pin	TBD		mV
I_{PD}	Powerdown Mode Current		TBD	mA
T_{LLYV}	ALE Low to READY Setup		$T_{OSC} - 65$	ns
T_{LLGV}	ALE Low to BUSWIDTH Setup		$T_{OSC} - 60$	ns
T_{AVDV}	Address Valid to Input Data Valid		$3\,T_{OSC} - 60$	ns
T_{RLDV}	$\overline{RD}$ Low to Input Data Valid		$T_{OSC} - 25$	ns
T_{LHLL}	ALE High Period	$T_{OSC} - 12$	$T_{OSC} + 12$	ns
T_{RHAX}	AD_8–AD_{15} Hold after $\overline{RD}$ Rising	$T_{OSC} - 50$		ns
T_{HALAZ}	$\overline{HLDA}$ Low to Address Float		−25	ns
T_{HALBZ}	$\overline{HLDA}$ Low to $\overline{BHE}$, INST, RD, $\overline{WR}$ Float		−30	ns
T_{HAHBV}	$\overline{HLDA}$ High to $\overline{BHE}$, INST, RD, $\overline{WR}$ Valid	−25		ns
A/D Absolute Error	Absolute Error		±6	LSBs

FUNCTIONAL DEVIATIONS

1. The DJNZW instruction is not guaranteed to be functional. The instruction, if encountered, will not cause an unimplemented opcode interrupt. (The opcode for DJNZW is 0E1 Hex.) The DJNZ (byte) instruction works correctly and should be used instead.
2. The CDE function is not guaranteed to work. The CDE pin must be directly connected to V_{SS}.
3. The HSI unit has two errata: one dealing with resolution and the other with first entries into the FIFO.

   The HSI resolution is 9 states instead of 8 states. Events on the same line may be lost if they occur faster than once every 9 state times.

   There is a mismatch between the 9 state time HSI resolution and the 8 state time timer. This causes one time value to be unused every 9 timer counts. Events may receive a time-tag one count later than expected because of this "skipped" time value.

   If the first two events into an empty FIFO (not including the Holding Register) occur in the same internal phase, both are recorded with one time-tag. Otherwise, if the second event occurs within 9 states after the first, its time-tag is one count later than the first's. If this is the "skipped" time value, the second event's time-tag is 2 counts later than the first's.

   If the FIFO and Holding Register are empty, the first event will transfer into the Holding Register after 8 state times, leaving the FIFO empty again. If the second event occurs after this time, it will act as a new first event into an empty FIFO.
4. The serial port Framing Error flag fails to indicate an error if the bit preceding the stop bit is a 1. This is the case in both the 8-bit and 9-bit modes. False framing errors are never generated.
5. The serial port RI flag is not generated after the first byte is received. The problem does not occur if the baud rate is reloaded after each reception.
6. If the unsigned divide instruction (byte or word) is the last instruction in the queue as HOLD or READY is asserted, the result may be incorrect.

DATA SHEET REVISION HISTORY

This data sheet (270918-002) is valid for devices marked with a "B" at the end of the top side tracking number. Data sheets are changed as new device information becomes available. Verify with your local Intel sales office that you have the latest version before finalizing a design or ordering devices.

The following differences exist between this data sheet and the previous version (-001).

1. The commercial and Express (extended temperature and extended burn-in) devices were combined in this data sheet. The Express only data sheet (270780-002) is now obsolete.
2. The EPROM devices were removed from this data sheet. They are now in a separate data sheet (270909).
3. The 80C196KB devices were removed from this data sheet. Only the 80C196KB10, 83C196KB10, 80C196KB12 and 83C196KB12 devices are now covered.
4. Changes were made to the format of the data sheet and the SFR descriptions were removed.
5. Two errata were added: the serial port RI flag and the DIVIDE during HOLD/READY.
6. Three specifications for the extended temperature and extended burn-in devices were changed: V_{IH2} Min was changed from 2.4V to 2.6V, T_{XHCH} Min was changed from 35 ns to 40 ns, and T_{HVCH} Min was changed from 90 ns to 85 ns.

The -001 data sheet integrated the 87C196KB (order number 270590-003) and the 83C196KB/80C196KB (order number 270634-003) data sheets. The following differences exist between the -001 data sheet and each of the above mentioned data sheets.

1. The status of the data sheet was upgraded from ADVANCE INFORMATION to PRELIMINARY.
2. The warning about the ABSOLUTE MAXIMUM RATINGS was reworded and a notice of disclaimer was added to the electrical specifications section.
3. V_{IH2} was increased from 2.2V to 2.6V.
4. I_{IL1} was increased from -950 μA to -1.2 mA. This change was documented in the previous revision of the data sheets but the DC Characteristics table did not reflect the change.
5. Maximum I_{PD} specification was added to the DC table and I_{PD} note was deleted.

80C198/83C198/80C194/83C194 COMMERCIAL/EXPRESS CHMOS MICROCONTROLLER

83C198 — 8 Kbytes of Factory Mask-Programmed ROM
80C198 — ROMless

- **232 Byte Register File**
- **Register-to-Register Architecture**
- **28 Interrupt Sources/16 Vectors**
- **2.3 μs 16 x 16 Multiply (12 MHz)**
- **4.0 μs 32/16 Divide (12 MHz)**
- **Powerdown and Idle Modes**
- **16-Bit Watchdog Timer**
- **8-Bit External Bus**
- **Extended Temperature Available**
- **Full Duplex Serial Port**
- **High Speed I/O Subsystem**
- **16-Bit Timer**
- **16-Bit Counter**
- **Pulse-Width-Modulated Output**
- **Four 16-Bit Software Timers**
- **10-Bit A/D Converter with Sample/Hold**
- **Extended Burn-In Available**

The 80C198 is the low cost member of the CHMOS MCS®-96 microcontroller family. Intel's CHMOS process provides a high performance processor along with low power consumption. To further reduce power requirements, the processor can be placed into Idle or Powerdown Mode.

The 83C198 is an 80C198 with 8 Kbytes on-chip ROM. In this document, the 80C198 will also refer to the 83C198, 80C194 and 83C194 unless otherwise stated. Bit, byte, word and some 32-bit operations are available on the 80C198. With a 12 MHz oscillator a 16-bit addition takes 0.66 μs, and the instruction times average 0.5 μs to 1.5 μs in typical applications.

Four high-speed capture inputs are provided to record times when events occur. Six high-speed outputs are available for pulse or waveform generation. The high-speed output can also generate four software timers or start an A/D conversion. Events can be based on the timer or counter. Also provided on-chip are an A/D converter, serial port, watchdog timer and a pulse-width-modulated output signal.

The 80C194 and 83C194 do not have the on-chip A/D converter.

With the commercial (standard) temperature option, operational characteristics are guaranteed over the temperature range of 0°C to +70°C. With the extended temperature range option, operational characteristics are guaranteed over the temperature range of −40°C to +85°C. With the extended burn-in option, the burn-in is dynamic for a minimum time of 160 hours at 125°C with V_{CC} = 5.5V ±0.5V, following the guidelines in MIL-STD-883, Method 1015. The specifications which are different for the extended temperature and extended burn-in devices are listed in this data sheet. Otherwise, the commercial specifications apply for both.

October 1991
Order Number: 270815-003

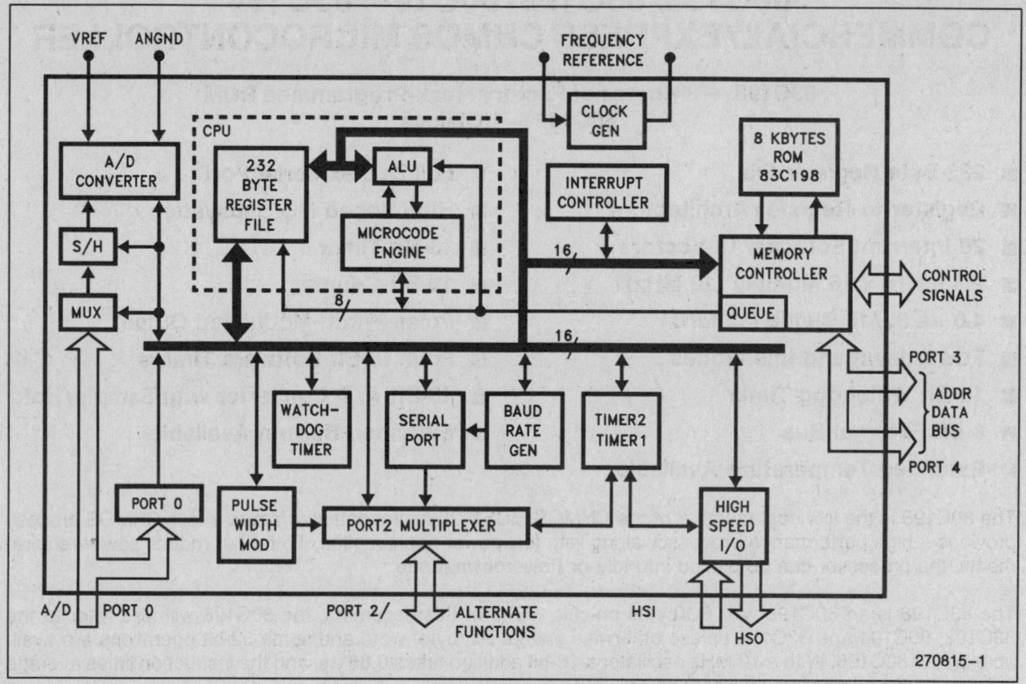

Figure 1. 80C198/80C198 Block Diagram

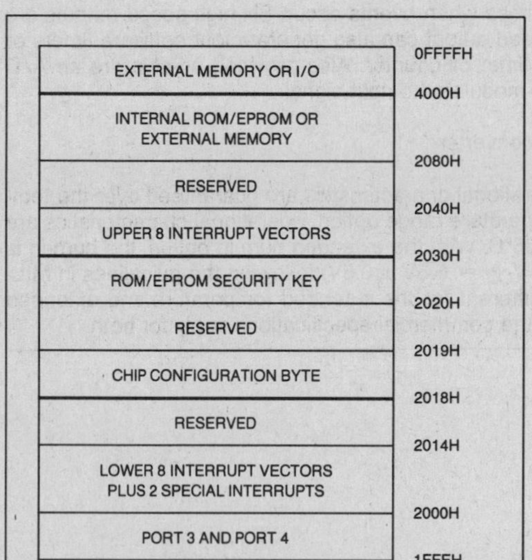

Figure 2. Memory Map

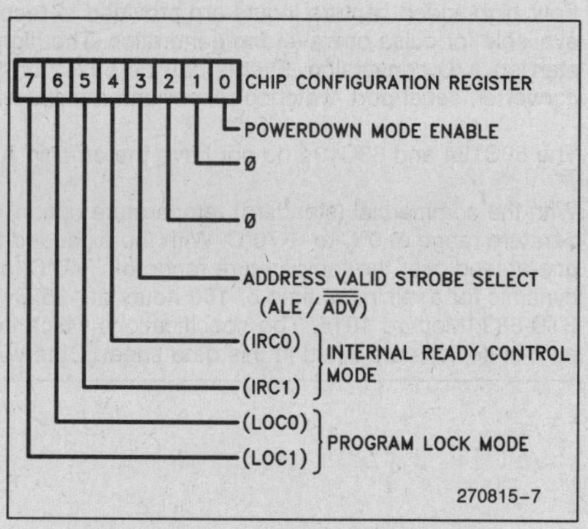

Figure 3. Chip Configuration (2018H)

PACKAGING

The 80C198 and 83C198 are available in a 52-pin PLCC package and an 80-pin QFP package. Contact your local sales office to determine the exact ordering code for the part desired.

Package Designators:
N = 52-pin PLCC
S = 80-pin QFP

Prefix Designators:
T = Extended temperature
L = Extended temperature and extended burn-in

Package Type	θ_{ja}	θ_{jc}
PLCC	35°C/W	12°C/W
QFP	85°C/W	—

Figure 4. 8XC198 Thermal Characteristics

All thermal impedance data is approximate for static air conditions at 1W of power dissipation. Values will change depending on operating conditions and application. See the Intel *Packaging Handbook* (Order Number 240800) for a description of Intel's thermal impedance test methodology.

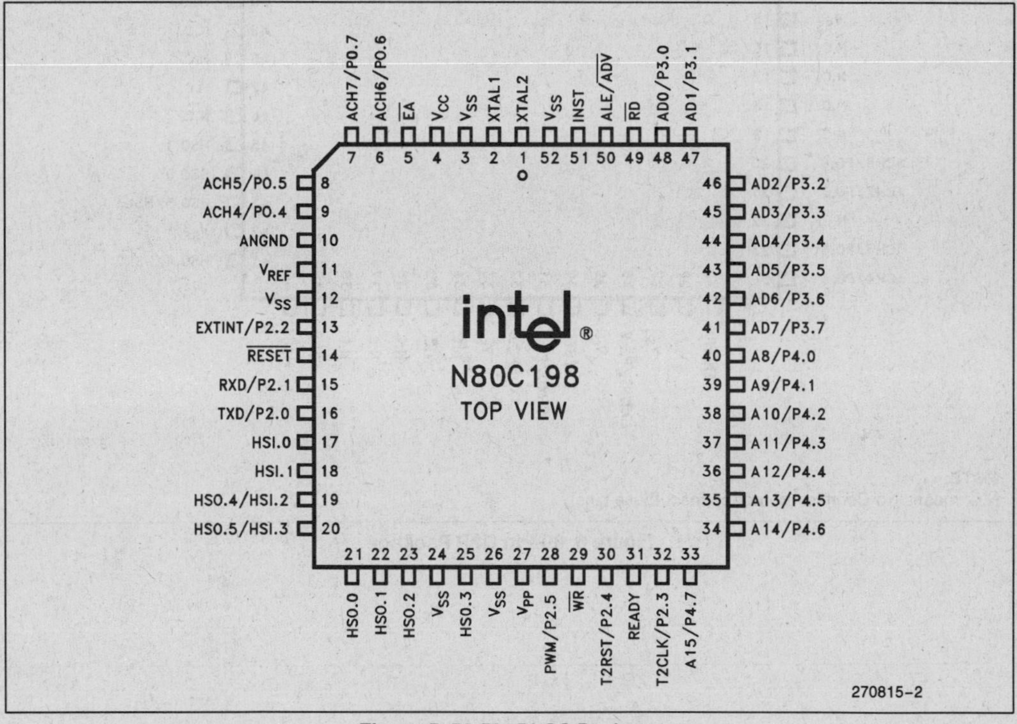

Figure 5. 52-Pin PLCC Package

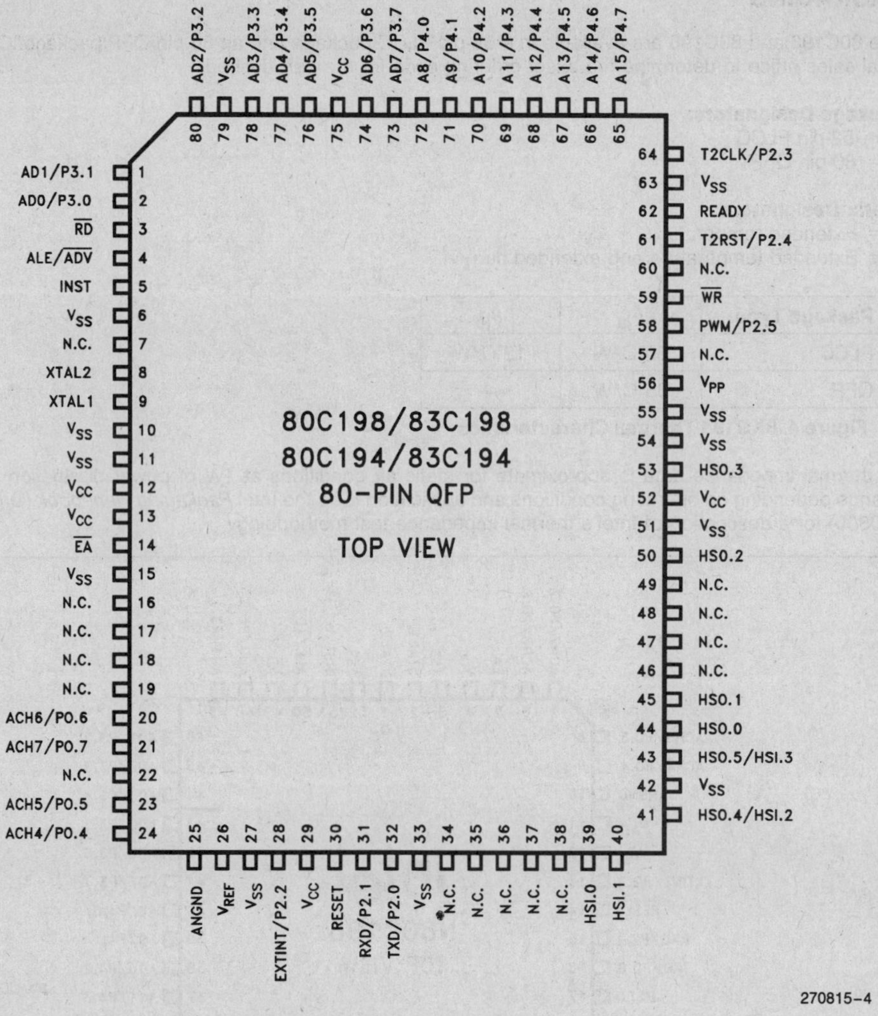

NOTE:
N.C. means No Connect (do not connect these pins).

Figure 6. 80-Pin QFP Package

PIN DESCRIPTIONS

Symbol	Name and Function
V_{CC}	Main supply voltage (5V).
V_{SS}	The PLCC package has 5 V_{SS} pins and the QFP package has 12 V_{SS} pins. All must be connected to circuit ground.
V_{REF}	Reference voltage for the A/D converter (5V). V_{REF} is also the supply voltage to the analog portion of the A/D converter and the logic used to read Port 0. Must be connected for A/D and Port 0 to function.
ANGND	Reference ground for the A/D converter. Must be held at nominally the same potential as V_{SS}.
V_{PP}	Timing pin for the return from powerdown circuit. Connect this pin with a 1 μF capacitor to V_{SS}. If this function is not used, connect to V_{CC}. This pin is the programming voltage on the EPROM device.
XTAL1	Input of the oscillator inverter and of the internal clock generator.
XTAL2	Output of the oscillator inverter.
$\overline{RESET}$	Reset input to and open-drain output from the chip. Input low for at least 4 state times to reset the chip. The subsequent low-to-high transition commences the Reset Sequence in which the PSW is cleared, a byte read from 2018H loads CCR, and a jump to location 2080H is executed. Input high for normal operation. RESET has an internal pullup.
INST	Output high during an external memory read indicates the read is an instruction fetch. INST is valid throughout the bus cycle. INST is activated only during external memory accesses and output low for a data fetch.
$\overline{EA}$	Input for memory select (External Access). $\overline{EA}$ equal to a TTL-high causes memory accesses to locations 2000H through 3FFFH to be directed to on-chip ROM/EPROM. $\overline{EA}$ equal to a TTL-low causes accesses to these locations to be directed to off-chip memory.
ALE/$\overline{ADV}$	Address Latch Enable or Address Valid output, as selected by CCR. Both pin options provide a latch to demultiplex the address from the address/data bus. When the pin is $\overline{ADV}$, it goes inactive high at the end of the bus cycle. $\overline{ADV}$ can be used as a chip select for external memory. ALE/$\overline{ADV}$ is activated only during external memory accesses.
$\overline{RD}$	Read signal output to external memory. $\overline{RD}$ is activated only during external memory reads.
$\overline{WR}$	Write output to external memory. $\overline{WR}$ will go low for every external write.
READY	Ready input to lengthen external memory cycles, for interfacing to slow or dynamic memory, or for bus sharing. If the pin is high, CPU operation continues in a normal manner. When the external memory is not being used, READY has no effect. Internal control of the number of wait states inserted into a bus cycle held not ready is available through configuration of CCR.
HSI	Inputs to High Speed Input Unit. Four HSI pins are available: HSI.0, HSI.1, HSI.2 and HSI.3. Two of them (HSI.2 and HSI.3) are shared with the HSO Unit.
HSO	Outputs from High Speed Output Unit. Six HSO pins are available: HSO.0, HSO.1, HSO.2, HSO.3, HSO.4 and HSO.5. Two of them (HSO.4 and HSO.5) are shared with the HSI Unit.
Port 0	4-bit high impedance input-only port. These pins can be used as digital inputs and/or as analog inputs to the on-chip A/D converter. These pins set the Programming Mode on the EPROM device.
Port 2	Multi-functional port. All of its pins are shared with other functions in the 80C198.

4

PIN DESCRIPTIONS (Continued)

Symbol	Name and Function
Ports 3 and 4	8-bit bidirectional I/O ports with open drain outputs. These pins are shared with the multiplexed address/data bus which has strong internal pullups. Available as I/O only on the ROM and EPROM devices.
TxD	The TxD pin is used for serial port transmission in Modes 1, 2 and 3. The TxD function is enabled by setting IOC1 5. In mode 0 the pin is used as the serial clock output.
RxD	Serial Port Receive pin used for serial port reception. The RxD function is enabled by setting SPCON.3. In mode 0 the pin functions as input or output data.
EXTINT	A positive transition on the EXTINT pin will generate an external interrupt. EXTINT is selected as the external interrupt source by setting IOC1.1 high.
T2CLK	The T2CLK pin is the Timer2 clock input or the serial port baud rate generator input.
T2RST	A rising edge on the T2RST pin will reset Timer2. The external reset function is enabled by setting IOCO.03 T2RST is enabled as the reset source by clearing IOCO.5.
PWM	Port 2.5 can be enabled as a PWM output by setting IOC1.O The duty cycle of the PWM is determined by the value loaded into the PWM-CONTROL register (17H).

ELECTRICAL CHARACTERISTICS
ABSOLUTE MAXIMUM RATINGS*

Ambient Temperature
Under Bias .... −55°C to +125°C
Storage Temperature .... −65°C to +150°C
Voltage On Any Pin to V_{SS}........ −0.5V to +7.0V
Power Dissipation(1) 1.5W

NOTE:
1. Power dissipation is based on package heat transfer, not device power consumption.

NOTICE: This data sheet contains preliminary information on new products in production. It is valid for the devices indicated in the revision history. The specifications are subject to change without notice.

**WARNING:* *Stressing the device beyond the "Absolute Maximum Ratings" may cause permanent damage. These are stress ratings only. Operation beyond the "Operating Conditions" is not recommended and extended exposure beyond the "Operating Conditions" may affect device reliability.*

OPERATING CONDITIONS

(All characteristics in this data sheet apply to these operating conditions unless otherwise noted.)

Symbol	Description	Min	Max	Units
T_A	Ambient Temperature Under Bias **Commercial** Temp.	0	+70	°C
T_A	Ambient Temperature Under Bias **Extended** Temp.	−40	+85	°C
V_{CC}	Digital Supply Voltage	4.50	5.50	V
V_{REF}	Analog Supply Voltage	4.50	5.50	V
F_{OSC}	Oscillator Frequency	3.5	12	MHz

NOTE:
ANGND and V_{SS} should be nominally at the same potential.

4

DC CHARACTERISTICS

Symbol	Description	Min	Max	Units	Test Conditions
V_{IL}	Input Low Voltage	−0.5	0.8	V	
V_{IH}	Input High Voltage (Note 1)	0.2 V_{CC} + 1.0	V_{CC} + 0.5	V	
V_{IH1}	Input High Voltage on XTAL 1	0.7 V_{CC}	V_{CC} + 0.5	V	
V_{IH2}	Input High Voltage on RESET	2.6	V_{CC} + 0.5	V	
V_{OL}	Output Low Voltage		0.3 0.45 1.5	V V V	I_{OL} = 200 μA I_{OL} = 32 mA I_{OL} = 7 mA
V_{OH}	Output High Voltage (Standard Outputs)	V_{CC} − 0.3 V_{CC} − 0.7 V_{CC} − 1.5		V V V	I_{OH} = −200 μA I_{OH} = −3.2 mA I_{OH} = −7 mA
I_{LI}	Input Leakage Current (Std. Inputs)		±10	μA	$0 < V_{IN} < V_{CC} - 0.3V$
I_{LI1}	Input Leakage Current (Port 0)		+3	μA	$0 < V_{IN} < V_{REF}$
I_{IL1}	Logical 0 Input Current in Reset (Note 2) (ALE, $\overline{RD}$, $\overline{WR}$, INST, P2.0)		−1.2	mA	V_{IN} = 0.45 V
Hyst	Hysteresis on RESET Pin	300		mV	

NOTE:
1. All pins except RESET and XTAL1.
2. Holding these pins below V_{IH} in Reset may cause the part to enter test modes.

DC CHARACTERISTICS (Continued)

Symbol	Description	Min	Typ(6)	Max	Units	Test Conditions
I_{CC}	Active Mode Current in Reset		40	55	mA	XTAL1 = 12 MHz $V_{CC} = V_{PP} = V_{REF} = 5.5V$
I_{REF}	A/D Converter Reference Current		2	5	mA	
I_{IDLE}	Idle Mode Current		10	22	mA	
I_{CC1}	Active Mode Current		15	22	mA	XTAL1 = 3.5 MHz
I_{PD}	Powerdown Mode Current		5	50	μA	$V_{CC} = V_{PP} = V_{REF} = 5.5V$
R_{RST}	Reset Pullup Resistor	6K		65K	Ω	
C_S	Pin Capacitance (Any Pin to V_{SS})			10	pF	$F_{TEST} = 1.0$ MHz

NOTES:
(Notes apply to all specifications)
1. Standard Outputs include AD0–15, $\overline{RD}$, $\overline{WR}$, ALE, INST, HSO pins, PWM/P2.5, RESET, Ports 3 and 4, TXD/P2.0 and RXD (in serial mode 0). The V_{OH} specification is not valid for RESET. Ports 3 and 4 are open-drain outputs.
2. Standard Inputs include HSI pins, $\overline{EA}$, READY, RXD/P2.1, EXTINT/P2.2, T2CLK/P2.3 and T2RST/P2.4.
3. Maximum current per pin must be externally limited to the following values if V_{OL} is held above 0.45V or V_{OH} is held below V_{CC} − 0.7V:
   I_{OL} on Output pins: 10 mA
   I_{OH} on Standard Output pins: 10 mA
4. Maximum current per bus pin (data and control) during normal operation is ±3.2 mA.
5. During normal (non-transient) conditions the following total current limits apply:

HSO, P2.0, RXD, $\overline{RESET}$	I_{OL}: 29 mA	I_{OH}: 26 mA
P2.5, $\overline{WR}$	I_{OL}: 13 mA	I_{OH}: 11 mA
AD0–AD15	I_{OL}: 52 mA	I_{OH}: 52 mA
$\overline{RD}$, ALE, INST	I_{OL}: 13 mA	I_{OH}: 13 mA

6. Typicals are based on a limited number of samples and are not guaranteed. The values listed are at room temperature and $V_{REF} = V_{CC} = 5V$.

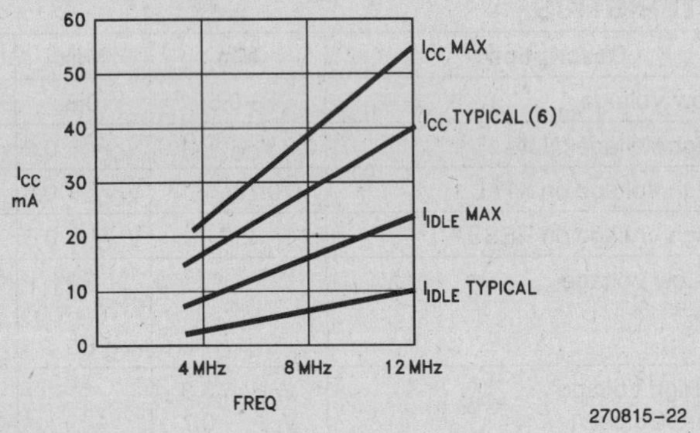

I_{CC} Max = 3.88 × FREQ + 8.43
I_{IDLE} Max = 1.65 × FREQ + 2.2

Figure 7. I_{CC} and I_{IDLE} vs Frequency

AC CHARACTERISTICS

Test Conditions: Capacitive load on all pins = 100 pF, Rise and fall times = 10 ns, F_{OSC} = 12 MHz

The system must meet these specifications to work with the 80C198/83C198:

Symbol	Description	Min	Max	Units	Notes
T_{AVYV}	Address Valid to Ready Setup		$2\,T_{OSC} - 85$	ns	
T_{LLYV}	ALE Low to READY Setup		$T_{OSC} - 70$	ns	
T_{YLYH}	Non READY Time	No upper limit		ns	
T_{LLYX}	READY Hold after ALE Low	$T_{OSC} - 15$	$2\,T_{OSC} - 40$	ns	(Note 1)
T_{AVDV}	Address Valid to Input Data Valid		$3\,T_{OSC} - 60$	ns	(Note 2)
T_{RLDV}	$\overline{RD}$ Active to Input Data Valid		$T_{OSC} - 23$	ns	(Note 2)
T_{RHDZ}	End of $\overline{RD}$ to Input Data Float		$T_{OSC} - 20$	ns	
T_{RXDX}	Data Hold after $\overline{RD}$ Inactive	0		ns	

NOTES:
1. If max is exceeded, additional wait states will occur.
2. When using wait states, add $2\,T_{OSC} \times n$, where n = number of wait states.

The 80C198/83C198 will meet these specifications:

Symbol	Description	Min	Max	Units	Notes
F_{XTAL}	Frequency on $XTAL_1$	3.5	12	MHz	(Note 1)
T_{OSC}	I/F_{XTAL}	83	286	ns	
T_{LHLH}	ALE Cycle Time	$4\,T_{OSC}$		ns	(Note 4)
T_{LHLL}	ALE High Period	$T_{OSC} - 10$	$T_{OSC} + 10$	ns	
T_{AVLL}	Address Setup to ALE Falling Edge	$T_{OSC} - 20$		ns	
T_{LLAX}	Address Hold after ALE Falling Edge	$T_{OSC} - 40$		ns	
T_{LLRL}	ALE Falling Edge to $\overline{RD}$ Falling Edge	$T_{OSC} - 30$		ns	
T_{RLRH}	$\overline{RD}$ Low Period	$T_{OSC} - 5$	$T_{OSC} + 25$	ns	(Note 4)
T_{RHLH}	$\overline{RD}$ Rising Edge to ALE Rising Edge	T_{OSC}	$T_{OSC} + 25$	ns	(Note 3)
T_{RLAZ}	$\overline{RD}$ Low to Address Float		10	ns	
T_{LLWL}	ALE Falling Edge to $\overline{WR}$ Falling Edge	$T_{OSC} - 10$		ns	
T_{QVWH}	Data Stable to $\overline{WR}$ Rising Edge	$T_{OSC} - 23$		ns	(Note 4)
T_{WLWH}	$\overline{WR}$ Low Period	$T_{OSC} - 30$	$T_{OSC} + 5$	ns	(Note 4)
T_{WHQX}	Data Hold after $\overline{WR}$ Rising Edge	$T_{OSC} - 2.5$		ns	
T_{WHLH}	$\overline{WR}$ Rising Edge to ALE Rising Edge	$T_{OSC} - 10$	$T_{OSC} + 15$	ns	(Note 3)
T_{WHBX}	INST Hold after $\overline{WR}$ Rising Edge	$T_{OSC} - 10$		ns	
T_{RHBX}	INST Hold after $\overline{RD}$ Rising Edge	$T_{OSC} - 10$		ns	
T_{WHAX}	AD8–15 Hold after $\overline{WR}$ Rising Edge	$T_{OSC} - 50$		ns	
T_{RHAX}	AD8–15 Hold after $\overline{RD}$ Rising Edge	$T_{OSC} - 25$		ns	

NOTES:
1. Testing performed at 3.5 MHz. However, the part is static by design and will typically operate below 1 Hz.
2. Typical specification, not guaranteed.
3. Assuming back-to-back bus cycles.
4. When using wait states, add $2\,T_{OSC} \times n$, where n = number of wait states.

System Bus Timings

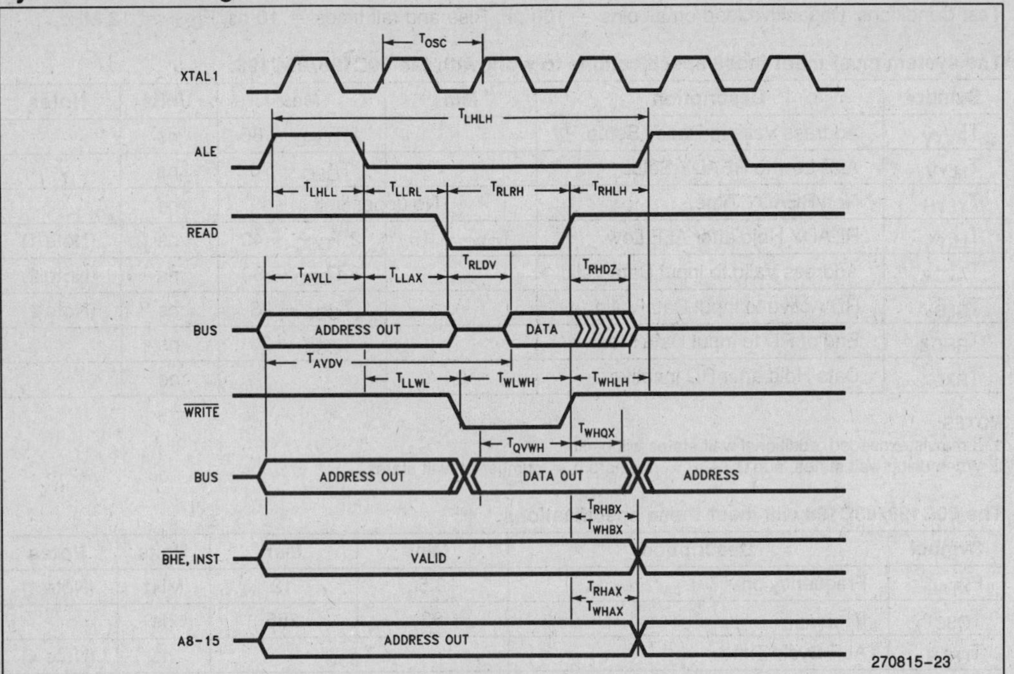

270815-23

READY Timings (One Wait State)

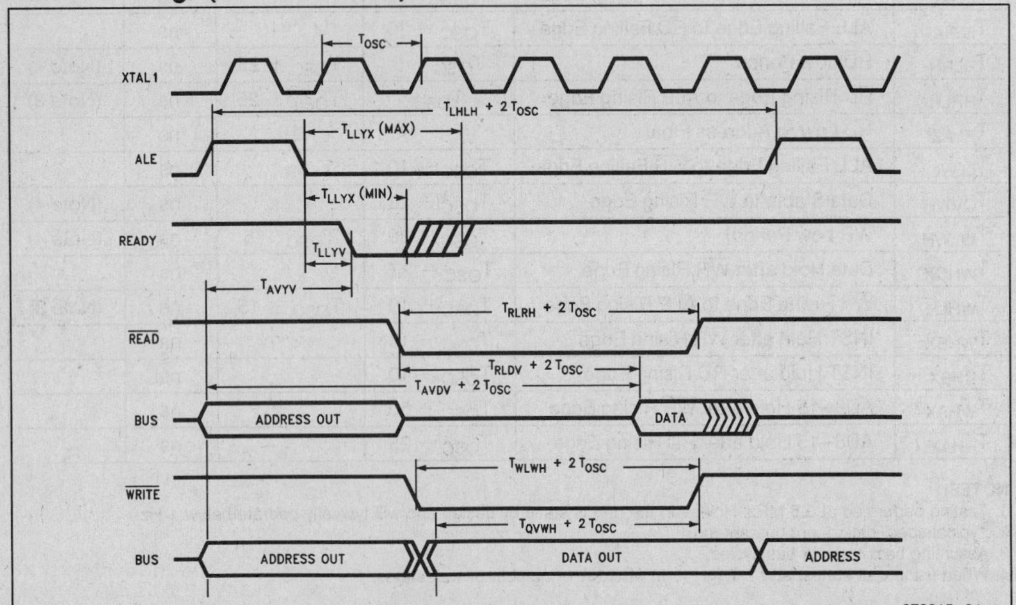

270815-24

EXTERNAL CLOCK DRIVE

Symbol	Parameter	Min	Max	Units
$1/T_{XLXL}$	Oscillator Frequency	3.5	12.0	MHz
T_{XLXL}	Oscillator Period	83	286	ns
T_{XHXX}	High Time	32		ns
T_{XLXX}	Low Time	32		ns
T_{XLXH}	Rise Time		10	ns
T_{XHXL}	Fall Time		10	ns

EXTERNAL CLOCK DRIVE WAVEFORMS

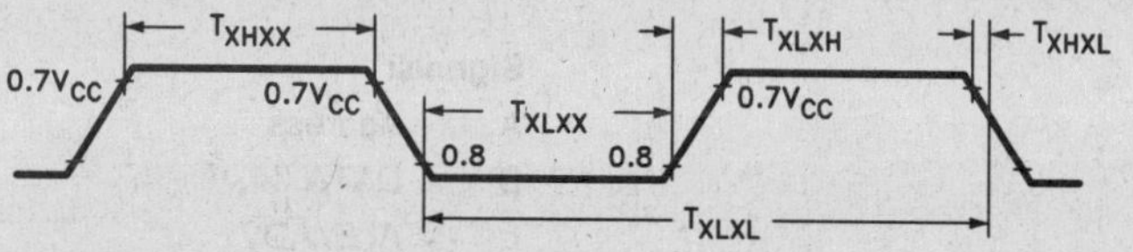

270815-25

An external oscillator may encounter as much as a 100 pF load at XTAL1 when it starts-up. This is due to interaction between the amplifier and its feedback capacitance. Once the external signal meets the V_{IL} and V_{IH} specifications the capacitance will not exceed 20 pF.

EXTERNAL CRYSTAL CONNECTIONS

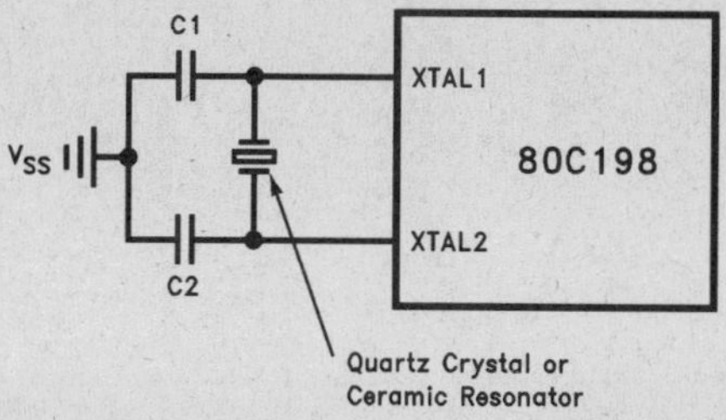

270815-29

NOTE:
Keep oscillator components close to chip and use short, direct traces to XTAL1, XTAL2 and V_{SS}. When using crystals, C1 = 20 pF and C2 = 20 pF. When using ceramic resonators, consult manufacturer for recommended capacitor values.

EXTERNAL CLOCK CONNECTIONS

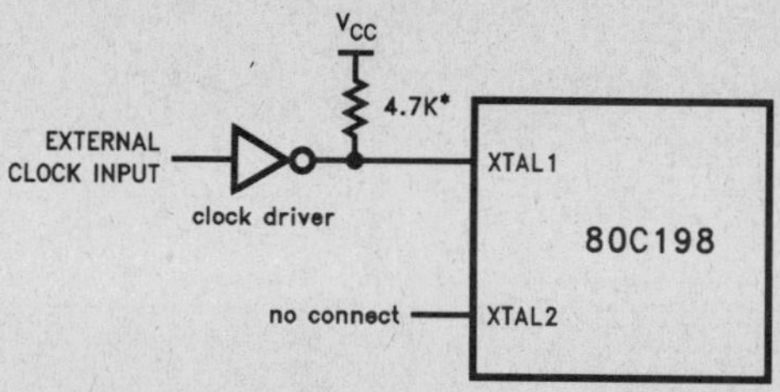

270815-30

*Required if TTL driver used.
Not needed if CMOS driver is used.

AC TESTING INPUT, OUTPUT WAVEFORMS

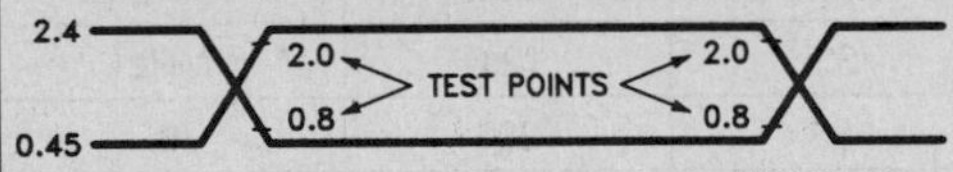

270815-26

AC Testing inputs are driven at 2.4V for a Logic "1" and 0.45V for a Logic "0" Timing measurements are made at 2.0V for a Logic "1" and 0.8V for a Logic "0".

FLOAT WAVEFORMS

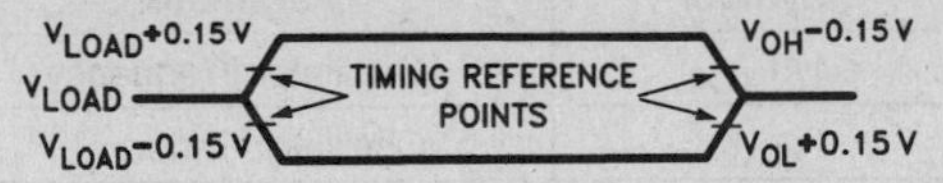

270815-27

For Timing Purposes a Port Pin is no Longer Floating when a 150 mV change from Load Voltage Occurs and Begins to Float when a 150 mV change from the Loaded V_{OH}/V_{OL} Level occurs $I_{OL}/I_{OH} = \pm 15$ mA.

EXPLANATION OF AC SYMBOLS

Each symbol is two pairs of letters prefixed by "T" for time. The characters in a pair indicate a signal and its condition, respectively. Symbols represent the time between the two signal/condition points.

Conditions:

H - High
L - Low
V - Valid
X - No Longer Valid
Z - Floating

Signals:

A - Address
D - DATA IN
L - ALE/$\overline{ADV}$
Q - DATA OUT
R - $\overline{RD}$
W - $\overline{WR}$
X - XTAL1
Y - READY

AC CHARACTERISTICS—SERIAL PORT—SHIFT REGISTER MODE

SERIAL PORT TIMING—SHIFT REGISTER MODE

Symbol	Parameter	Min	Max	Units
T_{XLXL}	Serial Port Clock Period (BRR ≥ 8002H)	6 T_{OSC}		ns
T_{XLXH}	Serial Port Clock Falling Edge to Rising Edge (BRR ≥ 8002H)	4 T_{OSC} − 50	4 T_{OSC} + 50	ns
T_{XLXL}	Serial Port Clock Period (BRR = 8001H)	4 T_{OSC}		ns
T_{XLXH}	Serial Port Clock Falling Edge to Rising Edge (BRR = 8001H)	2 T_{OSC} − 50	2 T_{OSC} + 50	ns
T_{QVXH}	Output Data Setup to Clock Rising Edge	2 T_{OSC} −50		ns
T_{XHQX}	Output Data Hold after Clock Rising Edge	2 T_{OSC} −50		ns
T_{XHQV}	Next Output Data Valid after Clock Rising Edge		2 T_{OSC} +50	ns
T_{DVXH}	Input Data Setup to Clock Rising Edge	T_{OSC} +50		ns
T_{XHDX}	Input Data Hold after Clock Rising Edge	0		ns
T_{XHQZ}	Last Clock Rising to Output Float		1 T_{OSC}	ns

WAVEFORM—SERIAL PORT—SHIFT REGISTER MODE

SERIAL PORT WAVEFORM—SHIFT REGISTER MODE

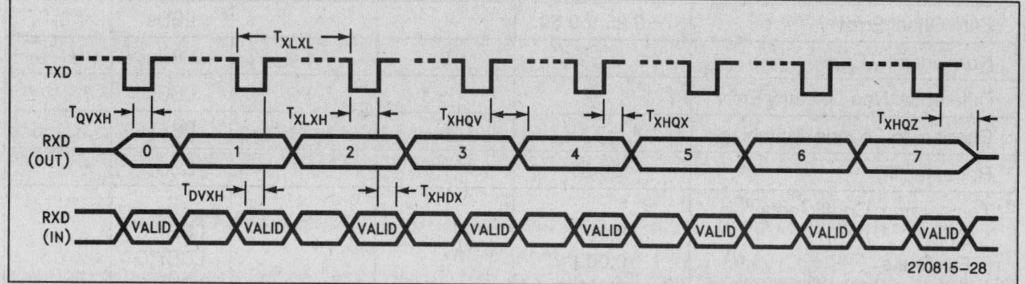

270815-28

A TO D CHARACTERISTICS

There are two modes of A/D operation: with or without clock prescaler. The speed of the A/D converter can be adjusted by setting a clock prescaler on or off. At high frequencies more time is needed for the comparator to settle. The maximum frequency with the clock prescaler disabled is 8 MHz. The conversion times with the prescaler turned on or off is shown in the table below.

The converter is ratiometric, so the absolute accuracy is directly dependent on the accuracy and stability of V_{REF}. V_{REF} must be close to V_{CC} since it supplies both the resistor ladder and the digital section of the converter.

The 80C194/83C194 does not have an A/D converter.

See the MCS-96 A/D Quick Reference for definition of A/D terms.

Conversion Time

Clock Prescaler On IOC2.4 = 0	Clock Prescaler Off IOC2.4 = 1
158 States 26.33 μs @ 12 MHz	91 States 22.75 μs @ 8 MHz

A/D CONVERTER SPECIFICATIONS

Parameter	Typical(1)	Minimum	Maximum	Units*	Notes
Resolution		512 9	1024 10	Levels Bits	
Absolute Error		0	±4	LSBs	
Full Scale Error	0.25 ±0.50			LSBs	
Zero Offset Error	−0.25 ±0.50			LSBs	
Non-Linearity Error	1.5 ±2.5	0	±4	LSBs	
Differential Non-Linearity Error		> −1	+2	LSBs	
Channel-to-Channel Matching	±0.1	0	±1	LSBs	
Repeatability	±0.25			LSBs	
Temperature Coefficients: Offset Full Scale Differential Non-Linearity	 0.009 0.009 0.009			 LSB/°C LSB/°C LSB/°C	
Off Isolation		−60		dB	2, 3
Feedthrough	−60			dB	2
V_{CC} Power Supply Rejection	−60			dB	2
Input Series Resistance		1K	5K	Ω	
DC Input Leakage		0	3.0	μA	4
Sample Time: Prescaler On Prescaler Off	15 8			States States	
Sampling Capacitor	3			pF	

NOTES:
*An "LSB", as used here, has a value of approximately 5 mV.
1. Typical values are expected for most devices at 25°C but are not tested or guaranteed.
2. DC to 100 KHz.
3. Multiplexer Break-Before-Make Guaranteed.
4. Resistance from device pin, through internal MUX, to sample capacitor.

EXTENDED TEMPERATURE/EXTENDED BURN-IN ONLY SPECIFICATIONS

Symbol	Description	Min	Max	Units
RESET Hysteresis	Hysteresis on RESET Pin	TBD		mV
IPD	Powerdown Mode Current		TBD	mA
T_{LLYV}	ALE Low to READY Setup		$T_{OSC} - 65$	ns
T_{AVDV}	Address Valid to Input Data Valid		$3\ T_{OSC} - 60$	ns
T_{RLDV}	$\overline{RD}$ Low to Input Data Valid		$T_{OSC} - 25$	ns
T_{LHLL}	ALE High Period	$T_{OSC} - 12$	$T_{OSC} + 12$	ns
T_{RHAX}	AD8–15 Hold after $\overline{RD}$ Rising	$T_{OSC} - 50$		ns
A/D Absolute Error	Absolute Error		+6	LSBs

FUNCTIONAL DEVIATIONS

1. The DJNZW instruction is not guaranteed to be functional. The instruction, if encountered, will not cause an unimplemented opcode interrupt. (The opcode for DJNZW is 0E1 Hex.) The DJNZ (byte) instruction works correctly and should be used instead.
2. The HSI unit has two errata: one dealing with resolution and the other with first entries into the FIFO.

   The HSI resolution is 9 states instead of 8 states. Events on the same line may be lost if they occur faster than once every 9 state times.

   There is a mismatch between the 9 state time HSI resolution and the 8 state time timer. This causes one time value to be unused every 9 timer counts. Events may receive a time-tag one count later than expected because of this "skipped" time value.

   If the first two events into an empty FIFO (not including the Holding Register) occurs within 9 states after the first, its time-tag is one count later than the first's. If this is the "skipped" time value, the second event's time-tag is 2 counts later than the first's.

   If the FIFO and Holding Register are empty, the first event will transfer into the Holding Register after 8 state times, leaving the FIFO empty again. If the second event occurs after this time, it will act as a new first event into an empty FIFO.
3. The serial port Framing Error flag fails to indicate an error if the bit preceding the stop bit is a 1. This is the case in both the 8-bit and 9-bit modes. False framing errors are never generated.
4. The serial port RI flag is not generated after the first byte is received. The problem does not occur if the baud rate is reloaded after each reception.
5. If the unsigned divide instruction (byte or word) is the last instruction in the queue as HOLD or READY is asserted, the result may be incorrect.

REVISION HISTORY

This data sheet (270815-003) is valid for devices marked with a "B" at the end of the top side tracking number. Data sheets are changed as new device information becomes available. Verify with your local Intel sales office that you have the latest version before finalizing a design or ordering devices.

The following differences exist between this data sheet and the previous version (-002).

1. The Express (extended temperature and extended burn-in) devices were added to this data sheet.
2. Changes were made to the format of the data sheet and the SFR descriptions were removed.
3. Four errata were added: the CDE pin, the HSI resolution, the serial port framing error flag, the serial port RI flag and the DIVIDE during HOLD/READY.
4. One specification for the extended temperature and extended burn-in devices was changed: V_{IH2} Min was changed from 2.4V to 2.6V.

Differences between -002 and the -001 version of the 80C198 data sheet.

1. V_{SS} pin description was altered to reflect the correct number of pins.
2. V_{IH2} Min was changed from 2.2V to 2.6V.
3. Max I_{PD} was added and the I_{PD} note was deleted.

For more detailed information on the 80C198, refer to the *80C196KB User's Guide*, order number 270651. The *80C196KB User's Guide* applies to the 80C198 except for the design considerations listed above. Because the 80C198 is a reduced pin count version, some 80C196KB features are not available and are listed here:

1. PORT 1. PORT1 is a quasi-bidirectional port.
   A. HOLD/HLDA. This feature is multiplexed on PORT1.5–.7 and is not available.
2. The A/D converter loses four of its input channels, ACH0–3.
3. T2CAPTURE (P2.7) Timer2 Capture feature is not available.
4. T2UP/DN (P2.6) The Timer2 UP/DOWN feature is not available.
5. CLKOUT
6. NMI
7. BUSWIDTH
8. BHE
9. PACT

8XC196KB/8XC196KB16
COMMERCIAL/EXPRESS CHMOS MICROCONTROLLER

- **8 Kbytes of On-Chip ROM/OTP Available**
- **232 Byte Register File**
- **Register-to-Register Architecture**
- **28 Interrupt Sources/16 Vectors**
- **1.75 μs 16 x 16 Multiply (16 MHz)**
- **3.0 μs 32/16 Divide (16 MHz)**
- **Powerdown and Idle Modes**
- **Five 8-Bit I/O Ports**
- **16-Bit Watchdog Timer**
- **12 MHz and 16 MHz Available**
- **Dedicated 15-Bit Baud Rate Generator**
- **Dynamically Configurable 8-Bit or 16-Bit Buswidth**
- **Full Duplex Serial Port**
- **High Speed I/O Subsystem**
- **16-Bit Timer**
- **16-Bit Up/Down Counter with Capture**
- **Pulse-Width-Modulated Output**
- **Four 16-Bit Software Timers**
- **10-Bit A/D Converter with Sample/Hold**
- **$\overline{\text{HOLD}}$/$\overline{\text{HLDA}}$ Bus Protocol**
- **Extended Temperature Available**

The 8XC196KB is a 16-bit microcontroller available in three different memory varieties: ROMless (80C196KB), 8K ROM (83C196KB) and 8K OTP (One Time Programmable—87C196KB). The 8XC196KB is a high performance member of the MCS® 96 microcontroller family. The 8XC196KB has the same peripheral set as the 8096BH and has a true superset of the 8096BH instructions. Intel's CHMOS process provides a high performance processor along with low power consumption. To further reduce power requirements, the processor can be placed into Idle or Powerdown Mode.

4

Bit, byte, word and some 32-bit operations are available on the 80C196KB. With a 16 MHz oscillator a 16-bit addition takes 0.50 μs, and the instruction times average 0.37 μs to 1.1 μs in typical applications.

Four high-speed capture inputs are provided to record times when events occur. Six high-speed outputs are available for pulse or waveform generation. The high-speed output can also generate four software timers or start an A/D conversion. Events can be based on the timer or up/down counter. Also provided on-chip are an A/D converter, serial port, watchdog timer and a pulse-width-modulated output signal.

The 8XC196KB has a maximum guaranteed frequency of 12 MHz. The 8XC196KB16 has a maximum guaranteed frequency of 16 MHz. All references to the 80C196KB also refer to the 80C196KB16; 83C196KB, Rxxx; 87C196KB and 87C196KB16 unless otherwise noted. The ROM device does not have a speed indicator at the end of the device name. Instead it has a ROM code number.

With the commercial (standard) temperature option, operational characteristics are guaranteed over the temperature range of 0°C to +70°C. With the extended temperature range option, operational characteristics are guaranteed over the temperature range of −40°C to +85°C.

Package Designators: N = 68-pin PLCC, S = 80-pin QFP (commercial only). Prefix Designators: T = Extended Temperature.

July 1994
Order Number: 270909-006

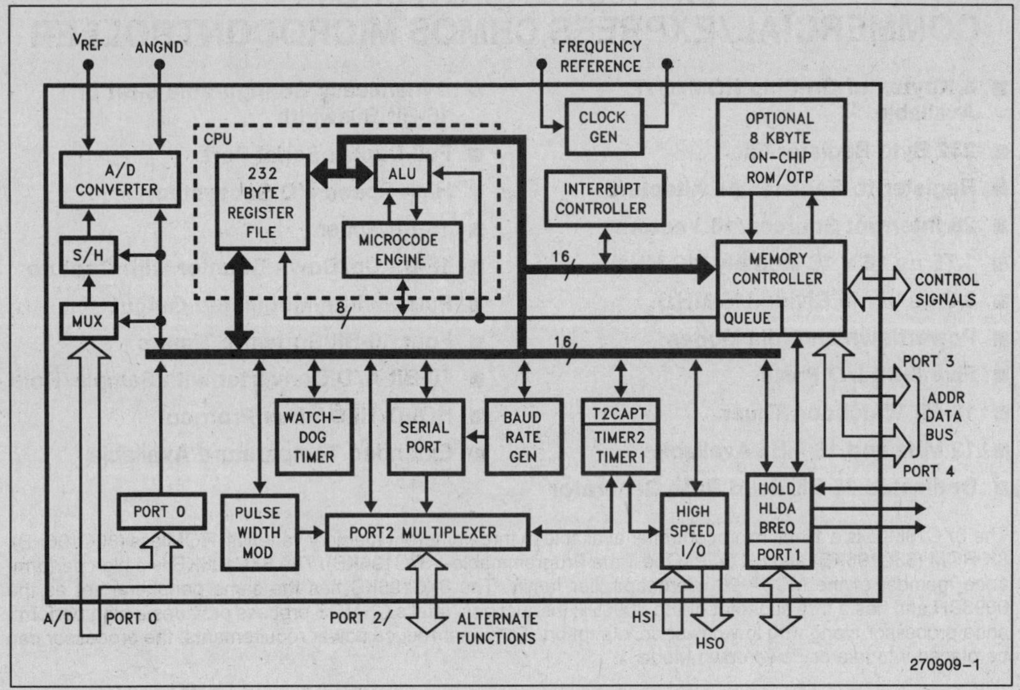

Figure 1. 8XC196KB Block Diagram

PROCESS INFORMATION

This device is manufactured on P629.0 and 629.1, a CHMOS III-E process. Additional process and reliability information is available in Intel's *Components Quality and Reliability Handbook,* Order Number 210997.

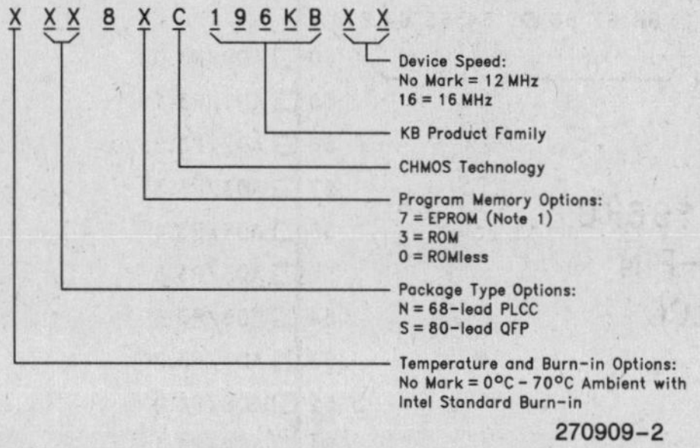

EXAMPLE: N87C196KB16 is 68-Lead PLCC OTPROM, 16 MHz.
For complete package dimensional data, refer to the Intel Packaging Handbook (Order Number 240800).

NOTE:
1. EPROMs are available as One Time Programmable (OTPROM) only.

Figure 2. The 8XC196KB Nomenclature

Table 1. Thermal Characteristics

Package Type	θ_{ja}	θ_{jc}
PLCC	35°C/W	13°C/W
QFP	70°C/W	4°C/W

All thermal impedance data is approximate for static air conditions at 1W of power dissipation. Values will change depending on operation conditions and application. See the Intel *Packaging Handbook* (order number 240800) for a description of Intel's thermal impedance test methodology.

Table 2. 8XC196KB Memory Map

Description	Address
External Memory or I/O	0FFFFH 04000H
Internal ROM/EPROM or External Memory (Determined by $\overline{EA}$)	3FFFH 2080H
Reserved. Must contain FFH. (Note 5)	207FH 2040H
Upper Interrupt Vectors	203FH 2030H
ROM/EPROM Security Key	202FH 2020H
Reserved. Must contain FFH. (Note 5)	201FH 201AH
Reserved. Must Contain 20H (Note 5)	2019H
CCB	2018H
Reserved. Must contain FFH. (Note 5)	2017H 2014H
Lower Interrupt Vectors	2013H 2000H
Port 3 and Port 4	1FFFH 1FFEH
External Memory	1FFDH 0100H
232 Bytes Register RAM (Note 1)	00FFH 0018H
CPU SFR's (Notes 1, 3)	0017H 0000H

NOTES:
1. Code executed in locations 0000H to 00FFH will be forced external.
2. Reserved memory locations must contain 0FFH unless noted.
3. Reserved SFR bit locations must contain 0.
4. Refer to 8XC196KB quick reference for SFR descriptions.
5. **WARNING:** Reserved memory locations must not be written or read. The contents and/or function of these locations may change with future revisions of the device. Therefore, a program that relies on one or more of these locations may not function properly.

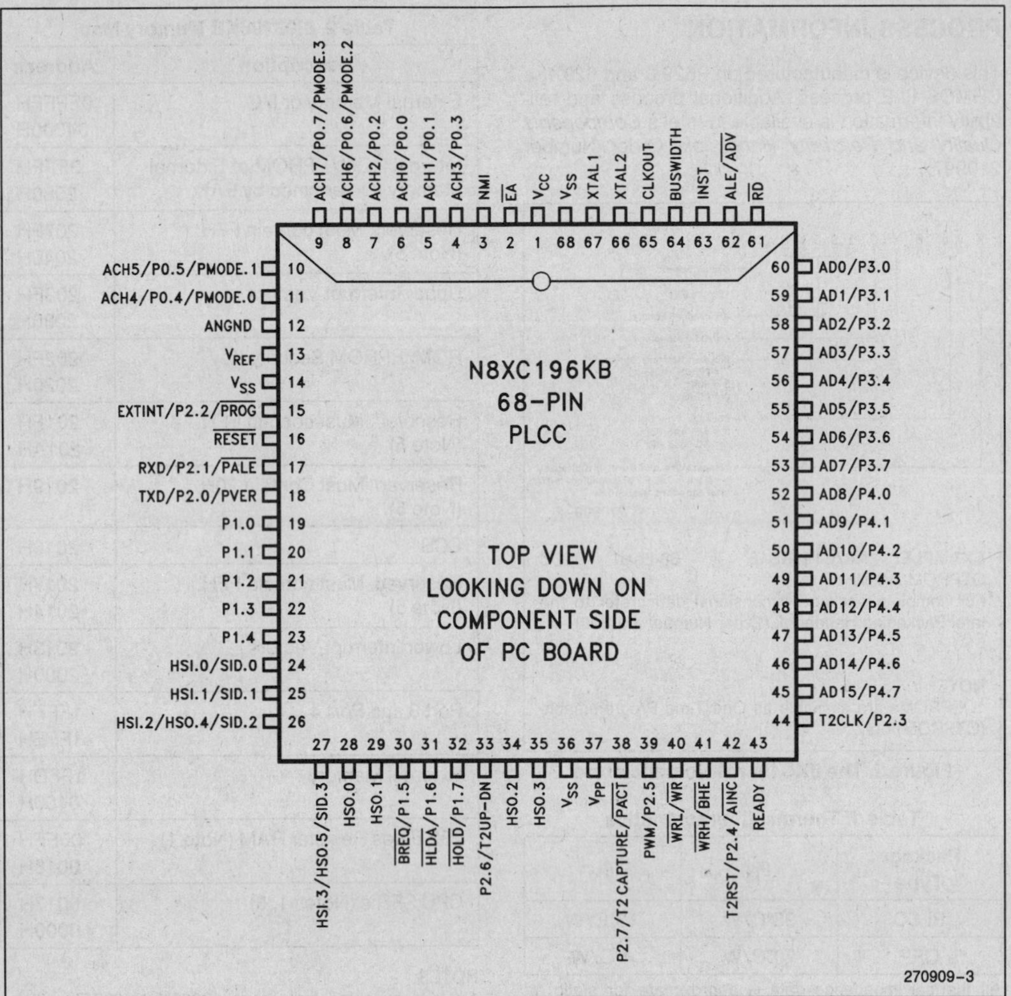

Figure 3. 68-Pin Package (PLCC Top View)

NOTE:
The above pin out diagram applies to the OTP (87C196KB) device. The OTP device uses all of the programming pins shown above. The ROM (83C196KB) device only uses programming pins: AINC, PALE, PMODE.n, and PROG. The ROMless (80C196KB) doesn't use any of the programming pins.

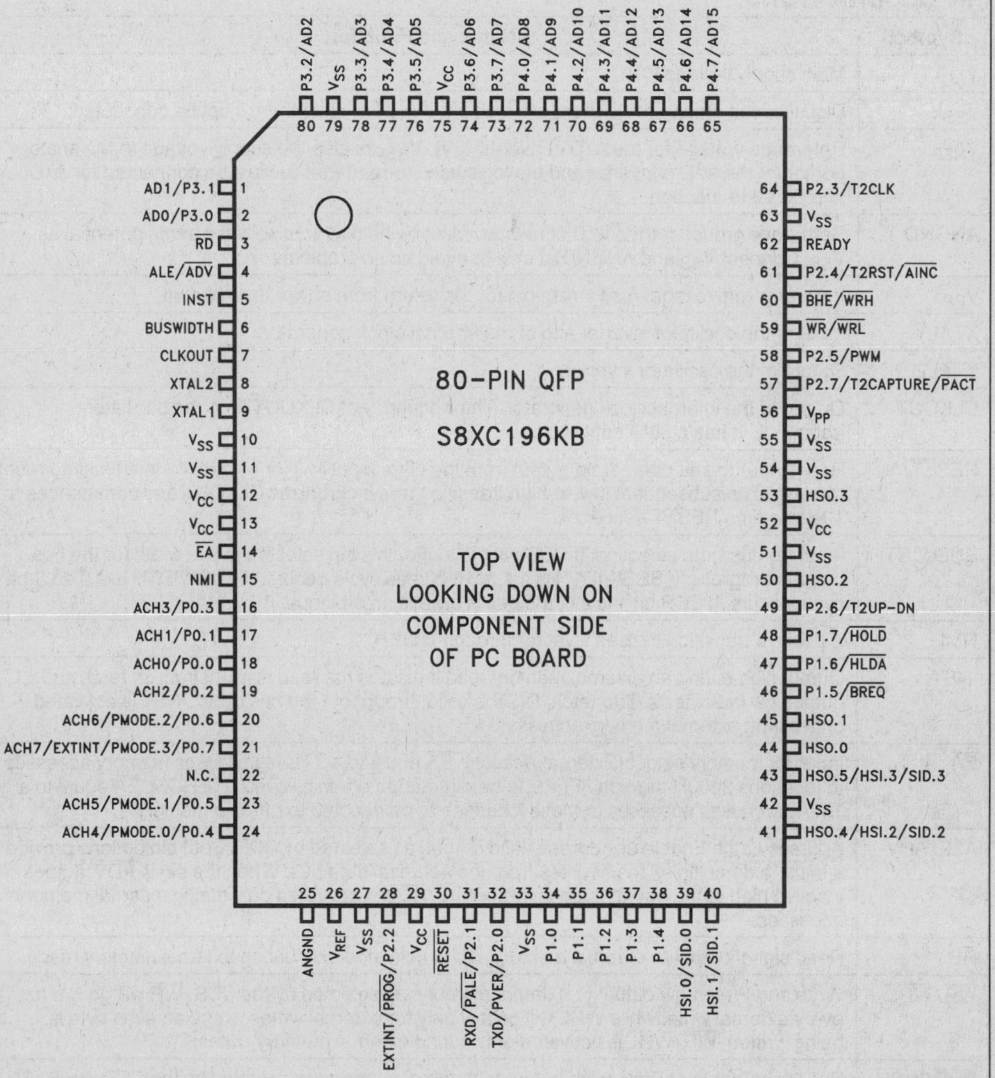

NOTE:
N.C. means No Connect (do not connect these pins).

Figure 4. 80-Pin QFP Package

NOTE:
The above pin out diagram applies to the OTP (87C196KB) device. The OTP device uses all of the programming pins shown above. The ROM (83C196KB) device only uses programming pins: AINC, PALE, PMODE.n, and PROG. The ROMless (80C196KB) doesn't use any of the programming pins.

PIN DESCRIPTIONS

Symbol	Name and Function
V_{CC}	Main supply voltage (5V).
V_{SS}	Digital circuit ground (0V). There are multiple V_{SS} pins, all of them must be connected.
V_{REF}	Reference voltage for the A/D converter (5V). V_{REF} is also the supply voltage to the analog portion of the A/D converter and the logic used to read Port 0. Must be connected for A/D and Port 0 to function.
ANGND	Reference ground for the A/D converter. Must be held at nominally the same potential as V_{SS}. Connect V_{SS} and ANGND at chip to avoid noise problems.
V_{PP}	Programming voltage. Also timing pin for the return from power down circuit.
XTAL1	Input of the oscillator inverter and of the internal clock generator.
XTAL2	Output of the oscillator inverter.
CLKOUT	Output of the internal clock generator. The frequency of CLKOUT is ½ the oscillator frequency. It has a 50% duty cycle.
$\overline{RESET}$	Reset input to and open-drain output from the chip. Input low for at least 4 state times to reset the chip. The subsequent low-to-high transition re-synchronizes CLKOUT and commences a 10-state-time RESET sequence.
BUSWIDTH	Input for buswidth selection. If CCR bit 1 is a one, this pin selects the bus width for the bus cycle in progress. If BUSWIDTH is a 1, a 16-bit bus cycle occurs. If BUSWIDTH is a 0 an 8-bit cycle occurs. If CCR bit 1 is a 0, the bus is always an 8-bit bus.
NMI	A positive transition causes a vector through 203EH.
INST	Output high during an external memory read indicates the read is an instruction fetch and output low indicates a data fetch. INST is valid throughout the bus cycle. INST is activated only during external memory accesses.
$\overline{EA}$	Input for memory select (External Access). $\overline{EA}$ equal to a TTL-high causes memory accesses to locations 2000H through 3FFFH to be directed to on-chip ROM/OTPROM. $\overline{EA}$ equal to a TTL-low causes accesses to these locations to be directed to off-chip memory.
ALE/$\overline{ADV}$	Address Latch Enable or Address Valid output, as selected by CCR. Both pin options provide a latch to demultiplex the address from the address/data bus. When the pin is $\overline{ADV}$, it goes inactive high at the end of the bus cycle. ALE/$\overline{ADV}$ is activated only during external memory accesses.
$\overline{RD}$	Read signal output to external memory. $\overline{RD}$ is activated only during external memory reads.
$\overline{WR}$/$\overline{WRL}$	Write and Write Low output to external memory, as selected by the CCR. $\overline{WR}$ will go low for every external write, while $\overline{WRL}$ will go low only for external writes where an even byte is being written. $\overline{WR}$/$\overline{WRL}$ is activated only during external memory writes.
$\overline{BHE}$/$\overline{WRH}$	Bus High Enable or Write High output to external memory, as selected by the CCR. $\overline{BHE}$ will go low for external writes to the high byte of the data bus. $\overline{WRH}$ will go low for external writes where an odd byte is being addressed. $\overline{BHE}$/$\overline{WRH}$ is activated only during external memory writes.
READY	Ready input to lengthen external memory cycles. If the pin is low prior to the falling edge of CLKOUT, the memory controller goes into a wait mode until the next positive transition in CLKOUT occurs with READY high. When the external memory is not being used, READY has no effect. Internal control of the number of wait states inserted into a bus cycle (held not ready) is available in the CCR.
HSI	Inputs to High Speed Input Unit. Four HSI pins are available: HSI.0, HSI.1, HSI.2 and HSI.3. Two of them (HSI.2 and HSI.3) are shared with the HSO Unit.
HSO	Outputs from High Speed Output Unit. Six HSO pins are available: HSO.0, HSO.1, HSO.2, HSO.3, HSO.4 and HSO.5. Two of them (HSO.4 and HSO.5) are shared with the HSI Unit.

PIN DESCRIPTIONS (Continued)

Symbol	Name and Function
Port 0	8-bit high impedance input-only port. Three pins can be used as digital inputs and/or as analog inputs to the on-chip A/D converter.
Port 1	8-bit quasi-bidirectional I/O port. These pins are shared with $\overline{\text{HOLD}}$, $\overline{\text{HLDA}}$ and $\overline{\text{BREQ}}$.
Port 2	8-bit multi-functional port. All of its pins are shared with other functions in the 87C196KB. Pins P2.6 and P2.7 are quasi-bidirectional.
Ports 3 and 4	8-bit bidirectional I/O ports with open drain outputs. These pins are shared with the multiplexed address/data bus, which has strong internal pullups.
$\overline{\text{HOLD}}$	Bus Hold input requesting control of the bus. Enabled by setting WSR.7.
$\overline{\text{HLDA}}$	Bus Hold acknowledge output indicating release of the bus. Enabled by setting WSR.7.
$\overline{\text{BREQ}}$	Bus Request output activated when the bus controller has a pending external memory cycle. Enabled by setting WSR.7.
TxD	The TxD pin is used for serial port transmission in Modes 1, 2 and 3. In Mode 0 the pin is used as the serial clock output.
RxD	Serial Port Receive pin used for serial port reception. In Mode 0 the pin functions as input or output data.
EXTINT	A rising edge on the EXTINT pin will generate an external interrupt.
T2CLK	The T2CLK pin is the Timer2 clock input or the serial port baud rate generator input.
T2RST	A rising edge on the T2RST pin will reset Timer2.
PWM	The pulse width modulator output.
T2UP-DN	The T2UPDN pin controls the direction of Timer2 as an up or down counter.
T2CAPTURE	A rising edge on P2.7 will capture the value of Timer2 in the T2CAPTURE register.
PMODE	Programming Mode Select. Determines the EPROM programming algorithm that is performed. PMODE is sampled after a chip reset and should be static while the part is operating.
SID	Slave ID Number. Used to assign each slave a pin of Port 3 or 4 to use for passing programming verification acknowledgement.
$\overline{\text{PALE}}$	Programming ALE Input. Accepted by the 87C196KB when it is in Slave Programming Mode. Used to indicate that Ports 3 and 4 contain a command/address.
$\overline{\text{PROG}}$	Programming. Falling edge indicates valid data on PBUS and the beginning of programming. Rising edge indicates end of programming.
$\overline{\text{PACT}}$	Programming Active. Used in the Auto Programming Mode to indicate when programming activity is complete.
$\overline{\text{PVAL}}$	Program Valid. This signal indicates the success or failure of programming in the Auto Programming Mode. A zero indicates successful programming.
PVER	Program Verification. Used in Slave Programming and Auto CLB Programming Modes. Signal is low after rising edge of PROG if the programming was not successful.
$\overline{\text{AINC}}$	Auto Increment. Active low signal indicates that the auto increment mode is enabled. Auto Increment will allow reading or writing of sequential EPROM locations without address transactions across the PBUS for each read or write.
Ports 3 and 4 (Programming Mode)	Address/Command/Data Bus. Used to pass commands, addresses, and data to and from slave mode 87C196KBs. Used by chips in Auto Programming Mode to pass command, addresses and data to slaves. Also used in the Auto Programming Mode as a regular system bus to access external memory. Should have pullups to V_{CC} when used in slave programming mode.

4

ELECTRICAL CHARACTERISTICS
ABSOLUTE MAXIMUM RATINGS*

Ambient Temperature
Under Bias .... −55°C to +125°C

Storage Temperature .... −65°C to +150°C

Voltage On Any Pin to V_{SS} .... −0.5V to +7.0V

Power Dissipation(1) 1.5W

NOTE:
1. Power dissipation is based on package heat transfer limitations, not device power consumption.

NOTICE: This data sheet contains preliminary information on new products in production. The specifications are subject to change without notice. Verify with your local Intel Sales office that you have the latest data sheet before finalizing a design.

**WARNING:* *Stressing the device beyond the "Absolute Maximum Ratings" may cause permanent damage. These are stress ratings only. Operation beyond the "Operating Conditions" is not recommended and extended exposure beyond the "Operating Conditions" may affect device reliability.*

OPERATING CONDITIONS

(All characteristics in this data sheet apply to these operating conditions unless otherwise noted.)

Symbol	Description	Min	Max	Units
T_A	Ambient Temperature Under Bias	0	+70	°C
V_{CC}	Digital Supply Voltage	4.50	5.50	V
V_{REF}	Analog Supply Voltage	4.50	5.50	V
F_{OSC}	Oscillator Frequency **12 MHz**	3.5	12	MHz
F_{OSC}	Oscillator Frequency **16 MHz**	3.5	16	MHz

NOTE:
ANGND and V_{SS} should be nominally at the same potential.

DC CHARACTERISTICS

Symbol	Description	Min	Max	Units	Test Conditions
V_{IL}	Input Low Voltage	−0.5	0.8	V	
V_{IH}	Input High Voltage (All Pins except XTAL1 and RESET)	0.2 V_{CC} + 0.9	V_{CC} + 0.5	V	
V_{IH1}	Input High Voltage on XTAL 1	0.7 V_{CC}	V_{CC} + 0.5	V	
V_{IH2}	Input High Voltage on RESET	2.6	V_{CC} + 0.5	V	
V_{OL}	Output Low Voltage		0.3 0.45 1.5	V V V	I_{OL} = 200 μA I_{OL} = 3.2 mA I_{OL} = 7 mA
V_{OH}	Output High Voltage (Standard Outputs)(2)	V_{CC} − 0.3 V_{CC} − 0.7 V_{CC} − 1.5		V V V	I_{OH} = −200 μA I_{OH} = −3.2 mA I_{OH} = −7 mA
V_{OH1}	Output High Voltage (Quasi-bidirectional Outputs)(1)	V_{CC} − 0.3 V_{CC} − 0.7 V_{CC} − 1.5		V V V	I_{OH} = −10 μA I_{OH} = −30 μA I_{OH} = −60 μA
I_{LI}	Input Leakage Current (Std. Inputs)(3)		±10	μA	$0 < V_{IN} < V_{CC} - 0.3V$
I_{LI1}	Input Leakage Current (Port 0)		+3	μA	$0 < V_{IN} < V_{REF}$
I_{TL}	1 to 0 Transition Current (QBD Pins)(1)		−800	μA	V_{IN} = 2.0V
I_{IL}	Logical 0 Input Current (QBD Pins)(1)		−50	μA	V_{IN} = 0.45V

DC CHARACTERISTICS (Continued)

Symbol	Description	Min	Typ(7)	Max	Units	Test Conditions
I_{IL1}	Logical 0 Input Current in Reset $\overline{BHE}$, $\overline{WR}$, P2.0			−850	μA	$V_{IN} = 0.45V$
I_{IL2}	Logical 0 Input Current in Reset ALE, $\overline{RD}$, INST			−7	mA	$V_{IN} = 0.45V$
I_{IH1}	Logical 1 Input Current on NMI Pin			100	μA	$V_{IN} = 2.0V$
Hyst.	Hysteresis on RESET Pin	300			mV	
I_{CC}	Active Mode Current in Reset		50	60	mA	XTAL1 = 16 MHz $V_{CC} = V_{PP} = V_{REF} = 5.5V$
I_{REF}	A/D Converter Reference Current		2	5	mA	
I_{IDLE}	Idle Mode Current		10	25	mA	
I_{PD}	Powerdown Mode Current		5	30	μA	$V_{CC} = V_{PP} = V_{REF} = 5.5V$
R_{RST}	Reset Pullup Resistor	6K		50K	Ω	
C_S	Pin Capacitance (Any Pin to V_{SS})			10	pF	$F_{TEST} = 1.0$ MHz

NOTES: (Notes apply to all specifications)

1. QBD (Quasi-bidirectional) pins include Port 1, P2.6 and P2.7.
2. Standard Outputs include AD0–15, $\overline{RD}$, $\overline{WR}$, ALE, $\overline{BHE}$, INST, HSO pins, PWM/P2.5, CLKOUT, RESET, Ports 3 and 4, TXD/P2.0 and RXD (in serial mode 0). The V_{OH} specification is not valid for RESET. Ports 3 and 4 are open-drain outputs.
3. Standard Inputs include HSI pins, $\overline{EA}$, READY, BUSWIDTH, NMI, RXD/P2.1, EXTINT/P2.2, T2CLK/P2.3 and T2RST/P2.4.
4. Maximum current per pin must be externally limited to the following values if V_{OL} is held above 0.45V or V_{OH} is held below V_{CC} − 0.7V:
   - I_{OL} on Output pins: 10 mA
   - I_{OH} on quasi-bidirectional pins: self limiting
   - I_{OH} on Standard Output pins: 10 mA
5. Maximum current per bus pin (data and control) during normal operation is ±3.2 mA.
6. During normal (non-transient) conditions the following total current limits apply:

Port 1, P2.6	I_{OL}: 29 mA	I_{OH} is self limiting
HSO, P2.0, RXD, $\overline{RESET}$	I_{OL}: 29 mA	I_{OH}: 26 mA
P2.5, P2.7, $\overline{WR}$, $\overline{BHE}$	I_{OL}: 13 mA	I_{OH}: 11 mA
AD0–AD15	I_{OL}: 52 mA	I_{OH}: 52 mA
$\overline{RD}$, ALE, INST–CLKOUT	I_{OL}: 13 mA	I_{OH}: 13 mA

7. Typicals are based on a limited number of samples and are not guaranteed. The values listed are at room temperature and $V_{REF} = V_{CC} = 5V$.

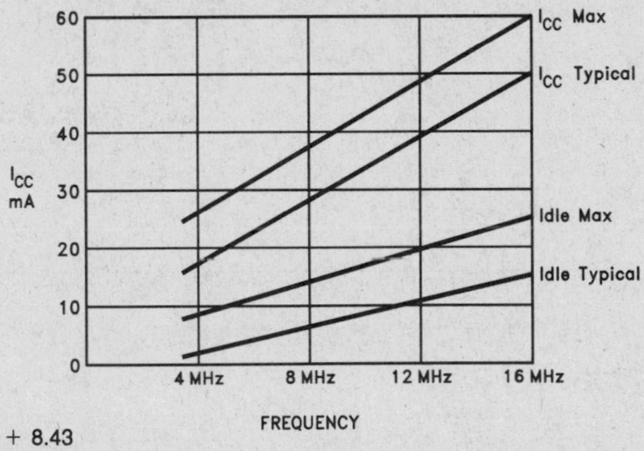

I_{CC} Max = 3.88 × FREQ + 8.43
I_{IDLE} Max = 1.65 × FREQ + 2.2

270909–5

Figure 6. I_{CC} and I_{IDLE} vs Frequency

AC CHARACTERISTICS

Test Conditions: Capacitive load on all pins = 100 pF, Rise and fall times = 10 ns, F_{OSC} = 12/16 MHz

The system must meet these specifications to work with the 87C196KB:

Symbol	Description	Min	Max	Units	Notes
T_{AVYV}	Address Valid to READY Setup		$2 T_{OSC} - 75$	ns	
T_{YLYH}	NonREADY Time	No upper limit		ns	
T_{CLYX}	READY Hold after CLKOUT Low	0	$T_{OSC} - 30$	ns	(Note 1)
T_{LLYX}	READY Hold after ALE Low	$T_{OSC} - 15$	$2 T_{OSC} - 40$	ns	(Note 1)
T_{AVGV}	Address Valid to Buswidth Setup		$2 T_{OSC} - 75$	ns	
T_{CLGX}	Buswidth Hold after CLKOUT Low	0		ns	
T_{AVDV}	Address Valid to Input Data Valid		$3 T_{OSC} - 55$	ns	(Note 2)
T_{RLDV}	$\overline{RD}$ Active to Input Data Valid		$T_{OSC} - 23$	ns	(Note 2)
T_{CLDV}	CLKOUT Low to Input Data Valid		$T_{OSC} - 50$	ns	
T_{RHDZ}	End of $\overline{RD}$ to Input Data Float		$T_{OSC} - 20$	ns	
T_{RXDX}	Data Hold after $\overline{RD}$ Inactive	0		ns	

NOTES:
1. If max is exceeded, additional wait states will occur.
2. When using wait states, add $2 T_{OSC} \times n$ where n = number of wait states.

AC CHARACTERISTICS (Continued)

Test Conditions: Capacitive load on all pins = 100 pF, Rise and fall times = 10 ns, F_{OSC} = 12/16 MHz

The 87C196KB will meet these specifications:

Symbol	Description	Min	Max	Units	Notes
F_{XTAL}	Frequency on XTAL1 **12 MHz**	3.5	12.0	MHz	(Note 2)
F_{XTAL}	Frequency on XTAL1 **16 MHz**	3.5	16.0	MHz	(Note 2)
T_{OSC}	1/F_{XTAL} **12 MHz**	83.3	286	ns	
T_{OSC}	1/F_{XTAL} **16 MHz**	62.5	286	ns	
T_{XHCH}	XTAL1 High to CLKOUT High or Low	+20	+110	ns	
T_{CLCL}	CLKOUT Cycle Time	2 T_{OSC}		ns	
T_{CHCL}	CLKOUT High Period	T_{OSC} − 10	T_{OSC} + 10	ns	
T_{CLLH}	CLKOUT Falling Edge to ALE Rising	−10	+10	ns	
T_{LLCH}	ALE Falling Edge to CLKOUT Rising	−15	+15	ns	
T_{LHLH}	ALE Cycle Time	4 T_{OSC}		ns	(Note 3)
T_{LHLL}	ALE High Period	T_{OSC} − 10	T_{OSC} + 10	ns	
T_{AVLL}	Address Setup to ALE Falling Edge	T_{OSC} − 20		ns	
T_{LLAX}	Address Hold after ALE Falling Edge	T_{OSC} − 40		ns	
T_{LLRL}	ALE Falling Edge to $\overline{RD}$ Falling Edge	T_{OSC} − 35		ns	
T_{RLCL}	$\overline{RD}$ Low to CLKOUT Falling Edge	+4	+25	ns	
T_{RLRH}	$\overline{RD}$ Low Period	T_{OSC} − 5	T_{OSC} + 25	ns	(Note 3)
T_{RHLH}	$\overline{RD}$ Rising Edge to ALE Rising Edge	T_{OSC}	T_{OSC} + 25	ns	(Note 1)
T_{RLAZ}	$\overline{RD}$ Low to Address Float		+5	ns	
T_{LLWL}	ALE Falling Edge to $\overline{WR}$ Falling Edge	T_{OSC} − 10		ns	
T_{CLWL}	CLKOUT Low to $\overline{WR}$ Falling Edge	0	+25	ns	
T_{QVWH}	Data Stable to $\overline{WR}$ Rising Edge	T_{OSC} − 23		ns	(Note 3)
T_{CHWH}	CLKOUT High to $\overline{WR}$ Rising Edge	−5	+15	ns	
T_{WLWH}	$\overline{WR}$ Low Period	T_{OSC} − 15	T_{OSC} + 5	ns	(Note 3)
T_{WHQX}	Data Hold after $\overline{WR}$ Rising Edge	T_{OSC} − 15		ns	
T_{WHLH}	$\overline{WR}$ Rising Edge to ALE Rising Edge	T_{OSC} − 15	T_{OSC} + 10	ns	(Note 1)
T_{WHBX}	$\overline{BHE}$, INST HOLD after $\overline{WR}$ Rising Edge	T_{OSC} − 15		ns	
T_{RHBX}	$\overline{BHE}$, INST HOLD after $\overline{RD}$ Rising Edge	T_{OSC} − 10		ns	
T_{WHAX}	AD8–15 hold after $\overline{WR}$ Rising Edge	T_{OSC} − 30		ns	
T_{RHAX}	AD8–15 hold after $\overline{RD}$ Rising Edge	T_{OSC} − 25		ns	

NOTES:
1. Assuming back-to-back bus cycles.
2. Testing performed at 3.5 MHz, however, the device is static by design and will typically operate below 1 Hz.
3. When using wait states, all 2 T_{OSC} + n where n = number of wait states.

4

System Bus Timings

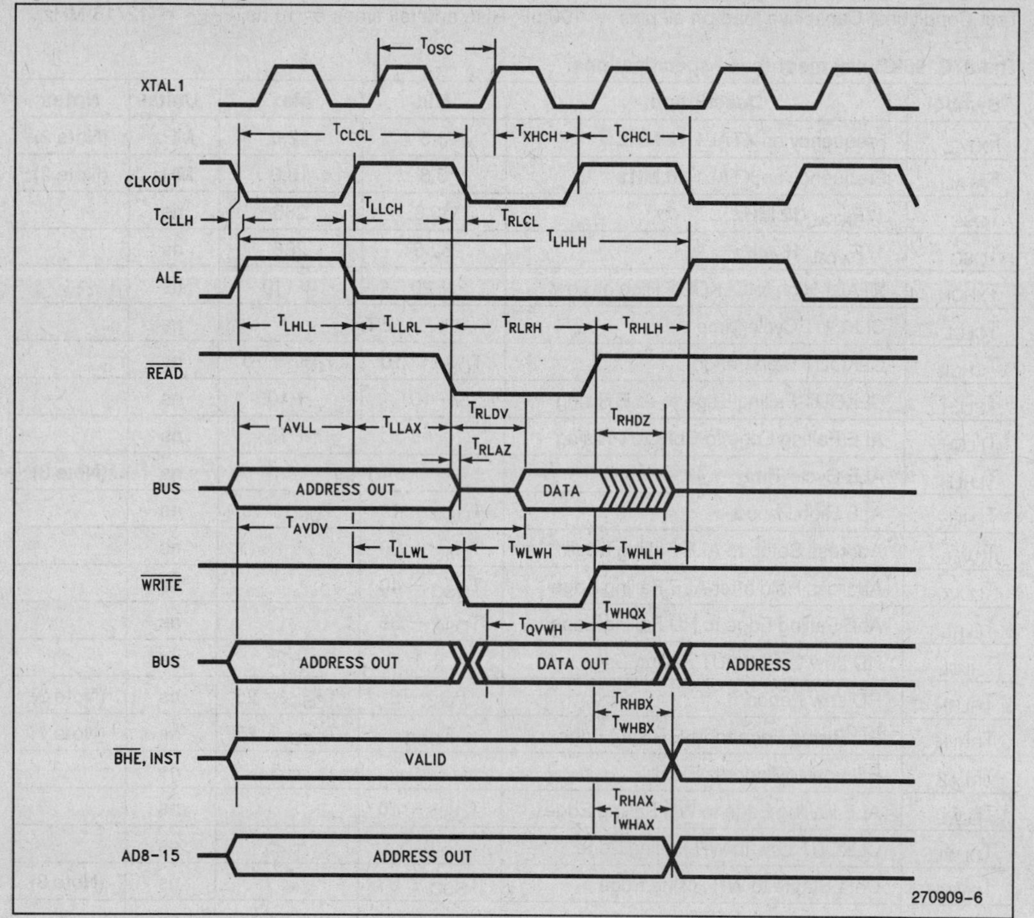

READY Timings (One Wait State)

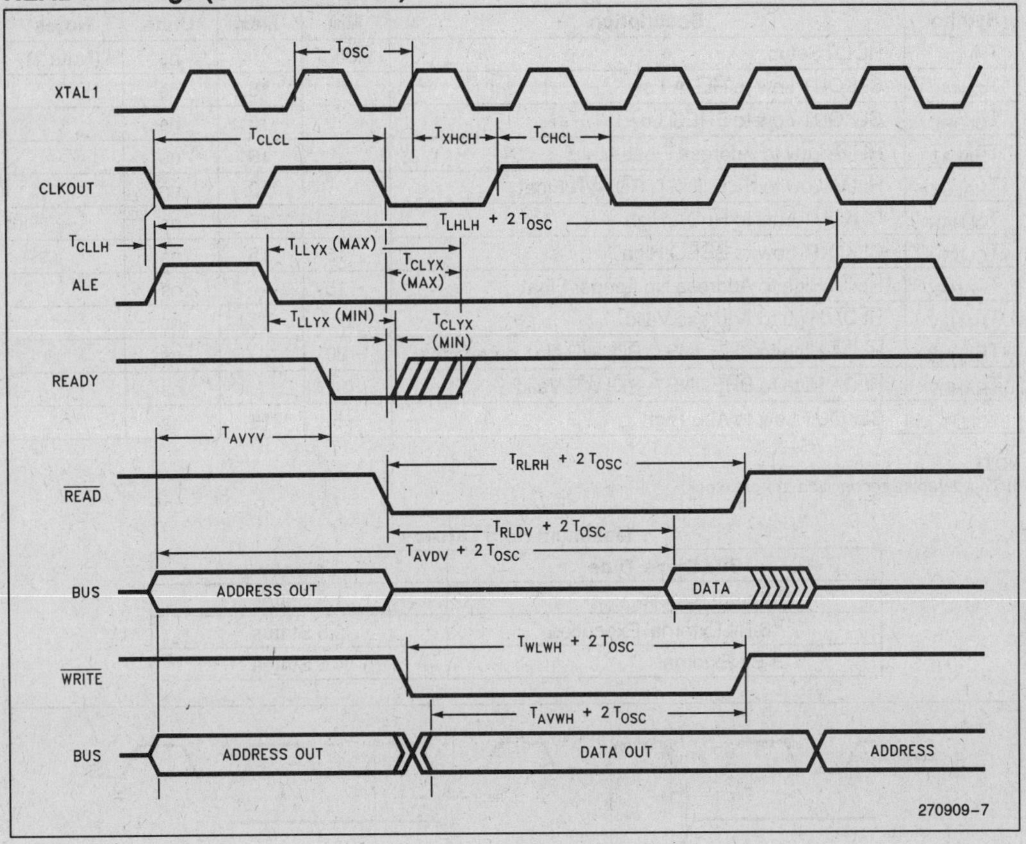

270909-7

Buswidth Bus Timings

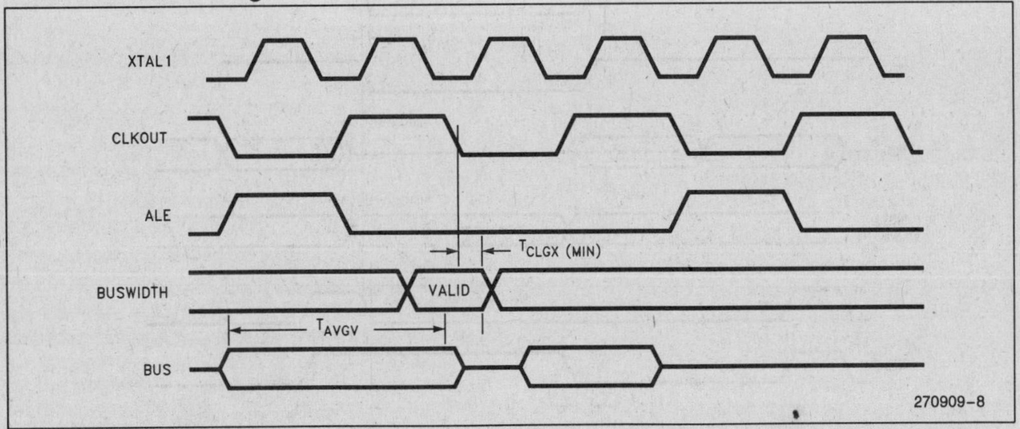

270909-8

$\overline{\text{HOLD}}$/$\overline{\text{HLDA}}$ Timings

Symbol	Description	Min	Max	Units	Notes
T_{HVCH}	$\overline{\text{HOLD}}$ Setup	55		ns	(Note 1)
T_{CLHAL}	CLKOUT Low to $\overline{\text{HLDA}}$ Low		15	ns	
T_{CLBRL}	CLKOUT Low to $\overline{\text{BREQ}}$ Low		15	ns	
T_{HALAZ}	$\overline{\text{HLDA}}$ Low to Address Float		10	ns	
T_{HALBZ}	$\overline{\text{HLDA}}$ Low to $\overline{\text{BHE}}$, INST, $\overline{\text{RD}}$, $\overline{\text{WR}}$ Float		10	ns	
T_{CLHAH}	CLKOUT Low to $\overline{\text{HLDA}}$ High	−15	15	ns	
T_{CLBRH}	CLKOUT Low to $\overline{\text{BREQ}}$ High	−15	15	ns	
T_{HAHAX}	$\overline{\text{HLDA}}$ High to Address No Longer Float	−15		ns	
T_{HAHAV}	$\overline{\text{HLDA}}$ High to Address Valid	0		ns	
T_{HAHBX}	$\overline{\text{HLDA}}$ High to $\overline{\text{BHE}}$, INST, $\overline{\text{RD}}$, $\overline{\text{WR}}$ No Longer Float	−20		ns	
T_{HAHBV}	$\overline{\text{HLDA}}$ High to $\overline{\text{BHE}}$, INST, $\overline{\text{RD}}$, $\overline{\text{WR}}$ Valid	0		ns	
T_{CLLH}	CLKOUT Low to ALE High	−5	15	ns	

NOTE:
1. To guarantee recognition at next clock.

Maximum Hold Latency

Bus Cycle Type	Latency
Internal Access	1.5 States
16-Bit External Execution	2.5 States
8-Bit External	4.5 States

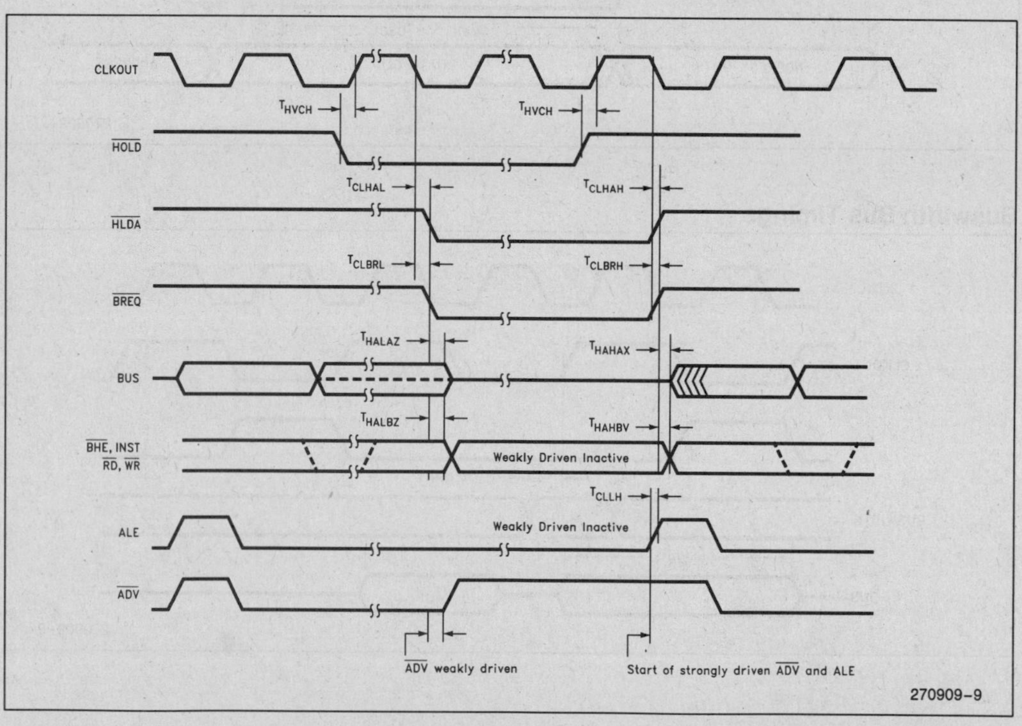

EXTERNAL CLOCK DRIVE

Symbol	Parameter	Min	Max	Units
$1/T_{XLXL}$	Oscillator Frequency **12 MHz**	3.5	12.0	MHz
$1/T_{XLXL}$	Oscillator Frequency **16 MHz**	3.5	16	MHz
T_{XLXL}	Oscillator Period **12 MHz**	83.3	286	ns
T_{XLXL}	Oscillator Period **16 MHz**	62.5	286	ns
T_{XHXX}	High Time	21.25		ns
T_{XLXX}	Low Time	21.25		ns
T_{XLXH}	Rise Time		10	ns
T_{XHXL}	Fall Time		10	ns

EXTERNAL CLOCK DRIVE WAVEFORMS

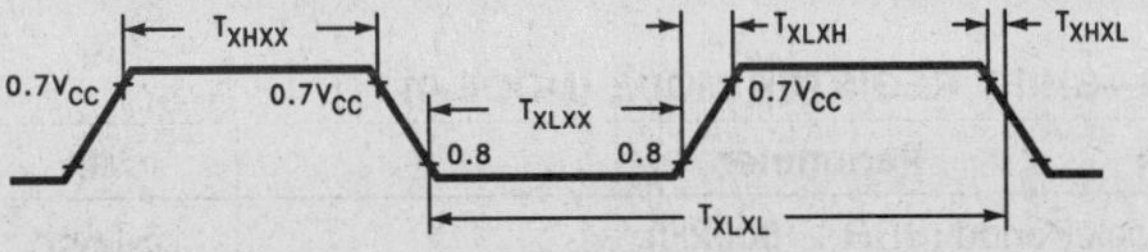

270909-10

An external oscillator may encounter as much as a 100 pF load at XTAL1 when it starts-up. This is due to interaction between the amplifier and its feedback capacitance. Once the external signal meets the V_{IL} and V_{IH} specifications, the capacitance will not exceed 20 pF.

EXTERNAL CRYSTAL CONNECTIONS

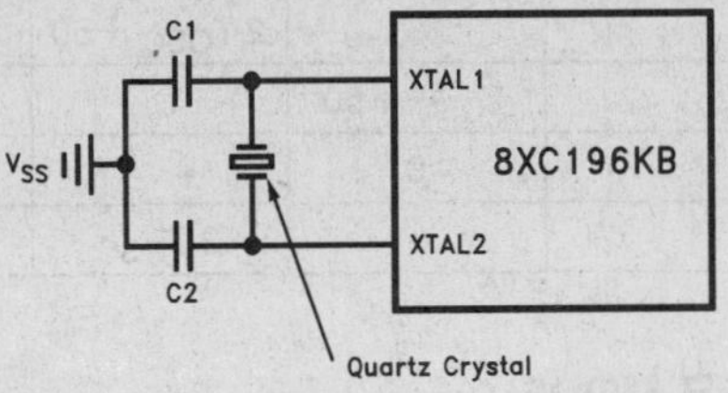

270909-11

NOTE:
Keep oscillator components close to chip and use short, direct traces to XTAL1, XTAL2 and V_{SS}. When using crystals, C1 = 20 pF, C2 = 20 pF. When using ceramic resonators, consult manufacturer for recommended circuitry.

EXTERNAL CLOCK CONNECTIONS

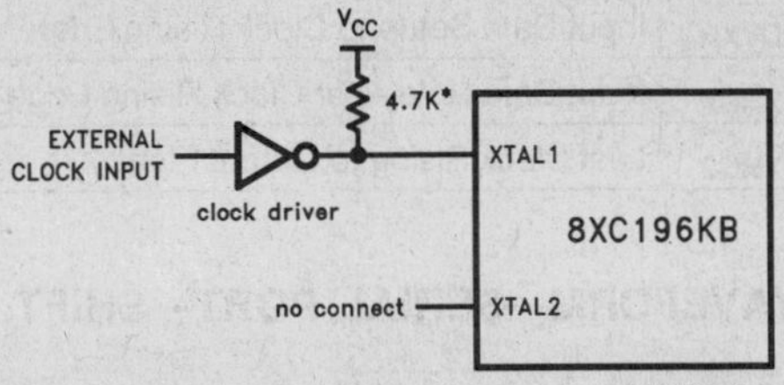

270909-12

* Required if open-collector TTL driver used
Not needed if CMOS driver is used.

AC TESTING INPUT, OUTPUT WAVEFORMS

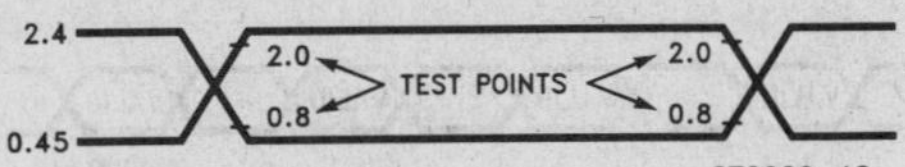

270909-13

AC Testing inputs are driven at 2.4V for a Logic "1" and 0.45V for a Logic "0" Timing measurements are made at 2.0V for a Logic "1" and 0.8V for a Logic "0".

FLOAT WAVEFORMS

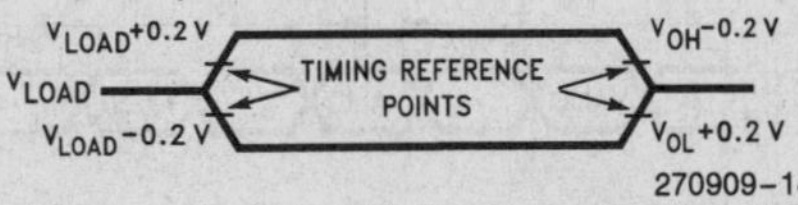

270909-14

For Timing Purposes a Port Pin is no Longer Floating when a 200 mV change from Load Voltage Occurs and Begins to Float when a 200 mV change from the Loaded V_{OH}/V_{OL} Level occurs; $I_{OL}/I_{OH} = \pm 15$ mA.

EXPLANATION OF AC SYMBOLS

Each symbol is two pairs of letters prefixed by "T" for time. The characters in a pair indicate a signal and its condition, respectively. Symbols represent the time between the two signal/condition points.

Conditions:

Signals:

H - High	A - Address	G - Buswidth	R - $\overline{RD}$
L - Low	B - $\overline{BHE}$	H - $\overline{HOLD}$	W - $\overline{WR}/\overline{WRH}/\overline{WRL}$
V - Valid	BR - $\overline{BREQ}$	HA - $\overline{HLDA}$	X - XTAL1
X - No Longer Valid	C - CLKOUT	L - ALE/$\overline{ADV}$	Y - READY
Z - Floating	D - DATA IN	Q - DATA OUT	

AC CHARACTERISTICS—SERIAL PORT—SHIFT REGISTER MODE

SERIAL PORT TIMING—SHIFT REGISTER MODE (MODE 0)

Symbol	Parameter	Min	Max	Units
T_{XLXL}	Serial Port Clock Period (BRR ≥ 8002H)	6 T_{OSC}		ns
T_{XLXH}	Serial Port Clock Falling Edge to Rising Edge (BRR ≥ 8002H)	4 T_{OSC} − 50	4 T_{OSC} + 50	ns
T_{XLXL}	Serial Port Clock Period (BRR = 8001H)	4 T_{OSC}		ns
T_{XLXH}	Serial Port Clock Falling Edge to Rising Edge (BRR = 8001H)	2 T_{OSC} − 50	2 T_{OSC} + 50	ns
T_{QVXH}	Output Data Setup to Clock Rising Edge	2 T_{OSC} −50		ns
T_{XHQX}	Output Data Hold after Clock Rising Edge	2 T_{OSC} −50		ns
T_{XHQV}	Next Output Data Valid after Clock Rising Edge		2 T_{OSC} +50	ns
T_{DVXH}	Input Data Setup to Clock Rising Edge	T_{OSC} +50		ns
T_{XHDX}	Input Data Hold after Clock Rising Edge	0		ns
T_{XHQZ}	Last Clock Rising to Output Float		2 T_{OSC}	ns

WAVEFORM—SERIAL PORT—SHIFT REGISTER MODE

SERIAL PORT WAVEFORM—SHIFT REGISTER MODE (MODE 0)

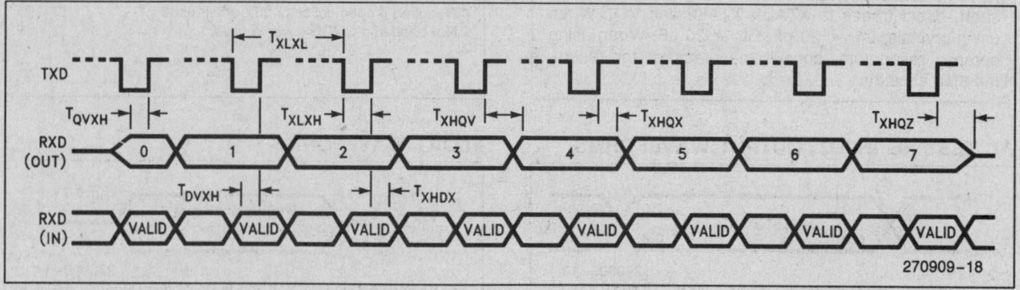

PRELIMINARY

10-BIT A/D CHARACTERISTICS

At a clock speed of 6 MHz or less, the clock prescaler should be disabled. This is accomplished by setting IOC2.4 = 1.

At higher frequencies (greater than 6 MHz) the clock prescaler should be enabled (IOC2.4 = 0) to allow the comparator to settle.

The table below shows two different clock speeds and their corresponding A/D conversion and sample times.

State times are calculated as follows:

$$\text{state time} = \frac{2}{\text{XTAL1}}$$

The converter is ratiometric, so the absolute accuracy is directly dependent on the accuracy and stability of V_{REF}. V_{REF} must be close to V_{CC} since it supplies both the resistor ladder and the digital section of the converter.

See the MCS-96 A/D Converter Quick Reference for definition of A/D terms.

Example Sample and Conversion Times

A/D Clock Prescaler	Clock Speed (MHz)	Sample Time (States)	Sample Time at Clock Speed (μs)	Conversion Time (States)	Conversion Time at Clock Speed (μs)
IOC2.4 = 0 → ON	16	15	1.875	156.5	19.6
IOC2.4 = 1 → OFF	6	8	2.667	89.5	29.8

A/D CONVERTER SPECIFICATIONS

Parameter	Typical(1)	Minimum	Maximum	Units*	Notes
Resolution		1024 10	1024 10	Levels Bits	
Absolute Error		0	±3	LSBs	
Full Scale Error	0.25 ±0.50			LSBs	
Zero Offset Error	0.25 ±0.50			LSBs	
Non-Linearity Error	1.5 ±2.5	0	±3	LSBs	
Differential Non-Linearity Error		>−1	+2	LSBs	
Channel-to-Channel Matching	±0.1	0	±1	LSBs	
Repeatability	±0.25			LSBs	
Temperature Coefficients: Offset Full Scale Differential Non-Linearity	 0.009 0.009 0.009			 LSB/°C LSB/°C LSB/°C	
Off Isolation		−60		dB	2, 3
Feedthrough	−60			dB	2
V_{CC} Power Supply Rejection	−60			dB	2
Input Series Resistance		750	1.2K	Ω	4
DC Input Leakage		0	±3.0	μA	
Sampling Capacitor	3			pF	

NOTES:
*An "LSB", as used here, has a value of approximately 5 mV.
1. Typical values are expected for most devices at 25°C.
2. DC to 100 KHz.
3. Multiplexer Break-Before-Make Guaranteed.
4. Resistance from device pin, through internal MUX, to sample capacitor.

OTPROM SPECIFICATIONS

OTPROM PROGRAMMING OPERATING CONDITIONS

Symbol	Parameter	Min	Max	Units
T_A	Ambient Temperature During Programming	20	30	C
V_{CC}, V_{PD}, V_{REF}(1)	Supply Voltages During Programming	4.5	5.5	V
V_{EA}	Programming Mode Supply Voltage	12.50	13.0	V(2)
V_{PP}	EPROM Programming Supply Voltage	12.50	13.0	V(2)
V_{SS}, ANGND(3)	Digital and Analog Ground	0	0	V
F_{OSC}	Oscillator Frequency **12 MHz**	6.0	12.0	MHz
F_{OSC}	Oscillator Frequency **16 MHz**	6.0	16.0	MHz

NOTES:
1. V_{CC}, V_{PD} and V_{REF} should nominally be at the same voltage during programming.
2. V_{EA} and V_{PP} must never exceed the maximum voltage for any amount of time or the device may be damaged.
3. V_{SS} and ANGND should nominally be at the same voltage (0V) during programming.

AC OTPROM PROGRAMMING CHARACTERISTICS

Symbol	Description	Min	Max	Units
T_{SHLL}	Reset High to First $\overline{PALE}$ Low	1100		T_{OSC}
T_{LLLH}	$\overline{PALE}$ Pulse Width	40		T_{OSC}
T_{AVLL}	Address Setup Time	0		T_{OSC}
T_{LLAX}	Address Hold Time	50		T_{OSC}
T_{LLVL}	$\overline{PALE}$ Low to PVER Low		60	T_{OSC}
T_{PLDV}	$\overline{PROG}$ Low to Word Dump Valid		50	T_{OSC}
T_{PHDX}	Word Dump Data Hold		50	T_{OSC}
T_{DVPL}	Data Setup Time	0		T_{OSC}
T_{PLDX}	Data Hold Time	50		T_{OSC}
T_{PLPH}	$\overline{PROG}$ Pulse Width	40		T_{OSC}
T_{PHLL}	$\overline{PROG}$ High to Next $\overline{PALE}$ Low	120		T_{OSC}
T_{LHPL}	$\overline{PALE}$ High to $\overline{PROG}$ Low	220		T_{OSC}
T_{PHPL}	$\overline{PROG}$ High to Next $\overline{PROG}$ Low	120		T_{OSC}
T_{PHIL}	$\overline{PROG}$ High to AINC Low	0		T_{OSC}
T_{ILIH}	$\overline{AINC}$ Pulse Width	40		T_{OSC}
T_{ILVH}	PVER Hold after $\overline{AINC}$ Low	50		T_{OSC}
T_{ILPL}	$\overline{AINC}$ Low to $\overline{PROG}$ Low	170		T_{OSC}
T_{PHVL}	$\overline{PROG}$ High to PVER Low		90	T_{OSC}

DC OTPROM PROGRAMMING CHARACTERISTICS

Symbol	Description	Min	Max	Units
I_{PP}	V_{PP} Supply Current (When Programming)		100	mA

NOTE:
Do not apply V_{PP} until V_{CC} is stable and within specifications and the oscillator/clock has stabilized or the device may be damaged.

OTPROM PROGRAMMING WAVEFORMS

SLAVE PROGRAMMING MODE DATA PROGRAM MODE WITH SINGLE PROGRAM PULSE

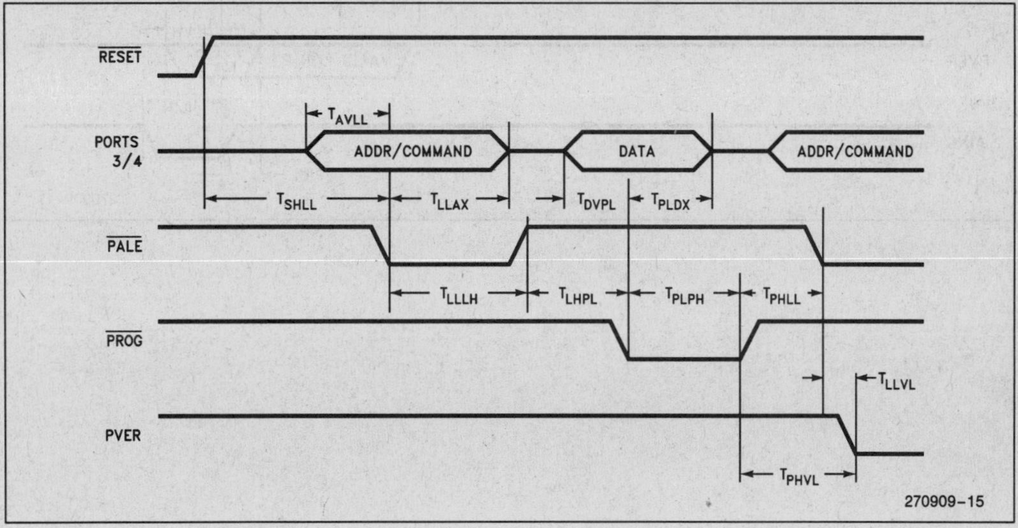

270909–15

SLAVE PROGRAMMING MODE IN WORD DUMP OR DATA VERIFY MODE WITH AUTO INCREMENT

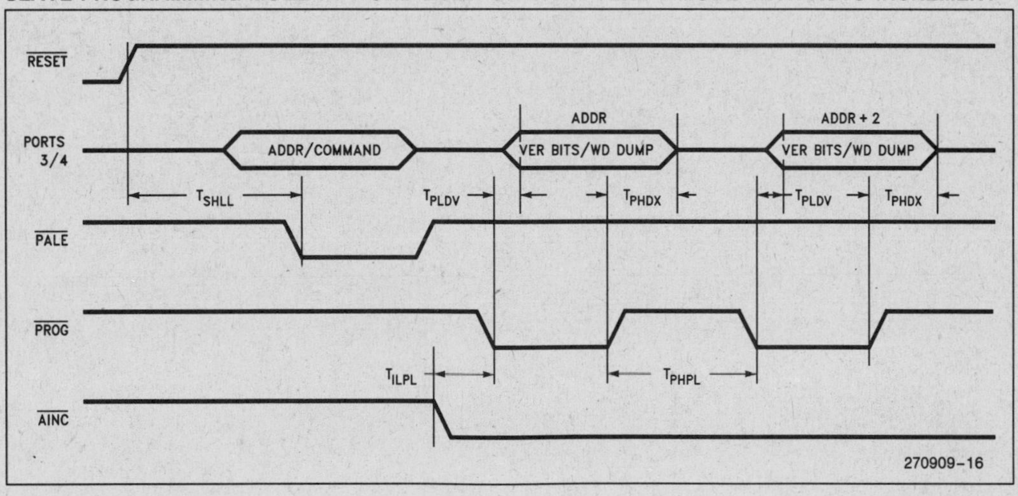

270909–16

4

SLAVE PROGRAMMING MODE TIMING IN DATA PROGRAM MODE WITH REPEATED PROG PULSE AND AUTO INCREMENT

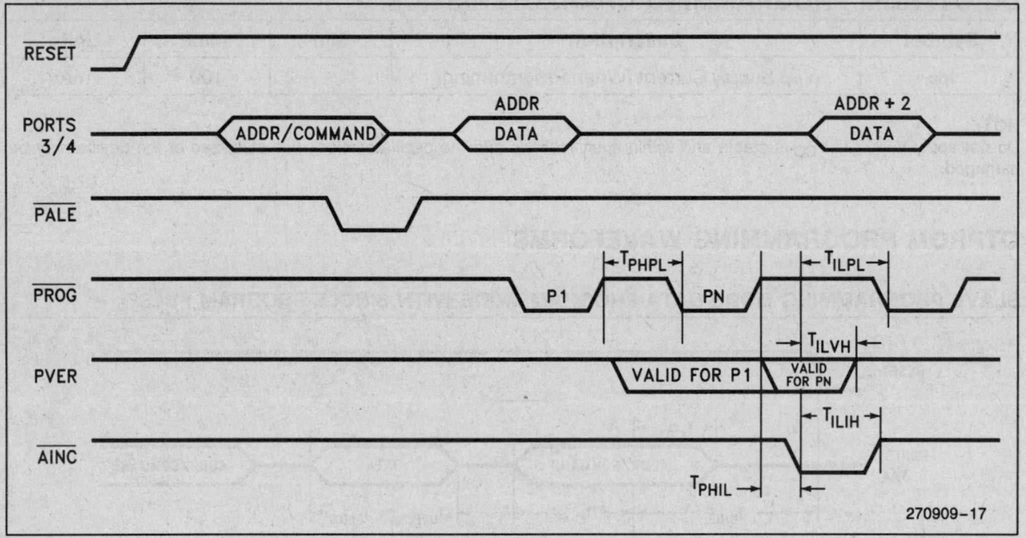

FUNCTIONAL DEVIATIONS

Devices marked with an "E", "F" or "G" have the following errata.

1. Missed Interrupt on P0.7, EXTINT

Interrupts occurring on P0.7 could be missed since the INT_PEND EXTINT bit may not be set. See techbit MC0893.

2. HSI_MODE Divide-by-Eight

See Faxback #2192

REVISION HISTORY

This data sheet (270909-006) is valid for devices with an "E", "F" or "G" at the end of the top side tracking number. Data sheets are changed as new device information becomes available. Verify with your local Intel sales office that you have the latest version before finalizing a design or ordering devices.

The following differences exist between this data sheet (270909-006) and (270909-005):

1. Removed "Word Addressable Only" from Port 3 and 4 in Table 2.
2. Removed ICC1, active mode current at 3.5 MHz. This specification is not longer required.
3. Removed TLLYV and TLLGV from waveform diagrams.
4. The HSI errata and CMPL with R0 were removed as this is now considered normal operation.
5. The HSI_MODE divide-by-eight errata was added to the known errata section.

The following differences exist between this data sheet (270909-005) and (270909-004):

1. I_{TL} MAX was $-650\ \mu A$ (270909-004). Now I_{TL} MAX is $-800\ \mu A$ (270909-005).
2. I_{IL2} was named I_{IL1} (270909-004). Now I_{IL2} is correctly named (270909-005).
3. I_{IL1} was omitted (270909-004). I_{IL1} MAX was added. I_{IL1} MAX is $-850\ \mu A$ (270909-005).
4. T_{LLYV} and T_{LLGV} (270909-004) were removed. These timings are not required in high-speed system designs.
5. An errata was added to the known errata section. There is a possibility to miss an external interrupt on P0.7 EXTINT.

The following differences exist between this data sheet (270909-004) and (270909-003):

1. The ROM (80C196KB), and ROMless (83C196KB) were combined with this data sheet resulting in no specification differences.
2. The description of the prescalar bit for the A/D has been enhanced.
3. T_{HAHBV}MIN was -15 ns (270909-003). Now T_{HAHBV}MIN is -20 ns (270909-004).
4. T_{XHQZ}MAX was 1 TOSC (270909-003). Now T_{XHQZ}MAX is 2 TOSC (270909-004). This should have no impact on designs using synchronous serial mode 0.
5. The change indicators for the 80C196KB are "E", "F" and "G". Previously there was only one change indicator "E". The change indicator is used for tracking purposes. The change indicator is the last character in the FPO number. The FPO number is the second line on the top side of the device.

The following differences exist between (-003) and version (-002).

1. The 12 MHz and 16 MHz devices were combined in this data sheet. The 87C196KB 12 MHz only data sheet (272035-001) is now obsolete.
2. Changes were made to the format of the data sheet and the SFR descriptions were removed.
3. The -002 version of this data sheet was valid for devices marked with a "B" or a "D" at the end of the top side tracking number.
4. The OSCILLATOR errata was removed.
5. An errata was not documented in the -002 data sheet for devices marked with a "B" or a "D". This is the DIVIDE DURING HOLD/READY errata. When HOLD or READY is active and DIV/DIVB is the last instruction in the queue, the divide result may be incorrect.
6. T_{XCH} was changed from Min = 40 ns to Min = 20 ns.
7. T_{RLCL} was changed from Min = 5 ns to Min = 4 ns.
9. I_{IL1} was changed from Max = −6 mA to Max = −7 mA.
10. T_{HAHBV} was changed from Min = −10 ns to Min = −15 ns.

Differences between the -002 and -001 data sheets.

1. The -001 version of this data sheet was valid for devices marked with a "C" at the end of the top side tracking number.
2. Added 64L SDIP and 80L QFP packages.
3. Added IIH1.
4. Changed T_{CHWH} Min from − 10 ns to − 5 ns.
5. Changed T_{CHWH} Max from + 10 ns to + 15 ns.
6. Changed T_{WLWH} Min from T_{OSC} − 20 ns to T_{OSC} − 15 ns.
7. Changed T_{WHQX} Min from T_{OSC} − 10 ns to T_{OSC} − 15 ns.
8. Changed T_{WHLH} Min from T_{OSC} − 10 ns to T_{OSC} − 15 ns.
9. Changed T_{WHLH} Max from T_{OSC} + 15 ns to T_{OSC} + 10 ns.
10. Changed T_{WHBX} Min from T_{OSC} − 10 ns to T_{OSC} − 15 ns.
11. Changed T_{HVCH} Min from 85 ns to 55 ns.
12. Remove T_{HVCH} Max.
13. Changed T_{CLHAL} Min from − 10 ns to − 15 ns.
14. Changed T_{CLHAL} Max from 20 ns to 15 ns.
15. Changed T_{CLBRL} Min from − 10 ns to − 15 ns.
16. Changed T_{CLBRL} Max from 20 ns to 15 ns.
17. Changed T_{HAHAX} Min from − 10 ns to − 15 ns.
18. Added HSI description to Functional Deviations.
19. Added Oscillator description to Functional Deviations.

8XC198
COMMERCIAL/EXPRESS CHMOS MICROCONTROLLER

8 Kbytes of OTPROM

- 8 Kbytes of On-Chip OTPROM or ROM
- 232 Byte Register File
- Register-to-Register Architecture
- 28 Interrupt Sources/16 Vectors
- 1.75 μs 16 x 16 Multiply (16 MHz)
- 3.0 μs 32/16 Divide (16 MHz)
- Powerdown and Idle Modes
- 16-Bit Watchdog Timer
- 8-Bit External Bus
- 16 MHz Standard
- Full Duplex Serial Port
- High Speed I/O Subsystem
- 16-Bit Timer
- 16-Bit Counter
- Pulse-Width-Modulated Output
- Four 16-Bit Software Timers
- 10-Bit A/D Converter with Sample/Hold
- Extended Temperature Available

The 8XC198 family offers low-cost entry into Intel's powerful MCS®-96 16-bit microcontroller architecture. Intel's CHMOS process provides a high performance processor along with low power consumption. To further reduce power requirements, the processor can be placed into Idle or Powerdown Mode.

The 8XC198 is the 8-bit bus version of the 8XC196KB. The prefixes mean: 80 (ROMless), 83 (ROM), 87 (OTP) One Time Programmable. The ROM and OTP are available in 8 Kbytes.

Bit, byte, word and some 32-bit operations are available on the 8XC198. With a 16 MHz oscillator a 16-bit addition takes 0.50 μs, and the instruction times average 0.37 μs to 1.1 μs in typical applications.

Four high-speed capture inputs are provided to record times when events occur. Six high-speed outputs are available for pulse or waveform generation. The high-speed output can also generate four software timers or start an A/D conversion. Events can be based on the timer or counter. Also provided on-chip are an A/D converter, serial port, watchdog timer and a pulse-width-modulated output signal.

With the commercial (standard) temperature option, operational characteristics are guaranteed over the temperature range of 0°C to +70°C. Wth the extended temperature range option, operational characteristics are guaranteed over the temperature range of −40°C to +85°C.

MCS®-96 is a registered trademark of Intel Corporation.

October 1992
Order Number: 272034-003

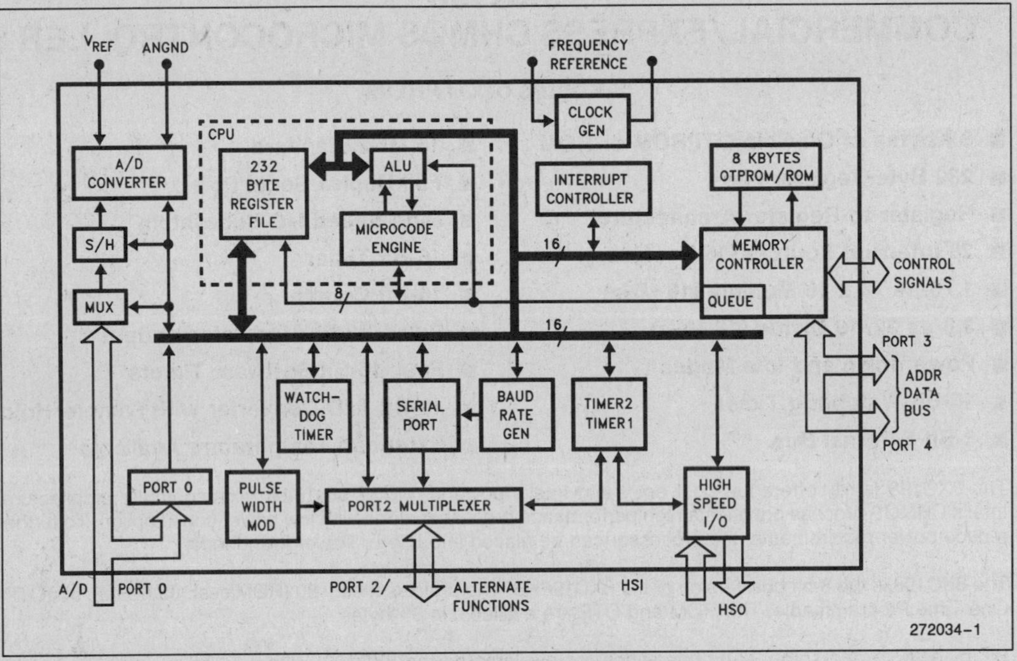

Figure 1. 87C198 Block Diagram

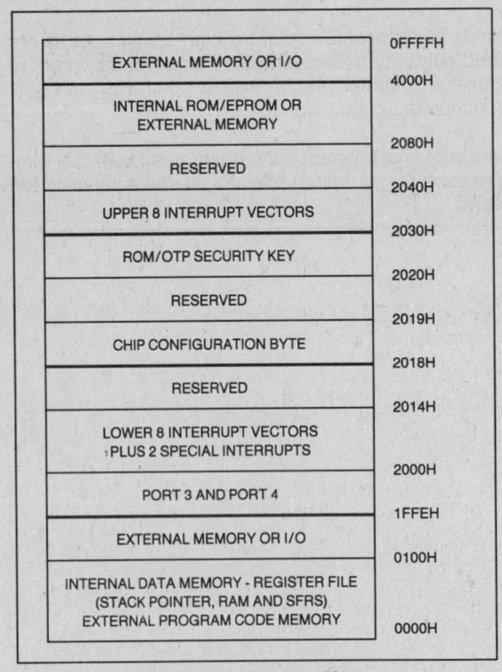

Figure 2. Memory Map

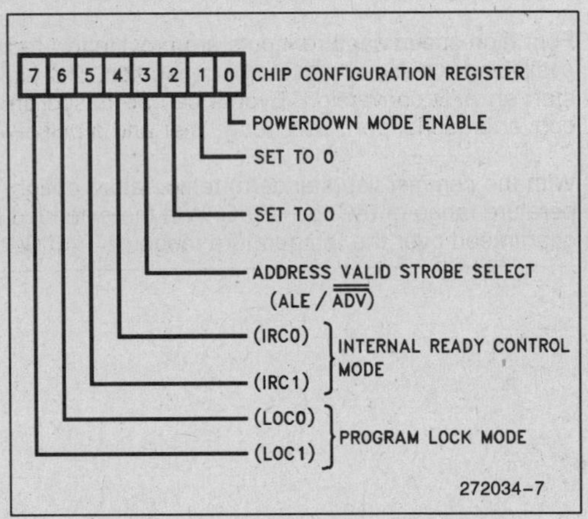

Figure 3. Chip Configuration (2018H)

WARNING:
Reserved memory locations must not be written or read. The contents and/or function of these locations may change with future revisions of the device. Therefore, a program that relies on one or more of these locations may not function properly.

PACKAGING

The 8XC198 is available in a 52-pin PLCC package and an 80-pin QFP package. Contact your local sales office to determine the exact ordering code for the part desired.

Package Designators:
N = 52-pin PLCC
S = 80-pin QFP

Thermal Characteristics

Package Type	θ_{ja}	θ_{jc}
PLCC	40°C/W	
QFP	70°C/W	4°C/W

All thermal impedance data is approximate for static air conditions at 1W of power dissipation. Values will change depending on operating conditions and application. See the Intel *Packaging Handbook* (Order Number 240800) for a description of Intel's thermal impedance test methodology.

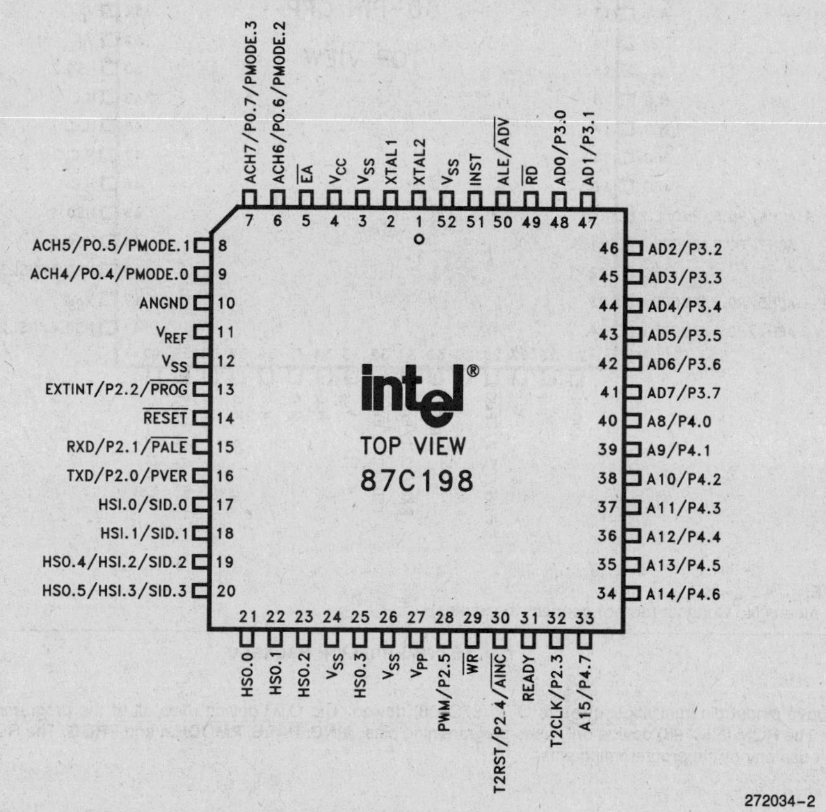

Figure 4. 52-Pin PLCC Package

NOTE:
The above pinout diagram applies to the OTP (87C198) device. The OTP device uses all of the programming pins shown above. The ROM (83C198) device only uses programming pins: **AINC, PALE, PMODE.n** and **PROG**. The ROMless (80C198) doesn't use any of the programming pins.

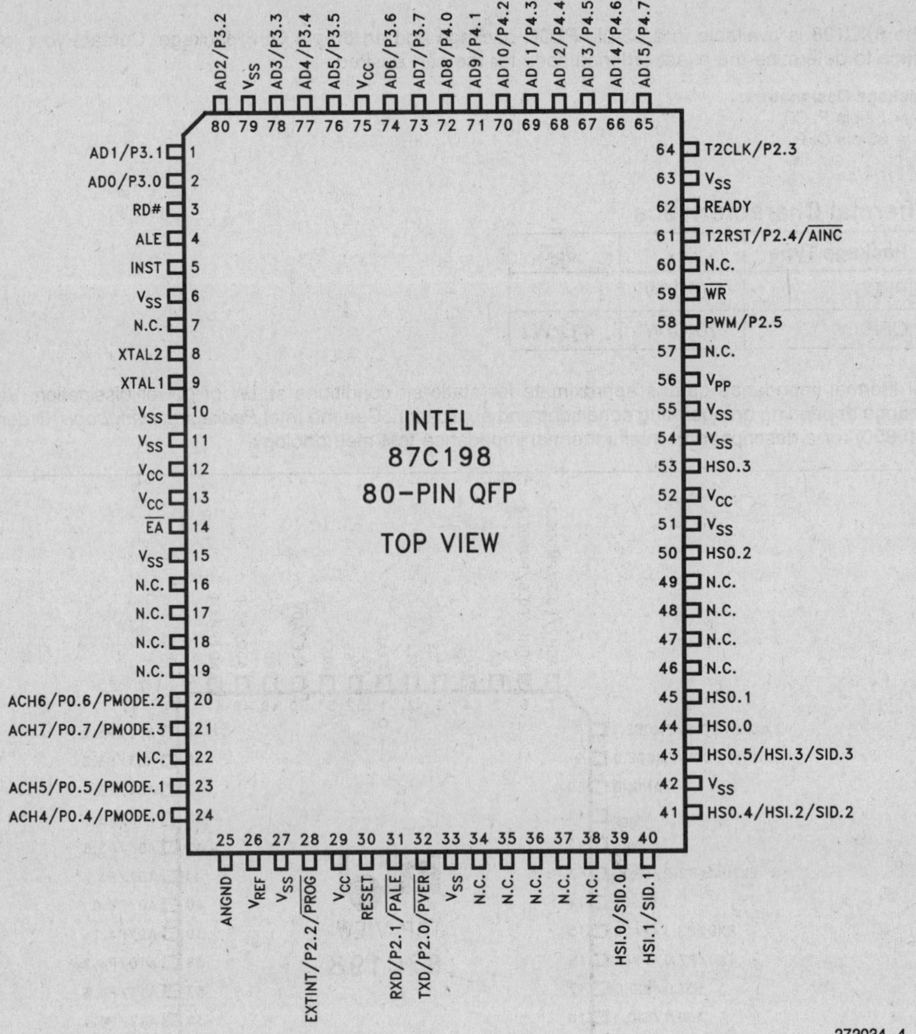

NOTE:
N.C. means No Connect (do not connect these pins).

Figure 5. 80-Pin QFP Package

NOTE:
The above pinout diagram applies to the OTP (87C198) device. The OTP device uses all of the programming pins shown above. The ROM (83C198) device only uses programming pins: **AINC, PALE, PMODE.n** and **PROG**. The ROMless (80C198) doesn't use any of the programming pins.

PIN DESCRIPTIONS

Symbol	Name and Function
V_{CC}	Main supply voltage (5V).
V_{SS}	The PLCC package has 5 V_{SS} pins and the QFP package has 12 V_{SS} pins. All must be connected to digital ground.
V_{REF}	Reference voltage for the A/D converter (5V). V_{REF} is also the supply voltage to the analog portion of the A/D converter and the logic used to read Port 0. Must be connected for A/D and Port 0 to function.
ANGND	Reference ground for the A/D converter. Must be held at nominally the same potential as V_{SS}.
V_{PP}	Programming Voltage. Also, timing pin for the return from powerdown circuit.
XTAL1	Input of the oscillator inverter and of the internal clock generator.
XTAL2	Output of the oscillator inverter.
$\overline{RESET}$	Reset input to and open-drain output from the chip. Input low for at least 4 state times to reset the chip. The subsequent low-to-high transition commences the 10-state Reset Sequence.
INST	Output high during an external memory read indicates the read is an instruction fetch. INST is valid throughout the bus cycle. INST is activated only during external memory accesses and output low for a data fetch.
$\overline{EA}$	Input for memory select (External Access). $\overline{EA}$ equal to a TTL-high causes memory accesses to locations 2000H through 3FFFH to be directed to on-chip ROM/EPROM. $\overline{EA}$ equal to a TTL-low causes accesses to these locations to be directed to off-chip memory.
ALE/$\overline{ADV}$	Address Latch Enable or Address Valid output, as selected by CCR. Both pin options provide a latch to demultiplex the address from the address/data bus. When the pin is $\overline{ADV}$, it goes inactive high at the end of the bus cycle. ALE/$\overline{ADV}$ is activated only during external memory accesses.
$\overline{RD}$	Read signal output to external memory. $\overline{RD}$ is activated only during external memory reads.
$\overline{WR}$	Write output to external memory. $\overline{WR}$ will go low for every external write.
READY	Ready input to lengthen external memory cycles. When the external memory is not being used, READY has no effect. Internal control of the number of wait states inserted into a bus cycle held not ready is available through configuration of CCR.
HSI	Inputs to High Speed Input Unit. Four HSI pins are available: HSI.0, HSI.1, HSI.2 and HSI.3. Two of them (HSI.2 and HSI.3) are shared with the HSO Unit.
HSO	Outputs from High Speed Output Unit. Six HSO pins are available: HSO.0, HSO.1, HSO.2, HSO.3, HSO.4 and HSO.5. Two of them (HSO.4 and HSO.5) are shared with the HSI Unit.
Port 0	4-bit high impedance input-only port. These pins can be used as digital inputs and/or as analog inputs to the on-chip A/D converter. These pins set the Programming Mode on the EPROM device.

PIN DESCRIPTIONS (Continued)

Symbol	Name and Function
Port 2	Multi-functional port. All of its pins are shared with other functions in the 80C198.
Ports 3 and 4	8-bit bidirectional I/O ports with open drain outputs. These pins are shared with the multiplexed address/data bus which has strong internal pullups. Available as I/O only on the ROM and EPROM devices.
TxD	The TxD pin is used for serial port transmission in Modes 1, 2 and 3. In mode 0 the pin is used as the serial clock output.
RxD	Serial Port Receive pin used for serial port reception. In mode 0 the pin functions as input or output data.
EXTINT	A positive transition on the EXTINT pin will generate an external interrupt.
T2CLK	The T2CLK pin is the Timer2 clock input or the serial port baud rate generator input.
T2RST	A rising edge on the T2RST pin will reset Timer2.
PWM	The PWM output.
PMODE	Programming Mode Select. Determines the EPROM programming algorithm that is performed. PMODE is sampled after a chip reset and should be static while the part is operating.
SID	Slave ID Number. Used to assign each slave a pin of Port 3 or 4 to use for passing programming verification acknowledgement.
$\overline{\text{PALE}}$	Programming ALE Input. Accepted by the 87C196KB when it is in Slave Programming Mode. Used to indicate that Ports 3 and 4 contain a command/address.
$\overline{\text{PROG}}$	Programming. Falling edge indicates valid data on PBUS and the beginning of programming. Rising edge indicates end of programming.
$\overline{\text{PVAL}}$	Program Valid. This signal indicates the success or failure of programming in the Auto Programming Mode. A zero indicates successful programming.
PVER	Program Verification. Used in Slave Programming and Auto CLB Programming Modes. Signal is low after rising edge of PROG if the programming was not successful.
$\overline{\text{AINC}}$	Auto Increment. Active low signal indicates that the auto increment mode is enabled. Auto Increment will allow reading or writing of sequential EPROM locations without address transactions across the PBUS for each read or write.
PORTS 3 and 4 (when programming)	Address/Command/Data Bus. Used to pass commands, addresses, and data to and from slave mode 87C196KBs. Used by chips in Auto Programming Mode to pass command, addresses and data to slaves. Also used in the Auto Programming Mode as a regular system bus to access external memory. Should have pullups to V_{CC} (15 kΩ).

ELECTRICAL CHARACTERISTICS

ABSOLUTE MAXIMUM RATINGS*

Ambient Temperature under Bias .... −55°C to +125°C

Storage Temperature .... −65°C to +150°C

Voltage on V_{PP} or $\overline{EA}$ to V_{SS} or ANGND .... −0.3V to +13.0V

Voltage on Any Other Pin to V_{SS} .. −0.5V to +7.0V

Power Dissipation(1) 1.5W

NOTE:
1. Power dissipation is based on package heat transfer limitations, not device power consumption.

NOTICE: This data sheet contains preliminary information on new products in production. It is valid for the devices indicated in the revision history. The specifications are subject to change without notice.

***WARNING:** *Stressing the device beyond the "Absolute Maximum Ratings" may cause permanent damage. These are stress ratings only. Operation beyond the "Operating Conditions" is not recommended and extended exposure beyond the "Operating Conditions" may affect device reliability.*

OPERATING CONDITIONS

(All characteristics in this data sheet apply to these operating conditions unless otherwise noted.)

Symbol	Description	Min	Max	Units
T_A	Ambient Temperature Under Bias	0	+70	°C
V_{CC}	Digital Supply Voltage	4.50	5.50	V
V_{REF}	Analog Supply Voltage	4.50	5.50	V
F_{OSC}	Oscillator Frequency **16 MHz**	3.5	16	MHz

NOTE:
ANGND and V_{SS} should be nominally at the same potential.

DC CHARACTERISTICS

Symbol	Description	Min	Max	Units	Test Conditions
V_{IL}	Input Low Voltage	−0.5	0.8	V	
V_{IH}	Input High Voltage (1)	0.2 V_{CC} + 0.9	V_{CC} + 0.5	V	
V_{IH1}	Input High Voltage on XTAL1	0.7 V_{CC}	V_{CC} + 0.5	V	
V_{IH2}	Input High Voltage on RESET	2.6	V_{CC} + 0.5	V	
V_{OL}	Output Low Voltage		0.3 0.45 1.5	V V V	I_{OL} = 200 μA I_{OL} = 32 mA I_{OL} = 7 mA
V_{OH}	Output High Voltage (Standard Outputs)	V_{CC} − 0.3 V_{CC} − 0.7 V_{CC} − 1.5		V V V	I_{OH} = −200 μA I_{OH} = −3.2 mA I_{OH} = −7 mA
I_{LI}	Input Leakage Current (Std. Inputs)		±10	μA	$0 < V_{IN} < V_{CC} - 0.3V$
I_{LI1}	Input Leakage Current (Port 0)		+3	μA	$0 < V_{IN} < V_{REF}$
I_{IL1}	Logical 0 Input Current in Reset (ALE, $\overline{RD}$, INST)		−6	mA	V_{IN} = 0.45 V
Hyst	Hysteresis on RESET Pin	300		mV	

NOTE:
1. All pins except RESET and XTAL1.

DC CHARACTERISTICS (Continued)

Symbol	Description	Min	Typ(6)	Max	Units	Test Conditions
I_{CC}	Active Mode Current in Reset		50	60	mA	XTAL1 = 16 MHz $V_{CC} = V_{PP} = V_{REF} = 5.5V$
I_{REF}	A/D Converter Reference Current		2	5	mA	
I_{IDLE}	Idle Mode Current		10	25	mA	
I_{CC1}	Active Mode Current		15	25	mA	XTAL1 = 3.5 MHz
I_{PD}	Powerdown Mode Current		5	30	μA	$V_{CC} = V_{PP} = V_{REF} = 5.5V$
R_{RST}	Reset Pullup Resistor	6K		50K	Ω	
C_S	Pin Capacitance (Any Pin to V_{SS})			10	pF	$F_{TEST} = 1.0$ MHz

NOTES:
(Notes apply to all specifications)
1. Standard Outputs include AD0–15, $\overline{RD}$, $\overline{WR}$, ALE, INST, HSO pins, PWM/P2.5, RESET, Ports 3 and 4, TXD/P2.0 and RXD (in serial mode 0). The V_{OH} specification is not valid for RESET. Ports 3 and 4 are open-drain outputs.
2. Standard Inputs include HSI pins, $\overline{EA}$, READY, RXD/P2.1, EXTINT/P2.2, T2CLK/P2.3 and T2RST/P2.4.
3. Maximum current per pin must be externally limited to the following values if V_{OL} is held above 0.45V or V_{OH} is held below $V_{CC} - 0.7V$:
   I_{OL} on Output pins: 10 mA
   I_{OH} on Standard Output pins: 10 mA
4. Maximum current per bus pin (data and control) during normal operation is ±3.2 mA.
5. During normal (non-transient) conditions the following total current limits apply:

HSO, P2.0, RXD, $\overline{RESET}$	I_{OL}: 29 mA	I_{OH}: 26 mA
P2.5, $\overline{WR}$	I_{OL}: 13 mA	I_{OH}: 11 mA
AD0–AD15	I_{OL}: 52 mA	I_{OH}: 52 mA
$\overline{RD}$, ALE, INST	I_{OL}: 13 mA	I_{OH}: 13 mA

6. Typicals are based on a limited number of samples and are not guaranteed. The values listed are at room temperature and $V_{REF} = V_{CC} = 5V$.

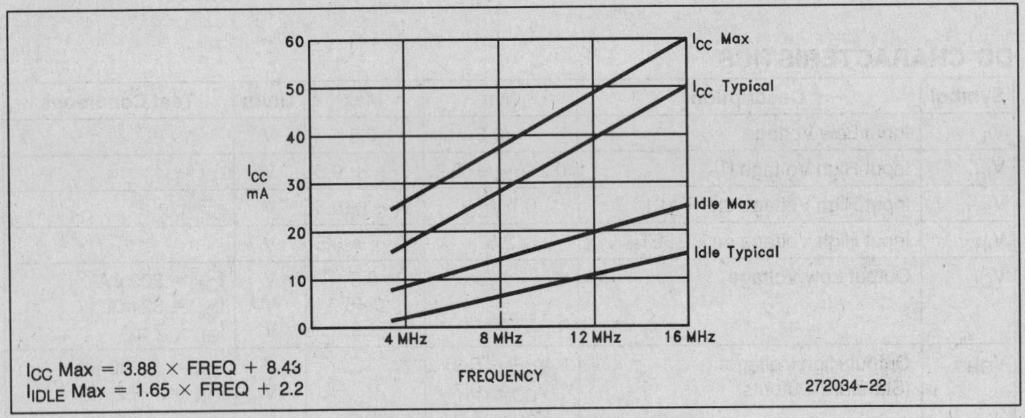

Figure 8. I_{CC} and I_{IDLE} vs Frequency

AC CHARACTERISTICS

Test Conditions: Capacitive load on all pins = 100 pF, Rise and fall times = 10 ns, F_{OSC} = 12/16 MHz

The system must meet these specifications to work with the 87C198:

Symbol	Description	Min	Max	Units	Notes
T_{AVYV}	Address Valid to Ready Setup		$2 T_{OSC} - 75$	ns	
T_{YLYH}	Non READY Time	No upper limit		ns	
T_{LLYX}	READY Hold after ALE Low	$T_{OSC} - 15$	$2 T_{OSC} - 40$	ns	(Note 1)
T_{AVDV}	Address Valid to Input Data Valid		$3 T_{OSC} - 55$	ns	(Note 2)
T_{RLDV}	$\overline{RD}$ Active to Input Data Valid		$T_{OSC} - 23$	ns	(Note 2)
T_{RHDZ}	End of $\overline{RD}$ to Input Data Float		$T_{OSC} - 20$	ns	
T_{RXDX}	Data Hold after $\overline{RD}$ Inactive	0		ns	

NOTES:
1. If max is exceeded, additional wait states will occur.
2. When using wait states, add $2 T_{OSC} \times n$, where n = number of wait states.

AC CHARACTERISTICS

Test Conditions: Capacitive load on all pins = 100 pF, Rise and fall times = 10 ns, F_{OSC} = 12/16 MHz

The 87C198 will meet these specifications:

Symbol	Description	Min	Max	Units	Notes
F_{XTAL}	Frequency on XTAL1 **12 MHz**	3.5	12	MHz	(Note 1)
F_{XTAL}	Frequency on XTAL1 **16 MHz**	3.5	16	MHz	(Note 1)
T_{OSC}	$1/F_{XTAL}$ **12 MHz**	83.3	286	ns	
T_{OSC}	$1/F_{XTAL}$ **16 MHz**	62.5	286	ns	
T_{LHLH}	ALE Cycle Time	4 T_{OSC}		ns	(Note 3)
T_{LHLL}	ALE High Period	T_{OSC} − 10	T_{OSC} + 10	ns	
T_{AVLL}	Address Setup to ALE Falling Edge	T_{OSC} − 20		ns	
T_{LLAX}	Address Hold after ALE Falling Edge	T_{OSC} − 40		ns	
T_{LLRL}	ALE Falling Edge to $\overline{RD}$ Falling Edge	T_{OSC} − 35		ns	
T_{RLRH}	$\overline{RD}$ Low Period	T_{OSC} − 5	T_{OSC} + 25	ns	(Note 3)
T_{RHLH}	$\overline{RD}$ Rising Edge to ALE Rising Edge	T_{OSC}	T_{OSC} + 25	ns	(Note 2)
T_{RLAZ}	$\overline{RD}$ Low to Address Float		5	ns	
T_{LLWL}	ALE Falling Edge to $\overline{WR}$ Falling Edge	T_{OSC} − 10		ns	
T_{QVWH}	Data Stable to $\overline{WR}$ Rising Edge	T_{OSC} − 23		ns	(Note 3)
T_{WLWH}	$\overline{WR}$ Low Period	T_{OSC} − 15	T_{OSC} + 5	ns	(Note 3)
T_{WHQX}	Data Hold after $\overline{WR}$ Rising Edge	T_{OSC} − 15		ns	
T_{WHLH}	$\overline{WR}$ Rising Edge to ALE Rising Edge	T_{OSC} − 15	T_{OSC} + 10	ns	(Note 2)
T_{WHBX}	INST Hold after $\overline{WR}$ Rising Edge	T_{OSC} − 15		ns	
T_{LLBX}	INST Hold after ALE Rising Edge	T_{OSC} − 10		ns	
T_{RHBX}	INST Hold after $\overline{RD}$ Rising Edge	T_{OSC} − 10		ns	
T_{WHAX}	AD8–15 Hold after $\overline{WR}$ Rising Edge	T_{OSC} − 30		ns	
T_{RHAX}	AD8–15 Hold after $\overline{RD}$ Rising Edge	T_{OSC} − 25		ns	

NOTES:
1. Testing performed at 3.5 MHz. However, the part is static by design and will typically operate below 1 Hz.
2. Assuming back-to-back bus cycles.
3. When using wait states, add 2 T_{OSC} × n, where n = number of wait states.

System Bus Timings

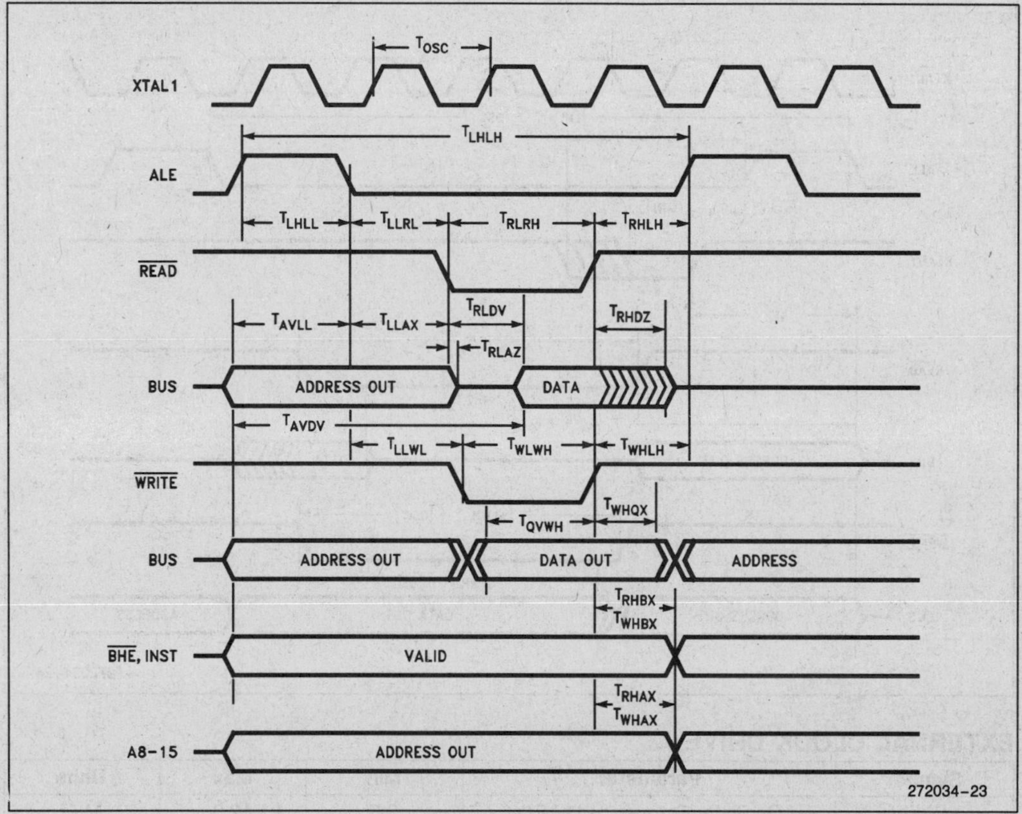

4

READY Timings (One Wait State)

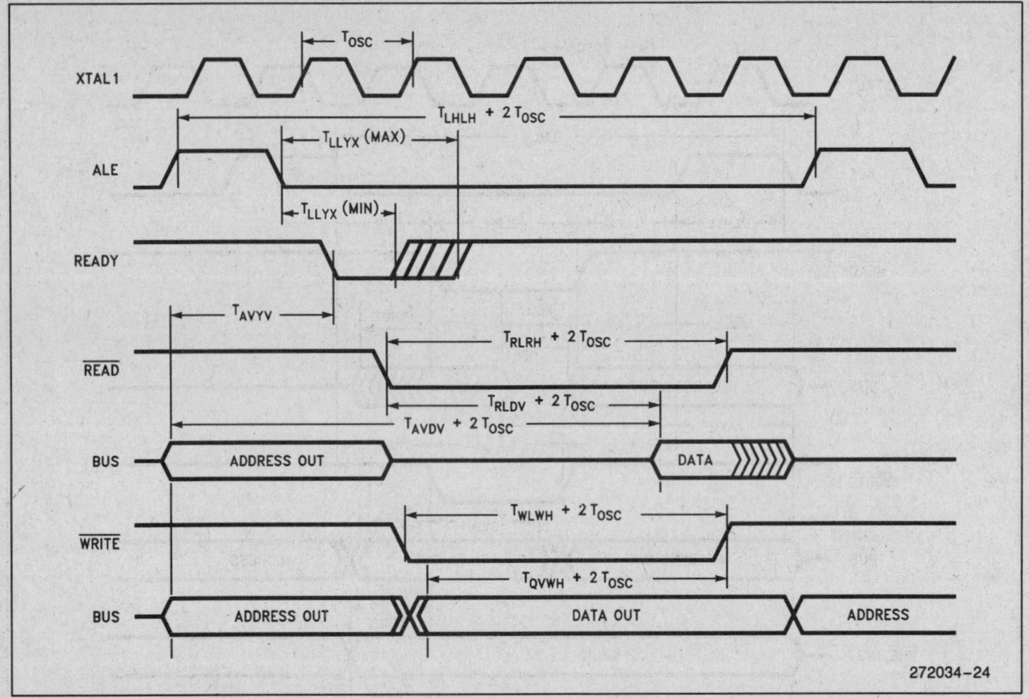

272034-24

EXTERNAL CLOCK DRIVE

Symbol	Parameter	Min	Max	Units
$1/T_{XLXL}$	Oscillator Frequency **12 MHz**	3.5	12.0	MHz
$1/T_{XLXL}$	Oscillator Frequency **16 MHz**	3.5	16.0	MHz
T_{XLXL}	Oscillator Period **12 MHz**	83.3	286	ns
T_{XLXL}	Oscillator Period **16 MHz**	62.5	286	ns
T_{XHXX}	High Time	21.25		ns
T_{XLXX}	Low Time	21.25		ns
T_{XLXH}	Rise Time		10	ns
T_{XHXL}	Fall Time		10	ns

EXTERNAL CLOCK DRIVE WAVEFORMS

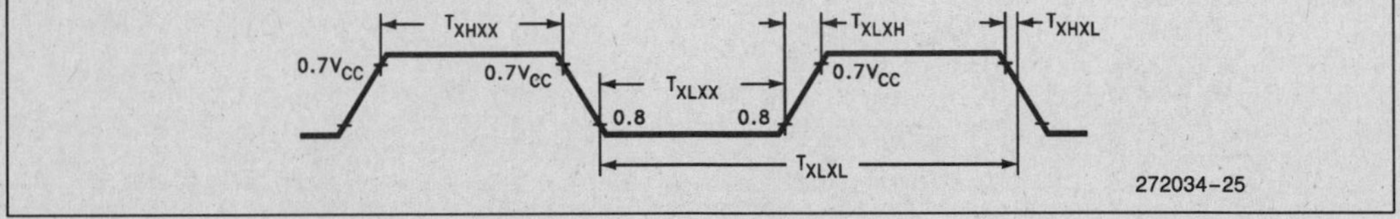

272034-25

An external oscillator may encounter as much as a 100 pF load at XTAL1 when it starts-up. This is due to interaction between the amplifier and its feedback capacitance. Once the external signal meets the V_{IL} and V_{IH} specifications the capacitance will not exceed 20 pF.

EXTERNAL CRYSTAL CONNECTIONS

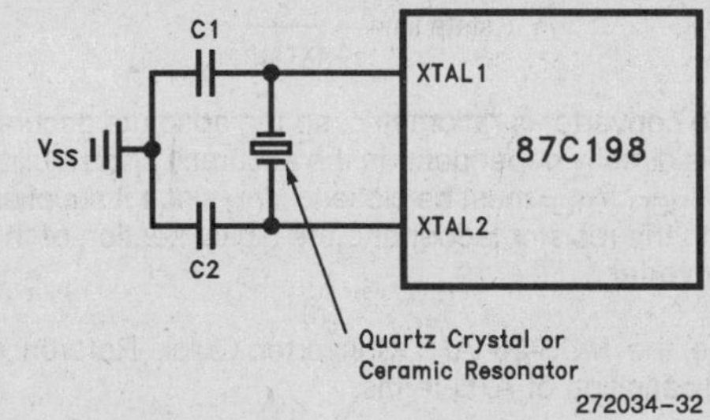

272034-32

NOTE:
Keep oscillator components close to chip and use short direct traces to XTAL1, XTAL2 and V_{SS}. When using crystals, C1 = 20 pF, C2 = 20 pF. When using ceramic resonators consult manufacturer for recommended capacitor values.

EXTERNAL CLOCK CONNECTIONS

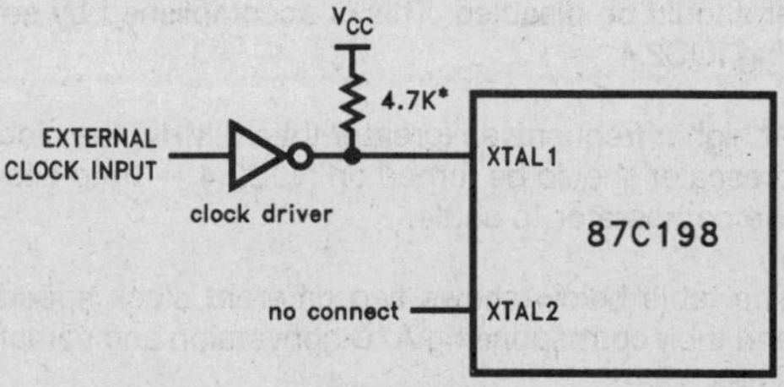

272034-33

NOTE:
*Required if open collector TTL driver used. Not needed if CMOS driver is used.

AC TESTING INPUT, OUTPUT WAVEFORMS

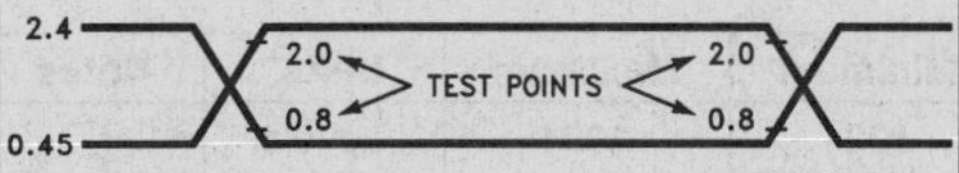

272034-26

AC Testing inputs are driven at 2.4V for a Logic "1" and 0.45V for a Logic "0" Timing measurements are made at 2.0V for a Logic "1" and 0.8V for a Logic "0".

FLOAT WAVEFORMS

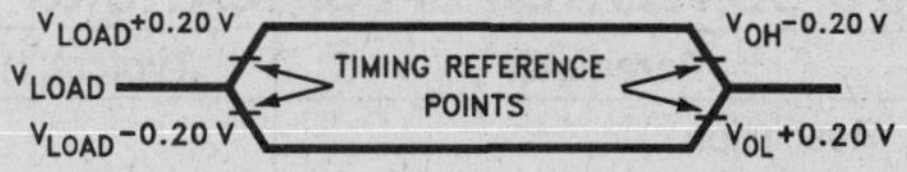

272034-27

For Timing Purposes a Port Pin is no Longer Floating when a 200 mV change from Load Voltage Occurs and Begins to Float when a 200 mV change from the Loaded V_{OH}/V_{OL} Level occurs I_{OL}/I_{OH} = ±15 mA.

EXPLANATION OF AC SYMBOLS

Each symbol is two pairs of letters prefixed by "T" for time. The characters in a pair indicate a signal and its condition, respectively. Symbols represent the time between the two signal/condition points.

Conditions:

H - High
L - Low
V - Valid
X - No Longer Valid
Z - Floating

Signals:

A - Address
D - DATA IN
L - ALE/$\overline{ADV}$
Q - DATA OUT
R - $\overline{RD}$
W - $\overline{WR}$
X - XTAL1
Y - READY

10-BIT AID CHARACTERISTICS

At a clock speed of 6 MHz or less, the clock prescaler should be disabled. This is accomplished by setting IOC2.4 = 1.

At higher frequencies (greater than 6 MHz) the clock prescaler should be turned on (IOC2.4 = 0) to allow the comparator to settle.

The table below shows two different clock speeds and their corresponding A/D conversion and sample times.

State times are calculated as follows:

$$\text{state time} = \frac{2}{f_{XTAL1}}$$

The converter is ratiometric, so the absolute accuracy is directly dependent on the accuracy and stability of V_{REF}. V_{REF} must be close to V_{CC} since it supplies both the resistor ladder and the digital section of the converter.

See the MCS-96 A/D Converter Quick Reference for definition of A/D terms.

Example Sample and Conversion Times

AID Clock Prescaler	Clock Speed (MHz)	Sample Time (States)	Sample Time at Clock Speed (μs)	Conversion Time (States)	Conversion Time at Clock Speed (μs)
IOC2.4=0 → ON	16	15	1.875	156.5	19.6
IOC2.4=1 → OFF	6	8	2.667	89.5	29.8

A/D CONVERTER SPECIFICATIONS

Parameter	Typical(1)	Minimum	Maximum	Units*	Notes
Resolution		1024 10	1024 10	Levels Bits	
Absolute Error		0	±3	LSBs	
Full Scale Error	0.25 ±0.50			LSBs	
Zero Offset Error	−0.25 ±0.50			LSBs	
Non-Linearity Error	1.5 ±2.5	0	±3	LSBs	
Differential Non-Linearity Error		> −1	+2	LSBs	
Channel-to-Channel Matching	±0.1	0	±1	LSBs	
Repeatability	±0.25			LSBs	
Temperature Coefficients: Offset Full Scale Differential Non-Linearity	 0.009 0.009 0.009			 LSB/°C LSB/°C LSB/°C	
Off Isolation		−60		dB	2, 3
Feedthrough	−60			dB	2
V_{CC} Power Supply Rejection	−60			dB	2
Input Series Resistance		750	1.2K	Ω	4
DC Input Leakage		0	3.0	μA	
Sample Time: Prescaler On Prescaler Off	15 8			States States	
Sampling Capacitor	3			pF	

NOTES:
*An "LSB", as used here, has a value of approximately 5 mV.
1. Typical values are expected for most devices at 25°C but are not tested or guaranteed.
2. DC to 100 KHz.
3. Multiplexer Break-Before-Make Guaranteed.
4. Resistance from device pin, through internal MUX, to sample capacitor.

EPROM SPECIFICATIONS

EPROM PROGRAMMING OPERATING CONDITIONS

Symbol	Parameter	Min	Max	Units
T_A	Ambient Temperature during Programming	20	30	°C
V_{CC}, V_{PD}, V_{REF}(1)	Supply Voltages during Programming	4.5	5.5	V
V_{EA}	Programming Mode Supply Voltage	12.50	13.0	V(2)
V_{PP}	EPROM Programming Supply Voltage	12.50	13.0	V(2)
V_{SS}, ANGND(3)	Digital and Analog Ground	0	0	V
F_{OSC}	Oscillator Frequency **16 MHz**	6.0	16.0	MHz

NOTES:
1. V_{CC}, V_{PD} and V_{REF} should nominally be at the same voltage during programming.
2. V_{EA} and V_{PP} must never exceed the maximum voltage for any amount of time or the device may be damaged.
3. V_{SS} and ANGND should nominally be at the same voltage (0V) during programming.

AC EPROM PROGRAMMING CHARACTERISTICS

Symbol	Description	Min	Max	Units
T_{SHLL}	Reset High to First $\overline{PALE}$ Low	1100		T_{OSC}
T_{LLLH}	$\overline{PALE}$ Pulse Width	40		T_{OSC}
T_{AVLL}	Address Setup Time	0		T_{OSC}
T_{LLAX}	Address Hold Time	50		T_{OSC}
T_{LLVL}	$\overline{PALE}$ Low to PVER Low		60	T_{OSC}
T_{PLDV}	$\overline{PROG}$ Low to Word Dump Valid		50	T_{OSC}
T_{PHDX}	Word Dump Data Hold		50	T_{OSC}
T_{DVPL}	Data Setup Time	0		T_{OSC}
T_{PLDX}	Data Hold Time	50		T_{OSC}
T_{PLPH}	$\overline{PROG}$ Pulse Width	40		T_{OSC}
T_{PHLL}	$\overline{PROG}$ High to Next $\overline{PALE}$ Low	120		T_{OSC}
T_{LHPL}	$\overline{PALE}$ High to $\overline{PROG}$ Low	220		T_{OSC}
T_{PHPL}	$\overline{PROG}$ High to Next $\overline{PROG}$ Low	120		T_{OSC}
T_{PHIL}	$\overline{PROG}$ High to AINC Low	0		T_{OSC}
T_{ILIH}	$\overline{AINC}$ Pulse Width	40		T_{OSC}
T_{ILVH}	PVER Hold after $\overline{AINC}$ Low	50		T_{OSC}
T_{ILPL}	$\overline{AINC}$ Low to $\overline{PROG}$ Low	170		T_{OSC}
T_{PHVL}	$\overline{PROG}$ High to PVER Low		90	T_{OSC}

DC EPROM PROGRAMMING CHARACTERISTICS

Symbol	Description	Min	Max	Units
I_{PP}	V_{PP} Supply Current (When Programming)		100	mA

EPROM PROGRAMMING WAVEFORMS

SLAVE PROGRAMMING MODE DATA PROGRAM MODE WITH SINGLE PROGRAM PULSE

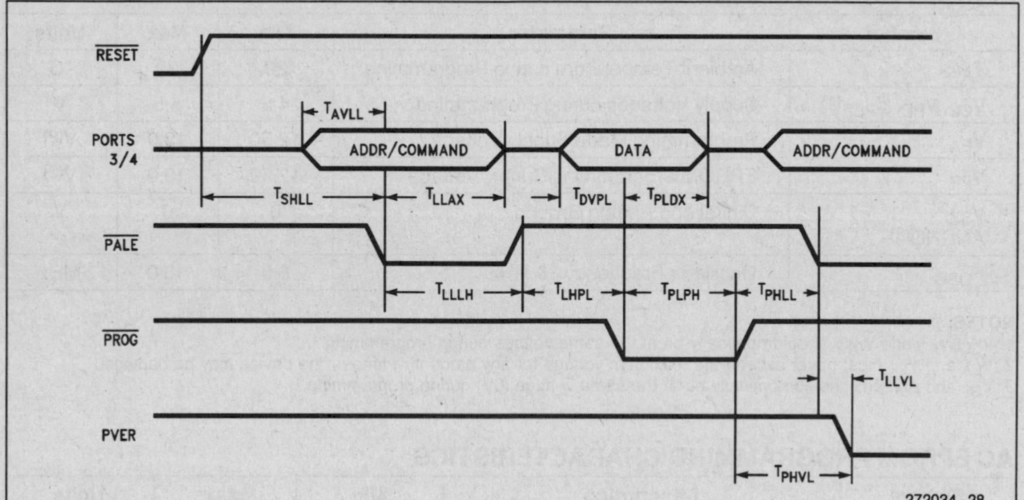

272034-28

SLAVE PROGRAMMING MODE IN WORD DUMP OR DATA VERIFY MODE WITH AUTO INCREMENT

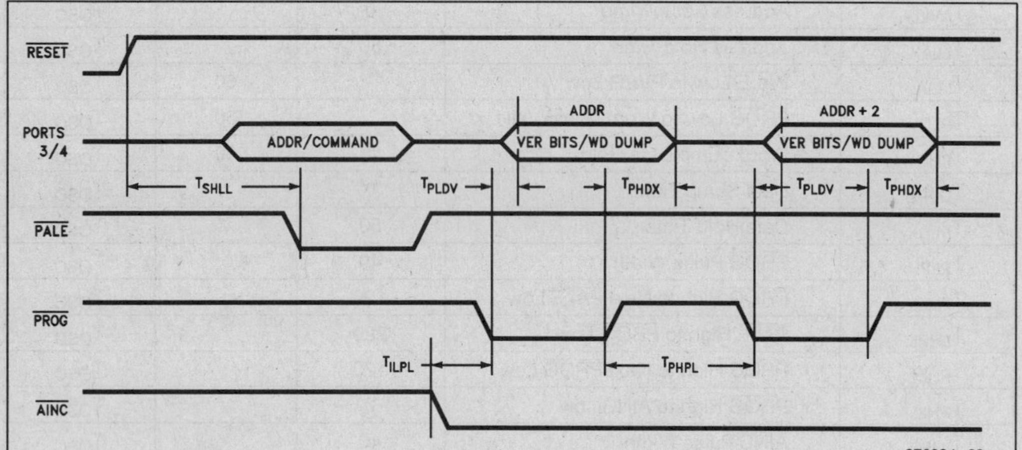

272034-29

SLAVE PROGRAMMING MODE TIMING IN DATA PROGRAM MODE WITH REPEATED PROG PULSE AND AUTO INCREMENT

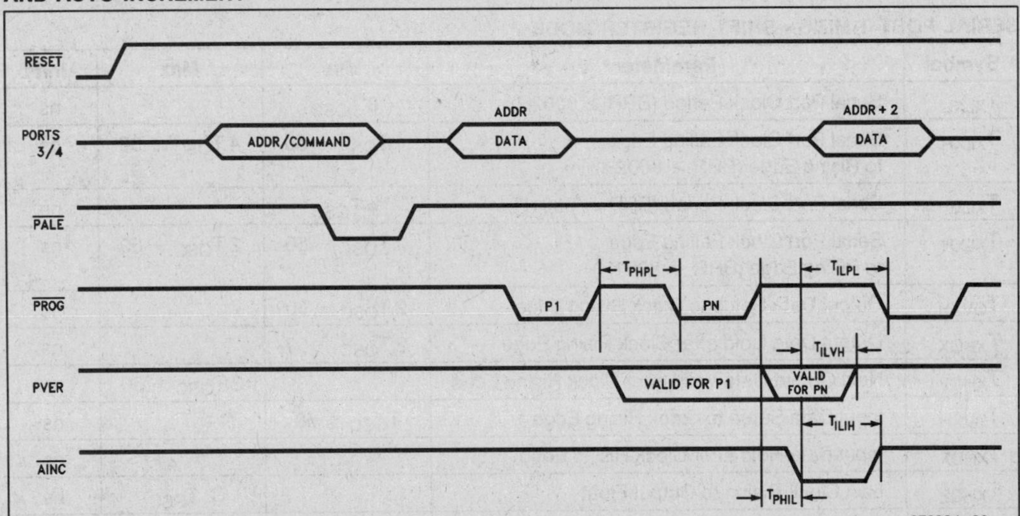

272034–30

AC CHARACTERISTICS—SERIAL PORT—SHIFT REGISTER MODE

SERIAL PORT TIMING—SHIFT REGISTER MODE

Symbol	Parameter	Min	Max	Units
T_{XLXL}	Serial Port Clock Period (BRR ≥ 8002H)	6 T_{OSC}		ns
T_{XLXH}	Serial Port Clock Falling Edge to Rising Edge (BRR ≥ 8002H)	4 T_{OSC} − 50	4 T_{OSC} + 50	ns
T_{XLXL}	Serial Port Clock Period (BRR = 8001H)	4 T_{OSC}		ns
T_{XLXH}	Serial Port Clock Falling Edge to Rising Edge (BRR = 8001H)	2 T_{OSC} − 50	2 T_{OSC} + 50	ns
T_{QVXH}	Output Data Setup to Clock Rising Edge	2 T_{OSC} − 50		ns
T_{XHQX}	Output Data Hold after Clock Rising Edge	2 T_{OSC} − 50		ns
T_{XHQV}	Next Output Data Valid after Clock Rising Edge		2 T_{OSC} + 50	ns
T_{DVXH}	Input Data Setup to Clock Rising Edge	T_{OSC} + 50		ns
T_{XHDX}	Input Data Hold after Clock Rising Edge	0		ns
T_{XHQZ}	Last Clock Rising to Output Float		2 T_{OSC}	ns

WAVEFORM—SERIAL PORT—SHIFT REGISTER MODE

SERIAL PORT WAVEFORM—SHIFT REGISTER MODE

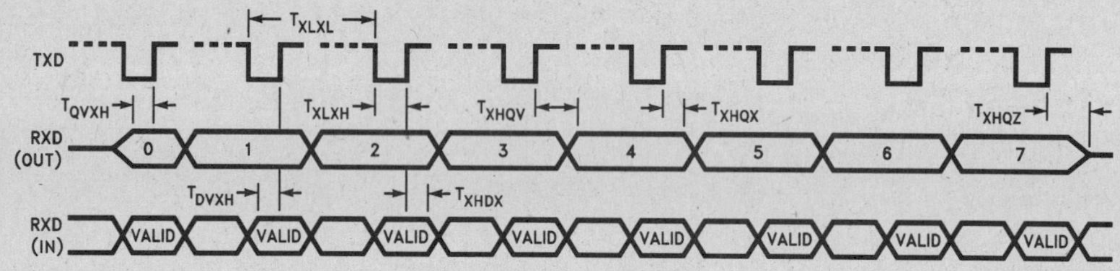

272034-31

FUNCTIONAL DEVIATIONS

Devices marked with an "E", "F", or "G" have the following errata.

1. HIGH SPEED INPUTS

The High Speed Input (HSI) has three deviations from the specifications.

NOTE:
"Events" are defined as one or more pin transitions. "Entries" are defined as the recording of one or more events.

A. The resolution is nine states instead of eight states. Events occurring on the same pin more frequently than once every nine states may be lost.

B. A mismatch between the nine state HSI resolution and the eight state hardware timer causes one time-tag value to be skipped every nine timer counts. Events may receive a time-tag one count later than expected.

C. If the FIFO and Holding Register are empty, the first event will transfer into the Holding Register, leaving the FIFO empty again. The next event that occurs will be the first event loaded into the empty FIFO. If the first two events into an empty FIFO (not counting the Holding Register) occur coincident with each other, both are recorded as one entry with one time-tag. If the second event occurs within 9 states after the first, the events will be entered separately with time-tags at least one count apart. If the second event enters the FIFO coincident with the "skipped" time-tag situation (see B above) the time-tags will be at least two counts apart.

2. CMPL with R0

Using CMPL with register 0 can set incorrect flags. Don't use register 0 with the compare long instruction. Use another long word register and set it equal to zero. See Techbit MC0692.

REVISION HISTORY

This data sheet (272034-003) is valid for devices marked with an "E", "F", or "G" at the end of the top side tracking number. Data sheets are changed as new device information becomes available. Verify with your local Intel sales office that you have the latest version before finalizing a design or ordering devices.

The following differences exist between this data sheet and the previous version (-002).

1. This data sheet added the ROMless and ROM devices 80C198 and 83C198 respectively.
2. The description of the A/D converter prescalar bit was improved.

intel®

8XC196KC/8XC196KC20 COMMERCIAL/EXPRESS CHMOS MICROCONTROLLER

87C196KC—16 Kbytes of On-Chip OTPROM
83C196KC—16 Kbytes ROM
80C196KC—ROMless

- 16 and 20 MHz Available
- 488 Byte Register RAM
- Register-to-Register Architecture
- 28 Interrupt Sources/16 Vectors
- Peripheral Transaction Server
- 1.4 μs 16 x 16 Multiply (20 MHz)
- 2.4 μs 32/16 Divide (20 MHz)
- Powerdown and Idle Modes
- Five 8-Bit I/O Ports
- 16-Bit Watchdog Timer
- Extended Temperature Available
- Dynamically Configurable 8-Bit or 16-Bit Buswidth
- Full Duplex Serial Port
- High Speed I/O Subsystem
- 16-Bit Timer
- 16-Bit Up/Down Counter with Capture
- 3 Pulse-Width-Modulated Outputs
- Four 16-Bit Software Timers
- 8- or 10-Bit A/D Converter with Sample/Hold
- $\overline{HOLD}/\overline{HLDA}$ Bus Protocol
- OTPROM One-Time Programmable Version

The 80C196KC 16-bit microcontroller is a high performance member of the MCS® 96 microcontroller family. The 80C196KC is an enhanced 80C196KB device with 488 bytes RAM, 16 and 20 MHz operation and an optional 16 Kbytes of ROM/OTPROM. Intel's CHMOS III process provides a high performance processor along with low power consumption.

The 87C196KC is an 80C196KC with 16 Kbytes on-chip OTPROM. The 83C196KC is an 80C196KC with 16 Kbytes factory programmed ROM. In this document, the 80C196KC will refer to all products unless otherwise stated.

Four high-speed capture inputs are provided to record times when events occur. Six high-speed outputs are available for pulse or waveform generation. The high-speed output can also generate four software timers or start an A/D conversion. Events can be based on the timer or up/down counter.

With the commercial (standard) temperature option, operational characteristics are guaranteed over the temperature range of 0°C to +70°C. With the extended (Express) temperature range option, operational characteristics are guaranteed over the temperature range of −40°C to +85°C. Unless otherwise noted, the specifications are the same for both options.

See the Packaging information for extended temperature designators.

November 1994
Order Number: 270942-005

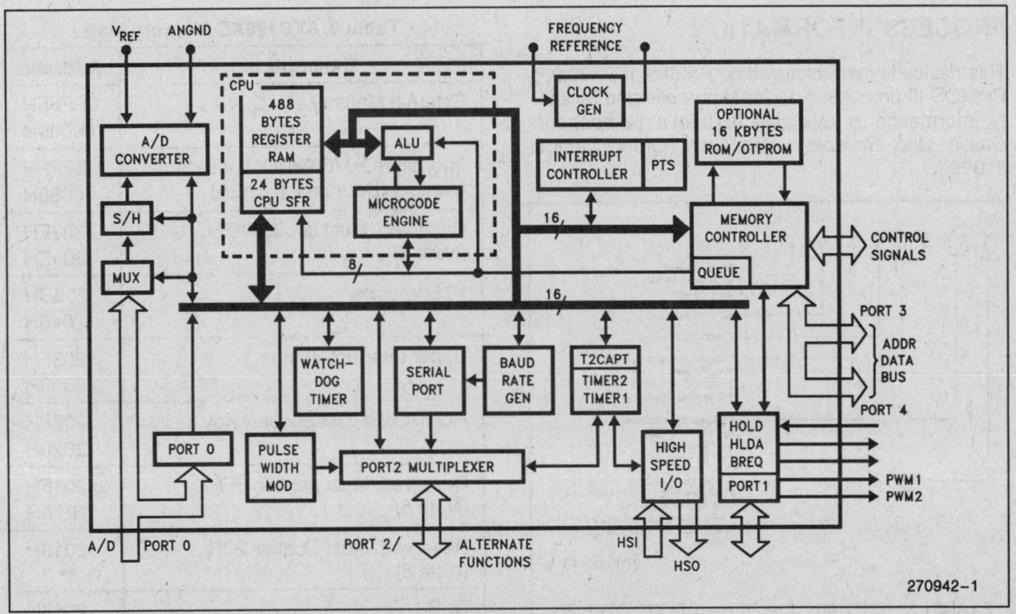

Figure 1. 8XC196KC Block Diagram

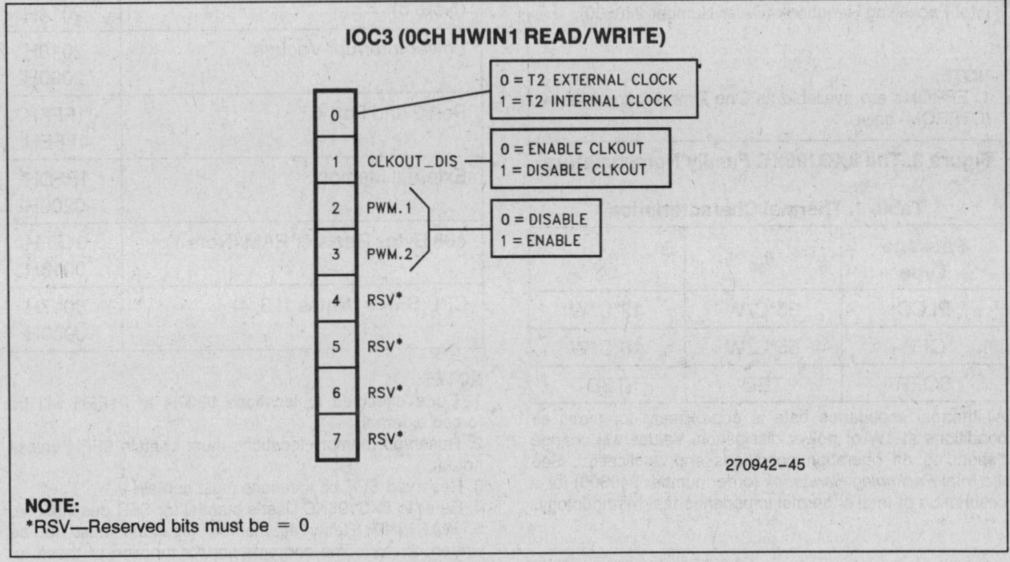

Figure 2. 8XC196KC New SFR Bit (CLKOUT Disable)

PROCESS INFORMATION

This device is manufactured on PX29.5 or PX29.9, a CHMOS III process. Additional process and reliability information is available in Intel's *Components Quality and Reliability Handbook,* Order Number 210997.

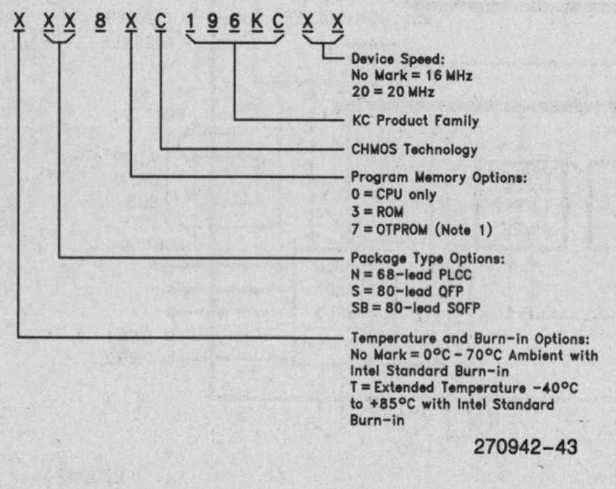

EXAMPLE: N87C196KC is 68-Lead PLCC OTPROM, 16 MHz.
For complete package dimensional data, refer to the Intel Packaging Handbook (Order Number 240800).

NOTE:
1. EPROMs are available as One Time Programmable (OTPROM) only.

Figure 3. The 8XC196KC Family Nomenclature

Table 1. Thermal Characteristics

Package Type	θ_{ja}	θ_{jc}
PLCC	35°C/W	13°C/W
QFP	55°C/W	16°C/W
SQFP	TBD	TBD

All thermal impedance data is approximate for static air conditions at 1W of power dissipation. Values will change depending on operation conditions and application. See the Intel *Packaging Handbook* (order number 240800) for a description of Intel's thermal impedance test methodology.

Table 2. 8XC196KC Memory Map

Description	Address
External Memory or I/O	0FFFFH 06000H
Internal ROM/OTPROM or External Memory (Determined by $\overline{EA}$)	5FFFH 2080H
Reserved. Must contain FFH. (Note 5)	207FH 205EH
PTS Vectors	205DH 2040H
Upper Interrupt Vectors	203FH 2030H
ROM/OTPROM Security Key	202FH 2020H
Reserved. Must contain FFH. (Note 5)	201FH 201AH
Reserved. Must Contain 20H (Note 5)	2019H
CCB	2018H
Reserved. Must contain FFH. (Note 5)	2017H 2014H
Lower Interrupt Vectors	2013H 2000H
Port 3 and Port 4	1FFFH 1FFEH
External Memory	1FFDH 0200H
488 Bytes Register RAM (Note 1)	01FFH 0018H
CPU SFR's (Notes 1, 3, 4)	0017H 0000H

NOTES:
1. Code executed in locations 0000H to 01FFH will be forced external.
2. Reserved memory locations must contain 0FFH unless noted.
3. Reserved SFR bit locations must contain 0.
4. Refer to 8XC196KC User's manual for SFR descriptions.
5. **WARNING:** Reserved memory locations must not be written or read. The contents and/or function of these locations may change with future revisions of the device. Therefore, a program that relies on one or more of these locations may not function properly.

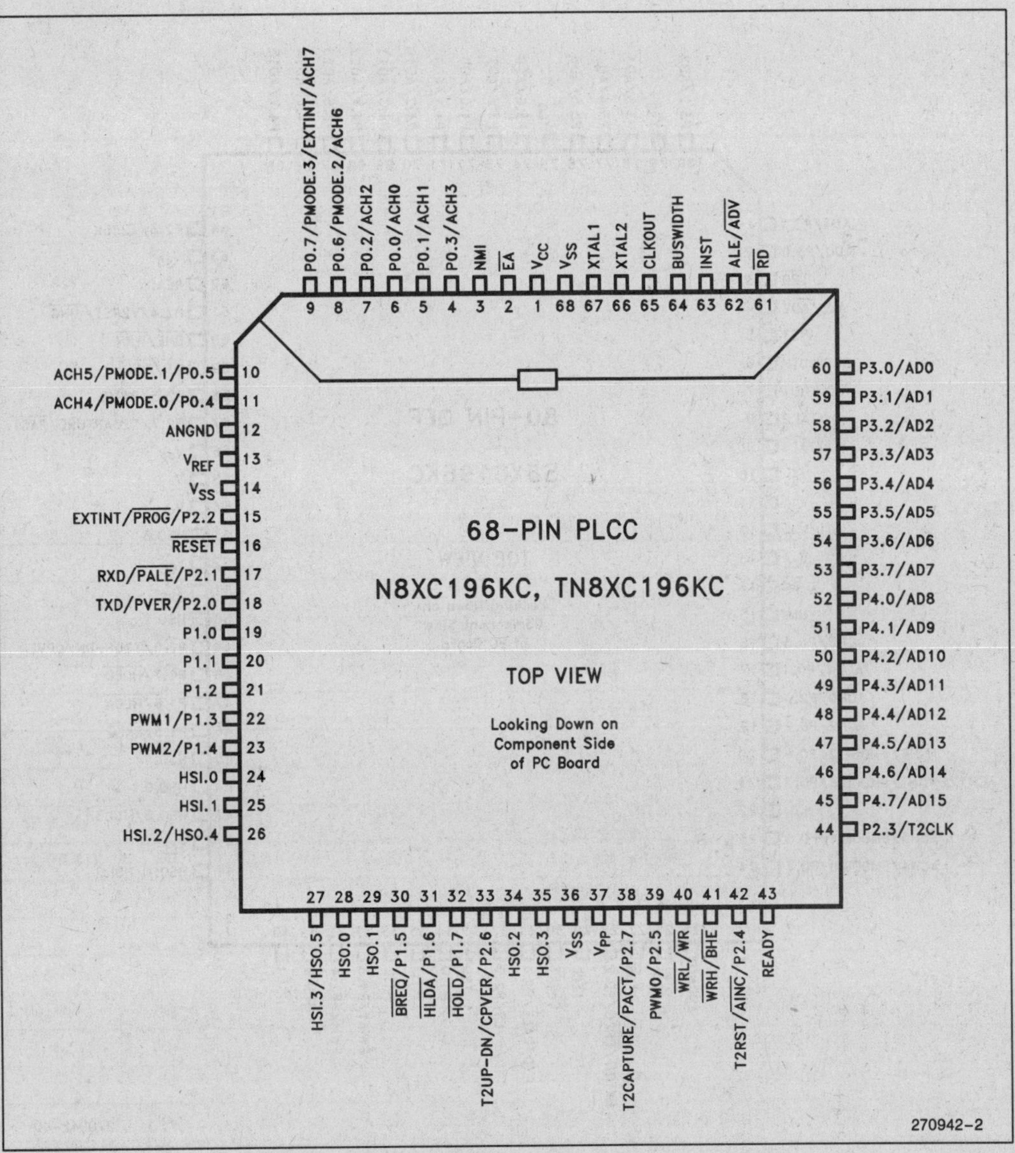

Figure 4. 68-Lead PLCC Package

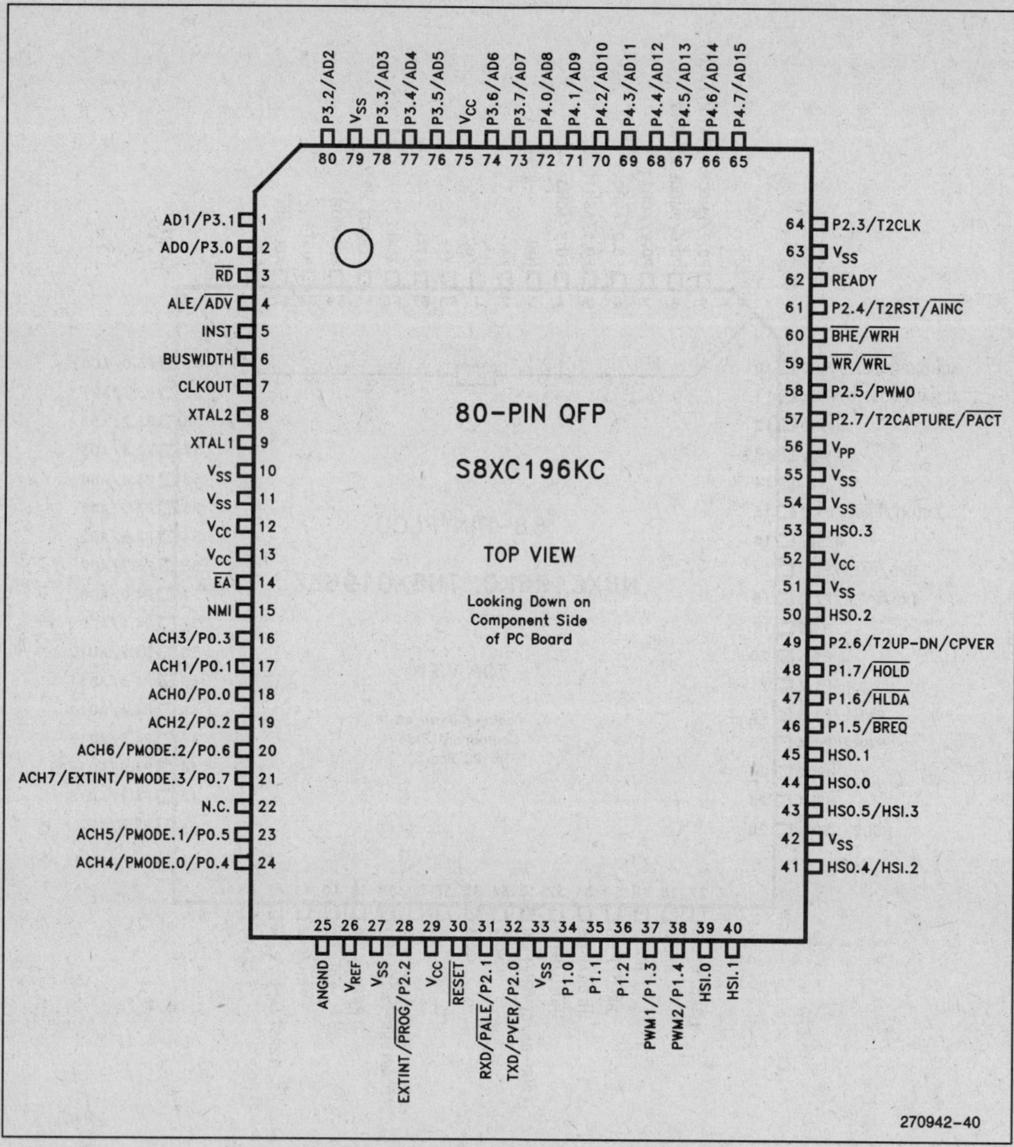

Figure 5. S8XC196KC 80-Pin QFP Package

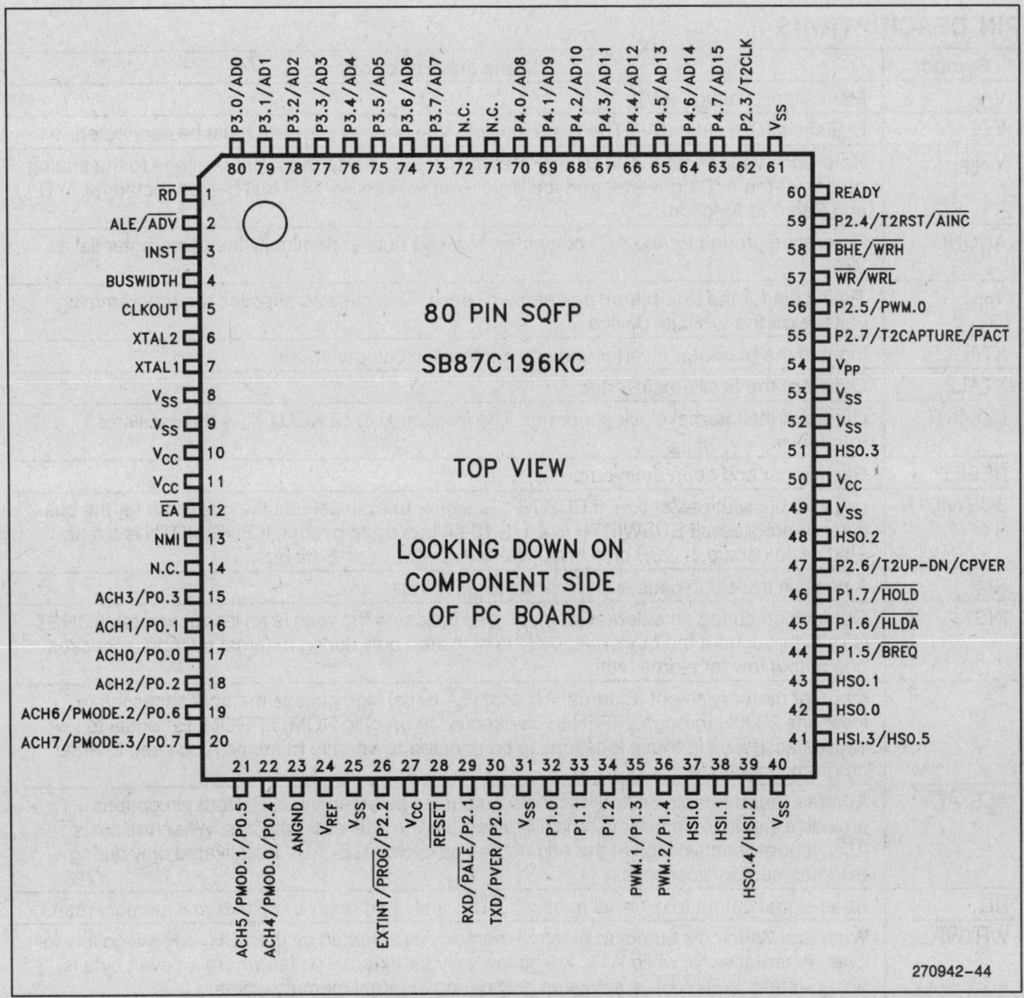

Figure 6. 80-Pin SQFP Package

4

PIN DESCRIPTIONS

Symbol	Name and Function
V_{CC}	Main supply voltage (5V).
V_{SS}	Digital circuit ground (0V). There are multiple V_{SS} pins, all of which must be connected.
V_{REF}	Reference voltage for the A/D converter (5V). V_{REF} is also the supply voltage to the analog portion of the A/D converter and the logic used to read Port 0. Must be connected for A/D and Port 0 to function.
ANGND	Reference ground for the A/D converter. Must be held at nominally the same potential as V_{SS}.
V_{PP}	Timing pin for the return from powerdown circuit. This pin also supplies the programming voltage on the EPROM device.
XTAL1	Input of the oscillator inverter and of the internal clock generator.
XTAL2	Output of the oscillator inverter.
CLKOUT	Output of the internal clock generator. The frequency of CLKOUT is ½ the oscillator frequency.
$\overline{RESET}$	Reset input and open drain output.
BUSWIDTH	Input for buswidth selection. If CCR bit 1 is a one, this pin selects the bus width for the bus cycle in progress. If BUSWIDTH is a 1, a 16-bit bus cycle occurs. If BUSWIDTH is a 0 an 8-bit cycle occurs. If CCR bit 1 is a 0, the bus is always an 8-bit bus.
NMI	A positive transition causes a vector through 203EH.
INST	Output high during an external memory read indicates the read is an instruction fetch. INST is valid throughout the bus cycle. INST is activated only during external memory accesses and output low for a data fetch.
$\overline{EA}$	Input for memory select (External Access). $\overline{EA}$ equal high causes memory accesses to locations 2000H through 5FFFH to be directed to on-chip ROM/EPROM. $\overline{EA}$ equal to low causes accesses to those locations to be directed to off-chip memory. Also used to enter programming mode.
ALE/$\overline{ADV}$	Address Latch Enable or Address Valid output, as selected by CCR. Both pin options provide a signal to demultiplex the address from the address/data bus. When the pin is $\overline{ADV}$, it goes inactive high at the end of the bus cycle. ALE/$\overline{ADV}$ is activated only during external memory accesses.
$\overline{RD}$	Read signal output to external memory. $\overline{RD}$ is activated only during external memory reads.
$\overline{WR}$/$\overline{WRL}$	Write and Write Low output to external memory, as selected by the CCR. $\overline{WR}$ will go low for every external write, while $\overline{WRL}$ will go low only for external writes where an even byte is being written. $\overline{WR}$/$\overline{WRL}$ is activated only during external memory writes.
$\overline{BHE}$/$\overline{WRH}$	Bus High Enable or Write High output to external memory, as selected by the CCR. $\overline{BHE}$ will go low for external writes to the high byte of the data bus. $\overline{WRH}$ will go low for external writes where an odd byte is being written. $\overline{BHE}$/$\overline{WRH}$ is activated only during external memory writes.
READY	Ready input to lengthen external memory cycles, for interfacing to slow or dynamic memory, or for bus sharing. When the external memory is not being used, READY has no effect.
HSI	Inputs to High Speed Input Unit. Four HSI pins are available: HSI.0, HSI.1, HSI.2 and HSI.3. Two of them (HSI.2 and HSI.3) are shared with the HSO Unit.
HSO	Outputs from High Speed Output Unit. Six HSO pins are available: HSO.0, HSO.1, HSO.2, HSI.3, HSO.4 and HSO.5. Two of them (HSO.4 and HSO.5) are shared with the HSI Unit.
Port 0	8-bit high impedance input-only port. These pins can be used as digital inputs and/or as analog inputs to the on-chip A/D converter.
Port 1	8-bit quasi-bidirectional I/O port.
Port 2	8-bit multi-functional port. All of its pins are shared with other functions in the 80C196KC. Pins 2.6 and 2.7 are quasi-bidirectionai.

PIN DESCRIPTIONS (Continued)

Symbol	Name and Function
Ports 3 and 4	8-bit bidirectional I/O ports with open drain outputs. These pins are shared with the multiplexed address/data bus which has strong internal pullups.
$\overline{HOLD}$	Bus Hold input requesting control of the bus.
$\overline{HLDA}$	Bus Hold acknowledge output indicating release of the bus.
$\overline{BREQ}$	Bus Request output activated when the bus controller has a pending external memory cycle.
PMODE	Determines the EPROM programming mode.
$\overline{PACT}$	A low signal in Auto Programming mode indicates that programming is in process. A high signal indicates programming is complete.
CPVER	Cummulative Program Output Verification. Pin is high if all locations have programmed correctly since entering a programming mode.
$\overline{PALE}$	A falling edge in Slave Programming Mode and Auto Configuration Byte Programming Mode indicates that ports 3 and 4 contain valid programming address/command information (input to slave).
$\overline{PROG}$	A falling edge in Slave Programming Mode indicates that ports 3 and 4 contain valid programming data (input to slave).
PVER	A high signal in Slave Programmig Mode and Auto Configuration Byte Programming Mode indicates the byte programmed correctly.
$\overline{AINC}$	Auto Increment. Active low input signal indicates that the auto increment mode is enabled. Auto Increment will allow reading or writing of sequential EPROM locations without address transactions across the PBUS for each read or write.

ELECTRICAL CHARACTERISTICS
ABSOLUTE MAXIMUM RATINGS*

Ambient Temperature
Under Bias .... −55°C to +125°C

Storage Temperature .... −65°C to +150°C

Voltage On Any Pin to V_{SS}...... −0.5V to +7.0V(1)

Voltage from $\overline{EA}$ or
V_{PP} to V_{SS} or ANGND................ +13.00V

Power Dissipation ....1.5W(2)

NOTE:
1. This includes V_{PP} and $\overline{EA}$ on ROM or CPU only devices.
2. Power dissipation is based on package heat transfer limitations, not device power consumption.

NOTICE: This is a production data sheet. It is valid for the devices indicated in the revision history. The specifications are subject to change without notice.

**WARNING: Stressing the device beyond the "Absolute Maximum Ratings" may cause permanent damage. These are stress ratings only. Operation beyond the "Operating Conditions" is not recommended and extended exposure beyond the "Operating Conditions" may affect device reliability.*

OPERATING CONDITIONS

Symbol	Description	Min	Max	Units
T_A	Ambient Temperature Under Bias **Commercial** Temp.	0	+70	°C
T_A	Ambient Temperature Under Bias **Extended** Temp.	−40	+85	°C
V_{CC}	Digital Supply Voltage	4.50	5.50	V
V_{REF}	Analog Supply Voltage	4.00	5.50	V
ANGND	Analog Ground Voltage	V_{SS} − 0.4	V_{SS} + 0.4	V(1)
F_{OSC}	Oscillator Frequency **(8XC196KC)**	8	16	MHz
F_{OSC}	Oscillator Frequency **(8XC196KC20)**	8	20	MHz

NOTE:
1. ANGND and V_{SS} should be nominally at the same potential.

DC CHARACTERISTICS (Over Specified Operating Conditions)

Symbol	Description	Min	Typ	Max	Units	Test Conditions
V_{IL}	Input Low Voltage	−0.5		0.8	V	
V_{IH}	Input High Voltage (Note 1)	0.2 V_{CC} + 1.0		V_{CC} + 0.5	V	
V_{IH1}	Input High Voltage on XTAL 1	0.7 V_{CC}		V_{CC} + 0.5	V	
V_{IH2}	Input High Voltage on RESET	2.2		V_{CC} + 0.5	V	
V_{HYS}	Hysteresis on $\overline{RESET}$	300			mV	V_{CC} = 5.0V
V_{OL}	Output Low Voltage			0.3 0.45 1.5	V V V	I_{OL} = 200 μA I_{OL} = 2.8 mA I_{OL} = 7 mA
V_{OL1}	Output Low Voltage in RESET on P2.5 (Note 2)			0.8	V	I_{OL} = +0.4 mA
V_{OH}	Output High Voltage (Standard Outputs)	V_{CC} − 0.3 V_{CC} − 0.7 V_{CC} − 1.5			V V V	I_{OH} = −200 μA I_{OH} = −3.2 mA I_{OH} = −7 mA

DC CHARACTERISTICS (Over Specified Operating Conditions) (Continued)

Symbol	Description	Min	Typ	Max	Units	Test Conditions
V_{OH1}	Output High Voltage (Quasi-bidirectional Outputs)	$V_{CC} - 0.3$ $V_{CC} - 0.7$ $V_{CC} - 1.5$			V V V	$I_{OH} = -10\ \mu A$ $I_{OH} = -30\ \mu A$ $I_{OH} = -60\ \mu A$
I_{OH1}	Logical 1 Output Current in Reset. on P2.0. Do not exceed this or device may enter test modes.	−0.8			mA	$V_{IH} = V_{CC} - 1.5V$
I_{IL2}	Logical 0 Input Current in Reset on P2.0. Maximum current that must be sunk by external device to ensure test mode entry.			TBD	mA	$V_{IN} = 0.45V$
I_{IH1}	Logical 1 Input Current. Maximum current that external device must source to initiate NMI.			+200	μA	$V_{IN} = V_{CC} = 2.4V$
I_{LI}	Input Leakage Current (Std. Inputs)			±10	μA	$0 < V_{IN} < V_{CC} - 0.3V$
I_{LI1}	Input Leakage Current (Port 0)			±3	μA	$0 < V_{IN} < V_{REF}$
I_{TL}	1 to 0 Transition Current (QBD Pins)			−650	μA	$V_{IN} = 2.0V$
I_{IL}	Logical 0 Input Current (QBD Pins)			−70	μA	$V_{IN} = 0.45V$
I_{IL1}	Ports 3 and 4 in Reset			−70	μA	$V_{IN} = 0.45V$
I_{CC}	Active Mode Current in Reset **(8XC196KC)**		65	75	mA	XTAL1 = 16 MHz $V_{CC} = V_{PP} = V_{REF} = 5.5V$
I_{CC}	Active Mode Current in Reset **(8XC196KC20)**		80	92	mA	XTAL1 = 20 MHz $V_{CC} = V_{PP} = V_{REF} = 5.5V$
I_{IDLE}	Idle Mode Current **(8XC196KC)**		17	25	mA	XTAL1 = 16 MHz $V_{CC} = V_{PP} = V_{REF} = 5.5V$
I_{IDLE}	Idle Mode Current **(8XC196KC20)**		21	30	mA	XTAL1 = 20 MHz $V_{CC} = V_{PP} = V_{REF} = 5.5V$
I_{PD}	Powerdown Mode Current		8	15	μA	$V_{CC} = V_{PP} = V_{REF} = 5.5V$
I_{REF}	A/D Converter Reference Current		2	5	mA	$V_{CC} = V_{PP} = V_{REF} = 5.5V$
R_{RST}	Reset Pullup Resistor	6K		65K	Ω	$V_{CC} = 5.5V, V_{IN} = 4.0V$
C_S	Pin Capacitance (Any Pin to V_{SS})			10	pF	

NOTES:
1. All pins except RESET and XTAL1.
2. Violating these specifications in Reset may cause the part to enter test modes.
3. Commercial specifications apply to express parts except where noted.
4. QBD (Quasi-bidirectional) pins include Port 1, P2.6 and P2.7.
5. Standard Outputs include AD0–15, $\overline{RD}$, $\overline{WR}$, ALE, $\overline{BHE}$, INST, HSO pins, PWM/P2.5, CLKOUT, RESET, Ports 3 and 4, TXD/P2.0 and RXD (in serial mode 0). The V_{OH} specification is not valid for RESET. Ports 3 and 4 are open-drain outputs.
6. Standard Inputs include HSI pins, READY, BUSWIDTH, RXD/P2.1, EXTINT/P2.2, T2CLK/P2.3 and T2RST/P2.4.
7. Maximum current per pin must be externally limited to the following values if V_{OL} is held above 0.45V or V_{OH} is held below $V_{CC} - 0.7V$:
   - I_{OL} on Output pins: 10 mA
   - I_{OH} on quasi-bidirectional pins: self limiting
   - I_{OH} on Standard Output pins: 10 mA
8. Maximum current per bus pin (data and control) during normal operation is ±3.2 mA.
9. During normal (non-transient) conditions the following total current limits apply:

Port 1, P2.6	I_{OL}: 29 mA	I_{OH} is self limiting
HSO, P2.0, RXD, $\overline{RESET}$	I_{OL}: 29 mA	I_{OH}: 26 mA
P2.5, P2.7, $\overline{WR}$, $\overline{BHE}$	I_{OL}: 13 mA	I_{OH}: 11 mA
AD0–AD15	I_{OL}: 52 mA	I_{OH}: 52 mA
$\overline{RD}$, ALE, INST–CLKOUT	I_{OL}: 13 mA	I_{OH}: 13 mA

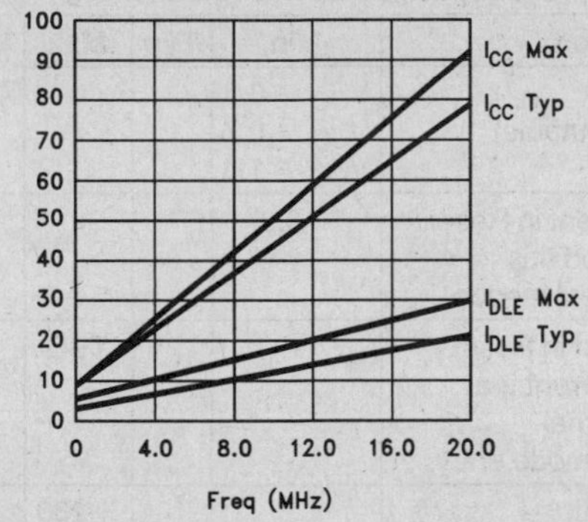

270942–17

I_{CC} Max = 4.13 × Frequency + 9 mA
I_{CC} Typ = 3.50 × Frequency + 9 mA
I_{IDLE} Max = 1.25 × Frequency + 5 mA
I_{IDLE} Typ = 0.88 × Frequency + 3 mA

NOTE:
Frequencies below 8 MHz are shown for reference only; no testing is performed.

Figure 7. I_{CC} and I_{IDLE} vs Frequency

AC CHARACTERISTICS

For use over specified operating conditions.

Test Conditions: Capacitive load on all pins = 100 pF, Rise and fall times = 10 ns, F_{OSC} = 16 MHz

The system must meet these specifications to work with the 80C196KC:

Symbol	Description	Min	Max	Units	Notes
T_{AVYV}	Address Valid to READY Setup		$2\ T_{OSC} - 68$	ns	
T_{YLYH}	Non READY Time	No upper limit		ns	
T_{CLYX}	READY Hold after CLKOUT Low	0	$T_{OSC} - 30$	ns	(Note 1)
T_{LLYX}	READY Hold after ALE Low	$T_{OSC} - 15$	$2\ T_{OSC} - 40$	ns	(Note 1)
T_{AVGV}	Address Valid to Buswidth Setup		$2\ T_{OSC} - 68$	ns	
T_{CLGX}	Buswidth Hold after CLKOUT Low	0		ns	
T_{AVDV}	Address Valid to Input Data Valid		$3\ T_{OSC} - 55$	ns	(Note 2)
T_{RLDV}	$\overline{RD}$ Active to Input Data Valid		$T_{OSC} - 22$	ns	(Note 2)
T_{CLDV}	CLKOUT Low to Input Data Valid		$T_{OSC} - 45$	ns	
T_{RHDZ}	End of $\overline{RD}$ to Input Data Float		T_{OSC}	ns	
T_{RXDX}	Data Hold after $\overline{RD}$ Inactive	0		ns	

NOTES:
1. If max is exceeded, additional wait states will occur.
2. If wait states are used, add $2\ T_{OSC}$ * N, where N = number of wait states.

AC CHARACTERISTICS (Continued)

For user over specified operating conditions.

Test Conditions: Capacitive load on all pins = 100 pF, Rise and fall times = 10 ns, F_{OSC} = 16 MHz

The 80C196KC will meet these specifications:

Symbol	Description	Min	Max	Units	Notes
F_{XTAL}	Frequency on XTAL1 **(8XC196KC)**	8	16	MHz	(Note 1)
F_{XTAL}	Frequency on XTAL1 **(8XC196KC20)**	8	20	MHz	(Note 1)
T_{OSC}	I/F_{XTAL} **(8XC196KC)**	62.5	125	ns	
T_{OSC}	I/F_{XTAL} **(8XC196KC20)**	50	125	ns	
T_{XHCH}	XTAL1 High to CLKOUT High or Low	+20	+110	ns	
T_{CLCL}	CLKOUT Cycle Time	2 T_{OSC}		ns	
T_{CHCL}	CLKOUT High Period	T_{OSC} − 10	T_{OSC} + 15	ns	
T_{CLLH}	CLKOUT Falling Edge to ALE Rising	−5	+15	ns	
T_{LLCH}	ALE Falling Edge to CLKOUT Rising	−20	+15	ns	
T_{LHLH}	ALE Cycle Time	4 T_{OSC}		ns	(Note 4)
T_{LHLL}	ALE High Period	T_{OSC} − 10	T_{OSC} + 10	ns	
T_{AVLL}	Address Setup to ALE Falling Edge	T_{OSC} − 15			
T_{LLAX}	Address Hold after ALE Falling Edge	T_{OSC} − 35		ns	
T_{LLRL}	ALE Falling Edge to $\overline{RD}$ Falling Edge	T_{OSC} − 30		ns	
T_{RLCL}	$\overline{RD}$ Low to CLKOUT Falling Edge	+4	+30	ns	
T_{RLRH}	$\overline{RD}$ Low Period	T_{OSC} − 5		ns	(Note 4)
T_{RHLH}	$\overline{RD}$ Rising Edge to ALE Rising Edge	T_{OSC}	T_{OSC} + 25	ns	(Note 2)
T_{RLAZ}	$\overline{RD}$ Low to Address Float		+5	ns	
T_{LLWL}	ALE Falling Edge to $\overline{WR}$ Falling Edge	T_{OSC} − 10		ns	
T_{CLWL}	CLKOUT Low to $\overline{WR}$ Falling Edge	0	+25	ns	
T_{QVWH}	Data Stable to $\overline{WR}$ Rising Edge	T_{OSC} − 23			(Note 4)
T_{CHWH}	CLKOUT High to $\overline{WR}$ Rising Edge	−5	+15	ns	
T_{WLWH}	$\overline{WR}$ Low Period	T_{OSC} − 20		ns	(Note 4)
T_{WHQX}	Data Hold after $\overline{WR}$ Rising Edge	T_{OSC} − 25		ns	
T_{WHLH}	$\overline{WR}$ Rising Edge to ALE Rising Edge	T_{OSC} − 10	T_{OSC} + 15	ns	(Note 2)
T_{WHBX}	$\overline{BHE}$, INST after $\overline{WR}$ Rising Edge	T_{OSC} − 10		ns	
T_{WHAX}	AD8–15 HOLD after $\overline{WR}$ Rising	T_{OSC} − 30		ns	(Note 3)
T_{RHBX}	$\overline{BHE}$, INST after $\overline{RD}$ Rising Edge	T_{OSC} − 10		ns	
T_{RHAX}	AD8–15 HOLD after $\overline{RD}$ Rising	T_{OSC} − 25		ns	(Note 3)

NOTES:
1. Testing performed at 8 MHz. However, the device is static by design and will typically operate below 1 Hz.
2. Assuming back-to-back bus cycles.
3. 8-Bit bus only.
4. If wait states are used, add 2 T_{OSC} * N, where N = number of wait states.

System Bus Timings

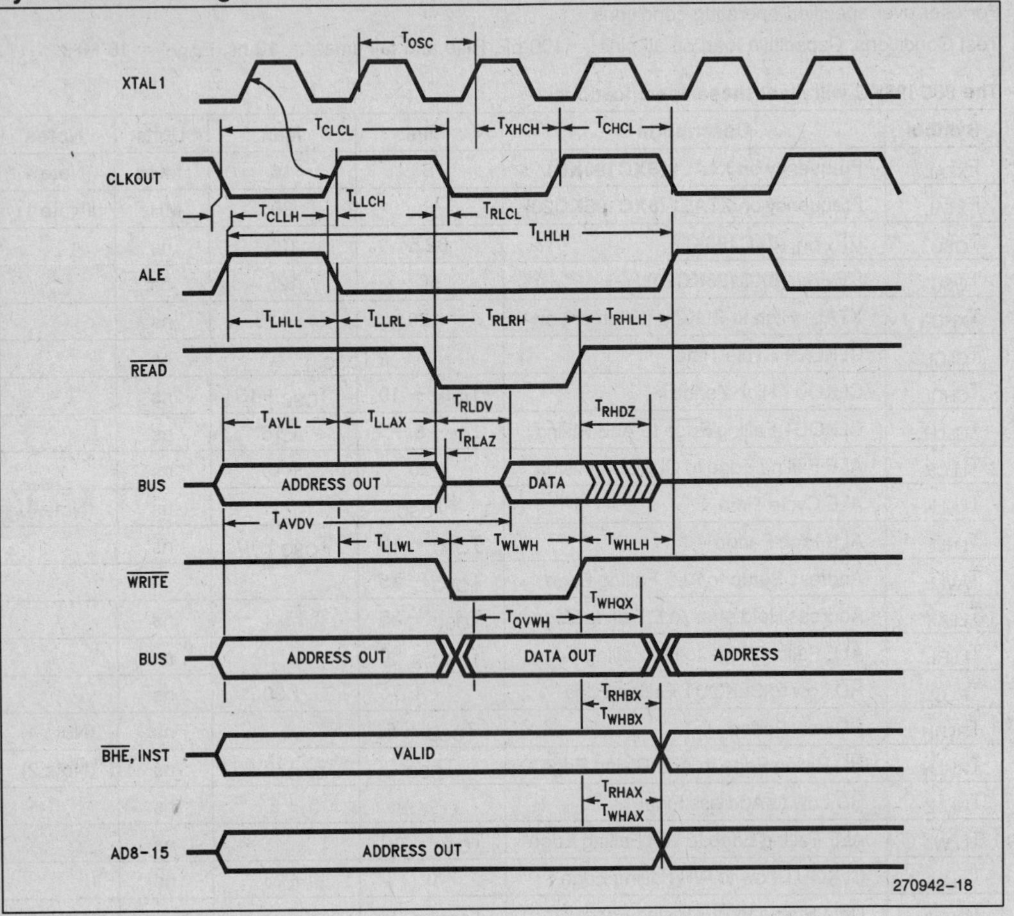

READY Timings (One Wait State)

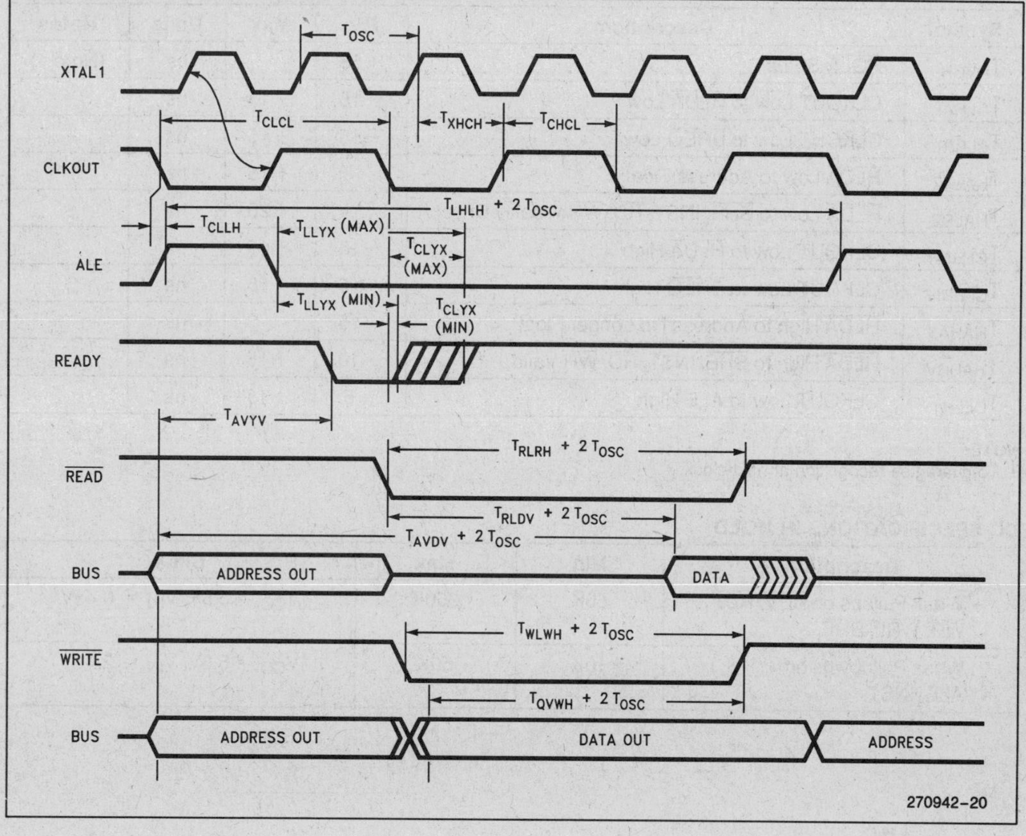

270942-20

Buswidth Timings

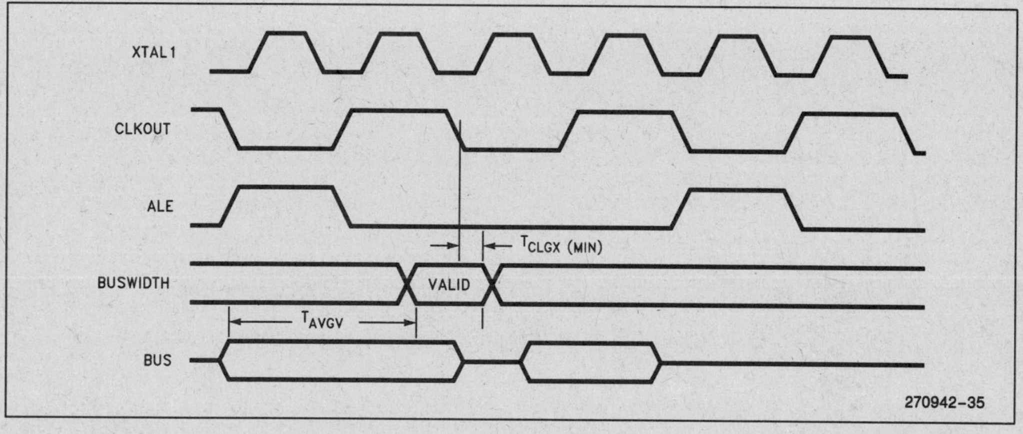

270942-35

$\overline{HOLD}$/$\overline{HLDA}$ Timings

Symbol	Description	Min	Max	Units	Notes
T_{HVCH}	$\overline{HOLD}$ Setup	+55		ns	(Note 1)
T_{CLHAL}	CLKOUT Low to $\overline{HLDA}$ Low	−15	+15	ns	
T_{CLBRL}	CLKOUT Low to $\overline{BREQ}$ Low	−15	+15	ns	
T_{HALAZ}	$\overline{HLDA}$ Low to Address Float		+15	ns	
T_{HALBZ}	$\overline{HLDA}$ Low to $\overline{BHE}$, INST, $\overline{RD}$, $\overline{WR}$ Weakly Driven		+20	ns	
T_{CLHAH}	CLKOUT Low to $\overline{HLDA}$ High	−15	+15	ns	
T_{CLBRH}	CLKOUT Low to $\overline{BREQ}$ High	−15	+15	ns	
T_{HAHAX}	$\overline{HLDA}$ High to Address No Longer Float	−15		ns	
T_{HAHBV}	$\overline{HLDA}$ High to $\overline{BHE}$, INST, $\overline{RD}$, $\overline{WR}$ Valid	−10	+15	ns	
T_{CLLH}	CLKOUT Low to ALE High	−5	+15	ns	

NOTE:
1. To guarantee recognition at next clock.

DC SPECIFICATIONS IN HOLD

Description	Min	Max	Units
Weak Pullups on $\overline{ADV}$, $\overline{RD}$, $\overline{WR}$, $\overline{WRL}$, $\overline{BHE}$	50K	250K	V_{CC} = 5.5V, V_{IN} = 0.45V
Weak Pulldowns on ALE, INST	10K	50K	V_{CC} = 5.5V, V_{IN} = 2.4

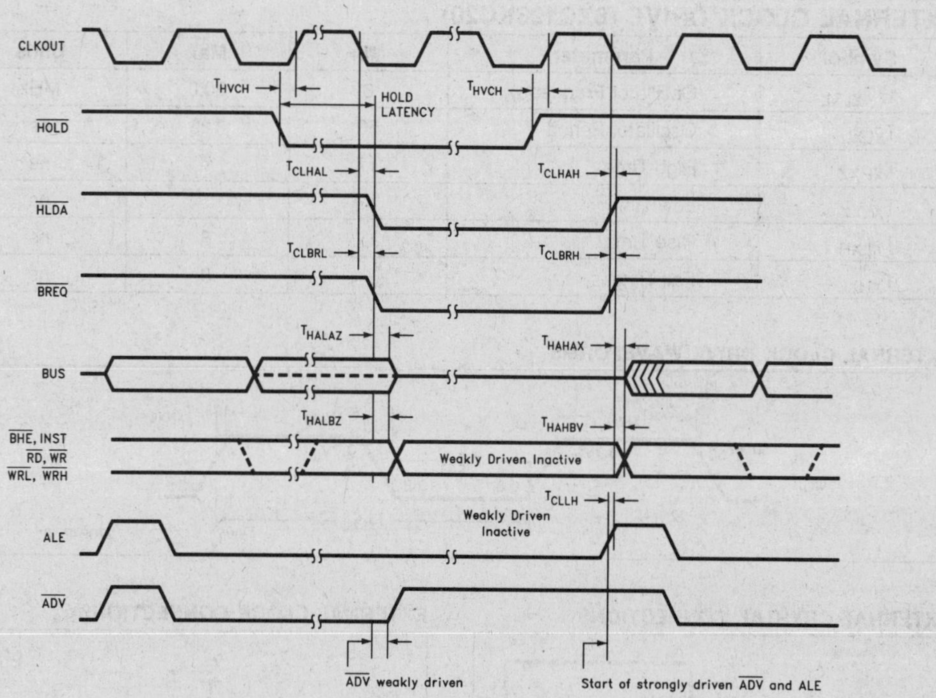

270942–36

Maximum Hold Latency

Bus Cycle Type	
Internal Execution	1.5 States
16-Bit External Execution	2.5 States
8-Bit External Execution	4.5 States

EXTERNAL CLOCK DRIVE (8XC196KC)

Symbol	Parameter	Min	Max	Units
$1/T_{XLXL}$	Oscillator Frequency	8	16.0	MHz
T_{XLXL}	Oscillator Period	62.5	125	ns
T_{XHXX}	High Time	20		ns
T_{XLXX}	Low Time	20		ns
T_{XLXH}	Rise Time		10	ns
T_{XHXL}	Fall Time		10	ns

EXTERNAL CLOCK DRIVE (8XC196KC20)

Symbol	Parameter	Min	Max	Units
$1/T_{XLXL}$	Oscillator Frequency	8	20.0	MHz
T_{XLXL}	Oscillator Period	50	125	ns
T_{XHXX}	High Time	17		ns
T_{XLXX}	Low Time	17		ns
T_{XLXH}	Rise Time		8	ns
T_{XHXL}	Fall Time		8	ns

EXTERNAL CLOCK DRIVE WAVEFORMS

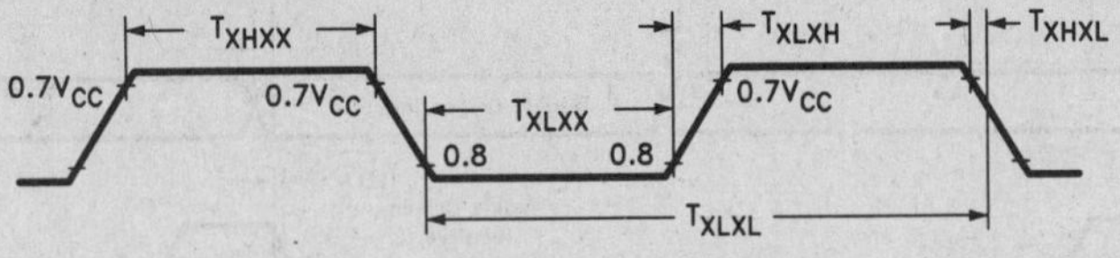

270942-21

EXTERNAL CRYSTAL CONNECTIONS

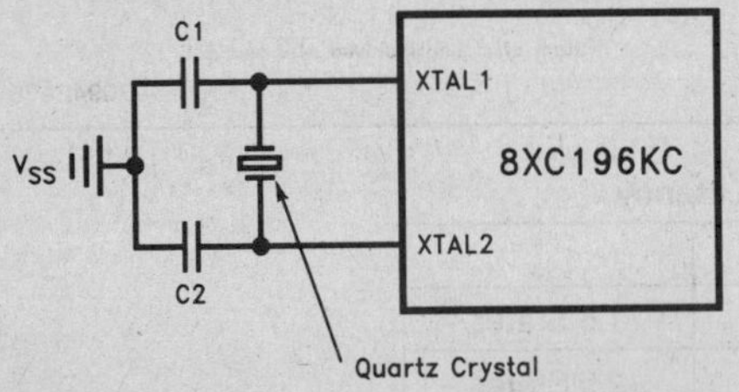

270942-41

NOTE:
Keep oscillator components close to chip and use short, direct traces to XTAL1, XTAL2 and V_{SS}. When using crystals, C1 = C2 ≈ 20 pF. When using ceramic resonators, consult manufacturer for recommended circuitry.

EXTERNAL CLOCK CONNECTIONS

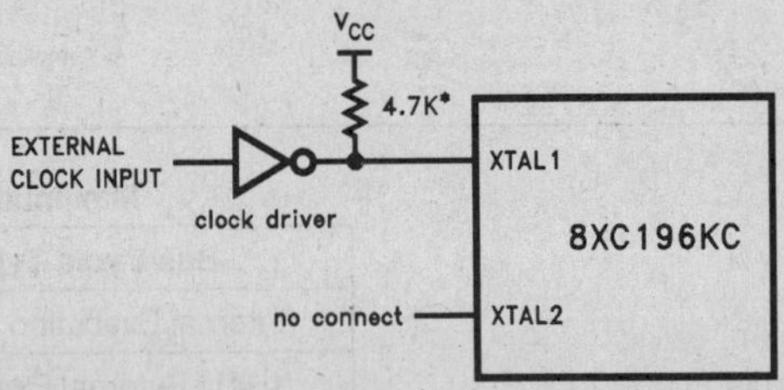

270942-42

NOTE:
*Required if TTL driver used.
Not needed if CMOS driver is used.

AC TESTING INPUT, OUTPUT WAVEFORMS

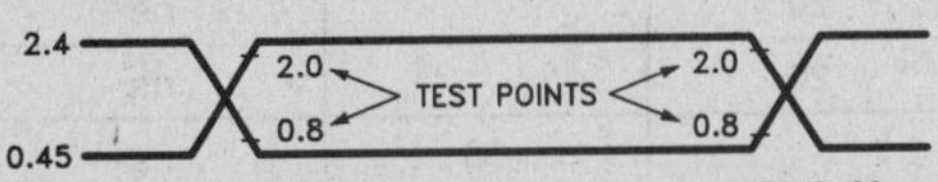

270942-22

AC Testing inputs are driven at 2.4V for a Logic "1" and 0.45V for a Logic "0" Timing measurements are made at 2.0V for a Logic "1" and 0.8V for a Logic "0".

FLOAT WAVEFORMS

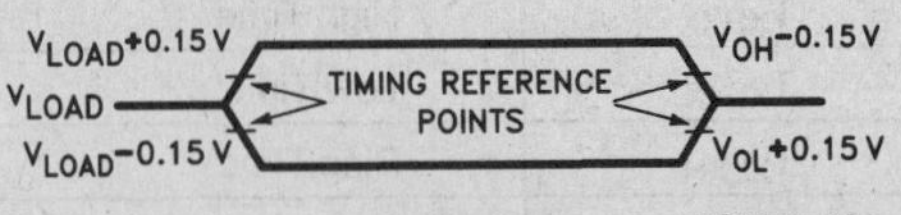

270942-23

For Timing Purposes a Port Pin is no Longer Floating when a 150 mV change from Load Voltage Occurs and Begins to Float when a 150 mV change from the Loaded V_{OH}/V_{OL} Level occurs; I_{OL}/I_{OH} = ±15 mA.

EXPLANATION OF AC SYMBOLS

Each symbol is two pairs of letters prefixed by "T" for time. The characters in a pair indicate a signal and its condition, respectively. Symbols represent the time between the two signal/condition points.

Conditions:
H— High
L— Low
V— Valid
X— No Longer Valid
Z— Floating

Signals:
A— Address
B— $\overline{BHE}$
C— CLKOUT
D— DATA
G— Buswidth
H— $\overline{HOLD}$
HA— $\overline{HLDA}$
L— ALE/$\overline{ADV}$
BR— $\overline{BREQ}$
R— $\overline{RD}$
W— $\overline{WR}/\overline{WRH}/\overline{WRL}$
X— XTAL1
Y— READY
Q— Data Out

AC CHARACTERISTICS—SERIAL PORT—SHIFT REGISTER MODE

SERIAL PORT TIMING—SHIFT REGISTER MODE (MODE 0)

Symbol	Parameter	Min	Max	Units
T_{XLXL}	Serial Port Clock Period (BRR ≥ 8002H)	6 T_{OSC}		ns
T_{XLXH}	Serial Port Clock Falling Edge to Rising Edge (BRR ≥ 8002H)	4 T_{OSC} −50	4 T_{OSC} +50	ns
T_{XLXL}	Serial Port Clock Period (BRR = 8001H)	4 T_{OSC}		ns
T_{XLXH}	Serial Port Clock Falling Edge to Rising Edge (BRR = 8001H)	2 T_{OSC} −50	2 T_{OSC} +50	ns
T_{QVXH}	Output Data Setup to Clock Rising Edge	2 T_{OSC} − 50		ns
T_{XHQX}	Output Data Hold after Clock Rising Edge	2 T_{OSC} − 50		ns
T_{XHQV}	Next Output Data Valid after Clock Rising Edge		2 T_{OSC} + 50	ns
T_{DVXH}	Input Data Setup to Clock Rising Edge	T_{OSC} +50		ns
T_{XHDX}	Input Data Hold after Clock Rising Edge	0		ns
T_{XHQZ}	Last Clock Rising to Output Float		1 T_{OSC}	ns

WAVEFORM—SERIAL PORT—SHIFT REGISTER MODE

SERIAL PORT WAVEFORM—SHIFT REGISTER MODE (MODE 0)

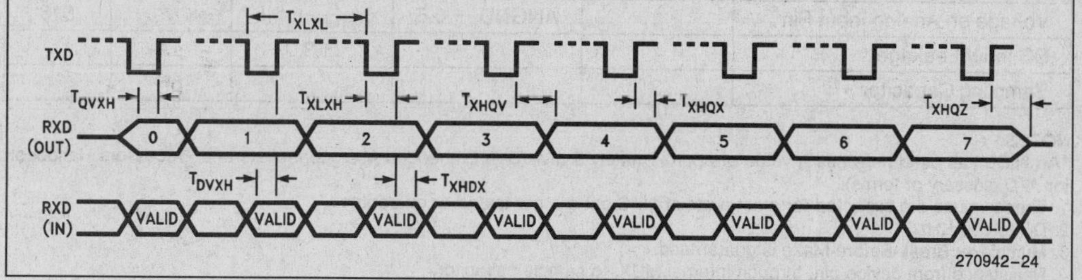

A to D CHARACTERISTICS

The A/D converter is ratiometric, so absolute accuracy is dependent on the accuracy and stability of V_{REF}.

10-BIT MODE A/D OPERATING CONDITIONS

Symbol	Description	Min	Max	Units
T_A	Ambient Temperature **Commercial** Temp.	0	+70	°C
T_A	Ambient Temperature **Extended** Temp.	−40	+85	°C
V_{CC}	Digital Supply Voltage	4.50	5.50	V
V_{REF}	Analog Supply Voltage	4.00	5.50	V
T_{SAM}	Sample Time	1.0		μs(1)
T_{CONV}	Conversion Time	10	20	μs(1)
F_{OSC}	Oscillator Frequency **(8XC196KC)**	8.0	16.0	MHz
F_{OSC}	Oscillator Frequency **(8XC196KC20)**	8.0	20.0	MHz

NOTE:
ANGND and V_{SS} should nominally be at the same potential, 0.00V.
1. The value of AD_TIME is selected to meet these specifications.

10-BIT MODE A/D CHARACTERISTICS (Over Specified Operating Conditions)

Parameter	Typical(1)	Minimum	Maximum	Units*	Notes
Resolution		1024 10	1024 10	Levels Bits	
Absolute Error		0	±3	LSBs	
Full Scale Error	0.25 ± 0.5			LSBs	
Zero Offset Error	0.25 ± 0.5			LSBs	
Non-Linearity	1.0 ± 2.0	0	±3	LSBs	
Differential Non-Linearity Error		> −1	+2	LSBs	
Channel-to-Channel Matching	±0.1	0	±1	LSBs	
Repeatability	±0.25			LSBs	
Temperature Coefficients: Offset Full Scale Differential Non-Linearity	 0.009 0.009 0.009			 LSB/°C LSB/°C LSB/°C	
Off Isolation		−60		dB	1, 2
Feedthrough	−60			dB	1
V_{CC} Power Supply Rejection	−60			dB	1
Input Series Resistance		750	1.2K	Ω	4
Voltage on Analog Input Pin		ANGND − 0.5	V_{REF} + 0.5	V	5, 6
DC Input Leakage		0	±3.0	μA	
Sampling Capacitor	3			pF	

NOTES:
*An "LSB" as used here has a value of approxiimately 5 mV. (See Embedded Microcontrollers and Processors Handbook for A/D glossary of terms).
1. These values are expected for most parts at 25°C but are not tested or guaranteed.
2. DC to 100 KHz.
3. Multiplexer Break-Before-Make is guaranteed.
4. Resistance from device pin, through internal MUX, to sample capacitor.
5. These values may be exceeded if the pin current is limited to ±2 mA.
6. Applying voltages beyond these specifications will degrade the accuracy of all channels being converted.
7. All conversions performed with processor in IDLE mode.

8-BIT MODE A/D OPERATING CONDITIONS

Symbol	Description	Min	Max	Units
T_A	Ambient Temperature **Commercial** Temp.	0	+70	°C
T_A	Ambient Temperature **Extended** Temp.	−40	+85	°C
V_{CC}	Digital Supply Voltage	4.50	5.50	V
V_{REF}	Analog Supply Voltage	4.00	5.50	V
T_{SAM}	Sample Time	1.0		μs(1)
T_{CONV}	Conversion Time	7	20	μs(1)
F_{OSC}	Oscillator Frequency **(8XC196KC)**	8.0	16.0	MHz
F_{OSC}	Oscillator Frequency **(8XC196KC20)**	8.0	20.0	MHz

NOTE:
ANGND and V_{SS} should nominally be at the same potential, 0.00V.
1. The value of AD_TIME is selected to meet these specifications.

8-BIT MODE A/D CHARACTERISTICS (Over Specified Operating Conditions)

Parameter	Typical	Minimum	Maximum	Units*	Notes
Resolution		256 8	256 8	Levels Bits	
Absolute Error		0	±1	LSBs	
Full Scale Error	±0.5			LSBs	
Zero Offset Error	±0.5			LSBs	
Non-Linearity		0	±1	LSBs	
Differential Non-Linearity Error		> −1	+1	LSBs	
Channel-to-Channel Matching			±1	LSBs	
Repeatability	±0.25			LSBs	
Temperature Coefficients: Offset Full Scale Differential Non-Linearity	 0.003 0.003 0.003			 LSB/°C LSB/°C LSB/°C	
Off Isolation		−60		dB	2, 3
Feedthrough	−60			dB	2
V_{CC} Power Supply Rejection	−60			dB	2
Input Series Resistance		750	1.2K	Ωs	4
Voltage on Analog Input Pin		V_{SS} − 0.5	V_{REF} + 0.5	V	5, 6
DC Input Leakage		0	±3.0	μA	
Sampling Capacitor	3			pF	

NOTES:
*An "LSB" as used here has a value of approximately 20 mV. (See Embedded Microcontrollers and Processors Handbook for A/D glossary of terms).
1. These values are expected for most parts at 25°C but are not tested or guaranteed.
2. DC to 100 KHz.
3. Multiplexer Break-Before-Make is guaranteed.
4. Resistance from device pin, through internal MUX, to sample capacitor.
5. These values may be exceeded if pin current is limited to ±2 mA.
6. Applying voltages beyond these specifications will degrade the accuracy of all channels being converted.
7. All conversions performed with processor in IDLE mode.

EPROM SPECIFICATIONS

OPERATING CONDITIONS DURING PROGRAMMING

Symbol	Description	Min	Max	Units
T_A	Ambient Temperature During Programming	20	30	C
V_{CC}	Supply Voltage During Programming	4.5	5.5	V(1)
V_{REF}	Reference Supply Voltage During Programming	4.5	5.5	V(1)
V_{PP}	Programming Voltage	12.25	12.75	V(2)
V_{EA}	EA Pin Voltage	12.25	12.75	V(2)
F_{OSC}	Oscillator Frequency During Auto and Slave Mode Programming	6.0	8.0	MHz
F_{OSC}	Oscillator Frequency During Run-Time Programming **(8XC196KC)**	6.0	16.0	MHz
F_{OSC}	Oscillator Frequency During Run-Time Programming **(8XC196KC20)**	6.0	20.0	MHz

NOTES:
1. V_{CC} and V_{REF} should nominally be at the same voltage during programming.
2. V_{PP} and V_{EA} must never exceed the maximum specification, or the device may be damaged.
3. V_{SS} and ANGND should nominally be at the same potential (0V).
4. Load capacitance during Auto and Slave Mode programming = 150 pF.

AC EPROM PROGRAMMING CHARACTERISTICS

Symbol	Description	Min	Max	Units
T_{SHLL}	Reset High to First $\overline{\text{PALE}}$ Low	1100		T_{OSC}
T_{LLLH}	$\overline{\text{PALE}}$ Pulse Width	50		T_{OSC}
T_{AVLL}	Address Setup Time	0		T_{OSC}
T_{LLAX}	Address Hold Time	100		T_{OSC}
T_{PLDV}	$\overline{\text{PROG}}$ Low to Word Dump Valid		50	T_{OSC}
T_{PHDX}	Word Dump Data Hold		50	T_{OSC}
T_{DVPL}	Data Setup Time	0		T_{OSC}
T_{PLDX}	Data Hold Time	400		T_{OSC}
T_{PLPH}(1)	$\overline{\text{PROG}}$ Pulse Width	50		T_{OSC}
T_{PHLL}	$\overline{\text{PROG}}$ High to Next $\overline{\text{PALE}}$ Low	220		T_{OSC}
T_{LHPL}	$\overline{\text{PALE}}$ High to $\overline{\text{PROG}}$ Low	220		T_{OSC}
T_{PHPL}	$\overline{\text{PROG}}$ High to Next $\overline{\text{PROG}}$ Low	220		T_{OSC}
T_{PHIL}	$\overline{\text{PROG}}$ High to AINC Low	0		T_{OSC}
T_{ILIH}	$\overline{\text{AINC}}$ Pulse Width	240		T_{OSC}
T_{ILVH}	PVER Hold after $\overline{\text{AINC}}$ Low	50		T_{OSC}
T_{ILPL}	$\overline{\text{AINC}}$ Low to $\overline{\text{PROG}}$ Low	170		T_{OSC}
T_{PHVL}	$\overline{\text{PROG}}$ High to $\overline{\text{PVER}}$ Valid		220	T_{OSC}

NOTE:
1. This specification is for the Word Dump Mode. For programming pulses, use the Modified Quick Pulse Algorithm. See user's manual for further information.

DC EPROM PROGRAMMING CHARACTERISTICS

Symbol	Description	Min	Max	Units
I_{PP}	V_{PP} Supply Current (When Programming)		100	mA

NOTE:
Do not apply V_{PP} until V_{CC} is stable and within specifications and the oscillator/clock has stabilized or the device may be damaged.

EPROM PROGRAMMING WAVEFORMS

SLAVE PROGRAMMING MODE DATA PROGRAM MODE WITH SINGLE PROGRAM PULSE

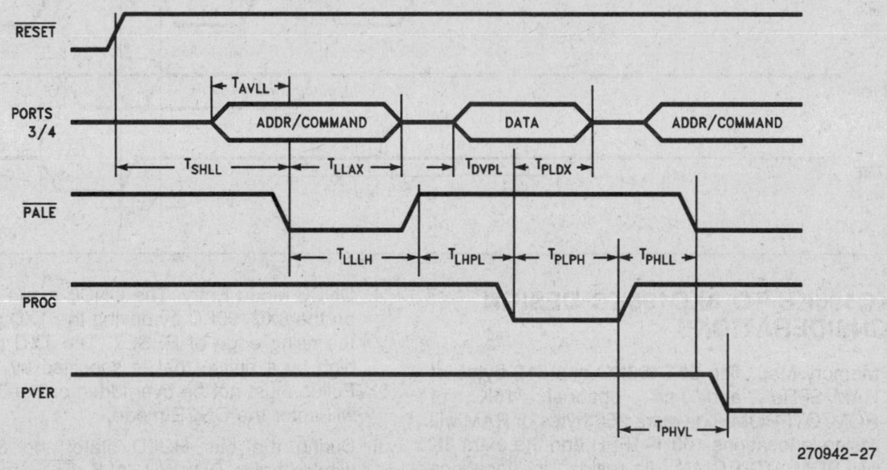

NOTE:
P3.0 must be high ("1")

SLAVE PROGRAMMING MODE IN WORD DUMP WITH AUTO INCREMENT

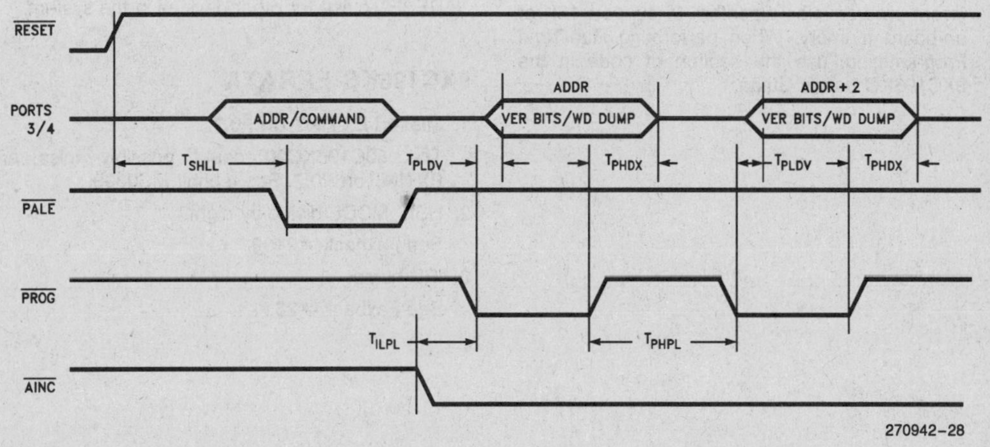

NOTE:
P3.0 must be low ("0")

SLAVE PROGRAMMING MODE TIMING IN DATA PROGRAM WITH REPEATED PROG PULSE AND AUTO INCREMENT

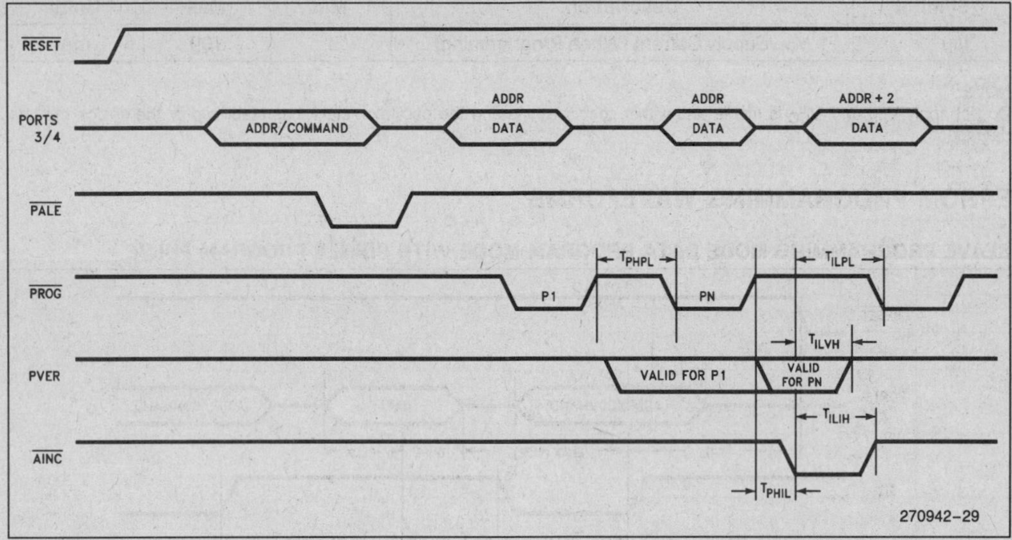

8XC196KB TO 8XC196KC DESIGN CONSIDERATIONS

1. Memory Map. The 8XC196KC has 512 bytes of RAM/SFRs and an optional 16K of ROM/OTPROM. The extra 256 bytes of RAM will reside in locations 100H–1FFH and the extra 8K of ROM/OTPROM will reside in locations 4000H–5FFFH. These locations are external memory on the 8XC196KB.
2. The CDE pin on the KB has become a V_{SS} pin on the KC to support 16/20 MHz operation.
3. EPROM programming. The 8XC196KC has a different programming algorithm to support 16K of on-board memory. When performing Run-Time Programming, use the section of code in the 8XC196KC User's Guide.
4. ONCE Mode Entry. The ONCE mode is entered on the 8XC196KC by driving the TXD pin low on the rising edge of RESET. The TXD pin is held high by a pullup that is specified by I_{OH1}. This Pullup must not be overridden or the 8XC196KC will enter the ONCE mode.
5. During the bus HOLD state, the 8XC196KC weakly holds $\overline{RD}$, $\overline{WR}$, ALE, $\overline{BHE}$ and INST in their inactive states. The 8XC196KB only holds ALE in its inactive state.
6. A RESET pulse from the 8XC196KC is 16 states rather than 4 states as on the 8XC196KB (i.e., a watchdog timer overflow). This provides a longer RESET pulse for other devices in the system.

8XC196KC ERRATA

1. **Missed EXTINT on P0.7.**

   The 80C196KC20 could possibly miss an EXTINT on P0.7. See techbit MC0893.
2. **HSI_MODE divide-by-eight.**

   See Faxback #2192.
3. **IPD hump.**

   See Faxback #2311.

DATA SHEET REVISION HISTORY

This data sheet is valid for devices with a "H", "L" or "M" at the end of the topside tracking number. The topside tracking number consists of nine characters and is the second line on the top side of the device. Data sheets are changed as new device information becomes available. Verify with your local Intel sales office that you have the latest version before finalizing a design or ordering devices.

The following are differences between the 270942-004 and 270942-005 datasheets:

1. Removed "Word Addressable Only" from Port 3 and 4 in Table 2.
2. Renamed PVAL to CPVER.
3. Removed T_{LLYV} and T_{LLGV} from the waveform diagrams.
4. Added HSI_MODE divide-by-eight and IPD hump to 8XC196KC errata.

The following are important differences between the 270942-002 and 270942-004 data sheets:

1. NMI during PTS, QBD port glitch and Divide HOLD/READY erratas were fixed and have been removed from the data sheet. The HSI errata is also removed as this is now considered normal operation.
2. Combined 16 and 20 MHz data sheets. Data sheet 270924-001 (20 MHz) is now obsolete.
3. Added 80-lead SQFP package pinout.
4. Added documentation for CLKOUT disable bit.
5. θ_{JA} for QFP package was changed to 55°C/W from 42°C/W.
6. θ_{JC} for QFP package was changed to 16°C/W from TBD°C/W.
7. T_{SAM} (MIN) in 10-bit mode was changed to 1.0 μs from 3.0 μs.
8. T_{SAM} (MIN) in 8-bit mode was changed to 1.0 μs from 2.0 μs.
9. I_{IL1} specification for port 2.0 was renamed I_{IL2}.
10. I_{IL2} (MAX) is changed to TBD from − 6 mA.
11. I_{IH1} (MAX) is changed to +200 μA from +100 μA.
12. I_{IH1} test condition changes to V_{IN} = 2.4V from V_{IN} = 5.5V.
13. V_{HYS} is changed to 300 mV from 150 mV.
14. I_{CC} (TYP) at 16 MHz is changed to 65 mA from 50 mA.
15. I_{CC} (MAX) at 16 MHz is changed to 75 mA from 70 mA.
16. I_{CC} (TYP) at 20 MHz is changed to 80 mA from 60 mA.
17. I_{CC} (MAX) at 20 MHz is changed to 92 mA from 86 mA.
18. I_{IDLE} (TYP) at 16 MHz is changed to 17 mA from 15 mA.
19. I_{IDLE} (MAX) at 16 MHz is changed to 25 mA from 30 mA.
20. I_{IDLE} (TYP) at 20 MHz is changed to 21 mA from 15 mA.
21. I_{IDLE} (MAX) at 20 MHz is changed to 30 mA from 35 mA.
22. I_{PD} (TYP) at 16 MHz is changed to 8 μA from 15 μA.
23. I_{PD} (MAX) at 16 MHz is changed to 15 μA from TBD.
24. I_{PD} (TYP) at 20 MHz is changed to 8 μA from 18 μA.
25. I_{PD} (MAX) at 20 MHz is changed to 15 μA from TBD.
26. T_{CLDV} (MAX) is changed to T_{OSC} −45 ns from T_{OSC} − 50 ns.
27. T_{LLAX} (MIN) is changed to T_{OSC} −35 ns from T_{OSC} − 40 ns.
28. T_{CHWH} (MIN) is changed to −5 ns from −10 ns.
29. T_{RHAX} (MIN) is changed to T_{OSC} − 25 ns from T_{OSC} − 30 ns.
30. T_{HALAZ} (MAX) is changed to +15 ns from +10 ns.
31. T_{HALBZ} (MAX) is changed to +20 ns from +15 ns.

32. T_{HAHBV} (MAX) is now specified at +15 ns, was formerly unspecified.
33. The T_{LLYV} and T_{LLGV} specifications were removed. These specifications are not required in high-speed systems designs.
34. Added EXTINT, P0.7 errata to Errata section.

The following are the important differences between the -001 and -002 versions of data sheet 270942.

1. Express and Commercial devices are combined into one data sheet. The Express only data sheet 270794-001 is obsolete.
2. Removed KB/KC feature set differences, pin definition table, and SFR locations and bitmaps.
3. Added programming pin function to package drawings and pin descriptions.
4. Changed absolute maximum temperature under bias from 0°C to +70°C to −55°C to +125°C.
5. Replaced V_{OH2} specification with I_{OH1} and I_{IL1} specifications.
6. Added I_{IH1} specification for NMI pulldown resistors.
7. Added maximum hold latency table.
8. Added external oscillator and external clock circuit drawings.
9. Changed Clock Drive T_{XHXX} and T_{XLXX} Min spec to 20 ns.
10. Fixed Serial Port T_{XLXH} specification.
11. Added 8- and 10-bit mode A/D operating conditions tables.
12. Specified operating range for sample and convert times.
13. Added specification for voltage on analog input pin.
14. Put operating conditions for EPROM programming into tabular format.

8XC196KD/8XC196KD20
COMMERCIAL CHMOS MICROCONTROLLER

87C196KD—32 Kbytes of On-Chip OTPROM
83C196KD—32 Kbytes of ROM

- **16 MHz and 20 MHz Available**
- **1000 Byte Register RAM**
- **Register-to-Register Architecture**
- **28 Interrupt Sources/16 Vectors**
- **Peripheral Transaction Server**
- **1.4 μs 16 x 16 Multiply (20 MHz)**
- **2.4 μs 32/16 Divide (20 MHz)**
- **Powerdown and Idle Modes**
- **Five 8-Bit I/O Ports**
- **16-Bit Watchdog Timer**
- **Dynamically Configurable 8-Bit or 16-Bit Buswidth**
- **Full Duplex Serial Port**
- **High Speed I/O Subsystem**
- **16-Bit Timer**
- **16-Bit Up/Down Counter with Capture**
- **3 Pulse-Width-Modulated Outputs**
- **Four 16-Bit Software Timers**
- **8- or 10-Bit A/D Converter with Sample/Hold**
- **$\overline{\text{HOLD}}$/$\overline{\text{HLDA}}$ Bus Protocol**
- **OTP One-Time Programmable Version**

The 8XC196KD 16-bit microcontroller is a high performance member of the MCS® 96 microcontroller family. The 8XC196KD is an enhanced 80C196KC device with 1000 bytes RAM, 16 MHz operation and an optional 32 Kbytes of ROM/EPROM. Intel's CHMOS III process provides a high performance processor along with low power consumption.

The 8XC196KD has a maximum guaranteed frequency of 16 MHz. The 8XC196KD20 has a maximum guaranteed frequency of 20 MHz. Unless otherwise noted, all references to the 8XC196KD also refer to the 8XC196KD20.

Four high-speed capture inputs are provided to record times when events occur. Six high-speed outputs are available for pulse or waveform generation. The high-speed output can also generate four software timers or start an A/D conversion. Events can be based on the timer or up/down counter.

November 1994
Order Number: 272145-003

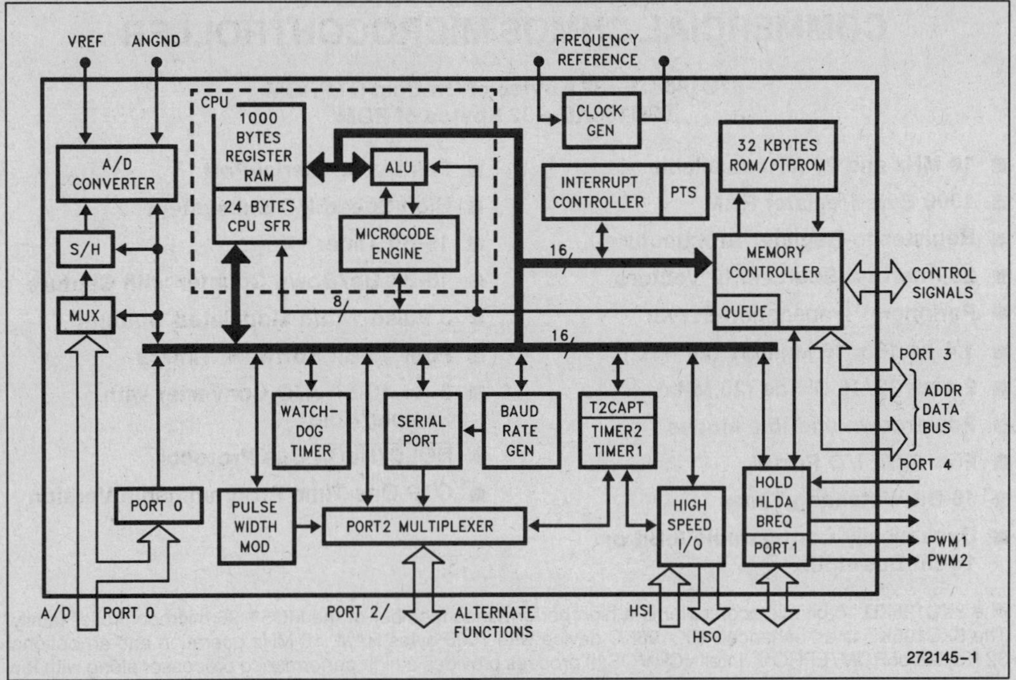

Figure 1. 8XC196KD Block Diagram

87C196KD ENHANCED FEATURE SET OVER THE 87C196KC

1. The 87C196KD has twice the RAM and twice the OTPROM space of the 87C196KC.
2. The vertical windowing scheme has been extended to allow all 1000 bytes of register RAM to be windowed into the lower register file.

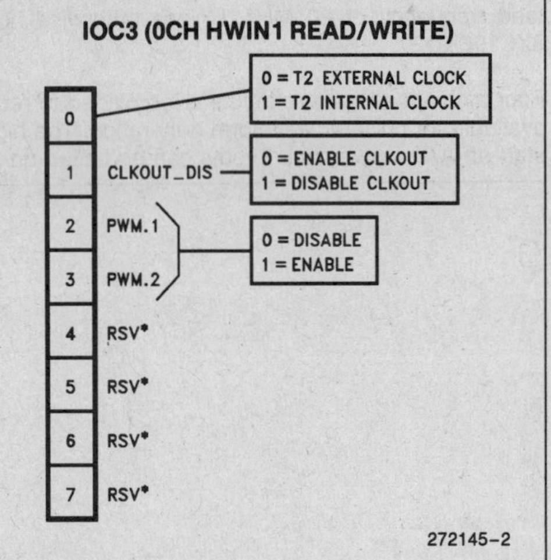

NOTE:
*RSV—Reserved bits must be = 0

Figure 2. 87C196KD New SFR Bit (CLKOUT Disable)

8XC196KD VERTICAL WINDOWING MAP

Table 1. 128-Byte Windows

Address to Remap	Device Series	WSR Contents
0380H	KD	X001 0111B = 17H
0300H	KD	X001 0110B = 16H
0280H	KD	X001 0101B = 15H
0200H	KD	X001 0100B = 14H
0180H	KC, KD	X001 0011B = 13H
0100H	KC, KD	X001 0010B = 12H
0080H	KC, KD	X001 0001B = 11H
0000H	KC, KD	X001 0000B = 10H

Window in Lower Register File: 80H–FFH

Table 2. 64-Byte Windows

Address to Remap	Device Series	WSR Contents
03C0H	KD	X010 1111B = 2FH
0380H	KD	X010 1110B = 2EH
0340H	KD	X010 1101B = 2DH
0300H	KD	X010 1100B = 2CH
02C0H	KD	X010 1011B = 2BH
0280H	KD	X010 1010B = 2AH
0240H	KD	X010 1001B = 29H
0200H	KD	X010 1000B = 28H
01C0H	KC, KD	X010 0111B = 27H
0180H	KC, KD	X010 0110B = 26H
0140H	KC, KD	X010 0101B = 25H
0100H	KC, KD	X010 0100B = 24H
00C0H	KC, KD	X010 0011B = 23H
0080H	KC, KD	X010 0010B = 22H
0040H	KC, KD	X010 0001B = 21H
0000H	KC, KD	X010 0000B = 20H

Window in Lower Register File: C0H–FFH

Table 3. 32-Byte Windows

Address to Remap	Device Series	WSR Contents
03E0H	KD	X101 1111B = 5FH
03C0H	KD	X101 1110B = 5EH
03A0H	KD	X101 1101B = 5DH
0380H	KD	X101 1100B = 5CH
0360H	KD	X101 1011B = 5BH
0340H	KD	X101 1010B = 5AH
0320H	KD	X101 1001B = 59H
0300H	KD	X101 1000B = 58H
02E0H	KD	X101 0111B = 57H
02C0H	KD	X101 0110B = 56H
02A0H	KD	X101 0101B = 55H
0280H	KD	X101 0100B = 54H
0260H	KD	X101 0011B = 53H
0240H	KD	X101 0010B = 52H
0220H	KD	X101 0001B = 51H
0200H	KD	X101 0000B = 50H
01E0H	KC, KD	X100 1111B = 4FH
01C0H	KC, KD	X100 1110B = 4EH
01A0H	KC, KD	X100 1101B = 4DH
0180H	KC, KD	X100 1100B = 4CH
0160H	KC, KD	X100 1011B = 4BH
0140H	KC, KD	X100 1010B = 4AH
0120H	KC, KD	X100 1001B = 49H
0100H	KC, KD	X100 1000B = 48H
00E0H	KC, KD	X100 0111B = 47H
00C0H	KC, KD	X100 0110B = 46H
00A0H	KC, KD	X100 0101B = 45H
0080H	KC, KD	X100 0100B = 44H
0060H	KC, KD	X100 0011B = 43H
0040H	KC, KD	X100 0010B = 42H
0020H	KC, KD	X100 0001B = 41H
0000H	KC, KD	X100 0000B = 40H

Window in Lower Register File: E0H–FFH

PROCESS INFORMATION

This device is manufactured on PX29.5 or PX29.9, a CHMOS III process. Additional process and reliability information is available in Intel's *Components Quality and Reliability Handbook,* Order Number 210997.

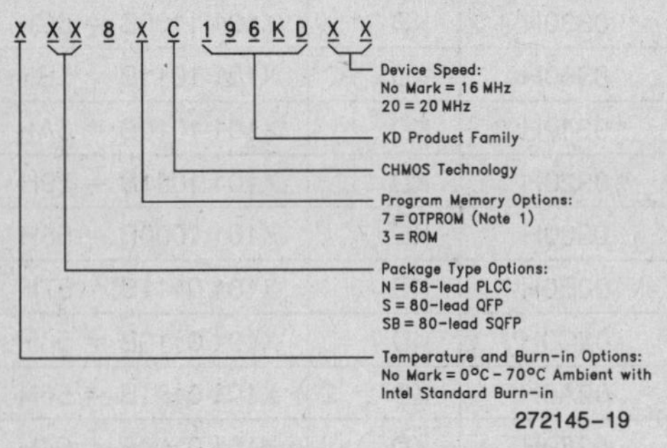

EXAMPLE: N87C196KD20 is 68-Lead PLCC OTPROM, 20 MHz.
For complete package dimensional data, refer to the Intel Packaging Handbook (Order Number 240800).

NOTE:
1. EPROMs are available as One Time Programmable (OTPROM) only.

Figure 3. The 8XC196KD Family Nomenclature

Table 4. Thermal Characteristics

Package Type	θ_{ja}	θ_{jc}
PLCC	35°C/W	13°C/W
QFP	56°C/W	12°C/W
SQFP	68°C/W	15.5°C/W

All thermal impedance data is approximate for static air conditions at 1W of power dissipation. Values will change depending on operation conditions and application. See the Intel *Packaging Handbook* (order number 240800) for a description of Intel's thermal impedance test methodology.

Table 5. 8XC196KD Memory Map

Description	Address
External Memory or I/O	0FFFFH 0A000H
Internal ROM/OTPROM or External Memory (Determined by $\overline{EA}$)	9FFFH 2080H
Reserved. Must contain FFH. (Note 5)	207FH 205EH
PTS Vectors	205DH 2040H
Upper Interrupt Vectors	203FH 2030H
ROM/OTPROM Security Key	202FH 2020H
Reserved. Must contain FFH. (Note 5)	201FH 201AH
Reserved. Must Contain 20H (Note 5)	2019H
CCB	2018H
Reserved. Must contain FFH. (Note 5)	2017H 2014H
Lower Interrupt Vectors	2013H 2000H
Port 3 and Port 4	1FFFH 1FFEH
External Memory	1FFDH 0400H
1000 Bytes Register RAM (Note 1)	03FFH 0018H
CPU SFR's (Notes 1, 3)	0017H 0000H

NOTES:
1. Code executed in locations 0000H to 03FFH will be forced external.
2. Reserved memory locations must contain 0FFH unless noted.
3. Reserved SFR bit locations must contain 0.
4. Refer to 8XC196KC for SFR descriptions.
5. **WARNING:** Reserved memory locations must not be written or read. The contents and/or function of these locations may change with future revisions of the device. Therefore, a program that relies on one or more of these locations may not function properly.

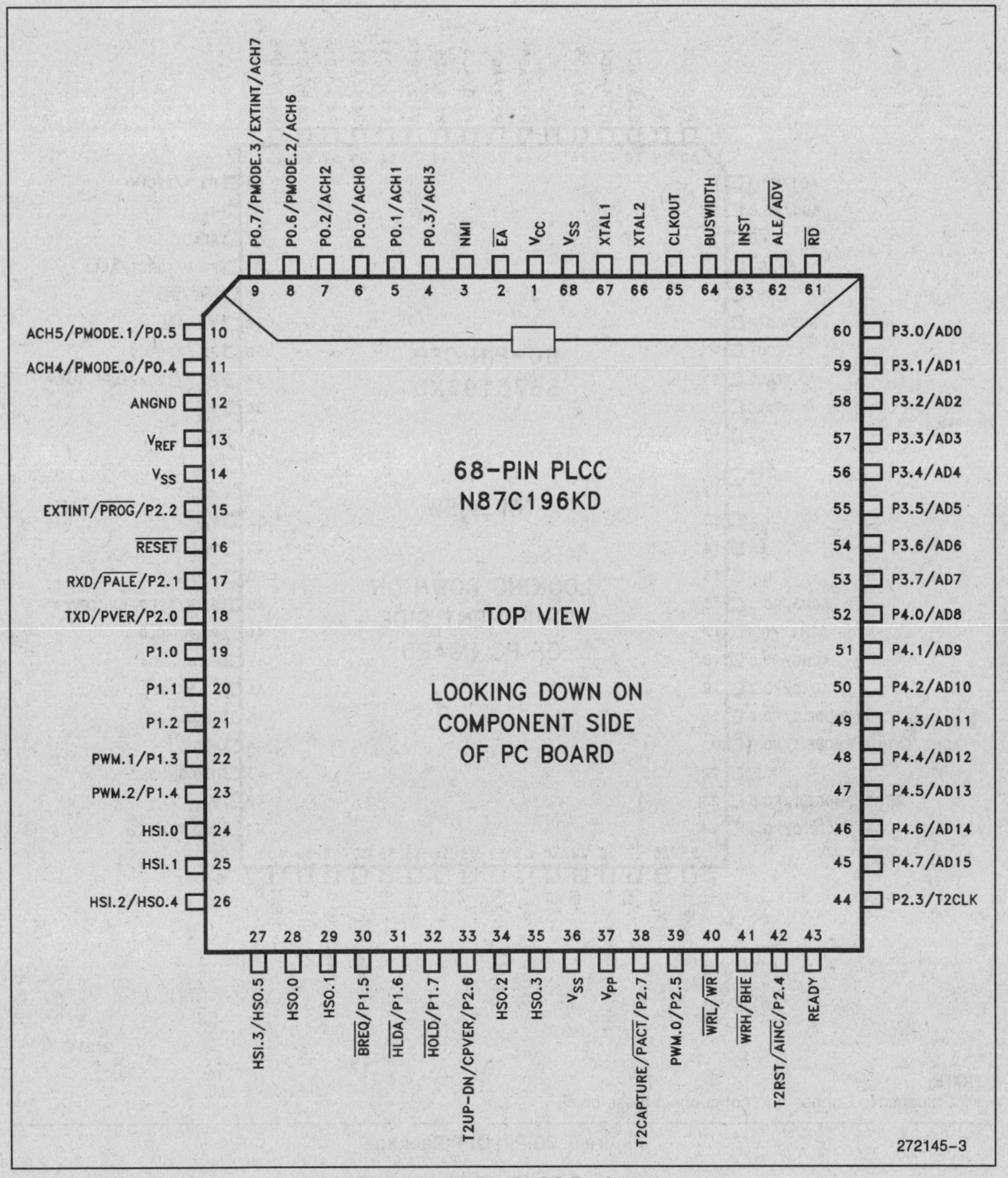

Figure 4. 68-Pin PLCC Package

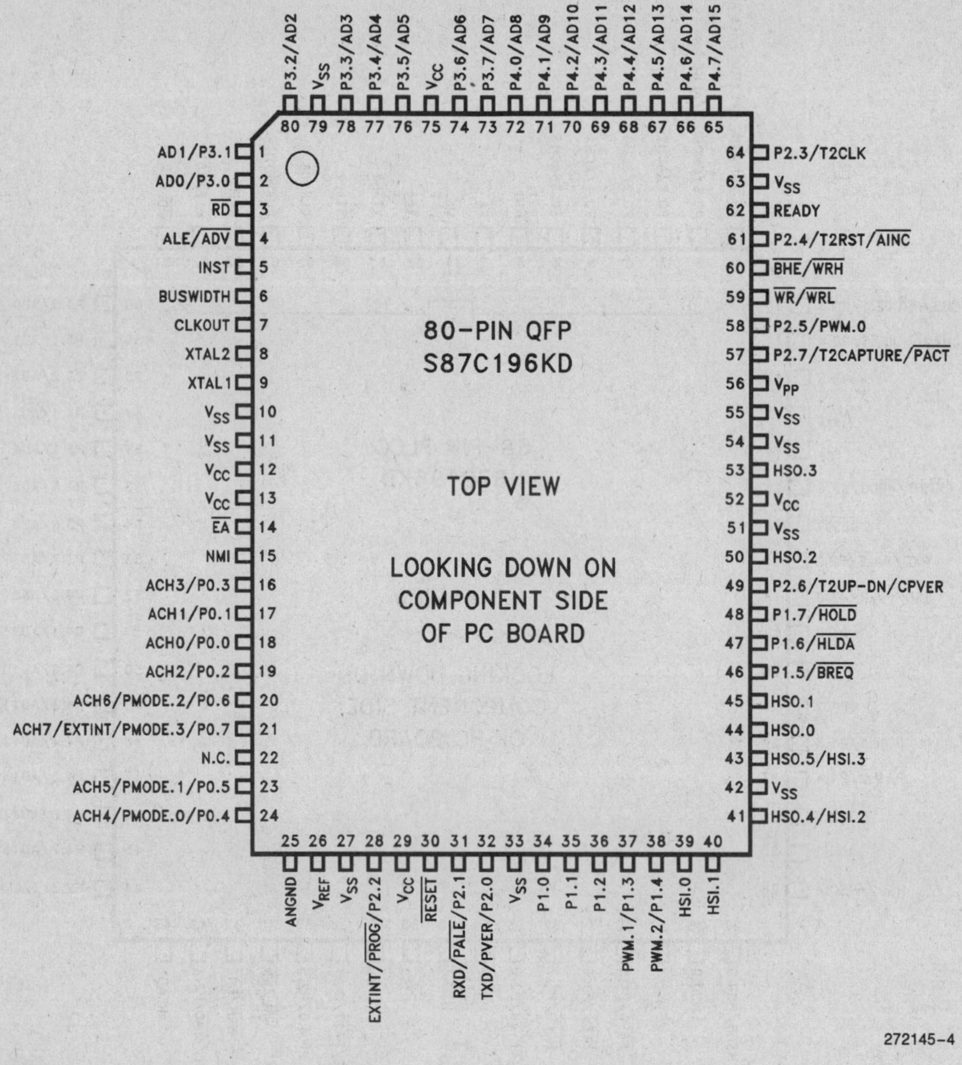

NOTE:
N.C. means No Connect (do not connect these pins).

Figure 5. 80-Pin QFP Package

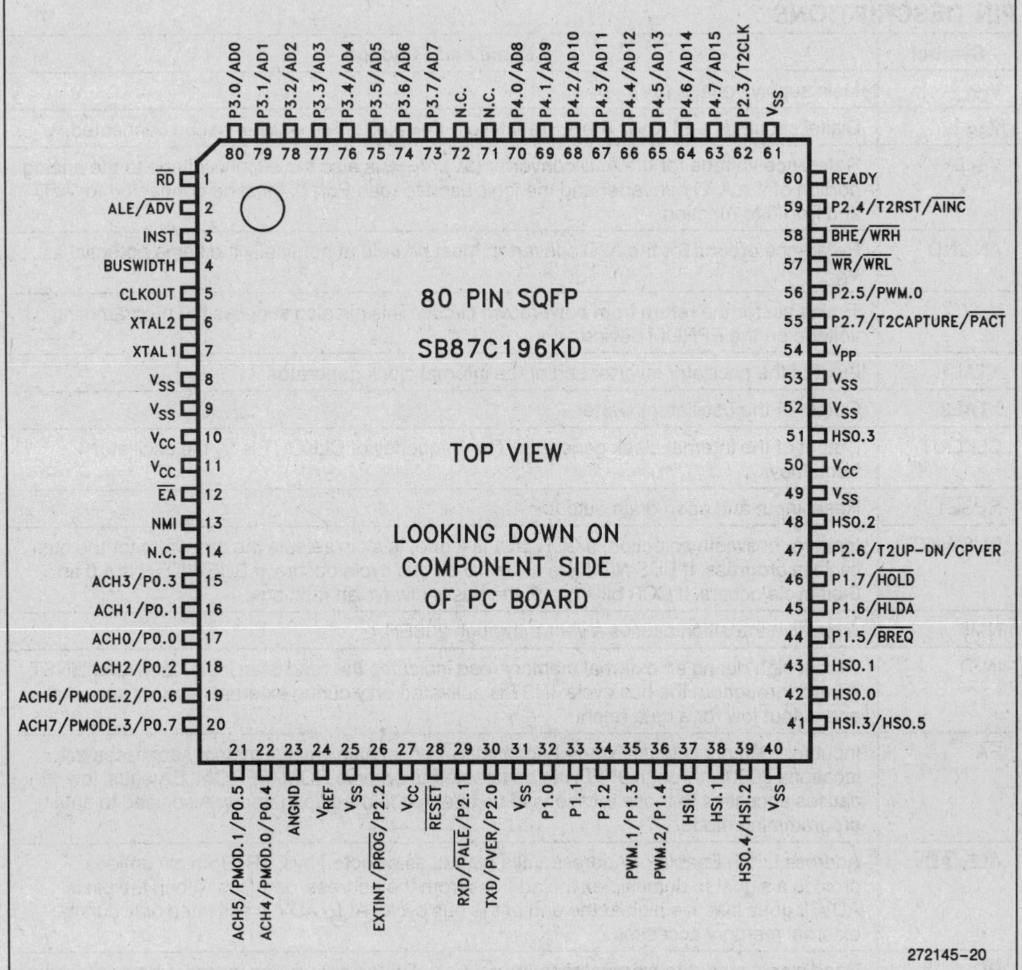

NOTE:
N.C. means No Connect (do not connect these pins).

Figure 6. 80-Pin SQFP Package

PIN DESCRIPTIONS

Symbol	Name and Function
V_{CC}	Main supply voltage (5V).
V_{SS}	Digital circuit ground (0V). There are multiple V_{SS} pins, all of which must be connected.
V_{REF}	Reference voltage for the A/D converter (5V). V_{REF} is also the supply voltage to the analog portion of the A/D converter and the logic used to read Port 0. Must be connected for A/D and Port 0 to function.
ANGND	Reference ground for the A/D converter. Must be held at nominally the same potential as V_{SS}.
V_{PP}	Timing pin for the return from powerdown circuit. This pin also supplies the programming voltage on the EPROM device.
XTAL1	Input of the oscillator inverter and of the internal clock generator.
XTAL2	Output of the oscillator inverter.
CLKOUT	Output of the internal clock generator. The frequency of CLKOUT is ½ the oscillator frequency.
$\overline{RESET}$	Reset input and open drain output.
BUSWIDTH	Input for buswidth selection. If CCR bit 1 is a one, this pin selects the bus width for the bus cycle in progress. If BUSWIDTH is a 1, a 16-bit bus cycle occurs. If BUSWIDTH is a 0 an 8-bit cycle occurs. If CCR bit 1 is a 0, the bus is always an 8-bit bus.
NMI	A positive transition causes a vector through 203EH.
INST	Output high during an external memory read indicates the read is an instruction fetch. INST is valid throughout the bus cycle. INST is activated only during external memory accesses and output low for a data fetch.
$\overline{EA}$	Input for memory select (External Access). $\overline{EA}$ equal high causes memory accesses to locations 2000H through 9FFFH to be directed to on-chip ROM/EPROM. $\overline{EA}$ equal low causes accesses to those locations to be directed to off-chip memory. Also used to enter programming mode.
ALE/$\overline{ADV}$	Address Latch Enable or Address Valid output, as selected by CCR. Both pin options provide a signal to demultiplex the address from the address/data bus. When the pin is $\overline{ADV}$, it goes inactive high at the end of the bus cycle. ALE/$\overline{ADV}$ is activated only during external memory accesses.
$\overline{RD}$	Read signal output to external memory. $\overline{RD}$ is activated only during external memory reads.
$\overline{WR}$/$\overline{WRL}$	Write and Write Low output to external memory, as selected by the CCR. $\overline{WR}$ will go low for every external write, while $\overline{WRL}$ will go low only for external writes where an even byte is being written. $\overline{WR}$/$\overline{WRL}$ is activated only during external memory writes.
$\overline{BHE}$/$\overline{WRH}$	Bus High Enable or Write High output to external memory, as selected by the CCR. $\overline{BHE}$ will go low for external writes to the high byte of the data bus. $\overline{WRH}$ will go low for external writes where an odd byte is being written. $\overline{BHE}$/$\overline{WRH}$ is activated only during external memory writes.
READY	Ready input to lengthen external memory cycles, for interfacing to slow or dynamic memory, or for bus sharing. When the external memory is not being used, READY has no effect.
HSI	Inputs to High Speed Input Unit. Four HSI pins are available: HSI.0, HSI.1, HSI.2 and HSI.3. Two of them (HSI.2 and HSI.3) are shared with the HSO Unit.
HSO	Outputs from High Speed Output Unit. Six HSO pins are available: HSO.0, HSO.1, HSO.2, HSI.3, HSO.4 and HSO.5. Two of them (HSO.4 and HSO.5) are shared with the HSI Unit.

PIN DESCRIPTIONS (Continued)

Symbol	Name and Function
Port 0	8-bit high impedance input-only port. These pins can be used as digital inputs and/or as analog inputs to the on-chip A/D converter.
Port 1	8-bit quasi-bidirectional I/O port.
Port 2	8-bit multi-functional port. All of its pins are shared with other functions in the 8XC196KD. Pins 2.6 and 2.7 are quasi-bidirectional.
Ports 3 and 4	8-bit bidirectional I/O ports with open drain outputs. These pins are shared with the multiplexed address/data bus which has strong internal pullups.
$\overline{HOLD}$	Bus Hold input requesting control of the bus.
$\overline{HLDA}$	Bus Hold acknowledge output indicating release of the bus.
$\overline{BREQ}$	Bus Request output activated when the bus controller has a pending external memory cycle.
PMODE	Determines the EPROM programming mode.
$\overline{PACT}$	A low signal in Auto Programming mode indicates that programming is in process. A high signal indicates programming is complete.
$\overline{PALE}$	A falling edge in Slave Programming Mode and Auto Configuration Byte Programming Mode indicates that ports 3 and 4 contain valid programming address/command information (input to slave).
$\overline{PROG}$	A falling edge in Slave Programming Mode indicates that ports 3 and 4 contain valid programming data (input to slave).
PVER	A high signal in Slave Programming Mode and Auto Configuration Byte Programming Mode indicates the byte programmed correctly.
CPVER	Cummulative Program Output Verification. Pin is high if all locations have programmed correctly since entering a programming mode.
$\overline{AINC}$	Auto Increment. Active low input enables the auto increment mode. Auto increment allows reading or writing sequential EPROM locations without address transactions across the PBUS for each read or write.

ELECTRICAL CHARACTERISTICS

ABSOLUTE MAXIMUM RATINGS*

Ambient Temperature
Under Bias .... −55°C to +125°C

Storage Temperature .... −65°C to +150°C

Voltage On Any Pin to V_{SS}
Except EA and V_{PP} .... −0.5V to +7.0V(1)

Voltage from $\overline{EA}$ or
V_{PP} to V_{SS} or ANGND .... −0.5V to +13.00V

Power Dissipation .... 1.5W(2)

NOTES:
1. This includes V_{PP} and $\overline{EA}$ on ROM or CPU only devices.
2. Power dissipation is based on package heat transfer limitations, not device power consumption.

NOTICE: This data sheet contains information on products in the sampling and initial production phases of development. It is valid for the devices indicated in the revision history. The specifications are subject to change without notice.

**WARNING: Stressing the device beyond the "Absolute Maximum Ratings" may cause permanent damage. These are stress ratings only. Operation beyond the "Operating Conditions" is not recommended and extended exposure beyond the "Operating Conditions" may affect device reliability.*

OPERATING CONDITIONS

Symbol	Description	Min	Max	Units
T_A	Ambient Temperature Under Bias **Commercial** Temp.	0	+70	°C
V_{CC}	Digital Supply Voltage	4.50	5.50	V
V_{REF}	Analog Supply Voltage	4.00	5.50	V
ANGND	Analog Ground Voltage	V_{SS} − 0.4	V_{SS} + 0.4	V(1)
F_{OSC}	Oscillator Frequency **(8XC196KD)**	8	16	MHz
F_{OSC}	Oscillator Frequency **(8XC196KD20)**	8	20	MHz

NOTE:
1. ANGND and V_{SS} should be nominally at the same potential.

DC CHARACTERISTICS (Over Specified Operating Conditions)

Symbol	Description	Min	Max	Units	Test Conditions
V_{IL}	Input Low Voltage	−0.5	0.8	V	
V_{IH}	Input High Voltage (Note 1)	0.2 V_{CC} + 1.0	V_{CC} + 0.5	V	
V_{HYS}	Hysteresis on $\overline{RESET}$	300		mV	V_{CC} = 5.0V
V_{IH1}	Input High Voltage on XTAL 1	0.7 V_{CC}	V_{CC} + 0.5	V	
V_{IH2}	Input High Voltage on RESET	2.2	V_{CC} + 0.5	V	
V_{OL}	Output Low Voltage		0.3 0.45 1.5	V V V	I_{OL} = 200 μA I_{OL} = 2.8 mA I_{OL} = 7 mA
V_{OL1}	Output Low Voltage in RESET on P2.5 (Note 2)		0.8	V	I_{OL} = +0.4 mA
V_{OH}	Output High Voltage (Standard Outputs) (Note 4)	V_{CC} − 0.3 V_{CC} − 0.7 V_{CC} − 1.5		V V V	I_{OH} = −200 μA I_{OH} = −3.2 mA I_{OH} = −7 mA
V_{OH1}	Output High Voltage (Quasi-bidirectional Outputs) (Note 3)	V_{CC} − 0.3 V_{CC} − 0.7 V_{CC} − 1.5		V V V	I_{OH} = −10 μA I_{OH} = −30 μA I_{OH} = −60 μA

DC CHARACTERISTICS (Over Specified Operating Conditions) (Continued)

Symbol	Description	Min	Typ	Max	Units	Test Conditions
I_{OH1}	Logical 1 Output Current in Reset on P2.0. Do not exceed this or device may enter test modes.	−0.8			mA	$V_{IH} = V_{CC} - 1.5V$
I_{IL2}	Logical 0 Input Current in Reset on P2.0. Maximum current that must be sunk by external device to ensure test mode entry.			−12.0	mA	$V_{IN} = 0.45V$
I_{IH1}	Logical 1 Input Current. Maximum current that external device must source to initiate NMI.			+200	μA	$V_{IN} = 2.4V$
I_{LI}	Input Leakage Current (Std. Inputs) (Note 5)			±10	μA	$0 < V_{IN} < V_{CC} - 0.3V$
I_{LI1}	Input Leakage Current (Port 0)			±3	μA	$0 < V_{IN} < V_{REF}$
I_{TL}	1 to 0 Transition Current (QBD Pins)			−650	μA	$V_{IN} = 2.0V$
I_{IL}	Logical 0 Input Current (QBD Pins)			−70	μA	$V_{IN} = 0.45V$
I_{IL1}	AD Bus in Reset			−70	μA	$V_{IN} = 0.45V$
I_{CC}	Active Mode Current in Reset **(8XC196KD)**		65	75	mA	XTAL1 = 16 MHz $V_{CC} = V_{PP} = V_{REF} = 5.5V$
I_{CC}	Active Mode Current in Reset **(8XC196KD20)**		80	92	mA	XTAL1 = 20 MHz $V_{CC} = V_{PP} = V_{REF} = 5.5V$
I_{IDLE}	Idle Mode Current **(8XC196KD)**		17	25	mA	XTAL1 = 16 MHz $V_{CC} = V_{PP} = V_{REF} = 5.5V$
I_{IDLE}	Idle Mode Current **(8XC196KD20)**		21	30	mA	XTAL1 = 20 MHz $V_{CC} = V_{PP} = V_{REF} = 5.5V$
I_{PD}	Powerdown Mode Current		8	15	μA	$V_{CC} = V_{PP} = V_{REF} = 5.5V$
I_{REF}	A/D Converter Reference Current		2	5	mA	$V_{CC} = V_{PP} = V_{REF} = 5.5V$
R_{RST}	Reset Pullup Resistor	6K		65K	Ω	$V_{CC} = 5.5V, V_{IN} = 4.0V$
C_S	Pin Capacitance (Any Pin to V_{SS})			10	pF	

NOTES:
1. All pins except RESET and XTAL1.
2. Violating these specifications in Reset may cause the part to enter test modes.
3. QBD (Quasi-bidirectional) pins include Port 1, P2.6 and P2.7.
4. Standard Outputs include AD0–15, $\overline{RD}$, $\overline{WR}$, ALE, $\overline{BHE}$, INST, HSO pins, PWM/P2.5, CLKOUT, RESET, Ports 3 and 4, TXD/P2.0 and RXD (in serial mode 0). The V_{OH} specification is not valid for RESET. Ports 3 and 4 are open-drain outputs.
5. Standard Inputs include HSI pins, READY, BUSWIDTH, RXD/P2.1, EXTINT/P2.2, T2CLK/P2.3 and T2RST/P2.4.
6. Maximum current per pin must be externally limited to the following values if V_{OL} is held above 0.45V or V_{OH} is held below V_{CC} − 0.7V:
   I_{OL} on Output pins: 10 mA
   I_{OH} on quasi-bidirectional pins: self limiting
   I_{OH} on Standard Output pins: 10 mA
7. Maximum current per bus pin (data and control) during normal operation is ±3.2 mA.
8. During normal (non-transient) conditions the following total current limits apply:

Port 1, P2.6	I_{OL}: 29 mA	I_{OH} is self limiting
HSO, P2.0, RXD, $\overline{RESET}$	I_{OL}: 29 mA	I_{OH}: 26 mA
P2.5, P2.7, $\overline{WR}$, $\overline{BHE}$	I_{OL}: 13 mA	I_{OH}: 11 mA
AD0–AD15	I_{OL}: 52 mA	I_{OH}: 52 mA
$\overline{RD}$, ALE, INST–CLKOUT	I_{OL}: 13 mA	I_{OH}: 13 mA

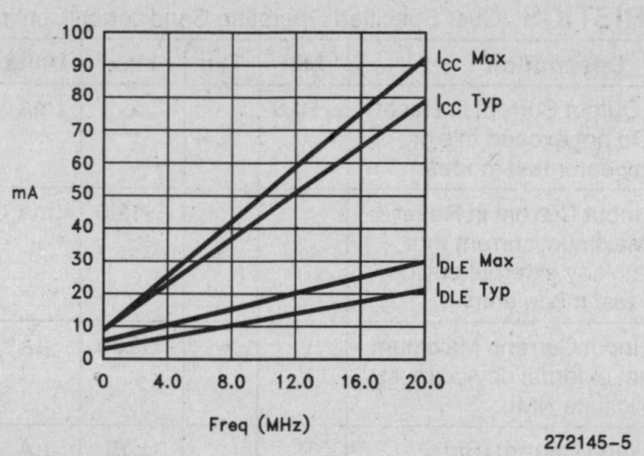

I_{CC} Max = 4.13 × Frequency + 9 mA
I_{CC} Typ = 3.50 × Frequency + 9 mA
I_{IDLE} Max = 1.25 × Frequency + 5 mA
I_{IDLE} Typ = 0.88 × Frequency + 3 mA

NOTE:
Frequencies below 8 MHz are shown for reference only; no testing is performed.

Figure 7. I_{CC} and I_{IDLE} vs Frequency

AC CHARACTERISTICS

For use over specified operating conditions.

Test Conditions: Capacitive load on all pins = 100 pF, Rise and fall times = 10 ns, F_{OSC} = 16/20 MHz

The system must meet these specifications to work with the 80C196KD:

Symbol	Description	Min	Max	Units	Notes
T_{AVYV}	Address Valid to READY Setup		2 T_{OSC} − 68	ns	
T_{YLYH}	Non READY Time	No upper limit		ns	
T_{CLYX}	READY Hold after CLKOUT Low	0	T_{OSC} − 30	ns	(Note 1)
T_{LLYX}	READY Hold after ALE Low	T_{OSC} − 15	2 T_{OSC} − 40	ns	(Note 1)
T_{AVGV}	Address Valid to Buswidth Setup		2 T_{OSC} − 68	ns	
T_{CLGX}	Buswidth Hold after CLKOUT Low	0		ns	
T_{AVDV}	Address Valid to Input Data Valid		3 T_{OSC} − 55	ns	(Note 2)
T_{RLDV}	$\overline{RD}$ Active to Input Data Valid		T_{OSC} − 22	ns	(Note 2)
T_{CLDV}	CLKOUT Low to Input Data Valid		T_{OSC} − 45	ns	
T_{RHDZ}	End of $\overline{RD}$ to Input Data Float		T_{OSC}	ns	
T_{RXDX}	Data Hold after $\overline{RD}$ Inactive	0		ns	

NOTES:
1. If max is exceeded, additional wait states will occur.
2. If wait states are used, add 2 T_{OSC} * N, where N = number of wait states.

AC CHARACTERISTICS (Continued)

For use over specified operating conditions.

Test Conditions: Capacitive load on all pins = 100 pF, Rise and fall times = 10 ns, F_{OSC} = 16/20 MHz

The 80C196KD will meet these specifications:

Symbol	Description	Min	Max	Units	Notes
F_{XTAL}	Frequency on XTAL1 **(8XC196KD)**	8	16	MHz	(Note 1)
F_{XTAL}	Frequency on XTAL1 **(8XC196KD20)**	8	20	MHz	(Note 1)
T_{OSC}	I/F_{XTAL} **(8XC196KD)**	62.5	125	ns	
T_{OSC}	I/F_{XTAL} **(8XC196KD20)**	50	125	ns	
T_{XHCH}	XTAL1 High to CLKOUT High or Low	+20	+110	ns	
T_{CLCL}	CLKOUT Cycle Time	2 T_{OSC}		ns	
T_{CHCL}	CLKOUT High Period	T_{OSC} − 10	T_{OSC} + 15	ns	
T_{CLLH}	CLKOUT Falling Edge to ALE Rising	−5	+15	ns	
T_{LLCH}	ALE Falling Edge to CLKOUT Rising	−20	+15	ns	
T_{LHLH}	ALE Cycle Time	4 T_{OSC}		ns	(Note 4)
T_{LHLL}	ALE High Period	T_{OSC} − 10	T_{OSC} + 10	ns	
T_{AVLL}	Address Setup to ALE Falling Edge	T_{OSC} − 15			
T_{LLAX}	Address Hold after ALE Falling Edge	T_{OSC} − 35		ns	
T_{LLRL}	ALE Falling Edge to $\overline{RD}$ Falling Edge	T_{OSC} − 30		ns	
T_{RLCL}	$\overline{RD}$ Low to CLKOUT Falling Edge	+4	+30	ns	
T_{RLRH}	$\overline{RD}$ Low Period	T_{OSC} − 5		ns	(Note 4)
T_{RHLH}	$\overline{RD}$ Rising Edge to ALE Rising Edge	T_{OSC}	T_{OSC} + 25	ns	(Note 2)
T_{RLAZ}	$\overline{RD}$ Low to Address Float		+5	ns	
T_{LLWL}	ALE Falling Edge to $\overline{WR}$ Falling Edge	T_{OSC} − 10		ns	
T_{CLWL}	CLKOUT Low to $\overline{WR}$ Falling Edge	0	+25	ns	
T_{QVWH}	Data Stable to $\overline{WR}$ Rising Edge	T_{OSC} − 23			(Note 4)
T_{CHWH}	CLKOUT High to $\overline{WR}$ Rising Edge	−5	+15	ns	
T_{WLWH}	$\overline{WR}$ Low Period	T_{OSC} − 20		ns	(Note 4)
T_{WHQX}	Data Hold after $\overline{WR}$ Rising Edge	T_{OSC} − 25		ns	
T_{WHLH}	$\overline{WR}$ Rising Edge to ALE Rising Edge	T_{OSC} − 10	T_{OSC} + 15	ns	(Note 2)
T_{WHBX}	$\overline{BHE}$, INST after $\overline{WR}$ Rising Edge	T_{OSC} − 10		ns	
T_{WHAX}	AD8–15 HOLD after $\overline{WR}$ Rising	T_{OSC} − 30		ns	(Note 3)
T_{RHBX}	$\overline{BHE}$, INST after $\overline{RD}$ Rising Edge	T_{OSC} − 10		ns	
T_{RHAX}	AD8–15 HOLD after $\overline{RD}$ Rising	T_{OSC} − 25		ns	(Note 3)

NOTES:
1. Testing performed at 8 MHz. However, the device is static by design and will typically operate below 1 Hz.
2. Assuming back-to-back bus cycles.
3. 8-Bit bus only.
4. If wait states are used, add 2 T_{OSC} * N, where N = number of wait states.

System Bus Timings

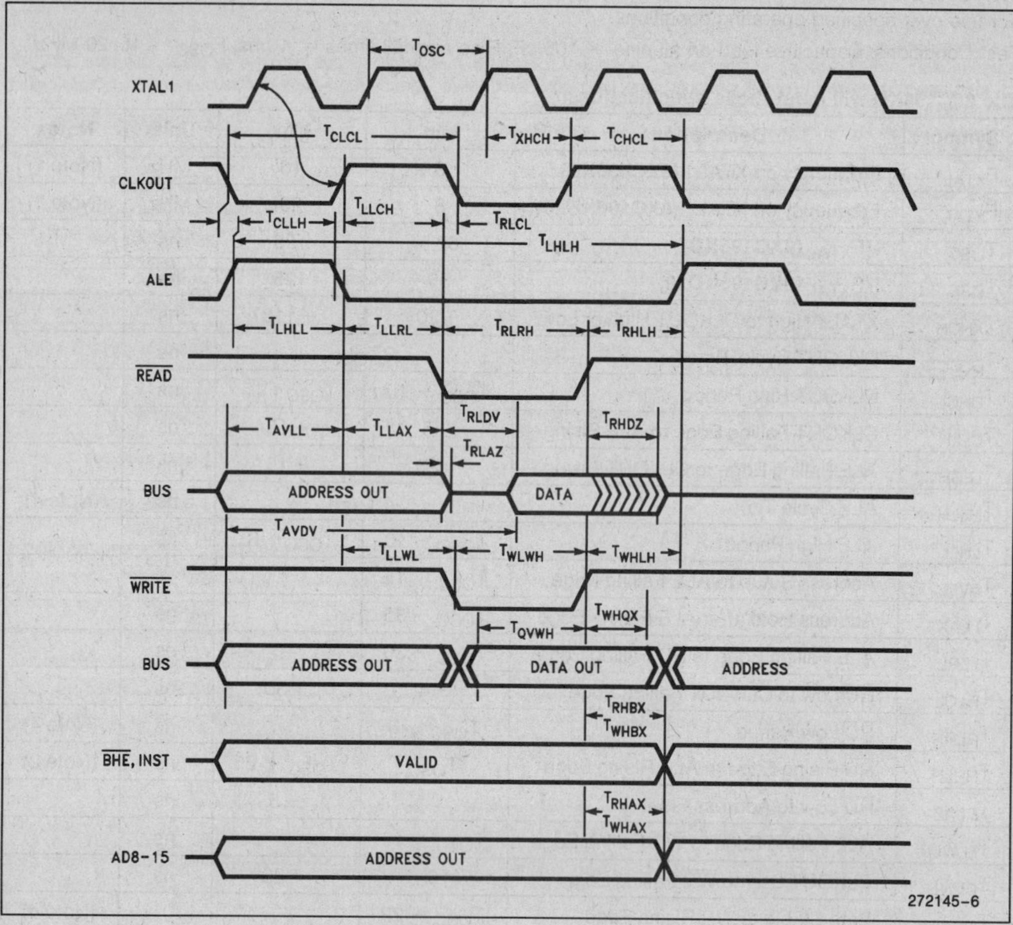

READY Timings (One Wait State)

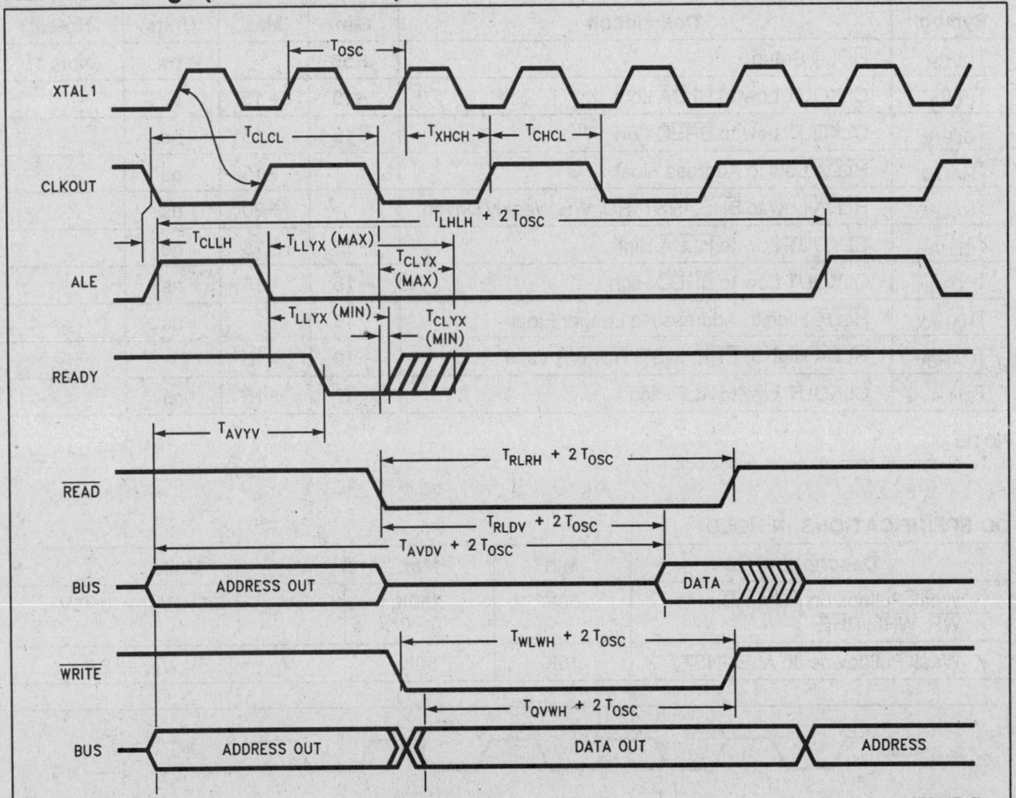

272145-7

Buswidth Timings

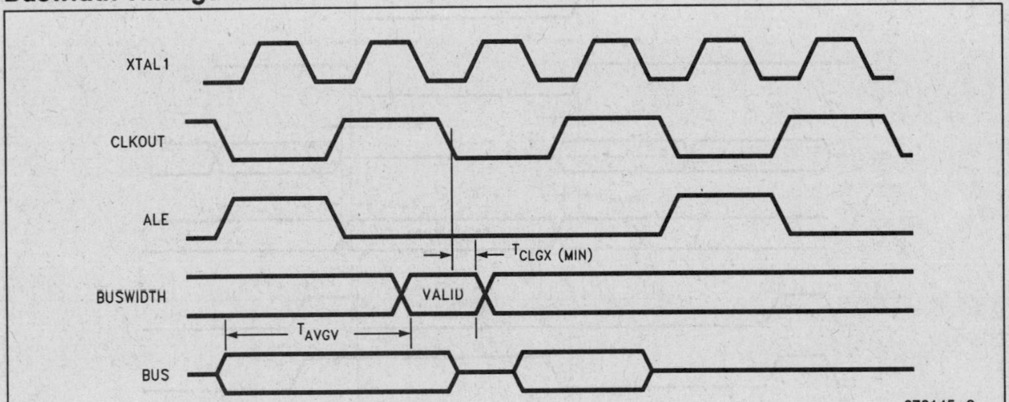

272145-8

$\overline{HOLD}/\overline{HLDA}$ TIMINGS

Symbol	Description	Min	Max	Units	Notes
T_{HVCH}	$\overline{HOLD}$ Setup	+55		ns	(Note 1)
T_{CLHAL}	CLKOUT Low to $\overline{HLDA}$ Low	−15	+15	ns	
T_{CLBRL}	CLKOUT Low to $\overline{BREQ}$ Low	−15	+15	ns	
T_{HALAZ}	$\overline{HLDA}$ Low to Address Float		+15	ns	
T_{HALBZ}	$\overline{HLDA}$ Low to $\overline{BHE}$, INST, $\overline{RD}$, $\overline{WR}$ Weakly Driven		+20	ns	
T_{CLHAH}	CLKOUT Low to $\overline{HLDA}$ High	−15	+15	ns	
T_{CLBRH}	CLKOUT Low to $\overline{BREQ}$ High	−15	+15	ns	
T_{HAHAX}	$\overline{HLDA}$ High to Address No Longer Float	−15		ns	
T_{HAHBV}	$\overline{HLDA}$ High to $\overline{BHE}$, INST, $\overline{RD}$, $\overline{WR}$ Valid	−10	+15	ns	
T_{CLLH}	CLKOUT Low to ALE High	−5	+15	ns	

NOTE:
1. To guarantee recognition at next clock.

DC SPECIFICATIONS IN HOLD

Description	Min	Max	Units
Weak Pullups on $\overline{ADV}$, $\overline{RD}$, $\overline{WR}$, $\overline{WRL}$, $\overline{BHE}$	50K	250K	V_{CC} = 5.5V, V_{IN} = 0.45V
Weak Pulldowns on ALE, INST	10K	50K	V_{CC} = 5.5V, V_{IN} = 2.4

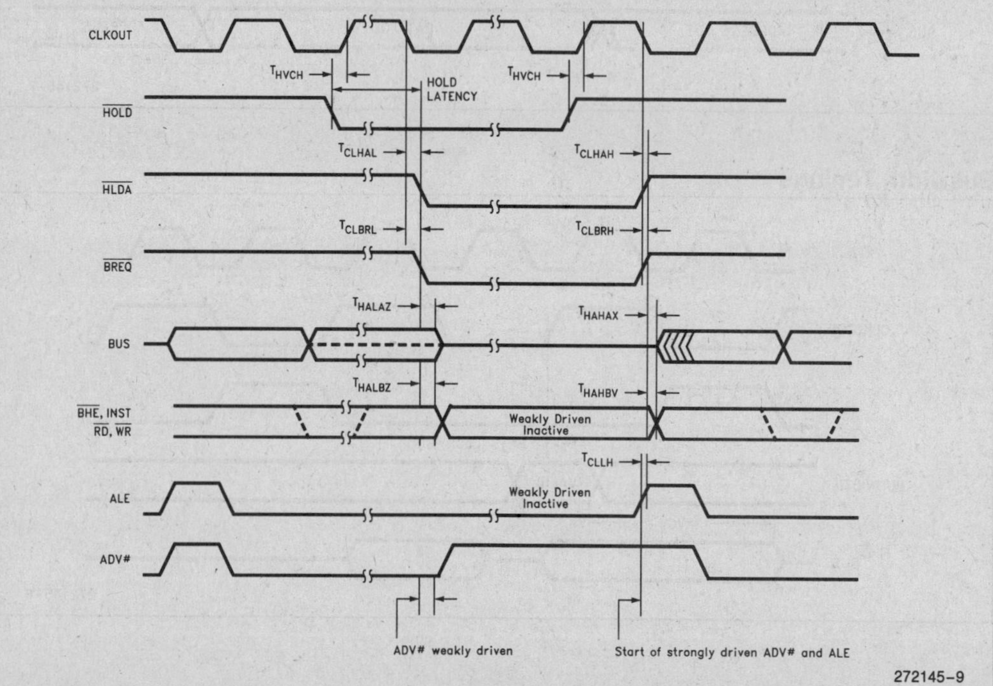

272145-9

MAXIMUM HOLD LATENCY

Bus Cycle Type	
Internal Execution	1.5 States
16-Bit External Execution	2.5 States
8-Bit External Execution	4.5 States

EXTERNAL CLOCK DRIVE (8XC196KD)

Symbol	Parameter	Min	Max	Units
$1/T_{XLXL}$	Oscillator Frequency	8	16.0	MHz
T_{XLXL}	Oscillator Period	62.5	125	ns
T_{XHXX}	High Time	20		ns
T_{XLXX}	Low Time	20		ns
T_{XLXH}	Rise Time		10	ns
T_{XHXL}	Fall Time		10	ns

EXTERNAL CLOCK DRIVE (8XC196KD20)

Symbol	Parameter	Min	Max	Units
$1/T_{XLXL}$	Oscillator Frequency	8	20.0	MHz
T_{XLXL}	Oscillator Period	50	125	ns
T_{XHXX}	High Time	17		ns
T_{XLXX}	Low Time	17		ns
T_{XLXH}	Rise Time		8	ns
T_{XHXL}	Fall Time		8	ns

EXTERNAL CLOCK DRIVE WAVEFORMS

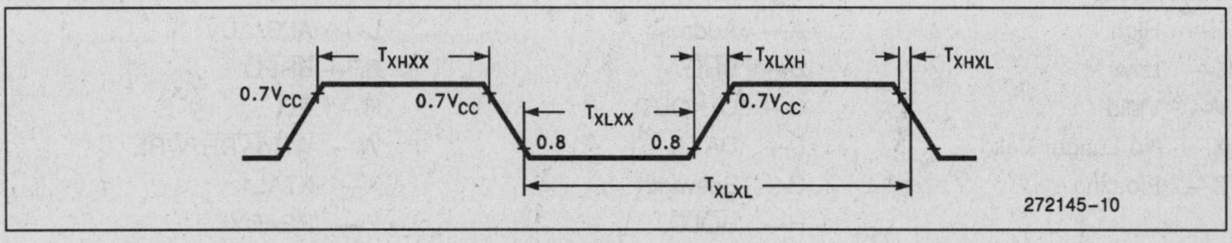

EXTERNAL CRYSTAL CONNECTIONS

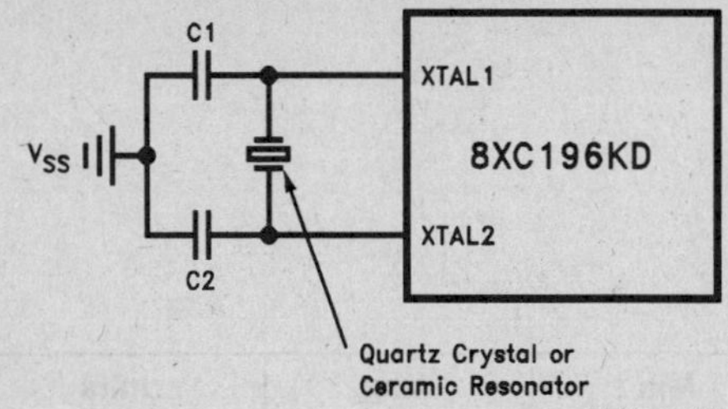

272145-13

NOTE:
Keep oscillator components close to chip and use short, direct traces to XTAL1, XTAL2 and V_{SS}. When using ceramic crystals, C1 = 20 pF, C2 = 20 pF. When using ceramic resonators consult manufacturer for recommended capacitor values.

EXTERNAL CLOCK CONNECTIONS

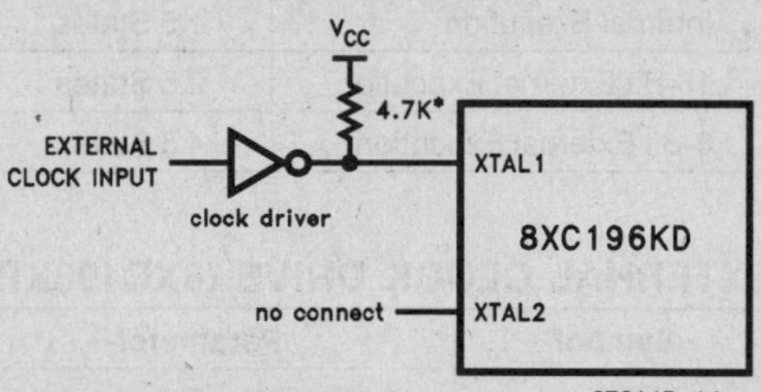

272145-14

NOTE:
*Required if TTL driver used.
Not needed if CMOS driver is used.

AC TESTING INPUT, OUTPUT WAVEFORMS

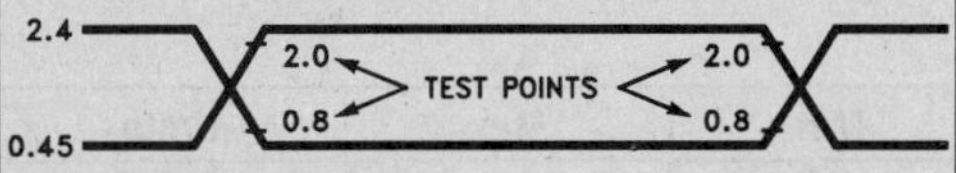

272145-11

AC Testing inputs are driven at 2.4V for a Logic "1" and 0.45V for a Logic "0" Timing measurements are made at 2.0V for a Logic "1" and 0.8V for a Logic "0".

FLOAT WAVEFORMS

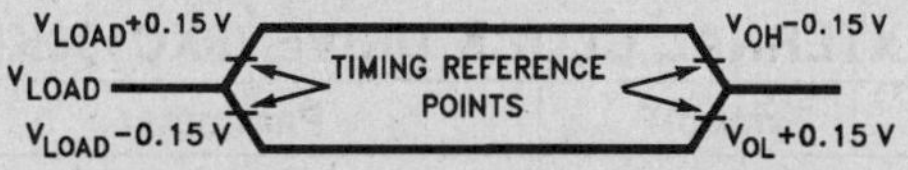

272145-12

For Timing Purposes a Port Pin is no Longer Floating when a 150 mV change from Load Voltage Occurs, and Begins to Float when a 150 mV change from the Loaded V_{OH}/V_{OL} Level occurs; I_{OL}/I_{OH} = ±15 mA.

EXPLANATION OF AC SYMBOLS

Each symbol is two pairs of letters prefixed by "T" for time. The characters in a pair indicate a signal and its condition, respectively. Symbols represent the time between the two signal/condition points.

Conditions:

H— High
L— Low
V— Valid
X— No Longer Valid
Z— Floating

Signals:

A— Address
B— $\overline{\text{BHE}}$
C— CLKOUT
D— DATA
G— Buswidth
H— $\overline{\text{HOLD}}$
HA— $\overline{\text{HLDA}}$
L— ALE/$\overline{\text{ADV}}$
BR— $\overline{\text{BREQ}}$
R— $\overline{\text{RD}}$
W— $\overline{\text{WR}}$/$\overline{\text{WRH}}$/$\overline{\text{WRL}}$
X— XTAL1
Y— READY
Q— Data Out

AC CHARACTERISTICS—SERIAL PORT—SHIFT REGISTER MODE

SERIAL PORT TIMING—SHIFT REGISTER MODE (MODE 0)

Symbol	Parameter	Min	Max	Units
T_{XLXL}	Serial Port Clock Period (BRR ≥ 8002H)	6 T_{OSC}		ns
T_{XLXH}	Serial Port Clock Falling Edge to Rising Edge (BRR ≥ 8002H)	4 T_{OSC} −50	4 T_{OSC} +50	ns
T_{XLXL}	Serial Port Clock Period (BRR = 8001H)	4 T_{OSC}		ns
T_{XLXH}	Serial Port Clock Falling Edge to Rising Edge (BRR = 8001H)	2 T_{OSC} −50	2 T_{OSC} +50	ns
T_{QVXH}	Output Data Valid to Clock Rising Edge	2 T_{OSC} − 50		ns
T_{XHQX}	Output Data Hold after Clock Rising Edge	2 T_{OSC} − 50		ns
T_{XHQV}	Next Output Data Valid after Clock Rising Edge		2 T_{OSC} + 50	ns
T_{DVXH}	Input Data Setup to Clock Rising Edge	T_{OSC} +50		ns
T_{XHDX}	Input Data Hold after Clock Rising Edge	0		ns
T_{XHQZ}	Last Clock Rising to Output Float		1 T_{OSC}	ns

WAVEFORM—SERIAL PORT—SHIFT REGISTER MODE

SERIAL PORT WAVEFORM—SHIFT REGISTER MODE (MODE 0)

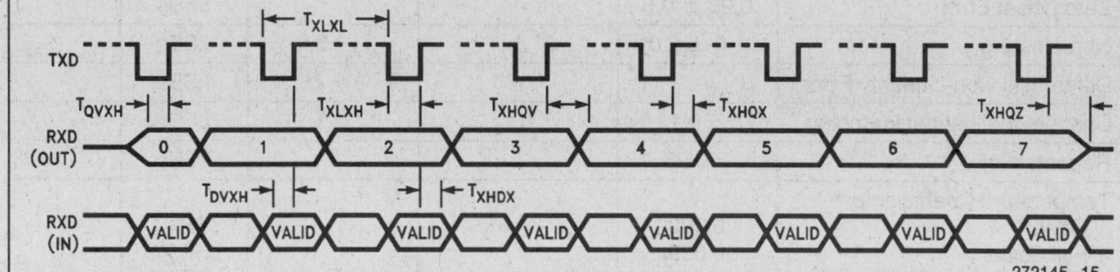

272145–15

A to D CHARACTERISTICS

The A/D converter is ratiometric, so absolute accuracy is dependent on the accuracy and stability of V_{REF}.

10-BIT MODE A/D OPERATING CONDITIONS

Symbol	Description	Min	Max	Units
T_A	Ambient Temperature **Commercial** Temp.	0	+70	°C
V_{CC}	Digital Supply Voltage	4.50	5.50	V
V_{REF}	Analog Supply Voltage	4.00	5.50	V
ANGND	Analog Ground Voltage	V_{SS} − 0.40	V_{CC} + 0.40	V
T_{SAM}	Sample Time	1.0		μs(1)
T_{CONV}	Conversion Time	10	20	μs(1)
F_{OSC}	Oscillator Frequency **(8XC196KD)**	8.0	16.0	MHz
F_{OSC}	Oscillator Frequency **(8XC196KD20)**	8.0	20.0	MHz

NOTE:
1. The value of AD_TIME is selected to meet these specifications.

10-BIT MODE A/D CHARACTERISTICS (Over Specified Operating Conditions)

Parameter	Typical(1)	Minimum	Maximum	Units*	Notes
Resolution		1024 10	1024 10	Levels Bits	
Absolute Error		0	±3	LSBs	
Full Scale Error	0.25 ± 0.5			LSBs	
Zero Offset Error	0.25 ± 0.5			LSBs	
Non-Linearity	1.0 ± 2.0	0	±3	LSBs	
Differential Non-Linearity Error		> −1	+2	LSBs	
Channel-to-Channel Matching	±0.1	0	±1	LSBs	
Repeatability	±0.25			LSBs	
Temperature Coefficients: Offset Full Scale Differential Non-Linearity	 0.009 0.009 0.009			 LSB/°C LSB/°C LSB/°C	
Off Isolation		−60		dB	2, 3
Feedthrough	−60			dB	2
V_{CC} Power Supply Rejection	−60			dB	2
Input Series Resistance		750	1.2K	Ω	4
Voltage on Analog Input Pin		ANGND − 0.5	V_{REF} + 0.5	V	5, 6
DC Input Leakage		0	±3.0	μA	
Sampling Capacitor	3			pF	

NOTES:
*An "LSB" as used here has a value of approxiimately 5 mV. (See Embedded Microcontrollers and Processors Handbook for A/D glossary of terms.)
1. These values are expected for most parts at 25°C but are not tested or guaranteed.
2. DC to 100 KHz.
3. Multiplexer Break-Before-Make is guaranteed.
4. Resistance from device pin, through internal MUX, to sample capacitor.
5. These values may be exceeded if the pin current is limited to ±2 mA.
6. Applying voltages beyond these specifications will degrade the accuracy of other channels being converted.
7. All conversions performed with processor in IDLE mode.

8-BIT MODE A/D OPERATING CONDITIONS

Symbol	Description	Min	Max	Units
T_A	Ambient Temperature **Commercial** Temp.	0	+70	°C
V_{CC}	Digital Supply Voltage	4.50	5.50	V
V_{REF}	Analog Supply Voltage	4.00	5.50	V
ANGND	Analog Ground Voltage	V_{SS} −0.40	V_{SS} + 0.40	V
T_{SAM}	Sample Time	1.0		μs(1)
T_{CONV}	Conversion Time	7	20	μs(1)
F_{OSC}	Oscillator Frequency **(8XC196KD)**	8.0	16.0	MHz
F_{OSC}	Oscillator Frequency **(8XC196KD20)**	8.0	20.0	MHz

NOTE:
1. The value of AD_TIME is selected to meet these specifications.

8-BIT MODE A/D CHARACTERISTICS (Over Specified Operating Conditions)

Parameter	Typical(1)	Minimum	Maximum	Units*	Notes
Resolution		256 8	256 8	Levels Bits	
Absolute Error		0	±1	LSBs	
Full Scale Error	±0.5			LSBs	
Zero Offset Error	±0.5			LSBs	
Non-Linearity		0	±1	LSBs	
Differential Non-Linearity Error		> −1	+1	LSBs	
Channel-to-Channel Matching			±1	LSBs	
Repeatability	±0.25			LSBs	
Temperature Coefficients: Offset Full Scale Differential Non-Linearity	 0.003 0.003 0.003			 LSB/°C LSB/°C LSB/°C	
Off Isolation		−60		dB	2, 3
Feedthrough	−60			dB	2
V_{CC} Power Supply Rejection	−60			dB	2
Input Series Resistance		750	1.2K	Ω	4
Voltage on Analog Input Pin		V_{SS} − 0.5	V_{REF} + 0.5	V	5, 6
DC Input Leakage		0	±3.0	μA	
Sampling Capacitor	3			pF	

NOTES:
*An "LSB" as used here has a value of approximately 20 mV. (See Embedded Microcontrollers and Processors Handbook for A/D glossary of terms).
1. These values are expected for most parts at 25°C but are not tested or guaranteed.
2. DC to 100 KHz.
3. Multiplexer Break-Before-Make is guaranteed.
4. Resistance from device pin, through internal MUX, to sample capacitor.
5. These values may be exceeded if pin current is limited to ±2 mA.
6. Applying voltages beyond these specifications will degrade the accuracy of other channels being converted.
7. All conversions performed with processor in IDLE mode.

OTPROM SPECIFICATIONS

OPERATING CONDITIONS

Symbol	Description	Min	Max	Units
T_A	Ambient Temperature During Programming	20	30	C
V_{CC}	Supply Voltage During Programming	4.5	5.5	V(1)
V_{REF}	Reference Supply Voltage During Programming	4.5	5.5	V(1)
V_{PP}	Programming Voltage	12.25	12.75	V(2)
V_{EA}	EA Pin Voltage	12.25	12.75	V(2)
F_{OSC}	Oscillator Frequency during Auto and Slave Mode Programming	6.0	8.0	MHz
F_{OSC}	Oscillator Frequency during Run-Time Programming **(8XC196KD)**	6.0	16.0	MHz
F_{OSC}	Oscillator Frequency during Run-Time Programming **(8XC196KD20)**	6.0	20.0	MHz

NOTES:
1. V_{CC} and V_{REF} should nominally be at the same voltage during programming.
2. V_{PP} and V_{EA} must never exceed the maximum specification, or the device may be damaged.
3. V_{SS} and ANGND should nominally be at the same potential (0V).
4. Load capacitance during Auto and Slave Mode programming = 150 pF.

AC OTPROM PROGRAMMING CHARACTERISTICS (SLAVE MODE)

Symbol	Description	Min	Max	Units
T_{SHLL}	Reset High to First $\overline{PALE}$ Low	1100		T_{OSC}
T_{LLLH}	$\overline{PALE}$ Pulse Width	50		T_{OSC}
T_{AVLL}	Address Setup Time	0		T_{OSC}
T_{LLAX}	Address Hold Time	100		T_{OSC}
T_{PLDV}	$\overline{PROG}$ Low to Word Dump Valid		50	T_{OSC}
T_{PHDX}	Word Dump Data Hold		50	T_{OSC}
T_{DVPL}	Data Setup Time	0		T_{OSC}
T_{PLDX}	Data Hold Time	400		T_{OSC}
T_{PLPH}(1)	$\overline{PROG}$ Pulse Width	50		T_{OSC}
T_{PHLL}	$\overline{PROG}$ High to Next $\overline{PALE}$ Low	220		T_{OSC}
T_{LHPL}	$\overline{PALE}$ High to $\overline{PROG}$ Low	220		T_{OSC}
T_{PHPL}	$\overline{PROG}$ High to Next $\overline{PROG}$ Low	220		T_{OSC}
T_{PHIL}	$\overline{PROG}$ High to AINC Low	0		T_{OSC}
T_{ILIH}	$\overline{AINC}$ Pulse Width	240		T_{OSC}
T_{ILVH}	PVER Hold after $\overline{AINC}$ Low	50		T_{OSC}
T_{ILPL}	$\overline{AINC}$ Low to $\overline{PROG}$ Low	170		T_{OSC}
T_{PHVL}	$\overline{PROG}$ High to $\overline{PVER}$ Valid		220	T_{OSC}

NOTE:
1. This specification is for the Word Dump Mode. For programming pulses, use the Modified Quick Pulse Algorithm.

DC OTPROM PROGRAMMING CHARACTERISTICS

Symbol	Description	Min	Max	Units
I_{PP}	V_{PP} Supply Current (When Programming)		100	mA

NOTE:
Do not apply V_{PP} until V_{CC} is stable and within specifications and the oscillator/clock has stabilized or the device may be damaged.

OTPROM PROGRAMMING WAVEFORMS

SLAVE PROGRAMMING MODE DATA PROGRAM MODE WITH SINGLE PROGRAM PULSE

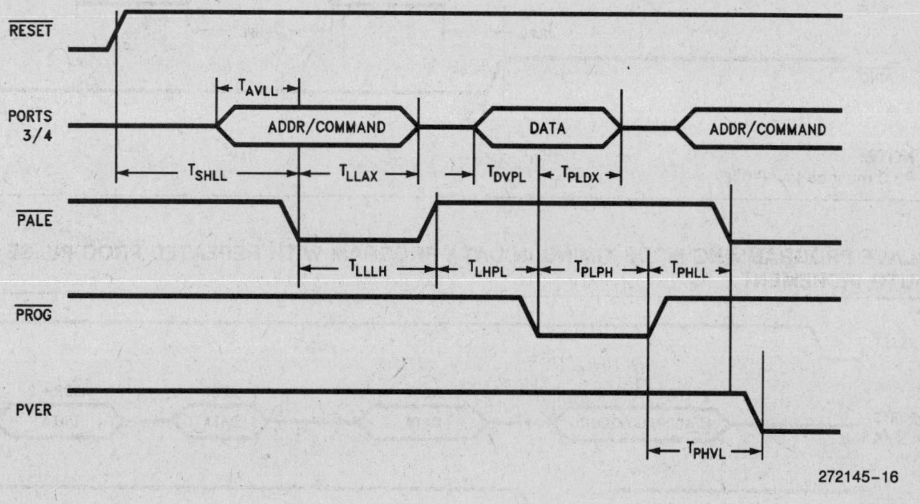

NOTE:
P3.0 must be high ("1")

SLAVE PROGRAMMING MODE IN WORD DUMP WITH AUTO INCREMENT

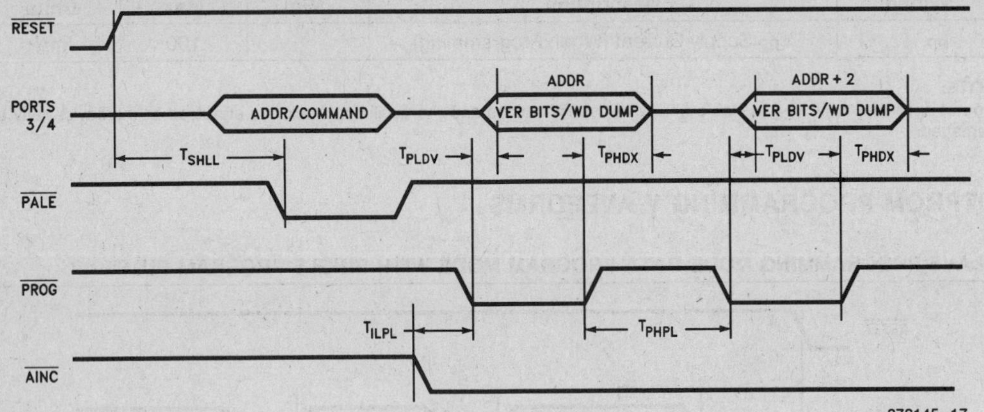

272145-17

NOTE:
P3.0 must be low ("0")

SLAVE PROGRAMMING MODE TIMING IN DATA PROGRAM WITH REPEATED PROG PULSE AND AUTO INCREMENT

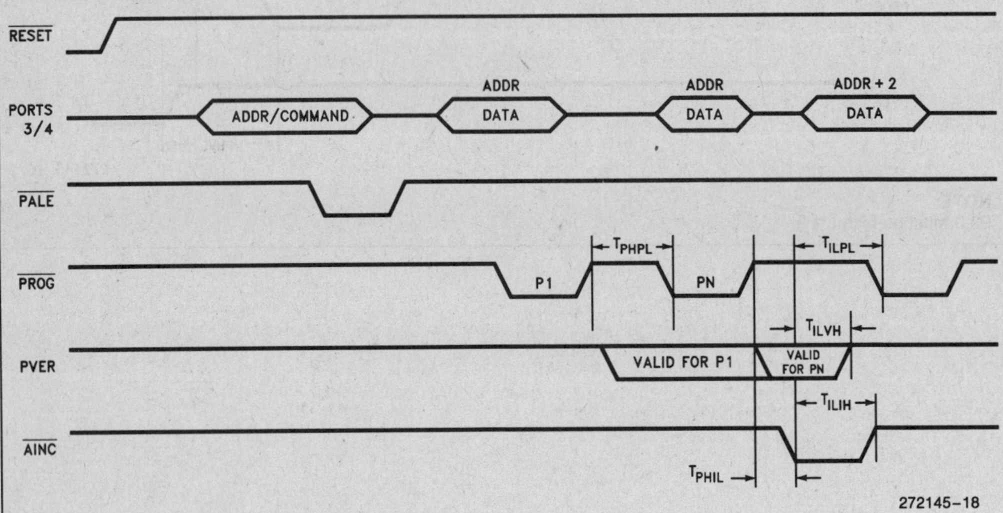

272145-18

ADVANCE INFORMATION

8XC196KC TO 8XC196KD DESIGN CONSIDERATIONS

1. Memory Map. The 8XC196KD has 1024 bytes of RAM/SFRs and 32K of OTPROM. The extra 512 bytes of RAM reside in locations 0200H to 03FFH, and the extra 16 Kbytes of OTPROM reside in locations 6000H to 9FFFH. On the 87C196KC these locations are always external, so KC code may have to be modified to run on the KD.
2. The vertical window scheme has been extended to include all on-chip RAM.
3. IOC3.1 controls the CLKOUT signal. This bit must be 0 to enable CLKOUT.
4. The 87C196KD has a different autoprogramming algorithm to support 32K of on-chip OTPROM.

8XC196KD ERRATA

1. 83C196KD can possibly miss interrupts on P0.7. See techbit MC0893.

DATA SHEET REVISION HISTORY

This data sheet is valid for devices with a "D" and "E" at the end of the topside tracking number. Data sheets are changed as new device information becomes available. Verify with your local Intel sales office that you have the latest version before finalizing a design or ordering devices.

The following are important differences between the 272145-002 and 272145-003 data sheets:

1. I_{IL1} specification (logic 0 input current in reset) was misnamed. It is renamed I_{IL2}.
2. T_{LLYV} and T_{LLGV} were removed. These specifications are not necessary for high-speed system designs.
3. An errata with 83C196KD P0.7 EXTINT was added to the errata section.

The following are important differences between the 272145-001 and 272145-002 data sheets:

1. Added 20 MHz specifications.
2. Added 80-lead SQFP package pinout.
3. Changed QFP Package θ_{JA} to 56°C/W from 42°C/W.
4. Changed V_{HYS} to 300 mV from 150 mV.
5. Changed I_{CC} Typical specification at 16 MHz to 65 mA from 50 mA.
6. Changed I_{CC} Maximum specification at 16 MHz to 75 mA from 70 mA.
7. Changed I_{IDLE} Typical specification to 17 mA from 15 mA.
8. Changed I_{IDLE} Maximum specification to 25 mA from 30 mA.
9. Changed I_{PD} Typical specification to 8 μA from 15 μA.
10. Added I_{PD} Maximum specification.
11. Changed T_{CLDV} Maximum specification to $T_{OSC} - 45$ from $T_{OSC} - 50$.
12. Changed T_{LLAX} Minimum specification to $T_{OSC} - 35$ from $T_{OSC} - 40$.
13. Changed T_{CHWH} Minimum specification to -5 from -10.
14. Changed T_{RHAX} Minimum specification to $T_{OSC} - 25$ from $T_{OSC} - 30$.
15. Changed T_{HALAZ} Maximum specification to $+15$ from $+10$.
16. Changed T_{HALBZ} Maximum specification to $+20$ from $+15$.
17. Added T_{HAHBV} Maximum specification.
18. Changed T_{SAM} for 10-bit mode to 1 μs from 3 μs.
19. Changed T_{SAM} for 8-bit mode to 1 μs from 2 μs.
20. Changed I_{IH1} test condition to $V_{IN} = 2.4V$ from 5.5V.
21. Changed I_{IH1} maximum specification to $+200$ μA from $+100$ μA.
22. Removed NMI from list of standard inputs.
23. Updated I_{CC} and I_{IDLE} vs frequency graph.
24. Updated note under DC EPROM Programming Characteristics.
25. Changed I_{LI1} maximum specification to -12 mA from -6 mA.

4

8XC196KR/KQ/JR/JQ COMMERCIAL/EXPRESS CHMOS MICROCONTROLLER

87C196KR/KQ/JR/JQ—16 Kbytes of On-Chip OTPROM
80C196KR/KQ/JR/JQ—ROMless

- High Performance CHMOS 16-Bit CPU
- 16 MHz Operating Frequency
- Up to 488 Bytes of On-Chip Register RAM
- 256 Bytes of Additional RAM (Code or Data RAM)
- Register-Register Architecture
- 8 Channel/10-Bit A/D with Sample/Hold
- 37 Prioritized Interrupt Sources
- Up to Seven 8-Bit (56) I/O Ports
- Full Duplex Serial I/O Port (SIO) and Full Duplex Synchronous Serial I/O Port (SSIO) with Dedicated Baud Rate Generators
- Interprocessor Communication Slave Port
- Watchdog Timer
- High-Speed Peripheral Transaction Server (PTS)
- Two Programmable 16-Bit Timer/Counters with Prescale, Cascading, Standard and Quadrature Counting Inputs
- 10 High-Speed Capture/Compare (EPA)
- Two Dedicated High Speed Compare Registers
- Two Flexible 16-Bit Timer/Counters
- Quadrature Counting Inputs
- Flexible 8-/16-Bit External Bus
- Programmable Bus (HOLD/HLDA)
- 1.75 μs 16 x 16 Multiply
- 3 μs 32/16 Divide
- Extended Temperature Available
- 68-Pin and 52-Pin PLCC Packages

Device	Pins/Package	OTPROM	Reg RAM	Internal RAM	I/O	EPA	SIO	SSIO	A/D
87C196KR	68 p PLCC	16K	512	256	56	10	Y	Y	8
87C196KQ	68 p PLCC	12K	384	128	56	10	Y	Y	8
87C196JR	52 p PLCC	16K	512	256	41	6	Y	Y	6
87C196JQ	52 p PLCC	12K	384	128	41	6	Y	Y	6
80C196KR	68 p PLCC	0	512	256	56	10	Y	Y	8
80C196KQ	68 p PLCC	0	384	128	56	10	Y	Y	8
80C196JR	52 p PLCC	0	512	256	41	6	Y	Y	6
80C196JQ	52 p PLCC	0	384	128	41	6	Y	Y	6

The 87C196KR/KQ/JR/JQ devices represent the 4th generation of MCS® 96 products implemented on Intel's advanced 1 micron process technology. These products are members of the 80C196 family of devices and the instruction set is the same as that of the 80C196KC. The 87C196JR is a 52-lead version of the 87C196KR device, while the 87C196KQ/JQ are memory scalars of the 87C196KR/JR.

The MCS 96 microcontroller family members are all high-performance microcontrollers with a 16-bit CPU. The 87C196KR is composed of the high-speed (16 MHz) core as well as the following peripherals: up to 16 Kbytes of on-chip EPROM, up to 512 bytes of Register RAM, 256 bytes of Code RAM, an eight-channel 10-bit analog to digital converter, an (8096 compatible) asynchronous/synchronous serial I/O port, an additional synchronous serial I/O port, 10 modularized multiplexed capture and compare channels (called the Event Processor Array), a sophisticated prioritized interrupt structure with the programmable Peripheral Transaction Server (PTS).

November 1994
Order Number: 270912-004

Additional register space is allocated for the EPA and can be windowed into the lower Register RAM area.

With the commercial (standard) temperature option, operational characteristics are guaranteed over the temperature range of 0°C to +70°C. With the extended (**Express**) temperature range option, operational characteristics are guaranteed over the temperature range of −40°C to +85°C. Unless otherwise noted, the specifications are the same for both options.

See the prefix identification for extended temperature designators.

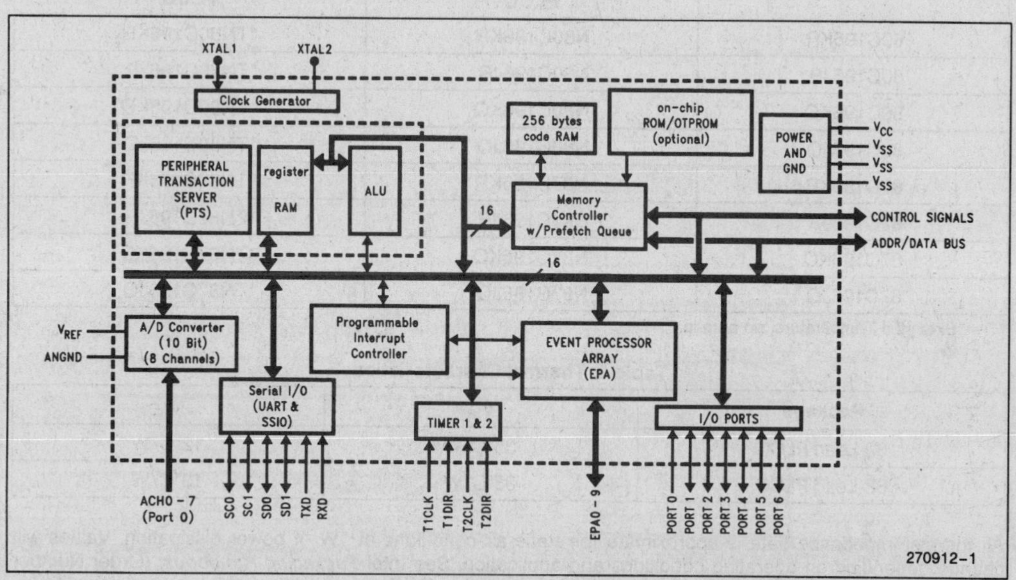

Figure 1. Block Diagram

0FFFFH 06000H	External Memory
05FFFH 02080H	Internal ROM/EPROM or External Memory
0207FH 0205EH	Reserved
0205DH 02040H	PTS Vectors
0203FH 02030H	Interrupt Vectors (upper)
0202FH 02020H	ROM/EPROM Security Key
0201FH	Reserved
0201BH	Reserved (must contain 20H)
0201AH	CCB1
02019H	Reserved (must contain 20H)

02018H	CCB0
02017H 02014H	Reserved
02013H 02000H	Interrupt Vectors (lower)
01FFFH 01F00H	Internal SFRs
01EFFH 00500H	External Memory
004FFH 00400H	Internal RAM
003FFH 00200H	External Memory
001FFH 18H	Register File
17H 00H	CPU SFR's

NOTES:
1. Reserved memory locations must contain 0FFH unless noted.
2. Reserved SFR bit locations must contain 0H unless noted.
3. **WARNING:** Reserved memory locations must not be written or read. The contents and/or function of these locations may change with future revisions of the device. Therefore, a program that relies on one or more of these locations may not function properly.

Process Information

The 8XC196KR/JR/KQ/JQ is manufactured on PX29.5, a CHMOS IV process. Additional process and reliability information is available in Intel's *Components Quality and Reliability Handbook,* Order Number 210997.

Table 1. Prefix Identification

Device	Commercial PLCC	Express PLCC
80C196KR	N80C196KR	*TN80C196KR
80C196JR	N80C196JR	*TN80C196JR
80C196KQ	N80C196KQ	*TN80C196KQ
80C196JQ	N80C196JQ	*TN80C196JQ
87C196KR	N87C196KR	*TN87C196KR
87C196JR	N87C196JR	*TN87C196JR
87C196KQ	N87C196KQ	*TN87C196KQ
87C196JQ	N87C196JQ	*TN87C196JQ

*T = Extended Temperature, no burn-in.

Table 2. Thermal Characteristics

Package	θ_{ja}	θ_{jc}
52-Lead PLCC	35°C/W	12°C/W
68-Lead PLCC	35°C/W	13°C/W

All thermal impedance data is approximate for static air conditions at 1W of power dissipation. Values will change depending on operating conditions and application. See Intel *Packaging Handbook,* (Order Number 240800) for a description of Intel's thermal impedance test methodology.

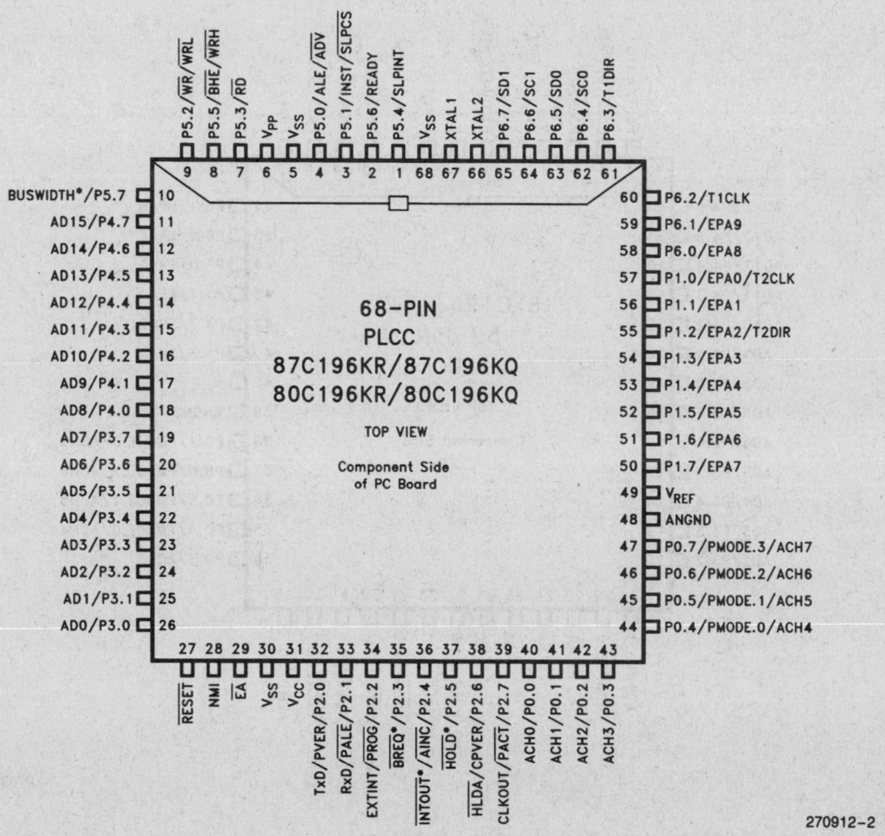

NOTE:
*In earlier versions of documentation these pins were referred to as:
INTOUT → INTINTOUT
BREQ → INTB
BUSWIDTH → BUSW
HOLD → HLD

Figure 2. Package Diagrams

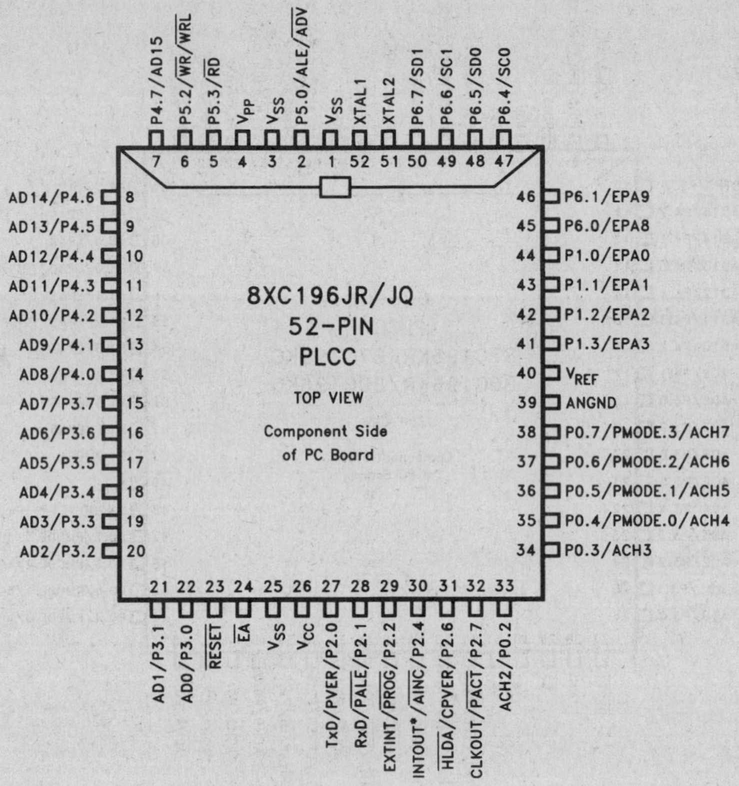

270912-3

NOTE:
*In earlier versions of documentation these pins were referred to as:
INTOUT → INTINTOUT
BREQ → INTB
BUSWIDTH → BUSW
HOLD → HLD

Figure 2. Package Diagrams (Continued)

ADVANCE INFORMATION

PIN DESCRIPTIONS

Symbol	Name and Function
V_{CC}	Main supply voltage (+5V).
V_{SS}	Digital circuit ground (0V). There are three V_{SS} pins, all of which MUST be connected.
V_{REF}	Reference for the A/D converter (+5V). V_{REF} is also the supply voltage to the analog portion of the A/D converter and the logic used to read Port 0. Must be connected for A/D and Port 0 to function.
ANGND	Reference ground for the A/D converter. Must be held at nominally the same potential as V_{SS}.
V_{PP}	Programming voltage for the OTPROM parts. It should be +12.5V for programming. It is also the timing pin for the return from powerdown circuit. If this function is not used, V_{PP} must be tied to V_{CC}.
ACH0–ACH7/PORT0	Analog inputs to the on-chip A/D converter.
$\overline{AINC}$	Input to automatically increment the address when in Programming mode.
ALE/$\overline{ADV}$/P5.0	Address Latch Enable or Address Valid output, as selected by CCR. Both pin options allow a latch to demultiplex the address/data bus. When the pin is $\overline{ADV}$, it goes inactive (high) at the end of the bus cycle. When the pin is ALE, the address can be latched on the falling edge. ALE/$\overline{ADV}$ is active only during external memory accesses. Can be used as standard I/O when not used as ALE.
$\overline{BHE}$/$\overline{WRH}$/P5.5	Byte High Enable or Write High output, as selected by the CCR. $\overline{BHE}$ = 0 selects the bank of memory that is connected to the high byte of the data bus. If the $\overline{WRH}$ function is selected, the pin will go low if the bus cycle is writing to an odd memory location. $\overline{BHE}$/$\overline{WRH}$ is only valid during 16-bit external memory cycles. Can be used as standard I/O when not used as $\overline{BHE}$/$\overline{WRH}$.
$\overline{BREQ}$/P2.3	Bus Request output activated when the bus controller has a pending external memory cycle. Can be used as standard I/O when not used as $\overline{BREQ}$.
BUSWIDTH/P5.7	Input for bus width selection. If CCR bit 1 = 1 and CCR1 bit 2 = 1, this pin dynamically controls the Bus width of the bus cycle in progress. If BUSWIDTH is low, an 8-bit cycle occurs. If BUSWIDTH is high, a 16-bit cycle occurs. Can be used as standard I/O when not used as BUSWIDTH.
CLKOUT/P2.7	Output of the internal clock generator. The frequency is ½ the oscillator frequency. It has a 50% duty cycle. Can be used as standard I/O when not used as CLKOUT.
CPVER	Cumulative Program Verify output. Indicates when all EPROM locations program correctly.
$\overline{EA}$	Input for memory select (External Access). $\overline{EA}$ = 1 causes memory accesses from locations 2000H to 5FFFH to be directed to on-chip EPROM/ROM. $\overline{EA}$ = 0 causes all memory accesses to be directed to off-chip memory. $\overline{EA}$ = +12.5V causes execution to begin in the Programming Mode. $\overline{EA}$ is latched at reset.
EPA0–7/PORT1 EPA8-9/P6.0–6.1	Event Processor Array pin for High Speed capture and compare. EPA0 and EPA2 also function as T2CLK and T2DIR. Can be used as standard I/O when not used as EPA or T2 clock functions.
EXTINT/P2.2	A positive transition on this pin causes a maskable interrupt vector through memory location 203CH. May be used as standard I/O if not used as EXTINT.
$\overline{HLDA}$/P2.6	Bus Hold Acknowledge output indicating release of the bus. Can be used as standard I/O when not used as $\overline{HLDA}$.
$\overline{HOLD}$/P2.5	Bus Hold input requesting control of the bus. Can be used as standard I/O when not used as $\overline{HOLD}$.

PIN DESCRIPTIONS (Continued)

Symbol	Name and Function
INST/P5.1	Output high during an external memory instruction fetch. INST is valid throughout the bus cycle. INST is low otherwise. Can be used as standard I/O when not used as INST.
$\overline{\text{INTOUT}}$/P2.4	Interrupt output indicating that a pending interrupt requires use of the external bus. Can be used as standard I/O if not used as $\overline{\text{INTOUT}}$.
NMI	A positive transition causes a non-maskable interrupt vector through memory location 203EH. If not used, this pin should be tied to V_{SS}. May be used by Intel Evaluation boards.
$\overline{\text{PACT}}$	Output that indicates when the device is currently programming itself. Not active during slave programming.
$\overline{\text{PALE}}$	Input to latch the address during programming modes.
PMODE.0–PMODE.3	Programming mode select inputs.
PORT0	8-bit high impedance input-only port. Also used as A/D converter inputs. Port 0 pins should not be left floating. These pins are also used as inputs by EPROM parts to select the Programming Mode.
PORT1	8-bit bidirectional standard I/O port. All of its pins are shared with the EPA.
PORT2	8-bit bidirectional standard I/O port. All of its pins are shared with other functions (TxD, RxD, EXTINT, $\overline{\text{BREQ}}$, $\overline{\text{INTOUT}}$, $\overline{\text{HOLD}}$, $\overline{\text{HLDA}}$, CLKOUT).
PORT3 PORT4	8-bit bidirectional standard I/O with open drain outputs. These pins are shared with the multiplexed address/data bus which uses strong internal pullups.
PORT5	8-bit bidirectional standard I/O port. All of its pins are shared with other functions (ALE/$\overline{\text{ADV}}$, INST, $\overline{\text{WR}}$/$\overline{\text{WRL}}$, $\overline{\text{RD}}$, SLPINT, $\overline{\text{BHE}}$/$\overline{\text{WRH}}$, READY, BUSWIDTH).
PORT6	8-bit bidirectional standard I/O port. All of its pins are shared with other functions (EPA8, EPA9, T1CLK, T1DIR, SC0, SD0, SC1, SD1).
$\overline{\text{PROG}}$	Programming mode enable input.
PVER	Program Verify output. Goes high after a byte/word is programmed to indicate a successful operation.
$\overline{\text{RD}}$/P5.3	Read signal output to external memory. $\overline{\text{RD}}$ is low only during external memory reads. Can be used as standard I/O when not used as $\overline{\text{RD}}$.
READY/P5.6	Ready input to lengthen external memory cycles. If READY = 1, CPU operation continues in a normal manner. If READY = 0 with the appropriate timings, the memory controller inserts wait states until the next positive transition of CLKOUT occurs with READY = 1. Can be used as standard I/O when not used as READY.
$\overline{\text{RESET}}$	Reset input to and output from the chip. Held low for at least 16 state times to reset the chip. The subsequent low to high transition resynchronizes CLKOUT and commences a 10-state time sequence. Input high for normal operation. $\overline{\text{RESET}}$ has an internal pullup.
RXD/P2.1	Receive data input pin for the Serial I/O port. Can be used as standard I/O if not used as RXD.
$\overline{\text{SLPCS}}$	Slave port chip select input pin. Can be used as standard I/O if not used as $\overline{\text{SLPCS}}$.
SLPINT/P5.4	Slave Port Interrupt Output pin. Can be used as standard I/O when not used as SLPINT.
SSIO/P6.4–6.7 (SC0, SD0, SC1, SD1)	Synchronous Serial I/O pins. SC0/SC1 serve as clock pins and SD0/SD1 are data pins. Can be used as standard I/O if not used for serial I/O.

PIN DESCRIPTIONS (Continued)

Symbol	Name and Function
T1CLK/P6.2	TIMER1 Clock input. The timer increments or decrements on both positive and negative edges. Can be used as standard I/O when not used as T1CLK.
T1DIR/P6.3	TIMER1 Direction input. The timer increments when this pin is high and decrements when this pin is low. Can be used as standard I/O when not used as T1DIR.
T2CLK/P1.0	TIMER2 Clock input. The timer increments or decrements on both positive and negative edges. Can be used as standard I/O when not used as T2CLK.
T2DIR/P1.2	TIMER2 Direction input. The timer increments when this pin is high and decrements when this pin is low. Can be used as standard I/O when not used as T2DIR.
TXD/P2.0	Transmit data output pins for the Serial I/O port. Can be used as standard I/O if not used as TXD.
$\overline{WR}$/$\overline{WRL}$/P5.2	Write and Write Low output to external memory. $\overline{WR}$ will go low for every external write. $\overline{WRL}$ will go low only for external writes where an even byte is being written. $\overline{WR}$/$\overline{WRL}$ is active only during external memory writes. Can be used as standard I/O when not used as $\overline{WR}$/$\overline{WRL}$.
XTAL1	Input of the oscillator inverter and the internal clock generator. This pin should be used when using an external clock source.
XTAL2	Output of the oscillator inverter.

ELECTRICAL CHARACTERISTICS
ABSOLUTE MAXIMUM RATINGS**

Storage Temperature .... −60°C to +150°C

Ambient Temperature under Bias .... −55°C to +125°C

Voltage from V_{PP} or $\overline{EA}$ to V_{SS} or ANGND .... −0.5V to +13.0V

Voltage from Any Other Pin to V_{SS} or ANGND .... −0.5V to +7.0V
This includes V_{PP} on ROM and CPU devices.

Power Dissipation.......................... 1.0W
(based on PACKAGE heat transfer limitations, not device power consumption)

NOTICE: This data sheet contains information on products in the sampling and initial production phases of development. It is valid for the devices indicated in the revision history. The specifications are subject to change without notice.

**WARNING: Stressing the device beyond the "Absolute Maximum Ratings" may cause permanent damage. These are stress ratings only. Operation beyond the "Operating Conditions" is not recommended and extended exposure beyond the "Operating Conditions" may affect device reliability.*

OPERATING CONDITIONS

Symbol	Description	Min	Max	Units
T_A	Ambient Temperature under Bias **Commercial** Temp.	0	+70	°C
T_A	Ambient Temperature under Bias **Extended** Temp.	−40	+85	°C
V_{CC}	Digital Supply Voltage	4.50	5.50	V
V_{REF}	Analog Supply Voltage	4.50	5.50	V
F_{OSC}	Oscillator Frequency	4	16	MHz(4)

NOTE:
ANGND and V_{SS} should be nominally at the same potential.

DC CHARACTERISTICS (Over Specified Operating Conditions)(9)

Symbol	Parameter	Min	Typ(6)	Max	Units	Test Conditions
V_{IL}	Input Low Voltage (All Pins)	−0.5V		0.3 V_{CC}	V	
V_{IH}	Input High Voltage	0.7 V_{CC}		V_{CC} + 0.5	V	
V_{OL}	Output Low Voltage (Outputs Configured as Push/Pull)			0.3 0.45 1.5	V V V	I_{OL} = 200 μA(3, 5) I_{OL} = 3.2 mA I_{OL} = 7.0 mA
V_{OH}	Output High Voltage (Outputs Configured as Push/Pull)	V_{CC} − 0.3 V_{CC} − 0.7 V_{CC} − 1.5			V V V	I_{OH} = −200 μA(3, 5, 8) I_{OH} = −3.2 mA I_{OH} = −7.0 mA
V_{OH2}	Output High Voltage in RESET	V_{CC} − 1V			V	I_{OH} = −15 μA(1, 7)
I_{LI}	Input Leakage Current (Std. Inputs)			±10	μA	$V_{SS} < V_{IN} < V_{CC} - 0.3V$(2)
I_{LI1}	Input Leakage Current (Port 0—A/D Inputs)		±1	±3	μA	$V_{SS} < V_{IN} < V_{REF}$
I_{IH}	Input High Current (NMI)			+175	μA	$V_{SS} < V_{IN} < V_{CC} - 0.3V$(10)

DC CHARACTERISTICS (Over Specified Operating Conditions)(9) (Continued)

Symbol	Parameter	Min	Typ(6)	Max	Units	Test Conditions
I_{CC}	V_{CC} Supply Current		60	75	mA	XTAL1 = 16 MHz, $V_{CC} = V_{PP} = V_{REF} = 5.5V$ (While Device in Reset)
I_{REF}	A/D Reference Supply Current		2	5	mA	
I_{IDLE}	Idle Mode Current		15	30	mA	XTAL1 = 16 MHz, $V_{CC} = V_{PP} = V_{REF} = 5.5V$
I_{PD}	Powerdown Mode Current(6)		50	TBD	μA	$V_{CC} = V_{PP} = V_{REF} = 5.5V$
R_{RST}	Reset Pullup Resistor	6K		65K	Ω	
C_S	Pin Capacitance (Any Pin to V_{SS})			10	pF	$F_{TEST} = 1.0$ MHz
R_{WPU}	Weak Pullup Resistance (Approx)		150K		Ω	(6)

NOTES:
1. All BD (Bidirectional) pins except INST and CLKOUT. BD pins include Port1, Port2, Port3, Port4, Port5 (as a port), and Port6.
2. Standard Input pins include XTAL1, $\overline{EA}$, $\overline{RESET}$, and Port 1/2/3/4/5/6 when setup as inputs.
3. All Bidirectional I/O pins when configured as Outputs (Push/Pull).
4. Device is Static and should operate below 1 Hz, but only tested down to 4 MHz.
5. Maximum I_{OL}/I_{OH} currents per pin will be characterized and published at a later date.
6. Typicals are based on limited number of samples and are not guaranteed. The values listed are at room temperature and $V_{REF} = V_{CC} = 5.0V$.
7. Violating these specifications in reset may cause the device to enter test modes (P5.4 and P2.6).
8. This specification applies to P3/4 only when used as an address bus supplying the address.
9. All voltages are referenced relative to V_{SS}. When used, V_{SS} refers to the device pin.
10. Worst case is at upper limit of test conditions.

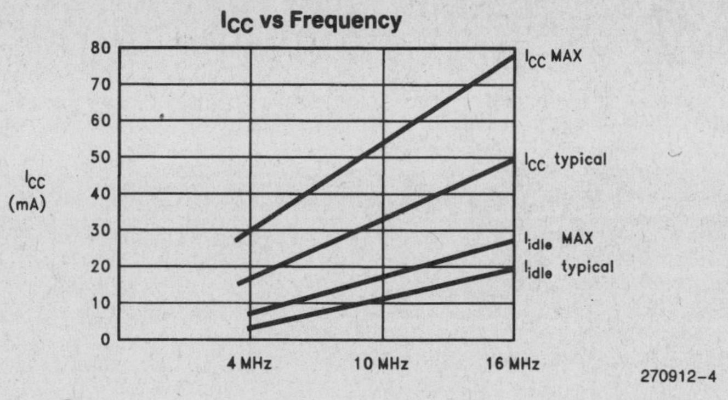

NOTES:
I_{CC} Max = 3.88 × FREQ + 13.43
I_{idle} Max = 1.65 × FREQ + 2.2

4

AC CHARACTERISTICS (Over Specified Operating Conditions)

Test Conditions: Capacitance Load on All Pins = 100 pF, Rise and Fall Times = 10 ns.

The system must meet these specifications to work with the 87C196KR/KQ/JR/JQ:

Symbol	Parameter	Min	Max	Units
T_{AVYV}	Address Valid to READY Setup		$2\ T_{OSC} - 75$	ns(2)
T_{LLYV}	ALE Low to READY Setup		$T_{OSC} - 70$	ns(2)
T_{YLYH}	Non READY Time	No Upper Limit		ns
T_{CLYX}	READY Hold after CLKOUT Low	0	$T_{OSC} - 30$	ns(1, 2)
T_{LLYX}	READY Hold after ALE Low	$T_{OSC} - 15$	$2\ T_{OSC} - 40$	ns(1, 2)
T_{AVGV}	Address Valid to Buswidth Setup		$2\ T_{OSC} - 75$	ns(2)
T_{LLGV}	ALE Low to Buswidth Setup		$T_{OSC} - 60$	ns(2)
T_{CLGX}	Buswidth Hold after CLKOUT Low	0		ns(2)
T_{AVDV}	Address Valid to Input Data Valid		$3\ T_{OSC} - 55$	ns
T_{RLDV}	$\overline{RD}$ Active to Input Data Valid		$T_{OSC} - 22$	ns
T_{CLDV}	CLKOUT Low to Input Data Valid		$T_{OSC} - 50$	ns
T_{RHDZ}	End of $\overline{RD}$ to Input Data Float		T_{OSC}	ns
T_{RXDX}	Data Hold after $\overline{RD}$ Inactive	0		ns

NOTE:
1. If max is exceeded, additional wait states will occur.
2. Does not apply to JR/JQ.

AC CHARACTERISTICS (Over Specified Operating Conditions)

Test Conditions: Capacitance Load on All Pins = 100 pF, Rise and Fall Times = 10 ns.

The 87C196KR/KQ/JR/JQ will meet these specifications.

Symbol	Parameter	Min	Max	Units
F_{XTAL}	Oscillator Frequency	4.0	16.0	MHz(1)
T_{OSC}	Oscillator Period (1/Fxtal)	62.5	250	ns
T_{XHCH}	XTAL1 High to CLKOUT High or Low	20	110	ns(2)
T_{CLCL}	CLKOUT Period	2 T_{OSC}		ns
T_{CHCL}	CLKOUT High Period	$T_{OSC}-10$	$T_{OSC}+15$	ns
T_{CLLH}	CLKOUT Falling Edge to ALE Rising	−10	15	ns
T_{LLCH}	ALE/$\overline{ADV}$ Falling Edge to CLKOUT Rising	−20	15	ns
T_{LHLH}	ALE/$\overline{ADV}$ Cycle Time	4 T_{OSC}		ns(5)
T_{LHLL}	ALE/$\overline{ADV}$ High Period	$T_{OSC}-10$	$T_{OSC}+10$	ns
T_{AVLL}	Address Setup to ALE/$\overline{ADV}$ Falling Edge	$T_{OSC}-15$		ns
T_{LLAX}	Address Hold after ALE/$\overline{ADV}$ Falling Edge	$T_{OSC}-40$		ns
T_{LLRL}	ALE/$\overline{ADV}$ Falling Edge to $\overline{RD}$ Falling Edge	$T_{OSC}-30$		ns
T_{RLCL}	$\overline{RD}$ Low to CLKOUT Falling Edge	4	30	ns
T_{RLRH}	$\overline{RD}$ Low Period	$T_{OSC}-5$		ns(5)
T_{RHLH}	$\overline{RD}$ Rising Edge to ALE/$\overline{ADV}$ Rising Edge	T_{OSC}	$T_{OSC}+25$	ns(3)
T_{RLAZ}	$\overline{RD}$ Low to Address Float		5	ns
T_{LLWL}	ALE/$\overline{ADV}$ Falling Edge to $\overline{WR}$ Falling Edge	$T_{OSC}-10$		ns
T_{CLWL}	CLKOUT Low to $\overline{WR}$ Falling Edge	−5	25	ns
T_{QVWH}	Data Stable to $\overline{WR}$ Rising Edge	$T_{OSC}-23$		ns
T_{CHWH}	CLKOUT High to $\overline{WR}$ Rising Edge	−10	15	ns
T_{WLWH}	$\overline{WR}$ Low Period	$T_{OSC}-30$		ns(5)
T_{WHQX}	Data Hold after $\overline{WR}$ Rising Edge	$T_{OSC}-25$		ns
T_{WHLH}	$\overline{WR}$ Rising Edge to ALE/$\overline{ADV}$ Rising Edge	$T_{OSC}-10$	$T_{OSC}+15$	ns(3)
T_{WHBX}	$\overline{BHE}$, INST Hold after $\overline{WR}$ Rising Edge	$T_{OSC}-10$		ns(6)
T_{WHAX}	AD8–15 Hold after $\overline{WR}$ Rising Edge	$T_{OSC}-30$(4)		ns
T_{RHBX}	$\overline{BHE}$, INST Hold after $\overline{RD}$ Rising Edge	$T_{OSC}-10$		ns(6)
T_{RHAX}	AD8–15 Hold after $\overline{RD}$ Rising Edge	$T_{OSC}-30$(4)		ns
T_{BVLL}	BHE Valid to ALE Falling Edge	$T_{OSC}-15$		ns(6)

NOTES:
1. Testing performed at 4.0 MHz, however, the device is static by design and will typically operate below 1 Hz.
2. Typical specifications, not guaranteed.
3. Assuming back-to-back bus cycles.
4. 8-bit bus only.
5. If wait states are used, add 2 $T_{OSC} \times n$, where n = number of wait states.
6. Does not apply to JR/JQ.

System Bus Timing

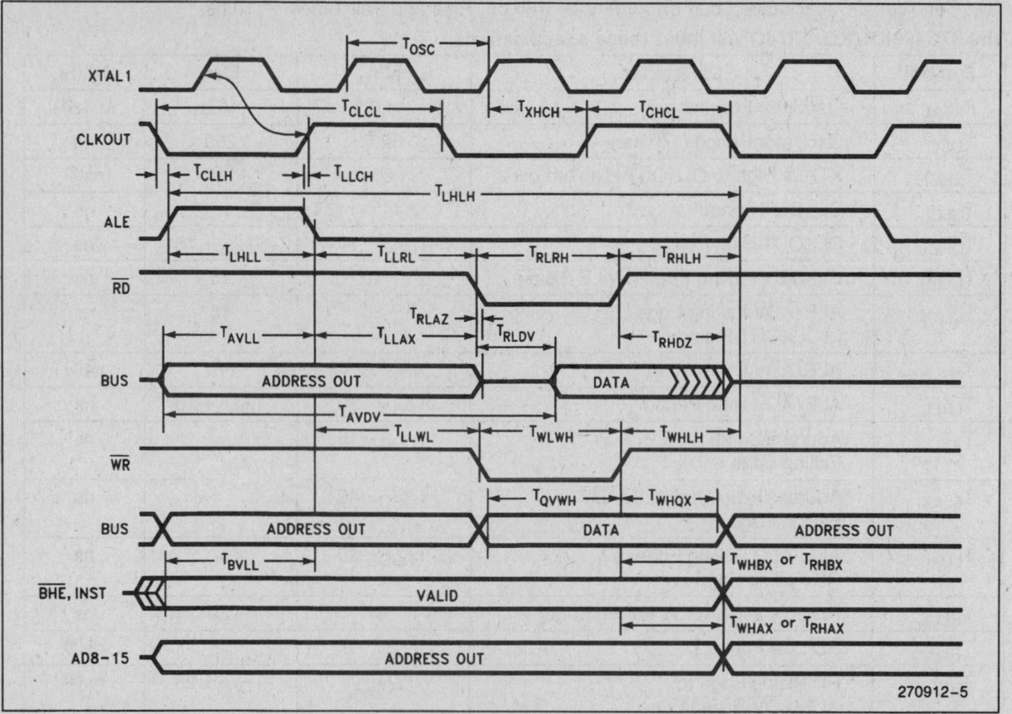

270912–5

Buswidth Timings

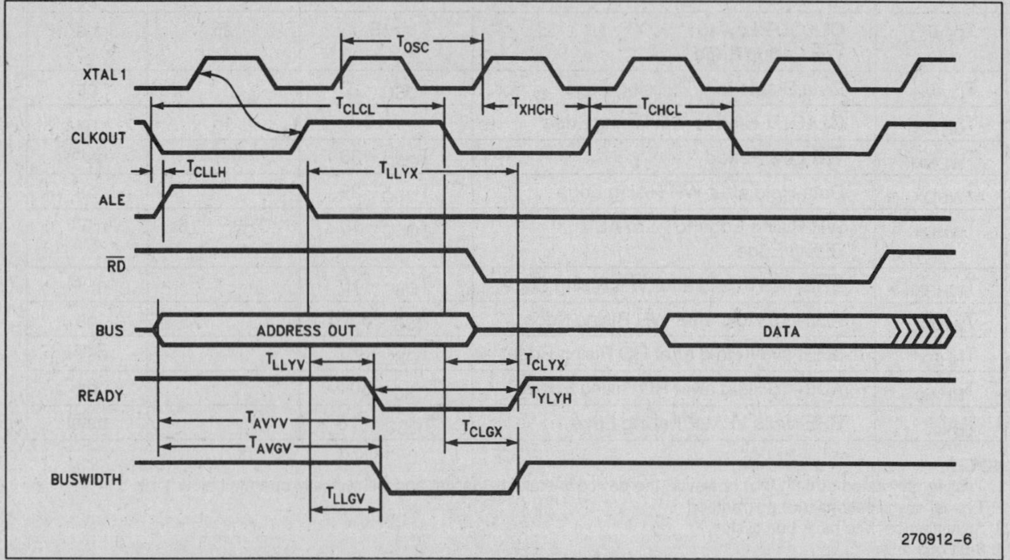

270912–6

$\overline{HOLD}$/$\overline{HLDA}$ Timings

Symbol	Description	Min	Max	Units	Notes
T_{HVCH}	$\overline{HOLD}$ Setup	65		ns	(1, 2)
T_{CLHAL}	CLKOUT Low to $\overline{HLDA}$ Low	−15	15	ns	(2)
T_{CLBRL}	CLKOUT Low to $\overline{BREQ}$ Low	−15	15	ns	(2)
T_{AZHAL}	$\overline{HLDA}$ Low to Address Float		25	ns	(2)
T_{BZHAL}	$\overline{HLDA}$ Low to $\overline{BHE}$, INST, $\overline{RD}$, $\overline{WR}$ Weakly Driven		25	ns	(2)
T_{CLHAH}	CLKOUT Low to $\overline{HLDA}$ High	−15	15	ns	(2)
T_{CLBRH}	CLKOUT Low to $\overline{BREQ}$ High	−15	15	ns	(2)
T_{HAHAX}	$\overline{HLDA}$ High to Address No Longer Float	−15		ns	(2)
T_{HAHBV}	$\overline{HLDA}$ High to BHE, INST, RD, WR Valid	−10		ns	(2)
T_{CLLH}	CLKOUT Low to ALE High	−10	15	ns	

NOTE:
1. To guarantee recognition at next clock.
2. Does not apply to JR/JQ.

HOLD LATENCY

	Max
Internal Access	1.5 States
16-Bit External Execution	2.5 States
8-Bit External Execution	4.5 States

DC SPECIFICATIONS IN HOLD

Parameter	Min	Max	Units
Weak Pullups on $\overline{ADV}$, $\overline{RD}$, $\overline{WR}$, $\overline{WRL}$, $\overline{BHE}$	50K	250K	V_{CC} = 5.5V, V_{IN} = 0.45V
Weak Pulldowns on ALE, INST	10K	50K	V_{CC} = 5.5V, V_{IN} = 2.4V

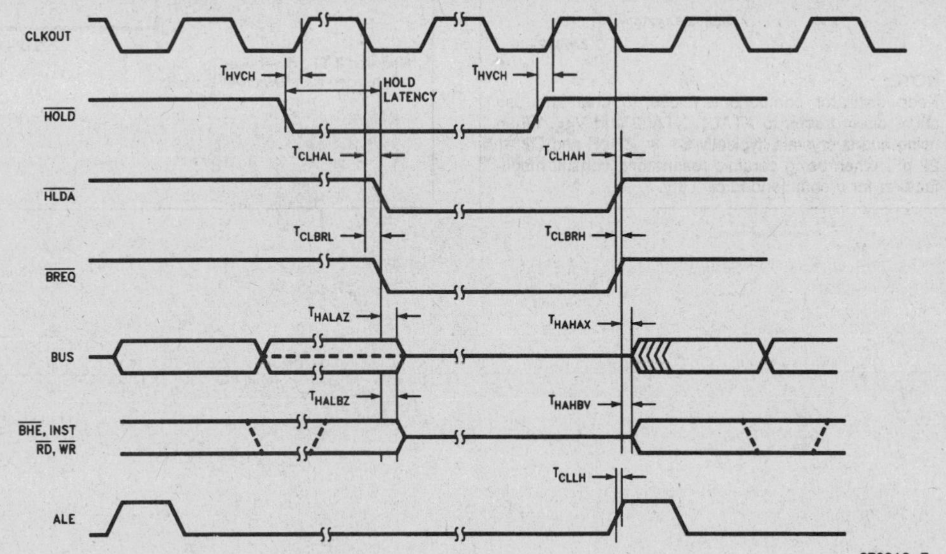

270912-7

EXTERNAL CLOCK DRIVE

Symbol	Parameter	Min	Max	Units
$1/T_{XLXL}$	Oscillator Frequency	4.0	16	MHz
T_{XLXL}	Oscillator Period (T_{OSC})	62.5	250	ns
T_{XHXX}	High Time	0.35 T_{OSC}	0.65 T_{OSC}	ns
T_{XLXX}	Low Time	0.35 T_{OSC}	0.65 T_{OSC}	ns
T_{XLXH}	Rise Time		10	ns
T_{XHXL}	Fall Time		10	ns

EXTERNAL CLOCK DRIVE WAVEFORMS

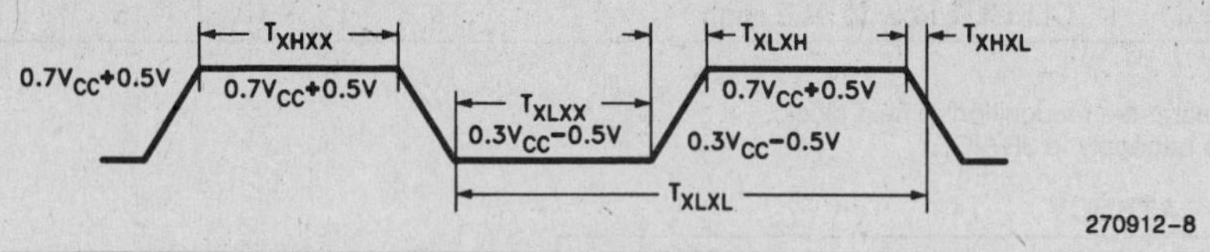

270912-8

EXTERNAL CRYSTAL CONNECTIONS

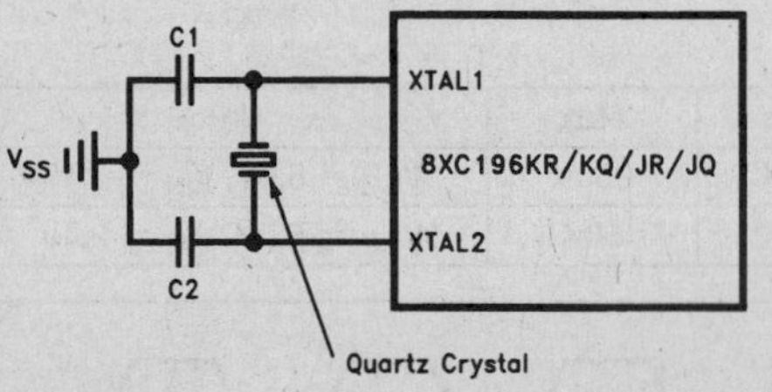

270912-16

NOTE:
Keep oscillator components close to chip and use short, direct traces to XTAL1, XTAL2 and V_{SS}. When using quartz crystals, typically C1 ≈ 22 pF and C2 ≈ 22 pF. When using ceramic resonators, consult manufacturer for recommended circuitry.

EXTERNAL CLOCK CONNECTIONS

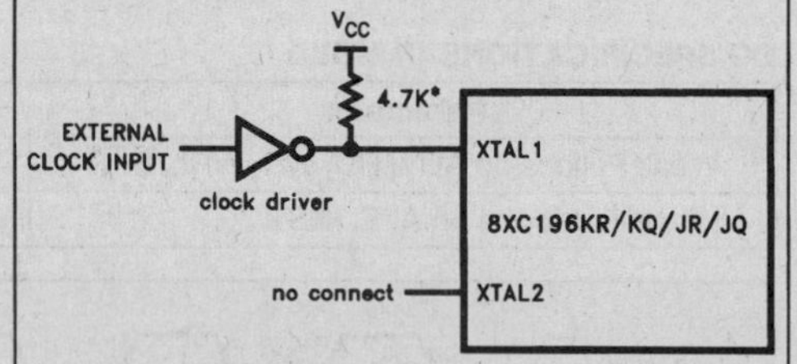

270912-17

*Required if TTL driver used.
Not needed if CMOS driver is used.

AC TESTING INPUT, OUTPUT WAVEFORMS

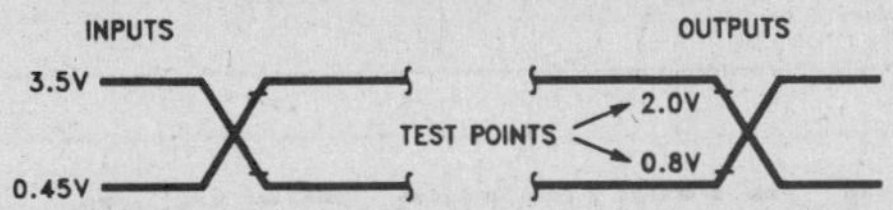

270912-9

NOTE:
AC Testing Inputs are driven at 3.5V for a logic "1" and 0.45V for a logic "0". Timing measurements are made at 2.0V for a logic "1" and 0.8V for logic "0".

FLOAT WAVEFORMS

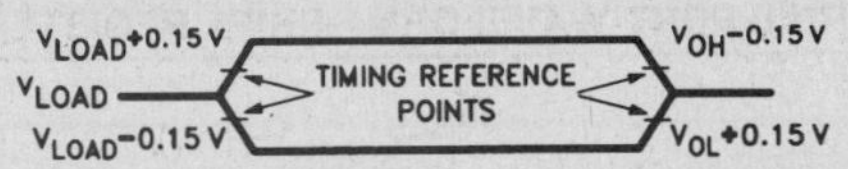

270912-10

NOTE:
For Timing Purposes a Port Pin is no longer floating when a 150 mV change from Load Voltage Occurs and Begins to Float when a 150 mV change from the Loading V_{OH}/V_{OL} Level occurs $I_{OL}/I_{OH} \leq 15$ mA.

EXPLANATION OF AC SYMBOLS

Each symbol is two pairs of letters prefixed by "T" for time. The characters in a pair indicate a signal and its condition, respectively. Symbols represent the time between the two signal/condition points.

Conditions:

H— High
L— Low
V— Valid
X— No Longer Valid
Z— Floating

Signals:

A— Address
B— $\overline{BHE}$
BR— $\overline{BREQ}$
C— CLKOUT
D— DATA
G— Buswidth
H— $\overline{HOLD}$
HA— $\overline{HLDA}$
L— ALE/$\overline{ADV}$
Q— Data Out
R— $\overline{RD}$
W— $\overline{WR}$/$\overline{WRH}$/$\overline{WRL}$
X— XTAL1
Y— READY

AC CHARACTERISTICS—SERIAL PORT—SHIFT REGISTER MODE (MODE 0)

SERIAL PORT TIMING—SHIFT REGISTER MODE (Over Specified Operating Conditions)
Test Conditions: Load Capacitance = 100 pF

Symbol	Parameter	Min	Max	Units
T_{XLXL}	Serial Port Clock Period	8 T_{OSC}		ns
T_{XLXH}	Serial Port Clock Falling Edge to Rising Edge	4 T_{OSC} − 50	4 T_{OSC} + 50	ns
T_{QVXH}	Output Data Setup to Clock Rising Edge	3 T_{OSC}		ns
T_{XHQX}	Output Data Hold after Clock Rising Edge	2 T_{OSC} − 50		ns
T_{XHQV}	Next Output Data Valid after Clock Rising Edge		2 T_{OSC} + 50	ns
T_{DVXH}	Input Data Setup to Clock Rising Edge	2 T_{OSC} + 200		ns
T_{XHDX}	Input Data Hold after Clock Rising Edge	0		ns
T_{XHQZ}	Last Clock Rising to Output Float		5 T_{OSC}	ns

4

WAVEFORM—SERIAL PORT—SHIFT REGISTER MODE (MODE 0)

SERIAL PORT WAVEFORM—SHIFT REGISTER MODE

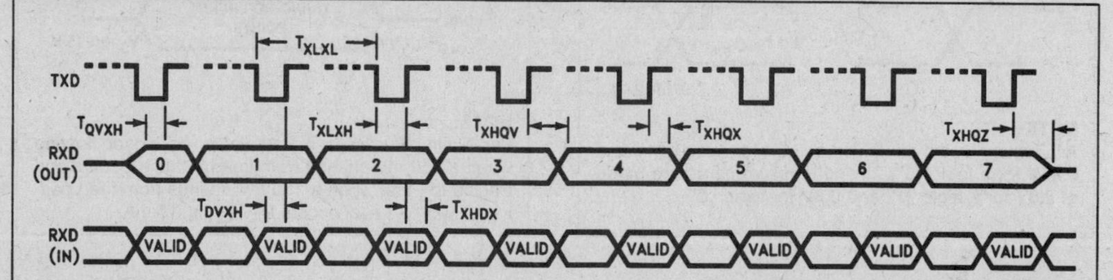

270912-14

A TO D

The speed of the A/D converter in the 10-bit or 8-bit modes can be adjusted by setting the AD_TIME special function register to the appropriate value. The AD_TIME register only programs the speed at which the conversions are performed, not the speed it can convert correctly.

The converter is ratiometric, so absolute accuracy is dependent on the accuracy and stability of V_{REF}.

A/D CONVERTER SPECIFICATION

After a conversion is started, the device is placed in the IDLE mode until the conversion is complete. Testing is performed at $V_{REF} = 5.12V$.

There is an AD_TEST register that allows for conversion on ANGND and V_{REF} as well as zero offset adjustment. The Absolute Error listed is WITHOUT doing any adjustments.

10-BIT A/D OPERATING CONDITIONS(1)

Symbol	Description	Min	Max	Units
T_A	Ambient Temperature **Commercial** Temp.	0	+70	°C
T_A	Ambient Temperature **Extended** Temp.	−40	+85	°C
V_{CC}	Digital Supply Voltage	4.50	5.50	V
V_{REF}	Analog Supply Voltage	4.50	5.50(2)	V
T_{SAM}	Sample Time	2.0		μs(3)
T_{CONV}	Conversion Time	16.5	19.5	μs(3)
F_{OSC}	Oscillator Frequency	4	16	MHz

NOTES:
1. ANGND and V_{SS} should nominally be at the same potential.
2. V_{REF} must not exceed V_{CC} by more than +0.5V.
3. The value of AD_TIME is selected to meet these specifications.

10-BIT MODE A/D CHARACTERISTICS (Over Specified Operating Conditions)

Parameter	Typical (1)	Min	Max	Units*
Resolution		1024 10	1024 10	Level Bits
Absolute Error		0	±3	LSBs
Full Scale Error	0.25 ±0.5			LSBs
Zero Offset Error	0.25 ±0.5			LSBs
Non-Linearity	1.0 ±2.0		±3	LSBs
Differential Non-Linearity		> −0.5	+0.5	LSBs
Channel-to-Channel Matching	±0.1	0	±1	LSBs
Repeatability	±0.25	0		LSBs
Temperature Coefficients: Offset Fullscale Differential Non-Linearity	 0.009 0.009 0.009			 LSB/C LSB/C LSB/C
Off Isolation		−60		dB(2, 3)
Feedthrough	−60			dB(2)
V_{CC} Power Supply Rejection	−60			dB(2)
Input Series Resistance		750	1.2K	Ω(4)
Voltage on Analog Input Pin		ANGND − 0.5	V_{REF} + 0.5	V
Sampling Capacitor	2			pF
DC Input Leakage		0	±3	μA

NOTES:
*An "LSB", as used here, has a value of approximately 5 mV. (See Embedded Microcontrollers and Processors Handbook for A/D glossary of terms).
1. These values are expected for most parts at 25°C but are not tested or guaranteed.
2. DC to 100 KHz.
3. Multiplexer Break-Before-Make Guaranteed.
4. Resistance from device pin, through internal multiplexer, to sample capacitor.

8-BIT A/D OPERATING CONDITIONS(1)

Symbol	Description	Min	Max	Units
T_A	Ambient Temperature **Commercial** Temp.	0	+70	°C
T_A	Ambient Temperature **Extended** Temp.	−40	+85	°C
V_{CC}	Digital Supply Voltage	4.50	5.50	V
V_{REF}	Analog Supply Voltage	4.50	5.50(2)	V
T_{SAM}	Sample Time	2.0		μs(3)
T_{CONV}	Conversion Time	16.5	19.5	μs(3)
F_{OSC}	Oscillator Frequency	4	16	MHz

NOTES:
1. ANGND and V_{SS} should nominally be at the same potential.
2. V_{REF} must not exceed V_{CC} by more than +0.5V.
3. The value of AD_TIME is selected to meet these specifications.

8-BIT MODE A/D CHARACTERISTICS (Over Specified Operating Conditions)

The 8-bit mode trades off resolution for a faster conversion time. The AD_TIME register must be used when performing an 8-bit conversion.

Parameter	Typ(1)	Minimum	Maximum	Units*	Notes
Resolution		256 8	256 8	Levels Bits	
Absolute Error		0	±2	LSBs	
Full Scale Error	±0.5			LSBs	
Zero Offset Error	±0.5			LSBs	
Non-Linearity		0	±2	LSBs	
Differential Non-Linearity Error		> −1	+1	LSBs	
Channel-to-Channel Matching			±1	LSBs	
Repeatability	±0.25			LSBs	
Temperature Coefficients: Offset Full Scale Differential Non-Linearity	 0.003 0.003 0.003			 LSB/°C LSB/°C LSB/°C	
Off Isolation		−60		dB(2, 3)	
Feedthrough	−60			dB(2)	
V_{CC} Power Supply Rejection	−60			dB(2)	
Input Series Resistance		750	1.2K	Ω	
Voltage on Analog Input Pin		ANGND − 0.5	V_{REF} + 0.5	V	
Sampling Capacitor	2			pF	
DC Input Leakage		0	±3	μA	

NOTES:
*An "LSB", as used here, has a value of approximately 20 mV.
1. Typical values are expected for most devices at 25°C.
2. DC to 100 KHz.
3. Multiplexer Break-Before-Make Guaranteed.
4. Resistance from device pin, through internal multiplexer, to sample capacitor.

OTPROM PROGRAMMING

OPERATING CONDITIONS DURING PROGRAMMING(3)

Symbol	Description	Min	Max	Units
T_A	Ambient Temperature during Programming	20	30	°C
V_{CC}	Supply Voltage during Programming	4.5	5.5	V(1)
V_{REF}	Reference Supply Voltage during Programming	4.5	5.5	V(1)
V_{PP}	Programming Voltage	12.25	12.75	V(2)
V_{EA}	$\overline{EA}$ Pin Voltage	12.25	12.75	V(2)
F_{OSC}	Oscillator Frequency during Auto and Slave Mode Programming	6.0	8.0	MHz
F_{OSC}	Oscillator Frequency during Run-Time Programming	6.0	12.0	MHz

NOTES:
1. V_{CC} and V_{REF} should nominally be at the same voltage during programming.
2. V_{PP} and V_{EA} must never exceed the maximum specification, or the device may be damaged.
3. V_{SS} and ANGND should nominally be at the same potential (0V).

AC OTPROM PROGRAMMING CHARACTERISTICS

Symbol	Description	Min	Max	Units
T_{AVLL}	Address Setup Time	0		T_{OSC}
T_{LLAX}	Address Hold Time	100		T_{OSC}
T_{DVPL}	Data Setup Time	0		T_{OSC}
T_{PLDX}	Data Hold Time	400		T_{OSC}
T_{LLLH}	$\overline{PALE}$ Pulse Width	50		T_{OSC}
T_{PLPH}	$\overline{PROG}$ Pulse Width(1)	50		T_{OSC}
T_{LHPL}	$\overline{PALE}$ High to $\overline{PROG}$ Low	220		T_{OSC}
T_{PHLL}	$\overline{PROG}$ High to Next $\overline{PALE}$ Low	220		T_{OSC}
T_{PHDX}	Word Dump Hold Time		50	T_{OSC}
T_{PHPL}	$\overline{PROG}$ High to Next $\overline{PROG}$ Low	220		T_{OSC}
T_{LHPL}	$\overline{PALE}$ High to $\overline{PROG}$ Low	220		T_{OSC}
T_{PLDV}	$\overline{PROG}$ Low to Word Dump Valid		50	T_{OSC}
T_{SHLL}	$\overline{RESET}$ High to First $\overline{PALE}$ Low	1100		T_{OSC}
T_{PHIL}	$\overline{PROG}$ High to $\overline{AINC}$ Low	0		T_{OSC}
T_{ILIH}	$\overline{AINC}$ Pulse Width	240		T_{OSC}
T_{ILVH}	PVER Hold after $\overline{AINC}$ Low	50		T_{OSC}
T_{ILPL}	$\overline{AINC}$ Low to $\overline{PROG}$ Low	170		T_{OSC}
T_{PHVL}	$\overline{PROG}$ High to $\overline{PVER}$ Valid		220	T_{OSC}

NOTE:
1. This specification is for the word dump mode. For programming pulses use 100 μs.

DC OTPROM PROGRAMMING CHARACTERISTICS

Symbol	Parameter	Min	Max	Units
I_{PP}	V_{PP} Programming Supply Current		100	mA

NOTE:
Do not apply V_{PP} until V_{CC} is stable and within specifications and the oscillator/clock has stabilized or the device may be damaged.

OTPROM PROGRAMMING WAVEFORMS

SLAVE PROGRAMMING MODE DATA PROGRAM MODE WITH SINGLE PROGRAM PULSE

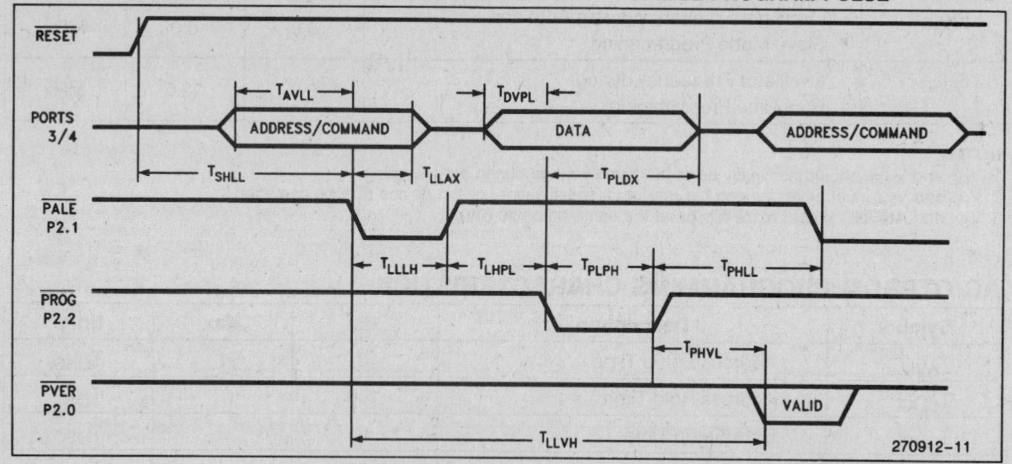

270912-11

SLAVE PROGRAMMING MODE IN WORD DUMP OR DATA VERIFY MODE WITH AUTO INCREMENT

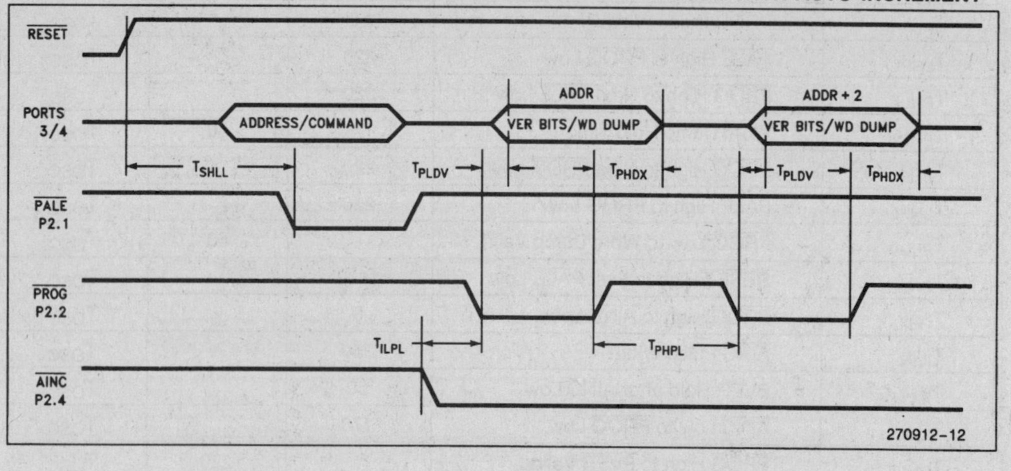

270912-12

SLAVE PROGRAMMING MODE TIMING IN DATA PROGRAM MODE WITH REPEATED PROG PULSE AND AUTO INCREMENT

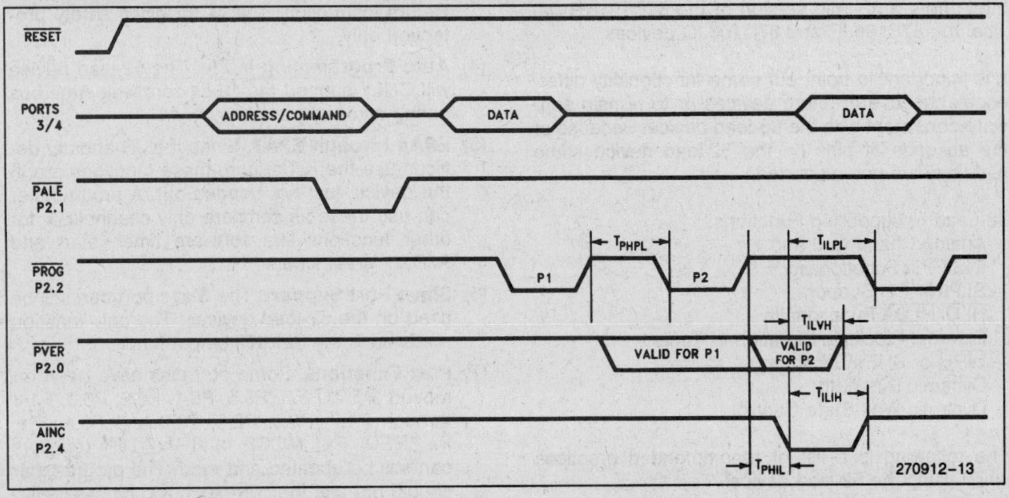

87C196KR/87C196JR ERRATA

1. I_{OH2}

   Current devices do not meet the test condition for V_{OH2} of $-15\ \mu A$. Instead the devices are guaranteed to source a minimum of $-6\ \mu A$.

87C196KR/87C196JR DESIGN CONSIDERATIONS

1. **EPA Timers**

   Special care must be taken when resetting/writing the EPA timers. This is more of a software technique than a device errata. For example: The EPA timers do not generate a "time valid" signal when the counter is either reset or written. This means that if a compare event is programmed in the EPA/Compare channel for a value of "0000H" (when reset) or equal to a written value, the compared event will NOT happen. However, if the timers are allowed to increment/decrement to that value, that compare event WILL occur.

2. **Port 6.4, 6.5, 6.6, 6.7**

   The user is not allowed to modify the P6_REG register when these pins are configured as Special Function P6_MODE.x = 1). During software manipulation of these registers, it is a good practice to first change the P6_MODE register, then modify the P6_REG register when switching from SF to LSIO.

3. **P2.7 (CLKOUT)**

   Port 2.7 (CLKOUT) does not operate in open drain mode.

4. Current versions of the 8XC196KQ/JQ are fabricated with 16K of internal OTPROM, 512 bytes of register RAM, and 256 bytes of internal RAM. The memory map of the 8XC196KQ/JQ is identical to the 8XC196KR/JR. However, the extra memory locations are not tested and should not be used. Intel may disable this extra memory on future versions of the 8XC196KQ/JQ. Any software that relies on reading or writing these locations may not function correctly on future devices.

   Two steps the user should always incorporate to insure future compatibility are:

   A) The program must contain a jump to a location greater than 16K before the 12K boundary is reached. This is necessary only if greater than 12K of program memory is required and portions of the program executes from internal OTPROM.

   B) Use program memory from 12K to 16K only if $\overline{EA}$ is tied to ground. Never use data memory from 180H to 1FFH or from 480H to 4FFH.

52-LEAD DEVICES

Intel offers a 52-lead version of the 87C196KR device: the 87C196JR and 87C196JQ devices.

It is important to point out some functionality differences because of future devices or to remain software consistent with the 68-lead device. Because of the absence of pins on the 52-lead device some functions are not supported.

52-Lead Unsupported Functions:

- Analog Channels 0 and 1
- INST Pin Functionality
- SLPINT Pin Support
- $\overline{\text{HLD}}/\overline{\text{HLDA}}$ Functionality
- External Clocking/Direction of Timer1
- $\overline{\text{WRH}}$ or BHE Functions
- Dynamic Buswidth
- Dynamic Wait State Control

The following is a list of recommended practices when using the 52-lead device:

(1) **External Memory.** Use an 8-bit bus mode only. There is neither a $\overline{\text{WRH}}$ or BUSWIDTH pin. The bus cannot dynamically switch from 8- to 16-bit or vice versa. Set the CCB bytes to an 8-bit only mode, using $\overline{\text{WR}}$ function only.

(2) **Wait State Control.** Use the CCB bytes to configure the maximum number of wait states. If the READY pin is selected to be a system function, the device will lockup waiting for READY. If the READY pin is configured as LSIO (default after $\overline{\text{RESET}}$), the internal logic will receive a logic "0" level and insert the CCB defined number of wait states in the bus cycle. DON'T USE IRC = "111".

(3) **NMI Support.** The NMI is not bonded out. Make the NMI vector at location 203Eh vector to a Return instruction. This is for glitch safety protection only.

(4) **Auto-Programming Mode.** The 52-lead device will ONLY support the 16-bit zero wait state bus during auto-programming.

(5) **EPA4 through EPA7.** Since the JR and JQ devices use the KR silicon, these functions are in the device, just not bonded out. A programmer can use these as compare only channels or for other functions like software timer, start and A/D, or reset timers.

(6) **Slave Port Support.** The Slave port can still be used on the 52-lead devices. The only function removed is the SLPINT output function.

(7) **Port Functions.** Some port pins have been removed. P5.7, P5.6, P5.5, P5.1, P6.2, P6.3, P1.4 through P1.7, P2.3, P2.5, P0.0 and P0.1. The Px_REG, Px_MODE, and Px_DIR registers can still be updated and read. The programmer should not use the corresponding bits associated with the removed port pins to conditionally branch in software. Treat these bits as RESERVED.

Additionally, these port pins should be setup internally by software as follows:

1. Written to Px_REG as "1" or "0".
2. Configured as Push/Pull, Px_DIR as "0".
3. Configured as LSIO.

This configuration will effectively strap the pin either high or low. *DO NOT Configure as Open Drain output "1", or as an Input pin. This device is CMOS.*

REVISION HISTORY

This data sheet (270912-003) supercedes 270912-002 and is valid for devices with a "C" at the end of the topside tracking number. Data sheets are changed as new device information becomes available. Verify that you have the latest version before finalizing a design or ordering devices.

1. Removed the following errata:
   - Slave Programming Mode
   - EPA_MASK1/EPA_PEND1
   - BMOVI
   - PTS and Other Interrupts
   - Serial Port Framing Error
   - Remap Mode on EPA3
   - A/D Abort
   - PTS/NMI Conflict
   - Data Output Register Cleared
   - Divide Error during HOLD/READY
   - SIO Mode 0
   - EPAIPV Multiplied by Two
   - (These were fixed on the C-step)
2. Moved the following from Errata to Design Considerations:
   - EPA Timers
   - Port 6.4, 6.5, 6.6, 6.7 (and reworded)
   - P2.7 (CLKOUT)
   - Oscillator Noise Sensitivity
3. Added New Errata:
   - I_{OH2} (also existed on A-step)
4. Added SLPCS to Package Diagrams and Pin Descriptions
5. Added T_{BVLL}
6. Added I_{IH} for NMI
7. Added notes to AC Characteristics identifying specifications that do not apply to JR/JQ
8. Changed T_{CLLH} from −5 ns to −10 ns under HOLD/HLDA Timings
9. Changed T_{HVCH} from 55 ns to 65 ns
10. Changed T_{AZHAL} from 10 ns to 25 ns
11. Changed T_{BZHAL} from 10 ns to 25 ns
12. Changed I_{CC} (max) from 70 mA to 75 mA
13. Changed I_{CC} formula from (3.88 × Freq + 8.43) to (3.88X Freq + 13.43)
14. Changed V_{OH2} test point from −50 µA to −15 µA
15. Changed Note 1 in DC parameters
16. Changed External Clock min/max, high/low times from percentage to ratio of T_{OSC}
17. Removed NMI from standard inputs (Note 2 under DC Characteristics)
18. Removed V_{OL1} spec
19. Removed T_{CLBV}
20. Added JQ/KQ design consideration

Data sheet 270912-002 supercedes 270912-001 and is valid for devices with an "A" at the end of the topside tracking number.

1. Removed:
   - CPU features descriptions
   - Peripheral features descriptions
   - SFR Operation (placed in Quick Reference)
   - SFR Maps (placed in Quick Reference)
   - SFR Bit Maps (placed in Quick Reference)
   - I_{IL} in DC Characteristics
   - T_{CLHAL} Max and T_{CLBRH} Max
   - Incorrect Sample and Convert time table from 8-bit A/D
2. Added:
   - Express options
   - Bullets on front page
   - Memory Map
   - Process Information
   - Prefix Identification
   - Thermal Characteristics
   - Programming functions to pin-out and pin descriptions
   - Ambient Temperature under Bias to Absolute Maximum Ratings
   - Note relating to Power Dissipation in Absolute Maximum Rating
   - Notes 8 and 9 to DC Characteristics
   - T_{CLBV} to AC Characteristics
   - Title to Buswidth timing diagram
   - T_{YLYH} to Buswidth timing diagram
   - Hold latency spec
   - External Crystal Connection diagram
   - External Clock Connection diagram
   - 10-bit A/D Operating Conditions Table
   - Title to 10-bit and 8-bit mode A/D Characteristics
   - Voltage on Analog Input Pin specification
   - Sampling Capacitor typical value
   - Note 4 to 10-bit and 8-bit A/D Specifications
   - 8-bit A/D Operating Conditions Table

Off Isolation, Feedthrough, V_{CC} Power Supply Rejection, Input Series Resistance, Voltage on Analog Input Pin, Sampling Capacitor and DC Input Leakage to 8-bit A/D specifications

Notes 1, 2 and 3 to EPROM Programming Conditions

New Errata (Items 10 to 16)

3. Changed:

Title of data sheet from "8XC196KR/KQ/JR/JQ 16-BIT HIGH PERFORMANCE CHMOS MICROCONTROLLER" to "8XC196KR/KQ/JR/JQ COMMERCIAL/EXPRESS CHMOS MICROCONTROLLER"

Several bullets on cover sheet

Register RAM numbers in table on front page to match device

Operating conditions to tabular format

Note 1 in DC Characteristics to include Ports 3 and 4

0 to V_{SS} for I_{LI} and I_{LI1} in DC Characteristics

Format of symbols in AC Characteristics

T_{CLCH} to T_{CLLH} in System Bus Timing diagram

T_{XLXL} Max from 286 ns to 250 ns

T_{XHXX} from T_{OSC} − 44 ns to 35%/65%

T_{XLXX} from T_{OSC} − 44 ns to 35%/65%

T_{XLXH} from T_{OSC} − 50 ns to 10 ns

T_{XHXL} from T_{OSC} − 50 ns to 10 ns

AC Testing Input, Output Waveform

Introductory text on A to D Characteristics and Converter Specification

DC Input Leakage from ±1 μA to ±3 μA in A/D Specifications.

Power Dissipation from 0.5W to 1.0W.

Wording in Float Waveform from 100 mV to 150 mV.

EPROM Programming Characteristics to Operating Conditions table.

Data sheet (270912-001) is valid for devices with an "A" at the end of the topside tracking number. This is the first version of the data sheet.

8XC196KT
COMMERCIAL CHMOS MICROCONTROLLER

- High Performance CHMOS 16-Bit CPU
- Up to 32 Kbytes of On-Chip EPROM
- Up to 1 Kbyte of On-Chip Register RAM
- Up to 512 Bytes of Additional RAM (Internal RAM)
- Register-Register Architecture
- 8 Channel/10-Bit A/D with Sample/Hold
- 37 Prioritized Interrupt Sources
- Up to Seven 8-Bit (56) I/O Ports
- Full Duplex Serial I/O Port
- Dedicated Baud Rate Generator
- Interprocessor Communication Slave Port
- Selectable Bus Timing Modes for Flexible Interfacing
- Oscillator Fail Detection Circuitry
- High Speed Peripheral Transaction Server (PTS)
- Two Dedicated 16-Bit High-Speed Compare Registers
- 10 High Speed Capture/Compare (EPA)
- Full Duplex Synchronous Serial I/O Port (SSIO)
- Two Flexible 16-Bit Timer/Counters
- Quadrature Counting Inputs
- Flexible 8-/16-Bit External Bus (Programmable)
- Programmable Bus (HOLD/HLDA)
- 1.75 μs 16 x 16 Multiply
- 3 μs 32/16 Divide
- 68-Pin PLCC Package

Device	Pins/Package	EPROM	Reg RAM	Code RAM	Address Space	I/O	EPA	A/D
8XC196KT	68-Pin PLCC	32K	1K	512	64 Kbyte	56	10	8

X = 7 EPROM Device
X = 0 ROMless Device

The 8XC196KT 16-bit microcontroller is a high performance member of the MCS®-96 microcontroller family. The 8XC196KT is an enhanced 8XC196KR device with 1000 byte register RAM, 512 bytes internal RAM, 16 MHz operation and an optional 32 Kbytes of ROM/EPROM. Intel's CHMOS III-E process provides a high performance processor along with low power consumption.

The 8XC196KT has a maximum guaranteed frequency of 16 MHz.

Ten high-speed capture/compare modules are provided. As capture modules event times with 250 ns resolution can be recorded and generate interrupts. As compare modules events such as toggling of a port pin, starting an A/D conversion, pulse width modulation, and software timers can be generated. Events can be based on the timer or up/down counter.

October 1993
Order Number: 272266-002

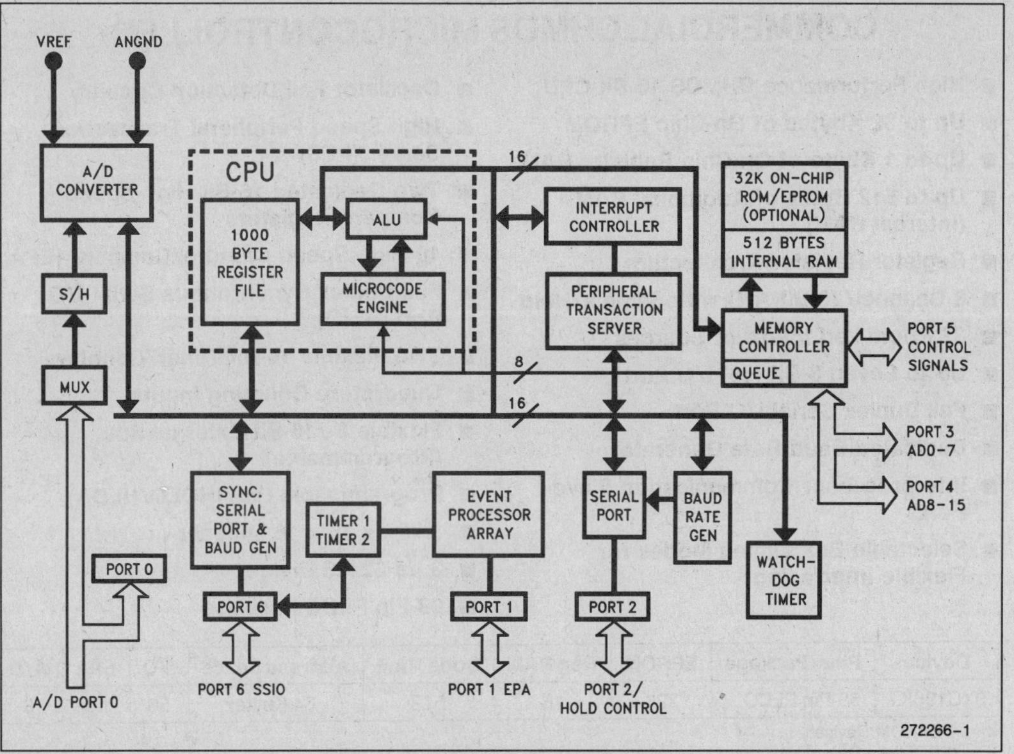

Figure 1. 8XC196KT Block Diagram

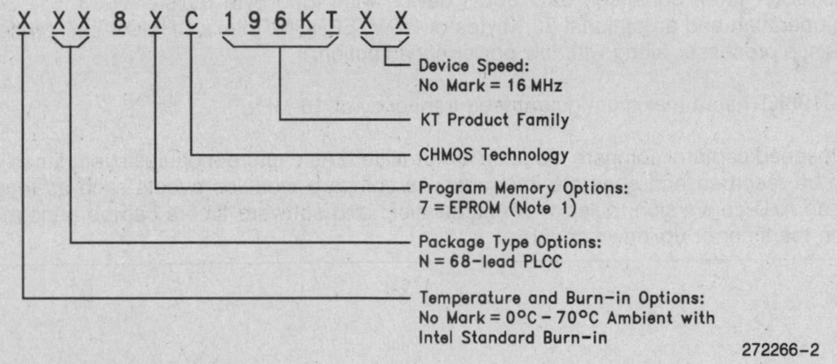

EXAMPLE: N87C196KT is 68-Lead PLCC OTPROM.
For complete package dimensional data, refer to the Intel Packaging Handbok (Order Number 240800).

NOTE:
1. EPROMs are available as One Time Programmable (OTPROM) only.

Figure 2. The 8XC196KT Family Nomenclature

PROCESS INFORMATION

This device is manufactured on PX29.5, a CHMOS III-E process. Additional process and reliability information is available in Intel's *Components Quality and Reliability Handbook*, Order Number 210997.

Table 1. Thermal Characteristics

Package Type	θ_{ja}	θ_{jc}
PLCC	36.5°C/W	13°C/W

All thermal impedance data is approximate for static air conditions at 1W of power dissipation. Values will change depending on operation conditions and application. See the Intel *Packaging Handbook* (Order Number 240800) for a description of Intel's thermal impedance test methodology.

8XC196KT Memory Map

FFFFH A000H	24 Kbytes External Memory	
9FFFH 2080H	Internal EPROM or External Memory	
207FH 2000H	Reserved Memory (Internal EPROM or External Memory)(4, 5)	
1FFFH 1FE0H	Memory Mapped Special Function Registers (SFR's)	
1FDFH 1F00H	Internal Special Function Registers (SFR's)(4)	
1EFFH 0600H	External Memory	
05FFH 0400H	Internal Code or Data RAM (Address with Indirect or Indexed Modes)	
03FFH 0100H	Register RAM	Upper Register File (Address with Indirect or Indexed Modes or through Windows)(1)
00FFH 0018H	Register RAM	Lower Register File (Address with Direct, Indirect, or Indexed Modes)(1, 3)
0017H 0000H	CPU SFR's	

NOTES:
1. Code executed in locations 0000H to 03FFH will be forced external.
2. Reserved memory locations must contain 0FFH unless noted.
3. Reserved SFR bit locations must be written with 0.
4. Refer to 8XC196KT for SFR descriptions.
5. **WARNING:** The contents or functions of reserved memory locations may change with future revisions of the device. Therefore, a program that relies on one or more of these locations may not function properly.

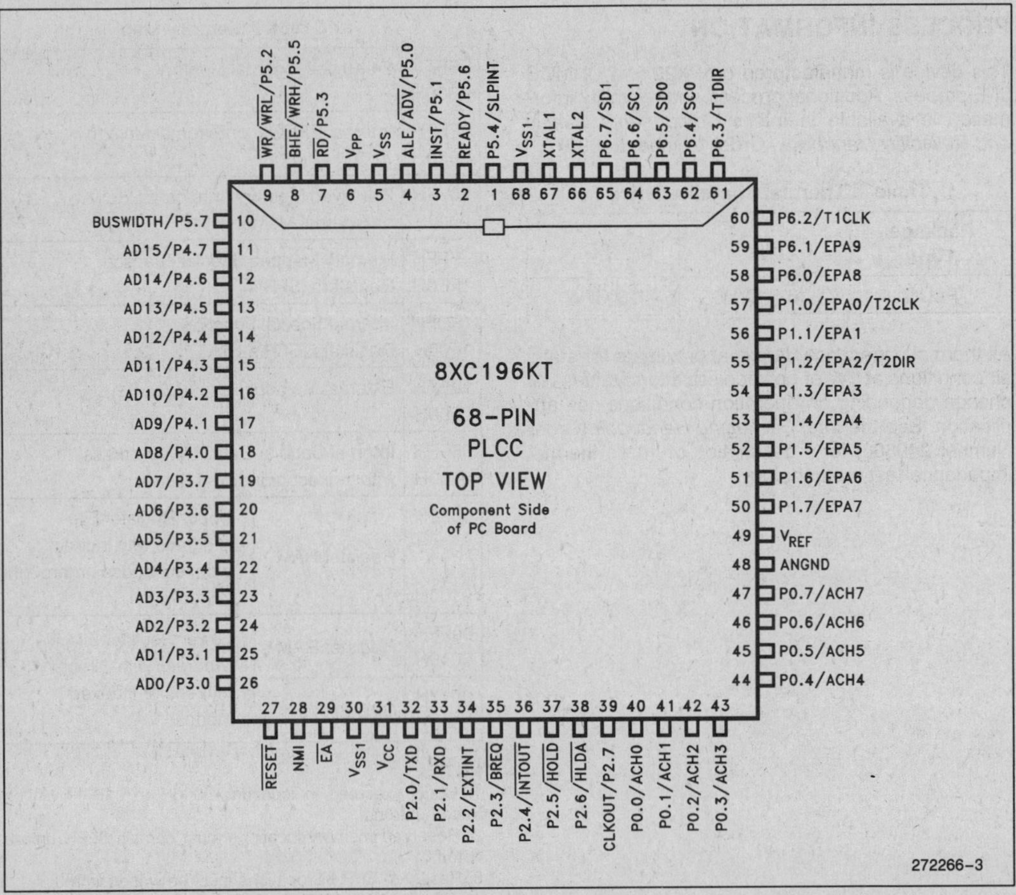
WR/WRL/P5.2
BHE/WRH/P5.5
RD/P5.3
VPP
VSS
ALE/ADV/P5.0
INST/P5.1
READY/P5.6
P5.4/SLPINT
VSS1
XTAL1
XTAL2
P6.7/SD1
P6.6/SC1
P6.5/SD0
P6.4/SC0
P6.3/T1DIR
BUSWIDTH/P5.7
AD15/P4.7
AD14/P4.6
AD13/P4.5
AD12/P4.4
AD11/P4.3
AD10/P4.2
AD9/P4.1
AD8/P4.0
AD7/P3.7
AD6/P3.6
AD5/P3.5
AD4/P3.4
AD3/P3.3
AD2/P3.2
AD1/P3.1
AD0/P3.0
8XC196KT
68-PIN
PLCC
TOP VIEW
Component Side
of PC Board
P6.2/T1CLK
P6.1/EPA9
P6.0/EPA8
P1.0/EPA0/T2CLK
P1.1/EPA1
P1.2/EPA2/T2DIR
P1.3/EPA3
P1.4/EPA4
P1.5/EPA5
P1.6/EPA6
P1.7/EPA7
VREF
ANGND
P0.7/ACH7
P0.6/ACH6
P0.5/ACH5
P0.4/ACH4
RESET
NMI
EA
VSS1
VCC
P2.0/TXD
P2.1/RXD
P2.2/EXTINT
P2.3/BREQ
P2.4/INTOUT
P2.5/HOLD
P2.6/HLDA
CLKOUT/P2.7
P0.0/ACH0
P0.1/ACH1
P0.2/ACH2
P0.3/ACH3
272266-3

PIN DESCRIPTIONS

Symbol	Name and Function
V_{CC}	Main supply voltage (+5V).
V_{SS}, V_{SSI}, V_{SSI}	Digital circuit ground (0V). There are multiple V_{SS} pins, all of which MUST be connected.
V_{REF}	Reference for the A/D converter (+5V). V_{REF} is also the supply voltage to the analog portion of the A/D converter and the logic used to read Port 0. Must be connected for A/D and Port 0 to function.
V_{PP}	Programming voltage for the EPROM parts. It should be +12.5V for programming. It is also the timing pin for the return from powerdown circuit.
ANGND	Reference ground for the A/D converter. Must be held at nominally the same potential as V_{SS}.
XTAL1	Input of the oscillator inverter and the internal clock generator.
XTAL2	Output of the oscillator inverter.
P.27/CLKOUT	Output of the internal clock generator. The frequency is ½ the oscillator frequency. It has a 50% duty cycle. Also LSIO pin.
$\overline{\text{RESET}}$	Reset input to and open-drain output from the chip.
P5.7/BUSWIDTH	Input for bus width selection. If CCR bit 1 is a one and CCR1 bit 2 is a one, this pin dyamically controls the Buswidth of the bus cycle in progress. If BUSWIDTH is low, an 8-bit cycle occurs, if BUSWIDTH is high, a 16-bit cycle occurs. If CCR bit 1 is "0" and CCR1 bit 2 is "1", all bus cycles are 8-bit, if CCR bit 1 is "1" and CCR1 bit 2 is "0", all bus cycles are 16-bit. CCR bit 1 = "0" and CCR1 bit 2 = "0" is illegal. Also an LSIO pin when not used as BUSWIDTH.
NMI	A positive transition causes a non maskable interrupt vector through memory location 203EH.
P5.1/INST	Output high during an external memory read indicates the read is an instruction fetch. INST is valid throughout the bus cycle. INST is active only during external memory fetches, during internal EPROM fetches INST is held low. Also LSIO when not INST.
$\overline{\text{EA}}$	Input for memory select (External Access). $\overline{\text{EA}}$ equal to a high causes memory accesses to locations 2000H through 9FFFH to be directed to on-chip EPROM/ROM. $\overline{\text{EA}}$ equal to a low causes accesses to these locations to be directed to off-chip memory. $\overline{\text{EA}}$ = +12.5V causes execution to begin in the Programming Mode. $\overline{\text{EA}}$ is latched at reset.
$\overline{\text{HOLD}}$	Bus Hold input requesting control of the bus.
$\overline{\text{HLDA}}$	Bus hold acknowledge output indicating release of the bus.
$\overline{\text{BREQ}}$	Bus request output activated when the bus controller has a pending external memory cycle.
P5.0/ALE/$\overline{\text{ADV}}$	Address Latch Enable or Address Valid output, as selected by CCR. Both pin options provide a latch to demultiplex the address from the address/data bus. When the pin is $\overline{\text{ADV}}$, it goes inactive (high) at the end of the bus cycle. $\overline{\text{ADV}}$ can be used as a chip select for external memory. ALE/$\overline{\text{ADV}}$ is active only during external memory accesses. Also LSIO when not used as ALE.
P5.3/$\overline{\text{RD}}$	Read signal output to external memory. $\overline{\text{RD}}$ is active only during external memory reads or LSIO when not used as $\overline{\text{RD}}$.
P5.2/$\overline{\text{WR}}$/$\overline{\text{WRL}}$	Write and Write Low output to external memory, as selected by the CCR, $\overline{\text{WR}}$ will go low for every external write, while $\overline{\text{WRL}}$ will go low only for external writes where an even byte is being written. $\overline{\text{WR}}$/$\overline{\text{WRL}}$ is active during external memory writes. Also an LSIO pin when not used as $\overline{\text{WR}}$/$\overline{\text{WRL}}$.

PIN DESCRIPTIONS (Continued)

Symbol	Name and Function
P5.5/$\overline{BHE}$/$\overline{WRH}$	Byte High Enable or Write High output, as selected by the CCR. $\overline{BHE}$ = 0 selects the bank of memory that is connected to the high byte of the data bus. A0 = 0 selects that bank of memory that is connected to the low byte. Thus accesses to a 16-bit wide memory can be to the low byte only (A0 = 0, $\overline{BHE}$ = 1), to the high byte only (A0 = 1, $\overline{BHE}$ = 0) or both bytes (A0 = 0, $\overline{BHE}$ = 0). If the $\overline{WRH}$ function is selected, the pin will go low if the bus cycle is writing to an odd memory location. $\overline{BHE}$/$\overline{WRH}$ is only valid during 16-bit external memory write cycles. Also an LSIO pin when not BHE/$\overline{WRH}$.
P5.6/READY	Ready input to lengthen external memory cycles, for interfacing with slow or dynamic memory, or for bus sharing. If the pin is high, CPU operation continues in a normal manner. If the pin is low prior to the falling edge of CLKOUT, the memory controller goes into a wait state mode until the next positive transition in CLKOUT occurs with READY high. When external memory is not used, READY has no effect. The max number of wait states inserted into the bus cycle is controlled by the CCR/CCR1. Also an LSIO pin when READY is not selected.
P5.4/SLPINT	Dual function I/O pin. As a bidirectional port pin or as a system function. The system function is a Slave Port Interrupt Output Pin.
P6.2/T1CLK	Dual function I/O pin. Primary function is that of a bidirectional I/O pin, however, it may also be used as a TIMER1 Clock input. The TIMER1 will increment or decrement on both positive and negative edges of this pin.
P6.3/T1DIR	Dual function I/O pin. Primary function is that of a bidirectional I/O pin, however, it may also be used as a TIMER1 Direction input. The TIMER1 will increment when this pin is high and decrements when this pin is low.
PORT1/EPA0–7 P6.0–6.1/EPA8–9	Dual function I/O port pins. Primary function is that of bidirectional I/O. System function is that of High Speed capture and compare. EPA0 and EPA2 have yet another function of T2CLK and T2DIR of the TIMER2 timer/counter.
PORT 0/ACH0–7	8-bit high impedance input-only port. These pins can be used as digital inputs and/or as analog inputs to the on-chip A/D converter. These pins are also used as inputs to EPROM parts to select the Programming Mode.
P6.3–6.7/SSIO	Dual function I/O ports that have a system function as Synchronous Serial I/O. Two pins are clocks and two pins are data, providing full duplex capability.
PORT 2	8-bit multi-functional port. All of its pins are shared with other functions.
PORT 3 and 4	8-bit bidirectional I/O ports with open drain outputs. These pins are shared with the multiplexed address/data bus which has strong internal pullups.

ABSOLUTE MAXIMUM RATINGS*

Storage Temperature .... −60°C to +150°C
Voltage from V_{PP} or $\overline{EA}$ to V_{SS} or ANGND .... −0.5V to +13.0V
Voltage from Any Other Pin to V_{SS} or ANGND .... −0.5 to +7.0V
This includes V_{PP} on ROM and CPU devices.
Power Dissipation.......................... 0.5W

NOTICE: This data sheet contains preliminary information on new products in production. The specifications are subject to change without notice. Verify with your local Intel Sales office that you have the latest data sheet before finalizing a design.

**WARNING: Stressing the device beyond the "Absolute Maximum Ratings" may cause permanent damage. These are stress ratings only. Operation beyond the "Operating Conditions" is not recommended and extended exposure beyond the "Operating Conditions" may affect device reliability.*

OPERATING CONDITIONS

Symbol	Parameter	Min	Max	Units
T_A	Ambient Temperature Under Bias	0	+70	°C
V_{CC}	Digital Supply Voltage	4.50	5.50	V
V_{REF}	Analog Supply Voltage	4.50	5.50	V
F_{OSC}	Oscillator Frequency	4	16	MHz (Note 4)

NOTE:
ANGND and V_{SS} should be nominally at the same potential.

DC CHARACTERISTICS (Under Listed Operating Conditions)

Symbol	Parameter	Min	Typ	Max	Units	Test Conditions
I_{CC}	V_{CC} Supply Current			90	mA	XTAL1 = 16 MHz, $V_{CC} = V_{PP} = V_{REF} = 5.5V$ (While device in Reset)
I_{REF}	A/D Reference Supply Current			5	mA	
I_{IDLE}	Idle Mode Current			40	mA	XTAL1 = 16 MHz, $V_{CC} = V_{PP} = V_{REF} = 5.5V$
I_{PD}	Powerdown Mode Current		50	75	μA	$V_{CC} = V_{PP} = V_{REF} = 5.5V$(6, 11)
V_{IL}	Input Low Voltage (all pins)	−0.5V		0.3 V_{CC}	V	For PORT0(10)
V_{IH}	Input High Voltage	0.7 V_{CC}		V_{CC} + 0.5	V	For PORT0(10)
V_{IH1}	Input High Voltage XTAL1	0.7 V_{CC}		V_{CC} + 0.5	V	XTAL1 Input Pin Only(1)
V_{IH2}	Input High Voltage on RESET	0.7 V_{CC}		V_{CC} + 0.5	V	RESET input pin only
V_{OL}	Output Low Voltage (Outputs Configured as Complementary)			0.3 0.45 1.5	V V V	I_{OL} = 200 μA(3,5) I_{OL} = 3.2 mA I_{OL} = 7.0 mA
V_{OH}	Output High Voltage (Outputs Configured as Complementary)	V_{CC} − 0.3 V_{CC} − 0.7 V_{CC} − 1.5			V V V	I_{OH} = −200μA(3,5) I_{OH} = −3.2 mA I_{OH} = −7.0 mA
I_{LI}	Input Leakage Current (Std. Inputs)			±10	μA	$V_{SS} < V_{IN} < V_{CC}$
I_{LI1}	Input Leakage Current (Port 0)			±3	μA	$V_{SS} < V_{IN} < V_{REF}$
I_{IL}	Logical 0 Input Current			−70	μA	V_{IN} = 0.45V(1)

DC CHARACTERISTICS (Under Listed Operating Conditions) (Continued)

Symbol	Parameter	Min	Typ	Max	Units	Test Conditions
V_{OH1}	SLPINT (P5.4) and HLDA (P2.6) Output High Voltage in RESET	2.0			V	$I_{OH} = 0.8$ mA(7)
V_{OH2}	Output High Voltage in RESET	$V_{CC} - 1V$			V	$I_{OH} = -6\ \mu A$(1)
C_S	Pin Capacitance (Any pin to V_{SS})			10	pF	$f_{test} = 1.0$ MHz(6)
R_{WPU}	Weak Pullup Resistance		150K		Ω	(Note 6)

NOTES:
1. All BD (bidirectional) pins except INST and CLKOUT. INST and CLKOUT are excluded due to their not being weakly pulled high in reset. BD pins include Port1, Port2, Port3, Port4, Port5 and Port6 except SPLINT (P5.4) and HLDA (P2.6).
2. Standard input pins include XTAL1, $\overline{EA}$, RESET, and Port 1/2/5/6 when setup as inputs.
3. All bidirectional I/O pins when configured as Outputs (Push/Pull).
4. Device is static and should operate below 1 Hz, but only tested down to 4 MHz.
5. Maximum I_{OL}/I_{OH} currents per pin will be characterized and published at a later date.
6. Typicals are based on limited number of samples and are not guaranteed. The values listed are at room temperature and $V_{REF} = V_{CC} = 5.0V$.
7. Violating these specifications in reset may cause the device to enter test modes (P5.4 and P2.6).
8. When P0 is used as analog inputs, refer to A/D specifications for this characteristic.
9. For temperatures $<100°C$ typical is 10 μA.

8XC196KT ADDITIONAL BUS TIMING MODES

The 8XC196KT devices have 3 additional bus timing modes for external memory interfacing.

MODE 3:

Mode 3 is the standard timing mode.

MODE 0:

Mode 0 is the standard timing mode, but 1 (minimum) wait state is always inserted in external bus cycles.

MODE 1:

Mode 1 is the long R/W mode. This mode advances $\overline{RD}$ and $\overline{WR}$ signals by 1 T_{OSC} creating a 2 T_{OSC} $\overline{RD}/\overline{WR}$ low time. ALE is also advanced by 0.5 T_{OSC} but ALE high time remains 1 T_{OSC}.

MODE 2:

Mode 2 is the long R/W mode with Early Address. Mode 2 is similar to Mode 1 with respect to RD, WR, and ALE signals. Additionally, the address is output on the bus 0.5 T_{OSC} earlier in the bus cycle.

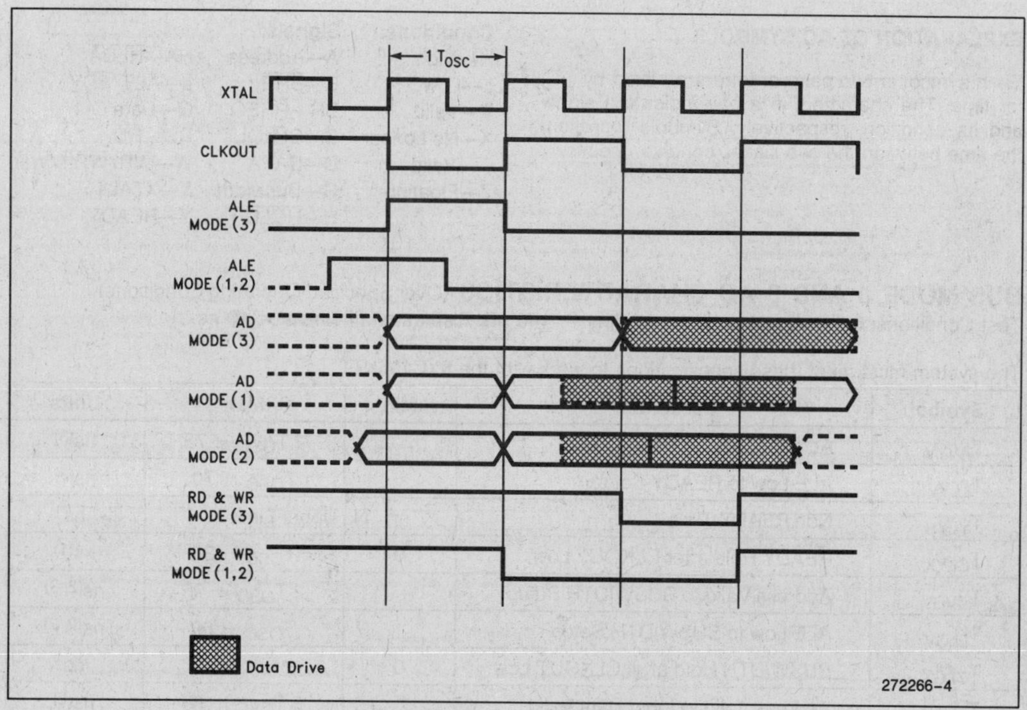

Detailed MODE 1, 2, 3, Comparison

EXPLANATION OF AC SYMBOLS

Each symbol is two pairs of letters prefixed by "T" for time. The characters in a pair indicate a signal and its condition, respectively. Symbols represent the time between the two signal/condition points.

Conditions:

H—High
L—Low
V—Valid
X—No Longer Valid
Z—Floating

Signals:

A—Address
B—$\overline{BHE}$
BR—$\overline{BREQ}$
C—CLKOUT
D—DATA
G—Buswidth
H—$\overline{HOLD}$
HA—$\overline{HLDA}$
L—ALE/$\overline{ADV}$
Q—Data Out
RD—$\overline{RD}$
W—$\overline{WR}$/$\overline{WRH}$/$\overline{WRL}$
X—XTAL1
Y—READY

BUS MODE 0 AND 3 AC CHARACTERISTICS (Over Specified Operating Conditions)

Test Conditions: Capacitance Load on All Pins = 100 pF, Rise and Fall Times = 10 ns.

The system must meet these specifications to work with the 8XC196KT.

Symbol	Parameter	Min	Max	Units
T_{AVYV}	Address Valid to Ready Setup		2 T_{OSC} − 75	ns(3)
T_{LLYV}	ALE Low to READY Setup		T_{OSC} − 70	ns(3)
T_{YLYH}	Non READY Time	No Upper Limit		ns
T_{CLYX}	READY Hold after CLKOUT Low	0	T_{OSC} − 30	ns(1)
T_{AVGV}	Address Valid to BUSWIDTH Setup		2 T_{OSC} − 75	ns(2, 3)
T_{LLGV}	ALE Low to BUSWIDTH Setup		T_{OSC} − 60	ns(2, 3)
T_{CLGX}	BUSWIDTH Hold after CLKOUT Low	0		ns
T_{AVDV}	Address Valid to Input Data Valid		3 T_{OSC} − 55	ns(2)
T_{RLDV}	RD active to input Data Valid		T_{OSC} − 30	ns(2)
T_{CLDV}	CLKOUT Low to Input Data Valid		T_{OSC} − 60	ns
T_{RHDZ}	End of RD to Input Data Float		T_{OSC}	ns
T_{RHDX}	Data Hold after RD High	0		ns

NOTES:
1. If Max is exceeded, additional wait states will occur.
2. If wait states are used, add 2 Tosc × n, where n = number of wait states.
3. If mode 0 is selected, one wait state minimum is always added. If additional wait states are required, add 2 Tosc to the specification.

BUS MODE 0 AND 3 AC CHARACTERISTICS (Over Specified Operating Conditions) (Continued)

Test Conditions: Capacitance Load on All Pins = 100 pF, Rise and Fall Times = 10 ns.

The 8XC196KT will meet these specifications

Symbol	Parameter	Min	Max	Units
T_{XTAL} → F_{XTAL}	Frequency on XTAL1	4.0	16.0	MHz(1)
T_{OSC}	XTAL1 Period ($1/F_{XTAL}$)	62.5	250	ns
T_{XHCH}	XTAL1 High to CLKOUT High or Low	+20	110	ns
T_{OFD}	Clock Failure to Reset Pulled Low(6)	4	40	μs
T_{CLCL}	CLKOUT Period	2 T_{OSC}		ns
T_{CHCL}	CLKOUT High Period	T_{OSC} − 10	T_{OSC} + 30	ns
T_{CLLH}	CLKOUT Low to ALE/ADV High	−10	+15	ns
T_{LLCH}	ALE/ADV Low to CLKOUT High	−25	+15	ns
T_{LHLH}	ALE/ADV Cycle Time	4 T_{OSC}		ns(5)
T_{LHLL}	ALE/ADV High Time	T_{OSC} − 10	T_{OSC} + 10	ns
T_{AVLL}	Address Valid to ALE Low	T_{OSC} − 15		ns
T_{LLAX}	Address Hold After ALE/ADV Low	T_{OSC} − 40		ns
T_{LLRL}	ALE/ADV Low to RD Low	T_{OSC} − 40		ns
T_{RLCL}	RD Low to CLKOUT Low	+4	+35	ns
T_{RLRH}	RD Low Period	T_{OSC} − 5		ns(5)
T_{RHLH}	RD High to ALE/ADV High	T_{OSC}	T_{OSC} + 25	ns(3)
T_{RLAZ}	RD Low to Address Float		+5	ns
T_{LLWL}	ALE/ADV Low to WR Low	T_{OSC} − 10		ns
T_{CLWL}	CLKOUT Low to WR Low	−10	+25	ns
T_{QVWH}	Data Valid before WR High	T_{OSC} − 23		ns
T_{CHWH}	CLKOUT High to WR High	−10	+15	ns
T_{WLWH}	WR Low Period	T_{OSC} − 30		ns(5)
T_{WHQX}	Data Hold after WR High	T_{OSC} − 30		ns
T_{WHLH}	WR High to ALE/ADV High	T_{OSC} − 10	T_{OSC} + 15	ns(3)
T_{WHBX}	BHE, INST Hold after WR High	T_{OSC} − 10		ns
T_{WHAX}	AD8–15 Hold after WR High	T_{OSC} − 30		ns(4)
T_{RHBX}	BHE, INST Hold after RD High	T_{OSC} − 10		ns
T_{RHAX}	AD8–15 Hold after RD High	T_{OSC} − 30		ns(4)

NOTES:
1. Testing performed at 4.0 MHz, however, the device is static by design and will typically operate below 1 Hz.
2. Typical specifications, not guaranteed.
3. Assuming back-to-back bus cycles.
4. 8-bit bus only.
5. If wait states are used, add 2 Tosc × n, where n = number of wait states. If mode 0 (1 automatic wait state added) operation is selected, add 2 T_{OSC} to specification.
6. T_{OFD} is the time for the oscillator fail detect circuit (OFD) to react to a clock failure. The OFD circuitry is enabled by programming the UPROM location 0778H with the value 0004H. KT customer QROM codes need to equate location 2016H to the value 0CDEH if the oscillator fail detect (OFD) function is desired. Intel manufacturing uses location 2016H as a flag to determine whether or not to program the Clock Detect Enable (CDE) bit. Programming the CDE bit enables oscillator fail detection.

BUS MODE 0 AND 3 8XC196KT SYSTEM BUS TIMING

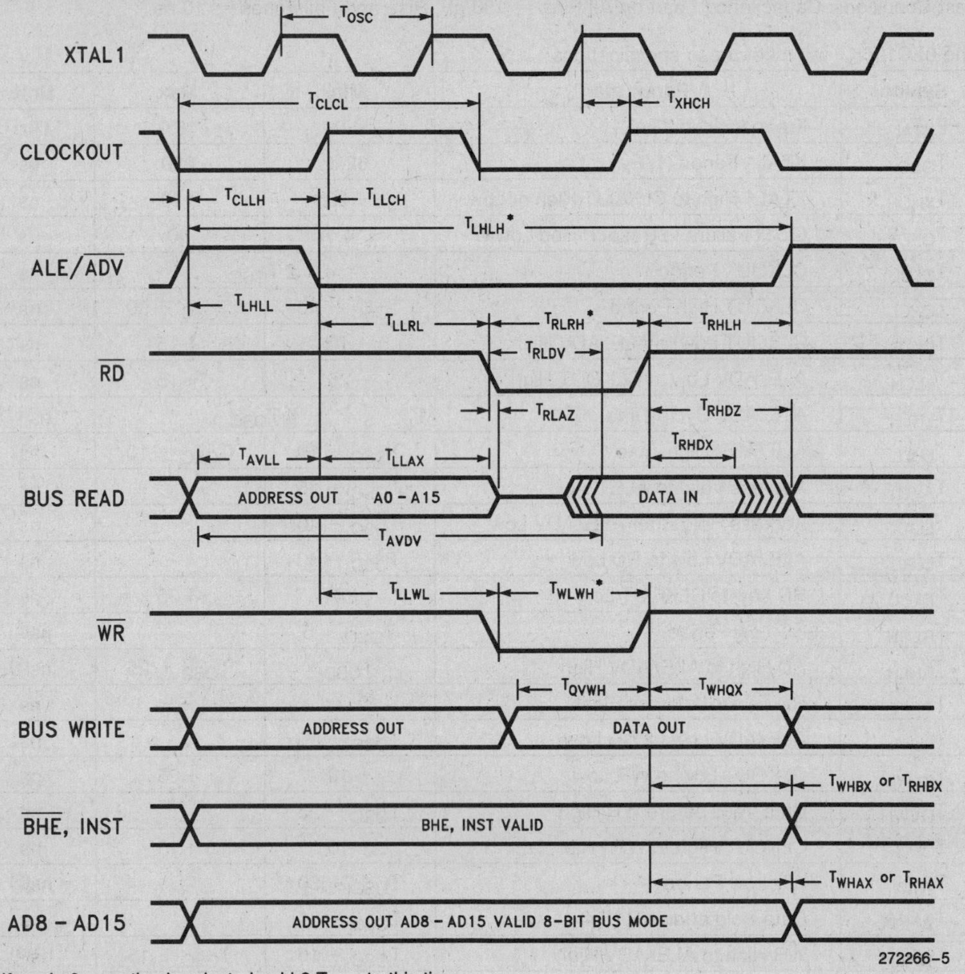

* If mode 0 operation is selected, add 2 T_{OSC} to this time.

MODE 0 AND 3 8XC196KT READY TIMINGS (ONE WAIT STATE)

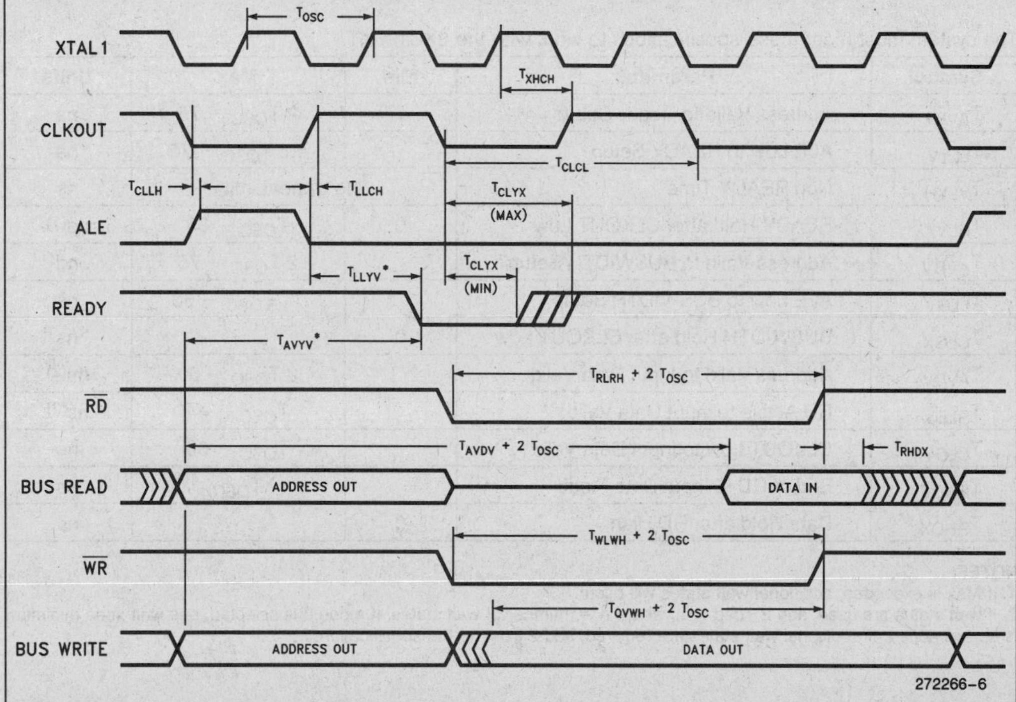

*If mode 0 selected, one wait state is always added. If additional wait states are required, add 2 Tosc to these specifications.

8XC196KT BUSWIDTH TIMINGS

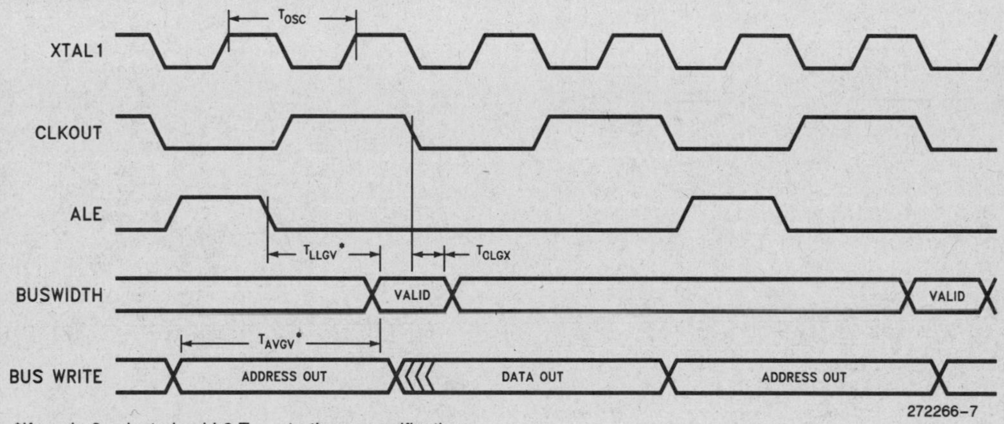

*If mode 0 selected, add 2 T_{OSC} to these specifications.

BUS MODE 1—AC CHARACTERISTICS (Over Specified Operating Conditions)

Test Conditions: Capacitance Load on All Pins = 100 pF, Rise and Fall Times = 10 ns.

The system must meet these specifications to work with the 8XC196KT.

Symbol	Parameter	Min	Max	Units
T_{AVYV}	Address Valid to Ready Setup		$2\ T_{OSC} - 75$	ns
T_{LLYV}	ALE Low to READY Setup		$1.5\ T_{OSC} - 70$	ns
T_{YLYH}	Non READY Time	No Upper Limit		ns
T_{CLYX}	READY Hold after CLKOUT Low	0	$T_{OSC} - 30$	ns(1)
T_{AVGV}	Address Valid to BUSWIDTH Setup		$2\ T_{OSC} - 75$	nd
T_{LLGV}	ALE Low to BUSWIDTH Setup		$T_{1.5\ OSC} - 60$	ns
T_{CLGX}	BUSWIDTH Hold after CLKOUT Low	0		ns
T_{AVDV}	Address Valid to Input Data Valid		$3\ T_{OSC} - 60$	ns(2)
T_{RLDV}	RD Active to input Data Valid		$T_{OSC} - 44$	ns(2)
T_{CLDV}	CLKOUT Low to Input Data Valid		$T_{OSC} - 60$	ns
T_{RHDZ}	End of RD to Input Data Float		T_{OSC}	ns
T_{RHDX}	Data Hold after RD High	0		ns

NOTES:
1. If Max is exceeded, additional wait states will occur.
2. If wait states are used, add 2 Tosc × n, where n = number of wait states. If mode 0 is selected, one wait state minimum is always added. If additional wait states are required, add 2 Tosc to the specification.

BUS MODE 1—AC CHARACTERISTICS (Over Specified Operating Conditions)

Test Conditions: Capacitance Load on All Pins = 100 pF, Rise and Fall Times = 10 ns.

The 8XC196KT will meet these specifications

Symbol	Parameter	Min	Max	Units
F_{XTAL}	Frequency on XTAL1	4.0	16.0	MHz(1)
T_{OSC}	XTAL1 Period ($1/F_{XTAL}$)	62.5	250	ns
T_{XHCH}	XTAL1 High to CLKOUT High or Low	+20		ns
T_{CLCL}	CLKOUT Period	$2\ T_{OSC}$		ns
T_{CHCL}	CLKOUT High Period	$T_{OSC} - 10$	$T_{OSC} + 27$	ns
T_{CHLH}	CLKOUT High to ALE/ADV High	$0.5\ T_{OSC} - 15$	$0.5\ T_{OSC} + 15$	ns
T_{CLLL}	CLKOUT Low to ALE/ADV Low	$0.5\ T_{OSC} - 25$	$0.5\ T_{OSC} + 15$	ns
T_{LHLH}	ALE/ADV Cycle Time	$4\ T_{OSC}$		ns(5)
T_{LHLL}	ALE/ADV High Time	$T_{OSC} - 10$	$T_{OSC} + 10$	ns
T_{AVLL}	Address Valid to ALE Low	$0.5\ T_{OSC} - 15$		ns
T_{LLAX}	Address Hold After ALE/ADV Low	$0.5\ T_{OSC} - 20$		ns
T_{LLRL}	ALE/ADV Low to RD Low	$0.5\ T_{OSC} - 15$		ns
T_{RLCL}	RD Low to CLKOUT Low	$T_{OSC} - 10$	$T_{OSC} + 30$	ns
T_{RLRH}	RD Low Period	$2T_{OSC} - 20$		ns(5)
T_{RHLH}	RD High to ALE/ADV High	$0.5\ T_{OSC}$	$0.5\ T_{OSC} + 25$	ns(3)
T_{RLAZ}	RD Low to Address Float		+5	ns
T_{LLWL}	ALE/ADV Low to WR Low	$0.5\ T_{OSC} - 10$		ns
T_{CLWL}	CLKOUT Low to WR Low	$T_{OSC} - 15$	$T_{OSC} + 25$	ns
T_{QVWH}	Data Valid before WR High	$2\ T_{OSC} - 23$		ns
T_{CHWH}	CLKOUT High to WR High	−10	+15	ns
T_{WLWH}	WR Low Period	$2\ T_{OSC} - 15$		ns(5)
T_{WHQX}	Data Hold after WR High	$0.5\ T_{OSC} - 25$		ns
T_{WHLH}	WR High to ALE/ADV High	$0.5\ T_{OSC} - 10$	$0.5\ T_{OSC} + 10$	ns(3)
T_{WHBX}	BHE Hold after WR High	$T_{OSC} - 15$		ns
T_{WHIX}	INST Hold after WR High	$0.5\ T_{OSC} - 15$		
T_{WHAX}	AD8–15 Hold after WR High	$0.5\ T_{OSC} - 30$		ns(4)
T_{RHBX}	BHE Hold after RD High	$T_{OSC} - 32$		ns
T_{RHIX}	INST Hold after RD High	$0.5\ T_{OSC} - 32$		
T_{RHAX}	AD8–15 Hold after RD High	$0.5\ T_{OSC} - 30$		ns(4)

NOTES:
1. Testing performed at 8.0 MHz, however, the device is static by design and will typically operate below 1 Hz.
2. Typical specifications, not guaranteed.
3. Assuming back-to-back bus cycles.
4. 8-bit bus only.
5. If wait states are used, add $2\ T_{OSC} \times n$, where n = number of wait states. If mode 0 (1 automatic wait state added) operation is selected, add $2\ T_{OSC}$ to specification.

MODE 1—8XC196KT SYSTEM BUS TIMING

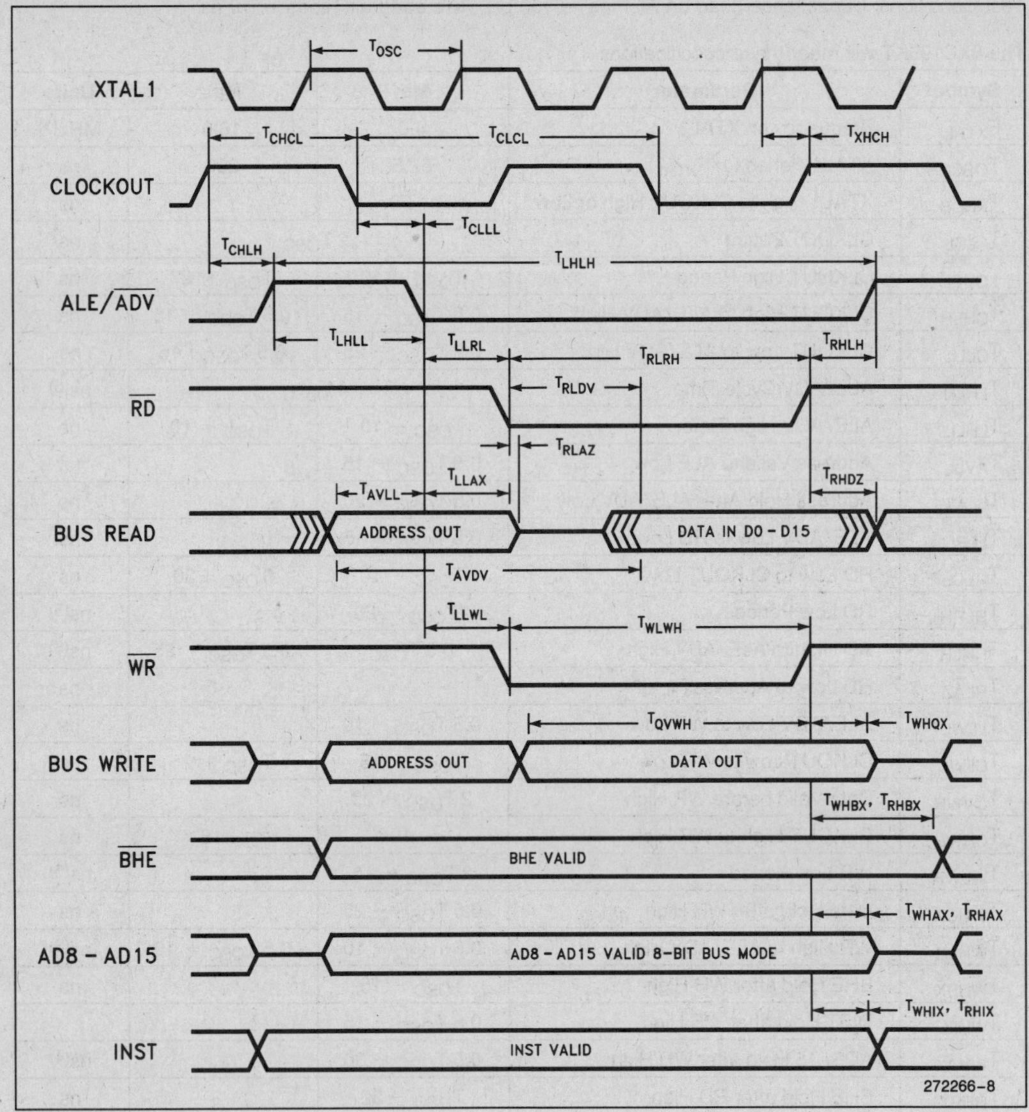

272266-8

MODE 1—8XC196KT READY TIMINGS (ONE WAIT STATE)

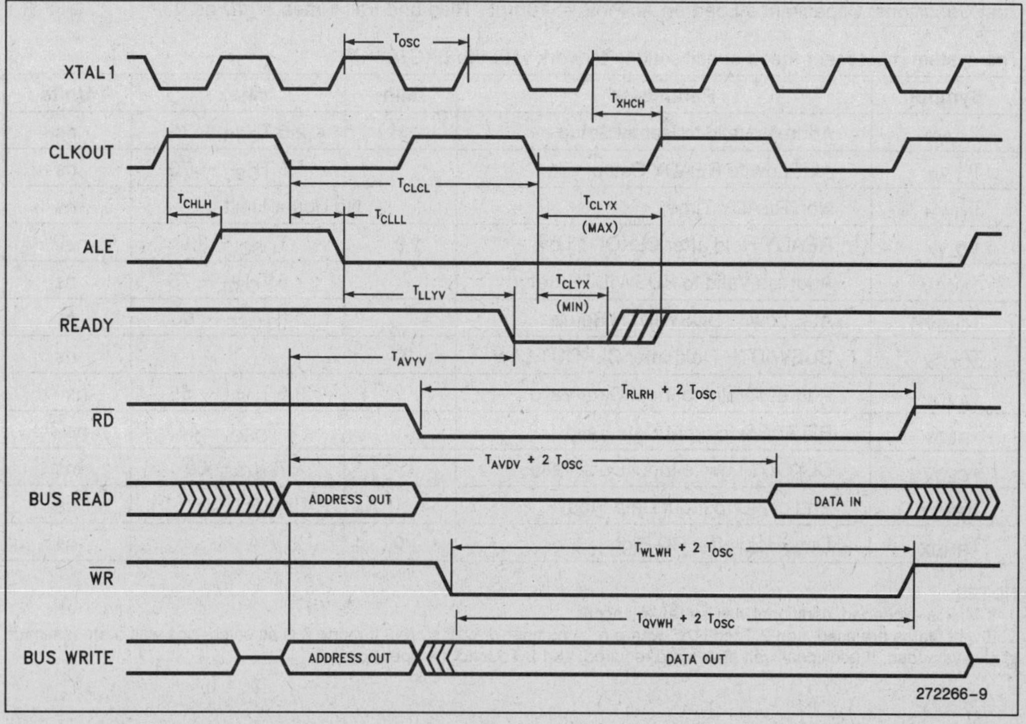

MODE 1—8XC196KT BUSWIDTH TIMINGS

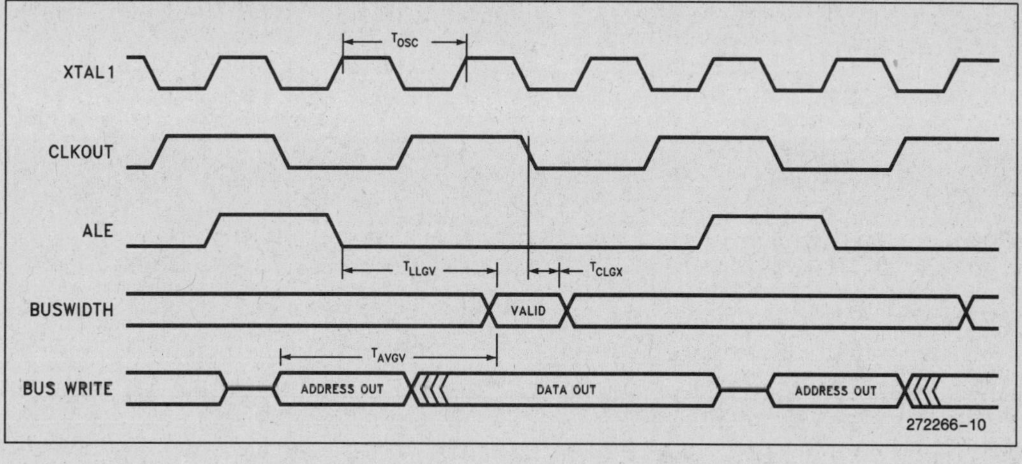

4

BUS MODE 2—AC CHARACTERISTICS (Over Specified Operating Conditions)

Test Conditions: Capacitance Load on All Pins = 100 pF, Rise and Fall Times = 10 ns.

The system must meet these specifications to work with the 8XC196KT.

Symbol	Parameter	Min	Max	Units
T_{AVYV}	Address Valid to Ready Setup		$2.5\ T_{OSC} - 75$	ns
T_{LLYV}	ALE Low to READY Setup		$1.5\ T_{OSC} - 70$	ns
T_{YLYH}	Non READY Time	No Upper Limit		ns
T_{CLYX}	READY Hold after CLKOUT Low	0	$T_{OSC} - 30$	ns(1)
T_{AVGV}	Address Valid to BUSWIDTH Setup		$2.5\ T_{OSC} - 75$	ns
T_{LLGV}	ALE Low to BUSWIDTH Setup		$T_{1.5\ OSC} - 60$	ns
T_{CLGX}	BUSWIDTH Hold after CLKOUT Low	0		ns
T_{AVDV}	Address Valid to Input Data Valid		$3.5\ T_{OSC} - 55$	ns(2)
T_{RLDV}	RD Active to Input Data Valid		$2\ T_{OSC} - 44$	ns(2)
T_{CLDV}	CLKOUT Low to Input Data Valid		$T_{OSC} - 60$	ns
T_{RHDZ}	End of RD to Input Data Float		$0.5\ T_{OSC}$	ns
T_{RHDX}	Data Hold after RD High	0		ns

NOTES:
1. If Max is exceeded, additional wait states will occur.
2. If wait states are used, add 2 Tosc × n, where n = number of wait states. If mode 0 is selected, one wait state minimum is always added. If additional wait states are required, add 2 Tosc to the specification.

BUS MODE 2—AC CHARACTERISTICS (Over Specified Operating Conditions)

Test Conditions: Capacitance Load on All Pins = 100 pF, Rise and Fall Times = 10 ns.

The 8XC196KT will meet these specifications

Symbol	Parameter	Min	Max	Units
F_{XTAL}	Frequency on XTAL1	8.0	16.0	MHz(1)
T_{OSC}	XTAL1 Period ($1/F_{XTAL}$)	62.5	125	ns
T_{XHCH}	XTAL1 High to CLKOUT High or Low	+20	+85	ns
T_{CLCL}	CLKOUT Period	2 T_{OSC}		ns
T_{CHCL}	CLKOUT High Period	$T_{OSC} - 10$	$T_{OSC} + 27$	ns
T_{CHLH}	CLKOUT High to ALE/ADV High	$0.5\ T_{OSC} - 15$	$0.5\ T_{OSC} + 15$	ns
T_{CLLL}	CLKOUT Low to ALE/ADV Low	$0.5\ T_{OSC} - 25$	$0.5\ T_{OSC} + 15$	ns
T_{LHLH}	ALE/ADV Cycle Time	4 T_{OSC}		ns(5)
T_{LHLL}	ALE/ADV High Time	$T_{OSC} - 10$	$T_{OSC} + 10$	ns
T_{AVLL}	Address Valid to ALE Low	$T_{OSC} - 15$		ns
T_{LLAX}	Address Hold After ALE/ADV Low	$0.5\ T_{OSC} - 20$		ns
T_{LLRL}	ALE/ADV Low to RD Low	$0.5\ T_{OSC} - 15$		ns
T_{RLCL}	RD Low to CLKOUT Low	$T_{OSC} - 10$	$T_{OSC} + 30$	ns
T_{RLRH}	RD Low Period	$2T_{OSC} - 20$		ns(5)
T_{RHLH}	RD High to ALE/ADV High	$0.5\ T_{OSC} - 5$	$0.5\ T_{OSC} + 25$	ns(3)
T_{RLAZ}	RD Low to Address Float		+5	ns
T_{LLWL}	ALE/ADV Low to WR Low	$0.5\ T_{OSC} - 10$		ns
T_{CLWL}	CLKOUT Low to WR Low	$T_{OSC} - 22$	$T_{OSC} + 25$	ns
T_{QVWH}	Data Valid before WR High	$2\ T_{OSC} - 25$		ns
T_{CHWH}	CLKOUT High to WR High	−10	+15	ns
T_{WLWH}	WR Low Period	$2\ T_{OSC} - 20$		ns(5)
T_{WHQX}	Data Hold after WR High	$0.5\ T_{OSC} - 25$		ns
T_{WHLH}	WR High to ALE/ADV High	$0.5\ T_{OSC} - 10$	$0.5\ T_{OSC} + 10$	ns(3)
T_{WHBX}	BHE Hold after WR High	$T_{OSC} - 15$		ns
T_{WHIX}	INST Hold after WR High	$0.5\ T_{OSC} - 15$		
T_{WHAX}	AD8–15 Hold after WR High	$0.5\ T_{OSC} - 30$		ns(4)
T_{RHBX}	BHE Hold after RD High	$T_{OSC} - 32$		ns
T_{RHIX}	INST Hold after RD High	$0.5\ T_{OSC} - 32$		
T_{RHAX}	AD8–15 Hold after RD High	$0.5\ T_{OSC} - 30$		ns(4)

NOTES:
1. Testing performed at 8.0 MHz, however, the device is static by design and will typically operate below 1 Hz.
2. Typical specifications, not guaranteed.
3. Assuming back-to-back bus cycles.
4. 8-bit bus only.
5. If wait states are used, add 2 $T_{OSC} \times n$, where n = number of wait states. If mode 0 (1 automatic wait state added) operation is selected, add 2 T_{OSC} to specification.

MODE 2—8XC196KT SYSTEM BUS TIMING

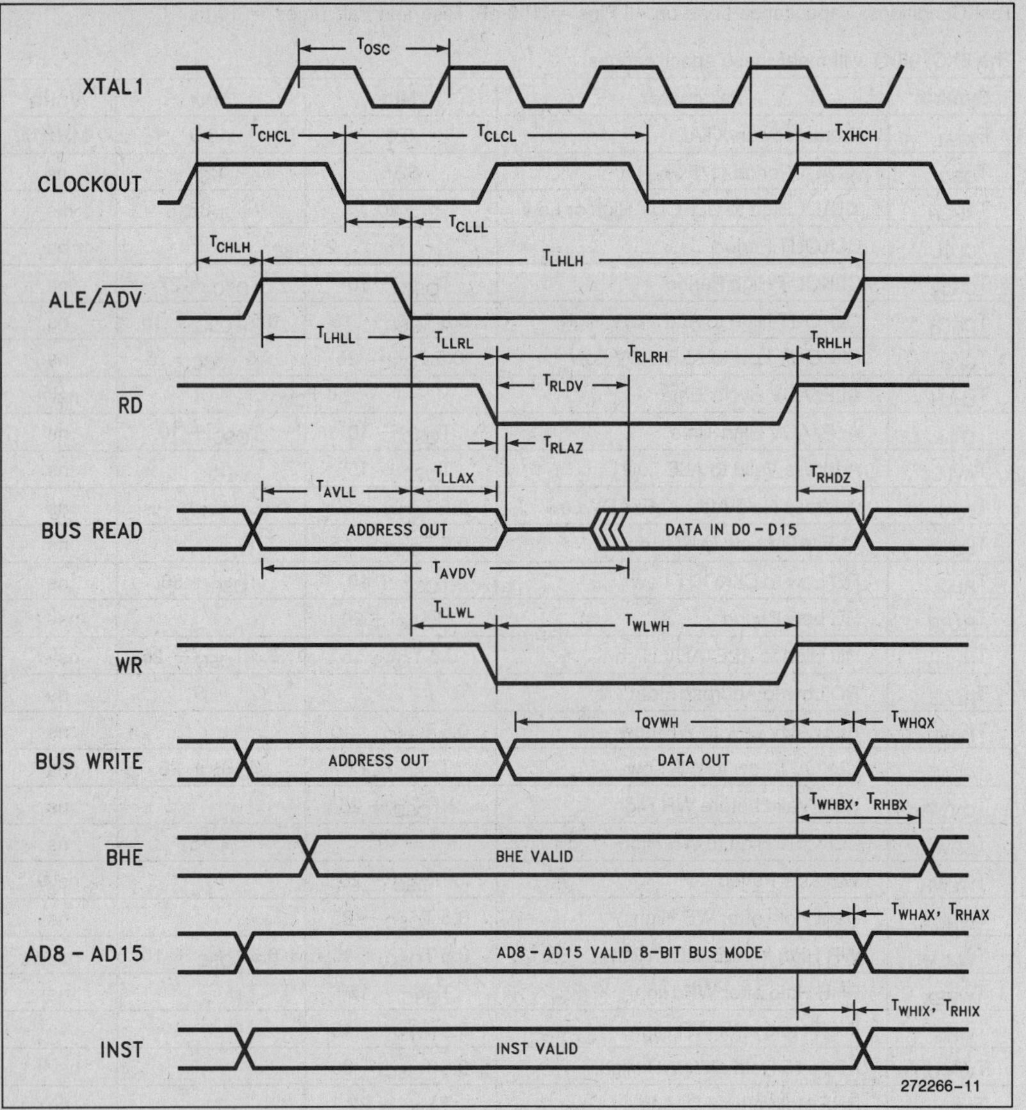

MODE 2—8XC196KT READY TIMINGS (ONE WAIT STATE)

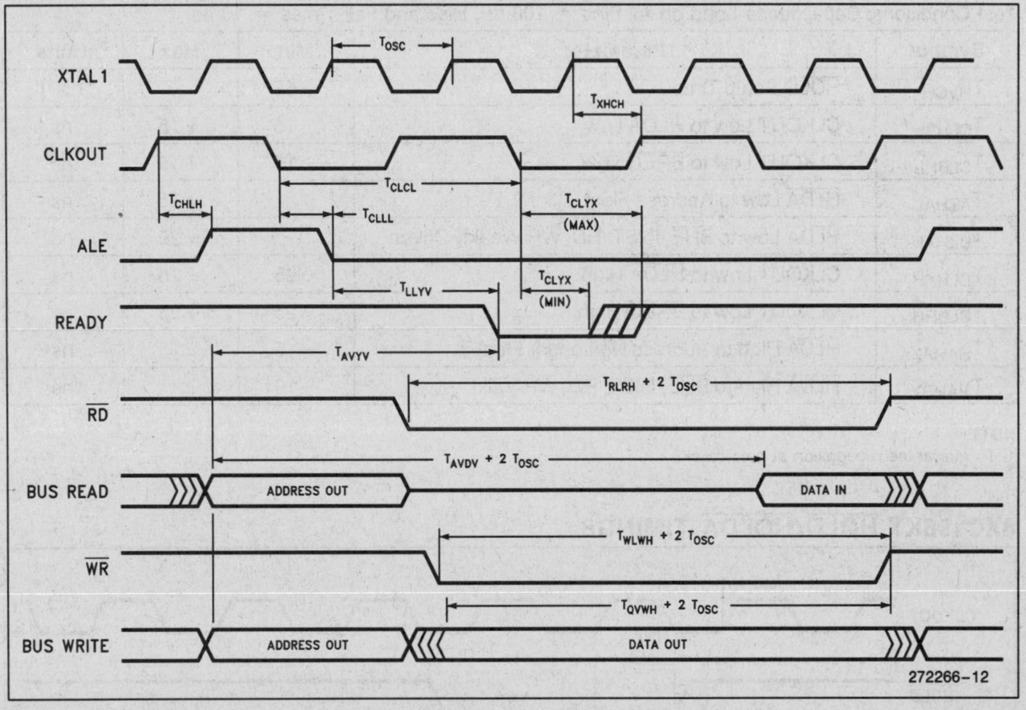

MODE 2—8XC196KT BUSWIDTH TIMINGS

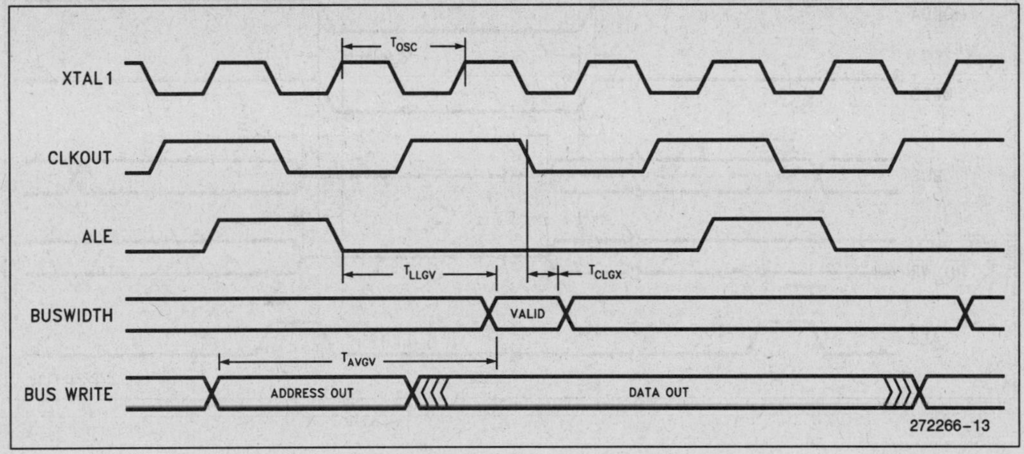

BUS MODE 0, 1, 2, and 3 HOLD/HOLDA TIMINGS (Over Specified Operation Conditions)

Test Conditions: Capacitance Load on All Pins = 100 pF, Rise and Fall Times = 10 ns.

Symbol	Parameter	Min	Max	Units
T_{HVCH}	$\overline{HOLD}$ Setup Time	+65		ns(1)
T_{CLHAL}	CLKOUT Low to $\overline{HLDA}$ Low	−15	+15	ns
T_{CLBRL}	CLKOUT Low to $\overline{BREQ}$ Low	−15	+15	ns
T_{AZHAL}	$\overline{HLDA}$ Low to Address Float		+25	ns
T_{BZHAL}	$\overline{HLDA}$ Low to $\overline{BHE}$, INST, $\overline{RD}$, $\overline{WR}$ Weakly Driven		+25	ns
T_{CLHAH}	CLKOUT Low to $\overline{HLDA}$ High	−25	+15	ns
T_{CLBRH}	CLKOUT Low to $\overline{BREQ}$ High	−25	+25	ns
T_{HAHAX}	$\overline{HLDA}$ High to Address No Longer Float	−15		ns
T_{HAHBV}	$\overline{HLDA}$ High to $\overline{BHE}$, INST, $\overline{RD}$, $\overline{WR}$ Valid	−10		ns

NOTE:
1. To guarantee recognition at next clock.

8XC196KT $\overline{HOLD}$/$\overline{HOLDA}$ TIMINGS

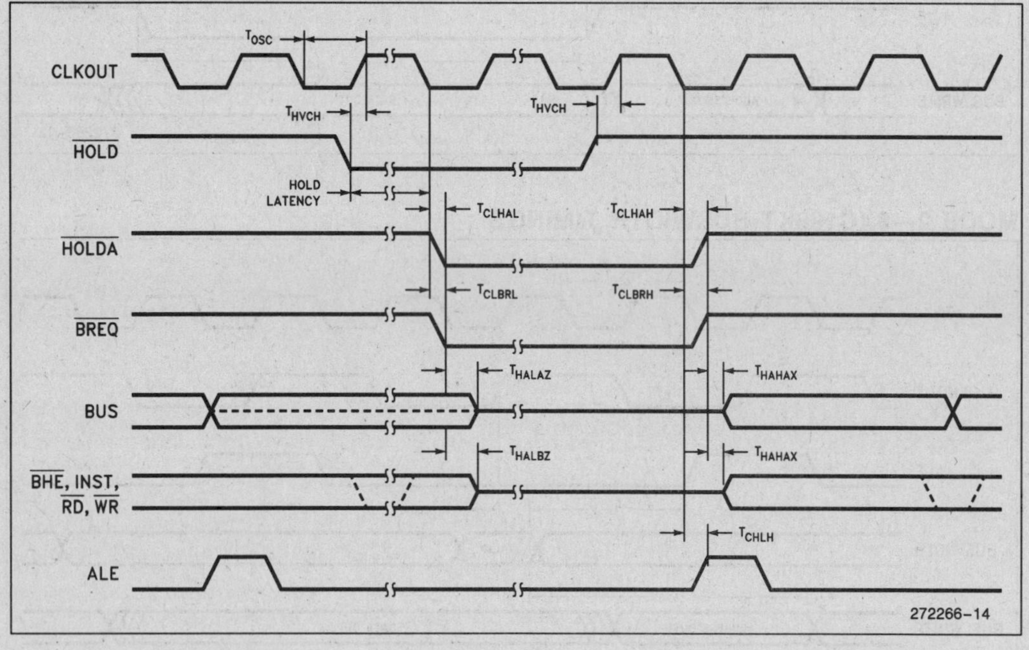

AC CHARACTERISTICS—SLAVE PORT

SLAVE PORT WAVEFORM—(SLPL = 0)

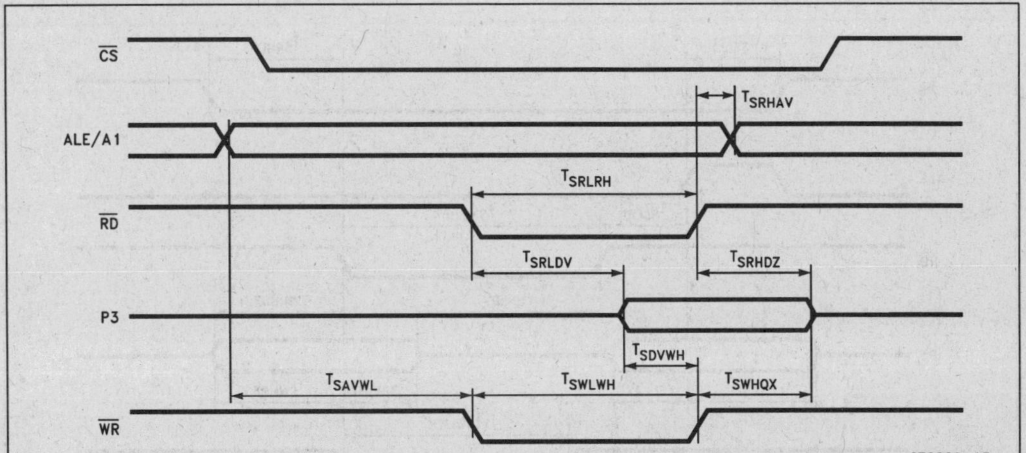

272266-15

SLAVE PORT TIMING—(SLPL = 0, 1, 2, 3)

Symbol	Parameter	Min	Max	Units
T_{SAVWL}	Address Valid to $\overline{WR}$ Low	50		ns
T_{SRHAV}	$\overline{RD}$ High to Address Valid	60		ns
T_{SRLRH}	$\overline{RD}$ Low Period	T_{OSC}		ns
T_{SWLWH}	$\overline{WR}$ Low Period	T_{OSC}		ns
T_{SRLDV}	$\overline{RD}$ Low to Output Data Valid		60	ns
T_{SDVWH}	Input Data Setup to $\overline{WR}$ High	20		ns
T_{SWHQX}	$\overline{WR}$ High to Data Invalid	30		ns
T_{SRHDZ}	$\overline{RD}$ High to Data Float	15		ns

NOTES:
1. Test Conditions: F_{OSC} = 16 MHz, T_{OSC} = 60 ns. Rise/Fall Time = 10 ns. Capacitive Pin Load = 100 pF.
2. These values are not tested in production, and are based upon theoretical estimates and/or laboratory tests.
3. Specifications above are advanced information and are subject to change.

AC CHARACTERISTICS—SLAVE PORT (Continued)

SLAVE PORT WAVEFORM—(SLPL = 1)

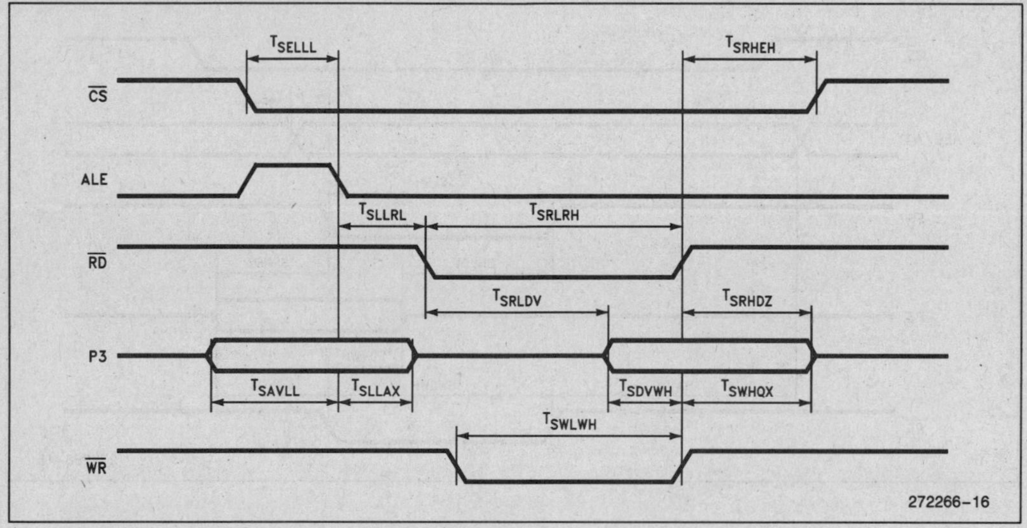

SLAVE PORT TIMING—(SLPL = 1, 2, 3)

Symbol	Parameter	Min	Max	Units
T_{SELLL}	$\overline{CS}$ Low to ALE Low	20		ns
T_{SRHEH}	$\overline{RD}$ or $\overline{WR}$ High to $\overline{CS}$ High	60		ns
T_{SLLRL}	ALE Low to $\overline{RD}$ Low	T_{OSC}		ns
T_{SRLRH}	$\overline{RD}$ Low Period	T_{OSC}		ns
T_{SWLWH}	$\overline{WR}$ Low Period	T_{OSC}		ns
T_{SAVLL}	Address Valid to ALE Low	20		ns
T_{SLLAX}	ALE Low to Address Invalid	20		ns
T_{SRLDV}	$\overline{RD}$ Low to Output Data Valid		60	ns
T_{SDVWH}	Input Data Setup to $\overline{WR}$ High	20		ns
T_{SWHQX}	$\overline{WR}$ High to Data Invalid	30		ns
T_{SRHDZ}	$\overline{RD}$ High to Data Float	15		ns

NOTES:
1. Test Conditions: F_{OSC} = 16 MHz, T_{OSC} = 60 ns. Rise/Fall Time = 10 ns. Capacitive Pin Load = 100 pF.
2. These values are not tested in production, and are based upon theoretical estimates and/or laboratory tests.
3. Specifications above are advanced information and are subject to change.

EXTERNAL CLOCK DRIVE

Symbol	Parameter	Min	Max	Units
$1/T_{XLXL}$	Oscillator Frequency	4	16	MHz
T_{XLXL}	Oscillator Period (T_{OSC})	62.5	250	ns
T_{XHXX}	High Time	$0.35 \times T_{OSC}$	$0.65\ T_{OSC}$	ns
T_{XLXX}	Low Time	$0.35 \times T_{OSC}$	$0.65\ T_{OSC}$	ns
T_{XLXH}	Rise Time		10	ns
T_{XHXL}	Fall Time		10	ns

EXTERNAL CLOCK DRIVE WAVEFORMS

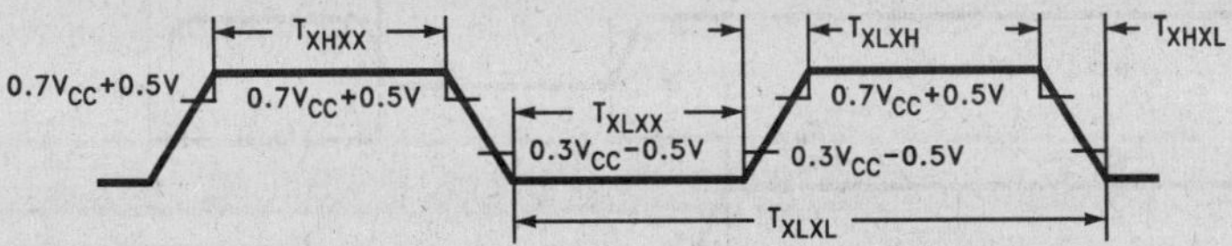

272266–17

AC TESTING INPUT, OUTPUT WAVEFORMS

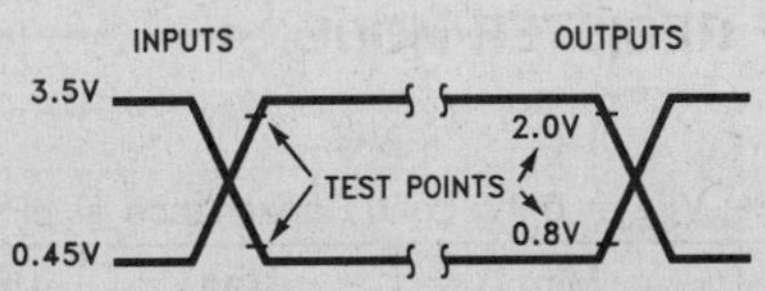

272266–18

AC Testing inputs are driven at 3.5V for a logic "1" and 0.45V for a logic "0". Timing measurements are made at 2.0V for a logic "1" and 0.8V for logic "0".

FLOAT WAVEFORMS

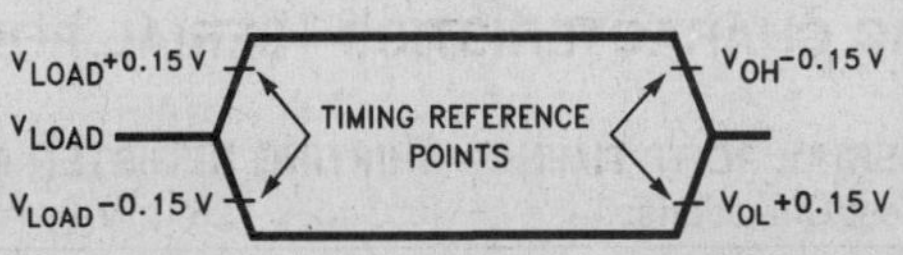

272266–19

For timing purposes a Port Pin is no longer floating when a 150 mV change from load voltage occurs and begins to float when a 150 mV change from the loading V_{OH}/V_{OL} level occurs $I_{OL}/I_{OH} \leq 15$ mA.

WAVEFORM—SERIAL PORT—SHIFT REGISTER MODE

SERIAL PORT WAVEFORM—SHIFT REGISTER MODE

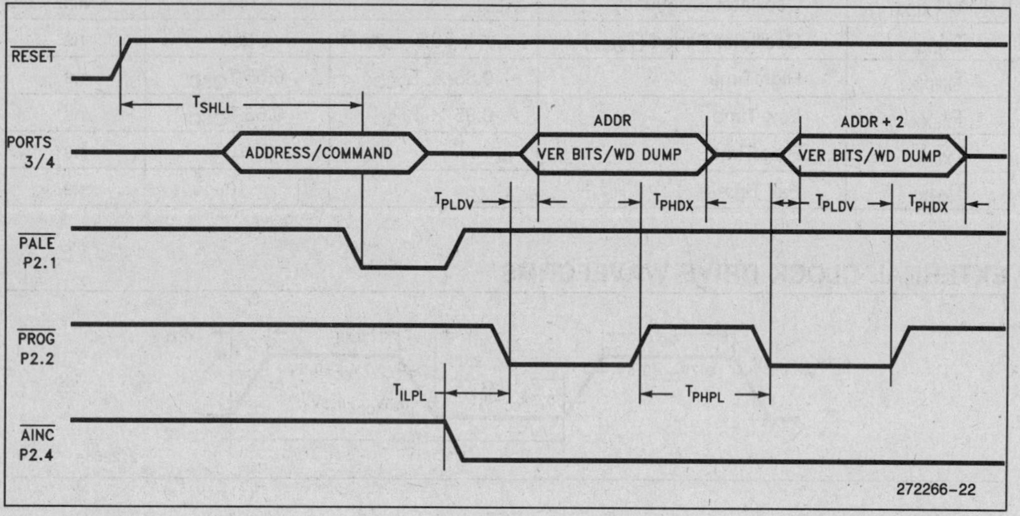

AC CHARACTERISTICS—SERIAL PORT-SHIFT REGISTER MODE

SERIAL PORT TIMING—SHIFTING REGISTER MODE

Test Conditions: $T_A = -40°C$ to $+125°C$; $V_{CC} = 5.0V \pm 10\%$; $V_{SS} = 0.0V$; Load Capacitance = pF

Symbol	Parameter	Min	Max	Units
T_{XLXL}	Serial Port Clock Period	8 T_{OSC}		ns
T_{XLXH}	Serial Port Clock Falling Edge to Rising Edge	4 T_{OSC} − 50	T_{OSC} + 50	ns
T_{QVXH}	Output Data Setup to Clock Rising Edge	3 T_{OSC}		ns
T_{XHQX}	Output Data Hold after Clock Rising Edge	2 T_{OSC} − 50		ns
T_{XHQV}	Next Output Data Valid after Clock Rising Edge		2 T_{OSC} + 50	ns
T_{DVXH}	Input Data Setup to Clock Rising Edge	2 T_{OSC} + 200		ns
T_{XHDX}(1)	Input Data Hold after Clock Rising Edge	0		ns
T_{XHQZ}(1)	Last Clock Rising to Output Float		5 T_{OSC}	ns

NOTE:
1. Parameters not tested.

A TO D CHARACTERISTICS

The A/D converter is ratiometric, so absolute accuracy is dependent on the accuracy and stability of V_{REF}.

10-BIT MODE A/D OPERATING CONDITIONS

Symbol	Description	Min	Max	Units
T_A	Ambient Temperature	0	+70	°C
V_{CC}	Digital Supply Voltage	4.50	5.50	V
V_{REF}	Analog Supply Voltage	4.50	5.50	V(1)
T_{SAM}	Sample Time	1.0		μs(2)
T_{CONV}	Conversion Time	10	15	μs(2)
F_{OSC}	Oscillator Frequency	4.0	16.0	MHz

NOTES:
1. V_{REF} must be within 0.5V of V_{CC}.
2. The value of AD_TIME is selected to meet these specifications.

10-BIT MODE A/D CHARACTERISTICS (Using Above Operating Conditions)(6)

Parameter	Typ*(1)	Min	Max	Units*
Resolution		1024 10	1024 10	Level Bits
Absolute Error		0	±3.0	LSBs
Full Scale Error	0.25 ±0.5			LSBs
Zero Offset Error	0.25 ±0.5			LSBs
Non-Linearity	1.0 ±2.0		±3.0	LSBs
Differential Non-Linearity		−0.75	+0.75	LSBs
Channel-to-Channel Matching	±0.1	0	±1.0	LSBs
Repeatability	±0.25	0		LSBs(1)
Temperature Coefficients: Offset Full Scale Differential Non-Linearity	 0.009 0.009 0.009			 LSB/C(1) LSB/C(1) LSB/C(1)
Off Isolation		−60		dB(1,2,3)
Feedthrough	−60			dB(1,2)
V_{CC} Power Supply Rejection	−60			dB(1,2)
Input Resistance		750	1.2K	Ω(4)
DC Input Leakage	±1.0	0	±3.0	μA
Voltage on Analog Input Pin		ANGND − 0.5	V_{REF} + 0.5	V(5)
Sampling Capacitor	3.0			pF

*An "LSB" as used here has a value of approximately 20 mV.

NOTES:
1. These values are expected for most parts at 25°C, but are not tested or guaranteed.
2. DC to 100 KHz.
3. Multiplexer break-before-make is guaranteed.
4. Resistance from device pin, through internal MUX, to sample capacitor.
5. Applying voltages beyond these specifications will degrade the accuracy of other channels being converted.
6. All conversions performed with processor in IDLE mode.

8-BIT MODE A/D OPERATING CONDITIONS

Symbol	Description	Min	Max	Units
T_A	Ambient Temperature	0	+70	°C
V_{CC}	Digital Supply Voltage	4.50	5.50	V
V_{REF}	Analog Supply Voltage	4.50	5.50	V(1)
T_{SAM}	Sample Time	1.0		μs(2)
T_{CONV}	Conversion Time	7	20	μs(2)
F_{OSC}	Oscillator Frequency	4.0	16.0	MHz

NOTES:
1. V_{REF} must be within 0.5V of V_{CC}.
2. The value of AD_TIME is selected to meet these specifications.

8-BIT MODE A/D CHARACTERISTICS (Using Above Operating Conditions)(6)

Parameter	Typ*(1)	Min	Max	Units*
Resolution		256 8	256 8	Level Bits
Absolute Error		0	±1.0	LSBs
Full Scale Error	±0.5			LSBs
Zero Offset Error	±0.5			LSBs
Non-Linearity		0	±1.0	LSBs
Differential Non-Linearity		−0.5	+0.5	LSBs
Channel-to-Channel Matching		0	±1.0	LSBs
Repeatability	±0.25	0		LSBs(1)
Temperature Coefficients: Offset Full Scale Differential Non-Linearity	 0.003 0.003 0.003			 LSB/C(1) LSB/C(1) LSB/C(1)
Off Isolation		−60		dB(1,2,3)
Feedthrough	−60			dB(1,2)
V_{CC} Power Supply Rejection	−60			dB(1,2)
Input Resistance		750	1.2K	Ω(4)
DC Input Leakage	±1.0	0	±3.0	μA
Voltage on Analog Input Pin		ANGND − 0.5	V_{REF} + 0.5	V(5)
Sampling Capacitor	3.0			pF

*An "LSB" as used here has a value of approximately 20 mV.

NOTES:
1. These values are expected for most parts at 25°C, but are not tested or guaranteed.
2. DC to 100 KHz.
3. Multiplexer break-before-make is guaranteed.
4. Resistance from device pin, through internal MUX, to sample capacitor.
5. Applying voltage beyond these specifications will degrade the accuracy of other channels being converted.
6. All conversions performed with processor in IDLE mode.

EPROM SPECIFICATIONS

OPERATING CONDITIONS

Symbol	Description	Min	Max	Units
T_A	Ambient Temperature during Programming	20	30	°C
V_{CC}	Supply Voltage during Programming	4.5	5.5	V(1)
V_{REF}	Reference Supply Voltage during Programming	4.5	5.5	V(1)
V_{PP}	Programming Voltage	12.25	12.75	V(2)
V_{EA}	EA Pin Voltage	12.25	12.75	V(2)
F_{OSC}	Oscillator Frequency during Auto and Slave Mode Programming	6.0	8.0	MHz
F_{OSC}	Oscillator Frequency during Run-Time Programming	6.0	16.0	MHz

NOTES:
1. V_{CC} and V_{REF} should nominally be at the same voltage during programming.
2. V_{PP} and V_{EA} must never exceed the maximum specification, or the device may be damaged.
3. V_{SS} and ANGND should nominally be at the same potential (0V).
4. Load capacitance during Auto and Slave Mode programming = 150 pF.

AC EPROM PROGRAMMING CHARACTERISTICS (SLAVE MODE)

Symbol	Parameter	Min	Max	Units
T_{AVLL}	Address Setup Time	0		T_{OSC}
T_{LLAX}	Address Hold Time	100		T_{OSC}
T_{DVPL}	Data Setup Time	0		T_{OSC}
T_{PLDX}	Data Hold Time	400		T_{OSC}
T_{LLLH}	$\overline{PALE}$ Pulse Width	50		T_{OSC}
T_{PLPH}	$\overline{PROG}$ Pulse Width(2)	50		T_{OSC}
T_{LHPL}	$\overline{PALE}$ High to $\overline{PROG}$ Low	220		T_{OSC}
T_{PHLL}	$\overline{PROG}$ High to next $\overline{PALE}$ Low	220		T_{OSC}
T_{PHDX}	Word Dump Hold Time		50	T_{OSC}
T_{PHPL}	$\overline{PROG}$ High to next $\overline{PROG}$ Low	220		T_{OSC}
T_{LHPL}	$\overline{PALE}$ High to $\overline{PROG}$ Low	220		T_{OSC}
T_{PLDV}	$\overline{PROG}$ Low to Word Dump Valid		50	T_{OSC}
T_{SHLL}	$\overline{RESET}$ High to First $\overline{PALE}$ Low	1100		T_{OSC}
T_{PHIL}	$\overline{PROG}$ High to $\overline{AINC}$ Low	0		T_{OSC}
T_{ILIH}	$\overline{AINC}$ Pulse Width	240		T_{OSC}
T_{ILVH}	PVER Hold after $\overline{AINC}$ Low	50		T_{OSC}
T_{ILPL}	$\overline{AINC}$ Low to $\overline{PROG}$ Low	170		T_{OSC}
T_{PHVL}	$\overline{PROG}$ High to $\overline{PVER}$ Valid		220	T_{OSC}

NOTES:
1. Run-time programming is done with Fosc = 6.0 MHz to 10.0 MHz, V_{CC}, V_{PD}, V_{REF} = 5V ±0.5V, T_C = 25°C ±5°C and V_{PP} = 12.5V ±0.25V. For run-time programming over a full operating range, contact factory.
2. Programming specifications are not tested, but guaranteed by design.
3. This specification is for the word dump mode. For programming pulses use Modified Quick Pulse Algorithm.

DC EPROM PROGRAMMING CHARACTERISTICS

Symbol	Parameter	Min	Max	Units
I_{PP}	V_{PP} Programming Supply Current		200	mA

NOTE:
Don not apply V_{PP} until V_{CC} is stable and within specifications and the oscillator/clock has stabilized or the device may be damaged.

EPROM PROGRAMMING WAVEFORMS

SLAVE PROGRAMMING MODE DATA PROGRAM MODE WITH SINGLE PROGRAM PULSE

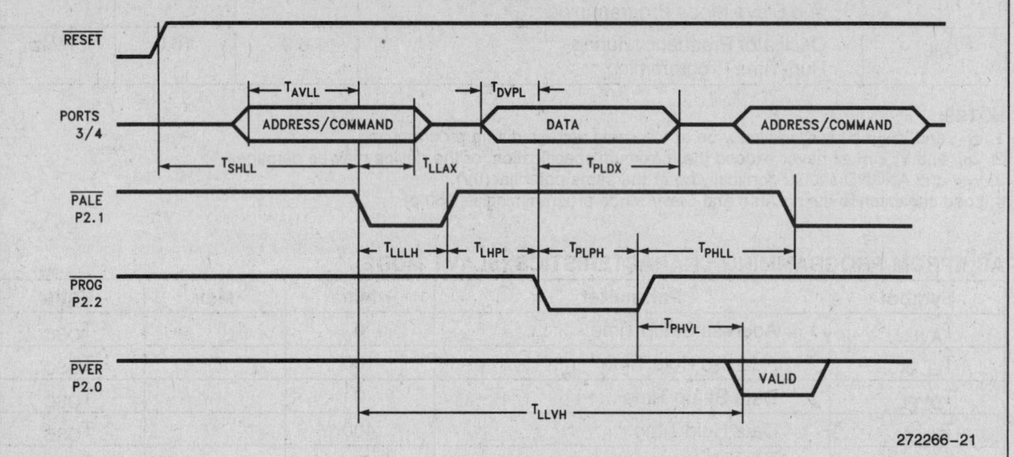

NOTE:
P3.0 must be HIGH ("1")

SLAVE PROGRAMMING MODE IN WORD DUMP MODE WITH AUTO INCREMENT

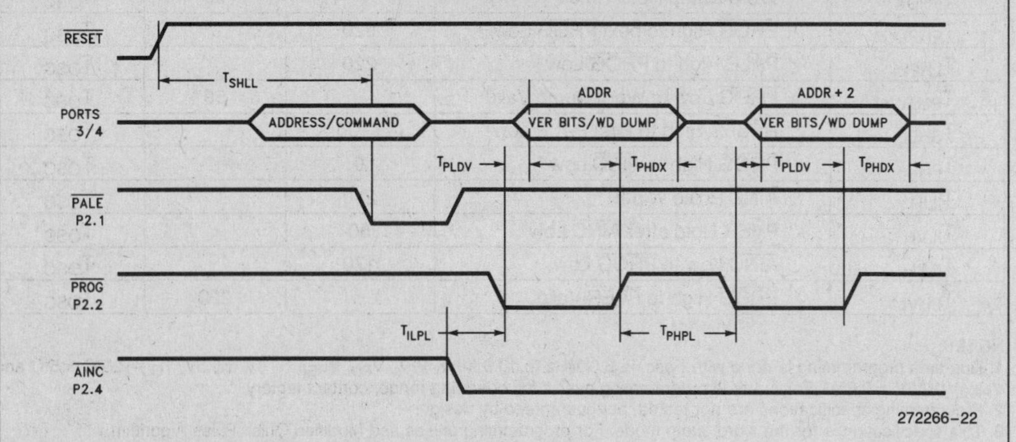

NOTE:
P3.0 must be LOW ("0")

SLAVE PROGRAMMING MODE TIMING IN DATA PROGRAM MODE WITH REPEATED PROG PULSE AND AUTO INCREMENT

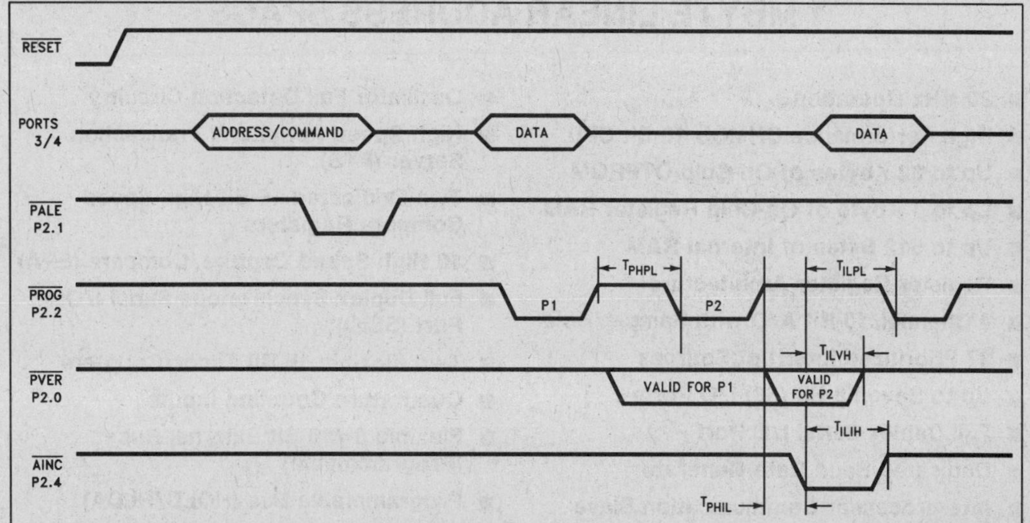

272266-23

8XC196KT ERRATA

The following is a list of all known functional deviations for 8XC196KT devices. C-step devices can be identified by a special mark following the eight digit FPO number on the top of the package. For C-step devices, this mark is a "C".

1. The following reserved op-codes do not generate the unimplemented op-code interrupt: 1Ch, 1Dh, 1Eh, 1Fh, E3h, E4h, E6h, E8h, E9h, EAh, EBh and F1h.

8XC196NT
CHMOS MICROCONTROLLER WITH 1 MBYTE LINEAR ADDRESS SPACE

- 20 MHz Operation
- High Performance CHMOS 16-Bit CPU
- Up to 32 Kbytes of On-Chip OTPROM
- Up to 1 Kbyte of On-Chip Register RAM
- Up to 512 Bytes of Internal RAM
- Register-Register Architecture
- 4 Channel/10-Bit A/D with Sample/Hold
- 37 Prioritized Interrupt Sources
- Up to Seven 8-Bit (56) I/O Ports
- Full Duplex Serial I/O Port
- Dedicated Baud Rate Generator
- Interprocessor Communication Slave Port
- Selectable Bus Timing Modes for Flexible External Memory Interfacing
- Oscillator Fail Detection Circuitry
- High Speed Peripheral Transaction Server (PTS)
- Two Dedicated 16-Bit High-Speed Compare Registers
- 10 High Speed Capture/Compare (EPA)
- Full Duplex Synchronous Serial I/O Port (SSIO)
- Two Flexible 16-Bit Timer/Counters
- Quadrature Counting Inputs
- Flexible 8-/16-Bit External Bus (Programmable)
- Programmable Bus (HOLD/HLDA)
- 1.4 μs 16 x 16 Multiply
- 2.4 μs 32/16 Divide
- 68-Pin Package

Device	Pins/Package	OTPROM	Reg RAM	Code RAM	Address Space	I/O	EPA	A/D
8XC196NT	68P PLCC	32K	1K	512	1 Mbyte	56	10	4

X = 7 OTPROM Device
X = 0 ROMLESS

The 8XC196NT 16-bit microcontroller is a high performance member of the MCS® 96 microcontroller family. The 8XC196NT is an enhanced 8XC196KR device with 1 Mbyte of linear address space, 1000 bytes of register RAM, 512 bytes of internal RAM, 20 MHz operation and an optional 32 Kbytes of OTPROM. Intel's CHMOS III-E process provides a high performance processor along with low power consumption.

Ten high-speed capture/compare modules are provided. As capture modules event times with 200 ns resolution can be recorded and generate interrupts. As compare modules events such as toggling of a port pin, starting an A/D conversion, pulse width modulation, and software timers can be generated. Events can be based on the timer or up/down counter.

September 1994
Order Number: 272267-004

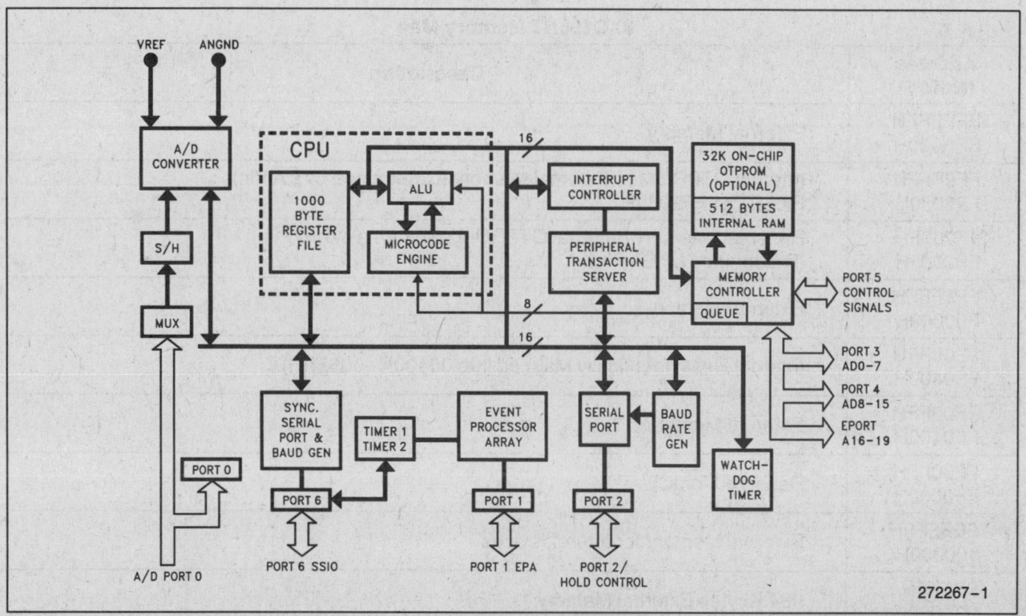

Figure 1. 8XC196NT Block Diagram

PROCESS INFORMATION

This device is manufactured on P629.5, a CHMOS III-E process. Additional process and reliability information is available in Intel's *Components Quality and Reliability Handbook*, Order Number 210997.

Table 1. Thermal Characteristics

Package Type	θ_{JA}	θ_{JC}
PLCC	36.5°C/W	13°C/W

All thermal impedance data is approximate for static air conditions at 1W of power dissipation. Values will change depending on operation conditions and application. See the Intel *Packaging Handbook* (order number 240800) for a description of Intel's thermal impedance test methodology.

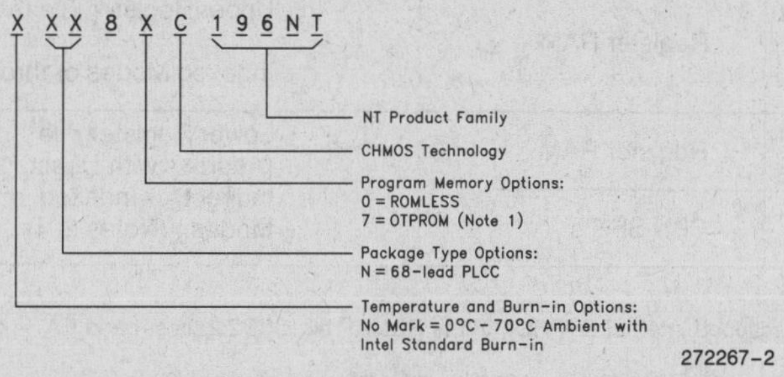

EXAMPLE: N87C196NT is 68-Lead PLCC OTPROM.
For complete package dimensional data, refer to the Intel Packaging Handbook (Order Number 240800).

Figure 2. The 8XC186NT Familiy Nomenclature

8XC196NT Memory Map

Address (Note 7)	Description	
FFFFFFH FFA000H	External Memory	
FF9FFFH FF2080H	Internal OTPROM or External Memory (Determined by $\overline{EA}$ Pin) RESET at FF2080H	
FF207FH FF2000H	Reserved Memory (Internal OTPROM or External Memory) (Determined by $\overline{EA}$ Pin)	
FF1FFFH FF0600H	External Memory	
FF05FFH FF0400H	Internal RAM (Identically Mapped into 00400H–005FFH)	
FF03FFH FF0100H	External Memory	
FF00FFH FF0000H	Reserved for ICE	
FEFFFFH 100000H	External Memory for future devices	
FFFFFH 00A000H	984 Kbytes External Memory	
009FFFH 002080H	Internal OTPROM or External Memory (Note 1)	
00207FH 002000H	Reserved Memory (Internal OTPROM or External Memory) (Notes 1, 3, and 6)	
001FFFH 001FE0H	Memory Mapped Special Function Registers (SFR's)	
001FDFH 001F00H	Internal Special Function Registers (SFR's) (Note 5)	
001EFFH 000600H	External Memory	
0005FFH 000400H	Internal RAM (Address with Indirect or Indexed Modes)	
0003FFH 000100H	Register RAM	Upper Register File (Address with Indirect or Indexed Modes or through Windows.) (Note 2)
0000FFH 000018H	Register RAM	Lower Register File (Address with Direct, Indirect, or Indexed Modes.) (Notes 2, 4)
000017H 000000H	CPU SFR's	

NOTES:
1. These areas are mapped internal OTPROM if the REMAP bit (CCB2.2) is set and $\overline{EA}$ = 5V. Otherwise they are external memory.
2. Code executed in locations 00000H to 003FFH will be forced external.
3. Reserved memory locations must contain 0FFH unless noted.
4. Reserved SFR bit locations must be written with 0.
5. Refer to 8XC196NT User's Guide and Quick Reference for SFR descriptions.
6. **WARNING:** The contents or functions of reserved memory locations may change with future revisions of the device. Therefore, a program that relies on one or more of these locations may not function properly.
7. The 8XC196NT internally uses 24 bit address, but only 20 address lines are bonded out allowing 1 Mbyte external address space.

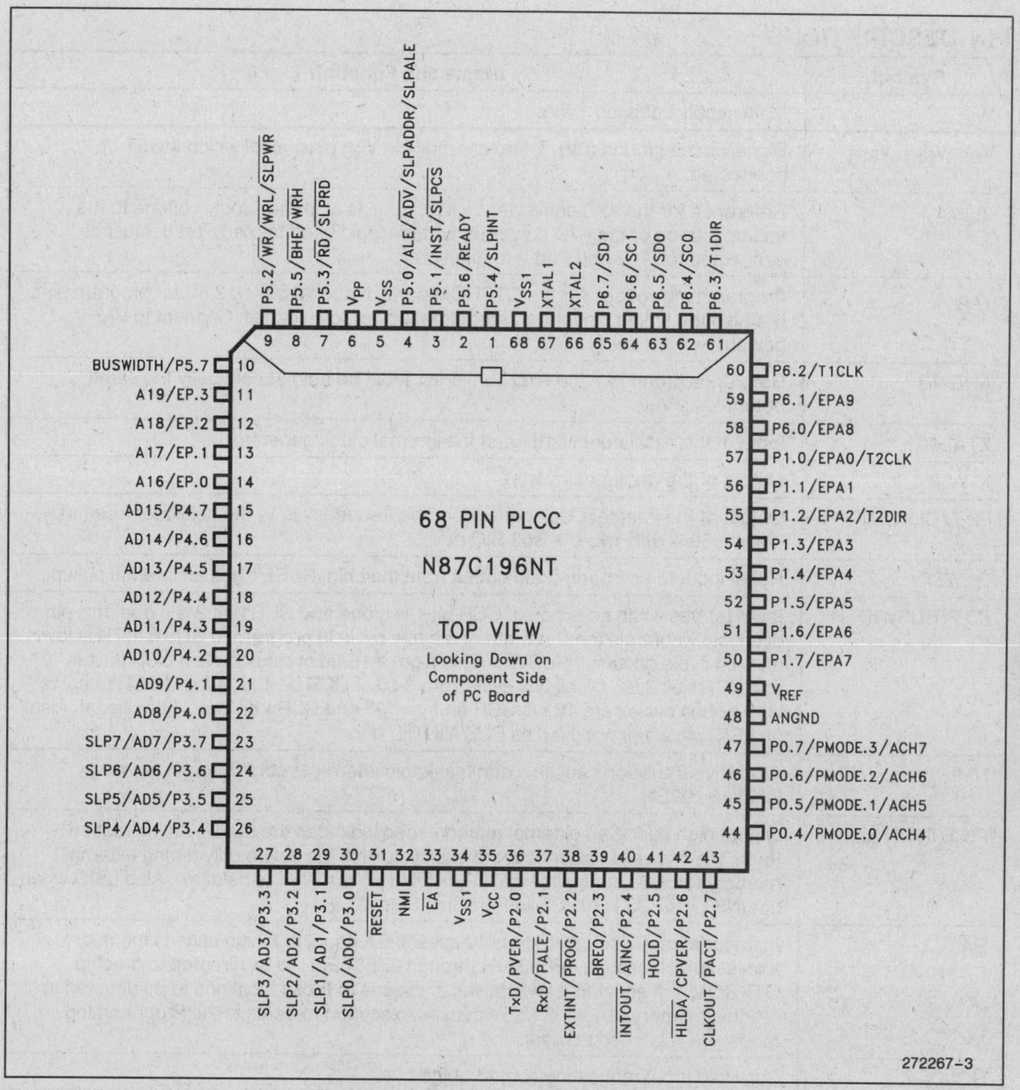

Figure 3. 68-Pin PLCC Package Diagram

4

PIN DESCRIPTIONS

Symbol	Name and Function
V_{CC}	Main supply voltage (+5V).
V_{SS}, V_{SS1}, V_{SS1}	Digital circuit ground (0V). There are multiple V_{SS} pins, all of which MUST be connected.
V_{REF}	Reference for the A/D converter (+5V). V_{REF} is also the supply voltage to the analog portion of the A/D converter and the logic used to read Port 0. Must be connected for A/D and Port 0 to function.
V_{PP}	Programming voltage for the OTPROM parts. It should be +12.5V for programming. It is also the timing pin for the return from powerdown circuit. Connect to V_{CC} if powerdown not being used.
ANGND	Reference ground for the A/D converter. Must be held at nominally the same potential as V_{SS}.
XTAL1	Input of the oscillator inverter and the internal clock generator.
XTAL2	Output of the oscillator inverter.
P2.7/CLKOUT	Output of the internal clock generator. The frequency is ½ the oscillator frequency. It has a 50% duty cycle. Also LSIO pin.
$\overline{\text{RESET}}$	Reset input to and open-drain output from the chip. $\overline{\text{RESET}}$ has an internal pullup.
P5.7/BUSWIDTH	Input for bus width selection. If CCR bit 1 is a one and CCR1 bit 2 is a one, this pin dyamically controls the Buswidth of the bus cycle in progress. If BUSWIDTH is low, an 8-bit cycle occurs, if BUSWIDTH is high, a 16-bit cycle occurs. If CCR bit 1 is "0" and CCR1 bit 2 is "1", all bus cycles are 8-bit, if CCR bit 1 is "1" and CCR1 bit 2 is "0", all bus cycles are 16-bit. CCR bit 1 = "0" and CCR1 bit 2 = "0" is illegal. Also an LSIO pin when not used as BUSWIDTH.
NMI	A positive transition causes a non maskable interrupt vector through memory location 203EH.
P5.1/INST/$\overline{\text{SLPCS}}$	Output high during an external memory read indicates the read is an instruction fetch. INST is valid throughout the bus cycle. INST is active only during external memory fetches, during internal OTPROM fetches INST is held low. Also LSIO when not INST. $\overline{\text{SLPCS}}$ is the Slave Port Chip Select.
$\overline{\text{EA}}$	Input for memory select (External Access). $\overline{\text{EA}}$ equal to a high causes memory accesses to locations 0FF2000H through 0FF9FFFH to be directed to on-chip OTPROM. $\overline{\text{EA}}$ equal to a low causes accesses to these locations to be directed to off-chip memory. $\overline{\text{EA}}$ = +12.5V causes execution to begin in the Programming Mode. $\overline{\text{EA}}$ is latched at reset.
$\overline{\text{HOLD}}$	Bus Hold Input requesting control of the bus.
$\overline{\text{HLDA}}$	Bus Hold acknowledge output indicating release of the bus.
$\overline{\text{BREQ}}$	Bus Request output activated when the bus controller has a pending external memory cycle.
P5.0/ALE/$\overline{\text{ADV}}$/SLPADDR/SLPALE	Address Latch Enable or Address Valid output, as selected by CCR. Both pin options provide a latch to demultiplex the address from the address/data bus. When the pin is $\overline{\text{ADV}}$, it goes inactive (high) at the end of the bus cycle. $\overline{\text{ADV}}$ can be used as a chip select for external memory. ALE/$\overline{\text{ADV}}$ is active only during external memory accesses. Also LSIO when not used as ALE. SLPADDR is the Slave Port Address Control Input and SLPALE is the Slave Port Address Latch Enable Input.
P5.3/$\overline{\text{RD}}$/$\overline{\text{SLPRD}}$	Read signal output to external memory. $\overline{\text{RD}}$ is active only during external memory reads or LSIO when not used as $\overline{\text{RD}}$. $\overline{\text{SLPRD}}$ is the Slave Port Read Control Input.

PIN DESCRIPTIONS (Continued)

Symbol	Name and Function
P5.2/$\overline{WR}$/$\overline{WRL}$/$\overline{SLPWR}$	Write and Write Low output to external memory, as selected by the CCR, $\overline{WR}$ will go low for every external write, while $\overline{WRL}$ will go low only for external writes where an even byte is being written. $\overline{WR}$/$\overline{WRL}$ is active during external memory writes. Also an LSIO pin when not used as $\overline{WR}$/$\overline{WRL}$. $\overline{SLPWR}$ is the Slave Port Write Control Input
P5.5/$\overline{BHE}$/$\overline{WRH}$	Byte High Enable or Write High output, as selected by the CCR. $\overline{BHE}$ = 0 selects the bank of memory that is connected to the high byte of the data bus. A0 = 0 selects that bank of memory that is connected to the low byte. Thus accesses to a 16-bit wide memory can be to the low byte only (A0 = 0, $\overline{BHE}$ = 1), to the high byte only (A0 = 1, $\overline{BHE}$ = 0) or both bytes (A0 = 0, $\overline{BHE}$ = 0). If the $\overline{WRH}$ function is selected, the pin will go low if the bus cycle is writing to an odd memory location. $\overline{BHE}$/$\overline{WRH}$ is only valid during 16-bit external memory read/write cycles. Also an LSIO pin when not BHE/$\overline{WRH}$.
P5.6/READY	Ready input to lengthen external memory cycles, for interfacing with slow or dynamic memory, or for bus sharing. If the pin is high, CPU operation continues in a normal manner. If the pin is low prior to the falling edge of CLKOUT, the memory controller goes into a wait state mode until the next positive transition in CLKOUT occurs with READY high. When external memory is not used, READY has no effect. The max number of wait states inserted into the bus cycle is controlled by the CCR/CCR1. Also an LSIO pin when READY is not selected.
P5.4/SLPINT	Dual function I/O pin. As a bidirectional port pin or as a system function. The system function is a Slave Port Interrupt Output Pin.
P6.2/T1CLK	Dual function I/O pin. Primary function is that of a bidirectional I/O pin, however, it may also be used as a TIMER1 Clock input. The TIMER1 will increment or decrement on both positive and negative edges of this pin.
P6.3/T1DIR	Dual function I/O pin. Primary function is that of a bidirectional I/O pin, however, it may also be used as a TIMER1 Direction input. The TIMER1 will increment when this pin is high and decrements when this pin is low.
PORT1/EPA0–7 P6.0–6.1/EPA8–9	Dual function I/O port pins. Primary function is that of bidirectional I/O. System function is that of High Speed capture and compare. EPA0 and EPA2 have yet another function of T2CLK and T2DIR of the TIMER2 timer/counter.
PORT 0/ACH4–7	4-bit high impedance input-only port. These pins can be used as digital inputs and/or as analog inputs to the on-chip A/D converter. These pins are also used as inputs to OTPROM parts to select the Programming Mode.
P6.3–6.7/SSIO	Dual function I/O ports that have a system function as Synchronous Serial I/O. Two pins are clocks and two pins are data, providing full duplex capability.
PORT 2	8-bit multi-functional port. All of its pins are shared with other functions.
PORT 3 and 4	8-bit bidirectional I/O ports with open drain outputs. These pins are shared with the multiplexed address/data bus which has strong internal pullups.
EPORT	8-bit bidirectional standard and I/O port. These bits are shared with the extended address bus, A16–A19. Pin function is selected on a per pin basis.
$\overline{INTOUT}$	Interrupt Output. This active-low output indicates that a pending interrupt requires use of the external bus.
SLP0–SLP7	Slave Port Address/Data Bus

ABSOLUTE MAXIMUM RATINGS*

Storage Temperature .... −60°C to +150°C

Voltage from V_{PP} or $\overline{EA}$ to V_{SS} or ANGND .... −0.5V to +13.0V

Voltage from Any Other Pin to V_{SS} or ANGND .... −0.5 to +7.0V

This includes V_{PP} on ROM and CPU devices.

Power Dissipation.......................... 0.5W

NOTICE: This data sheet contains information on products in the sampling and initial production phases of development. The specifications are subject to change without notice. Verify with your local Intel Sales office that you have the latest data sheet before finalizing a design.

**WARNING: Stressing the device beyond the "Absolute Maximum Ratings" may cause permanent damage. These are stress ratings only. Operation beyond the "Operating Conditions" is not recommended and extended exposure beyond the "Operating Conditions" may affect device reliability.*

OPERATING CONDITIONS

Symbol	Parameter	Min	Max	Units
T_A	Ambient Temperature Under Bias	0	+70	°C
V_{CC}	Digital Supply Voltage	4.50	5.50	V
V_{REF}	Analog Supply Voltage	4.50	5.50	V
F_{OSC}	Oscillator Frequency	4	20	MHz (Note 4)

NOTE:
ANGND and V_{SS} should be nominally at the same potential.

DC CHARACTERISTICS (Under Listed Operating Conditions)

Symbol	Parameter	Min	Typ	Max	Units	Test Conditions
I_{CC}	V_{CC} Supply Current			90	mA	XTAL1 = 20 MHz, $V_{CC} = V_{PP} = V_{REF} = 5.5V$ (While device in Reset)
I_{REF}	A/D Reference Supply Current			5	mA	
I_{IDLE}	Idle Mode Current			40	mA	XTAL1 = 20 MHz, $V_{CC} = V_{PP} = V_{REF} = 5.5V$
I_{PD}	Powerdown Mode Current(6)		50	75	μA	$V_{CC} = V_{PP} = V_{REF} = 5.5V$(11)
V_{IL}	Input Low Voltage (all pins)	−0.5V		0.3 V_{CC}	V	For PORT0(10)
V_{IH}	Input High Voltage	0.7 V_{CC}		V_{CC} + 0.5	V	For PORT0(10)
V_{IH1}	Input High Voltage XTAL1	0.7 V_{CC}		V_{CC} + 0.5	V	XTAL1 Input Pin Only(1)
V_{IH2}	Input High Voltage on RESET	0.7 V_{CC}		V_{CC} + 0.5	V	RESET input pin only
V_{OL}	Output Low Voltage (Outputs Configured as Complementary)			0.3 0.45 1.5	V V V	I_{OL} = 200 μA(3,5) I_{OL} = 3.2 mA I_{OL} = 7.0 mA
V_{OH}	Output High Voltage (Outputs Configured as Complementary)	V_{CC} − 0.3 V_{CC} − 0.7 V_{CC} − 1.5			V V V	I_{OH} = −200μA(3,5) I_{OH} = −3.2 mA I_{OH} = −7.0 mA
I_{LI}	Input Leakage Current (Std. Inputs)			±10	μA	$V_{SS} < V_{IN} < V_{CC}$
I_{LI1}	Input Leakage Current (Port 0)			±3	μA	$V_{CC} < V_{IN} < V_{REF}$
I_{IL}	Logical 0 Input Current			−70	μA	V_{IN} = 0.45V(1)

DC CHARACTERISTICS (Under Listed Operating Conditions) (Continued)

Symbol	Parameter	Min	Typ	Max	Units	Test Conditions
V_{OL1}	Output Low Voltage in RESET			0.8	V	(Note 7)
V_{OH1}	SLPINT (P5.4) and HLDA (P2.6) Output High Voltage in RESET	2.0			V	$I_{OH} = 0.8$ mA(7)
V_{OH2}	Output High Voltage in RESET	$V_{CC} - 1V$			V	$I_{OH} = -6\ \mu A$(1)
C_S	Pin Capacitance (Any pin to V_{SS})			10	pF	$f_{test} = 1.0$ MHz
R_{WPU}	Weak Pullup Resistance		150K		Ω	(Note 6)
R_{RST}	Reset Pullup	65K		180K	Ω	

NOTES:

1. All BD (bidirectional) pins except INST and CLKOUT. INST and CLKOUT are excluded due to their not being weakly pulled high in reset. BD pins include Port1, Port2, Port3, Port4, Port5, Port6 and EPORT except SPLINT (P5.4) and HLDA (P2.6).
2. Standard input pins include XTAL1, $\overline{EA}$, RESET, and Port 1/2/5/6 and EPORT when setup as inputs.
3. All bidirectional I/O pins when configured as Outputs (Push/Pull).
4. Device is static and should operate below 1 Hz, but only tested down to 4 MHz.
5. Maximum I_{OL}/I_{OH} currents per pin will be characterized and published at a later date.
6. Typicals are based on limited number of samples and are not guaranteed. The values listed are at room temperature and $V_{REF} = V_{CC} = 5.5V$.
7. Violating these specifications in reset may cause the device to enter test modes (P5.4 and P2.6).
8. TBD = To Be Determined.
9. Pullup present during return from powerdown condition.
10. When P0 is used as analog inputs, refer to A/D specifications.
11. For temperatures $<100°C$ typical is 10 μA.

4

8XC196NT ADDITIONAL BUS TIMING MODES

The 8XC196NT device has 3 additional bus timing modes for external memory interfacing.

MODE 3:

Mode 3 is the standard timing mode. Use this mode for systems that emulate the 8XC196KR bus timings.

MODE 0:

Mode 0 is the standard timing mode, but 1 (minimum) wait state is always inserted in external bus cycles.

MODE 1:

Mode 1 is the long R/W mode. This mode advances $\overline{RD}$ and $\overline{WR}$ signals by 1 T_{OSC} creating a 2 T_{OSC} $\overline{RD}/\overline{WR}$ low time. ALE is also advanced by 0.5 T_{OSC} but ALE high time remains 1 T_{OSC}.

MODE 2:

Mode 2 is the long R/W mode with Early Address. Mode 2 is similar to Mode 1 with respect to $\overline{RD}$, $\overline{WR}$, and ALE signals. Additionally, the address is output on the bus 0.5 T_{OSC} earlier in the bus cycle.

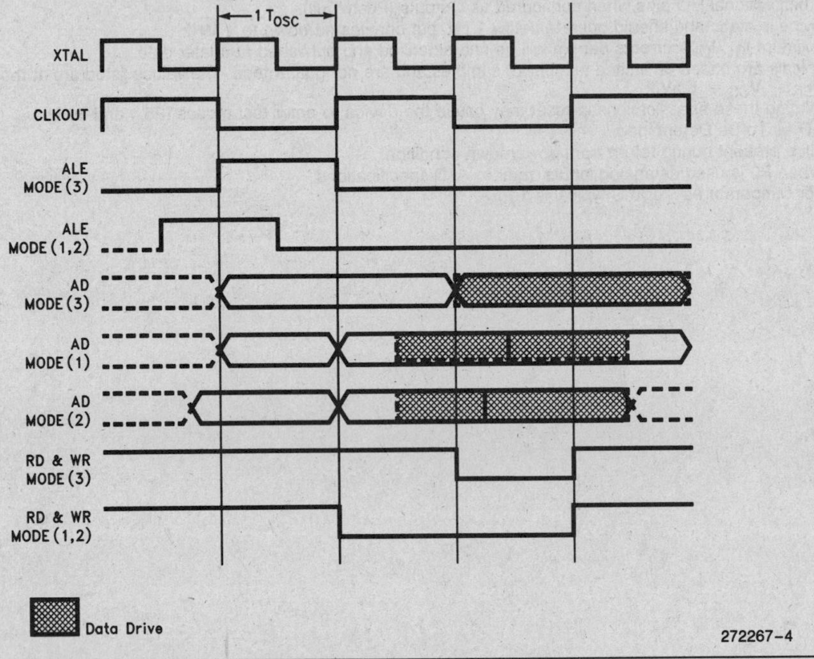

Figure 4. Detailed MODE 1, 2, 3, Comparison

EXPLANATION OF AC SYMBOLS

Each symbol is two pairs of letters prefixed by "T" for time. The characters in a pair indicate a signal and its condition, respectively. Symbols represent the time between the two signal/condition points.

Conditions:
H—High
L—Low
V—Valid
X—No Longer Valid
Z—Floating

Signals:
A—Address
B—$\overline{BHE}$
BR—$\overline{BREQ}$
C—CLKOUT
D—DATA
G—Buswidth
H—$\overline{HOLD}$
HA—$\overline{HLDA}$
L—ALE/$\overline{ADV}$
Q—Data Out
RD—$\overline{RD}$
W—$\overline{WR}$/$\overline{WRH}$/$\overline{WRL}$
X—XTAL1
Y—READY

BUS MODE 0 and 3—AC CHARACTERISTICS (Over Specified Operating Conditions)

Test Conditions: Capacitance Load on All Pins = 100 pF, Rise and Fall Times = 10 ns.

The system must meet these specifications to work with the 8XC196NT.

Symbol	Parameter	Min	Max	Units
T_{AVYV}	Address Valid to Ready Setup		$2T_{OSC} - 75$	ns(3)
T_{YLYH}	Non READY Time	No Upper Limit		ns
T_{CLYX}	READY Hold after CLKOUT Low	0	$T_{OSC} - 30$	ns(1)
T_{AVGV}	Address Valid to BUSWIDTH Setup		$2T_{OSC} - 75$	ns(2, 3)
T_{LLGV}	ALE Low to BUSWIDTH Setup		$T_{OSC} - 60$	ns(2, 3)
T_{CLGX}	BUSWIDTH Hold after CLKOUT Low	0		ns
T_{AVDV}	Address Valid to Input Data Valid		$3T_{OSC} - 55$	ns(2)
T_{RLDV}	$\overline{RD}$ active to input Data Valid		$T_{OSC} - 30$	ns(2)
T_{CLDV}	CLKOUT Low to Input Data Valid		$T_{OSC} - 60$	ns
T_{RHDZ}	End of $\overline{RD}$ to Input Data Float		T_{OSC}	ns
T_{RHDX}	Data Hold after $\overline{RD}$ High	0		ns

NOTES:
1. If Max is exceeded, additional wait states will occur.
2. If wait states are used, add $2T_{OSC} \times n$, where n = number of wait states.
3. If mode 0 is selected, one wait state minimum is always added. If additional wait states are required, add $2T_{OSC}$ to the specification.

BUS MODE 0 and 3—AC CHARACTERISTICS (Over Specified Operating Conditions)

Test Conditions: Capacitance Load on All Pins = 100 pF, Rise and Fall Times = 10 ns.

The 8XC196NT will meet these specifications

Symbol	Parameter	Min	Max	Units
F_{XTAL}	Frequency on XTAL1	4.0	20	MHz(1)
T_{OSC}	XTAL1 Period (1/F_{XTAL})	50	250	ns
T_{XHCH}	XTAL1 High to CLKOUT High or Low	+20	110	ns
T_{OFD}	Clock Failure to Reset Pulled Low(6)	4	40	μs
T_{CLCL}	CLKOUT Period	2 T_{OSC}		ns
T_{CHCL}	CLKOUT High Period	T_{OSC} − 10	T_{OSC} + 30	ns
T_{CLLH}	CLKOUT Low to ALE/$\overline{ADV}$ High	−10	+15	ns
T_{LLCH}	ALE/$\overline{ADV}$ Low to CLKOUT High	−25	+15	ns
T_{LHLH}	ALE/$\overline{ADV}$ Cycle Time	4 T_{OSC}		ns(5)
T_{LHLL}	ALE/$\overline{ADV}$ High Time	T_{OSC} − 10	T_{OSC} + 10	ns
T_{AVLL}	Address Valid to ALE Low	T_{OSC} − 15		ns
T_{LLAX}	Address Hold After ALE/$\overline{ADV}$ Low	T_{OSC} − 40		ns
T_{LLRL}	ALE/$\overline{ADV}$ Low to $\overline{RD}$ Low	T_{OSC} − 40		ns
T_{RLCL}	$\overline{RD}$ Low to CLKOUT Low	−5	+35	ns
T_{RLRH}	$\overline{RD}$ Low Period	T_{OSC} − 5		ns(5)
T_{RHLH}	$\overline{RD}$ High to ALE/$\overline{ADV}$ High	T_{OSC}	T_{OSC} + 25	ns(3)
T_{RLAZ}	$\overline{RD}$ Low to Address Float		+5	ns
T_{LLWL}	ALE/$\overline{ADV}$ Low to $\overline{WR}$ Low	T_{OSC} − 10		ns
T_{CLWL}	CLKOUT Low to $\overline{WR}$ Low	−10	+25	ns
T_{QVWH}	Data Valid before $\overline{WR}$ High	T_{OSC} − 23		ns
T_{CHWH}	CLKOUT High to $\overline{WR}$ High	−10	+15	ns
T_{WLWH}	$\overline{WR}$ Low Period	T_{OSC} − 30		ns(5)
T_{WHQX}	Data Hold after $\overline{WR}$ High	T_{OSC} − 35		ns
T_{WHLH}	$\overline{WR}$ High to ALE/$\overline{ADV}$ High	T_{OSC} − 10	T_{OSC} + 15	ns(3)
T_{WHBX}	$\overline{BHE}$, INST Hold after $\overline{WR}$ High	T_{OSC} − 10		ns
T_{WHAX}	AD8–15 Hold after $\overline{WR}$ High	T_{OSC} − 30		ns(4)
T_{RHBX}	$\overline{BHE}$, INST Hold after $\overline{RD}$ High	T_{OSC} − 10		ns
T_{RHAX}	AD8–15 Hold after $\overline{RD}$ High	T_{OSC} − 30		ns(4)

NOTES:
1. Testing performed at 8.0 MHz, however, the device is static by design and will typically operate below 1 Hz.
2. Typical specifications, not guaranteed.
3. Assuming back-to-back bus cycles.
4. 8-bit bus only.
5. If wait states are used, add 2 $T_{OSC} \times n$, where n = number of wait states. If mode 0 (1 automatic wait state added) operation is selected, add 2 T_{OSC} to specification.
6. T_{OFD} is the time for the oscillator fail detect circuit (OFD) to react to a clock failure. The OFD circuitry is enabled by programming the UPROM location 0778H with the value 0004H. NT/NQ customer QROM codes need to equate location 2016H to the value 0CDEH if the oscillator fail detect (OFD) function is desired. Intel manufacturing uses location 2016H as a flag to determine whether or not to program the Clock Detect Enable (CDE) bit. Programming the CDE bit enables oscillator fail detection.

BUS MODE 0 and 3—8XC196NT SYSTEM BUS TIMING

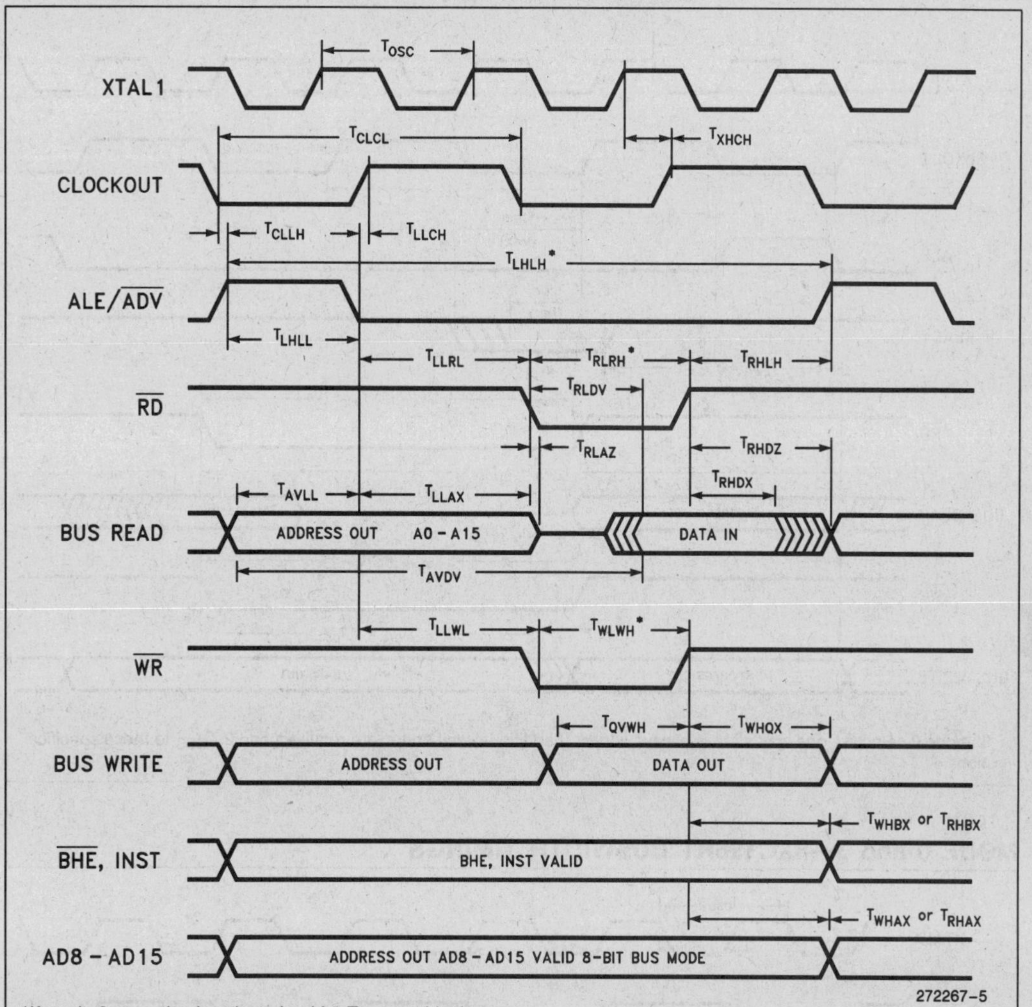

*If mode 0 operation is selected, add 2 T_{OSC} to this time.

4

8XC196NT MODE 0 and 3—READY TIMINGS (ONE WAIT STATE)

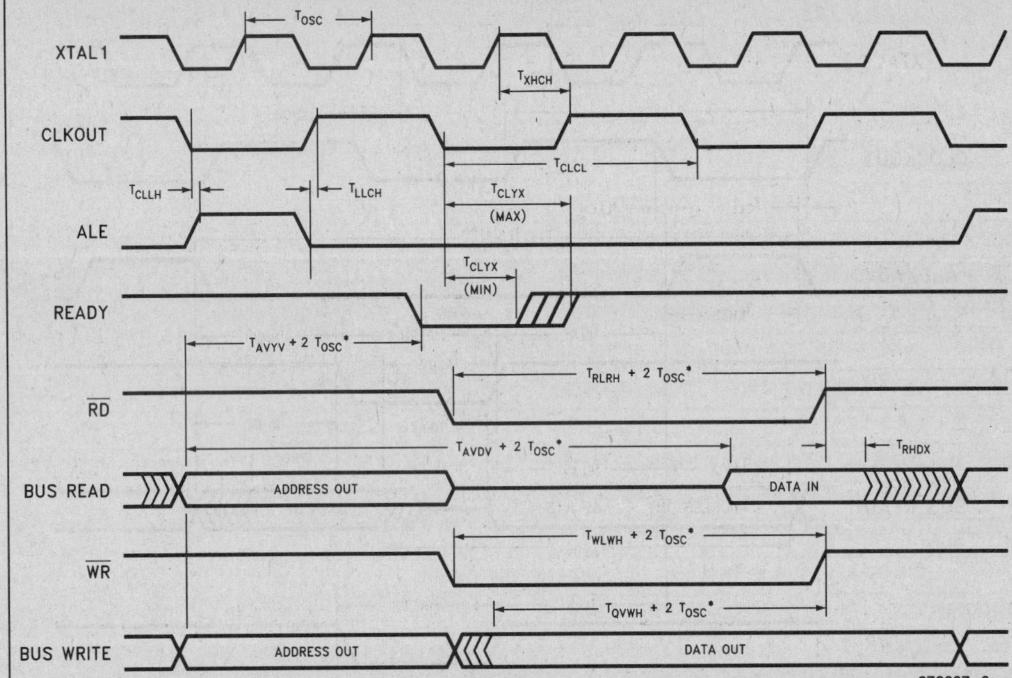

272267–6

*If mode 0 selected, one wait state is always added. If additional wait states are required, add 2 T_{OSC} to these specifications.

MODE 0 and 3—8XC196NT BUSWIDTH TIMINGS

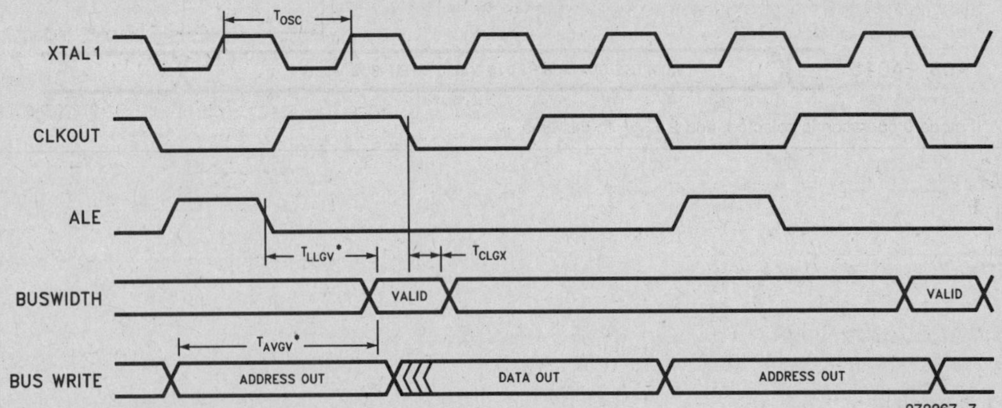

272267–7

*If mode 0 selected, add 2 T_{OSC} to these specifications.

BUS MODE 1—AC CHARACTERISTICS (Over Specified Operating Conditions)

Test Conditions: Capacitance Load on All Pins = 100 pF, Rise and Fall Times = 10 ns.

The system must meet these specifications to work with the 8XC196NT.

Symbol	Parameter	Min	Max	Units
T_{AVYV}	Address Valid to Ready Setup		$2\ T_{OSC} - 75$	ns
T_{YLYH}	Non READY Time		No Upper Limit	ns
T_{CLYX}	READY Hold after CLKOUT Low	0	$T_{OSC} - 30$	ns(1)
T_{AVGV}	Address Valid to BUSWIDTH Setup		$2\ T_{OSC} - 75$	ns
T_{LLGV}	ALE Low to BUSWIDTH Setup		$1.5\ T_{OSC} - 60$	ns
T_{CLGX}	BUSWIDTH Hold after CLKOUT Low	0		ns
T_{AVDV}	Address Valid to Input Data Valid		$3\ T_{OSC} - 60$	ns(2)
T_{RLDV}	$\overline{RD}$ active to input Data Valid		$2\ T_{OSC} - 44$	ns(2)
T_{CLDV}	CLKOUT Low to Input Data Valid		$T_{OSC} - 60$	ns
T_{RHDZ}	End of $\overline{RD}$ to Input Data Float		T_{OSC}	ns
T_{RHDX}	Data Hold after $\overline{RD}$ High	0		ns

NOTES:
1. If Max is exceeded, additional wait states will occur.
2. If wait states are used, add $2\ T_{OSC} \times n$, where n = number of wait states.

4

BUS MODE 1—AC CHARACTERISTICS (Over Specified Operating Conditions)

Test Conditions: Capacitance Load on All Pins = 100 pF, Rise and Fall Times = 10 ns.

The 8XC196NT will meet these specifications

Symbol	Parameter	Min	Max	Units
F_{XTAL}	Frequency on XTAL1	8.0	20	MHz(1)
T_{OSC}	XTAL1 Period ($1/F_{XTAL}$)	50	125	ns
T_{XHCH}	XTAL1 High to CLKOUT High or Low	+20	110	ns
T_{CLCL}	CLKOUT Period	$2\ T_{OSC}$		ns
T_{CHCL}	CLKOUT High Period	$T_{OSC} - 10$	$T_{OSC} + 27$	ns
T_{CHLH}	CLKOUT HIGH to ALE/$\overline{ADV}$ High	$0.5\ T_{OSC} - 15$	$0.5\ T_{OSC} + 15$	ns
T_{CLLL}	CLKOUT LOW to ALE/$\overline{ADV}$ Low	$0.5\ T_{OSC} - 25$	$0.5\ T_{OSC} + 15$	ns
T_{LHLH}	ALE/$\overline{ADV}$ Cycle Time	$4\ T_{OSC}$		ns(5)
T_{LHLL}	ALE/$\overline{ADV}$ High Time	$T_{OSC} - 20$	$T_{OSC} + 10$	ns
T_{AVLL}	Address Valid to ALE Low	$0.5\ T_{OSC} - 20$		ns
T_{LLAX}	Address Hold After ALE/$\overline{ADV}$ Low	$0.5\ T_{OSC} - 25$		ns
T_{LLRL}	ALE/$\overline{ADV}$ Low to $\overline{RD}$ Low	$0.5\ T_{OSC} - 15$		ns
T_{RLCL}	$\overline{RD}$ Low to CLKOUT Low	$T_{OSC} - 10$	$T_{OSC} + 30$	ns
T_{RLRH}	$\overline{RD}$ Low Period	$2\ T_{OSC} - 20$		ns(5)
T_{RHLH}	$\overline{RD}$ High to ALE/$\overline{ADV}$ High	$0.5\ T_{OSC}$	$0.5\ T_{OSC} + 25$	ns(3)
T_{RLAZ}	$\overline{RD}$ Low to Address Float		+5	ns
T_{LLWL}	ALE/$\overline{ADV}$ Low to $\overline{WR}$ Low	$0.5\ T_{OSC} - 10$		ns
T_{CLWL}	CLKOUT Low to $\overline{WR}$ Low	$T_{OSC} - 15$	$T_{OSC} + 25$	ns
T_{QVWH}	Data Valid before $\overline{WR}$ High	$2\ T_{OSC} - 23$		ns
T_{CHWH}	CLKOUT High to $\overline{WR}$ High	−10	+15	ns
T_{WLWH}	$\overline{WR}$ Low Period	$2\ T_{OSC} - 15$		ns(5)
T_{WHQX}	Data Hold after $\overline{WR}$ High	$0.5\ T_{OSC} - 12$		ns
T_{WHLH}	$\overline{WR}$ High to ALE/$\overline{ADV}$ High	$0.5\ T_{OSC} - 10$	$0.5\ T_{OSC} + 15$	ns(3)
T_{WHBX}	$\overline{BHE}$ Hold after $\overline{WR}$ High	$T_{OSC} - 15$		ns
T_{WHIX}	INST Hold after $\overline{WR}$ High	$0.5\ T_{OSC} - 15$		
T_{WHAX}	AD8–15 Hold after $\overline{WR}$ High	$0.5\ T_{OSC} - 30$		ns(4)
T_{RHBX}	$\overline{BHE}$ Hold after $\overline{RD}$ High	$T_{OSC} - 32$		ns
T_{RHIX}	INST Hold after $\overline{RD}$ High	$0.5\ T_{OSC} - 32$		
T_{RHAX}	AD8–15 Hold after $\overline{RD}$ High	$0.5\ T_{OSC} - 30$		ns(4)

NOTES:
1. Testing performed at 8.0 MHz, however, the device is static by design and will typically operate below 1 Hz.
2. Typical specifications, not guaranteed.
3. Assuming back-to-back bus cycles.
4. 8-bit bus only.
5. If wait states are used, add $2\ T_{OSC} \times n$, where n = number of wait states.

MODE 1—8XC196NT SYSTEM BUS TIMING

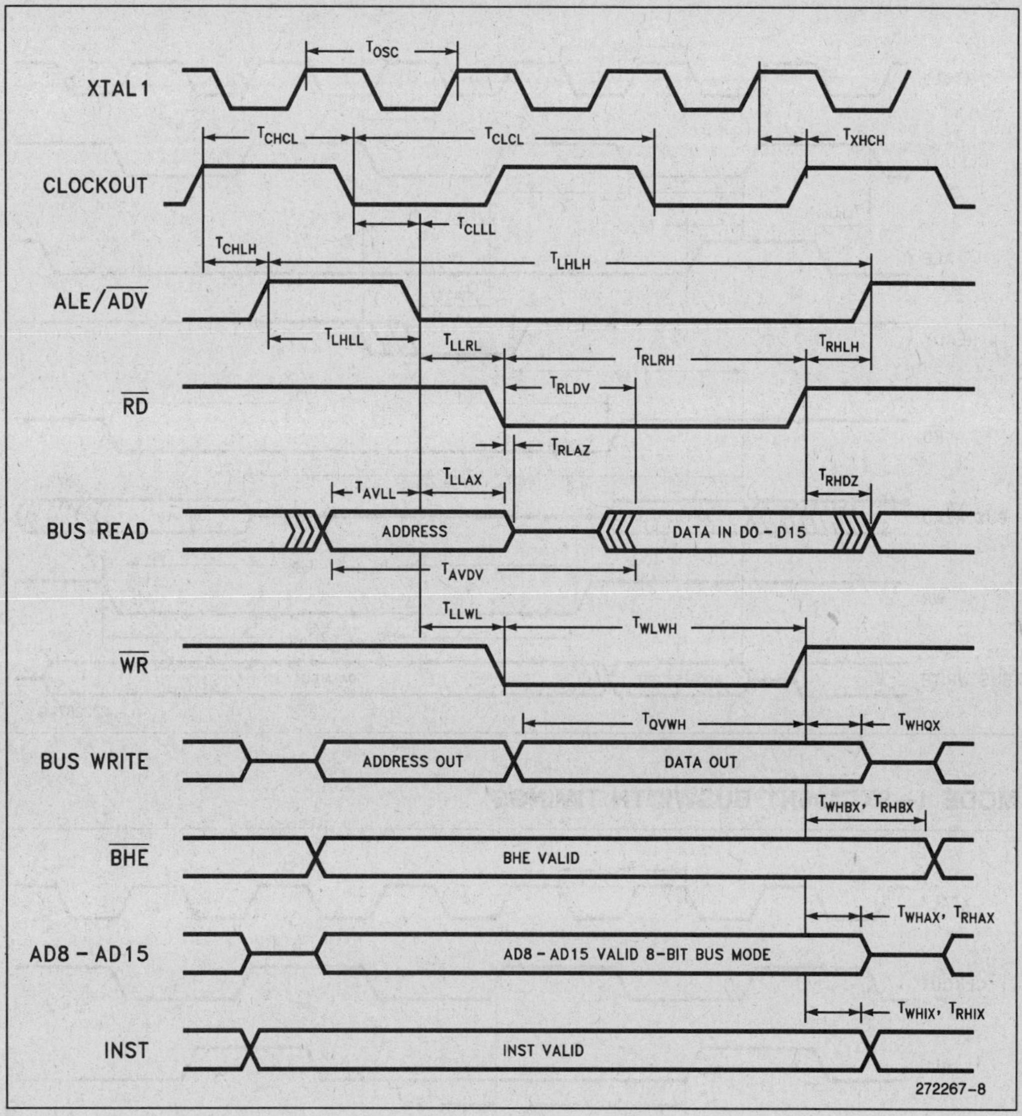

MODE 1—8XC196NT READY TIMINGS (ONE WAIT STATE)

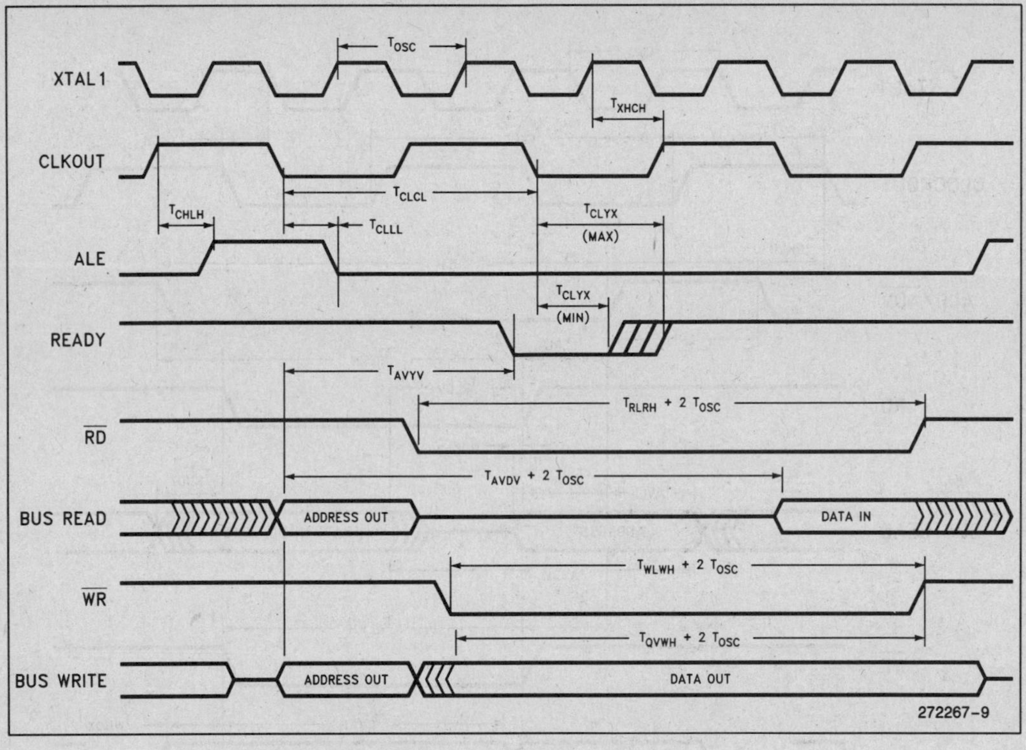

MODE 1—8XC196NT BUSWIDTH TIMINGS

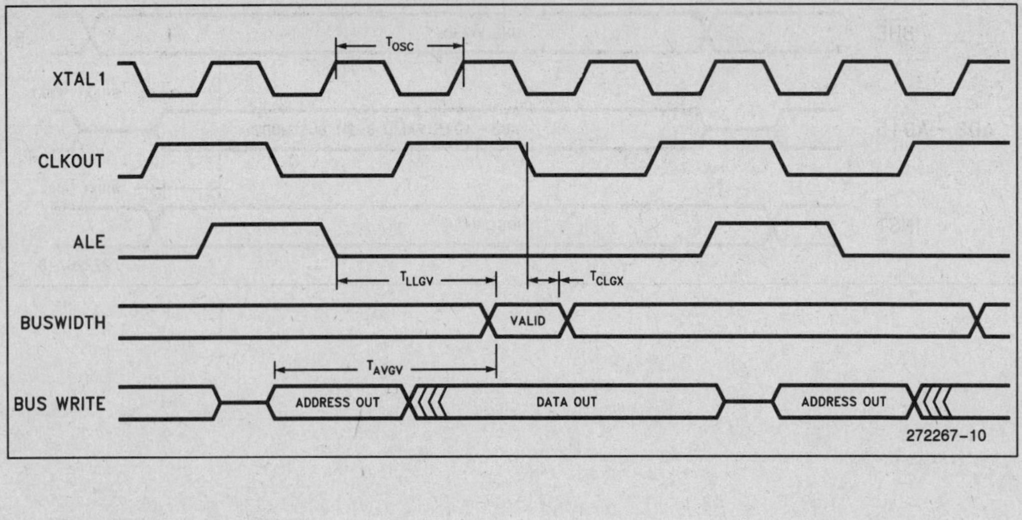

BUS MODE 2—AC CHARACTERISTICS (Over Specified Operating Conditions)

Test Conditions: Capacitance Load on All Pins = 100 pF, Rise and Fall Times = 10 ns.

The system must meet these specifications to work with the 8XC196NT.

Symbol	Parameter	Min	Max	Units
T_{AVYV}	Address Valid to Ready Setup		$2.5\ T_{OSC} - 75$	ns
T_{YLYH}	Non READY Time	No Upper Limit		ns
T_{CLYX}	READY Hold after CLKOUT Low	0	$T_{OSC} - 30$	ns(1)
T_{AVGV}	Address Valid to BUSWIDTH Setup		$2.5\ T_{OSC} - 75$	ns
T_{LLGV}	ALE Low to BUSWIDTH Setup		$1.5\ T_{OSC} - 60$	ns
T_{CLGX}	BUSWIDTH Hold after CLKOUT Low	0		ns
T_{AVDV}	Address Valid to Input Data Valid		$3.5\ T_{OSC} - 55$	ns(2)
T_{RLDV}	$\overline{RD}$ active to input Data Valid		$2\ T_{OSC} - 44$	ns(2)
T_{CLDV}	CLKOUT Low to Input Data Valid		$T_{OSC} - 60$	ns
T_{RHDZ}	End of $\overline{RD}$ to Input Data Float		$0.5\ T_{OSC}$	ns
T_{RHDX}	Data Hold after $\overline{RD}$ High	0		ns

NOTES:
1. If Max is exceeded, additional wait states will occur.
2. If wait states are used, add $2\ T_{OSC} \times n$, where n = number of wait states.

BUS MODE 2—AC CHARACTERISTICS (Over Specified Operating Conditions)

Test Conditions: Capacitance Load on All Pins = 100 pF, Rise and Fall Times = 10 ns.

The 8XC196NT will meet these specifications

Symbol	Parameter	Min	Max	Units
F_{XTAL}	Frequency on XTAL1	8.0	20	MHz(1)
T_{OSC}	XTAL1 Period ($1/F_{XTAL}$)	50	125	ns
T_{XHCH}	XTAL1 High to CLKOUT High or Low	+20	+85	ns
T_{CLCL}	CLKOUT Period	$2\ T_{OSC}$		ns
T_{CHCL}	CLKOUT High Period	$T_{OSC} - 10$	$T_{OSC} + 27$	ns
T_{CHLH}	CLKOUT HIGH to ALE/$\overline{ADV}$ High	$0.5\ T_{OSC} - 15$	$0.5\ T_{OSC} + 15$	ns
T_{CLLL}	CLKOUT LOW to ALE/$\overline{ADV}$ Low	$0.5\ T_{OSC} - 25$	$0.5\ T_{OSC} + 15$	ns
T_{LHLH}	ALE/$\overline{ADV}$ Cycle Time	$4\ T_{OSC}$		ns(5)
T_{LHLL}	ALE/$\overline{ADV}$ High Time	$T_{OSC} - 20$	$T_{OSC} + 10$	ns
T_{AVLL}	Address Valid to ALE Low	$T_{OSC} - 15$		ns
T_{LLAX}	Address Hold After ALE/$\overline{ADV}$ Low	$0.5\ T_{OSC} - 20$		ns
T_{LLRL}	ALE/$\overline{ADV}$ Low to $\overline{RD}$ Low	$0.5\ T_{OSC} - 15$		ns
T_{RLCL}	$\overline{RD}$ Low to CLKOUT Low	$T_{OSC} - 10$	$T_{OSC} + 30$	ns
T_{RLRH}	$\overline{RD}$ Low Period	$2\ T_{OSC} - 20$		ns(5)
T_{RHLH}	$\overline{RD}$ High to ALE/$\overline{ADV}$ High	$0.5\ T_{OSC} - 5$	$0.5\ T_{OSC} + 25$	ns(3)
T_{RLAZ}	$\overline{RD}$ Low to Address Float		+5	ns
T_{LLWL}	ALE/$\overline{ADV}$ Low to $\overline{WR}$ Low	$0.5\ T_{OSC} - 10$		ns
T_{CLWL}	CLKOUT Low to $\overline{WR}$ Low	$T_{OSC} - 22$	$T_{OSC} + 25$	ns
T_{QVWH}	Data Valid before $\overline{WR}$ High	$2\ T_{OSC} - 25$		ns
T_{CHWH}	CLKOUT High to $\overline{WR}$ High	−10	+15	ns
T_{WLWH}	$\overline{WR}$ Low Period	$2\ T_{OSC} - 20$		ns(5)
T_{WHQX}	Data Hold after $\overline{WR}$ High	$0.5\ T_{OSC} - 12$		ns
T_{WHLH}	$\overline{WR}$ High to ALE/$\overline{ADV}$ High	$0.5\ T_{OSC} - 10$	$0.5\ T_{OSC} + 10$	ns(3)
T_{WHBX}	$\overline{BHE}$ Hold after $\overline{WR}$ High	$T_{OSC} - 15$		ns
T_{WHIX}	INST Hold after $\overline{WR}$ High	$0.5\ T_{OSC} - 15$		
T_{WHAX}	AD8–15 Hold after $\overline{WR}$ High	$0.5\ T_{OSC} - 30$		ns(4)
T_{RHBX}	$\overline{BHE}$ Hold after $\overline{RD}$ High	$T_{OSC} - 32$		ns
T_{RHIX}	INST Hold after $\overline{RD}$ High	$0.5\ T_{OSC} - 32$		
T_{RHAX}	AD8–15 Hold after $\overline{RD}$ High	$0.5\ T_{OSC} - 30$		ns(4)

NOTES:
1. Testing performed at 8.0 MHz, however, the device is static by design and will typically operate below 1 Hz.
2. Typical specifications, not guaranteed.
3. Assuming back-to-back bus cycles.
4. 8-bit bus only.
5. If wait states are used, add $2\ T_{OSC} \times n$, where n = number of wait states.

MODE 2—8XC196NT SYSTEM BUS TIMING

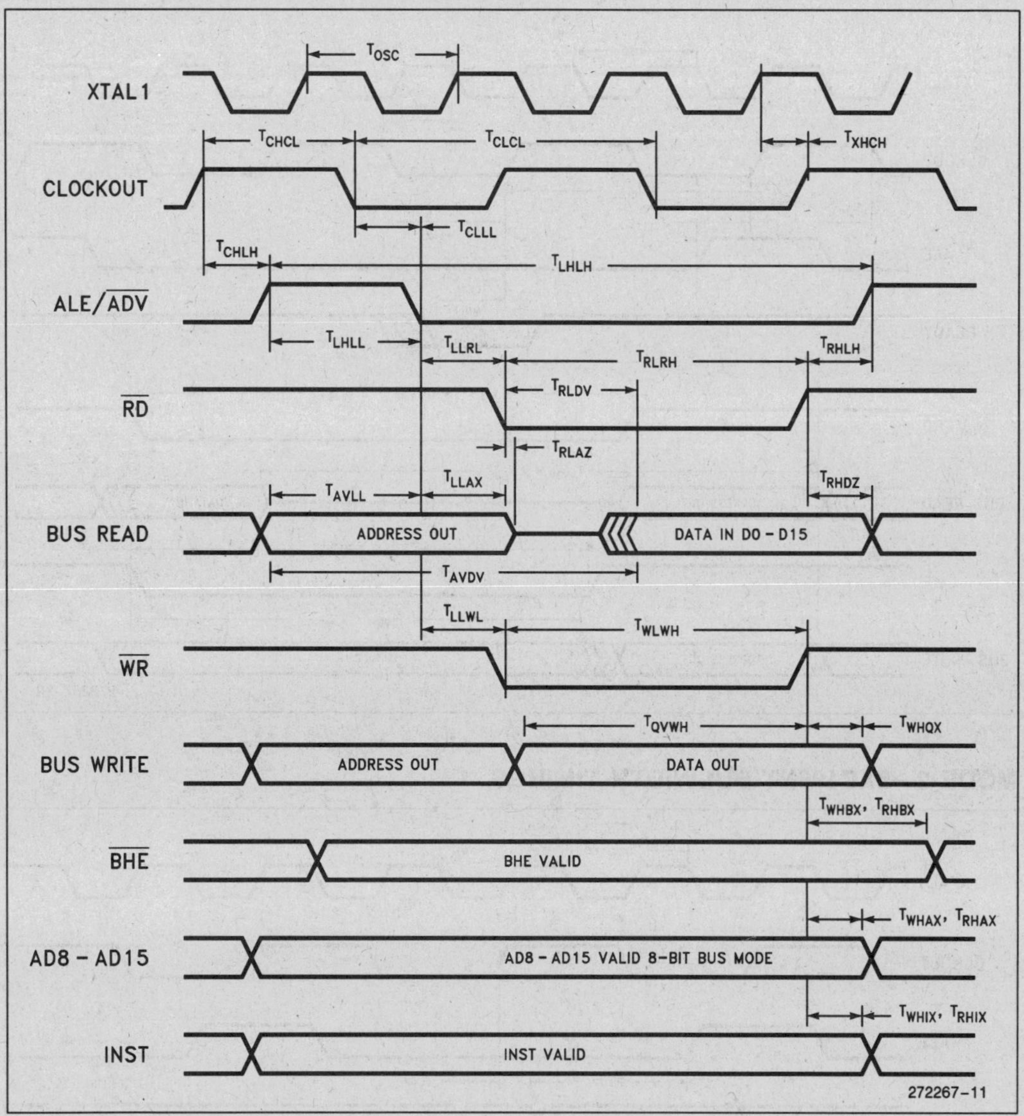

MODE 2—8XC196NT READY TIMINGS (ONE WAIT STATE)

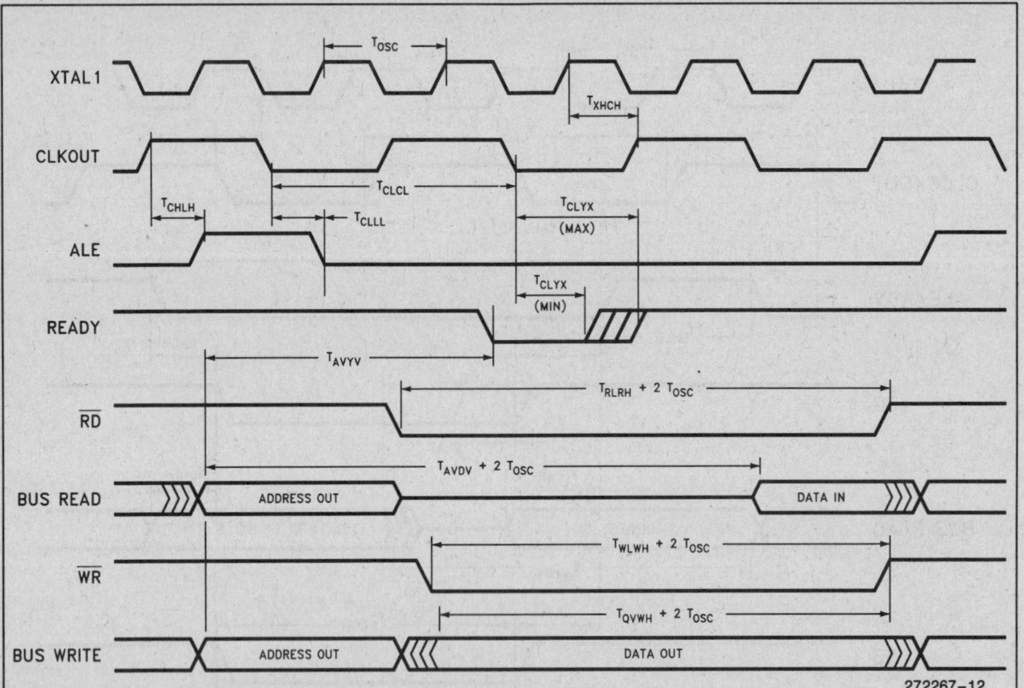

272267–12

MODE 2—8XC196NT BUSWIDTH TIMINGS

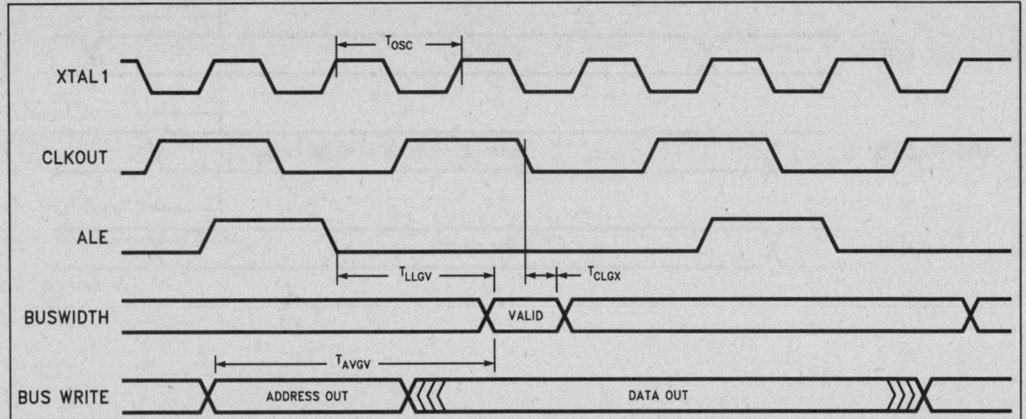

272267–13

BUS MODE 0, 1, 2, and 3—$\overline{HOLD}$/$\overline{HLDA}$ TIMINGS (Over Specified Operation Conditions)

Test Conditions: Capacitance Load on All Pins = 100 pF, Rise and Fall Times = 10 ns.

Symbol	Parameter	Min	Max	Units
T_{HVCH}	$\overline{HOLD}$ Setup Time	+65		ns(1)
T_{CLHAL}	CLKOUT Low to $\overline{HLDA}$ Low	−15	+15	ns
T_{CLBRL}	CLKOUT Low to $\overline{BREQ}$ Low	−15	+15	ns
T_{HALAZ}	$\overline{HLDA}$ Low to Address Float		+25	ns
T_{HALBZ}	$\overline{HLDA}$ Low to $\overline{BHE}$, INST, $\overline{RD}$, $\overline{WR}$ Weakly Driven		+25	ns
T_{CLHAH}	CLKOUT Low to $\overline{HLDA}$ High	−25	+15	ns
T_{CLBRH}	CLKOUT Low to $\overline{BREQ}$ High	−25	+25	ns
T_{HAHAX}	$\overline{HLDA}$ High to Address No Longer Float	−15		ns
T_{HAHBV}	$\overline{HLDA}$ High to $\overline{BHE}$, INST, $\overline{RD}$, $\overline{WR}$ Valid	−10		ns

NOTE:
1. To guarantee recognition at next clock.

8XC196NT $\overline{HOLD}$/$\overline{HLDA}$ TIMINGS

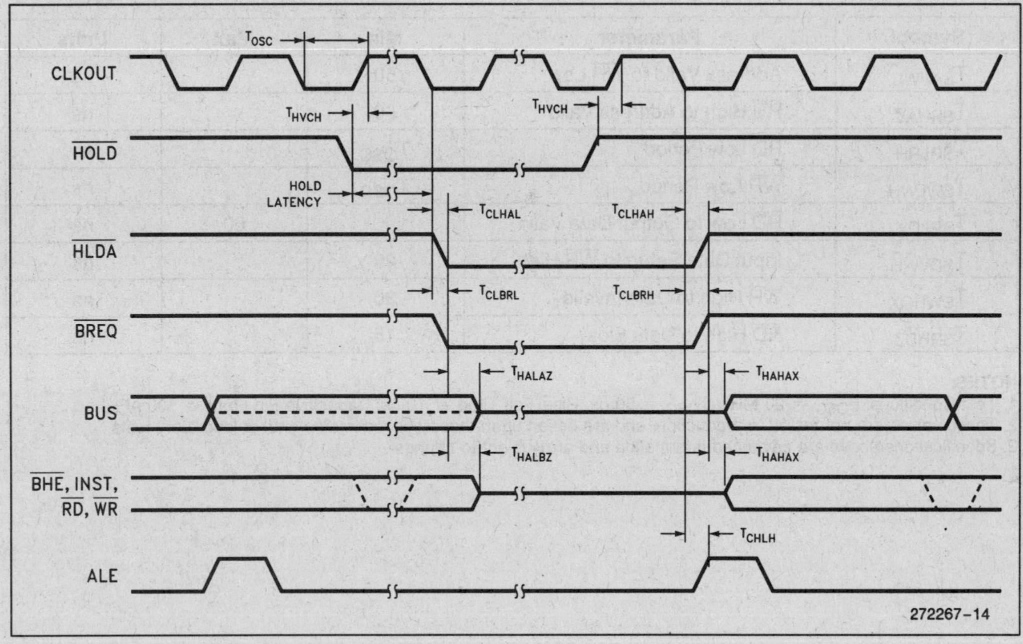

AC CHARACTERISTICS—SLAVE PORT

SLAVE PORT WAVEFORM—(SLPL = 0)

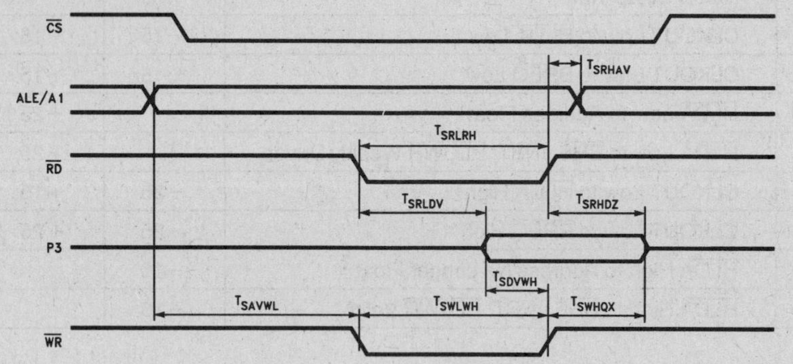

272267-15

SLAVE PORT TIMING—(SLPL = 0)

Symbol	Parameter	Min	Max	Units
T_{SAVWL}	Address Valid to $\overline{WR}$ Low	50		ns
T_{SRHAV}	$\overline{RD}$ High to Address Valid	60		ns
T_{SRLRH}	$\overline{RD}$ Low Period	T_{OSC}		ns
T_{SWLWH}	$\overline{WR}$ Low Period	T_{OSC}		ns
T_{SRLDV}	$\overline{RD}$ Low to Output Data Valid		60	ns
T_{SDVWH}	Input Data Setup to $\overline{WR}$ High	20		ns
T_{SWHQX}	$\overline{WR}$ High to Data Invalid	30		ns
T_{SRHDZ}	$\overline{RD}$ High to Data Float	15		ns

NOTES:
1. Test Conditions: F_{OSC} = 20 MHz, T_{OSC} = 50 ns. Rise/Fall Time = 10 ns. Capacitive Pin Load = 100 pF.
2. These values are not tested in production, and are based upon theoretical estimates and/or laboratory tests.
3. Specifications above are advanced information and are subject to change.

AC CHARACTERISTICS—SLAVE PORT (Continued)

SLAVE PORT WAVEFORM—(SLPL = 1)

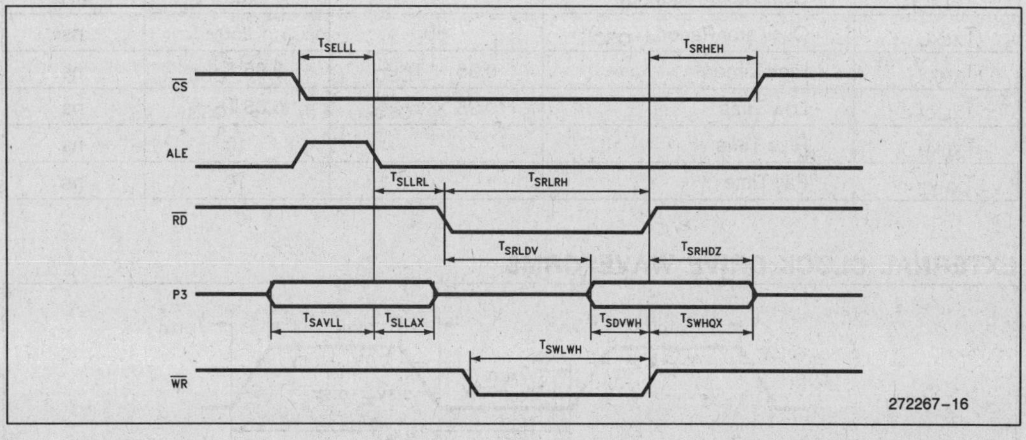

SLAVE PORT TIMING—(SLPL = 1)

Symbol	Parameter	Min	Max	Units
T_{SELLL}	$\overline{CS}$ Low to ALE Low	20		ns
T_{SRHEH}	$\overline{RD}$ or $\overline{WR}$ High to $\overline{CS}$ High	60		ns
T_{SLLRL}	ALE Low to $\overline{RD}$ Low	T_{OSC}		ns
T_{SRLRH}	$\overline{RD}$ Low Period	T_{OSC}		ns
T_{SWLWH}	$\overline{WR}$ Low Period	T_{OSC}		ns
T_{SAVLL}	Address Valid to ALE Low	20		ns
T_{SLLAX}	ALE Low to Address Invalid	20		ns
T_{SRLDV}	$\overline{RD}$ Low to Output Data Valid		60	ns
T_{SDVWH}	Input Data Setup to $\overline{WR}$ High	20		ns
T_{SWHQX}	$\overline{WR}$ High to Data Invalid	30		ns
T_{SRHDZ}	$\overline{RD}$ High to Data Float	15		ns

NOTES:
1. Test Conditions: F_{OSC} = 20 MHz, T_{OSC} = 50 ns. Rise/Fall Time = 10 ns. Capacitive Pin Load = 100 pF.
2. These values are not tested in production, and are based upon theoretical estimates and/or laboratory tests.
3. Specifications above are advanced information and are subject to change.

EXTERNAL CLOCK DRIVE

Symbol	Parameter	Min	Max	Units
$1/T_{XLXL}$	Oscillator Frequency	4	20	MHz
T_{XLXL}	Oscillator Period (T_{OSC})	50	250	ns
T_{XHXX}	High Time	$0.35 \times T_{OSC}$	$0.65\ T_{OSC}$	ns
T_{XLXX}	Low Time	$0.35 \times T_{OSC}$	$0.65\ T_{OSC}$	ns
T_{XLXH}	Rise Time		10	ns
T_{XHXL}	Fall Time		10	ns

EXTERNAL CLOCK DRIVE WAVEFORMS

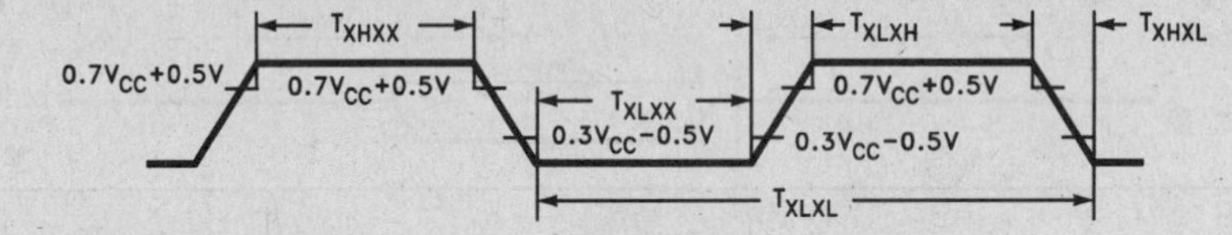

272267-17

AC TESTING INPUT, OUTPUT WAVEFORMS

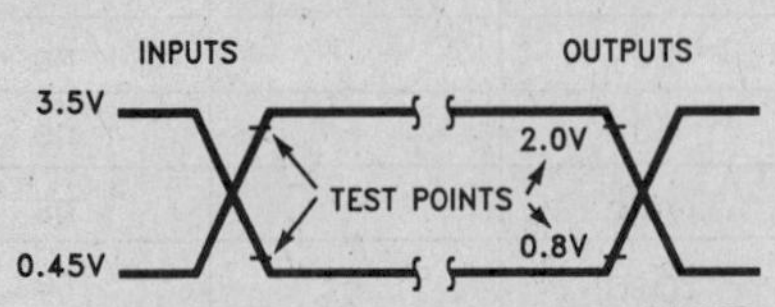

272267-18

AC Testing inputs are driven at 3.5V for a logic "1" and 0.45V for a logic "0". Timing measurements are made at 2.0V for a logic "1" and 0.8V for logic "0".

FLOAT WAVEFORMS

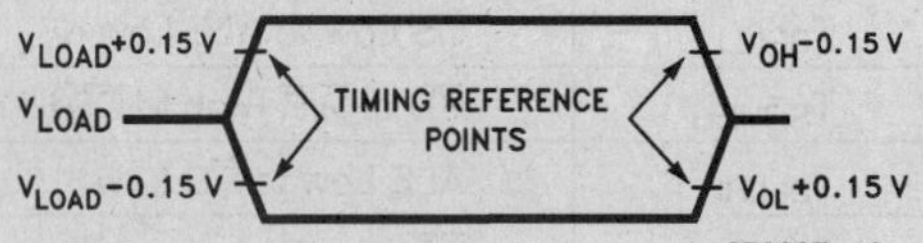

272267-19

For timing purposes a Port Pin is no longer floating when a 150 mV change from load voltage occurs and begins to float when a 150 mV change from the loading V_{OH}/V_{OL} level occurs $I_{OL}/I_{OH} \leq 15$ mA.

WAVEFORM—SERIAL PORT—SHIFT REGISTER MODE

SERIAL PORT WAVEFORM—SHIFT REGISTER MODE (MODE 0)

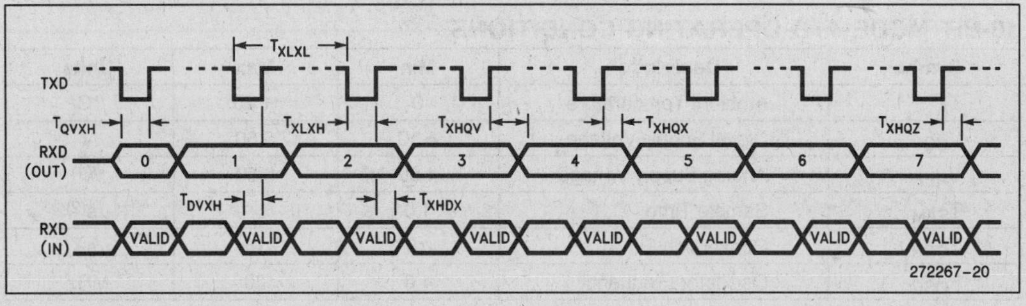

AC CHARACTERISTICS—SERIAL PORT-SHIFT REGISTER MODE

SERIAL PORT TIMING—SHIFT REGISTER MODE (MODE 0)

Test Conditions: $T_A = -40°C$ to $+125°C$; $V_{CC} = 5.0V \pm 10\%$; $V_{SS} = 0.0V$; Load Capacitance = pF

Symbol	Parameter	Min	Max	Units
T_{XLXL}(2)	Serial Port Clock Period (BRR ≥ 8002H) Receive Only	6 T_{OSC}		ns
T_{XLXH}(2)	Serial Port Clock Falling Edge to Rising Edge (BRR ≥ 8002H)	4 T_{OSC} − 50	4 T_{OSC} + 50	ns
T_{XLXL}(2)	Serial Port Clock Period (BRR = 8001H) Transmit Only	4 T_{OSC}		ns
T_{XLXH}(2)	Serial Port Clock Falling Edge to Rising Edge (BRR = 8001H)	2 T_{OSC} − 50	2 T_{OSC} + 50	ns
T_{QVXH}	Output Data Setup to Clock Rising Edge	3 T_{OSC}		ns
T_{XHQX}	Output Data Hold after Clock Rising Edge	2 T_{OSC} − 50		ns
T_{XHQV}	Next Output Data Valid after Clock Rising Edge		2 T_{OSC} + 50	ns
T_{DVXH}	Input Data Setup to Clock Rising Edge	2 T_{OSC} + 200		ns
T_{XHDX}(1)	Input Data Hold after Clock Rising Edge	0		ns
T_{XHQZ}(1)	Last Clock Rising to Output Float		5 T_{OSC}	ns

NOTES:
1. Parameters not tested.
2. The minimum baud rate register value for Receive is 8002H. The minimum baud rate register value for Transmit is 8001H.

A to D CHARACTERISTICS

The A/D converter is ratiometric, so absolute accuracy is dependent on the accuracy and stability of V_{REF}.

10-BIT MODE A/D OPERATING CONDITIONS

Symbol	Description	Min	Max	Units
T_A	Ambient Temperature	0	+70	°C
V_{CC}	Digital Supply Voltage	4.50	5.50	V
V_{REF}	Analog Supply Voltage	4.50	5.50	V(1)
T_{SAM}	Sample Time	1.0		μs(2)
T_{CONV}	Conversion Time	10	15	μs(2)
F_{OSC}	Oscillator Frequency	4.0	20	MHz

NOTES:
1. V_{REF} must be within 0.5V of V_{CC}.
2. The value of AD_TIME is selected to meet these specifications.

10-BIT MODE A/D CHARACTERISTICS (Using Above Operating Conditions)(6)

Parameter	Typ*(1)	Min	Max	Units*
Resolution		1024 10	1024 10	Level Bits
Absolute Error		0	±3.0	LSBs
Full Scale Error	0.25 ±0.5			LSBs
Zero Offset Error	0.25 ±0.5			LSBs
Non-Linearity	1.0 ±2.0		±3.0	LSBs
Differential Non-Linearity		−0.75	+0.75	LSBs
Channel-to-Channel Matching	±0.1	0	±1.0	LSBs
Repeatability	±0.25	0		LSBs(1)
Temperature Coefficients: Offset Full Scale Differential Non-Linearity	 0.009 0.009 0.009			 LSB/C(1) LSB/C(1) LSB/C(1)
Off Isolation		−60		dB(1,2,3)
Feedthrough	−60			dB(1,2)
V_{CC} Power Supply Rejection	−60			dB(1,2)
Input Resistance		750	1.2K	Ω(4)
DC Input Leakage	±1.0	0	±3.0	μA
Voltage on Analog Input Pin		ANGND − 0.5	V_{REF} + 0.5	V(5)
Sampling Capacitor	3.0			pF

*An "LSB" as used here has a value of approximately 5 mV.

NOTES:
1. These values are expected for most parts at 25°C, but are not tested or guaranteed.
2. DC to 100 KHz.
3. Multiplexer break-before-make is guaranteed.
4. Resistance from device pin, through internal MUX, to sample capacitor.
5. Applying voltages beyond these specifications will degrade the accuracy of other channels being converted.
6. All conversions performed with processor in IDLE mode.

8-BIT MODE A/D OPERATING CONDITIONS

Symbol	Description	Min	Max	Units
T_A	Ambient Temperature	0	+70	°C
V_{CC}	Digital Supply Voltage	4.50	5.50	V
V_{REF}	Analog Supply Voltage	4.50	5.50	V(1)
T_{SAM}	Sample Time	1.0		μs(2)
T_{CONV}	Conversion Time	7	20	μs(2)
F_{OSC}	Oscillator Frequency	4.0	20	MHz

NOTES:
1. V_{REF} must be within 0.5V of V_{CC}.
2. The value of AD_TIME is selected to meet these specifications.

8-BIT MODE A/D CHARACTERISTICS (Using Above Operating Conditions)(6)

Parameter	Typ*(1)	Min	Max	Units*
Resolution		256 8	256 8	Level Bits
Absolute Error		0	±1.0	LSBs
Full Scale Error	±0.5			LSBs
Zero Offset Error	±0.5			LSBs
Non-Linearity		0	±1.0	LSBs
Differential Non-Linearity		−0.5	+0.5	LSBs
Channel-to-Channel Matching		0	±1.0	LSBs
Repeatability	±0.25	0		LSBs(1)
Temperature Coefficients: Offset Full Scale Differential Non-Linearity	 0.003 0.003 0.003			 LSB/C(1) LSB/C(1) LSB/C(1)
Off Isolation		−60		dB(1,2,3)
Feedthrough	−60			dB(1,2)
V_{CC} Power Supply Rejection	−60			dB(1,2)
Input Resistance		750	1.2K	Ω(4)
DC Input Leakage	±1.0	0	±3.0	μA
Voltage on Analog Input Pin		ANGND − 0.5	V_{REF} + 0.5	V(5)
Sampling Capacitor	3.0			pF

*An "LSB" as used here has a value of approximately 5 mV.

NOTES:
1. These values are expected for most parts at 25°C, but are not tested or guaranteed.
2. DC to 100 KHz.
3. Multiplexer break-before-make is guaranteed.
4. Resistance from device pin, through internal MUX, to sample capacitor.
5. Applying voltage beyond these specifications will degrade the accuracy of other channels being converted.
6. All conversions performed with processor in IDLE mode.

OTPROM SPECIFICATIONS

OPERATING CONDITIONS

Symbol	Description	Min	Max	Units
T_A	Ambient Temperature During Programming	20	30	°C
V_{CC}	Supply Voltage During Programming	4.5	5.5	V(1)
V_{REF}	Reference Supply Voltage During Programming	4.5	5.5	V(1)
V_{PP}	Programming Voltage	12.25	12.75	V(2)
V_{EA}	EA Pin Voltage	12.25	12.75	V(2)
F_{OSC}	Oscillator Frequency during Auto and Slave Mode Programming	6.0	8.0	MHz
F_{OSC}	Oscillator Frequency during Run-Time Programming	6.0	20.0	MHz

NOTES:
1. V_{CC} and V_{REF} should nominally be at the same voltage during programming.
2. V_{PP} and V_{EA} must never exceed the maximum specification, or the device may be damaged.
3. V_{SS} and ANGND should nominally be at the same potential (0V).
4. Load capacitance during Auto and Slave Mode programming = 150 pF.

AC OTPROM PROGRAMMING CHARACTERISTICS (SLAVE MODE)

Symbol	Parameter	Min	Max	Units
T_{AVLL}	Address Setup Time	0		T_{OSC}
T_{LLAX}	Address Hold Time	100		T_{OSC}
T_{DVPL}	Data Setup Time	0		T_{OSC}
T_{PLDX}	Data Hold Time	400		T_{OSC}
T_{LLLH}	$\overline{PALE}$ Pulse Width	50		T_{OSC}
T_{PLPH}	$\overline{PROG}$ Pulse Width(2)	50		T_{OSC}
T_{LHPL}	$\overline{PALE}$ High to $\overline{PROG}$ Low	220		T_{OSC}
T_{PHLL}	$\overline{PROG}$ High to next $\overline{PALE}$ Low	220		T_{OSC}
T_{PHDX}	Word Dump Hold Time		50	T_{OSC}
T_{PHPL}	$\overline{PROG}$ High to next $\overline{PROG}$ Low	220		T_{OSC}
T_{LHPL}	$\overline{PALE}$ High to $\overline{PROG}$ Low	220		T_{OSC}
T_{PLDV}	$\overline{PROG}$ Low to Word Dump Valid		50	T_{OSC}
T_{SHLL}	$\overline{RESET}$ High to First $\overline{PALE}$ Low	1100		T_{OSC}
T_{PHIL}	$\overline{PROG}$ High to $\overline{AINC}$ Low	0		T_{OSC}
T_{ILIH}	$\overline{AINC}$ Pulse Width	240		T_{OSC}
T_{ILVH}	PVER Hold after $\overline{AINC}$ Low	50		T_{OSC}
T_{ILPL}	$\overline{AINC}$ Low to $\overline{PROG}$ Low	170		T_{OSC}
T_{PHVL}	$\overline{PROG}$ High to $\overline{PVER}$ Valid		220	T_{OSC}

NOTES:
1. Run-time programming is done with F_{OSC} = 6.0 MHz to 10.0 MHz, V_{CC}, V_{PD}, V_{REF} = 5V ±0.5V, T_C = 25°C ±5°C and V_{PP} = 12.5V ±0.25V. For run-time programming over a full operating range, contact factory.
2. This specification is for the word dump mode. For programming pulses use Modified Quick Pulse Algorithm.

DC OTPROM PROGRAMMING CHARACTERISTICS

Symbol	Parameter	Min	Max	Units
I_{PP}	V_{PP} Programming Supply Current		200	mA

NOTE:
Do not apply V_{PP} unti V_{CC} is stable and within specifications and the oscillator/clock has stabilized or the device may be damaged.

OTPROM PROGRAMMING WAVEFORMS

SLAVE PROGRAMMING MODE DATA PROGRAM MODE WITH SINGLE PROGRAM PULSE

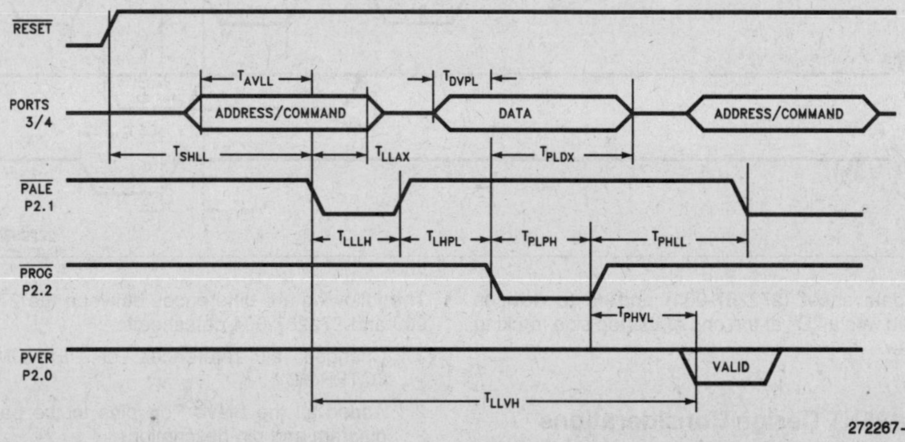

NOTE:
P3.0 must be high ("1")

SLAVE PROGRAMMING MODE IN WORD DUMP MODE WITH AUTO INCREMENT

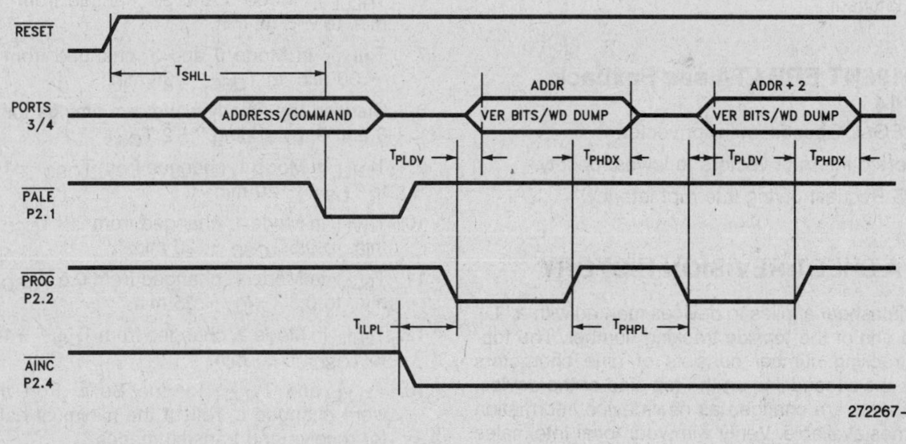

NOTE:
P3.0 must be low ("0")

SLAVE PROGRAMMING MODE TIMING IN DATA PROGRAM MODE WITH REPEATED PROG PULSE AND AUTO INCREMENT

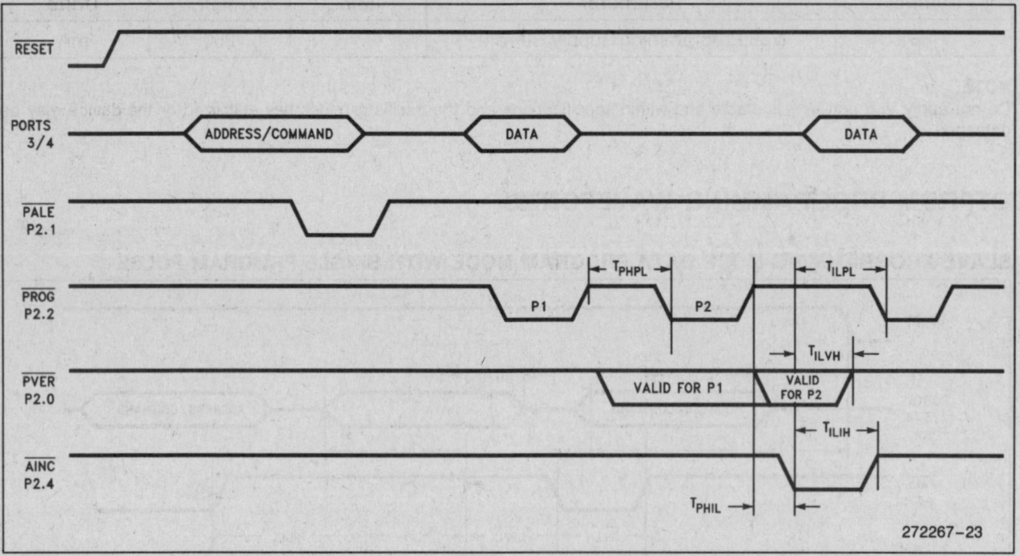

This data sheet (272267-004) applies to devices marked with a "D" at the end of the top side tracking number.

8XC196NT Design Considerations

1. When operating in bus timing modes 1 or 2, the upper and lower address/data lines must be latched. Even in 8-bit bus mode, the upper address lines must be latched. In modes 0 and 3, the upper address lines DO NOT NEED to be latched in 8-bit bus width mode. But in 16-bit buswidth mode the upper address lines need to be latched.

8XC196NT ERRATA see Faxback #2344

1. ILLEGAL Opcode interrupt vector.
2. Aborted Interrupt vectors to lowest priority.
3. PTS Request during Interrupt latency.

DATA SHEET REVISION HISTORY

This datasheet applies to devices marked with a "D" at the end of the topside tracking number. The topside tracking number consists of nine characters and is the second line on the top side of the device. Datasheets are changed as new device information becomes available. Verify with your local Intel sales office that you have the latest version before finalizing a design or ordering devices.

The following are differences between the 272267-003 and 272267-004 datasheets:

1. Changed all references of "EPROM" to "OTPROM".
2. Added all the Slave Port pins to the package diagram and pin descriptions.
3. Added $\overline{\text{INTOUT}}$ pin to pin descriptions.
4. Changed ILI1 (input leakage current for Port 0) from $\pm 1\ \mu A$ to $\pm 3\ \mu A$.
5. Removed T_{LLYV} from AC characterisics and waveform diagrams.
6. T_{RLCL} in Mode 0 and 3, changed from +4 ns min. to −5 ns min.
7. T_{WHQX} in Mode 0 and 3, changed from T_{OSC} − 30 min. to T_{OSC} − 35 min.
8. Clarified the Ready waveform timings for Mode 0 and 3, by adding "+2 T_{OSC}*".
9. T_{LHLL} in Mode 1, changed from T_{OSC} − 10 min. to T_{OSC} − 20 min.
10. T_{AVLL} in Mode 1, changed from 0.5 T_{OSC} − 15 min. to 0.5 T_{OSC} − 20 min.
11. T_{LLAX} in Mode 1, changed from 0.5 T_{OSC} − 20 min. to 0.5 T_{OSC} − 25 min.
12. T_{LHLL} in Mode 2, changed from T_{OSC} − 10 min. to T_{OSC} − 20 min.
13. T_{XLXL} and T_{XLXH} for the Serial Port timings were changed to reflect the minimum baudrate for receive and transmit modes.
14. Added the 8XC196NT ERRATA section.

8XC196NP COMMERCIAL CHMOS 16-BIT MICROCONTROLLER

- 25 MHz Operation at 4.5–5.5 Volts
  16 MHz Operation at 2.97–4.5 Volts
- 1 Mbyte of Linear Address Space
- Optional 4 Kbytes of ROM
- 1000 Bytes of Register RAM
- Register-register Architecture
- 33 I/O Port Pins
- 31 Prioritized Interrupt Sources
- 4 External Interrupt Pins and NMI Pin
- 2 Flexible 16-bit Timer/Counters with Quadrature Counting Capability
- 3 Pulse-width Modulator (PWM) Outputs with High Drive Capability
- Full-duplex Serial Port with Dedicated Baud-rate Generator
- Peripheral Transaction Server
- Chip-select Unit
  - 6 Chip-select Pins
  - Dynamic Demultiplexed/Multiplexed Address/Data Bus for Each Chip-select
  - Programmable Wait States (0, 1, 2, or 3) for Each Chip-select
  - Programmable Bus Width (8- or 16-bit) for Each Chip-select
  - Programmable Address Range for Each Chip-select
- Event Processor Array (EPA) with 4 High-speed Capture/Compare Channels
- 1.12 μs 16 x 16 Unsigned Multiplication
- 1.92 μs 32/16 Unsigned Division
- 100-pin SQFP or 100-pin QFP Package
- Complete System Development Support
- High-speed CHMOS Technology

4

The 8XC196NP is a member of Intel's 16-bit MCS® 96 microcontroller family. The device features 1 Mbyte of linear address space, a demultiplexed bus, and a chip-select unit. The external bus can dynamically switch between multiplexed and demultiplexed operation. When operating at 25 MHz in demultiplexed mode, the 8XC196NP can access a 100 ns memory device with zero wait states. The 8XC196NP is available without ROM (80C196NP) or with 4 Kbytes of ROM (83C196NP).

October 1994
Order Number: 272459-003

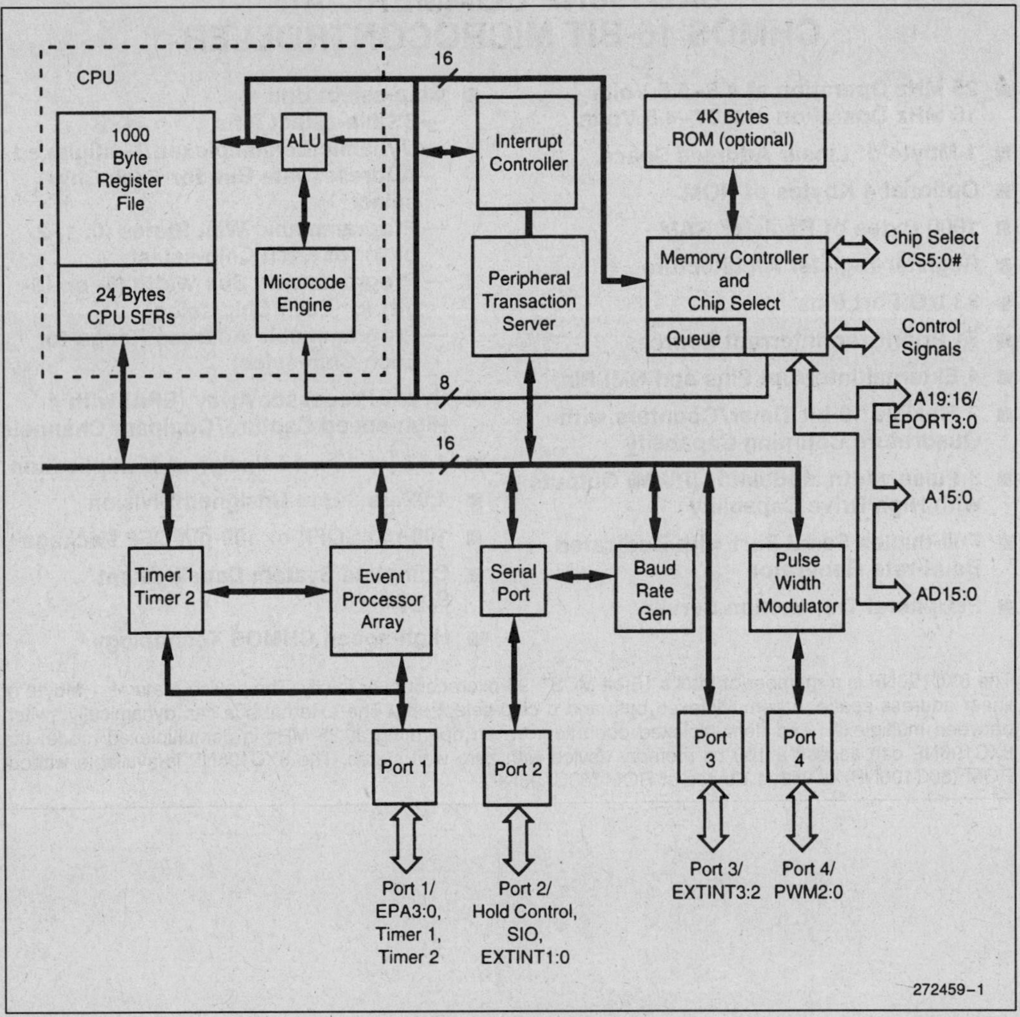

Figure 1. 8XC196NP Block Diagram

PROCESS INFORMATION

This device is manufactured on P648.2, a CHMOS IV process. Additional process and reliability information is available in Intel's *Components Quality and Reliability Handbook* (order number 210997).

All thermal impedance data is approximate for static air conditions at 1 watt of power dissipation. Values will change depending on operating conditions and the application. The Intel *Packaging Handbook* (order number 240800) describes Intel's thermal impedance test methodology.

Table 1. Thermal Characteristics

Package Type	θ_{JA}	θ_{JC}
100-Pin SQFP	55°C/W	14°C/W
100-Pin QFP	56°C/W	16°C/W

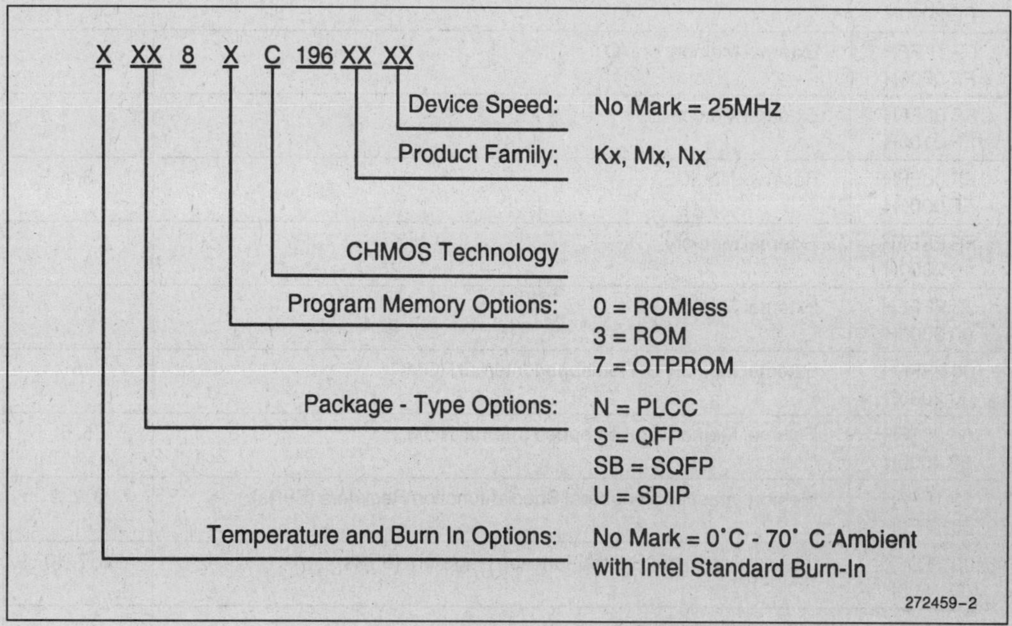

Figure 2. The 8XC196NP Family Nomenclature

Table 2. 8XC196NP Memory Map

Address (Note 1)	Description	Notes
FF FFFFH FF 3000H	External Memory or I/O	
FF 2FFFH FF 2080H	Internal ROM or External Memory (determined by EA# pin)	2
FF 207FH FF 2000H	Internal ROM or External Memory (determined by EA# pin)	3
FF 1FFFH FF 0600H	External Memory or I/O	
FF 05FFH FF 0100H	External Memory	
FF 00FFH FF 0000H	Reserved for ICE	3, 4
FE FFFFH 10 0000H	External Memory	
0F FFFFH 00 3000H	External Memory	
00 2FFFH 00 2080H	External Memory or Remapped Internal ROM	5
00 207FH 00 2000H	External Memory or Remapped Internal ROM	5, 6
00 1FFFH 00 1FE0H	Memory-mapped Peripheral Special-function Registers (SFRs)	3, 7, 9
00 1FDFH 00 1F00H	Internal Peripheral Special-function Registers (SFRs)	3, 7, 10
00 1EFFH 00 0400H	External Memory	
00 03FFH 00 0100H	Upper Register File (general-purpose register RAM)	8, 10
00 00FFH 00 0018H	Lower Register File (general-purpose register RAM and Stack Pointer)	8, 11
00 0017H 00 0000H	Lower Register File (CPU SFRs)	3, 7, 8, 11

NOTES:
1. Internally, there are 24 address bits (A23:0); however, only 20 address lines (A19:0) are bonded out. The external address space is 1 Mbyte (x00000H–xFFFFFH).
2. The 8XC196NP resets to FF2080H in internal ROM or xF2080H in external memory.
3. Unless otherwise noted, write 0FFH to reserved memory locations and write 0 to reserved SFR bits.
4. Locations xF0000H–xF00FFH are reserved for ICE.
5. These areas are mapped into internal ROM if the REMAP bit (CCB1.2) is set and EA# is at logic 1. Otherwise, they are mapped to external memory.
6. **WARNING:** The contents or functions of these memory locations may change with future device revisions, in which case a program that relies on one or more of these locations may not function properly.
7. Refer to the *8XC196NP User's Manual* or *8XC196NP Quick Reference* for SFR descriptions.
8. Code executed in locations 000000H to 0003FFH will be forced external.
9. Address with indirect, indexed, or extended modes.
10. Address with indirect, indexed, or extended modes or through register windows.
11. Address with direct, indirect, indexed, or extended modes or through register windows.

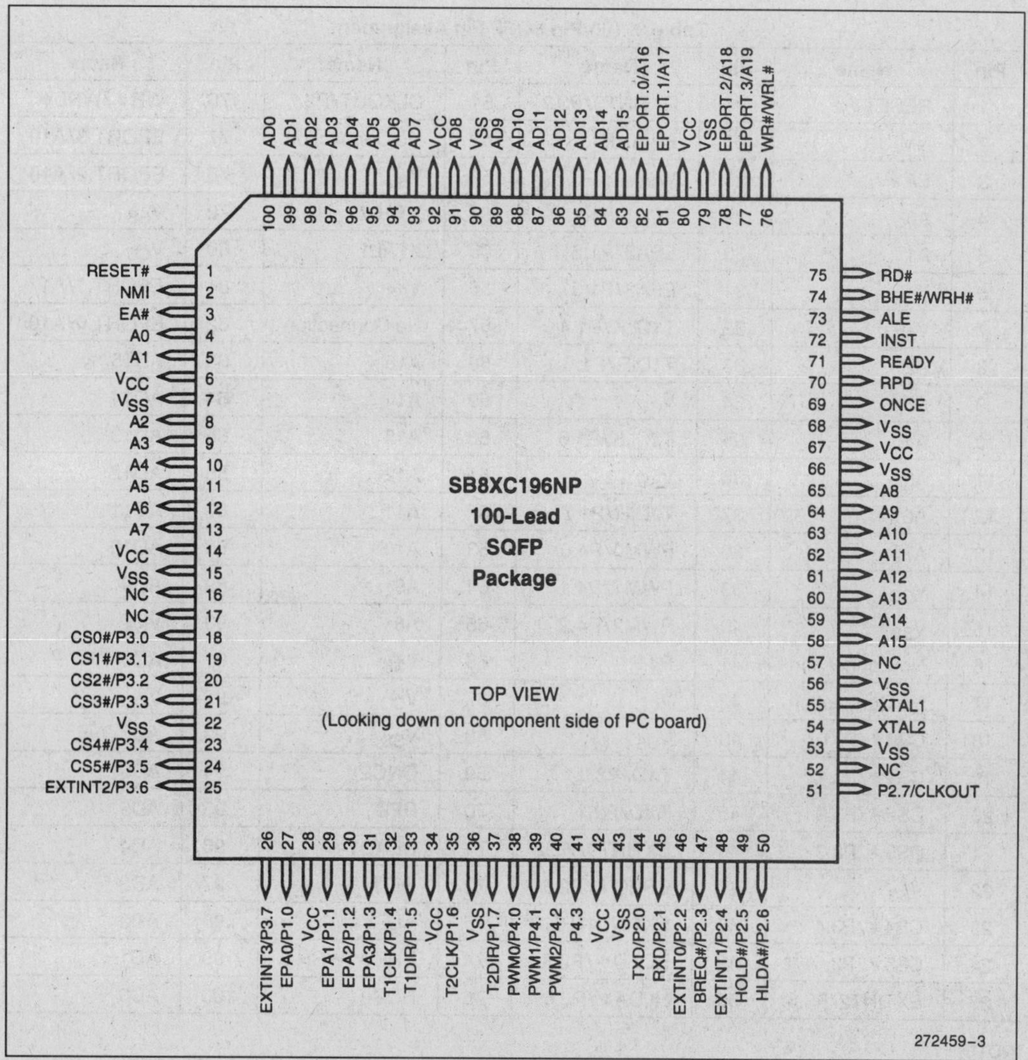

Figure 3. 8XC196NP 100-Pin SQFP Package

Table 3. 100-Pin SQFP Pin Assignment

Pin	Name	Pin	Name	Pin	Name	Pin	Name
1	RESET#	26	EXTINT3/P3.7	51	CLKOUT/P2.7	76	WR#/WRL#
2	NMI	27	EPA0/P1.0	52	No Connection	77	EPORT.3/A19
3	EA#	28	V_{CC}	53	V_{SS}	78	EPORT.2/A18
4	A0	29	EPA1/P1.1	54	XTAL2	79	V_{SS}
5	A1	30	EPA2/P1.2	55	XTAL1	80	V_{CC}
6	V_{CC}	31	EPA3/P1.3	56	V_{SS}	81	EPORT.1/A17
7	V_{SS}	32	T1CLK/P1.4	57	No Connection	82	EPORT.0/A16
8	A2	33	T1DIR/P1.5	58	A15	83	AD15
9	A3	34	V_{CC}	59	A14	84	AD14
10	A4	35	T2CLK/P1.6	60	A13	85	AD13
11	A5	36	V_{SS}	61	A12	86	AD12
12	A6	37	T2DIR/P1.7	62	A11	87	AD11
13	A7	38	PWM0/P4.0	63	A10	88	AD10
14	V_{CC}	39	PWM1/P4.1	64	A9	89	AD9
15	V_{SS}	40	PWM2/P4.2	65	A8	90	V_{SS}
16	No Connection	41	P4.3	66	V_{SS}	91	AD8
17	No Connection	42	V_{CC}	67	V_{CC}	92	V_{CC}
18	CS0#/P3.0	43	V_{SS}	68	V_{SS}	93	AD7
19	CS1#/P3.1	44	TXD/P2.0	69	ONCE	94	AD6
20	CS2#/P3.2	45	RXD/P2.1	70	RPD	95	AD5
21	CS3#/P3.3	46	EXTINT0/P2.2	71	READY	96	AD4
22	V_{SS}	47	BREQ#/P2.3	72	INST	97	AD3
23	CS4#/P3.4	48	EXTINT1/P2.4	73	ALE	98	AD2
24	CS5#/P3.5	49	HOLD#/P2.5	74	BHE#/WRH#	99	AD1
25	EXTINT2/P3.6	50	HLDA#/P2.6	75	RD#	100	AD0

NOTE:
To be compatible with future versions of the Nx family, tie the no connection (NC) pins as follows: Pin 57 = V_{SS}, Pin 16 = V_{CC}, Pin 17 = NC, and Pin 52 = V_{CC}.

Table 4. 100-Pin SQFP Pin Assignment Arranged by Functional Categories

Address & Data	
Name	**Pin**
A0	4
A1	5
A2	8
A3	9
A4	10
A5	11
A6	12
A7	13
A8	65
A9	64
A10	63
A11	62
A12	61
A13	60
A14	59
A15	58
A16	82
A17	81
A18	78
A19	77
AD0	100
AD1	99
AD2	98
AD3	97
AD4	96
AD5	95
AD6	94
AD7	93
AD8	91
AD9	89
AD10	88
AD11	87
AD12	86

Address & Data (cont)	
Name	**Pin**
AD13	85
AD14	84
AD15	83

Bus Control & Status	
Name	**Pin**
ALE	73
BHE#/WRH#	74
BREQ#	47
HOLD#	49
HLDA#	50
INST	72
RD#	75
READY	71
WR#/WRL#	76

Processor Control	
Name	**Pin**
CLKOUT	51
EA#	3
EXTINT0	46
EXTINT1	48
EXTINT2	25
EXTINT3	26
NMI	2
ONCE	69
RESET#	1
RPD	70
XTAL1	55
XTAL2	54

Input/Output	
Name	**Pin**
CS0#/P3.0	18
CS1#/P3.1	19
CS2#/P3.2	20
CS3#/P3.3	21
CS4#/P3.4	23
CS5#/P3.5	24
EPA0/P1.0	27
EPA1/P1.1	29
EPA2/P1.2	30
EAP3/P1.3	31
EPORT.0	82
EPORT.1	81
EPORT.2	78
EPORT.3	77
P2.2	46
P2.3	47
P2.4	48
P2.5	49
P2.6	50
P2.7	51
P3.6	25
P3.7	26
P4.3	41
PWM0/P4.0	38
PWM1/P4.1	39
PWM2/P4.2	40
RXD/P2.1	45
T1CLK/P1.4	32
T1DIR/P1.5	33
T2CLK/P1.6	35
T2DIR/P1.7	37
TXD/P2.0	44

Power & Ground	
Name	**Pin**
V_{CC}	6
V_{CC}	14
V_{CC}	28
V_{CC}	34
V_{CC}	42
V_{CC}	67
V_{CC}	80
V_{CC}	92
V_{SS}	7
V_{SS}	15
V_{SS}	22
V_{SS}	36
V_{SS}	43
V_{SS}	53
V_{SS}	56
V_{SS}	66
V_{SS}	68
V_{SS}	79
V_{SS}	90

No Connection	
Name	**Pin**
NC	16
NC	17
NC	52
NC	57

4

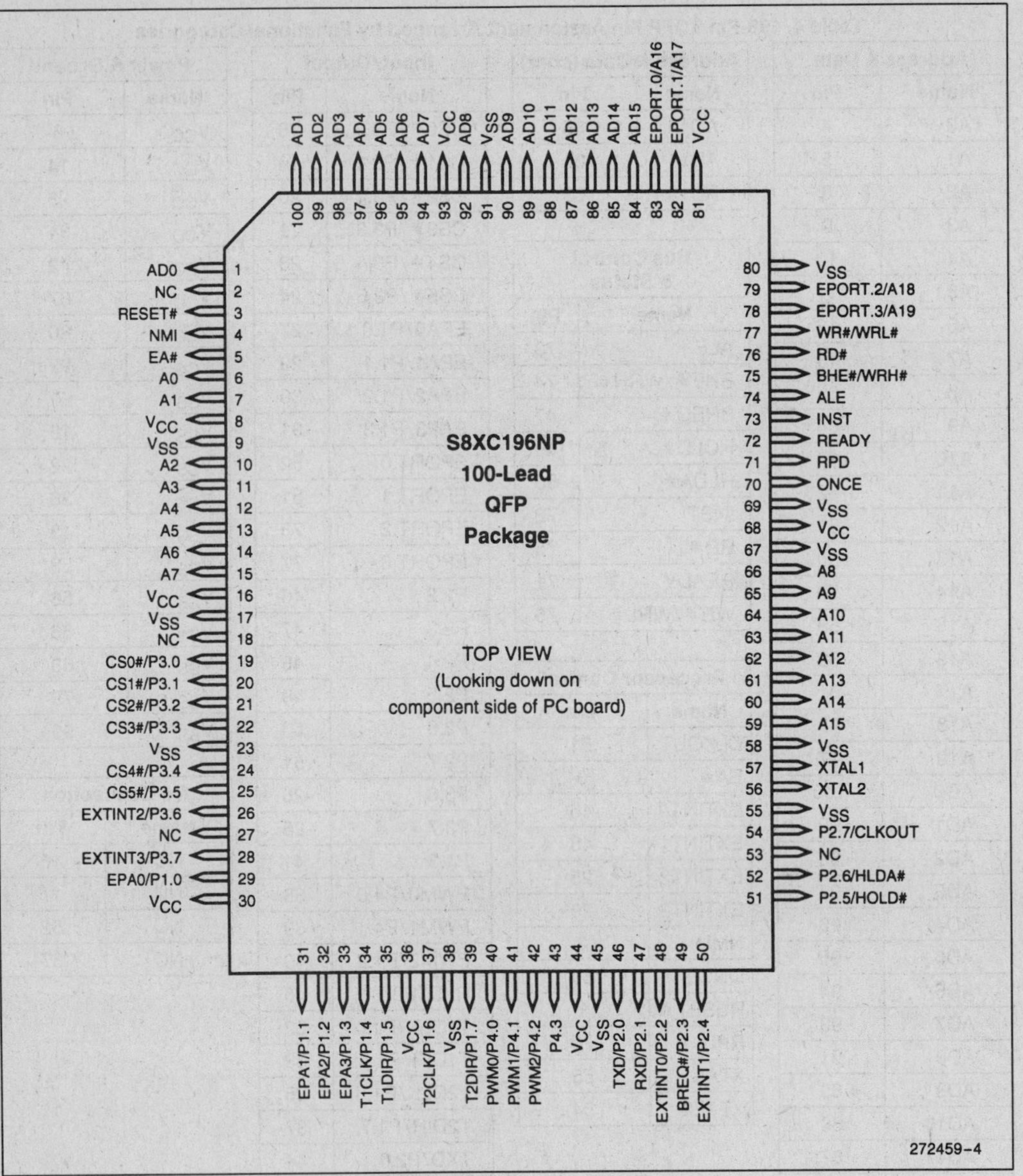

Figure 4. 8XC196NP 100-Pin QFP Package

Table 5. 100-Pin QFP Pin Assignment

Pin	Name	Pin	Name	Pin	Name	Pin	Name
1	AD0	26	EXTINT2/P3.6	51	HOLD#/P2.5	76	RD#
2	No Connection	27	No Connection	52	HLDA#/P2.6	77	WR#/WRL#
3	RESET#	28	EXTINT3/P3.7	53	No Connection	78	EPORT.3/A19
4	NMI	29	EPA0/P1.0	54	CLKOUT/P2.7	79	EPORT.2/A18
5	EA#	30	V_{CC}	55	V_{SS}	80	V_{SS}
6	A0	31	EPA1/P1.1	56	XTAL2	81	V_{CC}
7	A1	32	EPA2/P1.2	57	XTAL1	82	EPORT.1/A17
8	V_{CC}	33	EPA3/P1.3	58	V_{SS}	83	EPORT.0/A16
9	V_{SS}	34	T1CLK/P1.4	59	A15	84	AD15
10	A2	35	T1DIR/P1.5	60	A14	85	AD14
11	A3	36	V_{CC}	61	A13	86	AD13
12	A4	37	T2CLK/P1.6	62	A12	87	AD12
13	A5	38	V_{SS}	63	A11	88	AD11
14	A6	39	T2DIR/P1.7	64	A10	89	AD10
15	A7	40	PWM0/P4.0	65	A9	90	AD9
16	V_{CC}	41	PWM1/P4.1	66	A8	91	V_{SS}
17	V_{SS}	42	PWM2/P4.2	67	V_{SS}	92	AD8
18	No Connection	43	P4.3	68	V_{CC}	93	V_{CC}
19	CS0#/P3.0	44	V_{CC}	69	V_{SS}	94	AD7
20	CS1#/P3.1	45	V_{SS}	70	ONCE	95	AD6
21	CS2#/P3.2	46	TXD/P2.0	71	RPD	96	AD5
22	CS3#/P3.3	47	RXD/P2.1	72	READY	97	AD4
23	V_{SS}	48	EXTINT0/P2.2	73	INST	98	AD3
24	CS4#/P3.4	49	BREQ#/P2.3	74	ALE	99	AD2
25	CS5#/P3.5	50	EXTINT1/P2.4	75	BHE#/WRH#	100	AD1

NOTE:
To be compatible with future versions of the Nx family, tie the no connection (NC) pins as follows: Pin 2 = V_{SS}, Pin 18 = V_{CC}, Pin 27 = NC, and Pin 53 = V_{CC}.

4

Table 6. 100-pin QFP Pin Assignment Arranged by Functional Categories

Address & Data

Name	Pin
A0	6
A1	7
A2	10
A3	11
A4	12
A5	13
A6	14
A7	15
A8	66
A9	65
A10	64
A11	63
A12	62
A13	61
A14	60
A15	59
A16	83
A17	82
A18	79
A19	78
AD0	1
AD1	100
AD2	99
AD3	98
AD4	97
AD5	96
AD6	95
AD7	94
AD8	92
AD9	90
AD10	89
AD11	88
AD12	87

Address & Data (cont)

Name	Pin
AD13	86
AD14	85
AD15	84

Bus Control & Status

Name	Pin
ALE	74
BHE#/WRH#	75
BREQ#	49
HOLD#	51
HLDA#	52
INST	73
RD#	76
READY	72
WR#/WRL#	77

Processor Control

Name	Pin
CLKOUT	54
EA#	5
EXTINT0	48
EXTINT1	50
EXTINT2	26
EXTINT3	28
NMI	4
ONCE	70
RESET#	3
RPD	71
XTAL1	57
XTAL2	56

Input/Output

Name	Pin
CS0#/P3.0	19
CS1#/P3.1	20
CS2#/P3.2	21
CS3#/P3.3	22
CS4#/P3.4	24
CS5#/P3.5	25
EPA0/P1.0	29
EPA1/P1.1	31
EPA2/P1.2	32
EAP3/P1.3	33
EPORT.0	83
EPORT.1	82
EPORT.2	79
EPORT.3	78
P2.2	48
P2.3	49
P2.4	50
P2.5	51
P2.6	52
P2.7	54
P3.6	26
P3.7	28
P4.3	43
PWM0/P4.0	40
PWM1/P4.1	41
PWM2/P4.2	42
RXD/P2.1	47
T1CLK/P1.4	34
T1DIR/P1.5	35
T2CLK/P1.6	37
T2DIR/P1.7	39
TXD/P2.0	46

Power & Ground

Name	Pin
V_{CC}	8
V_{CC}	16
V_{CC}	30
V_{CC}	36
V_{CC}	44
V_{CC}	68
V_{CC}	81
V_{CC}	93
V_{SS}	9
V_{SS}	17
V_{SS}	23
V_{SS}	38
V_{SS}	45
V_{SS}	55
V_{SS}	58
V_{SS}	67
V_{SS}	69
V_{SS}	80
V_{SS}	91

No Connection

Name	Pin
NC	2
NC	18
NC	27
NC	53

PIN DESCRIPTIONS

Table 7. Pin Descriptions

Signal Name	Type	Description	Multiplexed With
A15:0	I/O	**System Address Bus.** These pins provide address bits 15–0 during the entire external memory cycle.	—
A19:16	I/O	**Address Lines 19–16.** These address lines provide address bits 19–16 during the entire external memory cycle, supporting extended addressing of the 1 Mbyte address space. **NOTE:** Internally, there are 24 address bits (A23:0); however, only 20 address lines (A19:0) are bonded out. The external address space is 1 Mbyte (x00000H–xFFFFFH). The 8XC196NP resets to FF2080H in internal ROM or xF2080H in external memory.	EPORT.3:0
AD15:0	I/O	**Address/Data Lines 15–0.** Multiplexed address/data bus in multiplexed bus mode. Data-only bus in demultiplexed bus mode.	—
ALE	O	**Address Latch Enable.** This active-high output signal is asserted only during external memory cycles. ALE signals the start of an external bus cycle and indicates that valid address information is available on the system address bus (A19:0). An external latch can use ALE to demultiplex the address from the address/data bus in multiplexed mode.	—
BHE#	O	**Byte High Enable.** During 16-bit bus cycles, this active-low output signal is asserted for word reads and writes and high-byte reads and writes to external memory. BHE# indicates that valid data is being transferred over the upper half of the system data bus. BHE#, in conjunction with A0, selects the memory byte to be accessed: **BHE# A0 Byte(s) Accessed** 0 0 both bytes 0 1 high byte only 1 0 low byte only	WRH#
BREQ#	O	**Bus Request.** This active-low output signal is asserted during a hold cycle when the bus controller has a pending external memory cycle. The device can assert BREQ# at the same time as or after it asserts HLDA#. Once it is asserted, BREQ# remains asserted until HOLD# is removed.	P2.3
CLKOUT	O	**Clock Output.** Output of the internal clock generator. The CLKOUT frequency is ½ the oscillator frequency input (XTAL1). CLKOUT has a 50% duty cycle.	P2.7
CS5:0#	O	**Chip-select Lines 5–0.** The active-low output CSx# is asserted during an external memory cycle when the address to be accessed is in the range programmed for chip select *x*. If the external memory address falls outside the ranges assigned to CS5:0#, the BUSCON values for CS5# are used by default. Immediately following reset, CS0# is automatically assigned to the range xF2000H–xF20FFH.	P3.5:0

4

Table 7. Pin Descriptions (Continued)

Signal Name	Type	Description	Multiplexed With
EA#	I	**External Access.** This active-low input signal directs memory accesses to on-chip or off-chip memory. If EA# is low, the memory access is off-chip. If EA# is high and the memory address is within FF2000H–FF2FFFH, the access is to on-chip ROM. Otherwise, an access with EA# high is to off-chip memory. EA# is **not** sampled on the rising edge of RESET#; it can be asserted/deasserted dynamically in normal operating mode. For devices without ROM, EA# must be tied low.	—
EPA3:0	I/O	**Event Processor Array (EPA) Input/Output pins.** High-speed input/output pins. Remapping provides higher resolution. With remapping enabled, EPA0 and EPA1 control the EPA1 output pin, EPA2 and EPA3 control the EPA3 output pin, and the EPA0 and EPA2 pins can be used as LSIO pins.	P1.3:0
EPORT.3:0	I/O	**Extended Addressing Port.** This is a 4-bit, bidirectional, memory-mapped I/O port. The pins are shared with the extended address bus, A19:16.	A19:16
EXTINT0 EXTINT1 EXTINT2 EXTINT3	I	**External Interrupt.** In normal operating mode, a positive transition on EXTINTx sets the EXTINTx interrupt pending flag. (EXTINTx is sampled during Phase 2 (CLKOUT high). The minimum high time is two state times.) If the chip is in Idle mode and if EXTINTx is enabled, a positive transition on EXTINTx brings the chip back to normal operation, where the first action is to execute the EXTINTx service routine. After completion of the service routine, execution resumes at the instruction following the one that put the chip into Idle mode. In Powerdown mode, applying a logical 1 to EXTINTx causes the chip to return to normal operating mode. If EXTINTx is enabled, the EXTINTx service routine is executed. Otherwise, execution continues at the instruction following the IDLPD instruction that put the chip into Powerdown mode. EXTINTx is normally a *sampled input*; however, the powerdown circuitry uses it as a level-sensitive input in Powerdown mode.	P2.2 P2.4 P3.6 P3.7
HOLD#	I	**Bus Hold Request.** This active-low input signal is used to request control of the bus if hold is enabled.	P2.5
HLDA#	O	**Bus Hold Acknowledge.** This active-low output indicates that the 8XC196NP has released the bus as the result of another device asserting HOLD#.	P2.6
INST	O	**Instruction Fetch.** This active-high output signal is valid only during external memory bus cycles. When high, INST indicates that an instruction is being fetched from external memory. The signal remains high during the entire bus cycle of an external instruction fetch. INST is low for data accesses, including interrupt vector fetches and chip configuration byte reads. INST is low during internal memory fetches. No LSIO function is associated with this pin.	—

Table 7. Pin Descriptions (Continued)

Signal Name	Type	Description	Multiplexed With
NMI	I	**Nonmaskable Interrupt.** In normal operating mode, a positive transition on NMI causes a vector through the NMI interrupt at location 0FF203EH. NMI must be asserted for greater than one state time to guarantee that it is recognized. If the chip is in Idle mode, a positive transition on NMI brings the chip back to normal operation, where the first action is to execute the NMI service routine. After completion of the service routine, execution resumes at the instruction following the one that put the chip into Idle mode. In Powerdown mode, NMI causes a return to normal operating mode only if it is tied to a masked EXTINTx.	—
ONCE	I	**On-circuit Emulation.** Asserting this pin during a return from reset places the device into on-circuit emulation (ONCE) mode. This mode puts all pins at a high-impedance state and allows for connection of an emulator. The device must be reset to exit ONCE mode. The value of ONCE is latched when the RESET# pin goes inactive. For normal execution connect the ONCE pin to V_{SS}.	—
P1.3:0 P1.4 P1.5 P1.6 P1.7	I/O	**Port 1.** This is an 8-bit, bidirectional, standard I/O port.	EPA3:0 T1CLK T1DIR T2CLK T2DIR
P2.0 P2.1 P2.2 P2.3 P2.4 P2.5 P2.6 P2.7	I/O	**Port 2.** This is an 8-bit, bidirectional, standard I/O port.	TXD RXD EXTINT0 BREQ# EXTINT1 HOLD# HLDA# CLKOUT
P3.5:0 P3.6 P3.7	I/O	**Port 3.** This is an 8-bit, bidirectional, memory-mapped I/O port.	CS5:0# EXTINT2 EXTINT3
P4.2: P4.3	I/O	**Port 4.** This is a 4-bit, bidirectional, memory-mapped I/O port with high-current drive capability.	PWM2:0
PWM2:0	O	**Pulse Width Modulator Outputs.** These PWM output pins have high-current drive capability.	P4.2:0
RD#	O	**Read.** Read-signal output to external memory. RD# is asserted only during external memory reads.	—
READY	I	**Ready Input.** This active-high input signal is used to lengthen external memory cycles for slow memory by generating wait states in addition to the programmed wait states. When READY is high, CPU operation continues in a normal manner with wait states inserted as programmed in the CCB0 or BUSCONx registers. READY is ignored for all internal memory accesses.	—

4

Table 7. Pin Descriptions (Continued)

Signal Name	Type	Description	Multiplexed With
RESET#	I/O	**Reset.** Reset input to and open-drain output from the chip. A falling edge on RESET# initiates the reset process. When RESET# is first asserted, the chip turns on a pull-down transistor connected to the RESET pin for 16 state times. This function can also be activated by execution of the RST instruction. In the Powerdown and Idle modes, asserting RESET# causes the chip to reset and return to normal operating mode. RESET# is a level-sensitive input.	—
RPD	O	**Return from Power Down.** This is the timing pin for the return-from-powerdown circuit. Connect a 1 μF capacitor from this pin to V_{SS} and connect a 1 MΩ resistor from this pin to V_{CC}. If powerdown mode is not to be used, leave this pin unconnected.	—
RXD	I/O	**Receive Serial Data.** In modes 1, 2, and 3, RXD is used to receive serial port data. In mode 0, it functions as an input or an open-drain output for data.	P2.1
T1CLK	I	**External Clock.** External clock for Timer 1. Timer 1 increments (or decrements) on both rising and falling edges of T1CLK. Also used in conjunction with T1DIR for quadrature counting mode. **and** External clock for the serial I/O baud-rate generator (program selectable).	P1.4
T2CLK	I	**Timer 2 External Clock.** External clock for Timer 2. Timer 2 increments (or decrements) on both rising and falling edges of T2CLK. Also used in conjunction with T2DIR for quadrature counting mode.	P1.6
T1DIR	I	**Timer 1 External Direction.** External direction (up/down) for Timer 1. Timer 1 increments when T1DIR is high and decrements when it is low. Also used in conjunction with T1CLK for quadrature counting mode.	P1.5
T2DIR	I	**Timer 2 External Direction.** External direction (up/down) for Timer 2. Timer 2 increments when T2DIR is high and decrements when it is low. Also used in conjunction with T2CLK for quadrature counting mode.	P1.7
TXD	O	**Transmit Serial Data.** In serial I/O modes 1, 2, and 3, TXD is used to transmit serial port data. In mode 0, it is used as the serial clock output.	P2.0
V_{CC}	PWR	**Digital Supply Voltage** (3.3 and 5 volts). There are multiple V_{CC} pins, all of which must be connected.	—
V_{SS}	GND	**Digital Circuit Ground** (0 volts). There are multiple V_{SS} pins, all of which must be connected to ground.	—
WR#	O	**Write.** This active-low output indicates that an external write is occurring. This signal is asserted only during external memory writes.	WRL#
WRH#	O	**Write High.** During 16-bit bus cycles, this active-low output signal is asserted for high-byte writes and word writes to external memory. During 8-bit bus cycles, WRH# is asserted for all write operations.	BHE#
WRL#	O	**Write Low.** During 16-bit bus cycles, this active-low output signal is asserted for low-byte writes and word writes. During 8-bit bus cycles, WRL# is asserted for all write operations.	WR#

Table 7. Pin Descriptions (Continued)

Signal Name	Type	Description	Multiplexed With
XTAL1	I	Input to the on-chip oscillator inverter and to the internal clock generator. If an external oscillator is used, XTAL1 also serves as the clock input (the XTAL1 V_{IH} specification must be met.) **and** Internal clock source for the serial I/O baud-rate generator, timers, and PWM unit (program selectable).	—
XTAL2	O	Output of the on-chip oscillator inverter. Leave XTAL2 floating unless it is connected to a crystal/resonator circuit.	—

ELECTRICAL CHARACTERISTICS

ABSOLUTE MAXIMUM RATINGS*

Storage Temperature .... −60°C to +150°C
Supply Voltage with Respect to V_{SS} ...... −0.5V to +7.0V
Power Dissipation 1.5W

OPERATING CONDITIONS*

T_A (Ambient Temperature Under Bias) .... 0°C to +70°C
V_{CC} (Digital Supply Voltage) .... 2.97V to 5.5V
F_{OSC1} (Oscillator Frequency for V_{CC} = 4.5–5.5 V) (Note 1, 2) .... 8 MHz to 25 MHz
F_{OSC2} (Oscillator Frequency for V_{CC} = 2.97–4.5 V) (Note 1, 2) .... 8 MHz to 16 MHz

NOTES:
1. This device is static and should operate below 1 Hz, but has been tested only down to 8 MHz.
2. This device functions across 2.97V to 5.5V, but is specified across two ranges, 2.97–4.5V and 4.5–5.5V.

NOTICE: This document contains information on products in the design phase of development. Do not finalize a design with this information. Revised information will be published when the product is available. Verify with your local Intel Sales Office that you have the latest data sheet before finalizing a design.

****WARNING:*** *Stressing the device beyond the "Absolute Maximum Ratings" may cause permanent damage. These are stress ratings only. Operation beyond the "Operating Conditions" is not recommended and extended exposure beyond the "Operating Conditions" may affect device reliability.*

4

DC Characteristics

Table 8. DC Characteristics at V_{CC} = 4.5–5.5V

Symbol	Parameter	Min	Typ	Max	Units	Test Conditions
I_{CC}	V_{CC} Supply Current		80	120	mA	XTAL1 = 25 MHz V_{CC} = 5.5V Device in Reset
I_{IDLE}	Idle Mode Current		24	36	mA	XTAL1 = 25 MHz V_{CC} = 5.5V
I_{PD}	Powerdown Mode Current (Note 1)		50	75	μA	V_{CC} = 5.5V (Note 2)
I_{LI}	Input Leakage Current (Standard Inputs)			±10	μA	$V_{SS} < V_{IN} < V_{CC}$
V_{IL}	Input Low Voltage (all pins)	−0.5		0.8	V	
V_{IH}	Input High Voltage	0.2 V_{CC} +1		V_{CC} +0.5	V	
V_{IL1}	Input Low Voltage XTAL1	−0.5		0.3 V_{CC}	V	
V_{IH1}	Input High Voltage XTAL1	0.7 V_{CC}		V_{CC} +0.5	V	
V_{OL}	Output Low Voltage (output configured as complementary) (Note 3)			0.3 0.45 1.5	V V V	I_{OL} = 200 μA I_{OL} = 3.2 mA I_{OL} = 7.0 mA
V_{OH}	Output High Voltage (output configured as complementary)	V_{CC} − 0.3 V_{CC} − 0.7 V_{CC} − 1.5			V V V	I_{OH} = −200 μA I_{OH} = −3.2 mA I_{OH} = −7.0 mA
V_{OL1}	Output Low Voltage on P4.x (output configured as complementary)			0.45 0.6	V V	I_{OL} = 10 mA I_{OL} = 15 mA
V_{OL2}	Output Low Voltage in RESET on ALE, INST, and NMI			0.45	V	I_{OL} = 3 μA
V_{OH1}	Output High Voltage in RESET (Note 4)	V_{CC} − 0.7			V	I_{OH} = 3 μA
V_{OL3}	Output Low Voltage in RESET for ONCE pin			0.45	V	I_{OL} = 30 μA
V_{OL4}	Output Low Voltage on XTAL2			0.3 0.45 1.5	V V V	I_{OL} = 100 μA I_{OL} = 700 μA I_{OL} = 3 mA

NOTES:
1. Typical values are based on a limited number of samples and are not guaranteed. The values listed are at room temperature with V_{CC} = 5.0V.
2. For temperatures below 100°C, typical is 10 μA.
3. For all pins except P4.3:0, which have higher drive capability (see V_{OL1}).
4. For all pins that were weakly pulled high during RESET. This **excludes** ALE, INST, and NMI, which were weakly pulled low (see V_{OL2}) and ONCE, which was pulled medium low (see V_{OL3}).

Table 8. DC Characteristics at V_{CC} = 4.5–5.5V (Continued)

Symbol	Parameter	Min	Typ	Max	Units	Test Conditions
V_{OH2}	Output High Voltage on XTAL2	$V_{CC} - 0.3$ $V_{CC} - 0.7$ $V_{CC} - 1.5$			V V V	$I_{OH} = -100\ \mu A$ $I_{OH} = -700\ \mu A$ $I_{OH} = -3$ mA
$V_{TH+} - V_{TH-}$	Hysteresis voltage width on RESET# pin (Note 1)		0.3		V	
C_S	Pin Capacitance (any pin to V_{SS})			10	pF	f_{TEST} = 1.0 MHz
R_{RST}	RESET Pull-up Resistor	9		95	kΩ	V_{CC} = 5.5 V, V_{IN} = 4.0V

NOTES:
1. Typical values are based on a limited number of samples and are not guaranteed. The values listed are at room temperature with V_{CC} = 5.0V.
2. For temperatures below 100°C, typical is 10 μA.
3. For all pins except P4.3:0, which have higher drive capability (see V_{OL1}).
4. For all pins that were weakly pulled high during RESET. This **excludes** ALE, INST, and NMI, which were weakly pulled low (see V_{OL2}) and ONCE, which was pulled medium low (see V_{OL3}).

4

Table 9. DC Characteristics at V_{CC} = 2.97–5.5V

Symbol	Parameter	Min	Typ	Max	Units	Test Conditions
V_{IL}	Input Low Voltage (all pins)	−0.5		0.4	V	
V_{IH}	Input High Voltage	0.2 V_{CC} +1.3		V_{CC} +0.5	V	
V_{IL1}	Input Low Voltage XTAL1	−0.5		0.3 V_{CC}	V	
V_{IH1}	Input High Voltage XTAL1	0.7 V_{CC}		V_{CC} + 0.5	V	
V_{OL}	Output Low Voltage (output configured as complementary) (Note 3)			0.3 0.45	V V	I_{OL} = 200 μA I_{OL} = 3.2 mA
V_{OH}	Output High Voltage (output configured as complementary)	V_{CC} − 0.3 V_{CC} − 0.7			V V	I_{OH} = −200 μA I_{OH} = −3.2 mA
V_{OH2}	Output High Voltage on XTAL2	V_{CC} − 0.3 $_{CC}$ − 0.7			V V	I_{OH} = −100 μA I_{OH} = −500 μA
V_{OL1}	Output Low Voltage on P4.x (output configured as complementary)			0.45 0.6	V V	I_{OL} = 8 mA I_{OL} = 10 mA
V_{OL2}	Output Low Voltage in RESET on ALE, INST, and NMI			0.45	V	I_{OL} = 2 μA
V_{OH1}	Output High Voltage in RESET (Note 4)	V_{CC} − 0.7			V	I_{OH} = 2 μA
V_{OL3}	Output Low Voltage in RESET for ONCE pin			0.45	V	I_{OL} = 30 μA
V_{OL4}	Output Low Voltage on XTAL2			0.3 0.45	V V	I_{OL} = 100 μA I_{OL} = 500 μA
$V_{TH+}-V_{TH-}$	Hysteresis voltage width on RESET# pin (Note 1)		0.3		V	

NOTES:
1. Typical values are based on a limited number of samples and are not guaranteed. The values listed are at room temperature and with V_{CC} = 3.3V.
2. For temperatures below 100°C, typical is 10 μA.
3. For all pins except P4.3:0, which have higher drive capability.
4. For all pins that were weakly pulled high during RESET. This **excludes** ALE, INST, and NMI, which were weakly pulled low (see V_{OL2}) and ONCE, which was pulled medium low (see V_{OL3}).

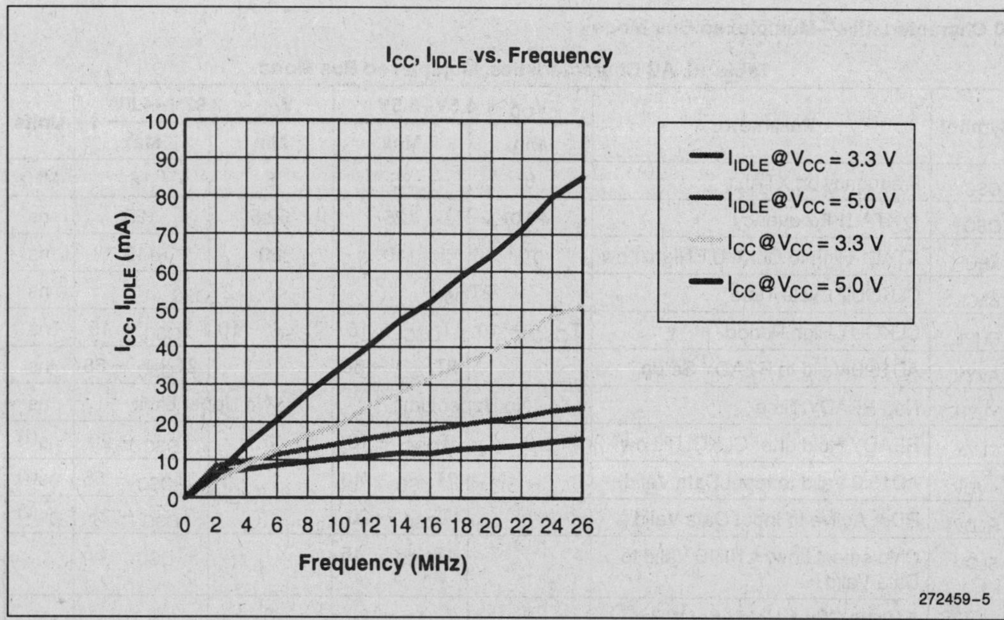

Figure 5. I_{CC}, I_{IDLE} versus Frequency

AC Characteristics—Multiplexed Bus Mode

Table 10. AC Characteristics, Multiplexed Bus Mode

Symbol	Parameter	V_{CC} = 4.5V–5.5V		V_{CC} = 2.97V–4.5V		Units
		Min	Max	Min	Max	
F_{OSC}	Frequency on XTAL1	8	25	8	16	MHz
T_{OSC}	1/XTAL1 Frequency	40.0	125	62.5	125	ns
T_{XHCH}	XTAL1 High to CLKOUT High/Low	10	110	20	110	ns
T_{CLCL}	CLKOUT Cycle Time	$2T_{OSC}$		$2T_{OSC}$		ns
T_{CHCL}	CLKOUT High Period	$T_{OSC} - 10$	$T_{OSC} + 10$	$T_{OSC} - 10$	$T_{OSC} + 15$	ns
T_{AVYV}	AD15:0 Valid to READY Setup		$2T_{OSC} - 50$		$2T_{OSC} - 68$	ns
T_{YLYH}	Non READY Time	No Upper Limit		No Upper Limit		ns
T_{CLYX}	READY Hold after CLKOUT Low	0	$T_{OSC} - 10$	0	$T_{OSC} - 20$	ns(1)
T_{AVDV}	AD15:0 Valid to Input Data Valid		$3T_{OSC} - 40$		$3T_{OSC} - 55$	ns(2)
T_{RLDV}	RD# Active to Input Data Valid		$T_{OSC} - 20$		$T_{OSC} - 25$	ns(2)
T_{SLDV}	Chip-select Low, A19:16 Valid to Data Valid		$4T_{OSC} - 45$		$4T_{OSC} - 65$	
T_{WHSH}	A19:16, CSx# Hold after WR# Rising Edge	0		0		
T_{RHSH}	A19:16, CSx# Hold after RD# Rising Edge	0		0		
T_{CLDV}	CLKOUT Low to Input Data Valid		$T_{OSC} - 35$		$T_{OSC} - 50$	ns
T_{RHDZ}	End of RD# to Input Data Float		$T_{OSC} - 5$		$T_{OSC} - 10$	ns
T_{RXDX}	Data Hold after RD# Inactive	0		0		ns
T_{CLLH}	CLKOUT Low to ALE High	−10	10	−12	10	ns
T_{LLCH}	ALE Low to CLKOUT High	−15	10	−10	15	ns
T_{LHLH}	ALE Cycle Time	$4T_{OSC}$		$4T_{OSC}$		ns(2)
T_{LHLL}	ALE High Period	$T_{OSC} - 10$	$T_{OSC} + 10$	$T_{OSC} - 15$	$T_{OSC} + 5$	ns
T_{AVLL}	AD15:0 Valid to ALE Low	$T_{OSC} - 15$		$T_{OSC} - 18$		ns
T_{LLAX}	AD15:0 Hold after ALE Low	$T_{OSC} - 25$		$T_{OSC} - 25$		ns
T_{LLRL}	ALE Low to RD# Low	$T_{OSC} - 15$		$T_{OSC} - 30$		ns
T_{RLCL}	RD# Low to CLKOUT Low	0	20	10	30	ns
T_{RLRH}	RD# Low Period	T_{OSC}		$T_{OSC} - 10$		ns(2)
T_{RHLH}	RD# High to ALE High	$T_{OSC} - 5$	$T_{OSC} + 15$	$T_{OSC} - 5$	$T_{OSC} + 20$	ns(3)

NOTES:
1. Exceeding the maximum specification causes additional wait states.
2. If wait states are used, add $2T_{OSC} \times n$, where n = number of wait states.
3. Assuming back-to-back bus cycles.
4. 8-bit bus only.

Table 10. AC Characteristics, Multiplexed Bus Mode (Continued)

Symbol	Parameter	V_{CC} = 4.5V–5.5V		V_{CC} = 2.97V–4.5V		Units
		Min	Max	Min	Max	
T_{RLAZ}	RD# Low to Address Float		5		5	ns
T_{LLWL}	ALE Low to WR# Low	$T_{OSC} - 15$		$T_{OSC} - 30$		ns
T_{CLWL}	CLKOUT Low to WR# Low	−15	10	−18	10	ns
T_{QVWH}	Data Valid before WR# High	$T_{OSC} - 15$		$T_{OSC} - 23$		ns(2)
T_{CHWH}	CLKOUT High to WR# High	−10	10	−10	10	ns
T_{WLWH}	WR# Low Period	$T_{OSC} - 5$		$T_{OSC} - 10$		ns(2)
T_{WHQX}	Data Hold after WR# High	$T_{OSC} - 20$		$T_{OSC} - 30$		ns
T_{WHLH}	WR# High to ALE High	$T_{OSC} - 12$	$T_{OSC} + 20$	$T_{OSC} - 12$	$T_{OSC} + 20$	ns(3)
T_{WHBX}	BHE#, INST Hold after WR# High	$T_{OSC} - 10$		$T_{OSC} - 10$		ns
T_{WHAX}	A15:8 Hold after WR# High	$T_{OSC} - 10$		$T_{OSC} - 30$		ns(4)
T_{RHBX}	BHE#, INST Hold after RD# High	$T_{OSC} - 10$		$T_{OSC} - 10$		ns
T_{RHAX}	A15:8 Hold after RD# High	$T_{OSC} - 10$		$T_{OSC} - 25$		ns(4)

NOTES:
1. Exceeding the maximum specification causes additional wait states.
2. If wait states are used, add $2T_{OSC} \times n$, where n = number of wait states.
3. Assuming back-to-back bus cycles.
4. 8-bit bus only.

SYSTEM BUS TIMINGS, MULTIPLEXED BUS

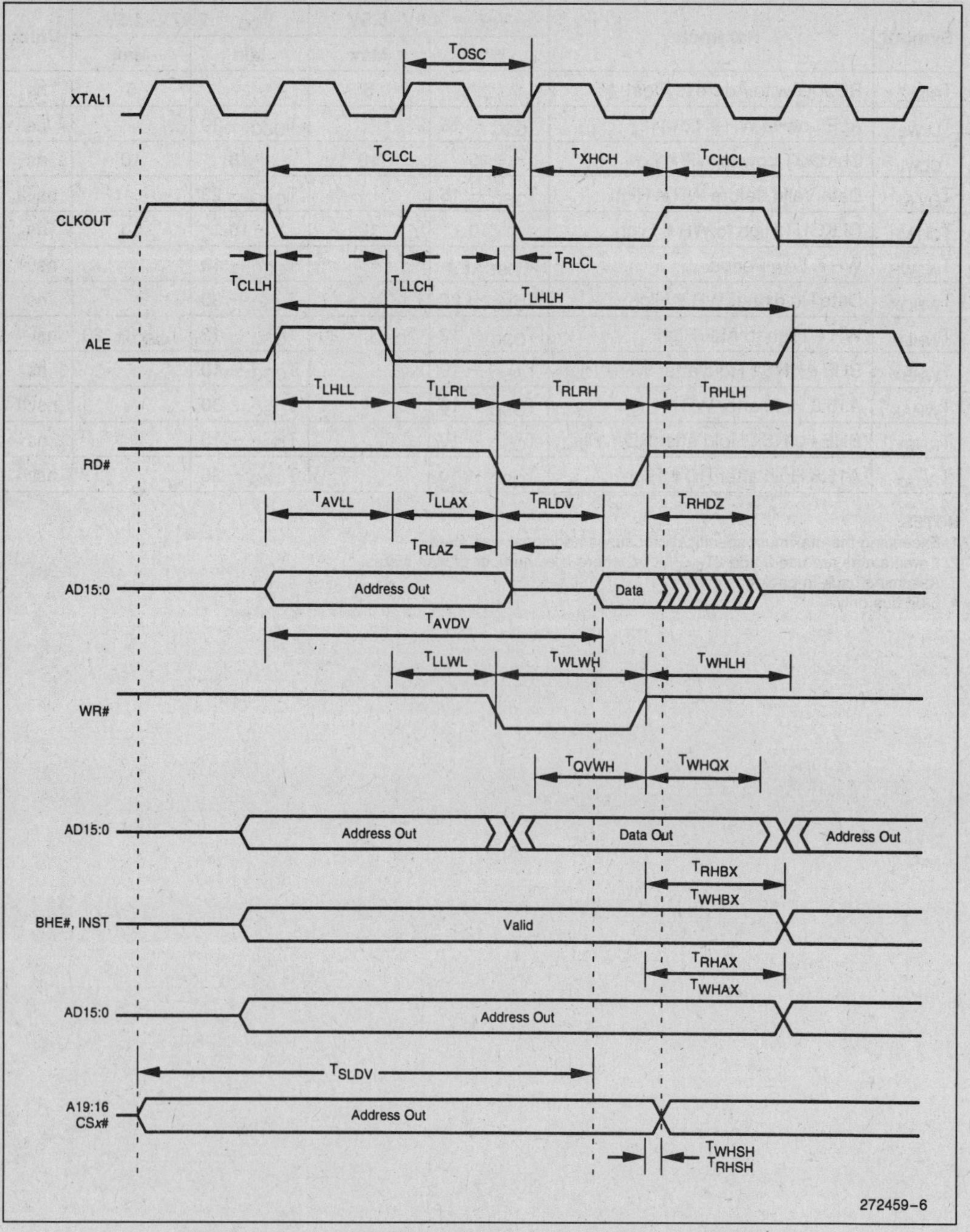

Figure 6. System Bus Timing Diagram (Multiplexed Bus Mode)

AC Characteristics—Demultiplexed Bus Mode

Table 11. AC Characteristics, Demultiplexed Bus Mode

Symbol	Parameter	V_{CC} = 4.5V–5.5V		V_{CC} = 2.97V–4.5V		Units
		Min	Max	Min	Max	
F_{OSC}	Frequency on XTAL1	8	25	8	16	MHz
T_{OSC}	1/XTAL1 Frequency	40.0	125	62.5	125	ns
T_{XHCH}	XTAL1 High to CLKOUT High/Low	10	110	20	110	ns
T_{CLCL}	CLKOUT Cycle Time	$2T_{OSC}$		$2T_{OSC}$		ns
T_{CHCL}	CLKOUT High Period	$T_{OSC} - 10$	$T_{OSC} + 10$	$T_{OSC} - 10$	$T_{OSC} + 15$	ns
T_{AVYV}	A19:0, CSx# Valid to READY Setup		$3T_{OSC} - 60$		$3T_{OSC} - 85$	ns
T_{YLYH}	Non READY Time	No Upper Limit		No Upper Limit		ns
T_{CLYX}	READY Hold after CLKOUT Low	0	$T_{OSC} - 10$		$T_{OSC} - 30$	ns(1)
T_{AVDV}	A19:0, CSx# Valid to Input Data Valid		$4T_{OSC} - 50$		$4T_{OSC} - 72$	ns(2)
T_{RLDV}	RD# Active to Input Data Valid		$2T_{OSC} - 25$		$2T_{OSC} - 33$	ns(2)
T_{CLDV}	CLKOUT Low to Input Data Valid		$T_{OSC} - 35$		$T_{OSC} - 50$	ns
T_{RHDZ}	End of RD# to Input Data Float		$T_{OSC} - 5$		$T_{OSC} - 5$	ns
T_{RXDX}	Data Hold after RD# Inactive	0		0		ns
T_{CLLH}	CLKOUT Low to ALE High	−10	10	−12	10	ns
T_{LLCH}	ALE Low to CLKOUT High	−15	10	−15	15	ns
T_{LHLH}	ALE Cycle Time	$4T_{OSC}$		$4T_{OSC}$		ns(2)
T_{LHLL}	ALE High Period	$T_{OSC} - 10$	$T_{OSC} + 10$	$T_{OSC} - 12$	$T_{OSC} + 10$	ns
T_{AVLL}	Address Valid to ALE Low	NA		NA		ns
T_{LLAX}	Address Hold after ALE Low	NA		NA		ns
T_{LLRL}	ALE Low to RD# Low	NA		NA		ns
T_{RLCH}	RD# Low to CLKOUT High	0	15	−5	20	ns
T_{RLRH}	RD# Low Period	$2T_{OSC} - 10$		$2T_{OSC} - 10$		ns(2)
T_{RHLH}	RD# High to ALE High	$T_{OSC} - 5$	$T_{OSC} + 20$	$T_{OSC} - 5$	$T_{OSC} + 20$	ns(3)
T_{RLAZ}	RD# Low to Address Float		NA		NA	ns
T_{LLWL}	ALE Low to WR# Low	NA		NA		ns
T_{WLCH}	WR# Low to CLKOUT High	−5	10	−10	10	ns
T_{QVWH}	Data Valid before WR# High	$3T_{OSC} - 37$		$3T_{OSC} - 55$		ns(2)
T_{CHWH}	CLKOUT High to WR# High	−15	5	−15	5	ns
T_{WLWH}	WR# Low Period	$2T_{OSC} - 10$		$2T_{OSC} - 13$		ns(2)

NOTES:
1. Exceeding the maximum specification causes additional wait states.
2. If wait states are used, add 2Tosc × *n*, where *n* = number of wait states.
3. Assuming back-to-back bus cycles.

Table 11. AC Characteristics, Demultiplexed Bus Mode (Continued)

Symbol	Parameter	V_{CC} = 4.5V–5.5V		V_{CC} = 2.97V–4.5V		Units
		Min	Max	Min	Max	
T_{WHQX}	Data Hold after WR# High	$T_{OSC} - 20$		$T_{OSC} - 25$		ns
T_{WHLH}	WR# High to ALE High	$T_{OSC} - 5$	$T_{OSC} + 20$	$T_{OSC} - 10$	$T_{OSC} + 20$	ns(3)
T_{WHBX}	BHE#, INST Hold after WR# High	$T_{OSC} - 10$		$T_{OSC} - 10$		ns
T_{WHAX}	A19:0, CSx# Hold after WR# High	0		0		ns
T_{RHBX}	BHE#, INST Hold after RD# High	$T_{OSC} - 10$		$T_{OSC} - 10$		ns
T_{RHAX}	A19:0, CSx# Hold after RD# High	0		0		ns

NOTES:
1. Exceeding the maximum specification causes additional wait states.
2. If wait states are used, add 2Tosc × *n*, where *n* = number of wait states.
3. Assuming back-to-back bus cycles.

SYSTEM BUS TIMINGS, DEMULTIPLEXED BUS

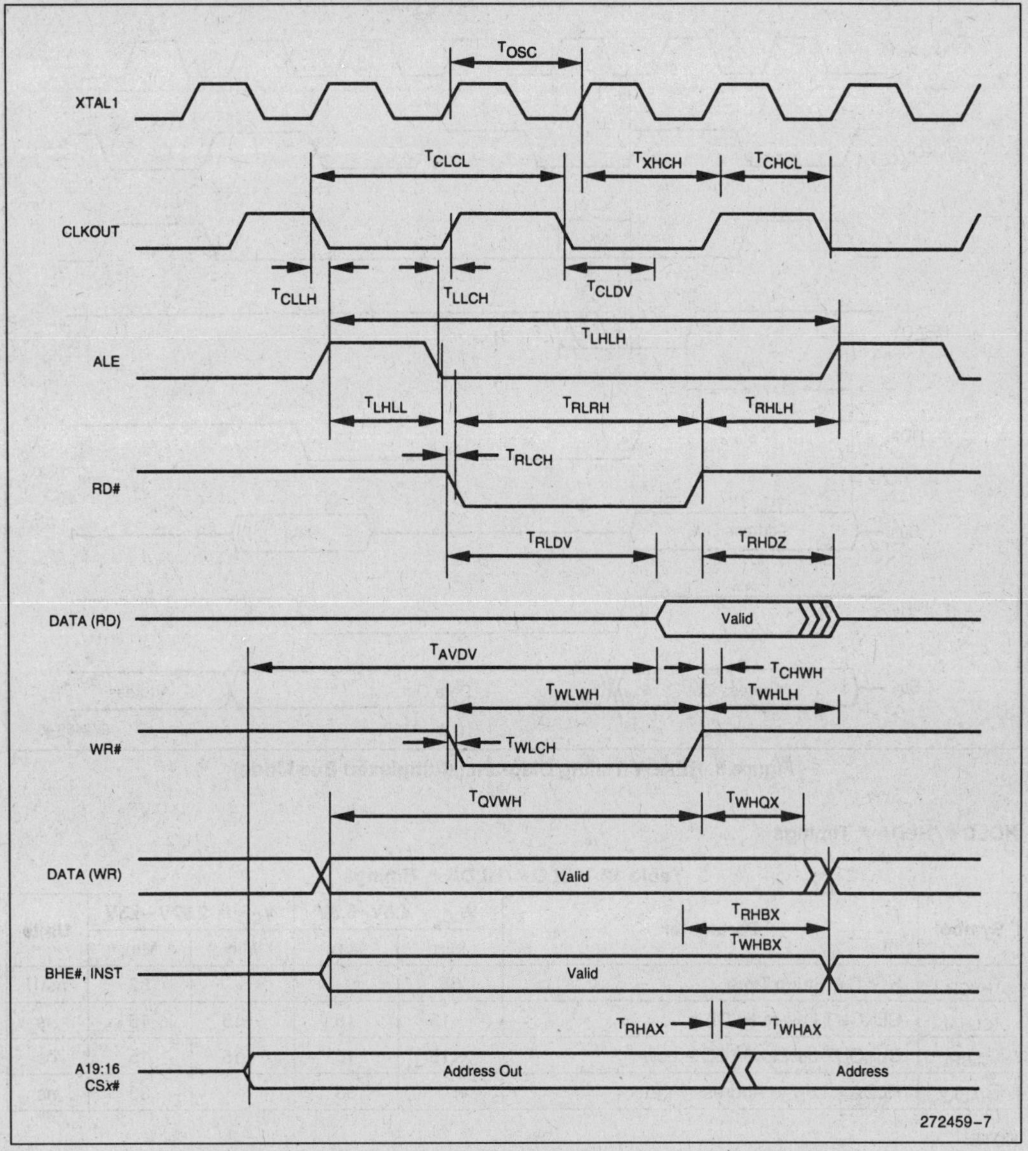

Figure 7. System Bus Timing Diagram (Demultiplexed Bus Mode)

READY TIMING, MULTIPLEXED BUS

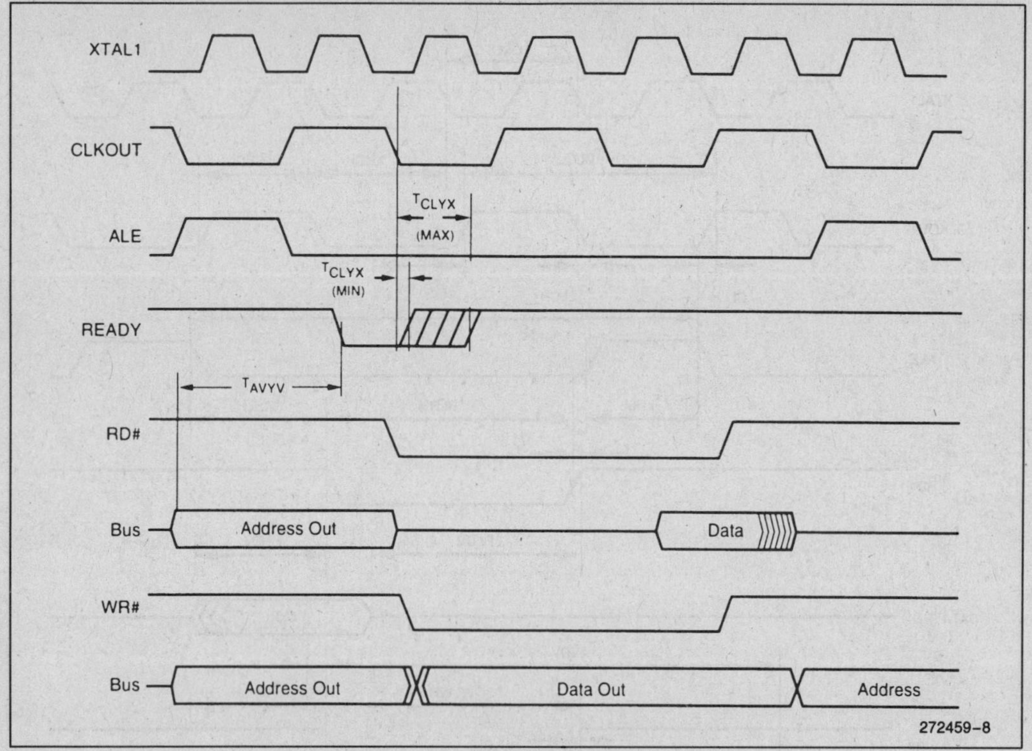

Figure 8. READY Timing Diagram (Multiplexed Bus Mode)

HOLD#/HLDA# Timings

Table 12. HOLD#/HLDA# Timings

Symbol	Parameter	V_{CC} = 4.5V–5.5V		V_{CC} = 2.97V–4.5V		Units
		Min	Max	Min	Max	
T_{HVCH}	HOLD# Setup Time	65			83	ns(1)
T_{CLHAL}	CLKOUT Low to HLDA# Low	−15	15	−15	15	ns
T_{CLBRL}	CLKOUT Low to BREQ# Low	−15	15	−15	15	ns
T_{HALAZ}	HLDA# Low to Address Float		33		33	ns

NOTE:
1. To guarantee recognition at next clock.

Table 12. HOLD#/HLDA# Timings (Continued)

Symbol	Parameter	V_{CC} = 4.5V–5.5V		V_{CC} = 2.97V–4.5V		Units
		Min	Max	Min	Max	
T_{HALBZ}	HLDA# Low to BHE#, INST, RD#, WR# Weakly Driven		25		25	ns
T_{CLHAH}	CLKOUT Low to HLDA# High	−25	15	−25	15	ns
T_{CLBRH}	CLKOUT Low to BREQ# High	−25	25	−25	25	ns
T_{HAHAX}	HLDA# High to Address No Longer Float	−20		−20		ns
T_{HAHBV}	HLDA# High to BHE#, INST, RD#, WR# Valid	−20		−20		ns

NOTE:
1. To guarantee recognition at next clock.

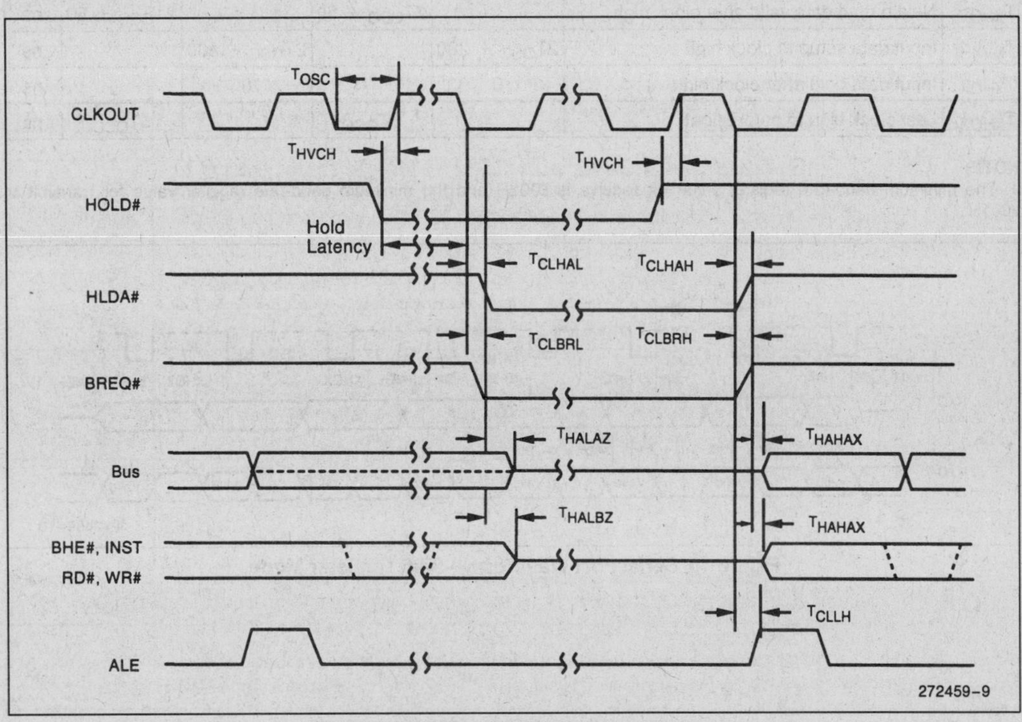

Figure 9. HOLD#/HLDA# Timing Diagram

AC Characteristics—Serial Port, Shift Register Mode

Table 13. Serial Port Timing—Shift Register Mode

Symbol	Parameter	V_{CC} = 4.5V–5.5V		V_{CC} = 2.97V–4.5V		Units
		Min	Max	Min	Max	
T_{XLXL}	Serial Port Clock period (BRR ≥ x002H) (BRR = x001H) (Note 1)	 $6T_{OSC}$ $4T_{OSC}$		 $6T_{OSC}$ $4T_{OSC}$		 ns ns
T_{XLXH}	Serial Port Clock falling edge to rising edge (BRR ≥ x002H) (BRR = x001H) (Note 1)	 $4T_{OSC} - 50$ $2T_{OSC} - 50$	 $T_{OSC} + 50$ $2T_{OSC} + 50$	 $4T_{OSC} - 50$ $2T_{OSC} - 50$	 $T_{OSC} + 50$ $2T_{OSC} + 50$	 ns ns
T_{QVXH}	Output data setup to clock high	$3T_{OSC}$		$3T_{OSC}$		ns
T_{XHQX}	Output data hold after clock high	$2T_{OSC} - 50$		$2T_{OSC} - 50$		ns
T_{XHQV}	Next output data valid after clock high		$2T_{OSC} + 50$		$2T_{OSC} + 50$	ns
T_{DVXH}	Input data setup to clock high	$2T_{OSC} + 200$		$2T_{OSC} + 200$		ns
T_{XHDX}	Input data hold after clock high	0		0		ns
T_{XHQZ}	Last clock high to output float		$5T_{OSC}$		$5T_{OSC}$	ns

NOTE:
1. The minimum baud-rate register value for receive is x002H and the minimum baud-rate register value for transmit is x001H.

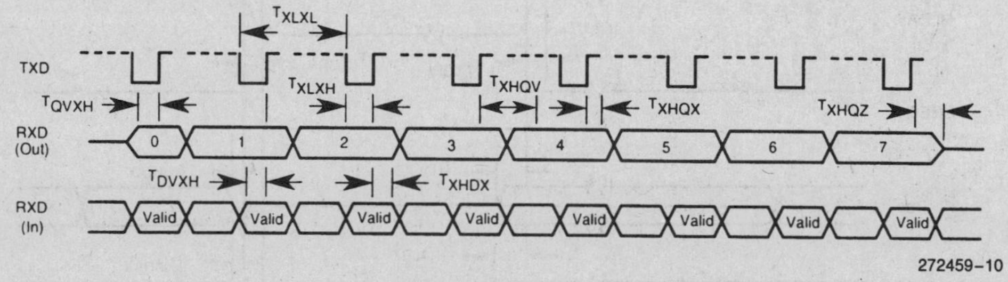

Figure 10. Serial Port Waveform—Shift Register Mode

External Clock Drive

Table 14. External Clock Drive

Symbol	Parameter	Min	Max	Units
$1/T_{XLXL}$	Oscillator Frequency	8	16/25	MHz
T_{XLXL}	Oscillator Period (T_{OSC})	62.5/40.0	125	ns
T_{XHXX}	High Time	0.35 T_{OSC}	0.65 T_{OSC}	ns
T_{XLXX}	Low Time	0.35 T_{OSC}	0.65 T_{OSC}	ns
T_{XLXH}	Rise Time		10	ns
T_{XHXL}	Fall Time		10	ns

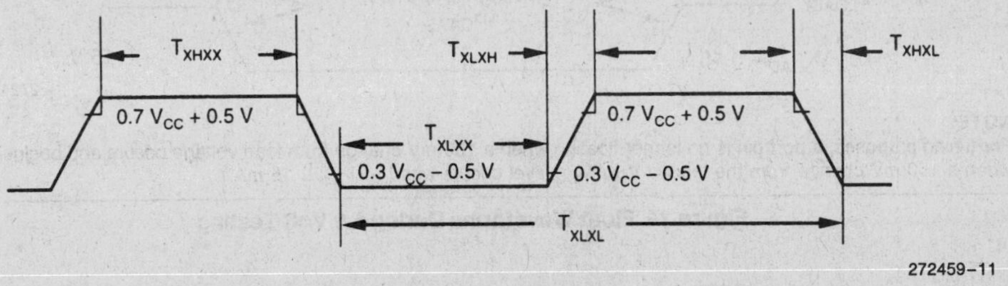

Figure 11. External Clock Drive Waveforms

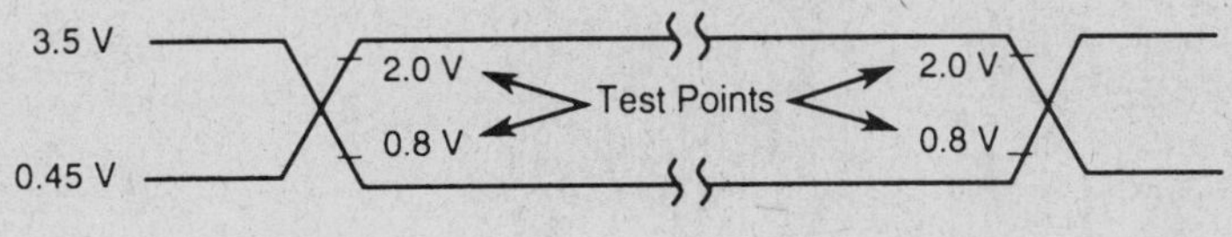

NOTE:
AC testing inputs are driven at 3.5V for a logic "1" and 0.45V for a logic "0". Timing measurements are made at 2.0V for a logic "1" and 0.8V for a logic "0".

Figure 12. AC Testing Output Waveforms During 5.0 Volt Testing

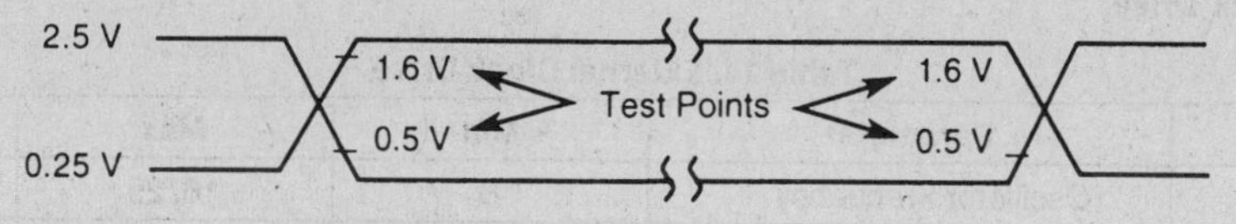

NOTE:
AC testing inputs are driven at 2.5V for a logic "1" and 0.25V for a logic "0". Timing measurements are made at 1.6V for a logic "1" and 0.5V for a logic "0".

Figure 13. AC Testing Output Waveforms During 3.3 Volt Testing

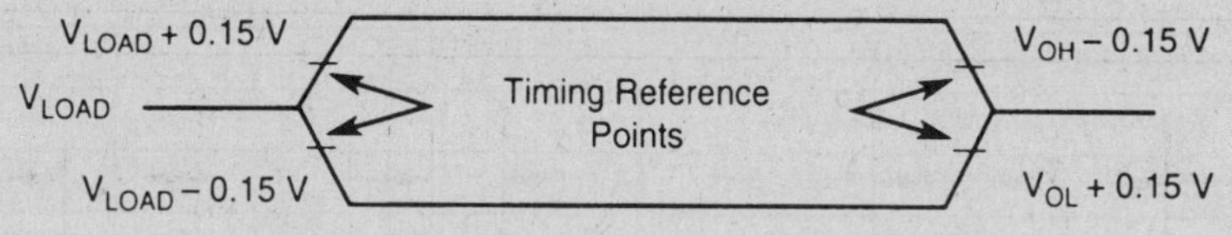

NOTE:
For timing purposes, a port pin is no longer floating when a 150 mV change from load voltage occurs and begins to float when a 150 mV change from the loaded V_{OH}/V_{OL} level occurs with $I_{OL}/I_{OH} \leq 15$ mA.

Figure 14. Float Waveforms During 5.0 Volt Testing

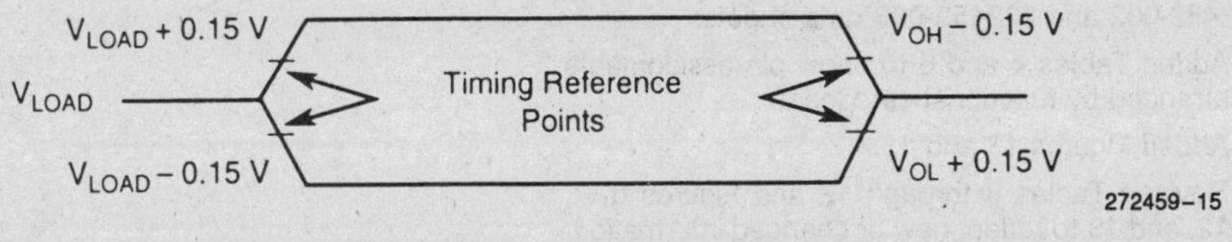

NOTE:
For timing purposes, a port pin is no longer floating when a 150 mV change from load voltage occurs and begins to float when a 150 mV change from the loaded V_{OH}/V_{OL} level occurs with $I_{OL}/I_{OH} \leq$ 10 mA.

Figure 15. Float Waveforms During 3.3 Volt Testing

8XC196NP ERRATA

Change identifiers have been used on embedded products since 1990. The change identifier is the last character in the FPO number. The FPO number is typically a nine character number located on the second line of the topside package mark. Below is listed the errata applicable to the A-step (denoted by an "A" at the end of the topside tracking number).

1. The illegal opcode interrupt vector is not taken when an illegal opcode is encountered. A branch to an unknown location occurs.
2. (1-Mbyte mode only.) If an interrupt is aborted, intentionally or unintentionally, an undesired branch to the lowest priority interrupt vector (FF2000H) may occur even if the lowest priority interrupt is not enabled. This may occur if any bit in the INT_MASK, INT_MASK1, INT_PEND, or INT_PEND1 register is cleared after the corresponding INT_PEND or INT_PEND1 bit is set.
3. (1-Mbyte mode only.) If a standard interrupt occurs at approximately the same time (this time is code dependent and therefore cannot be stated as an exact number of state times) as a PTS serviced interrupt, the PTS interrupt may be processed as a standard interrupt. The standard interrupt service routine for a PTS serviced interrupt (End-of-PTS) is typically used to modify the PTS control block and re-enable the PTS by setting the corresponding bit in the PTSSEL register. When this anomaly occurs, the End-of-PTS service routine will execute regardless of the value in PTSCOUNT. As a result, an undetermined number of PTS cycles will not occur. This applies to all PTS interrupts.

DATA SHEET REVISION HISTORY

This data sheet is valid for devices with an "A" at the end of the topside tracking number. Data sheets are changed as new device information becomes available. Verify with your local Intel sales office that you have the latest version before finalizing a design or ordering devices.

The following are important differences between the 272459-001 and 272459-002 data sheets:

1. Added 25 MHz specification data.
2. Added pin connections for "no connection" (NC) pins for future compatibility.

The following are important differences between the 272459-002 and 272459-003 data sheets:

1. Added Tables 4 and 6 to show pin assignments arranged by functional categories.
2. Added Figures 13 and 15.
3. Revised Tables 8 through 12 and Figures 6, 7, 12, and 13 to reflect new or changed information.

8XC196MC INDUSTRIAL MOTOR CONTROL MICROCONTROLLER

87C196MC 16 Kbytes of On-Chip OTPROM*
87C196MC, ROM 16 Kbytes of On-Chip Factory-Programmed OTPROM
80C196MC ROMless

- **High-Performance CHMOS 16-Bit CPU**
- **16 Kbytes of On-Chip OTPROM/ Factory-Programmed OTPROM**
- **488 bytes of On-Chip Register RAM**
- **Register to Register Architecture**
- **Up to 53 I/O Lines**
- **Peripheral Transaction Server (PTS) with 11 Prioritized Sources**
- **Event Processor Array (EPA)**
  - **4 High Speed Capture/Compare Modules**
  - **4 High Speed Compare Modules**
- **Extended Temperature Standard**
- **Two 16-Bit Timers with Quadrature Decoder Input**
- **3-Phase Complementary Waveform Generator**
- **13 Channel 8/10-Bit A/D with Sample/ Hold with Zero Offset Adjustment H/W**
- **14 Prioritized Interrupt Sources**
- **Flexible 8-/16-Bit External Bus**
- **1.75 μs 16 x 16 Multiply**
- **3 μs 32/16 Divide**
- **Idle and Power Down Modes**

4

The 8XC196MC is a 16-bit microcontroller designed primarily to control 3 phase AC induction and DC brushless motors. The 8XC196MC is based on Intel's MCS® 96 16-bit microcontroller architecture and is manufactured with Intel's CHMOS process.

The 8XC196MC has a three phase waveform generator specifically designed for use in "Inverter" motor control applications. This peripheral allows for pulse width modulation, three phase sine wave generation with minimal CPU intervention. It generates 3 complementary non-overlapping PWM pulses with resolutions of 0.125 μs (edge trigger) or 0.250 μs (centered).

The 8XC196MC has 16 Kbytes on-chip OTPROM/ROM and 488 bytes of on-chip RAM. It is available in three packages; PLCC (84-L), SDIP (64-L) and EIAJ/QFP (80-L).

Note that the 64-L SDIP package does not include P1.4, P2.7, P5.1 and the CLKOUT pins.

Operational characteristics are guaranteed over the temperature range of −40°C to +85°C.

The 87C196MC contains 16 Kbytes on-chip OTPROM. The 83C196MC contains 16 Kbytes on-chip ROM. All references to the 80C196MC also refers to the 83C196MC and 87C196MC unless noted.

---

*OTPROM (One Time Programmable Read Only Memory) is the same as EPROM but it comes in an unwindowed package and cannot be erased. It is user programmable.

April 1994
Order Number: 270946-005

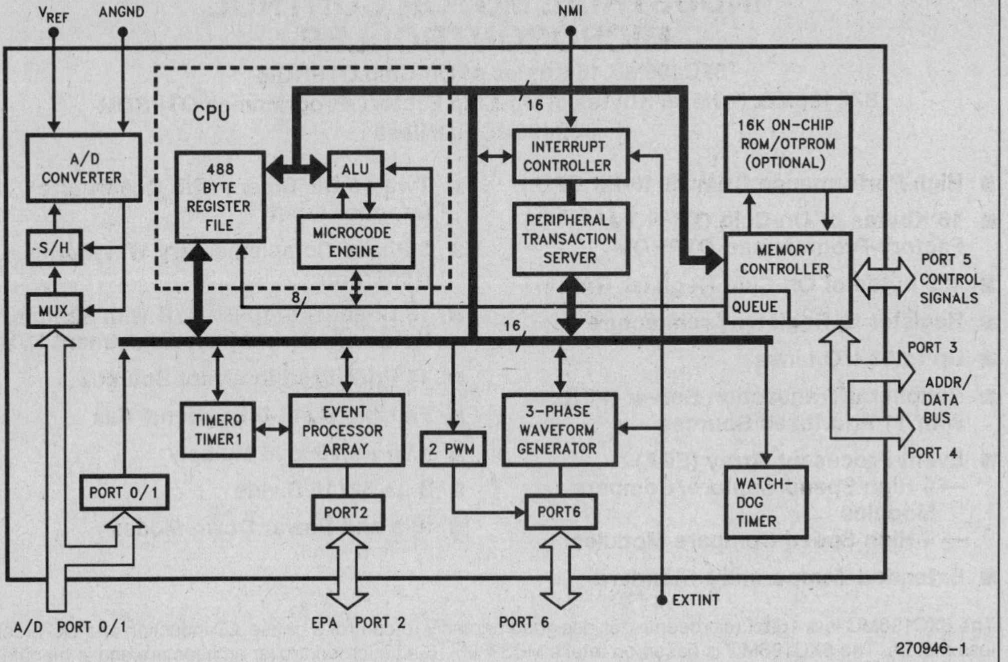

NOTE:
Connections between the standard I/O ports and the bus are not shown.

Figure 1. 87C196MC Block Diagram

PROCESS INFORMATION

This device is manufactured on PX29.5, a CHMOS III-E process. Additional process and reliability information is available in Intel's *Components Quality and Reliability Handbook,* Order Number 210997.

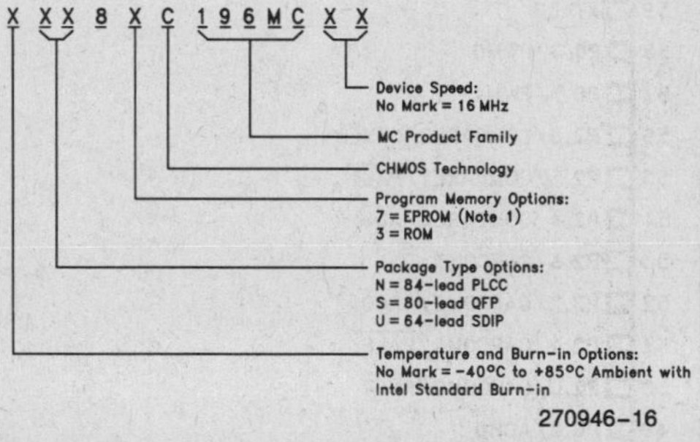

EXAMPLE: N87C196MC is 84-Lead PLCC OTPROM, 16 MHz.
For complete package dimensional data, refer to the Intel Packaging Handbook (Order Number 240800).

NOTE:
1. EPROMs are available as One Time Programmable (OTPROM) only.

Figure 3. The 8XC196MC Family Nomenclature

Thermal Characteristics

Package Type	θ_{ja}	θ_{jc}
PLCC	35°C/W	13°C/W
QFP	56°C/W	12°C/W
SDIP	TBD	TBD

All thermal impedance data is approximate for static air conditions at 1W of power dissipation. Values will change depending on operation conditions and application. See the Intel *Packaging Handbook* (order number 240800) for a description of Intel's thermal impedance test methodology.

8XC196MC Memory Map

Description	Address
External Memory or I/O	0FFFFH 06000H
Internal ROM/EPROM or External Memory (Determined by $\overline{EA}$)	5FFFH 2080H
Reserved. Must contain FFH. (Note 5)	207FH 205EH
PTS Vectors	205DH 2040H
Upper Interrupt Vectors	203FH 2030H
ROM/EPROM Security Key	202FH 2020H
Reserved. Must contain FFH. (Note 5)	201FH 201CH
Reserved. Must Contain 20H (Note 5)	201BH
CCB1	201AH
Reserved. Must Contain 20H (Note 5)	2019H
CCB0	2018H
Reserved. Must contain FFH. (Note 5)	2017H 2014H
Lower Interrupt Vectors	2013H 2000H
SFR's	1FFFH 1F00H
External Memory	1EFFH 0200H
488 Bytes Register RAM (Note 1)	01FFH 0018H
CPU SFR's (Notes 1, 3)	0017H 0000H

NOTES:
1. Code executed in locations 0000H to 03FFH will be forced external.
2. Reserved memory locations must contain 0FFH unless noted.
3. Reserved SFR bit locations must contain 0.
4. Refer to 8XC196KC for SFR descriptions.
5. **WARNING:** Reserved memory locations must not be written or read. The contents and/or function of these locations may change with future revisions of the device. Therefore, a program that relies on one or more of these locations may not function properly.

4

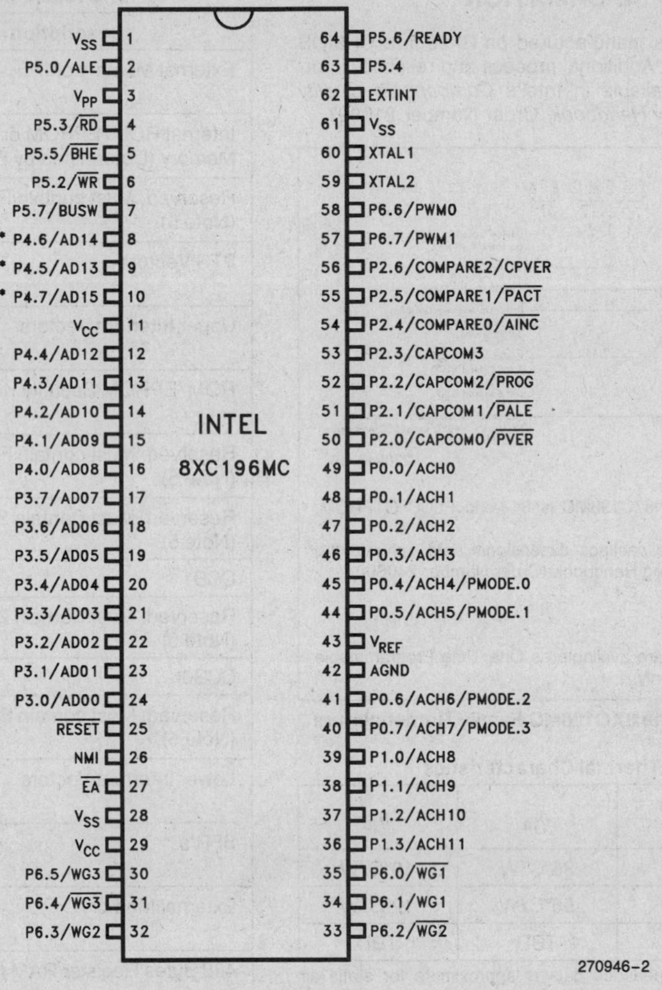

NOTE:
*The pin sequence is correct.
The 64-Lead SDIP package does not include the following pins: P1.4/ACH12, P2.7/COMPARE3, P5.1/INST, CLKOUT.

Figure 2. 64-Lead Shrink DIP (SDIP) Package

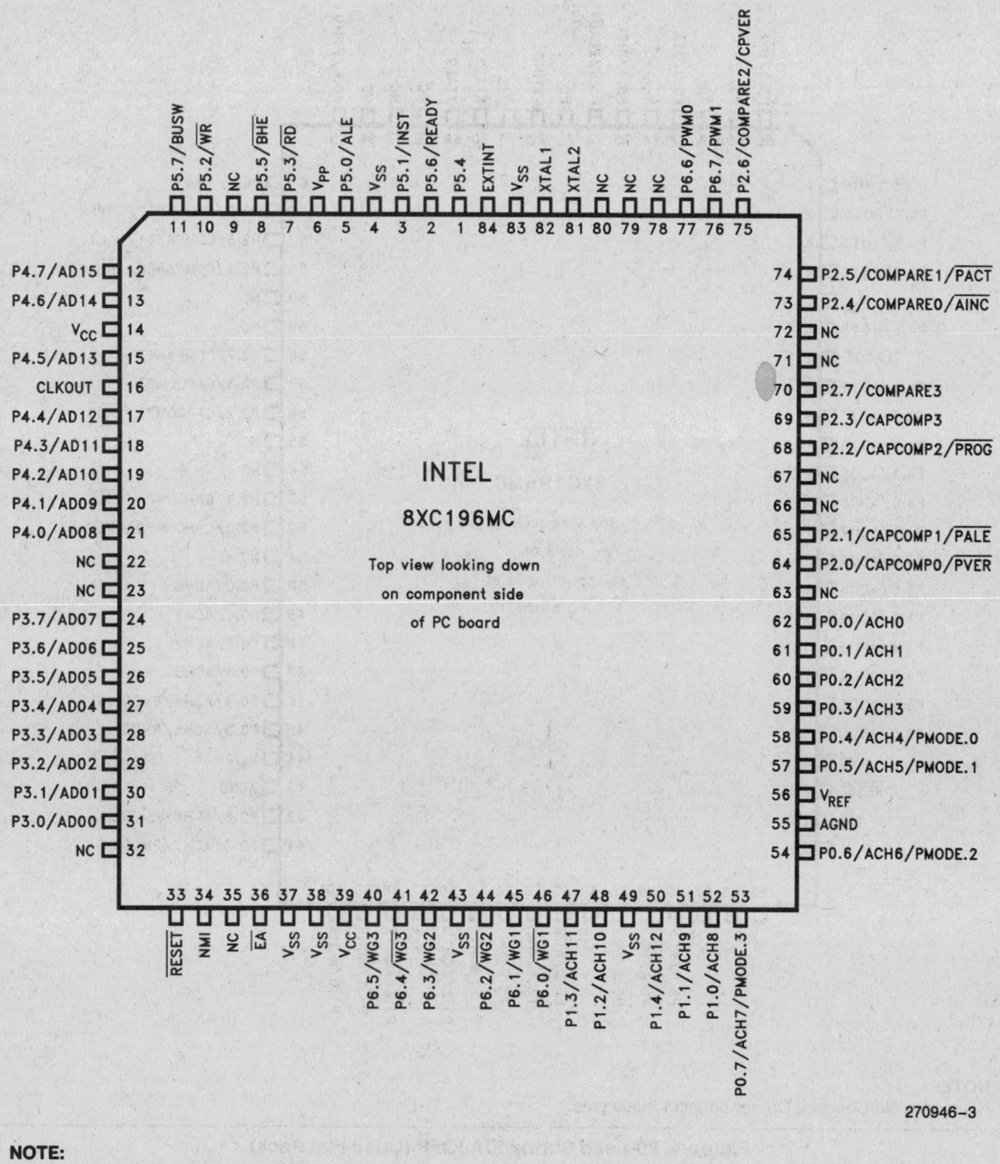

NOTE:
NC means No Connect. Do not connect these pins.

Figure 3. 84-Lead PLCC Package

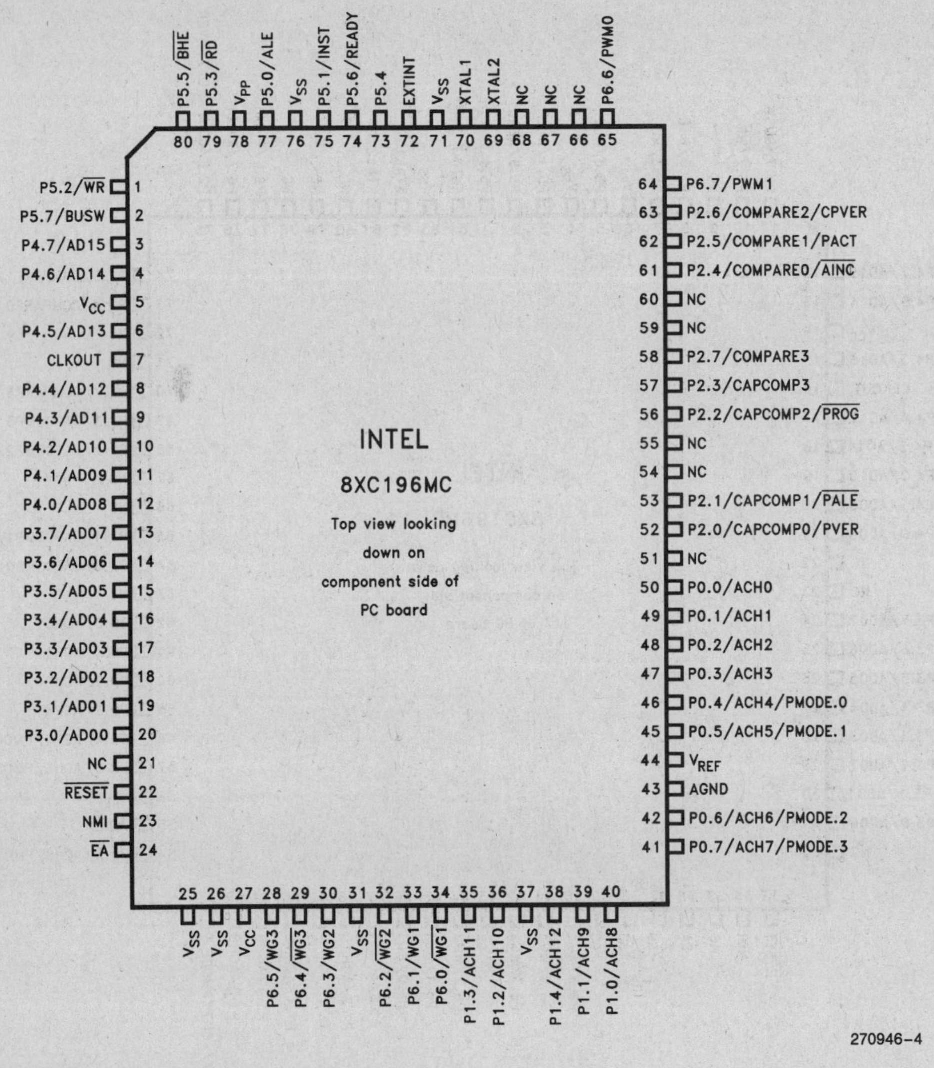

NOTE:
NC means No Connect. Do not connect these pins.

Figure 4. 80-Lead Shrink EIAJQFP (Quad Flat Pack)

PIN DESCRIPTIONS (Alphabetically Ordered)

Symbol	Function
ACH0–ACH12 (P0.0–P0.7, P1.0–P1.4)	Analog inputs to the on-chip A/D converter. ACH0–7 share the input pins with P0.0–7 and ACH8–12 share pins with P1.0–4. If the A/D is not used, the port pins can be used as standard input ports.
ANGND	Reference ground for the A/D converter. Must be held at nominally the same potential as V_{SS}.
ALE/$\overline{ADV}$(P5.0)	Address Latch Enable or Address Valid output, as selected by CCR. Both options allow a latch to demultiplex the address/data bus on the signal's falling edge. When the pin is $\overline{ADV}$, it goes inactive (high) at the end of the bus cycle. ALE/$\overline{ADV}$ is active only during external memory accesses. Can be used as standard I/O when not used as ALE/ADV.
$\overline{BHE}$/$\overline{WRH}$ (P5.5)	Byte High Enable or Write High output, as selected by the CCR. $\overline{BHE}$ will go low for external writes to the high byte of the data bus. $\overline{WRH}$ will go low for external writes where an odd byte is being written. $\overline{BHE}$/$\overline{WRH}$ is activated only during external memory writes.
BUSWIDTH (P5.7)	Input for bus width selection. If CCR bits 1 and 2 = 1, this pin dynamically controls the bus width of the bus cycle in progress. If BUSWIDTH is low, an 8-bit cycle occurs. If it is high, a 16-bit cycle occurs. This pin can be used as standard I/O when not used as BUSWIDTH.
CAPCOMP0–CAPCOMP3 (P2.0–P2.3)	The EPA Capture/Compare pins. These pins share P2.0–P2.3. If not used for the EPA, they can be configured as standard I/O pins.
CLKOUT	Output of the internal clock generator. The frequency is ½ of the oscillator frequency. It has a 50% duty cycle.
COMPARE0–COMPARE3 (P2.4–P2.7)	The EPA Compare pins. These pins share P2.4–P2.7. If not used for the EPA, they can be configured as standard I/O pins.
$\overline{EA}$	External Access enable pin. $\overline{EA}$ = 0 causes all memory accesses to be external to the chip. $\overline{EA}$ = 1 causes memory accesses from location 2000H to 5FFFH to be from the on-chip OTPROM/QROM. $\overline{EA}$ = 12.5V causes execution to begin in the programming mode. $\overline{EA}$ is latched at reset.
EXTINT	A programmable input on this pin causes a maskable interrupt vector through memory location 203CH. The input may be selected to be a positive/negative edge or a high/low level using WG_PROTECT (1FCEH).
INST (P5.1)	INST is high during the instruction fetch from the external memory and throughout the bus cycle. It is low otherwise. This pin can be configured as standard I/O if not used as INST.
NMI	A positive transition on this pin causes a non-maskable interrupt which vectors to memory location 203EH. If not used, it should be tied to V_{SS}. May be used by Intel Evaluation boards.
PORT0	8-bit high impedance input-only port. Also used as A/D converter inputs. Port0 pins should not be left floating. These pins also used to select programming modes in the OTPROM devices.
PORT1	5-bit high impedance input-only port. P1.0–P1.4 are also used as A/D converter inputs. In addition, P1.2 and P1.3 can be used as Timer 1 clock input and direction select respectively.
PORT2	8-bit bidirectional I/O port. All of the Port2 pins are shared with the EPA I/O pins (CAPCOMP0–3 and COMPARE0–3).
PORT3 PORT4	8-bit bidirectional I/O ports with open drain outputs. These pins are shared with the multiplexed address/data bus which uses strong internal pullups.
PORT5	8-bit bidirectional I/O port. 7 of the pins are shared with bus control signals ($\overline{ALE}$, INST, $\overline{WR}$, $\overline{RD}$, $\overline{BHE}$, READY, BUSWIDTH). Can be used as standard I/O.

4

PIN DESCRIPTIONS (Alphabetically Ordered) (Continued)

Symbol	Function
PORT6	8-bit output port. P6.6 and P6.7 output PWM, the others are used as the Wave Form Generator outputs. Can be used as standard output ports.
PWM0, PWM1 (P6.6, P6.7)	Programmable duty cycle, Programmable frequency Pulse Width Modulator pins. The duty cycle has a resolution of 256 steps, and the frequency can vary from 122 Hz to 31 KHz (16 MHz input clock). Pins may be configured as standard output if PWM is not used.
$\overline{RD}$ (P5.3)	Read signal output to external memory. $\overline{RD}$ is low only during external memory reads. Can be used as standard I/O when not used as $\overline{RD}$.
READY (P5.6)	Ready input to lengthen external memory cycles. If READY = 0, the memory controller inserts wait states until the next positive transition of CLKOUT occurs with READY = 1. Can be used as standard I/O when not used as READY.
$\overline{RESET}$	Reset input to and open-drain output from the chip. Held low for at least 16 state times to reset the chip. Input high for normal operation. $\overline{RESET}$ has an Ohmic internal pullup resistor.
T1CLK (P1.2)	Timer 0 Clock input. This pin has two other alternate functions: ACH10 and P1.2.
T1DIR (P1.3)	Timer 0 Direction input. This pin has two other alternate functions: ACH11 and P1.3.
V_{PP}	The programming voltage is applied to this pin. It is also the timing pin for the return from Power Down circuit. Connect this pin with a 1 μF capacitor to V_{SS} and a 1 MΩ resistor to V_{CC}. If the Power Down feature is not used, connect the pin to V_{CC}.
WG1–WG3/$\overline{WG1}$–$\overline{WG3}$ (P6.0–P6.5)	3 phase output signals and their complements used in motor control applications. The pins can also be configured as standard output pins.
$\overline{WR}$/$\overline{WRL}$ (P5.2)	Write and Write Low output to external memory. $\overline{WR}$ will go low every external write. $\overline{WRL}$ will go low only for external writes to an even byte. Can be used as standard I/O when not used as $\overline{WR}$/$\overline{WRL}$.
XTAL1	Input of the oscillator inverter and the internal clock generator. This pin should be used when using an external clock source.
XTAL2	Output of the oscillator inverter.
PMODE (P0.4–7)	Determines the EPROM programming mode.
$\overline{PACT}$ (P2.5)	A low signal in Auto Programming mode indicates that programming is in process. A high signal indicates programming is complete.
$\overline{PALE}$ (P2.1)	A falling edge in Slave Programming Mode and Auto Configuration Byte Programming Mode indicates that ports 3 and 4 contain valid programming address/command information (input to slave).
$\overline{PROG}$ (P2.2)	A falling edge in Slave Programming Mode begins programming. A rising edge ends programming.
PVER (P2.0)	A high signal in Slave Programming Mode and Auto Configuration Byte Programming Mode indicates the byte programmed correctly.
CPVER (P2.6)	Cumulative Program Verification. Pin is high if all locations since entering a programming mode have programmed correctly.
$\overline{AINC}$ (P2.4)	Auto Increment. Active low input enables the auto increment mode. Auto increment will allow reading or writing of sequential EPROM locations without address transactions across the PBUS for each read or write.

ABSOLUTE MAXIMUM RATINGS

Ambient Temperature
Under Bias .... −40°C to +85°C

Storage Temperature .... −65°C to +150°C

Voltage from $\overline{EA}$ or V_{PP}
to V_{SS} or ANGND.................... +13.00V

Voltage on V_{PP} or $\overline{EQ}$
to V_{SS} or ANGND .... −0.5V to 13.0V

Voltage on Any Other Pin
to V_{SS} or ANGND .... −0.5V to +7.0V(1)

Power Dissipation ....1.5W(2)

NOTES:
1. This includes V_{PP} and $\overline{EA}$ on ROM or CPU only devices.
2. Power dissipation is based on package heat transfer limitations, not device power consumption.

NOTICE: This data sheet contains preliminary information on new products in production. The specifications are subject to change without notice. Verify with your local Intel Sales office that you have the latest data sheet before finalizing a design.

**WARNING: Stressing the device beyond the "Absolute Maximum Ratings" may cause permanent damage. These are stress ratings only. Operation beyond the "Operating Conditions" is not recommended and extended exposure beyond the "Operating Conditions" may affect device reliability.*

OPERATING CONDITIONS

Symbol	Description	Min	Max	Units
T_A	Ambient Temperature Under Bias	−40	+85	°C
V_{CC}	Digital Supply Voltage	4.50	5.50	V
V_{REF}	Analog Supply Voltage	4.00	5.50	V
F_{OSC}	Oscillator Frequency	8	16	MHz

NOTE:
ANGND and V_{SS} should be nominally at the same potential. Also V_{SS} and V_{SS1} must be at the same potential.

4

DC ELECTRICAL CHARACTERISTICS (Over Specified Operating Conditions)

Symbol	Parameter	Min	Max	Units	Test Conditions
V_{IL}	Input Low Voltage	−0.5	0.3 V_{CC}	V	
V_{IH}	Input High Voltage	0.7 V_{CC}	V_{CC} + 0.5	V	
V_{OL}	Output Low Voltage Port 2 and 5, P6.6, P6.7, CLKOUT		0.3 0.45 1.5	V V V	I_{OL} = 200 μA I_{OL} = 3.2 mA I_{OL} = 7 mA
V_{OL1}	Output Low Voltage on Port 3/4		1.0	V	I_{OL} = 15 mA
V_{OL2}	Output Low Voltage on Port 6.0–6.5		0.45	V	I_{OL} = 10 mA
V_{OH}	Output High Voltage	V_{CC} − 0.3 V_{CC} − 0.7 V_{CC} − 1.5		V V V	I_{OH} = −200 μA I_{OH} = −3.2 mA I_{OH} = −7 mA
$V_{th+} - V_{th-}$	Hysteresis Voltage Width on $\overline{RESET}$	0.2		V	Typical

DC ELECTRICAL CHARACTERISTICS (Over Specified Operating Conditions) (Continued)

Symbol	Parameter	Min	Typ	Max	Units	Test Conditions
I_{LI}	Input Leakage Current on All Input Only Pins			±10	μA	0V < V_{IN} < V_{CC}−0.3V (in RESET)
I_{LI1}	Input Leakage Current on Port0 and Port1			±3	μA	0V < V_{IN} < V_{REF}
I_{IL}	Input Low Current on BD Ports (Note 1)			−70	μA	V_{IN} = 0.3 V_{CC}
I_{IL1}	Input Low Current on P5.4 and P2.6 during Reset			−7	mA	0.2 V_{CC}
I_{OH}	Output High Current on P5.4 and P2.6 during Reset	−2			mA	0.7 V_{CC}
I_{CC}	Active Mode Current in Reset		50	70	mA	XTAL1 = 16 MHz, V_{CC} = V_{PP} = V_{REF} = 5.5V
I_{REF}	A/D Conversion Reference Current		2	5	mA	
I_{IDL}	Idle Mode Current		15	30	mA	
I_{PD}	Power-Down Mode Current		5	50	μA	V_{CC} = V_{PP} = V_{REF} = 5.5V
R_{RST}	$\overline{\text{RESET}}$ Pin Pullup Resistor	6k		65k	Ω	
C_S	Pin Capacitance (Any Pin to V_{SS})			10	pF	F_{TEST} = 1.0 MHz

NOTES:

1. BD (Bidirectional ports) include:
   P2.0–P2.7, except P2.6
   P3.0–P3.7
   P4.0–P4.7
   P5.0–P5.3
   P5.5–P5.7
2. During normal (non-transient) conditions, the following total current limits apply:

P6.0–P6.5	I_{OL}: 40 mA	I_{OH}: 28 mA
P3	I_{OL}: 90 mA	I_{OH}: 42 mA
P4	I_{OL}: 90 mA	I_{OH}: 42 mA
P5, CLKOUT	I_{OL}: 35 mA	I_{OH}: 35 mA
P2, P6.6, P6.7	I_{OL}: 63 mA	I_{OH}: 63 mA

EXPLANATION OF AC SYMBOLS

Each symbol is two pairs of letters prefixed by "T" for time. The characters in a pair indicate a signal and its condition, respectively. Symbols represent the time between the two signal/condition points.

Conditions:

H — High
L — Low
V — Valid
X — No Longer Valid
Z — Floating

Signals:

A — Address
B — $\overline{BHE}$
C — CLKOUT
D — DATA
G — Buswidth
H — $\overline{HOLD}$
HA — $\overline{HLDA}$
L — ALE/$\overline{ADV}$
BR — $\overline{BREQ}$
R — $\overline{RD}$
W — $\overline{WR}$/$\overline{WRH}$/$\overline{WRL}$
X — XTAL1
Y — READY
Q — Data Out

AC ELECTRICAL CHARACTERISTICS (Over Specified Operating Conditions)

Test Conditions: Capacitive load on all pins = 100 pF, Rise and fall times = 10 ns, F_{OSC} = 16 MHz.

The system must meet the following specifications to work with the 87C196MC:

Symbol	Parameter	Min	Max	Units	Notes
F_{XTAL}	Frequency on XTAL1	8	16	MHz	3
T_{OSC}	1/F_{XTAL}	62.5	125	ns	
T_{AVYV}	Address Valid to READY Setup		$2\ T_{OSC} - 75$	ns	
T_{LLYV}	ALE Low to READY Setup		$T_{OSC} - 70$	ns	4
T_{YLYH}	Not READY Time	No Upper Limit		ns	
T_{CLYX}	READY Hold after CLKOUT Low	0	$T_{OSC} - 30$	ns	1
T_{LLYX}	READY Hold after ALE Low	$T_{OSC} - 15$	$2\ T_{OSC} - 40$	ns	1
T_{AVGV}	Address Valid to BUSWIDTH Setup		$2\ T_{OSC} - 75$	ns	
T_{LLGV}	ALE Low to BUSWIDTH Setup		$T_{OSC} - 60$	ns	4
T_{CLGX}	Buswidth Hold after CLKOUT Low	0		ns	
T_{AVDV}	Address Valid to Input Data Valid		$3\ T_{OSC} - 55$	ns	2
T_{RLDV}	$\overline{RD}$ Active to Input Data Valid		$T_{OSC} - 22$	ns	2
T_{CLDV}	CLKOUT Low to Input Data Valid		$T_{OSC} - 50$	ns	
T_{RHDZ}	End of $\overline{RD}$ to Input Data Float		T_{OSC}	ns	
T_{RXDX}	Data Hold after $\overline{RD}$ Inactive	0		ns	

NOTES:
1. If Max is exceeded, additional wait states will occur.
2. If wait states are used, add 2 T_{OSC} * N, where N = number of wait states.
3. Testing performed at 8 MHz. However, the device is static by design and will typically operate below 1 Hz.
4. These timings are included for compatibility with older −90 and BH products. They should not be used for newer high-speed designs.

AC ELECTRICAL CHARACTERISTICS (Continued)

Test Conditions: Capacitive load on all pins = 100 pF, Rise and fall times = 10 ns, F_{OSC} = 16 MHz.

The 87C196MC will meet the following timing specifications:

Symbol	Parameter	Min	Max	Units	Notes
T_{XHCH}	XTAL1 to CLKOUT High or Low	30	110	ns	
T_{CLCL}	CLKOUT Cycle Time	2 T_{OSC}		ns	
T_{CHCL}	CLKOUT High Period	T_{OSC} − 10	T_{OSC} + 15	ns	
T_{CLLH}	CLKOUT Falling Edge to ALE Rising	−5	15	ns	
T_{LLCH}	ALE Falling Edge to CLKOUT Rising	−20	15	ns	
T_{LHLH}	ALE Cycle Time	4 T_{OSC}		ns	3
T_{LHLL}	ALE High Period	T_{OSC} − 10	T_{OSC} + 10	ns	
T_{AVLL}	Address Setup to ALE Falling Edge	T_{OSC} − 15		ns	
T_{LLAX}	Address Hold after ALE Falling	T_{OSC} − 40		ns	
T_{LLRL}	ALE Falling Edge to $\overline{RD}$ Falling	T_{OSC} − 30		ns	
T_{RLCL}	$\overline{RD}$ Low to CLKOUT Falling Edge	4	30	ns	
T_{RLRH}	$\overline{RD}$ Low Period	T_{OSC} − 5	T_{OSC} + 25	ns	3
T_{RHLH}	$\overline{RD}$ Rising Edge to ALE Rising Edge	T_{OSC}	T_{OSC} + 25	ns	1
T_{RLAZ}	$\overline{RD}$ Low to Address Float		5	ns	
T_{LLWL}	ALE Falling Edge to $\overline{WR}$ Falling	T_{OSC} − 10		ns	
T_{CLWL}	CLKOUT Low to $\overline{WR}$ Falling Edge	0	25	ns	
T_{QVWH}	Data Stable to $\overline{WR}$ Rising Edge	T_{OSC} − 23		ns	
T_{CHWH}	CLKOUT High to $\overline{WR}$ Rising Edge	−10	15	ns	
T_{WLWH}	$\overline{WR}$ Low Period	T_{OSC} − 30		ns	3
T_{WHQX}	Data Hold after $\overline{WR}$ Rising Edge	T_{OSC} − 25		ns	
T_{WHLH}	$\overline{WR}$ Rising Edge to ALE Rising Edge	T_{OSC} − 10	T_{OSC} + 15	ns	1
T_{WHBX}	$\overline{BHE}$, INST Hold after $\overline{WR}$ Rising	T_{OSC} − 10		ns	
T_{WHAX}	AD8–15 Hold after $\overline{WR}$ Rising	T_{OSC} − 30		ns	2
T_{RHBX}	$\overline{BHE}$, INST Hold after $\overline{RD}$ Rising	T_{OSC} − 10		ns	
T_{RHAX}	AD8–15 Hold after $\overline{RD}$ Rising	T_{OSC} − 30		ns	2

NOTES:
1. Assuming back to back cycles.
2. 8-bit bus only.
3. If wait states are used, add 2 T_{OSC}*N, where N = number of wait states.

SYSTEM BUS TIMINGS

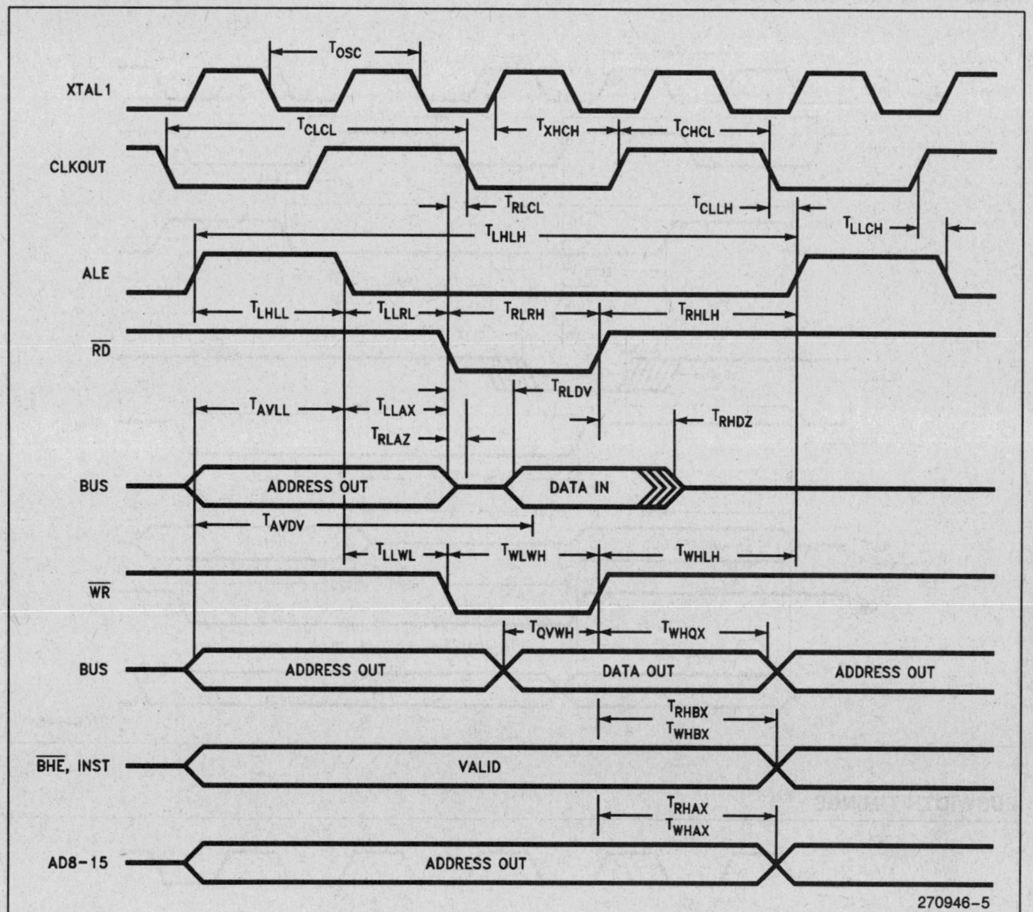

270946-5

READY TIMINGS (One Wait State)

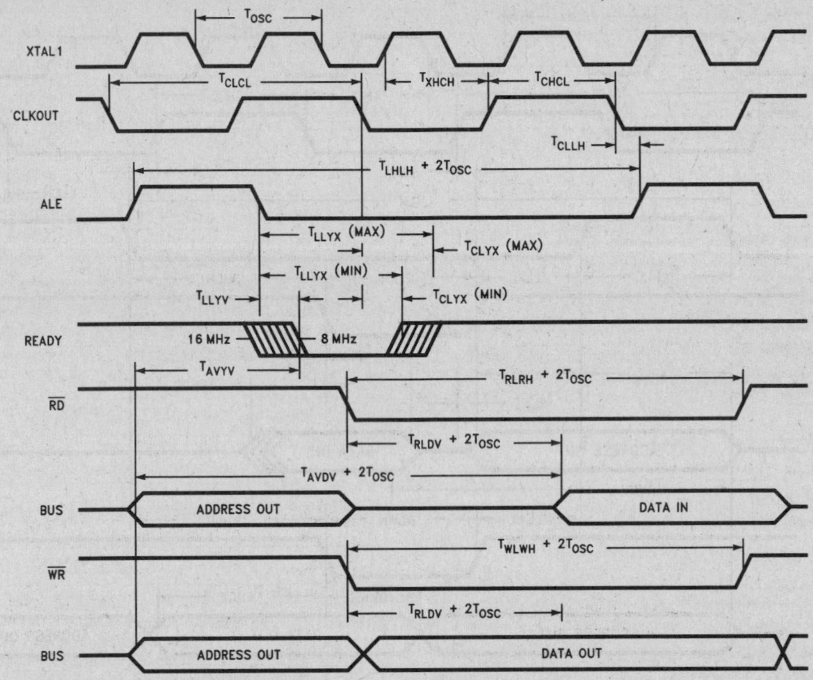

270946-6

BUSWIDTH TIMINGS

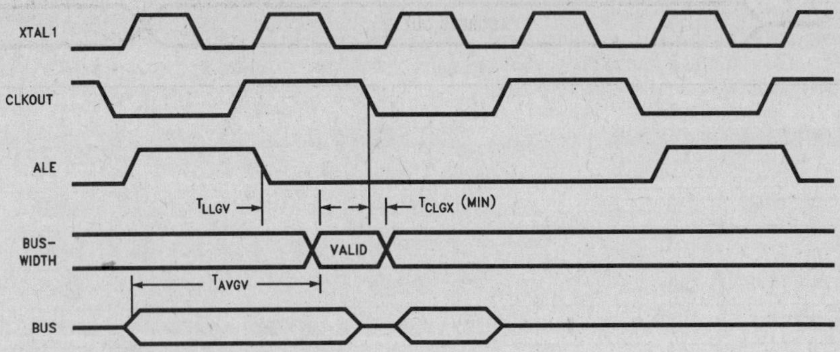

270946-7

PRELIMINARY

EXTERNAL CLOCK DRIVE

Symbol	Parameter	Min	Max	Units
$1/T_{XLXL}$	Oscillator Frequency	8	16.0	MHz
T_{XLXL}	Oscillator Period	62.5	125	ns
T_{XHXX}	High Time	22		ns
T_{XLXX}	Low Time	22		ns
T_{XLXH}	Rise Time		10	ns
T_{XHXL}	Fall Time		10	ns

EXTERNAL CRYSTAL CONNECTIONS

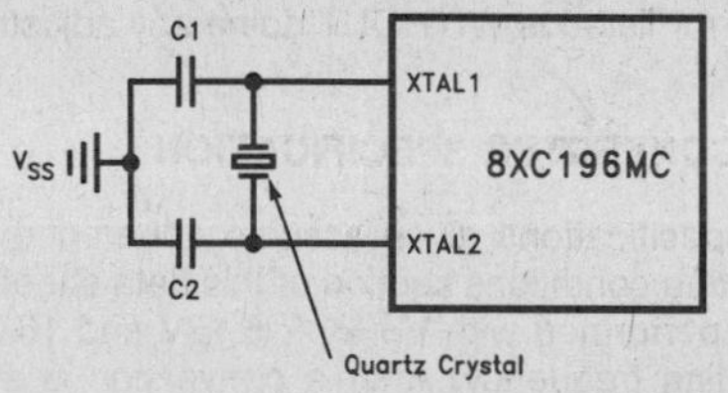

270946-14

NOTE:
Keep oscillator components close to chip and use short, direct traces to XTAL1, XTAL2 and V_{SS}. When using crystals, C1 = 20 pF, C2 = 20 pF. When using ceramic resonators, consult manufacturer for recommended circuitry.

EXTERNAL CLOCK CONNECTIONS

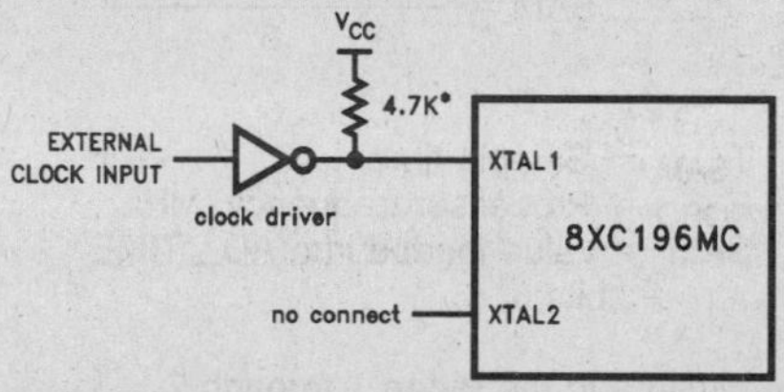

270946-15

* Required if TTL driver used.
Not needed if CMOS driver is used.

EXTERNAL CLOCK DRIVE WAVEFORMS

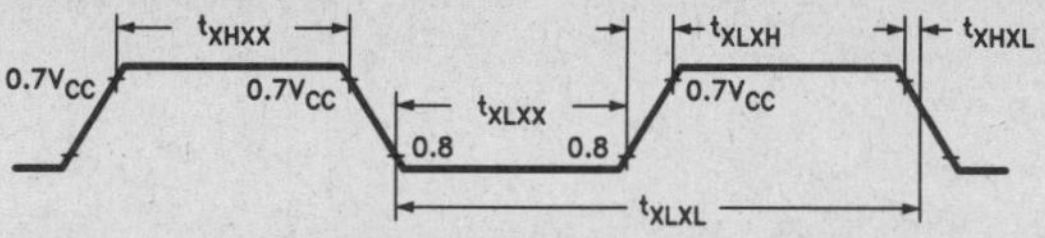

270946-8

An external oscillator may encounter as much as a 100 pF load at XTAL1 when it starts-up. This is due to interaction between the amplifier and its feedback capacitance. Once the external signal meets the V_{IL} and V_{IH} specifications the capacitance will not exceed 20 pF.

AC TESTING INPUT, OUTPUT WAVEFORMS

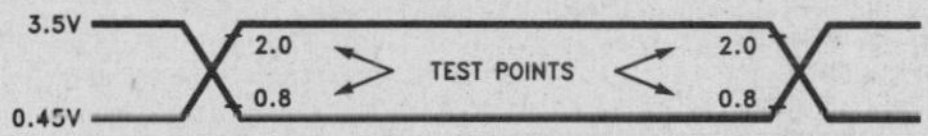

270946-9

AC Testing inputs are driven at 3.5V for a Logic "1" and 0.45V for a Logic "0". Timing measurements are made at 2.0V for a Logic "1" and 0.8V for a Logic "0".

FLOAT WAVEFORMS

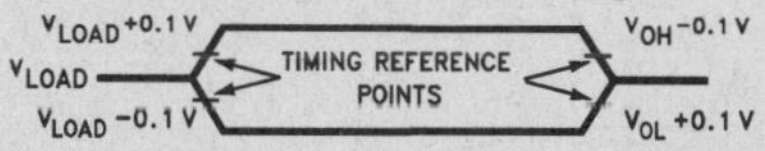

270946-10

For Timing Purposes a Port Pin is no Longer Floating when a 100 mV change from Load Voltage Occurs and Begins to Float when a 100 mV change from the Loaded V_{OH}/V_{OL} Level occurs $I_{OL}/I_{OH} = \leq \pm 15$ mA.

A TO D CHARACTERISTICS

The sample and conversion time of the A/D converter in the 8-bit or 10-bit modes is programmed by loading a byte into the AD_TIME Special Function Register. This allows optimizing the A/D operation for specific applications. The AD_TIME register is functional for all possible values, but the accuracy of the A/D converter is only guaranteed for the times specificed in the operating conditions table.

The value loaded into AD_TIME bits 5, 6, 7 determines the sample time, T_{SAM}, and is calculated using the following formula:

$$SAM = \frac{(T_{SAM} \times F_{OSC}) - 2}{8}$$

T_{SAM} = Sample time, µs
F_{OSC} = Processor frequency, MHz
SAM = Value loaded into AD_TIME bits 5, 6, 7

SAM must be in the range 1 through 7.

The value loaded into AD_TIME bits 0–5 determines the conversion time, T_{CONV}, and is calculated using the following formula:

$$CONV = \frac{(T_{CONV} \times F_{OSC}) - 3}{2B} - 1$$

T_{CONV} = Conversion time, µs
F_{OSC} = Processor frequency, MHz
B = 8 for 8-bit conversion
B = 10 for 10-bit conversion
CONV = Value loaded into AD_TIME bits 0–5

CONV must be in the range 2 through 31.

The converter is ratiometric, so absolute accuracy is dependent on the accuracy and stability of V_{REF}. V_{REF} must be close to V_{CC} since it supplies both the resistor ladder and the analog portion of the converter and input port pins. There is also an AD_TEST SFR that allows for conversion on ANGND and V_{REF} as well as adjusting the zero offset. The absolute error listed is WITHOUT doing any adjustments.

A/D CONVERTER SPECIFICATION

The specifications given assume adherence to the operating conditions section of this data sheet. Testing is performed with V_{REF} = 5.12V and 16.0 MHz operating frequency. After a conversion is started, the device is placed in the IDLE mode until the conversion is complete.

10-BIT MODE A/D OPERATING CONDITIONS

Symbol	Description	Min	Max	Units
T_A	Ambient Temperature	−40	+85	°C
V_{CC}	Digital Supply Voltage	4.50	5.50	V
V_{REF}	Analog Supply Voltage	4.00	5.50	V(1)
T_{SAM}	Sample Time	1.0		μs(2)
T_{CONV}	Conversion Time	10.0	20.0	μs(2)
F_{OSC}	Oscillator Frequency	8.0	16.0	MHz

NOTES:
ANGND and V_{SS} should nominally be at the same potential.
1. V_{REF} must be within 0.5V of V_{CC}.
2. The value of AD_TIME is selected to meet these specifications.

10-BIT MODE A/D CHARACTERISTICS (Over Specified Operating Conditions)

Parameter	Typical(1)	Min	Max	Units*
Resolution		1024 10	1024 10	Levels Bits
Absolute Error		0	±4	LSBs
Full Scale Error	0.25 ±0.5			LSBs
Zero Offset Error	0.25 ±0.5			LSBs
Non-Linearity	1.0 ±2.0		±4	LSBs
Differential Non-Linearity		>−1	+2	LSBs
Channel-to-Channel Matching	±0.1	0	±1.0	LSBs
Repeatability	±0.25	0		LSBs
Temperature Coefficients: Offset Full Scale Differential Non-Linearity	 0.009 0.009 0.009			 LSB/C LSB/C LSB/C
Off Isolation		−60		dB(2, 3)
Feedthrough	−60			dB(2)
V_{CC} Power Supply Rejection	−60			dB(2)
Input Series Resistance		750	2K	Ω(4)
Voltage on Analog Input Pin		ANGND − 0.5	V_{REF} + 0.5	V(5, 6)
Sampling Capacitor	3			pF
DC Input Leakage	±1	0	±3.0	μA

NOTES:
*An "LSB", as used here has a value of approximately 5 mV. (See Embedded Microcontrollers and Processors Handbook for A/D glossary of terms).
1. These values are expected for most parts at 25°C but are not tested or guaranteed.
2. DC to 100 KHz.
3. Multiplexer Break-Before-Make is guaranteed.
4. Resistance from device pin, through internal MUX, to sample capacitor.
5. These values may be exceeded if the pin current is limited to ±2 mA.
6. Applying voltages beyond these specifications will degrade the accuracy of other channels being converted.
7. All conversions performed with processor in IDLE mode.

8-BIT MODE A/D OPERATING CONDITIONS

Symbol	Description	Min	Max	Units
T_A	Ambient Temperature	−40	+85	°C
V_{CC}	Digital Supply Voltage	4.50	5.50	V
V_{REF}	Analog Supply Voltage	4.00	5.50	V(1)
T_{SAM}	Sample Time	1.0		μs(2)
T_{CONV}	Conversion Time	7.0	20.0	μs(2)
F_{OSC}	Oscillator Frequency	8.0	16.0	MHz

NOTES:
ANGND and V_{SS} should nominally be at the same potential.
1. V_{REF} must be within 0.5V of V_{CC}.
2. The value of AD_TIME is selected to meet these specifications.

8-BIT MODE A/D CHARACTERISTICS (Over the Above Operating Conditions)

Parameter	Typical(1)	Min	Max	Units*
Resolution		256 8	256 8	Level Bits
Absolute Error		0	±1	LSBs
Full Scale Error	±0.5			LSBs
Zero Offset Error	±0.5			LSBs
Non-Linearity		0	±1	LSBs
Differential Non-Linearity		> −1	+1	LSBs
Channel-to-Channel Matching		0	±1.0	LSBs
Repeatability	±0.25			LSBs
Temperature Coefficients: Offset Full Scale Differential Non-Linearity	 0.003 0.003 0.003			 LSB/C LSB/C LSB/C
Off Isolation		−60		dB(2, 3)
Feedthrough	−60			dB(2)
V_{CC} Power Supply Rejection	−60			dB(2)
Input Series Resistance		750	2K	Ω(4)
Voltage on Analog Input Pin		V_{SS} − 0.5	V_{REF} + 0.5	V(5, 6)
Sampling Capacitor	3			pF
DC Input Leakage	±1	0	±3.0	μA

NOTES:
*An "LSB" as used here, has a value of approximately 20 mV. (See Embedded Microcontrollers and Processors Handbook for A/D glossary of terms).
1. These values are expected for most parts at 25°C but are not tested or guaranteed.
2. DC to 100 KHz.
3. Multiplexer Break-Before-Make is guaranteed.
4. Resistance from device pin, through internal MUX, to sample capacitor.
5. These values may be exceeded if the pin current is limited to ±2 mA.
6. Applying voltages beyond these specifications will degrade the accuracy of other channels being converted.
7. All conversions performed with processor in IDLE mode.

EPROM SPECIFICATIONS

OPERATING CONDITIONS

Symbol	Description	Min	Max	Units
T_A	Ambient Temperature during Programming	20	30	°C
V_{CC}	Supply Voltage during Programming	4.5	5.5	V(1)
V_{REF}	Reference Supply Voltage during Programming	4.5	5.5	V(1)
V_{PP}	Programming Voltage	12.25	12.75	V(2)
V_{EA}	EA Pin Voltage	12.25	12.75	V(2)
F_{OSC}	Oscillator Frequency during Auto and Slave Mode Programming	6.0	8.0	MHz
T_{OSC}	Oscillator Frequency during Run-Time Programming	6.0	12.0	MHz

NOTES:
1. V_{CC} and V_{REF} should nominally be at the same voltage during programming.
2. V_{PP} and V_{EA} must never exceed the maximum specification, or the device may be damaged.
3. V_{SS} and ANGND should nominally be at the same potential (0V).
4. Load capacitance during Auto and Slave Mode programming = 150 pF.

AC EPROM PROGRAMMING CHARACTERISTICS

Symbol	Parameter	Min	Max	Units
T_{SHLL}	Reset High to First $\overline{PALE}$ Low	1100		T_{OSC}
T_{LLLH}	$\overline{PALE}$ Pulse Width	50		T_{OSC}
T_{AVLL}	Address Setup Time	0		T_{OSC}
T_{LLAX}	Address Hold Time	100		T_{OSC}
T_{PLDV}	$\overline{PROG}$ Low to Word Dump Valid		50	T_{OSC}
T_{PHDX}	Word Dump Data Hold		50	T_{OSC}
T_{DVPL}	Data Setup Time	0		T_{OSC}
T_{PLDX}	Data Hold Time	400		T_{OSC}
T_{PLPH}(1)	$\overline{PROG}$ Pulse Width	50		T_{OSC}
T_{PHLL}	$\overline{PROG}$ High to Next $\overline{PALE}$ Low	220		T_{OSC}
T_{LHPL}	$\overline{PALE}$ High to $\overline{PROG}$ Low	220		T_{OSC}
T_{PHPL}	$\overline{PROG}$ High to Next $\overline{PROG}$ Low	220		T_{OSC}
T_{PHIL}	PROG High to AINC Low	0		T_{OSC}
T_{ILIH}	$\overline{AINC}$ Pulse Width	240		T_{OSC}
T_{ILVH}	PVER Hold after $\overline{AINC}$ Low	50		T_{OSC}
T_{ILPL}	$\overline{AINC}$ Low to $\overline{PROG}$ Low	170		T_{OSC}
T_{PHVL}	$\overline{PROG}$ High to $\overline{PVER}$ Valid		220	T_{OSC}

NOTE:
1. This specification is for the Word Dump Mode. For programming pulses, use the Modified Quick Pulse Algorithm.

DC EPROM PROGRAMMING CHARACTERISTICS

Symbol	Parameter	Min	Max	Units
I_{PP}	V_{PP} Supply Current (When Programming)		100	mA

NOTE:
Do not apply V_{PP} until V_{CC} is stable and within specifications and the oscillator/clock has stabilized or the device may be damaged.

SLAVE PROGRAMMING MODE DATA PROGRAM MODE WITH SINGLE PROGRAM PULSE

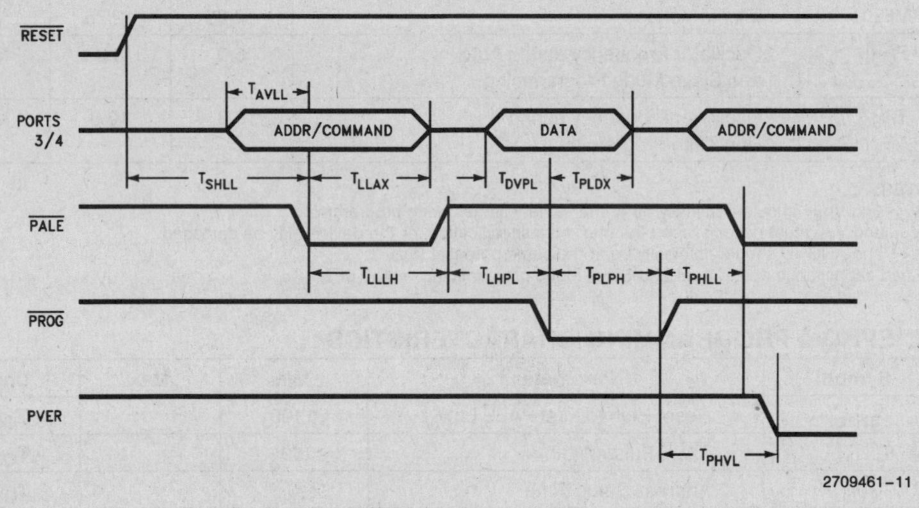

NOTE:
P3.0 must be high ("1")

SLAVE PROGRAMMING MODE IN WORD DUMP WITH AUTO INCREMENT

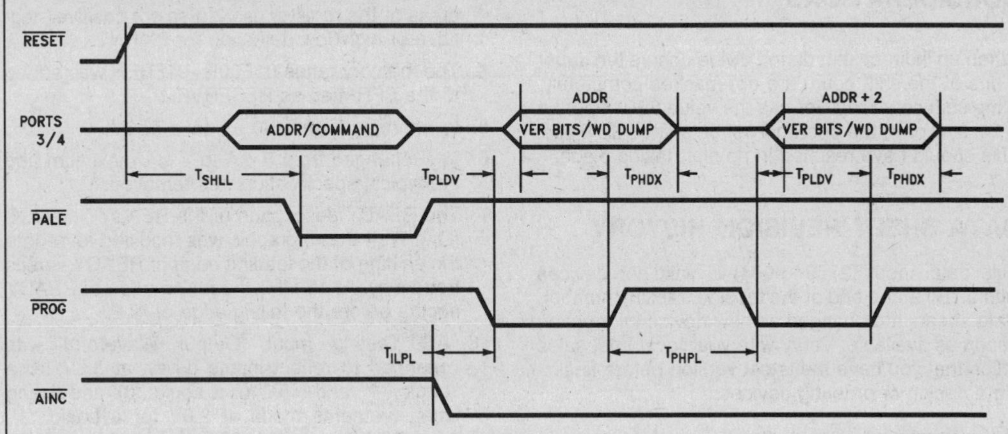

270946–12

NOTE:
P3.0 must be low ("0")

SLAVE PROGRAMMING MODE TIMING IN DATA PROGRAM WITH REPEATED PROG PULSE AND AUTO INCREMENT

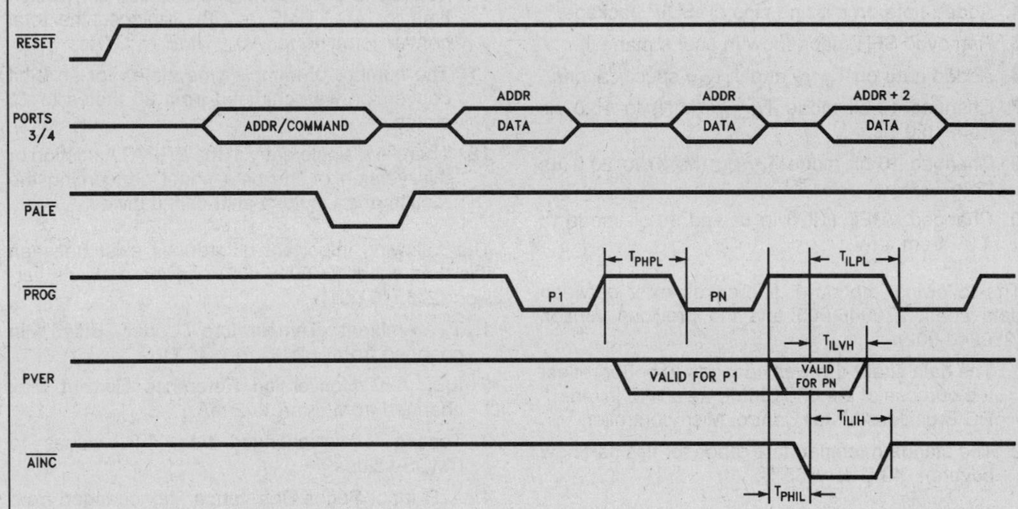

270946–13

87C196MC DESIGN CONSIDERATIONS

When an indirect shift during divide occurs the upper 3 bits of the shift count are not masked completely. If the shift count register has the value 32∗n where n = 1, 3, 5 or 7, the operand will be shifted 32 times. This should have resulted in no shift taking place.

DATA SHEET REVISION HISTORY

This data sheet (270946-004) is valid for devices with a "B" at the end of the topside tracking number. Data sheets are changed as new device information becomes available. Verify with your local Intel sales office that you have the latest version before finalizing a design or ordering devices.

The following important differences exist between this data sheet (270946-002) and the previous version (270946-003):

1. The data sheet was reorganized to standard format.
2. Added 83C196MC device.
3. Added package thermal characteristics.
4. Added note on missing pins on SDIP package.
5. Removed SFR maps (now in user's manual).
6. Added note on T_{LLYV} and T_{LLGV} specifications.
7. Changed 10-bit mode T_{CONV} (MIN) to 10.0 μs from 15.0 μs.
8. Changed 10-bit mode T_{CONV} (MAX) to 20.0 μs from 18.0 μs.
9. Changed VREF (MIN) in 8- and 10-bit mode to 4.0V from 4.5V.

The following important differences exist between data sheet 270946-003 and the previous version (270946-002):

1. The data sheet title was changed to better reflect the purpose of the 87C196MC as an AC Inverter/DC Brushless Motor Control Microcontroller.
2. The standard temperature range for this part now covers −40°C to +85°C.
3. EXTINT function description now includes WG_PROTECT (1FCEH) as the name and address of the register used to select positive/negative or high/low detection for EXTINT.
4. The memory range 01F00H–01FBFH was added to the SFR map as RESERVED.
5. I_{IL} changed from −60 μA to −70 μA.
6. I_{REF} changed from 5 mA to 2 mA maximum and the typical specification was removed.
7. The READY description of the READY TIMINGS (One Wait State) graphic was modified to denote the shifting of the leading edge of READY versus frequency. At 16 MHz the falling edge of READY occurs before the falling edge of ALE.
8. AC Testing Input, Output Waveform was changed to reflect inputs driven at 3.5V for a Logic "1" and .45V for a Logic "0" and timing measurements made at 2.0V for a Logic "1" and 0.8V for a Logic "0".
9. Float Waveform was changed from I_{OL}/I_{OH} = ±15 mA to $I_{OL}/I_{OH} \leq \pm 15$ mA
10. AD_TIME register for 10-bit conversions was changed from 0C7H to 0D8H. The number of sample time states was changed from 24 to 25 states, the conversion time states was changed from 80 to 240 states, and the total conversion time for AD_TIME = D8H replaced the total conversion time for AD_TIME = C7H.
11. The number of sample time states for an 8-bit conversion was changed from 20 states to 21 states.
12. There is a single entry in the ERRATA section of this version of the data sheet concerning the results of an indirect shift during divide.

The following important differences exist between this data sheet (270946-002) and the previous version (270946-001):

1. T_A Ambient Temperature Under Bias Min changed from −20°C to −40°C.
2. I_{REF} A/D Conversion Reference Current Max changed from 5 mA to 2 mA.
3. Testing levels changed from TTL values to CMOS values.
4. A/D Input Series Resistance Max changed from 1.2 KΩ to 2 KΩ.

8XC196MD INDUSTRIAL MOTOR CONTROL MICROCONTROLLER

87C196MD 16 Kbytes of On-Chip OTPROM*
87C196MD, ROM 16 Kbytes of On-Chip Factory-Programmed OTPROM
80C196MD ROMless

- **High-Performance CHMOS 16-Bit CPU**
- **16 Kbytes of On-Chip OTPROM/ Factory-Programmed OTPROM**
- **488 bytes of On-Chip Register RAM**
- **Register to Register Architecture**
- **Up to 64 I/O Lines**
- **Peripheral Transaction Server (PTS) with 17 Prioritized Sources**
- **Event Processor Array (EPA)**
  - **6 High Speed Capture/Compare Modules**
  - **6 High Speed Compare Modules**
- **Extended Temperature Standard**
- **Programmable Frequency Generator**
- **Two 16-Bit Timers with Quadrature Counting Input**
- **3-Phase Complementary Waveform Generator**
- **14 Channel 8/10-Bit A/D with Sample/ Hold with Zero Offset Adjustment H/W**
- **18 Prioritized Interrupt Sources**
- **Flexible 8-/16-Bit External Bus**
- **1.75 μs 16 x 16 Multiply**
- **3 μs 32/16 Divide**
- **Idle and Power Down Modes**

4

The 8XC196MD is a 16-bit microcontroller designed primarily to control 3 phase AC induction and DC brushless motors. The 8XC196MD is based on Intel's MCS® 96 16-bit microcontroller architecture and is manufactured with Intel's CHMOS process.

The 8XC196MD has a three phase waveform generator specifically designed for use in "Inverter" motor control applications. This peripheral allows for pulse width modulation, three phase sine wave generation with minimal CPU intervention. It generates 3 complementary non-overlapping PWM pulses with resolutions of 0.125 μs (edge trigger) or 0.250 μs (centered).

The 8XC196MD has 16 Kbytes on-chip OTPROM/ROM and 488 bytes of on-chip RAM. It is available in two packages; PLCC (84-L) and EIAJ/QFP (80-L).

Operational characteristics are guaranteed over the temperature range of −40°C to +85°C.

The 87C196MD contains 16 Kbytes on-chip OTPROM. The 83C196MD contains 16 Kbytes on-chip ROM. All references to the 80C196MD also refers to the 83C196MD and 87C196MD unless noted.

*OTPROM (One Time Programmable Read Only Memory) is the same as EPROM but it comes in an unwindowed package and cannot be erased. It is user programmable.

April 1994
Order Number: 272323-002

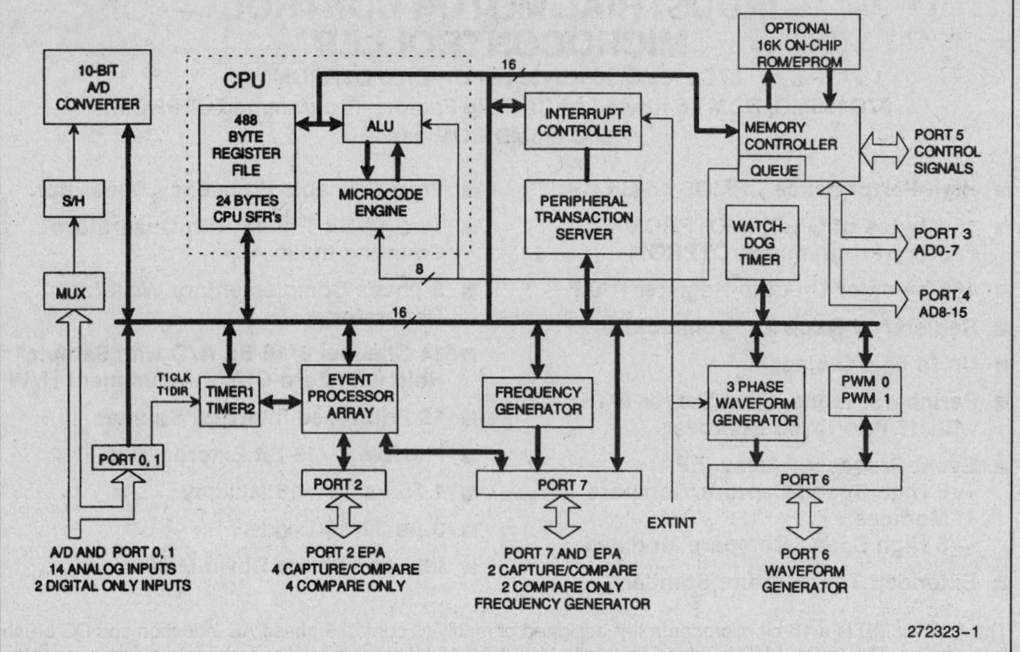

NOTE:
Connections between the standard I/O ports and the bus are not shown.

Figure 1. 87C196MD Block Diagram

PROCESS INFORMATION

This device is manufactured on PX29.5, a CHMOS III-E process. Additional process and reliability information is available in Intel's *Components Quality and Reliability Handbook,* Order Number 210997.

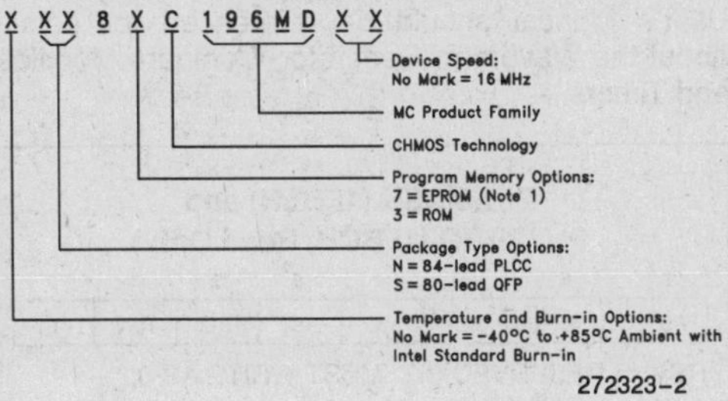

EXAMPLE: N87C196MD is 84-Lead PLCC OTPROM, 16 MHz.
For complete package dimensional data, refer to the Intel Packaging Handbook (Order Number 240800).

NOTE:
EPROMs are available as One Time Programmable (OTPROM) only.

Figure 2. The 8XC196MD Family Nomenclature

Table 1. Thermal Characteristics

Package Type	θ_{ja}	θ_{jc}
PLCC	35°C/W	13°C/W
QFP	56°C/W	12°C/W

All thermal impedance data is approximate for static air conditions at 1W of power dissipation. Values will change depending on operation conditions and application. See the Intel *Packaging Handbook* (order number 240800) for a description of Intel's thermal impedance test methodology.

Table 2. 8XC196MD Memory Map

Description	Address
External Memory or I/O	0FFFFH 06000H
Internal ROM/EPROM or External Memory (Determined by $\overline{EA}$)	5FFFH 2080H
Reserved. Must contain FFH. (Note 5)	207FH 205EH
PTS Vectors	205DH 2040H
Upper Interrupt Vectors	203FH 2030H
ROM/EPROM Security Key	202FH 2020H
Reserved. Must contain FFH. (Note 5)	201FH 201CH
Reserved. Must Contain 20H (Note 5)	201BH
CCB1	201AH
Reserved. Must Contain 20H (Note 5)	2019H
CCB0	2018H
Reserved. Must contain FFH. (Note 5)	2017H 2014H
Lower Interrupt Vectors	2013H 2000H
SFR's	1FFFH 1F00H
External Memory	1EFFH 0200H
488 Bytes Register RAM (Note 1)	01FFH 0018H
CPU SFR's (Notes 1, 3)	0017H 0000H

NOTES:
1. Code executed in locations 0000H to 01FFH will be forced external.
2. Reserved memory locations must contain 0FFH unless noted.
3. Reserved SFR bit locations must contain 0.
4. Refer to 8XC196MC for SFR descriptions.
5. **WARNING:** Reserved memory locations must not be written or read. The contents and/or function of these locations may change with future revisions of the device. Therefore, a program that relies on one or more of these locations may not function properly.

8XC196MC AND 8XC196MD DIFFERENCES

INT_MASK1/INT_PEND1 Registers

There are some differences between the 8XC196MC and 8XC196MD INT_MASK1/ INT_PEND1 registers. The 8XC196MD interrupt mask and pending registers are shown below. Notice that the CAPCOM5, COMP4, and CAPCOM4 bits are reserved bits on the 8XC196MC. The PI bit of the INT_PEND1 register will be set when a Waveform Generator or Compare Module 5 event occurs and the corresponding bit in the PI_MASK register is set. The PI interrupt vector can be taken when the PI bit in the INT_MASK1 register is set. The 8XC196MC User's Manual should be referenced for details about the interrupts.

INT_MASK1 (0031H) and INT_PEND1 (0012H)

7	6	5	4	3	2	1	0
RSV	EXTINT	PI	CAPCOM5*	COMP4*	CAPCOM4*	COMP3	CAPCOM3

RSV = RESERVED BIT. MUST WRITE AS 0
* = THIS BIT RESERVED ON 8XC196MC.

Figure 3. Interrupt Mask and Status Registers

PTSSRV and PTSSEL Register

Similarly, there are differences between 8XC196MC and 8XC196MD PTS registers. The 8XC196MD PTS registers are shown below. Notice the CAPCOM5, COMP4, and CAPCOM4 bits are reserved bits on the 8XC196MC. The PI bit in the PTSSRV will be set when a Waveform Generator or Compare Module 5 end of PTS interrupt occurs and the corresponding bit in the PI_MASK register is set. The PI PTS vector can be used when the PI bit in the PTSSEL register is set. The 8XC196MC User's Manual should be referenced for details about the PTS.

PTSSEL (0004H) and PTSSRV (0006H)

15	14	13	12	11	10	9	8
RSV	EXTINT	PI	CAPCOM5*	COMP4*	CAPCOM4*	COMP3	CAPCOM3

7	6	5	4	3	2	1	0
COMP2	CAPCOM2	COMP1	CAPCOM1	COMP0	CAPCOM0	AD_DONE	TOVF

RSV = RESERVED BIT. MUST WRITE AS 0
* = THIS BIT RESERVED ON 8XC196MC.

Figure 4. PTS Select and Service Registers

PI_MASK and PI_PEND Registers

The PI_MASK/PI_PEND registers contain the bits for the Compare Module 5 (COMP5) Waveform Generator (WG), Timer 1 Overflow (TFI), and Timer 2 Overflow (TF2) mask/status flag. The diagram below shows the registers. Notice that the COMP5 bit is a reserved bit on the 8XC196MC. The 8XC196MC User's Manual should be referenced for details about the Waveform Generator, Compare Modules, and Timers.

PI_MASK (1FBEH) and PI_PEND (1FBCH, Read Only)

7	6	5	4	3	2	1	0
RSV	COMP5*	RSV	WG	RSV	TF2	RSV	TF1

RSV = RESERVED BIT. MUST WRITE AS 0, READ AS 1.
* = THIS BIT RESERVED ON 8XC196MC.

Figure 5. Peripheral Interrupt Mask and Status Registers

The PI bit in the INT_PEND1 register is set if a Waveform Generator event or Compare Module 5 event occurs and the corresponding PI_MASK bit is set. For either of these events to cause an interrupt, the PI bit in the INT_MASK1 register and the corresponding event bit in the PI_MASK register must be set.

Similarly, the TOVF bit in the INT_PEND register is set if Timer 1 or Timer 2 overflow and the corresponding bit in the PI_MASK register is set. For either of these two events to cause an interrupt, the TOVF bit in the INT_MASK register and the corresponding event bit in the PI_MASK must be set.

Upon a PI and/or a TOVF interrupt, it may be necessary to check if the Compare Module 5, the Waveform Generator, Timer 1, or Timer 2 event caused the interrupt. The PI_PEND will give this information. However, it should be noted that reading the PI_PEND register will clear the register. So the individual bits in the PI_PEND register must be read by loading PI_PEND into another "shadow" register, then checking the "shadow" register to see what event occurred.

Table 3. Interrupt Sources, Vectors and Priorities

Interrupt Source	Symbol	Interrupt Service			PTS Service		
		Name	Vector	Priority	Name	Vector	Priority
Capture/Compare5	CAPCOMP5	INT12	2038H	12	PTS12	2058H	27
Compare4	COMP4	INT11	2036H	11	PTS11	2056H	26
Capture/Compare4	CAPCOMP4	INT10	2034H	10	PTS10	2054H	25

Interrupt and PTS Vectors

The 8XC196MD has three new interrupt and PTS vectors which are Capture/Compare5, Compare 4, and Capture/Compare4. Table 3 shows these interrupt vectors and priorities. These are shown as reserved vectors in the 8XC196MC User's Manual.

Frequency Generator

The Frequency Generator (FG) Peripheral which was not available on the 8XC196MC device, is available on the 8XC196MD device. The FG outputs a programmable-frequency 50% duty cycle waveform on the FREQOUT pin (P7.7). There are two 8-bit registers which control the FG peripheral:

— Frequency Generator Control Register (FG_CON) at 1FB8h

— Frequency Generator Period Count Register (FG_COUNT) at 1FBAh.

The FG_CON can be read or written. This register is loaded with a value which determines the number of counts necessary for toggling the output. The following equation should be used to calculate the FG_CON value:

$$\text{FG\_CON value} = \frac{F_{XTAL}}{16 * (\text{FG Frequency})} - 1$$

where FG Frequency is from 4 kHz to 1 MHz.

The FG_COUNT is loaded with the FG_CON register value. The FG_COUNT register is decremented every eighth state time. When it reaches 00h, the FG_COUNT register will send a signal to toggle the output pin and reload the FG_COUNT register with the value in the FG_CON register. The FG_COUNT can only be read, not written.

The FREQOUT pin (P7.7) must be configured for a special function to use it for the Frequency Generator feature.

Port 7

Port 7 is an additional bidirectional port that was not available on the 8XC196MC device. Port 7 can be used as I/O or some of the pins have special functions. The pins are listed below followed by their special functions.

Table 4. Port 7 Special Function Pins

Pin	Special Function
P7.0	CAPCOMP4
P7.1	CAPCOMP5
P7.2	CAPCOMP4
P7.3	CAPCOMP5
P7.4	
P7.5	
P7.6	
P7.7	FREQOUT

The special functions of the pins are selected in the Port 7 SFRs. The Port 2 I/O Port section of the 8XC196MC User's Manual can be referenced when setting up the Port 7 SFRs. Port 7 SFRs are located in the following locations:

Table 5. Port 7 Special Function Registers

SFR	Address
P7_MODE	1FD1h
P7_DIR	1FD3h
P7_REG	1FD5h
P7_PIN	1FD7h

4

Port 1

There are three additional Port 1 input pins (P1.5–P1.7) that were not available on the 8XC196MC. These pins are listed below followed by their function:

Table 6. New 8XC196MD Port 1 Pins

Pin	Description
P1.5	Digital or Analog Input
P1.6	Digital Input
P1.7	Digital Input

NOTE:
P1.5 was a V_{SS} pin on the 8XC196MC device. If P1.5 and P1.6 are not being used these pins can remain connected to V_{SS}.

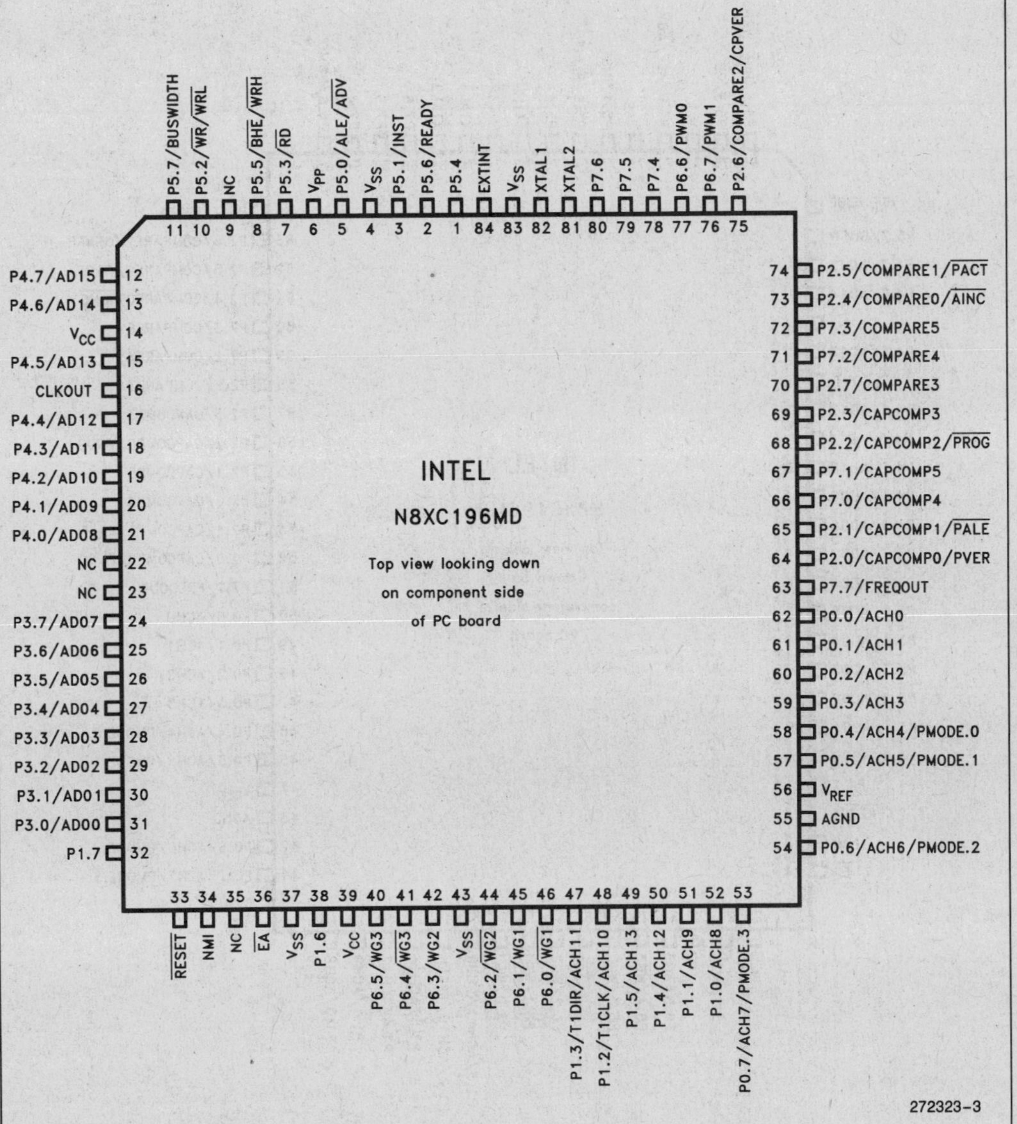

NOTE:
NC means No Connect. Do not connect these pins.

Figure 6. 84-Lead PLCC Package

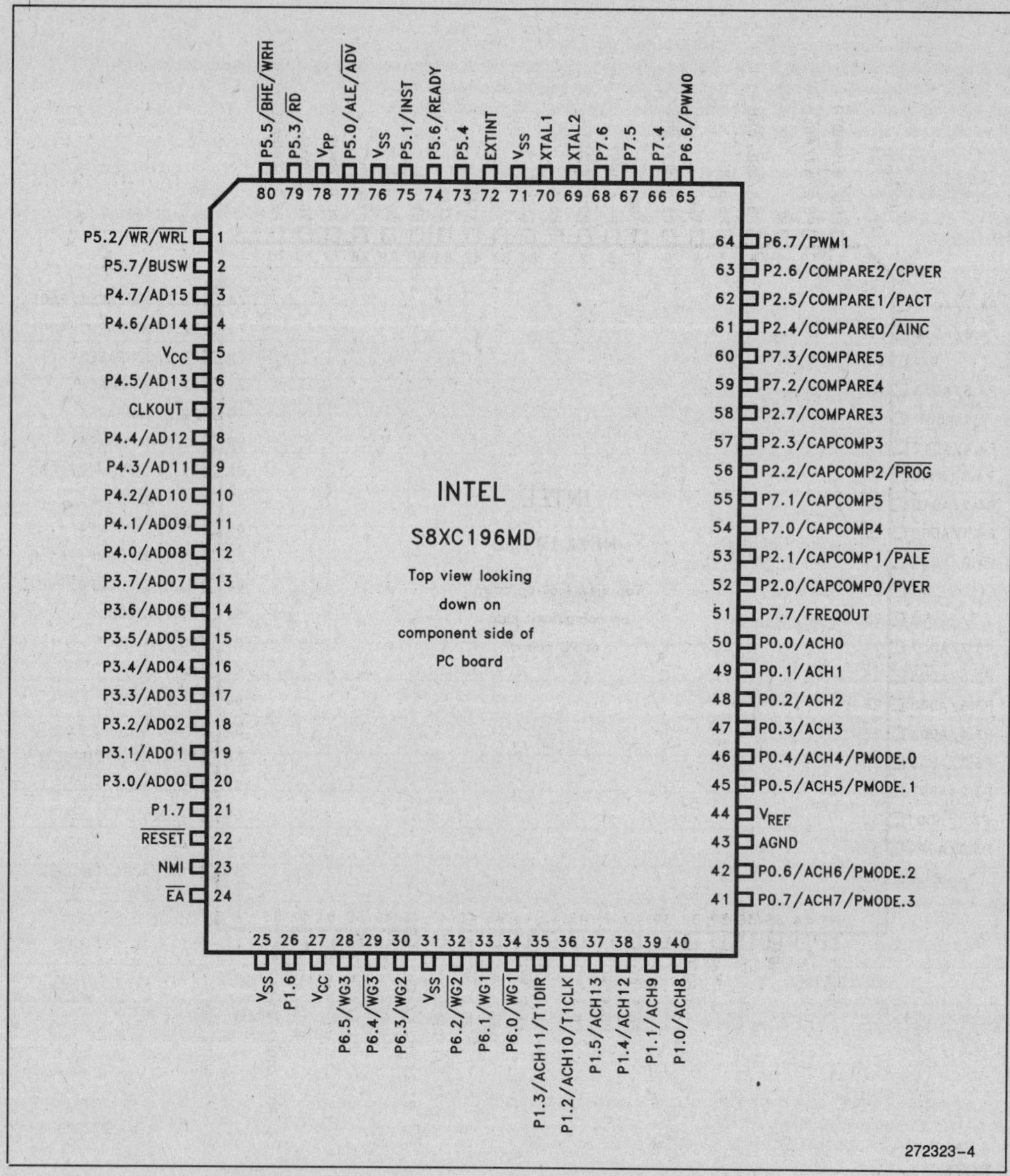

Figure 7. 80-Lead Shrink EIAJQFP (Quad Flat Pack)

PIN DESCRIPTIONS (Alphabetically Ordered)

Symbol	Function
ACH0–ACH13 (P0.0–P0.7, P1.0–P1.5)	Analog inputs to the on-chip A/D converter. ACH0–7 share the input pins with P0.0–7 and ACH8–13 share pins with P1.0–5. If the A/D is not used, the port pins can be used as standard input ports.
ANGND	Reference ground for the A/D converter. Must be held at nominally the same potential as V_{SS}.
ALE/$\overline{ADV}$(P5.0)	Address Latch Enable or Address Valid output, as selected by CCR. Both options allow a latch to demultiplex the address/data bus on the signal's falling edge. When the pin is $\overline{ADV}$, it goes inactive (high) at the end of the bus cycle. ALE/$\overline{ADV}$ is active only during external memory accesses. Can be used as standard I/O when not used as ALE/ADV.
$\overline{BHE}$/$\overline{WRH}$ (P5.5)	Byte High Enable or Write High output, as selected by the CCR. $\overline{BHE}$ will go low for external writes to the high byte of the data bus. $\overline{WRH}$ will go low for external writes where an odd byte is being written. $\overline{BHE}$/$\overline{WRH}$ is activated only during external memory writes.
BUSWIDTH (P5.7)	Input for bus width selection. If CCR bits 1 and 2 = 1, this pin dynamically controls the bus width of the bus cycle in progress. If BUSWIDTH is low, an 8-bit cycle occurs. If it is high, a 16-bit cycle occurs. This pin can be used as standard I/O when not used as BUSWIDTH.
CAPCOMP0–CAPCOMP5 (P2.0–P2.3, P7.0–P7.1)	The EPA Capture/Compare pins. CAPCOMP0–3 share the pins with P2.0–P2.3. CAPCOMP4–5 share the pins with P7.0–P7.1. If not used for the EPA, they can be configured as standard I/O pins.
CLKOUT	Output of the internal clock generator. The frequency is ½ of the oscillator frequency. It has a 50% duty cycle.
COMPARE0–COMPARE5 (P2.4–P2.7, P7.2–P7.3)	The EPA Compare pins. COMPARE0–3 share the pins with P2.4–P2.7. COMPARE4–5 share the pins with P7.2–P7.3. If not used for the EPA, they can be configured as standard I/O pins.
$\overline{EA}$	External Access enable pin. $\overline{EA}$ = 0 causes all memory accesses to be external to the chip. $\overline{EA}$ = 1 causes memory accesses from location 2000H to 5FFFH to be from the on-chip OTPROM/ROM. $\overline{EA}$ = 12.5V causes execution to begin in the programming mode. $\overline{EA}$ is latched at reset.
EXTINT	A programmable input on this pin causes a maskable interrupt vector through memory location 203CH. The input may be selected to be a positive/negative edge or a high/low level using WG_PROTECT (1FCEH).
FREQOUT	Programmable frequency output pin. The frequency can vary from 4 KHz to 1 MHz (16 MHz input clock). It has a 50% duty cycle. Pin may be configured as standard I/O if FREQOUT is not used.
INST (P5.1)	INST is high during the instruction fetch from the external memory and throughout the bus cycle. It is low otherwise. This pin can be configured as standard I/O if not used as INST.
NMI	A positive transition on this pin causes a non-maskable interrupt which vectors to memory location 203EH. If not used, it should be tied to V_{SS}. May be used by Intel Evaluation boards.
PORT0	8-bit high impedance input-only port. Also used as A/D converter inputs. Port0 pins should not be left floating. These pins also used to select programming modes in the OTPROM devices.
PORT1	8-bit high impedance input-only port. P1.0–P1.5 are also used as A/D converter inputs. In addition, P1.2 and P1.3 can be used as Timer 1 clock input and direction select respectively. P1.6–P1.7 can be used as input-only pins.

PIN DESCRIPTIONS (Alphabetically Ordered) (Continued)

Symbol	Function
PORT2	8-bit bidirectional I/O port. All of the Port2 pins are shared with the EPA I/O pins (CAPCOMP0–3 and COMPARE0–3).
PORT3 PORT4	8-bit bidirectional I/O ports with open drain outputs. These pins are shared with the multiplexed address/data bus which uses strong internal pullups.
PORT5	8-bit bidirectional I/O port. 7 of the pins are shared with bus control signals ($\overline{ALE}$, INST, $\overline{WR}$, $\overline{RD}$, $\overline{BHE}$, READY, BUSWIDTH). Can be used as standard I/O.
PORT6	8-bit output port. P6.6 and P6.7 output PWM, the others are used as the Wave Form Generator outputs. Can be used as standard output ports.
PORT7	8-bit bidirectional I/O port. P7.0–P7.3 can be used as EPA I/O pins (CAPCOMP4–5 and COMPARE4–5). P7.7 can be used as FREQOUT output pin. P7.4–P7.6 are standard I/O pins.
PWM0, PWM1 (P6.6, P6.7)	Programmable duty cycle, Programmable frequency Pulse Width Modulator pins. The duty cycle has a resolution of 256 steps, and the frequency can vary from 122 Hz to 31 KHz (16 MHz input clock). Pins may be configured as standard output if PWM is not used.
$\overline{RD}$ (P5.3)	Read signal output to external memory. $\overline{RD}$ is low only during external memory reads. Can be used as standard I/O when not used as $\overline{RD}$.
READY (P5.6)	Ready input to lengthen external memory cycles. If READY = 0, the memory controller inserts wait states until the next positive transition of CLKOUT occurs with READY = 1. Can be used as standard I/O when not used as READY.
$\overline{RESET}$	Reset input to and open-drain output from the chip. Held low for at least 16 state times to reset the chip. Input high for normal operation. $\overline{RESET}$ has an Ohmic internal pullup resistor.
T1CLK (P1.2)	Timer 1 Clock input. This pin has two other alternate functions: ACH10 and P1.2.
T1DIR (P1.3)	Timer 1 Direction input. This pin has two other alternate functions: ACH11 and P1.3.
V_{PP}	The programming voltage is applied to this pin. It is also the timing pin for the return from Power Down circuit. Connect this pin with a 1 μF capacitor to V_{SS} and a 1 MΩ resistor to V_{CC}. If the Power Down feature is not used, connect the pin to V_{CC}.
WG1–WG3/$\overline{WG1}$–$\overline{WG3}$ (P6.0–P6.5)	3 phase output signals and their complements used in motor control applications. The pins can also be configured as standard output pins.
$\overline{WR}$/$\overline{WRL}$ (P5.2)	Write and Write Low output to external memory. $\overline{WR}$ will go low every external write. $\overline{WRL}$ will go low only for external writes to an even byte. Can be used as standard I/O when not used as $\overline{WR}$/$\overline{WRL}$.
XTAL1	Input of the oscillator inverter and the internal clock generator. This pin should be used when using an external clock source.
XTAL2	Output of the oscillator inverter.
PMODE (P0.4–7)	Determines the EPROM programming mode.
$\overline{PACT}$ (P2.5)	A low signal in Auto Programming mode indicates that programming is in process. A high signal indicates programming is complete.

PIN DESCRIPTIONS (Alphabetically Ordered) (Continued)

Symbol	Function
$\overline{PALE}$ (P2.1)	A falling edge in Slave Programming Mode and Auto Configuration Byte Programming Mode indicates that ports 3 and 4 contain valid programming address/command information (input to slave).
$\overline{PROG}$ (P2.2)	A falling edge in Slave Programming Mode begins programming. A rising edge ends programming.
PVER (P2.0)	A high signal in Slave Programming Mode and Auto Configuration Byte Programming Mode indicates the byte programmed correctly.
CPVER (P2.6)	Cumulative Program Verification. Pin is high if all locations since entering a programming mode have programmed correctly.
$\overline{AINC}$ (P2.4)	Auto Increment. Active low input enables the auto increment mode. Auto increment will allow reading or writing of sequential EPROM locations without address transactions across the PBUS for each read or write.

ABSOLUTE MAXIMUM RATINGS

Ambient Temperature
Under Bias .... −40°C to +85°C

Storage Temperature .... −65°C to +150°C

Voltage from $\overline{EA}$ or V_{PP}
to V_{SS} or ANGND .... −0.5V to +13.00V

Voltage on Any Other Pin
to V_{SS} or ANGND .... −0.5V to +7.0V(1)

Power Dissipation .... 1.5W(2)

NOTES:
1. This includes V_{PP} and $\overline{EA}$ on ROM or CPU only devices.
2. Power dissipation is based on package heat transfer limitations, not device power consumption.

NOTICE: This data sheet contains preliminary information on new products in production. The specifications are subject to change without notice. Verify with your local Intel Sales office that you have the latest data sheet before finalizing a design.

**WARNING: Stressing the device beyond the "Absolute Maximum Ratings" may cause permanent damage. These are stress ratings only. Operation beyond the "Operating Conditions" is not recommended and extended exposure beyond the "Operating Conditions" may affect device reliability.*

OPERATING CONDITIONS

Symbol	Description	Min	Max	Units
T_A	Ambient Temperature Under Bias	−40	+85	°C
V_{CC}	Digital Supply Voltage	4.50	5.50	V
V_{REF}	Analog Supply Voltage	4.00	5.50	V
F_{OSC}	Oscillator Frequency	8	16	MHz

NOTE:
ANGND and V_{SS} should be nominally at the same potential. Also V_{SS} and V_{SS1} must be at the same potential.

DC ELECTRICAL CHARACTERISTICS (Over Specified Operating Conditions)

Symbol	Parameter	Min	Max	Units	Test Conditions
V_{IL}	Input Low Voltage	−0.5	0.3 V_{CC}	V	
V_{IH}	Input High Voltage	0.7 V_{CC}	V_{CC} + 0.5	V	
V_{OL}	Output Low Voltage Port 2, 5, and 7, P6.6, P6.7, CLKOUT		0.3 0.45 1.5	V V V	I_{OL} = 200 μA I_{OL} = 3.2 mA I_{OL} = 7 mA
V_{OL1}	Output Low Voltage on Port 3/4		1.0	V	I_{OL} = 15 mA
V_{OL2}	Output Low Voltage on Port 6.0–6.5		0.45	V	I_{OL} = 10 mA
V_{OH}	Output High Voltage	V_{CC} − 0.3 V_{CC} − 0.7 V_{CC} − 1.5		V V V	I_{OH} = −200 μA I_{OH} = −3.2 mA I_{OH} = −7 mA
$V_{th+}-V_{th-}$	Hysteresis Voltage Width on $\overline{RESET}$	0.2		V	Typical

DC ELECTRICAL CHARACTERISTICS (Over Specified Operating Conditions) (Continued)

Symbol	Parameter	Min	Typ	Max	Units	Test Conditions
I_{LI}	Input Leakage Current on All Input Only Pins			±10	μA	0V < V_{IN} < V_{CC}−0.3V (in RESET)
I_{LI1}	Input Leakage Current on Port0 and Port1			±3	μA	0V < V_{IN} < V_{REF}
I_{IL}	Input Low Current on BD Ports (Note 1)			−70	μA	V_{IN} = 0.3 V_{CC}
I_{IL1}	Input Low Current on P5.4 and P2.6 during Reset (Note 3)			−10	mA	0.2 V_{CC}
I_{OH}	Output High Current on P5.4 and P2.6 during Reset (Note 4)	−2			mA	0.7 V_{CC}
I_{CC}	Active Mode Current in Reset		50	70	mA	XTAL1 = 16 MHz, V_{CC} = V_{PP} = V_{REF} = 5.5V
I_{REF}	A/D Conversion Reference Current		2	5	mA	
I_{IDL}	Idle Mode Current		15	30	mA	
I_{PD}	Power-Down Mode Current		5	50	μA	V_{CC} = V_{PP} = V_{REF} = 5.5V
R_{RST}	$\overline{\text{RESET}}$ Pin Pullup Resistor	6k		65k	Ω	
C_S	Pin Capacitance (Any Pin to V_{SS})			10	pF	F_{TEST} = 1.0 MHz

NOTES:
1. BD (Bidirectional ports) include:
   P2.0–P2.7, except P2.6
   P3.0–P3.7
   P4.0–P4.7
   P5.0–P5.3
   P5.5–P5.7
   P7.0–P7.7
2. During normal (non-transient) conditions, the following total current limits apply:

P6.0–P6.5	I_{OL}: 40 mA	I_{OH}: 28 mA
P3	I_{OL}: 90 mA	I_{OH}: 42 mA
P4	I_{OL}: 90 mA	I_{OH}: 42 mA
P5, CLKOUT	I_{OL}: 35 mA	I_{OH}: 35 mA
P2, P6.6, P6.7, P7	I_{OL}: 63 mA	I_{OH}: 63 mA

3. Maximum current that must be sunk by external device to ensure test mode entry.
4. Do not exceed minimum current or device may enter test mode.

4

EXPLANATION OF AC SYMBOLS

Each symbol is two pairs of letters prefixed by "T" for time. The characters in a pair indicate a signal and its condition, respectively. Symbols represent the time between the two signal/condition points.

Conditions:

H — High
L — Low
V — Valid
X — No Longer Valid
Z — Floating

Signals:

A — Address
B — $\overline{BHE}$
C — CLKOUT
D — DATA
G — Buswidth
H — $\overline{HOLD}$
HA — $\overline{HLDA}$
L — ALE/$\overline{ADV}$
BR — $\overline{BREQ}$
R — $\overline{RD}$
W — $\overline{WR}/\overline{WRH}/\overline{WRL}$
X — XTAL1
Y — READY
Q — Data Out

AC ELECTRICAL CHARACTERISTICS (Over Specified Operating Conditions)

Test Conditions: Capacitive load on all pins = 100 pF, Rise and fall times = 10 ns, F_{OSC} = 16 MHz.

The system must meet the following specifications to work with the 87C196MD:

Symbol	Parameter	Min	Max	Units	Notes
F_{XTAL}	Frequency on XTAL1	8	16	MHz	3
T_{OSC}	$1/F_{XTAL}$	62.5	125	ns	
T_{AVYV}	Address Valid to READY Setup		$2 T_{OSC} - 75$	ns	
T_{LLYV}	ALE Low to READY Setup		$T_{OSC} - 70$	ns	4
T_{YLYH}	Not READY Time	No Upper Limit		ns	
T_{CLYX}	READY Hold after CLKOUT Low	0	$T_{OSC} - 30$	ns	1
T_{LLYX}	READY Hold after ALE Low	$T_{OSC} - 15$	$2 T_{OSC} - 40$	ns	1
T_{AVGV}	Address Valid to BUSWIDTH Setup		$2 T_{OSC} - 75$	ns	
T_{LLGV}	ALE Low to BUSWIDTH Setup		$T_{OSC} - 60$	ns	4
T_{CLGX}	Buswidth Hold after CLKOUT Low	0		ns	
T_{AVDV}	Address Valid to Input Data Valid		$3 T_{OSC} - 55$	ns	2
T_{RLDV}	$\overline{RD}$ Active to Input Data Valid		$T_{OSC} - 22$	ns	2
T_{CLDV}	CLKOUT Low to Input Data Valid		$T_{OSC} - 50$	ns	
T_{RHDZ}	End of $\overline{RD}$ to Input Data Float		T_{OSC}	ns	
T_{RXDX}	Data Hold after $\overline{RD}$ Inactive	0		ns	

NOTES:
1. If Max is exceeded, additional wait states will occur.
2. If wait states are used, add $2 T_{OSC} * N$, where N = number of wait states.
3. Testing performed at 8 MHz. However, the device is static by design and will typically operate below 1 Hz.
4. These timings are included for compatibility with older −90 and BH products. They should not be used for newer high-speed designs.

AC ELECTRICAL CHARACTERISTICS (Continued)

Test Conditions: Capacitive load on all pins = 100 pF, Rise and fall times = 10 ns, F_{OSC} = 16 MHz.

The 87C196MD will meet the following timing specifications:

Symbol	Parameter	Min	Max	Units	Notes
T_{XHCH}	XTAL1 to CLKOUT High or Low	30	110	ns	
T_{CLCL}	CLKOUT Cycle Time	2 T_{OSC}		ns	
T_{CHCL}	CLKOUT High Period	$T_{OSC} - 10$	$T_{OSC} + 15$	ns	
T_{CLLH}	CLKOUT Falling Edge to ALE Rising	−5	15	ns	
T_{LLCH}	ALE Falling Edge to CLKOUT Rising	−20	15	ns	
T_{LHLH}	ALE Cycle Time	4 T_{OSC}		ns	3
T_{LHLL}	ALE High Period	$T_{OSC} - 10$	$T_{OSC} + 10$	ns	
T_{AVLL}	Address Setup to ALE Falling Edge	$T_{OSC} - 15$		ns	
T_{LLAX}	Address Hold after ALE Falling	$T_{OSC} - 40$		ns	
T_{LLRL}	ALE Falling Edge to $\overline{RD}$ Falling	$T_{OSC} - 30$		ns	
T_{RLCL}	$\overline{RD}$ Low to CLKOUT Falling Edge	4	30	ns	
T_{RLRH}	$\overline{RD}$ Low Period	$T_{OSC} - 5$	$T_{OSC} + 25$	ns	3
T_{RHLH}	$\overline{RD}$ Rising Edge to ALE Rising Edge	T_{OSC}	$T_{OSC} + 25$	ns	1
T_{RLAZ}	$\overline{RD}$ Low to Address Float		5	ns	
T_{LLWL}	ALE Falling Edge to $\overline{WR}$ Falling	$T_{OSC} - 10$		ns	
T_{CLWL}	CLKOUT Low to $\overline{WR}$ Falling Edge	0	25	ns	
T_{QVWH}	Data Stable to $\overline{WR}$ Rising Edge	$T_{OSC} - 23$		ns	
T_{CHWH}	CLKOUT High to $\overline{WR}$ Rising Edge	−10	15	ns	
T_{WLWH}	$\overline{WR}$ Low Period	$T_{OSC} - 30$		ns	3
T_{WHQX}	Data Hold after $\overline{WR}$ Rising Edge	$T_{OSC} - 25$		ns	
T_{WHLH}	$\overline{WR}$ Rising Edge to ALE Rising Edge	$T_{OSC} - 10$	$T_{OSC} + 15$	ns	1
T_{WHBX}	$\overline{BHE}$, INST Hold after $\overline{WR}$ Rising	$T_{OSC} - 10$		ns	
T_{WHAX}	AD8–15 Hold after $\overline{WR}$ Rising	$T_{OSC} - 30$		ns	2
T_{RHBX}	$\overline{BHE}$, INST Hold after $\overline{RD}$ Rising	$T_{OSC} - 10$		ns	
T_{RHAX}	AD8–15 Hold after $\overline{RD}$ Rising	$T_{OSC} - 30$		ns	2

NOTES:
1. Assuming back to back cycles.
2. 8-bit bus only.
3. If wait states are used, add 2 T_{OSC}*N, where N = number of wait states.

SYSTEM BUS TIMINGS

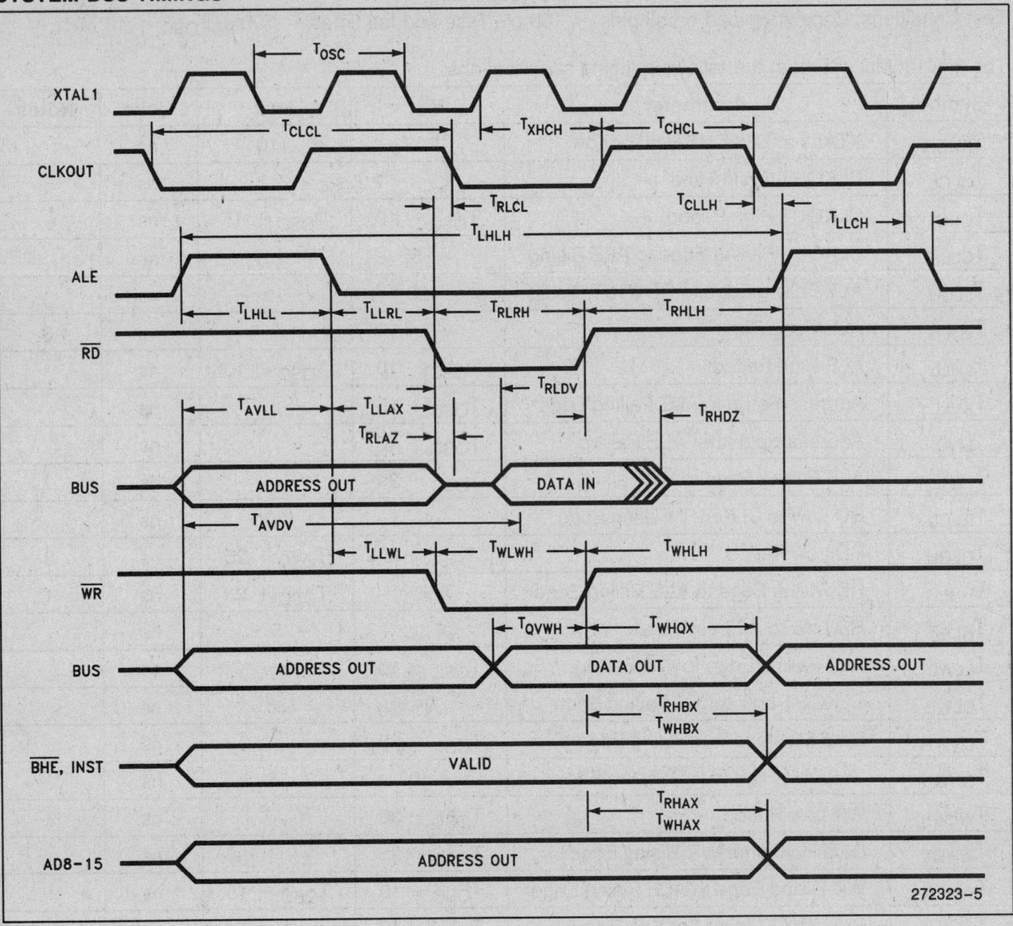

READY TIMINGS (One Wait State)

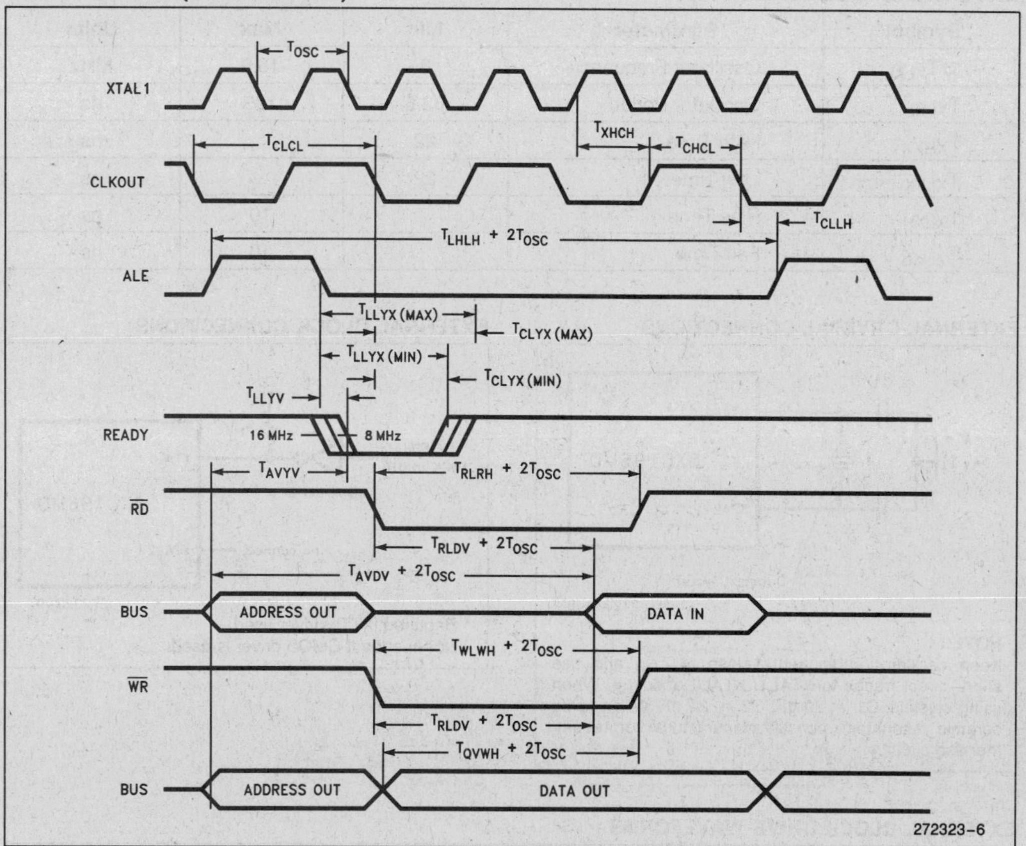

272323–6

BUSWIDTH TIMINGS

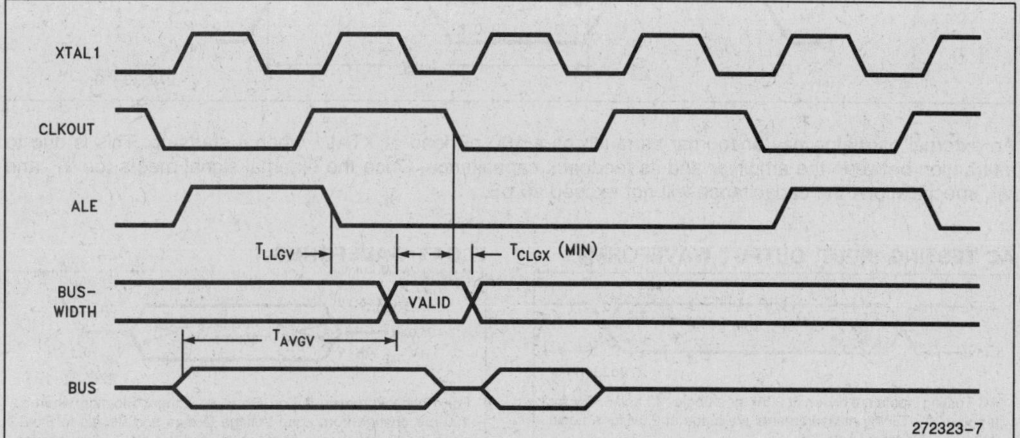

272323–7

EXTERNAL CLOCK DRIVE

Symbol	Parameter	Min	Max	Units
$1/T_{XLXL}$	Oscillator Frequency	8	16.0	MHz
T_{XLXL}	Oscillator Period	62.5	125	ns
T_{XHXX}	High Time	22		ns
T_{XLXX}	Low Time	22		ns
T_{XLXH}	Rise Time		10	ns
T_{XHXL}	Fall Time		10	ns

EXTERNAL CRYSTAL CONNECTIONS

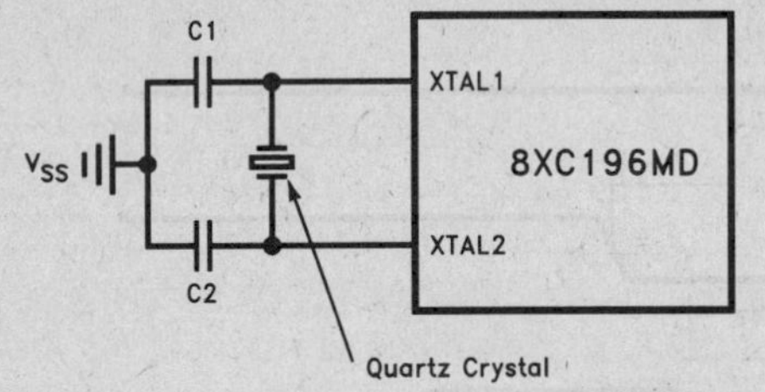

272323-8

NOTE:
Keep oscillator components close to chip and use short, direct traces to XTAL1, XTAL2 and V_{SS}. When using crystals, C1 = 20 pF, C2 = 20 pF. When using ceramic resonators, consult manufacturer for recommended circuitry.

EXTERNAL CLOCK CONNECTIONS

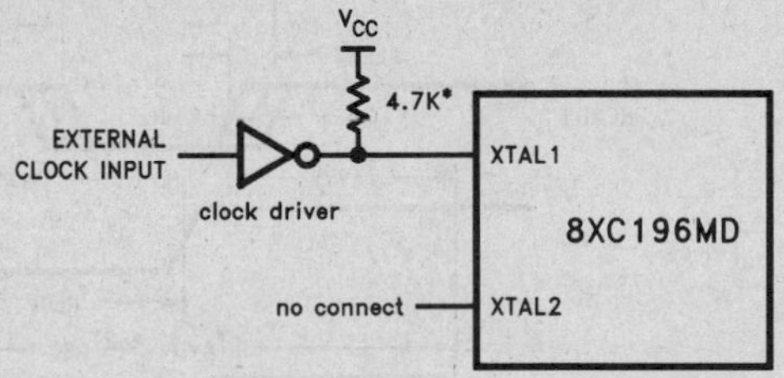

272323-9

* Required if TTL driver used.
Not needed if CMOS driver is used.

EXTERNAL CLOCK DRIVE WAVEFORMS

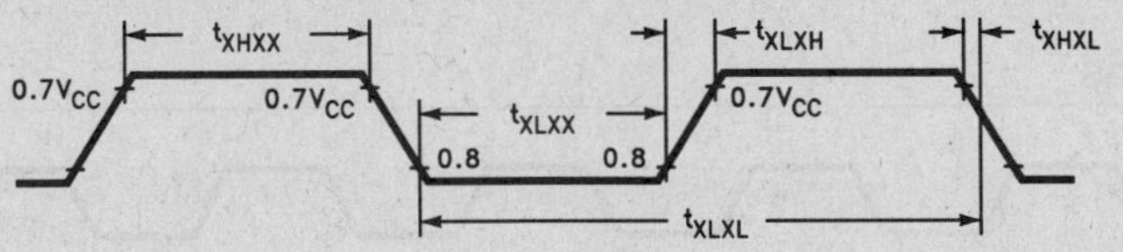

272323-10

An external oscillator may encounter as much as a 100 pF load at XTAL1 when it starts-up. This is due to interaction between the amplifier and its feedback capacitance. Once the external signal meets the V_{IL} and V_{IH} specifications the capacitance will not exceed 20 pF.

AC TESTING INPUT, OUTPUT WAVEFORMS

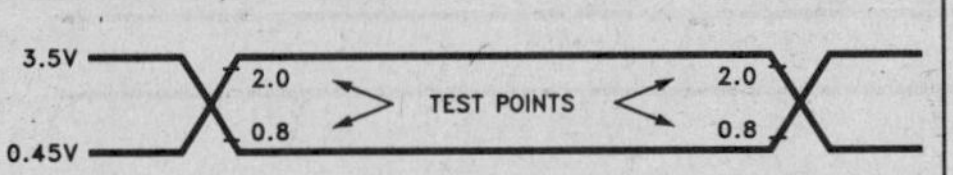

272323-11

AC Testing inputs are driven at 3.5V for a Logic "1" and 0.45V for a Logic "0". Timing measurements are made at 2.0V for a Logic "1" and 0.8V for a Logic "0".

FLOAT WAVEFORMS

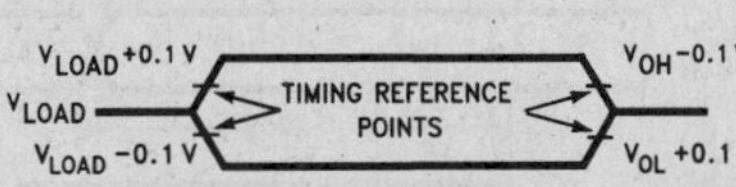

272323-12

For Timing Purposes a Port Pin is no Longer Floating when a 100 mV change from Load Voltage Occurs and Begins to Float when a 100 mV change from the Loaded V_{OH}/V_{OL} Level occurs $I_{OL}/I_{OH} = \pm 15$ mA.

A TO D CHARACTERISTICS

The sample and conversion time of the A/D converter in the 8-bit or 10-bit modes is programmed by loading a byte into the AD_TIME Special Function Register. This allows optimizing the A/D operation for specific applications. The AD_TIME register is functional for all possible values, but the accuracy of the A/D converter is only guaranteed for the times specificed in the operating conditions table.

The value loaded into AD_TIME bits 5, 6, 7 determines the sample time, T_{SAM}, and is calculated using the following formula:

$$SAM = \frac{(T_{SAM} \times F_{OSC}) - 2}{8}$$

T_{SAM} = Sample time, μs
F_{OSC} = Processor frequency, MHz
SAM = Value loaded into AD_TIME bits 5, 6, 7

SAM must be in the range 1 through 7.

The value loaded into AD_TIME bits 0–5 determines the conversion time, T_{CONV}, and is calculated using the following formula:

$$CONV = \frac{(T_{CONV} \times F_{OSC}) - 3}{2B} - 1$$

T_{CONV} = Conversion time, μs
F_{OSC} = Processor frequency, MHz
B = 8 for 8-bit conversion
B = 10 for 10-bit conversion
CONV = Value loaded into AD_TIME bits 0–5

CONV must be in the range 2 through 31.

The converter is ratiometric, so absolute accuracy is dependent on the accuracy and stability of V_{REF}. V_{REF} must be close to V_{CC} since it supplies both the resistor ladder and the analog portion of the converter and input port pins. There is also an AD_TEST SFR that allows for conversion on ANGND and V_{REF} as well as adjusting the zero offset. The absolute error listed is WITHOUT doing any adjustments.

A/D CONVERTER SPECIFICATION

The specifications given assume adherence to the operating conditions section of this data sheet. Testing is performed with V_{REF} = 5.12V and 16.0 MHz operating frequency. After a conversion is started, the device is placed in the IDLE mode until the conversion is complete.

4

10-BIT MODE A/D OPERATING CONDITIONS

Symbol	Description	Min	Max	Units
T_A	Ambient Temperature	−40	+85	°C
V_{CC}	Digital Supply Voltage	4.50	5.50	V
V_{REF}	Analog Supply Voltage	4.00	5.50	V(1)
T_{SAM}	Sample Time	1.0		μs(2)
T_{CONV}	Conversion Time	10.0	20.0	μs(2)
F_{OSC}	Oscillator Frequency	8.0	16.0	MHz

NOTES:
ANGND and V_{SS} should nominally be at the same potential.
1. V_{REF} must be within 0.5V of V_{CC}.
2. The value of AD_TIME is selected to meet these specifications.

10-BIT MODE A/D CHARACTERISTICS (Over Specified Operating Conditions)

Parameter	Typical(1)	Min	Max	Units*
Resolution		1024 10	1024 10	Levels Bits
Absolute Error		0	±4	LSBs
Full Scale Error	0.25 ±0.5			LSBs
Zero Offset Error	0.25 ±0.5			LSBs
Non-Linearity	1.0 ±2.0		±4	LSBs
Differential Non-Linearity		> −1	+2	LSBs
Channel-to-Channel Matching	±0.1	0	±1.0	LSBs
Repeatability	±0.25	0		LSBs
Temperature Coefficients: Offset Full Scale Differential Non-Linearity	 0.009 0.009 0.009			 LSB/C LSB/C LSB/C
Off Isolation		−60		dB(2, 3)
Feedthrough	−60			dB(2)
V_{CC} Power Supply Rejection	−60			dB(2)
Input Series Resistance		750	2K	Ω(4)
Voltage on Analog Input Pin		ANGND − 0.5	V_{REF} + 0.5	V(5, 6)
Sampling Capacitor	3			pF
DC Input Leakage	±1	0	±3.0	μA

NOTES:
*An "LSB", as used here has a value of approximately 5 mV. (See Embedded Microcontrollers and Processors Handbook for A/D glossary of terms).
1. These values are expected for most parts at 25°C but are not tested or guaranteed.
2. DC to 100 KHz.
3. Multiplexer Break-Before-Make is guaranteed.
4. Resistance from device pin, through internal MUX, to sample capacitor.
5. These values may be exceeded if the pin current is limited to ±2 mA.
6. Applying voltages beyond these specifications will degrade the accuracy of other channels being converted.
7. All conversions performed with processor in IDLE mode.

8-BIT MODE A/D OPERATING CONDITIONS

Symbol	Description	Min	Max	Units
T_A	Ambient Temperature	−40	+85	°C
V_{CC}	Digital Supply Voltage	4.50	5.50	V
V_{REF}	Analog Supply Voltage	4.00	5.50	V(1)
T_{SAM}	Sample Time	1.0		μs(2)
T_{CONV}	Conversion Time	7.0	20.0	μs(2)
F_{OSC}	Oscillator Frequency	8.0	16.0	MHz

NOTES:
ANGND and V_{SS} should nominally be at the same potential.
1. V_{REF} must be within 0.5V of V_{CC}.
2. The value of AD_TIME is selected to meet these specifications.

8-BIT MODE A/D CHARACTERISTICS (Over the Above Operating Conditions)

Parameter	Typical(1)	Min	Max	Units*
Resolution		256 8	256 8	Level Bits
Absolute Error		0	±1	LSBs
Full Scale Error	±0.5			LSBs
Zero Offset Error	±0.5			LSBs
Non-Linearity		0	±1	LSBs
Differential Non-Linearity		> −1	+1	LSBs
Channel-to-Channel Matching		0	±1.0	LSBs
Repeatability	±0.25			LSBs
Temperature Coefficients: Offset Full Scale Differential Non-Linearity	 0.003 0.003 0.003			 LSB/C LSB/C LSB/C
Off Isolation		−60		dB(2, 3)
Feedthrough	−60			dB(2)
V_{CC} Power Supply Rejection	−60			dB(2)
Input Series Resistance		750	2K	Ω(4)
Voltage on Analog Input Pin		V_{SS} − 0.5	V_{REF} + 0.5	V(5, 6)
Sampling Capacitor	3			pF
DC Input Leakage	±1	0	±3.0	μA

NOTES:
*An "LSB" as used here, has a value of approximately 20 mV. (See Embedded Microcontrollers and Processors Handbook for A/D glossary of terms).
1. These values are expected for most parts at 25°C but are not tested or guaranteed.
2. DC to 100 KHz.
3. Multiplexer Break-Before-Make is guaranteed.
4. Resistance from device pin, through internal MUX, to sample capacitor.
5. These values may be exceeded if the pin current is limited to ±2 mA.
6. Applying voltages beyond these specifications will degrade the accuracy of other channels being converted.
7. All conversions performed with processor in IDLE mode.

EPROM SPECIFICATIONS

OPERATING CONDITIONS DURING PROGRAMMING

Symbol	Description	Min	Max	Units
T_A	Ambient Temperature during Programming	20	30	°C
V_{CC}	Supply Voltage during Programming	4.5	5.5	V(1)
V_{REF}	Reference Supply Voltage during Programming	4.5	5.5	V(1)
V_{PP}	Programming Voltage	12.25	12.75	V(2)
V_{EA}	EA Pin Voltage	12.25	12.75	V(2)
F_{OSC}	Oscillator Frequency during Auto and Slave Mode Programming	6.0	8.0	MHz
T_{OSC}	Oscillator Frequency during Run-Time Programming	6.0	12.0	MHz

NOTES:
1. V_{CC} and V_{REF} should nominally be at the same voltage during programming.
2. V_{PP} and V_{EA} must never exceed the maximum specification, or the device may be damaged.
3. V_{SS} and ANGND should nominally be at the same potential (0V).
4. Load capacitance during Auto and Slave Mode programming = 150 pF.

AC EPROM PROGRAMMING CHARACTERISTICS (SLAVE MODE)

Symbol	Parameter	Min	Max	Units
T_{SHLL}	Reset High to First $\overline{\text{PALE}}$ Low	1100		T_{OSC}
T_{LLLH}	$\overline{\text{PALE}}$ Pulse Width	50		T_{OSC}
T_{AVLL}	Address Setup Time	0		T_{OSC}
T_{LLAX}	Address Hold Time	100		T_{OSC}
T_{PLDV}	$\overline{\text{PROG}}$ Low to Word Dump Valid		50	T_{OSC}
T_{PHDX}	Word Dump Data Hold		50	T_{OSC}
T_{DVPL}	Data Setup Time	0		T_{OSC}
T_{PLDX}	Data Hold Time	400		T_{OSC}
T_{PLPH}(1)	$\overline{\text{PROG}}$ Pulse Width	50		T_{OSC}
T_{PHLL}	$\overline{\text{PROG}}$ High to Next $\overline{\text{PALE}}$ Low	220		T_{OSC}
T_{LHPL}	$\overline{\text{PALE}}$ High to $\overline{\text{PROG}}$ Low	220		T_{OSC}
T_{PHPL}	$\overline{\text{PROG}}$ High to Next $\overline{\text{PROG}}$ Low	220		T_{OSC}
T_{PHIL}	PROG High to AINC Low	0		T_{OSC}
T_{ILIH}	$\overline{\text{AINC}}$ Pulse Width	240		T_{OSC}
T_{ILVH}	PVER Hold after $\overline{\text{AINC}}$ Low	50		T_{OSC}
T_{ILPL}	$\overline{\text{AINC}}$ Low to $\overline{\text{PROG}}$ Low	170		T_{OSC}
T_{PHVL}	$\overline{\text{PROG}}$ High to PVER Valid		220	T_{OSC}

NOTE:
1. This specification is for the Word Dump Mode. For programming pulses, use the Modified Quick Pulse Algorithm.

DC EPROM PROGRAMMING CHARACTERISTICS

Symbol	Parameter	Min	Max	Units
I_{PP}	V_{PP} Supply Current (When Programming)		100	mA

NOTE:
Do not apply V_{PP} until V_{CC} is stable and within specifications and the oscillator/clock has stabilized or the device may be damaged.

SLAVE PROGRAMMING MODE DATA PROGRAM MODE WITH SINGLE PROGRAM PULSE

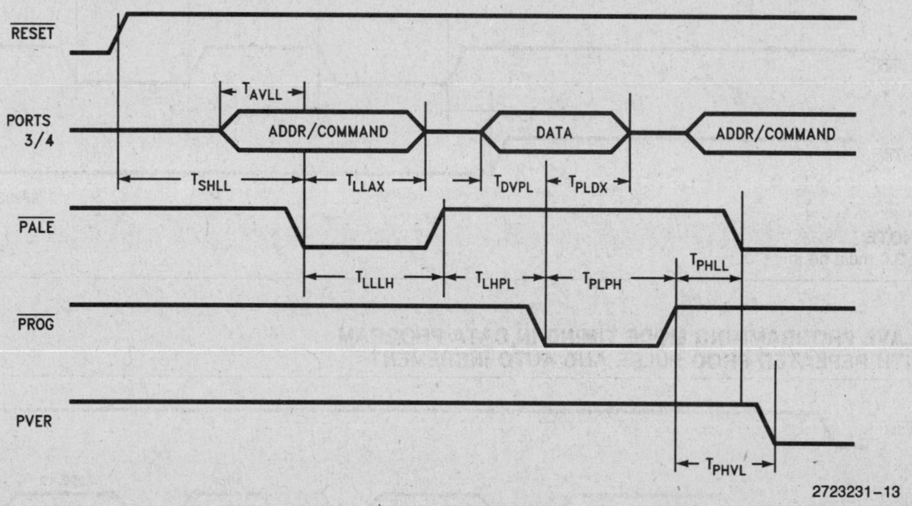

NOTE:
P3.0 must be high ("1")

SLAVE PROGRAMMING MODE IN WORD DUMP WITH AUTO INCREMENT

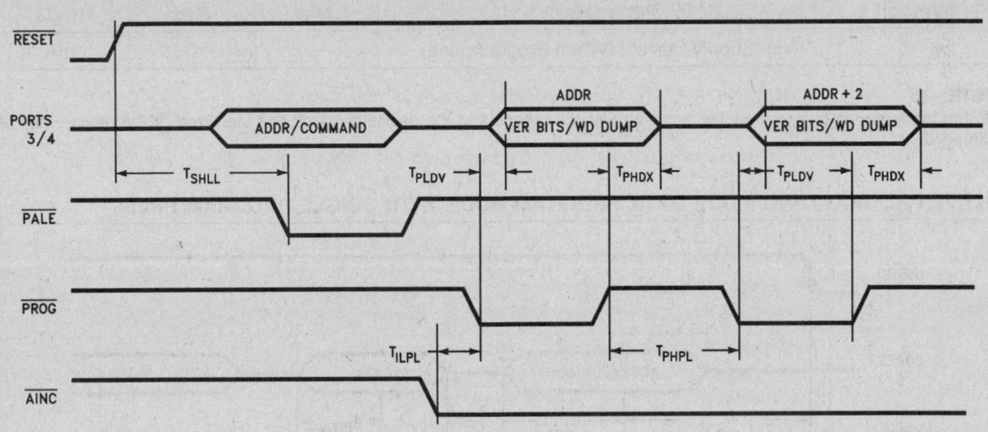

272323-14

NOTE:
P3.0 must be low ("0")

SLAVE PROGRAMMING MODE TIMING IN DATA PROGRAM WITH REPEATED PROG PULSE AND AUTO INCREMENT

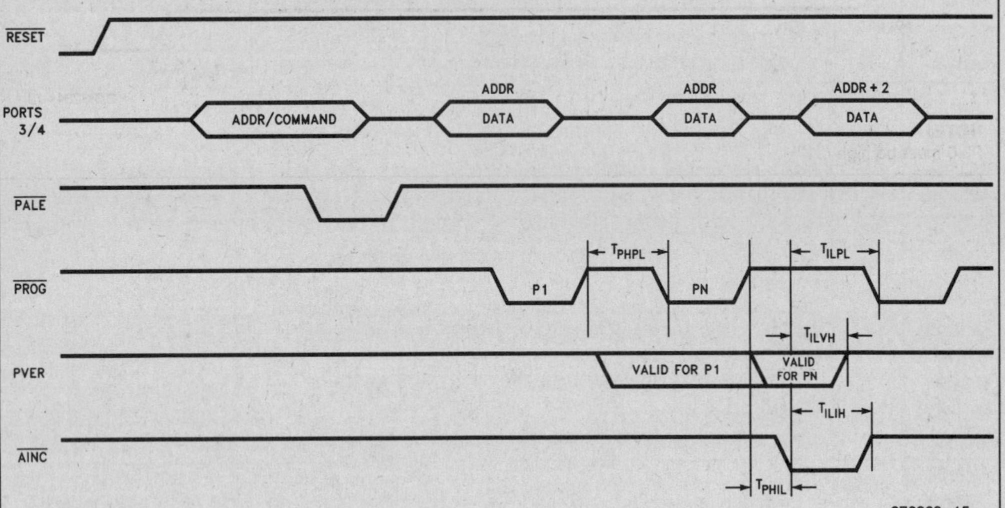

272323-15

87C196MD DESIGN CONSIDERATIONS

When an indirect shift during divide occurs the upper 3 bits of the shift count are not masked completely. If the shift count register has the value 32 * n where n = 1, 3, 5 or 7, the operand will be shifted 32 times. This should have resulted in no shift taking place.

8XC196MC to 8XC196MD Design Considerations

8XC196MC and 8XC196MD are pin compatible. However, there were several pins that were not connected (NC) on the 8XC196MC that are I/O pins on the 8XC196MD. Port 7 is a bidirectional port added to the 8XC196MD. Port 1 has one additional analog or digital input that was connected to V_{SS} on the 8XC196MC. Port 1 also has two additional digital inputs. See 8XC196MC and 8XC196MD Differences Section of this data sheet.

DATA SHEET REVISION HISTORY

This is the initial data sheet (272323-001). It is valid for devices with a "B" at the end of the topside tracking number. Data sheets are changed as new device information becomes available. Verify with your local Intel sales office that you have the latest version before finalizing a design or ordering devices.

8XC196MH INDUSTRIAL MOTOR CONTROL CHMOS MICROCONTROLLER

- High Performance CHMOS 16-bit CPU
- 16 MHz Operating Frequency
- 32 Kbytes of On-chip OTPROM/ROM
- 744 Bytes of On-chip Register RAM
- Register-to-register Architecture
- 16 Prioritized Interrupt Sources
- Peripheral Transaction Server (PTS) with 15 Prioritized Sources
- Up to 52 I/O Lines
- 3-phase Complementary Waveform Generator
- 8-channel 8- or 10-bit A/D with Sample and Hold
- 2-channel UART
- Event Processor Array (EPA) with 2 High-speed Capture/Compare Modules and 4 High-speed Compare-only Modules
- Two Programmable 16-bit Timers with Quadrature Counting Inputs
- Two Pulse-width Modulator (PWM) Outputs with High Drive Capability
- Flexible 8- or 16-bit External Bus
- 1.75 μs 16 x 16 Multiply
- 3 μs 32/16 Divide
- Extended Temperature Available
- Idle and Powerdown Modes
- Watchdog Timer

The 8XC196MH is a member of Intel's family of 16-bit MCS® 96 microcontrollers. It is designed primarily to control three-phase AC induction and DC brushless motors. It features an enhanced three-phase waveform generator specifically designed for use in "inverter" motor-control applications. This peripheral provides pulse-width modulation and three-phase sine wave generation with minimal CPU intervention. It generates three complementary non-overlapping PWM pulses with resolutions of 0.125 μs (edge triggered) or 0.250 μs (centered).

The 8XC196MH has two dedicated serial port peripherals, allowing less software overhead. The watchdog timer can be programmed with one of four time options.

The 8XC196MH is available without internal memory (80C196MH), with 32 Kbytes of factory programmed OTPROM* (83C196MH), or with 32 Kbytes of user programmable OTPROM (87C196MH). It is available in three packages: 84-lead PLCC, 80-lead Shrink EIAJ/QFP, and 64-lead SDIP. The 64-lead package does not contain pins for the P5.1/INST and P6.7/PWM1 signals.

Operational characteristics are guaranteed over the temperature range of −40°C to +85°C.

---

*One-Time Programmable Read-Only Memory (OTPROM) is similar to EPROM but comes in an unwindowed package and cannot be erased. It is user programmable.

November 1994
Order Number: 272543-001

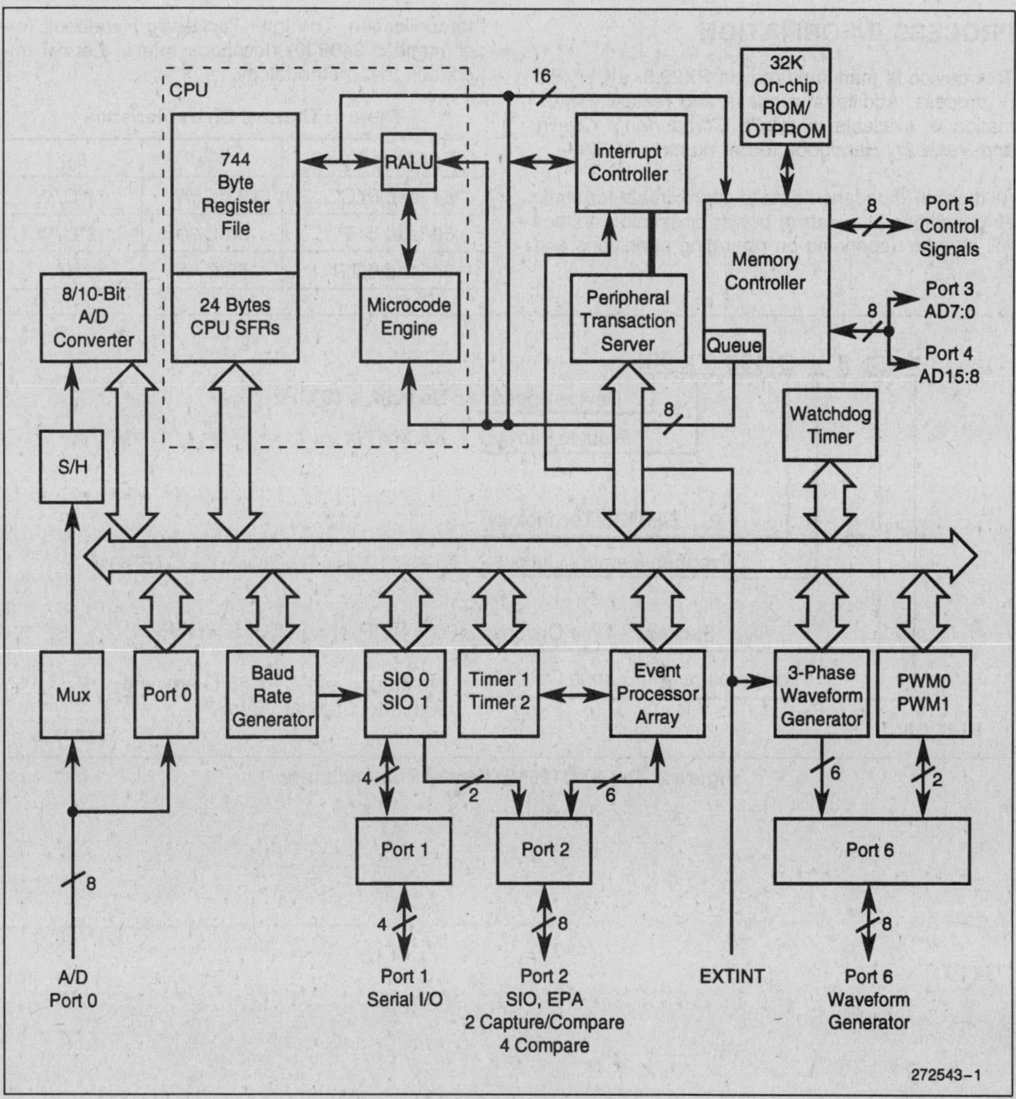

Figure 1. 8XC196MH Block Diagram

PROCESS INFORMATION

This device is manufactured on PX29.5, a CHMOS IV process. Additional process and reliability information is available in Intel's *Components Quality and Reliability* Handbook (order number 210997).

All thermal impedance data is approximate for static air conditions at 1 watt of power dissipation. Values will change depending on operating conditions and the application. The Intel *Packaging* Handbook (order number 240800) describes Intel's thermal impedance test methodology.

Table 1. Thermal Characteristics

Package Type	θ_{JA}	θ_{JC}
84-lead PLCC	33°C/W	11°C/W
80-lead QFP	56°C/W	12°C/W
64-lead SDIP	56°C/W	N/A

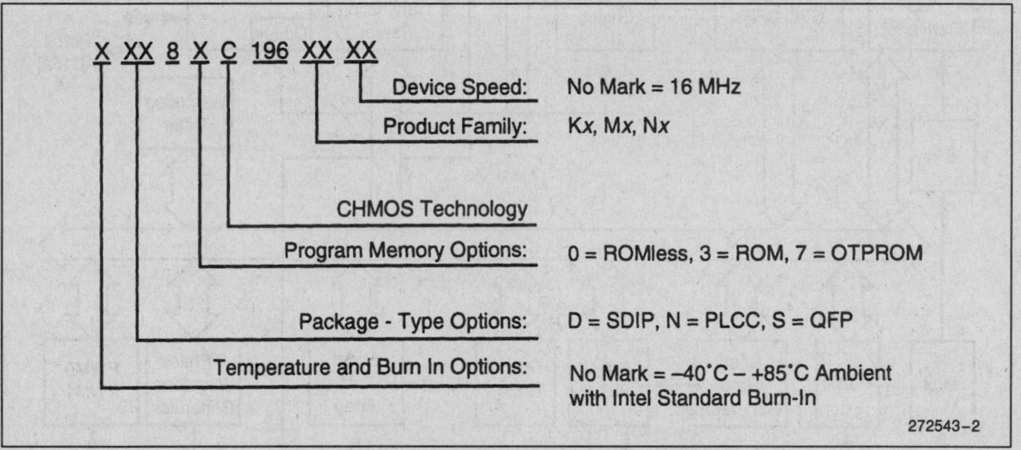

Figure 2. The 8XC196MH Family Nomenclature

Table 2. 8XC196MH Memory Map

Address (1)	Description	Notes
0FFFFH 0A000H	External Memory	
09FFFH 02080H	Internal ROM/OTPROM or External Memory	
0207FH 0205EH	Reserved	1, 2
0205DH 02040H	PTS Vectors	
0203FH 02030H	Interrupt Vectors (upper)	
0202FH 02020H	ROM/OTPROM Security Key	
0201FH 0201CH	Reserved	1, 2
0201BH	Reserved (must contain 20H)	
0201AH	CCB1	
02019H	Reserved (must contain 20H)	
02018H	CCB0	
02017H 02014H	Reserved	
02013H 02000H	Interrupt Vectors (lower)	
01FFFH 01F00H	Internal SFRs	1
1EFFH 300H	External Memory	
2FFH 18H	Register RAM	3
17H 00H	CPU SFRs	1

NOTES:
1. Unless otherwise noted, write 0FFH to reserved memory locations and write 0 to reserved SFR bits.
2. **WARNING:** The contents and/or function of reserved locations may change with future revisions of the device.
3. Code executed in locations 0000H to 02FFH will be forced external.

Table 3. Signals Arranged by Functional Categories

Address & Data
AD15:0

Bus Control & Status
ALE/ADV#
BHE#/WRH#
BUSWIDTH
INST
READY
RD#
WR#/WRL#

Power & Ground
ANGND
V_{CC}
V_{PP}
V_{REF}
V_{SS}

Programming Control
AINC#
CPVER
PACT#
PALE#
PBUS15:0
PMODE.3:0
PROG#
PVER

Processor Control
EA#
EXTINT
NMI
ONCE#
RESET#
XTAL1
XTAL2

Input/Output
P0.0/ACH0
P0.1/ACH1
P0.2/ACH2
P0.3/ACH3
P0.4/ACH4
P0.5/ACH5
P0.6/ACH6/T1CLK
P0.7/ACH7/T1DIR
P1.0/TXD0
P1.1/RXD0
P1.2/TXD1
P1.3/RXD1
P2.0/EPA0
P2.1/SCLK0#/BCLK0
P2.2/EPA1
P2.3/COMP3
P2.4/COMP0

Input/Output (Cont'd)
P2.5/COMP1
P2.6/COMP2
P2.7/SCLK1#/BCLK1
P3.7:0
P4.7:0
P5.7:0
P6.0/WG1#
P6.1/WG1
P6.2/WG2#
P6.3/WG2
P6.4/WG3#
P6.5/WG3
P6.6/PWM0
P6.7/PWM1

NOTE:
The following signals are not available in the 64-pin package: P5.1, P6.7, INST, and PWM1.

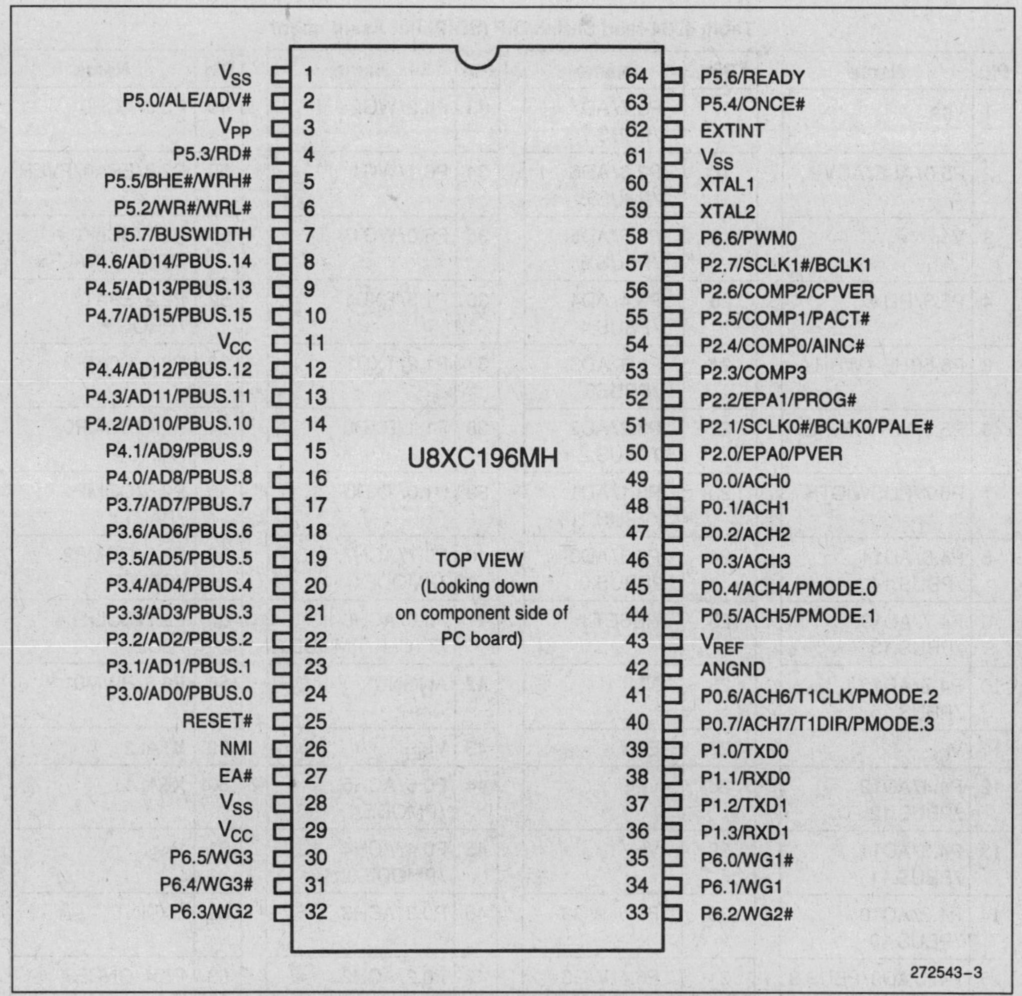

Figure 3. 8XC196MH 64-lead Shrink DIP (SDIP) Package

4

Table 4. 64-lead Shrink DIP (SDIP) Pin Assignment

Pin	Name	Pin	Name	Pin	Name	Pin	Name
1	V_{SS}	17	P3.7/AD7 /PBUS.7	33	P6.2/WG2#	49	P0.0/ACH0
2	P5.0/ALE/ADV#	18	P3.6/AD6 /PBUS.6	34	P6.1/WG1	50	P2.0/EPA0/PVER
3	V_{PP}	19	P3.5/AD5 /PBUS.5	35	P6.0/WG1#	51	P2.1/SCLK0# /BCKL0/PALE#
4	P5.3/RD#	20	P3.4/AD4 /PBUS.4	36	P1.3/RXD1	52	P2.2/EPA1 /PROG#
5	P5.5BHE#WRH#	21	P3.3/AD3 /PBUS3	37	P1.2/TXD1	53	P2.3/COMP3
6	P5.2/WR#/WRL#	22	P3.2/AD2 /PBUS.2	38	P1.1/RXD0	54	P2.4/COMP0 /AINC#
7	P5.7/BUSWIDTH	23	P3.1/AD1 /PBUS.1	39	P1.0/TXD0	55	P2.5/COMP1 /PACT#
8	P4.6/AD14 /PBUS.14	24	P3.0/AD0 /PBUS.0	40	P0.7/ACH7/T1DIR /PMODE.3	56	P2.6/COMP2 /CPVER
9	P4.7/AD13 /PBUS.13	25	RESET#	41	P0.6/ACH6 /T1CLK/PMODE.2	57	P2.7/SCLK1# /BCLK1
10	P4.7/AD15 /PBUS.15	26	NMI	42	ANGND	58	P6.6/PWM0
11	V_{CC}	27	EA#	43	V_{REF}	59	XTAL2
12	P4.4/AD12 /PBUS.12	28	V_{SS}	44	P0.5/ACH5 /PMODE.1	60	XTAL1
13	P4.3/AD11 /PBUS.11	29	V_{CC}	45	P0.4/ACH4 /PMODE.0	61	V_{SS}
14	P4.2/AD10 /PBUS.10	30	P6.5/WG3	46	P0.3/ACH3	62	EXTINT
15	P4.1/AD9/PBUS.9	31	P6.4/WG3#	47	P0.2/ACH2	63	P5.4/ONCE#
16	P4.0/AD8/PBUS.8	32	P6.3WG2	48	P0.1/ACH1	64	P5.6/READY

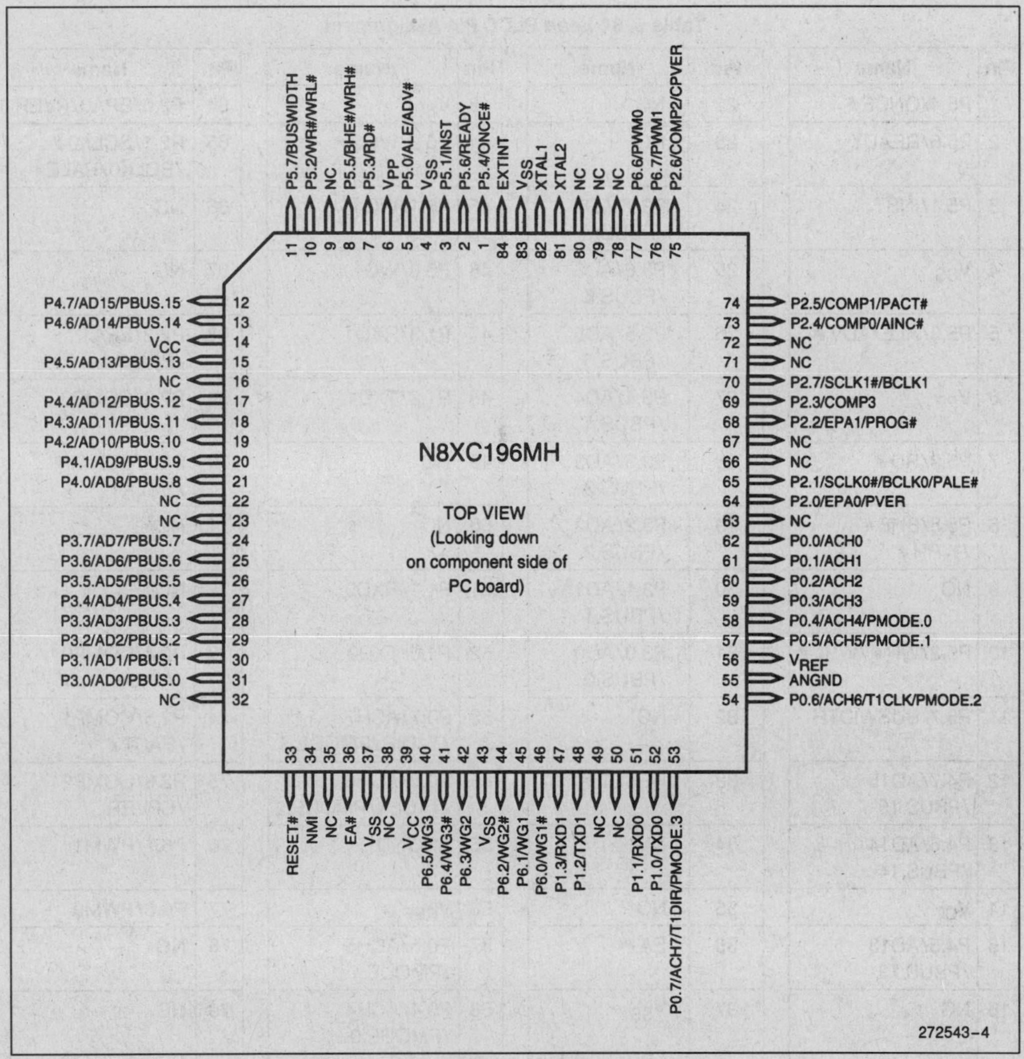

Figure 4. 8XC196MH 84-lead PLCC Package

Table 5. 84-Lead PLCC Pin Assignment

Pin	Name	Pin	Name	Pin	Name	Pin	Name
1	P5.4/ONCE#	22	NC	43	V_{SS}	64	P2.0/EPA0/PVER
2	P5.6/READY	23	NC	44	P6.2/WG2#	65	P2.1/SCLK0# /BCLK0/PALE#
3	P5.1/INST	24	P3.7/AD7 /PBUS.7	45	P6.1/WG1	66	NC
4	V_{SS}	25	P3.6/AD6 /PBUS.6	46	P6.0/WG1#	67	NC
5	P5.0/ALE/ADV#	26	P3.5/AD5 /PBUS.5	47	P1.3/RXD1	68	P2.2/EPA1 /PROG#
6	V_{PP}	27	P3.4/AD4 /PBUS.4	48	P1.2/TXD1	69	P2.3/COMP3
7	P5.3/RD#	28	P3.3/AD3 /PBUS.3	49	NC	70	P2.7/SCLK1# /BCLK1
8	P5.5/BHE# /WRH#	29	P3.2/AD2 /PBUS.2	50	NC	71	NC
9	NC	30	P3.1/AD1 /PBUS.1	51	P1.1/RXD0	72	NC
10	P5.2/WR#/WRL#	31	P3.0/AD0 /PBUS.0	52	P1.0/TXD0	73	P2.4/COMP0 /AINC#
11	P5.7/BUSWIDTH	32	NC	53	P0.7/ACH7 /T1DIR/PMODE.3	74	P2.5/COMP1 /PACT#
12	P4.7/AD15 /PBUS.15	33	RESET#	54	P0.6/ACH6 /T1CLK/PMODE.2	75	P2.6/COMP2 /CPVER
13	P4.6/AD14 /PBUS.14	34	NMI	55	ANGND	76	P6.7/PWM1
14	V_{CC}	35	NC	56	V_{REF}	77	P6.6/PWM0
15	P4.5/AD13 /PBUS.13	36	EA#	57	P0.5/ACH5 /PMODE.1	78	NC
16	NC	37	V_{SS}	58	P0.4/ACH4 /PMODE.0	79	NC
17	P4.4/AD12 /PBUS.12	38	NC	59	P0.3/ACH3	80	NC
18	P4.3/AD11 /PBUS.11	39	V_{CC}	60	P0.2/ACH2	81	XTAL2
19	P4.2/AD10 /PBUS.10	40	P6.5/WG3	61	P0.1/ACH1	82	XTAL1
20	P4.1/AD9/PBUS.9	41	P6.4/WG3#	62	P0.0/ACH0	83	V_{SS}
21	P4.0/AD8/PBUS.8	42	P6.3/WG2	63	NC	84	EXTINT

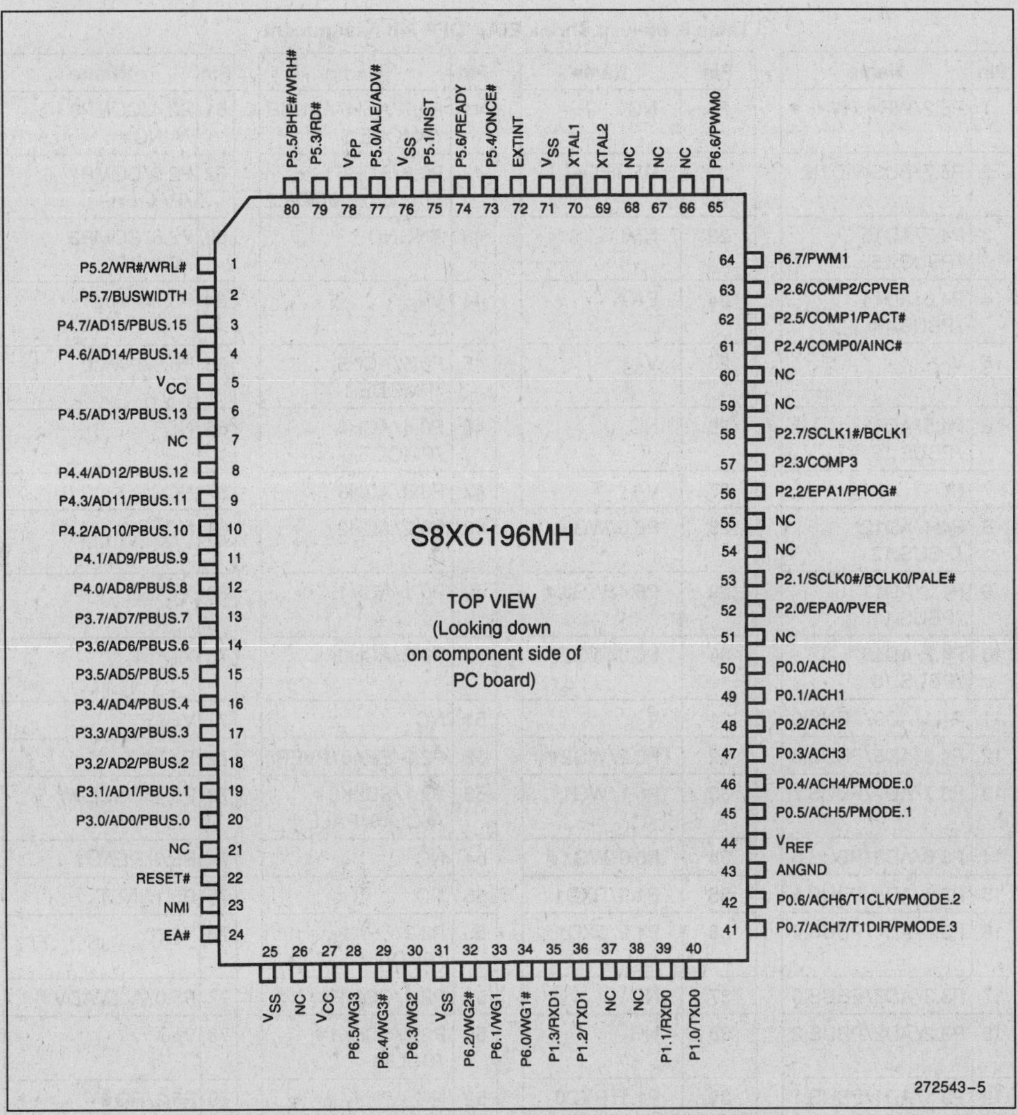

Figure 5. 8XC196MH 80-lead Shrink EIAJ/QFP Package

Table 6. 80-lead Shrink EIAJ/QFP Pin Assignment

Pin	Name	Pin	Name	Pin	Name	Pin	Name
1	P5.2/WR#/WRL#	21	NC	41	P0.7/ACH7/T1DIR /PMODE.3	61	P2.4/COMP0 /AINC#
2	P5.7/BUSWIDTH	22	RESET#	42	P0.6/ACH6 /T1CLK/PMODE.2	62	P2.5/COMP1 /PACT#
3	P4.7/AD15 /PBUS.15	23	NMI	43	ANGND	63	P2.6/COMP2 /CPVER
4	P4.6/AD14 /PBUS.14	24	EA#	44	V_{REF}	64	P6.7/PWM1
5	V_{CC}	25	V_{SS}	45	P0.5/ACH5 /PMODE.1	65	P6.6/PWM0
6	P4.5/AD13 /PBUS.13	26	NC	46	P0.4/ACH4 /PMODE.0	66	NC
7	NC	27	V_{CC}	47	P0.3/ACH3	67	NC
8	P4.4/AD12 /PBUS.12	28	P6.5/WG3	48	P0.2/ACH2	68	NC
9	P4.3/AD11 /PBUS.11	29	P6.4/WG3#	49	P0.1/ACH1	69	XTAL2
10	P4.2/AD10 /PBUS.10	30	P6.3/WG2	50	P0.0/ACH0	70	XTAL1
11	P4.1/AD9/PBUS.9	31	V_{SS}	51	NC	71	V_{SS}
12	P4.0/AD8/PBUS.8	32	P6.2/WG2#	52	P2.0/EPA0/PVER	72	EXTINT
13	P3.7/AD7/PBUS.7	33	P6.1/WG1	53	P2.1/SCLK0# /BCLK0/PALE#	73	P5.4/ONCE#
14	P3.6/AD6/PBUS.6	34	P6.0/WG1#	54	NC	74	P5.6/READY
15	P3.5/AD5/PBUS.5	35	P1.3/RXD1	55	NC	75	P5.1/INST
16	P3.4/AD4/PBUS.4	36	P1.2/TXD1	56	P2.2/EPA1 /PROG#	76	V_{SS}
17	P3.3/AD3/PBUS.3	37	NC	57	P2.3/COMP3	77	P5.0/ALE/ADV#
18	P3.2/AD2/PBUS.2	38	NC	58	P2.7/SCLK1# /BCLK1	78	V_{PP}
19	P3.1/AD1/PBUS.1	39	P1.1/RXD0	59	NC	79	P5.3/RD#
20	P3.0/AD0/PBUS.0	40	P1.0/TXD0	60	NC	80	P5.5/BHE#/WRH#

PIN DESCRIPTIONS

Table 7. Signal Descriptions

Signal Name	Type	Description	Multiplexed With
ACH7 ACH6 ACH5 ACH4 ACH3:0	1	**ANALOG CHANNELS.** These pins are analog inputs to the A/D converter. These pins are multiplexed with the port 0 pins. While it is possible for the pins to function simultaneously as analog and digital inputs, this is not recommended because reading the port while a conversion is in process can produce unreliable conversion results. The ANGND and V_{REF} pins must be connected for the A/D converter and the multiplexed port pins to function.	P0.7/T1DIR/PMODE.3 P0.6/T1CLK/PMODE.2 P0.5/PMODE.1 P0.4/PMODE.0 P0.3:0
AD15:8 AD7:0	I/O	**ADDRESS/DATA LINES.** These pins provide a multiplexed address and data bus. During the address phase of the bus cycle, address bits 0–15 are presented on the bus and can be latched using ALE or ADV#. During the data phase, 8- or 16-bit data is transferred.	P4.7:0/PBUS.15:8 P3.7:0/PBUS.7:0
ADV#	O	**ADDRESS VALID.** This active-low output signal is asserted only during external memory accesses. ADV# indicates that valid address information is available on the system address/data bus. The signal remains low while a valid bus cycle is in progress and is returned high as soon as the bus cycle completes. An external latch can use the ADV# signal to demultiplex the address from the address/data bus. Used with a decoder, ADV# can generate chip-selects for external memory.	P5.0/ALE
AINC#	I	**AUTO INCREMENT.** In slave programming mode, this active-low input signal enables the autoincrement mode. Auto increment allows reading from or writing to sequential OTPROM locations without requiring address transactions across the programming bus for each read or write.	P2.4/COMP0
ALE	O	**ADDRESS LATCH ENABLE.** This active-high output signal is asserted only during external memory cycles. ALE signals the start of an external bus cycle and indicates that valid address information is available on the system address/data bus. ALE differs from ADV# in that it is not returned high until a new bus cycle is to begin. An external latch can use ALE to demultiplex the address from the address/data bus.	P5.0/ADV#
ANGND	GND	**ANALOG GROUND.** Reference ground for the A/D converter and the logic used to read port 0. ANGND must be held at nominally the same potential as V_{SS}.	—
BCLK1 BCLK0	I	**SERIAL COMMUNICATIONS BAUD CLOCK 0 AND 1.** BCLK0 and 1 are alternate clock sources for the serial ports. The maximum input frequency is $F_{OSC}/4$.	P2.7/SCLK1# P2.1/SCLK0#/PALE#

Table 7. Signal Descriptions (Continued)

Signal Name	Type	Description	Multiplexed With
BHE#	O	**BYTE HIGH ENABLE.** During 16-bit bus cycles, this active-low output signal is asserted for word reads and writes and for high-byte reads and writes to external memory. BHE# indicates that valid data is being transferred over the upper half of the system address/data bus. BHE#, in conjunction with A0, selects the memory byte to be accessed: **BHE# A0 Byte(s) Accessed** 0 0 both bytes 0 1 high byte only 1 0 low byte only	P5.5/WRH#
BUSWIDTH	I	**BUS WIDTH.** When enabled in the chip configuration register, this active-high input signal dynamically selects the bus width of the bus cycle in progress. When BUSWIDTH is high, a 16-bit bus cycle occurs; when BUSWIDTH is low, an 8-bit bus cycle occurs. BUSWIDTH is active during a CCR fetch.	P5.7
COMP3 COMP2 COMP1 COMP0	O	**EVENT PROCESSOR ARRAY (EPA) COMPARE PINS.** These signals are the output of the EPA compare modules. These pins are multiplexed with other signals and may be configured as standard I/O.	P2.3 P2.6/CPVER P2.5/PACT# P2.4/AINC#
CPVER	O	**CUMULATIVE PROGRAM VERIFICATION.** This active-high output signal indicates whether any verify errors have occurred since the device entered programming mode. CPVER remains high until a verify error occurs, at which time it is driven low. Once an error occurs, CPVER remains low until the device exits programming mode. When high, CPVER indicates that all locations have programmed correctly since the device entered programming mode.	P2.6/COMP2
EA#	I	**EXTERNAL ACCESS.** This active-low input signal directs memory accesses to on-chip or off-chip memory. If EA# is low, the memory access is off-chip. If EA# is high and the memory address is within 2000H–2FFFH, the access is to on-chip ROM or OTPROM. Otherwise, an access with EA# high is to off-chip memory. EA# is sampled **only** on the rising edge of RESET#. If EA# = V_{EA} on the rising edge of RESET#. the device enters the programming mode selected by PMODE.3:0. For devices without ROM, EA# must be tied low.	—
EPA1 EPA0	I/O	**EVENT PROCESSOR ARRAY (EPA) INPUT/OUTPUT PINS.** These are the high-speed input/output pins for the EPA capture/compare modules. These pins are multiplexed with other signals and may be configured as standard I/O.	P2.2/PROG# P2.0/PVER

Table 7. Signal Descriptions (Continued)

Signal Name	Type	Description	Multiplexed With
EXTINT	I	**EXTERNAL INTERRUPT.** This programmable interrupt is controlled by the WG_PROTECT register. This register controls whether the interrupt is edge triggered or sampled and whether a rising edge/high level or falling edge/low level activates the interrupt. This interrupt vectors through memory location 203CH. If the chip is in idle mode and if EXTINT is enabled, a valid EXTINT interrupt brings the chip back to normal operation, where the first action is to execute the EXTINT service routine. After completion of the service routine, execution resumes at the instruction following the one that put the chip into idle mode. In powerdown mode, a valid EXTINT interrupt causes the chip to return to normal operating mode. If EXTINT is enabled, the EXTINT service routine is executed. Otherwise, execution continues at the instruction following the IDLPD instruction that put the chip into powerdown mode.	—
INST	O	**INSTRUCTION FETCH.** This active-high output signal is valid only during external memory bus cycles. When high, INST indicates that an instruction is being fetched from external memory. The signal remains high during the entire bus cycle of an external instruction fetch. INST is low for data accesses, including interrupt vector fetches and chip configuration byte reads. INST is low during internal memory fetches.	P5.1
NMI	I	**NONMASKABLE INTERRUPT.** In normal operating mode, a rising edge on NMI causes a vector through the NMI interrupt at location 203EH. NMI must be asserted for greater than one state time to guarantee that it is recognized. In idle mode, a rising edge on NMI brings the chip back to normal operation, where the first action is to execute the NMI service routine. After completion of the service routine, execution resumes at the instruction following the one that put the chip into idle mode. In powerdown mode, NMI causes a return to normal operating mode only if it is tied to EXTINT.	—
ONCE#	I	**ON-CIRCUIT EMULATION.** Holding this pin low while the RESET# signal transitions from a low to a high places the device into on-circuit emulation (ONCE) mode. ONCE mode isolates the device from other components in the system to allow the use of a clip-on emulator for system debugging. This mode puts all pins except XTAL1 and XTAL2 into a high-impedance state. To exit ONCE mode, reset the device by pulling the RESET# signal low.	P5.4

4

Table 7. Signal Descriptions (Continued)

Signal Name	Type	Description	Multiplexed With
P0.7 P0.6 P0.5 P0.4 P0.3:0	I	**PORT 0.** This is a high-impedance, input-only port. Port 0 pins should not be left floating. These pins may individually be used as analog inputs (ACHx) or digital inputs (P0.x). While it is possible for the pins to function simultaneously as analog and digital inputs, this is not recommended because reading port 0 while a conversion is in process can produce unreliable conversion results. ANGND and V_{REF} must be connected for port 0 and the A/D converter to function.	ACH7/T1DIR/PMODE.3 ACH6/T1CLK/PMODE.2 ACH5/PMODE.1 ACH4/PMODE.0 ACH3:0
P1.3 P1.2 P1.1 P1.0	I	**PORT 1.** This is a 4-bit, bidirectional, standard I/O port that is multiplexed with individually selectable special-function signals. (Used as PBUS.15:12 in Auto-programming Mode.)	RXD1 TXD1 RXD1 TXD0
P2.7 P2.6 P2.5 P2.4 P2.3 P2.2 P2.1 P2.0	I/O	**PORT 2.** This is an 8-bit, bidirectional, standard I/O port that is multiplexed with individually selectable special-function signals. P2.6 is multiplexed with a special test mode function. To prevent accidental entry into test modes, always configure P2.6 as an output.	SCLK1#/BCLK1 COMP2/CPVER COMP1/PACT# COMP0/AINC# COMP3 EPA1/PROG# SCLK0#/BCLK0/PALE# EPA0/PVER
P3.7:0	I/O	**PORT 3.** This is an 8-bit, bidirectional, memory-mapped I/O port with open-drain outputs. The pins are shared with the multiplexed address/data bus, which has complementary drivers. In programming modes, port 3 serves as the low byte of the programming bus (PBUS).	AD7:0/PBUS.7:0
P4.7:0	I/O	**PORT 4.** This is an 8-bit, bidirectional, memory-mapped I/O port with open-drain outputs. The pins are shared with the multiplexed address/data bus, which has complementary drivers. In programming modes, port 4 serves as the high byte of the programming bus (PBUS).	AD15:8/PBUS.15:8
P5.7 P5.6 P5.5 P5.4 P5.3 P5.2 P5.1 P5.0	I/O	**PORT 5.** This is an 8-bit, bidirectional, standard I/O port that is multiplexed with individually selectable control signals. Because P5.4 is multiplexed with the ONCE# function, always configure it as an output to prevent accidental entry into ONCE mode.	BUSWIDTH READY BHE#/WRH# ONCE# RD# WR#/WRL# INST ALE/ADV#

Table 7. Signal Descriptions (Continued)

Signal Name	Type	Description	Multiplexed With
P6.7 P6.6 P6.5 P6.4 P6.3 P6.2 P6.1 P6.0	O	**PORT 6.** This is an 8-bit output port that is multiplexed with the special functions of the waveform generator and PWM peripherals. The WG_OUT register configures the pins, establishes the output polarity, and controls whether changes to the outputs are synchronized with an event or take effect immediately.	PWM1 PWM0 WG3 WG3# WG2 WG2# WG1 WG1#
PACT#	O	**PROGRAMMING ACTIVE.** In auto-programming mode, PACT# low indicates that programming activity is occurring.	P2.5/COMP1
PALE#	I	**PROGRAMMING ALE.** In slave programming mode, this active-low input indicates that ports 3 and 4 contain a command/address. When PALE# is asserted, data and commands on ports 3 and 4 are read into the device.	P2.1/SCLK0#/BCLK0
PBUS.15:8 PBUS.7:0	I/O	**PROGRAMMING BUS.** In programming modes, used as a bidirectional port with open-drain outputs to pass commands, addresses, and data to or from the device. Used as a regular system bus to access external memory during auto-programming mode. When using slave programming mode, the PBUS is used in open-drain I/O port mode (not as a system bus). In slave programming mode, you must add external pull-up resistors to read data from the device during the dump word routine.	P4.7:0/AD15:8 P3.7:0/AD7:0
PMODE.3 PMODE.2 PMODE.1 PMODE.0	I	**PROGRAMMING MODE SELECT.** Determines the OTPROM programming algorithm that is to be performed. PMODE is sampled after a device reset when EA# = V_{EA} and must be stable while the device is operating.	P0.7/ACH7/T1DIR P0.6/ACH6/T1CLK P0.5/ACH5 P0.4/ACH4
PROG#	I	**PROGRAMMING START.** This active-low input is valid only in slave programming mode. The rising edge of PROG# latches data on the PBUS and begins programming. The falling edge of PROG# ends programming.	P2.2/EPA1
PVER	O	**PROGRAM VERIFICATION.** In programming modes, this active-high output signal is asserted to indicate that the word has programmed correctly. (PVER low after the rising edge of PROG# indicates an error.)	P2.0/EPA0
PWM1:0	O	**PULSE WIDTH MODULATOR OUTPUTS.** These are PWM output pins with high-current drive capability. The duty cycle and frequency-pulse-widths are programmable.	P6.7:6
RD#	O	**READ.** Read-signal output to external memory. RD# is asserted only during external memory reads.	P5.3

4

Table 7. Signal Descriptions (Continued)

Signal Name	Type	Description	Multiplexed With
READY	I	**READY INPUT.** This active-high input signal is used to lengthen external memory cycles for slow memory by generating wait states. When READY is high, CPU operation continues in a normal manner. If READY is low, the memory controller inserts wait states until the READY signal goes high or until the number of wait states is equal to the number programmed into the chip configuration register. READY is ignored for all internal memory accesses.	P5.6
RESET #	I/O	**RESET.** Reset input to and open-drain output from the chip. A falling edge on RESET # initiates the reset process. When RESET # is first asserted, the chip turns on a pull-down transistor connected to the RESET pin for 16 state times. This function can also be activated by execution of the RST instruction. In the powerdown and idle modes, asserting RESET # causes the chip to reset and return to normal operating mode. RESET # is a level-sensitive input.	—
RXD1 RXD0	I/O	**RECEIVE SERIAL DATA 0 AND 1.** In modes 1, 2, and 3, RXD0 and 1 are used to receive serial port data. In mode 0, they function as either inputs or open-drain outputs for data.	P1.3 P1.1
SCLK1 # SCLK0 #	I/O	**SYNCHRONOUS CLOCK PIN 0 AND 1.** In mode 4, these are the bidirectional, shift clock signals that synchronize the serial data transfer. Data is transferred 8 bits at a time with the LSB first. The DIR bit (SP_CONx.7) controls the direction of SCLKx signal. DIR = 0 The internal shift clock is output on SCLKx. DIR = 1 An external shift clock is input on SCLKx.	P2.7/BCLK1 P2.1/BCLK0
T1CLK	I	**EXTERNAL CLOCK.** External clock for timer 1. Timer 1 increments (or decrements) on both rising and falling edges of T1CLK. Also used in conjunction with T1DIR for quadrature counting mode.	P0.6/ACH6/PMODE.2
T1DIR	I	**TIMER 1 EXTERNAL DIRECTION.** External direction (up/down) for timer 1. Timer 1 increments when T1DIR is high and decrements when it is low. Also used in conjunction with T1CLK for quadrature counting mode.	P0.7/ACH7/PMODE.3
TXD1 TXD0		**TRANSMIT SERIAL DATA 0 AND 1.** In serial I/O modes 1, 2, and 3, TXD0 and 1 are used to transmit serial port data. In mode 0, they are used as the serial clock output.	P1.2 P1.0
V_{CC}	PWR	**DIGITAL SUPPLY VOLTAGE.** Connect each V_{CC} pin to the digital supply voltage.	—
V_{PP}	PWR	**PROGRAMMING VOLTAGE.** Set to 12.5V when programming the on-chip OTPROM. Also the timing pin for the "return from power-down" circuit.	—

Table 7. Signal Descriptions (Continued)

Signal Name	Type	Description	Multiplexed With
V_{REF}	PWR	**REFERENCE VOLTAGE FOR THE A/D CONVERTER.** V_{REF} is also the supply voltage to the analog portion of the A/D converter and the logic used to read Port 0. V_{REF} must be connected for the A/D and port 0 to function.	—
V_{SS}	GND	**DIGITAL CIRCUIT GROUND** (0 volts). Connect each V_{SS} pin to ground.	—
WG3 WG2 WG1	O	**WAVEFORM GENERATOR PHASE 1–3 POSITIVE OUTPUTS.** 3-phase output signals used in motion-control applications.	P6.5 P6.3 P6.1
WG3# WG2# WG1#	O	**WAVEFORM GENERATOR PHASE 1–3 NEGATIVE OUTPUTS.** Complementary 3-phase output signals used in motion-control applications.	P6.4 P6.2 P6.0
WR#	O	**WRITE.** This active-low output indicates that an external write is occurring. This signal is asserted only during external memory writes.	P5.2/WRL#
WRH#	0	**WRITE HIGH.** During 16-bit bus cycles, this active-low output signal is asserted for high-byte writes and word writes to external memory. During 8-bit bus cycles, WRH# is asserted for all write operations.	P5.5/BHE#
WRL#	O	**WRITE LOW.** During 16-bit bus cycles, this active-low output signal is asserted for low-byte writes and word writes. During 8-bit bus cycles, WRL# is asserted for all write operations.	P5.2/WR#
XTAL1	I	**CLOCK/OSCILLATOR INPUT.** Input to the on-chip oscillator inverter and the internal clock generator. Also provides the clock input for the serial I/O baud-rate generator, timers, and PWM unit. If an external oscillator is used, connect the external clock input signal to XTAL1 and ensure that the XTAL1 V_{IH} specification is met.	—
XTAL2	O	**OSCILLATOR OUTPUT.** Output of the on-chip oscillator inverter. When using the on-chip oscillator, connect XTAL2 to an external crystal or resonator. When using an external clock source, let XTAL2 float.	—

4

ELECTRICAL CHARACTERISTICS

NOTICE: This data sheet contains preliminary information on new products in production. It is valid for the devices indicated in the revision history. The specifications are subject to change without notice.

ABSOLUTE MAXIMUM RATINGS*

Storage Temperature .... −65°C to +150°C

Ambient Temperature
Under Bias .... −40°C to +85°

Voltage from V_{PP} or EA# to
V_{SS} or ANGND (Note 1) ...... −0.5V to +13.0V

Voltage with respect to
V_{SS} or ANGND (Note 1)........ −0.5V to +7.0V
(This includes V_{PP} on ROM and CPU devices.)

Power Dissipation........................1.5W
(based on package heat transfer limitations, not device power consumption)

WARNING: Stressing the device beyond the "Absolute Maximum Ratings" may cause permanent damage. These are stress ratings only. Operation beyond the "Operating Conditions" is not recommended and extended exposure beyond the "Operating Conditions" may affect device reliability.

OPERATING CONDITIONS*

T_A (Ambient Temperature
Under Bias).................. −40°C to +85°C

V_{CC} (Digital Supply Voltage) .... 4.50V to 5.50V

V_{REF} (Analog Supply Voltage) ...... 4.50V to 5.50V

F_{OSC} (Oscillator Frequency)
(Note 2) .... 8 MHz to 16 MHz

NOTES:

1. ANGND and V_{SS} should be at nominally the same potential.
2. Testing is performed down to 8 MHz, although the device is static by design and will typically operate below 1 Hz.

DC CHARACTERISTICS

Table 8. DC Characteristics over Specified Operating Conditions

Symbol	Parameter	Min	Typ (4)	Max	Units	Test Conditions
V_{IL}	Input Low Voltage (standard inputs (1))	−0.5		0.3 V_{CC}	V	
V_{IL1}	Input Low Voltage (RESET#, ports 3, 4, and 5)	−0.5		0.8	V	
V_{IH}	Input High Voltage (standard inputs (1))	0.7 V_{CC}		V_{CC} + 0.5	V	
V_{IH1}	Input High Voltage (RESET#, ports 3, 4, and 5)	0.2 V_{CC} + 1.0		V_{CC} + 0.5	V	
V_{OL}	Output Low Voltage (RESET#, ports 1, 2, 5, P6.6, P6.7, and XTAL2)			0.3 0.45 1.5	V V V	I_{OL} = 200 μA I_{OL} = 3.2 mA I_{OL} = 7.0 mA
V_{OL1}	Output Low Voltage (ports 3, 4)			1.0	V	I_{OL} = 7 mA
V_{OL2}	Output Low Voltage (P6.5:0)			0.45	V	I_{OL} = 10 mA
V_{OH}	Output High Voltage (output pins and I/O configured as push/pull outputs)	V_{CC} − 0.3 V_{CC} − 0.7 V_{CC} − 1.5			V V V	I_{OH} = − 200 μA I_{OH} = − 3.2 mA I_{OH} = − 7.0 mA
V_{TH+} − V_{TH-}	Hysteresis Voltage Width on RESET# pin	0.2			V	
I_{LI}	Input Leakage Current (standard inputs (1))			±10	μA	V_{SS} < V_{IN} < V_{CC} − 0.3V
I_{LI1}	Input Leakage Current (port 0 − A/D inputs)			±3	μA	V_{SS} < V_{IN} < V_{REF}
I_{IH}	Input High Current (NMI)			300	μA	V_{IN} = 0.7 V_{CC}
I_{IL}	Input Low Current (port 2, except P2.6)			−70	μA	V_{IN} = 0.3 V_{CC}

NOTES:
1. Standard input pins include XTAL1, EA#, and Ports 1 and 2 when configured as inputs.
2. Maximum current that an external device must sink to ensure test mode entry.
3. Violating these specifications during reset may cause the device to enter test modes.
4. Typical values are based on a limited number of samples and are not guaranteed. Operating conditions for typical values are room temperature and V_{REF} = V_{CC} = 5.5V.
5. Testing is performed down to 8 MHz, although the device is static by design and will typically operate below 1 Hz.
6. All voltages are referenced relative to V_{SS}. When used, V_{SS} refers to the device pin.
7. Table 9 lists the total current limits during normal (non-transient conditions). The total current listed is the sum of the pins listed for each specification value.

Table 8. DC Characteristics over Specified Operating Conditions (Continued)

Symbol	Parameter	Min	Typ (4)	Max	Units	Test Conditions
I_{IL1}	Input Low Current (P5.4 and P2.6 during reset) (2)			−10	mA	$V_{IN} = 0.8V$
I_{IL2}	Input Low Current (ports 3, 4, and 5, except P5.4)			−300	μA	$V_{IN} = 0.8V$
I_{IL3}	Input Low Current (port 1)			−300	μA	$V_{IN} = 0.3\ V_{CC}$
I_{OH}	Output High Current (P5.4 and P2.6 during reset) (3)	−0.2			mA	$0.7\ V_{CC}$
I_{OH1}	Output High Current (P6.5:0 during reset)	−6		−40	μA	$0.7\ V_{CC}$
I_{CC}	V_{CC} Supply Current		50	70	mA	XTAL1 = 16 MHz $V_{CC} = 5.5V$ $V_{PP} = 5.5V$ $V_{REF} = 5.5V$
I_{REF}	A/D Reference Supply Current		2	5	mA	
I_{IDLE}	Idle Mode Current		15	30	mA	
I_{PD}	Powerdown Mode Current (4)		5	50	μA	
R_{RST}	Reset Pull-up Resistor	6		65	kΩ	
C_S	Pin Capacitance (any pin to V_{SS})			10	pF	$F_{TEST} = 1.0$ MHz

NOTES:
1. Standard input pins include XTAL1, EA#, and Ports 1 and 2 when configured as inputs.
2. Maximum current that an external device must sink to ensure test mode entry.
3. Violating these specifications during reset may cause the device to enter test modes.
4. Typical values are based on a limited number of samples and are not guaranteed. Operating conditions for typical values are room temperature and $V_{REF} = V_{CC} = 5.5V$.
5. Testing is performed down to 8 MHz, although the device is static by design and will typically operate below 1 Hz.
6. All voltages are referenced relative to V_{SS}. When used, V_{SS} refers to the device pin.
7. Table 9 lists the total current limits during normal (non-transient conditions). The total current listed is the sum of the pins listed for each specification value.

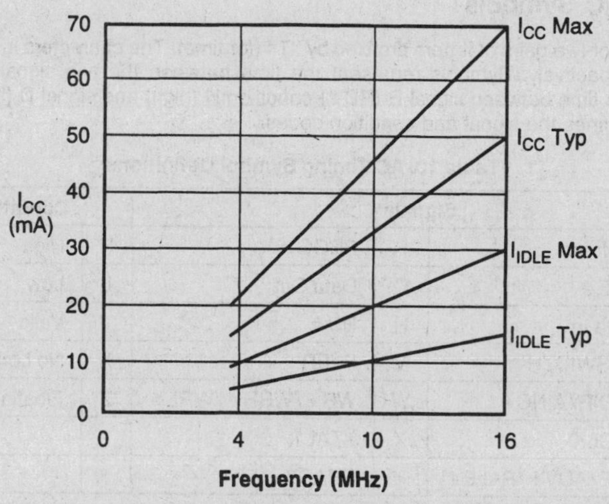

Figure 6. I_{CC}, I_{IDLE} versus Frequency

Table 9. Total Current Limits During Normal (Non-transient) Conditions

Signal Names	Maximum I_{OL} Limits	Maximum I_{OH} Limits
Port 1	25 mA	−25 mA
Port 2, P6.6, P6.7	40 mA	−40 mA
Port 3	40 mA	−30 mA
Port 4	40 mA	−30 mA
Port 5	40 mA	−30 mA
P6.5:0	40 mA	−30 mA

Explanation of AC Symbols

Each symbol consists of two pairs of letters prefixed by "T" (for time). The characters in a pair indicate a signal and its condition, respectively. Symbols represent the time between the two signal/condition points. For example, T_{RHDZ} is the time between signal R (RD#) condition H (high) and signal D (Input Data) condition Z (floating). Table 10 defines the signal and condition codes.

Table 10. AC Timing Symbol Definitions

Signals				Conditions	
A	Address	P	PROG#	H	High
B	BHE#	Q	Data Out	L	Low
D	Data In	R	RD#	V	Valid
G	BUSWIDTH	V	PVER	X	No Longer Valid
I	T1DIR/AINC#	W	WR#/WRH#/WRL#	Z	Floating
K	T1CLK	X	XTAL1		
L	ALE/ADV#/PALE#	Y	READY		

AC CHARACTERISTICS (OVER SPECIFIED OPERATION CONDITIONS)

Table 11 defines the AC timing specifications that the external memory system must meet and those that the 8XC196MH will provide.

Table 11. AC Timing Definitions (1)

Symbol	Parameter	Min	Max	Units	Notes
F_{OSC}	Frequency on XTAL1	8	16	MHz	4
T_{OSC}	$1/F_{OSC}$	62.5	125	ns	
The External Memory System Must Meet These Specifications					
T_{AVYV}	Address Valid to READY Setup		$2T_{OSC} - 75$	ns	
T_{LLYV}	ALE/ADV# Low to READY Setup		$T_{OSC} - 70$	ns	
T_{YLYH}	Non READY Time	No Upper Limit		ns	
T_{LLYX}	READY Hold after ALE/ADV# Low	$T_{OSC} - 15$	$2T_{OSC} - 40$	ns	2
T_{AVGV}	Address Valid to BUSWIDTH Setup		$2T_{OSC} - 75$	ns	
T_{LLGV}	ALE/ADV# Low to BUSWIDTH Setup		$T_{OSC} - 60$	ns	

NOTES:
1. Test Conditions: Capacitive load on all pins = 100 pF, rise and fall times = 10 ns, F_{OSC} = 16 MHz.
2. Exceeding the maximum specification causes additional wait states.
3. If wait states are used, add $2T_{OSC} \times n$, where n = number of wait states.
4. Testing is performed down to 8 MHz, although the device is static by design and will typically operate below 1 Hz.
5. Assuming back-to-back bus cycles.
6. 8-bit bus only.

Table 11. AC Timing Definitions (1)

Symbol	Parameter	Min	Max	Units	Notes
The External Memory System Must Meet These Specifications (Continued)					
T_{LLGX}	BUSWIDTH Hold after ALE/ADV# Low	T_{OSC}		ns	
T_{LHDV}	ALE/ADV# High to Input Data Valid		$3T_{OSC} - 55$	ns	
T_{AVDV}	Address Valid to Input Data Valid		$3T_{OSC} - 55$	ns	3
T_{RLDV}	RD# Active to Input Data Valid		$T_{OSC} - 30$	ns	3
T_{RHDZ}	End of RD# to Input Data Float		T_{OSC}	ns	
T_{RXDX}	Data Hold after RD# Inactive	0		ns	
The 8XC196MH will Meet These Specifications					
T_{XHLH}	XTAL1 Rising Edge to ALE Rising	20	110	ns	
T_{XHLL}	XTAL1 Rising Edge to ALE Falling	20	110	ns	
T_{LHLH}	ALE/ADV# Cycle Time	$4T_{OSC}$		ns	3
T_{LHLL}	ALE/ADV# High Period	$T_{OSC} - 10$	$T_{OSC} + 10$	ns	
T_{AVLH}	Address Valid to ALE/ADV# High	$T_{OSC} - 17$		ns	
T_{AVLL}	Address Valid to ALE/ADV# Low	$T_{OSC} - 17$		ns	
T_{LLAX}	Address Hold after ALE/ADV# Low	$T_{OSC} - 40$		ns	
T_{LLRL}	ALE/ADV# Low to RD# Low	$T_{OSC} - 30$		ns	
T_{RLRH}	RD# Low Period	$T_{OSC} - 5$	$T_{OSC} + 25$	ns	3
T_{RHLH}	RD# High to ALE/ADV# High	T_{OSC}	$T_{OSC} + 25$	ns	5
T_{RLAZ}	RD# Low to Address Float		5	ns	
T_{LLWL}	ALE/ADV# Low to WR# Low	$T_{OSC} - 10$		ns	
T_{QVWH}	Data Valid before WR# High	$T_{OSC} - 23$		ns	
T_{WLWH}	WR# Low Period	$T_{OSC} - 30$		ns	3
T_{WHQX}	Data Hold after WR# High	$T_{OSC} - 25$		ns	
T_{WHLH}	WR# High to ALE/ADV# High	$T_{OSC} - 10$	$T_{OSC} + 15$	ns	5
T_{WHBX}	BHE#, INST Hold after WR# High	$T_{OSC} - 10$		ns	
T_{WHAX}	A15:8 Hold after WR# High	$T_{OSC} - 30$		ns	6
T_{RHBX}	BHE#, INST Hold after RD# High	$T_{OSC} - 10$		ns	
T_{RHAX}	A15:8 Hold after RD# High	$T_{OSC} - 30$		ns	6

NOTES:
1. Test Conditions: Capacitive load on all pins = 100 pF, rise and fall times = 10 ns, F_{OSC} = 16 MHz.
2. Exceeding the maximum specification causes additional wait states.
3. If wait states are used, add $2T_{OSC} \times n$, where n = number of wait states.
4. Testing is performed down to 8 MHz, although the device is static by design and will typically operate below 1 Hz.
5. Assuming back-to-back bus cycles.
6. 8-bit bus only.

SYSTEM BUS TIMINGS

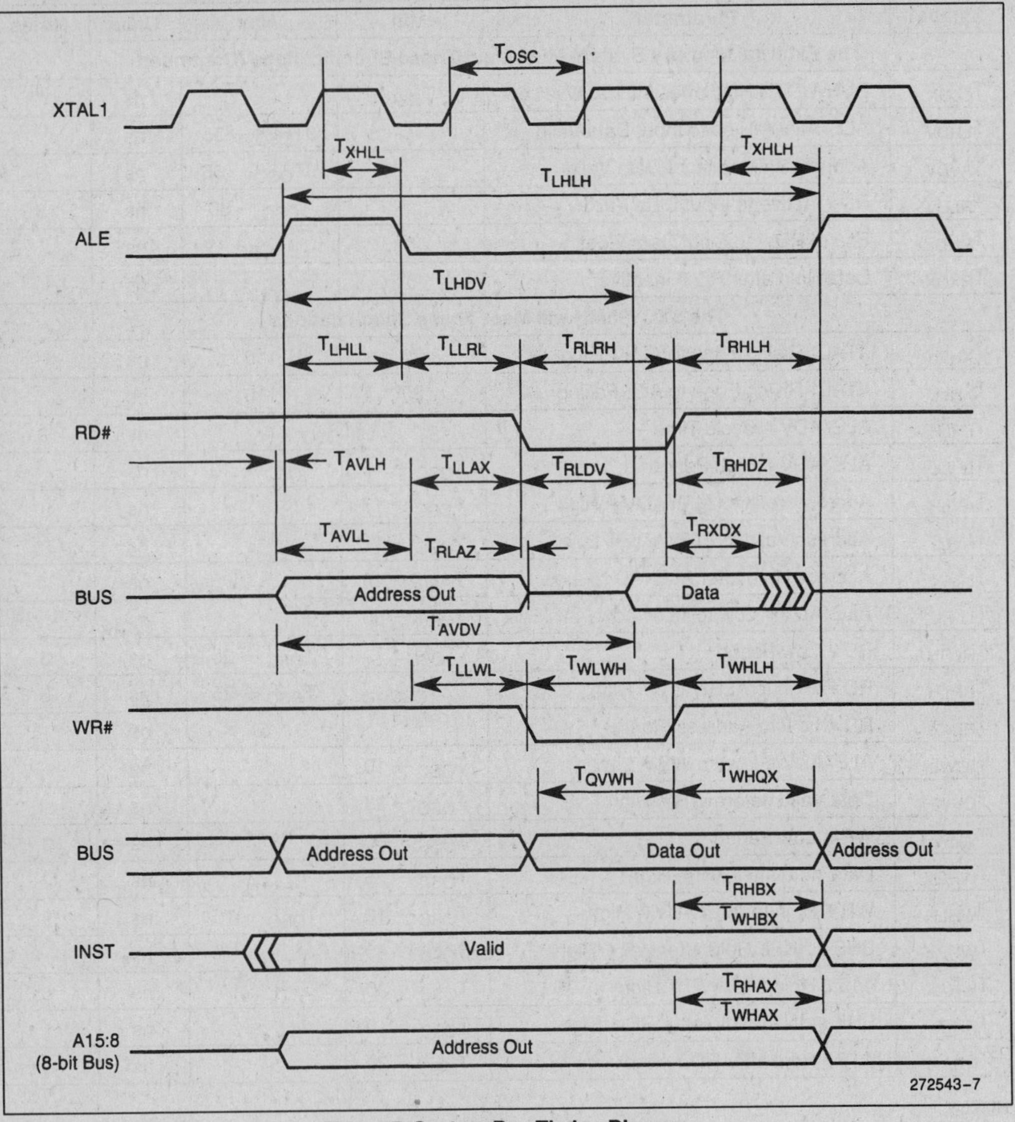

Figure 7. System Bus Timing Diagram

READY TIMING (ONE WAIT STATE)

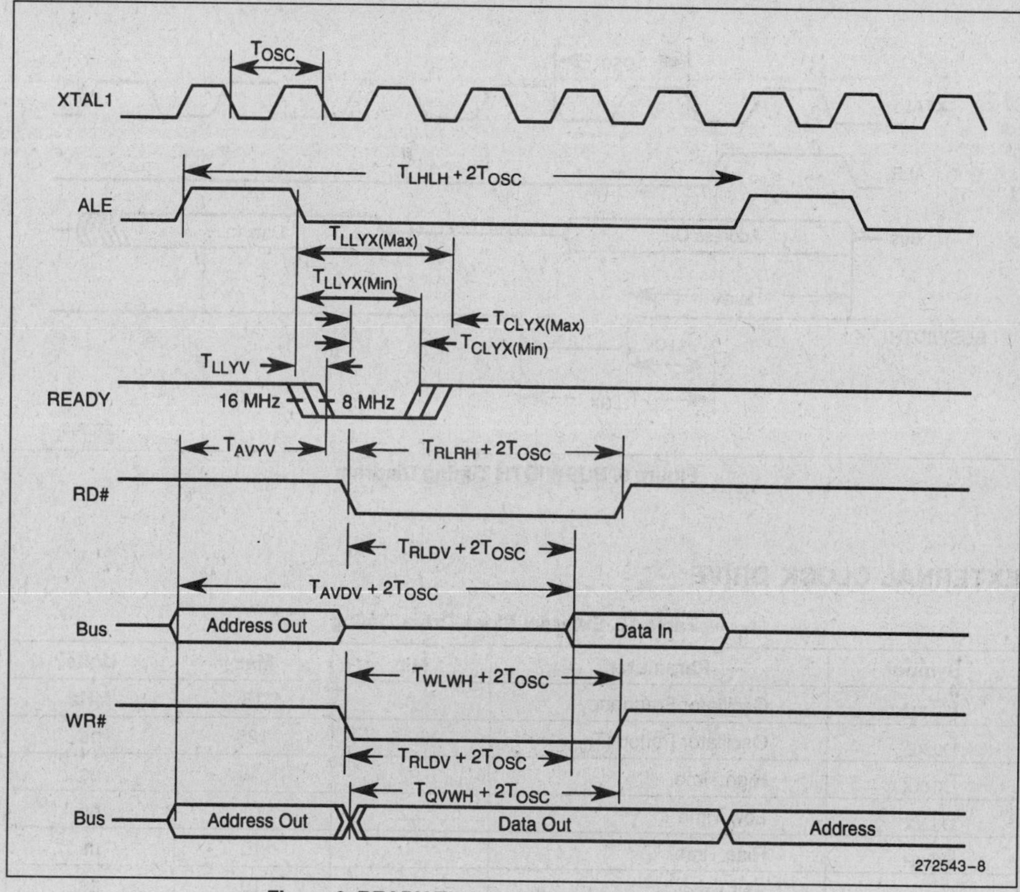

Figure 8. READY Timing Diagram (One Wait State)

BUSWIDTH TIMING

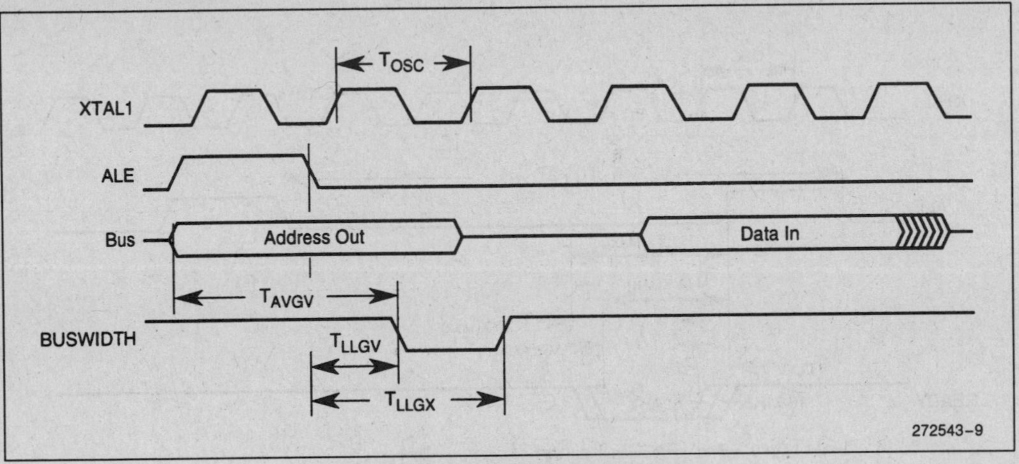

Figure 9. BUSWIDTH Timing Diagram

EXTERNAL CLOCK DRIVE

Table 12. External Clock Drive Timing

Symbol	Parameter	Min	Max	Units
$1/T_{XLXL}$	Oscillator Frequency	8	16	MHz
T_{XLXL}	Oscillator Period (T_{OSC})	62.5	125	ns
T_{XHXX}	High Time	22		ns
T_{XLXX}	Low Time	22		ns
T_{XLXH}	Rise Time		10	ns
T_{XHXL}	Fall Time		10	ns

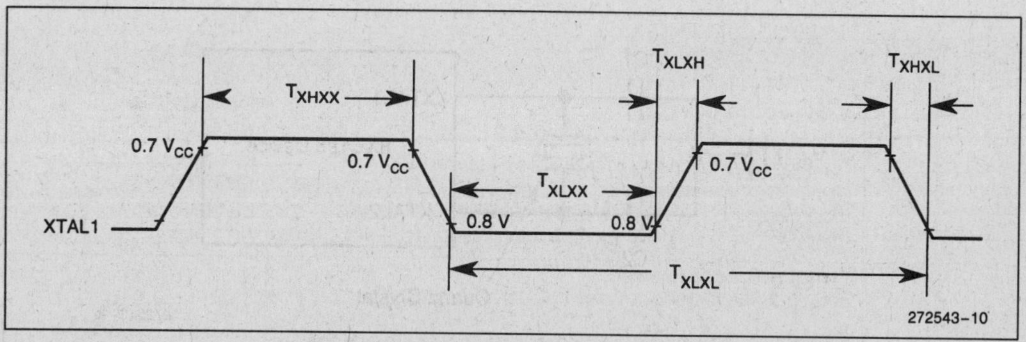

Figure 10. External Clock Drive Waveforms

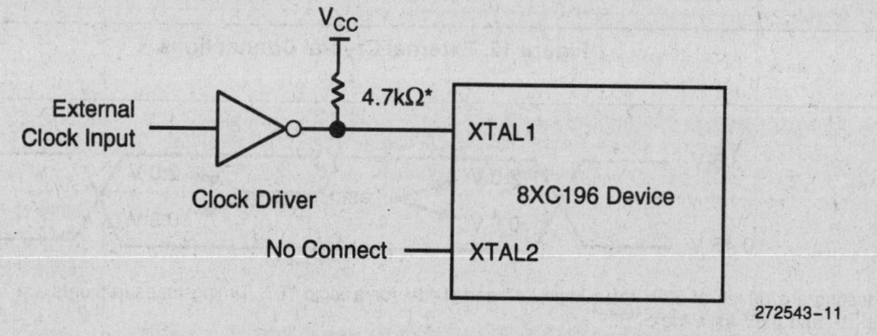

NOTE:
*Required if TTL driver is used. Not needed if CMOS driver is used.

Figure 11. External Clock Connections

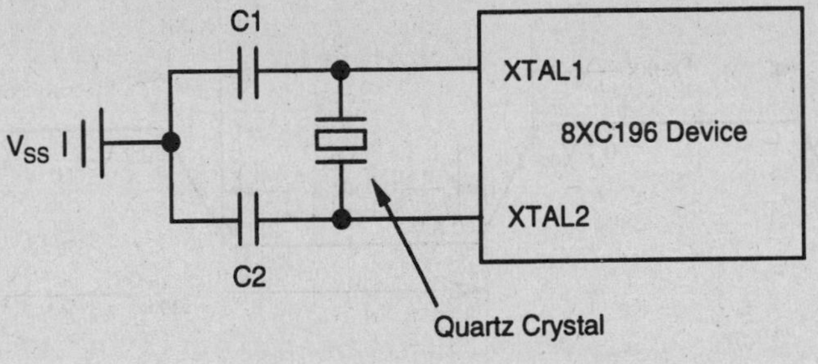

NOTE:
Keep oscillator components close to the chip and use short, direct traces to XTAL1, XTAL2, and V_{SS}. When using crystals, C1 = C2 = 20 pF. When using ceramic resonators, consult the manufacturer for recommended oscillator circuitry.

Figure 12. External Crystal Connections

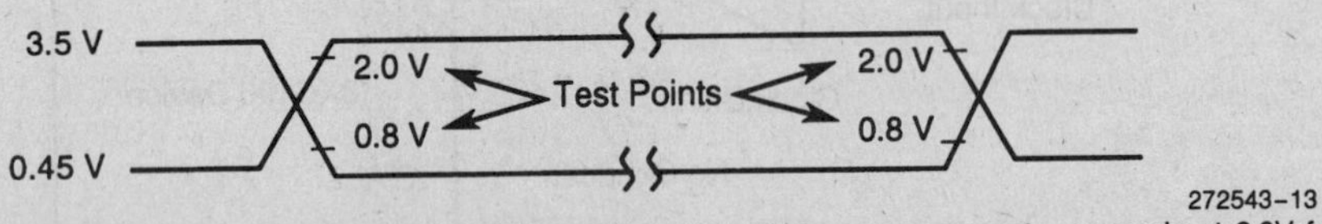

AC testing are driven at 3.5V for a logic "1" and 0.45V for a logic "0". Timing measurements are made at 2.0V for a logic "1" and 0.8V for a logic "0".

Figure 13. AC Testing Input, Output Waveforms

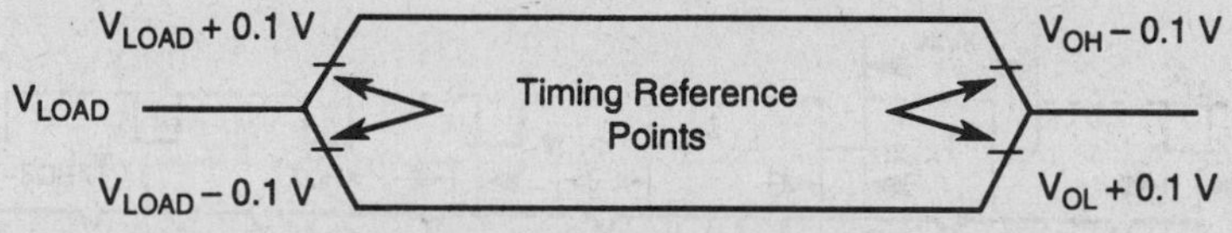

For timing purposes, a port pin is no longer floating when a 100 mV change from load voltage occurs and begins to float when a 100 mV change from the loading V_{OH}/V_{OL} level occurs with $I_{OL}/I_{OH} \leq$ 15 mA.

Figure 14. Float Waveforms

AC CHARACTERISTICS—SERIAL PORT, SHIFT REGISTER MODE

Table 13. Serial Port Timing—Shift Register Mode (Mode 0)

Symbol	Parameter	Min	Max	Units	Notes
T_{XLXL}	Serial Port Clock Period (Baud-rate *n* ≥ 8002H) (Baud-rate *n* = 8001H)	 $6T_{OSC}$ $4T_{OSC}$		 ns ns	 1, 2
T_{XLXH}	Serial Port Clock Low Period (Baud-rate *n* ≥ 8002H) (Baud-rate *n* = 8001H)	 $4T_{OSC} - 50$ $2T_{OSC} - 50$	 $4T_{OSC} + 50$ $2T_{OSC} + 50$	 ns ns	 1, 2
T_{QVXH}	Output Data Setup to Clock High	$2T_{OSC} - 50$		ns	
T_{XHQX}	Output Data Hold after Clock High	$2T_{OSC} - 50$		ns	
T_{XHQV}	Next Output Data Valid after Clock High		$2T_{OSC} + 50$	ns	
T_{DVXH}	Input Data Setup to Clock High	$T_{OSC} + 50$		ns	
T_{XHDX}	Input Data Hold after Clock High	0		ns	
T_{XHQZ}	Last Clock High to Output Float		T_{OSC}	ns	

NOTES:
1. n for Baud-rate n signifies Serial Port 0 or 1.
2. Maximum Serial Port Mode 0 reception is with Baud-rate n ≥ 8002H.

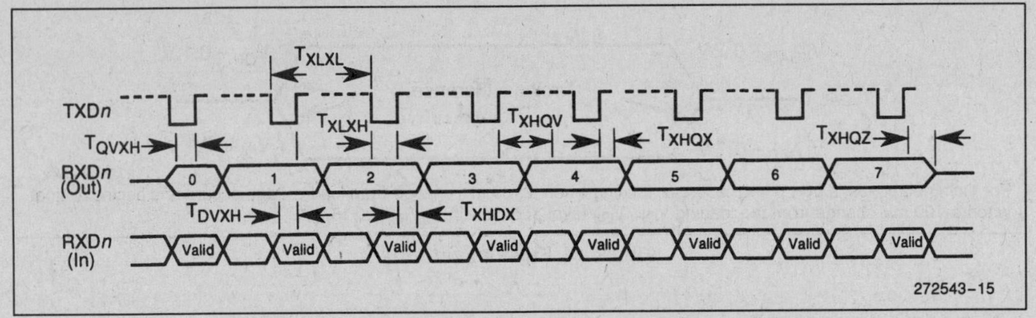

Figure 15. Serial Port Waveform—Shift Register Mode (Mode 0)

Table 14. Serial Port Timing—Mode 4

Symbol	Parameter	Min	Max	Units
T_{XLXL}	Serial Port Clock Period (DIR = 0)	$16T_{OSC}$	$131072T_{OSC}$	ns
T_{XLXX}	Serial Port Clock Low Period (DIR = 0/1)	$(T_{XLXL}/2) - 30$		ns
T_{XHXX}	Serial Port Clock High Period (DIR = 0/1)	$(T_{XLXL}/2) - 30$		ns
T_{XLXL}	Serial Port Clock Period (DIR = 1)	$16T_{OSC}$		ns
T_{XHXL}	Serial Clock Falling Time (DIR = 1)	0	20	ns
T_{XLXH}	Serial Clock Rising Time (DIR = 1)	0	20	ns
T_{XLQV}	Clock Low to Output Data Setup		$7.5T_{OSC} - 50$	ns
T_{XLQX}	Output Data Hold after Clock Low	0		ns
T_{XHQX}	Last Output Data Hold after Clock High (DIR = 1)	$13.7T_{OSC}$		ns
T_{DVXX}	Input Data Setup to Clock Low Invalid	0		ns
T_{XHDH}	Input Data Hold after Clock High	$6T_{OSC}$		ns

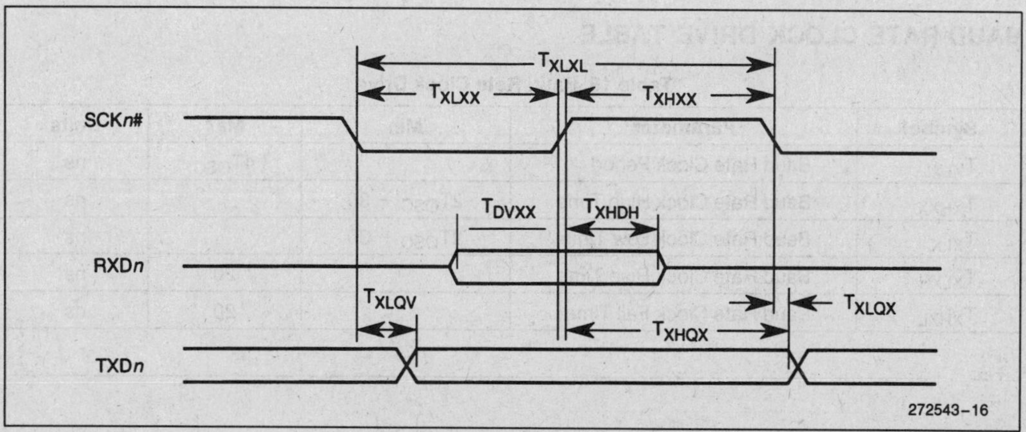

Figure 16. Serial Port Waveform—Mode 4

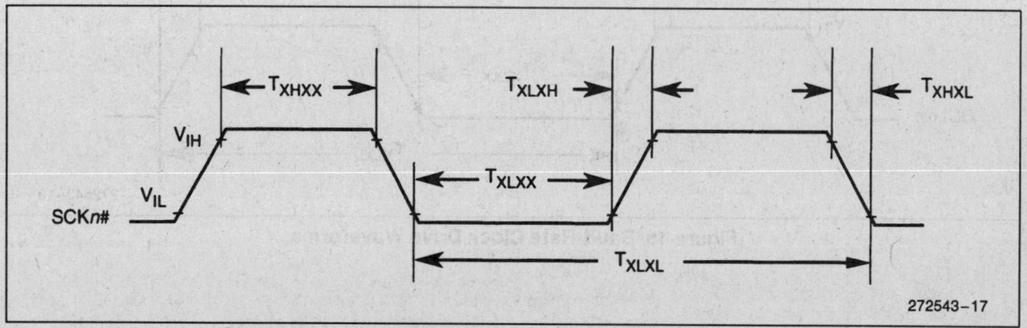

Figure 17. Serial Port Waveform—Clock Drive (DIR = 1)

BAUD-RATE CLOCK DRIVE TABLE

Table 15. Baud Rate Clock Drive

Symbol	Parameter	Min	Max	Units
T_{XLXL}	Baud Rate Clock Period		$4T_{OSC}$	ns
T_{XHXX}	Baud Rate Clock High Time	$2T_{OSC} - 30$		ns
T_{XLXX}	Baud Rate Clock Low Time	$2T_{OSC} - 30$		ns
T_{XLXH}	Baud Rate Clock Rise Time		20	ns
T_{XHXL}	Baud Rate Clock Fall Time		20	ns

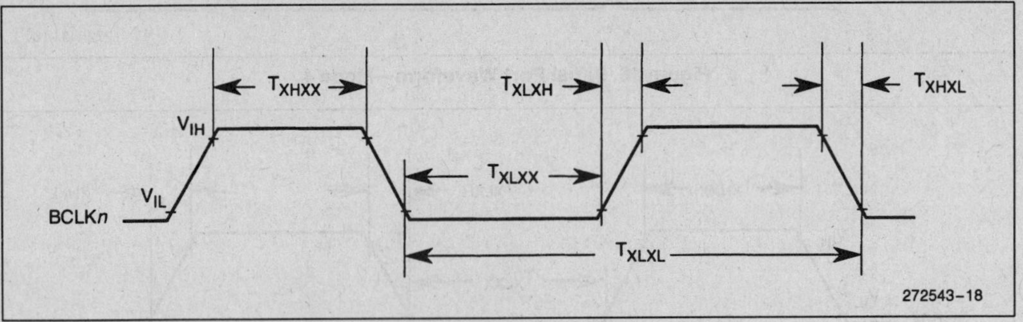

Figure 18. Baud-Rate Clock Drive Waveforms

A/D SAMPLE AND CONVERSION TIMES

Two parameters, sample time and conversion time, control the time required for an A/D conversion. The sample time is the length of time that the analog input voltage is actually connected to the sample capacitor. If this time is too short, the sample capacitor will not charge completely. If the sample time is too long, the input voltage may change and cause conversion errors. The conversion time is the length of time required to convert the analog input voltage stored on the sample capacitor to a digital value. The conversion time must be long enough for the comparator and circuitry to settle and resolve the voltage. Excessively long conversion times allow the sample capacitor to discharge, degrading accuracy.

The AD_TIME register programs the A/D sample and conversion times. Use the T_{SAM} and T_{CONV} specifications in Tables 16 and 18 to determine appropriate values for SAM and CONV; otherwise, erroneous conversion results may occur.

Use the following formulas to determine the SAM and CONV values:

$$SAM = \frac{T_{SAM} \times F_{OSC} - 2}{8}$$

$$CONV = \left[\frac{T_{CONV} \times F_{OSC} - 3}{2 \times B}\right] - 1$$

where:

SAM = 1 to 7
CONV = 2 to 31
T_{SAM} is the sample time, in μsec (Tables 16 and 18)
T_{CONV} is the conversion time, in μsec (Tables 16 and 18)
F_{OSC} is the XTAL1 frequency, in MHz
B is the number of bits to be converted (8 or 10)

When the SAM and CONV values are known, write them to the AD_TIME register. Do not write to this register while a conversion is in progress; the results are unpredictable.

4

AC Characteristics—A/D Converter

Table 16. 10-bit A/D Operating Conditions (1)

Symbol	Description	Min	Max	Units	Notes
T_A	Ambient Temperature	−40	+85	°C	
V_{CC}	Digital Supply Voltage	4.50	5.50	V	
V_{REF}	Analog Supply Voltage	4.50	5.50	V	2
T_{SAM}	Sample Time	1.0		μs	3
T_{CONV}	Conversion Time	10.0	20.0	μs	3
F_{OSC}	Oscillator Frequency	8	16	MHz	

NOTES:
1. ANGND and V_{SS} should nominally be at the same potential.
2. V_{REF} must not exceed V_{CC} by more than + 0.5 V because V_{REF} supplies both the resistor ladder and the analog portion of the converter and input port pins.
3. Program the AD_TIME register to meet the T_{SAM} and T_{CONV} specifications.

Table 17. 10-bit Mode A/D Characteristics Over Specified Operating Conditions (1)

Parameter	Typical (3)	Min	Max	Units (2)	Notes
Resolution		1024 10	1024 10	Levels Bits	
Absolute Error		0	±3	LSBs	
Full-scale Error	0.25 ± 0.5			LSBs	
Zero Offset Error	0.25 ± 0.5			LSBs	
Nonlinearity	1.0 ± 2.0		±3	LSBs	
Differential Nonlinearity		−0.75	+0.75	LSBs	
Channel-to-channel Matching	± 0.1	0	±1	LSBs	
Repeatability	±0.25	0		LSBs	

NOTES:
1. Testing is performed with $V_{REF} = 5.12V$ and $F_{OSC} = 16$ MHz.
2. An LSB, as used here, has a value of approximately 5 mV.
3. Typical values are based on a limited number of samples and are not guaranteed. Operating conditions for typical values are room temperature and $V_{REF} = V_{CC} = 5.5V$.
4. DC to 100 KHz.
5. Multiplexer break-before-make guaranteed.
6. Resistance from device pin, through internal multiplexer, to sample capacitor.
7. These values may be exceeded if the pin current is limited to ±2 mA.
8. Applying voltage beyond these specifications will degrade the accuracy of other channels being converted.
9. All conversions were performed with processor in idle mode.

Table 17. 10-bit Mode A/D Characteristics Over Specified Operating Conditions (1) (Continued)

Parameter	Typical (3)	Min	Max	Units (2)	Notes
Temperature Coefficients: Offset Full-scale Differential Nonlinearity	 0.009 0.009 0.009			 LSB/C LSB/C LSB/C	
Off-isolation		−60		dB	4, 5
Feedthrough	−60			dB	4
V_{CC} Power Supply Rejection	−60			dB	6
Input Series Resistance		750	1.2K	Ω	4
Voltage on Analog Input Pin		ANGND − 0.5	V_{REF} + 0.5	V	7, 8
Sampling Capacitor	3			pF	
DC Input Leakage	±1.0	0	±3	μA	

NOTES:
1. Testing is performed with V_{REF} = 5.12V and F_{OSC} = 16 MHz.
2. An LSB, as used here, has a value of approximately 5 mV.
3. Typical values are based on a limited number of samples and are not guaranteed. Operating conditions for typical values are room temperature and V_{REF} = V_{CC} = 5.5V.
4. DC to 100 KHz.
5. Multiplexer break-before-make guaranteed.
6. Resistance from device pin, through internal multiplexer, to sample capacitor.
7. These values may be exceeded if the pin current is limited to ±2 mA.
8. Applying voltage beyond these specifications will degrade the accuracy of other channels being converted.
9. All conversions were performed with processor in idle mode.

Table 18. 8-bit A/D Operating Conditions (1)

Symbol	Description	Min	Max	Units	Notes
T_A	Ambient Temperature	−40	+85	°C	
V_{CC}	Digital Supply Voltage	4.50	5.50	V	
V_{REF}	Analog Supply Voltage	4.50	5.50	V	2
T_{SAM}	Sample Time	1.0		μs	3
T_{CONV}	Conversion Time	7.0	20.0	μs	3
F_{OSC}	Oscillator Frequency	8	16	MHz	

NOTES:
1. ANGND and V_{SS} should nominally be at the same potential.
2. V_{REF} must not exceed V_{CC} by more than +0.5 V because V_{REF} supplies both the resistor ladder and the analog portion of the converter and input port pins.
3. Program the AD_TIME register to meet the T_{SAM} and T_{CONV} specifications.

Table 19. 8-bit Mode A/D Characteristics Over Specified Operating Conditions (1)

Parameter	Typical (3)	Min	Max	Units (2)	Notes
Resolution		256 8	256 8	Levels Bits	
Absolute Error		0	±1	LSBs	
Full-scale Error	±0.5			LSBs	
Zero Offset Error	±0.5			LSBs	
Nonlinearity		0	±1	LSBs	
Differential Nonlinearity		−0.5	+ 0.5	LSBs	
Channel-to-channel Matching		0	±1	LSBs	
Repeatability	±0.25	0		LSBs	
Temperature Coefficients: Offset Full-scale Differential Nonlinearity	 0.003 0.003 0.003			 LSB/°C LSB/°C LSB/°C	
Off Isolation		−60		dB	4, 5
Feedthrough	−60			dB	4
V_{CC} Power Supply Rejection	−60			dB	4
Input Series Resistance		750	1.2K	Ω	6
Voltage on Analog Input Pin		ANGND − 0.5	V_{REF} + 0.5	V	7, 8
Sampling Capacitor	3			pF	
DC Input Leakage	±1	0	±3	μA	

NOTES:
1. Testing is performed with V_{REF} = 5.12V and F_{OSC} = 16 MHz.
2. An LSB, as used here, has a value of approximately 20 mV.
3. Typical values are based on a limited number of samples and are not guaranteed. Operating conditions for typical values are room temperature and $V_{REF} = V_{CC}$ = 5.5V.
4. DC to 100 KHz.
5. Multiplexer break-before-make guaranteed.
6. Resistance from device pin, through internal multiplexer, to sample capacitor.
7. These values may be exceeded if the pin current is limited to ±2 mA.
8. Applying voltage beyond these specifications will degrade the accuracy of other channels being converted.
9. All conversions were performed with processor in idle mode.

OTPROM SPECIFICATIONS

Table 20. Programming Operating Conditions

Symbol	Description	Min	Max	Units	Notes
T_A	Ambient Temperature	20	30	°C	
V_{CC}	Supply Voltage During Programming	4.50	5.50	V	3
V_{REF}	Reference Supply Voltage During Programming	4.50	5.50V	V	3
V_{PP}	Programming Voltage	12.25	12.75	V	2
V_{EA}	EA Pin Voltage	12.25	12.75	V	2
F_{OSC}	Oscillator Frequency During Auto and Slave Mode Programming	6	8	MHz	
	Oscillator Frequency During Run-Time Programming	6	12	MHz	

NOTES:
1. V_{CC} and V_{REF} should be at nominally the same voltage during programming.
2. If V_{PP} and V_{EA} exceed the maximum specification, the device may be damaged.
3. V_{SS} and ANGND should be at nominally the same potential (0 volts).
4. Load capacitance during auto and slave mode programming = 150 pF.

Table 21. AC OTPROM Programming Characteristics

Symbol	Description	Min	Max	Units
T_{AVLL}	Address Setup Time	0		T_{OSC}
T_{LLAX}	Address Hold Time	100		T_{OSC}
T_{DVPL}	Data Setup Time	0		T_{OSC}
T_{PLDX}	Data Hold Time	400		T_{OSC}
T_{LLLH}	PALE# Pulse Width	50		T_{OSC}
T_{PLPH}	PROG# Pulse Width (1)	50		T_{OSC}
T_{PHLL}	PROG# High to Next PALE# Low	220		T_{OSC}
T_{PHDX}	Word Dump Hold Time		50	T_{OSC}
T_{PHPL}	PROG# High to Next PROG# Low	220		T_{OSC}
T_{LHPL}	PALE# High to PROG# Low	220		T_{OSC}
T_{PLDV}	PROG# Low to Word Dump Valid		50	T_{OSC}
T_{SHLL}	RESET# High to First PALE# Low	1100		T_{OSC}
T_{PHIL}	PROG# High to AINC# Low	0		T_{OSC}
T_{ILIH}	AINC# Pulse Width	240		T_{OSC}

NOTE:
1. This specification is for Word Dump Mode. For programming pulses, use the Modified Quick Pulse Algorithm explained in the User's Manual.

Table 21. AC OTPROM Programming Characteristics (Continued)

Symbol	Description	Min	Max	Units
T_{ILVH}	PVER Hold after AINC# Low	50		T_{OSC}
T_{ILPL}	AINC# Low to PROG# Low	170		T_{OSC}
T_{PHVL}	PROG# High to PVER Valid		220	T_{OSC}

NOTE:
1. This specification is for Word Dump Mode. For programming pulses, use the Modified Quick Pulse Algorithm explained in the User's Manual.

Table 22. DC OTPROM Programming Characteristics

Symbol	Parameter	Min	Max	Units
I_{PP}	V_{PP} Supply Current (when programming)		100	mA

NOTE:
Do not apply V_{PP} until V_{CC} is stable and within specifications and the oscillator/clock has stabiliized. Otherwise, the device may be damaged.

OTPROM POGRAMMING WAVEFORMS

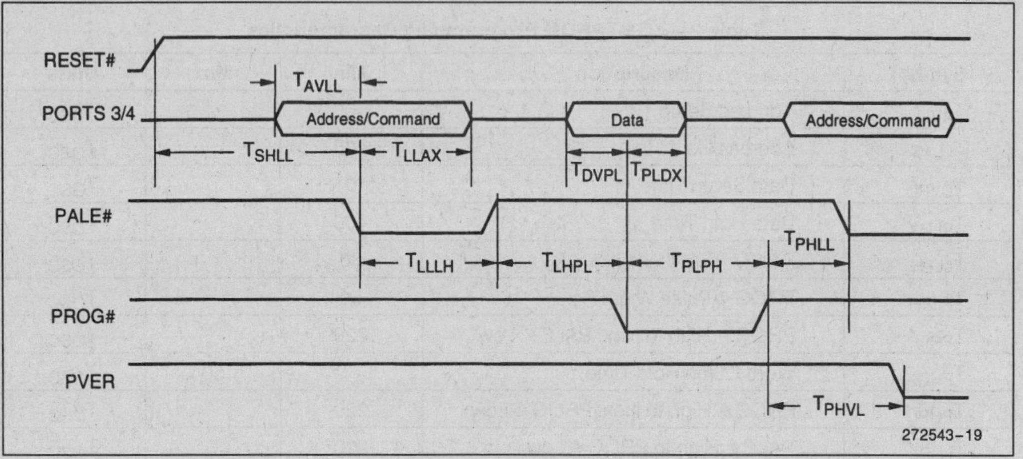

Figure 19. Slave Programming Mode Data Program Mode with Single Program Pulse

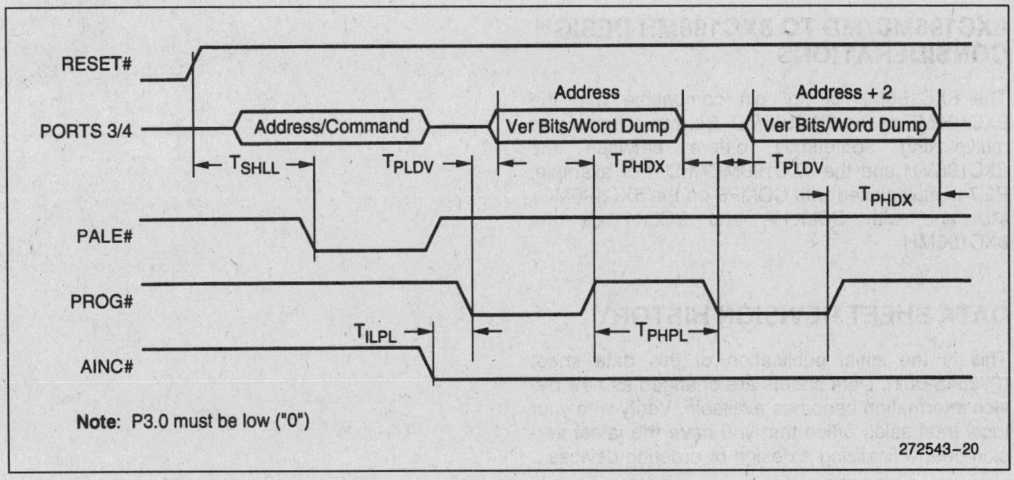

Figure 20. Slave Programming Mode in Word Dump with Autoincrement Timing

SLAVE PROGRAMMING MODE IN WORD DUMP WITH AUTOINCREMENT

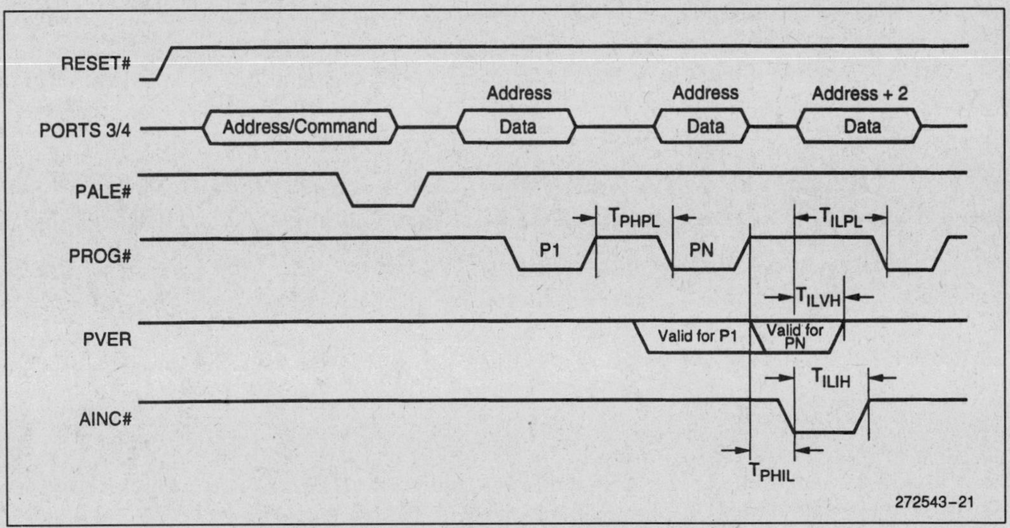

Figure 21. Slave Programming Mode in Data Program with Repeated Program Pulse and Autoincrement

8XC196MC/MD TO 8XC196MH DESIGN CONSIDERATIONS

The 8XC196MH is not pin compatible with the 8XC196MC or the 8XC196MD. Be aware that signal multiplexing sometimes differs between the 8XC196MH and the 8XC196MC/MD. For example, P2.7 is multiplexed with COMP3 on the 8XC196MC/MD and with SCLK1# and BCLK1 on the 8XC196MH.

DATA SHEET REVISION HISTORY

This is the initial publication of this data sheet (272543-001). Data sheets are changed as new device information becomes available. Verify with your local Intel sales office that you have the latest version before finalizing a design or ordering devices.

MCS®-51 and MCS®-96 Packaging Information

January 1994

Order Number: 272118-001

MCS-51 and MCS-96 Packaging Information

CONTENTS

Intel Product Identification Codes

_ _

Up to 15 Alphanumeric Characters for Device Types

Up to 6 Alphanumeric Characters to Show Customer Specific Requirements

Package Type

- A – Ceramic Pin Grid Array
- C – Ceramic Dual In-Line Package
- D – Cerdip Dual In-Line Package
- KU – Plastic Quad Flatpack Package, Fine Pitch, Die Up
- N – Plastic Leaded Chip Carrier
- P – Plastic Dual In-Line Package
- R – Ceramic Leadless Chip Carrier
- S – Quad Flatpack Package
- U – Plastic Dual In-Line Package (Shrink)

- L – Indicates extended operating temperature range (−40°C to +85°C) express product with 160 ± 8 hrs. dynamic burn-in.
- Q – Indicates commercial temperature range (0°C to 70°C) express product with 160 ± 8 hrs. dynamic burn-in
- T – Indicates extended temperature range (−40°C to +85°C) express product without burn-in.

272118–12

EXAMPLES:

N80C196KR	PLCC, 16 MHz, Commercial Temperature Range
LN87C54	PLCC, 12 MHz, Extended Temperature Range (Express)

40-LEAD PLASTIC DUAL IN-LINE PACKAGE (TYPE N)

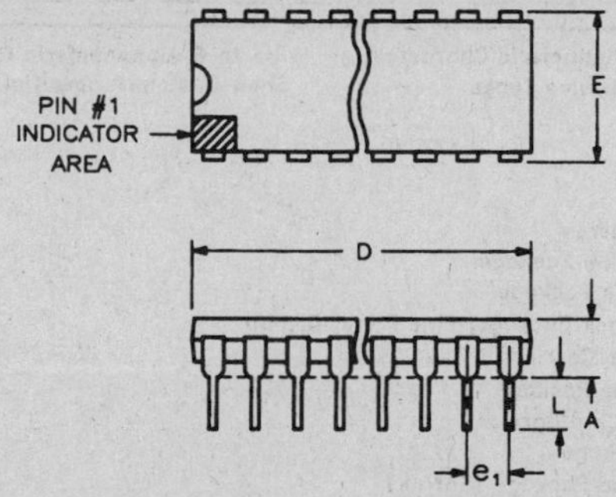

272118-1

Family: Plastic Dual In-Line Package		
Symbol	**Millimeters Approx***	**Inches Approx***
A	5	0.2
D	53	2.1
E	16	0.6
e_1	2.5	0.10
L	3	0.1

*For exact dimensions consult the Packaging Handbook (#240800).

40-LEAD CERDIP DUAL IN-LINE PACKAGE (TYPE D)

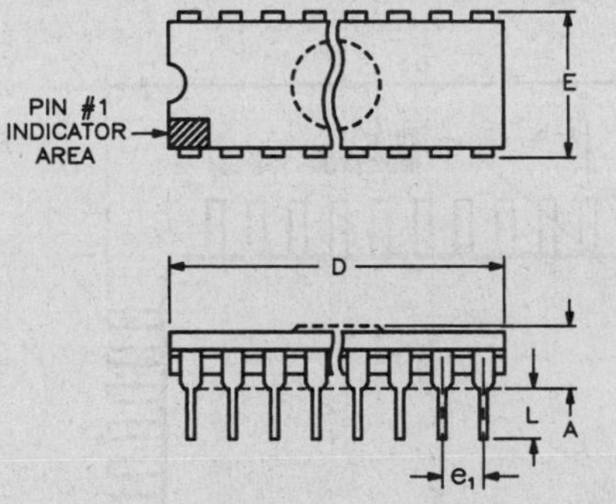

272118-2

Family: Cerdip Dual In-Line Package		
Symbol	Millimeters Approx*	Inches Approx*
A	5.8	0.2
D	53	2.1
E	16	0.6
e_1	2.5	0.10
L	3	0.1

*For exact dimensions consult the Packaging Handbook (#240800).

4

44-LEAD QUAD FLATPACK PACKAGE (TYPE S) VARIATION: SQUARE

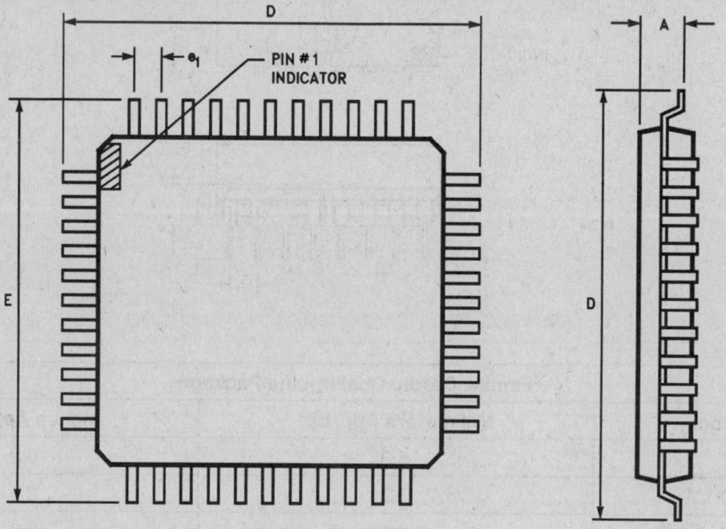

Family: Quad Flatpack Package	
Symbol	**Millimeters Approx***
A	3
D	13
E	13
e_1	0.8

*For exact dimensions consult the Packaging Handbook (#240800).

48-LEAD PLASTIC DUAL IN-LINE PACKAGE (TYPE N)

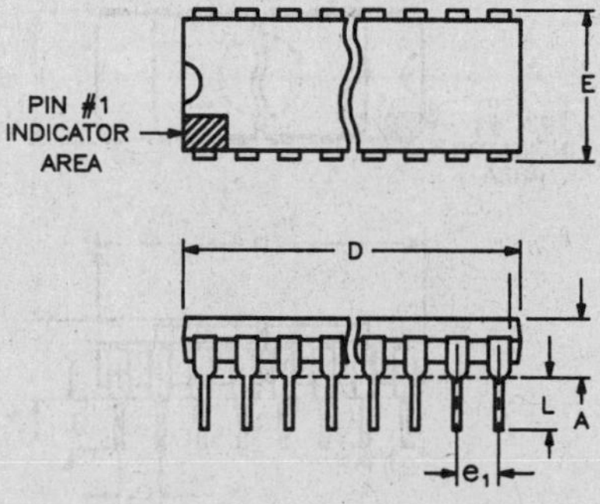

272118–4

Family: Plastic Dual In-Line Package		
Symbol	**Millimeters Approx***	**Inches Approx***
A	5	0.2
D	62	2.5
E	16	0.6
e_1	2.5	0.1
L	3	0.1

*For exact dimensions consult the Packaging Handbook (#240800).

4

48-LEAD CERAMIC DUAL IN-LINE PACKAGE (TYPE C)

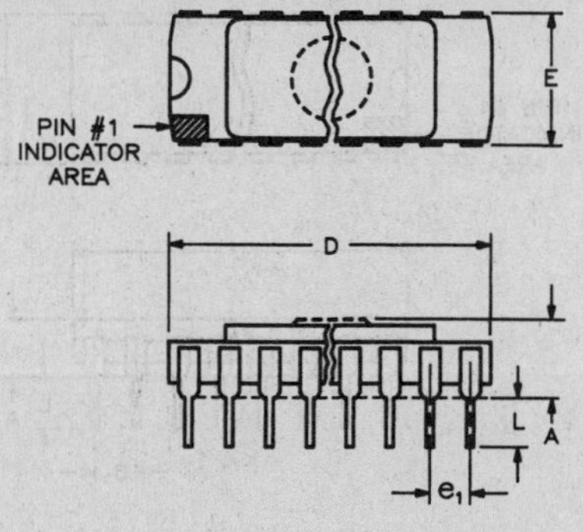

272118-5

Family: Ceramic Side Braze Dual In-Line		
Symbol	**Millimeters Approx***	**Inches Approx***
A	6(1)	0.2(1)
A	7(2)	0.3(2)
D	62	2.5
E	16	0.6
e_1	2.5	0.1
L	3	0.1

NOTES:
1. Solid LID
2. EPROM LID
*For exact dimensions consult the Packaging Handbook (#240800).

64-LEAD PLASTIC DUAL IN-LINE PACKAGE (SHRINK) (TYPE U)

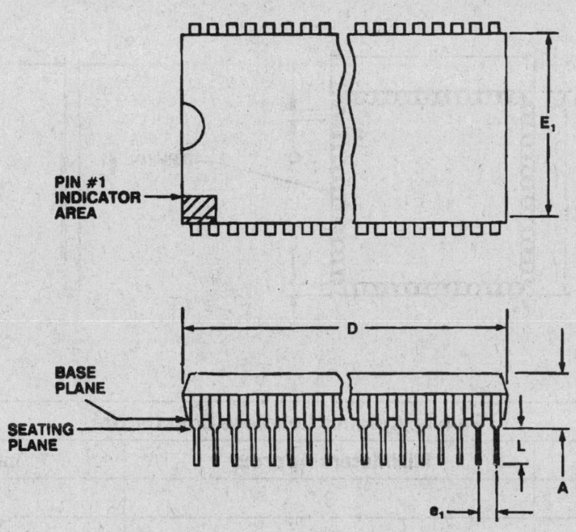

272118-6

Family: Plastic Dual In-Line Package		
Symbol	**Millimeters Approx***	**Inches Approx***
A	6	0.3
D	59	2.3
E_1	18	0.7
e_1	1.8	0.07
L	3	0.1

*For exact dimensions consult the Packaging Handbook (#240800).

68-CERAMIC LEADLESS CHIP CARRIER (TYPE R) VARIATION: B

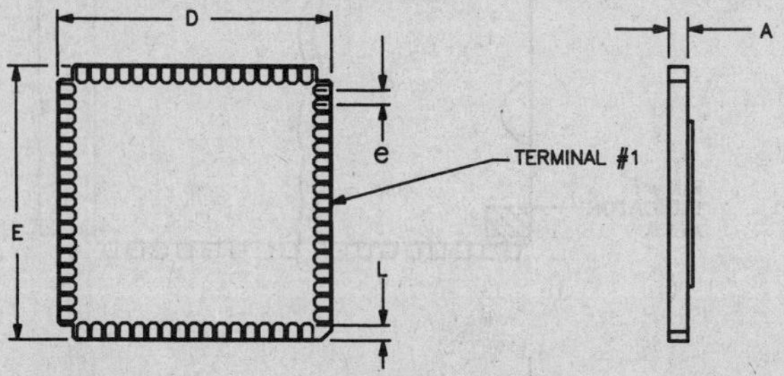

272118-7

Family: Ceramic Leadless Chip Carrier		
Symbol	**Millimeters Approx***	**Inches Approx***
A	3	0.1
D	25	1.0
E	25	1.0
e	1.3	0.05
L	1	0.1

*For exact dimensions consult the Packaging Handbook (#240800).

68-LEAD CERAMIC PIN GRID ARRAY PACKAGE (TYPE A)

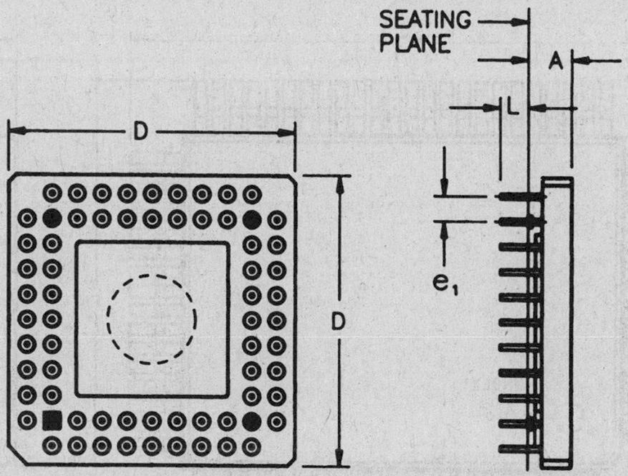

272118-8

Family: Ceramic Pin Grid Array Package		
Symbol	**Millimeters Approx***	**Inches Approx***
A	5	0.2
D	30	1.2
e_1	2.5	0.1
L	2	0.1

*For exact dimensions consult the Packaging Handbook (#240800).

4

80-LEAD QUAD FLATPACK PACKAGE (TYPE S)
VARIATION: RECTANGULAR

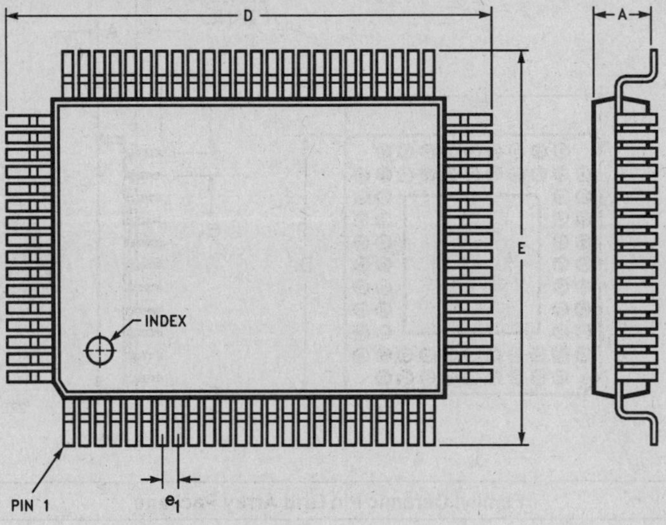

272118–9

Family: Quad Flatpack Package	
Symbol	Millimeters Approx*
A	3
D	25
E	19
e_1	0.8

*For exact dimensions consult the Packaging Handbook (#240800).

100-LEAD PLASTIC QUAD FLATPACK PACKAGE (TYPE KU)

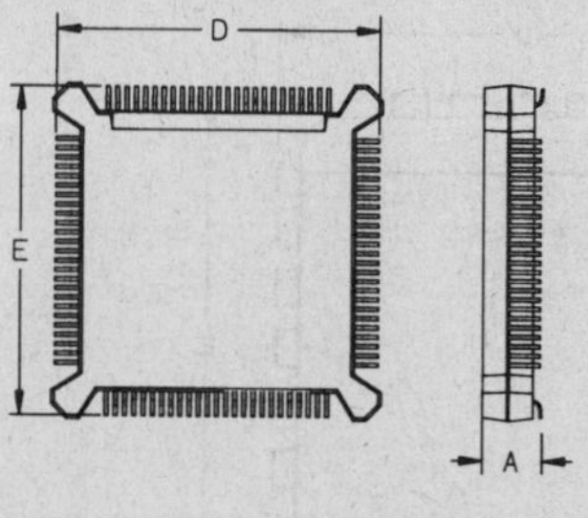

Family: Plastic Quad Flatpack (0.025 Inch (0.635mm) Pitch)			
Symbol	**Description**	**Inches Approx***	**Millimeters Approx***
A	Package Height	0.2	5
D, E	Terminal Dimension	0.9	23

*For exact dimensions consult the Packaging Handbook (#240800).

4

44/52/68/84-LEAD PLASTIC LEADED CHIP CARRIER (TYPE N)

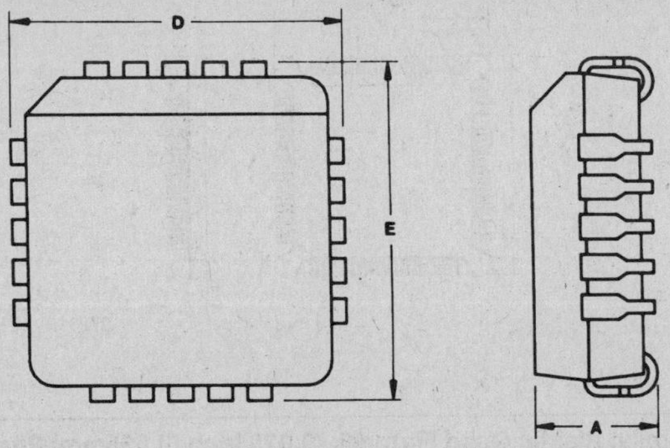

272118-11

Family: Plastic Leaded Chip Carrier—Square		
Symbol	44-Lead	
	Millimeters Approx*	Inches Approx*
A	5	0.2
D	18	0.7
E	18	0.7

Family: Plastic Leaded Chip Carrier—Square						
Symbol	Millimeters Approx*			Inches Approx*		
	68-Lead	52-Lead	84-Lead	68-Lead	52-Lead	84-Lead
A	5	5	5	0.2	0.2	0.2
D	26	21	31	1.0	0.8	1.2
E	26	21	31	1.0	0.8	1.2

*For exact dimensions consult the Packaging Handbook (#240800).

intel®

5

The RUPI Family

The RUPI-44 Family: Microcontroller with On-Chip Communication Controller

October 1988

Order Number: 296163-001

THE RUPI-44 FAMILY: MICROCONTROLLER WITH ON-CHIP COMMUNICATION CONTROLLER

CONTENTS

PAGE

INTRODUCTION

The RUPI-44 family is designed for applications requiring local intelligence at remote nodes, and communication capability among these distributed nodes. The RUPI-44 integrates onto a single chip Intel's highest performance microcontroller, the 8051-core, with an intelligent and high performance Serial communication controller, called the Serial Interface Unit, or SIU. See Figure 1. This dual controller architecture allows complex control and high speed data communication functions to be realized cost effectively.

The RUPI-44 family consists of three pin compatible parts:

- 8344—8051 Microcontroller with SIU
- 8044—An 8344 with 4K bytes of on-chip ROM program memory
- 8744—An 8344 with 4K bytes of on-chip EPROM program memory

1.0 ARCHITECTURE OVERVIEW

The 8044's dual controller architecture enables the RUPI to perform complex control tasks and high speed communication in a distributed network environment.

The 8044 microcontroller is the 8051-core, and maintains complete software compatibility with it. The microcontroller contains a powerful CPU with on-chip peripherals, making it capable of serving sophisticated real-time control applications such as instrumentation, industrial control, and intelligent computer peripherals. The microcontroller features on-chip peripherals such as two 16-bit timer/counters and 5 source interrupt capability with programmable priority levels. The microcontroller's high performance CPU executes most instructions in 1 microsecond, and can perform an 8 × 8 multiply in 4 microseconds. The CPU features a Boolean processor that can perform operations on 256 directly addressable bits. 192 bytes of on-chip data RAM can be extended to 64K bytes externally. 4K bytes of on-chip program ROM can be extended to 64K bytes externally. The CPU and SIU run concurrently. See Figure 2.

The SIU is designed to perform serial communications with little or no CPU involvement. The SIU supports data rates up to 2.4 Mbps, externally clocked, and 375 Kbps self clocked (i.e., the data clock is recovered by an on-chip digital phase locked loop). SIU hardware supports the HDLC/SDLC protocol: zero bit insertion/deletion, address recognition, cyclic redundancy check, and frame number sequence check are automatically performed.

The SIU's Auto mode greatly reduces communication software overhead. The AUTO mode supports the SDLC Normal Response Mode, by performing secondary station responses in hardware without any CPU involvement. The Auto mode's interrupt control and frame sequence numbering capability eliminates software overhead normally required in conventional systems. By using the Auto mode, the CPU is free to concentrate on real time control of the application.

5

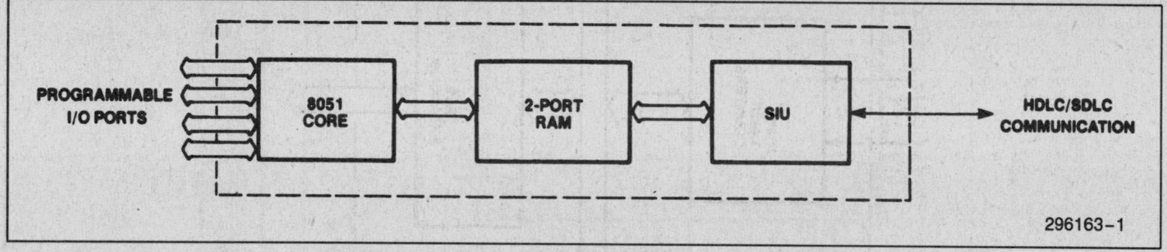

Figure 1. RUPI-44 Dual Controller Architecture

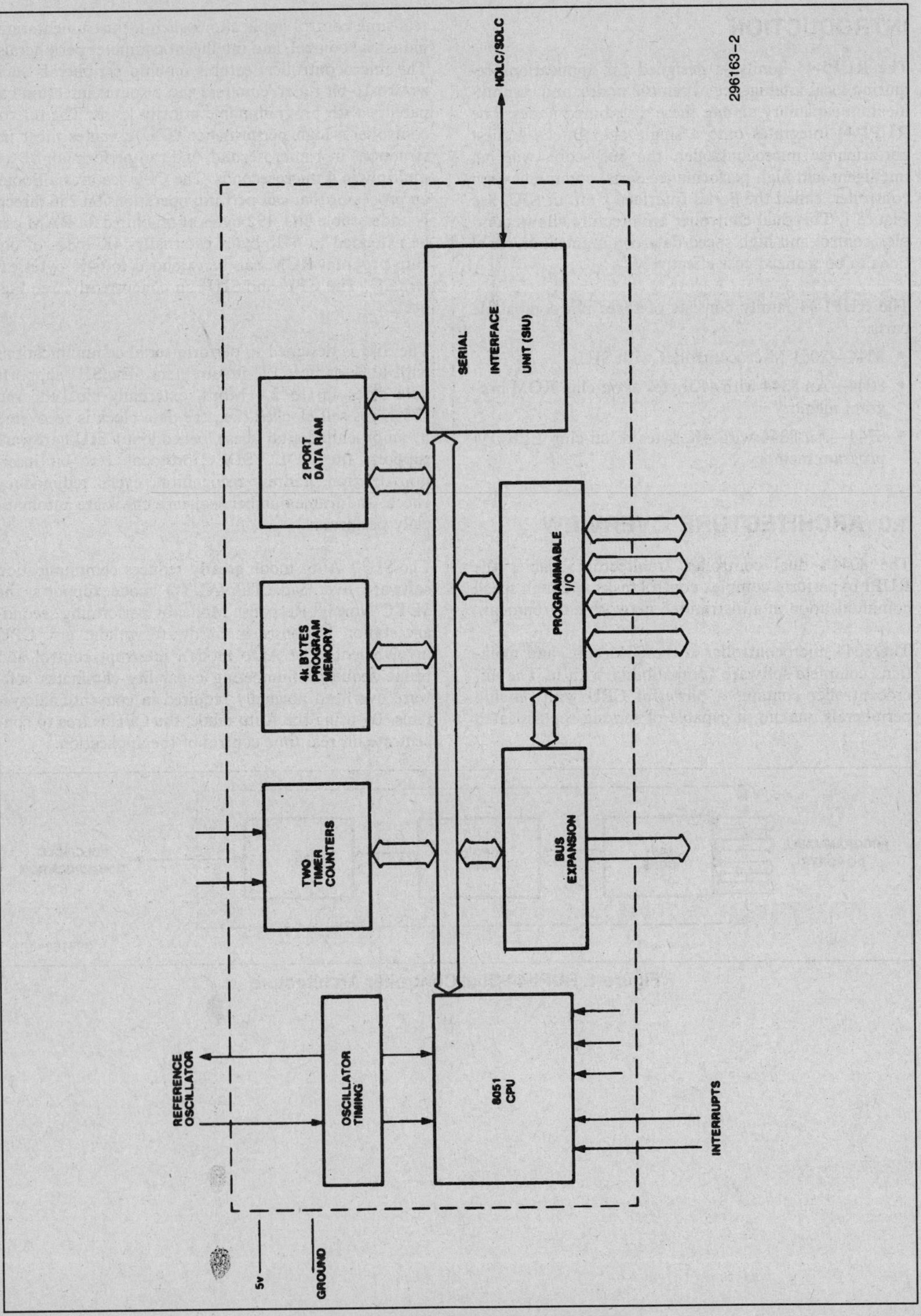

Figure 2. Simplified 8044 Block Diagram

2.0 THE HDLC/SDLC PROTOCOLS

2.1 HDLC/SDLC Advantages over Async

The High Level Data Link Control, HDLC, is a standard communication link control established by the International Standards Organization (ISO). SDLC is a subset of HDLC.

HDLC and SDLC are both well recognized standard serial protocols. The Synchronous Data Link Control, SDLC, is an IBM standard communication protocol. IBM originally developed SDLC to provide efficient, reliable and simple communication between terminals and computers.

The major advantages of SDLC/HDLC over Asynchronous communications protocol (Async):

- SIMPLE: Data Transparency
- EFFICIENT: Well Defined Message-Level Operation
- RELIABLE: Frame Check Sequence and Frame Numbering

The SDLC reduces system complexity. HDLC/SDLC are "data transparent" protocols. Data transparency means that an arbitrary data stream can be sent without concern that some of the data could be mistaken for a protocol controller. Data transparency relieves the communication controller having to detect special characters.

SDLC/HDLC provides more data throughout than Async. SDLC/HDLC runs at Message-level Operation which transmits multiple bytes within the frame, whereas Async is based on character-level operation. Async transmits or receives a character at a time. Since Async requires start and stop bits in every transmission, there is a considerable waste of overhead compared to SDLC/HDLC.

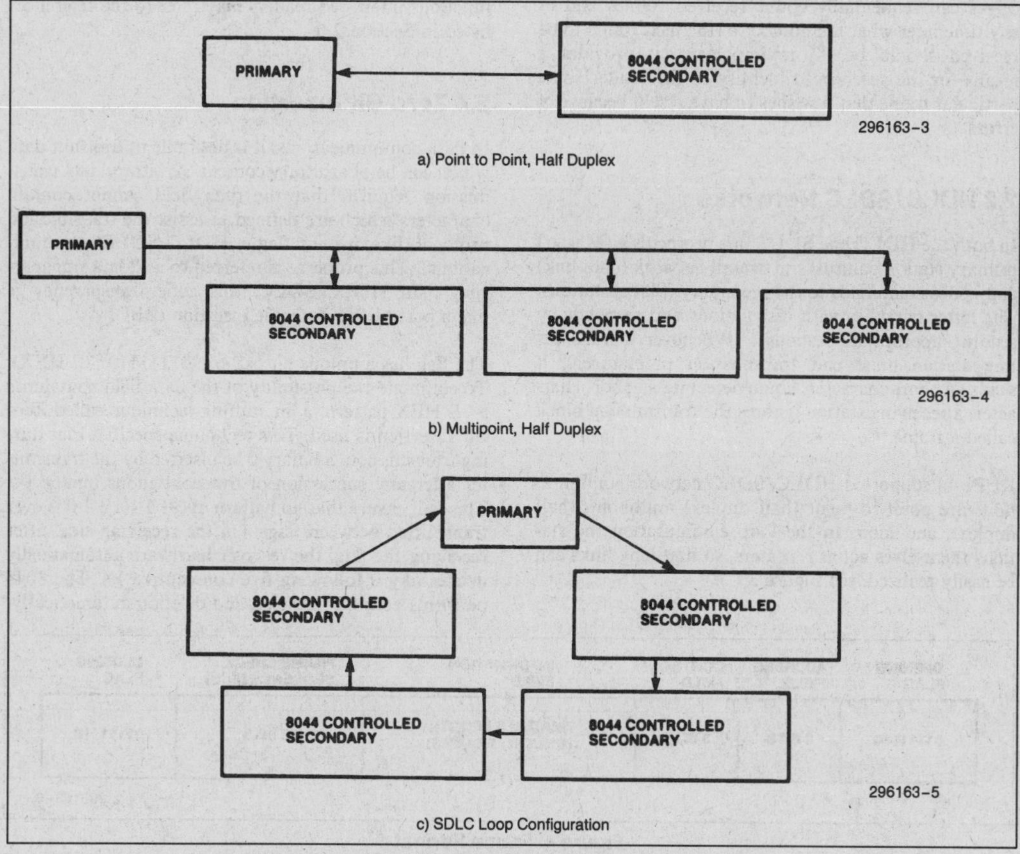

Figure 3. RUPI-44 Supported Network Configurations

Due to SDLC/HDLC's well delineated field (see Figure 4) the CPU does not have to interpret character by character to determine control field and information field. In the case of Async, CPU must look at each character to interpret what it means. The practical advantage of such feature is straight forward use of DMA for information transfer.

In addition, SDLC/HDLC further improves Data throughput using implied Acknowledgement of transferred information. A station using SDLC/HDLC may acknowledge previously received information while transmitting different information in the same frame. In addition, up to 7 messages may be outstanding before an acknowledgement is required.

The HDLC/SDLC protocol can be used to realize reliable data links. Reliable Data transmission is ensured at the bit level by sending a frame check sequence, cyclic redudancy checking, within the frame. Reliable frame transmission is ensured by sending a frame number identification with each frame. This means that a receiver can sequentially count received frames and at any time infer what the number of the next frame to be received should be. More important, it provides a means for the receiver to identify to the sender some particular frame that it wishes to have resent because of errors.

2.2 HDLC/SDLC Networks

In both the HDLC and SDLC line protocols a (Master) primary station controls the overall network (data link) and issues commands to the secondary (Slave) stations. The latter complies with instructions and responds by sending appropriate responses. Whenever a transmitting station must end transmission prematurely, it sends an abort character. Upon detecting an abort character, a receiving station ignores the transmission block called a frame.

RUPI-44 supported HDLC/SDLC network configurations are point to point (half duplex) multipoint (half duplex), and loop. In the loop configuration the stations themselves act as repeaters, so that long links can be easily realized, see Figure 3.

2.3 Frames

An HDLC/SDLC frame consists of five basic fields: Flag, Address, Control, Data and Error Detection. A frame is bounded by flags—opening and closing flags. An address field is 8 bits wide in SDLC, extendable to 2 or more bytes in HDLC. The control field is also 8 bits wide, extendable to two bytes in HDLC. The SDLC data field or information field may be any number of bytes. The HDLC data field may or may not be on an 8 bit boundary. A powerful error detection code called Frame Check Sequence contains the calculated CRC (Cycle Redundancy Code) for all the bits between the flags. See Figure 4.

In HDLC and SDLC are three types of frames; an Information Frame is used to transfer data, a Supervisory Frame is used for control purposes, and a Nonsequenced Frame is used for initialization and control of the secondary stations.

For a more detailed discussion of higher level protocol functions interested readers may refer to the references listed in Section 2.6.

2.4 Zero Bit Insertion

In data communications, it is desirable to transmit data which can be of arbitrary content. Arbitrary data transmission requires that the data field cannot contain characters which are defined to assist the transmission protocol (like opening flag in HDLC/SDLC communications). This property is referred to as "data transparency". In HDLC/SDLC, this code transparency is made possible by Zero Bit Insertion (ZBI).

The flag has a unique bit pattern: 01111110 (7E HEX). To eliminate the possibility of the data field containing a 7E HEX pattern, a bit stuffing technique called Zero Bit Insertion is used. This technique specifies that during transmission, a binary 0 be inserted by the transmitter after any succession of five contiguous binary 1's. This will ensure that no pattern of 0 1 1 1 1 1 1 0 is ever transmitted between flags. On the receiving side, after receiving the flag, the receiver hardware automatically deletes any 0 following five consecutive 1's. The 8044 performs zero bit insertion and deletion automatically.

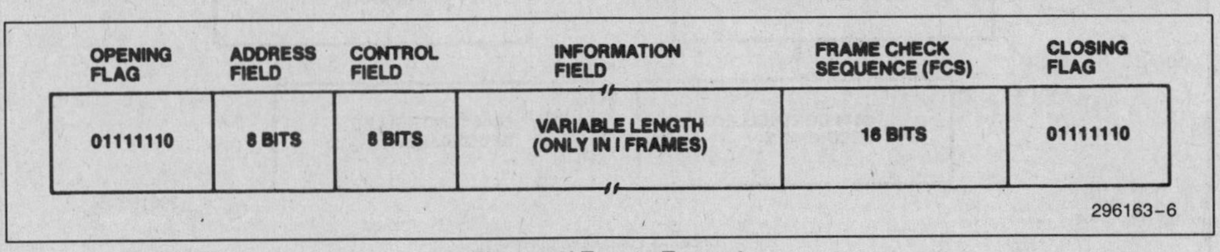

Figure 4. Frame Format

2.5 Non-return to Zero Inverted (NR21)

NRZI is a method of clock and data encoding that is well suited to the HDLC/SDLC protocol. It allows HDLC/SDLC protocols to be used with low cost asynchronous modems. NRZI coding is done at the transmitter to enable clock recovery from the data at the receiver terminal by using standard digital phase locked loop (DPLL) techniques. NRZI coding specifies that the signal condition does not change for transmitting a 1, while a 0 causes a change of state. NRZI coding ensures that an active data line will have a transition at least every 5-bit times (recall Zero Bit Insertion), while contiguous 0's will cause a change of state. Thus, ZBI and NRZI encoding makes it possible for the 8044's on-chip DPLL to recover a receive clock (from received data) synchronized to the received data and at the same time ensure data transparency.

2.6 References

1. *IBM Synchronous Data Link Control General Information GA27-3093-2 File No. GENL-09.*
2. *Standard Network Access Protocol Specification, DATAPAC Trans-Canada Telephone System CCG111.*
3. *IBM 3650 Retail Store System Loop Interface OEM Information, IBM, GA27-3098-0.*
4. *Guidebook to Data Communications, Training Manual, Hewlett-Packard 5955-1715.*
5. *"Serial Backplane Suits Multiprocessor Architectures", Mike Webb, Computer Design, July 1984, pp. 85–96.*
6. *"Serial Bus Simplifies Distributed Control", P.D. MacWilliams, Control Engineering, June 1984, pp. 101–104.*
7. *"Chips Support Two Local Area Networks", Bob Dahlberg, Computer Design, May 1984, pp. 107–114.*
8. *"Build a VLSI-based Workstation for the Ethernet Environment", Mike Webb, EDN, 23 February 1984, pp. 297–307.*
9. *"Networking With the 8044", Young Sohn & Charles Gopen, Digital Design, May 1984, pp. 136–137.*

3.0 RUPI-44 DESIGN SUPPORT

3.1 Design Tool Support

A critical design consideration is time to market. Intel provides a sophisticated set of design tools to speed hardware and software development time of 8044 based products. These include ICE-44, ASM-51, PL/M-51, and EMV-44.

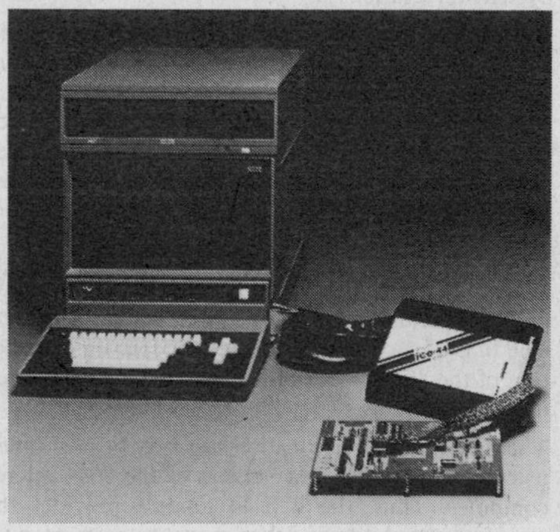

296163–7

Figure 5. RUPI-44 Development Support Configuration Intellec System, ICE™-44 Buffer Box, and ICE-44 Module Plugged into a User Prototype Board

A primary tool is the 8044 In Circuit Emulator, called ICE-44. See Figure 5. In conjunction with Intel's Intellec Microprocessor Development System, the ICE-44 emulator allows hardware and software development to proceed interactively. This approach is more effective than the traditional method of independent hardware and software development followed by system integration. With the ICE-44 module, prototype hardware can be added to the system as it is designed. Software and hardware integration occurs while the product is being developed.

The ICE-44 emulator assists four stages of development:

1) Software Debugging

   It can be operated without being connected to the user's system before any of the user's hardware is available. In this stage ICE-44 debugging capabilities can be used in conjunction with the Intellec text editor and 8044 macroassembler to facilitate program development.

2) Hardware Development

   The ICE-44 module's precise emulation characteristics and full-speed program RAM make it a valuable tool for debugging hardware, including the time-critical SDLC serial port, parallel port, and timer interfaces.

5

3) System Integration

Integration of software and hardware can begin when any functional element of the user system hardware is connected to the 8044 socket. As each section of the user's hardware is completed, it is added to the prototype. Thus, each section of the hardware and software is system tested in real-time operation as it becomes available.

4) System Test

When the user's prototype is complete, it is tested with the final version of the user system software. The ICE-44 module is then used for real-time emulation of the 8044 to debug the system as a completed unit.

The final product verification test may be performed using the 8744 EPROM version of the 8044 microcomputer. Thus, the ICE-44 module provides the user with the ability to debug a prototype or production system at any stage in its development.

A conversion kit, ICE-44 CON, is available to upgrade an ICE-51 module to ICE-44.

Intel's ASM-51 Assembler supports the 8044 special function registers and assembly program development. PL/M-51 provides designers with a high level language for the 8044. Programming in PL/M can greatly reduce development time, and ensure quick time to market.

These tools have recently been expanded with the addition of the EMV-44CON. This conversion kit allows you to convert an EMV-51 into an EMV-44 emulation vehicle. The resultant low cost emulator is designed for use with an iPDS Personal Development System, which also supports the ASM-51 assembler and PL/M-51. See Figure 6.

Emulation support is similar to the ICE-44 with support for Software and Hardware Development, System

296163–8

Figure 6. RUPI-44 iPDS Personal Development System, EMV-44 Buffer Box, and EMV-44 Module Plugged into a User Prototype Board

Integration, and System Test. The iPDS's rugged portability and ease of use also make it an ideal system for production tests and field service of your finished design. In addition, the iPDS offers EPROM programming module for the 8744, and direct communications with the 8044-based BITBUS via an optional iSBX-344 distributed control module.

3.2 8051 Workshop

Intel provides 8051 training to its customers through the 5-day 8051 workshop. Familiarity with the 8051 and 8044 is achieved through a combination of lecture and laboratory exercises.

For designers not familiar with the 8051, the workshop is an effective way to become proficient with the 8051 architecture and capabilities.

8044 Architecture

October 1988

Order Number: 296164-001

8044 ARCHITECTURE

CONTENTS

PAGE

GENERAL

The 8044 is based on the 8051 core. The 8044 replaces the 8051's serial port with an intelligent HDLC/SDLC controller called the Serial Interface or SIU. Thus the differences between the two result from the 8044's increased on-chip RAM (192 bytes) and additional special function registers necessary to control the SIU. Aside from the increased memory, the SIU itself, and differences in 5 pins (for the serial port), the 8044 and 8051 are compatible.

This chapter describes the differences between the 8044 and 8051. Information pertaining to the 8051 core, eg. instruction set, port operation, EPROM programming, etc. is located in the 8051 sections of this manual.

A block diagram of the 8044 is shown in Figure 1. The pinpoint is shown on the inside front cover.

1.0 MEMORY ORGANIZATION OVERVIEW

The 8044 maintains separate address spaces for Program Memory and Data Memory. The Program Memory can be up to 64K bytes long, of which the lowest 4K bytes are in the on-chip ROM.

If the $\overline{EA}$ pin is held high, the 8044 executes out of internal ROM unlwess the Program Counter exceeds 0FFFH. Fetches from locations 1000H through FFFFH are directed to external Program Memory.

If the $\overline{EA}$ pin is held low, the 8044 fetches all instructions from external Program Memory.

The Data Memory consists of 192 bytes of on-chip RAM, plus 35 Special Function Registers, in addition to which the device is capable of accessing up to 64K bytes of external data memory.

The Program Memory uses 16-bit addresses. The external Data Memory can use erither 8-bit or 16-bit addresses. The internal Data Memory uses 8-bit addresses, which provide a 256-location address space. The lower 192 addresses access the on-chip RAM. The Special Function Registers occupy various locations in the upper 128 bytes of the same address space.

The lowest 32 bytes in the internal RAM (locations 00 through 1FH) are divided into 4 banks of registers, each bank consisting of 8 bytes. Any one of these banks can be selected to be the "working registers" of the CPU, and can be accessed by a 3-bit address in the same byte as the opcode of an instruction. Thus, a large number of instructions are one-byte instructions.

The next higher 16 bytes of the internal RAM (locations 20H through 2FH) have individually addressable bits. These are provided for use as software flags or for one-bit (Boolean) processing. This bit-addressing capability is an important feature of the 8044. In addition to the 128 individually addressable bits in RAM, twelve of the Special Function Registers also have individually addressable bits.

A memory map is shown in Figure 2.

1.1 Special Function Registers

The Special Function Registers are as follows:

*	ACC	Accumulator (A Register)
*	B	B Register
*	PSW	Program Status Word
	SP	Stack Pointer
	DPTR	Data Pointer (consisting of DPH AND DPL)
*	P0	Port 0
*	P1	Port 1
*	P2	Port 2
*	P3	Port 3
*	IP	Interrupt Priority
*	IE	Interrupt Enable
	TMOD	Timer/Counter Mode
*	TCON	Timer/Counter Control
	TH0	Timer/Counter 0 (high byte)
	TL0	Timer/Counter 0 (low byte)
	TH1	Timer/Counter 1 (high byte)
	TL1	Timer/Counter 1 (low byte)
	SMD	Serial Mode
*	STS	Status/Command
*	NSNR	Send/Receive Count
	STAD	Station Address
	TBS	Transmit Buffer Start Address
	TBL	Transmit Buffer Length
	TCB	Transmit Control Byte
	RBS	Receive Buffer Start Address
	RBL	Receive Buffer Length
	RFL	Received Field Length
	RCB	Received Control Byte
	DMA CNT	DMA Count
	FIFO	FIFO (three bytes)
	SIUST	SIU State Counter
	PCON	Power Control

The registers marked with * are both byte- and bit-addressable.

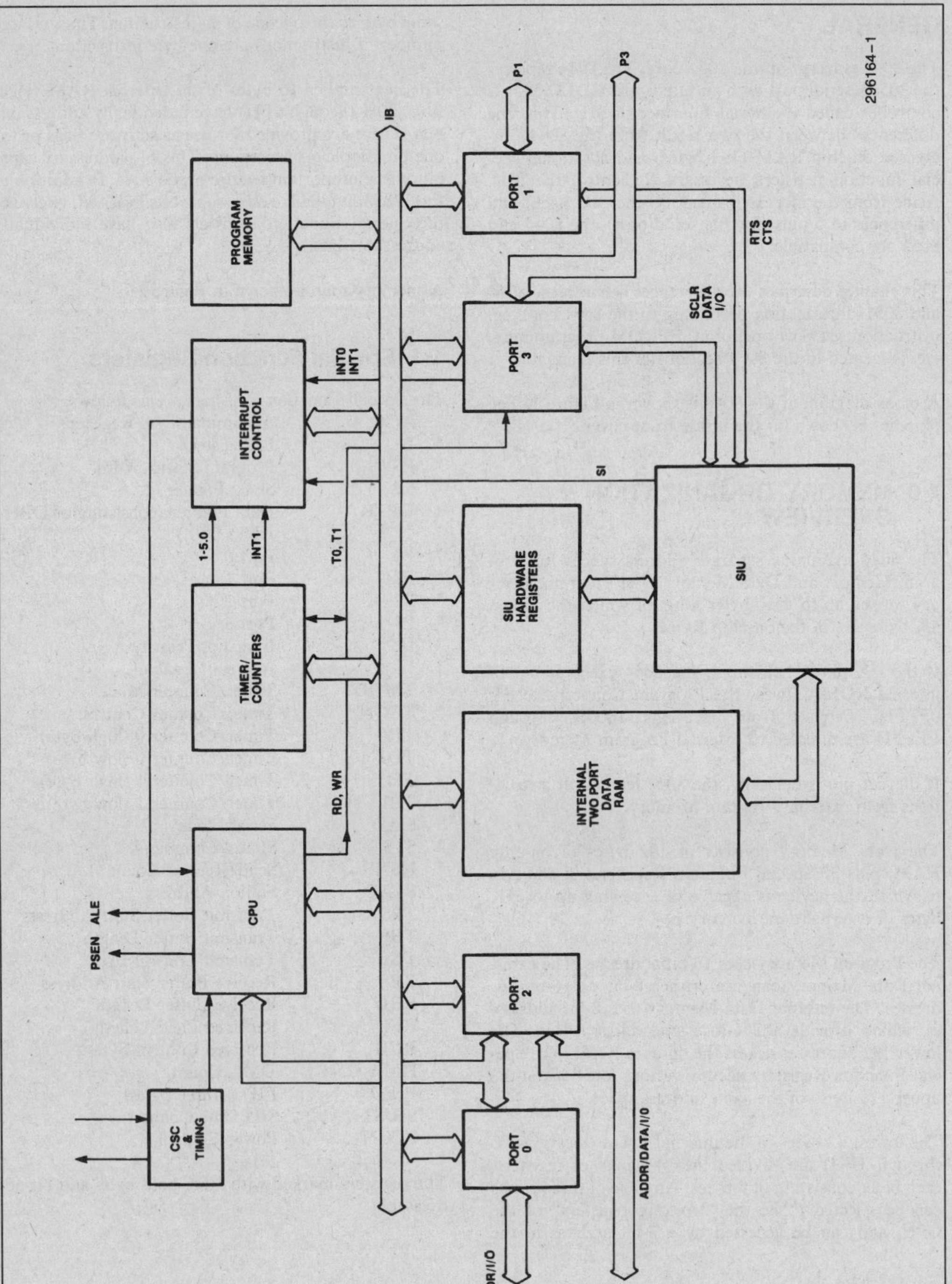

Figure 1. RUPI Block Diagram

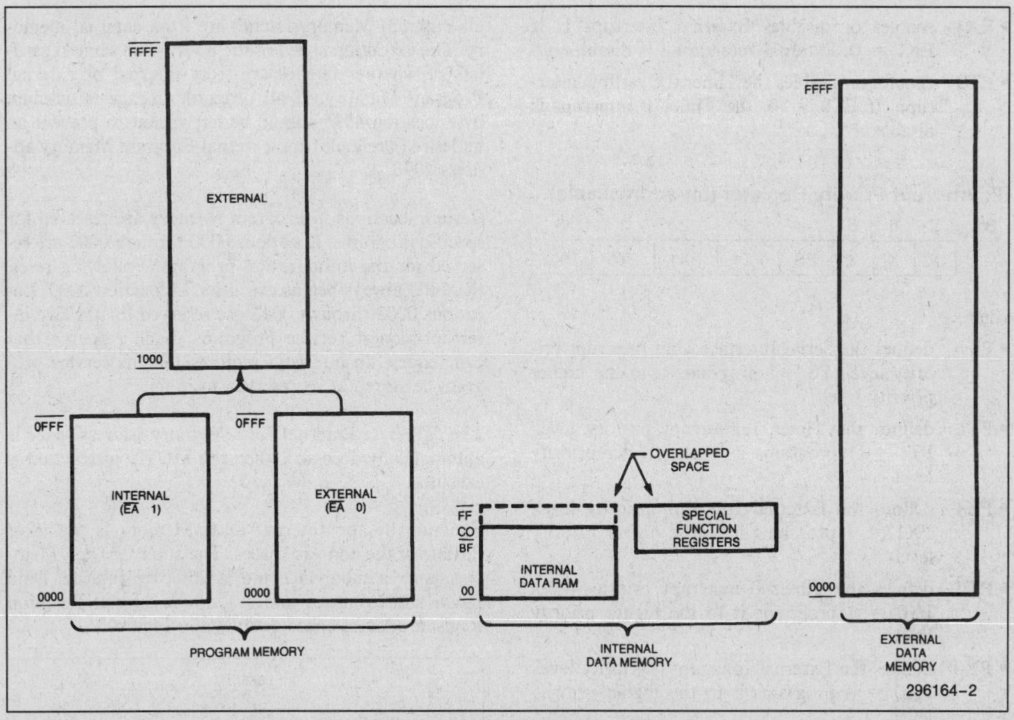

Figure 2. RUPI-44 Memory Map

Stack Pointer

The Stack Pointer is 8 bits wide. The stack can reside anywhere in the 192 bytes of on-chip RAM. When the 8044 is reset, the stack pointer is initialized to 07H. When executing a PUSH or a CALL, the stack pointer is incremented before data is stored, so the stack would begin at location 08H.

1.2 Interrupt Control Registers

The Interrupt Request Flags are as listed below:

Source	*Request Flag*	*Location*
External Interrupt 0	$\overline{INT0}$, if IT0 = 0 IE0, if IT0 = 1	P3.2 TCON.1
Timer 0 Overflow	TF0	TCON.5
External Interrupt 1	$\overline{INT1}$, if IT1 = 0 IE1, if IT1 = 1	P3.3 TCON.3
Timer 1 Overflow	TF1	TCON.7
Serial Interface Unit	SI	STS.4

External Interrupt control bits IT0 and IT1 are in TCON.0 and TCON.2, respectively. Reset leaves all flags inactive, with IT0 and IT1 cleared.

All the interrupt flags can be set or cleared by software, with the same effect as by hardware.

The Enable and Priority Control Registers are shown below. All of these control bits are set or cleared by software. All are cleared by reset.

IE: Interrupt Enable Register (bit-addressable)

Bit:	7	6	5	4	3	2	1	0
	EA	X	X	ES	ET1	EX1	ET0	EX0

where:

- EA — disables all interrupts. If EA = 0, no interrupt will be acknowledged. If EA = 1, each interrupt source is individually enabled or disabled by setting or clearing its enable bit.
- ES — enables or disables the Serial Interface Unit interrupt. If ES = 0, the Serial Interface Unit interrupt is disabled.
- ET1 — enables or disables the Timer 1 Overflow interrupt. If ET1 = 0, the Timer 1 interrupt is disabled.

- EX1 enables or disables External Interrupt 1. If EX1 = 0, External Interrupt 1 is disabled.
- ET0 enables or disables the Timer 0 Overflow interrupt. If ET0 = 0, the Timer 0 interrupt is disabled.

IP: Interrupt Priority Register (bit-addressable)

Bit:	7	6	5	4	3	2	1	0
	X	X	X	PS	PT1	PX1	PT0	PX0

where:

- PS defines the Serial Interface Unit interrupt priority level. PS = 1 programs it to the higher priority level.
- PT1 defines the Timer 1 interrupt priority level. PT1 = 1 programs it to the higher priority level.
- PX1 defines the External Interrupt priority level. PX1 = 1 programs it to the higher priority level.
- PT0 defines the Timer 0 interrupt priority level. PT0 = 1 programs it to the higher priority level.
- PX0 defines the External Interrupt 0 priority level. PX0 = 1 programs it to the higher priority level.

2.0 MEMORY ORGANIZATION DETAILS

In the 8044 family the memory is organized over three address spaces and the program counter. The memory spaces shown in Figure 2 are the:

- 64K-byte Program Memory address space
- 64K-byte External Data Memory address space
- 320-byte Internal Data Memory address space

The 16-bit Program Counter register provides the 8044 with its 64K addressing capabilities. The Program Counter allows the user to execute calls and branches to any location within the Program Memory space. There are no instructions that permit program execution to move from the Program Memory space to any of the data memory spaces.

In the 8044 and 8744 the lower 4K of the 64K Program Memory address space is filled by internal ROM and EPROM, respectively. By tying the $\overline{EA}$ pin high, the processor can be forced to fetch from the internal ROM/EPROM for Program Memory addresses 0 through 4K. Bus expansion for accessing Program Memory beyond 4K is automatic since external instruction fetches occur automatically when the Program Counter increases above 4095. If the $\overline{EA}$ pin is tied low all Program Memory fetches are from external memory. The execution speed of the 8044 is the same regardless of whether fetches are from internal or external Program Memory. If all program storage is on-chip, byte location 4095 should be left vacant to prevent an undesired prefetch from external Program Memory address 4096.

Certain locations in Program Memory are reserved for specific programs. Locations 0000 through 0002 are reserved for the initialization program. Following reset, the CPU always begins execution at location 0000. Locations 0003 through 0042 are reserved for the five interrupt-request service programs. Each resource that can request an interrupt requires that its service program be stored at its reserved location.

The 64K-byte External Data Memory address space is automatically accessed when the MOVX instruction is executed.

Functionally the Internal Data Memory is the most flexible of the address spaces. The Internal Data Memory space is subdivided into a 256-byte Internal Data RAM address space and a 128-byte Special Function Register address space as shown in Figure 3.

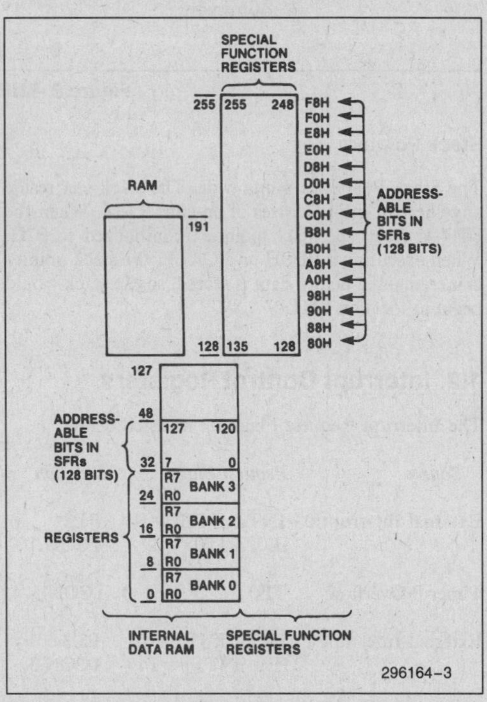

Figure 3. Internal Data Memory Address Space

The Internal Data RAM address space is 0 to 255. Four 8-Register Banks occupy locations 0 through 31. The stack can be located anywhere in the Internal Data RAM address space. In addition, 128 bit locations of the on-chip RAM are accessible through Direct Addressing. These bits reside in Internal Data RAM at byte locations 32 through 47. Currently locations 0 through 191 of the Internal Data RAM address space are filled with on-chip RAM.

The stack depth is limited only by the available Internal Data RAM, thanks to an 8-bit reloadable Stack Pointer. The stack is used for storing the Program Counter during subroutine calls and may be used for passing parameters. Any byte of Internal Data RAM or Special Function Register accessible though Direct Addressing can be pushed/popped.

The Special Function Register address space is 128 to 255. All registers except the Program Counter and the four 8-Register Banks reside here. Memory mapping the Special Function Registers allows them to be accessed as easily as internal RAM. As such, they can be operated on by most instructions. In the overlapping memory space (address 128-191), indirect addressing is used to access RAM, and direct addressing is used to access the SFR's. The SFR's at addresses 192–255 are also accessed using direct addressing. The Special Function Registers are listed in Figure 4. Their mapping in the Special Function Register address space is shown in Figures 5 and 6.

Performing a read from a location of the Internal Data memory where neither a byte of Internal Data RAM (i.e., RAM addresses 192-255) nor a Special Function Register exists will access data of indeterminable value.

Architecturally, each memory space is a linear sequence of 8-bit wide bytes. By Intel convention the storage of multi-byte address and data operands in program and data memories is the least significant byte at the low-order address and the most significant byte at the high-order address. Within byte X, the most significant bit is represented by X.7 while the least significant bit is X.0. Any deviation from these conventions will be explicitly stated in the text.

2.1 Operand Addressing

There are five methods of addressing source operands. They are Register Addressing, Direct Addressing, Register-Indirect Addressing, Immediate Addressing

ARITHMETIC REGISTERS:
Accumulator*, B register*,
Program Status Word*
POINTERS:
Stack Pointer, Data Pointer (high & low)
PARALLEL I/O PORTS:
Port 3*, Port 2*, Port 1*, Port 0*
INTERRUPT SYSTEM:
Interrupt Priority Control*,
Interrupt Enable Control*
TIMERS:
Timer Mode, Timer Control*, Timer 1
(high & low), Timer 0 (high & low)
SERIAL INTERFACE UNIT:
Transmit Buffer Start,
Transmit Buffer Length,
Transmit Control Byte,
Send Count Receive Count*,
DMA Count,
Station Address
Receive Field Length
Receive Buffer Start
Receive Buffer Length
Receive Control Byte,
Serial Mode,
Status Register.*

*Bits in these registers are bit addressable.

Figure 4. Special Function Registers

ARITHMETIC REGISTERS:
Accumulator*, B register*,
Program Status Word*
POINTERS:
Stack Pointer, Data Pointer (high & low)
PARALLEL I/O PORTS:
Port 3*, Port 2*, Port 1*, Port 0*
INTERRUPT SYSTEM:
Interrupt Priority Control*,
Interrupt Enable Control*
TIMERS:
Timer Mode, Timer Control*, Timer 1
(high & low), Timer 0 (high & low)
SERIAL INTERFACE UNIT:
Serial Mode, Status/Command*,
Send/Receive Count*, Station Address,
Transmit Buffer Start Address,
Transmit Buffer Length,
Transmit Control Byte,
Receive Buffer Start Address,
Receive Buffer Length,
Receive Field Length,
Receive Control Byte,
DMA Count,
FIFO (three bytes),
SIU Controller State Counter

*Bits in these registers are bit-addressable.

Figure 5. Mapping of Special Function Registers

and Base-Register-plus Index-Register-Indirect Addressing. The first three of these methods can also be used to address a destination operand. Since operations in the 8044 require 0 (NOP only), 1, 2, 3 or 4 operands, these five addressing methods are used in combinations to provide the 8044 with its 21 addressing modes.

Most instructions have a "destination, source" field that specifies the data type, addressing methods and operands involved. For operations other than moves, the destination operand is also a source operand. For example, in "subtract-with-borrow A,#5" the A register receives the result of the value in register A minus 5, minus C.

Most operations involve operands that are located in Internal Data Memory. The selection of the Program Memory space or External Data Memory space for a second operand is determined by the operation mnemonic unless it is an immediate operand. The subset of the Internal Data Memory being addressed is determined by the addressing method and address value. For example, the Special Function Registers can be accessed only through Direct Addressing with an address of 128-255. A summary of the operand addressing methods is shown in Figure 6. The following paragraphs describe the five addressing methods.

2.2 Register Addressing

Register Addressing permits access to the eight registers (R7-R0) of the selected Register Bank (RB). One of the four 8-Register Banks is selected by a two-bit field in the PSW. The registers may also be accessed through Direct Addressing and Register-Indirect Addressing, since the four Register Banks are mapped into the lowest 32 bytes of internal Data RAM as shown in Figures 9 and 10. Other Internal Data Memory locations that are addressed as registers are A, B, C, AB and DPTR.

2.3 Direct Addressing

Direct Addressing provides the only means of accessing the memory-mapped byte-wide Special Function Registers and memory mapped bits within the Special Function Registers and Internal Data RAM. Direct Addressing of bytes may also be used to access the lower 128 bytes of Internal Data RAM. Direct Addressing of bits gains access to a 128 bit subset of the Special Function Registers as shown in Figures 5, 6, 9, and 10.

REGISTER NAMES	SYMBOLIC ADDRESS	BIT ADDRESS	BYTE ADDRESS
B REGISTER	B	247 through 240	240 (F0H) ←
ACCUMULATOR	ACC	231 through 224	224 (E0H) ←
*THREE BYTE FIFO	FIFO		223 (DFH)
	FIFO		222 (DEH)
	FIFO		221 (DDH)
TRANSMIT BUFFER START	TBS		220 (DCH)
TRANSMIT BUFFER LENGTH	TBL		219 (DBH)
TRANSMIT CONTROL BYTE	TCB		218 (DAH)
*SIU STATE COUNTER	SIUST		217 (D9H)
SEND COUNT RECEIVE COUNT	NSNR	223 through 216	216 (D8H) ←
PROGRAM STATUS WORD	PSW	215 through 208	208 (D0H) ←
*DMA COUNT	DMA CNT		207 (CFH)
STATION ADDRESS	STAD		206 (CEH)
RECEIVE FIELD LENGTH	RFL		205 (CDH)
RECEIVE BUFFER START	RBS		204 (CCH)
RECEIVE BUFFER LENGTH	RBL		203 (CBH)
RECEIVE CONTROL BYTE	RCB		202 (CAH)
SERIAL MODE	SMD		201 (C9H)
STATUS REGISTER	STS	207 through 200	200 (C8H) ←
INTERRUPT PRIORITY CONTROL	IP	191 through 184	184 (B8H) ←
PORT 3	P3	183 through 176	176 (B0H) ←
INTERRUPT ENABLE CONTROL	IE	175 through 168	168 (A8H) ←
PORT 2	P2	167 through 160	160 (A0H) ←
PORT 1	P1	151 through 144	144 (90H) ←
TIMER HIGH 1	TH1		141 (8DH)
TIMER HIGH 0	TH0		140 (8CH)
TIMER LOW 1	TL1		139 (8BH)
TIMER LOW 0	TL0		138 (8AH)
TIMER MODE	TMOD		137 (89H)
TIMER CONTROL	TCON	143 through 136	136 (88H) ←
DATA POINTER HIGH	DPH		131 (83H)
DATA POINTER LOW	DPL		130 (82H)
STACK POINTER	SP		129 (81H)
PORT 0	P0	135 through 128	128 (80H) ←

← SFR's CONTAINING DIRECT ADDRESSABLE BITS

296164-4

Figure 6. Mapping of Special Function Registers

Direct Byte Address	(MSB)		Bit Address					(LSB)	Hardware Register Symbol
240	F7	F6	F5	F4	F3	F2	F1	F0	8
224	E7	E6	E5	E4	E3	E2	E1	E0	ACC
	NS2	NS1	NS0	SES	NR2	NR1	NR0	SER	
216	DF	DE	DD	DC	DB	DA	D9	D8	NSNR
	CY	AC	FO	RS1	RSO	OV		P	
204	D7	D6	D5	D4	D3	D2	D1	D0	PSW
	TBF	RE	RTS	SI	BV	CPB	AM	RBP	
200	CF	CE	CD	CC	CB	CA	C9	C8	STS
				PS	PT1	PX1	PT0	PX0	
184	—	—	—	BC	BB	BA	B9	B8	1P
176	B7	B6	B5	B4	B3	B2	B1	B0	P3
	EA			E5	ET1	EX1	ET0	EX0	
168	AF	—	—	AC	AB	AA	A9	A8	1E
160	A7	A6	A5	A4	A3	A2	A1	A0	P2
144	97	96	95	94	93	92	91	90	P1
	TF1	TR1	TF0	TR0	IE1	IT1	IE0	IT0	
136	8F	8E	8D	8C	8B	8A	89	88	TCON
128	87	86	85	84	83	82	81	80	P0

Figure 7. Special Function Register Bit Address

- **Register Addressing**
  - — **R7-R0**
  - — **A, B, C (bit), AB (two bytes), DPTR (double byte)**
- **Direct Addressing**
  - — **Lower 128 bytes of Internal Data RAM**
  - — **Special Function Registers**
  - — **128 bits in subset of Special Function Register address space**
- **Register-Indirect Addressing**
  - — **Internal Data RAM [@R1, @R0, @SP (PUSH and POP only)]**
  - — **Least Significant Nibbles in Internal Data RAM (@R1, @R0)**
  - — **External Data Memory (@R1, @R0, @DPTR)**
- **Immediate Addressing**
  - — **Program Memory (in-code constant)**
- **Base-Register-plus Index-Register-Indirect Addressing**
  - — **Program Memory (@ DPTR + A, @ PC + A)**

Figure 8. Operand Addressing Methods

RAM BYTE	(MSB)							(LSB)	
BFH									191
2FH	7F	7E	7D	7C	7B	7A	79	78	47
2EH	77	76	75	74	73	72	71	70	46
2DH	6F	6E	6D	6C	6B	6A	69	68	45
2CH	67	66	65	64	63	62	61	60	44
2BH	5F	5E	5D	5C	5B	5A	59	58	43
2AH	57	56	55	54	53	52	51	50	42
29H	4F	4E	4D	4C	4B	4A	49	48	41
28H	47	46	45	44	43	42	41	40	40
27H	3F	3E	3D	3C	3B	3A	39	38	39
26H	37	36	35	34	33	32	31	30	38
25H	2F	2E	2D	2C	2B	2A	29	28	37
24H	27	26	25	24	23	22	21	20	36
23H	1F	1E	1D	1C	1B	1A	19	18	35
22H	17	16	15	14	13	12	11	10	34
21H	0F	0E	0D	0C	0B	0A	09	08	33
20H	07	06	05	04	03	02	01	00	32
									31
1FH	Bank 3								
18H									24
17H	Bank 2								23
10H									16
0FH	Bank 1								15
08H									8
07H	Bank 0								7
00H									0

296164-5

Figure 9. RAM Bit Address

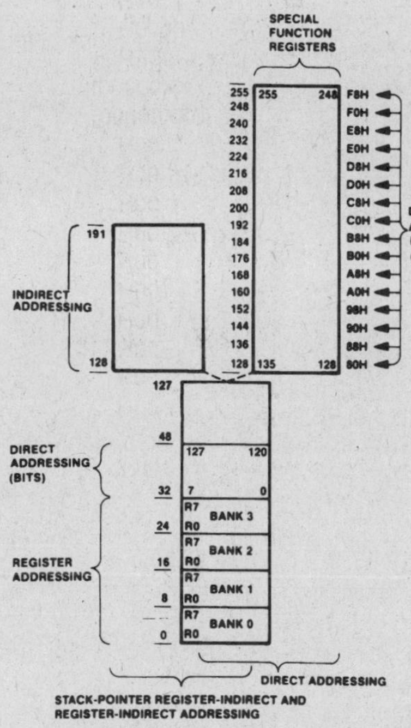

296164-6

Figure 10. Addressing Operands in Internal Data Memory

5

Register-Indirect Addressing using the content of R1 or R0 in the selected Register Bank, or using the content of the Stack Pointer (PUSH and POP only), addresses the Internal Data RAM. Register-Indirect Addressing is also used for accessing the External Data Memory. In this case, either R1 or R0 in the selected Register Bank may be used for accessing locations within a 256-byte block. The block number can be preselected by the contents of a port. The 16-bit Data Pointer may be used for accessing any location within the full 64K external address space.

3.0 RESET

Reset is accomplished by holding the RST pin high for at least two machine cycles (24 oscillator periods) *while the oscillator is running.* The CPU responds by executing an internal reset. It also configures the ALE and PSEN pins as inputs. (They are quasi-bidirectional.) The internal reset is executed during the second cycle in which RST is high and is repeated every cycle until RST goes low. It leaves the internal registers as follows:

Register	Content
PC	0000H
A	00H
B	00H
PSW	00H
SP	07H
DPTR	0000H
P0-P3	0FFH
IP	(XXX00000)
IE	(0XX00000)
TMOD	00H
TCON	00H
TH0	00H
TL0	00H
TH1	00H
TL1	00H
SMD	00H
STS	00H
NSNR	00H
STAD	00H
TBS	00H
TBL	00H
TCB	00H
RBS	00H
RBL	00H
RFL	00H
RCB	00H
DMA CNT	00H
FIFO1	00H
FIFO2	00H
FIFO3	00H
SIUST	01H
PCON	(0XXXXXXX)

The internal RAM is not affected by reset. When VCC is turned on, the RAM content is indeterminate unless VPD was applied prior to VCC being turned off (see Power Down Operation.)

4.0 RUPI-44 FAMILY PIN DESCRIPTION

VSS: Circuit ground potential.

VCC: Supply voltage during programming (of the 8744), verification (of the 8044 or 8744), and normal operation.

Port 0: Port 0 is an 8-bit open drain bidirectional I/O port. It is also the multiplexed low-order address and data bus during accessses to external memory (during which accesses it activates internal pullups). It also outputs instruction bytes during program verification. (External pullups are required during program verification.) Port 0 can sink eight LS TTL inputs.

Port 1: Port 1 is an 8-bit bidirectional I/O port with internal pullups. It receives the low-order address byte during program verification in the 8044 or 8744. Port 1 can sink/source four LS TTL inputs, It can drive MOS inputs without external pullups.

Two of the Port 1 pins serve alternate functions, as listed below:

Port Pin	*Alternate Function*
P1.6	$\overline{\text{RTS}}$ (Request to Send). In a non-loop configuration, $\overline{\text{RTS}}$ signals that the 8044 is ready to transmit data.

8044AH/8344AH/8744H
HIGH PERFORMANCE 8-BIT MICROCONTROLLER WITH ON-CHIP SERIAL COMMUNICATION CONTROLLER

- **8044AH—Includes Factory Mask Programmable ROM**
- **8344AH—For Use with External Program Memory**
- **8744H—Includes User Programmable/Eraseable EPROM**

8051 MICROCONTROLLER CORE

- **Optimized for Real Time Control 12 MHz Clock, Priority Interrupts, 32 Programmable I/O Lines, Two 16-bit Timer/Counters**
- **Boolean Processor**
- **4K × 8 ROM, 192 × 8 RAM**
- **64K Accessible External Program Memory**
- **64K Accessible External Data Memory**
- **4 μs Multiply and Divide**

SERIAL INTERFACE UNIT (SIU)

- **Serial Communication Processor that Operates Concurrently to CPU**
- **2.4 Mbps Maximum Data Rate**
- **375 Kbps using On-Chip Phase Locked Loop**
- **Communication Software in Silicon:**
  — Complete Data Link Functions
  — Automatic Station Response
- **Operates as an SDLC Primary or Secondary Station**

The RUPI-44 family integrates a high performance 8-bit Microcontroller, the Intel 8051 Core, with an Intelligent/high performance HDLC/SDLC serial communication controller, called the Serial Interface Unit (SIU). See Figure 1. This dual architecture allows complex control and high speed data communication functions to be realized cost effectively.

Specifically, the 8044's Microcontroller features: 4K byte On-Chip program memory space; 32 I/O lines; two 16-bit timer/event counters; a 5-source; 2-level interrupt structure; a full duplex serial channel; a Boolean processor; and on-chip oscillator and clock circuitry. Standard TTL and most byte-oriented MCS-80 and MCS-85 peripherals can be used for I/O amd memory expansion.

The Serial Interface Unit (SIU) manages the interface to a high speed serial link. The SIU offloads the On-Chip 8051 Microcontroller of communication tasks, thereby freeing the CPU to concentrate on real time control tasks.

The RUPI-44 family consists of the 8044, 8744, and 8344. All three devices are identical except in respect of on-chip program memory. The 8044 contains 4K bytes of mask-programmable ROM. User programmable EPROM replaces ROM in the 8744. The 8344 addresses all program memory externally.

The RUPI-44 devices are fabricated with Intel's reliable +5 volt, silicon-gate HMOSII technology and packaged in a 40-pin DIP.

The 8744H is available in a hermetically sealed, ceramic, 40-lead dual in-line package which includes a window that allows for EPROM erasure when exposed to ultraviolet light (See Erasure Characteristics). During normal operation, ambient light may adversely affect the functionality of the chip. Therefore applications which expose the 8744H to ambient light may require an opaque label over the window.

8044's Dual Controller Architecture

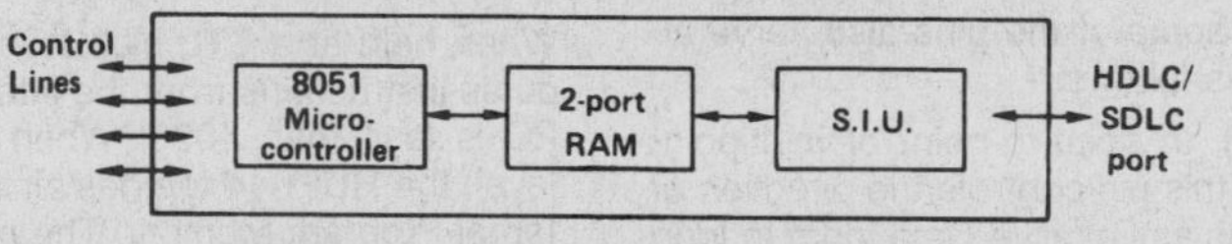

231663-1

Figure 1. Dual Controller Architecture

Table 1. RUPI-44 Family Pin Description

VSS

Circuit ground potential.

VCC

+5V power supply during operation and program verification.

PORT 0

Port 0 is an 8-bit open drain bidirectional I/O port. It is also the multiplexed low-order address and data bus when using external memory. It is used for data output during program verification. Port 0 can sink/source eight LS TTL loads (six in 8744).

PORT 1

Port 1 is an 8-bit quasi-bidirectional I/O port. It is used for the low-order address byte during program verification. Port 1 can sink/source four LS TTL loads.

In non-loop mode two of the I/O lines serve alternate functions:

- $\overline{RTS}$ (P1.6). Request-to-Send output. A low indicates that the RUPI-44 is ready to transmit.
- $\overline{CTS}$ (P1.7) Clear-to-Send input. A low indicates that a receiving station is ready to receive.

PORT 2

Port 2 is an 8-bit quasi-bidirection I/O port. It also emits the high-order address byte when accessing external memory. It is used for the high-order address and the control signals during program verification. Port 2 can sink/source four LS TTL loads.

PORT 3

Port 3 is an 8-bit quasi-bidirectional I/O port. It also contains the interrupt, timer, serial port and RD and WR pins that are used by various options. The output latch corresponding to a secondary function must be programmed to a one (1) for that function to operate. Port 3 can sink/source four LS LTT loads.

In addition to I/O, some of the pins also serve alternate functions as follows:

- I/$\overline{O}$ RxD (P3.0). In point-to-point or multipoint configurations, this pin controls the direction of pin P3.1. Serves as Receive Data input in loop and diagnostic modes.
- DATA TxD (P3.1) In point-to-point or multipoint configurations, this pin functions as data input/output. In loop mode, it serves as transmit pin. A '0' written to this pin enables diagnostic mode.
- $\overline{INT0}$ (P3.2). Interrupt 0 input or gate control input for counter 0.
- $\overline{INT1}$ (P3.3). Interrupt 1 input or gate control input for counter 1.
- TO (P3.4). Input to counter 0.
- SCLK T1 (P3.5). In addition to I/O, this pin provides input to counter 1 or serves as SCLK (serial clock) input.
- $\overline{WR}$ (P3.6). The write control signal latches the data byte from Port 0 into the External Data Memory.
- $\overline{RD}$ (P3.7). The read control signal enables External Data Memory to Port 0.

RST

A high on this pin for two machine cycles while the oscillator is running resets the device. A small external pulldown resistor ($\approx$8.2KΩ) from RST to V_{SS} permits power-on reset when a capacitor ($\approx$10μf) is also connected from this pin to V_{CC}.

ALE/$\overline{PROG}$

Provides Address Latch Enable output used for latching the address into external memory during normal operation. It is activated every six oscillator periods except during an external data memory access. It also receives the program pulse input for programming the EPROM version.

$\overline{PSEN}$

The Program Store Enable output is a control signal that enables the external Program Memory to the bus during external fetch operations. It is activated every six oscillator periods, except during external data memory accesses. Remains high during internal program execution.

$\overline{EA}$/VPP

When held at a TTL high level, the RUPI-44 executes instructions from the internal ROM when the PC is less than 4096. When held at a TTL low level, the RUPI-44 fetches all instructions from external Program Memory. The pin also receives the 21V EPROM programming supply voltage on the 8744.

Table 1. RUPI-44 Family Pin Description (Continued)

XTAL 1	XTAL 2
Input to the oscillator's high gain amplifier. Required when a crystal is used. Connect to VSS when external source is used on XTAL 2.	Output from the oscillator's amplifier. Input to the internal timing circuitry. A crystal or external source can be used.

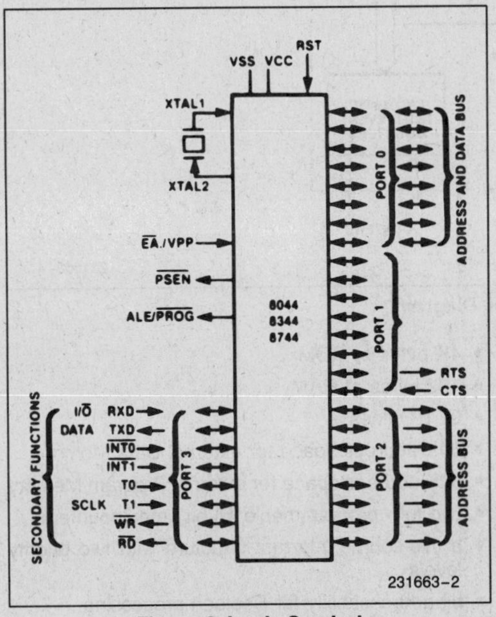

Figure 2. Logic Symbol

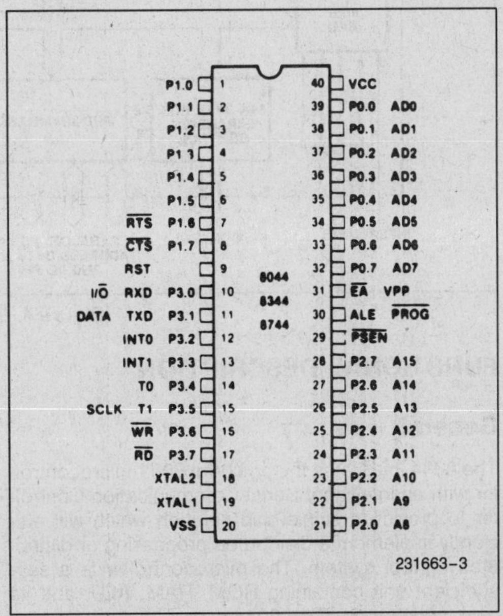

Figure 3A. DIP Pin Configuration

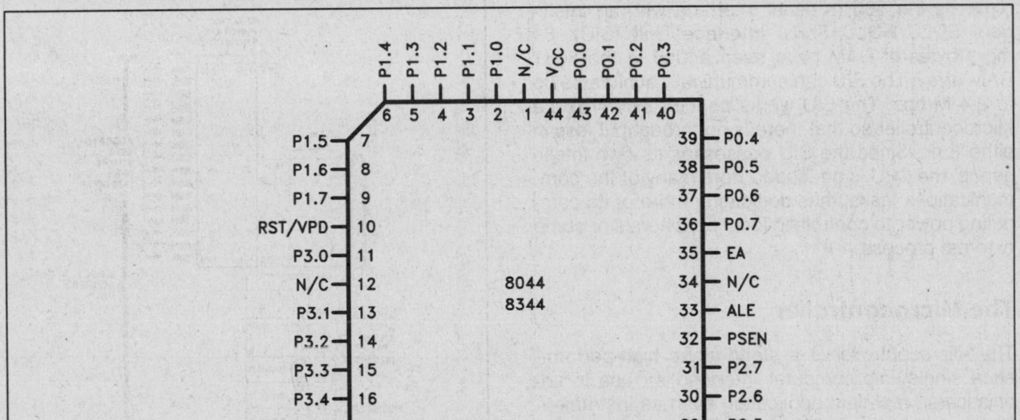

Figure 3B. PLCC Pin Configuration

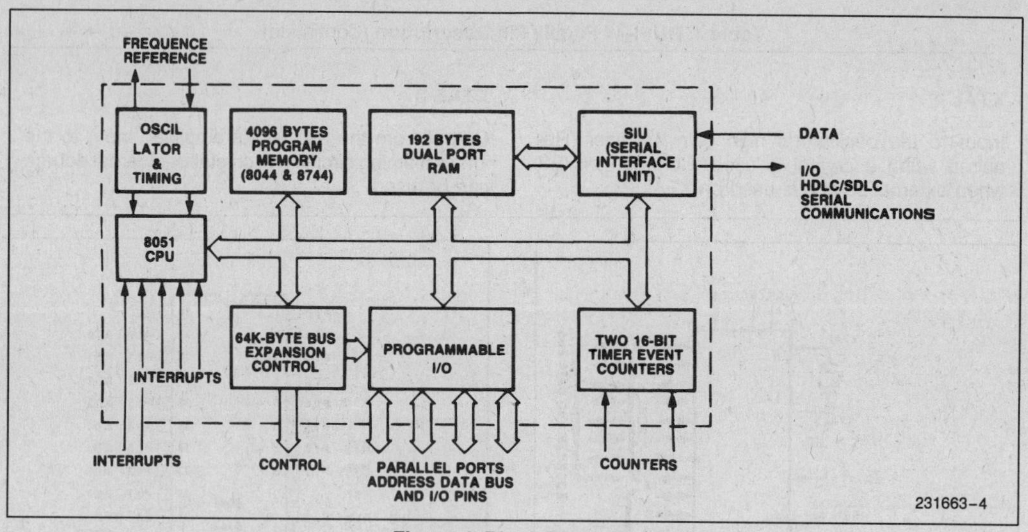

Figure 4. Block Diagram

FUNCTIONAL DESCRIPTION

General

The 8044 integrates the powerful 8051 microcontroller with an intelligent Serial Communication Controller to provide a single-chip solution which will efficiently implement a distributed processing or distributed control system. The microcontroller is a self-sufficient unit containing ROM, RAM, ALU, and its own peripherals. The 8044's architecture and instruction set are identical to the 8051's. The 8044 replaces the 8051's serial interface with an intelligent SDLC/HDLC Serial Interface Unit (SIU). 64 more bytes of RAM have been added to the 8051 RAM array. The SIU can communicate at bit rates up to 2.4 M bps. The SIU works concurrently with the Microcontroller so that there is no throughput loss in either unit. Since the SIU possesses its own intelligence, the CPU is off-loaded from many of the communications tasks, thus dedicating more of its computing power to controlling local peripherals or some external process.

The Microcontroller

The microcontroller is a stand-alone high-performance single-chip computer intended for use in sophisticated real-time application such as instrumentation, industrial control, and intelligent computer peripherals.

The major features of the microcontroller are:

- 8-bit CPU
- on-chip oscillator
- 4K bytes of ROM
- 192 bytes of RAM
- 32 I/O lines
- 64K address space for external Data Memory
- 64K address space for external Program Memory
- two fully programmable 16-bit timer/counters
- a five-source interrupt structure with two priority levels
- bit addressability for Boolean processing

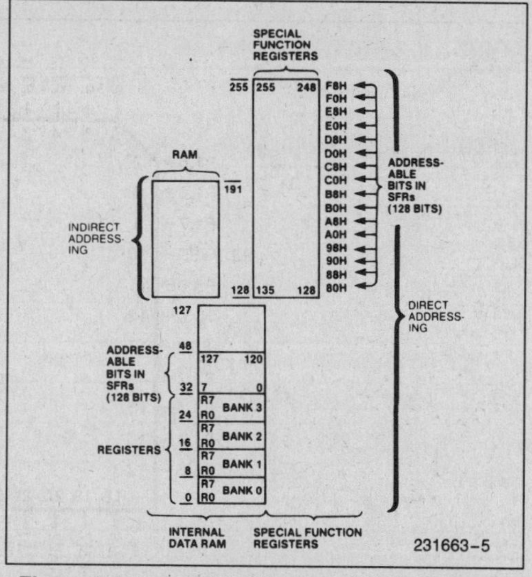

Figure 5. Internal Data Memory Address Space

- 1 μs instruction cycle time for 60% of the instructions 2 μs instruction cycle time for 40% of the instructions
- 4 μs cycle time for 8 by 8 bit unsigned Multiply/Divide

INTERNAL DATA MEMORY

Functionally the Internal Data Memory is the most flexible of the address spaces. The Internal Data Memory space is subdivided into a 256-byte Internal Data RAM address space and a 128-bit Special Function Register address space as shown in Figure 5.

The Internal Data RAM address space is 0 to 255. Four 8-Register Banks occupy locations 0 through 31. The stack can be located anywhere in the Internal Data RAM address space. In addition, 128 bit locations of the on-chip RAM are accessible through Direct Addressing. These bits reside in Internal Data RAM at byte locations 32 through 47. Currently locations 0 through 191 of the Internal Data RAM address space are filled with on-chip RAM.

Parallel I/O

The 8044 has 32 general-purpose I/O lines which are arranged into four groups of eight lines. Each group is called a port. Hence there are four ports; Port 0, Port 1, Port 2, and Port 3. Up to five lines from Port 3 are dedicated to supporting the serial channel when the SIU is invoked. Due to the nature of the serial port, two of Port 3's I/O lines (P3.0 and P3.1) do not have latched outputs. This is true whether or not the serial channel is used.

Port 0 and Port 2 also have an alternate dedicated function. When placed in the external access mode, Port 0 and Port 2 become the means by which the 8044 communicates with external program memory. Port 0 and Port 2 are also the means by which the 8044 communicates with external data memory. Peripherals can be memory mapped into the address space and controlled by the 8044.

Table 2. MCS®-51 Instruction Set Description

Mnemonic		Description	Byte	Cyc
ARITHMETIC OPERATIONS				
ADD	A,Rn	Add register to Accumulator	1	1
ADD	A,direct	Add direct byte to Accumulator	2	1
ADD	A,@Ri	Add indirect RAM to Accumulator	1	1
ADD	A,#data	Add immediate data to Accumulator	2	1
ADDC	A,Rn	Add register to Accumulator with Carry	1	1
ADDC	A,direct	Add direct byte to A with Carry flag	2	1
ADDC	A,@Ri	Add indirect RAM to A with Carry flag	1	1
ADDC	A,#data	Add immediate data to A wlth Carry flag	2	1
SUBB	A,Rn	Subtract register from A with Borrow	1	1
SUBB	A,direct	Subtract direct byte from A with Borrow	2	1
ARITHMETIC OPERATIONS (Continued)				
SUBB	A,@Ri	Subtract indirect RAM from A with Borrow	1	1
SUBB	A,#data	Subtract immed data from A with Borrow	2	1
INC	A	Increment Accumulator	1	1
INC	Rn	Increment register	1	1
INC	direct	Increment direct byte	2	1
INC	@Ri	Increment indirect RAM	1	1
INC	DPTR	Increment Data Pointer	1	2
DEC	A	Decrement Accumulator	1	1
DEC	Rn	Decrement register	1	1
DEC	direct	Decrement direct byte	2	1
DEC	@Ri	Decrement indirect RAM	1	1
MUL	AB	Multiply A & B	1	4
DIV	AB	Divide A by B	1	4
DA	A	Decimal Adjust Accumulator	1	1

5

Table 2. MCS®-51 Instruction Set Description (Continued)

Mnemonic		Description	Byte	Cyc
LOGICAL OPERATIONS				
ANL	A,Rn	AND register to Accumulator	1	1
ANL	A,direct	AND direct byte to Accumulator	2	1
ANL	A,@RI	AND indirect RAM to Accumulator	1	1
ANL	A,#data	AND immediate data to Accumulator	2	1
ANL	direct,A	AND Accumulator to direct byte	2	1
ANL	direct,#data	AND immediate data to direct byte	3	2
ORL	A,Rn	OR register to Accumulator	1	1
ORL	A,direct	OR direct byte to Accumulator	2	1
ORL	A,@Ri	OR indirect RAM to Accumulator	1	1
ORL	A,#data	OR immediate data to Accumulator	2	1
ORL	direct,A	OR Accumulator to direct byte	2	1
ORL	direct,#data	OR immediate data to direct byte	3	2
XRL	A,Rn	Exclusive-OR register to Accumulator	1	1
XRL	A,direct	Exclusive-OR direct byte to Accumulator	2	1
XRL	A,@RI	Exclusive-OR indirect RAM to A	1	1
XRL	A,#data	Exclusive-OR immediate data to A	2	1
XRL	direct,A	Exclusive-OR Accumulator to direct byte	2	1
XRL	direct,#data	Exclusive-OR immediate data to direct	3	2
CLR	A	Clear Accumulator	1	1
CPL	A	Complement Accumulator	1	1

Mnemonic		Description	Byte	Cyc
LOGICAL OPERATIONS (Continued)				
RL	A	Rotate Accumulator Left	1	1
RLC	A	Rotate A Left through the Carry flag	1	1
RR	A	Rotate Accumulator Right	1	1
RRC	A	Rotate A Right through Carry flag	1	1
SWAP	A	Swap nibbles within the Accumulator	1	1
DATA TRANSFER				
MOV	A,Rn	Move register to Accumulator	1	1
MOV	A,direct	Move direct byte to Accumulator	2	1
MOV	A,@RI	Move indirect RAM to Accumulator	1	1
MOV	A,#data	Move immediate data to Accumulator	2	1
MOV	Rn,A	Move Accumulator to register	1	1
MOV	Rn,direct	Move direct byte to register	2	2
MOV	Rn,#data	Move immediate data to register	2	1
MOV	direct,A	Move Accumulator to direct byte	2	1
MOV	direct,Rn	Move register to direct byte	2	2
MOV	direct,direct	Move direct byte to direct	3	2
MOV	direct,@Ri	Move indirect RAM to direct byte	2	2
MOV	direct,#data	Move immediate data to direct byte	3	2
MOV	@Ri,A	Move Accumulator to indirect RAM	1	1
MOV	@Ri,direct	Move direct byte to indirect RAM	2	2

Table 2. MCS®-51 Instruction Set Description (Continued)

Mnemonic		Description	Byte	Cyc
DATA TRANSFER (Continued)				
MOV	@Ri, #data	Move immediate data to indirect RAM	2	1
MOV	DPTR, #data16	Load Data Pointer with a 16-bit constant	3	2
MOVC	A,@A + DPTR	Move Code byte relative to DPTR to A	1	2
MOVC	A,@A + PC	Move Code byte relative to PC to A	1	2
MOVX	A,@Ri	Move External RAM (8-bit addr) to A	1	2
MOVX	A,@DPTR	Move External RAM (16-bit addr) to A	1	2
MOVX	@Ri,A	Move A to External RAM (8-bit addr)	1	2
MOVX	@DPTR,A	Move A to External RAM (16-bit) addr	1	2
PUSH	direct	Push direct byte onto stack	2	2
POP	direct	Pop direct byte from stack	2	2
XCH	A,Rn	Exchange register with Accumulator	1	1
XCH	A,direct	Exchange direct byte with Accumulator	2	1
XCH	A,@Ri	Exchange indirect RAM with A	1	1
XCHD	A,@Ri	Exchange low-order Digit ind RAM w A	1	1
BOOLEAN VARIABLE MANIPULATION				
CLR	C	Clear Carry flag	1	1
CLR	bit	Clear direct bit	2	1
SETB	C	Set Carry Flag	1	1
SETB	bit	Set direct Bit	2	1
CPL	C	Complement Carry Flag	1	1
CPL	bit	Complement direct bit	2	1
ANL	C,bit	AND direct bit to Carry flag	2	2
BOOLEAN VARIABLE MANIPULATION (Continued)				
ANL	C,/bit	AND complement of direct bit to Carry	2	2
ORL	C/bit	OR direct bit to Carry flag	2	2
ORL	C,/bit	OR complement of direct bit to Carry	2	2
MOV	C,/bit	Move direct bit to Carry flag	2	1
MOV	bit,C	Move Carry flag to direct bit	2	2
PROGRAM AND MACHINE CONTROL				
ACALL	addr11	Absolute Subroutine Call	2	2
LCALL	addr16	Long Subroutine Call	3	2
RET		Return from subroutine	1	2
RETI		Return from interrupt	1	2
AJMP	addr11	Absolute Jump	2	2
LJMP	addr16	Long Jump	3	2
SJMP	rel	Short Jump (relative addr)	2	2
JMP	@A + DPTR	Jump indirect relative to the DPTR	1	2
JZ	rel	Jump if Accumulator is Zero	2	2
JNZ	rel	Jump if Accumulator is Not Zero	2	2
JC	rel	Jump if Carry flag is set	2	2
JNC	rel	Jump if No Carry flag	2	2
JB	bit,rel	Jump if direct Bit set	3	2
JNB	bit,rel	Jump if direct Bit Not set	3	2
JBC	bit,rel	Jump if direct Bit is set & Clear bit	3	2
CJNE	A,direct,rel	Compare direct to A & Jump if Not Equal	3	2
CJNE	A, #data,rel	Comp, immed, to A & Jump if Not Equal	3	2

5

Table 2. MCS®-51 Instruction Set Description (Continued)

Mnemonic	Description	Byte	Cyc
PROGRAM AND MACHINE CONTROL (Continued)			
CJNE Rn,#data,rel	Comp, immed, to reg & Jump if Not Equal	3	2
CJNE @Ri,#data, rel	Comp, immed, to ind. & Jump if Not Equal	3	2
DJNZ Rn,rel	Decrement register & Jump if Not Zero	2	2
DJNZ direct,rel	Decrement direct & Jump if Not Zero	3	2
NOP	No operation	1	1

Notes on data addressing modes:

Rn — Working register R0-R7

direct — 128 internal RAM locations, any I/O port, control or status register

@Ri — Indirect internal RAM location addressed by register R0 or R1

Notes on data addressing modes:

(Continued)

#data — 8-bit constant included in instruction

#data16 — 16-bit constant included as bytes 2 & 3 of instruction

bit — 128 software flags, any I/O pin, controll or status bit

Notes on program addressing modes:

addr16 — Destination address for LCALL & LJMP may be anywhere within the 64-K program memory address space

Addr11 — Destination address for ACALL & AJMP will be within the same 2-K page of program memory as the first byte of the following instruction

rel — SJMP and all conditional jumps include an 8-bit offset byte, Range is +127 −128 bytes relative to first byte of the following instruction

Timer/Counters

The 8044 contains two 16-bit counters which can be used for measuring time intervals, measuring pulse widths, counting events, generating precise periodic interrupt requests, and clocking the serial communications. Internally the Timers are clocked at 1/12 of the crystal frequency, which is the instruction cycle time. Externally the counters can run up to 500 KHz.

Interrupt System

External events and the real-time driven on-chip peripherals require service by the CPU asynchronous to the execution of any particular section of code. To tie the asynchronous activities of these functions to normal program execution, a sophisticated multiple-source, two priority level, nested interrupt system is provided. Interrupt response latency ranges from 3 μsec to 7 μsec when using a 12 MHz clock.

All five interrupt sources can be mapped into one of the two priority levels. Each interrupt source can be enabled or disabled individually or the entire interrupt system can be enabled or disabled. The five interrupt sources are: Serial Interface Unit, Timer 1, Timer 2, and two external interrupts. The external interrupts can be either level or edge triggered.

Serial Interface Unit (SIU)

The Serial Interface Unit is used for HDLC/SDLC communications. It handles Zero Bit Insertion/Deletion, Flags automatic access recognization, and a 16-bit cyclic redundancy check. In addition it implements in hardware a subset of the SDLC protocol certain applications it is advantageous to have the CPU control the reception or transmission of every single frame. For this reason the SIU has two modes of operation: "AUTO" and "FLEXIBLE" (or "NON-AUTO"). It is in the AUTO mode that the SIU responds to SDLC frames without CPU intervention; whereas, in the FLEXIBLE mode the reception or transmission of every single frame will be under CPU control.

There are three control registers and eight parameter registers that are used to operate the serial interface. These registers are shown in Figure 5 and Figure 6. The control register set the modes of operation and provide status information. The eight parameter registers buffer the station address, receive and transmit control bytes, and point to the on-chip transmit and receive buffers.

Data to be received or transmitted by the SIU must be buffered anywhere within the 192 bytes of on-chip RAM. Transmit and receive buffers are not allowed to "wrap around" in RAM; a "buffer end" is generated after address 191 is reached.

REGISTER NAMES	SYMBOLIC ADDRESS	BIT ADDRESS	BYTE ADDRESS
B REGISTER	B	247 through 240	240 (F0H)
ACCUMULATOR	ACC	231 through 224	224 (E0H)
*THREE BYTE FIFO	FIFO		223 (DFH)
	FIFO		222 (DEH)
	FIFO		221 (DDH)
TRANSMIT BUFFER START	TBS		220 (DCH)
TRANSMIT BUFFER LENGTH	TBL		219 (DBH)
TRANSMIT CONTROL BYTE	TCB		218 (DAH)
* SIU STATE COUNTER	SIUST		217 (D9H)
SEND COUNT RECEIVE COUNT	NSNR	223 through 216	216 (D8H)
PROGRAM STATUS WORD	PSW	215 through 208	208 (D0H)
*DMA COUNT	DMA CNT		207 (CFH)
STATION ADDRESS	STAD		206 (CEH)
RECEIVE FIELD LENGTH	RFL		205 (CDH)
RECEIVE BUFFER START	RBS		204 (CCH)
RECEIVE BUFFER LENGTH	RBL		203 (CBH)
RECEIVE CONTROL BYTE	RCB		202 (CAH)
SERIAL MODE	SMD		201 (C9H)
STATUS REGISTER	STS	207 through 200	200 (C8H)
INTERRUPT PRIORITY CONTROL	IP	191 through 184	184 (B8H)
PORT 3	P3	183 through 176	176 (B0H)
INTERRUPT ENABLE CONTROL	IE	175 through 168	168 (A8H)
PORT 2	P2	167 through 160	160 (A0H)
PORT 1	P1	151 through 144	144 (90H)
TIMER HIGH 1	TH1		141 (8DH)
TIMER HIGH 0	TH0		140 (8CH)
TIMER LOW 1	TL1		139 (8BH)
TIMER LOW 0	TL0		138 (8AH)
TIMER MODE	TMOD		137 (89H)
TIMER CONTROL	TCON	143 through 136	136 (88H)
DATA POINTER HIGH	DPH		131 (83H)
DATA POINTER LOW	DPL		130 (82H)
STACK POINTER	SP		129 (81H)
PORT 0	P0	135 through 128	128 (80H)

SFR's CONTAINING DIRECT ADDRESSABLE BITS

231663-6

NOTE:
*ICE Support Hardware registers. Under normal operating conditions there is no need for the CPU to access these registers.

Figure 5. Mapping of Special Function Registers

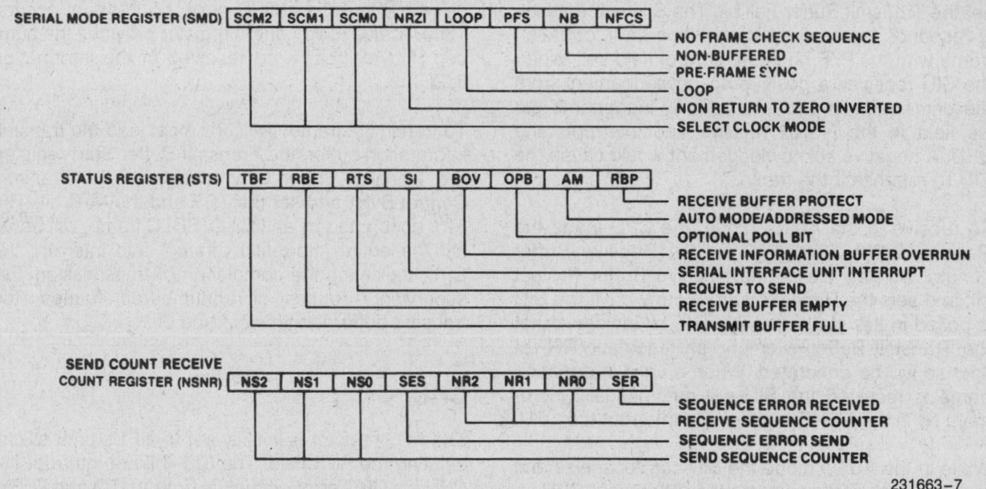

Figure 6. Serial Interface Unit Control Registers

5

With the addition of only a few bytes of code, the 8044's frame size is not limited to the size of its internal RAM (192 bytes), but rather by the size of external buffer with no degradation of the RUPI's features (e.g. NRZI, zero bit insertion/deletion, address recognition, cyclic redundancy check). There is a special function register called SIUST whose contents dictates the operation of the SIU. At low data rates, one section of the SIU (the Byte Processor) performs no function during known intervals. For a given data rate, these intervals (stand-by mode) are fixed. The above characteristics make it possible to program the CPU to move data to/from external RAM and to force the SIU to perform some desired hardware tasks while transmission or reception is taking place. With these modifications, external RAM can be utilized as a transmit and received buffer instead of the internal RAM.

AUTO Mode

In the AUTO mode the SIU implements in hardware a subset of the SDLC protocol such that it responds to many SDLC frames without CPU intervention. All AUTO mode responses to the primary station will comform to IBM's SDLC definition. The advantages of the AUTO mode are that less software is required to implement a secondary station, and the hardware generated response to polls is much faster than doing it in software. However, the Auto mode can not be used at a primary station.

To transmit in the AUTO mode the CPU must load the Transmit Information Buffer, Transmit Buffer Start register, Transmit Buffer Length register, and set the Transmit Buffer Full bit. The SIU automatically responds to a poll by transmitting an information frame with the P/F bit in the control field set. When the SIU receives a positive acknowledgement from the primary station, it automatically increments the Ns field in the NSNR register and interrupts the CPU. A negative acknowledgement would cause the SIU to retransmit the frame.

To receive in the AUTO mode, the CPU loads the Receive Buffer Start register, the Receive Buffer Length register, clears the Receive Buffer Protect bit, and sets the Receive Buffer Empty bit. If the SIU is polled in this state, and the TBF bit indicates that the Transmit Buffer is empty, an automatic RR response will be generated. When a valid information frame is received the SIU will automatically increment Nr in the NSNR register and interrupt the CPU.

While in the AUTO mode the SIU can recognize and respond to the following commands without CPU intervention: I (Information), RR (Receive Ready), RNR (Receive Not Ready), REJ (Reject), and UP (Unnumbered Poll). The SIU can generate the following responses without CPU intervention: I (Information), RR (Receive Ready), and RNR (Receive Not Ready).

When the Receive Buffer Empty bit (RBE) indicates that the Receive Buffer is empty, the receiver is enabled, and when the RBE bit indicates that the Receive Buffer is full, the receiver is disabled. Assuming that the Receiver Buffer is empty, the SIU will respond to a poll with an I frame if the Transmit Buffer is full. If the Transmit Buffer is empty, the SIU will respond to a poll with a RR command if the Receive Buffer Protect bit (RBP) is cleared, or an RNR command if RBP is set.

FLEXIBLE (or NON-AUTO) Mode

In the FLEXIBLE mode all communications are under control of the CPU. It is the CPU's task to encode and decode control fields, manage acknowledgements, and adhere to the requirements of the HDLC/SDLC protocols. The 8044 can be used as a primary or a secondary station in this mode.

To receive a frame in the FLEXIBLE mode, the CPU must load the Receive Buffer Start register, the Receive Buffer Length register, clear the Receive Buffer Protect bit, and set the Receive Buffer Empty bit. If a valid opening flag is received and the address field matches the byte in the Station Address register or the address field contains a broadcast address, the 8044 loads the control field in the receive control byte register, and loads the I field in the receive buffer. If there is no CRC error, the SIU interrupts the CPU, indicating a frame has just been received. If there is a CRC error, no interrupt occurs. The Receive Field Length register provides the number of bytes that were received in the information field.

To transmit a frame, the CPU must load the transmit information buffer, the Transmit Buffer Start register, the Transmit Buffer Length register, the Transmit Control Byte, and set the TBF and the RTS bit. The SIU, unsolicited by an HDLC/SDLC frame, will transmit the entire information frame, and interrupt the CPU, indicating the completion of transmission. For supervisory frames or unnumbered frames, the transmit buffer length would be 0.

CRC

The FCS register is initially set to all 1's prior to calculating the FCS field. The SIU will not interrupt the CPU if a CRC error occurs (in both AUTO and FLEXIBLE modes). The CRC error is cleared upon receiving of an opening flag.

Frame Format Options

In addition to the standard SDLC frame format, the 8044 will support the frames displayed in Figure 7. The standard SDLC frame is shown at the top of this figure. For the remaining frames the information field will incorporate the control or address bytes and the frame check sequences; therefore these fields will be stored in the Transmit and Receive buffers. For example, in the non-buffered mode the third byte is treated as the beginning of the information field. In the non-addressed mode, the information field begins after the opening flag. The mode bits to set the frame format options are found in the Serial Mode register and the Status register.

FRAME OPTION	NFCS	NB	AM[1]	FRAME FORMAT
Standard SDLC NON-AUTO Mode	0	0	0	F | A | C | I | FCS | F
Standard SDLC AUTO Mode	0	0	1	F | A | C | I | FCS | F
Non-Buffered Mode NON-AUTO Mode	0	1	1	F | A | I | FCS | F
Non-Addressed Mode NON-AUTO Mode	0	1	0	F | I | FCS | F
No FCS Field NON-AUTO Mode	1	0	0	F | A | C | I | F
No FCS Field AUTO Mode	1	0	1	F | A | C | I | F
No FCS Field Non-Buffered Mode NON-AUTO Mode	1	1	1	F | A | I | F
No FCS Field Non-Addressed Mode NON-AUTO Mode	1	1	0	F | I | F

Mode Bits:
AM — "AUTO" Mode/Addressed Mode
NB — Non-Buffered Mode
NFCS — No FCS Field Mode

Key to Abbreviations:
F = Flag (01111110)
A = Address Field
C = Control Field
I = Information Field
FCS = Frame Check Sequence

Note 1:
The AM bit function is controlled by the NB bit. When NB = 0, AM becomes AUTO mode select, when NB = 1, AM becomes Address mode select.

Figure 7. Frame Format Options

Extended Addressing

To realize an extended control field or an extended address field using the HDLC protocol, the FLEXIBLE mode must be used. For an extended control field, the SIU is programmed to be in the non-buffered mode. The extended control field will be the first and second bytes in the Receive and Transmit Buffers. For extended addressing the SIU is placed in the non-addressed mode. In this mode the CPU must implement the address recognition for received frames. The addressing field will be the initial bytes in the Transmit and Receive buffers followed by the control field.

The SIU can transmit and receive only frames which are multiples of 8 bits. For frames received with other than 8-bit multiples, a CRC error will cause the SIU to reject the frame.

SDLC Loop Networks

The SIU can be used in an SDLC loop as a secondary or primary station. When the SIU is placed in the Loop mode it receives the data on pin 10 and transmits the data one bit time delayed on pin 11. It can also recognize the Go ahead signal and change it into a flag when it is ready to transmit. As a secondary station the SIU can be used in the AUTO or FLEXIBLE modes. As a primary station the FLEXIBLE mode is used; however, additional hardware is required for generating the Go Ahead bit pattern. In the Loop mode the maximum data rate is 1 Mbps clocked or 375 Kpbs self-clocked.

SDLC Multidrop Networks

The SIU can be used in a SDLC non-loop configuration as a secondary or primary station. When the SIU is placed in the non-loop mode, data is received and transmitted on pin 11, and pin 10 drives a tri-state buffer. In non-loop mode, modem interface pins, RTS and CTS, become available.

Data Clocking Options

The 8044's serial port can operate in an externally clocked or self clocked system. A clocked system provides to the 8044 a clock synchronization to the data. A self-clocked system uses the 8044's on-chip Digital Phase Locked Loop (DPLL) to recover the clock from the data, and clock this data into the Serial Receive Shift Register.

In this mode, a clock synchronized with the data is externally fed into the 8044. This clock may be generated from an External Phase Locked Loop, or possibly supplied along with the data. The 8044 can transmit and receive data in this mode at rates up to 2.4 Mbps.

This self clocked mode allows data transfer without a common system data clock. An on-chip Digital Phase Locked Loop is employed to recover the data clock which is encoded in the data stream. The DPLL will converge to the nominal bit center within eight bit transitions, worst case. The DPLL requires a reference clock of either 16 times (16x) or 32 times (32x) the data rate. This reference clock may be externally applied or internally generated. When internally generated either the 8044's internal logic clock (crystal frequency divided by two) or the timer 1 overflow is used as the reference clock. Using the internal timer 1 clock the data rates can vary from 244 to 62.5 Kbps. Using the internal logic clock at a 16x sampling rate, receive data can either be 187.5 Kbps, or 375 Kbps. When the reference clock for the DPLL is externally applied the data rates can vary from 0 to 375 Kbps at a 16x sampling rate.

To aid in a Phase Locked Loop capture, the SIU has a NRZI (Non Return to Zero Inverted) data encoding and decoding option. Additionally the SIU has a preframe sync option that transmits two bytes of alternating 1's and 0's to ensure that the receive station DPLL will be synchronized with the data by the time it receives the opening flag.

Control and Status Registers

There are three SIU Control and Status Registers:

Serial Mode Register (SMD)

Status/Command Register (STS)

Send/Receive Count Register (NSNR)

The SMD, STS, and NSNR, registers are all cleared by system reset. This assures that the SIU will power up in an idle state (neither receiving nor transmitting).

These registers and their bit assignments are described below.

SMD: Serial Mode Register (byte-addressable)

Bit 7:	6	5	4	3	2	1	0
SCM2	SCM1	SCM0	NRZI	LOOP	PFS	NB	NFCS

The Serial Mode Register (Address C9H) selects the operational modes of the SIU. The 8044 CPU can both read and write SMD. The SIU can read SMD but cannot write to it. To prevent conflict between CPU and SIU access to SMD, the CPU should write SMD only when the Request To Send (RTS) and

Receive Buffer Empty (RBE) bits (in the STS register) are both false (0). Normally, SMD is accessed only during initialization.

The individual bits of the Serial Mode Register are as follows:

Bit #	Name	Description
SMD.0	NFCS	No FCS field in the SDLC frame.
SMD.1	NB	Non-Buffered mode. No control field in the SDLC frame.
SMD.2	PFS	Pre-Frame Sync mode. In this mode, the 8044 transmits two bytes before the first flag of a frame, for DPLL synchronization. If NRZI is enabled, 00H is sent; otherwise, 55H is sent. In either case, 16 preframe transitions are guaranteed.
SMD.3	LOOP	Loop configuration.
SMD.4	NRZI	NRZI coding option. If bit = 1, NRZI coding is used. If bit = 0, then it is straight binary (NRZ).
SMD.5	SCM0	Select Clock Mode—Bit 0
SMD.6	SCM1	Select Clock Mode—Bit 1
SMD.7	SCM2	Select Clock Mode—Bit 2

The SCM bits decode as follows:

SCM			Clock Mode	Data Rate
2	1	0		(Bits/sec)*
0	0	0	Externally clocked	0-2.4M**
0	0	1	Reserved	
0	1	0	Self clocked, timer overflow	244-62.5K
0	1	1	Reserved	
1	0	0	Self clocked, external 16x	0-375K
1	0	1	Self clocked, external 32x	0-187.5K
1	1	0	Self clocked, internal fixed	375K
1	1	1	Self clocked, internal fixed	187.5K

NOTES:
*Based on a 12 Mhz crystal frequency
**0-1 M bps in loop configuration

STS: Status/Command Register (bit-addressable)

Bit:	7	6	5	4	3	2	1	0
	TBF	RBE	RTS	SI	BOV	OPB	AM	RBP

The Status/Command Register (Address C8H) provides operational control of the SIU by the 8044 CPU, and enables the SIU to post status information for the CPU's access. The SIU can read STS, and can alter certain bits, as indicated below. The CPU can both read and write STS asynchronously. However, 2-cycle instructions that access STS during both cycles ('JBC/B, REL' and 'MOV/B, C.') should not be used, since the SIU may write to STS between the two CPU accesses.

The individual bits of the Status/Command Register are as follows:

Bit #	Name	Description
STS.0	RBP	Receive Buffer Protect. Inhibits writing of data into the receive buffer. In AUTO mode, RBP forces an RNR response instead of an RR.
STS.1	AM	AUTO Mode/Addressed Mode. Selects AUTO mode where AUTO mode is allowed. If NB is true, (= 1), the AM bit selects the addressed mode. AM may be cleared by the SIU.
STS.2	OPB	Optional Poll Bit. Determines whether the SIU will generate an AUTO response to an optional poll (UP with P = 0). OPM may be set or cleared by the SIU.
STS.3	BOV	Receive Buffer Overrun. BOV may be set or cleared by the SIU.
STS.4	SI	SIU Interrupt. This is one of the five interrupt sources to the CPU. The vector location = 23H. SI may be set by the SIU. It should be cleared by the CPU before returning from an interrupt routine.
STS.5	RTS	Request To Send. Indicates that the 8044 is ready to transmit or is transmitting. RTS may be read or written by the CPU. RTS may be read by the SIU, and in AUTO mode may be written by the SIU.
STS.6	RBE	Receive Buffer Empty. RBE can be thought of as Receive Enable. RBE is set to one by the CPU when it is ready to receive a frame, or has just read the buffer, and to zero by the SIU when a frame has been received.
STS.7	TBF	Transmit Buffer Full. Written by the CPU to indicate that it has filled the transmit buffer. TBF may be cleared by the SIU.

NSNR: Send/Receive Count Register (bit-addressable)

Bit:	7	6	5	4	3	2	1	0
	NS2	NS1	NS0	SES	NR2	NR1	NR0	SER

The Send/Receive Count Register (Address D8H) contains the transmit and receive sequence numbers, plus tally error indications. The SIU can both read and write NSNR. The 8044 CPU can both read and write NSNR asynchronously. However, 2-cycle instructions that access NSNR during both cycles ('JBC /B, REL,' and 'MOV /B,C') should not be used, since the SIU may write to NSMR between the two 8044 CPU accesses.

The individual bits of the Send/Receive Count Register are as follows:

Bit #	Name	Description
NSNR.0	SER	Receive Sequence Error: NS (P) ≠ NR (S)
NSNR.1	NR0	Receive Sequence Counter—Bit 0
NSNR.2	NR1	Receive Sequence Counter—Bit 1
NSNR.3	NR2	Receive Sequence Counter—Bit 2
NSNR.4	SES	Send Sequence Error: NR (P) ≠ NS (S) and NR (P) ≠ NS (S) + 1
NSNR.5	NS0	Send Sequence Counter—Bit 0
NSNR.6	NS1	Send Sequence Counter—Bit 1
NSNR.7	NS2	Send Sequence Counter—Bit 2

Parameter Registers

There are eight parameter registers that are used in connection with SIU operation. All eight registers may be read or written by the 8044 CPU. RFL and RCB are normally loaded by the SIU.

The eight parameter registers are as follows:

STAD: Station Address Register (byte-addressable)

The Station Address register (Address CEH) contains the station address. To prevent acess conflict, the CPU should access STAD only when the SIU is idle (RTS = 0 and RBE = 0). Normally, STAD is accessed only during initialization.

TBS: Transmit Buffer Start Address Register (byte-addressable)

The Transmit Buffer Start address register (Address DCH) points to the location in on-chip RAM for the beginning of the I-field of the frame to be transmitted. The CPU should access TBS only when the SIU is not transmitting a frame (when TBF = 0).

TBL: Transmit Buffer Length Register (byte = addressable)

The Transmit Buffer Length register (Address DBH) contains the length (in bytes) of the I-field to be transmitted. A blank I-field (TBL = 0) is valid. The CPU should access TBL only when the SIU is not transmitting a frame (when TBF = 0).

NOTE:
The transmit and receive buffers are not allowed to "wrap around" in the on-chip RAM. A "buffer end" is automatically generated if address 191 (BFH) is reached.

TCB: Transmit Control Byte Register (byte-addressable)

The Transmit Control Byte register (Address DAH) contains the byte which is to be placed in the control field of the transmitted frame, during NON-AUTO mode transmission. The CPU should access TCB only when the SIU is not transmitting a frame (when TBF = 0). The $N_S$and N_R counters are not used in the NON-AUTO mode.

RBS: Receive Buffer Start Address Register (byte-addressable)

The Receive Buffer Start address register (Address CCH) points to the location in on-chip RAM where the beginning of the I-field of the frame being received is to be stored. The CPU should write RBS only when the SIU is not receiving a frame (when RBE = 0).

RBL: Receive Buffer Length Register (byte-addressable)

The Receive Buffer Length register (Address CBH) contains the length (in bytes) of the area in on-chip RAM allocated for the received I-field. RBL=0 is valid. The CPU should write RBL only when RBE=0.

RFL: Receive Field Length Register (byte-addressable)

The Receive Field Length register (Address CDH) contains the length (in bytes) of the received I-field that has just been loaded into on-chip RAM. RFL is loaded by the SIU. RFL = 0 is valid. RFL should be accessed by the CPU only when RBE = 0.

RCB: Receive Control Byte Register (byte-addressable)

The Received Control Byte register (Address CAH) contains the control field of the frame that has just been received. RCB is loaded by the SIU. The CPU can only read RCB, and should only access RCB when RBE = 0.

ICE Support

The 8044 In-Circuit Emulator (ICE-44) allows the user to exercise the 8044 application system and monitor the execution of instructions in real time.

The emulator operates with Intel's Intellec development system. The development system interfaces with the user's 8044 system through an in-cable buffer box. The cable terminates in a 8044 pin-compatible plug, which fits into the 8044 socket in the user's system. With the emulator plug in place, the user can excercise his system in real time while collecting up to 255 instruction cycles of real-time data. In addition, he can single-step the program.

Static RAM is available (in the in-cable buffer box) to emulate the 8044 internal and external program memory and external data memory. The designer can display and alter the contents of the replacement memory in the buffer box, the internal data memory, and the internal 8044 registers, including the SFR's.

SIUST: SIU State Counter (byte-addressable)

The SIU State Counter (Address D9H) reflects the state of the internal logic which is under SIU control. Therefore, care must be taken not to write into this register. This register provides a useful means for debugging 8044 receiver problem.

ABSOLUTE MAXIMUM RATINGS*

Ambient Temperature Under Bias ...... 0°C to 70°C
Storage Temperature .... −65°C to −150°C
Voltage on $\overline{EA}$, VPP Pin to VSS ... −0.5V to −21.5V
Voltage on Any Other Pin to VSS .... −0.5V to −7V
Power Dissipation 2W

NOTICE: This is a production data sheet. The specifications are subject to change without notice.

**WARNING: Stressing the device beyond the "Absolute Maximum Ratings" may cause permanent damage. These are stress ratings only. Operation beyond the "Operating Conditions" is not recommended and extended exposure beyond the "Operating Conditions" may affect device reliability.*

D.C. CHARACTERISTICS T_A = 0°C to 70°C, VCC = 5V = 10%, VSS = 0V

Symbol	Parameter		Min	Max	Unit	Test Conditions
VIL	Input Low Voltage (Except $\overline{EA}$ Pin of 8744H)		−0.5	0.8	V	
VIL1	Input Low Voltage to $\overline{EA}$ Pin of 8744H		0	0.8	V	
VIH	Input High Voltage (Except XTAL2, RST)		2.0	VCC + 0.5	V	
VIH1	Input High Voltage to XTAL2, RST		2.5	VCC + 0.5	V	XTAL1 = VSS
VOL	Output Low Voltage (Ports 1, 2, 3)*			0.45	V	IOL = 1.6mA
VOL1	Output Low Voltage (Port 0,ALE,$\overline{PSEN}$)*					
		8744H		0.60 0.45	V V	IOL = 3.2 mA IOL = 2.4 mA
		8044AH/8344AH		0.45	V	IOL = 3.2 mA
VOH	Output High Voltage (Ports 1, 2, 3)		2.4		V	IOH = −80 μA
VOH1	Output High Voltage (Port 0 in External Bus Mode, ALE, $\overline{PSEN}$)		2.4		V	IOH = −400 μA
IIL	Logical 0 Input Current (Ports 1, 2, 3)			−500	μA	Vin = 0.45V
IIL1	Logical 0 Input Current to $\overline{EA}$ Pin of 8744H only			−15	mA	
IIL2	Logical 0 Input Current (XTAL2)			−4.0	mA	Vin = 0.45V
ILI	Input Leakage Current (Port 0) 8744H 8044AH/8344AH			 ±100 ±10	 μA μA	 0.45 < Vin < VCC 0.45 < Vin < VCC
IIH	Logical 1 Input Current to $\overline{EA}$ Pin of 8744H			500	μA	
IIH1	Input Current to RST to Activate Reset			500	μA	Vin < (VCC − 1.5V)
ICC	Power Supply Current: 8744H 8044AH/8344AH			 285 200	 mA mA	All Outputs Disconnected: $\overline{EA}$ = VCC
CIO	Pin Capacitance			10	pF	Test Freq. = 1MHz(1)

***NOTES:**
1. Sampled not 100% tested. T_A = 25°C.
2. Capacitive loading on Ports 0 and 2 may cause spurious noise pulses to be superimposed on the VOLs of ALE and Ports 1 and 3. The noise is due to external bus capacitance discharging into the Port 0 and Port 2 pin when these pins make 1-to-0 transitions during bus operations. In the worst cases (capacitive loading > 100 pF), the noise pulse on the ALE line may exceed 0.8V. In such cases it may be desirable to qualify ALE with a Schmitt Trigger, or use an address latch with a Schmitt Trigger STROBE input.

A.C. CHARACTERISTICS

T_A = 0°C to +70°C, VCC = 5V ± 10%, VSS = 0V, Load Capacitance for Port 0, ALE, and PSEN = 100 pF, Load Capacitance for All Other Outputs = 80 pF

EXTERNAL PROGRAM MEMORY CHARACTERISTICS

Symbol	Parameter	12 MHz Osc		Variable Clock 1/TCLCL = 3.5 MHz to 12 MHz		Unit
		Min	Max	Min	Max	
TLHLL	ALE Pulse Width	127		2TCLCL-40		ns
TAVLL	Address Valid to ALE Low	43		TCLCL-40		ns
TLLAX[1]	Address Hold After ALE Low	48		TCLCL-35		ns
TLLIV	ALE Low to Valid Instr in 8744H 8044AH/8344AH		 183 233		 4TCLCL-150 4TCLCL-100	ns
TLLPL	ALE Low to $\overline{PSEN}$ Low	58		TCLCL-25		ns
TPLPH	$\overline{PSEN}$ Pulse Width 8744H 8044AH/8344AH	 190 215		 3TCLCL-60 3TCLCL-35		 ns ns
TPLIV	$\overline{PSEN}$ Low to Valid Instr in 8744H 8044AH/8344AH		 100 125		 3TCLCL-150 3TCLCL-125	 ns ns
TPXIX	Input Instr Hold After $\overline{PSEN}$	0		0		ns
TPXIZ[2]	Input Instr Float After $\overline{PSEN}$		63		TCLCL-20	ns
TPXAV[2]	$\overline{PSEN}$ to Address Valid	75		TCLCL-8		ns
TAVIV	Address to Valid Instr in 8744H 8044AH/8344AH		 267 302		 5TCLCL-150 5TCLCL-115	 ns ns
TAZPL	Address Float to $\overline{PSEN}$	−25		−25		ns

NOTES:
1. TLLAX for access to program memory is different from TLLAX for data memory.
2. Interfacing RUPI-44 devices with float times up to 75ns is permissible. This limited bus contention will not cause any damage to Port 0 drivers.

5

EXTERNAL DATA MEMORY CHARACTERISTICS

Symbol	Parameter	12 MHz Osc		Variable Clock 1/TCLCL = 3.5 MHz to 12 MHz		Unit
		Min	Max	Min	Max	
TRLRH	$\overline{RD}$ Pulse Width	400		6TCLCL-100		ns
TWLWH	$\overline{WR}$ Pulse Width	400		6TCLCL-100		ns
TLLAX	Address Hold after ALE	48		TCLCL-35		ns
TRLDV	$\overline{RD}$ Low to Valid Data in		252		5TCLCL-165	ns
TRHDX	Data Hold After $\overline{RD}$	0		0		ns
TRHDZ	Data Float After $\overline{RD}$		97		2TCLCL-70	ns
TLLDV	ALE Low to Valid Data In		517		8TCLCL-150	ns
TAVDV	Address to Valid Data In		585		9TCLCL-165	ns
TLLWL	ALE Low to $\overline{RD}$ or $\overline{WR}$ Low	200	300	3TCLCL-50	3TLCLCL+50	ns
TAVWL	Address to $\overline{RD}$ or $\overline{WR}$ Low	203		4TCLCL-130		ns
TQVWX	Data Valid to $\overline{WR}$ Transition					
	8744H	13		TCLCL-70		ns
	8044AH/8344AH	23		TCLCL-60		ns
TQVWH	Data Setup Before $\overline{WR}$ High	433		7TCLCL-150		ns
TWHQX	Data Held After $\overline{WR}$	33		TCLCL-50		ns
TRLAZ	$\overline{RD}$ Low to Address Float		25		25	ns
TWHLH	$\overline{RD}$ or $\overline{WR}$ High to ALE High					
	8744H	33	133	TCLCL-50	TCLCL+50	ns
	8044AH/8344AH	43	123	TCLCL-40	TCLCL+50	ns

NOTE:
1. TLLAX for access to program memory is different from TLLAX for access data memory.

Serial Interface Characteristics

Symbol	Parameter	Min	Max	Unit
TDCY	Data Clock	420		ns
TDCL	Data Clock Low	180		ns
TDCH	Data Clock High	100		ns
tTD	Transmit Data Delay		140	ns
tDSS	Data Setup Time	40		ns
tDHS	Data Hold Time	40		ns

WAVEFORMS

Memory Access

PROGRAM MEMORY READ CYCLE

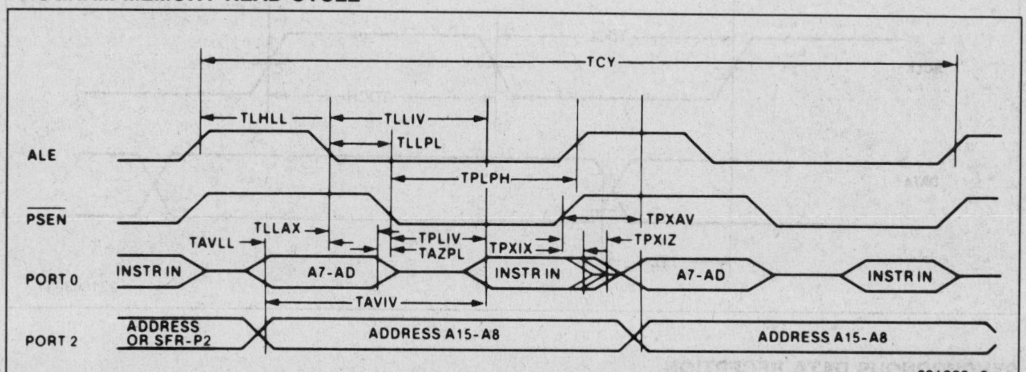

231663-8

DATA MEMORY READ CYCLE

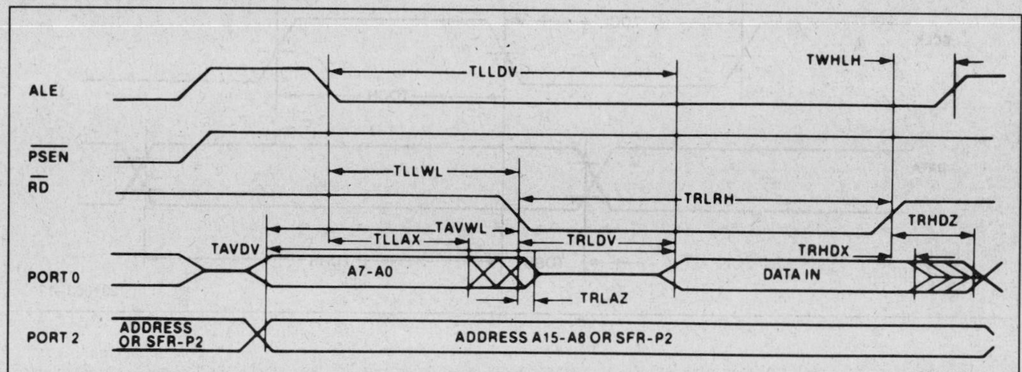

231663-9

DATA MEMORY WRITE CYCLE

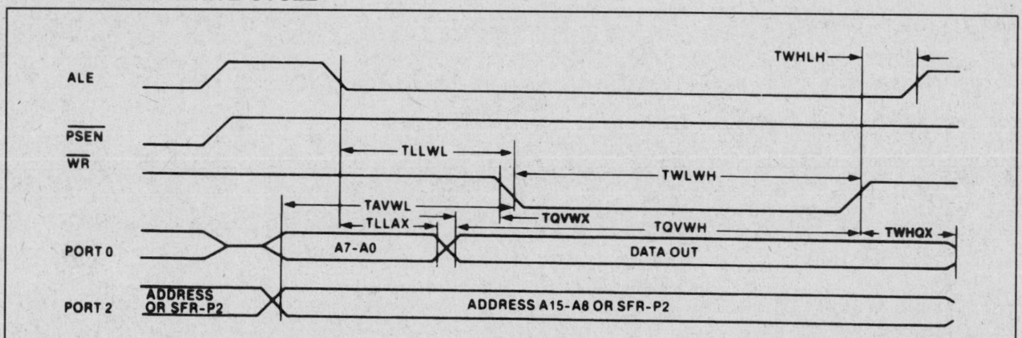

231663-10

SERIAL I/O WAVEFORMS

SYNCHRONOUS DATA TRANSMISSION

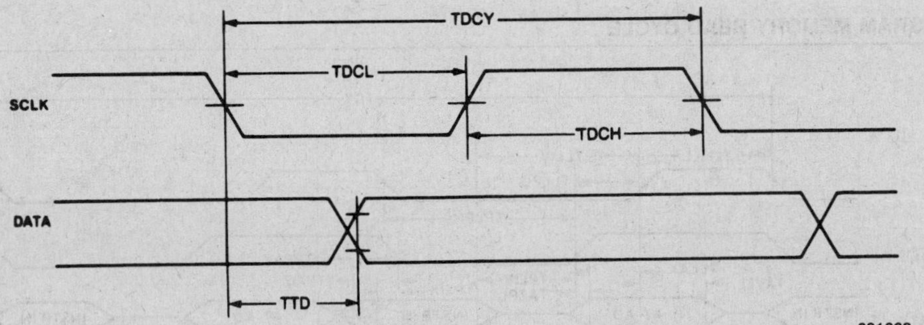

231663–11

SYNCHRONOUS DATA RECEPTION

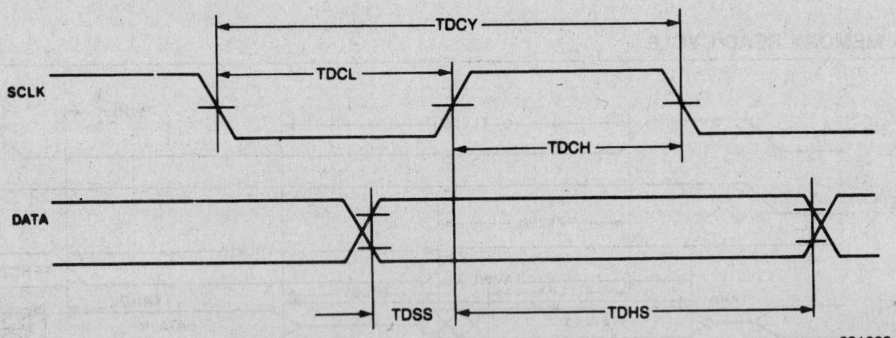

231663–12

AC TESTING INPUT, OUTPUT, FLOAT WAVEFORMS

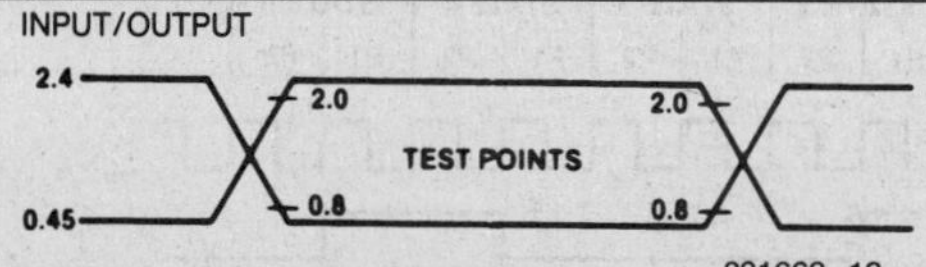

231663-13

AC testing inputs are driven at 2.4V for a Logic "1" and 0.45V for a Logic "0" Timing measurements are made at 2.0V for a Logic "1" and 0.8V for a Logic "0".

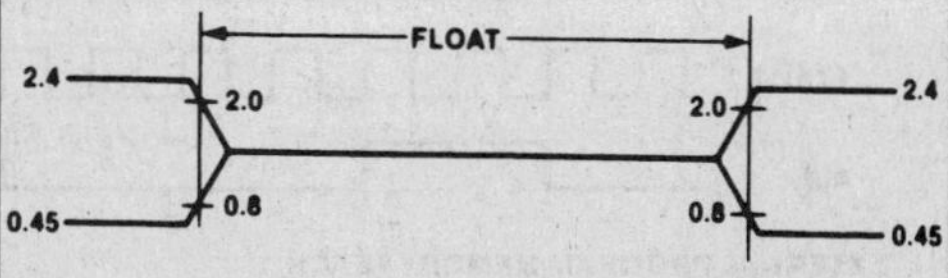

231663-14

EXTERNAL CLOCK DRIVE XTAL2

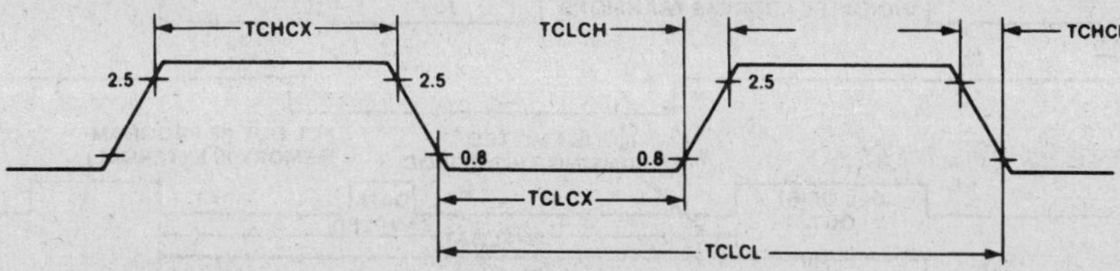

231663-15

Symbol	Parameter	Variable Clock Freq = 3.5 MHz to 12 MHz		Unit
		Min	Max	
TCLCL	Oscillator Period	83.3	285.7	ns
TCHCX	High Time	20	TCLCL-TCLCX	ns
TCLCX	Low Time	20	TCLCL-TCHCX	ns
TCLCH	Rise Time		20	ns
TCHCL	Fall Time		20	ns

5

CLOCK WAVEFORMS

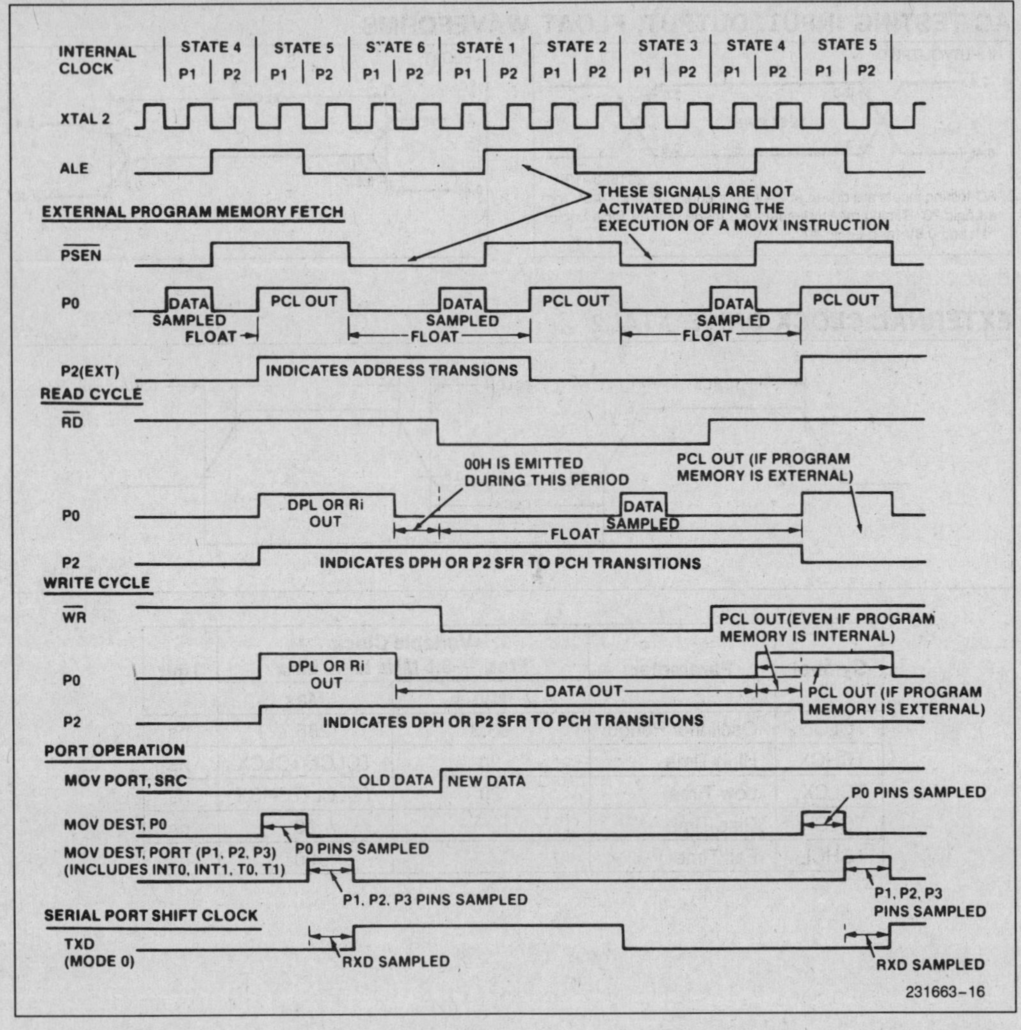

This diagram indicates when signals are clocked internally. The time it takes the signals to propagate to the pins, however, ranges from 25 to 125 ns. This propagation delay is dependent on variables such as temperature and pin loading. Propagation also varies from output to output and component to component. Typically though, (T_A = 25°C, fully loaded) RD and WR propagation delays are approximately 50 ns. The other signals are typically 85 ns. Propagation delays are incorporated in the AC specifications.

8744H EPROM CHARACTERISTICS

Erasure Characteristics

Erasure of the 8744H Program Memory begins to occur when the chip is exposed to light with wavelengths shorter than approximately 4,000 Ångstroms. Since sunlight and fluorescent lighting have wavelengths in this range, constant exposure to these light sources over an extended period of time (about 1 week in sunlight, or 3 years in room-level fluorescent lighting) could cause unintentional erasure. If an application subjects the 8744H to this type of exposure, it is suggested that an opaque label be placed over the window.

The recommended erasure procedure is exposure to ultraviolet light (at 2537 Ångstroms) to an integrated dose of at least 15 W-sec/cm^2 rating for 20 to 30 minutes, at a distance of about 1 inch, should be sufficient.

Erasure leaves the array in an all 1s state.

Programming the EPROM

To be programmed, the 8744H must be running with a 4 to 6 MHz oscillator. (The reason the oscillator needs to be running is that the internal bus is being used to transfer address and program data to appropriate registers.) The address of an EPROM location to be programmed is applied to Port 1 and pins P2.0-P2.3 of Port 2, while the data byte is applied to Port 0. Pins P2.4-P2.6 and $\overline{\text{PSEN}}$ should be held low, and P2.7 and RST high. (These are all TTL levels except RST, which requires 2.5V for high.) $\overline{\text{EA}}$/VPP is held normally high, and is pulsed to +21V. While EA/VPP is at 21V, the ALE/$\overline{\text{PROG}}$ pin, which is normally being held high, is pulsed low for 50 msec. Then EA/VPP is returned to high. This is illustrated in Figure 8. Detailed timing specifications are provided in the EPROM Programming and Verification Characteristics section of this data sheet.

Program Memory Security

The program memory security feature is developed around a "security bit" in the 8744H EPROM array. Once this "hidden bit" is programmed, electrical access to the contents of the entire program memory array becomes impossible. Activation of this feature is accomplished by programming the 8744H as described in "Programming the EPROM" with the exception that P2.6 is held at a TTL high rather than a TTL low. In addition, Port 1 and P2.0-P2.3 may be in any state. Figure 9 illustrates the security bit programming configuration. Deactivating the security feature, which again allows programmability of the EPROM, is accomplished by exposing the EPROM to ultraviolet light. This exposure, as described in "Erasure Characteristics," erases the entire EPROM array. Therefore, attempted retrieval of "protected code" results in its destruction.

Program Verification

Program Memory may be read only when the "security feature" has not been activated. Refer to Figure 10 for Program Verification setup. To read the Program Memory, the following procedure can be used. The unit must be running with a 4 to 6 MHz oscillator. The address of a Program Memory location to be read is applied to Port 1 and pins P2.0-P2.3 of Port 2. Pins P2.4-P2.6 and $\overline{\text{PSEN}}$ are held at TTL low, while the ALE/$\overline{\text{PROG}}$, RST, and $\overline{\text{EA}}$/VPP pins are held at TTL high. (These are all TTL levels except RST, which requires 2.5V for high.) Port 0 will be the data output lines. P2.7 can be used as a read strobe. While P2.7 is held high, the Port 0 pins float. When P2.7 is strobed low, the contents of the addressed location will appear at Port 0. External pullups (e.g., 10K) are required on Port 0 during program verification.

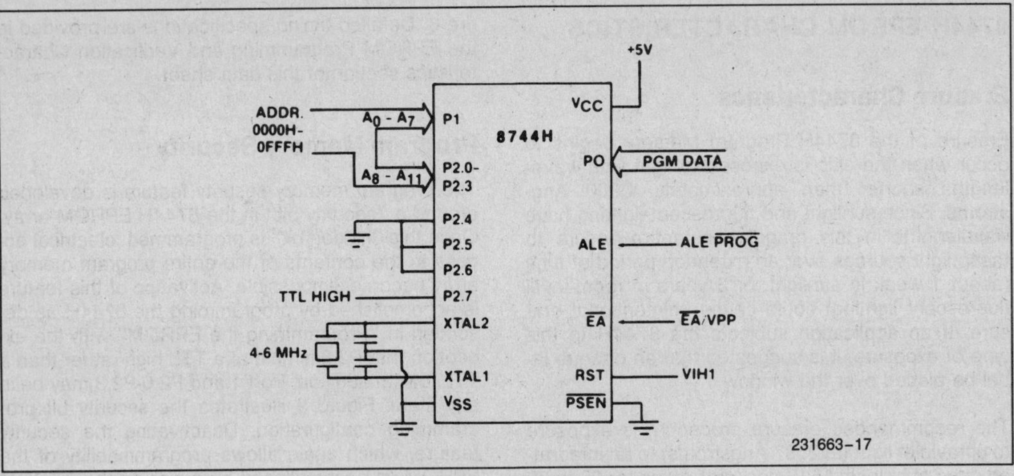

Figure 8. Programming Configuration

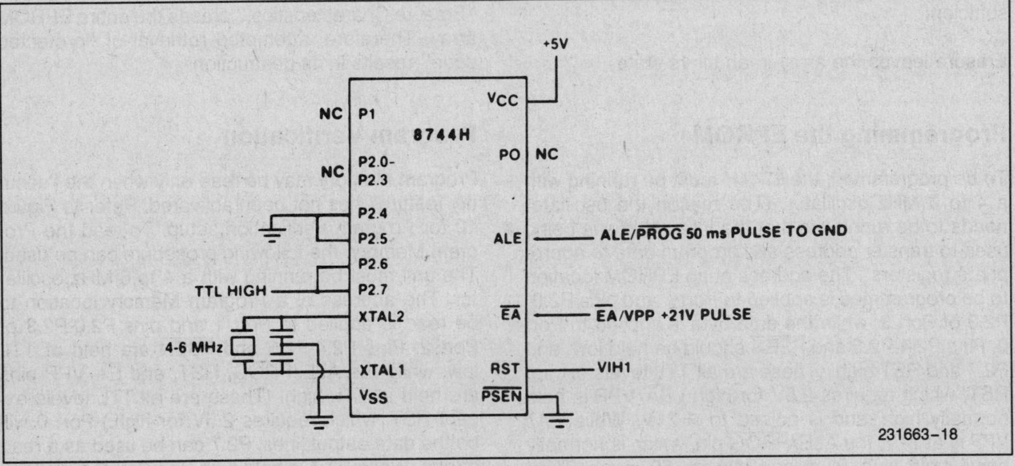

Figure 9. Security Bit Programming Configuration

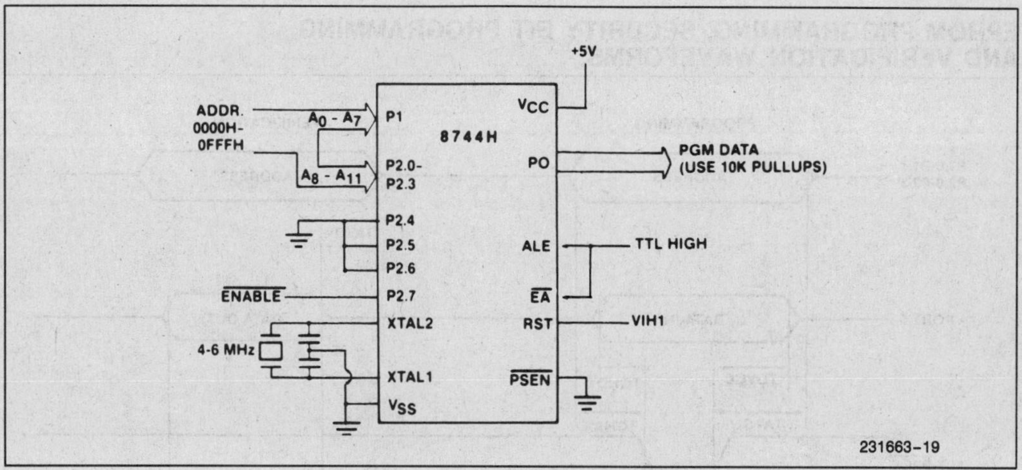

Figure 10. Program Verification Configuration

EPROM PROGRAMMING, SECURITY BIT PROGRAMMING AND VERIFICATION CHARACTERISTICS

TA = 21°C to 27°C, V_{CC} = 4.5V to 5.5V, V_{SS} = 0V

Symbol	Parameter	Min	Max	Units
V_{PP}	Programming Supply Voltage	20.5	21.5	V
IPP	Programming Current		30	mA
1/TCLCL	Oscillator Frequency	4	6	MHz
TAVGL	Address Setup to $\overline{PROG}$	48TCLCL		
TGHAX	Address Hold after $\overline{PROG}$	48TCLCL		
TDVGL	Data Setup to $\overline{PROG}$	48TCLCL		
TGHDX	Data Hold after $\overline{PROG}$	48TCLCL		
TEHSH	$\overline{ENABLE}$ High to Vpp	48TCLCL		
TSHGL	V_{PP} Setup to $\overline{PROG}$	10		μsec
TGHSL	V_{PP} Hold after $\overline{PROG}$	10		μsec
TGLGH	$\overline{PROG}$ Width	45	55	msec
TAVQV	Address to Data Valid		48TCLCL	
TELQV	$\overline{ENABLE}$ to Data Valid		48TCLCL	
TEHQZ	Data Float after $\overline{ENABLE}$	0	48TCLCL	

EPROM PROGRAMMING, SECURITY BIT PROGRAMMING AND VERIFICATION WAVEFORMS

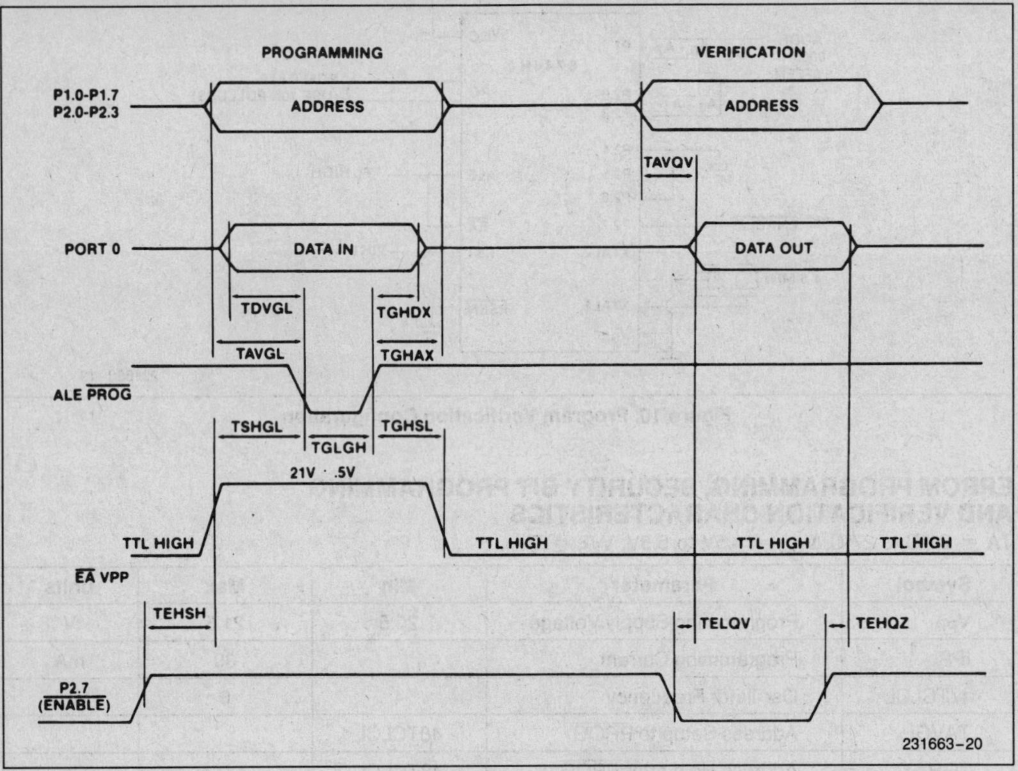

intel®

6

CAN 82527 Controller

6

82527
SERIAL COMMUNICATIONS CONTROLLER
CONTROLLER AREA NETWORK PROTOCOL

Automotive

- **Supports CAN Specification 2.0**
  - **Standard Data and Remote Frames**
  - **Extended Data and Remote Frames**
- **Programmable Global Mask**
  - **Standard Message Identifier**
  - **Extended Message Identifier**
- **15 Message Objects of 8-Byte Data Length**
  - **14 Tx/Rx Buffers**
  - **1 Rx Buffer with Programmable Mask**
- **Flexible CPU Interface**
  - **8-Bit Multiplexed**
  - **16-Bit Multiplexed**
  - **8-Bit Non-Multiplexed (Synchronous/Asynchronous)**
  - **Serial Interface**
- **Programmable Bit Rate**
- **Programmable Clock Output**
- **Flexible Interrupt Structure**
- **Flexible Status Interface**
- **Configurable Output Driver**
- **Configurable Input Comparator**
- **Two 8-Bit Bidirectional I/O Ports**
- **44-Lead PLCC Package**
- **Pinout Compatibility with the 82526**

The 82527 serial communications controller is a highly integrated device that performs serial communication according to the CAN protocol. It performs all serial communication functions such as transmission and reception of messages, message filtering, transmit search, and interrupt search with minimal interaction from the host microcontroller, or CPU.

The 82527 is Intel's first device to support the standard and extended message frames in CAN Specification 2.0 Part B. It has the capability to transmit, receive, and perform message filtering on extended message frames. Due to the backwardly compatible nature of CAN Specification 2.0, the 82527 also fully supports the standard message frames in CAN Specification 2.0 Part A.

The 82527 features a powerful CPU interface that offers flexibility to directly interface to many different CPUs. It can be configured to interface with CPUs using an 8-bit multiplexed, 16-bit multiplexed, or 8-bit non-multiplexed address/data bus for Intel and non-Intel architectures. A flexible serial interface (SPI) is also available when a parallel CPU interface is not required.

6

The 82527 provides storage for 15 message objects of 8-byte data length. Each message object can be configured as either transmit or receive except for the last message object. The last message object is a receive-only buffer with a special mask design to allow select groups of different message identifiers to be received.

The 82527 also implements a global masking feature for message filtering. This feature allows the user to globally mask any identifier bits of the incoming message. The programmable global mask can be used for both standard and extended messages.

The 82527 offers hardware, or pinout, compatibility with the 82526. It is pin-to-pin compatible with the 82526 except for pins 9, 30, and 44. These pins are used as chip selects on the 82526 and are used as CPU interface mode selection pins on the 82527.

The 82527 is fabricated using Intel's reliable CHMOS III 5V technology and is available in a 44-lead PLCC for the automotive temperature range (−40°C to +125°C).

NOTICE:

This is an ADVANCE INFORMATION DATA SHEET. The A.C. and D.C. parameters contained within this data sheet may change after full automotive temperature characterization of the device has been performed. Contact your local sales office before finalizing the Timing and D.C. characteristics of a design to verify you have the latest information.

August 1993
Order Number: 272250-003

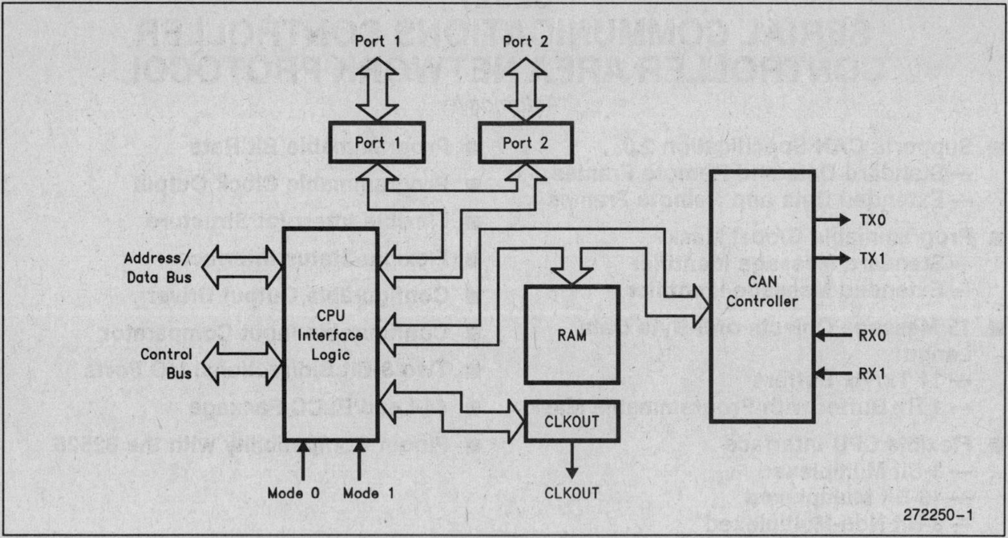

Figure 1. 82527 Block Diagram

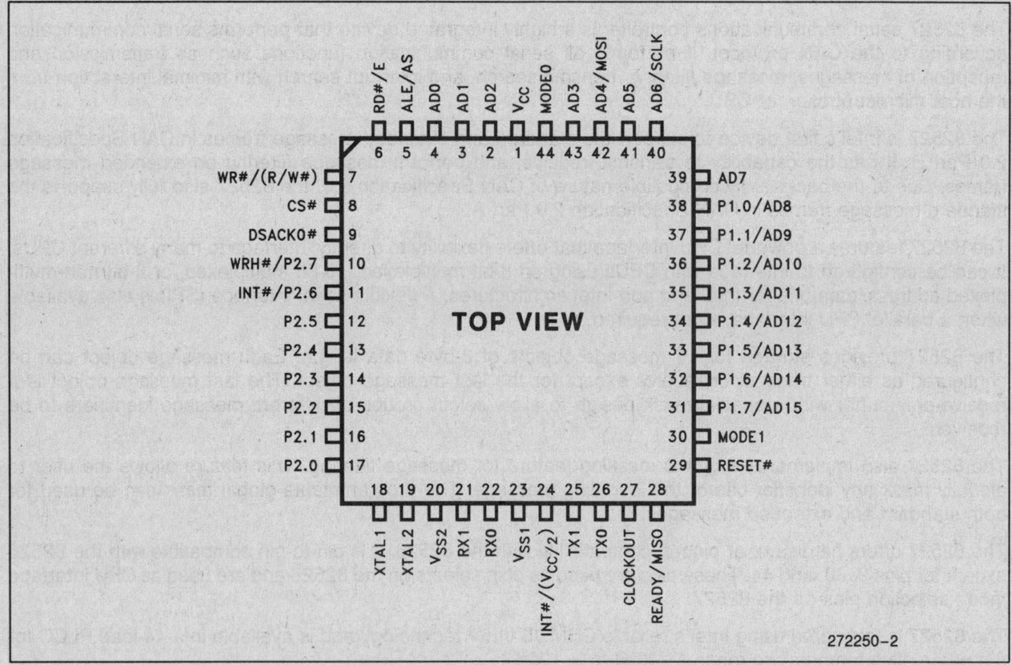

Figure 2. 44-Pin PLCC Package

PIN DESCRIPTION

The 82527 pins are described in this section. Table 1 presents the legend for interpreting the pin types.

Table 1. Pin Type Legend

Symbol	Description
I	Input only pin
O	Output only pin
I/O	Pin can be either input or output

PIN DESCRIPTIONS

Pin #	Pin Name	Pin Type	Pin Description
23	V_{SS1}	Ground	GROUND connection must be shorted externally to a V_{SS} board plane. Provides digital ground.
20	V_{SS2}	Ground	GROUND connection must be shorted externally to a V_{SS} board plane. Provides ground for analog comparator.
1	V_{CC}	Power	POWER connection must be shorted externally to +5V DC. Provides power for entire device.
18	XTAL1	I	Input for an external clock. XTAL1 (along with XTAL2) are the crystal connections to an internal oscillator.
19	XTAL2	O	Push-pull output from the internal oscillator. XTAL2 (along with XTAL1) are the crystal connections to an internal oscillator. If an external oscillator is used XTAL2 must be floated, or not be connected. XTAL2 must not be used as a clock output to drive other CPUs.
27	CLKOUT	O	Programmable clock output. This output may be used to drive the oscillator of the host microcontroller.
29	RESET#	I	A falling edge (high-to-low) transition causes a hardware reset.
8	CS#	I	A low level on this pin enables CPU access to the 82527 device.
24	INT# ($V_{CC}/2$)	O O	The interrupt pin is an open-drain output to the host microcontroller. $V_{CC}/2$ is the power supply for the ISO low speed physical layer. The function of this pin is determined by the MUX bit in the CPU Interface Register (Address 02H) as follows: MUX = 1: pin 24 = $V_{CC}/2$, pin 11 = INT# MUX = 0: pin 24 = INT#
22 21	RX0 RX1	I I	Inputs from the CAN bus line(s) to the input comparator. A recessive level is read when RX0 > RX1. A dominant level is read when RX1 > RX0. When the CoBy bit (Bus Configuration register) is programmed as a "1", the input comparator is bypassed and RX0 is the CAN bus line input.
26 25	TX0 TX1	O O	Serial data push-pull output to the CAN bus line. During a recessive bit TX0 is high and TX1 is low. During a dominant bit TX0 is low and TX1 is high.

6

Pin #	Pin Name	Pin Type	Pin Description
4 3 2 43 42 41 40 39	AD0/A0/ICP AD1/A1/CP AD2/A2/CSAS AD3/A3/STE AD4/A4/MOSI AD5/A5 AD6/A6/SCLK AD7/A7	I/O-I-I I/O-I-I I/O-I-I I/O-I I/O-I-I I/O-I I/O-I-I I/O-I	Address/Data bus in 8-bit multiplexed mode. Address bus in 8-bit non-multiplexed mode. Low byte of A/D bus in 16-bit multiplexed mode. In Serial Interface mode, the following pins have the following meaning: AD0: ICP Idle Clock Polarity AD1: CP Clock Phase AD2: CSAS Chip Select Active State AD3: STE Sync Transmit Enable AD6: SCLK Serial Clock Input AD4: MOSI Serial Data Input
38 37 36 35 34 33 32 31	AD8/D0/P1.0 AD9/D1/P1.1 AD10/D2/P1.2 AD11/D3/P1.3 AD12/D4/P1.4 AD13/D5/P1.5 AD14/D6/P1.6 AD15/D7/P1.7	I/O-O-I/O I/O-O-I/O I/O-O-I/O I/O-O-I/O I/O-O-I/O I/O-O-I/O I/O-O-I/O I/O-O-I/O	High byte of A/D bus in 16-bit multiplexed mode. Data bus in 8-bit non-multiplexed mode. Low speed I/O port. P1 pins in 8-bit multiplexed mode and serial mode. Port pins have weak pullups until the port is configured by writing to 9FH and AFH.
17 16 15 14 13 12 11 10	P2.0 P2.1 P2.2 P2.3 P2.4 P2.5 P2.6/INT# P2.7/WRH#	I/O I/O I/O I/O I/O I/O I/O-O I/O-I	P2 in all modes. P2.6 is INT# when MUX = 1 and is open-drain. P2.7 is WRH# in 16-bit multiplexed mode.
44 30	Mode0 Mode1	I I	These pins select one of the four parallel interfaces. These pins are weakly held low during reset. **Mode1 Mode0** 0 0 8-bit multiplexed — Intel 0 0 Serial Interface mode entered when RD# = 0, WR# = 0 upon reset. 0 1 16-bit multiplexed — Intel 1 0 8-bit multiplexed — non-Intel 1 1 8-bit non-multiplexed
5	ALE/AS	I-I	ALE used for Intel modes. AS used for non-Intel modes.
6	RD# E	I I	RD# used for Intel modes. E used for non-Intel modes.
7	WR# R/W#	I I	WR# used for Intel modes. R/W# used for non-Intel modes.
28	READY MISO	O O	READY is an output to synchronize accesses from the host microcontroller to the 82527. READY is an open-drain output to the host microcontroller. MISO is the serial data output for the serial interface mode.
9	DSACK0#	O	DSACK0# is an open-drain output to synchronize accesses from the host microcontroller to the 82527.

ELECTRICAL CHARACTERISTICS

ABSOLUTE MAXIMUM RATINGS*

Storage Temperature .... −60°C to +150°C

Voltage from Any Pin to V_{SS} .... −0.5V to +7.0V

NOTICE: This data sheet contains information on products in the sampling and initial production phases of development. The specifications are subject to change without notice. Verify with your local Intel Sales office that you have the latest data sheet before finalizing a design.

**WARNING: Stressing the device beyond the "Absolute Maximum Ratings" may cause permanent damage. These are stress ratings only. Operation beyond the "Operating Conditions" is not recommended and extended exposure beyond the "Operating Conditions" may affect device reliability.*

D.C. Characteristics V_{CC} = 5V ±10%; T_A = −40°C to +125°C

Symbol	Parameter	Min	Max	Conditions
V_{IL}	Input Low Voltage (All except XTAL1, XTAL2, RX0, RX1, AD0–AD7 in Mode 3)	−0.5V	0.8V	
V_{IL1}	Input Low Voltage for AD0–AD7 in Mode 3	−0.5V	0.5V	
V_{IL2}	Input Low Voltage (RX0) for Comparator Bypass Mode		0.5V	
V_{IH}	Input High Voltage (All except XTAL1, XTAL2, RX0, RX1, RESET#)	3.0V	V_{CC} + 0.5V	
V_{IH1}	Input High Voltage (RESET#)	3.0V	V_{CC} + 0.5V	
V_{IH1}	Input High Voltage (RESET#) Hysteresis on RESET#	3.0V 200 mV	V_{CC} + 0.5V	
V_{IH2}	Input High Voltage (RX0) for Comparator Bypass Mode	4.0V		
V_{OL}	Output Low Voltage (All Outputs except TX0, TX1)		0.45V	I_{OL} = 1.6 mA
V_{OH}	Output High Voltage (All Outputs except TX0, TX1, CLOCKOUT)	V_{CC} − 0.8V		I_{OH} = −200 μA
V_{OHR1}	Output High Voltage (CLOCKOUT)	0.8 V_{CC}		I_{OH} = −80 μA
I_{LK}	Input Leakage Current		±10 μA	$V_{SS} < V_{IN} < V_{CC}$
C_{IN}	PIN Capacitance**		10 pF	f_{XTAL} = 1 KHz
I_{CC}	Supply Current(1)		50 mA	f_{XTAL} = 16 MHz
I_{SLEEP}	Sleep Current(1) with V_{CC}/2 Output Enabled, No Load with V_{CC}/2 Output Disabled		 700 μA 100 μA	
I_{PD}	Powerdown Current(1)		25 μA	XTAL1 Clocked

NOTE:
**Typical value based on characterization data.
Port pins are weakly held after reset until the port configuration registers are written (9FH, AFH).
1. All pins are driven to V_{SS} or V_{CC} including RX0 and RX1.

6

PHYSICAL LAYER SPECIFICATIONS Load Condition: 100 pF

D.C. Characteristics $V_{CC} = 5V \pm 10\%$; $T_A = -40°C$ to $+125°C$

RX0/RX1 and TX0/TX1	Min	Max	Conditions
Input Voltage	$-0.5V$	$V_{CC} + 0.5V$	
Common Mode Range	$V_{SS} + 1V$	$V_{CC} - 1V$	
Differential Input Threshold		± 100 mV	
Internal Delay 1: Sum of the Comparator Input Delay and the TX0/TX1 Output Driver Delay		60 ns	Load on TX0, TX1 = 100 pF, +100 mV to −100 mV RX0/RX1 differential
Internal Delay 2: Sum of the RX0 Pin Delay (if the Comparator is Bypassed) and the TX0/TX1 Output Driver Delay		50 ns	Load on TX0, TX1 = 100 pF
Source Current on Each TX0, TX1	−10 mA		$V_{OUT} = V_{CC} - 1.0V$
Sink Current on Each TX0, TX1	10 mA		$V_{OUT} = 1.0V$
$V_{CC}/2$			
$V_{CC}/2$	2.38V	2.62V	$I_{OUT} \leq 75\ \mu A$, $V_{CC} = 5V$

CLOCKOUT SPECIFICATIONS

Load Condition: 50 pF

Parameter	Min	Max
CLOCKOUT Frequency	XTAL/15	XTAL

A.C. Characteristics for 8/16-Bit Multiplexed Intel Modes (Modes 0, 1)

Conditions: V_{CC} = 5V ±10%, V_{SS} = 0V, T_A = −40°C to +125°C, C_L = 100 pF

Symbol	Parameter	Min	Max	Conditions
$1/t_{XTAL}$	Oscillator Frequency	8 MHz	16 MHz	
$1/t_{SCLK}$	System Clock Frequency	4 MHz	10 MHz	
$1/t_{MCLK}$	Memory Clock Frequency	2 MHz	8 MHz	
t_{AVLL}	Address Valid to ALE Low	20 ns		
t_{LLAX}	Address Hold after ALE Low	20 ns		
t_{LHLL}	ALE High Time	30 ns		
t_{LLRL}	ALE Low to RD# Low	20 ns		
t_{CLLL}	CS# Low to ALE Low	20 ns		
t_{QVWH}	Data Setup to WR# High	30 ns		
t_{WHQX}	Input Data Hold after WR# High	20 ns		
t_{WLWH}	WR# Pulse Width	40 ns		
t_{WHLH}	WR# High to Next ALE High	8 ns		
t_{WHCH}	WR# High to CS# High	0 ns		
t_{RLRH}	RD# Pulse Width This time is long enough to initiate a double read cycle by loading the High Speed Registers (04H, 05H), but is too short to READ from 04H and 05H (See t_{RLDV})	40 ns		
t_{RLDV}	RD# Low to Data Valid (Only for Registers 02H, 04H, 05H)	0 ns	45 ns	
t_{RLDV1}	RD# Low Data to Data Valid (for Registers except 02H, 04H, 05H) for Read Cycle without a Previous Write(1) for Read Cycle with a Previous Write(1)		 1.5 t_{MCLK} + 100 ns 3.5 t_{MCLK} + 100 ns	
t_{RHDZ}	Data Float after RD# High	0 ns	45 ns	
t_{CLYV}	CS# Low to READY Setup Condition: Load Capacitance on the READY Output: 50 pF		32 ns 40 ns	V_{OL} = 1.0V V_{OL} = 0.45V
t_{WLYZ}	WR# Low to READY Float for a Write Cycle if No Previous Write is Pending(2)		145 ns	
t_{WHYZ}	End of Last Write to READY Float for a Write Cycle if a Previous Write Cycle is Active(2)		2 t_{MCLK} + 100 ns	
t_{RLYZ}	RD# Low to READY Float (for registers except 02H, 04H, 05H) for Read Cycle without a Previous Write(1) for Read Cycle with a Previous Write(1)		 2 t_{MCLK} + 100 ns 4 t_{MCLK} + 100 ns	
t_{WHDV}	WR# High to Output Data Valid on Port 1/2	t_{MCLK}	2 t_{MCLK} + 500 ns	

NOTES:
References to WR# also pertain to WRH#.
1. Definition of "read cycle without a previous write": The time between the rising edge of WR#/WRH# (for the previous write cycle) and the falling edge of RD# (for the current read cycle) is greater than 2 t_{MCLK}.
2. Definition of "write cycle with a previous write": The time between the rising edge of WR#/WRH# (for the previous write cycle) and the rising edge of WR#/WRH# (for the current write cycle) is less than 2 t_{MCLK}.

A.C. Characteristics for 8/16-Bit Multiplexed Intel Modes (Modes 0, 1)

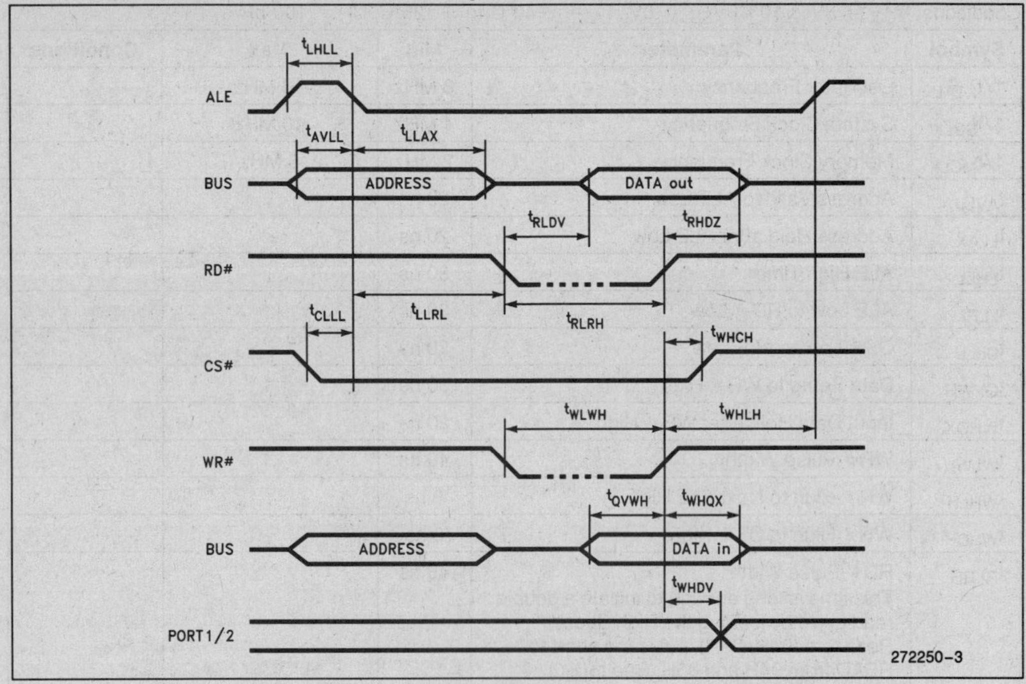

A.C. Characteristics for 8/16-Bit Multiplexed Intel Modes (Modes 0, 1)

Ready Output Timing for a Write Cycle if No Previous Write is Pending

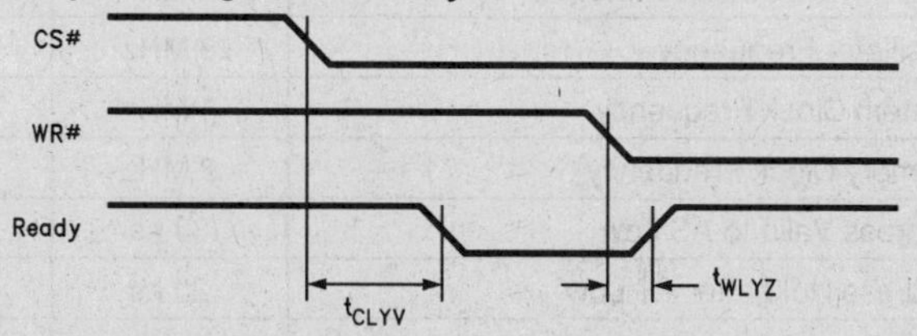

272250-4

Ready Output Timing for a Write Cycle if a Previous Write Cycle is Active

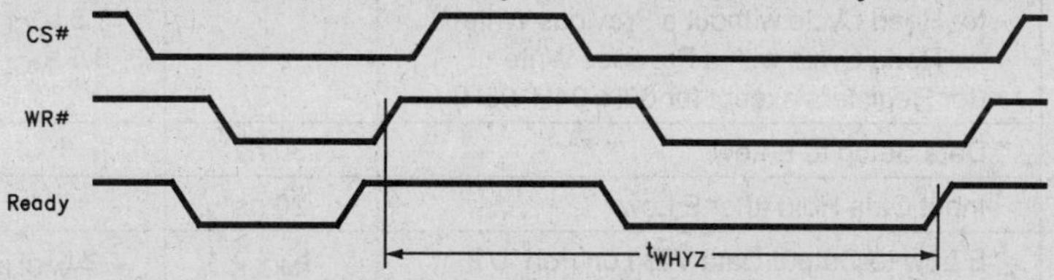

272250-5

Ready Output Timing for a Read Cycle

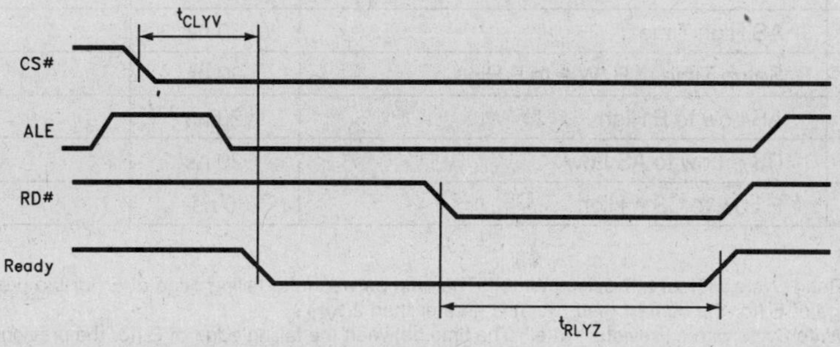

272250-6

6

A.C. Characteristics for 8-Bit Multiplexed Non-Intel Mode (Mode 2)

Conditions: $V_{CC} = 5V \pm 10\%$, $V_{SS} = 0V$, $T_A = -40°C$ to $+125°C$, $C_L = 100$ pF

Symbol	Parameter	Min	Max
$1/t_{XTAL}$	Oscillator Frequency	8 MHz	16 MHz
$1/t_{SCLK}$	System Clock Frequency	4 MHz	10 MHz
$1/t_{MCLK}$	Memory Clock Frequency	2 MHz	8 MHz
t_{AVSL}	Address Valid to AS Low	33 ns	
t_{SLAX}	Address Hold after AS Low	20 ns	
t_{ELDZ}	Data Float after E Low	0 ns	45 ns
t_{EHDV}	E High to Data Valid for Registers 02H, 04H, 05H	0 ns	45 ns
	for Read Cycle without a Previous Write(1) for Read Cycle with a Previous Write (for Registers except for 02H, 04H, 05H)		$1.5\ t_{MCLK} + 100$ ns $3.5\ t_{MCLK} + 100$ ns
t_{QVEL}	Data Setup to E Low	30 ns	
t_{ELQX}	Input Data Hold after E Low	20 ns	
t_{ELDV}	E Low to Output Data Valid on Port 1/2	t_{MCLK}	$2\ t_{MCLK} + 500$ ns
t_{EHEL}	E High Time	45 ns	
t_{ELEL}	End of Previous Write (Last E Low) to E Low for a Write Cycle	$2\ t_{MCLK}$	
t_{SHSL}	AS High Time	30 ns	
t_{RSEH}	Setup Time of R/W# to E High	30 ns	
t_{SLEH}	AS Low to E High	20 ns	
t_{CLSL}	CS# Low to AS Low	20 ns	
t_{ELCH}	E Low to CS# High	0 ns	

NOTES:
1. Definition of "Read Cycle without a Previous Write": The time between the falling edge of E (for the previous write cycle) and the rising edge of E (for the current read cycle) is greater than 2 t_{MCLK}.
2. Definition of "Write Cycle with a Previous Write": The time between the falling edge of E (for the previous write cycle) and the falling edge of E (for the current write cycle) is less than 2 t_{MCLK}.

A.C. Characteristics for 8-Bit Multiplexed Non-Intel Mode (Mode 2)

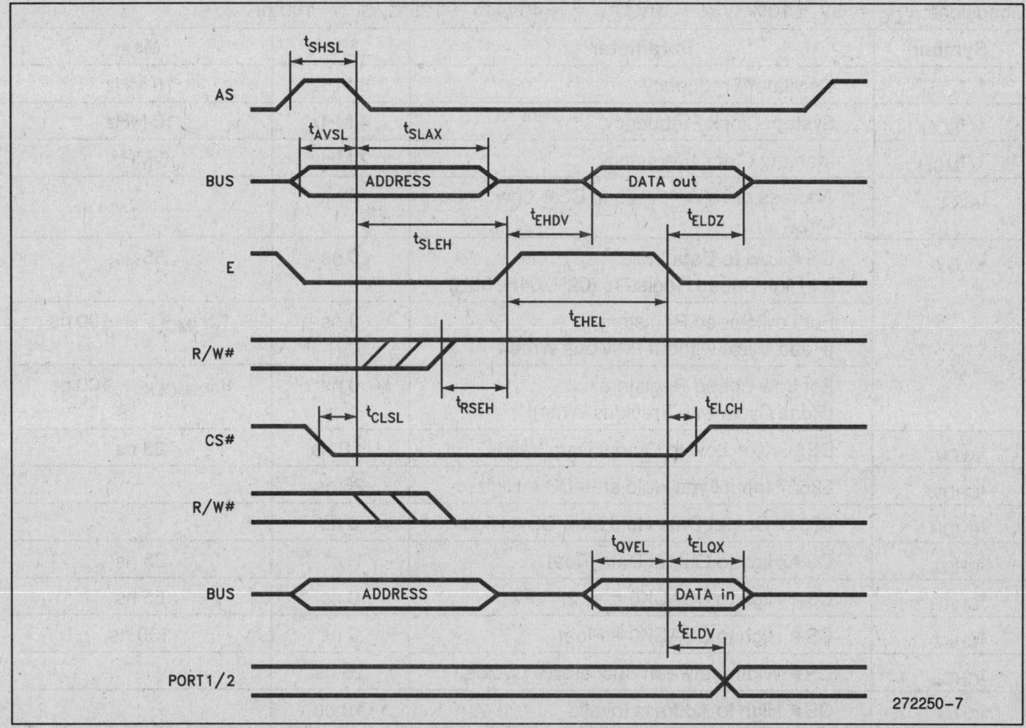

6

A.C. Characteristics for 8-Bit Non-Multiplexed Asynchronous (Mode 3)

Conditions: V_{CC} = 5V ±10%, V_{SS} = 0V, T_A = −40°C to +125°C, C_L = 100 pF

Symbol	Parameter	Min	Max
$1/t_{XTAL}$	Oscillator Frequency	8 MHz	16 MHz
$1/t_{SCLK}$	System Clock Frequency	4 MHz	10 MHz
$1/t_{MCLK}$	Memory Clock Frequency	2 MHz	8 MHz
t_{AVCL}	Address or R/W# Valid to CS# Low Setup	3 ns	
t_{CLDV}	CS# Low to Data Valid for High Speed Registers (02H, 04H, 05H)	0 ns	55 ns
	For Low Speed Registers (Read Cycle without Previous Write)(1)	0 ns	1.5 t_{MCLK} + 100 ns
	For Low Speed Registers (Read Cycle with Previous Write)(1)	0 ns	3.5 t_{MCLK} + 100 ns
t_{KLDV}	DSACK0# Low to Output Data Valid	<0 ns	23 ns
t_{CHDV}	82527 Input Data Hold after CS# High	25 ns	
t_{CHDH}	82527 Output Data Hold after CS# High	0 ns	
t_{CHDZ}	CS# High to Output Data Float		35 ns
t_{CHKH}	CS# High to DSACK0# = 2.4V(3)	0 ns	55 ns
t_{CHKZ}	CS# High to DSACK0# Float	0 ns	100 ns
t_{CHCL}	CS# Width between Successive Cycles	25 ns	
t_{CHAI}	CS# High to Address Invalid	10 ns	
t_{CHRI}	CS# High to R/W# Invalid	5 ns	
t_{CLCH}	CS# Width Low	65 ns	
t_{DVCH}	CPU Write Data Valid to CS# High	32 ns	
t_{CLKL}	CS# Low to DSACK0# Low for High Speed Registers and Low Speed Registers Write Access without Previous Write(2)	0 ns	65 ns
t_{CHKL}	End of Previous Write (CS# High) to DSACK0# Low for a Write Cycle with a Previous Write(2)	0 ns	2 t_{MCLK} + 145 ns

NOTES:

E and AS must be tied high in this mode.

1. Definition of "Read Cycle without a Previous Write": The time between the rising edge of CS# (for the previous write cycle) and the falling edge of CS# (for the current read cycle) is greater than 2 t_{MCLK}.
2. Definition of "Write Cycle without a Previous Write": The time between the rising edge of CS# (for the previous write cycle) and the rising edge of CS# (for the current write cycle) is greater than 2 t_{MCLK}.
3. An on-chip pullup will drive DSACK0# to approximately 2.4V. An external pullup is required to drive this signal to a higher voltage.

A.C. Characteristics for 8-Bit Non-Multiplexed Asynchronous Mode (Mode 3)

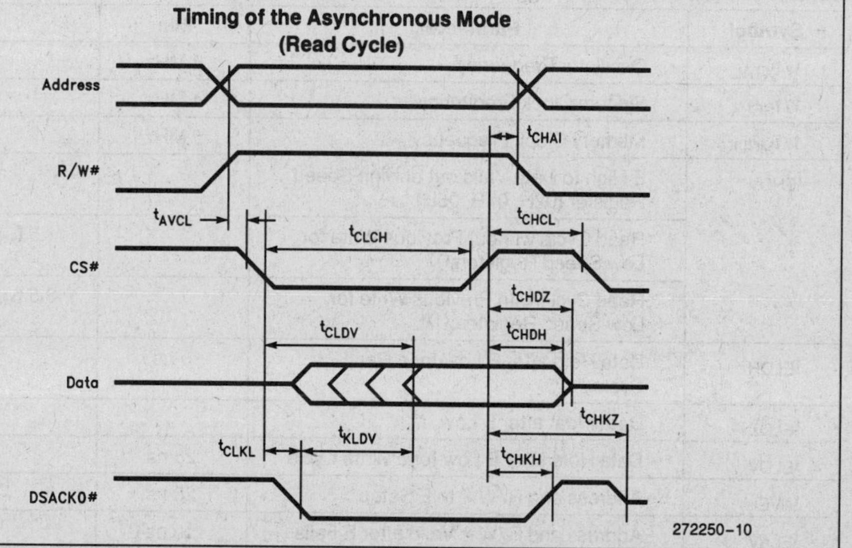

272250–10

A.C. Characteristics for 8-Bit Non-Multiplexed Asynchronous Mode (Mode 3)

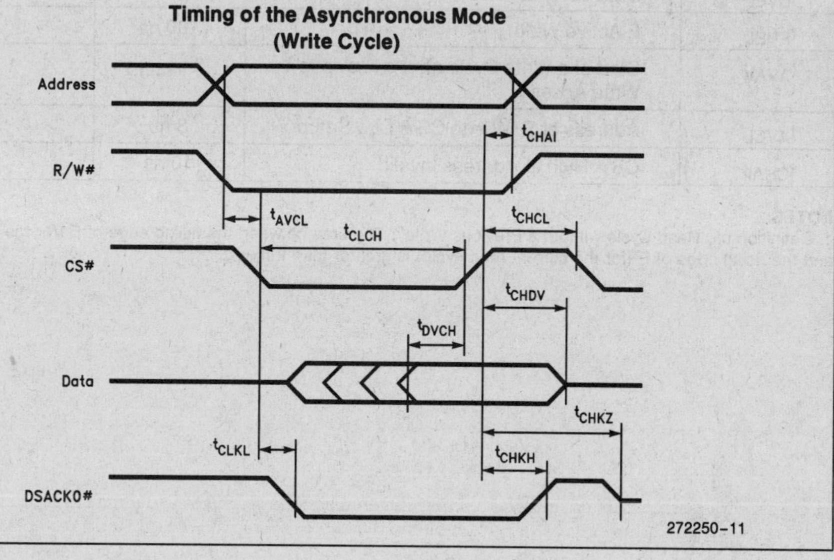

272250–11

A.C. Characteristics for 8-Bit Non-Multiplexed Synchronous Mode (Mode 3)

Conditions: V_{CC} = 5V ±10%, V_{SS} = 0V, T_A = −40°C to +125°C, C_L = 100 pF

Symbol	Parameter	Min	Max
$1/t_{XTAL}$	Oscillator Frequency	8 MHz	16 MHz
$1/t_{SCLK}$	System Clock Frequency	4 MHz	10 MHz
$1/t_{MCLK}$	Memory Clock Frequency	2 MHz	8 MHz
t_{EHDV}	E High to Data Valid out of High Speed Register (02H, 04H, 05H)		55 ns
	Read Cycle without Previous Write for Low Speed Registers(1)		1.5 t_{MCLK} + 100 ns
	Read Cycle with Previous Write for Low Speed Registers(1)		3.5 t_{MCLK} + 100 ns
t_{ELDH}	Data Hold after E Low for a Read Cycle	5 ns	
t_{ELDZ}	Data Float after E Low		35 ns
t_{ELDV}	Data Hold after E Low for a Write Cycle	25 ns	
t_{AVEH}	Address and R/W# to E Setup	25 ns	
t_{ELAV}	Address and R/W# Valid after E Falls	15 ns	
t_{CVEH}	CS# Valid to E High	0 ns	
t_{ELCV}	CS# Valid after E Low	0 ns	
t_{DVEL}	Data Setup to E Low	55 ns	
t_{EHEL}	E Active Width	100 ns	
t_{AVAV}	Start of a Write Cycle after a Previous Write Access	2 t_{MCLK}	
t_{AVCL}	Address or R/W# to CS# Low Setup	3 ns	
t_{CHAI}	CS# High to Address Invalid	10 ns	

NOTES:

1. Definition of "Read Cycle without a Previous Write": The time between the falling edge of E (for the previous write cycle) and the rising edge of E (for the current read cycle) is greater than 2 t_{MCLK}.

A.C. Characteristics for 8-Bit Non-Multiplexed Synchronous Mode (Mode 3)

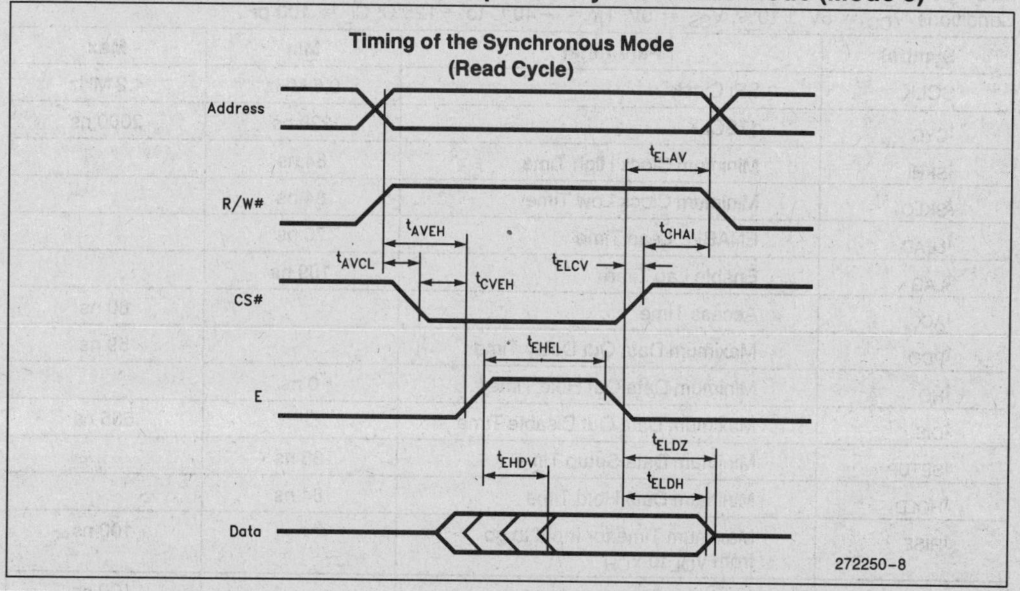

A.C. Characteristics for 8-Bit Non-Multiplexed Synchronous Mode (Mode 3)

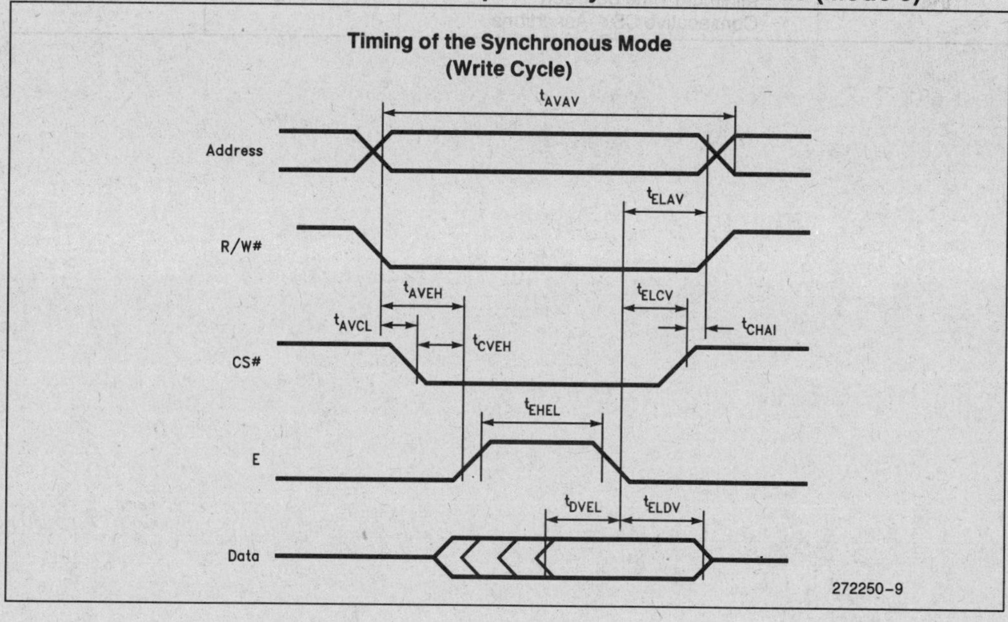

A.C. Characteristics for Serial Interface Mode

Conditions: V_{CC} = 5V ±10%, V_{SS} = 0V, T_A = −40°C to +125°C, C_L = 100 pF

Symbol	Parameter	Min	Max
SCLK	SPI Clock	0.5 MHz	4.2 MHz
t_{CYC}	1/SCLK	238 ns	2000 ns
t_{SKHI}	Minimum Clock High Time	84 ns	
t_{SKLO}	Minimum Clock Low Time	84 ns	
t_{LEAD}	ENABLE Lead Time	70 ns	
t_{LAG}	Enable Lag Time	109 ns	
t_{ACC}	Access Time		60 ns
t_{PDO}	Maximum Data Out Delay Time		59 ns
t_{HO}	Minimum Data Out Hold Time	0 ns	
t_{DIS}	Maximum Data Out Disable Time		665 ns
t_{SETUP}	Minimum Data Setup Time	35 ns	
t_{HOLD}	Minimum Data Hold Time	84 ns	
t_{RISE}	Maximum Time for Input to go from V_{OL} to V_{OH}		100 ns
t_{FALL}	Maximum Time for Input to go from V_{OH} to V_{OL}		100 ns
t_{CS}	Minimum Time between Consecutive CS# Assertions	670 ns	

A.C. Characteristics for Serial Interface Mode

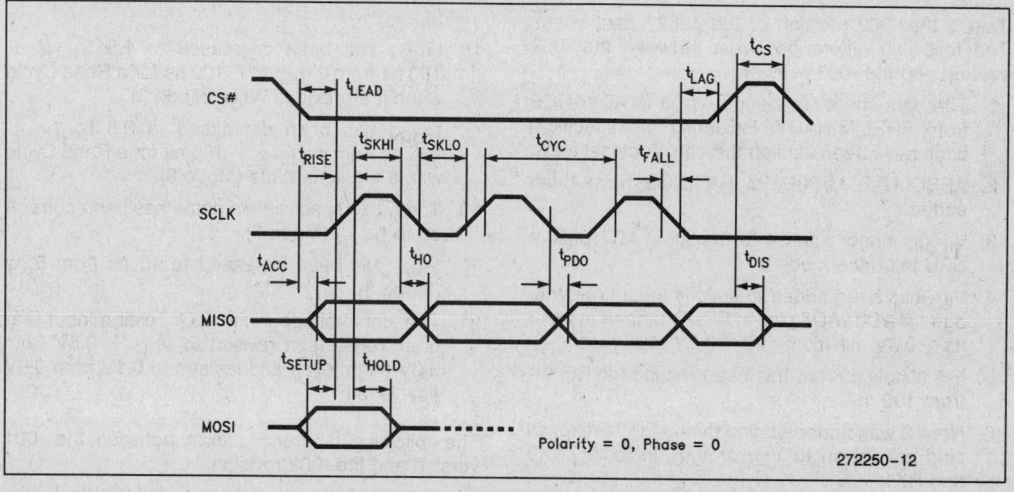

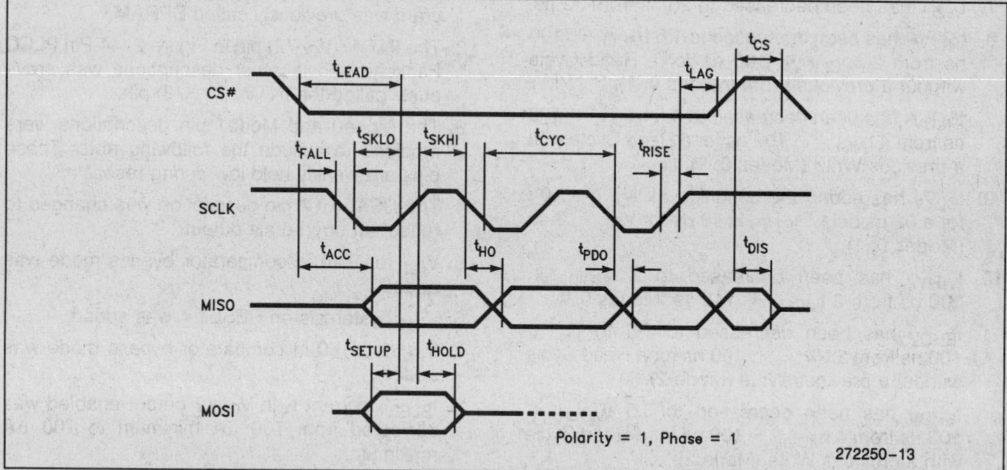

A.C. TESTING INPUT

Input, Output Waveforms

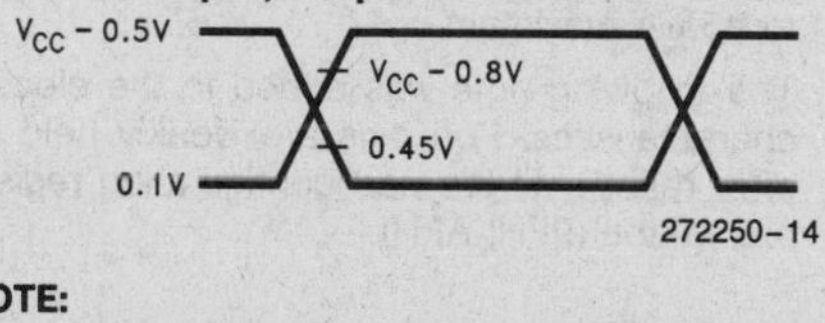

NOTE:
AC Inputs during testing are driven at V_{CC} − 0.5V for a Logic "1" and 0.1V for a Logic "0". Timing measurements are made at V_{OH} Min for a Logic "1" and V_{OL} Max for a Logic "0".

6

DATA SHEET REVISION HISTORY

This is the -003 revision of the 82527 data sheet. The following differences exist between the -002 version and the -003 revision.

1. The data sheet has been revised to ADVANCE from PRELIMINARY, indicating the specifications have been verified through electrical tests.
2. ABSOLUTE MAXIMUM RATINGS have been added.
3. V_{IL} no longer applies to the AD0–AD7 pins in CPU Interface mode 3.
4. V_{IL1} has been added to specify Input Low Voltage for AD0–AD7 pins in CPU Interface mode 3 as −0.5V minimum and +0.5V maximum.
5. I_{CC} supply current has been reduced to 50 mA from 100 mA.
6. Note 2 was added stating during I_{PD} testing, all pins are driven to V_{SS} or V_{CC}, including RX0 and RX1.
7. t_{AVLL} has been decreased to 20 ns from 33 ns.
8. t_{RLDV1} has been decreased to 1.5 t_{MCLK} + 100 ns from 2 t_{MCLK} + 100 ns for a Read Cycle without a previous Write (Modes 0, 1).

   t_{RLDV1} has been decreased to 3.5 t_{MCLK} + 100 ns from 4 t_{MCLK} + 100 ns for a Read Cycle with a previous Write (Modes 0, 1).
9. t_{CLYV} has added the condition of V_{OL} = 1.0V for a 32 ns delay. t_{CLYV} is 40 ns for V_{OL} = 0.45 (Modes 0, 1).
10. t_{WHYZ} has been decreased to 2 t_{MCLK} + 100 ns from 2 t_{MCLK} + 145 ns (Modes 0, 1).
11. t_{EHDV} has been decreased to 1.5 t_{MCLK} + 100 ns from 2 t_{MCLK} + 100 ns for a Read Cycle without a previous Write (Mode 2).

    t_{EHDV} has been decreased to 3.5 t_{MCLK} + 100 ns from 4 t_{MCLK} + 100 ns for a Read Cycle with a previous Write (Mode 2).
12. t_{ELEL} has been decreased to 2 t_{MCLK} from 2 t_{MCLK} + 145 ns (Mode 2).
13. t_{CLDV} has been decreased to 55 ns from 65 ns (Mode 3).
14. t_{CHKH} is specified for V_{IH} = 2.4V, decreased from V_{IH} = 3.0V. Note 3 has been added which states an on-chip pullup will drive DSACK0# to approximately 2.4V. An external pullup is required to drive this signal to a higher voltage (Mode 3).
15. t_{CHAI} has been increased to 10 ns from 5 ns. t_{CHAI} no longer includes CS# High to R/W# Invalid (Mode 3).
16. t_{CHRI} = 5 ns has been added to specify CS# High to R/W# Invalid (Mode 3).
17. t_{EHDV} has been decreased to 55 ns from 65 ns for Reads of the High Speed Registers (Mode 3).
18. t_{EHDV} has been decreased to 1.5 t_{MCLK} + 100 ns from 2 t_{MCLK} + 100 ns for a Read Cycle without a previous Write (Mode 3).

    t_{EHDV} has been decreased to 3.5 t_{MCLK} + 100 ns from 4 t_{MCLK} + 100 ns for a Read Cycle with a previous Write (Mode 3).
19. The t_{AVAL} specification name has been corrected to t_{AVAV} (Mode 3).
20. t_{CHAI} has been increased to 10 ns from 5 ns (Mode 3).
21. The input voltage in the A.C. Testing Input Diagram have been revised to V_{CC} − 0.5V from 3.0V (high level) and revised to 0.1V from 0.8V (low level).

The following differences exist between the -001 version and the -002 revision.

1. The RAM block in Figure 1. 82527 Block Diagram was previously called DPRAM.
2. The INT#/(V_{CC}/2) pin in Figure 2. 44-Pin PLCC Package and in other descriptions was previously called the INT#/(V_{DD}/2) pin.
3. The Mode0 and Mode1 pin descriptions were modified to include the following note: These pins are weakly held low during reset.
4. The DSACK0# pin description was changed to reflect an open-drain output.
5. V_{IL1} for RX0 in comparator bypass mode was added.
6. V_{IH1} hysteresis on RESET# was added.
7. V_{IH2} for RX0 in comparator bypass mode was added.
8. I_{SLEEP} current with V_{CC}/2 output enabled was corrected from 700 μA minimum to 700 μA maximum.
9. I_{SLEEP} current with V_{CC}/2 output disabled was corrected from 100 μA minimum to 100 μA maximum.
10. I_{PD} current was changed from 10 μA minimum to 25 μA maximum.
11. The following note was added to the electrical characteristics: Port pins are weakly held high after reset until the port configuration registers are written (9FH, AFH).

12. The following D.C. Characteristics Specifications have been removed and replaced by the Internal Delay 1 and Internal Delay 2 specifications. These specifications reflect the production test methodology which requires these two delays to be tested together.
    a. Delay Dominant to Recessive
    b. Delay Recessive to Dominant
    c. Input Delay with Comparator Bypassed
    d. Rise Time
    e. Fall Time
13. The following A.C. Characteristics for 8-Bit/16-Bit Multiplexed Intel Modes (Modes 0,1) have been changed:
    a. $1/t_{MCLK}$ has been increased to 8 MHz from 5 MHz.
    b. t_{LLAX} has been decreased to 20 ns from 22.5 ns.
    c. t_{LLRL} has been increased to 20 ns from 0 ns.
    d. t_{CLLL} has been added.
    e. t_{WHLH} has been increased to 8 ns from 0 ns.
    f. t_{WHCH} has been added.
    g. t_{RLDV1} has been added.
    h. t_{WLYH} has been changed to t_{WLYZ} to reflect the READY pin is an open-drain output.
    i. t_{WHYH} has been changed to t_{WHYZ} to reflect the READY pin is an open-drain output.
    j. t_{RLYH} has been changed to t_{RLYZ} to reflect the READY pin is an open-drain output.
    k. t_{WHDV} has been increased to 2 t_{MCLK} + 250 ns from 2 t_{MCLK} + 100 ns.
    l. The following note was added: References to WR# also pertain to WRH#.
    m. The following definition was added for a "read cycle without a previous write": The time between the rising edge of WR#/WRH# (for the previous write cycle) and the falling edge of RD# (for the current read cycle) is greater than 2 t_{MCLK}.
    n. The following definition was added for a "write cycle with a previous write": The time between the rising edge of WR#/WRH# (for the previous write cycle) and the next rising edge of WR#/WRH# (for the current write cycle) is less than 2 t_{MCLK}.
14. The timing diagrams for 8-Bit/16-Bit Multiplexed Intel Modes (Modes 0,1) have been changed to show ALE rising before CS# falls.
15. The following A.C. Characteristics for 8-Bit Multiplexed Non-Intel Modes (Modes 2) have been changed:
    a. $1/t_{MCLK}$ has been increased to 8 MHz from 5 MHz.
    b. t_{SLAX} has been decreased to 20 ns from 22.5 ns.
    c. t_{EVDV} has been decreased to (2, 4) t_{MCLK} + 100 ns from (2, 4) t_{MCLK} + 145 ns.
    d. t_{ELDV} minimum has been decreased to t_{MCLK} from t_{MCLK} + 100 ns.
    e. t_{ELDV} maximum has been increased to 2 t_{MCLK} + 500 ns from 2 t_{MCLK} + 100 ns.
    f. t_{EHEL} for registers except 02H, 04H, 05H has been renamed to t_{ELEL} and the specification has been decreased to 2 t_{MCLK} + 145 ns from 4 t_{MCLK} + 145 ns.
    g. t_{SLEH} has been increased to 20 ns from 0 ns.
    h. t_{CLSL} has been added.
    i. t_{ELCH} has been added.
    j. The following definition was added for a "read cycle without a previous write": The time between the falling edge of E (for the previous write cycle) and the rising edge of E (for the current read cycle) is greater than 2 t_{MCLK}.
    k. The following definition was added for a "write cycle with a previous write": The time between the falling edge of E (for the previous write cycle) and the next falling edge of E (for the current write cycle) is less than 2 t_{MCLK}.

16. The following A.C. Characteristics for 8-Bit Non-Multiplexed Asynchronous Mode (Mode 3) have been changed:
   a. $1/t_{MCLK}$ has been increased to 8 MHz from 5 MHz.
   b. t_{CLDV} has been decreased for low speed registers to (2, 4) t_{MCLK} + 100 ns from (2, 4) t_{MCLK} + 145 ns.
   c. t_{CHKH} comment "with 3.3 KΩ Pullup and 100 pF Load" has been removed since t_{CHKH} is tested with a current source.
   d. t_{CLKL} for a Write Access with a Previous Write has been renamed to t_{CHKL}.
   e. The note "E and AS must be tied high in this mode" has been added.
   f. The following definition was added for a "read cycle without a previous write": The time between the rising edge of CS# (for the previous write cycle) and the falling edge of CS# (for the current read cycle) is greater than 2 t_{MCLK}.
   g. The following definition was added for a "write cycle with a previous write": The time between the rising edge of CS# (for the previous write cycle) and the next rising edge of CS# (for the current write cycle) is less than 2 t_{MCLK}.
17. The following A.C. Characteristics for 8-Bit Non-Multiplexed Synchronous Mode (Mode 3) have been changed:
   a. $1/t_{MCLK}$ has been increased to 8 MHz from 5 MHz.
   b. t_{ELDZ} minimum has been removed.
   c. t_{AVCL} has been added.
   d. t_{CHAI} has been added.
   e. The following definition was added for a "read cycle without a previous write": The time between the falling edge of E (for the previous write cycle) and the rising edge of E (for the current read cycle) is greater than 2 t_{MCLK}.
   f. The following definition was added for a "write cycle with a previous write": The time between the falling edge of E (for the previous write cycle) and the next falling edge of E (for the current write cycle) is less than 2 t_{MCLK}.
18. The following A.C. Characteristics for Serial Interface Mode have been changed:
   a. t_{SKHI} has been decreased to 84 ns from 119 ns.
   b. t_{SKLO} has been decreased to 84 ns from 119 ns.
   c. t_{PDO} has been decreased to 59 ns from 84 ns.
   d. t_{SETUP} has been decreased to 35 ns from 59 ns.
   e. t_{HOLD} has been decreased to 84 ns from 109 ns.
19. The note in the A.C. Testing Input diagram referenced V_{OH} was previously named V_{IH}.

EV82527 EVALUATION KIT

272263-01

The EV82527 evaluation kit demonstrates the capabilities of the 82527 serial communications controller and the Controller Area Network (CAN) protocol. This evaluation kit represents a quick approach to learning the features of the 82527 device and CAN Specification 2.0 29-bit message identifiers.

DESCRIPTION

The EV82527 evaluation kit features the 82527 device and the CAN protocol, Specification 2.0. The 82527 device implements CAN Specification 2.0 and is optimized to allow the host microcontroller to remain dedicated to its application control function. The host microcontroller interface to the 82527 is analogous to that of a RAM. The transmission, reception and error confinement routines are hardwired in the 82527 and are transparent to the user.

FEATURES

The EV82527 evaluation kit consists of three boards: an EV87C196KR motherboard, an 82527 daughterboard and a DV82527 satellite board, plus a software monitor that assembles MCS®-96 code on-line.

- The motherboard is a fully functional evaluation board which contains the host microcontroller, the Intel 87C196KR.
- The daughterboard is configured with an RS-485 CAN bus interface and can be easily adapted to other physical layer implementations.
- The daughterboard communicates with a DV82527 satellite board acting as an additional network node. The satellite board requires no host-CPU programming and uses dip switches to choose various communication options.

BENEFITS

- Quick setup and installation
- Interfaces to high performance 16-bit host-CPU

6

October 1992
Order Number: 272263-001

intel.

EV82527 EVALUATION KIT

- Assists the development of CAN application software
- Demonstrates CAN Specification 2.0 protocol and features
- Uses standard Personal Computer host

87C196KR MOTHERBOARD

Assembly language programs for the 87C196KR motherboard may be downloaded to the microcontroller and executed. The monitor program has the following features:

- Program loading
- Program disassembly
- In-monitor assembler that allows program to be written on-line
- 16 breakpoints
- Single-stepping
- Specific commands to interrogate 82527 messages and status

The 87C916KR is a powerful 16-bit microcontroller with high speed I/O, an A/D converter, full duplex serial I/O (synchronous and asynchronous), 768 bytes of RAM and 16 Kbytes of EPROM.

82527 DAUGHTERBOARD

The interface between the 82527 CAN device and the 87C196KR microcontroller is completed by connecting the mother and daughterboards together. The 82527 device interfaces to the 87C196KR using either an 8- or 16-bit multiplexed address/data bus.

CAN bus communication utilizes the on-board RS-485 interface or connects to a user defined physical interface.

The two 8-bit I/O ports of the 82527 device connect to dip switches or LED displays allowing the user to change or monitor the operation of the 82527 and the 87C196KR devices.

DV82527 SATELLITE BOARD

The daughterboard connects to a satellite board via a cable serving as the CAN bus. It executes a series of fixed programs which are user-selected dip switches. The satellite board receives and transmits one-byte messages of either 11- or 29-bit message identifier format. Messages may use one of four possible message identifiers. The satellite sends remote frames as well.

The reception and transmission of satellite board messages is monitored on LED displays.

PERSONAL COMPUTER REQUIREMENTS

The EV82527 evaluation kit is hosted on an IBM PC AT, XT or BIOS-compatible clone. The PC must meet the following requirements:

- 512 Kbytes of memory
- One 1.2 Meg floppy Disk Drive
- MS-DOS 3.0 or later
- A serial (COM1 or COM2) at 9600 baud
- ASM-96, iC-96 or PL/M-96 or any 8096 Assembler/Compiler that generates Object Module Format code
- A text editor such as AEDIT